Engineering Physics

Second Edition

About the Authors

Hitendra K Malik is currently Professor of Physics at the Indian Institute of Technology Delhi, from where he received his PhD degree in the field of Plasma Physics, in 1995 at the age of 24. He has been a merit scholarship holder throughout his academic career. He is the recipient of Career Award from AICTE, Government of India, for his teaching and research, Outstanding Scientist Award from VIF, India, for his contributions to Science, and 2017 Albert Nelson Marquis Lifetime Achievement Award from USA. In addition, he received the prestigious Erasmus Mundus Visiting Fellowship from European Union (Germany and France), JSPS Fellowship (two times) from Japan, FRD Fellowship from South Africa and DAAD Fellowship from Germany. Owing to his worldwide recognition, his name has been included in 'Marquis Who's Who' in 2011, published from USA. Based on the survey conducted by ResearchGate (RG), his scientific score has been found within top 5% of the scientists and researchers all over the world.

Professor Malik is highly cited in India and abroad for his research work and books with h-index of 24 and i10-index of 70. Governments of India, Germany and France, through DST, CSIR, DRDO, AICTE, DAAD, CEFIPRA, etc., have provided him funding to accomplish 12 sponsored research projects. He is on the editorial board of 5 reputed research journals (including Springer). In recognition of his outstanding research and teaching contributions, he has been asked to deliver more than 50 keynote and invited talks in India, Japan, South Korea, USA, France, Germany, South Africa, and Turkey. Also, he has been chief guest in various universities, mentor of faculty colleagues of engineering institutions, and member of organizing and advisory committees of national and international conferences held in India and abroad.

He has guided 80 PhD, postgraduate and undergraduate theses, including 22 PhD theses in the area of laser/microwave plasma interactions, particle acceleration, solitons, Terahertz radiation, Hall thrusters, plasma material interaction, and nanotechnology. He has published more than 330 scientific papers in high impact factor journals and conferences, including 19 independent articles. He has been reviewer for 72 Journals of international repute, several sponsored research projects (Indian and Foreign agencies), and 18 PhD theses. He is an expert member of academic and administrative bodies of 14 different universities and institutions from 8 states of India including UGC.

Apart from this book, he has also authored another textbook on *Laser-Matter Interaction,* CRC Press, 3 Chapters in the Books *Wave Propagation*, InTechOpen Science, Croatia (featured as highly downloaded chapter), *Society, Sustainability and Environment*, Shivalik Prakashan, New Delhi, and *Plasma Science and Nanotechnology*, Apple Academic Press, exclusive worldwide distribution by CRC Press, a Taylor & Francis Group.

Ajay Kumar Singh has almost two decades of teaching experience in several engineering institutions across North India. Currently, he is Professor of Physics at the Department of Applied Sciences, Maharaja Surajmal Institute of Technology (MSIT), Janakpuri, New Delhi. He has also served as the Head of Department at MSIT. Earlier, he was associated as Professor (2003–2012) at the Department of Applied Science and Humanities, Dronacharya College of Engineering, Haryana.

Dr. Singh completed his PhD from Aligarh Muslim University in the year 1999. During his PhD, his work specifically focused on Uranium concentration in rock samples, soil samples and fly ash samples. He also investigated radon levels in low and high background areas. He has published more than 20 research papers and several articles in national and international journals and conferences. He has edited and co-authored several books on Environment, Water Resources, Nuclear Physics, and Engineering Physics. His book on Engineering Physics Practical and Tutorials has been highly appreciated by students. He is a life member of Plasma Science Society of India (PSSI).

Dr. Singh was the 'B.Tech First Year Syllabus Revision Committee' coordinator representing all the affiliated engineering colleges of Guru Gobind Singh Indraprastha University (GGSIPU). He is academic coordinator of PhD scholars enrolled at MSIT under University School of Information, Communication and Technology, GGSIPU, Dwarka, New Delhi. He is also supervisor of PhD students under USICT, which is a premier constituent institute of GGSIPU. Dr. Singh has also been teaching a special course on Nanotechnology for the USICT PhD students. He is also the teacher representative in the governing board of Maharaja Surajmal Institute of Technology.

Engineering Physics

Second Edition

Hitendra K Malik
Professor, Department of Physics
Indian Institute of Technology
Delhi

Ajay Kumar Singh
Professor, Department of Applied Sciences
Maharaja Surajmal Institute of Technology
(MSIT) New Delhi

McGraw Hill Education (India) Private Limited

Published by McGraw Hill Education (India) Private Limited,
Registered Office: Anjana Complex No: 5/90 A, Butt Road, St. Thomas Mount, Chennai – 600016, Tamil Nadu, India.

Engineering Physics, 2e

Tenth reprint 2024

Print Edition
ISBN-13: 978-93-5260-695-5
ISBN-10: 93-5260-695-7

E-Book Edition
ISBN-13: 978-93-5260-696-2
ISBN-10: 93-5260-696-5

11 12 13 14 15 RAJ 28 27 26 25 24

Typeset at The Composers, 260, C.A. Apt., Paschim Vihar, New Delhi 110063, and Printed and bound in India at Rajkamal Electric Press, Plot No. 2, Phase-IV, Kundli, Haryana

Visit us at: www.mheducation.co.in
Toll free in India: 1800 103 5875
Write to us at: support.india@mheducation.com
CIN: U80302TN2010PTC111532

Dedicated to

OMENDRA Bhaiya and all those moments that remain with me as a source of inspiration and help me to move ahead with great success, satisfaction and optimistic approach

Brief Contents

Contents

Foreword

It gives me immense pleasure to see the present textbook on "Engineering Physics" which covers almost the entire syllabus taught at undergraduate level at different engineering colleges and institutions throughout India. I complement the authors and appreciate their efforts in bringing out this book written in a very simple language. The text is comprehensive and the explanation of topics is commendable. I understand that this book carries all the elements required for a good presentation.

I have been a student of IIT Kharagpur and later on taught at IIT Delhi. Being a part of the IIT system, I recognise that the rigorous and enriching teaching experience at IITs originating from the interaction with the best engineering students and their strong feedback results in continuous evolution and refinement of the teachers. This spirit is reflected in the comprehensive and in-depth handling of important topics in a very simple manner in this book. I am happy to note that this textbook has been penned down by IITian and hope that it would serve to be a good textbook on the subject. Since this book also covers advanced topics, it will be an important learning resource for the teachers, and those students who wish to develop research skills and pursue higher studies. I hope that the book is well received in the academic world.

Prem Vrat

Professor Prem Vrat
Vice-Chancellor, U.P. Technical University, Lucknow
Founder Director, IIT Roorkee

Preface to the Second Edition

The first edition of the textbook was appreciated by the teachers and students of many universities, engineering colleges and institutes, including IIT's throughout India. Words of appreciation were also received from faculty colleagues from Japan, China, Taiwan, Russia, Canada, South Korea, Pakistan, Bangladesh, Turkey, Iran, South Africa, Germany, France, United Kingdom, and United States of America. Students preparing for GATE/CSIR competitive examinations also suggested for more examples in the book and inclusion of topics of postgraduate level. The students very enthusiastically informed us about the utility of the book for the preparation of interviews for admission in PhD programmes at IITs and other universities (including foreign universities) or to get government jobs in India.

In view of all the above points, we have come up with the second edition of the book, where we have used simple language for explaining each and every topic. We have included more physical insight, wherever required. Some chapters are thoroughly revised in terms of new topics and solved problems. We have also updated advanced topics keeping in mind the research going on in these fields. The solutions to the Objective-Type Questions are also provided at the end of the book.

In particular, **Chapter 4** includes details of the topic Population Inversion which covers various schemes for the same, i.e., two-level, three-level and four-level systems. In **Chapter 5**, a topic on Optical Fibres as a Dielectric Waveguide is included. After **Chapter 7** on Waves and Oscillations, a new **Chapter 8** on Simple Harmonic Motion and Sound Waves has been included that discusses standing waves, supersonic and shock waves, in addition to sound waves, Doppler effect and Lissajous figures. **Chapter 9** on Sound Waves and Acoustics of Buildings has been thoroughly revised. In this chapter, Recording and Reproduction of Sound has been withdrawn and other topics are revisited. New topics on ultrasonics have been included which talk about production of ultrasonic waves and their absorption, dispersion, detection and applications. In **Chapter 10** on Dielectrics, a topic Energy Stored in an Electrostatic Field is withdrawn as its concept is discussed in **Chapter 11** on Electromagnetism. Moreover, details of Clausius-Mosotti equation are revised with the inclusion of physical insight of this equation. The chapter on Electromagnetism has been thoroughly revised. For example, Section 11.21 has been rewritten in order to make the readers understand which form of the Maxwell's equations is appropriate for free space, dielectric medium and conducting medium and how are these equations modified in these media. Bound charges and bound currents are also discussed. The solution to wave equation in conducting medium is included as Section 11.28.1, where dispersion relation, skin depth and phase relationship of the electric and magnetic field vectors are discussed. New solved problems, objective-type questions and other practice problems are also included in order to provide an indepth knowledge on the electromagnetic fields and their propagation in different media.

In **Chapter 12** on Theory of Relativity, physical insight to two interesting topics, viz. Length Contraction and Time Dilation is provided. Several new solved problems on various topics are also provided for the readers. **Chapter 13** on Applied Nuclear Physics has been thoroughly revised and new topics are included on basic properties of nucleus, nuclear forces, binding energy of nucleus, nuclear stability and various nuclear models, in addition to more equations and problems, both solved and unsolved. Introduction part of **Chapter 16** on Quantum Mechanics has been revised. The topic on Thermionic Emission (Section 17.7) has been shortened but significance of Richardson's equation is included. The earlier **Chapter 21** on Photoconductivity and Photovoltaics has been withdrawn but its important topics, viz. photoconductivity, simple model of photoconductor and effect of traps, are included in **Chapter 18** on Bond Theory of Solids and Photoconductivity.

The much important **Chapter 22** on Nanophysics has been rewritten in view of recent advances in the field. Now, it is renamed as Nanoscience and Nanotechnology. Certain new topics are included to clarify how nanomaterials are different from bulk materials and to know the differences between nanoscience and nanotechnology. The chapter very systematically discusses the nanoscales in 1D, 2D, 3D and OD. Particularly, nanowires, carbon nanotubes, inorganic nanotubes, biopolymers, nanoparticles, buckyballs/fullerenes and quantum dots are discussed in detail along with the methods of their synthesis, properties and their applications. Finally, the applications, limitations and disadvantages of nanotechnology are also discussed.

The text is supported with online supplements for instructors and students. Instructor resources are available through *https://connect.mheducation.com* while student resources are available through *https://www.mheducation.co.in* (see the back cover for more details).

For Instructors

- Chapter-wise Power Point slides with diagrams and notes for effective lecture presentations

For Students

- An e-guide to aid last minute revision need

We believe the readers shall find the second edition of the book more beneficial in terms of syllabus covered, quality of topics, large number of solved problems aimed at providing physical insight to various topics, and teaching various methods of solving difficult problems. The systematic approach adopted in the present book shall certainly help the teachers and students providing for crystal clear understanding of the topics and carrying out research in the related fields. This edition will be vital in enhancing the self confidence of our UG and PG students which will help them in advancing their careers.

Finally, we look forward to receive feedback from the teachers and students on the recent edition of the book.

H K Malik
Ajay K Singh

Publisher's Note:

McGraw Hill Education (India) invites suggestions and comments, all of which can be sent to info.india@mheducation.com (kindly mention the title and author name in the subject line).

Piracy-related issues may also be reported.

Preface to the First Edition

Physics is a mandatory subject for all engineering students, where almost all the important elements of the subject are covered. Finally, these evolve as different branches of the engineering course. The book entitled Engineering Physics has been written keeping in mind the need of undergraduate students from various engineering and science colleges of all Indian universities. It caters to the complete syllabus for both–Physics-I and Physics-II papers in the first year Engineering Physics course.

The aim of writing this book has been to present the material in a concise and very simple way so that even weak students can grasp the fundamentals. In view of this, every chapter starts with a simple introduction and then related topics are covered with a detailed description along with the help of figures. Particularly the solved problems (compiled from University Question Papers) are at the end of each chapter. These problems are not merely numerical; many of them focus on reasoning and require thoughtful analysis. Finally, the chapters carry unsolved questions based on which the students would be able to test their knowledge as to what they have acquired after going through various chapters. *A chapter-end summary and list of important formulae will be helpful to students for a quick review during examinations. The rich pedagogy consists of solved examples (450), objective-type questions (230), short-answer questions (224) and practice problems (617).* The manuscript has been formulated in such a way that students shall grasp the subject easily and save their time as well. Since the complete syllabus is covered in a single book, it would be highly convenient to both.

The manuscript contains 22 chapters which have been prepared as per the syllabus taught in various colleges and institutions. In particular, the manuscript discusses optics, lasers, holography, fibre optics, waves, acoustics of buildings, electromagnetism, theory of relativity, nuclear physics, solid state physics, quantum physics, magnetic properties of solids, superconductivity, photoconductivity and photovoltaic, X-rays and nanophysics in a systematic manner. We have discussed advanced topics such as laser cooling, Bose-Einstein condensation, scanning electron microscope (SEM), scanning tunnelling microscope (STM), controlled fusion including plasma, Lawson criterion, inertial confinement fusion (ICF), plasma based accelerators, namely, plasma wake field accelerator, plasma beat wave accelerator, laser wake field accelerator and self-modulated laser wake field accelerator, and nanophysics with special emphasis on properties of nanoparticles, carbon nanotubes, synthesis of nanoparticles and applications of nanotechnology. These will be of interest to the teachers who are involved in teaching postgraduate courses at the universities and the students who opt for higher studies and research as their career. Moreover, a series of review questions and problems at the end of each chapter together with the solved questions would serve as a question bank for the students preparing for various competitive examinations. They will get an opportunity to learn the subject and test their knowledge on the same platform.

The structuring of the book provides in-depth coverage of all topics. **Chapter 1** discusses Interference. **Chapter 2** is on Diffraction. **Chapter 3** is devoted to Polarization. Coherence and Lasers are described in

Chapter 4. **Chapter 5** discusses Fibre Optics and its Applications, while Electron Optics is dealt with in **Chapter 6**. **Chapter 7** describes Waves and Oscillations. **Chapter 8** is on Sound Waves and Acoustics. **Chapter 9** is on Dielectrics. Electromagnetic Wave Propagation is described in **Chapter 10**. **Chapter 11** discusses the Theory of Relativity.

Chapter 12 is devoted to Nuclear Physics. Crystal Structure is described in **Chapter 13**. **Chapter 14** deals with the Development of Quantum Physics, while **Chapter 15** is on Quantum Mechanics. **Chapter 16** discusses Free Electron Theory. Band Theory of Solids is explained in **Chapter 17**. **Chapter 18** describes the Magnetic Properties of Solids. **Chapter 19** is on Superconductivity. **Chapter 20** explains X-rays in detail while **Chapter 21** is on Photoconductivity and Photovoltaics. Finally, **Chapter 22** discusses Nanophysics in great detail. The manuscript has been organised such that it provides a link between different topics of a chapter. In order to make it simpler, all the necessary mathematical steps have been given and the physical feature of the mathematical expressions is discussed as and when required.

This book is enriched with a number of resources for instructors and students.

We would like to thank the entire team of Tata McGrawHill Education *specifically Vibha Mahajan, Shalini Jha, Tina Jajoriya, Dipika Dey, Sohini Mukherji, Priyanka Negi and Baldev Raj* for bringing out this book in a very short time span. The reviewers of the book also deserve a special mention for taking out time to review the book. Their names are given below.

A K Jain	*IIT Roorkee*
Dhirendra Kumar	*Meerut Institute of Engineering and Technology, Uttar Pradesh*
Vinay Kumar	*SRMS CET, Bareilly*
Prerna Garg	*Meerut Institute of Technology, Uttar Pradesh*
Amit Kumar Srivastava	*Aryavrat Institute of Technology and Management, Lucknow*
Shyam Singh	*Aryavart Institute of Technology and Management, Lucknow*
R S Tiwari	*Apollo Institute of Engineering, Kanpur*
Kamlesh Pathak	*SVNIT, Surat, Gujarat*
Kanti Jotania	*M S University, Baroda, Gujarat*
Vijayalakshmi Sanyal	*Bharathiyar College of Engineering and Technology, Karaikal, Tamil Nadu*
A K Meikap	*NIT, Durgapur, West Bengal*
K Sivakumar	*Anna University, Chennai*

H K Malik

Ajay K Singh

Interference

1

LEARNING OBJECTIVES

After reading this chapter you will be able to

LO1 Explain interference through Young's double slit experiment

LO2 Describe the concept of wave and Huygen's principle

LO3 Illustrate phase and path difference

LO4 Explain coherence, its various types and coherent sources

LO5 Discuss analytical treatment of interference and conditions for sustained interference

LO6 Examine multiple beam superposition and interference by division of wavefront and amplitude

LO7 Review engineering/scientific applications of interferences including homodyne and heterodyne detection

Introduction

You would have seen beautiful colours in soap films or patch of oil floating on the surface of water. Moreover, the colour gets changed when you watch it from different angles. Did you ever try to find out the reason? In scientific language, this takes place due to the phenomenon of interference. The phenomenon of interference of light tells us about the wave nature of the light. In optics, the interference means the superposition of two or more waves which results in a new wave pattern. Here, we are talking about the interaction of waves emerging from the same source or when the frequencies of these waves are the same. In the context of light, which is an electromagnetic wave, we say that when the light from two different sources moves in the same direction, then these light wave trains superimpose upon each other. This results in the modification of distribution of intensity of light. According to the principle of superposition, this is called the interference of light. More precisely the interference can be defined as the interaction between two or more waves of the same or very close frequencies emitted from coherent sources (defined later), where the wavefronts are combined according to the principle of superposition. The resulting variation in the disturbances produced by the waves is called the interference pattern. Thomas Young, in 1802, explained the interference successfully in his double slit experiment.

1.1 YOUNG'S DOUBLE SLIT EXPERIMENT LO1

The phenomenon of interference may be better understood by taking two point light sources S_1 and S_2 which produce similar waves (Fig. 1.1). Let the sources S_1 and S_2 be at equal distances from the main source S while being close to each other. Since the sources emit waves in all the directions, the spherical waves first pass through S and then S_1 and S_2. Finally these waves expand into the space. The crests of the waves are represented by complete arcs and the troughs by dotted arcs. It is seen that constructive interference takes place at the points where the crests due to one source meet the crests due to another source or where their troughs meet each other. In this case, the resultant amplitude will be the sum of the amplitudes of the separate waves and hence the intensity of the light will be maximum at these points. Similarly, at those points where crests due to one source meet the troughs due to another source or vice-versa, the resultant amplitude will be the difference of the amplitudes of the separate waves. At these points the intensity of the waves (or light) will be minimum. Therefore, due to the intersection of these lines, an alternate bright and dark fringes are observed on the screen placed at the right side of the sources S_1 and S_2. These fringes are obtained due to the phenomenon of interference of light.

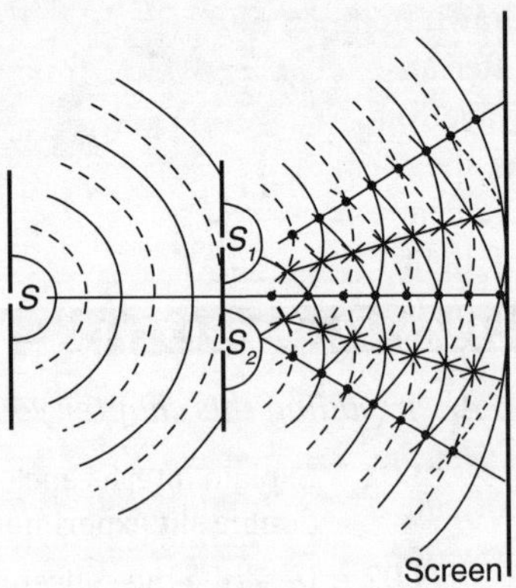

FIGURE 1.1

1.2 CONCEPT OF WAVES AND HUYGENS' PRINCIPLE LO2

A wave is a disturbance that propagates through space and time, usually with the transference of energy from one point to another without any particle of the medium being permanently displaced. Under this situation, the particles only oscillate about their equilibrium positions. If the oscillations of the particles are in the direction of wave propagation, then the wave is called longitudinal wave. However, if these oscillations take place in perpendicular direction with the direction of wave propagation, the wave is said to be transverse in nature. In electromagnetic waves, such as light waves, it is the changes in electric and magnetic fields which represent the wave disturbance. The progress of the wave propagation is described by the passage of a waveform through the medium with a certain velocity called the phase velocity or wave velocity. However, the energy is transferred at the group velocity of the waves making the waveform.

The wave theory of the light was proposed in 1678 by *Huygens*, a Dutch scientist. On the basis of his wave theory, he explained satisfactorily the phenomena of reflections, refraction etc. In the beginning, Huygens' supposed that these waves are longitudinal waves but later he came to know that these waves are transverse in nature. Huygens' gave a hypothesis for geometrical construction of the position of a common wavefront at any instant when the propagation of waves takes place in a medium. The wavefront is an imaginary surface joining the points of constant phase in a wave propagated through the medium. The way in which the wavefront is propagated further in the medium is given by Huygens' principle. This principle is based on the following assumptions:

(i) Each point on the given wavefront acts as a source of secondary wavelets.

(ii) The secondary wavelets from each point travel through space in all the directions with velocity of light.

(iii) A surface touching the secondary wavelets tangentially in the forward direction at any given time constructs the new wavefront at that instant. This is known as secondary wavefront.

In order to demonstrate the Huygens' principle, we consider the propagation of a spherical wavefront (Fig. 1.2a) or plane wavefront (Fig. 1.2b) in an isotropic (uniform) medium (for example, ether) emerging from a source of light S. At any time, suppose PQ is a section of the primary wavefront drawn from the source S. To find the position of the wavefront after an interval t, we take points 1, 2, 3, ... on the primary wavefront PQ. As per Huygens' principle, these points act as the source of secondary wavelets. Taking each point as the centre, we draw spheres of radii ct, where c is the speed of light. These spherical surfaces represent the position of secondary wavelets at time t. Further, we draw a surface P_1Q_1 that touches tangentially all these secondary wavelets in the forward direction. This surface P_1Q_1 is the secondary wavefront. Another surface P_2Q_2 in the backward direction is not called the secondary wavefront as there is no backward flow of the energy during the propagation of the light waves.

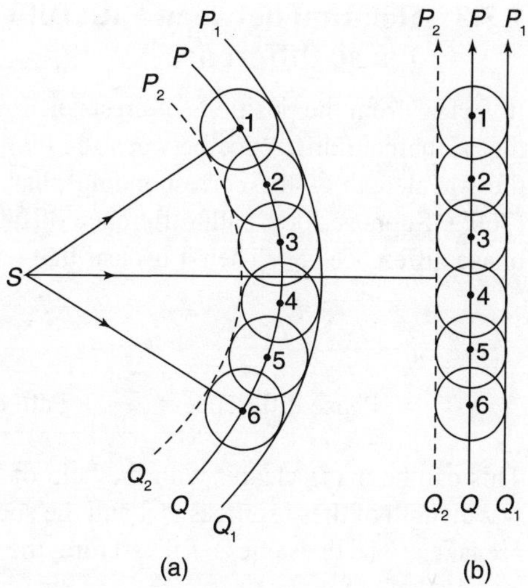

FIGURE 1.2

1.3 PHASE DIFFERENCE AND PATH DIFFERENCE — LO3

As mentioned, the interference pattern is obtained when the two or more waves superimpose each other. In order to understand this pattern it is very important to know about the path and phase differences between the interfering waves.

1.3.1 Phase Difference

Two waves that have the same frequencies and different phases are known to have a phase difference and are said to be out of phase, with each other. If the phase difference is 180°, then the two waves are said to be in antiphase and if it is 0°, then they are in phase as shown in Fig. 1.3(a and b). If the two interfering waves meet at a point where they are in antiphase, then the destructive interference occurs. However, if these two waves meet at a point where they are in the same phase, then the constructive interference takes place.

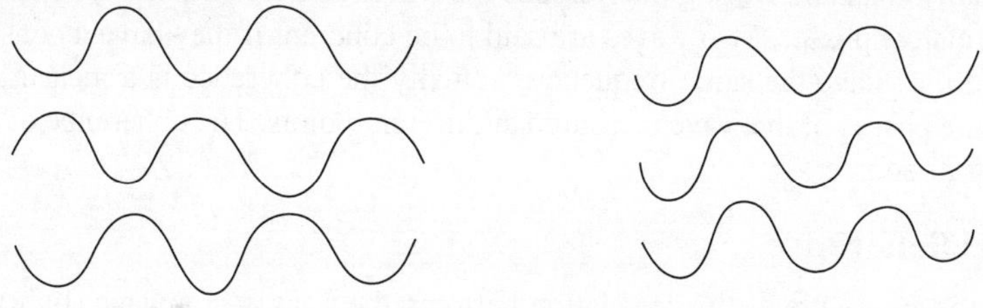

(a) (b)

FIGURE 1.3

1.3.2 Path Difference

In Fig. 1.4, while the two wave crests are traveling a different distance from their sources, they meet at a point P in such a way that a crest meets a crest. For this particular location on the pattern, the difference in distance traveled is known as **path difference**.

1.3.3 Relation between Path Difference and Phase Difference

It is clear from the positions of crests or troughs of the waves that if the path difference between the two waves is equal to the wavelength λ, the corresponding phase difference is 2π (360°). Suppose for a path difference of δ the corresponding phase difference is ϕ. Then it is clear that

$$\phi = \frac{2\pi}{\lambda}\delta$$

$$\text{Phase difference} = \frac{2\pi}{\lambda} \times \text{Path difference} \qquad \text{(i)}$$

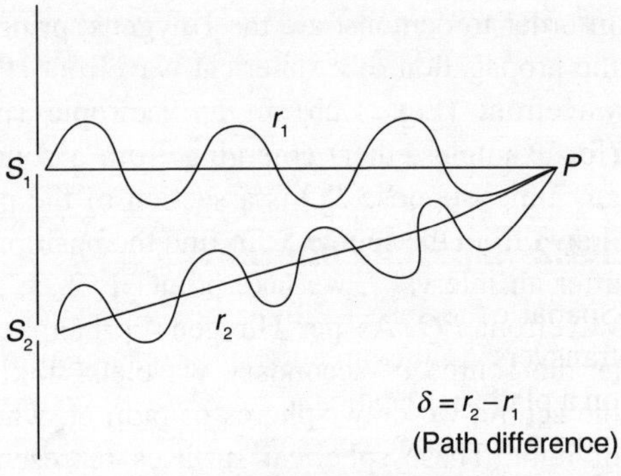

FIGURE 1.4

This can be made clearer with the help of Fig. 1.4, where two sources of waves S_1 and S_2 are shown. The wavelength of these sources is λ and the sources are in phase at S_1 and S_2. The frequencies of both the waves are taken to be the same as f. Therefore, the angular frequency $\omega = 2\pi f$. They travel at the same speed and the propagation constant for them is $k = \frac{2\pi}{\lambda}$. We can write the wave equations for both the waves at point P as

$y_1 = a\cos(\omega t - kr_1)$ for the wave emerging from source S_1 and

$y_2 = a\cos(\omega t - kr_2)$ for the wave emerging from source S_2

Here $(\omega t - kr_1)$ is the phase ϕ_1 and $(\omega t - kr_2)$ is the phase ϕ_2. Therefore, the phase difference between them is $\phi_1 - \phi_2$, given by $\phi_1 - \phi_2 = \omega t - kr_1 - \omega t + kr_2 = k(r_2 - r_1)$.

Using Eq. (i) and $k = \frac{2\pi}{\lambda}$, the path difference is obtained as

Path difference $\delta = r_2 - r_1$.

1.4 COHERENCE — LO4

Coherence is a property of waves that helps in getting stationary interference, i.e., the interference which is temporally and spatially constant. During interference the waves add constructively or subtract destructively, depending on their relative phases. Two waves are said to be coherent if they have a constant relative phase. This also means that they have the same frequency. Actually the coherence is a measure of the correlation that exists between the phases of the wave measured at different points. The coherence of a wave depends on the characteristics of its source.

1.4.1 Temporal Coherence

Temporal coherence is a measure of the correlation between the phases of a wave (light) at different points along the direction of wave propagation. If the phase difference of the wave crossing the two points lying along the direction of wave propagation is independent of time, then the wave is said to have temporal coherence. Temporal coherence is also known as **longitudinal coherence**. This tells us how monochromatic a source is. In Fig. 1.5A, a wave traveling along the positive x-direction is shown, where two points A and B are lying on the x-axis. Let the phases of the wave at these points at any instant t be ϕ_A and ϕ_B, respectively, and at a later time t' they be ϕ'_A and ϕ'_B. Under this situation, if the phase difference $\phi_B - \phi_A = \phi'_B - \phi'_A$, then the wave is said to have temporal coherence.

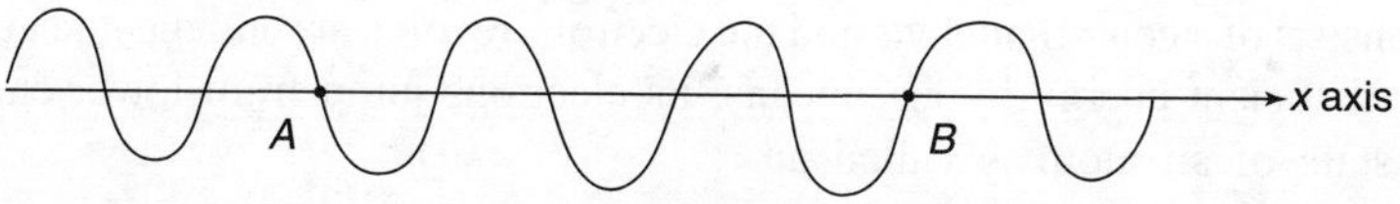

FIGURE 1.5A

1.4.2 Spatial Coherence

Spatial coherence is a measure of the correlation between the phases of a wave (light) at different points transverse to the direction of propagation. If the phase difference of the waves crossing the two points lying on a plane perpendicular to the direction of wave propagation is independent of time, then the wave is said to have spatial coherence. This tells us how uniform the phase of the wavefront is. In Fig. 1.5B, a wave traveling along the positive x-direction is shown, where $PQRS$ is a transverse plane and A and B are the two points situated on this plane within the waveforms. Let the waves crossing these points at any time t have the same phase ϕ and at a later time t' the phases of the waves are again the same but equal to ϕ'. Under this situation, the waves are said to have spatial coherence.

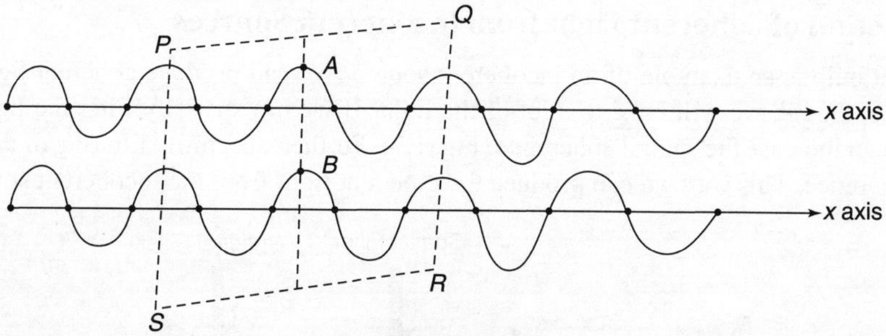

FIGURE 1.5B

1.4.3 Coherence Time and Coherence Length

A monochromatic source of light emits radiation of a single frequency (or wavelength). In practice, however, even the best source of light emits radiations with a finite range of wavelengths. For a single frequency wave, the time interval over which the phase remains constant is called the **coherence time**. The coherence time is generally represented by Δt. In a monochromatic sinusoidal wave the coherence time is infinity because the phase remains constant throughout. However, practically the coherence time exists and the distance traveled by the light pulses during this coherence time is known as coherence length ΔL. The coherence length is also called the **spatial interval**, which is the length over which the phase of the wave remains constant. The coherence length and coherence time are related to each other according to the following formula

$$\Delta L = c\Delta t$$

1.5 COHERENT SOURCES LO4

Two sources of light are said to be coherent, if they emit waves of the same frequency (or wavelength), nearly the same amplitude and maintain a constant phase difference between them. *Laser* is a good example of coherent source. In actual practice, it is not possible to have two independent sources which are coherent. This can be explained as follows. A source of light consists of large number of atoms. According to the atomic theory, each atom consists of a central nucleus and the electrons revolve around the nucleus in different orbits. When an atom gets sufficient energy by any means, its electrons jump from lower energy level to higher energy level. This state of an atom is called an **excited state**. The electron lives in this state only for about 10^{-8} seconds. After this interval of time the electrons fall back to the inner orbits. During this process, the atoms radiate energy in the form of light. Out of the large number of atoms some of them emit light at any instant of time and at the next instant other atoms do so and so on. This results in the emission of light waves with different phases. So, it is obvious that it is difficult to get coherent light from different parts of the same source (Fig. 1.6). Therefore, two independent sources of light can never act as coherent sources.

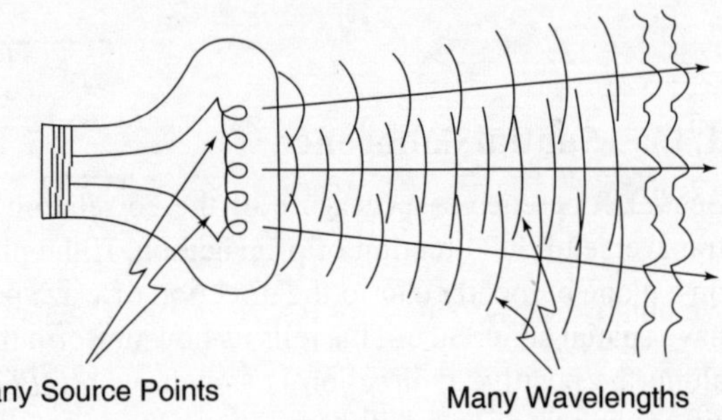

FIGURE 1.6

1.5.1 Production of Coherent Light from Incoherent Sources

An ordinary light bulb is an example of an incoherent source. We can produce coherent light from such an incoherent source, though we will have to a lot of the light. If we use spatially filter the light coming from such source, we can increase the spatial coherence (Fig. 1.7). Further, spectrally filtering of the light increases the temporal coherence. This way we can produce the coherent light from the incoherent source.

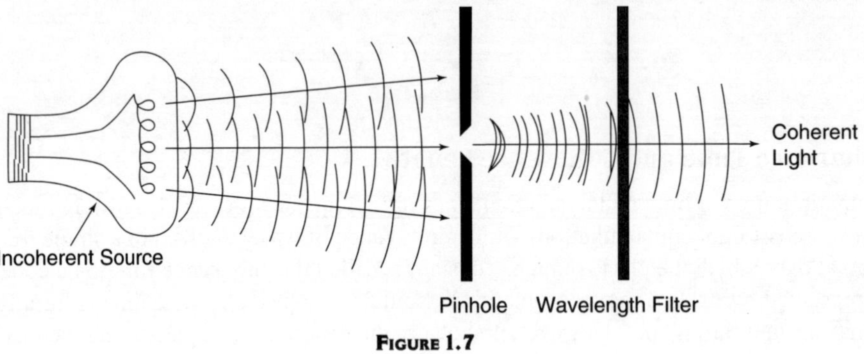

FIGURE 1.7

1.6 ANALYTICAL TREATMENT OF INTERFERENCE

LO5

Let us consider the superposition of two waves of same frequency ω and a constant phase difference ϕ traveling in the same direction. Their amplitudes are taken as a_1 and a_2, respectively. The displacement due to one wave at any instant is given by

$$\psi_1 = a_1 \sin \omega t \tag{i}$$

and the displacement due to another wave at the same instant is given by

$$\psi_2 = a_2 \sin (\omega t + \phi) \tag{ii}$$

According to the principle of superposition, the resultant displacement (ψ_R) is given by

$$\psi_R = \psi_1 + \psi_2 \tag{iii}$$

$$= a_1 \sin \omega t + a_2 \sin (\omega t + \phi)$$

$$= a_1 \sin \omega t + a_2 \sin \omega t \cos \phi + a_2 \cos \omega t \sin \phi$$

$$= (a_1 + a_2 \cos \phi) \sin \omega t + a_2 \sin \phi \cos \omega t \tag{iv}$$

Assuming $\quad a_1 + a_2 \cos \phi = A \cos \theta \quad$ (v)

$$a_2 \sin \phi = A \sin \theta \tag{vi}$$

We obtain using Eq. (iv) – (vi)

$$\psi_R = A \sin (\omega t + \theta) \tag{vii}$$

On squaring and adding Eqs. (v) and (vi), we have

$$A^2 (\sin^2\theta + \cos^2\theta) = a_2^2 \sin^2\phi + a_1^2 + 2a_1a_2 \cos \phi + a_2^2 \cos^2\phi$$

$$A^2 = a_1^2 + a_2^2 (\sin^2\phi + \cos^2\phi) + 2a_1a_2 \cos \phi \tag{viii}$$

The resultant intensity is therefore given by

$$I = A^2 = a_1^2 + a_2^2 + 2a_1a_2 \cos \phi \tag{ix}$$

The angle θ can be calculated from Eqs. (v) and (vi) as

$$\tan \theta = \frac{a_2 \sin \phi}{a_1 + a_2 \cos \phi} \tag{x}$$

1.6.1 Condition for Constructive Interference

It is clear from Eq. (ix) that the intensity, I will be maximum at points where the values of $\cos \phi$ are +1, i.e, phase difference ϕ be $2n\pi$, with $n = 0, 1, 2, 3, ...$ Then the maximum intensity is obtained from Eq. (ix) as

$$I_{max} = (a_1 + a_2)^2 \tag{xi}$$

In other words, the intensity will be maximum when the phase difference is an integral multiple of 2π. In this case,

$$I_{max} > a_1^2 + a_2^2$$

Thus, the resultant intensity will be greater than the sum of the individual intensities of the waves.

If $a_1 = a_2 = a$, then

$$I_{max} = 4a^2$$

1.6.2 Condition for Destructive Interference

It is clear from Eq. (ix) that the intensity I will be minimum at points where $\cos \phi = -1$. i.e., where phase difference $\phi = (2n + 1)\pi$, with $n = 0, 1, 2, 3,....$ Then Eq. (ix) gives

$$I_{min} = (a_1 - a_2)^2 \tag{xii}$$

Therefore, it is clear that in destructive interference the intensity will be minimum when the phase difference ϕ is an odd multiple of π.

If $a_1 = a_2$, then $I_{min} = 0$

If $a_1 \neq a_2$, then $I_{min} \neq 0$

$$I_{min} < a_1^2 + a_2^2$$

Thus, in the case of destructive interference the resultant intensity will be less than the sum of the individual intensities of the waves.

Figure 1.8 represents the intensity variation with phase differences ϕ graphically (for $a_1 = a_2 = a$).

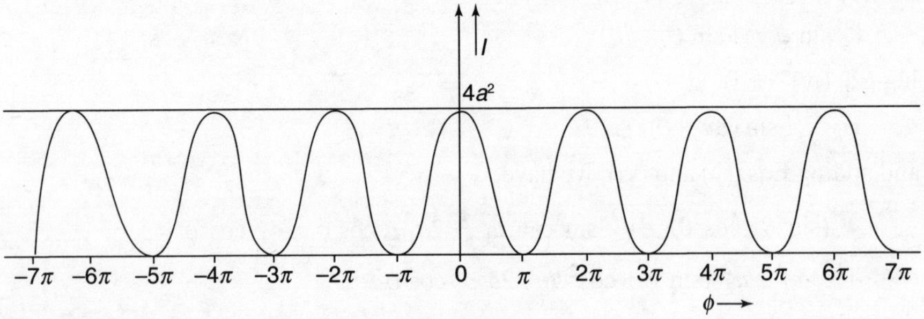

FIGURE 1.8

1.6.3 Conservation of Energy

The resultant intensity due to the interference of two waves a_1 sin ωt and a_2 sin $(\omega t + \phi)$ is given by Eq. (ix), reproduced below

$$I = a_1^2 + a_2^2 + 2a_1a_2 \cos \phi$$

$\therefore$ $$I_{max} = a_1^2 + a_2^2 + 2a_1a_2 = (a_1 + a_2)^2$$

and $$I_{min} = a_1^2 + a_2^2 - 2a_1a_2 = (a_1 - a_2)^2$$

If $a_1 = a_2 = a$ then

$$I_{max} = 4a^2 \text{ and } I_{min} = 0$$

Therefore, average intensity (I_{av}) will be obtained as

$$I_{av} = 2a^2$$

For unequal amplitudes a_1 and a_2 the average intensity would be $(a_1^2 + a_2^2)$. Thus, in interference only some part of energy is transferred from the position of minima to the position of maxima, and the average intensity or energy remains constant. This shows that the phenomenon of interference is in accordance with the law of conservation of energy.

1.7 CONDITIONS FOR SUSTAINED INTERFERENCE LO5

Sustained interference means a constant interference of light waves. In order to obtain such interference, the following conditions must be satisfied

(i) The two sources should emit waves of the same frequency (wavelength). If it is not so, then the positions of maxima and minima will change with time.

(ii) The waves from the two sources should propagate along the same direction with equal speeds.

(iii) The phase difference between the two interfering waves should be zero or it should remain constant. It means the sources emitting these waves must be coherent.

(iv) The two coherent sources should be very close to each other, otherwise the interference fringes will be very close to each other due to the large path difference between the interfering waves. For the large separation of the sources, the fringes may even overlap and the maxima and minima will not appear distinctly.

(v) A reasonable distance between the sources and screen should be kept, as the maxima and minima appear quite close if this distance is smaller. On the other hand, the large distance of the screen reduces the intensity.

(vi) In order to obtain distinct and clear maxima and minima, the amplitudes of the two interfering waves must be equal or nearly equal.

(vii) If the source is not narrow, it may act as a multi source. This will lead to a number of interference patterns. Therefore, the coherent sources must be narrow.

(viii) In order to obtain the pattern with constant fringe width and good intensity fringes, the sources should be monochromatic and the background should be dark.

1.7.1 Condition of Relative Phase Shift

This is regarding the introduction of additional phase change between the interfering waves when they emerge after reflecting from two different surfaces. In most of the situations, the reflection takes place when the beam propagates from the medium of lower refractive index to the medium of higher refractive index or vice-versa. When the reflection occurs with light going from a lower index toward a higher index, the condition is called **internal reflection**. However, when the reflection occurs for light going from a higher index toward a lower index, the condition is referred to as **external reflection**. A relative phase shift of π takes place between the externally and internally reflected beams so that an additional path difference of $\lambda/2$ is introduced between the two beams. If both the interfering beams get either internally or externally reflected, no phase shift takes place between them.

1.8 MULTIPLE BEAM SUPERPOSITION LO6

In Section 1.6, we have given theoretical analysis of the interference due to the superposition of two waves of the same frequency and the constant phase difference. The intensity of the interference pattern showed its dependence on the amplitudes of the interfering waves. However, now we consider a large number of waves of the same frequency and amplitude, which propagate in the same direction. The amount by which each wave train is ahead or lags behind the other is a matter of chance. Based on the amplitude and intensity of the resultant wave, we can examine the interference. We assume n number of wave trains whose individual amplitudes are equal ($= a$, say). The amplitude of the resultant wave can be understood as the amplitude of motion of a particle undergoing n simple harmonic motions (each of amplitude a) at once. In this case, if all these motions are in the same phase, the resultant wave will have an amplitude equal to na and the intensity would be n^2a^2, i.e., n^2 times that of one wave. However, in our case, the phases are distributed purely at random, as shown in Fig. 1.9 as per graphical method of compounding amplitudes. Here, the phases ϕ_1, ϕ_2, ϕ_3, ... take arbitrary values between 0 and 2π. The intensity due to the superposition of such waves can be calculated by the square of the resultant amplitude A. In order to find A^2, we should square the sum of the projections of all vectors a along the x-direction and add it to the square of the corresponding sum along the y-direction. The summation of projections along x-direction are given by the following expression

$$a(\cos\phi_1 + \cos\phi_2 + \cos\phi_3 + \ldots + \cos\phi_n)$$

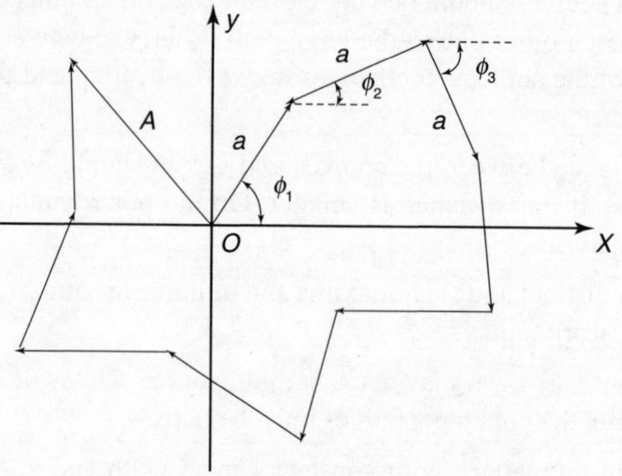

FIGURE 1.9

The square of quantity in the parentheses gives the terms of the form $\cos^2\phi_1$, $2\cos\phi_1\cos\phi_2$, etc. It is seen that the sum of these cross product terms increases approximately in proportion to number n. So we do not obtain a definite result with one given array of arbitrarily distributed waves. For a large number of such arrays, we find their average effect in computing the intensity in any physical problem. Under this situation, it is safe to conclude that these cross product terms will average to zero. So we consider only the $\cos^2\phi$ terms. Similarly, for the y projections of the vectors we obtain $\sin^2\phi$ terms. With this we have

$$I \approx A^2 = a^2(\cos^2\phi_1 + \cos^2\phi_2 + \cos^2\phi_3 + \ldots + \cos^2\phi_n) + a^2(\sin^2\phi_1 + \sin^2\phi_2 + \sin^2\phi_3 + \ldots + \sin^2\phi_n).$$

Using the identity $\sin^2\phi_p + \cos^2\phi_p = 1$, the above expression reduces to $I \approx a^2 \times n$.

Since a^2 is the intensity due to a single wave, the above relation shows that the average intensity resulting from the superposition of n waves with arbitrary phases is n times of a single wave. It means the resultant amplitude A increases in proportion with in length as n gets increased.

1.9 INTERFERENCE BY DIVISION OF WAVEFRONT LO6

This method uses multiple slits, lenses, prisms or mirrors for dividing a single wavefront laterally to form two smaller segments that can interfere with each other. In the division of a wavefront, the interfering beams of radiation that left the source in different directions and some optical means is used to bring the beams back together. This method is useful with small sources. Double slit experiment is an excellent example of interference by division of wavefront. Fresnel's biprism is also used for getting interference pattern based on this method.

1.9.1 Fresnel's Biprism

Fresnel's Biprism is a device by which we can obtain two virtual coherent sources of light to produce sustained interference. It is the combination of two acute angled prisms which are joined with their bases in such a way that one angle becomes obtuse angle θ' of about 179° and remaining two angles are acute angles each of about 1/2°, as shown in Fig. 1.10.

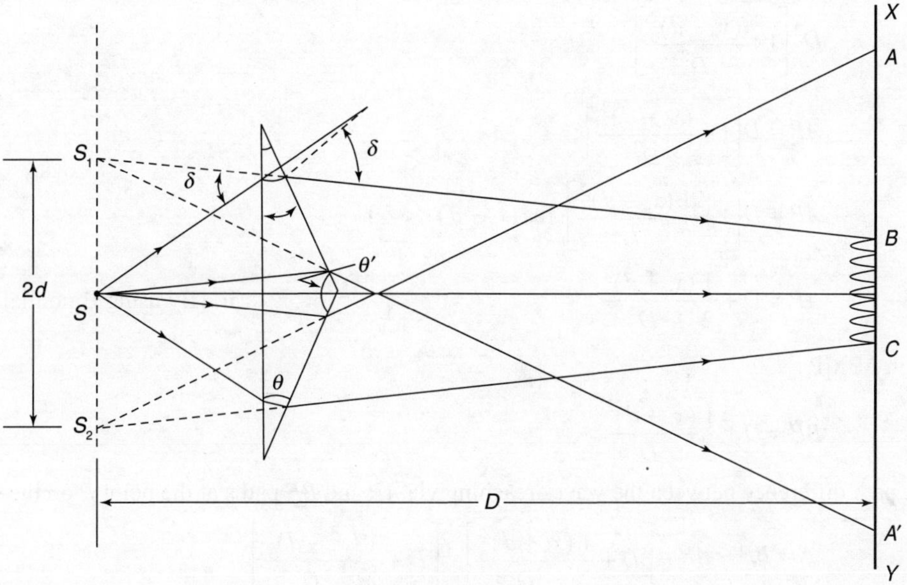

FIGURE 1.10

Let monochromatic light from slit S fall on the biprism, placed at a small distance from S. When the light falls on upper part of the biprism, it bends downward and appears to come from source S_1. Similarly, the other part of the light when falls on the lower part of the biprism, bends upward and appears to come from source S_2. Here, the images S_1 and S_2 act as two virtual coherent sources of light (Fig. 1.10). Coherent sources are the one that have a constant or zero phase difference throughout. In the situation, on placing the screen XY on right side of the biprism, we obtain an alternate bright and dark fringes in the overlapping region BC.

1.9.1.1 Theory of Fringes

Let A and B be two virtual coherent sources of light separated by a distance $2d$. The screen XY, on which the fringes are obtained, is separated by a distance D from the two coherent sources, as shown in Fig. 1.11. The point C on the screen is equidistant from A and B. Therefore, the path difference between the two waves from sources A and B at point C is zero. Thus the point C will be the centre of a bright fringe. On both sides of C, alternately bright and dark fringes are produced.

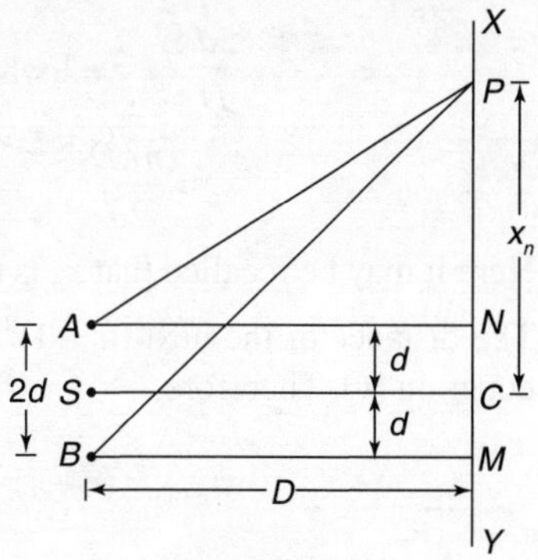

FIGURE 1.11

Draw perpendiculars AN and BM from A and B on the screen. Let the distance of a point P on the screen from the central bright fringe at C be x_n.

From geometry, we have

$$NP = x_n - d; \qquad MP = x_n + d$$

In right angled ΔANP,

$$AP^2 = AN^2 + NP^2 \tag{i}$$

$$= D^2 + (x_n - d)^2$$

$$D^2\left[1+\frac{(x_n-d)^2}{D^2}\right]$$

$$AP = D\left[1+\frac{(x_n-d)^2}{D^2}\right]^{1/2}$$

$$AP = D\left[1+\frac{1}{2}\frac{(x_n-d)^2}{D^2}\right], \text{[as } (x_n-d) << D]$$

$$AP = D+\frac{1}{2}\frac{(x_n+d)^2}{D} \qquad \text{(ii) [By using Binomial Theorem]}$$

Similarly, in ΔBMP,

$$BP = D+\frac{1}{2}\frac{(x_n+d)^2}{D} \qquad \text{(iii)}$$

Hence, the path difference between the waves reaching via *AP* and *BP* paths at the point *P* on the screen

$$\Delta = BP - AP = \left[D+\frac{1}{2}\frac{(x_n+d)^2}{D}\right]-\left[D+\frac{1}{2}\frac{(x_n-d)^2}{D}\right]$$

$$=\frac{4x_nd}{2D}$$

$$\Delta = \frac{2d}{D}x_n \qquad \text{(iv)}$$

Condition for Bright Fringes: In order to interfere constructively and produce bright fringes, the two rays should arrive at points *P* in phase. This is possible if the path difference is an integral multiple of λ. Therefore,

$$\Delta = n\lambda$$

$$\frac{2d}{D}x_n = n\lambda \text{ where } n = 0, 1, 2 \ldots\ldots$$

$$x_n = \frac{n\lambda D}{2d} \qquad \text{(v)}$$

Here it may be recalled that x_n is the distance of the nth order bright fringe from the central maxima.

The distance of the next $(n + 1)^{\text{th}}$ maximum from the point *C* can be calculated by replacing *n* by $n + 1$ in equation (v). Therefore,

$$x_{(n+1)} = (n+1)\frac{\lambda D}{2d}$$

The separation between two consecutive maxima gives the fringe width β, as follows

$$\beta = x_{n+1} - x_n$$

or fringe width

$$\beta = \frac{\lambda D}{2d} \qquad \text{(vi)}$$

Condition for Dark Fringes: In order to interfere destructively and produce dark fringe at point P, the two rays should arrive at this point in out of phase (phase difference of π). This is possible, if the path difference is an odd multiple of $\frac{\lambda}{2}$. Therefore,

$$\Delta = \left(n + \frac{1}{2}\right)\lambda \text{, where } n = 0, 1, 2, \ldots$$

From Eq. (iv)

$$\Delta = \frac{2d}{D} x_n = (2n+1)\frac{\lambda}{2} \tag{vii}$$

$$x_n = \frac{(2n+1)\lambda D}{4d} \tag{viii}$$

Equation (viii) gives the distance of nth order dark fringe from the point C. The distance of the next $(n+1)^{\text{th}}$ minimum from the point C will be

$$x_{(n+1)} = \frac{[2(n+1)+1]\lambda D}{4d}$$

$$= \frac{(2n+3)\lambda D}{4d} \tag{ix}$$

Hence, the fringe width between two consecutive minima would be

$$\beta = x_{(n+1)} - x_n = \frac{(2n+3)\lambda D}{4d} - \frac{(2n+1)\lambda D}{4d}$$

$$\beta = \frac{\lambda D}{2d} \tag{x}$$

It is clear from Eqs. (vi) and (x) that the bright and dark fringes are of equal width.

1.9.1.2 Experimental Method for Determination of Wavelength of Light

The experimental setup used for the determination of wavelength of light consists of a good quality heavy optical bench of about 1.5 meter length fitted with scale. It has four uprights that carry an adjustable slit S, a biprism, a convex lens and a micrometer eyepiece, respectively. These components are shown in Fig. 1.12. Each upright can be moved along the length of the optical bench and screws are provided to rotate the slit and biprism in their own planes and the eyepiece can also move at right angle to the length of the optical bench.

To obtain well defined and sharp interference fringes, the following adjustments are necessary:

(i) Labeled optical bench by using spirit level and leveling screws.

(ii) Adjust all uprights to the same height.

(iii) Illuminate the vertical slit by monochromatic source of light. Make the slit narrow.

(iv) Now place the biprism on the second upright and try to adjust its edge parallel to the slit until two equally bright virtual sources A and B are observed.

(v) Shift the micrometer eyepiece on the bench away from the slit and also move it at right angle to the length of optical bench until the fringes are observed in the field of view.

(vi) In order to get fine fringes, change the position of the biprism slowly in its own plane such that its edge remains parallel to the slit.

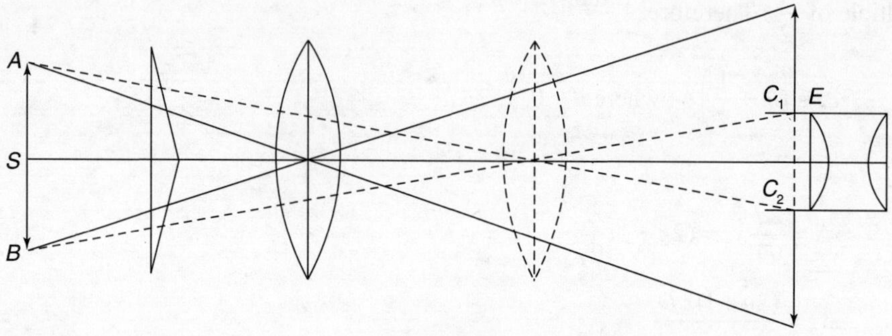

FIGURE 1.12

Lateral shift and its removal: On moving the micrometer eye piece on the bench towards the biprism, if the fringes appear to shift at right angle to the optical bench then it is known as **lateral shift** (Fig. 1.13(a)). However, if the principle axis and axis of optical bench become parallel, then no lateral shift remains, as shown in Fig. 1.13(b).

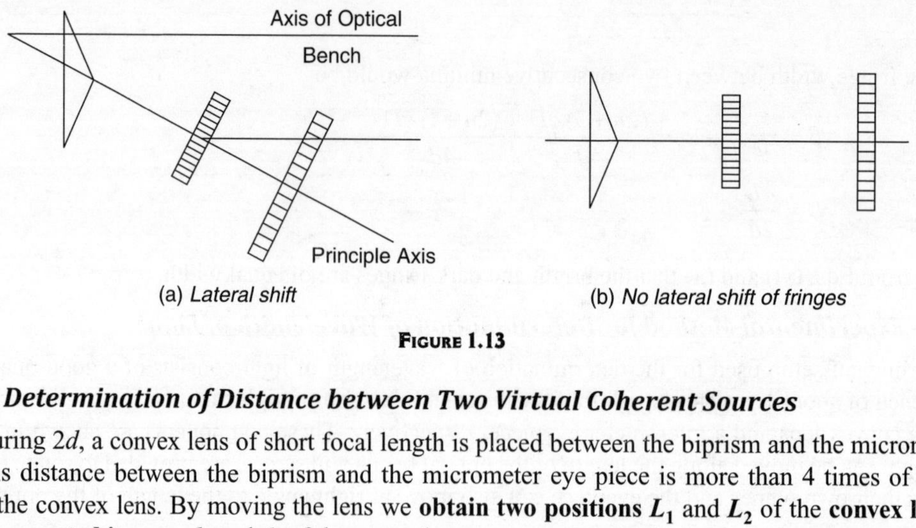

(a) *Lateral shift* (b) *No lateral shift of fringes*

FIGURE 1.13

1.9.1.3 Determination of Distance between Two Virtual Coherent Sources

For measuring $2d$, a convex lens of short focal length is placed between the biprism and the micrometer eye piece. This distance between the biprism and the micrometer eye piece is more than 4 times of the focal length of the convex lens. By moving the lens we **obtain two positions L_1 and L_2 of the convex lens such that two separated images d_1 and d_2 of the two coherent sources respectively can be observed,** as shown in Fig. 1.14.

For the first position of lens, L_1, the magnification is given as

$$\frac{v}{u} = \frac{d_1}{2d} \tag{i}$$

and for second position of the lens, the magnification is

$$\frac{u}{v} = \frac{d_2}{2d} \tag{ii}$$

Then from Eqs. (i) and (ii), we get

$$\frac{v}{u}\times\frac{u}{v}=\frac{d_1d_2}{(2d)^2} \text{ or } (2d)^2 = d_1d_2$$

or $$2d=\sqrt{d_1d_2} \quad \text{(iii)}$$

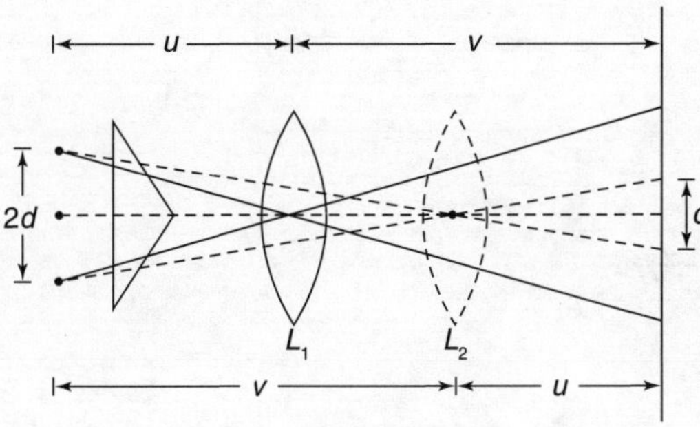

FIGURE 1.14

Therefore, the measurement of positions of images d_1 and d_2 will determine the distance $2d$ between the sources. The wavelength λ of monochromatic light can be calculated when we substitute the values of β, D and $2d$ in the formula $\lambda = \beta(2d/D)$, derived in the previous section.

1.9.1.4 Determination of Thickness of Thin Transparent Sheet (Displacement of Fringes)

Let A and B be two virtual coherent sources of light. The point C_0 on the screen is equidistant from both the sources (Fig. 1.15). When a transparent material plate G of thickness t and having refractive index μ, it is placed in the path of one of the light wave, we observe that the fringe which was originally at C_0 shifts to another position P, as shown in Fig. 1.15.

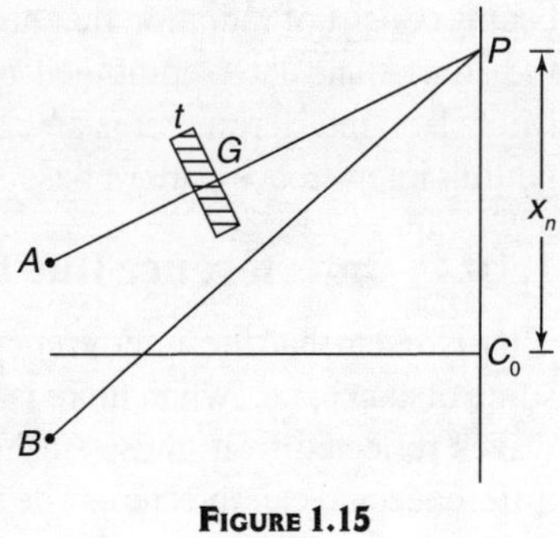

FIGURE 1.15

The time taken by the light wave from A to P partly through air and partly through the plate is the same as the time taken by the other light wave from B to P in air. If c and v be the velocity of light in air and in the plate, respectively, then

$$\frac{BP}{c}=\frac{AP-t}{c}+\frac{t}{v}$$

or $$\frac{BP}{c}=\frac{AP-t}{c}+\frac{\mu t}{c} \qquad \left[\because \mu=\frac{c}{v}\right]$$

or $$BP=(AP-t)+\mu t$$

or $$BP-AP=(\mu-1)t \quad \text{(i)}$$

Here $BP - AP$ is the path difference between the two interfering waves.

If the point P is originally occupied by the nth order bright fringe, then the path difference between the two interfering waves will be

$$BP-AP=n\lambda,$$

$$(\mu-1)\,t=n\lambda \quad \text{(ii)}$$

The distance x_n through which the fringe is shifted to point P from the central maximum C_0 is given by

$$x_n = \frac{n\lambda D}{2d} \quad \text{(iii)}$$

where, $\frac{\lambda D}{2d} = \beta =$ fringe width.

From Eq. (iii), we get

$$\frac{x_n \cdot 2d}{D} = n\lambda \quad \text{(iv)}$$

From Eqs. (ii) and (iv), we get

$$(\mu - 1)t = \frac{x_n \cdot 2d}{D}$$

or

$$t = \frac{x_n \cdot 2d}{(\mu - 1)D} \quad \text{(v)}$$

Therefore, by knowing x_n, $2d$, D and μ, we can calculate thickness t of the glass plate by using Eq. (v).

1.10 INTERFERENCE BY DIVISION OF AMPLITUDE LO6

The method, which is used to produce two coherent sources from a common source, is called **division of amplitude** that maintains the same width but reduced amplitude. After following different paths the two waves of reduced amplitudes are combined to produce an interference pattern. In this method, the interfering beams consist of radiation that has left the source in the same direction. This radiation is divided after leaving the source and later combined to produce interference. This method can be used with extended sources. Michelson interferometer is an example of interference by division of amplitude. Thin films are also used for getting interference pattern based on this method.

1.10.1 Interference Due to Thin Films

This is clear that the interference takes place when the two waves superimpose each other after traveling some distance, i.e., when there is a path difference between them. Since the thin film has its two surfaces, the waves reflected from these surfaces can attain a path difference and can interfere. The same may be applied on the waves that transmit through the film.

1.10.1.1 Thin Film of Uniform Thickness

Consider a uniform transparent film having thickness t and a refractive index μ. A ray of light AB incident at an angle i on the upper surface of the film is partly reflected along BC and partly refracted along BF at an angle r. At point F the wave BF is again partly reflected from the second surface along FD and partly emerges out along FK and so on. In this situation, the interference occurs between reflected waves BC and DE and also between the transmitted waves FK and GL (Fig. 1.16).

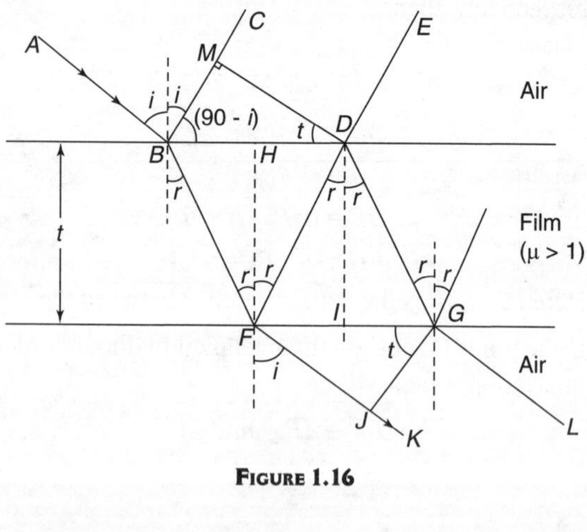

FIGURE 1.16

The path difference between the reflected rays

$$\Delta = (BF + FD)_{\text{in film}} - (BM)_{\text{in air}}$$

$$\Delta = \mu\,(BF + FD) - BM$$

$\because$ $$BF = FD$$

$\therefore$ $$\Delta = 2\,\mu BF - BM \qquad \text{(i)}$$

In the right angled ΔBFH,

$$\cos r = \frac{t}{BF} \text{ or } BF = \frac{t}{\cos r} \qquad \text{(ii)}$$

and $$\tan r = \frac{BH}{t} \text{ or } BH = t \tan r$$

$$BD = 2 \times BH$$

$\therefore$ $$BD = 2t \tan r \qquad \text{(iii)}$$

In the ΔBMD,

$$\sin i = \frac{BM}{BD} \text{ or } BM = BD \sin i$$

$\therefore$ $$BM = 2t \tan r \sin i \qquad \text{(iv)}$$

From Eqs. (i), (ii) and (iv), we get

$$\Delta = 2\mu\frac{t}{\cos r} - 2t \tan r \sin i \qquad \text{(v)}$$

$\because$ $$\mu = \frac{\sin i}{\sin r} \text{ or } \sin i = \mu \sin r \qquad \text{(vi)}$$

$\therefore$ $$\Delta = \frac{2\mu t}{\cos r} - 2t\frac{\sin r}{\cos r}\mu \sin r = \frac{2\mu t}{\cos r}[1 - \sin^2 r]$$

$$\Delta = 2\mu t \cos r \qquad \text{(vii)}$$

Equation (vii) represents only the apparent path difference and does not represent the effective total path difference. When the light is reflected from the surface of an optically denser medium in case of rad BC, a phase change of πa equivalent to path difference of $\lambda/2$ is introduced. Therefore, the total path difference between BC and DE will be

$$\Delta = 2\mu t \cos r + \lambda/2 \qquad \text{(viii)}$$

Condition for Maxima: To have a maximum at a particular point, the two rays should arrive there in phase. So the path difference must contain a whole number of wavelength, i.e.,

$$\Delta = n\lambda,\ n = 0, 1, 2..., \qquad \text{(ix)}$$

From Eq. (viii) and (ix), we get

$$2\mu t \cos r + \lambda/2 = n\lambda$$

$$2\mu t \cos r = n\lambda - \lambda/2$$

$$2\mu t \cos r = (2n - 1)\lambda/2 \qquad \text{(x)}$$

Condition for Minima: To have a minimum at a particular point, the two rays should arrive there in out of phase (odd multiple of π) for which the path difference must contain a half odd integral number of wavelength, i.e,

$$\Delta = \left(n + \frac{1}{2}\right)\lambda \qquad \text{(xi)}$$

Using Eq. (viii), we obtain

$$2\mu t \cos r = n\lambda \text{ where, } n = 0, 1, 2, 3, \ldots \qquad \text{(xii)}$$

It should be noted that the interference pattern will not be perfect because the intensities of the rays BC and DE are not the same and their amplitudes are different.

In order to obtain the interference between the transmitted waves, we calculate the path difference between the waves, FK and GL as under

$$\Delta = (FD + DG)_{\text{in film}} - (FJ)_{\text{in air}}$$

$$\Delta = \mu[FD + DG] - FJ$$

$\because$ $$FD = DG$$

$\therefore$ $$\Delta = 2\mu FD - FJ \qquad \text{(xiii)}$$

In ΔFDI, $$\cos r = \frac{DI}{FD} = \frac{t}{FD} \text{ or } FD = \frac{t}{\cos r} \qquad \text{(xiv)}$$

and $$\tan r = \frac{FI}{DI} = \frac{FI}{t} \text{ or } FI = t \tan r$$

$$FG = 2t \tan r \qquad \text{(xv)}$$

In right angled ΔFJG,

$$\sin i = \frac{FJ}{FG} \text{ or } FJ = FG \sin i$$

$\therefore$ $$FJ = 2t \tan r \sin i \qquad \text{(xvi)}$$

From Eq. (xiii), (xiv) and (xvi), we get

$$\Delta = \frac{2\mu t}{\cos r} - 2t \tan r \sin i$$

$$= \frac{2\mu t}{\cos r} - 2t\frac{\sin r}{\cos r}\mu \sin r \qquad \left[\mu = \frac{\sin i}{\sin r}\right]$$

$$= \frac{2\mu t}{\cos r}[1 - \sin^2 r] = 2\mu t \cos r$$

Since these two waves are emerging from the same medium, the additional phase difference (or path difference) will not be introduced. Therefore, the total path difference

$$\Delta = 2\mu t \cos r \qquad \text{(xvii)}$$

Condition for Maxima: As discussed, it is possible when

$$\Delta = n\lambda \qquad \text{(xviii)}$$

From Eqs. (xvii) and (xviii), we get

$$2\mu t \cos r = n\lambda \text{ where, } n = 0, 1, 2, 3, \ldots \qquad \text{(xix)}$$

Condition for Minima: For obtaining minimum intensity, we should have

$$\Delta = \left(n + \frac{1}{2}\right)\lambda$$

which gives $$2\mu t \cos r = \left(n + \frac{1}{2}\right)\lambda \text{ where, } n = 0, 1, 2, 3, \qquad \text{...(xx)}$$

Thus, the conditions for interference with transmitted light are obviously opposite to those obtained with reflected light. Hence, if the film appears dark in the reflected light, it will appear bright in the transmitted light and vice-versa. This shows that the interference pattern in the reflected and transmitted lights are complimentary to each other.

(i) Necessity of an Extended Source of Light for Interference in Thin Films

When a thin transparent film is exposed to white light and seen in the reflected light, different colours are seen in the film. These colours arise due to the interference of the light waves reflected from the top and bottom surfaces of the film. The path difference between the reflected rays depends upon the thickness t, refractive index μ of the film and the angle θ of inclination of the incident rays. The light which comes from any point from the surface of the film will include the colour whose wavelength satisfies the equation $2\mu t \cos r = (2n - 1)\ \lambda/2$ and only this colour will be present with the maximum intensity in the reflected light.

When the transparent film of a large thickness as compared to the wavelength of the light, is illuminated by white light, the path difference at any point of the film will be zero. In the case of such a thick film, at a given point, the condition of constructive interference is satisfied by a large number of wavelengths, as $<< t$. The condition of destructive interference is also satisfied at the same point for the large number of wavelengths. Therefore, consequently that point receives an average intensity due to the light of all wavelengths and no colours are observed.

In the context of realization of above phenomena it is always needed to use a broad power of light that will enable the eye to see whole of the film simultaneously.

If we use a point source, then we observe that different parts of reflected light cannot reach the eye due to small size of the pupil, as shown in Fig. 1.17(a). The reflected rays only from a small portion of the film can enter the eye. Hence, the whole of the film cannot be seen by the eye placed in a fixed position. However, if a

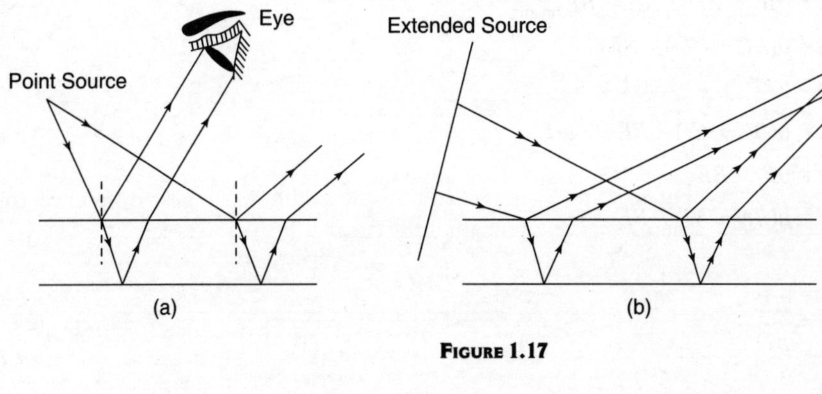

FIGURE 1.17

broad source of light is used to illuminate a thin film, the light reflected from each part of the film reaches the eye placed in a fixed position, as shown in Fig. 1.17(b). Hence, one can seen the entire film simultaneously by employing an extended source of light.

1.10.1.2 Non-uniform Thickness Film (Wedge Shaped Film)

Consider two plane surfaces *OM* and *OM′* inclined at an angle θ enclosing a wedge shaped air film of increasing thickness, as shown in Fig 1.18. A beam of monochromatic light is incident on the upper surface of the film and the interference occurs between the rays reflected at its upper and lower surfaces. The interference occurs between the reflected rays *BK* and *DL*, both of which are obtained from the same incident ray of light *AB*.

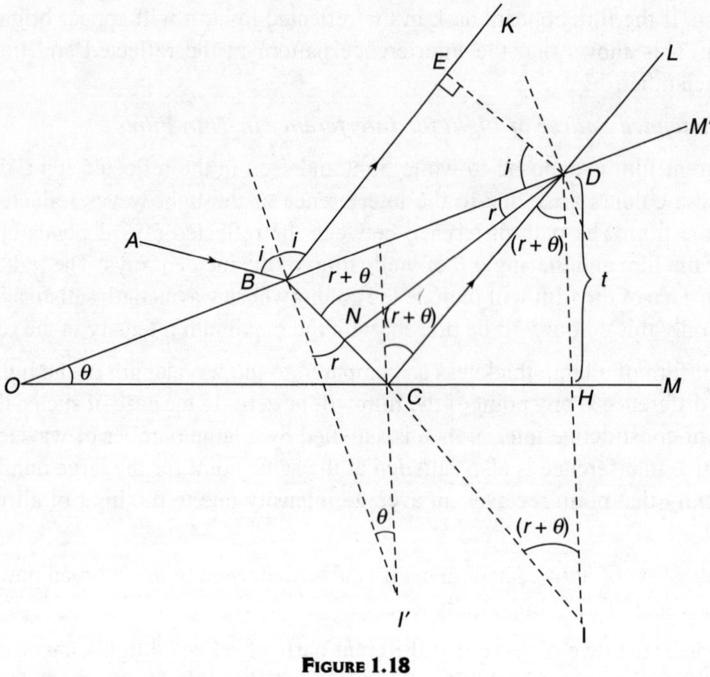

FIGURE 1.18

The path difference between the two reflected rays

$$\Delta = [BC + CD]_{\text{in film}} - [BE]_{\text{in air}}$$

$$\Delta = \mu(BC + CD) - BE$$

$$\because \quad CD = CI$$

$$\Delta = \mu(BC + CI) - BE$$

$$= \mu BI - BE$$

$$= \mu(BN + NI) - BE \tag{i}$$

In right angled ΔBED,

$$\sin i = \frac{BE}{BD} \tag{ii}$$

In right angled ΔBND,

$$\sin r = \frac{BN}{BD} \tag{iii}$$

Dividing Eq. (ii) by Eq. (iii), we get

$$\frac{\sin i}{\sin r} = \frac{BE}{BN} \text{ or } \frac{\sin i}{\sin r} = \mu = \frac{BE}{BN}$$

or
$$BE = \mu BN \tag{iv}$$

From Eqs. (i) and (iv), we get

$$\Delta = \mu(BN + NI) - \mu BN$$

or
$$\Delta = \mu NI \tag{v}$$

In right angled ΔDNI,

$$\cos(r + \theta) = \frac{NI}{DI}$$

$\therefore$
$$DI = DH + HI = t + t = 2t$$

$$\cos(r + \theta) = \frac{NI}{DI} \text{ or } NI = 2t\cos(r + \theta) \tag{vi}$$

From Eqs. (v) and (vi), we get

$$\Delta = 2\mu t\cos(r + \theta) \tag{vii}$$

Equation (vii), in the case of reflected light, does not represent the effective total path difference, as a phase difference of π (Stokes phase change) has been introduced through the reflection of wave BK. Therefore, the total path difference between the reflected rays,

$$\Delta = 2\mu t\cos(r + \theta) + \lambda/2 \tag{viii}$$

Equation (viii) shows that the path difference Δ depends on the thickness t. However, t is not uniform and it is different at different positions.

At $t = 0$, Eq. (viii) reads

$$\Delta = \lambda/2$$

which is the condition for darkness. Therefore, the edge of the film appears to be dark. This is called zero order band.

For normal incidence, $i = 0$ and $r = 0$. Then, the path difference

$$\Delta = 2\mu t\cos\theta + \lambda/2 \tag{ix}$$

Condition for Maxima: As explained earlier, the constructive interference takes place when

$$\Delta = n\lambda \quad \text{where, } n = 0, 1, 2, 3, \ldots \tag{x}$$

From Eqs. (ix) and (x), we get

$$2\mu t\cos\theta + \lambda/2 = n\lambda$$

$$2\mu t\cos\theta = (2n - 1)\lambda/2 \tag{xi}$$

Condition for Minima: In order to get destructive interference, the path difference

$$\Delta = \left(n + \frac{1}{2}\right)\lambda \tag{xii}$$

or $$2\mu t \cos\theta + \lambda/2 = \left(n + \frac{1}{2}\right)\lambda$$

$$2\mu t \cos\theta = n\lambda \text{ where, } n = 0, 1, 2, 3, \ldots \tag{xiii}$$

(i) Nature of Fringes

For normal incidence of the light waves or a parallel incident beam, the incident angle remains constant and hence the angle of refraction. If the light is monochromatic, then λ is also fixed. Therefore, the change in path difference will take place due to μt or thickness t of the film only. As we move outwards from the point of contact O, the thickness of the film increases. However, at a particular place along a line parallel to the edge, t has only one value. Since the locii of the points of constant thickness of the film are straight lines parallel to the edge, straight bright and dark fringes parallel to the edge will be formed in the reflected light. If we use the white light in place of monochromatic light, coloured fringes will be observed.

(ii) Derivation for Fringe Width

For a wedge shaped film the conditions of maxima and minima are reproduced below.

$$2\mu t \cos(r + \theta) = (2n - 1)\lambda/2$$

$$2\mu t \cos(r + \theta) = n\lambda$$

For normal incidence and small values of θ the above conditions read as

$$2\mu t = (2n - 1)\lambda/2 \tag{xiv}$$

and $$2\mu t = n\lambda \tag{xv}$$

If points A and C (Fig. 1.19) represent positions of two consecutive dark fringes corresponding to film thicknesses $AB = t_1$ and $CD = t_2$ respectively, then the fringe width (w) will be equal to BE. Now from Eq. (xv), we get the following condition corresponding to the points A and C.

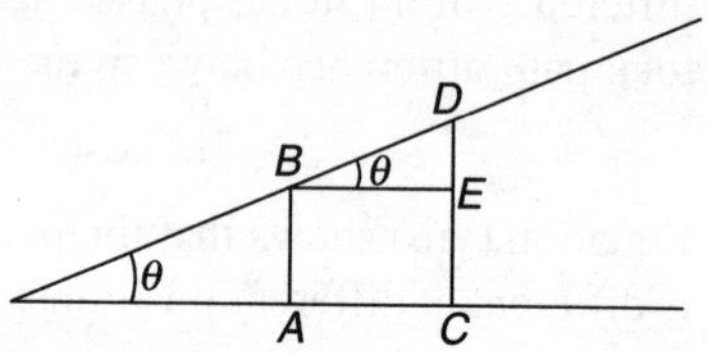

FIGURE 1.19

$$2\mu t_1 = n\lambda \text{ and } 2\mu t_2 = (n + 1)\lambda$$

or $$2\mu(t_2 - t_1) = \lambda$$

or $$2(CD - AB) = \lambda$$

or $$2(DE) = \lambda \tag{xvi}$$

But $$\tan\theta = DE/BE \text{ or } DE = BE\tan\theta \tag{xvii}$$

From Eqs. (xvi) and (xvii), we get

$$2\mu(BE\tan\theta) = \lambda$$

or $$BE = \frac{\lambda}{2\mu\tan\theta} = w$$

For smaller values of θ, $\tan\theta \simeq \theta$ and we get

$$w = \frac{\lambda}{2\mu\theta} \tag{xviii}$$

It is clear from (xviii) that the fringe width w is independent of thickness t for smaller angle θ. Therefore the fringes are equally spaced and of same width for fixed λ, μ and θ.

1.10.2 Newton's Rings

If a plano-convex lens is placed such that its curved surface lies on a glass plate, then an air film of gradually increasing thickness is formed between the two surfaces. When a beam of monochromatic (single wavelength) light is allowed to fall normally on this film and viewed as shown in Fig. 1.20, an alternating dark and bright circular fringes are observed. These circular fringes are formed because of the interference between the reflected waves from the top and the bottom surfaces of the air film. These fringes are circular since the air film has a circular symmetry and the thickness of the film corresponding to each fringe is same throughout the circle. The interference fringes so formed were first investigated by Newton and hence known as **Newton's rings**.

The path difference between the two reflected rays, can be obtained as done in the case of wedge shaped film. It is reproduced below as

$$\Delta = 2\mu t \cos(r + \theta) + \lambda/2 \qquad \text{(i)}$$

Where ($\lambda/2$) is due to **Stokes** phase change.

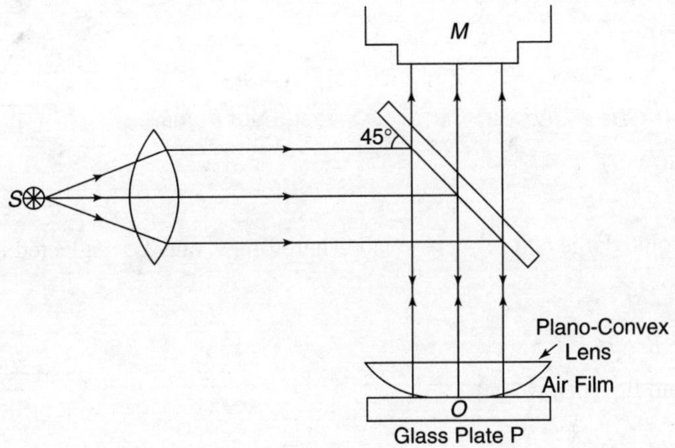

FIGURE 1.20

For normal incidence and an air film, $i = 0$, $r = 0$, $\mu = 1$. In addition, if θ is also very small, then $\cos\theta = 1$. Under this situation, the path differences becomes

$$\Delta = 2t + \frac{\lambda}{2} \qquad \text{(ii)}$$

Here t is the thickness of the air film at a particular point.

At the point of contact, $t = 0$

$$\Delta = \frac{\lambda}{2}$$

which is the condition of minimum intensity and hence, the central spot of the ring will be dark.

Condition for Maxima: For constructive interference

$$\Delta = n\lambda \quad \text{(iii)}$$

or
$$2t + \frac{\lambda}{2} = n\lambda$$

or
$$2t = (2n-1)\frac{\lambda}{2} \text{ where } n = 0, 1, 2, 3... \quad \text{(iv)}$$

Condition for Minima: For destructive interference

$$\Delta = \left(n + \frac{1}{2}\right)\lambda$$

or
$$2t + \frac{\lambda}{2} = n\lambda + \frac{\lambda}{2}$$

or
$$2t = n\lambda \text{ where } n = 0, 1, 2, 3... \quad \text{(v)}$$

Diameter of Dark and Bright Rings: Let us consider the thickness of the air film at point Q as t and r, as the radius of the fringe at that point together with R as the radius of curvature of the lens (Fig. 1.21).

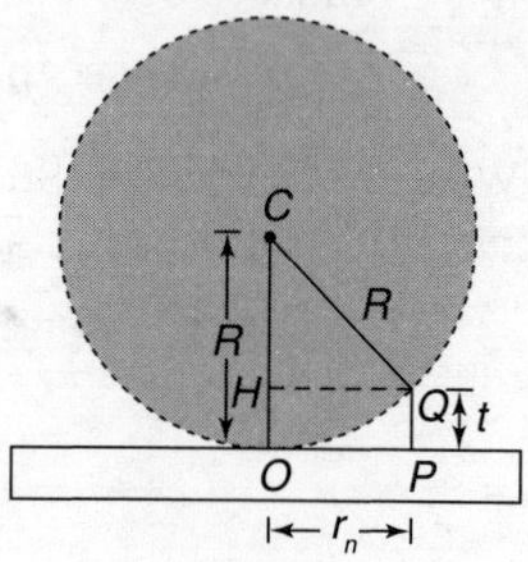

FIGURE 1.21

Hence,
$$OC = CQ = R, HQ = r_n$$
$$HC = R - t$$

In right angled ΔCHQ

$$CQ^2 = CH^2 + HQ^2$$
$$R^2 = (R-t)^2 + r_n^2$$

or
$$r_n^2 = 2Rt - t^2$$

In actual practice, R is quite large and t is very small. Therefore, t^2 may be neglected in comparison with 2Rt

$\therefore$
$$r_n^2 = 2Rt$$

or
$$r_n^2 = R \times 2t \quad \text{(vi)}$$

For Bright Rings: From Eq. (iii), we get

$$2t = (2n-1)\frac{\lambda}{2}$$

When we put this value of $2t$ in Eq. (vi), we get

$$r_n^2 = R \times (2n-1)\frac{\lambda}{2}$$

$$\left(\frac{D_n}{2}\right)^2 = R \times (2n-1)\frac{\lambda}{2}$$

or
$$D_n^2 = 2\lambda R(2n-1) \quad \text{(vii)}$$

The above equation gives the diameter D_n of nth order bright fringe as

$$D_n = \sqrt{2\lambda R(2n-1)}$$

$\therefore$
$$D_n \propto \sqrt{(2n-1)} \quad \text{(viii)}$$

Thus the diameter of the bright circular fringe(s) is proportional to the square root(s) of the odd natural numbers.

For Dark Rings: Applying the condition $2t = n\lambda$ for the dark rings, Eq. (vi) reads

$$r_n^2 = n\lambda R.$$

or

$$D_n^2 = 4n\lambda R$$

$\therefore$

$$D_n \propto \sqrt{n} \quad \text{(ix)}$$

Thus the diameter D_n of dark circular fringe(s) is proportional to the square root(s) of the natural numbers.

1.10.2.1 Determination of Wavelength of Light

We have seen that the diameter of nth order dark fringe in Newton's rings method is

$$D_n^2 = 4n\lambda R \quad \text{(x)}$$

From the above relation, the diameter of $(n+p)^{\text{th}}$ order dark fringe can be written as

$$D_{(n+p)}^2 = 4(n+p)\lambda R \quad \text{(xi)}$$

Subtracting Eq. (x) from equation (xi), we get

$$D_{(n+p)}^2 - D_n^2 = 4p\lambda R$$

or

$$\lambda = \frac{D_{(n+p)}^2 - D_n^2}{4pR}$$

Therefore, the measurement of diameters of the nth and $(n+p)^{\text{th}}$ dark fringes together with the radius of curvature of the lens gives us the wavelength of sodium light with the help of above formula.

1.10.2.2 Determination of Radius of Curvature of Plano Convex Lens

This is clear from the theory of Newton's rings that the measurement of diameters of nth order and $(n+p)^{\text{th}}$ order dark fringes play an important role in the determination of wavelength of monochromatic light. For this purpose, the following relation is used

$$\lambda = \frac{D_{(n+p)}^2 - D_n^2}{4pR}$$

Therefore, if we use the monochromatic source of light of known wavelength, it would be possible to determine the radius of curvature of the plano convex lens with the help of following formula

$$R = \frac{D_{(n+p)}^2 - D_n^2}{4p\lambda}$$

1.10.2.3 Determination of Refractive Index of a Liquid

The liquid whose refractive index is to be determined is placed between the lens and the glass plate and then we evaluate the diameters of the dark fringes.

The diameter of nth order dark fringe in air film is given by

$$D_n^2 = 4n\lambda R$$

Similarly, the diameter of nth order dark fringe in liquid film would be

$$[D_n^2]_{liquid} = \frac{4n\lambda R}{\mu}$$

where μ is the refractive index of the liquid and $D_{liquid} < D_{air}$

Therefore, the refractive index of the liquid can be calculated from the following formula once we are able to determine the diameters of dark fringes.

$$\mu = \frac{[D_n^2]_{air}}{[D_n^2]_{liquid}}$$

1.10.2.4 Newton's Rings in Transmitted Light

Newton's rings can be observed in reflected as well as in transmitted light. Figure 1.22 shows that the rays QA and HRB are the transmitted rays, which interfere. From the figure it is also clear that the ray QA suffers no reflection at a medium of higher index, so its phase does not change. However, the ray HRB encounters two reflections at the denser medium at Q and H. Since a phase change of π occurs at each reflection, the total phase change due to both reflections would be 2π. Therefore, there will not be any phase shift. In view of this, the path difference between the two transmitted rays QA and HRB would be

$$\Delta = 2\mu t \cos(r + \theta) \quad \text{(i)}$$

For air ($\mu = 1$), normal incidence ($r = 0$) and smaller angle θ ($\cos \theta = 1$), the path difference becomes

$$\Delta = 2t \quad \text{(ii)}$$

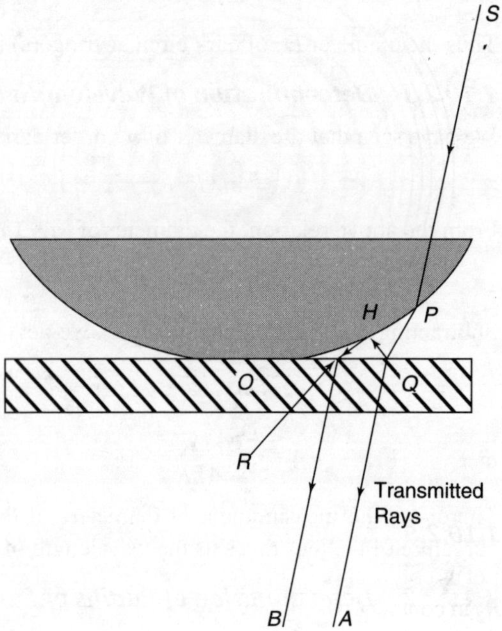

FIGURE 1.22

The above equations shows that at $t = 0$, the path difference between the two transmitted rays $\Delta = 0$. Therefore, at the centre, the bright fringe will appear.

From Eq. (ii), the conditions for maxima and minima can respectively be obtained as below

$$2t = n\lambda, n = 0, 1, 2, \ldots \quad \text{(iii)}$$

$$2t = (n + 1/2)\lambda, n = 0, 1, 2, \ldots \quad \text{(iv)}$$

Because of the same reason as discussed earlier, the fringes in the transmitted light will also be circular. The diameter of bright circular fringes can be obtained as

$$D_n = 2\sqrt{n\lambda R}$$

Thus the diameter of the bright fringes is proportional to the square root of natural numbers. When we calculate the diameter of dark circular fringes, it comes out to be

$$D_n = \sqrt{2(n+1)\lambda R}$$

This relation shows that the diameter of the dark fringes is proportional to the square root of odd natural numbers.

From the above relation, it is clear that the fringes observed in the transmitted light are exactly complementary to that of the reflected light. These fringes are much poorer in contrast as the transmitted rays emerge with lower intensity in comparison with the reflected rays. The Newton's rings obtained in the reflected as well as in the transmitted light are shown in Fig. 1.23a and b, respectively.

(a)

(b)

FIGURE 1.23

1.10.2.5 Newton's Rings formed between Two Curved Surfaces

Let us consider two curved surfaces of radii of curvature R_1 and R_2 in contact at the point O. A thin air film is enclosed between the two curved surfaces (Fig. 1.24). In this arrangement also, dark and bright rings are formed and can be seen with a traveling microscope.

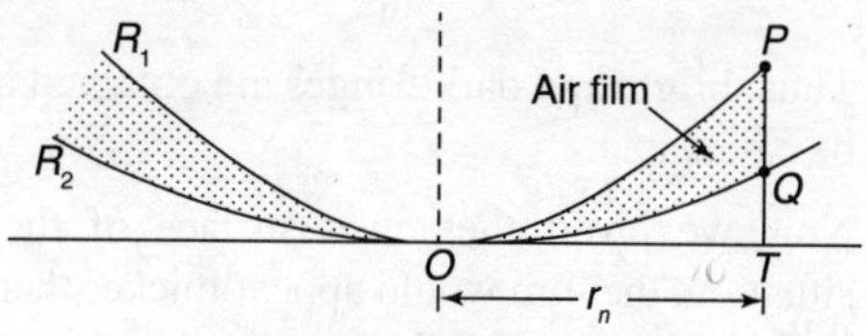

FIGURE 1.24

The thickness of the air film at P is

$$PQ = PT - QT$$

If the radius of nth dark ring be r_n, then from geometry,

$$PT = \frac{r_n^2}{2R_1} \text{ and } QT\frac{r_n^2}{2R_2}$$

$$\therefore \quad PQ = \frac{r_n^2}{2R_1} - \frac{r_n^2}{2R_2}$$

If we assume the thickness of the film as t, then

$$t = \frac{r_n^2}{2R_1} - \frac{r_n^2}{2R_2}$$

Now, this is clear from Fig. 1.24 that this type of film is similar to the wedge shaped film. Therefore, the path difference between the wave reflected from the upper and lower surfaces of the film would be

$$2\mu t \cos(r+\theta) + \frac{\lambda}{2}$$

For air ($\mu = 1$), normal incidence ($r = 0$) and the smaller angle θ, the path difference takes the form

$$2t + \frac{\lambda}{2}$$

Therefore, in case of reflected light, for n^{th} dark fringes

$$2t + \frac{\lambda}{2} = \left(n + \frac{1}{2}\right)\lambda$$

or $$2t = n\lambda$$

or $$2\left[\frac{r_n^2}{2R_1} - \frac{r_n^2}{2R_2}\right] = n\lambda$$

$$r_n^2\left[\frac{1}{R_1} - \frac{1}{R_2}\right] = n\lambda \text{ where } n = 0, 1, 2, 3\ldots \quad \text{(i)}$$

Similarly, for n^{th} bright fringe the path difference should satisfy the following condition

$$2t + \frac{\lambda}{2} = n\lambda$$

or $$2t = \left(n - \frac{1}{2}\right)\lambda$$

$$r_n^2\left[\frac{1}{R_1} - \frac{1}{R_2}\right] = \frac{(2n-1)\lambda}{2} \text{ where } n = 0, 1, 2, 3\ldots \quad \text{(ii)}$$

Thus, bright and dark fringes are obtained according to Eqs. (i) and (ii). The diameter of the fringes can also be calculated.

Now we invert the lower surface of the film. Under this situation, the film would appear thicker than the previous case (Fig. 1.25). The film thickness PQ in this case would be

$$PQ = PT + QT$$

$$t = \frac{r_n^2}{2R_1} + \frac{r_n^2}{2R_2}$$

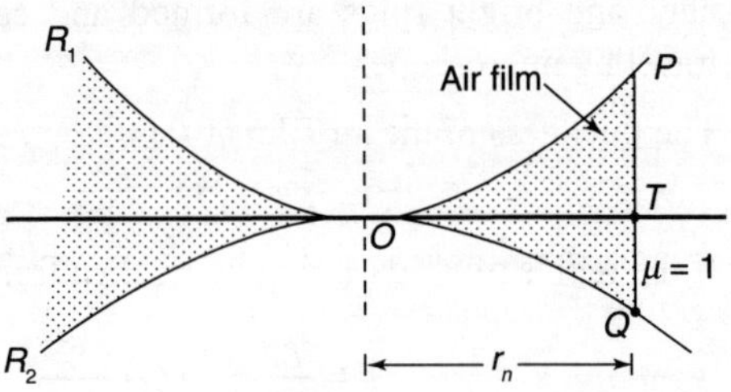

Figure 1.25

For the reasons explained in wedge-shaped film, the following condition should be satisfied in order to obtain nth order dark fringe of radius r_n

$$2t = n\lambda \text{(for air)}$$

or $$2r_n^2\left[\frac{1}{2R_1} + \frac{1}{2R_2}\right] = n\lambda$$

$$r_n^2\left[\frac{1}{R_1} + \frac{1}{R_2}\right] = n\lambda \text{ where } n = 0, 1, 2, 3\ldots \quad \text{(iii)}$$

For n^{th} bright fringe

$$2t = (2n-1)\frac{\lambda}{2}$$

or
$$2r_n^2\left[\frac{1}{2R_1}+\frac{1}{2R_2}\right] = (2n-1)\frac{\lambda}{2}$$

$$r_n^2\left[\frac{1}{R_1}+\frac{1}{R_2}\right] = (2n-1)\frac{\lambda}{2} \quad \text{where } n = 0, 1, 2, \text{....etc.} \qquad \text{(iv)}$$

A comparison of Eq. (i) with Eq. (iii) reveals that the diameter of dark fringes in the second case, where below curved surface looks like convex when viewed from above, would be smaller than the one in first case. This effect is similar to the situation as if we increase the width or thickness of the film. The same is the case for bright fringes.

1.10.3 Michelson's Interferometer

It consists of two highly polished mirrors M_1 and M_2 and two plane glass plates P and Q parallel to each other, as shown in Fig. 1.26. The glass plate P is half-silvered on its back surface and inclined at an angle of 45° to the beam of incident light. Another glass plate Q is such that P and Q are of equal thickness and of the same material. Two plane mirrors M_1 and M_2 are silvered on their front surfaces and mounted on two arms at right angle to each other. The position of the mirror M_1 can be changed with the help of a fine screw.

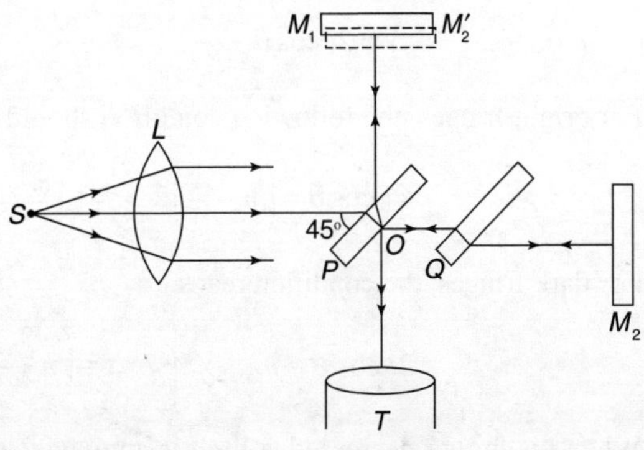

FIGURE 1.26

Light from a monochromatic source S, rendered parallel by a lens L, falls on the glass plate P. The semi-silvered plate P divides the incident light beam into two parts of nearly equal intensities, namely reflected and transmitted beams. The reflected beam moves towards mirror M_1 and falls normally on it and hence it is reflected back to P and enters the telescope T. The transmitted beam moves towards mirror M_2 and falls normally on it after passing through the plate Q. Therefore, it is reflected by the mirror M_2 and follows the same path. At P it is reflected to enter the telescope T. Since the beams entering the telescope have been derived from the same incident beam, these two rays are capable of giving the phenomenon of interference; thereby producing interference fringes.

Function of Plate Q: The beam going towards the mirror M_1 and reflected back, crosses the plate P twice, while the other beam in the absence of Q would travel wholly in air. Therefore, to compensate the additional path, the plate Q is used between the mirror M_2 and plate P. The light beam going towards the mirror M_2 and reflected back towards P also passes twice through the compensation plate Q. Therefore, the optical paths of the two rays in glass are the same.

Types of Fringes: The fringes in Michelson interferometer depend upon the inclination of M_1 and M_2. Let M_2' be the image of M_2 formed by the reflection at the half-silvered surface of the plate P so that $OM_2 = OM_2'$. The interference fringes may be regarded as formed by the light reflected from the surfaces of M_1 and M_2'. Thus, the arrangement is equivalent to an air-film enclosed between the reflecting surfaces M_1 and M_2'.

It is obvious that the path difference between the two beams produced by the reflecting surfaces M_1 and M_2' is equal to the twice of the thickness of the film $M_1 M_2'$. This path difference can be varied by moving M_1 backwards or forward parallel to itself. If we use monochromatic light, the pattern of bright and dark fringes will be formed. Here the shape of the fringes will depend upon the inclination of M_1 and M_2.

If M_1 and M_2 are exactly at right angles to each other, the reflecting surfaces M_1 and M_2' are parallel and hence air film between M_1 and M_2' is of constant thickness t so that we get circular fringes of equal inclination.

These fringes are called as **Haidinger's fringes** that can be seen in the field view of a telescope. When the distance between the mirrors M_1 and M_2 or between M_1 and M_2' is decreased, the circular fringes shrink and vanish at the centre. A ring disappears each time when the path $2t$ decreases by λ.

Since the vertical ray first gets reflected from the inner surface of P (internal reflection), and then from the front surface of the mirror M_1 (external reflection) a phase change of π takes place. The horizontal ray first gets reflected from the front surfaces of M_2 (external reflection) and then from the inner surface of glass plate P (external reflection), so there is no phase change. Therefore, the total path difference for normal incidence would be

$$\Delta = 2t\cos\theta + \frac{\lambda}{2}$$

For bright fringes, the following condition should be satisfied

$$2t\cos\theta = \left(n - \frac{1}{2}\right)\lambda \qquad [\because \Delta = \lambda] \tag{i}$$

For dark fringes, the condition reads

$$2t\cos\theta = n\lambda \qquad \left[\because \Delta = \left(n + \frac{1}{2}\right)\lambda\right] \tag{ii}$$

When t is further decreased, a limit is attained where M_1 and M_2' coincide and the path difference between the two rays becomes zero. Now the field of view is perfectly dark. When M_1 is further moved, the fringes appear again.

If M_1 and M_2 are not perfectly perpendicular, a wedge shaped film will be formed between M_1 and M_2' then we get almost straight line fringes of equal thickness in the field of view of telescope, as the radius of fringes is very large.

All the above discussed films are shown in Fig. 1.27.

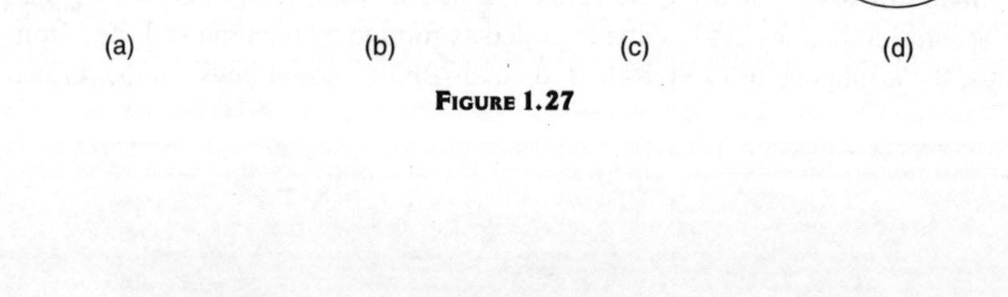

FIGURE 1.27

1.10.3.1 *Applications*

Michelson's interferometer uses the concept of interference that takes place with the help of two mirrors. The distance between one mirror and the image of another plays an important role in the formation of fringes. Michelson's interferometer has diverse applications, some of which are listed below.

(i) Determination of Wavelength of Light

First of all the Michelson's interferometer is set for circular fringes with central bright spot, which is possible when both the mirrors are parallel ($\theta = 0$). If t be the thickness of air film enclosed between the two mirrors (M_1 and M_2') and n be the order of the spot obtained, then for normal incidence $\cos r = 1$, we have

$$2t + \frac{\lambda}{2} = n\lambda$$

or
$$2t = \left(n - \frac{1}{2}\right)\lambda$$

If M_1 is moved $\frac{\lambda}{2}$ away from M_2', then an additional path difference of λ will be introduced and hence $(n+1)^{th}$ bright spot appears at the centre of the field. Thus each time M_1 moves through a distance $\frac{\lambda}{2}$, a new bright fringe appears. Therefore, if M_1 moves by a distance x (x_1 to x_2) and N new fringes appear at the centre of the field, then we have

$$x = x_2 - x_1 = N\frac{\lambda}{2}$$

or
$$\lambda = \frac{2(x_2 - x_1)}{N} = \frac{2x}{N} \qquad \lambda = \frac{2x}{N}$$

The difference $(x_2 - x_1)$ is measured with the help of micrometer screw and N is actually counted. The experiment is repeated for number of times and the mean value of λ is obtained.

(ii) Determination of Difference in Wavelengths

Michelson's interferometer is adjusted in order to obtain the circular fringes. Let the source be not monochromatic and have two wavelengths λ_1 and λ_2 ($\lambda_1 > \lambda_2$) which are very close to each other (as Sodium D lines). The two wavelengths form their separate fringe patterns but as λ_1 and λ_2 are very close to each other and thickness of air film is small, the two patterns practically coincide with each other. As the mirror M_1 is moved slowly, the two patterns separate slowly and when the thickness of air film is such that the dark fringe of λ_1 falls on bright fringe of λ_2, the result is maximum indistinctness. Now the mirror M_1 is further moved, say through a distance x, so that the next indistinct position is reached. In this position, if n fringes of λ_1 appear at the centre, then $(n+1)$ fringes of λ_2 should appear at the centre of the field of view. Hence

$$x = n\frac{\lambda_1}{2} \text{ and } x = (n+1)\frac{\lambda_2}{2}$$

or
$$n = \frac{2x}{\lambda_1} \qquad \text{(i)}$$

and
$$(n+1) = \frac{2x}{\lambda_2} \qquad \text{(ii)}$$

On subtracting Eq. (i) from Eq. (ii), we get

$$(n+1)-n=\frac{2x}{\lambda_2}-\frac{2x}{\lambda_1}$$

or
$$1=\frac{2x(\lambda_1-\lambda_2)}{\lambda_1\lambda_2}$$

or
$$(\lambda_1-\lambda_2)=\frac{\lambda_1\lambda_2}{2x}=\frac{\lambda_{av}^2}{2x} \text{ where } \lambda_1\lambda_2=\lambda_{av}^2 \text{ is the square of mean of } \lambda_1 \text{ and } \lambda_2.$$

Thus measuring the distance x moved by mirror M_1 between the two consecutive positions of maximum indistinctness, the difference between two wavelengths of the source can be determined, if λ_{av} is known.

(iii) Determination of Thickness and Refractive Index of a Thin Transparent Sheet

The Michelson's interferometer is adjusted for producing straight white light fringes and cross-wire is set up on the central bright fringe. Now insert thin transparent plate in the path of one of the interfering waves. On the inclusion of a plate of thickness t and refractive index μ, the path difference is increased by a factor of $2(\mu-1)t$. The fringes are therefore shifted. The mirror M_1 is now moved till the central fringe is again brought back to its initial position. The distance x traveled by the mirror M_1 is measured by micrometer. Therefore

$$2x=2(\mu-1)t \text{ or } t=\frac{x}{(\mu-1)} \tag{iii}$$

From Eq. (iii), we can write

$$\mu=\frac{x}{t}+1 \tag{iv}$$

Thus, by knowing the thickness of the transparent sheet and the distance x, we can calculate the refractive index of the sheet with the help of a Michelson's interferometer.

1.11 APPLICATIONS OF INTERFERENCE IN THE FIELD OF ENGINEERING LO7

The phenomenon of interference arises in many situations and the scientists and engineers have taken advantage of interference in designing and developing various instruments.

1.11.1 Testing of Optical Flatness of Surfaces

An example of the application of interference method is the testing of optical components for surface quality. The most important example is that of optical flats. However, the methods used for fl at surfaces can be adapted simply to test spherical surfaces.

1.11.1.1 Flatness Interferometers

With these interferometers we can compare the flatness of two surfaces by placing them in contact with slight wedge of air between them. This gives a tilt and thus the fringes start originating like that of Newton's ring between the two surfaces. To get half wavelength contours of the space between the surfaces, they should be viewed from infiity. Further, to avoid the risk of scratching, a desirable distance should be there between the two surfaces. Most common examples of flatness interferometers are *Fizeau* and *Twyman interferometers*.

(i) Fizeau Interferometer

In this type of interferometer, the sources and viewing point are kept at infinity (Fig. 1.28). This interferometer generates interference between the surface of a test sample and a reference surface that is brought close to the test sample. The interference images are recorded and analysed by an imaging optic system. However, the contrast and the shape of the interference signals depend on the reflectivity of the test samples.

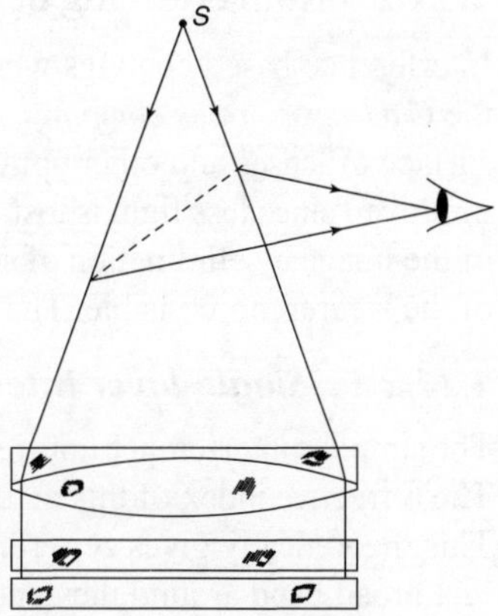

FIGURE 1.28

(ii) Twyman-Green Interferometer

This is an important instrument used to measure defects in optical components such as lenses, prisms, plane parallel windows, laser rods and plane mirrors. Twyman-Green interferometer, shown in Fig. 1.29 resembles Michelson interferometer in the beam splitter and mirror arrangement. However, the difference lies in the way of their illumination. In the case of Twyman-Green interferometer, we use a monochromatic point source which is located at the principal focus of a well-corrected lens whereas in Michelson interferometer an extended source is used. If the mirrors M_1 and M_2 are perpendicular to each other and the beam-splitter *BS* makes an angle of 45° with the normal of each mirror, then the interference is exactly analogous to thin film interference at normal incidence. Therefore, completely constructive interference is obtained when $d = m\lambda/2$, where d is the path difference between the two arms adjusted by translating the mirror M_1. The complete destructive interference is obtained when $d = (m + 1/2)\lambda/2$. With the help of rotation of mirror M_2 we can see fringes of equal thickness on the screen, as the angle of incidence is constant. This situation is analogous to interference pattern observed with collimated light and a thin film with varying thickness. In order to test the optical components, one of the mirrors is intentionally tilted to create fringes. Then the quality of the component can be determined from the change in the fringe pattern when the component is placed in the interferometer. Lens testing is specifically important for quantifying aberrations and measuring the focal length.

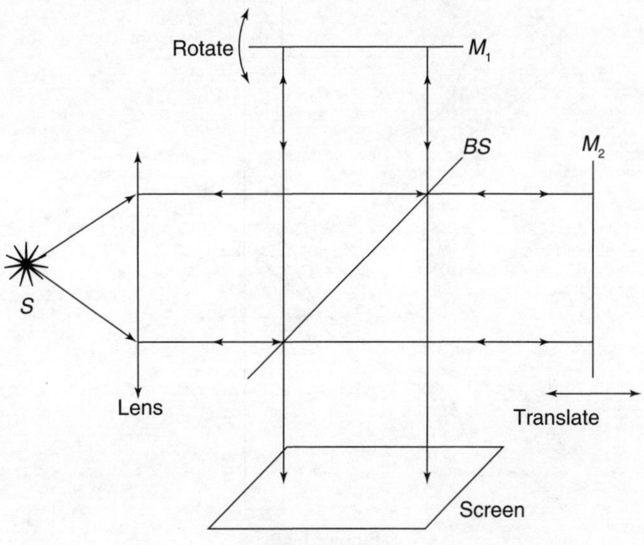

FIGURE 1.29

1.11.2 Nonreflecting or Antireflecting (AR) Coatings

Interference-based coatings were invented in November 1935 by *Alexander Smakula*, who was working for the *Carl Zeiss optics company*. Antireflecting coatings are a type of optical coatings. These are applied to the surface of lenses and other optical devices for reducing reflection. This way the efficiency of the system gets improved since less light is lost. For example, in a telescope the reduction in reflections improves the contrast of the image by elimination of stray light. In another applications a coating on eyeglass lenses makes the eyes of the wearer more visible. The anti-reflecting coatings can be mainly divided into three groups.

1.11.2.1 Single-layer Interference Coatings

The simplest interference non-reflecting coating consists of a single quarter-wave layer of transparent material. The refractive index of this material is taken to be equal to the square root of the substrate's refractive index. This theoretically gives zero reflectance at the center wavelength and decreased reflectance for wavelengths in a broad band around the center. The use of an intermediate layer to form an antireflection coating can be thought of as analogous to the technique of impedance matching of electrical signals. A similar method is used in fibre optic research where an index matching oil is sometimes used to temporarily defeat total internal reflection so that light may be coupled into or out of a fiber.

The antireflection coatings rely on an intermediate layer not only for its direct reduction of reflection coefficient, but also use the interference effect of a thin layer. If the layer thickness is controlled precisely and it is made exactly one quarter of the light's wavelength ($\lambda/4$), then it is called a quarter-wave coating (Fig. 1.30). In this case, the incident beam I, when reflected from the second interface, will travel exactly half its own wavelength further than the beam reflected from the first surface. The two reflected beams R_1 and R_2 will destructively interfere as they are exactly out of phase and cancel each other if their intensities are equal. Therefore, the reflection from the surface is suppressed and all the energy of the beam is propagated through the transmitted beam T. In the calculation of the reflection from a stack of layers, the transfer-matrix method can be used.

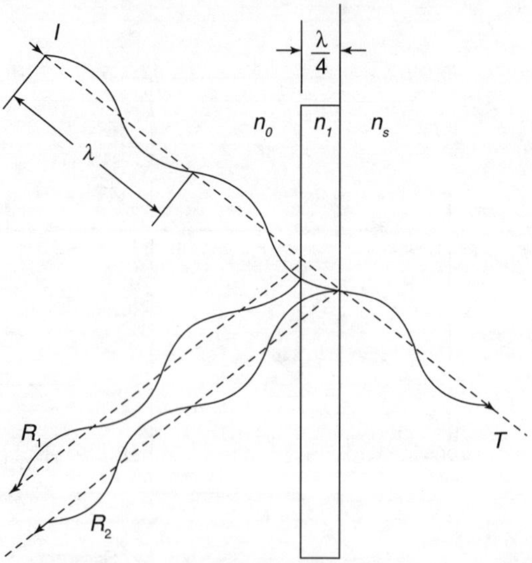

FIGURE 1.30

1.11.2.2 Multilayer Coatings or Multicoating

Multiple coating layers can also be used for *reflection reduction*. It is possible if we design them such that the reflections from the surfaces undergo maximum destructive interference. This can be done if we add a second quarter-wave thick higher-index layer between the low-index layer (for example, silica) and the substrate. Under this situation, the reflection from all three interfaces produces destructive interference and antireflection. Optical coatings can also be made with near-zero reflectance at multiple wavelengths or optimum performance at angles of incidence other than 0°.

1.11.2.3 Absorbing Antireflecting Coatings

Absorbing antireflecting coatings are an additional category of antireflection coatings. These coatings are useful in situations where *low reflectance* is required and high transmission through a surface is unimportant or undesirable. They can produce very low reflectance with few layers. They can often be produced more cheaply or at greater scale than standard non-absorbing anti-reflecting coatings. In sputter deposition system for such films, titanium nitride and niobium nitride are frequently used.

1.11.2.4 Practical Problems with AR Coatings

Real coatings do not reach perfect performance, though they are capable of reducing a surface's reflection coefficient to less than 0.1%. Practical details include correct calculation of the layer thickness. This is because the wavelength of the light is reduced inside a medium and this thickness will be $\lambda_0/4n_1$, where λ_0 is the vacuum wavelength and n_1 is the refractive index of the film. Finding suitable materials for use on ordinary glass is also another difficulty, since few useful substances have the required refractive index ($n \approx 1.23$) which will make both reflected rays exactly equal in intensity. Since magnesium fluoride (MgF_2) is hard-wearing and can be easily applied to substrates using *physical vapour deposition*, it is often used for this purpose even though its index is higher than desirable ($n = 1.38$).

1.12 SCIENTIFIC APPLICATIONS OF INTERFERENCE — LO7

In interferometry, we use the principle of superposition to combine different waves in a way that will cause the result of their combination to have some meaningful property, that is indicative of the original state of the waves. The phenomenon of interference is employed under various situations for its scientific applications. For a better understanding of the applications, we first need to know about the homodyne and heterodyne detections.

1.13 HOMODYNE AND HETERODYNE DETECTION — LO7

In standard interferometry, the interference occurs between the two beams at the same wavelength (or *carrier frequency*). The phase difference between the two beams results in a change in the intensity of the light on the detector. Measuring the resulting intensity of the light after the mixing of these two light beams is known as *homodyne detection*. In heterodyne detection, we modulate one of the two beams prior to detection, usually by a frequency shift. A special case of heterodyne detection is optical heterodyne detection, which detects the interference at the beat frequency.

1.13.1 Imaging Interferometry

In this interferometry, the pattern of radiation across a region can be represented as a function of position $i(x, y)$, i.e., an image and the pattern of incoming radiation $i(x, y)$ can be transformed into the Fourier domain

$f(u, v)$. A single detector measures information from a single point in (x, y) space. An interferometer measures the difference in phase between two points in the (x, y) domain. This corresponds to a single point in the (u, v) domain. An interferometer builds up a full picture by measuring multiple points in (u, v) space. The image $i(x, y)$ can then be restored by performing an inverse Fourier transform on the measured $f(u, v)$ data.

1.13.2 Holographic Interferometry (HI)

Holographic interferometry (HI) is a technique that enables static and dynamic displacements of objects with optically rough surfaces to be measured to optical interferometric precision, i.e., to fractions of a wavelength of light. These measurements can be applied to stress, strain and vibration analysis, as well as to nondestructive testing. It can also be used to detect optical path length variations in transparent media, which enables, for example, fluid flow to be visualised and analysed. It can also be used to generate contours representing the form of the surface. Holography interferometry is of two types.

(i) Live Holography Interferometry

Holography enables the light field scattered from an object to be recorded and replayed. If this recorded field is superimposed on the "live field" scattered from the object, then the two fields will be identical. However, if a small deformation is applied to the object, the relative phases of the two light fields will alter and it is possible to observe interference. This technique is known as live holographic interferometry.

(ii) Frozen-Fringe Holography

In this holography, it is also possible to obtain fringes by making two recordings of the light field scattered from the object on the same recording medium. The reconstructed light fields may then interfere to give fringes, which map out the displacement of the surface.

1.13.3 Electronic Speckle Pattern Interferometry

Electronic Speckle Pattern Interferometry (ESPI), also known as TV Holography, is a technique that uses laser light together with video detection, recording and processing to visualize static and dynamic displacements of components with optically rough surfaces. The visualisation is in the form of fringes on the image where each fringe normally represents a displacement of half a wavelength of the light used, i.e., quarter of a micrometre or so.

1.13.4 Angle Resolved Low Coherence Interferometry

Angle resolved low coherence interferometry is an emerging biomedical imaging technology that uses the properties of scattered light to measure the average size of cell structures, including the cell nuclei. The technology shows promise as a clinical tool for *in situ* detection of dysplatic or precancerous tissue.

1.13.5 Optical Coherence Tomography

This is a medical imaging technique based on low-coherence interferometry, where subsurface light reflections are resolved to give tomographic visualisation. Recent advances have struggled to combine the nanometre phase retrieval with the ranging capability of low-coherence interferometry.

1.13.6 Geodetic Standard Baseline Measurements

A famous use of white light interferometry is the precise measurement of geodetic standard baselines. Here the light path is split in two, and one leg is folded between a mirror pair 1 m apart. The other leg bounces once off a mirror 6 m away. The fringes will be seen only if the second path is precisely 6 times the first. Starting from a standard quartz gauge of 1 m length, it is possible to measure distances up to 864 m by repeated multiplication. Baselines thus established are used to calibrate geodetic distance measurement equipments. This leads to a metrologically traceable scale for geodetic networks measured by these instruments. More modern geodetic applications of laser interferometry are in calibrating the divisions on levelling staffs and in monitoring the free fall of a reflective prism within a ballistic or absolute gravimeter. This allows determination of gravity, i.e., the acceleration of free fall, directly from the physical definition at a few parts in a billion accuracy.

1.13.7 Interference Lithography

This is a technique for patterning regular arrays of fine features, without the use of complex optical systems or photo marks. The basic principle of this is the same as interferometry. An interference pattern between two or more coherent light waves is set up and recorded in a recording layer. This interference pattern consists of a periodic series of fringes of representing intensity maxima and minima. The benefit of using interference lithography is the quick generation of dense features over a wide area without loss of focus.

SUMMARY

We summarise the main outcome of the chapter as follows:

- We first discussed the phenomenon of interference and then explained it based on Young's double slits experiment.
- Concepts of wavefront and secondary wavelets were discussed based on Huygens' principle. Then secondary wavefront was introduced as the surface touching the secondary wavelets tangentially in the forward direction at any given time.
- Phase difference and path difference between the two waves play a key role for obtaining constructive or destructive interference. Therefore, phase and path differences were explained in detail together with their relation.
- For obtaining sustained interference pattern, the two sources should be coherent. So the concept of coherence, both temporal and spatial, was introduced and coherence time and coherence length were talked about.
- A short description of a technique for producing coherent light from incoherent sources was given.
- Analytical treatment of the interference was discussed where conditions were obtained for the constructive and destructive interferences.
- When a light wave gets reflected from a surface, a phase change may take place. Therefore, condition of relative phase shift was explained.
- Superposition was extended for n number of waves and it was observed that the resultant amplitude increases in proportion with $\sqrt{n}$ in length as n gets increased.
- Interfering waves can be produced either by division of wavefront or by division of amplitude. Therefore, the details of interference were discussed based on these two methods.
- Fresnel's biprism, which is used to create two virtual coherent sources, was discussed in detail for obtaining interference pattern and the related conditions for dark and bright fringes.

- Application of biprism for the determination of wavelength of light, distance between two virtual coherent sources and thickness of transparent sheet were discussed. The displacement of fringes by the introduction of thin transparent sheet in the path of one light wave was also explored.
- Thin films are used for the division of amplitude of light waves which superimpose each other. The interference pattern obtained by thin films of uniform and non-uniform thicknesses was investigated.
- When the air film is created between the curved surface of a plano-convex lens and the flat surface of a mirror, the interference takes place between the reflected as well as the transmitted light. Here the fringes are obtained in the form of rings known as Newton's rings.
- Newton's rings method was used for determination of the wavelength of light, radius of curvature of a plano-convex lens and the refractive index of liquid.
- The theory was extended to Newton's rings formed between two curved surfaces.
- Theory and practical applications of Michelson's interferometer were discussed. Clarification of path difference and the details of formation of fringes were given.
- Engineering applications of interference were included, particularly related to the testing of optical flatness of surfaces and nonreflecting or antireflecting coatings.
- Finally the scientific applications of interference were discussed related to various interferometry, tomography and lithography.

SOLVED EXAMPLES

EXAMPLE 1 If light of wavelength 660 nm has wave trains 13.2×10^{-6}m long, what would be the coherence time.

SOLUTION Given $\lambda = 6.6 \times 10^{-7}$m, coherence length $(\Delta L) = 1.32 \times 10^{-5}$m and coherence time (Δt) = ?

Formula used is $\Delta L = c \cdot \Delta t$

or
$$\Delta t = \frac{\Delta L}{c} = \frac{1.32 \times 10^{-5}}{3 \times 10^{8}} = \mathbf{4.4 \times 10^{-14}\ sec}$$

EXAMPLE 2 Coherence length of a light is 2.945×10^{-2} m and its wavelength is 5896 Å. Calculate the coherence time and number of oscillations corresponding to the coherence length.

SOLUTION Given $\Delta L = 2.945 \times 10^{-2}$m and $\lambda = 5.896 \times 10^{-7} \cdot \Delta t$ = ?

Formula used is $\Delta L = c\Delta t$

or
$$\Delta t = \frac{\Delta L}{c} = \frac{2.945 \times 10^{-2}}{3 \times 10^{8}} = \mathbf{9.816 \times 10^{-11}\ sec}$$

and number of oscillations in a length L,

$$n = \frac{\Delta L}{\lambda} = \frac{2.945 \times 10^{-2}}{5.896 \times 10^{-7}} = \mathbf{4.99 \times 10^{4}}$$

EXAMPLE 3 A coherent beam has band width of 1200 Hz. Obtain the coherence length.

SOLUTION Given $\Delta v = 1200$ Hz and $C = 3 \times 10^{8}$ m/s

By using the relation coherence length $L_C = \frac{C}{\Delta v}$

$$L_C = \frac{3.0 \times 10^8}{1200} = \mathbf{2.5 \times 10^5\ m}$$

EXAMPLE 4 Calculate the line-width, coherence time and frequency stability for a line of Krypton having a wavelength of 6.058×10^{-7}m and coherence length as 0.2 m.

SOLUTION Given $\lambda_{av} = 6.058 \times 10^{-7}$m, $\Delta L = 0.2$ m and $c = 3 \times 10^8$m/sec.

In Michelson's Interferometer we derived the following formula

$$1 = \frac{2x}{\lambda_2} - \frac{2x}{\lambda_1}$$

where x is the distance between two mirrors. The above expression can be written as

$$2x = \frac{\lambda_1 \lambda_2}{\Delta\lambda} = \frac{\lambda_{av}^2}{\Delta\lambda}$$

where λ_{av} is the mean wavelength of λ_1 and λ_2. Here $\Delta\lambda$ is called **line width**.

In view of the fact that the fringes are not observed if the path difference exceeds the coherence length ΔL, we can assume the beam to contain all wavelengths lying between λ and $(\lambda + d\lambda)$.

Therefore, $$2x = \Delta L = \frac{\lambda_{av}^2}{\Delta\lambda}$$

or $$\Delta\lambda = \frac{\lambda_{av}^2}{\Delta L}$$

$\because$ $$\text{frequency, } v = \frac{c}{\lambda}$$

$\therefore$ $$\left[\frac{\Delta v}{\Delta\lambda}\right] = \frac{c}{\lambda_{av}^2}$$

or $$\Delta v = \frac{c}{\lambda_{av}^2}\Delta\lambda = \frac{c}{\Delta L}$$

Here, Δv is called frequency spread of the line, which can be written in terms of Δt as follows.

$$\Delta v = \frac{1}{\Delta t} \text{ or } \Delta t = \frac{1}{\Delta v}$$

In addition, frequency stability is defined as the ratio of frequency spread and frequency of any spectral line, i.e.,

$$\text{frequency stability} = \frac{\Delta v}{v}$$

Line width $$\Delta\lambda = \frac{\lambda_{av}^2}{\Delta L} = \frac{(6.058 \times 10^{-7})^2}{0.2} = \mathbf{1.834 \times 10^{-12}\ m}$$

Frequency spread $$\Delta v = \frac{c}{\Delta L} = \frac{3 \times 10^8}{0.2} = 1.5 \times 10^9\ \text{Hz}$$

Frequency $$v = \frac{c}{\lambda} = \frac{3 \times 10^8}{6.058 \times 10^{-7}}$$

$$= \mathbf{4.952 \times 10^{14}\ Hz}$$

and Frequency stability

$$= \frac{\Delta v}{v} = \frac{1.5 \times 10^9}{4.952 \times 10^{14}} = \mathbf{3.0 \times 10^{-6}}$$

EXAMPLE 5 The Doppler width for an orange line of Krypton is 550×10^{-15} m. If the wavelength of light is 605.8 nm, calculate the coherent length.

SOLUTION Given Doppler line width $(\Delta\lambda) = 5.5 \times 10^{-13}$m.

$$\lambda = \lambda_{av} = 6.058 \times 10^{-7} \text{ m}, \Delta L = ?$$

Formula used is $$\Delta L = \frac{\lambda_{av}^2}{\Delta\lambda} = \frac{(6.058 \times 10^{-7})^2}{5.5 \times 10^{-13}} = \mathbf{0.6673\ m}$$

EXAMPLE 6 A mercury vapour lamp emits a light of wavelength 5461 Å with a band width of 6×10^8 Hz. Calculate the ratio of its coherence length with the coherence length of a He–Ne laser operating at a wavelength 6328 Å with a band width of 10^6 Hz.

SOLUTION For mercury vapour lamp, $\lambda_{av} = 5.461 \times 10^{-7}$ m, $= 6 \times 10^8$ Hz

Formula used is $$\left[\frac{\Delta v}{\Delta\lambda}\right] = \frac{c}{\lambda_{av}^2} \text{ or } \Delta\lambda = \frac{\lambda_{av}^2 \Delta v}{c}$$

or $$\Delta\lambda = \frac{(5.461 \times 10^{-7})^2 \times 6 \times 10^8}{3 \times 10^8}$$

$$= 5.964 \times 10^{-13} \text{m}$$

The coherence length is given by

$$\Delta L_1 = \frac{\lambda_{av}^2}{\Delta\lambda} = \frac{(5.641 \times 10^{-7})^2}{5.964 \times 10^{-13}} = 0.534 \text{ m}$$

For He – Ne laser,

Given $\lambda_{av} = 6.328 \times 10^{-7}$ m, $\Delta v = 10^6$ Hz

$$\Delta\lambda = \frac{\lambda_{av}^2 \Delta v}{c} = \frac{(6.328 \times 10^{-7})^2 \times 10^6}{3 \times 10^8}$$

$$\Delta\lambda = 1.335 \times 10^{-15} \text{ m}$$

Coherence length (for laser)

$$\Delta L_2 = \frac{\lambda_{av}^2}{\Delta\lambda} = \frac{(6.328 \times 10^{-7})^2}{1.335 \times 10^{-15}} = 299.952 \text{ m}$$

$$\therefore \quad \frac{\Delta L_1}{\Delta L_2} = \frac{0.534}{299.952} = \frac{1}{562} = \mathbf{1{:}562}$$

EXAMPLE 7 Find the coherence length of a laser beam for which the band width is 3000 Hz.

SOLUTION Given $\Delta v = 3000$ Hz.

Coherence length $(\Delta L) = c\Delta t$ and coherence time $(\Delta t) = \dfrac{1}{\Delta v}$

So $$\Delta t = \frac{1}{3000} = 3.333 \times 10^{-4} \text{ sec}$$

$$L = c \times t = 3 \times 10^8 \times 3.333 \times 10^{-4} = \mathbf{1.0 \times 10^5 m}$$

EXAMPLE 8 Calculate the resultant line-width, band width and coherence length assuming that we chop a continuous perfectly monochromatic beam of wavelength 6328 Å in 10^{-10} seconds using some sort of shutter.

SOLUTION Given $\lambda_{av} = 6.328 \times 10^{-7}$ m and $t = 10^{-10}$ sec

Coherence length, $\Delta L = c\Delta t = 3 \times 10^8 \times 10^{-10} = 3 \times 10^{-2}$ m

Bandwidth, $\Delta\gamma = \frac{1}{\Delta t} = \frac{1}{10^{-10}} = 10^{10}$ Hz

Line-width, $\Delta\lambda = \frac{\lambda^2 av}{C}\Delta\gamma = \frac{(6.328\times10^{-7})^2\times10^{10}}{3\times10^8}$

$= 1.335 \times 10^{-11}$ m

$= \mathbf{0.1335}$ **Å**

EXAMPLE 9 For a red cadmium line of wavelength 6438 Å and coherence length 38 cm deduce the order of magnitude of (a) coherence time and (b) spectral width of the line.

SOLUTION Given coherence length, $\Delta L = 0.38$ m and $\lambda_{av} = 6.43\ 8 \times 10^7$ m

Coherence time $\Delta t = ?$

Spectral line width $\Delta\lambda = ?$

$$\Delta L = c\,\Delta t \text{ or } \Delta t = \frac{\Delta L}{c}$$

$$\Delta t = \frac{0.38}{3\times10^8} = 1.266\times10^{-9}\text{ sec}$$

or $$\Delta L = \frac{\lambda_{av}^2}{\Delta\lambda} \text{ or } \Delta\lambda = \frac{\lambda_{av}^2}{\Delta L} = \frac{(6.438\times10^{-7})^2}{0.38}$$

$$\mathbf{\Delta\lambda = 1.09 \times 10^{-12}\ m}$$

EXAMPLE 10 The ratio of intensities of two waves that produce interference pattern is 16:1. Deduce the ratio of maximum to minimum intensities in fringe system.

SOLUTION Given $I_1 : I_2 = 16 : 1$

The intensity, $I \propto a^2$

∴ $$a_1^2 : a_2^2 = 16 : 1 \quad \text{or} \quad a_1 : a_2 = 4 : 1$$

or $$a_1 = 4a_2$$

$$\frac{I_{max}}{I_{min}} = \frac{(a_1+a_2)^2}{(a_1-a_2)^2} = \frac{(4a_2+a_2)^2}{(4a_2-a_2)^2} = \frac{25}{9}$$

i.e., $$I_{max} : I_{min} = \mathbf{25{:}9}$$

EXAMPLE 11 Two waves of same frequency with amplitudes 1.0 and 2.0 units, interfere at a point, where the phase difference is 60°. What is the resultant amplitude?

SOLUTION Given $a_1 = 1.0$ unit, $a_2 = 2.0$ unit and $\phi = 60°$

the resultant amplitude

$$R = \sqrt{a_1^2 + a_2^2 + 2a_1a_2\cos\phi}$$

$$= \sqrt{1+4+2} = \sqrt{7}$$

$= \mathbf{2.65\ units}$

EXAMPLE 12 Distance between two slits is 0.1 mm and the width of the fringes formed on the screen is 5 mm. If the distance between the screen and the slit is one meter, what would be the wavelength of light used?

SOLUTION Given $\beta = 5.0 \times 10^{-3}$ m, $2d = 1.0 \times 10^{-4}$ m and $D = 1.0$ m

Formula used is $\lambda = \dfrac{\beta\, 2d}{D} = \dfrac{5\times10^{-3}\times1.0\times10^{-4}}{1.0} = 5000$ Å

$\lambda = $ **5000 Å**

EXAMPLE 13 A biprism of angle 1° and refractive index 1.5 is at a distance of 40 cm from the slit. Find the fringe width at 60 cm from the biprism for sodium light of wavelength 5893 Å.

SOLUTION Given $a = .4$ m, $\mu = 1.5$, $\lambda = 5.893 \times 10^{-7}$ m and $D = 1.0$ m

From the Fig. 1.31 $\delta = \dfrac{d}{a}$ or $d = a\delta$

or $2d = 2a\delta$

The deviation produced in the incident light is given by

$$\delta = (\mu - 1)\,\alpha$$
$$2d = 2a\,(\mu - 1)\alpha$$

where α = angle of prism

$$1° = \frac{\pi}{180}\text{rad}$$

Fringe width $\beta = \dfrac{\lambda D}{2d} = \dfrac{5.893\times10^{-7}\times1.0}{2\times0.4(1.5-1)\pi/180} = \dfrac{5.893\times10^{-7}\times180}{2\times0.4\times0.5\times3.14} = \mathbf{0.0844\times10^{-3}\,m}$

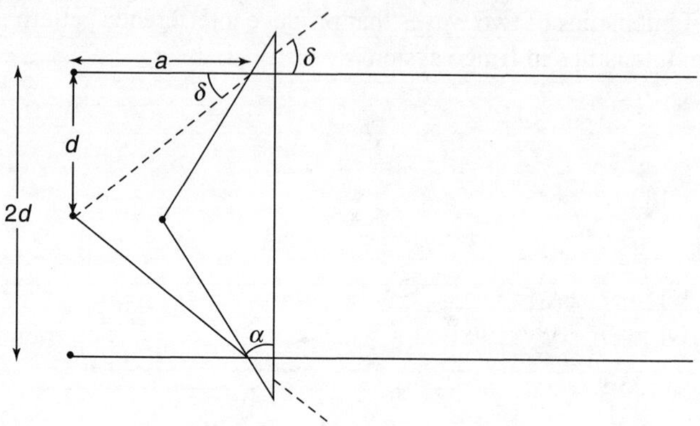

EXAMPLE 14 Interference fringes are produced by Fresnel's bi-prism on the focal plane of a reading microscope which is 1.0 m far from the slit. A lens interposed between the biprism and the microscope gives two images of the slit in two positions. If the images of the slits are 4.05 mm apart in one position and 2.90 mm apart in the other position and the wavelength of the sodium light is 5893 Å, find the distance between the consecutive interference bands?

SOLUTION Given $\lambda = 5.893 \times 10^{-7}$ m, $D = 1.0$ m, $d_1 = 4.05 \times 10^{-3}$ m and $d_2 = 2.90 \times 10^{-3}$ m

Formula used is $2d = \sqrt{d_1 d_2} = \sqrt{4.05\times2.90\times10^{-6}}$

or $2d = 3.427 \times 10^{-3}$ m

Now $\beta = \frac{\lambda D}{2d} = \frac{5.893 \times 10^{-7} \times 1.0}{3.427 \times 10^{-3}}$

$\beta = \mathbf{0.172\ mm}$

EXAMPLE 15 In a biprism experiment fringes were first obtained with the sodium light of wavelength 5890 Å and fringe width was measured to be 0.342 mm. Sodium light was then replaced with white light and central fringe was located. On introducing a thin glass sheet in half of the beam, the central fringe was shifted by 2.143 mm. Calculate the thickness of the glass sheet if the refractive index of glass is 1.542.

SOLUTION Given $\lambda = 5890 \times 10^{-10}$ m, = 1.542, $x_n = 2.143 \times 10^{-3}$ m

$\beta = 3.42 \times 10^{-4}$ m

Formula used is $x_n = n\beta$

or $n = \frac{x_n}{\beta} = \frac{2.143 \times 10^{-3}}{3.42 \times 10^{-4}} = 6.266$

$\therefore$ $n \approx 6$

$\because$ $(\mu - 1)t = n\lambda$

or $t = \frac{n\lambda}{(\mu - 1)} = \frac{6 \times 5890 \times 10^{-10}}{0.542}$

$t = \mathbf{6.52 \times 10^{-6} m.}$

EXAMPLE 16 Biprism is kept 10 cm away from the slit illuminated by monochromic light of $\lambda = 5896$ Å. The width of the fringes obtained on a screen placed at a distance of 90 cm from the biprism is 9.0×10^{-4} m. What is the distance between two coherent sources?

SOLUTION Given $a = 0.10$ m, $b = 0.90$ m and $D = a + b = 1.0$ m

$\lambda = 5.896 \times 10^{-7}$ m, $\beta = 9.0 \times 10^{-4}$ m

Formula used is $2d = \frac{\lambda D}{\beta} = \frac{5.896 \times 10^{-7} \times 1.0}{9 \times 10^{-4}}$

or $2d = \mathbf{6.55 \times 10^{-4} m}$

EXAMPLE 17 The distance between the slit and biprism and between biprism and screen are 50 cm each, Angle of biprism and refractive index are 179° and 1.5, respectively. Calculate the wavelength of light used if the distance between two successive fringes is 0.0135 m.

SOLUTION Given $\beta = 0.0135$ m, $a = b = 0.5$ m and $D = a + b = 1$

$\mu = 1.5, A = 179°, \alpha = \frac{180 - A}{2} = \left(\frac{1}{2}\right)^{\circ} \times \frac{\pi}{180} = \frac{\pi}{360}$ rad

Formula used is $l = \frac{2a(\mu - 1)\alpha}{D}\beta = \frac{2 \times 0.50 \times (1.5 - 1)}{1.0} \times \frac{\pi}{360} \times 0.0135$

$\lambda = \mathbf{5893\ Å}$

EXAMPLE 18 The distance between the slit and biprism and between biprism and eyepiece are 45 cm each. The obtuse angle of biprism is 178° and its refractive index is 1.5. If the fringe width is 15.6×10^{-3} cm, find the wavelength of light used.

SOLUTION $\beta = \dfrac{\lambda D}{2d}$ or $\lambda = \dfrac{\beta(2d)}{D}$

Given $a = 45$ cm, $D = 90$ cm, $\mu = 1.5$, $\alpha = 1° = \pi/180$ rad

$$\beta = 15.6 \times 10^{-3} \text{ cm}$$

$2d$ can be calculated by the relation

$$2d = 2a(\mu - 1)\alpha$$

$$= 2 \times 45 \times 0.5 \times (22/7) \times (1/180) = 0.786$$

$$\lambda = \frac{\beta(2d)}{0} = \frac{15.6 \times 10^{-3} \times 0.789}{90}$$

$$= \mathbf{13676\ Å}$$

EXAMPLE 19 In a biprism experiment, the eye piece was placed at a distance of 120 cm from the source. Calculate the wavelength of light, if the eye is required to move through a distance of 1.9 cm for 20 fringes and distance between two slits is 0.06 cm.

SOLUTION Given $x_n = 1.9$ cm, $n = 20$, $D = 120$ cm and $2d = 0.06$ cm.

Formula used is $b = \dfrac{x_n}{n} = \dfrac{1.9}{20} = 0.095$ cm and $\beta = \dfrac{\lambda D}{2d}$

or $\lambda = \dfrac{\beta 2d}{D} = \dfrac{0.095 \times 0.06}{120}$

$$\lambda = \mathbf{4750\ Å}$$

EXAMPLE 20 In a biprism experiment using light of wavelength 5890 Å, 40 fringes are observed in the field of view. If this light is replaced by light of wavelength 4358 Å. Calculate how many fringes are observed in the field of view.

SOLUTION Given $\lambda_1 = 5890$ Å, $N_1 = 40$ and $\lambda_2 = 4358$ Å, $N_2 = ?$

$$x = N_1\beta_1 = N_1\frac{\lambda_1 D}{2d} = N_2\frac{\lambda_2 D}{2d}$$

$\therefore$ $N_1\lambda_1 = N_2\lambda_2$

$$40 \times 5890 \times 10^{-10} = N_2 \times 4358 \times 10^{-10}$$

$$N_2 = \mathbf{54}$$

EXAMPLE 21 Light of wavelength 5893 Å is reflected at normal incidence from a soap film of refractive index 1.42. What is the least thickness of the film that will appear (a) bright and (b) dark?

SOLUTION Given $\lambda = 5.893 \times 10^{-7}$m, $i = r = 0$, $= 1.42$ and for smallest thickness $n = 1$

Condition for thin film to appear bright in reflected light is

$$2\mu t \cos r = (2n - 1)\lambda/2$$

or $t = \dfrac{(2n-1)\lambda/2}{2\mu \cos r} = \dfrac{(2-1) \times 5.893 \times 10^{-7}}{2 \times 1.42 \times 2 \times 1}$

$$= \mathbf{1.038 \times 10^{-4}\ mm}$$

Similarly condition for thin film to appear to dark in reflected light is

$$2\mu t \cos r = n\lambda$$

or $$t = \frac{n\lambda}{2\mu\cos r} = \frac{1\times 5.893\times 10^{-7}}{2\times 1.42\times 1}$$

$$= \mathbf{2.075\times 10^{-4}\,mm}$$

EXAMPLE 22 A parallel beam of light strikes an oil film ($\mu = 1.4$), fl floating on a surface of water ($\mu = 1.33$). When viewed at an angle of 30° from the normal 6th dark fringe is seen. Find the thickness of the film. (Given wavelength of light 589 nm).

SOLUTION Given $\lambda = 5.89\times 10^{-7}$ m, $\mu_{oil} = 1.4$, $i = 30°$ and $n = 6$

$$\mu_{oil} = \frac{\sin i}{\sin r} \text{ or } 1.4 = \frac{\sin 30°}{\sin r} \text{ or } \sin r = \frac{0.5}{1.4}$$

or $$\sin r = \mathbf{0.3571}$$

or $$r = \sin^{-1}(0.3571) = 20.92°$$

and $$\cos r = \cos 20.92 = 0.934$$

For n^{th} order dark fringe in reflected light, the condition is

$$2\mu t\cos r = n\lambda \text{ or } t = \frac{n\lambda}{2\mu\cos r}$$

$$t = \frac{6\times 5.89\times 10^{-7}}{2\times 1.4\times 0.934}$$

$$= \mathbf{1.351\times 10^{-3}\,mm}$$

EXAMPLE 23 Calculate the thickness of a soap film ($\mu = 1.463$) that will result in constructive interference in the reflected light, if the film is illuminated normally with light whose wavelength in free space is 6000 Å.

SOLUTION Given $\lambda = 6.0\times 10^{-7}$ m, $\mu = 1.463$, for normal incidence $i = r = 0°$ and for smallest thickness $n = 1$.

For constructive interference $2\,\mu t\cos r = (2n-1)\,\lambda/2$

$$t = \frac{(2n-1)\lambda}{2\times 2\times 1.463\times 1} = \frac{(2-1)\times 6.0\times 10^{-7}}{4\times 1.463}$$

$$= \mathbf{1.025\times 10^{-4}\,mm}$$

EXAMPLE 24 A parallel beam of sodium light ($\lambda = 5890$ Å) strikes a film of oil floating on water. When viewed at an angle of 30° from the normal, θ^{th} dark band is seen. Determine the thickness of the film. (Refractive index of oil = 1.46).

SOLUTION Given $\lambda = 5.89\times 10^{-7}$ m, $i = 30°$, $\mu = 1.46$ and $n = 8$

Condition for obtaining dark band is $2\mu t\cos r = n\lambda$ (i)

or $$t = \frac{n\lambda}{2\mu t\cos r}$$

As we know, $$\mu = \frac{\sin i}{\sin r} \quad \text{(ii)}$$

or $$r = \frac{\sin i}{\mu} \quad \text{(iii)}$$

or $$\sin r = \frac{\sin 30°}{1.46} = \frac{1}{2.92}$$

or $$\cos r = \sqrt{1-\sin^2 r}$$

By using Eq. (ii), we get $$= \sqrt{1-\left(\frac{1}{2.92}\right)^2}$$

$$= 0.94$$

$$t = \frac{8\times 5.89\times 10^{-7}}{2\times 1.46\times 0.94}$$

$$= \mathbf{1.72\times 10^{-3}\,mm}$$

Example 25 White light is reflected from an oil film of thickness 0.01 mm and refractive index 1.4 at an angle of 45° to the vertical. If the reflected light falls on the slit of a spectrometer, calculate the number of dark bands seen between wavelengths 4000 and 5000 Å.

Solution Given $t = 1.0\times 10^{-5}$ m, $\mu = 1.4$, $i = 45°$, $\lambda_1 = 4.0\times 10^{-7}$ m and $\lambda_2 = 5.0\times 10^{-7}$ m

Condition of dark bands in reflected light is

$$2\mu t \cos r = n\lambda \tag{i}$$

$$\mu = \frac{\sin i}{\sin r} \text{ or } \sin r = \frac{\sin i}{\mu}$$

or $$\sin r = \frac{\sin 45°}{1.4} = \frac{1/\sqrt{2}}{1.4} = \frac{1}{1.4\sqrt{2}}$$

$$\cos r = \sqrt{1-\sin^2 r} = \sqrt{1-\frac{1}{2\times 1.96}}$$

$$= 0.86$$

For wavelength λ_1, i.e, 4.0×10^{-7} m

$$2\mu t \cos r = n_1\lambda_1$$

$$n_1 = \frac{2\mu t\cos r}{\lambda_1} = \frac{2\times 1.4\times 1.0\times 10^{-5}\times 0.86}{4.0\times 10^{-7}} = 60.2$$

$$n_1 \approx 60$$

For wavelength λ_2, i.e, 5.0×10^{-7} m

$$n_2 = \frac{2\mu t\cos r}{\lambda_2} = \frac{2\times 1.4\times 1.0\times 10^{-5}\times 0.86}{5.0\times 10^{-7}}$$

$$= 48.16 \text{ or } n_2 \approx 48$$

$$n_1 - n_2 = 60 - 48 = \mathbf{12}$$

i.e, **12 dark bands are seen between wavelengths 4000 and 5000 Å.**

Example 26 A parallel beam of light of wavelength 5890 Å is incident on a glass plate having refractive index $\mu = 1.5$ such that the angle of refraction in the plate is 60°. Calculate the smallest thickness of glass plate which will appear dark by reflected light.

Solution Given $\lambda = 5.89\times 10^{-7}$ m, = 1.5 and $r = 60°$

Condition for the film to appear dark in reflected light is

$$2\mu t \cos r = n\lambda$$

For minimum thickness $n = 1$

$$t = \frac{\lambda}{2\mu \cos r} = \frac{5.89 \times 10^{-7}}{2 \times 1.5 \times 0.5}$$

$$= \mathbf{0.3927 \times 10^{-6} m}$$

EXAMPLE 27 A soap film of refractive index 1.333 is illuminated by white light incident at an angle of 45°. The light refracted by it is examined by a spectroscope and two consecutive bright bands are focused corresponding to the wavelength 6.1×10^{-5} cm and 6.0×10^{-5} cm. Find the thickness of the film.

SOLUTION Given $\mu = 1.333$, $i = 45°$, $\lambda_1 = 6.1 \times 10^{-7}$ m and $\lambda_2 = 6.0 \times 10^{-7}$ m

$$\mu = \frac{\sin i}{\sin r} \text{ or } \sin r = \frac{\sin 45°}{1.333} = \frac{1/\sqrt{2}}{1.333} = \frac{0.707}{1.333}$$

$$\sin r = 0.53$$

$$\cos r = \sqrt{1 - \sin^2 r} = \sqrt{1 - (0.53)^2} = 0.848$$

Condition of bright film to observe in transmitted case is

$$2\mu t \cos r = n\lambda_1 = (n + 1)\lambda_2$$

or

$$n \times 6.1 \times 10^{-7} = (n + 1) \times 6.0 \times 10^{7}$$

$$n = 60$$

and

$$t = \frac{n\lambda_1}{2\mu \cos r} = \frac{60 \times 6.1 \times 10^{-7}}{2 \times 1.333 \times 0.848}$$

$$= \mathbf{1.62 \times 10^{-5} mm}$$

EXAMPLE 28 A soap film suspended in air has thickness 5×10^{-5} cm viewed at an angle 35° to the normal. Find the wavelength of light in visible spectrum, which will be absent for a reflected light. The μ for the soap film is 1.33 and visible spectrum is in the range of 4000 to 7800 Å

SOLUTION Given $t = 5.0 \times 10^{-7}$ m, $i = 35°$ and $\mu = 1.33$

By using the relation

$$2\mu t \cos r = n\lambda \text{ and } \mu = \frac{\sin i}{\sin r}$$

$$\sin r = \frac{\sin i}{\sin \mu} \text{ and } \cos r = \sqrt{1 - \sin^2 r}$$

$$\cos r = \sqrt{1 - \left(\frac{\sin i}{\mu}\right)^2} = \sqrt{1 - \left(\frac{\sin 35}{1.33}\right)^2}$$

$$= 0.902$$

for first order ($n = 1$)

$$1 \cdot \lambda_1 = 2\mu + \cos r$$

$$\lambda_1 = 2 \times 1.33 \times 5.0 \times 10^{-7} \times 0.902$$

$$= 1.19 \times 10^{-6}$$

$$= 1200 \text{ Å}$$

In second order ($n = 2$)

$$2 \times \lambda_2 = 2\mu t + \cos r$$

or $$\lambda_2 = \mu + \cos r = 1.33 \times 5 \times 10^{-7} \times 0.902$$
$$= 6.0 \times 10^{-7} \text{ m (approx)}$$

In second order ($n = 3$)

$$3 \times \lambda_3 = 2\mu t + \cos r$$

or $$\lambda_3 = \frac{2 \times 1.33 \times 5.0 \times 10^{-7} \times 0.902}{3}$$
$$= 4000 \text{ Å (approx)}$$

In second order ($n = 4$)

$$4 \times \lambda_4 = 2\mu t + \cos r$$

or $$\lambda_4 = \frac{2 \times 1.33 \times 5.0 \times 10^{-7} \times 0.902}{4}$$
$$= \mathbf{3000 \text{ Å (approx)}}$$

Hence λ_2 and λ_3 wavelength of light in a visible spectrum will be absent.

Example 29 Calculate the thickness of a soap film ($\mu = 1.463$) that will result in constructive interference in the reflected light, if the film is illuminated normally with light whose wavelength in free space is 6000 Å.

Solution Given $\lambda = 6.0 \times 10^{-7}$ m, in this case $n = 1$, and for normal incidence $i = 0$ and $r = 0$

Condition for constructive interference $2\mu t \cos r = (2n - 1)\lambda/2$

$$t = \frac{(2n-1)\lambda}{4\mu \cos r} = \frac{1 \times 6.0 \times 10^{-7}}{4 \times 1.463 \times 1}$$
$$= \mathbf{1.025 \times 10^{-7} \text{ m.}}$$

Example 30 A thin film is illuminated by white light at an angle of incidence ($i = \sin^{-1}(4/5)$. In reflected light, two dark consecutive overlapping fringes are observed corresponding to wavelengths 6.1×10^{-7} m and $6.0 \times 1\ 0^{-7}$ m. The refractive index of the film is 4/3. Calculate the thickness of the film.

Solution Given $\lambda_1 = 6.1 \times 10^{-7}$ m, $\lambda_2 = 6.0 \times 10^{-7}$ m and $\mu = 4/3$

$$\mu = \frac{\sin i}{\sin r} \text{ or } \sin r = \frac{\sin i}{\mu} = \frac{4/5}{4/3} = \frac{3}{5}$$

and $$\cos r = \sqrt{1 - \sin^2 r} = \sqrt{1 - \frac{9}{25}} = \frac{4}{5} = 0.8$$

Condition for dark fringes is

$$2\mu t \cos r = n\lambda_1 = (n+1)\lambda_2$$

or $$n\lambda_1 = (n+1)\lambda_2$$
$$n \times 6.1 \times 10^{-7} = (n+1)\, 6.0 \times 10^{-7}$$
$$n(6.1 - 6.0) \times 10^{-7} = 6.0 \times 10^{-7}$$

or $$n = 60$$

and $$2\mu t \cos r = n\lambda_1$$

or $$t = \frac{60 \times 6.1 \times 10^{-7}}{2 \times \frac{4}{3} \times 0.8}$$

$$t = \mathbf{1.716 \times 10^{-2} \text{ mm}}$$

EXAMPLE 31 Two plane glass surfaces in contact along one edge are separated at the opposite edge by a thin wire. If 20 interference fringes are observed between these edges in sodium light at normal incidence. What is the thickness of wire?

SOLUTION Given $\lambda_{av} = 5893$ Å, $n = 20$, $i = r = 0$ and $\mu = 1$

$$w = \frac{\lambda}{2\mu\theta} = \frac{5.893 \times 10^{-7}\,\text{m}}{2 \times 1 \times \theta}$$

or

$$w\theta = \frac{5.893 \times 10^{-7}}{2}$$

From Fig. 1.32

$$\theta = \frac{t}{20\omega}$$

or

$$t = 20\,w\theta$$

$$= 20 \times \frac{5.893 \times 10^{-7}}{2}$$

$$\boldsymbol{t = 5.893 \times 10^{3}}\ \textbf{mm.}$$

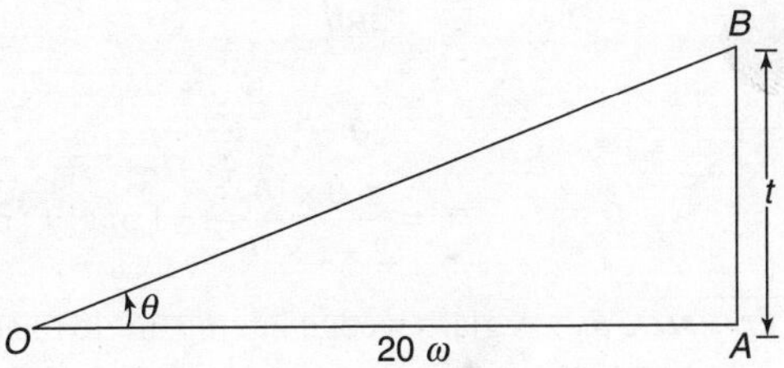

FIGURE 1.32

EXAMPLE 32 A wedge air film is enclosed between two glass plates touching at one edge and separated by a wire of 0.06×10^{-3} m diameter at a distance of 0.1 5 m from the edge. Calculate the fringe width. The light of wavelength 6.0×10^{-7} m from the broad source is allowed to fall normally on the film.

SOLUTION Given $\lambda = 6.0 \times 10^{-7}$ m and $\mu = 1$

$$w = \frac{\lambda}{2\mu\theta} \qquad \text{(i)}$$

From Fig. 1.33

$$\theta = \frac{6.0 \times 10^{-5}}{0.15} \qquad \text{(ii)}$$

$\therefore$

$$w = \frac{6.0 \times 10^{-7} \times 0.15}{2 \times 1 \times 6.0 \times 10^{-5}}$$

$$w = \textbf{0.75 mm.}$$

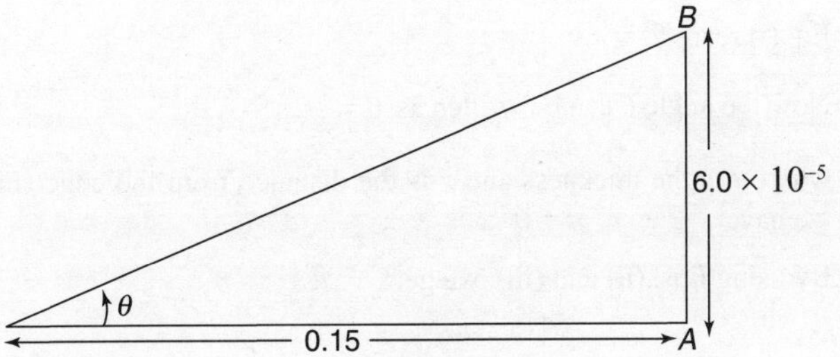

FIGURE 1.33

EXAMPLE 33 A wedge shaped film is illuminated by light of wavelength 4650 Å. The angle of wedge is 40″. Calculate the fringe separation between two consecutive fringes.

SOLUTION Given $\lambda = 4.56 \times 10^{-7}$ m and $\mu = 1$

$$\theta = 40'' = \frac{40}{3600} \times \frac{\pi}{180°}\,\text{rad}$$

$$= 1.9 \times 10^{-4}\,\text{rad}$$

$\therefore$

$$w = \frac{\lambda}{2\mu\alpha} = \frac{4.65 \times 10^{-7}}{2 \times 1 \times 1.9 \times 10^{-4}}$$

$$\textbf{w = 1.2 mm}$$

EXAMPLE 34 Two glass plates enclose a wedge-shaped air film touching at one edge are separated by a wire of 0.03 mm diameter at distance 15 cm from the edge. Monochromatic light ($\lambda = 6000$ Å) from a broad source falls normally on the film. Calculate the fringe-width.

SOLUTION Given $\lambda = 6.0 \times 10^{-7}$ m and $\mu = 1$

$$\left[\text{Angle } \theta = \frac{\text{Arc }(AB)}{\text{radius}}\right]$$

$$\theta = \frac{0.03 \times 10^{-3}}{0.15}$$

$$= 2.0 \times 10^{-4} \text{rad}$$

$$w = \frac{\lambda}{2\mu\theta}$$

$$= \frac{\lambda}{2\theta}$$

$$w = \frac{6.0 \times 10^{-7}}{2 \times 2 \times 10^{-4}} = 1.5 \times 10^{-3} \text{m} = \mathbf{1.5\ mm}$$

EXAMPLE 35 A glass wedge having angle 0.01 radian is illuminated normally by light of wavelength 5890 Å. At what distance from the edge of the wedge, will the 12th dark fringe be observed by reflected light?

SOLUTION Given $\lambda = 5.89 \times 10^{-7}$ m, $n = 12$, $\theta = 0.01$ rad and $\mu = 1$

Condition for obtaining dark fringe is

$$2\mu t \cos(r + \theta) = n\lambda \quad \text{(i)}$$

For normal incidence $i = r = 0$ and when θ is very small

$$\cos\theta \approx 1$$

Eq. (i) reads $2t = n\lambda$ (ii)

Now the angle θ can be written as $\theta = \dfrac{t}{x}$

where t is the thickness and x is the distance from the edge (Fig. 1.34) then we have

$$t = \theta \cdot x \quad \text{(iii)}$$

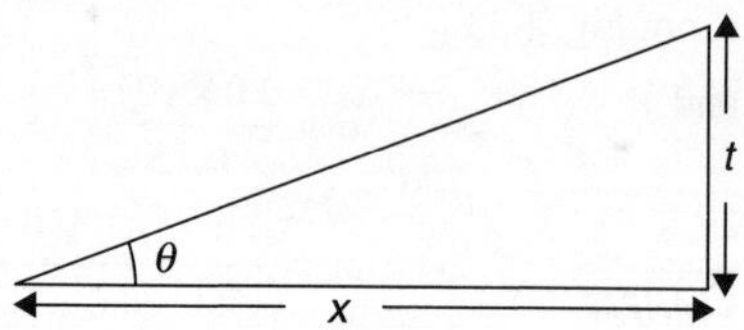

FIGURE 1.34

By using Eqs. (ii) and (iii), we get

$$2\theta \cdot x = n\lambda$$

or

$$x = \frac{n\lambda}{2\theta} = \frac{12 \times 5.89 \times 10^{-7}}{2 \times 0.01} = 3.5 \times 10^{-4} \text{ m}$$

$$\mathbf{x = 0.35\ mm.}$$

EXAMPLE 36 A glass wedge of angle 0.01 radian of illuminated by monochromatic light of wavelength 6000 Å falling normally on it. At what distance from the edge of the wedge will the 10th fringe be observed by reflected light?

SOLUTION Given $\alpha = 0.01$ radian, $\lambda = 10 \times \lambda = 6.0 \times 10^{-7}$ m the condition for dark fringe

$$= 2t = n\lambda \quad \text{(i)}$$

The angle of wedge $\alpha = \dfrac{t}{x}$

or

$$t = \alpha x \quad \text{(ii)}$$

Put the value of t from Eq. (ii) in Eq (i), we get

$$2\alpha x = n\lambda$$

$$x = \frac{n\lambda}{2\alpha} = \frac{10 \times 6.0 \times 10^{-7}}{2 \times 0.01} = \mathbf{3 \times 10^{-4}\ m}$$

EXAMPLE 37 Interference fringes are produced when monochromatic light is incident normally on a thin wedge-shaped film of refractive index 1.5. If the distance between two consecutive fringes is 0.02 mm. Find the angle of the film, the wavelength of light being 5.5×10^{-5} cm.

SOLUTION Given $\mu = 1.5$, $w = 0.02 \times 10^{-3}$ m and $\lambda = 5.5 \times 10^{7}$ m.

$$w = \frac{\lambda}{2\mu\theta} \text{ or } \theta = \frac{\lambda}{2\mu w} = \frac{5.5 \times 10^{-7}}{2 \times 1.5 \times 0.02 \times 10^{-3}}$$

$$= 0.009166 \text{ rad} = \mathbf{0.525°}$$

EXAMPLE 38 In Newton's rings experiment, the diameter of the 15th ring was found to be 0.59 cm and that of the 5th ring was 0.336 cm. If the radius of the plano convex lens is 100 cm, compute the wavelength of light used.

SOLUTION Given $D_{15} = 5.9 \times 10^{-3}$ m, $D_5 = 3.36 \times 10^{-3}$ m, $p = 10$ and $r = 1.0$ m.

Formula used is $$\lambda = \frac{D^2(n+p) - D_n^2}{4pR} = \frac{[(5.9)^2 - (3.36)^2] \times 10^{-6}}{4 \times 10 \times 1.0}$$

$$\lambda = \mathbf{5880 \text{ Å}}$$

EXAMPLE 39 In a Newton's rings experiment the radius of 10th and 20th rings are 0.2 and 0.3 cm, respectively, and the focal length of the plano-convex lens is 90 cm. Calculate the wavelength of light used in nanometers.

SOLUTION Given $f = 0.9$ m, $\mu = 1.5$, $D_{10} = 0.2$ cm and $D_{20} = 0.3$ cm. $p = 10$

Formula used is $$\frac{1}{f} = (\mu - 1)\left(\frac{1}{R_1} - \frac{1}{R_2}\right) = (1.5 - 1)\left[\frac{1}{R_2} - \frac{1}{\infty}\right], R = \infty$$

$$\frac{1}{0.9} = 0.5\left[\frac{1}{R_1}\right] \text{ or } R_1 = R = 0.45 \text{ m}$$

and $$\lambda = \frac{D_{(n+p)}^2 - D_n^2}{4pR} = \frac{D_{15}^2 - D_5^2}{4 \times 10 \times 0.45} = \frac{[(0.3)^2 - (0.2)^2] \times 10^{-4}}{4 \times 10 \times 0.45}$$

$$\lambda = \mathbf{277.8 \text{ nm}}$$

EXAMPLE 40 In a Newton's rings arrangement a thin convex lens of focal length 1.0 m. ($\mu = 1.5$) remains in contact with an optical flat and light of wavelength 5896×10^{-10} m is used. Newton's rings are observed normally by reflected light. What is the diameter of 7th bright ring?

SOLUTION Given $\mu = 1.5$, $f = 1.0$ m and $\lambda = 5.896 \times 10^{-7}$ m

$$R_1 = R \text{ and } R_2 = R$$

Formula used is $$\frac{1}{f} = (\mu - 1)\left(\frac{1}{R_1} - \frac{1}{R_2}\right)$$

or $$\frac{1}{1.0} = (1.5 - 1)\left[\frac{1}{R} + \frac{1}{R}\right]$$

or $$\frac{2}{R} = \frac{1}{0.5} \quad \text{or} \quad R = 1.0 \text{ m}$$

now $$D_n^2 = 4n\lambda R$$

for $n = 7$

$$D_7 = \sqrt{4 \times 7 \times 5.896 \times 10^{-7} \times 1.0}$$

$$D_7 = \mathbf{4.063 \times 10^{3} \text{ m}}$$

EXAMPLE 41 Light source emitting the light of wavelengths $\lambda_1 = 6.0 \times 10^{-7}$ m and $\lambda_2 = 4.8 \times 10^{-7}$ m is used to obtain Newton's rings in reflected light. It is found that the nth dark ring of λ_1 coincides with $(n+1)^{th}$ dark ring of λ_2. If the radius of curvature of the curved surface of the lens is 0.96 m. Calculate the diameter of $(n+1)^{th}$ dark ring of λ_2.

SOLUTION Given $\lambda_1 = 6.0 \times 10^{-7}$ m, $\lambda_2 = 4.8 \times 10^{-7}$ m and $R = 0.96$ n

The diameter of nth order dark ring of λ_1 is

$$D_n^2(\lambda_1) = 4n\lambda_1 R$$

Similarly, the diameter of $(n+1)^{th}$ order dark ring of λ_2

$$D_{(n+1)}^2(\lambda_2) = 4(n+1)\lambda_2 R$$

Since, $$D_n^2(\lambda_1) = D_{(n+1)}^2(\lambda_2)$$

$$4n\lambda_1 = 4(n+1)\lambda_2$$

or $$\frac{n+1}{n} = \frac{\lambda_1}{\lambda_2}$$

or $$1 + \frac{1}{n} = \frac{\lambda_1}{\lambda_2} \text{ or } \frac{1}{n} = \frac{\lambda_1 - \lambda_2}{\lambda_2}$$

or $$n = \frac{\lambda_2}{\lambda_1 - \lambda_2} = \frac{4.8 \times 10^{-7}}{(6.0 - 4.8) \times 10^{-7}} = 4.0$$

or $$n = 4$$

Hence, $$D_{(n+1)}^2(\lambda_2) = 4(n+1)\lambda_2 R = 4 \times 5 \times 4.8 \times 10^{-7} \times 0.96$$

$$D_{(n+1)} = 3.0358 \times 10^{-3}$$

or $$\mathbf{D_{(n+1)} = 3.04 \times 10^{-3} \text{ m.}}$$

EXAMPLE 42 In Newton's ring arrangement a source is emitting two wavelengths $\lambda_1 = 6.0 \times 10^{-7}$ m and $\lambda_2 = 5.9 \times 10^{-7}$ m. It is found that n^{th} dark ring due to one wavelength coincides with $(n+1)^{th}$ dark ring due to the other. Find the diameter of the nth dark ring if radius of curvature of the lens is 0.9 m.

SOLUTION Given $\lambda_1 = 6.0 \times 10^{-7}$ m, $\lambda_2 = 5.9 \times 10^{-7}$ m and $R = 0.9$ m.

The diameter of the nth order dark ring of λ_1 is

$$D_n^2(\lambda_1) = 4n\lambda_1 R$$

The diameter of the $(n+1)^{th}$ order dark ring of λ_2 is

$$D_{(n+1)}^2(\lambda_2) = 4_{(n+1)}\lambda_2 R$$

Since two rings coincide

$$4n\lambda_1 R = 4(n+1)\lambda_2 R$$

$$\frac{n+1}{n} = \frac{\lambda_1}{\lambda_2} \text{ or } n = \frac{\lambda_2}{\lambda_1 - \lambda_2}$$

$$n = \frac{5.9 \times 10^{-7}}{(6.0 - 5.9) \times 10^{-7}} = \mathbf{59}$$

Now $$D_n = \sqrt{4n\lambda_1 R} = \sqrt{4 \times 59 \times 6.0 \times 10^{-7} \times 0.9}$$

$$= 0.01128 \text{ m}$$

$$\mathbf{D_n = 0.0113 \text{ m}}$$

Example 43 Newton's rings are formed using light of wavelength 5896 Å in reflected light with a liquid placed between plane and curved surfaces. The diameter of 7th bright fringe is 0.4 cm and the radius of curvature is 1.0 m. Evaluate the refractive index of liquid.

Solution Given $D_7 = 4.0 \times 10^{-3}$ m, $\lambda = 5.896 \times 10^{-7}$ m, $R = 1.0$ m and $n = 7$.

$$D_n'^2 = \frac{2(2n-1)\lambda R}{\mu} \quad \text{or} \quad \mu = \frac{2(2n-1)\lambda_R}{D_n'^2}$$

$$\mu = \frac{2 \times 13 \times 5.896 \times 10^{-7} \times 1.0}{(4 \times 10^{-3})^2}$$

$$\boldsymbol{\mu = 0.96}$$

Example 44 If the diameter of nth dark ring in an arrangement giving Newton's ring changes from 0.3 cm and 0.25 cm as liquid is introduced between the lens and the plate, calculate the value of the refractive index of the liquid and also calculate the velocity of light in the liquid. Velocity of light in vacuum is 3×10^8 m/sec.

Solution Given $D_n = 3.0 \times 10^{-3}$ m, $D_n = 2.5 \times 10^{-3}$ m

Formula used is $D_n^2 = 4n\lambda R$

$$D_n^2 = \frac{4n\lambda R}{\mu}$$

$$\mu = \frac{D_n^2}{D_n^2} = \left[\frac{3.0 \times 10^{-3}}{2.5 \times 10^{-3}}\right]^2 = 1.44$$

$$\mu = 1.44$$

or $$V_{liq} = \frac{c}{\mu} = \frac{3 \times 10^8}{1.44}$$

$$\boldsymbol{V_{liq} = 2.08 \times 10^8 \text{ m/sec}}$$

Example 45 The Newton's rings are seen in reflected light of wavelength 5896Å. The radius of curvature of plano-convex lens is 1.0 meter. An air film is replaced by a liquid whose refractive index is to be calculated under the conditions if 16th ring is dark and its diameter is 5.1 mm.

Solution Given $D_{16} = 5.1 \times 10^{-3}$ m, $\lambda = 5.896 \times 10^{-7}$ m and $R = 1.0$ m

Formula used is $$D_n^2 = \frac{4n\lambda R}{\mu} \quad \text{or} \quad \frac{4n\lambda R}{D_n^2}$$

$$\mu = \frac{4 \times 16 \times 5.896 \times 10^{-7} \times 1.0}{(5.1 \times 10^{-3})^2}$$

$$\boldsymbol{\mu = 1.45}$$

Example 46 The Newton's rings are observed in reflected light of wavelength 6300 Å. A thin layer of liquid of refractive index 1.63 is formed between curved surface of plano-convex lens ($\mu = 1.69$) and plane glass plate ($\mu = 1.03$) and the radius of curvature of the curvex lens is 0.9 m. Find the radius of smallest dark ring.

Solution Given $\lambda = 6.3 \times 10^{-7}$ m, $\mu = 1.63$ and $R = 0.9$ m

Formula used is $$r_n'^2 = \frac{n\lambda R}{\mu} \quad (n = 1 \text{ for a smallest dark ring})$$

$$r_n'^2 = \frac{1 \times 6.3 \times 10^{-7} \times 0.9}{1.63} = 34.7853 \times 10^{-8} \text{m}$$

$$\boldsymbol{r_1^2 = 5.9 \times 10^{-4} \text{ m} = 0.59 \text{ mm}}$$

EXAMPLE 47 Newton's rings are observed with two different media between the glass surfaces. The nth rings have diameters as 10 : 7. Find the ratio of the refractive indices of the two media.

SOLUTION Given $Dn' : Dn'' = 10 : 7$

$$\because \quad D_n^2 = \frac{4n\lambda R}{\mu} \qquad \text{(i)}$$

For the first medium (μ_1)

$$D_n'^2 = \frac{4n\lambda R}{\mu_1} \qquad \text{(ii)}$$

For the first medium (μ_1)

$$D_n''^2 = \frac{4n\lambda R}{\mu_2} \qquad \text{(iii)}$$

$$\frac{\mu_1}{\mu_2} = \frac{D_n''^2}{D_n'^2} = \left[\frac{7}{10}\right]^2 = \frac{49}{100}$$

or $\quad \boldsymbol{\mu_1 : \mu_2 = 49 : 100}$

EXAMPLE 48 A combination of convex lens plane glass plate is illuminated by monochromatic light. The diameter of the 10th dark ring is measured in reflected light and is found to be 0.48 cm. Find the wavelength of light used. The radius of curvature of the lower face of the lens is 90 cm.

SOLUTION Given $R = 0.9$ m, $D_{10} = 4.8 \times 10^3$ m and $n = 10$

Formula used is $D_n^2 = 4n\lambda R$

or $$\lambda = \frac{D_n^2}{4nR}$$

or $$\lambda = \frac{(4.8 \times 10^{-3})^2}{4 \times 10 \times 0.9}$$

$$\boldsymbol{\lambda = 6400 \text{ Å}}$$

EXAMPLE 49 In Newton's rings experiment the diameter of 5th dark ring is reduced to half of its value after placing a liquid between plane glass plate and convex surface. Calculate the refractive index of liquid.

SOLUTION Given $D's = \dfrac{D_5}{2} \cdot \mu = ?$

Formula used is $D_n^2 = 4n\lambda R = D_5^2 = 4 \times 5 \times \lambda R$ or $D_5 = \sqrt{20\lambda R}$

$$\therefore \quad D_5'^2 = \frac{4 \times 5 \times \lambda R}{\mu} \quad \text{or} \quad D_5' = \sqrt{\frac{20\lambda R}{\mu}}$$

$$D_5' = \frac{D_5}{2} \quad \text{or} \quad \sqrt{\frac{20\lambda R}{\mu}} = \frac{\sqrt{20\lambda R}}{2}$$

or $$\frac{1}{\sqrt{\mu}} = \frac{1}{2} \text{ or } \mu = 4$$

or $\quad \boldsymbol{\mu = 4}$

EXAMPLE 50 Newton's rings by reflection are formed between two bi-convex lenses having equal radii of curvatures as 100 cm each. Calculate the distance between the 5th and 15th dark rings, using monochromatic light of wavelength 5400 Å.

SOLUTION Given $\lambda = 5.4 \times 10^{-7}$ m, $R_1 = R_2 = 1.0$ m

Formula used is $r_n^2 = \dfrac{n\lambda}{\left\{\dfrac{1}{R_1} + \dfrac{1}{R_2}\right\}}$ or $r_5 = \sqrt{\dfrac{5 \times 5.4 \times 10^{-7} \times 100}{\dfrac{1}{2} + \dfrac{1}{2}}} = 11.62 \times 10^{-4}\text{m}$

or $r_5 = 0.1162$ cm.

Similarly $r_{15} = \sqrt{\dfrac{15 \times 5.4 \times 10^{-7}}{2}} = 2.012 \times 10^{-3}$ m $= 0.2012$ cm

$\therefore$ Distance between 5th and 15th rings $= r_{15} - r_5 =$ **0.085 cm**

EXAMPLE 51 A Michelson interferometer is set for the white straight fringes. When a mica sheet of thickness 0.005 cm is put in front of the fixed mirror, then in order to bring back the coloured fringes to their original position, the movable mirror is moved by 0.0025 cm. Calculate the refractive index of mica.

SOLUTION Given $x = 2.5 \times 10^{-5}$ m and $t = 5.0 \times 10^{-5}$ m

Formula used is $2x = 2(\mu - 1)t$

where t is the thickness of mica sheet and is the refractive index

or $\mu = \dfrac{x}{t} + 1$

$$\mu = \frac{2.5 \times 10^{-5}\text{m}}{5.0 \times 10^{-5}} + 1 = 1.5$$

or $\mu = \mathbf{1.5}$

EXAMPLE 52 If a movable mirror of Michelson's interferometer is moved through a distance 0.06 mm, 200 fringes crossed the field of view. Find the wavelength of light.

SOLUTION Given $x = 6.0 \times 10^{-5}$ m and N = 200 fringes

Formula used is $x = N\lambda/2$

where x is the separation of movable mirror from the fixed mirror, then $6.0 \times 10^{-5} = 200 \times \lambda/2$ or = **6000 Å**

EXAMPLE 53 In Michelson's interferometer a thin plate is introduced in the path of one of the beams and it is found that 50 band crosses the line of observation. If the wavelength of light used is 5896 Å and $\mu = 1.4$, determine the thickness of the plate.

SOLUTION Given $n = 50$, $\lambda = 5896 \times 10^{-10}$ m and $\mu = 1.4$

Formula used is $2(\mu - 1)\, t = n\lambda$

or $= \dfrac{n\lambda}{2(\mu - 1)} = \dfrac{50 \times 5896 \times 10^{-10}}{2(1.4 - 1)}$

$t = \mathbf{3.68 \times 10^{-5}}$ **m**

EXAMPLE 54 Calculate the distance between successive positions of the movable mirror of Michelson's interferometer giving best fringes in case of a sodium source having wavelengths 5896 Å and 5890 Å. What will be the change in path difference between two successive reappearances of the interference pattern?

SOLUTION Given $\lambda_1 = 5.896$ m and $10^{-7}\lambda_2 = 5.89 \times 10^{-7}$ m

Formula used is $\Delta\lambda = (\lambda_1 - \lambda_2) = \dfrac{\lambda_1\lambda_2}{2x}$

or $$x=\frac{\lambda_1\lambda_2}{2(\lambda_1-\lambda_2)}=\frac{5.896\times 5.89\times 10^{-14}}{2\times 6\times 10^{-10}}$$

$$= 0.289 \text{ mm}$$

The path difference will be equal to

$$= 2x = 2\times 0.289 \text{ mm}$$

$$= \mathbf{0.5788\ mm}.$$

EXAMPLE 55 In Michelson's interferometer 100 fringes cross the field of view when the movable mirror is displaced through 0.02948 mm. Calculate the wavelength of monochromatic light used.

SOLUTION Given $x = 0.02948 \times 10^{-3}$ m and $n = 100$

Formula used is $$2x = n\lambda \text{ or } \lambda=\frac{2x}{n}=\frac{2\times 2.948\times 10^{-5}\text{ m}}{100}$$

or $$\lambda = 5.896 \times 10^{-7} \text{ m} = \mathbf{5896\ Å}$$

EXAMPLE 56 The wavelength of two components of D-lines of sodium are 5890 Å and 5896 Å. By how much distance one of the mirror of Michelson's interferometer be moved so as to obtain consecutive position of maximum distinctness.

SOLUTION Given $\lambda_1 = 5.896 \times 10^{-7}$ m and $\lambda_2 = 5.89 \times 10^{-7}$ m

Formula used $\Delta\lambda = \lambda_1 - \lambda_2 = \frac{\lambda_1\lambda_2}{2x}$

where x is distance through which the movable mirror is moved from one position of maxima to the next, then we have

$$x=\frac{\lambda_1\lambda_2}{2(\lambda_1-\lambda_2)}=\frac{5.896\times 5.89\times 10^{-14}}{2\times 6\times 10^{-10}}$$

$$\mathbf{x = 0.289\ mm}$$

EXAMPLE 57 In an experiment with Michelson's interferometer, the distance traveled by the mirror for two successive position of maximum distinctness was 0.2945 mm. If the mean wavelength for the two component of sodium D–line is 5893 Å, calculate the difference between the two wavelengths.

SOLUTION Given $x = 0.2945 \times 10^{-3}$ m and $\lambda_{av} = 5.893 \times 10^{-7}$ m

Formula used is $$\Delta\lambda = \lambda_1 - \lambda_2 = \frac{\lambda_{av}^2}{2x}=\frac{(5.893\times 10^{-7})^2}{2\times 0.2945\times 10^{-3}}$$

$$\Delta\lambda = \mathbf{5.896\ Å}.$$

EXAMPLE 58 In an experiment for determining the refractive index of a gas using Michelson's interferometer a shift of 140 fringes is observed when all the gas is removed from the tube. If the wavelength of light used is 5460 Å and the length of the tube is 20 cm, calculate the refractive index of the gas.

SOLUTION Given, $\lambda = 5.46 \times 10^{-7}$ m, $t = 0.2$ m and $n = 140$

Formula used $$2(\mu - 1)\, t = n\lambda$$

or $$\mu=\frac{n\lambda}{2t}+1=\frac{140\times 5.46\times 10^{-7}}{2\times 0.2}+1$$

$$= 0.00019 + 1$$

$$\mu = \mathbf{1.00019}$$

OBJECTIVE TYPE QUESTIONS

Q.1 Light waves are
(a) longitudinal waves (b) transverse waves
(c) both (a) and (b) (d) none of them

Q.2 Which of the following does not support the wave nature of light?
(a) interference (b) polarisation (c) compton effect (d) diffraction

Q.3 Colours in thin films are because of
(a) dispersion (b) diffraction (c) interference (d) none of them

Q.4 A phase difference π between two interfering beams is equivalent to path deference
(a) 2λ (b) λ (c) $\lambda/2$ (d) none of them

Q.5 The central fringe in Fresnel's biprism experiment in the reflected light is
(a) dark (b) non-uniform (c) bright (d) none of them

Q.6 In reflected light the central fringe of Newton's ring is
(a) dark (b) bright (c) non-uniform (d) none of them

Q.7 In Newton's rings experiment, the diameter of bright rings is proportional to
(a) odd natural numbers (b) natural numbers
(c) even natural numbers (d) square root of natural numbers

Q.8 Extended source is needed in
(a) Young's double slit experiment (b) Biprism experiment
(c) Newton's rings experiment (d) none of them

Q.9 Second glass plate in Michelson's interferometer is known as
(a) extra glass plate (b) compensating glass plate
(c) simple glass plate (d) none of them

Q.10 Intensity of light is proportional to
(a) double of its amplitude (b) square root of amplitude
(c) square of amplitude (d) none of these

Q.11 Which of the following device cannot produce coherent beam of light?
(a) Fresnel's biprism (b) Young's double slit
(c) Half-shade polarimeter (d) Lloyd's mirror

Q.12 Which of the following method(s) is/are based on division of amplitude?
(a) Newton's rings (b) Michelson interferometer
(c) (a) and (b) (d) None of these

Q.13 Which of the following method(s) is(are) based on division of wave front?
(a) Young's double slit (b) Newton's rings
(c) Fresnel's biprism (d) (a) and (c)

Q.14 What is the actual shape of interference fringes in Young's double slit experiment?
(a) circular (b) elliptical (c) parabolic (d) hyperbolic

Q.15 The phenomenon which cannot be explained by wave theory of light is
(a) Interference (b) Photoelectric effect
(c) Polarisation (d) Diffraction

Q.16 In which of the following the interference is produced by the division of amplitude?
(a) Lloyd's mirror (b) Newton's rings
(c) Young's double slit experiment (d) Fresnel's biprism

Q.17 When monochromatic light is replaced by white light in Fresnel's biprism arrangement, the central fringe is
(a) dark (b) coloured (c) white (d) none of these

Q.18 In Newton's rings arrangement, bright and dark rings are obtained using sodium yellow light. What happens if the top surface of the glass plate on which the lens is kept is highly silvered?
(a) fringes disappear (b) fringe width remains unchanged
(c) fringe width decreases (d) none of these

SHORT-ANSWER QUESTIONS

Q.1 What is a wavefront?
Q.2 What do you understand by phase difference and path difference?
Q.3 What do you understand by coherent sources?
Q.4 Discuss why two independent sources of light can never be coherent?
Q.5 How are two coherent sources obtained in practice?
Q.6 What are the conditions for interference of light?
Q.7 What are the conditions for maxima and minima in an interference pattern?
Q.8 Define fringe width.
Q.9 Explain fringe width obtained in Newton's rings experiment.
Q.10 Distinguish between division of wavefront and division of amplitude?
Q.11 What is Fresnel's biprism?
Q.12 Explain the formation of coherent sources by the use of Fresnel's biprism.
Q.13 Explain the effect of placing a very thin film in the path of one of the interfering beams.
Q.14 Explain the formation of colours when the white light is incident on a transparent thin film.
Q.15 Why a thick film shows no colours in reflected white light?
Q.16 Explain the necessity of extended source in interference with division of amplitude.
Q.17 What are Newton's rings?
Q.18 Explain why Newton's rings are circular.
Q.19 Discuss Michelson's interferometer.
Q.20 Give the applications of Michelson's interferometer.

PRACTICE PROBLEMS

Q.1 Briefly outline the wave theory of light. What is wavefront? How does it propagate?
Q.2 Explain clearly Huygens' principle for the propagation of light.
Q.3 What are coherent sources? What are the conditions for two sources to be coherent? How are they realised in practice? Can two independent sources become coherent?

Q.4 Define interference and explain the phenomenon of interference of light.

Q.5 Give the conditions for producing good interference fringes.

Q.6 State and explain in brief the conditions for (a) observance (b) good contrast of fringes and (c) stationary interference pattern.

Q.7 Discuss briefly the phenomenon of interference with relation to law of conservation of energy.

Q.8 Obtain an expression for fringe width in case of Young's double slit experiment. Prove that in this case of interference dark and bright bands are of equal width.

Q.9 A Young's double slits experiment is carried out with monochromatic light in air. What will be the change in wavelength and fringe width when the apparatus is immersed in water or the medium is replaced by an optically denser medium?

Q.10 Discuss how coherent sources are produced with the help of Fresnel's biprism. Explain with necessary theory how the wavelength of monochromatic light can be determined using a Fresnel's biprism?

Q.11 Describe Fresnel's biprism. Discuss in detail how the wavelength of monochromatic source of light can be determined with its help?

Q.12 Explain the formation of interference fringes by means of Fresnel's biprism when a monochromatic source of light is used, and derive the expression for the fringe width. How will you measure the wavelength of monochromatic light using biprism method?

Q.13 Describe the geometrical features of Fresnel's biprism. How can it be used to find the wavelength of light?

Q.14 Describe the construction, theory and working of Fresnel's biprism experiment to find the wavelength of light?

Q.15 Discuss the effect of introducing a thin mica sheet in the path of one of the interfering beams in a experiment. Deduce an expression for the displacement of the fringes. Describe how this method is used for finding the thickness of a thin glass plate?

Q.16 Explain the effect of introducing a thin plate of glass in the path of one of the interfering beams in the experiment. Calculate the displacement of fringes. Show how this method can be used for finding the refractive index, thickness of the plate and wavelength of light?

Q.17 Using an optical method how would you determine the thickness of a piece of transparent cello tape? Explain.

Q.18 Discuss the phenomenon of interference in thin films. Obtain the conditions for maxima and minima. Show that the interference pattern in the reflected and transmitted systems is complimentary. Why an extremely thin film appears black in reflected light?

Q.19 Why a broad source of light is necessary for observing colours in thin films?

Q.20 Explain analytically the colour of thin films. Why are the colours in reflected and transmitted light complementary?

Q.21 Describe the interference observed when a thin parallel shaped film is seen by reflected light normally.

Q.22 Explain how interference fringes are formed by a thin wedge shaped film when examined by normally reflected light. Find the expression for fringe-width. How will you estimate the difference of film thickness between two points?

Q.23 Explain the formation of Newton's rings in reflected light?

Q.24 What are Newton's rings? Explain can these be used to find the wavelength of light?

Q.25 Explain with theory, Newton's rings method to determine the wavelength of monochromatic light and discuss which source is preferred, point source or extended source?

Q.26 Discuss the formation of Newton's rings by reflected light. Describe the experimental arrangement and give necessary theory. Why are Newton's rings circular?

Q.27 Explain the phenomenon of interference in thin film and also explain with theory of Newton's rings experiment to find the wavelength of monochromatic light.

Q.28 With the help of a neat diagram show an experimental arrangement to produce Newton's rings by reflected sodium light. Prove that in reflected light the diameter of the dark rings is proportional to the square root of the natural number.

Q.29 Describe the principle of construction and working of Michelson's interferometer.

Q.30 Describe the principle, construction, theory and working of Michelson interferometer to find the wavelength and the difference in wavelength of a given light.

Q.31 Explain the working of Michelson interferometer. How will you produce circular fringes with it? How will you measure the difference in wavelength between D lines of sodium light?

Q.32 How do you obtain localised fringes in Michelson's interferometer? How straight fringes are obtained?

Q.33 How will you find the wavelength of monochromatic light with Michelson's interferometer?

Q.34 How will you use Michelson's interferometer to determine the thickness of a thin transparent sheet?

Q.35 Describe the Michelson's interferometer. How will you use it to calibrate a meter in terms of a standard wavelength?

Unsolved Questions

Q.1 Calculate the coherence length for a white light when its the wavelength ranges from 4000 Å to 7000 Å. **[Ans. 10^{-15} m]**

Q.2 One of the most ideal line of the Krypton has wavelength 6058 Å with line width of 0.0184 Å. Calculate the coherence length and band width. **[Ans. 0.20 m, 1.5×10^9 Hz]**

Q.3 The amplitudes of light waves emerging from two slits in Young's experiment are in the ratio of 1:2. Find the intensity ratio of the interference patterns. **[Ans. 9 : 1]**

Q.4 Two coherent sources whose intensity ratio is 81 : 1 produce interference fringes. Deduce the ratio of maximum to minimum intensity in fringe system. **[Ans: 25 : 16]**

Q.5 What is the separation between the slits of Young's double slit experiment that gives second order maxima at a distance of 5.0 mm from central maxima. The screen is at a distance of 2.0 m from the slit and the wavelengths of light is 500 nm. **[Ans: 0.4 mm].**

Q.6 The distance between the slit and biprism and that of between the biprism and screen are each 50 cm. The obtuse angle of biprism is 179° and its refractive index is 1.5. If the width of the fringes is 0.014 cm, calculate the wavelength of light. [**Hint:** $2d = 2a(\mu - 1)$] **[Ans: 6140Å]**

Q.7 Two narrow and parallel slits 0.1 cm apart are illuminated with a monochromatic light of wavelength 5893 Å. The interference pattern is observed at a distance of 25 cm from the slits. Calculate the fringe width. **[Ans: 0:147 nm].**

Q.8 On introduction of a thin sheet of mica, having thickness $1.2 \times 1\,0^{-4}$ cm, in the path of one of the interfering beams in a biprism experiment, the central fringe is shifted through a distance equal to the spacing between successive bright fringes. Calculate the refractive index of mica (Given $\lambda = 6 \times 10^{-7}$ m). **[Ans. 1.5]**

Q.9 In Fresnel's biprism experiment the fringes of 0.19 mm width are formed on the screen placed at a distance of 1.0 m from the slits. A convex lens is placed at a distance of 30 cm from the images of two coherent sources. The separation between the two images was found to be 0.70 cm. Calculate the wavelength of light used. **[Ans. 5550 Å]**

Q.10 The distance between two virtual images of a slit formed by a biprism is 0.3 mm. If fringes of width 0.59 mm are formed on a screen placed at a distance of 30 cm from the slit, calculate the wavelength of the light used. **[Ans. 5900 Å]**

Q.11 When a glass piece of thickness 3.6×10^{-4} cm is placed in the path of one of the interfering beams in a biprism experiment, it is found that the central bright fringe shifts through a distance equal to the width of four fringes. Calculate the refractive index of the piece of glass. Wavelength of light used is 5.4×10^{-5} cm. (**Hint**: $(\mu - 1)\, t = n\lambda$) **[Ans.: 1.6]**

Q.12 In a biprisms experiment fringe width is observed as 0.88 mm. What will it become if the distance between biprism and slit is reduced to 0.82 times its original distance. (**Hint**: $2D = \lambda\, D/\beta = 2a(\mu - 1)\alpha$) **[Ans. 1.07 mm]**

Q.13 When a thin soap film of refractive index 1.33 is seen by normally reflected light of sodium of wavelength 5893 Å, it appears to be black. Find the minimum thickness of the film.

[Ans: 2.215×10^{-7} m]

Q.14 A soap film of refractive index 1.43 is illuminated by a white light incident at an angle of 30°. The refracted light is examined by a spectroscope in which the dark band corresponding to wavelength 6000 Å is observed. Calculate the thickness of the film. **[Ans: 2.23×10^{-7} m]**

Q.15 A soap film of refractive index 1.33 is illuminated with light of different wavelengths at an angle of 45°. There is a complete darkness for wavelength 5890 Å. Calculate the thickness of the soap film. **[Ans: 2.614×10^{-7}m]**

Q.16 Using sodium light ($\lambda = 5893$ Å) interference fringes are formed by reflection from a thin air wedge. When viewed perpendicularly, 10 fringes are observed in a distance of 1.0 cm. Calculate the angle of the wedge. **[Ans: 2.95×10^{-4} rad]**

Q.17 Light of wavelength 6000 Å falls normally on the wedge-shaped film of refractive index 1.4 forming fringes that are 2.0 mm apart. Calculate the angle of the wedge. **[Ans: 1.07×10^{-4} rad]**

Q.18 A wedge-shaped air film having an angle of 40″ is illuminated by monochromatic light and fringes are observed vertically through a microscope. The distance between two consecutive bright fringes is 0.12 cm. Calculate the wavelength of light used. **[Ans: 4656 Å]**

Q.19 In Newton's rings experiment the diameter of 4th and 25th rings are 0.3 cm and 0.8 cm, respectively. Find the wavelength of light. Given $R = 100$ cm. **[Ans: 5875 Å]**

Q.20 In a Newton's rings experiment fringes are observed in reflected light of wavelength $5.9 \text{ m} \times 10^{-7}$ m The diameter of 10th dark fringe is 0.5cm. Find the radius of curvature of the lens and thickness of the air film. **[Ans. 1.059 m, 2.95×10^{-6} m]**

Q.21 In a Newton's ring experiment the radius of 10th and 20th rings are 0.2 and 0.3 cm, respectively, and the focal length of the plano-convex lens is 90 cm. Calculate the wavelength of light used in nanometre.

[Ans. 278 nm]

Q.22 In Newton's ring arrangement a source is emitting two wavelengths $\lambda_1 = 6 \times 10^{-7}$ and $\lambda_2 = 5.9 \times 10^{-7}$ m. It is found that n^{th} dark ring due to one wavelength coincides with $(n + 1)^{th}$ dark ring due to the other. Find the diameter of the nth dark ring if radius of curvature of the lens is 0.9 m.

[Ans. 1.1289×10^{2} m]

Q.23 Newton's ring are formed in reflected light of wavelength 6000 Å with a liquid between the plane and curved surfaces. If the diameter of sixth bright ring is 3.1 mm and the radius of curvature of the curved surface is 100 cm, calculate the refractive index of the liquid. **[Ans. 1.37]**

Q.24 Newton's rings are formed with reflected light of wavelength 5.895×10^{-7} m with a liquid between the plane and the curved surface. The diameter of the 5th dark is 0.3 cm and the radius of curvature of the curved surface is 1 m. Calculate the refractive index of the liquid. **[Ans. 1.31]**

Q.25 In Newton's rings experiment, the diameter of 7th dark ring is 3.4 mm.

(i) Calculate the diameter of 1 6th dark ring.

(ii) If a small amount of liquid is filled between the lens and glass plate, calculate the radius of 7th and 16th bright rings. (Given: $\mu_{liq} = 1.3$) **[Ans. (i) 5.25 mm; (ii) 3.0 mm, 4.6 mm]**

Q.26 Newton's rings are observed by reflection between two curved surfaces of radii of curvature 1.0 m and 1.5 m, which are in contact in same plane. Calculate the distance between the 8th and 1 8th dark rings using monochromatic light of wavelength 5890 Å. **[Ans: 1.88 mm]**

Q.27 A thin film of a material whose refractive index is 1.45 on being introduced in one of the arms of Michelson's interferometer causes a shift of 6 fringes. If the wavelength of the light used is 5890 Å, calculate the thickness of the film. **[Ans: 3.926 m]**

Q.28 A sheet of CaF_2 of refractive index 1.434 is inserted normally is one arm of a Michelson's interferometer. At $\lambda = 5896$ Å, 40 fringes are observed to be displaced. Calculate the thickness of the sheet.

[Ans: 27 m]

Q.29 A thin plate of refractive index 1.4 is introduced is the path of one of the beams of light in Michelson's interferometer and it is found that 50 fringes have crossed the line of observation. The wave length of light used is 5896 Å. Calculate the thickness of the plate. **[Ans: 36.85 m]**

Q.30 In Michelson's interferometer 500 fringes cross the field of view when the movable mirror is displaced through 0.147 mm. Calculate the wavelength of monochromatic light used. **[Ans. 5.88×10^{-7} m]**

Q.31 Michelson's interferometer illuminated by light of wavelength 6438 Å is used to measure the distance between two points. Calculate this distance if 239 fringes cross the reference mark when the mirror is moved from one point to the other. **[Ans: 7.69×10^{-5} m]**

Q.32 In a Michelson's interferometer 790 fringes cross the field of view when the movable mirror is displaced through 2.33×10^{-4} m. Calculate the wavelength of monochromatic light used.

[Ans. 5898 Å]

Diffraction

2

LEARNING OBJECTIVES

After reading this chapter you will be able to

LO1 Differentiate between diffraction and interference

LO2 Discuss types of diffraction and the differences between them

LO3 Illustrate Fresnel's Half-period zones

LO4 Analyse construction, theory and multi-focus behaviour of a zone plate

LO5 Explain Fresnel's diffraction by circular aperture

LO6 Demonstrate Fraunhofer diffraction by a single-slit/double slit/*n* slits in diffraction grating

LO7 Analyse the application of diffraction grating in determining wavelength of light

LO8 Explain resolving power of an optical instruments – telescope and microscope through Rayleigh criterion

LO9 Illustrate resolving and dispersive power of diffraction grating

Introduction

In the previous chapter, it was discussed that in order to produce an interference pattern, superposition of at least two beams or waves of light is necessary. For obtaining a sustained interference, these waves should be coherent and therefore, they were developed from a single source and were separated by the division of wavefront or amplitude. The same effect is implicated in the diffraction of light. The diffraction of light is described as the clear bending of waves around small obstacle and the spreading of waves to a certain extent into the region of geometrical shadow when a beam of light passes through a narrow slit. We can say that the diffraction is any deviation from geometrical optics resulting from the obstruction of a wavefront of light. Such effects are observed even if the obstacle is not opaque but causes local variations in the amplitude or phase of the wavefront of the transmitted light. For example, this effect can be seen when there is a modification in the properties of the medium through which the wave is traveling, like variation in the refractive index for light waves. Also, small bubbles or imperfections in a glass lens produce unwanted diffraction patterns when a monochromatic light is transmitted through it. This phenomenon can be suitably explained only by assuming the wave nature of light. The effects of diffraction are generally more prominent for the waves when the size of the diffracting object is of the order of the wavelength of the wave. The diffraction also has negative implications. For example, the edges of optical images are seen to be blurred by the diffraction. Therefore, the phenomenon of diffraction leads to a basic limitation in resolution of the instruments like camera, telescope, microscope, etc.

In addition to electromagnetic waves such as visible light, *x*-rays and radio waves, the diffraction occurs with other waves also including sound waves and water waves. Water or ocean waves diffract around jetties and other obstacles. Sound waves can diffract around objects. This is the reason we can still hear someone calling us even if we are hiding behind a wall.

2.1 YOUNG'S DOUBLE SLIT EXPERIMENT: DIFFRACTION OR INTERFERENCE? LO1

If you recall Young's double slit experiment, you will notice that in this experiment the light is blocked everywhere except at the two opening i.e., the wavefront is obstructed. In order to find the resulting interference fringe pattern, the two openings were treated as point sources. The same is true if long slits are used whose widths can be considered as points. However, by taking into account the finite size of the slits, a more complete analysis can be prepared. Then this problem becomes a problem of diffraction, where the interference pattern gets modified. In order to explain this, we apply Huygens' principle according to which every point of a given wavefront of light acts as a source of secondary spherical wavelets. Here, Fresnel added another assumption that the actual field at any point away from the wavefront can be found by a superposition of all these wavelets when we take into account both their amplitudes and phases. This is referred to as Huygens-Fresnel principle. In view of this, we consider every point of the wavefront emerging from each slit as a source of wavelets whose superposition generates the resultant field or diffraction pattern at some point on a screen. Therefore, we consider a continuous array of sources across both the slits for obtaining the diffraction pattern and do not take the slits as isolated point sources, as done in the case of interference.

2.2 DIFFERENCE BETWEEN DIFFRACTION AND INTERFERENCE

In simple words, the diffraction is the bending of light around an obstacle, whereas the interference is the meeting of two waves. In the phenomenon of diffraction, the interfering beams originate from a continuous distribution of sources as is clear from Huygens' principle. However, the interfering beams originate from a discrete number of sources in the phenomenon of interference. Interference pattern is obtained by the superposition of waves coming from two different wavefronts originating from the same source. However, the waves emerging from different parts of the same wavefront superimpose with each other to produce the diffraction pattern. The widths of the diffraction fringes are not equal, but the widths of the interference fringes may or may not be equal. If you focus on the points of minimum intensity, you will observe them as perfectly dark in the interference, but these points in the case of diffraction are not perfectly dark. Moreover, the bright fringes in the interference pattern are of uniform intensity but these are not of the same intensity in the diffraction pattern.

2.3 TYPES OF DIFFRACTION LO2

In order to obtain the diffraction pattern on a screen, we need a source of light, obstacle or aperture and the observation screen. Here, it is obvious that the distance of the source and the screen from the aperture will determine the diffraction pattern. Depending upon the distance of the source from the aperture, the wavefronts will reach the aperture either in the spherical form or in the plane form. The same is applicable to the wavefronts reaching the observation screen after emerging from the aperture. Based on these distances and hence the shapes of the wavefronts, the diffraction pattern is classified into two classes, namely Fraunhofer diffraction and Fresnel diffraction.

2.3.1 Fraunhofer Diffraction

We need to use plane wavefronts in order to obtain this type of diffraction. This is possible if both the source of light and the screen are effectively far enough from the aperture so that the wavefronts reaching the aperture and the observation screen can be considered plane. Then the source and the screen are said to be at infinite distances from the aperture. This condition can also be attained by using two convex lenses, out of which one makes the light from the source parallel before it falls on the aperture and the other helps focusing light after diffraction on the observation screen. This is clear that under the said arrangement the incident wavefront is plane and the secondary wavelets originating from the unblocked portions of the wavefront are in the same phase at every point in the plane of the aperture. Here, the diffraction is produced by the interference between parallel rays that are focused with the help of convex lens. This Fraunhofer type of diffraction is also called **far-field diffraction** and it is encountered in the case of a plane transmission grating (discussed later) or concave reflection grating.

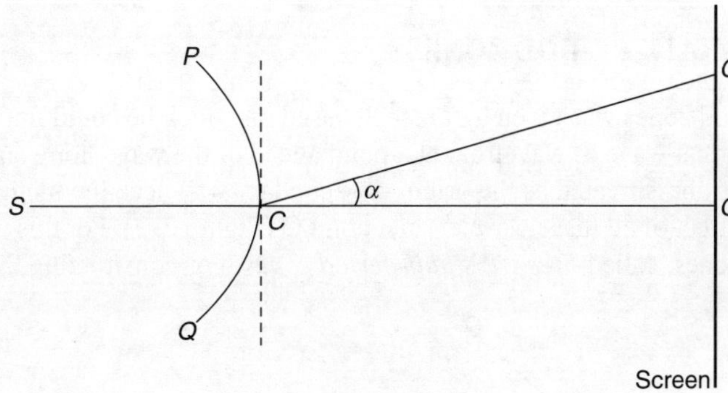

FIGURE 2.1

2.3.2 Fresnel Diffraction

If the source of light or the observation screen or both of them are at finite distances from the diffracting aperture, then the wavefronts falling on the aperture or reaching the screen will not be plane. These will be either spherical or cylindrical depending upon the situation. The diffraction obtained under this type of arrangement is called Fresnel diffraction for which the curvature of the wavefronts is important. This Fresnel type of diffraction is also called **near-field diffraction**. In this arrangement, lenses are not used to make the rays parallel or convergent. Therefore, the phase of secondary wavelets is not the same at all points in the plane of the diffracting aperture. Here the resulting field or the diffraction pattern is obtained by the superposition of these secondary wavelets emanating from different elements of unblocked portions of the wavefront. This was fantastically explained by Fresnel based on some assumptions. For example, he considered division of a wavefront into a large number of small area elements or zones called *Fresnel zones*. Under this situation, at any point O_1 on the screen (Figure 2.1), the resultant field will depend on the combined effect of all the secondary waves emanating from these zones. The effect of a particular zone at any point will depend on the distance between the point and the zone. Finally he considered the obliquity (angle α in the figure) of the point O_1 and took the obliquity factor as proportional to $(1 + \cos \alpha)$. Therefore, for an elementary wavefront at C, the resulting effect is maximum at O where $\alpha = 0°$ and $\cos \alpha = 1$. This effect becomes less significant when we move towards O_1, as the angle α gets increased. In the direction tangential to the wavefront (dashed line in the figure), this effect is one half of that at O, as the angle $\alpha = 90°$. Based on the obliquity factor, Fresnel could explain the non-existence of the wave in the backward direction, where $\alpha = 180°$ and hence the resulting effect is zero as $1 + \cos \alpha = 0$

2.3.3 Difference between Fresnel and Fraunhofer Diffractions

Fraunhofer diffraction is encountered in the case of gratings that contain number of slits. Fresnel type of diffraction is obtained when light suffers diffraction at a straight edge, a thin wire, a narrow slit, a small hole etc. These effects are the simplest to observe experimentally. For this reason Fresnel diffraction was historically the first type to be investigated. However, its explanation involves much more difficult mathematical theory than that required in handling the plane waves of Fraunhofer diffraction. The other difference between the Fraunhofer and Fresnel diffractions is related to the modification in the pattern when the observation screen is moved relative to the aperture. In the far-field approximation (Fraunhofer diffraction), when the screen is moved, the size of the diffraction pattern scales uniformly and the shape of the pattern does not change. However, in the near-field approximation (Fresnel diffraction), both the size and shape of the pattern depend on the distance between the diffracting aperture and the observation screen.

2.4 FRESNEL'S HALF-PERIOD ZONES LO3

The concept of half-period zones was given by Fresnel, based on which he could find the effect produced by a slightly divergent spherical wave or wavefront at a point ahead of the wave, for example on a screen. Since every point on the wavefront is treated as the origin of secondary wavelets, the method of half-period zones is used to calculate the effect of all these wavelets at a point O (Figure 2. 2). For this we divide the wavefront into a large number of zones, called *Fresnel 's half-period zones*. For constructing these zones, we consider

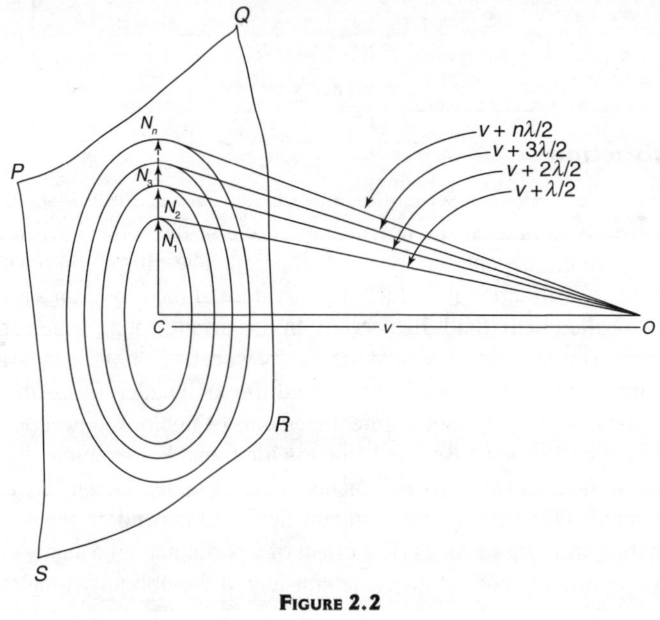

FIGURE 2.2

PQRS as a spherical wavefront of a monochromatic light of wavelength λ traveling toward the screen. We draw a perpendicular from the point O to the wavefront. This meets the wavefront at a point C at a distance of v. We draw a series of circles around C such that their distances from C are CN_1, CN_2, CN_3, ..., CN_n and each circle is a half wavelength farther from O. Therefore, the circles will be at distances $v + \lambda/2$, $v + 2\lambda/2$, $v + 3\lambda/2$,..., $v + n\lambda/2$ from O. This way the areas of the zones, i.e., the areas of the rings between successive circles, are equal (proved later). The area enclosed between CN_1, N_1N_2, N_2N_3 etc. are called first, second, third half-period zones, etc. respectively. Actually the difference of half a period in the vibrations from successive zones is the origin of the name half-period zones.

2.4.1 Radii and Areas of Half-Period Zones

In most of the cases the wavelength of light λ remains much smaller compared with the distance v. Under this condition i.e., when $\lambda << v$, we can calculate the radii and the areas of the zones. The radius of the first half-period zone

$$CN_1 \approx \sqrt{ON_1^2 - OC^2} = \sqrt{(v + \lambda/2)^2 - v^2} \approx \sqrt{v\lambda}$$

Here we have neglected a term $\lambda^2/4$ in view of the condition $\lambda << v$. Similarly the radius of the second half-period zone

$$CN_2 \approx \sqrt{ON_2^2 - OC^2} = \sqrt{(v + 2\lambda/2)^2 - v^2} \approx \sqrt{2v\lambda}$$

Thus the radius of nth half-period zone would be

$$CN_n \approx \sqrt{nv\lambda}$$

Based on the relations for the radii of the half-period zones, we can find the area of first, second, third half-period zone, etc. Since the radius of first half-period zone is CN_1, its area would be

$$a_1 = \pi CN_1^2 = \pi v\lambda$$

Similarly the area of second half-period zone is obtained as

$$a_2 = \pi(CN_2^2 - CN_1^2) = \pi v\lambda$$

This is clear from the above that the areas of all zones are the same under the approximation $\lambda << v$ and are independent of n. However, a more exact evaluation shows that the area gets increased very slowly with n.

2.4.2 Resultant Amplitude due to Whole Wavefront

As per Huygens' principle every point of the wavefront sends secondary wavelets in the same phase. However, these will not reach the point O in the same phase as they travel different distances. Since the average distance of O from any two consecutive zones differ by $\lambda/2$ so, the successive zones will produce resultants at O differing by a phase difference of π. If A_n be the resultant amplitude of the light due to n^{th} half-period zone, then its successive values will have alternate signs because of a phase change of π (or the reverse direction of amplitude vector) in moving from one zone to the next. It means if the amplitudes of the waves emerging from the first, second, third half-period zones, etc. are respectively A_1, A_2, A_3, etc., then the resultant amplitude A due to the whole wavefront would be

$$A = A_1 - A_2 + A_3 - ... - A_n \text{ (if } n \text{ is even)}$$

$$A = A_1 - A_2 + A_3 - ... + A_n \text{ (if } n \text{ is odd)}$$

or

$$A = A_1 - A_2 + A_3 - ... + (-1)^{n-1} A_n$$

Now as the magnitude of successive amplitudes goes on decreasing with the higher order of zones due to the increased average distance of the zone from O and the larger obliquity, the amplitude A_2 is slightly smaller than A_1 but slightly greater than A_3. Therefore, to the first approximation, we can assume $A_2 = \frac{A_1 + A_3}{2}$, $A_4 = \frac{A_3 + A_5}{2}$, etc. In view of this, we can expand the above series as

$$A = \frac{A_1}{2} + \left(-A_2 + \frac{A_1 + A_2}{2}\right) + \left(-A_4 + \frac{A_3 + A_5}{2}\right) + \ldots + \frac{A_{n-1}}{2} - A_n \text{ (if } n \text{ is even)}$$

$$A = \frac{A_1}{2} + \left(-A_2 + \frac{A_1 + A_3}{2}\right) + \left(-A_4 + \frac{A_3 + A_5}{2}\right) + \ldots + \frac{A_n}{2} \text{ (if } n \text{ is odd)}$$

If the number n is sufficiently large, then the effect due to n^{th} zone would become insignificant and the resultant amplitude due to the whole wavefront can be approximated as $A = \frac{A_1}{2}$ and hence the intensity $I = \frac{A_1^2}{4}$. Thus the amplitude due to whole wavefront at point O is just half of the one due to the first half-period and the intensity is equal to one forth of the intensity due to the first half-period zone a the point O.

2.5 ZONE PLATE LO4

A zone plate is an optical device that is used to verify the correctness of Fresnel's method of dividing a wavefront into half-period zones. It is a special diffracting screen designed to obstruct the light from the alternate half-period zones. It is a transparent plate on which a series of concentric circles are drawn with their radii proportional to the square root of natural numbers. This way the formed alternate annular zones are blocked. This type of plate behaves like a convex lens and produces an image of a source of light on the screen placed at a suitable distance.

2.5.1 Construction of Zone Plate

On a sheet of white paper we draw concentric circles with their radii proportional to the square root of natural numbers. The alternate zones are painted black and a significantly reduced photograph of this drawing is obtained on a glass plate. The negative, when kept in the light path from a distant point source, produces a large intensity at a point on its axis at a distance determined by the size of the zone and the wavelength of light used. We can construct two types of zone plates, namely positive zone plate and negative zone plate. If the odd zones are transparent and even zones are opaque on the zone plate, the plate is called *positive zone plate*. If the even zones are transparent and odd zones are opaque on the zone plate, the plate is called *negative zone plate* as shown in Fig. 2.3(A) (i) and (ii)).

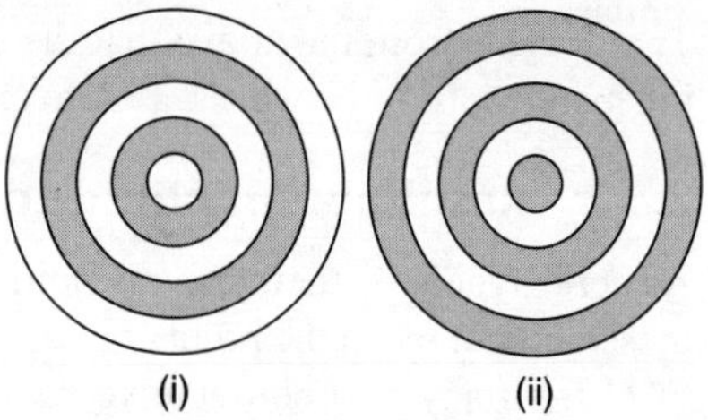

FIGURE 2.3A

2.5.2 Theory of Zone Plate

Let S be a point source of monochromatic light sending out spherical waves of wavelength λ and O is the position of screen for a bright image (Figure 2.3(B)). Consider an imaginary transparent plate PQ at a right angle to the line SO passing through the point C. Our aim is to determine the intensity of light at O due to the wavefront emerging from the source S. In order to find this, we divide the wavefront into half-period zones bounded by circles having centres at C and radii equal to CN_1, CN_2, CN_3, ..., CN_n as r_1, r_2, r_3,..., r_n. These radii divide the plate PQ into half-period zones such that from one zone to the next there is an increasing path difference of $\lambda/2$. So

$$SN_1 + N_1O = SC + CO + \lambda/2 \qquad (\because \lambda \text{ is small})$$
$$SN_2 + N_2O = SC + CO + 2\lambda/2, \ldots,$$
$$SN_n + N_nO = SC + CO + n\lambda/2.$$

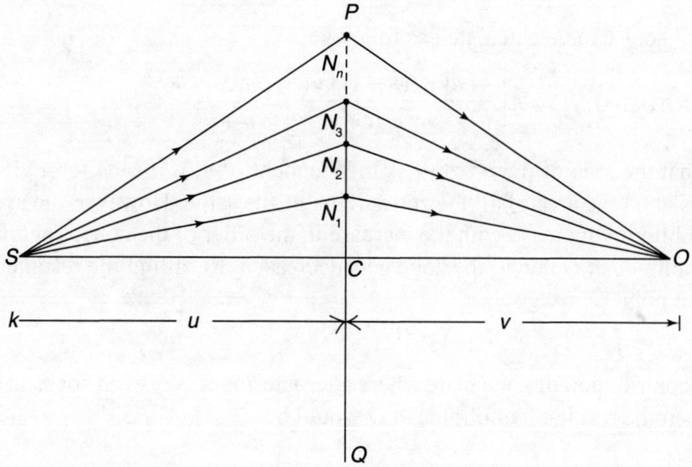

FIGURE 2.3B

To find the radius r_n of the nth circle or the nth zone, we have

$$SN_n + N_nO = SC + CO + n\lambda/2 \tag{i}$$

If we take $SC = u$ and $CO = v$, then

$$SN_n = \sqrt{(SC^2 + CN^2)} = \sqrt{(u^2 + r_n^{\,2})}$$
$$= u\left(1 + \frac{r_n^2}{u^2}\right)^{1/2} \approx u + \frac{r_n^2}{2u} \ (\because r_n << u)$$

Similarly,

$$N_nO \approx v + \frac{r_n^2}{2v}$$

Substituting these values in Eq. (i), we get

$$u + \frac{r_n^2}{2u} + v + \frac{r_n^2}{2v} = u + v + \frac{n\lambda}{2}$$

or
$$\frac{1}{u} + \frac{1}{v} = \frac{n\lambda}{r_n^2}$$

After applying the sign convention, the above formula takes the form

$$\frac{1}{v} + \frac{1}{u} = \frac{n\lambda}{r^n} \tag{ii}$$

or
$$r_n^2 = \frac{n\lambda uv}{c + v} \tag{iii}$$

Since u, v and λ are constants for a given light, object and image

$$r_n \propto \sqrt{n} \tag{iv}$$

It is clear from the above relation that the radii of zones are proportional to the square roots of natural numbers.

Now the area of the nth zone can be calculated as follows

$$a_n = \pi(r_n^2 - r_{n-1}^2) = \pi\left[\frac{n\lambda uv}{u+v} - \frac{(n-1)\lambda uv}{u+v}\right] = \frac{\pi\lambda uv}{u+v} \tag{v}$$

This relation shows that the area of the nth zone is independent of n. It means for a given object and image (for u and v to be constants) the area of all the zones remains the same. However, the average distance of the zone from O and the obliquity increase with the increase in the order of the zone. Therefore, as the order of a zone increases the amplitude at O due to the zone gets decreased. Recalling the resultant amplitude A due to the whole wavefront at point O, we get

$$A = A_1 - A_2 + A_3 - \ldots + (-1)^{n-1} A_n \tag{vi}$$

Now we focus on the contribution of zone plate where alternate zones, say even zones in case of positive zone plate, are blocked. Then the resultant amplitude at O would be

$$A = A_1 + A_3 + A_5 + \ldots \tag{vii}$$

Here it is clear that the resultant amplitude A is positive. However, A will be negative if odd zones are blocked (negative zone plate). Based on the sign of the resultant amplitude A, the zone plates were named as positive and negative zone plates. A comparison of Eq. (vii) with equation (vi) shows that the resultant amplitude produced by a zone plate (where the light is blocked by alternate half-period zones) is greater than that due to wholly unobstructed wavefront. Hence, the intensity at O is very much enhanced, i.e., the point O is extremely bright and can be said to be the image of S. This concentration of light at an axial point shows that the zone plate operates as a lens with O as a focal point. This explains the focusing action of the zone plate. In order to find the focal length of the zone plate, we concentrate on Eq. (ii) and observe that it is similar to the lens formula

$$\frac{1}{v} + \frac{1}{u} = \frac{1}{f} \tag{viii}$$

So a comparison of Eq. (viii) with Eq. (ii) gives

$$f_n = \frac{r_n^2}{n\lambda} \tag{ix}$$

This expression determines the focal length of a zone plate. Since the wavelength λ appears in the above expression, a zone plate will have severe chromatic aberrations. So, it behaves like a convergent lens.

2.5.3 Multi-focus Behaviour of a Zone Plate

For an object of infinity, i.e., when $u = \infty$, the radius of nth zone can be obtained from Eq. (iii) as

$$r_n = \sqrt{n\lambda v} \tag{x}$$

A comparison of Eq. (x) and Eq. (ix) infers that the distance v is the first focal length f_1 (when $n = 1$), given by

$$f_1 = \frac{r_1^2}{\lambda}$$

From Eq. (x) it is clear that for fixed r_n, the number n gets increased if we reduce the distance v. It means as the field point O is brought towards the zone plate along the axis, the same zonal area of radius r_1 will include more half-period zones. If the field point O is brought at a distance of $f_1/2$, $n = 2$ satisfied the relation (x) for the same zonal radius r_1. Therefore, each of the original zone in this case will now contain two half-period zones. For each original zone these two half-period zones contribute light at the focal point $v = f_1/2$ out of phase by π with each other. So they cancel and no light is focused by the zone plate at this focal point $f_1/2$. If we keep on moving the field point O towards the zone plate, we will find $n = 3$ when $v = f_1/3$ for the same zonal radius r_1. In this case now three half-period zones are contained in each of the original zones. Out of these three zones, the effect of two will be canceled due to a phase difference of π between them and the light will be focused at point $O(v = f_1/3)$ due to only one half-period zone. For the further movement of point O, we will find no light at $v = f_1/4$, light at $v = f_1/5$, etc. Therefore, we can conclude that a zone plate has multiple foci of focal lengths $f_1, f_1/3, f_1/5$, etc. For $v = f_1/3$ the contribution of each original zone is subdivided into three half-period zones at the observation point O. So the resultant amplitude will be

$$A = (A_1 - A_2 + A_3) + (-A_4 + A_5 - A_6) + (A_7 - A_8 + A_9) - \dots \tag{xi}$$

In the above expression the first parenthesis is due to the first zone, second parenthesis is due to the second zone, third parenthesis is due to the third zone, etc. For the zones that are reproduced on a smaller scale, the obliquity factor is not very important and we may estimate $A_j = A_1$ where $j = 2, 3, 4, \dots$ Therefore, the resultant amplitude at $v = f_1/3$ would be simply equal to A_1. However, at $v = f_1$ the amplitude will be three times of this amplitude. Thus, the amplitude at $v = f_1/3$, zone by zone, is reduced by a factor of 1/3 and hence, the intensity at this point is 1/9 that at $v = f_1$. This can be extended to point at $f_1/5$ also, where the original zone of radius r_1 will include five half-period zones. Thus the maximum intensity points along the axis and hence the foci of the zone plate can be found at $f_n = \dfrac{r_1^2}{n\lambda}$ with n as odd number.

2.5.4 Comparison between Zone Plate and Convex Lens

Based on the above discussion an understanding is developed that a zone plate operates as a convex lens. A zone plate has some similarities as well as some differences with a convex lens. For example, both show chromatic aberration as their focal lengths depend upon the wavelength λ. Also, the relation between the conjugate distances is similar for both of them. However, the differences between these two are listed below.

(i) In case of a zone plate the image is formed by diffraction whereas the rays in case of a lens are brought to focus by refraction.

(ii) The image due to a convex lens is more intense than due to a zone plate.

(iii) A convex lens has only one focus, whereas a zone plate has n number of foci of reduced intensity between the points O and C.

(iv) The focal length of a lens is given by the relation $\dfrac{1}{f} = (\mu - 1)\left(\dfrac{1}{R_1} - \dfrac{1}{R_2}\right)$, where μ is the refractive index of the material of lens and R_1 and R_2 are the radius of curvatures. However, the focal length of the zone plate is given by $\dfrac{1}{f} = \dfrac{n\lambda}{r_n^2}$

(v) The focal length of a lens is directly proportional to the wavelength λ as $f \propto \dfrac{1}{\mu} \Rightarrow f \propto \dfrac{v}{c} \Rightarrow f \propto \lambda$. However, the focal length of a zone plate is inversely proportional to λ.

(vi) Light takes the same time to go from S to O when passed through any part of the lens. However, in a zone plate, light from any transparent zone reaches the point O one period later than the light from the next inner zone.

2.5.5 Phase Reversal Zone Plate

If instead of blocking the alternate zones of the zone plate we introduce an additional path of $\lambda/2$ between the waves emanating from successive zones, the amplitudes from these successive zones will be in phase with each other. Thus we will get four times more intense image than an ordinary zone plate. Such type of zone plate is called phase reversal zone plate and was investigated by *Wood* in 1898.

2.6 FRESNEL'S DIFFRACTION BY A CIRCULAR APERTURE LO5

Let us consider Fig. 2.4 where the aperture AB is taken as circular. In order to study the diffraction through this circular aperture, Fresnel devised a method for calculating the contribution of various parts of the wavefront (arc PQ). He divided the aperture into zones with circular symmetry about the axis SCO along with C as the pole. As discussed earlier, the zones are nothing but the circles on the wavefront separated in such a way that each zone is $\lambda/2$ farther from the field point O than the previous zone. In the Fig. 2.4 XY is the screen at which the diffraction pattern is obtained and the shadow is seen above M and below N on the screen.

If we choose the size of the aperture AB such that it allows only the first half-period zone, then the resultant amplitude at point O will be A_1. However, if the whole wavefront is exposed and there are a large number of half-period zones, then the resultant amplitude at O will be

$$A = A_1 - A_2 + A_3 - A_4 + \ldots = A_1/2 \qquad \text{(i)}$$

In case of even and odd numbers of half-period zones, however, the resultant amplitude will be

$$A = \frac{A_1}{2} + \frac{A_{n-1}}{2} - A_n \ (\text{if } n \text{ is even}), \quad A = \frac{A_1}{2} + \frac{A_n}{2} \ (\text{if } n \text{ is odd}) \qquad \text{(ii)}$$

Therefore, we can conclude that if n is large as in the case of large aperture, A_n approaches to zero for both either n is odd or even and the resultant amplitude is half of that the first contributing zone.

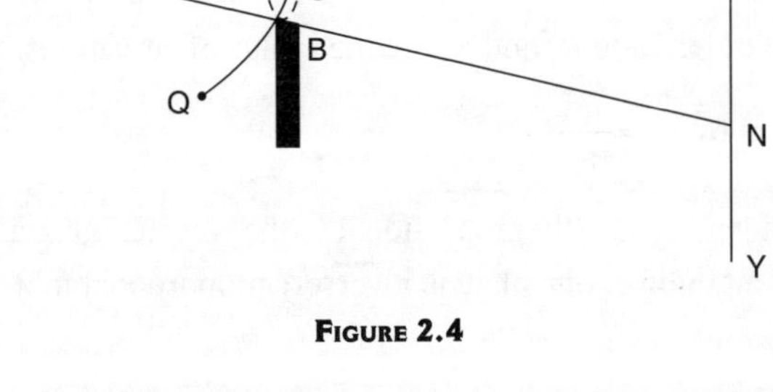

FIGURE 2.4

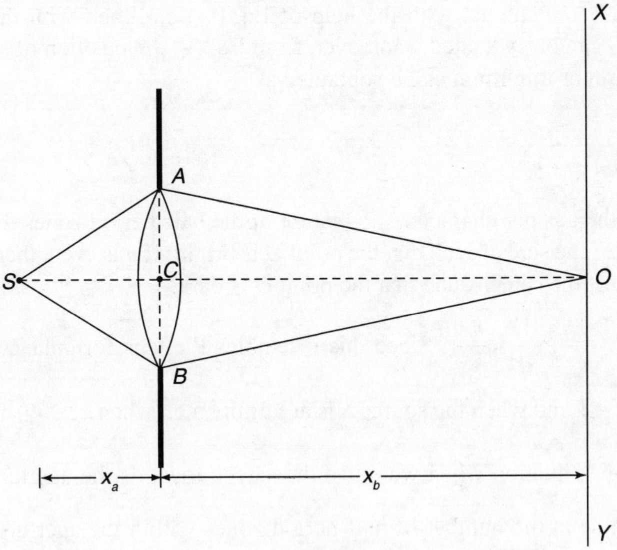

FIGURE 2.5

In order to find out the intensity at the point O due to the whole wavefront, we take help of Fig. 2.5. Here the distances $SC = x_a$ and $CO = x_b$ and the radius of the circular aperture AB is $CA = CB = r$. We can find the path difference between the waves reaching at O as follows

$$\delta = SA + AO - (SC + CO)$$

$$\delta = \sqrt{(x_a^2 + r^2)} + \sqrt{(x_b^2 + r^2)} - x_a - x_b$$

$$\delta = x_a\left(1 + \frac{r^2}{x_a^2}\right)^{1/2} + x_b\left(1 + \frac{r^2}{x_b^2}\right)^{1/2} - x_a - x_b$$

$$\delta = x_a\left(1 + \frac{r^2}{2x_a^2}\right) + x_b\left(1 + \frac{r^2}{2x_b^2}\right) - x_a - x_b$$

$$\delta = \frac{r^2}{2}\left(\frac{1}{x_a} + \frac{1}{x_b}\right) \tag{iii}$$

If the position of screen is fixed and the size of the aperture is such that it contains n number of half-period zones, then the path difference will be

$$\delta = \frac{n\lambda}{2} \tag{iv}$$

From Eqs. (iii) and (iv), we obtain

$$\frac{n\lambda}{2} = \frac{r^2}{2}\left(\frac{1}{x_a} + \frac{1}{x_b}\right) \tag{v}$$

$$n = \frac{r^2}{\lambda}\left(\frac{1}{x_a} + \frac{1}{x_b}\right) \tag{vi}$$

Now the number n can be calculated with the help of Eq. (vi) and then with the help of it the resultant amplitude at the point O can be evaluated. Moreover, from Eq. (v) the position of screen where the intensity would be either maximum or minimum can be obtained as

$$x_b = \frac{x_a r^2}{n\lambda x_a - r^2} \tag{vii}$$

If the number n is odd, then as per characteristic feature of the half-period zones the corresponding value of x_a will give the position of the screen such that the point O is bright. If n is even, then the corresponding value x_a will give the position of the screen such that the point O is dark.

From Eq. (v) we obtain $\frac{1}{x_a} + \frac{1}{x_b} = \frac{n\lambda}{r^2}$. Since this resembles the lens formula, we can find the first focal length as $f_1 = \frac{r^2}{\lambda}$ (for $n = 1$ and when the source S is at infinity, i.e., when $x_a = \infty$) and the radius $r = \sqrt{n\lambda x_b}$. Now we can analyze the situation when we move the screen towards the aperture. For the fixed width of the aperture (i.e., the value r) the number of half-period zones within the aperture will alternately be even and odd as per relation $r = \sqrt{n\lambda x_b}$. Therefore, the point O will alternately be dark (for even values of n) and bright (for odd values of n). However, if for some distance of the screen (value x_b) the aperture contains only a fraction of the first zone, then the light will spread to the geometrical shadow of the aperture. On the other hand, now we fix up the position of the screen and change the width of the aperture. When a circular aperture has the diameter same as the first half-period zone, then the resultant amplitude at the point O will be equal to A_1. If we make the aperture wider such that it has two half-period zones, there will be almost zero amplitude at the point O. Now if we remove all the opaque shields so that all zones of an unobstructed wavefront contribute, the resultant amplitude will become $A_1/2$ and hence the intensity one fourth of that due to the first zone aperture alone. These are some curious result because they are not so clear in ordinary experience. Another result of historic interest is achieved when a round obstacle or the disc is substituted such that it covers only the first half-period zone. So all the zones except the first will contribute to the resultant amplitude at O. As now the second zone is the first contributing zone, the light of amplitude $A_2/2$ focuses at the point O. Therefore, the intensity at the centre of the shadow of the obstacle will be almost the same as without disc!

2.6.1 Intensity at Nonaxial Point

In order to calculate the intensity at a nonaxial point such that O_1 on the screen XY at a distance of y from the central point O, we consider the circular aperture AB of radius r (Fig. 2.6). The screen is at a distance of x_b. We consider the source at an infinite distance from the aperture so that the incident wavefront is plane. As discussed earlier, the pole of the wavefront is defined as the point where the perpendicular from the point O meets the wavefront. So for the observation point O, the pole is C. However, in moving from point O to O_1, it can be noted from the figure that the pole of the wavefront C is shifted to the position C_1. In this situation, the aperture will not be symmetrical with respect to the half-period zones. This can be viewed as if the whole system of half-period zones is lifted up by the distance y as $CC_1 = OO_1 = y$. Thus the upper portions of the wavefront are now cut off by the obstacle between the points A and C_1. Obviously the distance AC_1 will determine the number of zones that are cut off and an increase in this distance shall lead more number of half-period zones to cut (Fig. 2.7). Under such situation, in addition to some fully exposed lower order half-period zones, the lower portion of higher order zones will contribute to the intensity at points O and O_1.

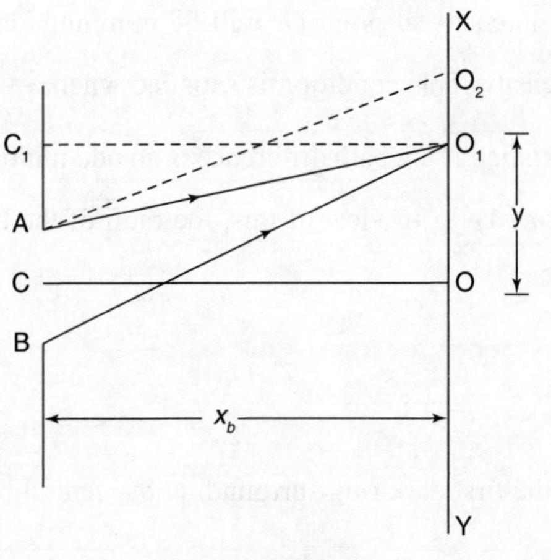

FIGURE 2.6

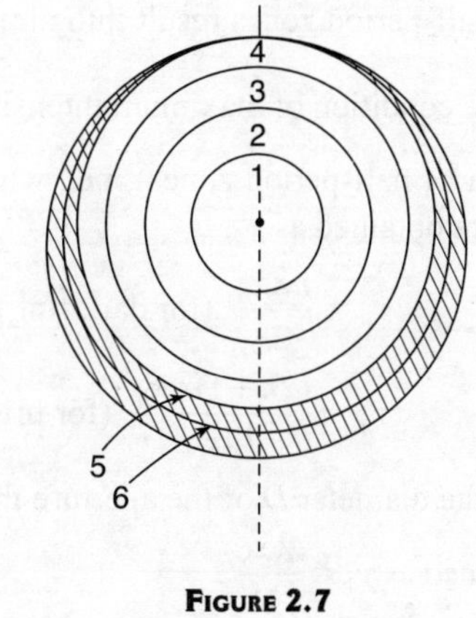

FIGURE 2.7

Let us find the maximum intensity at point O along the axis at a particular distance of the screen from the aperture (Fig. 2.6). Under this condition, let the aperture contain odd number of half-period zones (say five). As we move away from the axis from O to O_1 (nonaxial point), the pole C gets shifted to C_1. Suppose now first four half-period zones are fully exposed together with about half of the fifth and sixth zones. Then, the resultant amplitude at point O_1 would be

$$A = A_1 - A_2 + A_3 - A_4 + \frac{A_5}{2} - \frac{A_6}{2} = \frac{A_1}{2} + \left(\frac{A_1 + A_3}{2} - A_2\right) + \left(\frac{A_3 + A_5}{2} - A_4\right) - \frac{A_6}{2}$$

or $$A = \frac{A_1}{2} - \frac{A_6}{2} \qquad \text{(viii)}$$

Therefore, the point O_1 will have minimum intensity. Now we move further to point O_2 so that the first three half-period zones are fully exposed together with nearly half of the forth, fifth, sixth and seventh zones. Then the resultant amplitude will be maximum. Therefore, we obtain series of points along OX at which the intensity is alternately maximum and minimum. The same is true when we move toward Y, i.e., along OY. Therefore, we finally observe bright and dark rings of unequal widths about the point of observation O. If the aperture is large, these rings are seen only near the limits of the geometrical shadow and the intensity falls off rapidly within the shadow. However, in case of a small aperture which contains only a fraction of the first half-period zone, the observation point O will be bright and no rings within the geometrical image of the aperture will be seen.

The positions of the bright and dark rings can be obtained with the help of Fig. 2.6. Here the path difference between the secondary waves diffracted from A and B and focused on the point O_1 is

$$\delta = BO_1 - AO_1 = [(y + r)^2 + x_b^2]^{1/2} - [(y - r)^2 + x_b^2]^{1/2}$$

$$\delta = x_b\left[1 + \frac{(y + r)^2}{2x_b^2}\right] - x_b\left[1 + \frac{(y - r)^2}{2x_b^2}\right] \quad \text{(as the distance } y \ll x_b\text{)}$$

$$\delta = \frac{2ry}{x_b} \qquad \text{(ix)}$$

If the path difference δ is an even multiple of $\lambda/2$, the intensity at point O_1 will be minimum as the even number of half-period zones result into almost zero intensity. This condition is satisfied when $\frac{2ry}{x_b} = 2n\frac{\lambda}{2}$. However, the condition of maximum intensity will be satisfied if the path difference is an odd multiple of $\lambda/2$ (odd number of half-period zones), i.e., when $\frac{2ry}{x_b} = (2n+1)\frac{\lambda}{2}$. In view of this, the radii of the bright and dark rings are obtained as

$$y_n = \frac{n\lambda x_b}{2r} \text{ (for dark ring)} \tag{x}$$

$$y_n = \frac{(2n+1)\lambda x_b}{4r} \text{ (for bright ring)} \tag{xi}$$

In terms of the diameter D of the aperture the radius of the first dark ring surrounding the central bright ring can be obtained as $y_1 = \frac{\lambda x_b}{D}$.

This phenomenon of diffraction at a circular aperture has significance in the formation of images by telescopes and microscopes. For example, in a telescope the image of a star is seen to consist of bright central disc which is surrounded by dark and bright rings of gradually diminishing intensities.

2.7 FRAUNHOFER DIFFRACTION BY A SINGLE SLIT — LO6

Since in Fraunhofer diffraction the source is effectively at infinite distance, a collimated parallel beam of monochromatic light of wavelength λ can be taken as incident normally on a narrow slit AB of width b (Fig. 2.8). We divide this wavefront into a large number of points in each sending waves of equal amplitude a according to Huygens' principle. These waves get diffracted and then interfere to produce diffraction pattern on the screen. The secondary waves that travel along the direction of incident beam are focused at point C while those inclined at an angle θ with the direction of incident beam (due to diffraction) are focused at another point P (Fig. 2.8). To find out the resultant intensity at P, we draw a perpendicular AK on BK. It is clear from the figure that the optical paths of the waves traveled after the plane AK to the point P are equal. However, the optical paths of the waves originating from points on AB (from A toward B) and reaching the point P gradually increase. Hence, the phase difference between them gets larger. This is shown in Fig. 2.9(a) for n number of waves each of amplitude a. In this figure a phase difference of ϕ is taken between two successive waves. The resultant amplitude of these waves at point P is shown as R.

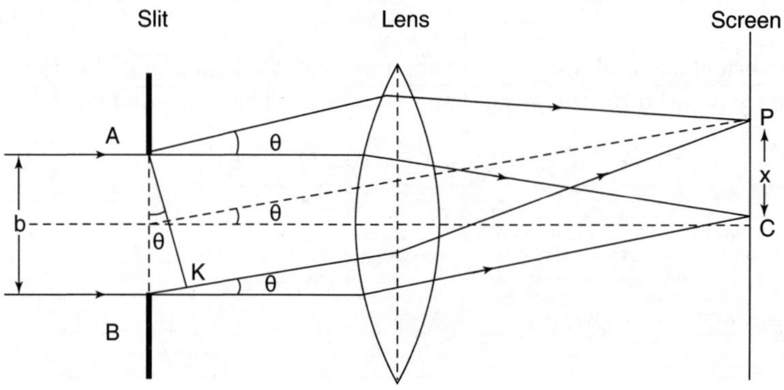

FIGURE 2.8

The total path difference between the waves originating from extreme points A and B is $BK = AB \sin\theta = b \sin\theta$. Therefore, the path difference between different waves originating from all the points of the slit AB vary between zero and $b \sin\theta$. The phase difference corresponding to path difference $b \sin\theta$ will be $(2\pi/\lambda)\, b \sin\theta$

Since the aperture is divided into n equal parts, the phase difference between any two consecutive parts will be $\frac{1}{n}\frac{2\pi}{\lambda} b \sin\theta\ (= \theta$, say).

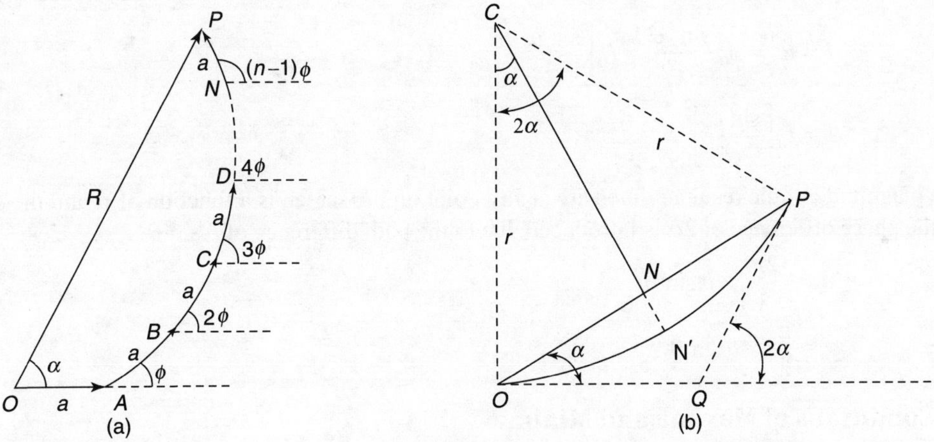

FIGURE 2.9

The resultant amplitude and intensity at point P due to all these secondary waves can be obtained by vector polygon method. Let α be the phase difference between the waves from the initial direction to the resultant (Fig. 2.9a), then 2α will be the total phase difference between the secondary waves originating from extreme points of the slit AB (Fig. 2.9b). Here, it is taken that all the amplitudes constitute an arc due to their large number and small phase difference between them. Because of the symmetry, we have $\angle O = \alpha$ and $\angle Q = 2\alpha$. The chord OP gives the resultant amplitude due to all the secondary waves at point P.

Then in the ΔOCN,

$$\sin\alpha = \frac{ON}{OC} = \frac{ON}{r}$$

or $$ON = r\sin\alpha \qquad \text{(i)}$$

where, r is the radius of the circular arc.

$\therefore$ Chord $OP = 2ON = 2r\sin\alpha$

$\therefore$ Chord OP = resultant amplitude

$\therefore$ $$R = 2r\sin\alpha \qquad \text{(ii)}$$

The length of the arc $ON'P = na$, where n is an integer number and a is the amplitude of each vibration (Fig. 2.9b)

We know that,

$$\angle PCO = 2\alpha = \frac{\text{Arc } ON'P}{\text{Radius}} = \frac{na}{r}$$

or $$2r = \frac{na}{\alpha} \qquad \text{(iii)}$$

Substituting the value of $2r$ in Eq. (ii), we get

$$R = na\frac{\sin\alpha}{\alpha}$$

$$R = A_0\frac{\sin\alpha}{\alpha} \quad \text{(iv)}$$

Here we have taken $na = A_0$. Thus, resultant intensity at point P on the screen is given by a measure of

$$I = R^2 = A_0^2\frac{\sin^2\alpha}{\alpha^2} = I_0\left[\frac{\sin\alpha}{\alpha}\right]^2$$

$$I = I_0\left[\frac{\sin\alpha}{\alpha}\right]^2 \quad \text{(v)}$$

Thus, the magnitude of the resultant intensity at any point on the screen is a function of α and the slit width b. Since the phase difference of 2α is introduced due to the path difference of $b \sin\theta$

$$\therefore \quad 2\alpha = \frac{2\pi}{\lambda} \times b\sin\theta$$

or

$$\alpha = \frac{\pi}{\lambda} \times b\sin\theta$$

2.7.1 Conditions of Maxima and Minima

It is clear from Eq. (iv) that the resultant amplitude R will be a maximum when

$$\frac{\sin\alpha}{\alpha} = 1, \text{ which s true when } \alpha \to 0, \text{ i.e.,}$$

$$\frac{\pi}{\lambda} b\sin\theta \to 0 \text{ or } \sin\theta \to 0$$

$$\Rightarrow \quad \theta = 0°$$

From Eq. (iv),

$$R = \frac{A_0}{\alpha}\sin a = \frac{A_0}{\alpha}\left[\alpha - \frac{\alpha^3}{\underline{|3}} + \frac{\alpha^5}{\underline{|5}} - \frac{\alpha^7}{\underline{|5}} + \ldots\right]$$

$$= A_0\left[1 - \frac{\alpha^2}{\underline{|3}} + \frac{\alpha^4}{\underline{|5}} + \frac{\alpha^6}{\underline{|7}} + \ldots\right]$$

$$R = A_0 \text{ for } \alpha = 0$$

Hence the intensity corresponding to $\alpha = 0$ is $I = R^2 = A_0^2 = I_0$. This is called as the intensity of central (principal) maximum.

In order to determine the position of maximum intensity, let us differentiate Eq (v) w.r.t. to α and equate it to zero, i.e.,

$$\frac{dI}{d\alpha} = 2I_0\frac{\sin\alpha}{\alpha}\left[\frac{\alpha\cos\alpha - \sin\alpha}{\alpha^2}\right] = 0$$

Since I_0 cannot be zero, either $\sin\alpha = 0$ or $\alpha\cos\alpha - \sin\alpha = 0$. The equation $\sin\alpha = 0$ determines the positions of minima [eq. (iv)] except when $\alpha = 0$ because it corresponds to the position of maximum. Therefore, the following condition should be satisfied for obtaining the minima

$$\sin\alpha = 0 \text{ or } = \alpha = \pm m\pi$$

or $$\frac{\pi}{\lambda} b \sin\theta = \pm m\pi$$

or $$b \sin\theta = \pm m\lambda, m = 1, 2, 3 \ldots$$

The position of maxima are given by

$$\alpha \cos\alpha - \sin\alpha = 0$$

or $$\alpha = \tan\alpha$$

This equation can be solved graphically by plotting the curves

$$y = \alpha$$

and $$y = \tan\alpha$$

The first relation $y = \alpha$ represents the equation of straight line passing through the origin making an angle 45° with the axis and the equation $y = \tan\alpha$ represents a discontinuous curve having a number of branches with asymptotes at the intervals of π (Fig. 2.10). The points of intersections of the these curves will give the values of α that will satisfy the relation $\alpha = \tan\alpha$.

Therefore, the maxima occur when

$$\alpha = \frac{3\pi}{2}, \frac{5\pi}{2}, \frac{7\pi}{2} \ldots \text{ or } \alpha = (2n+1)\frac{\pi}{2}, n = 1, 2, 3, \ldots$$

These are called points of secondary maxima. A measure of intensity of first secondary maxima is obtained from Eq. (v) with $\alpha = \frac{3\pi}{2}$, as

$$I_1 = A_0^2 \left[\frac{\sin\left(\frac{3\pi}{2}\right)}{\frac{3\pi}{2}}\right]^2 = \frac{4}{9\pi^2} I_0$$

Similarly, the intensity of second secondary maxima is

$$I_2 = A^2 \left[\frac{\sin\left(\frac{5\pi}{2}\right)}{\frac{5\pi}{2}}\right]^2 = \frac{4}{25\pi^2} I_0$$

Similarly, $$I_3 = A^2 \left[\frac{\sin\left(\frac{7\pi}{2}\right)}{\frac{7\pi}{2}}\right]^2 = \frac{4}{49\pi^2} I_0 \text{ and so on.}$$

Thus, the ratio of relative intensities of successive maxima are

$$1 : \frac{4}{9\pi^2} : \frac{4}{25\pi^2} : \frac{4}{49\pi^2} : \ldots$$

The intensity of the first secondary maxima $\frac{4}{9\pi^2}$, i.e., 4.5% that of principal maximum as shown in Fig. 2.11

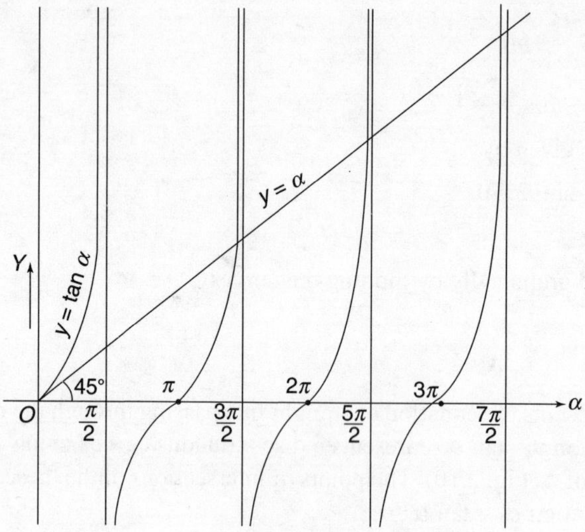

Figure 2.10

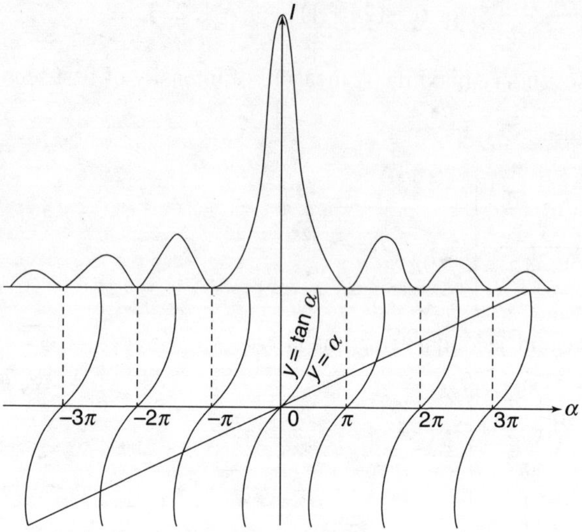

Figure 2.11

2.8 FRAUNHOFER DIFFRACTION BY DOUBLE SLITS

Let a parallel, collimated beam of monochromatic light of wavelength λ be incident normally on the two parallel slits *AB* and *ED* (Fig. 2.12) each having a width *b* and separated by an opaque distance *d*. The distance between the corresponding points of the two slits is $(b + d)$. Let the diffracted light be focused by a convex lens *L* on the screen *XY* placed in the focal plane of the lens. As discussed earlier, all the secondary waves traveling in the direction parallel to *MC* get focused at point *C*. Therefore, the point *C* corresponds to the position of the central bright maximum.

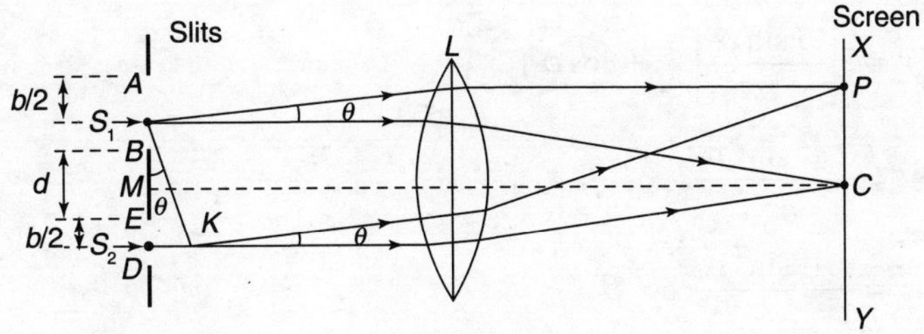

FIGURE 2.12

Further consider that the two slits are equivalent to two coherent sources placed at the middle points S_1 and S_2 of the slits AB and ED. Since the resultant amplitude due to a single slit is $A \sin \alpha/\alpha$ at any point P making an angle θ with MC, we may consider that each slit is sending a wave of amplitude $(A \sin \alpha/\alpha)$. the resultant amplitude due to interference of these two waves having a phase difference of ϕ' at point P can be calculated as follows. Take S_1K as the perpendicular drawn from S_1 on S_2K. Hence, the path difference between the rays at point P will be

$$S_2K = (b + d) \sin \theta \tag{i}$$

The phase difference between them will be

$$\phi' = \frac{2\pi}{\lambda} \times S_2K = \frac{2\pi}{\lambda}(b + d) \sin \theta \tag{ii}$$

The resultant amplitude R' at point P can be determined by using vector addition method of amplitudes, as shown in Fig. 2.13.

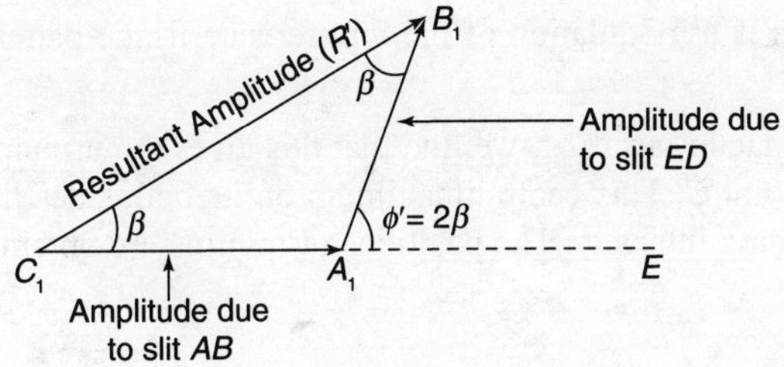

FIGURE 2.13

Since both the slits being of same size send the light of same amplitudes, we may write

$$C_1A_1 = A_1B_1 = A\frac{\sin \alpha}{\alpha} = R'$$

Now $\qquad \angle B_1A_1E = \phi' \; (= 2\beta, \text{say})$

$\therefore$ Resultant amplitude $= C_1B_1 = R'$ can be obtained as

$$C_1B_1^2 = C_1A_1^2 + A_1B_1^2 + 2C_1A_1 \cdot A_1B_1 \cos \phi'$$

$$(R')^2 = \left[\frac{A \sin \alpha}{\alpha}\right]^2 + \left[\frac{A \sin \alpha}{\alpha}\right]^2 + 2\left[\frac{A \sin \alpha}{\alpha}\right]\left[\frac{A \sin \alpha}{\alpha}\right] \cos \phi'$$

$$\text{or} \qquad (R')^2 = 2\left[\frac{A \sin \alpha}{\alpha}\right]^2 [1 + \cos \phi']$$

$$= \frac{2A^2 \sin^2 \alpha}{\alpha^2} 2\cos^2 \beta$$

$$= \frac{4A^2 \sin^2 \alpha}{\alpha^2} \cos^2 \beta \qquad \text{(iii)}$$

A measure of intensity can be obtained from the Eq. (iii) as

$$I = (R')^2 = \frac{4A^2 \sin^2 \alpha}{\alpha^2} \cos^2 \beta$$

$$\text{or} \qquad I = 4I_0 \frac{\sin^2 \alpha}{\alpha^2} \cos^2 \beta \qquad \text{(iv)}$$

Based on Eq. (iv), we conclude that the resultant intensity in the pattern depends upon two factors.

(i) The factor, $I_0 \dfrac{\sin^2 \alpha}{\alpha^2}$, which gives the diffraction pattern due to a single slit.

(ii) The factor $\cos^2 \beta$, which gives the interference pattern in the waves diffracted from the two slits.

From Eq. (iv) it is clear that the maximum intensity $I = 4I_0$, i.e., double slit provides the intensity four times of the obtained by single slit. Further, it is noted that the intensity I is a product of intensities obtained for double slit interference and single slit diffractions. Moreover, the expressions of α and β show that the factor $\cos^2\alpha$ varies more rapidly than the factor $\sin^2\alpha/\alpha^2$, as $d > b$. The product of sine and cosine factors proves that the double slit diffraction pattern is a modulation of the interference fringe pattern by a single slit diffraction envelope.

When we analyse diffraction factor $\sin^2\alpha/\alpha^2$, we find that this gives the principal maximum in the direction $\theta = 0°$ on the screen at the point C. This central maximum on its both sides has either alternate minima or subsidiary maxima of decreasing intensity. The positions of minima are obtained in the direction $\sin \alpha = 0$, when $\alpha \neq 0$. So

$$\alpha = \pm m\pi$$

$$\text{or} \qquad \frac{\pi}{\lambda} b \sin \theta = \pm m\pi$$

$$b \sin \theta = \pm m\lambda \qquad \text{(v)}$$

where $m = 1, 2, 3, \ldots$, etc. As mentioned $m = 0$ will give the position of maximum.

As discussed in the case of single slit diffraction the factor $\sin^2 \alpha/\alpha^2$ gives secondary maxima at the points

$\alpha = \dfrac{3\pi}{2}, \dfrac{5\pi}{2}, \dfrac{7\pi}{2}, \ldots$. Therefore, the positions of the secondary maxima are obtained in the direction

$$\alpha = (2n + 1)\frac{\pi}{2}$$

where $n = 1, 2, 3, \ldots$,

We can analyse the variation of intensity observed by the second factor, $\cos^2 \beta$, as follows.

For minimum intensity,

$$\cos^2 \beta = 0 \text{ or } \cos \beta = 0, \text{ i.e.,}$$

$$\beta = (2m + 1)\ \frac{\pi}{2},\ m = 0, 1, 2 \ldots$$

Using Eq. (ii), we get

$$\beta = \frac{\pi}{\lambda}\ (b + d) \sin \theta = (2m + 1) \frac{\pi}{2}$$

$$(b + d) \sin \theta = (2m + 1) \frac{\pi}{2} \qquad \text{(vi)}$$

Similarly, bright fringes or maxima are obtained for the values of θ for which

$$\cos^2 \beta = 1 \text{ or } \beta = \pm n\pi$$

or $$\frac{\pi}{\lambda}\ (b + d) \sin \theta = \pm n\pi$$

or $$(b + d) \sin \theta = \pm n\lambda \qquad \text{(vii)}$$

where $n = 0, 1, 2, 3, \ldots$

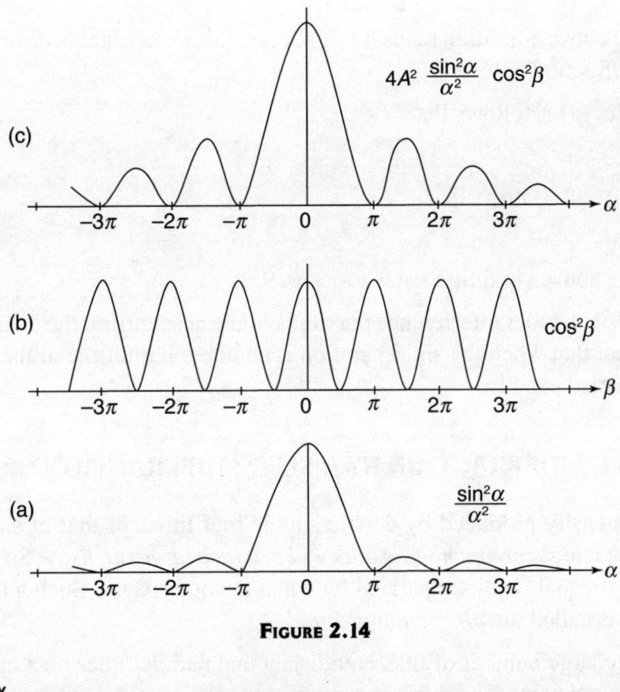

FIGURE 2.14

The variations of $\dfrac{\sin^2 \alpha}{\alpha^2}$ with α and $\cos^2 \beta$ is shown in fig. 2.14. The combined effect of these two, i.e., the resultant intensity in double slit Fraunhofer diffraction is shown in Fig. 2.14c.

2.8.1 Missing Orders in Diffraction Pattern

Here the directions of diffraction minima and interference maxima are respectively given by

$$b \sin \theta = \pm m\lambda, m = 1, 2, 3, \ldots \tag{viii}$$

and

$$(b + d) \sin \theta = \pm n\lambda, n = 0, 1, 2, \ldots \tag{ix}$$

Therefore, the number of interference maxima depends upon the relative values of width of the slits b and spacing between the two consecutive slits d. If 'b' is kept constant and 'd' is varied then certain number of interference maxima are found to be missing from the diffraction pattern. These are known as missing orders in double slit Fraunhofer diffraction pattern. This happens for a particular set of values of b and d, which simultaneously satisfy the conditions of diffractions minima [Eq. viii) and interference maxima Eq. (ix). Since diffraction minima dominate interference maxima the resultant intensity is obtained as zero. The condition for missing orders is obtained by dividing equation (ix) by equation (viii), i.e.,

$$\frac{b+d}{b} = \frac{n}{m} \tag{x}$$

Case-I: $d = b$, From Eq. (x)

$$\frac{b+d}{b} = \frac{n}{m} \text{ or } \frac{n}{m} = 2$$

or

$$n = 2m$$

Since, $m = 1, 2, 3, \ldots$, the above condition reads $n = 2, 4, 6, \ldots$ this shows that even order interference maxima will be missing in the diffraction pattern.

Case-II: $d = 2b$. From Eq. (x) it follows that

$$\frac{2b+b}{b} = \frac{n}{m} \text{ or } \frac{n}{m} = 3$$

or

$$n = 3m$$

Given $m = 1, 2, 3, \ldots$, the above condition reads $n = 3, 6, 9, \ldots$

Thus for $d = 2b$, 3rd, 6th, 9th, ... order interference maxima will be absent from the diffraction pattern. From the above discussion it is clear that when the slit separation is an integral multiple of the slit width, the condition for missing orders is satisfied exactly.

2.9 FRAUNHOFER DIFFRACTION BY *N* SLITS: DIFFRACTION GRATING

We have seen that the intensity produced by double slits is four times of that of single slit. Therefore, it is expected that *the device having more number of slits will produce large intensity*. So now we consider a large number of parallel slits of equal widths separated by equal opaque spaces. Such a device that makes use of multiple slit diffractions is called the *diffraction grating*.

It is constructed by ruling large number of fine, equidistant and parallel lines on a optically plane glass plate with the help of fine diamond point. The ruled lines are opaque to light while the space between any two lines is transparent to the light and acts as a slit. There are about 15,000 lines per inch in such a grating.

Let a parallel, collimated beam of monochromatic light of wavelength λ be incident normally on N-parallel slits (grating) each of width b and separated by a opaque distance d. The sum of b and d is known as *grating element*. The middle points in two consecutive slits separated by the distance $(b + d)$ are known as *corresponding points*. Let the diffracted light be focused by a convex lens L on the screen XY placed in the focal plane of the lens. All the secondary waves traveling in the direction parallel to the direction of incidence are brought to focus at a point C (Fig. 2.15a). The point C corresponds to the position of central bright maximum. The rays making an angle θ with the direction of incidence are focused at a point P (Fig. 2.15b).

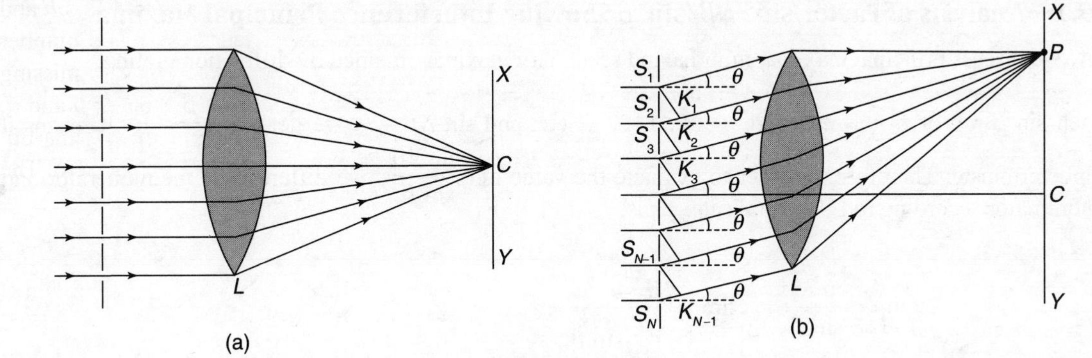

FIGURE 2.15

We may consider that the each slit in the grating is equivalent to an individual coherent source which is placed at the middle of each slit and sending a single wave of amplitude $\dfrac{A \sin \alpha}{\alpha}$ at angle θ with the direction of wave propagation. Here $\alpha = \dfrac{\pi}{\lambda} \times b \sin \theta$

If S_1K_1 be the perpendicular on S_2K_1, then the path difference between the waves originating from S_1 and S_2 is given by

$$S_2K_1 = (b + d) \sin \theta \tag{i}$$

The corresponding phase difference would be

$$\frac{2\pi}{\lambda}(b + d) \sin \theta = 2\beta \text{ (say)} \tag{ii}$$

It is clear from the above equation that the phase difference between two successive waves is constant and is equal to 2β (say). The phases increase in arithmetical progression (Fig. 2.16). Thus, we can construct the polygon of N amplitudes. The resultant amplitude and intensity at point P due to the waves from N slits can be obtained by vector polygon method and is given by

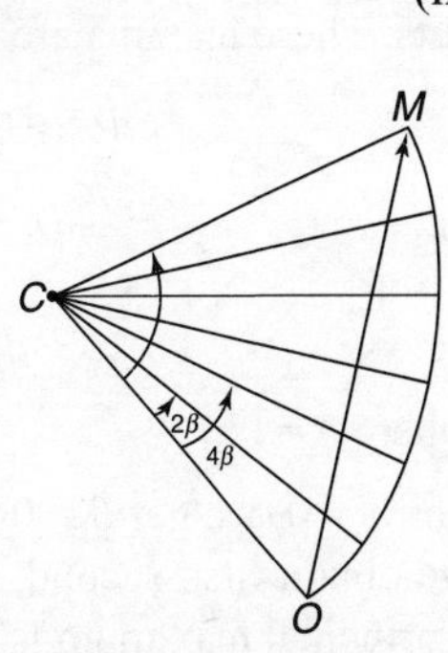

FIGURE 2.16

$$R = R_0 \frac{\sin N\beta}{\sin \beta} = \frac{A_0 \sin \alpha}{\alpha} \frac{\sin N\beta}{\sin \beta} \tag{iii}$$

A measure of the resultant intensity at point P is given by

$$I = R^2 = \frac{A_0^2 \sin^2 \alpha}{\alpha^2} \frac{\sin^2 N\beta}{\sin^2 \beta} \tag{iv}$$

The factor $\frac{A_0^2 \sin^2 \alpha}{\alpha^2}$ gives the intensity distribution in diffraction pattern due to a single slit, while the factor $\frac{\sin^2 N\beta}{\sin^2 \beta}$ yields the interference pattern due to N-slits.

2.9.1 Analysis of Factor $\sin^2 N\beta/\sin^2\beta$ Showing Interference Principal Maxima

Now we discuss principal maxima, minima and secondary maxima obtained by diffraction grating.

When $\sin \beta = 0$ or $\beta = \pm n\pi$, where $n = 0, 1, 2, \ldots$ etc. and $\sin N\beta = 0$, we get $\frac{\sin N\beta}{\sin \beta} = \frac{0}{0}$. It means it is indeterminate. Therefore, in order to evaluate the value of $\frac{\sin N\beta}{\sin \beta}$, we differentiate the numerator and denominator according to L'Hospital rule. Thus,

$$\lim_{\beta\to\pm n\pi} \frac{\sin N\beta}{\sin \beta} = \lim_{\beta\to\pm n\pi} \frac{\frac{d}{d\beta}(\sin N\beta)}{\frac{d}{d\beta}\sin \beta}$$

$$= \lim_{\beta\to\pm n\pi} \frac{N \cos N\beta}{\cos \beta} = \pm N$$

By substituting this value of $\frac{\sin N\beta}{\sin \beta}$ in Eq. (iv), we have

$$I = \frac{A_0^2 \sin^2 \alpha}{\alpha^2} N^2 \tag{v}$$

That is the resultant intensity of maxima becomes $\frac{A_0^2 \sin^2 \alpha}{\alpha^2} N^2$. Therefore, the resultant intensity of any of the principal maxima in the diffraction pattern can be obtained by multiplying N^2 to the factor $\frac{A_0^2 \sin^2 \alpha}{\alpha^2}$.

Being proportional to N^2, the brightness of the principal maxima increases with the increase of number of slits. These maxima are obtained in the direction given by

$$\beta = \pm n\pi$$

or $$\frac{\pi}{\lambda}(b+d)\sin\theta = \pm n\pi$$

or $$(b+d)\sin\theta = \pm n\lambda \tag{vi}$$

where $n = 0, 1, 2, \ldots$

For $n = 0$, we get $\theta = 0$ which gives zero order principal maximum. For the other values of n as 1, 2, 3, … we obtain first, second, third, … order principal maximum, respectively. The condition for the existence of a principal maximum is sometimes called the diffraction grating equation. The value of n gives the order the diffraction

Minima

The intensity expression (iv) shows that it is minimum when $\sin N\beta = 0$ but $\sin \beta \neq 0$.

Therefore, $\sin N\beta = 0$ or $N\beta = \pm m\pi$

or $$N(\pi/\lambda)\,(b+d)\sin\theta = \pm m\pi$$

or $$N(b+d)\sin\theta = \pm m\lambda \qquad \text{(vii)}$$

where m can have all integral values except 0, N, $2N$, $3N$ …, nN, because of these values of m, $\sin\beta = 0$, which gives different principal maxima.

It is clear from Eq. (vii) that $m = 0$ gives principal maximum of zero order. $m = 1, 2, 3, \ldots (N-1)$ give minima and $m = N$ gives again principal maximum of first order. Thus, there are $(N-1)$ equispaced minima between two consecutive principal maxima.

Secondary Maxima

As discussed there are $(N-1)$ minima between two successive principal maxima. In order to differentiate two consecutive minima there should be a maximum between them. Therefore, there would be $(N-2)$ maxima between $(N-1)$ minima. These maxima are known as *secondary maxima*. The positions of secondary maxima are obtained by differentiating equation (iv) w.r.t. β and putting it to zero.

$$\frac{dI}{d\beta} = \frac{A_0^2\sin^2\alpha}{\alpha^2}\,2\left[\frac{\sin N\beta}{\sin\beta}\right]\left[\frac{N\cos N\beta\sin\beta - \sin N\beta\cos\beta}{\sin^2\beta}\right] = 0$$

or $$N\cos N\beta\sin\beta - \sin N\beta\cos\beta = 0, \text{ as } \sin\beta = 0 \text{ gives principal maxima.}$$

or $$\tan N\beta = N\tan\beta$$

From this relation, we can draw the following triangle.

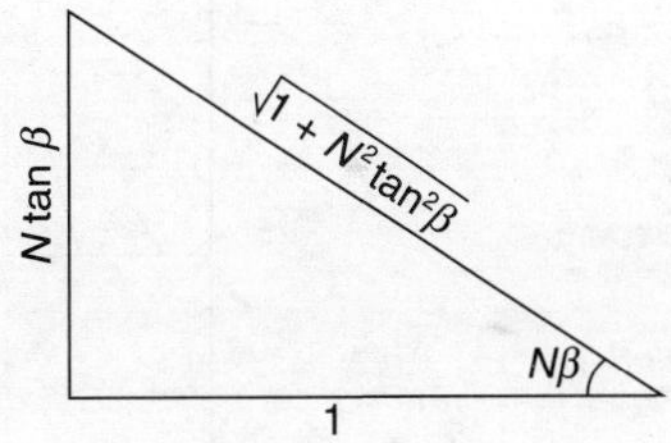

FIGURE 2.17

From Fig. 2.17, we have

$$\sin N\beta = \frac{N\tan\beta}{\sqrt{1+N^2\tan^2\beta}}$$

Hence, $$\frac{\sin^2 N\beta}{\sin^2\beta} = \frac{N^2\tan^2\beta/(\sqrt{1+N^2\tan^2\beta})^2}{\sin^2\beta}$$

$$= \frac{N^2}{\cos^2\beta(1+N^2\tan^2\beta)}$$

$$= \frac{N^2}{\cos^2\beta + N^2\sin^2\beta}$$

$$= \frac{N^2}{1-\sin^2\beta + N^2\sin^2\beta}$$

or $$\frac{\sin^2 N\beta}{\sin^2\beta} = \frac{N^2}{1+(N^2-1)\sin^2\beta} \qquad \text{(ix)}$$

From Eqs. (iv) and (ix), the intensity of secondary maxima is given by

$$I_s = \frac{A_0^2\sin^2\alpha}{\alpha^2}\cdot\frac{N^2}{1+(N^2-1)\sin^2\beta} \qquad \text{(x)}$$

It is clear from Eq. (x) that the intensity of secondary maxima is proportional to $\frac{N^2}{1+(N^2-1)\sin^2\beta}$. Since, the intensity of principal maxima is proportional to N^2.

$$\frac{\text{Intensity of secondary maxima}}{\text{Intensity of principal maxima}} = \frac{1}{1+(N^2-1)\sin^2\beta}$$

Hence, as N increases the intensity of secondary maxima decreases. In case of diffraction grating N is very large. Therefore, the secondary maxima are not visible in the spectrum and there is complete darkness between two successive principal maxima.

2.9.2 Diffraction Pattern

As mentioned earlier $\frac{\sin^2\alpha}{\alpha^2}$ is the diffraction factor and $\sin^2 N\beta/\sin^2\beta$ is the interference factor. In Fig. 2.18 we plot these two separately and also a combined effect (product) of them is shown. Thus the intensity distribution or the diffraction pattern due to N slits or diffraction grating is shown in Fig. 2.18c.

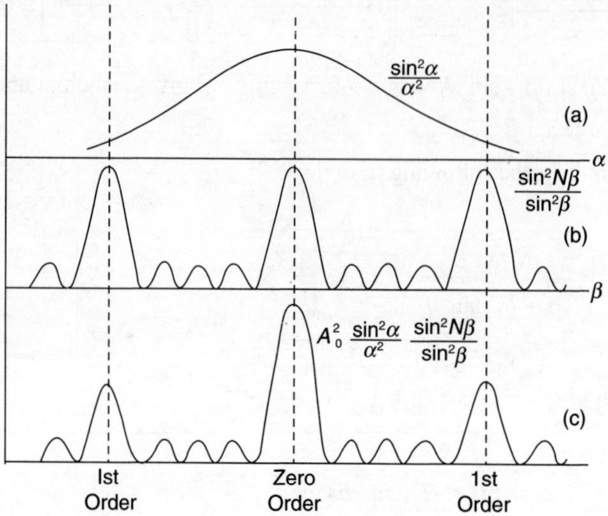

FIGURE 2.18

2.9.3 Missing Orders in Diffraction Pattern

As observed in double slit diffraction, for a particular set of values of slit width and slit separation, certain number of interference maxima are found to be missing. In the case of diffraction grating also, similar situation arises for certain values of b and d. It means the condition of missing order is met in the diffraction pattern of diffractions grating. For this the conditions of interference maxima and diffraction minima should be simultaneously satisfied. The conditions of interference maxima is given by equation (vi). For the diffraction minima, it can be noted from diffraction factor $\sin^2\alpha/\alpha^2$ that it should be zero. It means $\sin\alpha = 0$ but $\alpha \neq 0$.

Therefore, $\sin\left(\frac{\pi}{\lambda} b \sin\theta\right) = 0$. This would be true if

$$\frac{\pi}{\lambda} b \sin \theta = \pm m\pi$$

or $$b \sin \theta = \pm m\lambda.$$

Therefore, in order to meet the condition of missing orders following relations should be satisfied.

$$(b + d) \sin \theta = n\lambda, \boldsymbol{n} = 0, 1, 2, \ldots \qquad \text{(Interference maxima) (xi)}$$

$$b \sin \theta = m\lambda, m = 1, 2, 3, \ldots \qquad \text{(Diffraction minima) (xii)}$$

From Eqs. (xi) and (xii), we get

$$\frac{b+d}{b} = \frac{n}{m}$$

or $$n = \left(\frac{b+d}{b}\right) m$$

This is the condition of missing order of interference maxima in diffraction pattern. The absent orders are given below.

If $d = b$, then

$$n = \left(\frac{b+d}{b}\right) m = 2m$$

Therefore, for $m = 1, 2, 3, \ldots$ the 2nd, 4th, 6th, … order interference will be absent as $n = 2, 4, 6 \ldots$ If $d = 2b$, then, $n = 3, 6, 9, \ldots$ for $m = 1, 2, 3, \ldots$ Therefore the 3rd, 6th, 9th, … order interference will be absent from the diffraction pattern.

2.9.4 Angular Width of Principal Maxima

If θ_n be the direction of nth principal maxima and $(\theta_n + d\theta_n)$ and $(\theta_n - d\theta_n)$ the directions of first outer and inner sided minima adjacent to the nth maxima, then the value of angular width will be $2d\theta_n$ (Fig. 2.19).

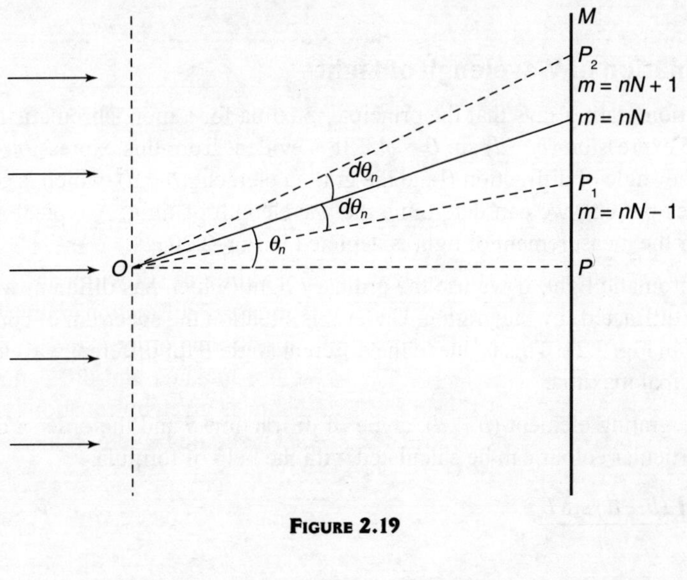

FIGURE 2.19

The directions of nth order principal maxima and minima are given as follows

$$(b + d) \sin \theta_n = n\lambda \quad \text{(xiii)}$$

and

$$N(b + d) \sin \theta_n = m\lambda \quad \text{(xiv)}$$

For the first order outer and inner sided minima adjacent to the nth maxima, θ_n should be replaced with $(\theta_n \pm d\theta_n)$ and $m = (nN \pm 1)$. Then from Eq. (xiv), we get

$$N(b + d) \sin (\theta_n \pm d\theta_n) = (nN \pm 1)\, \lambda \quad \text{(xv)}$$

or

$$N(b + d) (\sin \theta_n \cos d\theta_n \pm \cos \theta_n \sin d\theta_n) = (nN \pm 1)\lambda \quad \text{(xvi)}$$

For small values of $d\theta_n$, $\cos d\theta_n = 1$ and $\sin d\theta_n = d\theta_n$. With this equation (xvi) becomes

$$N(b + d) \sin \theta_n \pm N(b + d) \cos \theta_n\, d\theta_n = nN\lambda \pm \lambda \quad \text{(xvii)}$$

or

$$Nn\lambda \pm N(b + d) \text{ as on } d\theta_n = nNd \pm \lambda \quad \text{[with the help of Eqn (xiii)]}$$

With the help of Eq. (xiii), the above equation can be written as

$$N(b + d) \cos \theta_n\, d\theta_n = \lambda$$

or

$$d\theta_n = \frac{\lambda}{N(b + d) \cos \theta_n} \quad \text{(xviii)}$$

or

$$2d\theta_n = \frac{2\lambda}{N(b + d) \cos \theta_n} \quad \text{(xix)}$$

This is the expression for angular width of the nth order principal maxima, which shows that it depends on the total number of lines present on the grating and the wavelength of the light used in addition to the grating element.

2.10 APPLICATION OF DIFFRACTION GRATING LO7

In the previous chapter we learnt how to determine the wavelength of monochromatic light using mechanism of interference. We can also find the wavelength of light by using the concept of diffraction with the help of diffraction grating.

2.10.1 Determination of Wavelengh of Light

The theory of diffraction grating says that the principal maxima for a monochromatic light of wavelength λ is obtained as per the expression $(b + d) \sin \theta = n\lambda$. It is evident from this expression that if we are able to accurately measure the angle of diffraction θ and the grating element $(b + d)$, which is equal to the reciprocal of the number of lines per cm, we can determine the wavelength of light. A typical arrangement of using diffraction grating for the measurement of light is depicted in Fig. 2.20.

In place of a monochromatic light, if we use the ordinary light (which has different wavelengths), then the beam gets dispersed (diffracted) by the grating. Under this situation the spectrum of constituent wavelengths is obtained, as shown in Fig. 2.21. This is due to the different angle θ for different wavelength λ for satisfying the condition of principal maxima.

Thus by knowing the grating element $(b + d)$, angle of diffraction θ and the order n of the diffraction, the wavelength λ of a particular colour can be calculated with the help of formula

$$\lambda = \frac{(b + d) \sin \theta}{n}$$

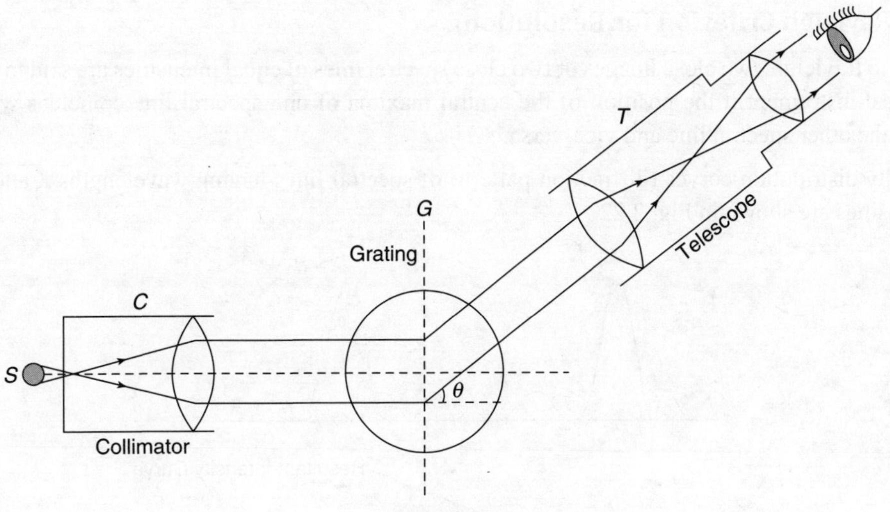

Figure 2.20

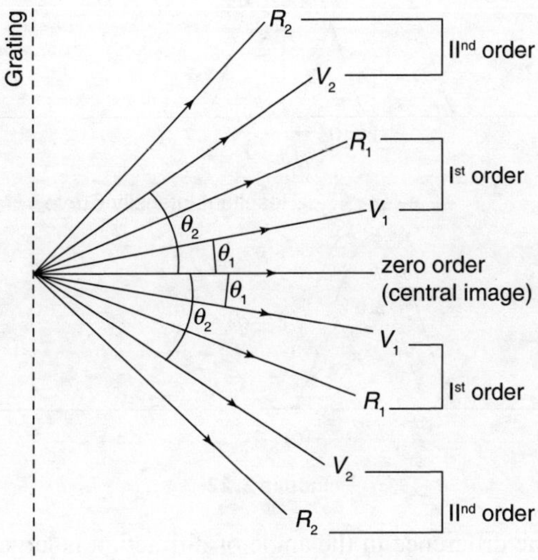

Figure 2.21

2.11 RESOLVING POWER OF AN OPTICAL INSTRUMENT LO8

If two images are very close to each other, they may appear as one and it is impossible for the eye to see them separately. Therefore, some optical instruments are used to resolve such images. Our eye can resolve two objects only when the angle subtended by them at the eye is greater than one minute $(1/60)°$. Here we say that the resolving limit of the normal eye is $(1/60)°$. The resolving power of an optical instrument is defined as its ability to just resolve the images of two close point sources or small object. It is the ability of instrument to measure the angular separation of two images that are close to each other.

2.11.1 Rayleigh Criterion for Resolution

According to Rayleigh, two close images or two close spectral lines of equal intensities are said to be resolved by an optical instrument if the position of the central maxima of one spectral line coincides with the first minima of the other spectral line and vice-versa.

The intensity distribution curves (diffraction pattern) of spectral lines having wavelengths λ and $\lambda + d\lambda$ of equal intensities are shown in Fig. 2.22.

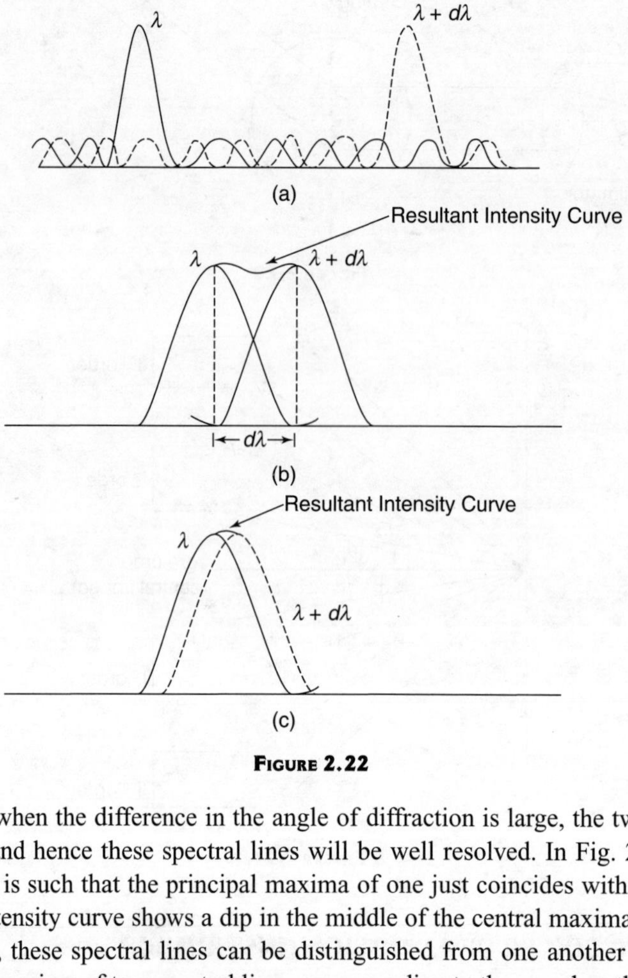

FIGURE 2.22

Figure 2.22a says that when the difference in the angle of diffraction is large, the two spectral lines can be seen as separate ones and hence these spectral lines will be well resolved. In Fig. 2.22b, the difference in the angle of diffraction is such that the principal maxima of one just coincides with the first minima of the other. Here resultant intensity curve shows a dip in the middle of the central maxima of these spectral lines. According to Rayleigh, these spectral lines can be distinguished from one another and are said to be just resolved. If the central maxima of two spectral lines corresponding to the wavelengths λ and $\lambda + d\lambda$ are very close to each other, as shown in Fig. 2.22c, then these two spectral lines overlap and they cannot be seen as separate ones.

According to Rayleigh criterion, two images or two close spectral lines of equal intensities are said to be just resolved when the resultant intensity at the dip is $(8/\pi^2)$ of the intensity of either central maxima. This can be proved as follows:

According to the theory of single slit Fraunhofer diffraction,

$$I = I_0 \frac{\sin^2 \alpha}{\alpha^2}$$

Since first minima is formed at an angle $\alpha = \pi$, the angle at the point of intersection will be $\pi/2$. In this case, intensity of each curve at the dip

$$I_1 = I_2 = I_0 \frac{\sin^2(\pi/2)}{(\pi/2)^2} = \frac{4}{\pi^2} I_0$$

The resultant intensity at the dip

$$I = I_1 + I_2 = \frac{4}{\pi^2} I_0 + \frac{4}{\pi^2} I_0$$

$$= \frac{8}{\pi^2} I_0$$

$$= 0.81 I_0$$

2.12 RESOLVING POWER OF A TELESCOPE (LO8)

Telescope is an optical instrument which is used to produce a magnified image of a distant object. The telescope can also form separate images of two close small objects situated at large distance. In this context it is important to investigate its resolving power.

Since a telescope consists of a system of lenses, we consider the diameter of the objective lens *AB* of the telescope as *a*. Further, we take two distant object *P* and *Q* such that they substend an angle θ on the objective lens of the telescope as shown in Fig. 2.23.

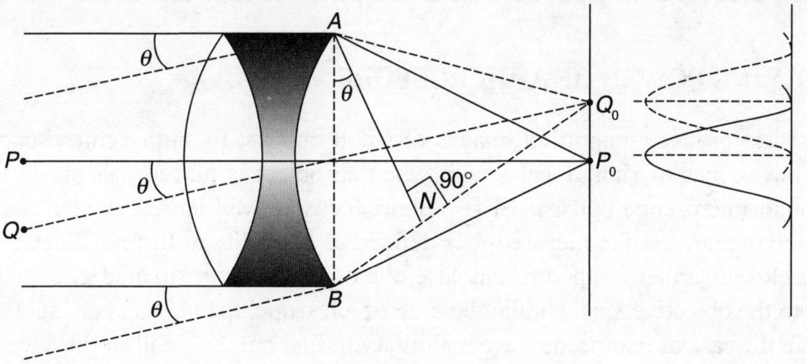

FIGURE 2.23

Now, we consider that a beam of light is incident on the objective lens of the telescope from these two neighbouring point sources. The image of each point source gives Fraunhofer diffraction pattern. Let P_0 and Q_0 be the positions of the central maximum of the images of *P* and *Q*. According to Rayleigh criterion, these two images are said to be resolved if the position of the central maximum of diffraction pattern of one coincides with first minimum of the other and vice-versa. All the secondary waves traveling in the direction AP_0 and BP_0 will meet on P_0 and the path difference between them will be zero. Thus, the point P_0 corresponds to the position of the central maximum of the first image. Similarly, the secondary waves traveling along AQ_0 and BQ_0 are met on Q_0 and the path difference between AQ_0 and BQ_0 is equal to BN.

$$BN = AB \sin \theta = a \sin \theta$$

or $$BN = a\theta \text{ (for small angle)}$$

If the path $BN = a\theta$ be equal to λ, then the position of the central maximum of Q_0 corresponds to the first minimum for the first image P_0 and hence this condition satisfies the Rayleigh criterion of resolution. Thus

$$a\theta = \lambda$$

or $$\theta = \lambda/a \tag{i}$$

The Eq. (i) holds goods for rectangular aperture.

For circular apertures, the modified from the Eq. (i) is written as

$$\theta = \frac{1.22\lambda}{a} \tag{ii}$$

where, λ is the wavelength of the light and θ refers to the limit of resolution and its reciprocal gives the resolving power of the telescope.

$$\text{Resolving power} = \frac{1}{\theta} = \frac{a}{1.22\lambda} \tag{iii}$$

The above equation says that the resolving power of a telescope would be higher, if the aperture a of the objective lens is taken larger.

Limit of Resolution of a Telescope

The limit of resolution is defined as an angle subtended at the objective lens of the telescope by two distant point objects which are just resolved when seen through the telescope. For the smaller values of this subtended angle on the objective, the resolving power of the telescope is said to be higher.

2.13 RESOLVING POWER OF A MICROSCOPE LO8

Microscope is used to produce magnified images of small objects. In simple microscope, a virtual image is formed by a convex lens of short focal length when an object is placed just inside its principal focus. However, compound microscope consists of two short focus convex lenses, named the objective and the eyepiece. The resolving power of a microscope is defined as its ability to form distinctly separate images of two objects lying close together. In the present case, the objects are not situated at large distance rather the objects are close to the objective lens. Unlike the case of telescope, the objects here subtend a large angle at the object plane. In the case of microscope, we mainly try to find out the smallest distance between two point objects which will produce images that are just resolved.

In Fig. 2.24, P and Q are two point objects separated by a distance d from each other and AB is the aperture of the objective lens of the microscope. P_0 and Q_0 are the corresponding Fraunhofer diffraction pattern of these two images. P_0 and Q_0 are the positions of the central maximum of P and Q that are surrounded by alternate dark and bright diffraction minima and maxima. According to Rayleigh criterion, these two images are said to be resolved if the position of the principal maximum of image P_0 corresponds to the first minimum of the image Q_0.

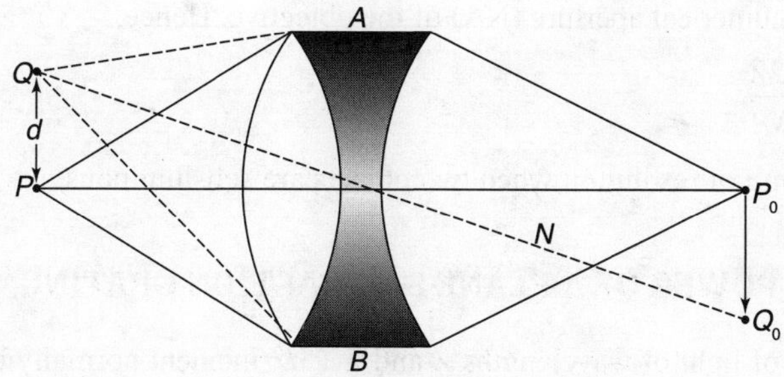

FIGURE 2.24

The path difference between the extreme rays originating from the point Q and reaching at P_0 is given by

$$\begin{aligned}\Delta &= (QB + BP_0) - (QA + AP_0)\\ &= QB - QA \qquad [\because BP_0 = AP_0]\\ &= (QM + MB) - (NA - NQ) \qquad \text{(Please refer to Fig. 2.25)}\\ &= (QM + PB) - (PA - NQ) \qquad [\text{For small distance between } P \text{ and } Q]\\ &= QM + NQ \qquad [\therefore PB = PA]\end{aligned}$$

From the triangles PNQ and PMQ in Fig. 2.25

$$QM = PQ \sin\theta = d \sin\theta$$

$$NQ = PQ \sin\theta = d \sin\theta$$

Therefore, the path difference = $QM + NQ = d\sin\theta + d\sin\theta = 2d\sin\theta$.

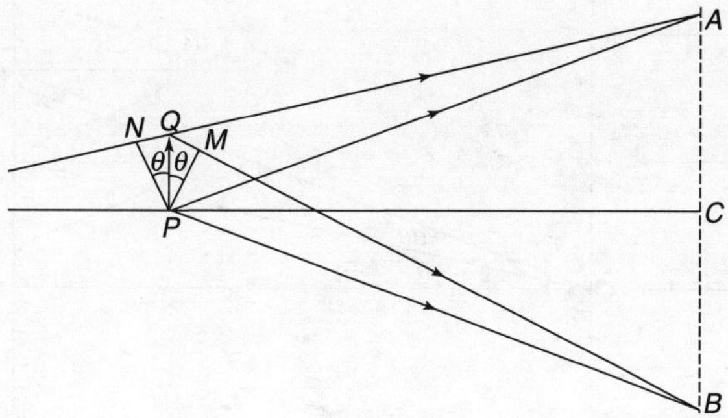

FIGURE 2.25

According to Airy, if the path difference is equal to 1.22λ (for circular aperture), then the maxima of image P_0 coincides with the minimum of the image Q_0. Therefore, these two images appear just resolved,. Thus,

$$2d\sin\theta = 1.22\lambda$$

$$d = \frac{1.22\lambda}{2\sin\theta} \qquad \text{(i)}$$

If the space between the object and objective lens is filled with an oil of refractive index μ, then

$$d = \frac{1.22\lambda}{2\mu\sin\theta} \qquad \text{(ii)}$$

Here μ sin θ is known as numerical aperture (NA) of the objective. Hence,

$$d = \frac{1.22\lambda}{2\,NA} \qquad \text{(iii)}$$

This equation gives the limit of resolution when two objects are self-luminous.

2.14 RESOLVING POWER OF A PLANE DIFFRACTION GRATING LO9

Consider a parallel beam of light of wavelengths λ and $\lambda + d\lambda$ incident normally on the plane transmission grating having grating element $(b + d)$ and total number of rulings N (Fig. 2.26). Then the resolving power of the grating is defined as the ratio of wavelength (λ) to the difference $d\lambda$ of the wavelength i.e., $\lambda/d\lambda$.

The separate diffraction pattern for λ and $\lambda + d\lambda$ is shown in the Fig. 2.26. According to Rayleigh criterion, these spectral lines are just resolved as the principal maxima of one line lies just on the first minima of the other.

Now the direction of nth principal maximum for a wavelength λ is given as

$$(b + d)\sin\theta = n\lambda \qquad \text{(i)}$$

The direction of nth principal maximum for a wavelength $\lambda + d\lambda$ is given by

$$(b + d)\sin(\theta + d\theta) = n(\lambda + d\lambda) \qquad \text{(ii)}$$

The equation of minima for wavelength λ is

$$N(b + d)\sin\theta = m\lambda \qquad \text{(iii)}$$

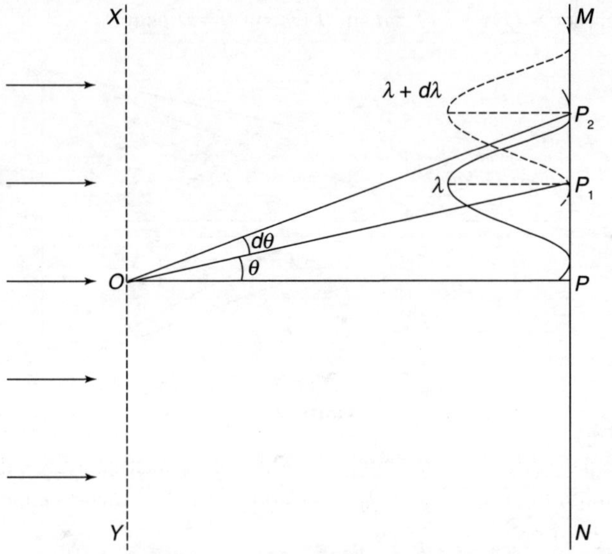

FIGURE 2.26

Here m has all the integral values except 0, N, $2N$, … nN, because for these values of m the condition for maxima is satisfied. Thus, first minimum adjacent to n^{th} principal maximum in the direction $(\theta + d\theta)$ can be obtained by substituting the values of m as $(nN + 1)$ in Eq. (iii). Therefore, first minima in the direction $(\theta + d\theta)$ is given by

$$N(b + d)\sin(\theta + d\theta) = (nN + 1)\lambda$$

or $$(b+d)\sin(\theta+d\theta)=\frac{(nN+1)\lambda}{N}$$

$$(b+d)\sin(\theta+d\theta)=n\lambda+\frac{\lambda}{N} \quad \text{(iv)}$$

A comparison of Eq. (iv) with Eq. (ii), i.e., the Rayleigh criterion for just resolution, gives

$$n(\lambda+d\lambda)=n\lambda+\frac{\lambda}{N}$$

or $$n\lambda+nd\lambda=n\lambda+\frac{\lambda}{N}$$

or $$nd\lambda=\frac{\lambda}{N}$$

or $$\frac{\lambda}{d\lambda}=nN$$

This is the required expression for the resolving power of the plane diffraction grating. This says that the number of lines per cm of a grating should be larger in order to increase its resolving power.

2.15 DISPERSIVE POWER OF A PLANE DIFFRACTION GRATING LO9

As we have seen, light of different wavelengths get dispersed/diffracted by the grating at different angles. In view of this, the angular dispersive power of a diffraction grating is defined as the rate of change of the angle of diffraction with the wavelength of light. It is denoted by $\frac{d\theta}{d\lambda}$.

For a plane transmission grating, the condition for principal maxima is

$$(b+d)\sin\theta=n\lambda \quad \text{(i)}$$

where $(b + d)$ is the grating element and θ is the angle of diffraction for n^{th} order principal maxima.

Differentiation of Eq. (i), w.r.t. λ gives

$$(b+d)\cos\theta\frac{d\theta}{d\lambda}=n$$

$\therefore$ $$\Rightarrow \text{Dispersive power}=\frac{d\theta}{d\lambda}=\frac{n}{(b+d)\cos\theta} \quad \text{(ii)}$$

Here $d\theta$ is the angular separation between the two lines having difference $d\lambda$ in the wavelengths.

We can make following observations from Eq. (ii).

(i) The dispersive power is directly proportional to n, i.e., to the order of diffraction. Thus, higher is the order, greater is the dispersive power. Hence the angular separation of two spectral lines is double in the second order diffraction as compared to that in the first order.

(ii) The dispersive power is inversely proportional to the grating element $(b + d)$. This means that the dispersive power is directly proportional to number of lines per cm of grating. Therefore, the angular dispersive power of two given lines is greater with a grating having larger number of lines per cm.

(iii) The dispersive power is inversely proportional to cos θ. thus, if the angle of diffraction $\theta = 0°$, cos $\theta = 1$ and hence the angular dispersion is minimum. Therefore, if θ is small, the value of cos θ may be taken as unity so the influence of cos θ may be neglected.

If we neglect the influence of the factor cos θ, then $d\theta \propto d\lambda$ for a given order, i.e., the angular dispersion of two spectral lines in a particular order is directly proportional to the difference in the wavelengths. Such spectrum is called a *normal spectrum*.

Linear Dispersive Power

If dx be the linear separation of two spectral lines differing in wavelengths by $d\lambda$ in the focal plane of a lens of focal length f, then we have

$$dx = f\,d\theta$$

Here, the linear dispersive power is defined as $\dfrac{dx}{d\lambda} = f\dfrac{d\theta}{d\lambda}$

$$= \frac{fn}{(b+d)\cos\theta} \qquad \text{(iii)}$$

SUMMARY

The main outcomes of this chapter are summarized as follows:

- Initially we discussed the phenomenon of diffraction.
- We clarified Young's double slit experiment with reference to both the phenomena of interference and diffraction.
- A clear distinction of the diffraction from the interference was given.
- Depending upon the distance of the source from the aperture, the incident wave can be realized either in the form of spherical wavefront or plane wavefront. The same is applicable to the wavefronts reaching the screen after emerging from the aperture. Based on these distances and hence the shapes of the wavefronts, the diffraction was divided into two classes, namely Fraunhofer diffraction and Fresnel diffraction.
- A concept of finding the resultant of a wavefront on the screen was given by Fresnel in terms of half-period zones. So Fresnel's half-period zones along with their construction were discussed.
- The concept of half-period zone was extended to zone plate, where alternative half-period zones are blocked and intense diffraction pattern is obtained.
- It was proved that a zone plate acts like a convex lens but it has multiple foci. Similarities and differences of a zone plate with a convex lens were summarized.
- Theory was given for Fresnel's diffraction by a circular aperture. Modification in the diffraction pattern at various nonaxial points on the screen was explained. The diffraction pattern was investigated when we move the screen toward the aperture by keeping the aperture diameter fixed. All these phenomena were discussed with the help of concept of half-period zones.
- Fraunhofer diffraction by a single slit and double slits was investigated by deriving the conditions of maxima and minima. It was discussed how certain number of interference maxima are found to be absent in the double slit experiment when we play with the slit width and slit separation. These conditions are called as the conditions for missing orders.
- Since the intensity in the case of double slit diffraction was found to be four times of the one obtained by the single slit, the need of an arrangement having a large number of slits was discussed and the

plane diffraction grating was introduced. The concepts of principal maxima and secondary maxima were brought in.

- The resolving power of an optical instrument is defined as its ability to just resolve the images of two close point sources or small objects. These two close images or two close spectral lines of equal intensities are said to be just resolved if the position of the central maxima of one spectral line coincides with the first minima of the other spectral line and vice-versa. This is called Rayleigh criterion for resolution.
- Expressions for resolving powers of a diffraction grating, telescope and microscope were obtained. Dispersive powers of a plane diffraction grating, both angular dispersive power and linear dispersive power, were discussed. Angular dispersion power is defined as the rate of change of the angle of diffraction with the wavelength of light.

Solved Examples

Example 1 A plane wavefront of light ($\lambda = 5000$ Å) is incident on an opening and is received on a screen at a distance of 100 cm from the opening. Find the radius of 80th half-period zone and the area of a half-period zone.

Solution Given $\lambda = 5.0 \times 10^{-7}$ m, $v = 100$ cm $= 1.0$ m and $n = 80$.

Radius of nth half-period zone

$$r_n = \sqrt{nv\lambda} = \sqrt{80 \times 1.0 \times 5 \times 10^{-7}}$$
$$= 6.32 \times 10^{-3} \text{ m}$$
$$= 0.632 \text{ cm.}$$

Area of half-period zone

$$= \pi v\lambda = 3.14 \times 100 \times 5 \times 10^{-5}$$
$$= \mathbf{0.0157\ cm^2}$$

Exmaple 2 Find the radius of the first half-period zone of a zone plate that behaves like a convex lens of focal length 60 cm. Given $\lambda = 6000$ Å.

Solution Given $f = 0.60$ m, $\lambda = 6.0 \times 10^{-7}$ m and $n = 1$.

Formula used is $\quad f_n = \dfrac{r_n^2}{n\lambda} \quad$ or $\quad f_1 = \dfrac{r_1^2}{\lambda}$

or $\quad r_1^2 = 0.6 \times 6.0 \times 10^{-7}$

$\quad r_1 = 0.6 \times 10^{-3}$ m

or $\quad \mathbf{r_1 = 0.6\ mm}$

Example 3 A parallel beam of light of wavelength 5×10^{-7} m falls on a circular aperture and the diffraction pattern is observed on a screen 0.30 m away. Find the radius of circular opening so that the intensity of light on the screen is 4 times the intensity in absence of opening.

Solution Given $v = 0.30$ m and $\lambda = 5 \times 10^{-7}$ m.

In the given case, the radius of the opening

= radius of 1st half-period zone

$$= \sqrt{v\lambda}$$

$$= \sqrt{0.30 \times 5 \times 10^{-7}}$$

$$= 0.387 \times 10^{-3} \text{ m.}$$

$$= \mathbf{0.0387\ mm}$$

EXAMPLE 4 Keeping the distance of observation point fixed as 0.50 m, calculate the number of half-period zones in a circular opening of radius (a) 2.0 mm (b) 20 mm, where the light of wavelength 6000Å is used.

SOLUTION Given $v = 0.50$ m, $r_1 = 2.0$ mm, $r_2 = 20$ mm and $\lambda = 6.0 \times 10^{-7}$ m.

Let us consider a_n as the area of circular opening, which contains n half-period zones.

$$a_n = n(\pi v\lambda) \qquad (\because \text{ opening contains } n \text{ half-period zones each of which has area} = \pi v\lambda)$$

$$a_n = \pi r^2$$

For $r_1 = 2.0$ mm $= 2.0 \times 10^{-3}$ m,

$$n = \frac{\pi r^2}{\pi v\lambda} = \frac{r_1^2}{v\lambda} = \frac{(2.0\times10^{-3})^2}{0.5\times6.0\times10^{-7}} = \frac{4.0\times10^{-6}}{3.0\times10^{-7}} = \frac{40}{3}$$

$$\mathbf{n = 13}$$

Similarly, for $r_2 = 20$ mm $= 0.02$ m

$$n = \frac{r_2^2}{v\lambda} = \frac{(0.02)^3}{0.5\times6.0\times10^{-7}}$$

$$= \frac{4.0\times10^{-4}}{3.0\times10^{-7}} = 1.333\times10^3$$

$$\mathbf{n = 1333}$$

EXAMPLE 5 The diameter of the first ring of a zone plate is 1.0 mm. If the plane waves of wavelength 5000 Å fall on the plate, find where a screen should be placed so that light is focused at the brightest spot.

SOLUTION Given $r_1 = ½$ mm $= 0.5 \times 10^{-3}$ m, $\lambda = 5.0 \times 10^{-7}$ m and $n = 1$.

Formula used is $f_n = \dfrac{r_n^2}{n\lambda}$

$\because \qquad n = 1$

$$\therefore \qquad f_1 = \frac{r_1^2}{\lambda} = \frac{(0.5\times10^{-3})^2}{5.0\times10^{-7}} = \frac{2.5\times10^{-7}}{5.0\times10^{-7}} = \mathbf{0.5\ m}$$

EXAMPLE 6 Find the radius of the first three transparent zones of a zone plate behaving like a convex lens of focal length 1.0 m for light of wavelength $\lambda = 5893$ Å.

SOLUTION Given $f = 1.0$ m and $\lambda = 5.893 \times 10^{-7}$ m.

Considering positive zone plate in which odd number of zone are transparent so that

$$r = \sqrt{fn\lambda}$$

Then, for 1st zone, $n = 1$

$$\therefore \qquad r_1 = \sqrt{1.0\times1\times5.893\times10^{-7}}$$

$$= 7.676 \times 10^{-4} \text{ m}$$

Similarly $r_3 = \sqrt{1.0 \times 3 \times 5.893 \times 10^{-7}}$

$= 1.329 \times 10^{-3}$ m.

and $r_5 = \sqrt{fn\lambda}$

$= \sqrt{1.0 \times 5 \times 5.893 \times 10^{-7}}$

$= 17.165 \times 10^{-4}$ m

$= \mathbf{1.716 \times 10^{-3}}$ **m.**

EXAMPLE 7 What is the radius of the tenth zone plate of focal length 20 cm for light of wavelength 5000 Å?

SOLUTION Given $f = 0.20$ m, $f = 5.0 \times 10^{-7}$ m and $n = 10$.

Formula used is $f_n = \dfrac{r_n^2}{n\lambda}$

or $r_{10}^2 = fn\lambda = 0.2 \times 10 \times 5.0 \times 10^{-7}$ m $= 10^{-6}$

or $r_{10} = 10^{-3}$ m

$\mathbf{r_{10} = 1.0}$ **mm.**

EXAMPLE 8 The image of a point source of light ($\lambda = 5890$ Å) at a distance 1.0 meter from the zone plate is observed at 2.0 meter on the other side. Calculate

(a) the focal length of the zone plate (b) power of a zero plate

(c) diameter of 1st zone.

SOLUTION Given $\lambda = 5.89 \times 10^{-7}$ m, $u = 1.0$ m and $v = 2.0$ m.

(a) Focal length

$$\frac{1}{f} = \frac{1}{u} + \frac{1}{v} = \frac{1}{1} + \frac{1}{2} = \frac{3}{2}$$

$$f = \frac{2}{3} = \mathbf{0.67\ m.}$$

(b) Power of zone plate

$$P = \frac{1}{f} \quad \text{[because zone plate acts as a convex lens]}$$

$$P = \frac{1}{2/3} = \frac{3}{2} = \mathbf{1.5\ D}$$

(c) Let r_1 be the radius of the first zone. Then,

$$r_1^2 = fn\lambda$$

$$r_1^2 = \frac{2}{3} \times 1 \times 5.893 \times 10^{-7}\,\text{m}^2$$

$$r_1^2 = 3.93 \times 10^{7}\,\text{m}^2$$

$$r_1 = 6.267 \times 10^{-4}\,\text{m}$$

Hence diameter of 1st zone

$= 2r_1 = 1.253 \times 10^{-3}$ m $=$ **1.253 mm**

EXAMPLE 9 If the focal length of the zone plate is 1.0 m for light of wavelength 6.0×10^{-7} m, what will be the focal length for the wavelength 5.0×10^{-7} m.

SOLUTION Given $\lambda_1 = 6.0 \times 10^{-7}$ m and $\lambda_2 = 5.0 \times 10^{-7}$ m.

Formula used is $f_n = \dfrac{r_n^2}{n\lambda}$

Thus, focal length f_1 and f_2 are

$$f_1 = \frac{r_n^2}{\lambda_1} \text{ and } f_2 = \frac{r_n^2}{2\lambda_2}$$

or $$\frac{f_2}{f_1} = \frac{\lambda_1}{\lambda_2} \quad \text{or} \quad f_2 = f_1 \frac{\lambda_1}{\lambda_2}$$

$$f_2 = 1.0 \times \frac{6.0 \times 10^{-7}}{5.0 \times 10^{-7}}$$

$$\boldsymbol{f = 1.2\ \text{m}}$$

EXAMPLE 10 An object is placed at 20 cm from a zone plate and the brightest image is situated at 20 cm from the zone plate with light of wavelength $\lambda = 4000$ Å. Calculate the number of Fresnel's zone in a radius of 1.0 cm of that plate.

SOLUTION Given $u = 0.20$ m, $v = 0.20$ m, $\lambda = 4.0 \times 10^{-7}$ m and $r = 0.01$ m.

Formula used is $f_n = \dfrac{r_n^2}{n\lambda}$ and $\dfrac{1}{f} = \dfrac{1}{u} + \dfrac{1}{v}$

$$\frac{1}{f} = \frac{1}{0.2} + \frac{1}{0.2} = \frac{2}{0.2} \text{ or } f = \frac{0.2}{2}\text{m} = 0.1\text{ m}$$

or $$f = 10\text{ cm}$$

So, the number of Fresnel's zone $(n) = \dfrac{r_n^2}{f_n\lambda}$

$$= \frac{(0.01)^2}{(0.1) \times 4 \times 10^{-7}}$$

$$= \frac{1.0 \times 10^{-4}}{4 \times 10^{-8}} = \frac{1.0}{4} \times 10^4 = 0.25 \times 10^4$$

$\therefore$ $$\boldsymbol{n = 2500}$$

EXAMPLE 11 The diameter of the central zone of a zone plate is 2.3 mm. If a point source of light ($\lambda = 5893$ Å) is placed at a distance of 6.0 m from it, calculate the position of the first image.

SOLUTION Given diameter $(d) = 2.3$ mm $= 2.3 \times 10^{-7}$ m, $r = d/2 = 1.15 \times 10^{-3}$ m, $\lambda = 5.893 \times 10^{-7}$ m and $n = 1$.

Formula used is $f_n = \dfrac{r_n^2}{n\lambda} = \dfrac{r^2}{\lambda} = \dfrac{(1.15 \times 10^{-3})^2}{5.893 \times 10^{-7}}$

$$= 2.244\text{ m}$$

$\therefore$ $$f = 2.2\text{ m}$$

$$\frac{1}{f} = \frac{1}{u} + \frac{1}{v} = \frac{1}{6} + \frac{1}{v}$$

$$\frac{1}{v}=\frac{1}{2.2}-\frac{1}{6}=\frac{6-2.2}{2.2\times 6}=\frac{3.8}{13.2}$$

$$v=\frac{13.2}{3.8}=3.47$$

$$\mathbf{v = 3.47\ m}$$

Hence, first image is formed at a distance of 3.47 m.

EXAMPLE 12 A zone plate is made by arranging the radii of the circles which define the zones such that they are the same as the radii of Newton's rings formed between a plane surface and the surface having radius of curvature 200 cm. Find the principal focal length of the zone plane.

SOLUTION Given $R = 2.0$ m.

Formula used is $\quad f_1=\dfrac{r_1^2}{\lambda} \qquad (\because n=1)$ (i)

By Newton's ring formula

$$r_n=\sqrt{n\lambda R}$$

For $n = 1$

$$r_1=\sqrt{\lambda R} \tag{ii}$$

By using Eqs. (i) and (ii), we have

$$f_1=\frac{\lambda R}{\lambda}=R=2.0\text{ m}$$

$$\mathbf{f_1 = 2.0\ m}$$

EXAMPLE 13 A single slit of width 1 mm is illuminated by light of wavelength 589 nm. Find the angular spread of the central maxima of diffraction pattern observed.

SOLUTION Given $\lambda = 5.89 \times 10^{-7}$ m and slit-width (b) 1.0×10^{-3} m.

Formula used is $\quad b \sin\theta = m\lambda$

For first minima, $\quad m = 1$

$\therefore \quad \sin\theta = \lambda/b$

$$\theta=\sin^{-1}[\lambda/b]=\sin^{-1}\left[\frac{5.89\times 10^{-7}}{1.0\times 10^{-3}}\right]$$

$$\mathbf{\theta = 0.03374°}$$

The angular spread of central maximum is 2θ

$\therefore \quad 2\theta = 2 \times 0.03374$

$$\mathbf{2\theta = 0.0675°}$$

Therefore, angular spread of central maximum is 0.0675°.

EXAMPLE 14 In Fraunhofer type diffraction at narrow slit of width 0.2 mm, a screen is placed 1.2 m away from the slit. In the fringe pattern, first minimum lie at 3.7 mm on either side of the central maximum. Find out the wavelength of light.

SOLUTION Given $b = 2 \times 10^{-4}$ m, $D = 1.2$ m and $x = 3.7 \times 10^{-3}$ m.

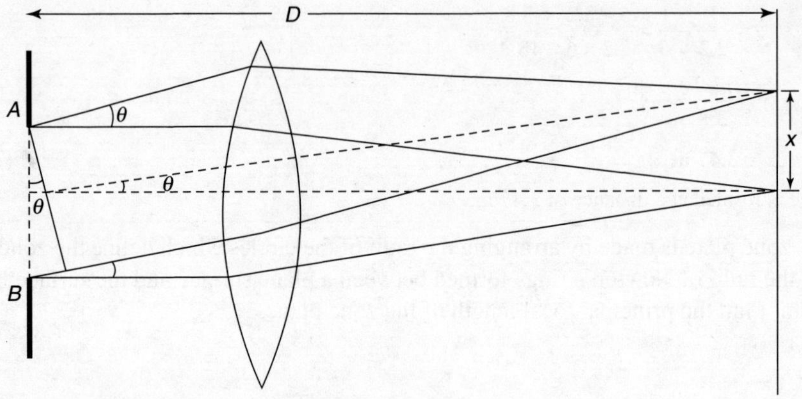

FIGURE 2.27

From the given figure, if θ is very small, then

$$\sin\theta = \frac{x}{D} \quad \text{(i)}$$

condition of minima $b\sin\theta = m\lambda$

For $m = 1$, then $\quad \sin\theta = \lambda/b \quad$ (ii)

By using Eqs. (i) and (ii), we get

$$\frac{\lambda}{b} = \frac{x}{D} \text{ or } \lambda = \frac{xb}{D} = \frac{3.7\times10^{-3}\times2\times10^{-4}}{1.2}$$

$$\boldsymbol{\lambda = 6167\ \text{Å}}$$

Example 15 A light of wavelength 550 nm falls normally on a slit of width 2.2 μm. Determine the angular position of second and third minima.

Solution Given $b = 2.2\times10^{-6}$ m and $\lambda = 5.5\times10^{-7}$ m.

The formula used is $b\sin\theta = m\lambda$

For angular position of second minima

$$\sin\theta_2 = \frac{2\lambda}{b} = \frac{2\times5.5\times10^{-7}}{2.2\times10^{-6}} = 0.5$$

or $\quad \theta_2 = \sin^{-1}(0.5) = 30°$

or $\quad \boldsymbol{\theta_2 = 30°}$

Similarly, for angular position of third minima

$$\sin\theta_3 = \frac{3\lambda}{b} = \frac{3\times5.5\times10^{-7}}{2.2\times10^{-6}} = 0.75$$

$$\boldsymbol{\theta_3 = 48.59°}$$

Example 16 In Fraunhofer diffraction at a slit of width 1.2×10^{-6} m, find the half-angular width of the central bright maximum if the slit is illuminated by light of wavelength 5890 Å.

Solution Given $b = 1.2\times10^{-6}$ m and $\lambda = 5.89\times10^{-7}$ m.

Formula used is $\quad b\sin\theta = m\lambda$

For first minima $m = 1$

$$\sin\theta = \lambda/b = \frac{5.89\times10^{-7}}{1.2\times10^{-6}} = 0.491$$

$$\theta = \sin^{-1}(0.491)$$

or $$\theta = \mathbf{29.41^\circ}$$

EXAMPLE 17 Parallel beam of light (5000 Å) is normally incident on a slit. The central maximum fans out at 30° on both sides of the direction of the incident light. Calculate the slit width. For what width of the slit the central maximum would spread out to 90° from the direction of the incident light?

SOLUTION Given $\lambda = 5.0 \times 10^{-7}$ m and $\theta = 30°$.

Formula used is $b \sin\theta = m\lambda$

$$b = \frac{\lambda}{\sin\theta} \quad (m = 1 \text{ for first minimum})$$

$$= \frac{5.0\times10^{-7}}{\sin 30^\circ} = 1.0\ \mu\text{m}$$

If $\theta = 90°, b = ?$

$$b = \frac{5.0\times10^{-7}}{\sin 90^\circ} = \mathbf{0.5\ \mu m}$$

EXAMPLE 18 A parallel beam of light (λ = 5890 Å) is incident perpendicularly on a slit of width 0.1 mm. Calculate angular width and linear width of central maximum formed on the screen 100 cm away.

SOLUTION Given $\lambda = 5.89 \times 10^{-7}$ m, b = 1.0×10^{-4} m and $D = 1.0$ m.

Formula used is $b \sin\theta = m\lambda$

For angular width of central maximum

$$\sin\theta = \frac{\lambda}{b} \quad (m = 1 \text{ for first minimum})$$

$$= \frac{5.89\times10^{-7}}{1.0\times10^{-4}} = 5.89 \times 10^{-3}$$

$$\theta = \sin^{-1}(0.00589)$$

$$= 0.3375^\circ$$

Therefore, the total angular spread of central maximum is 2θ, then

$$2\theta = 2\times0.3375^\circ$$

$$2\theta = 0.675^\circ$$

For linear width formula used is

$$\sin\theta = \frac{x}{D}$$ [Please see Fig. 2.27]

$$x = D\sin\theta = 1.0\times5.89\times10^{-3}$$

$$= 5.89 \times 10^{-3}\ \text{m}$$

Total linear separation = $2x = 2 \times 5.89 \times 10^{-3}$ m

$$= 0.01178\ \text{m}$$

$$= \mathbf{1.178\ cm}$$

Example 19 A single slit is illuminated by light composed of two wavelength λ_1 and λ_2. One observes that due to Fraunhofer diffraction the first minimum obtained for λ_1 coincides with the second diffraction minimum of λ_2. What is the relation between λ_1 and λ_2?

Solution Given $m = 1$ for λ_1 and $m = 2$ for λ_2.

Formula used is $b \sin\theta = m\lambda$

For λ_1, $m = 1$, then

$$b \sin\theta = \lambda_1 \qquad \text{(i)}$$

For λ_2, $m = 2$, then

$$b \sin\theta = 2\lambda_2 \qquad \text{(ii)}$$

From Eqs. (i) and (ii)

$$\lambda_1 = 2\lambda_2$$

i.e., λ_1 is double of λ_2

Example 20 Find the angular width of the central bright maximum in the Fraunhofer diffraction pattern of a slit of width 12×10^{-5} cm when the slit is illuminated by monochromatic light of wavelength 6000 Å.

Solution Given $b = 1.2 \times 10^{-6}$ m and $\lambda = 6.0 \times 10^{-7}$ m.

Formula used is $b \sin\theta = m\lambda$

or

$$\sin\theta = \frac{\lambda}{b} \qquad (m = 1)$$

$$= \frac{6.0 \times 10^{-7}}{1.2 \times 10^{-6}} = 0.5$$

$\therefore$

$$\theta = \sin^{-1}(0.5) = 30°$$

or

$$\theta = 30°$$

where θ is half-angular width.

Therefore, angular width of central maximum is

$$\mathbf{2\theta = 60°}$$

Example 21 Diffraction pattern of a single slit of width 0.5 cm is formed by a lens of focal length 40 cm. Calculate the distance between the first dark and next bright fringe from the axis. Wavelength is 4890 Å.

Solution Given $\lambda = 4890$ Å $= 4.89 \times 10^{-7}$ m, $f = 0.40$ m and $b = 0.005$ m.

Formula used is $b \sin\theta = m\lambda$ (i)

For first dark fringe, $m = 1$

$$\sin\theta = \frac{\lambda}{b} \qquad \text{(ii)}$$

and

$$\sin\theta = \frac{x}{f} \qquad \text{[Please see Fig. 2.27]} \qquad \text{(iii)}$$

By using Eqs. (iii) and (ii)

$$\frac{\lambda}{b} = \frac{x}{f}$$

$$x = \frac{\lambda f}{b} = \frac{4.89 \times 10^{-7} \times 0.40}{5.0 \times 10^{-3}}$$

$$= 3.912 \times 10^{-5}\text{m.}$$

For first secondary maxima

$$b \sin \theta = \frac{(2n+1)\lambda}{2}$$

$$\sin \theta = \frac{3\lambda}{2b} \quad (n = 1) \tag{iv}$$

and $$\sin \theta = \frac{x'}{f} \quad \text{[from Fig. 2.27]} \tag{v}$$

$$\frac{x'}{f} = \frac{3\lambda}{2b}$$

$$x' = \frac{3l}{2b} f = \frac{3 \times 4.89 \times 10^{-7} \times 0.40}{2 \times 5.0 \times 10^{-3}}$$

or $$x' = 5.868 \times 10^{-5} \text{ m}$$

Hence, the difference between 1st dark and next bright ring is

$$x_2 - x_1 = 5.868 \times 10^{-5} - 3.912 \times 10^{-5}$$

$$\mathbf{\Delta x = 1.956 \times 10^{-5} \text{ m}}$$

EXAMPLE 22 A plane of wavelength 5893 Å passes through a slit, which is 0.5 mm wide and forms a diffraction pattern on a screen placed on the focal plane of a lens of focal length 1.0 m. Calculate the separation of the dark band on either side of the central maximum.

SOLUTION Given $b = 0.5 \times 10^{-3}$ m and $\lambda = 5.893 \times 10^{-7}$ m.

Formula used is $b \sin \theta = m\lambda \ (m = 1)$

$$\sin \theta = \frac{\lambda}{b} = \frac{5.893 \times 10^{-7}}{5.0 \times 10^{-4}} = \frac{x}{f} \text{ or } x = 1.1786 \times 10^{-3} \times 1 = 1.1786 \times 10^{-3} \text{ m}$$

$$x = 1.1786 \text{ mm}$$

Hence the separation of dark band on either side

$$= \mathbf{2x = 2.357 \text{ mm.}}$$

EXAMPLE 23 Calculate the missing orders in a double slit Fraunhofer diffraction pattern, if the widths of slits are 0.08×10^{-3} m and they are 0.4×10^{-3} m apart.

SOLUTION Given $b = 0.08 \times 10^{-3}$ and $d = 0.4 \times 10^{-3}$ m.

The directions of interference maxima are given by

$$(b + d) \sin \theta = n\lambda \tag{i}$$

The directions of diffraction minima are given by

$$b \sin \theta = m\lambda \tag{ii}$$

Dividing Eq. (i) by Eq. (ii), we get

$$\frac{b+d}{b} = \frac{n}{m} \text{ or } \frac{(0.08 + 0.4) \times 10^{-3}}{0.08 \times 10^{-3}} = \frac{n}{m}$$

or $$\frac{6}{1} = \frac{n}{m}$$

$$n = 6m$$

$$\mathbf{= 6, 12, 18, \ldots \text{ etc.}} \qquad \mathbf{(m = 1, 2, 3, \ldots)}$$

Hence, 6th, 12th, 18th, … etc interference maxima will be missing in the diffraction pattern.

Example 24 In a double-slit Fraunhofer diffraction pattern the screen is 1.6 m away from the slits. The slit width are 0.2 mm and they are 0.4 mm. Calculate the wavelength of light if the fringe width is 2.5×10^{-3} m and also deduce the missing orders.

Solution Given slit width $b = 2.0 \times 10^{-4}$ m, $2d = 4.0 \times 10^{-4}$ m, $\beta = 2.5 \times 10^{-3}$ m and $D = 1.6$ m.

Formula used is $$\lambda = \frac{\beta 2d}{D} = \frac{2.5 \times 10^{-3} \times 4.0 \times 10^{-4}}{1.6} = 6250 \text{ Å}$$

The directions of interference maxima are

$$(b + d) \sin \theta = n\lambda \qquad \text{(i)}$$

The directions of diffraction minima are

$$b \sin \theta = m\lambda \qquad \text{(ii)}$$

then, from Eqs (i) and (ii), we get

$$\frac{b+d}{b} = \frac{n}{m} \text{ or } n = \frac{b+d}{b} m$$

$$n = \frac{(2.0 + 4.0) \times 10^{-4}}{2.0 \times 10^{-4}} m = 3m$$

$$\mathbf{= 3, 6, 9 \ldots \text{ etc. } (m = 1, 2, 3 \ldots)}$$

Hence 3rd, 6th, 9th, … interference maxima will be missing in the diffraction pattern.

Example 25 A parallel beam of sodium light is normally incident on a plane transmission grating having 4250 lines per cm and a second order spectral line is observed at an angle of 30°. Calculate the wavelength of light.

Solution Given $N = 4250$ lines per cm, $\theta = 30°$ and $n = 2$.

Formula used is $$(b + d) \sin \theta = n\lambda$$

$$\lambda = \frac{(b+d) \sin \theta}{n}$$

Now, $$(b+d) = \frac{1}{4250} \text{ cm}$$

$\therefore$ $$\lambda = \frac{1}{4250} \times \frac{\sin 30°}{2}$$

$$= 5882 \times 10^{-8} \text{ cm}$$

$$\lambda = \mathbf{5882 \text{ Å}}$$

Example 26 A parallel beam of monochromatic light is allowed to incident normally on a plane transmission grating having 5000 lines per cm and second order spectral line is found to be diffracted through 30°. Calculate the wavelength of light.

Solution Given $N = 5000$ lines per cm, $\theta = 30°$ and $n = 2$.

$$(b + d) = \frac{1}{N} = \frac{1}{5000} = 2.0 \times 10^{-4} \text{ cm}$$

Formula used is $$(b = d) \sin \theta = n\lambda \text{ or } \lambda = \frac{(b+d) \sin \theta}{n}$$

or $$\lambda = \frac{2.0 \times 10^{-4} \times \sin 30°}{2} = 5000 \times 10^{-8} \text{ cm}$$

$$\boldsymbol{\lambda} = \mathbf{5000 \text{ Å}}$$

EXAMPLE 27 In a grating spectrum, which spectral line in 4th order will overlap with 3rd order line of 5461 Å.

SOLUTION Given $n_1 = 4$, $\lambda_2 = 5461$ Å and $n_2 = 3$, $\lambda_1 = ?$

As per question

$$(b + d) \sin \theta = 4\, \lambda_1 = 3\, \lambda_2$$

or
$$\lambda_1 = \frac{3}{4}\lambda_2 = \frac{3}{4} \times 5461 = 4096$$

or
$$\boldsymbol{\lambda_1 = 4096 \text{ Å}}$$

EXAMPLE 28 In a plane transmission, grating the angle of diffraction for second order maxima for wavelength 5×10^{-5} cm is 30°. Calculate the number of lines in one centimeter of the grating surface.

SOLUTION Given $\lambda = 5 \times 10^{-5}$ cm, $\theta = 30°$ and $N = ?$, $n = 2$.

$$(b + d) \sin \theta = n\lambda$$

$$(b + d) = \frac{n\lambda}{\sin \theta} = \frac{2 \times 5 \times 10^{-5}}{\sin 30°} = \frac{10 \times 10^{-5}}{0.5}$$

$$= 2.0 \times 10^{-4} \text{ cm.}$$

The number of lines are

$$N = \frac{1}{(b+d)} = \frac{1}{2.0 \times 10^{-4}} = \frac{10^4}{2} \text{ per cm}$$

$$= \textbf{5000 lines/cm}$$

EXAMPLE 29 A plane grating has 15000 lines per inch. Find the angle of separation of the 5048 Å and 5016 Å lines of helium in second order spectrum.

SOLUTION Given $\lambda_1 = 5048 \times 10^{-8}$ cm, $\lambda_2 = 5016 \times 10^{-8}$ cm.

$$n = 2 \text{ and } b + d = \frac{2.54}{15000} = 1.693 \times 10^{-4} \text{ cm}$$

Formula used is $(b + d) \sin \theta = n\lambda$

For wavelength λ_1

$$(b + d) \sin \theta_1 = n\lambda_1$$

$$\sin \theta_1 = \frac{n\lambda_1}{b+d} = \frac{2 \times 5048 \times 10^{-8}}{1.693 \times 10^{-4}}$$

$$\sin \theta_1 = 0.5963$$

$$\theta_1 = 36.60°$$

Similarly,
$$\theta_2 = \sin^{-1} \frac{2 \times 5016 \times 10^{-8}}{1.693 \times 10^{-4}}$$

$$= 36.34°$$

Therefore, angle of separation $\Delta\theta = \theta_1 - \theta_2$

$$= \mathbf{0.26°}$$

EXAMPLE 30 A plane transmission grating having 6000 lines/cm is used to obtain a spectrum of light from a sodium lamp in the second order. Calculate the angular separation between the two sodium lines whose wavelengths are 5890 Å and 5896 Å.

Solution Given $N = 6000$ lines/cm, $a + b + \frac{1}{N} = \frac{1}{6000}$ cm, $= 1.67 \times 10^{-4}$ cm.

$\lambda_1 = 5890 \times 10^{-8}$ cm, $\lambda_2 = 5896 \times 10^{-8}$ cm, and $n = 2$

Formula used is $(b + d) \sin \theta = n\lambda$

For λ_1

$$(b + d) \sin \theta_1 = n\lambda_1$$

$$\sin \theta_1 = \frac{n\lambda_1}{(b+d)} = \frac{2 \times 5890 \times 10^{-8}}{1.67 \times 10^{-4}}$$

$$\sin \theta_1 = 0.7054$$

$$\theta_1 = \sin^{-1} (0.7054)$$

$$= 44.86°$$

Similarly, $$\theta_2 = \sin^{-1} \frac{2 \times 5896 \times 10^{-8}}{1.67 \times 10^{-4}}$$

$$= \sin^{-1} (0.7061)$$

$$= 44.92$$

Thus, the angular separation $(\Delta\theta) = \theta_2 = \theta_1$

$$= 44.92 - 44.86$$

$$\theta_2 - \theta_1 = \mathbf{0.06°}$$

Example 31 A diffraction grating used at normal incidence gives a line (5400 Å) in a certain order superposed on the violet line (4050 Å) of the next higher order. How many lines per cm are there in the grating if angle of diffraction is 30°?

Solution Given $\lambda_1 = 5400 \times 10^{-8}$ cm of nth order,

$\lambda_2 = 4050 \times 10^{-8}$ cm of $(n + 1)^{\text{th}}$ order

Formula used is $$(b + d) \sin \theta = n\lambda_1 = (n + 1)\lambda_2 \quad \text{(i)}$$

$$n = \frac{(b+d)\sin\theta}{\lambda_1} \quad \text{(ii)}$$

and $$n + 1 = \frac{(b+d)\sin\theta}{\lambda_2} \quad \text{(iii)}$$

Eliminating n by using Eqs (ii) and (iii), we get

$$\left(\frac{1}{\lambda_2} - \frac{1}{\lambda_1}\right) (b + d) \sin \theta = 1$$

$$(b + d) = \frac{\lambda_1\lambda_2}{\lambda_1 - \lambda_2} \frac{1}{\sin\theta} = \frac{5400 \times 10^{-8} \times 4050 \times 10^{-8}}{(5400 - 4050) \times 10^{-8} \times \sin 30°} = 32400 \times 10^{-8} \text{ cm}$$

or $$N = \frac{1}{b+d} = \mathbf{3086} \text{ lines/cm}$$

Example 32 A plane transmission grating produces an angular separation of 0.01 radian between two wavelengths observed at an angle of 30°. Given mean value of the wavelength as 5000 Å. Calculate the difference in two wavelengths if the spectrum is observed in the second order.

SOLUTION Given $d\theta = 0.01$ radian, $\theta = 30°$ and $\lambda = 5000 \times 10^{-8}$ cm.

Formula used is $(b + d) \sin \theta = n\lambda$ (i)

By differentiating Eq. (i)

$$(b + d) \cos \theta \, d\theta = n d\lambda \quad \text{(ii)}$$

Dividing Eq. (ii) by Eq. (i), we get

$$\frac{\cos \theta \, d\theta}{\sin \theta} = \frac{d\lambda}{\lambda} \quad \text{or } d\lambda = \lambda \cot \theta d\theta$$

$$d\lambda = 5000 \times 10^{-8} \cot 30° \times 0.01$$

$$\mathbf{d\lambda = 86.6 \text{ Å}}$$

EXAMPLE 33 How many orders will be visible if the wavelength of the incident radiation is 5000 Å and the number of lines on the grating is 2620 in one inch.

SOLUTION Given $N = 2620$ lines per inch and $\lambda = 5000 \times 10^{-8}$ cm.

$$b + d = \frac{1}{N} = 1 \text{ inch} = 2.54 \text{ cm} = 9.695 \times 10^{-4} \text{ cm}$$

Formula used is $(b + d) \sin \theta = n\lambda$

For maximum possible value $\sin \theta = 1$, then

$$\text{Order of spectrum } (n) = \frac{(b+d)}{\lambda} = \frac{9.695 \times 10^{-4}}{5.0 \times 10^{-5} \text{cm}}$$

$$= 19.38$$

$$= \mathbf{19}$$

That is 19th order will be visible.

EXAMPLE 34 What is the highest order spectrum which may be seen with monochromatic light of wavelength 5000Å by means of diffraction grating with 5000 lines/cm?

SOLUTION Given N= 5000 lines per cm and $\lambda = 5000 \times 10^{-8}$.

$$b + d = \frac{1}{N} = \frac{1}{5000} \text{cm}$$

Formula used is $(b + d) \sin \theta = n\lambda$ (i)

For highest order spectrum to be visible the value of $\sin \theta$ must be 1. Then Eq. (i) becomes

$$n = \frac{b+d}{\lambda} = \frac{1}{5000} \frac{1}{5 \times 10^{-5}} = \frac{100}{25} = \mathbf{4}$$

that is the highest order will be 4.

EXAMPLE 35 How many orders will be observed by a grating having 4000 lines per cm if it is illuminated by visible light in the range 4000 Å to 7000 Å.

SOLUTION Given $(b + d) = \dfrac{1}{4000}$ cm $= 2.5 \times 10^{-4}$ cm $= 2.5 \times 10^{-6}$ m.

Formula used is $(b + d) \sin \theta = n\lambda$

For $\lambda_1 = 4000$ Å

$$n_1 = \frac{(b+d)}{\lambda_1} \quad (\sin \theta = 1)$$

$$n_1 = \frac{2.5 \times 10^{-6}}{4.0 \times 10^{-7}} = 6.25$$

$$n_1 = 6.25$$

For $\lambda_2 = 7000$ Å

$$n_2 = \frac{2.5 \times 10^{-6}}{7.0 \times 10^{-7}} = \mathbf{3.57}$$

∴ The order of the spectrum varies from 3 to 6 depending upon the wavelength of the visible range.

EXAMPLE 36 A diffraction grating having 4000 lines/cm is illuminated normally by light of wavelength 5000 Å. Calculate its dispersive power in third order spectrum.

SOLUTION Given $\lambda = 5.0 \times 10^{-5}$ cm, $n = 3$, $N = 4000$ lines/cm and

$$(b + d) = \frac{1}{4000}\text{cm} = 2.5 \times 10^{-4} \text{ cm.}$$

Formula used is $$\frac{d\theta}{d\lambda} = \frac{n}{(b+d)\cos\theta} \quad \text{(i)}$$

or $$\frac{d\theta}{d\lambda} = \frac{n}{(b+d)\sqrt{1-\sin^2\theta}} \quad \text{(ii)}$$

As $(b + d)\sin\theta = n\lambda$ or $\sin\theta = \dfrac{n\lambda}{(b+d)}$ (iii)

By using Eqs (ii) and (iii), we have

$$\frac{d\theta}{d\lambda} = \frac{n}{(b+d)\sqrt{1-\left\{\dfrac{n\lambda}{(b+d)}\right\}^2}} = \frac{3}{\left(\dfrac{1}{4000}\right)\sqrt{1-\left(\dfrac{3\times5\times10^{-5}}{1/4000}\right)^2}}$$

$$= \frac{12000}{\sqrt{1-0.36}}$$

or $$\frac{d\theta}{d\lambda} = \mathbf{1.875 \times 10^4\ rad/cm}$$

EXAMPLE 37 Calculate the minimum number of lines in grating which will first resolve the lines of wavelengths 5890 Å and 5896 Å in the second order.

SOLUTION Given $\lambda_1 = 5.89 \times 10^{-5}$ cm, $\lambda_2 = 5.896 \times 10^{-5}$ cm, $n = 2$.

∴ $$\Delta\lambda = 6 \text{ Å} = 6 \times 10^{-8} \text{ cm}$$

$$\text{Resolving Power} = \frac{\lambda}{d\lambda} = nN$$

$$N = \frac{1}{n}\frac{\lambda}{\Delta\lambda} = \frac{5.890\times10^{-5}}{2\times6\times10^{-8}} = \frac{5890}{12} = \mathbf{490.8}$$

EXAMPLE 38 For proper resolution 491 lines are required. Calculate the minimum number of lines in a grating which will just resolve the sodium lines in the first order spectrum. The wavelengths are 5890 Å and 5896 Å.

SOLUTION Given $\lambda_1 = 5.89 \times 10^{-5}$ cm, $\lambda_2 = 5.896 \times 10^{-5}$ cm and

$\Delta\lambda = 6.0 \times 10^{-8}$ cm, $n = 1$.

Formula used is $\lambda/\Delta\lambda = n/N$

$$N = \frac{\lambda}{nd\lambda} = \frac{5.89 \times 10^{-5}}{1 \times 6 \times 10^{-8}} = \mathbf{981.66}$$

i.e., for proper resolution the minimum number of lines in grating will be 982.

EXAMPLE 39 In a grating the sodium doublet (5890 Å, 5896 Å) is viewed in third order at 30° to the normal and is resolved. Determine the grating spacing and the total width of the rulings.

SOLUTION Given $\theta = 30°$, $n = 3$, $\lambda_{mean} = 5.893 \times 10^{-5}$ cm.

Formula used is $(b + d) \sin \theta = n\lambda$

$$(b+d) = \frac{n\lambda}{\sin\theta} = \frac{3 \times 5.893 \times 10^{-5}\,\text{cm}}{\sin 30°}$$

$$= \frac{3 \times 5.893 \times 10^{-5}}{0.5}$$

$$= 3.536 \times 10^{-4}\,\text{cm}$$

$\because$ Resolving Power = n/N

$\therefore$ $$\frac{\lambda}{d\lambda} = nN$$

or $$N = \frac{\lambda}{nd\lambda} = \frac{5.893 \times 10^{-5}\,\text{cm}}{3 \times 6 \times 10^{-8}\,\text{cm}} = N = 327$$

Hence, the total width of the ruled surface is

$$N(b + d) = 327 \times 3.536 \times 10^{-4} = \mathbf{0.1156\ cm}$$

EXAMPLE 40 Find the separation of two points on the moon that can be resolved by a 500 cm telescope. The distance of the moon is 3.8×10^5 km. The eye is most sensitive to light of wavelength 5500 Å.

SOLUTION Given $\lambda = 5.5 \times 10^{-7}$ m, $a = 5.0$ m, $R = 3.8 \times 10^8$ m.

Limit of resolution of a telescope

$$\theta = \frac{1.22\lambda}{a} = \frac{1.22 \times 5.5 \times 10^{-7}}{5.0}$$

$$= 1.342 \times 10^{-7}\ \text{rad}$$

Let x be the distance between points, then

$$\theta = \frac{x}{R}$$

$$1.342 \times 10^{-7} = \frac{x}{3.8 \times 10^8}$$

or $\mathbf{x = 50.996\ m}$

EXAMPLE 41 What will be the diameter of a telescope objective which is required to resolve two stars separated by an angle of 10^{-3} degree? Assume $\lambda = 500$ Å.

SOLUTION Given $\lambda = 5.0 \times 10^{-7}$ m, $\theta = 10^{-3}$ deg $= \frac{\pi}{180} \times 10^{-3}$ rad.

Formula used is $$\theta = \frac{1.22\lambda}{a}$$

or $$a = \frac{1.22\lambda}{\theta} = \frac{1.22 \times 5.0 \times 10^{-7}}{\frac{\pi}{180} \times 10^{-3}}$$

$$= \frac{1.22 \times 1.8 \times 50 \times 10^{-3}}{3.14} = \mathbf{0.03495\ m}$$

EXAMPLE 42 Calculate the aperture of the objective of a telescope which may be used to resolve two stars separated by 2.44×10^{-6} radian for light of wavelength 6000 Å.

SOLUTION Given $\lambda = 6.0 \times 10^{-7}$ m and $\theta = 2.44 \times 10^{-6}$ rad.

Formula used is $$\theta = \frac{1.22\lambda}{a} \text{ or } a = \frac{1.22\lambda}{\theta}$$

or $$a = \frac{1.22 \times 6.0 \times 10^{-7}}{2.44 \times 10^{-6}} = 0.30 \text{ m}$$

or $$a = \mathbf{0.30\ m}$$

Hence, aperture of the objective is 0.30 m.

EXAMPLE 43 Two pin holes 1.5 mm apart are placed in front of a source of light of wavelength = 5.5×10^{-5} cm and seen through a telescope with its objective stopped down to a diameter of 0.4 cm. Find the maximum distance from the telescope at which the pin holes can be resolved.

SOLUTION Given $\lambda = 5.5 \times 10^{-5}$ m, $a = 0.004$ m and $x = 1.5 \times 10^{-3}$ m.

Formula used is $$\theta = \frac{1.22\lambda}{a} \quad \text{(i)}$$

and $$\theta = \frac{x}{R} \quad \text{(ii)}$$

$$\frac{x}{R} = \frac{1.22\lambda}{a}$$

or $$R = \frac{xa}{1.22\lambda} = \frac{1.5 \times 10^{-3} \times 4.0 \times 10^{-3}}{1.22 \times 5.5 \times 10^{-7}}$$

$$= \mathbf{8.9418\ m}$$

EXAMPLE 44 A microscope objective gathers light over a come of semi-angle 30° and uses visible light ($\lambda = 5500$ Å). Estimate its resolving limit.

SOLUTION Given $\lambda = 5.5 \times 10^{-7}$ m and $\theta = 30°$.

Formula used is $2d \sin\theta = 1.22\ \lambda$

or $$d = \frac{1.22\lambda}{2 \sin\theta}$$

$$d = \frac{1.22 \times 5.5 \times 10^{-7}}{2 \times \sin 30°} = \mathbf{6.76 \times 10^{-7}\ m}$$

Thus, the resolving limit of the microscope is 6.7×10^{-7} m.

EXAMPLE 45 A microscope is used to resolve two self-luminous objects separated by a distance of 4.0×10^{-5} cm. If the wavelength of light is 5461 Å, compute the numerical aperture of the objective.

SOLUTION Given $\lambda = 5.461 \times 10^{-7}$ m and $d = 4.0 \times 10^{-7}$ m.

Formula used is Numerical aperture $(NA) = \dfrac{1.22\lambda}{2d}$

$$NA = \frac{1.22 \times 5.461 \times 10^{-7}}{2 \times 4.0 \times 10^{-7}} = \mathbf{0.833}$$

Numerical aperture of the objective is 0.833.

EXAMPLE 46 A plane wave of light of wavelength 690 nm is incident on a vertical slit of width 10^{-4} m. Sketch the intensity distribution on a screen 3 m from the slit placed parallel to the slit aperture. At what distances from the central maximum do the first two zeroes occur?

SOLUTION Given $\lambda = 6.90 \times 10^{-7}$ m, $b = 10^{-4}$ m, $D = 3.0$ m.

By using the relation, $b \sin\theta = n\lambda$ and $\sin\theta = \dfrac{x}{D}$

we get $$x = \frac{n\lambda D}{b}$$

for $n = 1$

$$x = \frac{1 \times 6.90 \times 10^{-7} \times 3.0}{10^{-4}} = 2.07 \text{ cm}$$

for $n = 2$

$$x = \frac{2 \times 6.90 \times 10^{-7} \times 3.0}{10^{-4}} = \mathbf{4.14\ cm}$$

EXAMPLE 47 A grating having 15000 lines per inch produces spectra of a mercury arc. The green line of the mercury spectrum has a wavelength of 5461 Å. What is the angular separation between the first order and second order green line?

SOLUTION Given $N = 15000$ lines per inch, $\lambda = 5461 \times 10^{-8}$ cm.

By using the relation $(b + d)\sin\theta = n\lambda$

$$b + d = \frac{2.54}{15000} \text{cm} = 1.6933 \times 10^{-4} \text{ cm}$$

then $$\sin\theta = \frac{n\lambda}{b+d} = \frac{n \times 5461 \times 10^{-8}}{1.6933 \times 10^{-4}}$$

for $n = 1, \theta = \theta_1$

$$\sin\theta_1 = \frac{1 \times 5461 \times 10^{-8}}{1.6933 \times 10^{-4}} = 0.322506$$

$$\theta_1 = \sin^{-1}(0.322506) = 18.81°$$

for $n = 2, \theta = \theta_2$, then

$$\sin\theta_2 = \frac{n\lambda}{(b+d)} = \frac{2 \times 5461 \times 10^{-8}}{16933 \times 10^{-8}}$$

$$= 0.645013$$

$$\theta_2 = \sin^{-1}(0.645013) = 40.17°$$

Therefore, the angular separation of lines $(\Delta\theta)$

$$= \theta_2 - \theta_1$$

$$= 40.17° - 18.81° = \mathbf{21.36°}$$

EXAMPLE 48 Light is incident normally on a grating 0.5 cm wide with 2500 lines. Find the angle of the two sodium lines in the first order spectrum. Are the two lines resolved?

SOLUTION Given $b + d = \frac{0.5 \text{ cm}}{2500} = 2 \times 10^{-4}$ cm, $\lambda_1 = 5890$ Å.

and $\lambda_2 = 5896$ Å

$\theta - \theta_1$ for D_1, lines $(n = 0)$

$$\theta_1 = \sin^{-1}\left\{\frac{n\lambda}{(b+d)}\right\}$$

$$\theta_1 = \sin^{-1}\left[\frac{\lambda_1}{(b+d)}\right] = \sin^{-1}\left[\frac{5890 \times 10^{-8}}{20000 \times 10^{-8}}\right] = \sin^{-1}(0.2945)$$

$$\theta_1 = 17.13°$$

θ_2 for D_2 line $(n = 1)$

$$\theta_2 = \sin^{-1}\left[\frac{\lambda_2}{(b+d)}\right] = \sin^{-1}\left[\frac{5896 \times 10^{-8}}{20000 \times 10^{-8}}\right]$$

$$= \sin^{-1}[0.2948] = 17.15°$$

$$\Delta\theta = \theta_2 - \theta_1 = 0.02° = 0.02 \times 60' = 1.2' \times 60'' = 72''$$

Yes, these two lines will be just resolved.

OBJECTIVE TYPE QUESTIONS

Q.1 The diffraction phenomenon is
(a) bending of light around an obstacle (b) rectilinear propagation of light
(c) oscillation of light wave in one direction (d) none of them.

Q.2 The diameter of half-period zones are proportional to
(a) $1/\sqrt{n}$ (b) $\frac{1}{n}$ (c) $\sqrt{n}$ (d) $1/n$

Q.3 The ratio of areas of 1st to IInd half-period zones is
(a) 1 : 2 (b) 1: 1 (c) 2: 1 (d) 4 : 1

Q.4 To find prominent diffraction, the size of the diffracting object should be
(a) greater than the wavelength of light used (b) of the order of wavelength of light
(c) less than the wavelength of light (d) none of these

Q.5 Product of focal lengths of a zone plate and a convex lens
(a) is independent of wavelength of light
(b) is directly proportional to wavelength of light
(c) is inversely proportional to wavelength of light
(d) none of the above

Q.6 Zone plate has some similarities as well as some differences with a
(a) plano-convex lens (b) concave lens
(c) grating (d) convex lens

Q.7 In Fresnel diffraction
(a) source of light is kept at infinite distance from the aperture
(b) source of light is kept at finite distance from the aperture
(c) convex lens is used
(d) aperture width is selected so that it can act as a point source

Q.8 In Fraunhofer diffraction, the incident wavefront should be
(a) elliptical (b) plane (c) spherical (d) cylindrical

Q.9 In the diffraction pattern using circular aperture, when the screen is brought toward the aperture
(a) the intensity at the screen gradually increases
(b) the intensity at the screen gradually decreases
(c) the light is found to focus only to a fixed distance
(d) many points are observed where greater intensity is found

Q.10 Significant diffraction of x-rays can be obtained
(a) by a single slit (b) by a double slit
(c) by a diffraction (d) by an atomic crystal

Q.11 Maximum number of orders found with a grating is
(a) independent of grating element (b) directly proportional to the grating element
(c) proportional to the wavelength (d) none of these

Q.12 The resolving power of a grating increases
(a) as the number of ruled lines per cm of a grating increases
(b) as the number of ruled lines per cm of a grating decreases
(c) as grating element decreases
(d) none of these

Q.13 Which of the following statement is true in the case of dispersive power of a grating?
(a) it is directly proportional to number of lines per cm of grating
(b) it is inversely proportional to cos θ
(c) it is directly proportional to order of diffraction
(d) all of these

Q.14 In a plane diffraction grating, width of the principal maxima increases
(a) as number of slits decreases (b) as number of slits increases
(c) both (a) and (b) (d) none of these

Q.15 Which of the following statement is correct, if the width of the slit is reduced in single slit experiment?
(a) the intensity of fringes larged (b) the thickness of fringes reduced
(c) the thickness of fringes increased (d) none of these

Q.16 Which of the following colour (wavelength) will be diffracted more, when white light is used?
(a) red (b) green (c) violet (d) none of these

Q.17 The resolving limit of normal eye is
(a) $\left(\frac{1}{60}\right)^\circ$ (b) $\left(\frac{1}{30}\right)^\circ$ (c) $\left(\frac{1}{45}\right)^\circ$ (d) none of these

Q.18 Which of the following relation is true for the limit of resolution of telescope?
(a) $\frac{1}{\theta}$ (b) $\frac{a}{1.22\lambda}$ (c) $\frac{1.22\lambda}{a}$ (d) (a) and (b)

SHORT-ANSWER QUESTIONS

Q.1 What do you understand by diffraction of light?

Q.2 How will you distinguish between interference and diffraction?

Q.3 Discuss difference between Fraunhofer and Fresnel diffraction.

Q.4 Explain why I[st] secondary maximum corresponding to $\pi/2$ is not obtained in single and double slit diffraction pattern?

Q.5 What are Fresnel's half-period zones?

Q.6 What are the radii of zones of zone plate?

Q.7 In what respect a zone plate is similar or differ from a convex lens.

Q.8 What are the conditions of secondary maxima in diffraction due to single slit?

Q.9 What is a diffracting grating?

Q.10 On what factors does the dispersive power of a grating depend?

Q.11 What doe you mean by dispersive power?

Q.12 What is the ratio of intensity of secondary maximum and principal maximum?

Q.13 What is Rayleigh criterion of resolution?

Q.14 What do you mean by resolving power of a telescope?

Q.15 What do you mean by resolving power of a grating?

Q.16 What is the advantage of increasing the number of lines in a grating?

PRACTICE PROBLEMS

General Questions

Q.1 How many types of diffractions are there? Distinguish between Fresnel and Fraunhofer type of diffractions when the secondary wavelets are in the same phase at all points in the plane of the aperture.

Q.2 What is diffraction? Explain clearly the difference between interference and diffraction.

Q.3 What are Fresnel's half-period zones? Prove that the area of a half-period zone on a plane wavefront is independent of the order of the zone and that the amplitude due to a large wavefront at a point in front of it is just half that due to the first half-period zone acting alone. Hence give Fresnel's explanations of the rectilinear propagation of light.

Q.4 Explain the meaning of Fresnel's half-period zones. Why are they called so? What is the phase difference between wavelets from successive half-period zones? What are the factors on which the amplitude of the light waves from a half-period zone at the observation point depend?

Q.5 What is a zone plate and how is it made? Explain how a zone plate acts like a convergent lens having multiple foci. Derive an expression for its focal length.

Q.6 How is zone plate constructed?

Q.7 Compare the performance of zone plate with that of a converging lens.

Q.8 Give the theory of a zone plate. Show that a zone plate has multi foci. Compare the zone plate with a convex lens. What is meant by 'phase reversal zone plate'?

Q.9 Give the mathematical treatment of Fresnel's diffraction at a circular aperture.

Q.10 Describe the diffraction effects due to narrow circular aperture illuminated by monochromatic light from a point-source. Show that if the screen on which diffraction pattern is obtained be moving towards the aperture, the intensity at the centre would be alternately maximum and minimum.

Q.11 Derive an expression for the intensity distribution due to Fraunhofer diffraction at a single slit and show that the intensity of the first subsidiary maximum is about 4.5% of that of the principal maximum.

Q.12 Discuss the phenomena of Fraunhofer diffraction at a single slit and show that the relative intensities of the successive maximum are nearly
$1 : 4/9\pi^2 : 4/25\ \pi^2 : 4/49\ \pi^2$.

Q.13 Distinguish between interference and diffraction and explain the spectra formed by a grating and also discuss the cause of decrease of intensity of principal maxima with the increase of order.

Q.14 If the width of the opaque space of grating is doubled the width of the transparent space, which order of spectrum will be absent?

Q.15 Give the construction and theory of plane transmission grating and explain the formation of spectra by it.

Q.16 What is a grating? Explain the spectra, with theory, formed by a plane transmission diffraction grating. Show that the intensity is not uniformly distributed over all the maxima.

Q.17 Define resolving power and dispersive power of a grating. Obtain expressions for these in the case of plane transmission grating.

Q.18 Explain the phenomenon of Fraunhofer diffraction at a single slit.

Q.19 Explain how a chromatic beam gets dispersed while passing through a plane transmission diffraction grating. Derive an expression for the dispersive power.

Q.20 Write short note on plane transmission grating.

Q.21 Deduce expressions for dispersive and resolving power of a diffraction grating. Can we increase resolving power without affecting dispersive power? Discuss.

Q.22 Discuss two methods in detail for finding the wavelength of a given monochromatic light-one using interference of light and another based upon diffraction of light phenomenon.

UNSOLVED QUESTIONS

Q.1 A screen is placed at a distance of 100 cm from a circular hole illuminated by parallel beam of light of wavelength 6000 Å. Compute the radius of the fourth half-period zone. **[Ans: 1.6×10^{-3} m]**

Q.2 Find the first three focal lengths of a zone plate for which the radius of the first zone is 0.3 mm, for light of wavelength 5000 Å. **[Ans: 0.18 m, 0.06 m, 0.036 m]**

Q.3 What is the radius of the first zone in a zone plate of focal length 20 cm for light of wavelength 5000 Å. **[Ans: 3.16×10^{-3} m]**

Q.4 An object is placed at 20 cm from the zone plate and the brightest image is situated at 20 cm from the zone plate with light of $\lambda = 4000$ Å. Calculate the number of Fresnel's zones in the radius of 1.0 cm of the plate. **[Ans. 2500]**

Q.5 A zone plate is made such that the radii of the circles defining the zones are the same as the radii of Newton's rings formed between a plane surface and a surface whose radius of curvature is 150 cm. Find the primary focal length of the zone plate. **[Ans. 1.5 m]**

Q.6 When a circular aperture of diameter 2.0 mm is illuminated by a plane wave of light, the most intense point on the axis is at a distance of 200 cm from the aperture. Calculate the wavelength of light. **[Ans. 5000Å]**

Q.7 A light of wavelength 6000 Å passes through a narrow aperture of radius 0.09 cm. At what distance along the axis will first maximum intensity be observed? **[Ans. 1.35 m]**

Q.8 Calculate the half-angular width of the central bright maximum in Fraunhofer diffraction at a slit having width 1.23×10^{-6} m illuminated by monochromatic light of wavelength 5896 Å **[Ans. 28.61°]**

Q.9 A parallel beam of light of wavelength 6000 Å falls normally on a slit. In the diffraction pattern the first minimum lies at a distance 4.9 mm from the central maximum and the screen is placed 1.0 m from the slit. Calculate the width of the slit. **[Ans. 0.122 mm]**

Q.10 A single slit of width 0.14 mm is illuminated normally by monochromatic light and diffraction bands are observed on a screen 2.0 m away. If the centre of second dark band is 1.6 cm from the middle of the central bright band, deduce the wavelength of light used. **[Ans. 5600Å]**

Q.11 A screen is placed 2.0 m away from a narrow slit which is illuminated with light of wavelength 6000Å. If the first minimum lies 5 mm on either side of the central maximum. Calculate the slit width. **[Ans. 0.24 mm]**

Q.12 A parallel beam of monochromatic light is normally incident on a plane transmission grating having 12000 lines per cm. The second order spectral line is observed at an angle 45°. Find the wavelength of light used. **[Ans. 2946 Å]**

Q.13 A plane transmission grating having 5500 lines per cm is used to produce a spectrum of mercury light. What will be the angular separation of the two yellow lines 4770 Å and 5790 Å in the second order. **[Ans. 10′ of acc]**

Q.14 A parallel beam of monochromatic light is allowed to be incident normally. On a plane grating having 1250 lines per cm and a second order spectrum line is observed to be deviated through 30°. Calculate the wavelength of the spectral line. **[Ans. 2×10^{-4} cm]**

Q.15 Find the angular separation between two sodium lines 5890 Å and 5896 Å in the second order spectrum of a grating with 5000 lines/cm. The width of a grating is 0.5 cm. Can they be seen distinctly? **[Ans. 3 min, yes]**

Q.16 A plane transmission grating has 40,000 lines in all, with grating element 12.5×10^{-5} cm. Calculate the maximum resolving power for which it can be used in the range of wavelength 5000 Å. **[Ans. 80,000]**

Q.17 Calculate the aperture of the objective of a telescope which may be used to resolve stars separated by 4.88×10^{-6} radian for light of wavelength 6000 Å. **[Ans. 0.15 m]**

Q.18 A telescope objective has a focal length of 3.0 m and a diameter of 0.01 m. Find the distance between centres of the images of the two stars which are just resolved by it, assuming the wavelength of the light 5000 Å. **[Ans. 2.01×10^{-4} m]**

Q.19 Calculate the resolving power of a laboratory microscope if N.A. given on the objective is 0.12 and the wavelength of light used is 6000 Å. **[Ans. 4000]**

Q.20 A microscope is used to resolve two equally bright point objects separated by 5.55×10^{-7} m. Calculate the numerical aperture of the objective if light of wavelength 5460 Å is used. **[Ans. 0.6]**

Polarisation

3

LEARNING OBJECTIVES

After reading this chapter you will be able to

LO1 Explain transverse wave nature of polarisation

LO2 Illustrate the difference between unpolarised and polarised light

LO3 Enable to know the means of producing plane-polarised light–transmission, reflection, refraction and scattering

LO4 Discuss theory of production of plane, circularly and elliptically polarised light

LO5 Analyse optical activity and phenomenon of specific rotation

LO6 Demonstrate working of Half-shade polarimeter, Biquartz polarimeter and saccharimeter

LO7 Analyse photoelasticity

Introduction

You would have encountered problem in receiving signal when you watch TV or listen to your stereo system. In order to overcome this problem, you adjust the position of an antenna attached to the stereo system or align the TV antenna (receiving antenna) in the proper orientation. Did you ever think why this is necessary and what physics is involved in doing so? Actually this is required as some types of antennas, via the electrons, respond to the electric field of an electromagnetic wave (signal). If the orientation of the receiving antenna matches with the orientation of the electric field of the wave, the electric field causes the electrons to flow along the wires to generate a current. So the plane of the receiving antenna must be horizontal if the electric field of the signal broadcast by the station vibrates in a horizontal plane. If the field vibrates in a vertical plane, the orientation of the antenna should be changed to the vertical plane. So this is clear that by doing the adjustments of position or orientation of the antenna, we increase the strength of the signal, i.e., we improve the reception of the signal. The proper orientation of vibration of the electric field is nothing but the polarisation of the wave.

A light wave is an electromagnetic wave whose electric field and magnetic field vectors vibrate perpendicular to the direction of wave propagation. In order to completely identify the electromagnetic wave, it is sufficient to specify the electric field since the magnetic field can be determined once the electric field is known (*discussed later in the chapter on Electromagnetic Wave Propagation*). So a light wave whose electric field vector, also called as light vector, is vibrating in more than one plane is referred to as unpolarised light. The light emitted by the sun, by a lamp, or by a candle flame is unpolarised light. It is possible to convert unpolarised light into polarised light in which the vibrations occur only in a single plane. The process of converting unpolarised light into polarised light is known as polarisation. There are a variety of methods of polarising light. Any interaction of light with matter whose optical properties are asymmetrical along the directions transverse to the propagation vector provides a means of polarising light. Only transverse waves can be polarised. The polarisation of longitudinal waves such as sound waves is not possible as in these waves the vibrations occur only in the direction of wave propagation.

The phenomena of interference and diffraction discussed in the previous chapters show that the light travels in the form of waves. However, these phenomena do not tell us about the nature of light waves, i.e., whether the light waves are transverse or longitudinal or whether the vibrations are linear, circular or elliptical. Such important investigations represent the subject of polarisation of light.

3.1 MECHANICAL EXPERIMENT SHOWING POLARISATION OF TRANSVERSE WAVE LO1

In order to explain the polarisation of light, we consider a mechanical experiment by using a string whose two end points are P and Q and which is kept fixed with point Q. The string passes through two parallel slits S_1 and S_2 (Fig. 3.1). When the end point P is shaken up and down parallel to the slit S_1, a transverse wave is generated that travels towards the point Q and reaches there if the slit S_2 is parallel to the slit S_1 (Fig. 3.1a). Now we rotate the slit S_2 through 90° (Fig. 3.1b). Under this situation, the vibrations of the string are completely stopped by this slit and the string does not vibrate in the region between S_2 and Q. As is clear from this experiment, the vibrations of this wave are confined to a plane parallel to the axis of the slit S_1. Therefore, this wave is called *plane-polarised wave* or *linearly polarised wave*. If the end P of the string is moved in a circular manner, the

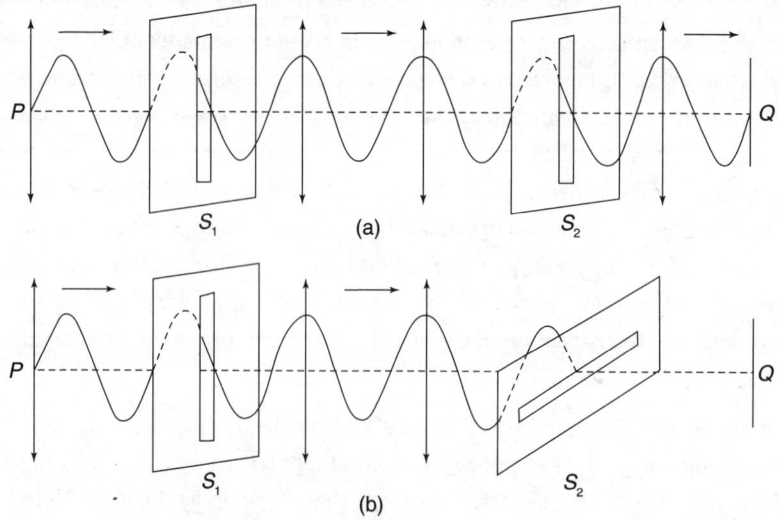

FIGURE 3.1

particles of the string will vibrate in the circle and the wave thus generated is called *circularly polarised wave*. If the end P of the string is moved in an elliptic manner, the particles of the string will vibrate in the ellipse and the wave thus generated is called *elliptically polarised wave*. Under these situations, only those vibrations can pass beyond the slit S_2 which are parallel to the axis of the slit S_2. However, the passage of longitudinal waves through the slit S_2 will always be possible in its any position with respect to the slit S_1.

3.2 DIFFERENCE BETWEEN UNPOLARISED LIGHT AND POLARISED LIGHT — LO2

An unpolarised light is symmetrical about the direction of propagation as the light vector vibrates along all possible directions perpendicular to the direction of propagation of light (Fig. 3.2a). In polarised light lack of symmetry is found about the direction of propagation of the light. Since light is an electromagnetic wave, the polarisation describes the orientation of the vibrations of the wave. The vibrations may be oriented in a single direction in linearly polarised light (Fig. 3.2b, c). In Fig. 3.2b, the vibrations of light vector are in vertical direction (in the plane of paper) whereas in Fig. 3.2c the vibrations are taking place in horizontal direction (perpendicular to the plane of paper). However, in circularly polarised light and elliptically polarised light, the direction of vibrations may rotate as the wave propagates. Circularly polarised light can rotate rightward or leftward in the direction of wave propagation. Which of these two rotations is present in a wave is called the *wave's chirality*. Chirality is a Greek word which means *handedness*. This is a property of asymmetry which is significantly useful in several branches of science.

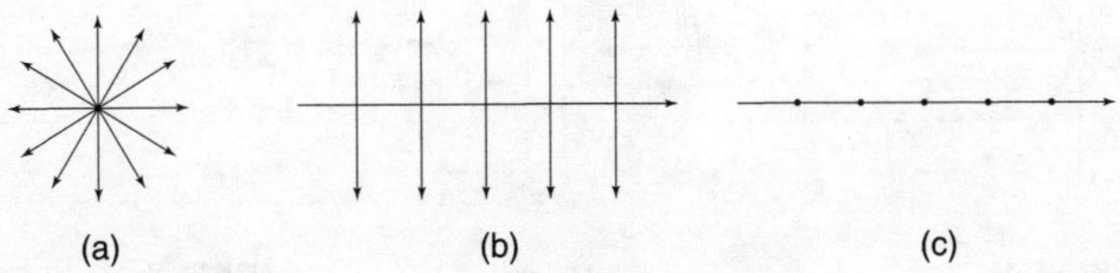

FIGURE 3.2

3.2.1 Plane of Polarisation and Plane of Vibration

As discussed earlier, the electric field and magnetic field vectors of the light waves are perpendicular to each other and to the direction of propagation of the wave. As per convention, the direction of electric field vector $\vec{E}$ is taken as the direction of polarisation of the wave. In view of this, the plane determined by the electric vector $\vec{E}$ and the direction of wave propagation (vector $\vec{k}$), is called the *plane of polarisation*, particularly for the plane (linearly) polarised light. This term is sometimes applied to the plane at right angles to this, i.e., to the plane containing the magnetic field vector. Then the plane containing the electric vector $\vec{E}$ and wave vector $\vec{k}$ is named as the plane of vibration. However, in a plane electromagnetic wave with elliptic polarisation and circular polarisation, the electric field vector $\vec{E}$ has components (out of phase) in both the directions perpendicular to $\vec{k}$. So the plane that contains only vibrations of $\vec{E}$ field can be called as the plane of polarisation. Hence, the plane of vibration and the plane of polarisation will be the same in this case.

3.3 MEANS OF PRODUCTION OF PLANE-POLARISED LIGHT — LO3

The different methods commonly used for the production of plane-polarised light may be classified under different categories: (a) polarisation by transmission, (b) polarisation by reflection, (c) polarisation by refraction, and (d) polarisation by scattering.

3.3.1 Polarisation by Transmission

It is possible to get the unpolarised light converted into a polarised light along with its vibrations in a single plane if the unpolarised light is passed through a Polaroid filter.

3.3.1.1 Polaroid Filter

The most general method of polarisation makes use of a Polaroid filter. The Polaroid filter has long chain molecules that are aligned in the same direction within the filter. The alignment of these molecules constitutes a polarisation axis that extends across the length of the filter. This axis allows electromagnetic waves to pass through whose vibrations are parallel to the axis. Thus any vibrations perpendicular to the polarisation axis are stopped by the filter. When an unpolarised light is passed through a Polaroid filter, it emerges with its vibrations in a single plane along with one half of its intensity. This way the emerging light is the polarised light. The relationship between the alignment of long chain molecules and the polarisation axis in a Polaroid filter is just opposite, i.e., a Polaroid filter with its long chain molecules aligned vertically will have a horizontally aligned polarisation axis. This type of a filter will stop all the vertical vibrations and allow only the horizontal vibrations to pass through. However, a Polaroid filter with its long chain molecules aligned horizontally will have a polarisation axis aligned vertically. So this filter will stop all the horizontal vibrations and allow only the vertical vibrations to pass through. It is clear that the two Polaroid filters oriented with their polarisation axes perpendicular to each other will stop all the light.

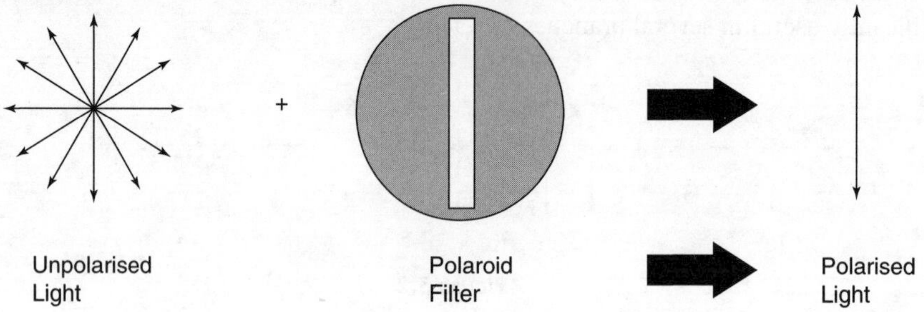

FIGURE 3.3

3.3.1.2 Malus' Law

Based on Malus' $\cos^2 \theta$ intensity law the intensity of transmitted light is described when there is a relative orientation between the directions of the polarisation of the incoming light and the polarisation axis of the filter. Therefore, the incoming light should be polarised. For this we first polarise the unpolarised light by passing it through a Polaroid filter, called the polariser. Then the polarised light is passed through a second Polaroid filter, called the *analyser*. The transmission axis of the analyser makes an angle θ with the plane of polarisation of the incident light (Fig. 3.4). This way the emerging light is polarised in the same direction as the transmission axis of the analyser and its intensity varies as the cosine square of the angle between the planes of the transmission of analyser and the polariser. This is called Malus' law.

In order to get the relation for resultant intensity, we consider A to be the amplitude of the plane-polarised light and θ as the angle between the planes of the polariser and the analyser (Fig. 3.5). The amplitude A can be resolved into two components $A \cos \theta$ and $A \sin \theta$. Only the component $A \cos \theta$ is transmitted through the analyser. Then a measure of the intensity of the transmitted light will be

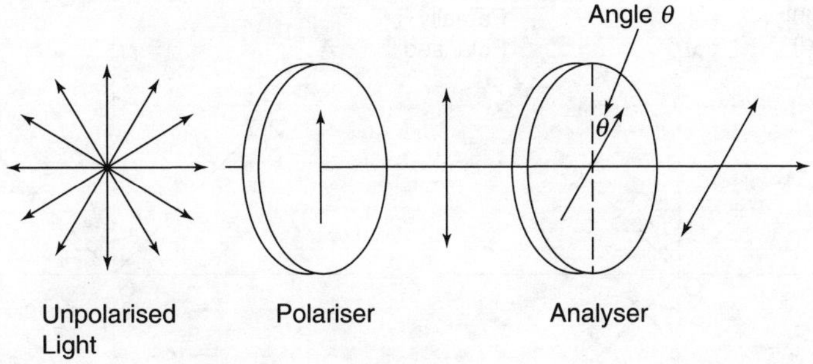

FIGURE 3.4

$$I = (A \cos \theta)^2$$
$$= A^2 \cos^2\theta = I_0 \cos^2\theta \qquad (I_0 = A^2)$$
$$I = I_0 \cos^2\theta$$

where I_0 is the maximum intensity. This equation is known as Malus' law, which gives the intensity of transmitted light. It is clear from this relation that the whole intensity is passed when the planes of the polariser and the analyser are parallel ($\theta = 0°$). However, the incident light is completely blocked when these axes are perpendicular to each other.

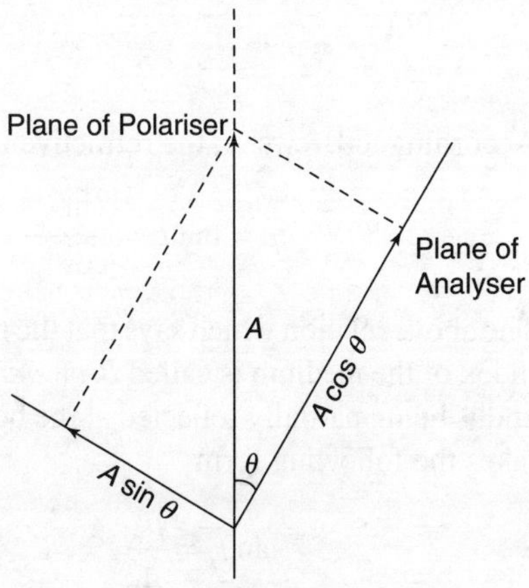

FIGURE 3.5

3.3.2 Polarisation by Reflection

We can obtain partially or sometimes fully polarised light when the light is reflected by the surface of an electrical insulator. In this case, the degree of polarisation depends on the angle of incidence of the light and the refractive index of the reflecting material. In some cases, the reflected light can be completely polarised parallel to the reflecting surface and perpendicular to the direction of the light propagation.

3.3.2.1 *Brewster's Law*

In 1808, *Malus* discovered a simple method for the polarisation of light by reflection. He found that when an ordinary light is reflected from the surface of a glass plate the reflected and refracted light beams are partially plane-polarised (Fig. 3.6a). This depends on the angle of incidence and at a particular angle of incidence (57.5° for a glass surface) the reflected light is completely plane-polarised while the transmitted light is partially polarised. This angle of incidence is known as *polarising angle*. *Brewster* also performed series of experiments in 1911 for studying the polarisation of light by reflection at the surfaces of different media. He also found that at a particular angle of incidence, the reflected light is completely polarised. The reflected light is the component of incident light polarised normal to the plane of incidence and therefore parallel to the surface (Fig. 3.6b) in view of the plane of incidence as the plane having vector $\vec{k}$ and unit vector normal to the surface. The incident angle at which the reflected light is completely polarised is known as *Brewster's angle* or angle of polarisation (i_p).

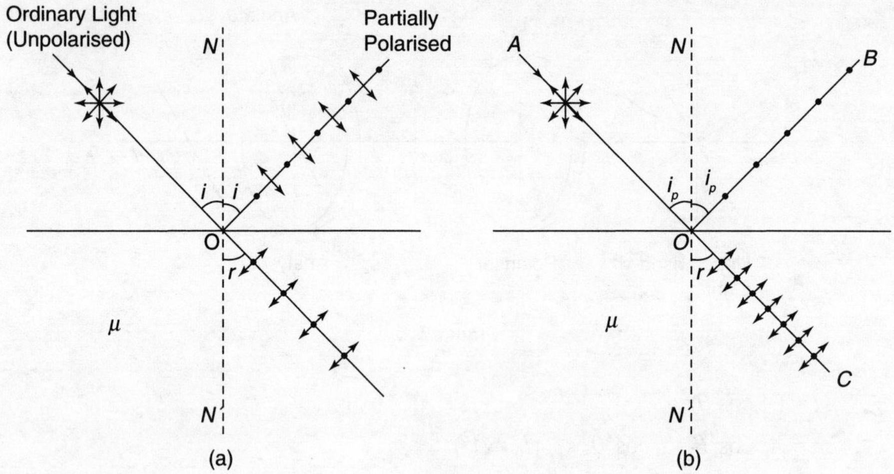

FIGURE 3.6

According to Brewster, the refractive index μ of the medium (Fig. 3.6b) is given by

$$\mu = \tan i_p = \frac{\sin i_p}{\cos i_p} \tag{i}$$

The above relation which says that the tangent of the angle of polarisation is numerically equal to the refractive index of the medium is called *Brewster's law*. If the light is propagating in a medium with refractive index μ_1 and is being partially reflected at the boundary with a medium of refractive index μ_2, Brewster's law however takes the following form

$$\tan i_p = \frac{\mu_2}{\mu_1}$$

The above polarising angle i_p is sometimes referred to as the Brewster angle of the material. As per Snell's law, for a glass surface in Fig. 3.6b we can write

$$\mu = \frac{\sin i_p}{\sin r} \tag{ii}$$

From Eqs. (i) and (ii), we get

$$\frac{\sin i_p}{\cos i_p} = \frac{\sin i_p}{\sin r} \tag{iii}$$

From Eq. (iii), we have

$$\cos i_p = \sin r = \cos\left(\frac{\pi}{2} - r\right)$$

$$i_p = \frac{\pi}{2} - r$$

$$i_p + r = \frac{\pi}{2}$$

As we know, $\angle NON' = \pi$, then we have,

$$\angle NOB + \angle BOC + \angle CON' = \pi$$

$$i_p + \angle BOC + r = \pi$$

$$\therefore \quad \angle BOC = \pi - (r + i_p)$$

$$= \pi - \frac{\pi}{2}$$

$$\angle BOC = \frac{\pi}{2}$$

i.e., at the polarising angle, the reflected and refracted rays are at right angle to each other.

3.3.2.2 *Biot's Polariscope*

Biot's polariscope is used for producing and detecting the plane-polarised light by reflection and is shown in Fig. 3.7. In the figure *A* and *B* are two parallel glass plates and their back surfaces are blackened to prevent the transmitted light (Fig. 3.7a). Let a ray of unpolarised light *PQ* be incident at the polarising angle on the polished face of glass plate *A* and it is reflected along *QR*. This ray *QR* is incident on the other glass plate *B* at the polarising angle and hence gets reflected along *RS*. It is seen that the ray *RS* is completely plane-polarised and its intensity is found to be maximum. The two glass plates *A* and *B* are known as the polariser and the analyser, respectively. When the glass plate *B* is gradually rotated about *QR* keeping the angle of incidence constant, the intensity of the reflected ray *RS* decreases and it becomes zero for 90° rotation of glass plate *B*, as shown in Fig. 3.7(b). If the glass plate *B* is further rotated, the intensity of the ray *RS* becomes maximum at 180°, minimum at 270° and again becomes maximum at 360°. By using this simple experiment, we can obtain and detect plane-polarised light.

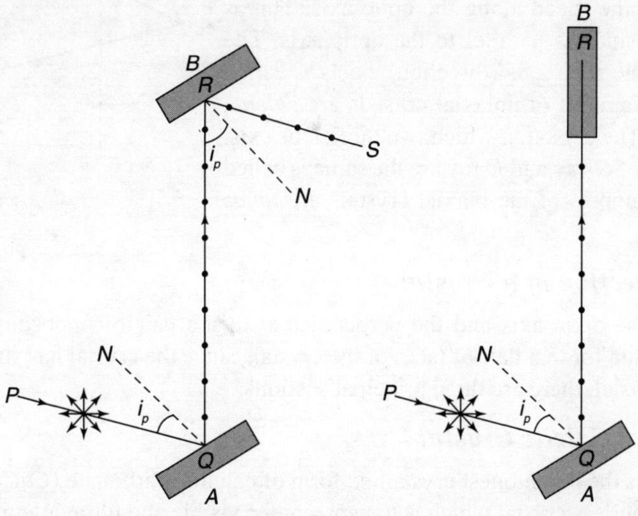

FIGURE 3.7

3.3.3 Polarisation by Refraction

Polarisation can also take place by the refraction of light, which occurs when a beam of light passes from one material into another material. Under this situation, the path of the light beam changes its direction at the surface of the two materials and then the refracted beam acquires some degree of polarisation. Mostly, the polarisation occurs in a plane perpendicular to the surface.

The light split into two beams upon entering the crystal and both the refracted light beams are polarised – one in a direction parallel to the surface and the other in a direction perpendicular to the surface. Since these two refracted rays are polarised with a perpendicular orientation, a polarising filter can be used to completely stop one of the images.

3.3.3.1 *Optic Axis*

You would have learnt in the experiments using lenses that there exists a line which passes through the centre of curvature of a lens surface such that the light rays are neither reflected nor refracted. This is called the *optic axis* or the principal axis. Similar situation arises for a particular type of crystal such as calcite crystal or tourmaline crystal (Fig. 3.8). For these substances there exists a specific direction within the crystal known as the optic axis or the principal axis, which is determined by the atomic configuration of the crystal. The optic axis of a calcite crystal is shown in Fig. 3.8 by the dotted line *AB*. Any ray of ordinary unpolarised light incident along the optic axis or parallel to this axis does not split up into two rays. The light ray gets split into two rays called as *ordinary ray* (*O*-ray) and *extraordinary ray* (*E*-ray) only when it makes an angle with the optic axis. It is observed that the ordinary and extraordinary rays propagate at the same speed along the optic axis. This is true for any direction which is parallel to the optic axis. The crystal in which only one such axis (direction) exists is called *uniaxial crystal*. The examples of uniaxial crystals are *calcite*, *tourmaline* and *quartz*. The crystal in which two directions exist along which the speeds of *O*-ray and *E*-ray are the same is called *biaxial crystal*. The examples of the biaxial crystals are *topaz* and *aragonite*.

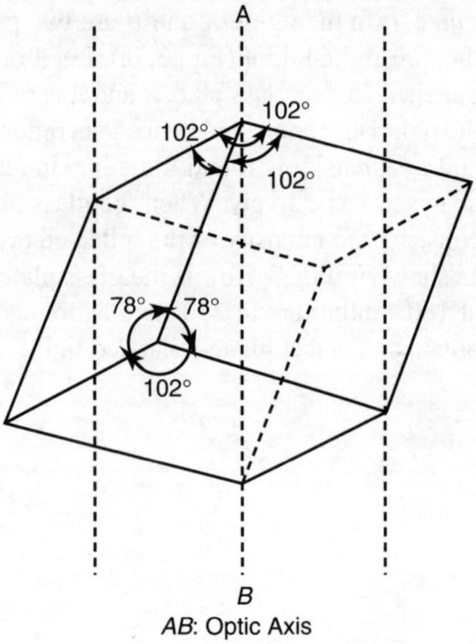

FIGURE 3.8

3.3.3.2 *Principal Section of a Crystal*

The plane containing the optic axis and the perpendicular to the pair of opposite faces of the crystal is known as principal section for that pair of faces of the crystal. Since the crystal has six faces, for each pair of opposite faces of the crystal, there are three principal sections.

3.3.3.3 *Geometry of Calcite Crystal*

The calcite or calcspar is the commonest crystalline form of calcium carbonate ($CaCO_3$). It is also known as *Iceland spar*. It is a colourless crystal which is transparent for visible and ultraviolet light. It occurs in nature in different variety of crystal form, for example in the rhombohedral class of the hexagonal system. It breaks readily into simple cleavage rhombohedrons, whose shape is shown in Fig. 3.8. It can be seen from the figure that its each face is a parallelogram with angles as 78° and 102°. An interesting feature of calcite is that each crystal can be made to slice or break along cleavage planes into two or more smaller crystals with faces that are parallelograms with angles 71° and 109° (Fig. 3.9).

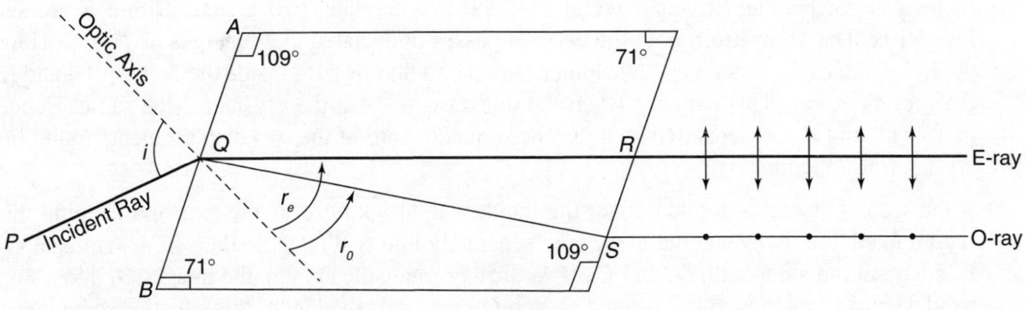

FIGURE 3.9

3.3.3.4 *Double Refraction*

Double refraction is an optical property in which a single ray of unpolarised light when passes through a uniaxial crystal gets split into two refracted rays, each propagating in a different direction. These two rays are named as *O*-ray and *E*-ray. Therefore, the phenomenon of splitting of light into two rays is known as *double refraction*.

If a narrow beam of unpolarised light *PQ* is incident on the face *AB* of the crystal at an angle *i*, it splits up into *O*-ray and *E*-ray inside the crystal. *O*-ray makes an angle of refraction r_o whereas the *E*-ray makes an angle r_e with the optic axis inside the crystal (Fig. 3.9). Both the rays are plane-polarised and their vibrations are at right angles to each other and to the direction of propagation. The refractive index for *O*-ray is observed to be constant in all the directions whereas the refractive index for *E*-ray varies according to the direction taken. This is because it has components that are both parallel and perpendicular to the optic axis of the crystal. Since the speed of light wave in a medium is the ratio of its speed in vacuum and the index of refraction for that wavelength, an *E*-ray can move either faster or slower than an *O*-ray. However, the velocity of *O*-ray inside the crystal is same in all the directions while the *E*-ray travels in the crystal with different velocities in different directions. For example, in a negative crystal such as calcite, where the refractive index for *E*-ray is smaller than that for *O*-ray ($\mu_e < \mu_0$; $\mu_0 = 1.6584$, $\mu_e = 1.4864$), the velocity of *E*-ray is larger than the velocity of *O*-ray in the crystal. On the other hand, in a positive crystal such as quartz, where the refractive index for *E*-ray is greater than that for *O*-ray ($\mu_e > \mu_0$), the velocity of *E*-ray is smaller than the velocity of *O*-ray in the crystal.

In view of Fig. 3.9, the refractive indices for *O*-ray and *E*-ray can be written as

$$\mu_0 = \frac{\sin i}{\sin r_o} \text{ and } \mu_e = \frac{\sin i}{\sin r_e}$$

Since the velocity of *O*-ray and hence the refractive index μ_0 inside the crystal is same in all the directions, this ray obeys Snell's Law. However, the *E*-ray does not obey Snell's Law as it travels in the crystal with different velocities in different directions, leading to different μ_e in different directions.

3.3.3.5 *Polarisation by Double Refraction*

The polarisation of light by double refraction in calcite is demonstrated in Fig. 3.10 where *AB* and *CD* are the principal sections of the two crystals. Here we rotate the second crystal and observe the following phenomena related to the *O*-ray and *E*-ray separated by the crystals.

(a) In the case of parallel principal sections of the two crystals, two images O and E are seen in Fig. 3.10a. The O-ray from both the crystals passes undeviated and emerges as O_1-ray. However, the E-ray passes the second crystal along a path parallel to its path inside the first crystal and finally emerges as E_1-ray. This happens when the thickness of both the crystals is the same. Hence, the images O_1 and E_1 are separated by a distance equal to sum of the two displacements found in each crystal, if used separately.

(b) If the second crystal is rotated about the incident light taking it as the axis and keeping the first crystal fixed, the O-ray and the E-ray split separately into two rays. So the two new images O_2 and E_2 are observed along with O_1 and E_1. If we further rotate the crystal, the images O_1 and O_2 remain fixed whereas E_1 and E_2 rotate around O_1 and O_2, respectively. Under this situation, the intensity of O_1 and E_1 decreases. When the principal section of the second crystal makes an angle of 45° with the principal section of the first crystal, the four images of equal intensities are seen. This is shown in Fig. 3.10b.

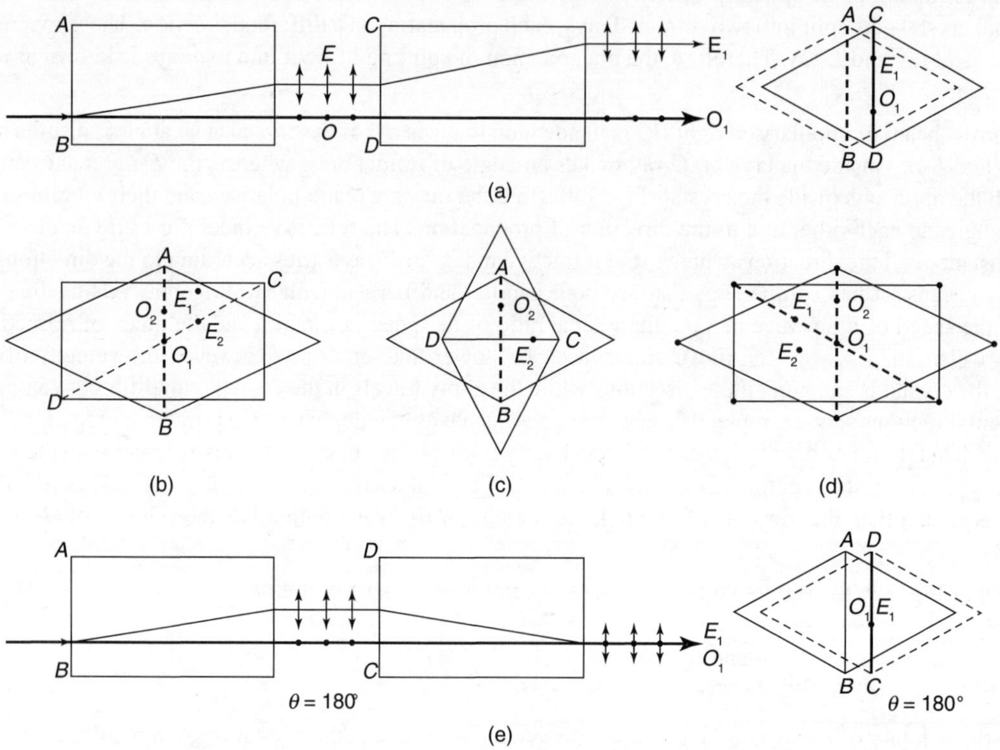

FIGURE 3.10

On continuing the rotation, the intensities of images O_1 and E_1 get decreased and the intensities of O_2 and E_2 get increased. At 90° rotation, the images O_1 and E_1 finally disappear and the new images O_2 and E_2 acquire maximum intensities Fig. 3.10c.

(c) For the further rotation of the second crystal, the images O_1 and E_1 again appear and the intensities of these images increase. Then the intensities of the images O_2 and E_2 decrease. At 135° angle of rotation, the intensities of four images become equal, as shown in Fig. 3.10d.

(d) At $\theta = 180°$, the principal sections of both the crystals are again parallel. However, their optic axes are oriented in the opposite directions (Fig. 3.10e). In this situation, the images O_2 and E_2 disappear and the images O_1 and E_1 superimpose with each other to form a single image that emerges from the second crystal.

Based on the above observations, this experiment demonstrates the polarisation of light. The first crystal produces plane-polarised vibrations whereas the second crystal analyses these vibrations.

Explanation Let the principal sections *AB* and *CD* of the first and second crystals, respectively, be inclined at an angle θ (Fig. 3.11). A ray of ordinary unpolarised light splits into two plane-polarised rays after emerging from the first crystal. When the *O*-ray vibrates perpendicular to the principal section *AB*, then the *E*-ray vibrations are along the principle section. Let *a* be the amplitude of each ray, represented by *NO* and *NE*, respectively. On entering the second crystal, each of the *O*- and *E*-rays is split into two components. The *O*-ray is split into two components as $O_1 = a \cos \theta$ and $E_2 = a \sin \theta$ whereas *E*-ray is split into two components as $E_1 = a \cos \theta$ and $O_2 = a \sin \theta$. Thus, a measure of the intensities of O_1 and E_1 is $(a \cos \theta)^2$ while that of O_2 and E_2 is $(a \sin \theta)^2$. Based on these expressions for the intensities, we can now discuss the different cases:

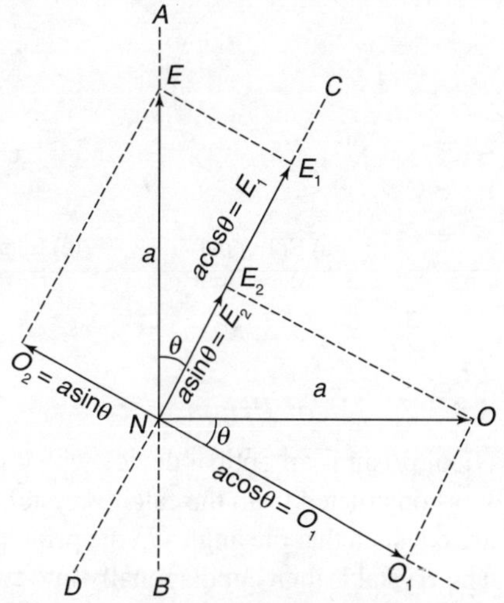

FIGURE 3.11

Case-1: At $\theta = 0°$ and $\theta = 180°$, we get $\cos^2\theta = 1$ and $\sin^2\theta = 0$. It means the intensities of O_1 and E_1 rays are maximum, while that of O_2 and E_2 is zero.

Case-2: If $\theta = 45°$ and $\theta = 135°$, we get $\cos^2\theta = ½$ and $\sin^2\theta = ½$. It means the intensities of O_1, E_1, O_2 and E_2 have the same values. Therefore, all the four images are equally bright.

Case-3: If $\theta = 90°$, we get $\cos^2\theta = 0$ and $\sin^2\theta = 1$. It means O_1 and E_1 vanish and O_2 and E_2 are the brightest.

It is clear from the above cases that the sum of the intensities of the two components is $a^2\cos^2\theta + a^2\sin^2\theta = a^2$, which is just equal to the intensity of the incident light beam.

3.3.3.6 *Huygens' Theory of Double Refraction*

Phenomenon of double refraction was explained by Huygens' for which he extended his principle of secondary wavelets and made the following assumptions.

(i) When a light wave strikes the surface of a doubly refracting crystal, each point of the crystal becomes the origin of two secondary wavelets, named as ordinary ray and extraordinary ray. These two wavelets spread out into the crystal.

(ii) The wavefront corresponding to ordinary ray is spherical as the velocity of ordinary ray remains the same in all the directions (Fig. 3.12a).

(iii) The wavefront corresponding to extraordinary ray is an ellipsoid of revolution with the optic axis as its axis of revolution (Fig. 3.12b). This is due to the fact that the velocity of *E*-ray is different in different directions in the crystal.

(iv) The two wavefronts corresponding to *O*-ray and *E*-ray touch each other along the optic axis since both the rays travel with the same velocity along the direction of optic axis.

(v) For negative uniaxial crystals (like calcite) in which the velocity of *O*-ray is less than the velocity of *E*-ray, sphere lies inside the ellipsoid (Fig. 3.12c). However, for positive uniaxial crystals (like quartz) the ellipsoid lies inside the sphere (Fig. 3.12d) since in this case the velocity of *O*-ray is greater than the velocity of *E*-ray.

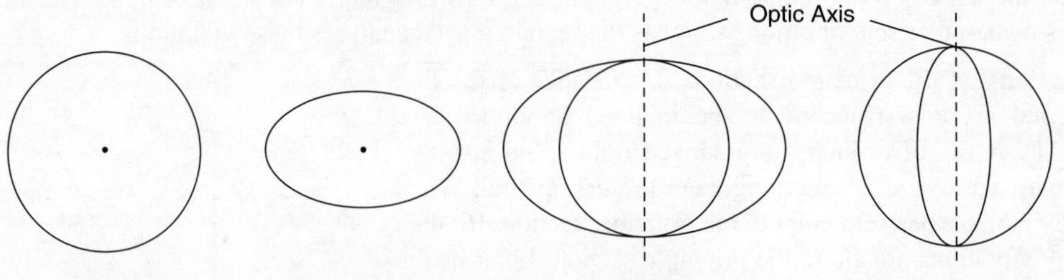

FIGURE 3.12

3.3.3.7 *Nicol Prism*

Nicol prism is an optical device which is used for producing and analysing plane-polarised light in practice. It is constructed from the calcite crystal *PQRS* having length three times of its width. Its end faces *PQ* and *RS* are cut such that the angles in the principal section become 68° and 112° in place of 71° and 109° (Fig. 3.13). The crystal is then cut diagonally into two parts. The surfaces of these parts are grinded to make optically flat and then these are polished. Thus polished surfaces are connected together with a special cement known as Canada Balsam.

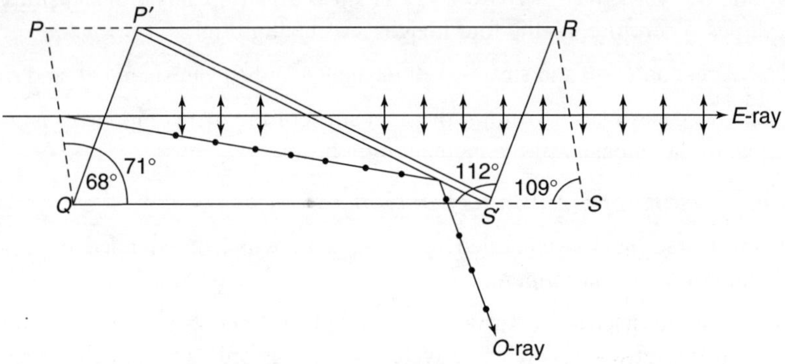

FIGURE 3.13

3.3.3.8 *Quarter-Wave Plate*

It is a plate of doubly refracting uniaxial crystal like quartz or calcite, whose refracting faces are cut parallel to the optic axis and its thickness is such that it introduces a phase change of $\pi/2$, i.e., a path change of $\lambda/4$ between the ordinary and extraordinary light waves. Let μ_0 and μ_e be the refractive indices for the ordinary and extraordinary light waves, respectively, and t the thickness of the plate. The path difference between the ordinary and extraordinary waves is given by

$$(\mu_0 - \mu_e)t = \frac{\lambda}{4} \text{ for negative crystal}$$

$$(\mu_e - \mu_0)t = \frac{\lambda}{4} \text{ for positive crystal}$$

$$t = \frac{\lambda}{4(\mu_0 - \mu_e)} \text{ for negative crystal}$$

Quarter-wave plate is used to produce circularly and elliptically polarised light.

3.3.3.9 *Half-Wave Plate*

It is a plate of doubly refracting uniaxial crystal like quartz or calcite, whose refracting faces are cut parallel to the optic axis and its thickness is such that it introduces a phase change of π, i.e., a path change of $\lambda/2$ between the ordinary and extraordinary light waves. For the refractive indices μ_0 and μ_e for ordinary and extraordinary light waves, the path difference is written as,

For Half-Wave Plate

$$(\mu_0 - \mu_e)t = \frac{\lambda}{2} \text{ for negative crystal}$$

$$(\mu_e - \mu_0)t = \frac{\lambda}{2} \text{ for positive crystal}$$

$$t = \frac{\lambda}{2(\mu_0 - \mu_e)} \text{ for negative crystal}$$

Half-wave plate is used to produce plane-polarised light. Quarter-wave plate and half-wave plate are known as phase retarding plates. The phase retardation can be calculated by using the following relation.

$$\delta = \frac{2\pi}{\lambda} \times \Delta x \text{ where } \Delta x \text{ is path difference.}$$

When a beam of unpolarised light is incident on the face $P'Q$, it gets split into two refracted rays, named O-ray and E-ray. These two rays are plane-polarised rays, whose vibrations are at right angles to each other. The refractive index of Canada balsam cement being 1.55 lies between those of ordinary and extraordinary rays. This is because the refractive indices of ordinary and extraordinary rays for calcite crystal are 1.6584 and 1.4864, respectively.

It is clear from the above discussion that Canada Balsam layer acts as an optically rarer medium for the ordinary ray and it acts as an optically denser medium for the extraordinary ray. When ordinary ray of light travels in the calcite crystal and enters the Canada balsam cement layer, it passes from denser to rarer medium. Moreover, the angle of incidence is greater than the critical angle, the incident ray is totally internally reflected from the crystal and only extraordinary ray is transmitted through the prism. Therefore, fully plane-polarised wave is generated with the help of Nicol prism.

Nicol Prism as a Polariser and an Analyser: In order to produce and analyse the plane-polarised light, we arrange two nicol prism as per Fig. 3.14. When a beam of unpolarised light is incident on the nicol prism, emergent beam from the prism is obtained as plane-polarised, and which has vibrations parallel to the principal section. This prism is therefore known as polariser. If this polarised beam falls on another parallel nicol prism P_2, whose principal section is parallel to that of P_1, then the incident beam will behave as E-ray

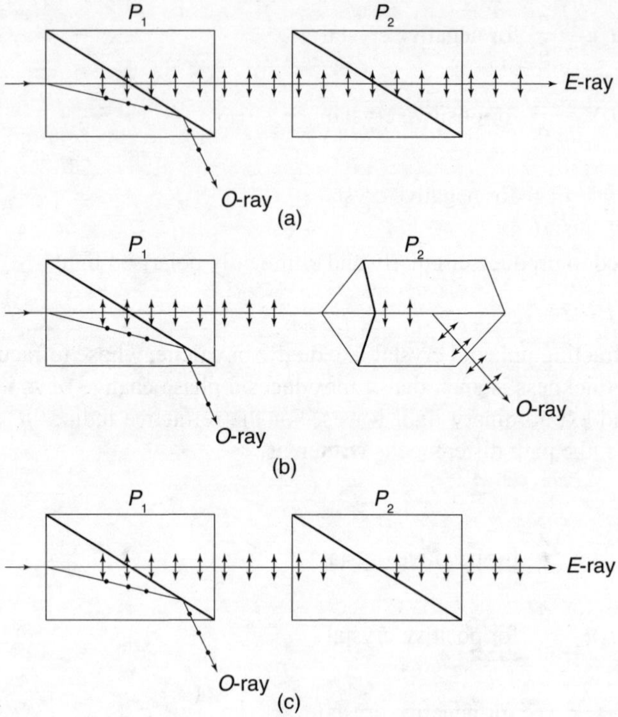

FIGURE 3.14

inside the nicol prism P_2 and gets completely transmitted through it (Fig. 3.14a). This way the intensity of emergent light will be maximum.

Now the nicol prism P_2 is rotated about its axis, then we note that the intensity of emerging light decreases and becomes zero at 90° rotation of the second prism (Fig. 3.14b). In this position, the vibrations of *E*-ray become perpendicular to the principal section of the analyser (nicol prism P_2). Hence, this ray behaves as *O*-ray for prism P_2 and it is totally internally reflected by Canada balsam layer. This fact can be used for detecting the plane-polarised light and the nicol prism P_2 acts as an analyser.

If the nicol prism P_2 is further rotated about its axis, the intensity of the light emerging from it increases and becomes maximum for the position when principal section of P_2 is again parallel to that of P_1 (Fig. 3.14c). Hence, the nicol prisms P_1 and P_2 acts as polariser and analyser, respectively.

3.3.4 Polarisation by Scattering

The scattering of light by the air molecules produces linearly polarised light in the plane perpendicular to the incident light. The scatterers can be imagined as tiny antennae which emit radiations (light) perpendicular to their axis of vibrations. If the charges in a molecule are vibrating along the *x*-axis, the radiation or light is not obtained along the *x*-axis rather the scattered light is found to be linearly polarised at 90° away from the beam direction. This leads the light to be partially polarised that undergoes Rayleigh scattering from the blue sky.

3.4 THEORY OF PRODUCTION OF PLANE, CIRCULARLY AND ELLIPTICALLY POLARISED LIGHT LO4

As discussed earlier, the light that has unidirectional vibrations is known as plane-polarised light or linearly polarised light. When two plane-polarised light waves are allowed to superimpose, and the resultant electric vector rotates in such a way that its tip traces a circle, the resultant light is known as circularly polarised light. However, if the magnitude of the resultant electric vector varies periodically during its rotation and its tip traces an ellipse, then the resultant light is called elliptically polarised light.

Let us consider a calcite crystal such that its refracting faces are cut parallel to its optic axis. Further we consider that the linear vibrations (amplitude A) in the incident light are along the direction PQ that makes an angle θ with the optic axis. (Fig. 3.15a). Under this situation, the incident plane-polarised light of amplitude A spits into two components $A\cos\theta$ (E-ray) and $A\sin\theta$ (O-ray). These components constitute E-ray and O-ray in view of their vibrations parallel to the optic axis and perpendicular to it, respectively. As per Huygens' theory, the E-ray and O-ray travel in the same direction (Fig. 3.15b) with different velocities. Since calcite is a negative crystal, the velocity of E-ray will be greater than that of O-ray. Hence, a phase difference is introduced between them after traveling through the plate.

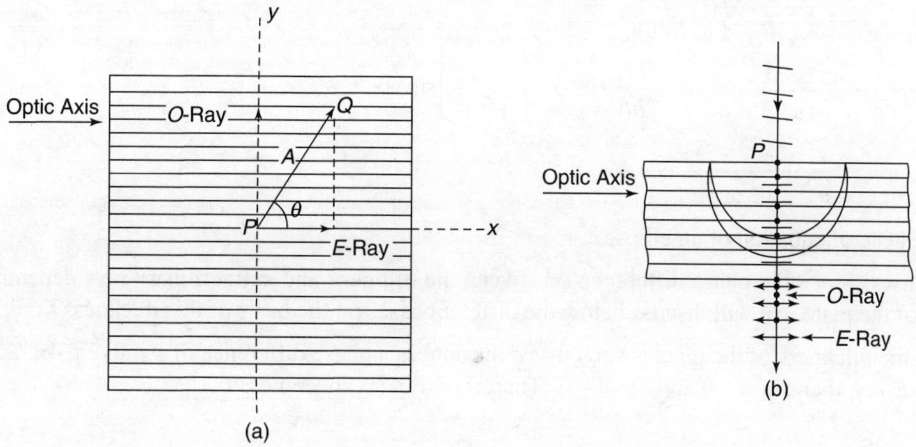

FIGURE 3.15

In view of the incident light wave as $A\sin\omega t$, we can represent the E-ray along the optic axis as

$$x = A\cos\theta\sin(\omega t + \phi) \tag{i}$$

Similarly, the O-ray along y-axis will be

$$y = A\sin\theta\sin\omega t \tag{ii}$$

Now assuming $A\cos\theta = a$ and $A\sin\theta = b$, we get

$$x = a\sin(\omega t + \phi) \tag{iii}$$

$$y = b\sin\omega t \tag{iv}$$

From Eq. (iv), we have

$$\sin\omega t = \frac{y}{b} \tag{v}$$

and $$\cos \omega t = \sqrt{1-\frac{y^2}{b^2}} \tag{vi}$$

Now from Eq. (iii), we get

$$\frac{x}{a} = \sin \omega t \cos \phi + \cos \omega t \sin \phi \tag{vii}$$

Putting the values of sin ωt and cos ωt from Eqs. (v) and (vi) in the above equation, we have

$$\frac{x}{a} = \frac{y}{b}\cos \phi + \sqrt{1-\frac{y^2}{b^2}}\sin \phi$$

or $$\frac{x}{a} - \frac{y}{b}\cos \phi = \sqrt{1-\frac{y^2}{b^2}}\sin \phi \tag{viii}$$

On squaring both sides, Eq. (viii) we get

$$\left[\frac{x}{a} - \frac{y}{b}\cos \phi\right]^2 = \left[1-\frac{y^2}{b^2}\right]\sin \phi$$

$$\frac{x^2}{a^2} + \frac{y^2}{b^2}\cos^2 \phi - \frac{2xy}{ab}\cos \phi = \left[1-\frac{y^2}{b^2}\right]\sin^2 \phi$$

or $$\frac{x^2}{a^2} + \frac{y^2}{b^2} - \frac{2xy}{ab}\cos \phi = \sin^2 \phi \tag{ix}$$

This is the general equation of an ellipse.

Special Cases: Since the phase difference ϕ between the ordinary and extraordinary rays depend upon the thickness of the plate, we will discuss below the different cases on the basis of this thickness t.

Case-I: If the thickness of the plate is such that it introduces a phase difference of $\phi = 0, 2\pi, 4\pi, \ldots$ between *O*-ray and *E*-ray, then sin $\phi = 0$ and cos $\phi = 1$. Therefore, Eq. (ix) becomes,

$$\frac{x^2}{a^2} + \frac{y^2}{b^2} - \frac{2xy}{ab} = 0$$

or $$\left[\frac{x}{a} - \frac{y}{b}\right]^2 = 0$$

or $$y = \frac{b}{a}x \tag{x}$$

This is the equation of straight line having the slope $\left(\frac{b}{a}\right)$ and passing through the origin Fig. 3.16(a). This concludes that the light emerging through the plate is plane-polarised.

Case-II: If the thickness of the plate is such that $\phi = \pi, 3\pi, 5\pi, \ldots$, then sin $\phi = 0$ and cos $\phi = -1$. Therefore, Eq. (ix) attains the form

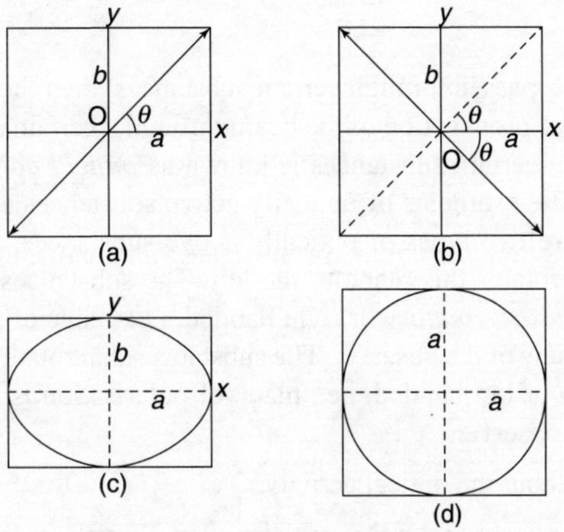

FIGURE 3.16

$$\frac{x^2}{a^2}+\frac{y^2}{b^2}+\frac{2xy}{ab}=0$$

or
$$\left[\frac{x}{a}+\frac{y}{b}\right]^2=0$$

$$y=\frac{-b}{a}x \tag{xi}$$

This is again an equation of straight line having the slope $\left(-\frac{b}{a}\right)$ (Fig. 3.16b). So we will have again the emergent light as plane-polarised light.

Case-III: If the thickness of the plate is such that $\phi=\frac{\pi}{2},\frac{3\pi}{2},\frac{5\pi}{2}\cdots$, then $\sin\phi=1$, $\cos\phi=0$ Eq. (ix) attains the form

$$\frac{x^2}{a^2}+\frac{y^2}{b^2}=1 \tag{xii}$$

This is the equation of an ellipse with its axis along x and y directions (Fig. 3.16c). Therefore, the emergent light will be elliptically polarised light.

Case-IV: If $a=b$ and ϕ satisfies the condition of Case-III

$$x^2+y^2=a^2$$

This is the equation of a circle of radius a. Thus, the emergent light will be circularly polarised light if the plate introduces a phase change of

$$\frac{\pi}{2},\frac{3\pi}{2},\frac{5\pi}{2}\text{ etc.}$$

From the above discussion it is clear that the plane and circularly polarised lights are the special cases of an elliptically polarised light which is obtained by the superposition of two plane-polarised lights.

3.5 OPTICAL ACTIVITY LO5

When a beam of plane-polarised light is passed through certain substances, then the plane of polarisation of the beam is rotated about the direction of propagation of the beam through a certain angle. This phenomenon of rotating the plane of polarisation by certain substances is known as *optical activity* and the material is known as optically active substance. The examples of optically active substances are *sugar crystals*, *sugar solution*, *sodium chlorate* etc. There are two types of optically active substances; one rotates the plane of polarisation to the right and the other rotates this plane to the left. The substances that rotate the plane of polarisation to the right are known as *dextro-rotatory* or right handed, i.e., plane of polarisation is rotated in clock wise direction from the point of view of the observer. The substances that rotate the plane of polarisation to the left are known as *leavo-rotatory* or left handed, i.e., plane of polarisation is rotated in anticlockwise direction from the point of view of the observer.

In 1815, *Biot* gave the following laws about the optical activity.

(i) The angle of rotation of a light beam of a given wavelength is directly proportional to the length of optical active substance traversed.

(ii) For a given path length, the angle of rotation is proportional to the concentration of the solution or vapour.

(iii) The rotation produced by optically active substances is equal to the algebraic sum of rotations made by the substances individually. The anticlockwise and clockwise rotations are taken with opposite signs.

(iv) The angle of rotation is inversly proportional to the square of wavelength. In case of quartz, the angle of rotation is given by

Angle of rotation = $A + B/\lambda^2$. where A and B are constants.

3.6 SPECIFIC ROTATION LO5

It is an intrinsic property of a pure material at a given wavelength and temperature. The plane of the linearly polarised light is rotated when it is passed through liquids containg an optically active substance, for example *sugar solution*, *sodium chlorate*, *cinnabar*, *camphor* in *alcohol,* etc.). The specific rotation S is defined as the observed angle of optical rotation θ when plane-polarised light is passed through a sample with a path length of 1 decimetre and a sample concentration of 1 gram per 1 millilitre. Therefore

$$S = \frac{\theta}{l \times c} = \frac{\theta}{l \times (m/V)} = \frac{\theta V}{l \times m}$$

In this equation, l is the path length in decimeters, and c is the concentration of the liquid in g/ml, for a sample at a temperature T(given in degrees Celsius) and wavelength λ (in nanometers). The formal unit for specific rotation is deg cm^2 g^{-1} but scientific literature uses just degrees.

3.7 LAURENT'S HALF-SHADE POLARIMETER LO6

It consists of two nicol prisms P_1 and P_2 as shown in Fig. 3.17. The nicol prism P_1 acts as a polariser and P_2 acts as an analyser. The half-wave plate H is placed between nicol prism P_1 and P_2. This plate consists of

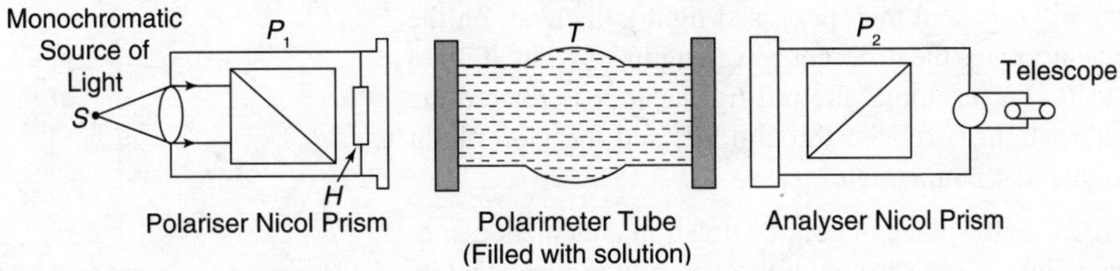

FIGURE 3.17

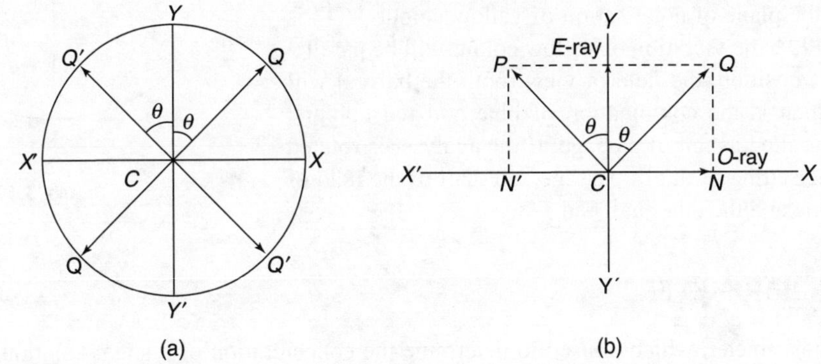

FIGURE 3.18

two semi-circular plates, one half of which is made of glass and the other half of quartz and both halves are attached together as shown in Fig. 3.18a. The thickness of the quartz plate is kept such that it introduces a phase difference of π between the ordinary and extraordinary vibrations.

The monochromatic light from the source S is incident on convex lens, from which it emerges as parallel beam and falls on the polarising nicol prism P_1. The light emerging from the polariser P_1 is plane-polarised and falls on the half-shade plate H and the on a polarimeter tube T filled with optically active solution. Finally the light emerged from tube falls on the analyser. Then we see emergent light with the help of a telescope. The analysing nicol prism can be rotated about its axis. Its rotations are measured in term of angle θ by using circular scale.

3.8 BIQUARTZ POLARIMETER LO6

The arrangement of biquartz polarimeter is the same as that of a Laurent's half-shade polarimeter. However, there are some different features of this polarimeter. For examples, it consists of a biquartz plate in place of Laurent's half-shade plate and it uses white light in place of monochromatic light.

A biquartz plate consists of two semicircular plates one of left-handed and the other of right-handed quartz, both cut perpendicular to optic axis and joined together so as to form a complete circular plate. The left-handed and right-handed quartz plates rotate the plane of polarisation of the incident beam in anti-clockwise and clockwise directions, respectively. The thickness of both half-plates is the same and is adjusted such that each rotates the plane of polarisation of yellow light through 90°. When a beam of white light is incident on the polarising nicol P_1 and then polarised light is incident on the Biquartz plate, it rotates the different wavelengths of white light to different extents. For example, the red light colour is rotated the least, the yellow light is rotated through 90° and the violet light rotates through a maximum angle.

In Fig. 3.19 XCX' is the plane of polarisation of the incident plane-polarised white light. Here, the analyser is set in such a way that its principal section is parallel to XCX'. In this figure, YCY' represents the plane of polarisation of yellow colour. Since with the use of biquartz plate, the plane of polarisation of yellow colour YCY' is rotated through 90°, the vibration of yellow colour will be parallel to XCX'. In this position, the field of view (both the halves) will appear grey which is the combination of blue and red colours. This position is called as tint of passage. If the analyser is rotated slightly from the setting of tint of passage, one-half of the field of view appears blue and the other half-red.

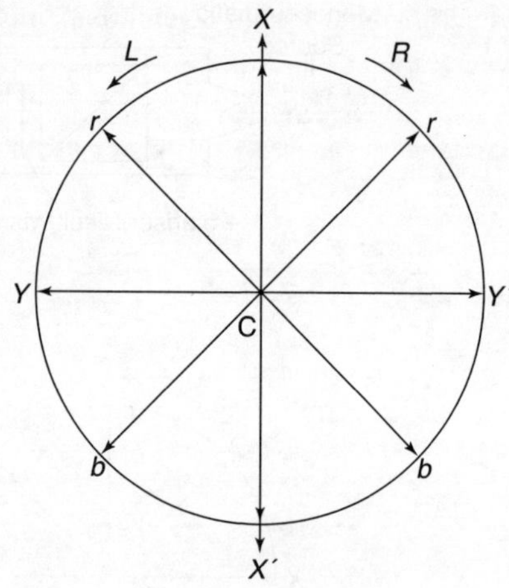

FIGURE 3.19

3.9 SACCHARIMETER

LO6

It is a type of polarimeter, which is used to determine the concentration of a sugar solution by measuring the angle of rotation of the plane of polarisation of the polarised light passing through a tube containing the solution. The concentration of the solution (m/V) is calculated by the formula obtained from the concept of specific rotation

$$\frac{m}{V} = \frac{\theta}{S \times l}$$

3.9.1 Laurent Saccharimeter

It consists of two Nicol prisms namely N_1 and N_2, as shown in Fig. 3.20a. Nicol N_1 is used to polarise the light, so it works as a polariser. Nicol N_2 works as an analyser. If N_1 and N_2 are kept parallel, the light can pass through them. Then these are said to be parallel. If N_1 and N_2 are not kept parallel so the light does not pass through them, then the Nicols are said to be crossed. Some substances like quartz, sugar solution, etc. possess the property of rotating the plane of polarisation of the light. The amount of this rotation can be measured by determining the angle through which Nicol N_2 is turned.

In order to determine the angle by which the Nicol N_2 is rotated, we keep a circular sheet made of quartz and glass just in front of polarising Nicol (Fig. 3.20b). Glass is of such thickness that it absorbs the same amount of light as the quartz does. Light gets separated into two components when it just reaches the quartz plate. These components pass through the plate with different velocities. Let these components be represented by OP and OE when vibrations at O take place in the direction OA (Fig. 3.20b). There will be a gradual change of phase between these components due to different transmission velocities. After a time this disturbance will reach a point in the plate where component displacement is along OP and along OE'. The resultant displacement is OA'. This difference is one half a period on leaving the quartz plate and the plate is said to be half-wave plate. Light passes through the glass undisturbed and its oscillations are still along the direction ED which is parallel to OA.

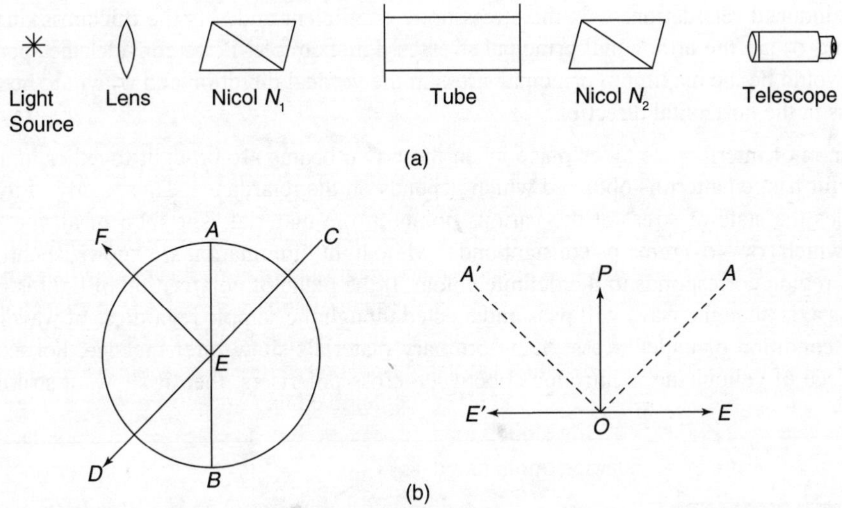

Figure 3.20

If Nicol N_2 is kept with its short diagonal at right angle to OA', this OA' component is not transmitted while OA passes through and the glass side appears illuminated. In fact, the light passes through from both sides of the plate, but both sides are not equally illuminated. When both sides present the same illumination, the principle plane of the Nicol is either along AB or normal to it. If both halves are equally dark, the Nicol is so placed that the smaller components are transmitted. If the Nicol is set for equal illumination on both sides and an active substance is interposed it will be necessary to rotate the Nicol to find the position of equal intensities. In this case, the amount of rotation determines the angle of rotation of the plane of polarisation.

3.10 PHOTOELASTICITY LO7

Photoelasticity is a phenomenon in which transparent isotropic substances such as plastic or glass could be made optically anisotropic by applying mechanical stress. It is an experimental technique for measuring and visualising stresses in structures by means of a phenomenon known as *birefringence*. The material becomes birefringent under the compression and tension. This induced birefringence in the material is proportional to the stress. It is found that the optic axis is induced in the direction of the stress in both the cases of compression and tension. If nonuniform stress is applied to the sample, the phenomena of birefringence or the retardation will not take place on a transmitted wave.

As discussed, two crossed polarisers ordinarily do not transmit light but when a photoelastic material is placed between them and the plane of polarisation of light is not parallel to the principal axis of the stress, some light will be transmitted. Under this situation, a complicated coloured fringe pattern is obtained, which is helpful to measure the internal mechanical stresses in the samples.

When a beam of plane-polarised light is passed through a specimen of birefringent material or photoelastic material, then the light is resolved along the two principal stress directions. Each component of the light experiences different refractive indices. Then relative phase retardation (retardance) at any point on the specimen material is directly proportional to the principal phase difference, i.e.,

$$R \propto (\sigma_1 - \sigma_2)$$

$$R = Cd(\sigma_1 - \sigma_2)$$

Here R is the induced retardation, C is the stress optic coefficient and d is the thickness of the specimen material, σ_1 and σ_2 are the orthogonal principal stresses. For example, if we consider the specimen to be a plate then σ_1 would be the maximum principal stress in the vertical direction and σ_2 would be the minimum principal stress in the horizontal direction.

The phenomenon of interference takes place when these two beams are brought together in a polariscope. Then a colourful fringe pattern is obtained which depends on the retardance. Thus, study of these colourful fringes provides the state of stress at the various points in the material. The locii of all the points on the specimen for which $(\sigma_1 - \sigma_2)$ remains constant under white light illumination are knows isochromatic region and each such region corresponds to the definite colour. If the plane of polarisation of light is parallel to the principal stress axis, then the wave will pass undeflected though the sample regardless of wavelength. Either under normal condition or under stress, many ordinary materials show birefringence. For example, when a crumpled piece of cellophane is introduced between cross polarisers, then it shows a striking variety of colours.

SUMMARY

Following important points can be noted related to the matter presented in this chapter.

- Starting with a general introduction of light as an electromagnetic wave, the concept of polarisation of light was introduced.
- Based on a mechanical experiment using a string, various types of polarisation of wave viz. linearly polarised wave, circularly polarised wave and elliptically polarised wave were discussed.
- Along with the appropriate figures, the difference between unpolarised light and polarised light was made clear.
- The important features of polarised light including the direction of polarisation, plane of polarisation and plane of vibration were discussed.
- Various means of production of plane-polarised light were introduced.
- Polaroid filter is commonly used for generating the plane-polarised light. So it was discussed in detail including the alignment of its long chain molecules and the polarisation axis.
- Malus' law of intensity was discussed related to the intensity of light emerging from a Polaroid filter (analyser) whose transmission axis make an angle with the axis of another Polaroid which is used to produce plane-polarised light.
- Concept of Brewster's angle was introduced for which the reflected light can be obtained as fully polarised light.
- Biot's polariscope was discussed in short. This can be used for producing and detecting the plane-polarised light by the reflection.
- Polarisation can also take place by the reflection of light which occurs when a beam of light passes from one material into another material. When the unpolarised light is passed through a particular type of crystal, the light gets split into two rays. These waves are polarised parallel and perpendicular to a particular direction in the crystal. This particular direction is called as optic axis. So optic axis was discussed in detail.

- The principal section of a crystal was also discussed.
- Geometry of calcite crystal was discussed with relation to the double refraction which is nothing but the splitting of light wave into two rays in the crystal. These two rays are referred to as ordinary ray (*O*-ray) and extraordinary ray (*E*-ray).
- Experimental observation of polarisation by double reflection was discussed in detail and an explanation was also given.
- Huygens' theory of double refraction was discussed qualitatively.
- Out of the two rays (*E*-ray and *O*-ray) obtained by double refraction, one ray can be suppressed with the use of Nicol prism. So it was discussed in detail.
- Since *E*-ray and *O*-ray are plane-polarised waves, their superposition can lead to the resultant wave as circularly polarised wave or elliptically polarised wave. In view of this, the concept of quarter-wave plate and half-wave plate for the inclusion of phase change of $\pi/2$ or π between *E*-ray and *O*-ray was given.
- Theory of production of plane-polarised light, circularly polarised light and elliptically polarised light was discussed in detail.
- There are some substances that rotate the plane of polarisation of light which is passed through them. These substances are called optically active substances and this property is called as optical activity. So this was discussed along with the inclusion of specific rotation.
- Polarimeters were discussed which are used to produce the polarised light and also to determine the rotation of its plane of polarisation when passed through an optically active substance. Specifically Laurent's half shade polarimeter and biquartz polarimeter were discussed.
- The concept of polarimeter was extended to Saccharimeter, which is used to determine the concentration of sugar solution based on the measurement of angle of rotation of the plane of polarisation of light.
- Lastly photoelasticity was discussed which is a phenomenon in which transparent isotropic substances such as plastic or glass could be made optically anisotropic by applying mechanical stress. It is an experimental technique for measuring and visualising stress in structures by means of a phenomenon known as birefringence. The material becomes birefringence under the compression and tension.

SOLVED EXAMPLES

EXAMPLE 1 Refractive index of glass is 1.5. Calculate Brewster's angle for it. Also calculate the angle of refraction.

SOLUTION Given $\mu = 1.5$

Brewster's law says, $\mu = \tan i_p$

or $$\tan i_p = \mu = 1.5 = \frac{3}{2}$$

$$i_p = 56.31°$$

$$r = 90 - i_p = 90 - 56.31 = 33.69°$$

$$\mathbf{r = 33.69°}$$

Example 2 The refractive index for water is 1.33. Calculate the polarising angle for water.

Solution Given $\mu = 1.33$

Formula used is $\tan i_p = \mu = 1.33$

$$i_p = \tan^{-1}(1.33) = 53.06°$$

$$\mathbf{i_p = 53.06°}$$

Example 3 The refractive indices of glass and water are 1.54 and 1.33, respectively. For which case polarising angle will be greater: for a beam incident from water to glass or for a beam incident from glass to water?

Solution Given $\mu_g = 1.54$ and $\mu_w = 1.33$

Formula used is $\tan i_p = \mu$

For water to glass ${}_w\mu_g = \dfrac{\mu_g}{\mu_w} = \dfrac{1.54}{1.33} = 1.16$

So $i_p = \tan^{-1}({}_w\mu_g) = \tan^{-1}(1.16)$

$$\mathbf{i_p = 49.23°}$$

For glass to water ${}_g\mu_w = \dfrac{\mu_w}{\mu_g} = \dfrac{1.33}{1.54} = 0.864$

So $i_p = \tan^{-1}(0.864) = \mathbf{40.82°}$

Hence, polarising angle (i_p) is greater for a beam incident from water to glass.

Example 4 If the polarising angle of a piece of glass for green light is 60°, calculate the angle of minimum deviation for a 60° prism made of same glass.

Solution Given $i_p = 60°$

$$\mu = \tan i_p = \tan 60°$$

or $\mu = 1.732$

In case of prism $\mu = \dfrac{\sin\left[\dfrac{A+\delta m}{2}\right]}{\sin\dfrac{A}{2}}$, where δm is the angle of minimum deviation and A is the prism angle

Here, $A = 60°$ and $\mu = 1.732$ $\delta m = ?$

$$\therefore \quad \frac{\sin\left[\dfrac{60° + \delta m}{2}\right]}{\sin\dfrac{60°}{2}} = 1.732$$

or $$\sin\left[\frac{60° + \delta m}{2}\right] = 1.732 \times \frac{1}{2} = 0.866$$

$$\frac{60° + \delta m}{2} = 60°$$

or $\delta m = \mathbf{60°}$

EXAMPLE 5 Determine the Brewster's angle for a glass of refractive index 1.5 when it is immersed in water of refractive index 1.33.

SOLUTION Given $\mu_g = 1.5$ and $\mu_w = 1.33$

Therefore, the refractive index of glass w.r.t water = $\dfrac{1.5}{1.33} = 1.128$

By Brewster's law, $\tan i_p = \mu$, where i_p = Brewster's angle

$$i_p = \tan^{-1}\mu = \tan^{-1}(1.128)$$

$$\mathbf{i_p = 48.4°}$$

EXAMPLE 6 A ray of light is incident on a glass plate of refractive index 1.732 at a polarising angle. Find the angle of incidence and angle of refraction.

SOLUTION Given $\mu = 1.732$, $\tan i_p = \mu$, where i_p = angle of polarisation = angle of incidence

$$i_p = \tan^{-1}(\mu) = \tan^{-1}(1.732) = 60°$$

Now $\quad i_p + r = 90°$ or $r = 90° - 60° = \mathbf{30}$

So the angle of incidence is 60° and angle of refraction is 30°.

EXAMPLE 7 If the angle between a polariser and analyser is 60°, what will be the intensity of transmitted light for original intensity of incident light as I_0?

SOLUTION Given $\theta = 60°$

According to Malus' law $I = I_0 \cos^2\theta$

$$I = I_0 \cos^2 60 = I_0(0.5)^2$$

$$\mathbf{I = 0.25I_0}$$

EXAMPLE 8 Unpolarised light is incident on two polarising sheets placed one on top of the other. What must be the angle between the characteristics direction of sheets if the intensity of transmitted light is (i) 1/3 of maximum intensity of the transmitted beam and (ii) 1/3 of intensity of incident beam. Assume that sheet reduces the intensity of unpolarised light by exactly 50%.

SOLUTION Consider intensity of unpolarised light as I_0. The intensity of polarised light transmitted by the first sheet would be

$$I = \frac{1}{2}I_0$$

Case-I: $\quad \theta_1 = ?$

$$I_1 = \frac{1}{3}I$$

$$I_1 = I\cos^2\theta_1 \quad \text{or} \quad \frac{1}{3}I = I\cos^2\theta_1$$

or $\quad \cos\theta_1 = \dfrac{1}{\sqrt{3}}$

or $\quad \mathbf{\theta_1 = 54.74°}$

Case-II: $\quad I_2 = \dfrac{1}{3}I_0 = \dfrac{2}{3}I$

$$I_2 = I\cos^2\theta_2 \quad \text{or} \quad \frac{2}{3}I = I\cos^2\theta_2$$

or $\cos\theta_2 = \dfrac{\sqrt{2}}{\sqrt{3}}$ or $\theta_2 = \mathbf{35.26°}$

Example 9 Two Nicols have parallel polarising directions so that the intensity of transmitted light is maximum. Through what angle must either Nicol be turned if intensity is to drop by one-fourth of its maximum value?

Solution The transmitted intensity will be $\frac{3}{4}$ of incident intensity, i.e., $\frac{3}{4} I_0$

Malus' Law $I = I_0 \cos^2\theta$

or $\frac{3}{4} I_0 = I_0 \cos^2\theta$

$$\cos\theta = \frac{\sqrt{3}}{2}$$

or $\theta = \mathbf{30°}$

Example 10 Two Nicol prisms are so arranged that the amount of light transmitted through them is maximum. What will be the percentage reduction in the intensity of the incident light when the analyser is rotated through (i) 30°, (ii) 45° (iii) 60° and (iv) 90°?

Solution

$$I = I_0 \cos^2\theta \text{ Malus Law} \tag{i}$$

$$\cos^2\theta = \frac{I}{I_0} \tag{ii}$$

where I_0 is the intensity of incident light and I is the intensity of transmitted light.

Therefore, percentage reduction in the intensity of incident light is

$$= \frac{I_0 - I}{I_0} \times 100 = \left(1 - \frac{I}{I_0}\right) \times 100 \tag{iii}$$

By using Eqs. (ii) and (iii), we get percentage reduction in intensity

$= (1 - \cos^2\theta) \times 100$

(i) For $\theta = \mathbf{30°}$, Percentage reduction in intensity

$= (1 - \cos^2 30°) \times 100$

$= [1 - (0.866)^2] \times 100$

$= \mathbf{25\%}$

(ii) For $\theta = \mathbf{45°}$, %reduction in intensity

$= (1 - \cos^2 45°) \times 100 = \mathbf{50\%}$

(iii) For $\theta = \mathbf{60°}$, % reduction in intensity

$= (1 - \cos^2 60) = \mathbf{75\%}$

and,

(iv) For $\theta = \mathbf{90°}$, % reduction in intensity

$= (1 - \cos^2 90) = \mathbf{100\%}$

Example 11 Two polaroids are adjusted so as to obtain maximum intensity. Through what angle should one polaroid be rotated to reduce the intensity to (i) half (ii) one fourth.

SOLUTION Formula used is $I = I_0 \cos^2\theta$ (i)

or $$\frac{I}{I_0} = \cos^2\theta \quad \text{(ii)}$$

Case (i) $$\frac{I_1}{I_0} = \frac{1}{2}$$

Then, by Eq. (i), we have $\cos^2\theta = \dfrac{1}{2}$

or $$\cos\theta = \frac{1}{\sqrt{2}}$$

or $$\boldsymbol{\theta = 45°}$$

Case (ii) $$\frac{I_1}{I_0} = \frac{1}{4}$$

$\therefore$ $$\cos^2\theta = \frac{1}{4} \quad \text{[using eqn (i)]}$$

$$\cos\theta = \frac{1}{2}$$

$$\boldsymbol{\theta = 60°}$$

EXAMPLE 12 Calculate the thickness of a half-wave plate of a quartz for a wavelength of 5000 Å. Here $\mu_e = 1.553$ and $\mu_0 = 1.544$

SOLUTION Given $\lambda = 5000$ Å, $\mu_e = 1.553$ and $\mu_0 = 1.544$

For Half-wave plate $$t = \frac{\lambda}{2(\mu_e - \mu_0)} = \frac{5.0 \times 10^{-7}}{2 \times (1.553 - 1.544)} = \mathbf{2.78 \times 10^{-5}\ m.}$$

EXAMPLE 13 Calculate the thickness of quarter-wave plate for light of wavelength 5893 Å, given refractive indices for ordinary ray and extraordinary ray as 1.544 and 1.533, respectively.

SOLUTION Given $\mu_o = 1.554$, $\mu_e = 1.533$ and $\lambda = 5.893 \times 10^{-7}$ m

Formula used is $$t = \frac{\lambda}{4(\mu_0 - \mu_e)} = \frac{5.893 \times 10^{-7}}{4 \times (1.554 - 1.533)} = 70.15 \times 10^{-7}$$

$$= 7.02 \times 10^{-6}\ \text{m.} = \mathbf{7.02\ \mu m}$$

EXAMPLE 14 Calculate the thickness of a quarter-wave plate of wavelength 5890 Å.
(i) Given $\mu_o = 1.55$ and $\mu_e = 1.50$ (ii) Given $\mu_o = 1.55$ and $\mu_e = 1.57$

SOLUTION Given $\lambda = 5.89 \times 10^{-7}$ m

(i) Formula used is $$t = \frac{\lambda}{4(\mu_0 - \mu_e)} = \frac{5.89 \times 10^{-7}}{4 \times (1.55 - 1.50)}$$

$$= 2.945 \times 10^{-6}\ \text{m}$$

$$\boldsymbol{t = 2.945\ \mu m}$$

(ii) $$t = \frac{\lambda}{4(\mu_e - \mu_0)} = \frac{5.89 \times 10^{-7}}{4 \times (1.57 - 1.55)}$$

$$= 7.36 \times 10^{-6}\ \text{m}$$

or $$\boldsymbol{t = 7.36\ \mu m}$$

EXAMPLE 15 Calculate the thickness of a calcite plate which would convert plane-polarised light into circularly-polarised light. The principal refractive indices are $\mu_o = 1.658$ and $\mu_e = 1.486$ at the wavelength 5890Å of light used.

SOLUTION Given $\mu_o = 1.658$, $\mu_e = 1.486$ and $\lambda = 5.89 \times 10^{-7}$ m

To convert plane-polarised light into circularly polarised light, path difference must be $\lambda/4$.

Hence,
$$t = \frac{\lambda}{4(\mu_0 - \mu_e)} = \frac{5.89 \times 10^{-7}}{4 \times (1.658 - 1.486)}$$
$$= \mathbf{8.56 \times 10^{-7}\ m}$$

EXAMPLE 16 Plane-polarised light passes through a quartz plate with its optic axis parallel to the faces. Calculate the least thickness of the plate for which the emergent beam will be plane-polarised. (Given $\mu_e = 1.5533$, $\mu_o = 1.5442$ and $\lambda = 5 \times 10^{-5}$ cm).

SOLUTION Given $\mu_e = 1.5533$, $\mu_0 = 1.5442$, and $\lambda = 5 \times 10^{-5}$ cm

$$t = \frac{\lambda}{2(\mu_e - \mu_0)} = \frac{5.0 \times 10^{-7}}{4 \times (1.5533 - 1.5442)}$$
$$= \mathbf{1.37 \times 10^{-5} m}$$

EXAMPLE 17 Find the thickness of a quarter-wave plate when the wavelength of light is equal to 5890Å $\mu_0 = 1.55$ and $\mu_e = 1.54$.

SOLUTION Given $\lambda = 5.89 \times 10^{-7}$ m, $\mu_o = 1.55$ and $\mu_e = 1.54$

Formula used is
$$t = \frac{\lambda}{4(\mu_0 - \mu_e)} = \frac{5.89 \times 10^{-7}}{4(1.55 - 1.54)}$$
$$\mathbf{t = 1.47 \times 10^{-5}\ m}$$

EXAMPLE 18 Quartz has refractive indices 1.553 and 1.544. Calculate the thickness of the quarter-wave plate for sodium light of wavelength 5890 Å.

SOLUTION Given $\mu_e = 1.553$ and $\mu_o = 1.544$

$$\lambda = 5.89 \times 10^{-7}\ \text{m}$$
$$t = \frac{5.89 \times 10^{-7}}{4 \times (1.553 - 1.544)}$$
$$\mathbf{t = 1.63 \times 10^{-5} m}$$

EXAMPLE 19 Plane-polarised light ($\lambda = 5 \times 10^{-7}$ m) is incident on a quartz plate cut parallel to the optic axis. Find the least thickness of the plate for which the ordinary and extraordinary rays combine to form a plane-polarised light on emergence. What multiples of this thickness would give the same result? The indices of the refraction of quartz are $\mu_e = 1.5533$ and $\mu_0 = 1.5542$.

SOLUTION Given $\mu_o = 1.5442$, $\mu_e = 1.5533$ and $\lambda = 5 \times 10^{-7}$ m

In the given case the quartz plate must act as a half-wave ($\lambda/2$) plate, then formula used is

$$t = \frac{\lambda}{2(\mu_e - \mu_0)} = \frac{5 \times 10^{-7}}{2(1.5533 - 1.5442)}$$

or $\quad t = \mathbf{2.75 \times 10^{-5}}$ **m**

The thickness that would give the similar result should be t, $3t$, $5t$ ... i.e., 2.75×10^{-5} m, 8.25×10^{-5} and so on.

EXAMPLE 20 On introducing a polarimeter tube 25 cm long and containing sugar solution of unknown strength, it is found that the plane of polarisation is rotated through 10°. Find the strength of the sugar solution in g/cm^3 (Given that the specific rotation of sugar solution is 60° per decimeter per unit concentration).

SOLUTION Given $\theta = 10°$, $S = 60°$ and $l = 25$ cm

Formula used is $s = \dfrac{\theta}{lc}$ where l is in decimeter

or $\quad s = \dfrac{10\theta}{lc}$ for l in cm

or $\quad c = \dfrac{10\theta}{ls} = \dfrac{10 \times 10}{25 \times 60}$

or $\quad \mathbf{c = 0.067}$ **g/cc**

EXAMPLE 21 Compute the specific rotation if the plane of polarisation is turned through 26.4°, traversing 20 cm length of 20% sugar solution.

SOLUTION Given $\theta = 26.4°$, $l = 20$ cm and $c = 20\% = 0.2$ g/cm^3

Formula used is $\quad S = \dfrac{10\theta}{lc} = \dfrac{10 \times 26.4}{20 \times 0.2}$

$\mathbf{S = 66°}$

EXAMPLE 22 The plane of polarisation of plane-polarised light is rotated through 6.5° in passing through a length of 2.0 decimeter of sugar solution of 5% concentration. Calculate the specific rotation of the sugar solution.

SOLUTION Given $\theta = 6.5°$, $l = 2$ dm and $c = 5\% = 0.05$ g/cc

Formula used is $\quad S = \dfrac{\theta}{lc} = \dfrac{6.5}{2 \times 0.05}$

$= \mathbf{65° \ (dm)^{-1} \ (gm/cc)^{-1}}$

EXAMPLE 23 80 gm of impure sugar when dissolved in a litre of water gives an optical rotation of 9.9° when placed in a tube of length 20 cm. If the specific rotation of sugar is 66°, find the percentage purity of the sugar sample.

SOLUTION Given $\theta = 9.9°$, $l = 20$ cm $= 2.0$ dm and $S = 66°$

Formula used $\quad S = \dfrac{\theta}{lc}$ or $c = \dfrac{\theta}{ls}$

or $\quad c = \dfrac{9.9}{2.0 \times 66} = 0.075$ g/cc

$= \mathbf{75}$ **gm/L [1 litre = $\mathbf{10^3}$ cc]**

80 g impure sugar is dissolved in one litre of water in which 75 g sugar is pure. Therefore, percentage of pure sugar is

$$\frac{75}{80} \times 100 = \mathbf{93.75\%}$$

EXAMPLE 24 A 20 cm long tube containing sugar solution rotates the plane of polarisation by 11°. If the specific rotation of sugar is 66°, determine the strength of the solution.

SOLUTION Given $\theta = 11°$, $l = 20$ cm and $S = 66°$

Formula used is $S = \dfrac{10\theta}{lc}$ or $c = \dfrac{10\theta}{lS} = \dfrac{10 \times 11}{20 \times 66}$

$$= \mathbf{0.0833\ g/cm^3}$$

EXAMPLE 25 Calculate the specific rotation if the plane of polarisation is turned through 26.4°, traversing 20 cm length of 20% sugar solution.

SOLUTION Given $\theta = 26.4°$, $l = 20$ cm, ad $c = 20\% = 0.2$ g/cc

Since l is given in cm, specific rotation $S = \dfrac{10\theta}{lc} = \dfrac{10 \times 26.4}{20 \times 0.2} = \mathbf{66°}$

EXAMPLE 26 A sugar solution in a tube of length 20 cm produces optical rotation of 13°. The solution is then diluted to one-third of its previous concentration. Find the optical rotation produced by 30 cm long tube containing the diluted solution.

SOLUTION $l = 20$ cm, $\theta = 13°$, $c' = \dfrac{c}{3}$ and $l' = 30$ cm

Formula used is $S = \dfrac{10 \times \theta}{lc}$

Therefore, $S = \dfrac{10\theta}{lc} = \dfrac{10\theta'}{l'c'}$

$$\theta' = \left(\frac{l'}{l}\right)\left(\frac{c'}{c}\right)\theta = \left(\frac{30}{20}\right)\left(\frac{c/3}{c}\right)13°$$

$$= \frac{3}{2} \times \frac{1}{3} \times 13° = \theta' = \mathbf{6.5°}$$

OBJECTIVE TYPE QUESTIONS

Q.1 Which of the following phenomenon tells about the transverse nature of light waves?
(a) interference (b) diffraction (c) polarisation (d) photoelectric effect

Q.2 Plane-polarised light has vibrations
(a) in one direction perpendicular to the direction of propagation
(b) along the direction of propagation
(c) in all directions perpendicular to the direction of propagation
(d) in two directions perpendicular to the direction of propagation

Q.3 Polarised light can be produced by
(a) reflection (b) refraction (c) double refraction (d) all of them

Q.4 At polarising angle, the reflected and refracted rays make angle
(a) 90° (b) 180° (c) 30° (d) none of these

Q.5 Brewster's law in terms of refractive index can be expressed as
(a) $\mu = \sin i_p$ (b) $\mu = \cos i_p$ (c) $\mu = \tan i_p$ (d) $\mu = \cot i_p$

Q.6 According to the Malus' Law, the intensity of polarised light emerging through the analyser varies as

(a) $I_0 \cos^2\theta$ (b) $I_0 \sin^2\theta$ (c) $I_0 \cos\theta$ (d) $\frac{I_0}{2}\cos^2\theta$

Q.7 Malus' discovered a simplest method for polarisation of light by reflection in the year

(a) 1808 (b) 1908 (c) 1856 (d) none of these

Q.8 Which one is the example of uniaxial crystal?

(a) calcite (b) tourmaline (c) quartz (d) all of them

Q.9 Which one is the example of biaxial crystal?

(a) sodium chloride (b) tourmaline (c) aragonite (d) none of them

Q.10 What happens if the ordinary unpolarised light is passed through a uniaxial crystal?

(a) light is split into two rays (b) light remains unaffected
(c) light is split into more than two rays (d) none of them

Q.11 What happens to O and E-rays if they travel along the optic axis?

(a) both rays travel with same velocity (b) O-ray travels faster than E-ray
(c) E-ray travels faster than O-ray (d) none of them.

Q.12 How many principal sections of uniaxial crystal has?

(a) 6 (b) 3 (c) 5 (d) 2

Q.13 At what angle of incidence of plane-polarised light with quarter-wave plate elliptically polarised light becomes circularly polarised?

(a) 90° (b) 45° (c) 60° (d) 30°

Q.14 How much phase change is introduced by a quarter-wave plate between ordinary and extraordinary rays?

(a) π (b) 2π (c) $\pi/2$ (d) $\pi/4$

Q.15 Dextrorotatory optically active substance rotates the plane of vibrations

(a) in clockwise direction (b) in anti-clockwise direction
(c) by 180° (d) none of them

Q.16 Which of the following relation is true for quartz crystal?

(a) $\mu_e > \mu_o$ (b) $\mu_o > \mu_e$ (c) $\mu_o = \mu_e$ (d) none of these

Q.17 Which of the following relation is true for quartz crystal?

(a) $v_o > v_e$ (b) $v_e > v_o$ (c) $v_e = v_o$ (d) none of these

Q.18 The substance that is capable of rotation of plane of vibration is known as

(a) optically active (b) optically inactive (c) both (a) and (b) (d) none of these

Q.19 If two polarising and analysing nicols are at 90° than, the emergent light passed through the analysing nicol becomes

(a) maximum (b) minimum (c) zero (d) none of these

SHORT-ANSWER QUESTIONS

Q.1 What do you understand by polarisation?

Q.2 Experimentally show that only the transverse waves can be polarised.

Q.3 What is the difference between polarised and unpolarised light?

Q.4 Distinguish between plane of vibration and plane of polarisation.
Q.5 What is Brewster's law?
Q.6 State Malus' law?
Q.7 What is the relation between angle of polarisation and refractive index?
Q.8 What do you understand by double refraction?
Q.9 What are ordinary and extraordinary rays?
Q.10 What is optic axis?
Q.11 What is principal section of the crystal?
Q.12 Differentiate between uniaxial and biaxial crystals?
Q.13 What is Nicol prism?
Q.14 Distinguish between linearly, circularly and elliptically polarised light.
Q.15 What are quarter-wave and half-wave plates?
Q.16 What do you mean by optical activity?
Q.17 What do you understand by specific rotation?
Q.18 What do you understand by photoelasticity?
Q.19 Write a short note on
(a) Polarised and unpolarised light,
(b) Double refraction,
(c) Nicol prism,
(d) Photoelasticity,
(e) Quarter and half-wave plate,
(f) Biquartz polarimeter,
(g) Half-shade polarimeter.

PRACTICE PROBLEMS

Q.1 Explain polarisation on the basis of electromagnetic theory of light.
Q.2 Explain the method and cause of production of plane-polarised light by reflection.
Q.3 State and derive Brewster's law. What does the law become when the rays of light travel from denser medium to rarer medium?
Q.4 By Brewster's law show that light incident on a transparent substance at polarising angle gives reflected and refracted rays at right angle to each other.
Q.5 How can you explain the transverse nature of a wave with the help of phenomenon of polarisation using mechanical and light experiments?
Q.6 Distinguish between polarised light and unpolarised light. Also define plane of vibration and plane of polarisation.
Q.7 How can you obtain linearly polarised light from a beam of a non-polarised light?
Q.8 How would you obtain plane-polarised light? Name various techniques and discuss one out of them which is based on double refraction phenomenon.
Q.9 What do you understand by double refraction? Explain how would you use the phenomenon to produce plane-polarised light and circularly polarised light?

Q.10 How do you use the phenomenon of double refraction to produce a plane-polarised light? Explain in detail.

Q.11 (a) Explain the phenomenon of double refraction in uniaxial crystal. (b) What are quarter-wave and half-wave plates? Explain their use in the study of different types of polarised light.

Q.12 Explain the phenomenon of polarisation of light. Describe the construction of a Nicol Prism, and show how it can be used as a polariser and as an analyser.

Q.13 Discuss the principle, construction and working of Nicol prism as polarised.

Q.14 Draw diagrams and discuss double refraction through uniaxial crystals due to a plane wave when

(i) optic axis is inclined to the upper face but lying in the plane of incidence.
(ii) optic axis is parallel to the upper face but lying in the plane of incidence.
(iii) optic axis is parallel to the upper face but perpendicular to the plane of incidence.
(iv) optic axis is perpendicular to the upper face.

Q.15 What do you understand by optical rotation? Explain Fresnel's theory of the rotation of the plane of polarisation. How would you increase the sensitivity of a pair of crossed Nicols?

Q.16 How would you distinguish between circularly polarised light and unpolarised light?

Q.17 How would you distinguish between plane, circularly and elliptically polarised light?

Q.18 What are plane-polarised circularly polarised and elliptically polarised light? Explain their production with the help of mathematical equations. Give the salient features of biquartz device.

Q.19 Give two differences between Laurent's half shade polarimeter and biquartz polarimeter.

Q.20 Define specific rotation. Describe the construction and working of Laurent's half-shade polarimeter. Discuss the relative merits of biquartz polarimeter and half-shade polarimeter.

Q.21 What is optical activity? Describe the construction, theory and working of biquartz polarimeter to find the optical rotation of a solution and also discuss the action of biquartz plate in it.

Q.22 What do you understand by a half and quarter-wave plate? Give the theory and construction of Laurent's half-shade polarimeter.

Q.23 What is specific rotation? Describe the construction and working of biquartz polarimeter to find the specific rotation of sugar solution and discuss the utility of biquartz plate in it.

UNSOLVED QUESTIONS

Q.1 A beam of light is incident on a glass plate at an angle of 58°6′ and the reflected beam is completely plane-polarised. Find the refractive index of glass. **[Ans: 1.6]**

Q.2 Refractive index of water is 1.33. Calculate the angle of polarisation for light reflected from the surface of a pond. **[Ans: 53.06°]**

Q.3 Critical angle for refraction for glass to air is 40°. Calculate the polarising angle for glass. **[Ans: 57.3°]**

Q.4 A beam of light traveling in water strikes a glass plate which is also immersed in water. When the angle of incidence is 51°, the reflected beam is found to be plane-polarised. Calculate the refractive index of glass. **[Ans: 1.235]**

Q.5 A polariser and an analyser are set in such a way that the intensity of the emergent light is maximum. What percentage of the maximum intensity of light is transmitted from the analyser if either is rotated by 30°, 45° and 60°? **[Ans: 75%, 50%, 25%]**

Q.6 Two Nicols have parallel polarising directions so that the intensity of transmitted light is maximum. Through what angle must either Nicol be turned if the intensity is to drop by one fourth of its maximum value? **[Ans: 30°]**

Q.7 An analysing Nicol examines two adjacent plane-polarised beams *A* and *B* whose planes of polarisaton are mutually perpendicular. In one postion of the analyser, beam *B* shows zero intensity. From this position a rotation of 30° shows the two beams as matched (i.e., of equal intensity). Deduce the intensity ratio I_A/I_B of the two beams **[Ans: 1/3]**

Q.8 Find the thickness of a quarter-wave plate for the wavelength 5890Å of light, when $\mu_0 = 1.55$ and $\mu_e = 1.54$. **[Ans: 1.4725×10^{-5} m]**

Q.9 Find the thickness of calcite plate which would convert plane-polarised light into circularly polarised light. The refractive indices are $\mu_0 = 1.658$ and $\mu_e = 1.486$ at the wavelength of light used as 5890Å. **[Ans: 8.56×10^{-9} m]**

Q.10 Calculate the thickness of a quarter-wave plate of quartz for sodium light of wavelength 5893Å. The refractive indices of quartz for ordinary and extra-ordinary rays are 1.5442 and 1.5533 respectively. **[Ans: 1.61×10^{-5} m]**

Q.11 Calculate the thickness of a doubly refracting crystal plate required to introduce a path difference of $\lambda/2$ between the ordinary and extraordinary rays when $\lambda = 6000$Å, $\mu_0 = 1.55$ and $\mu_e = 1.54$. **[Ans: 5×10^{-5} m]**

Q.12 Calculate the thickness of (i) a quarter-wave plate and (ii) a half-wave plate. Given that $\mu_e = 1.553$ and $\mu_0 = 1.544$ and $\lambda = 5000$Å. **[Ans: (i) 1.39×10^{-5} m (ii) 2.78×10^{-5} m]**

Q.13 A plane-polarised light is incident on a piece of quartz cut parallel to the axis. Find the least thickness for which the ordinary and extraordinary rays combine to form plane-polarised light. Given $\mu_0 = 1.5442$, $\mu_e = 1.55$ and $\lambda = 5 \times 10^{-5}$ cm. **[Ans: 2.75×10^{-4} m]**

Q.14 For calcite $\mu_0 = 1.658$ and $\mu_e = 1.486$ for sodium light. Calculate the minimum thickness of the quarter-wave plate for calcite. **[Ans: $t = 8.56 \times 10^{-7}$ m]**

Q.15 A 20 cm long tube containing sugar solution rotates the plane of polarisation by 11°. If the specific rotation of sugar is 66°, calculate the strength of the solution. **[Ans: 0.0833 g/cc]**

Q.16 A 200 mm long tube containing 48 cm^3 of sugar solution produces an optical rotation of 11° when placed in a saccharimeter. If the specific rotation of sugar solution is 66°, calculate the quantity of sugar contained in the tube in the form of a solution. **[Ans: 4.0 g]**

Q.17 A 20 cm long tube is filled with a solution of 15 g of cane sugar in 100 cc of water. Find the angle of rotation of the plane of polarisation of a beam of plane-polarised light when it passes through the solution. Specific rotation for cane sugar = 66.50 per dm/g per cm^3. **[Ans: 20°]**

Lasers and Holography

4

LEARNING OBJECTIVES

After reading this chapter you will be able to

LO1 Learn about absorption of radiation and different types of emissions

LO2 Understand the phenomenon of population inversion and characteristics of laser light

LO3 Know about the components and types of lasers.

LO4 Discuss the application of laser and laser cooling

LO5 Explain holography versus conventional photography, recording and reconstruction of image on a holograph

LO6 Illustrate types of holograms

LO7 Evaluate the applications of holography

Introduction

In the previous chapters, interesting phenomena of interference and diffraction of light including its polarisation have been investigated in detail. It was discussed that the interference has scientific as well as engineering applications. The concept of interference is applied to testing the surface quality of optical components and this led to the development of flatness interferometers. An exciting use of the concept of interference is made in the preparation of nonreflecting or antireflecting coatings that are applied to surfaces of lenses (for example, eye glass lenses) and other optical devices for reducing the reflections and hence in improving the efficiency of the system like telescope. However, you would have learnt that in order to realise the above mentioned phenomena in an efficient way there is a need of using the coherent and monochromatic sources as the phase of incoherent source (light) varies randomly with time and position. This need of monochromatic and coherent sources contributed to the birth of a special type of device that amplifies light and produces a highly intense and highly directional beam which mostly has a very pure wavelength. This device is called LASER. Lasers are available with power ranging roughly from 1 nW (= 10^{-9} W) to 10^5 PW (1 PW = 10^{-15} W) and with frequency ranging from 100 GHz (1 GHz = 10^9 Hz) to 100 PHz. Nowadays the lasers with pulse duration as short as ~ 1 fs (= 10^{-15} s) are available with their pulse energies as high as 10 kJ.

The name LASER is an acronym of Light Amplification by Stimulated Emission of Radiation. The immediate originator to the LASER is the MASER, formerly acronym of Microwave Amplification by Stimulated Emission of Radiation. Since the techniques have been extended to the infrared and optical regions, it has now come to stand for Molecular rather than Microwave amplification. A laser uses some processes that amplify light signals. These processes mainly include stimulated emission and optical feedback provided by mirrors. The stimulated emission takes place in amplifying medium contained by the laser. The application of set of mirrors is to feed the light back to the amplifying medium so that the developed beam is grown continuously. The key concept for realisation of the laser operation is the principle of coherence accompanying stimulated emission.

4.1 ABSORPTION AND EMISSION OF RADIATION — LO1

It is well known that an atom can be excited by supplying energy with an amount equal to the difference of its any two energy levels. Then after a very short duration of time the atom shall radiate energy when it comes down to its lower energy state. An electron undergoes a transition between two energy states E_1 and E_2 if the atom emits or absorbs a photon of appropriate energy as per the relation $E_2 - E_1 = h\nu$, where h is Planck constant and ν is the frequency of radiation.

4.1.1 Absorption of Radiation

At low temperatures, most of the atoms stay in lower energy states. If an atom is initially in the lower energy state $E_{1,}$ it can be raised to the higher energy state E_2 by the absorption of a photon of energy $h\nu$, as shown in Fig. 4.1. This is know as absorption of radiation and is represented by

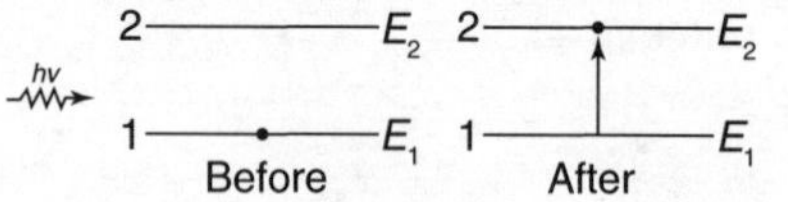

FIGURE 4.1

$$E_2 = E_1 + h\nu$$

$$\Rightarrow \quad E_2 - E_1 = \Delta E = h\nu \qquad \text{(i)}$$

The probability of occurrence of this absorption from state 1 to state 2 is proportional to the energy density $u(\nu)$ of the radiation

$$P_{12} = B_{12}\, u(\nu) \qquad \text{(ii)}$$

Where the proportionality constant B_{12} is known as the *Einstein's coefficient of absorption of radiation*.

Take an example of electron transition associated with visible and ultraviolet radiation interactions with matter. Here the absorption of a photon occurs only when the quantum energy of the photon precisely matches the energy gap between the initial and final states. In such interaction of radiation with matter, if there is no pair of energy state such that the photon energy can elevate the system from the lower to upper state, then the matter will be transparent to that radiation.

4.1.2 Spontaneous Emission

If an atom is initially in the upper state E_2, it can come down to lower state E_1 by emitting a photon of energy $h\nu$ as shown in Fig. 4.2. This is known as *spontaneous emission*. This is the natural radiation

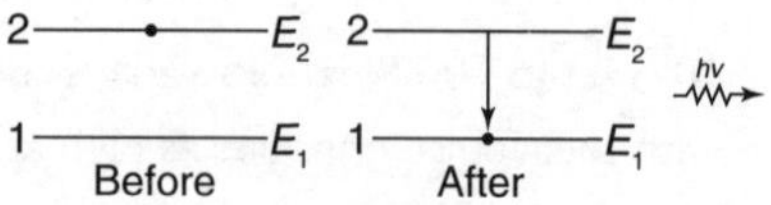

FIGURE 4.2

decay process that is inherent in all excited states of all materials. However, such emission is not always the dominant decay process.

The probability of occurrence of this spontaneous emission transition from state 2 to state 1 depends only on the properties of states 2 and 1 and is given by

$$P'_{12} = A_{21} \tag{iii}$$

Where A_{21} is known as the *Einstein's coefficient of spontaneous emission of radiation.*

4.1.3 Stimulated (Induced) Emission

Einstein was the first to point out a third possibility of induced emission, in which an incident photon of energy $h\nu$ causes a transition from upper state E_2 to the lower state E_1, as shown in Fig. 4.3. This occurs when

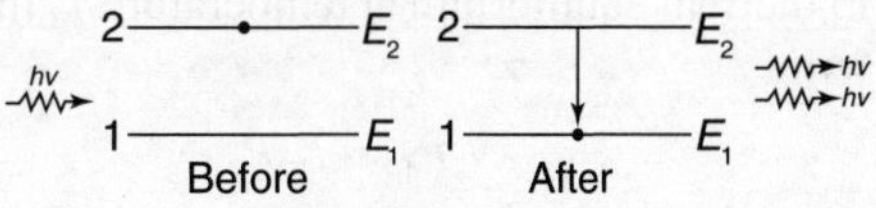

FIGURE 4.3

$$h\nu = \Delta E = E_2 - E_1$$

In a system of atoms in thermal equilibrium, the number of atoms in the ground state is generally much greater than in a higher energy state. This is known as *normal population* of atoms among the available energy states. A state in which the number of atoms in higher energy state is greater than that of lower energy state is known *population inversion.*

Therefore, the incoming photon stimulates the transition to the lower state and produces a second photon of the same energy, when a sizable population of electrons resides in upper level (Fig. 4.4). In this condition it is called a population inversion. This population inversion sets the stage for stimulated emission of multiple photons. This is the precondition for the light amplification in a laser. Since the emitted photons have a definite time and phase relation to each other, the light has a high degree of coherence. If these emitted photons are passed through an assembly of atoms, which fulfil the condition of population inversion, these are amplified. This amplification is very much clear from Fig. 4.4, which shows multiplication of photons emitted during the process of stimulated emission.

(energy of each photon is hv)

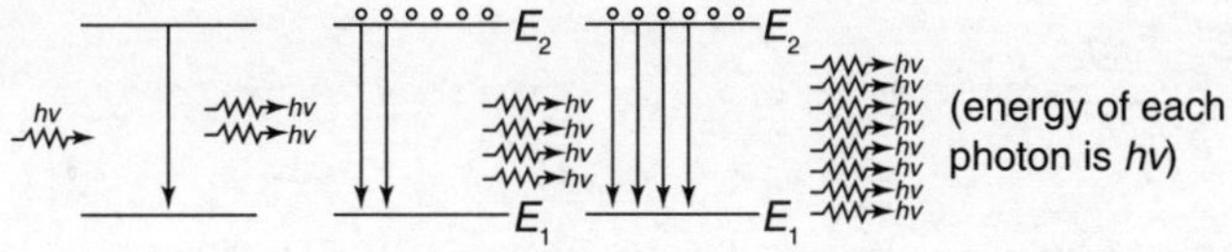

FIGURE 4.4

The probability of occurrence of stimulated emission transition from the upper level 2 to the lower level 1 is proportional to the energy density $u(\nu)$ of the radiation and is expressed as

$$P''_{21} = B_{21}\, u(\nu) \tag{iv}$$

Where B_{21} is the *Einstein's coefficient of stimulated emission of radiation.*

Thus, the total probability of emission transition from the upper level 2 to the lower level 1 is given by

$$P_{21} = P'_{21} + P''_{21}$$

or $$P_{21} = A_{21} + B_{21}\, u(\nu) \tag{v}$$

4.1.4 Relation between Einstein's Coefficients

Let N_1 and N_2 be the number of atoms at any instant in the state 1 and 2, respectively. The probability of absorption transition for number of atoms from state 1 to 2 per unit time is given by

$$N_1P_{12} = N_1B_{12}\, u\,(\nu) \tag{vi}$$

The total probability of transition for number of atoms from state 2 to 1, either by spontaneously or by stimulated emission per unit time is given by

$$N_2P_{21} = N_2[A_{21} + B_{21}\, u(\nu)] \tag{vii}$$

In thermal equilibrium at temperature T, the absorption and emission probabilities are equal and thus, we can write

$$N_1P_{12} = N_2P_{21}$$

or $$N_1B_{12}\, u(\nu) = N_2[A_{21} + B_{21}\, u(\nu)]$$

or $$u(\nu) = \frac{N_2A_{21}}{N_1B_{12} - N_2B_{21}}$$

or $$u(\nu) = \frac{A_{21}}{B_{21}}\frac{1}{(N_1/N_2)(B_{12}/B_{21}) - 1} \tag{viii}$$

But according to Einstein

$$B_{12} = B_{21} \tag{ix}$$

Then from Eq. (viii) and (ix), we get

$$u(\nu) = \frac{A_{21}}{B_{21}}\frac{1}{(N_1/N_2) - 1} \tag{x}$$

According to Boltzmann's law, the distribution of atoms among the energy states E_1 and E_2 at the thermal equilibrium at temperature T is given by

$$\frac{N_1}{N_2} = \frac{e^{-E_1/kT}}{e^{-E_2/kT}} = e^{(E_2 - E_1)/kT} \tag{xi}$$

or $$\frac{N_1}{N_2} = e^{h\nu/kT} \tag{xii}$$

Here k is the Boltzmann constant.

From Eq. (x), we can write

$$u(\nu) = \frac{A_{21}}{B_{21}}\frac{1}{e^{h\nu/kT} - 1} \tag{xiii}$$

Planck's radiation formula yields the energy density of radiation $u(\nu)$ as

$$u(\nu) = \frac{8\pi h\nu^3}{c^3}\frac{1}{e^{h\nu/kT} - 1} \tag{xiv}$$

Comparing Eq. (xiii) and (xiv), we get

$$\frac{A_{21}}{B_{21}} = \frac{8\pi h \nu^3}{c^3} \tag{xv}$$

Equation (xv) gives the relation between the probabilities of spontaneous and stimulated emissions. This is also known as the relation between the Einstein's coefficients *A* and *B*.

4.2 POPULATION INVERSION LO2

For a system with three energy states viz. E_1 (population N_1), E_2 (population N_2) and E_3 (population N_3) in equilibrium such that $E_1 < E_2 < E_3$, the uppermost level E_3 is populated least whereas the lowest level E_1 is populated most. Since the population in these states follow the trend $N_1 > N_2 > N_3$, the system shall absorb photons rather than emitting them. However, when a sizable population of electrons is achieved in the upper levels, the condition is known as population inversion (a non-equilibrium state). This condition sets the stage for stimulated emission of radiation, i.e., multiple photons, as the first few randomly emitted spontaneous photons trigger stimulated emission of more photons and those stimulated photons induce still more stimulated emissions, and so on.

The population inversion is the precondition for the light amplification occurring in LASER. In order to achieve this condition, a multilevel scheme is used. For example, the atoms are pumped into the highest level of the three levels. Then spontaneous de-excitation from this pumped level to the metastable level takes place and the laser emission occurs between the metastable level and the ground state. It is clear that energy has to be supplied to the laser medium in order to raise atoms from the lower level to the excited level and for maintaining population at the excited level at a value greater than that of the lower energy. One can think that heating the material can solve this purpose but this only increases the average energy of the atoms and does not enhance the population in the higher level. However, following schemes may be adopted to achieve the population inversion.

4.2.1 Schemes for Population Inversion

To discuss schemes for the population inversion in various energy level systems, we will first prove that the two-level system is not appropriate to achieve this condition of population inversion.

4.2.1.1 Two-level System

Consider the case of two-level system having energies E_1 and E_2 such that $E_2 > E_1$. We can easily find that the Einstein coefficients (or constants) for the upward (B_{12}) and downward (B_{21}) transitions are equal, i.e., $B_{12} = B_{21}$. It means, even with strong pumping, the population distribution in upper and lower levels can only be made equal, i.e., the optical pumping will at most only achieve equal population of a two-level system. This is due to the fact that the probabilities for raising an electron to the upper level and inducing the decay of an electron to the lower level (stimulated emission) are exactly the same. In other words, we can say that the numbers of electrons going up and coming down will be the same when both the levels are equally populated. So, we cannot achieve population inversion in the case of two energy levels system. Therefore, optical as well as any other pumping method needs either three or four level systems to attain population inversion. The solution is to use a third metastable level, where the electrons can stay for longer duration. Under this situation, the pumping will be between the other two levels and the electrons in the upper energy level will quickly decay into the metastable level, leaving the upper level practically unpopulated at all times. The transition from the metastable level to the ground level has a different frequency, which is the laser frequency.

The pumping frequency is between the upper level and the ground level. Thus the pumping is off-resonant to the laser transition and it will not trigger the stimulated emission.

4.2.1.2 *Three-Level System*

Bloembergen proposed a mechanism where atoms are pumped into an excited state by an external source of energy, for example by an electric pulse or an optical illumination. In addition to this excited state (say E_3), the system has a metastable state (say E_2) and the atoms from the upper level E_3 decays spontaneously to this metastable state and this transition is generally radiation less or non-radiative (the energy being given away to the lattice). The lifetime of the electrons in the metastable state E_2 is such that the rate of spontaneous decay from the upper level E_3 to the ground level (say E_1) is slower than the rate at which the atoms decay from the upper level to the metastable state, resulting in a population inversion between the metastable level and the ground state (Fig. 4.5). The population inversion can be achieved only by pumping into a higher lying level, followed by a rapid radiative or non-radiative transfer into the upper laser level. This is because in this way we can avoid the stimulated emission caused by the pump wave. The emitted photons here are confined to a laser cavity to stimulate further the emission from the excited atoms. Larger width of the excited level can make possible the absorption of a wider range of wavelengths to make pumping more effective. Ruby laser works on the principle of a three-level system.

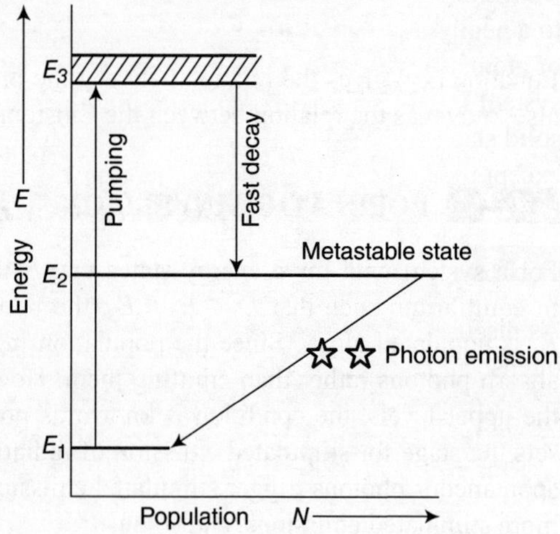

Fig. 4.5

Since the lower level involved in the lasing (population inversion) is the ground state of the atom, the three level system needs very high pumping power and yields low efficiency. Here more than half of the total number of the atoms have to be pumped to the excited state E_3 before achieving population inversion. The energy used to do this in each of the cycle is wasted. However, the pumping power can be reduced significantly if the lower level involved in the lasing is not the ground state. This will require at least a four-level system (Fig. 4.6). Here, the pumping will transfer atoms from the ground state E_1 to an excited state E_4, from where they decay rapidly into the metastable state E_3 to make population N_3 larger than population N_2 to achieve the condition of population inversion between E_3 and E_2 at moderate pumping.

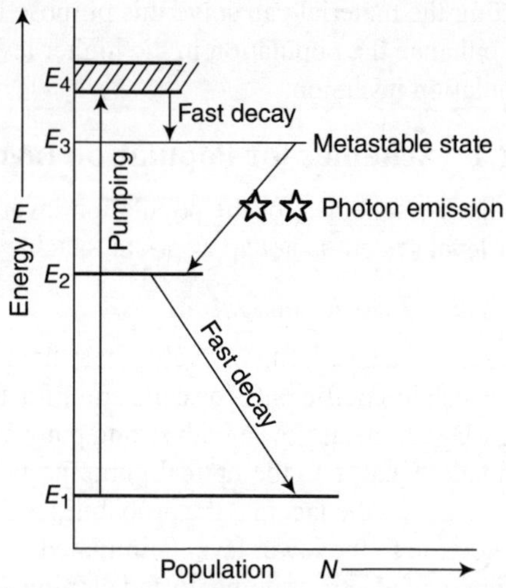

Fig. 4.6

4.2.1.3 *Four-Level System*

The schematic of four-level system is depicted in Fig. 4.6 where four energy levels having energies E_1, E_2, E_3 and E_4 with respective populations of N_1, N_2, N_3 and N_4 are shown. These energies follow the trend $E_4 > E_3 > E_2 > E_1$. Here an optical pumping excites the atoms from the ground state E_1 to the pump band E_4. The atoms from this level make a fast decay (radiationless transition) to the metastable energy level E_3. The population inversion of level E_3 with the level E_2 takes place when the lifetime of the transition from E_3

to E_2 is long compared to that of E_4 to E_3 (lasing level). The atoms in the metastable state E_3 relax and start to create laser transitions through spontaneous and stimulated emissions into energy level E_2. The transition from energy level E_2 to the ground state (level E_1) is fast just like level E_4. This quickly de-excited atom leads to a negligible population in the state E_2 and maintains the population inversion. Since only a small number of atoms need to be excited in the upper lasing level E_3 to form the population inversion, a four-level laser system is much more efficient and practical than the three-level laser system. The most popular four-level solid state gain medium is Nd:YAG. All lasers based on neodymium-doped gain media are four-level lasers except those operated on the ground state transition around 0.9–0.95 μm.

4.3 CHARACTERISTIC OF LASER LIGHT LO2

As discussed, laser radiation is achieved by the process of stimulated emission and the laser beam is highly intense and directional. This radiation of a very pure frequency has the following main characteristics.

(i) **Coherent:** In simple words, the meaning of coherent is highly ordered. The word coherent comes from another word "*Cohero*" which has the meaning "*to stick together*". In fact, different parts of the laser beam have a definite relationship to each other. This coherence is described in terms of temporal coherence (coherence in time) and spatial coherence (coherence in space) (Fig. 4.7) which are required to produce high quality interference.

Ordinary light is not coherent because it comes from independent atoms which emit on the time scale of 10^{-8} seconds. A train of incoherent photons is shown in (Fig. 4.8) from which it is clear that these photons are not in order, i.e., they do not have a definite relationship with each other. However, a degree of coherence can be found in sources like the mercury green line, but their coherence does not approach that of a laser.

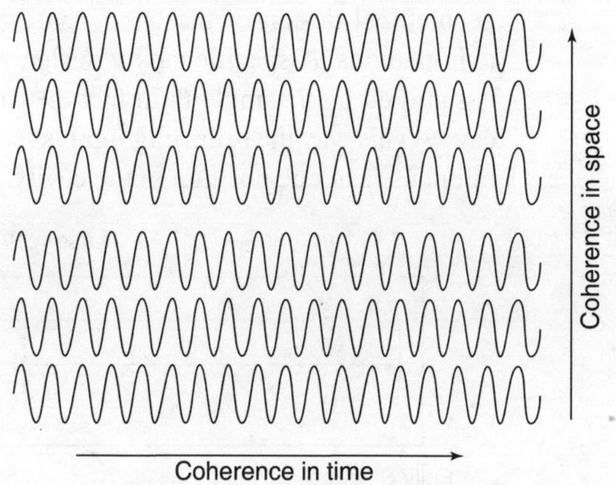

FIGURE 4.7

(ii) **Monochromatic:** The simple meaning of this word is that it is pure in colour or wavelength. The light from a laser typically comes from one atomic transition with a single precise wavelength. So the laser light has a single spectral colour and is almost the purest monochromatic light available. It means the laser light is not exactly monochromatic, but it has high degree of monochromaticity. The deviation from monochromaticity is due to the Doppler effect of the moving atoms or molecules from which the radiation originate.

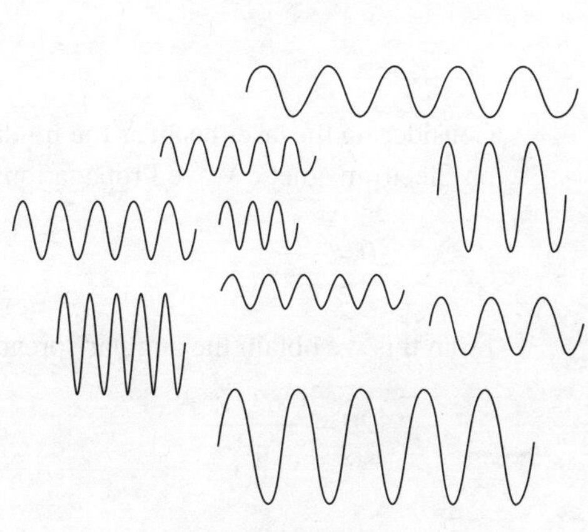

FIGURE 4.8

(iii) **Collimated:** Collimated means it does not spread out much. The light from a typical laser emerges in an extremely thin beam with very little divergence, i.e, the beam is highly collimated. The high degree of collimation arises from the fact that the cavity of the laser has very nearly parallel front and back mirrors as shown in (Fig. 4.9). Because of this the light attains a parallel path after reflections from these mirrors. As it is clear from the figure, the back mirror is made almost perfectly reflecting while the front mirror is about 99% reflecting. Thus about 1% beam comes out from it, which we see as the output beam. Under this process, however, the light passes back and forth between the mirrors many times in order to gain intensity by the stimulated emission of more photons at the same wavelength. If the light is a bit off axis, it will be lost from the beam.

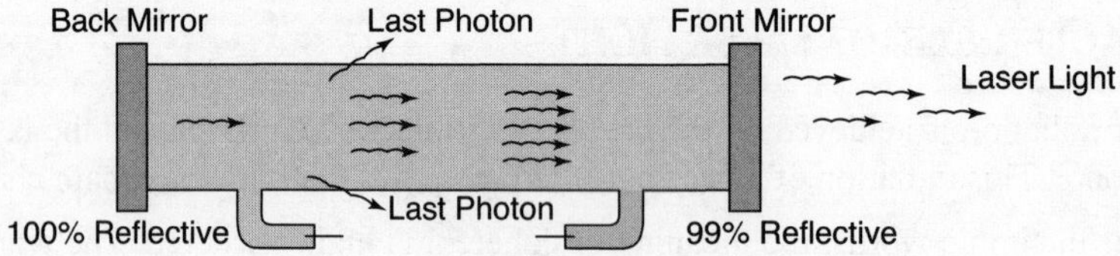

FIGURE 4.9

The high degree of collimation or the directionality of a laser beam (single mode) is due to the geometrical design of the laser cavity and to the fact that stimulated emission process produces twin photons. A specific cavity design is shown in Fig. 4.10, where the angular spread of a beam is signified by the angle θ. In fact the cavity mirrors are shaped with concave surfaces towards the cavity. This way the reflecting light is focused back into the cavity, which finally forms a beam waist of radius r_0 at one position in the cavity.

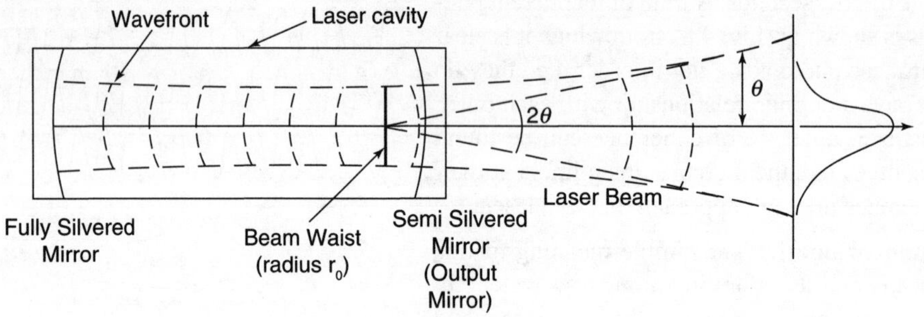

FIGURE 4.10

Considering the laser beam as the fundamental TEM_{00} mode (modes will be discussed in the chapter on Electromagnetic Wave Propagation), the half angle beam spread can be, written as

$$\theta = \frac{\lambda}{\pi r_0}$$

From this we obtain the angular spread as

$$2\theta = \frac{0.637\,\lambda}{r_0}$$

In addition to this, we can calculate the intensity, i.e., the power per unit area of a typical laser which is much greater than other sources of electromagnetic radiation. This is due to the directionality and compactness of the laser beam. In view of this, the intensity or irradiance of a laser beam in terms of its waist radius is given by the following relation

$$I = \frac{P}{A} = \frac{P}{\pi r_0^2}, \text{ where } P \text{ is the power.}$$

4.4 MAIN COMPONENTS OF LASER LO3

In order to understand the working principle of a laser, we should first know about the essential components of the laser. These are given below

(i) **Pumping:** The method of raising the molecules or atoms from their lower energy state to higher energy state is known as *optical pumping*. The optical pumping is needed for achieving population inversion which is precondition for stimulated emission. In this case, the rate of stimulated emission will exceed the rate of stimulated absorption. Hence, the intensity of light will increase during each pass through the medium.

(ii) **Active System:** A system in which the population inversion is to be achieved is called as active system or the gain medium for a laser. Laser systems are named based on the makeup of the gain medium, which may be a gas, liquid or solid. The energy levels in the gain medium, those participate in the radiation, determine the wavelength of laser radiation. Laser action has been observed in over half of the known elements. Two of the most popular transitions in gases are 632.8 nm visible radiation from neon and the 10.6 μm infrared radiation from the CO_2 molecule.

(iii) **Resonant Cavity:** In a laser, the active system or the gain medium is enclosed in an optical cavity (or resonant cavity) usually made up of two parallel surfaces, one of which is perfectly reflecting reflector and the other surface is partially reflecting reflector. In this resonant cavity, the intensity of photons is raised tremendously through stimulated emission process.

4.5 TYPES OF LASER LO3

Nowadays different kinds of lasers are available, the most common being in a digital communications. Virtually every house now has at least one – in their CD/DVD players and recorders. Some lasers can change colour – they are called tunable lasers. The lasers now operate from the infrared to the ultraviolet regions. Moreover, *X*-ray lasers are being developed using electron accelerators. The lasers now are available in the wide range viz, solid lasers, liquid lasers, gas lasers, semiconductor lasers, etc.

4.5.1 Ruby Laser: Solid State Laser

Ruby laser is a solid state laser, which consists of three main parts (i) working material (ii) optical resonant cavity and (iii) excitation source.

Working Material Ruby laser is made up of a crystal of ruby in the form of cylindrical rod having size 2 to 30 cm in length and 0.5 to 2 cm in diameter whose both ends are optically flat. One of the end is fully silvered and other is partially silvered, so that they can act as fully and partially reflecting surfaces, respectively, as shown in Fig. 4.11. Ruby rod is a crystal of Al_2O_3 in which chromium oxide is mixed as impurity so that some of the Al^{3+} ions are replaced by Cr^{3+} ions. These 'impurity' chromium ions give rise to the laser action.

The space between the two faces *A* and *B* is known as the resonant cavity, in which the light (photon) intensity can be built up by multiple reflections and through stimulated emission. The ruby rod is wound by a helical xenon flash light tube with an excitation source in the form of a power supply.

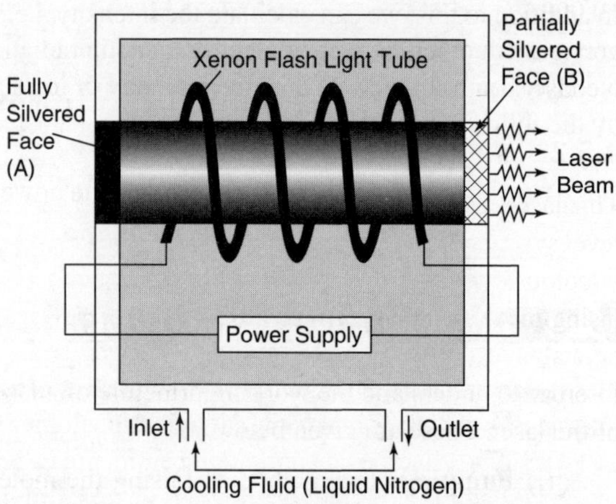

FIGURE 4.11

Working Principle of Ruby Laser In this laser, chromium ions are active centres which are responsible for the laser transition. A simplified energy level diagram of chromium ions in ruby crystal is shown in Fig. 4.12. In the normal state, most of the chromium ions are in the ground state E_1. When light from the flash tube of wavelength 5500 Å is made to fall upon the ruby rod, these incident photons are absorbed by the chromium ions that rise to the excited state E_3. Then they give a part of their energy to the crystal structure and reach the metastable state, i.e., the E_2 state. These ions in metastable state can remain for a longer duration 10^{-3} sec. Therefore, the number of ions in this state goes on increasing while at the same time number of ions in ground state goes on decreasing due to the optical pumping. Thus, the population inversion is established between the metastable state and the ground state.

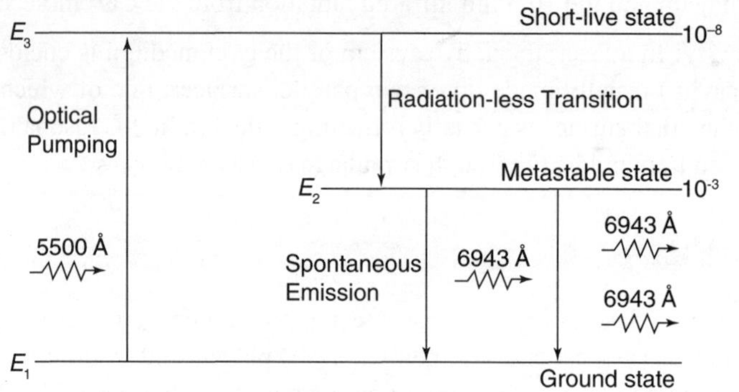

FIGURE 4.12

When an excited ion passes spontaneously from the metastable state to the ground state, it emits a photon of wavelength 6943 Å. This photon travels parallel to the axis of ruby rod and stimulates the surrounding ions present in the metastable state then by stimulated emission other photons are emitted, which are in the phase with the stimulating photons. By successive reflections of these photons at the ends of the rod, every time the stimulated emission is achieved, we obtain an intense, coherent and unidirectional laser beam from the partially silvered face B.

The ruby laser operates at about 1% efficiency. It may produces a laser beam of 1 mm to 25 mm in diameter. The beam obtained is in the form of pulses. However, on the advantage side, very strong beam as strong as

10,000 Watt in power is produced. Furthermore, the construction of this laser is simple and the operation is very easy. For this reason, this laser is also known as practical laser. Other examples of solid state lasers are Neodynium-YAG (N*d*-YAG), Neodynium-Glass (N*d*-Glass) and semiconductor lasers.

4.5.2 Nd-YAG Laser: Solid State Laser

This laser is capable of producing very high power emissions, as a result of its lasing medium operates as four level systems. The schematic of Nd-YAG laser is shown in Fig. 4.13. The lasing medium in the Nd-YAG laser is colourless, isotropic crystal called Yttrium aluminium garnet (YAG-$Y_2Al_5O_{12}$). The main dopant in the lasing medium is Neodymium (Nd^{3+}). When it is used in laser, Neodymium replaces 1% of Yttrium and the crystal takes a light blue colour. The YAG has a relatively high thermal conductivity, which improves thermal dissipation in thermal cavity. So continuous wave operation up to a few hundred watts is possible. Average power of up to 1 kW is available when it is operated in pulse mode.

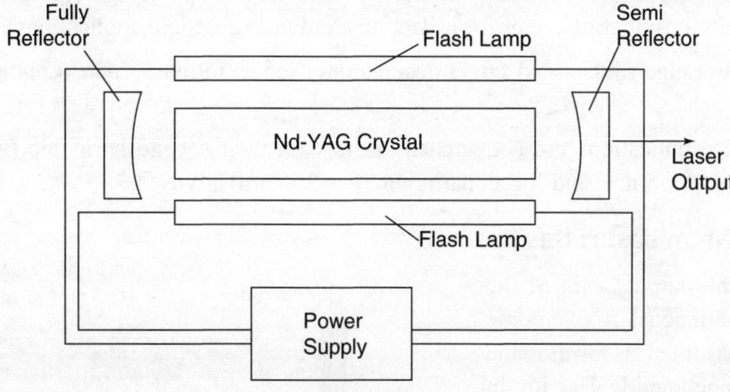

FIGURE 4.13

The energy level diagram for Nd-YAG is shown in Fig. 4.14. These levels arise from three inner shell 4f electrons of the Nd^{3+} ion, which are effectively screened by eight outer electrons ($5S^2$ and $5P^6$). For the operation of Nd-YAG lasers a cooling system is required. A Nd-YAG laser produces 30 times as much waste heat as laser output with an efficiency of about 3%. The waste heat must be removed in order to ensure proper laser operation by flooding the optical compartment with water. However optical distortion and image

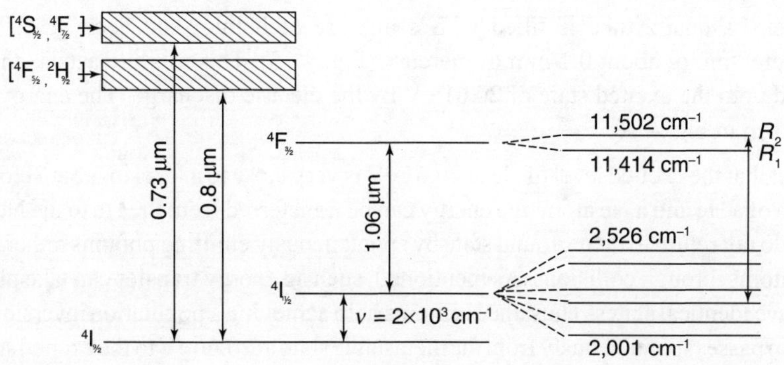

FIGURE 4.14

problem is created due to absorption of significant amount of flash lamp energy by water. This problem can be overcome by flowing water over the outside of the optical cavity and by encasing the lasing rod and flash lamp with transparent cooling jacket.

An advantage of Nd-YAG laser is that by using *Q*-switching, laser beam pulse frequency and shape can be tailored where a shutter moves rapidly in and out of the path of the beam. In this manner beam output is interrupted until a high level of population inversion and energy storage is achieved in the resonator. If the optical cavity is switched from no reflection (low *Q*) to near total reflection (high *Q*), the cycle can be optimised to build up the maximum population inversion before the pulse is generated. This way, we get a beam pulse with high energy up to 1 J and a short pulse period down to 10 ns is obtained.

Applications

(i) Nd-YAG is used in material processing such as welding and drilling.

(ii) It is also used in photo disruption of transparent membrane of pathological origin, which can appear in the interior chamber of eye or for iridectomy and in endoscopic applications.

(iii) It is used in range finders and target designators used in military context, which use *Q*-switched lasers.

(iv) In scientific applications the *Q*-switched lasers with their second harmonic (λ = 532 nm), third harmonic (λ = 355 nm) and fourth harmonic (λ = 266 nm) are used.

4.5.3 Helium-Neon Laser: Gas Laser

As we know that the output beam of the ruby laser is not continuous. To overcome this drawback, the gas filled laser was made by A. Javan, W. Bennett and D. Herriott in 1961. It consists of a quartz tube having the size about 1.5 cm in diameter and about 1 meter in length. The both ends of the tube are sealed by optically plane and parallel mirrors, one of them being partially silvered (90% reflective) and the other one is fully silvered (100% reflective).

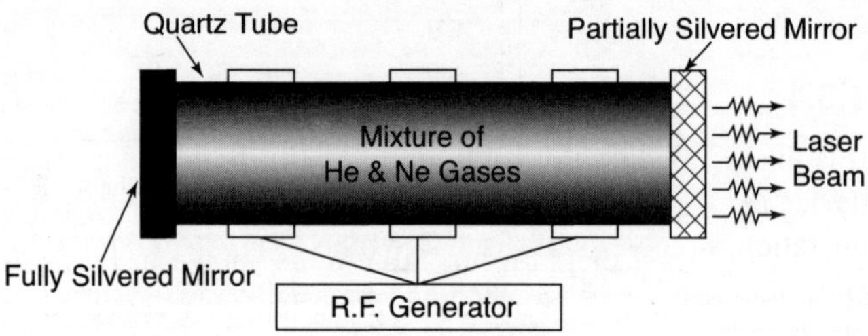

FIGURE 4.15

In this laser system, a quartz tube is filled with a mixture of helium and neon gases in the ratio 10:1 respectively, at a pressure of about 0.1 mm of mercury (Fig. 4.15). This mixture acts as the active medium. Helium is pumped upto the excited state of 20.61 eV by the electric discharge. The energy level diagram of He-Ne laser is shown in Fig. 4.16.

Here, it can be seen that the excited level of He at 20.61 eV is very close to a level in Ne at 20.66 eV. It is so close that upon collision of a He and a Ne atom, the energy can be transferred from the He to the Ne atoms. Thus, the excited He atoms do not return to their ground state by spontaneously emitting photons rather they transfer their energy to the Ne atoms through collisions. As mentioned, such an energy transfer can take place when the two colliding atoms have identical states. Thus, the He atoms help achieving a population inversion in the Ne atoms. An excited Ne atom passes spontaneously from the metastable state at 20.66 eV to the excited state at 18.70 eV by

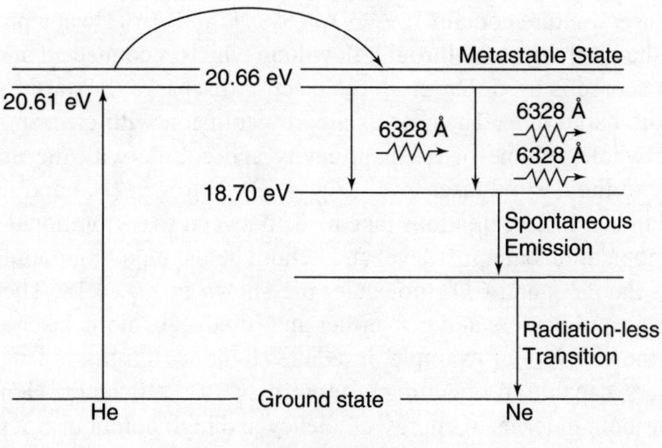

FIGURE 4.16

emitting a photon of wavelength 6328 Å. This photon travels through the gas mixture parallel to the axis of the tube and stimulates the surrounding Ne atoms present in the metastable state. This way we get other photons that are in the phase with the stimulating photons. These photons are reflected forth and back by the silvered ends and the number of photons gets amplified through stimulated emission every time. Finally, a portion of these intensified photons passes through the partially silvered end.

The He-Ne laser is the most common and inexpensive gas laser. Usually it is constructed to operate in the red light at 6328 Å and in the infrared at 15,230 Å. According to Garmire, an unfocused 1 mW – He Ne laser has a brightness equal to sunshine on a clear day (~ 0.1 W/cm^2) and is just as dangerous to stare at directly.

4.5.4 Carbon Dioxide Gas Laser

It is one of the earliest high power molecular gas laser that uses carbon dioxide gas molecule. This optical device is capable of continuous output powers above 10 kW. It is also capable of extremely high power pulse operation. It consists of discharge tube of size of about 2.5 cm in diameter and 5.0 cm in length. The both ends of the tube are sealed by optically plane and parallel mirrors, one of them being semi-silvered and other one is fully silvered. (Fig. 4.17).

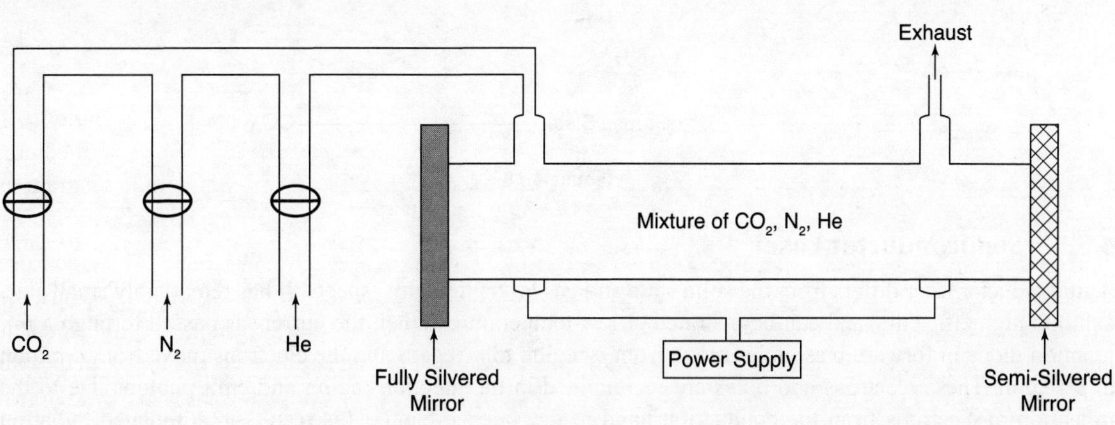

FIGURE 4.17

The carbon dioxide gas laser mixture contain 15% CO_2, 15% N_2 and 70% He at a pressure of few mm of Hg. This mixture is fed into the discharge tube through flow loop which is connected at one end of the discharge tube. The dc excitations source is used that produces electric discharge. In starting nitrogen molecules are allowed to enter in the discharge tube. They get excited by collision with electrons. Then excited nitrogen molecules flow into the whole volume of resonant cavity and collide with the unexcited CO_2 molecules and transfer their energy to the desired laser level (Fig. 4.18). Nitrogen (N_2) and helium (He) improve the efficiency of the laser action, while oscillations take place between two vibrational levels of CO_2. Nitrogen helps producing a large populations in upper level and helium helps removing population from lower energy level. Related energy levels of N_2 and CO_2 molecules are shown in Fig. 4.18. The radiated photons travel back and forth between the end mirrors and get further amplified. It exhibits laser action at several infrared frequencies but none in the visible. For example, it radiates light at 10.6 μm is far infrared region. It is one of the most efficient lasers, capable of operating at more than 30% efficiency. Hence, this laser is suitable for industrial applications both in terms of energy efficiency and high output beam; particularly it is used for welding and cutting.

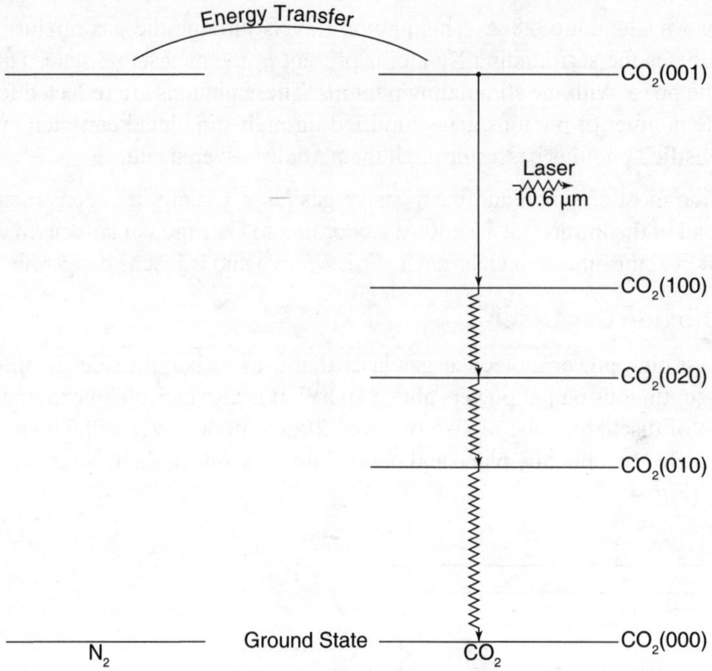

FIGURE 4.18

4.5.5 Semiconductor Laser

Semiconductor laser differs from the solid state and gas lasers in many aspects. It has remarkably small size, exhibits high efficiency and can be operated at low temperature. When the current is passed through a *p-n* junction diode in forward bias, holes move from *p*-region to *n*-region and the electrons move from *n*-region to *p*-region. These electrons and holes are recombined in the junction region and emit photons due to the transition of electrons from the conduction band to the valence band. This results in stimulated radiation coming from a very narrow region near the junction. The action is intensified by increasing the current and decreasing the junction thickness.

Semiconductor laser is made up of an active layer of gallium arsenide (GaAs) of thickness 0.2 microns. This is sandwiched in between a n-type GaAsAl and p-type GaAsAl layer as shown in Fig. 4.19. The resonant cavity is provided by polishing opposite faces of the GaAs crystal and the pumping occurs by passing electrical current from an ordinary source (Power Supply). From this system GaAs semiconductor, laser beams of wavelength ranging from 7000 Å to 30,000 Å can be produced.

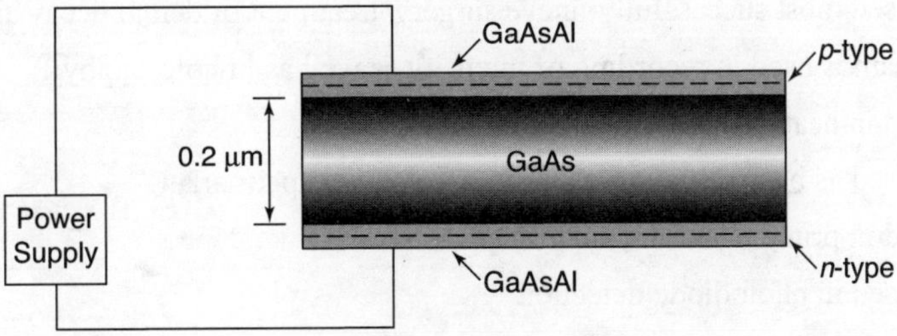

FIGURE 4.19

4.5.6 Advantages and Disadvantages of Ruby laser, He-Ne laser and Semiconductor laser

The merits and demerits of solid laser (Ruby), gas laser (He–Ne) and semiconductor laser are given in Table 4.1.

TABLE 4.1

	Ruby laser	*He-Ne laser*	*Semiconductor laser*
Advantages			
1	Easy to construct and operate	Easy to construct and operate	Easy in operation
2	Very strong and intense beam upto a power of 10 kW	Continuous beam	Long life, highly monochromatic, tunable and continuous beam
3	Beam diameter as large as 25 mm	Exceptionally monochromatic beam with high operation duration (10,000 hrs.)	Excellent efficiency with very high operation duration (20,000 hrs.)
Disadvantage			
	Its laser beam is only pulse like and its operation duration is very less (few hrs.)	It has got very low power about 0.5 – 5mW	It has got low power about 200 mW

4.6 APPLICATIONS OF LASERS LO4

Lasers have many applications in science, industry and medicine, some of which are listed below:

(i) Lasers have been used to measure long distances, so they are very useful in surveying and ranging. For this purpose, a fast laser pulse is sent to a corner reflector at the point to be measured and the time of reflection is measured to get the distance.

(ii) Lasers are electromagnetic waves of very high intensity and can be used to study the laws of interaction of atoms and molecules.

(iii) Lasers are suitable for communication and they have significant advantages because they are more nearly monochromatic. This allows the pulse shape to be maintained better over long distances. So communication can be sent at higher rates without overlap of the pulses.

(iv) Laser beams are highly intense and are used for welding, cutting of materials, machining and drilling holes, etc. Generally, carbon dioxide laser are used for such purposes, as it carries large power.

(v) Lasers are used most successfully in eye surgery, treatment of dental decay and skin diseases.

(vi) The laser beam is used in recording of intensity as well as in holography.

(vii) Laser is used in heat treatments for hardening.

(viii) Lasers are used as barcode scanners in library and in supermarket.

(ix) Laser is used in printers (Laser printers).

(x) Lasers are used in photodiode detection.

4.7 LASER COOLING LO4

So far you have learnt that the laser radiation is highly intense, highly coherent and highly monochromatic light. It is amazing that the field of laser radiation can be used to cool down atoms to very low temperature, for example up to 10^{-9} K. This can be understood based on an atom which is traveling toward a laser beam and absorbs a photon from the laser. In this situation, the atom will be slowed by the fact that the photon has momentum $p = E/c = h/\lambda$, where E is the energy, c is the speed of light, h is the Planck's constant and λ is the wavelength associated with the photon. If we assume that a number of sodium atoms are freely moving in a vacuum chamber at 300 K (room temperature), i.e., the rms velocity of the sodium atom is about 570 m/s. Then the momentum of the sodium atom can be reduced by the amount of the momentum of the photon, if a laser is tuned just below one of the sodium d-lines (5890 Å and 5896 Å, about 2.1 eV) and the sodium atom absorbs a laser photon when traveling toward the laser. It would take a large number of such absorptions to cool the sodium atoms to nearly 0 K. The change in speed from the absorption of one photon can be calculated from

$$\frac{\Delta p}{p} = \frac{p_{\text{photon}}}{mv} = \frac{\Delta v}{v} \Rightarrow \Delta v = \frac{p_{\text{photon}}}{m}$$

The above expression shows a lot of photons, but according to *Chu* a laser can induce on the order of 10^7 absorptions per second so that an atom could be stopped in a matter of milliseconds.

There is a conceptual problem that an absorption can also speed up an atom if it catches it from behind. So more absorptions from head-on photons are necessary to have. This is accomplished in practice by tuning the laser slightly below the resonance absorption of a stationary sodium atom. More precisely, with the opposing laser beams with perpendicular linear polarisations, atoms can be selectively driven or "optically pumped" into the lower energy levels. This method of cooling sodium atoms was proposed by *Theodore Hansch* and *Arthur Schawlow* at Stanford University in 1975 and was achieved by *Chu* at AT & T Bell Labs in 1985. Here sodium atoms were cooled from a thermal beam at 500K to about 240 μK.

4.8 HOLOGRAPHY LO5

Holography is one of the remarkable achievements of modern science and technology, which has been possible only because of the lasers. The word "holography" was originated from the Greek words "*holos*" and "*grapho*".

The meaning of "holos" is "*whole*" and of "grapho" is "*write*". So holography means complete record of the image. Holography is a three-dimensional (3D) laser photography. It is lensless photography in which an image is captured as an interference pattern. The image thus obtained is called a *hologram*, which is true 3D record of the object. Holography not only records the amplitude but also the phase of the light wave with the help of interferometric techniques. This recorded reference pattern contains more information than a focused image and enables the viewer to view a true 3D image, which exhibits parallax. The technique of holography was invented by *Gabor* in 1947.

4.8.1 Principle of Holography

In holography, there are two basic waves that come together to create the interference pattern. One wave is called *object wave* and another wave is called *reference wave*. When an object wave meets a reference wave, it creates a standing wave pattern of interference. This is then photographed, which we call a hologram.

4.8.2 Requirements of Holography

Following are some requirements for the absolute holography.

(i) Since holography is an interference phenomenon, there should not be a path difference between the object wave and the reference wave more than the coherence length. This is necessary to achieve stable interference fringes.

(ii) Spatial coherence is important so that the reference wave and the scattered object waves from different regions can interfere properly.

(iii) Since reconstructed image coordinates depend on wavelength as well as position of the reconstructing source, it is necessary that the source emits a narrow band of wavelength and it is not broad in the interest of obtaining good resolution in the reconstructed image.

(iv) In order to obtain aberrations free reconstructed image, it is necessary that the reconstructing source is of the same wavelength and is situated at the same position with respect to the hologram as the reference source.

(v) All recording arrangement like film, object, mirrors etc., must be motionless during the exposure.

4.9 HOLOGRAPHY VERSUS CONVENTIONAL PHOTOGRAPHY LO5

Holography represents a photographic process in a broad sense, but essentially it differs from a usual photo, as the phase of light waves scattered by the object carries the complete information about 3D structure of the object. A conventional photography is a 2D image of a 3D scene, which brings into focus every part of the scene that falls within the depth of the field of the lens. Due to this, a conventional photograph lacks the perception of the depth or the parallax with which we view a real life scene. Since a conventional photograph only records the intensity pattern, 3D character of the object scene is lost. Contrary to this, the hologram contains depth and parallax, which provides the ability to see around the object to objects placed behind. It gives information about amplitude as well as the phase of an object. So hologram preserves information about the object for latter observation.

In conventional photography, there is one to one relationship between object and image point as the light originating from a particular point of scene is collected by a lens focused on that particular point. However, in holography lens is not used and this is a complex interference pattern of microscopically spaced fringes. Hologram receives light from every point of a scene and hence there is no one to one relationship. This is a record of entire signal wave.

In conventional photography, radiated energy is recorded and phase relationship of wave arriving from different distances and directions is lost. However, in holography phase relationship is recorded by using the technique of interference of light waves.

4.10 RECORDING AND RECONSTRUCTION OF IMAGE ON HOLOGRAPH LO5

Recording of a hologram is a result of superpositions of the object wave and the reference wave, which is usually a plane wave. This interference pattern is recorded by a photographic plate that contains information about amplitude as well as phase of the object wave. In order to see the image, hologram is illuminated with another wave called the reconstruction wave which is identical to the reference wave in most of the cases. This is called reconstruction of image on the hologram.

4.10.1 Theory

If the object is a point scatterer and it is made of large number of such points, then the composite wave reflected by the object will be the vectorial sum of all object waves scattered from are these points. As mentioned earlier, holography records the object wave, particularly the phase (say φ) associated with it. So we can represent the object wave, which is due to the superposition of waves from point scatterers on the object, as

$$Y_1(x, z) = A_1(x, z) \cos (\phi - \omega t) \tag{i}$$

where ω is the frequency. The object wave represented by Eq. (i) lies in the plane of photographic plate at $y = 0$.

Now, we consider a reference wave propagating in the xy plane and inclined at an angle α from the y axis. In view of this, the field associated with the reference wave can be written as

$$\begin{aligned} Y_2(x, y, z) &= A_2 \cos (\vec{k} \cdot \vec{r} - \omega t) \\ &= A_2 \cos (kx \sin \alpha + ky \cos \alpha - \omega t) \end{aligned} \tag{ii}$$

At the photographic plate, i.e., at $y = 0$, this field becomes

$$Y_2(x, z) = A_2 \cos (kx \sin \alpha - \omega t).$$

Since the propagation constant $k = 2\pi/\lambda$, $kx \sin \alpha = \dfrac{\sin \alpha}{\lambda} 2\pi x$

Here $\sin \alpha/\lambda$ is defined as the spatial frequency (say ξ). So the field associated with the reference wave becomes

$$Y_2 (x, z) = A_2 \cos (2\pi\xi x - \omega t) \tag{iii}$$

A comparison of equation (iii) with equation (i) yields that the phase linearly varies with x.

Simple method of superposition enables us to calculate the total field at the photographic plate (at $y = 0$) as

$$Y = Y_1 + Y_2$$

$$Y(x, z, t) = A_1(x, z) \cos (\phi - \omega t) + A_2 \cos (2\pi\xi x - \omega t) \tag{iv}$$

In view of the response of photographic plate to the intensity we find below the measure of intensity pattern recorded by the photographic plate as

$$\begin{aligned} I(x, z) &= \text{Average value of } Y^2(x, z, t) \\ &= \langle Y^2 (x, z, t) \rangle \end{aligned}$$

or $$I(x, z) = A_1^2 (x, z) < \cos^2 (\phi - \omega t)> + A_2^2 < \cos^2 (2\pi\xi x - \omega t)$$
$$+ 2A_1(x, z) A_2 <\cos (\phi - \omega t) \cos (2\pi\xi x - \omega t> \quad \text{(v)}$$

As we know that $< \cos^2 (\phi - \omega t) > = ½$,

$$<\cos^2 (2\pi\xi x - \omega t) > = ½,$$
$$< 2\cos (\phi - \omega t) \cos (2\pi\xi x - \omega t) >$$
$$= ½ <\cos (\phi + 2\pi\xi x - 2\omega t) + \cos (\phi - 2\pi\xi x) >$$

[Using $2 \cos \theta_1 \cos \theta_2 = \cos (\theta_1 + \theta_2) + \cos (\theta_1 - \theta_2)$]

The average value of $\cos (\phi + 2\pi\xi x - 2\omega t)$ can be obtained by using simple integration

$$\frac{1}{2\pi}\int_0^{2\pi/\omega} \cos (\phi + 2\pi\xi x - 2\omega t)\, dt,$$

as the average value of $\cos \omega t$ over the period $T = 2\pi/\omega$

$$= \frac{1}{2\pi}\int_0^{2\pi/\omega} \cos \omega t\, dt.$$

So it comes out to be zero. With this, the intensity $I(x, z)$ is written as

$$I(x, y) = A_1^2 (x, z)/2 + A_2^2/2 + A_1(x, z) A_2 \cos (\phi - 2\pi\xi x) \quad \text{(vi)}$$

The above equation shows that the intensity I is the function of phase $\phi(x, z)$. It means the phase information of the object wave is recorded in the intensity pattern.

In order to obtain a hologram, we develop the photographic plate containing above intensity pattern. In this context, the ratio of the transmitted field to the incident field is defined as the transmittance of the hologram that depends on $I(x, z)$. Using a suitable developing process, the condition under which the transmittance is linearly related to $I(x, z)$ can be obtained. Under this condition, if $R_e(x, z)$ denotes the field of the reconstruction wave, at the hologram plane, then the transmitted field can be taken as

$$T_e(x, z)\ \alpha\ R_e(x, z)\, I(x, z).$$

Taking K_p as the proportionality coefficient and putting the value of $I(x, z)$ from equation (vi), we obtain

$$T_e(x, z) = K_p R_e(x, z)\left[\frac{A_1^2(x, z)}{2} + \frac{A_2^2}{2} + A_1(x, z)\, A_2 \cos (\phi - 2\pi\xi x)\right] \quad \text{(vii)}$$

In the case when the reconstruction wave is identical to the reference wave $Y_2(x, z)$, the above equation becomes

$$T_e(x, z) = K_p A_2\left[\frac{A_1^2(x, z)}{2} + \frac{A_2^2}{2}\right]\cos (2\pi\xi x - \omega t)$$
$$+ K_p A_2^2 A_1(x, z) \cdot \cos (2\pi\xi x - \omega t) \cos (\phi - 2\pi\xi x)$$

Again using $2 \cos \theta_1 \cos \theta_2 = \cos (\theta_1 + \theta_2) + \cos (\theta_1 - \theta_2)$, we get the following expression for $T_e(x, z)$

$$T_e(x, z) = K_p A_2\left[\frac{A_1^2(x, z)}{2} + \frac{A_2^2}{2}\right]\cos (2\pi\xi x - \omega t)$$
$$+ \frac{K_p A_2^2 A_1(x, z)}{2}\cos (4\pi\xi x - \phi - \omega t) \cdot \cos(\phi - \omega t) \quad \text{(viii)}$$

The above equation contains three terms, which may be analysed as follows.

(i) First term being proportional to A_2^2 represents the reconstruction wave whose amplitude is modulated by the term $A_1^2(x, z)$, i.e., by the amplitude of object wave. The factor cos $(2\pi\xi x - \omega t)$ shows that this part of the total field is traveling in the direction of the reference wave.

(ii) The second term is identical to equation (i) within a constant term. Hence, this represents the original object wave. Having appeared in transmitted field, it gives rise to a virtual image.

(iii) The third term carries the phase $\phi(x, z)$ in addition to the term $4\pi\xi x$, but with negative sign. It means this wave has a curvature opposite to the object wave, i.e., if the object wave is diverging spherical wave, then the last term (third term) shows a converging spherical wave. Hence, this wave forms a real image of the object contrary to the second term. This image can be photographed with the help of a film.

4.11 TYPES OF HOLOGRAMS LO6

In order to construct a hologram, we need two coherent light waves, one is the object wave carrying information about the object and the other is a plane wave that is called reference wave. There are various types of holograms, but the most common ones are the transmission hologram and the reflection holograms.

4.11.1 Transmission Hologram

This type of hologram is commonly used. If the object wave and the reference wave emerge from the same side of the holographic film, then the hologram is called transmission hologram (Fig. 4.20). Another characteristic of transmission hologram is the low diffraction efficiency and weak image reconstruction.

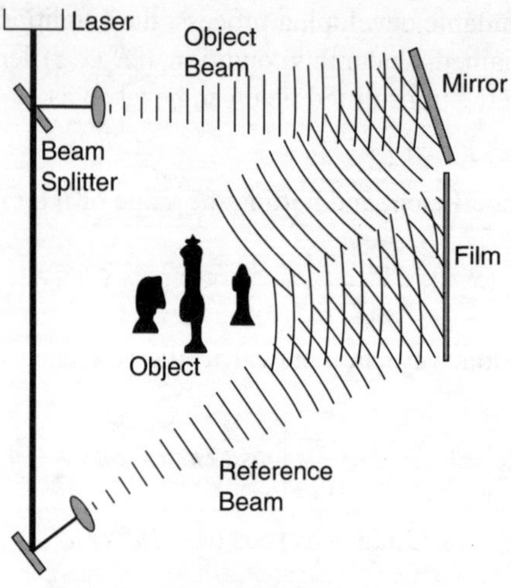

FIGURE 4.20

4.11.1.1 Recording Process

As mentioned, in order to make a hologram, two coherent light waves (laser light) are required (Fig. 4.21). The first one is called the object wave, which is reflected from the object and carries information about the object. The second one is called reference wave and is a plane wave without information. These two waves generate an interference pattern, which is recorded in the form of a hologram on film emulsion. For obtaining the stable interference patterns, absolutely stable conditions are required during the exposure of the film. This recorded hologram is called transmission hologram because the light passes through the holographic plate.

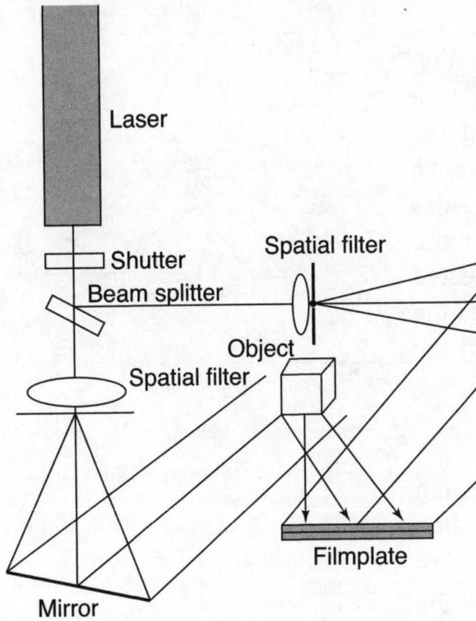

FIGURE 4.21

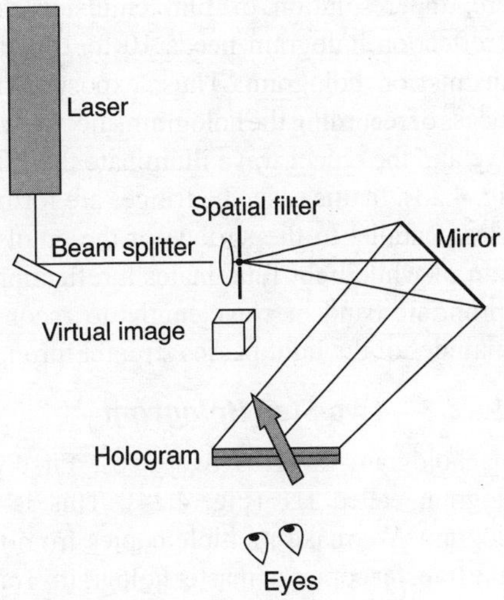

FIGURE 4.22

4.11.1.2 Reconstruction Process

We can reconstruct the holographic image by developing the hologram and then placing it in its original position in the reference beam as during its recording. If we look along the reconstructed object wave, we see a replica of the object and as we shift viewpoints we see object from different perspectives. Thus, the object appears to be three-dimensional. During the reconstruction of the transmission hologram, the light does not pass through the image, but it creates a wavefront that makes it appear as though the light had been generated in the position of the object. This image thus formed is called virtual image (Fig. 4.22). Contrary to this, an image having light actually passing through it is called a real image.

4.11.1.3 Properties

Some important properties of transmission hologram are as follows.

(i) When viewed with white light, the transmission holograms look like a blurry rainbow image.

(ii) These holograms are viewed as sharp images when we use the shining laser light through the hologram.

(iii) Less resolving power is needed in materials.

(iv) Transmission hologram can be formed in a simple setup.

(v) Greater depth of the scene is possible in transmission holograms.

4.11.2 Reflection Hologram

The holograms that are viewed with white light source on the same side as the viewer are known as reflection holograms. In such a hologram, a truly three-dimensional image is seen near its surface. This hologram is the most common type shown in galleries. The light is located on the viewer's side of the hologram at a specific angle and distance. The image thus formed consists of light reflected by the hologram. There are two types of reflection holograms.

4.11.2.1 One-Step Hologram

Here, the resolution of film emulsion is high, as the recording of reflection hologram needs 10 to 100 times much power than a transmission hologram. Thus, exposure time is long. During the process of recording the hologram, the two waves namely the reference wave and the object wave illuminate the film plate on opposite sides (Fig. 4.23). In this case, the fringes are formed in layers and are more or less parallel to the surface of the emulsion. If a highly directed beam of white light illuminates a reflection hologram, it selects the appropriate band of wavelengths to reconstruct the image and the remainder of the light passes straight through.

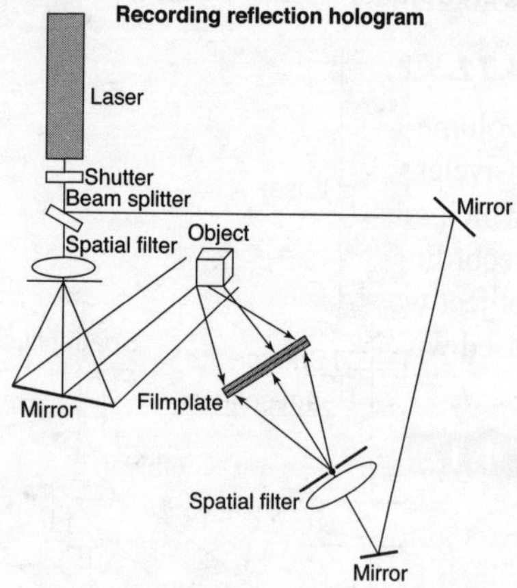

FIGURE 4.23

4.11.2.2 Two-Step Hologram

This hologram involves two steps. First we make a transmission hologram called H1 (Fig. 4.24). This is called a master or first hologram. We make multiple copies from the master hologram. We make transfer copies of master hologram. Transfer copy means making another hologram using the image on the master as the subject. These transfer holograms are either laser-visible transmission holograms or reflection holograms H2. Suppose we want any object in the final hologram just to appear half in front and half behind the recording plate. In such circumstances, the two step hologram is of great use.

4.11.2.3 Properties

Some important properties of reflection hologram are as follows.

(i) These holograms can be viewed in regular light.

(ii) The finished reflection hologram is monochromatic.

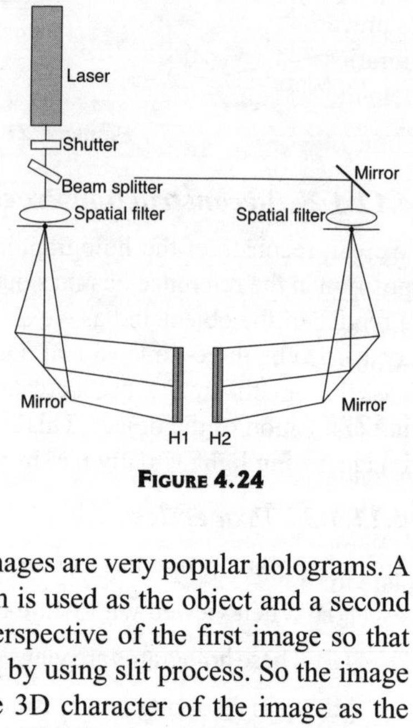

FIGURE 4.24

4.11.3 White Light Hologram: Rainbow Hologram

Rainbow holograms that can be viewed in white light and produce 3D images are very popular holograms. A double holographic process makes them, in which an ordinary hologram is used as the object and a second hologram is made through a slit. A horizontal slit limits the vertical perspective of the first image so that there is no vertical parallax. The coherence requirement can be removed by using slit process. So the image brightness is obtained from ordinary room light while maintaining the 3D character of the image as the viewer eye is moved horizontally. If viewer eye is moved vertically, no parallax is seen and the image colour sweeps through the rainbow spectrum from blue to red.

4.11.3.1 Embossed Hologram

Many variations of hologram can be made between the reflection and transmission types of holograms. Embossed hologram is one of such types of holograms, which is used widely in most security applications. These holograms offer an effective method of protection against any forms of manipulation, as they are too difficult to copy due to their complex structure. All credit cards and passports have embossed hologram. In this hologram, the original hologram is recorded in a photosensitive material called *photoresist*. These holograms are easily produced at large scale and also at a very low cost.

4.11.3.2 Volume Hologram

Volume holograms are produced when the thickness of the recording material is much larger than the light wavelength used for recording. These are transmission holograms and are also known as *thick holograms*, which are mainly considered as a high-density data storage technology. These are 3D holograms created by recording the interference pattern of two mutually coherent light waves. The angle of difference between the object wave and the reference wave is 90° to 180°. Due to certain unique properties, volume holograms are used widely in various spectroscopic and imaging applications.

4.12 APPLICATIONS OF HOLOGRAPHY

LO7

Holography represents examples of recombining of scattered radiation. It is a product of interference of light, which is used to measure very small optical path lengths with precision by using wavelength of light and interference. Now holography is being used in industry, communication and other engineering problems also. You would have seen hologram on tickets, original covers of software programs, credit cards etc. This is used to prevent *falsification*. Another important application is through bar code readers used in shops, warehouses, libraries and so on. In aircraft industry, holography technology is used through head up displays (HUD) which help the pilot to see instrument panel on to the windscreen.

Some other important applications of holography are given below.

4.12.1 Time Average Holographic Interferometry

This interferometry is very useful for determining or studying the modes of vibration of complex structures. Hologram is prepared using a long exposure time than the periods of vibrations being studied. This hologram freezes many images, mapping the motion of vibrating surface. Interference fringes pattern provides information about the relative vibrational amplitudes as a function of position on the surface.

4.12.2 Microscopy

A hologram contains many separate observations of microscopic particles. Image provided by hologram may be viewed by focusing on any depth of unchanging field. Microscopic hologram is made by illuminating the specimen by laser light, a part of which is split off outside the microscope and is routed to the photographic plate to rejoin the subject beam processed by the microscope. It can be shown that if $\lambda r > \lambda s$, where λr is the wavelength of reconstructing light and λs is the wavelength used in holography, then the magnification is

$$M = \left(\frac{v}{u}\right)\left(\frac{\lambda r}{\lambda s}\right)$$

Here u is the object distance from the film and v is the corresponding image distance from the hologram. However, these distances are equal, i.e., $u = v$, if the reference and reconstructed wavefronts are both plane wave.

4.12.3 Ultrasonic Hologram

As the words "ultrasonic holograms" suggest, the waves producing a hologram may not necessarily be electromagnetic in nature. Also, the holographic principles do not depend on the transverse nature of the radiation. Holograms generated with the help of ultrasonic waves are very useful because of the ability of such waves to penetrate the objects that are opaque to visible light. Holograms formed by ultrasonic waves are very useful to get 3D images inside the opaque bodies.

4.12.4 Holocameras

Hologram can be developed and viewed with the help of holocameras, which do not use photographic film. Thermoplastic recording material is used in holocameras and image development is done by electrical and thermal means. The image development does not need wet chemical processing. Also, it can be completed in a few seconds without repositioning the recording.

4.12.5 Holographic Data Storage

Data can be stored by holographic technique. It is very interesting that the data can be reduced to dimensions of the order of wavelength of light. Therefore, volume holograms can be useful to record vast quantities of information. Photosensitive crystal like potassium bromide with colour centres or lithium niobate are used in place of thick layered photoemulsion. Small rotation of crystal takes place of turning pages.

SUMMARY

The main topics discussed in this chapter are summarized below.

- Laser was introduced as a special type of device that amplifies light and produces a highly intense and highly directional beam which mostly has a very pure frequency.
- It was made clear the population inversion is the basic requirement for the operation of the laser.
- For achieving the laser radiation, the concept of stimulated emission was discussed in detail along with the inclusion of Einstein's coefficients.
- The main components of laser were discussed and based on the gain medium the lasers were classified as solid state laser, gas laser or semiconductor laser.
- Ruby laser, Nd-YAG laser, He-Ne laser, CO_2 laser and semiconductor laser were discussed in detail and the energy diagrams provided.
- It was mentioned that the lasers have diverse applications in different fields of science and technology. These applications were talked about in brief.
- A new concept of laser cooling was discussed in detail. It was shown how a highly intense and coherent light of laser can cool the sodium atoms to 10^{-6} K.
- Another exciting filed of holography was introduced and it was mentioned that with the help of lasers the holograms can be developed that give 3D picture of the objects.
- Principle and the requirements of the holography were discussed.
- The advance/additional features of holography from those of conventional photography were talked about.
- Detailed description of recording and reconstruction of image on holograph were discussed.
- Two types of holograms, namely transmission holograms and reflection holograms, were discussed in detail along with their recording and reconstruction processes and the properties.

- White light hologram was introduced, which is also known as rainbow hologram. Then the embossed and volume holograms were talked about.
- Various applications of holography were discussed including time average holographic interferometry, microscopy, ultrasonic holograms, holocameras and the holographic data storage.

SOLVED EXAMPLES

EXAMPLE 1 Determine the energy and momentum of a photon of a laser beam of wavelength 6328Å (Given: $h = 6.63 \times 10^{-34}$ J K sec. and $\boldsymbol{c} = 3.0 \times 10^8$ m/sec).

SOLUTION Given $\lambda = 6328 \times 10^{-10}$ m, $h = 6.63 \times 10^{-34}$ J K sec. and $c = 3 \times 10^8$ m/sec.

$$\text{Formula used } E = h\nu = \frac{hc}{\lambda}$$

$$= \frac{6.63\times10^{-34}\times3\times10^{8}}{6.328\times10^{-7}\text{ m}} = 1.05\times10^{-19}\text{ Joule}$$

$$E = 3.143 \text{ Joule}$$

$$\text{Momentum } p = \frac{E}{c} = \frac{h}{\lambda} = \frac{6.63\times10^{-34}}{6.328\times10^{-7}} = 1.05\times10^{-27}\text{ kg}\cdot\text{m/sec}$$

$$p = \mathbf{1.05\ kg\cdot m/sec.}$$

EXAMPLE 2 Calculate the energy of laser pulse in a ruby laser for 2.8×10^{19} Cr^{3+} ions. If the laser emits radiation of wavelength 6943Å.

SOLUTION Given: $\lambda = 6943 \times 10^{-10}$ m, $n = 2.8 \times 10^{19}$

The energy of a photon, $= h\nu$

and the total energy due to n Cr^{3+} ions is

$$E = nh\nu = n\frac{hc}{\lambda} = 2.8\times10^{19}\cdot\frac{6.63\times10^{-34}\times3\times10^{8}}{6.943\times10^{-7}}$$

$$= \mathbf{8.02\ J}$$

EXAMPLE 3 A three-level laser emits a light of wavelength of 5500 Å, What will be the ratio of population of upper level (E_2) to the lower energy level (E_1) if the optical pumping mechanism is shut off (Assume $T = 300$ K).

At what temperature for the conditions of (a) would the ratio of populations be 1/2?

SOLUTION Given $\lambda = 5500$ Å

Formula used is

$$E_2 - E_1 = h\nu = \frac{hc}{\lambda}$$

$$= \frac{(6.63\times10^{-34}\text{ J/sec})\times(3\times10^{8}\text{ m/sec})}{(5.5\times10^{-7}\text{ m})\times(1.6\times10^{-19}\text{ J/eV})}$$

$$= 2.26 \text{ eV}$$

and kT can be calculated as

$$kT = (8.62 \times 10^{-5}\text{ eV/K}) \times (300\text{ K})$$
$$= 0.0259\text{ eV}$$

The ratio of upper to the lower energy levels i.e.,

$$\frac{E_2}{E_1} = e^{(-E_2 - E_1)/kT} = e^{-2.26/0.0259}$$
$$= e^{-87.3}$$
$$\frac{E_2}{E_1} = \mathbf{1.3 \times 10^{-38}}$$

Given

$$E_2 / E_1 = \frac{1}{2}$$

Then by using above equation

$$\frac{E_2}{E_1} = \frac{1}{2} = e^{-(E_2 - E_1)/kT}$$

or $$e^{(E_2 - E_1)/kT} = 2 \quad \text{or} \quad \frac{E_2 - E_1}{kT} = \log_e^2$$

or $$T = \frac{E_2 - E_1}{K \log e\, 2} = \frac{2.26\text{ eV}}{\left(8.62 \times 10^{-5} \frac{\text{eV}}{K}\right) \times (0.693)} = 37832.75\text{ K}$$

or $$\mathbf{T = 37832\ K}$$

This temperature is much hotter than the sun.

Example 4 (a) A He-Ne laser of wavelength 6328 Å has an internal beam of radius 0.23 mm. What would be the beam divergence angle?

(b) What lower limit might be expected for the beam divergence, if we can control the beam-waist radius (r_0) by lower cavity design and selecting the wavelength. By what factor will the beam divergence decrease if we design a laser having a beam waist of 2.4 mm radius and wavelength 2000 Å?

Solution

(a) Given $\lambda = 6328$ Å, $r = 0.23$ mm

Formula used is

$$\theta = \frac{\lambda}{\pi r_0} = \frac{6328 \times 10^{-10}\text{m}}{3.14 \times 2.3 \times 10^{-4}\text{m}} = \mathbf{8.76 \times 10^{-4}\ rad.}$$

i.e., beam radius increases about 8.76 cm at every 100 m distance.

(b) Given, $\lambda = 2000$ Å, $r = 2.4$ mm

then

$$\theta = \frac{\lambda}{\pi r_0} = \frac{2000 \times 10^{-10}\text{m}}{3.14 \times 2.4 \times 10^{-3}\text{m}} = \mathbf{2.65 \times 10^{-5}\ rad.}$$

i.e., about 33 fold decrease in beam spread over the He-Ne laser described in part (a) and in this case beam radius increases about 2.65 mm at every 100 m distance.

Example 5 A pulsed ruby laser consists of ruby crystal in the form of a cylinder of size 6.0 cm in length and 1.0 cm in diameter. Ruby laser is made of Al_2O_3 crystal in our case one aluminium ion in every 3500 has been replaced by chromium ion Cr^{3+} ion and these same ions also produce laser light which occurs by three level mechanism at a wavelength of 6944 Å. [Given density (ρ) of Al_2O_3 = 3700 kg/m^3 and Molar mass = 0.102 kg/mol.]

Solution Given, length (l) = 6.0×10^{-2} m, diameter (D) = 1.0×10^{-2} m, λ = 6944 Å, density (ρ) of Al_2O_3 = 3700kg/m^3, Molar Mass M = 0.102 kg/Mol.

Formula used for no. of aluminium ions is

$$N_{\text{Al}} = \frac{2\text{N.m}}{M} = \frac{2N \cdot \rho \cdot V}{M}$$

where m is the mass of ruby cylinder and factor 2 accounts for two aluminium ions in each molecule of Al_2O_3. The volume V is given as

$$V = \pi r^2 l = \pi\left(\frac{D}{2}\right)^2 l = \frac{\pi}{4} D^2 l$$

$$= \frac{1}{4} \times 3.14 \times (1.0 \times 10^{-2})^2 \times 6.0 \times 10^{-2}$$

$$= 4.7 \times 10^{-6} \text{m}^3$$

Thus,

$$N_{\text{Al}} = \frac{2 \times (6.0 \times 10^{23} \text{per mol}) \times (3.7 \times 10^3 \text{kg/m}^3) \times 4.7 \times 10^{-6} \text{m}^3}{0.102}$$

$$= 2.1 \times 10^{23}$$

and the number of chromium ions Cr^{3+} ions is given by

$$N_{\text{cr}} = \frac{N_{\text{Al}}}{3500} = 6.0 \times 10^{19}$$

The energy of the stimulated emission photon is given by

$$E = h\nu = \frac{hc}{\lambda} = \frac{4.1 \times 10^{-15} \text{eV. sec} \times 3 \times 10^8 \text{m/sec}}{6.944 \times 10^{-7} \text{m}}$$

$$= 1.8 \text{ eV}.$$

And hence the total energy due to all the pulses is given by

$$E_{\text{total}} = N_{\text{cr}} \cdot E$$

$= 6.0 \times 10^{19} \times 1.8$ eV $\times 1.6 \times 10^{-19}$ J/eV = **17 Joules**

Example 6 Calculate the power per unit area delivered by a laser pulse of energy 4.0×10^{-3} Joule, the pulse length in time as 10^{-9} sec and when the pulse is focused on target to a very small spot of radius 1.5×10^{-5} m.

Solution Given $P = 4.0 \times 10^{-3}$ J, $r = 1.5 \times 10^{-5}$ m

Formula used for power delivered per unit area is given by

$$I = \frac{P}{A}, \text{ where } P = \frac{4.0 \times 10^{-3} \text{J}}{10^{-9} \text{ sec.}}$$

or $\qquad P = 4.0 \times 10^6$ W

and $A = \pi r^2 = 3.14 \times (1.5 \times 10^{-5})^2 = 7.065 \times 10^{-10}\ \text{m}^2$

so $$I = \frac{P}{A} = \frac{4.0 \times 10^6\ \text{W}}{7.065 \times 10^{-10}\ \text{m}^2} = 5.7 \times 10^{15}\ \text{W/m}^2$$

or $$\mathbf{I = 5.7 \times 10^{15}\ W/m^2}$$

EXAMPLE 7 A laser beam has wavelength of 7200 Å and aperture 5×10^{-3}. The laser beam is sent to moon at a distance 4×10^8 m from the earth. Determine (a) angular spread and (b) a real spread when it reaches the moon.

SOLUTION Given $\lambda = 7.2 \times 10^{-7}$ m,

$$\text{radius } r = \frac{d}{2} = 2.5 \times 10^{-3}\ \text{m}, D = 4.0 \times 10^8\ \text{m}$$

Formula used is

(a) Angular spread (θ)

$$= \frac{0.637\lambda}{r}$$

$$\theta = \frac{0.637 \times 7.2 \times 10^{-7}}{2.5 \times 10^{-3}}$$

or $$= 1.834 \times 10^{-4}\ \text{radian}$$

(b) Areal spread = $(\theta D)^2 = (4 \times 10^8 \times 1.834 \times 10^{-4})^2$

$$\mathbf{= 53.85 \times 10^8\ m^2}$$

EXAMPLE 8 A 0.1 W laser beam with an aperture of 5.0 mm emits a light of wavelength 6943 Å. Calculate the areal spread and intensity of the image when the beam is focused with a lens having focal length 100 mm.

SOLUTION Given:

$$\text{radius of aperture} = \frac{\text{diameter}}{2}$$

or $$r = 2.5 \times 10^{-3}\ \text{m}, \lambda = 6.943 \times 10^{-7}\ \text{m}, f = 0.1\ \text{m}, P = 0.1\ \text{W}$$

Formula used is

$$\text{Angular spread } (\theta) = \frac{0.637\lambda}{r}$$

$$\theta = \frac{0.637 \times 6.943 \times 10^{-7}\ \text{m}}{2.5 \times 10^{-3}}$$

or $$\theta = 1.769 \times 10^{-4}\ \text{radius}$$

$$\text{A real spread} = (\theta \cdot D)^2 = (\theta \cdot f)^2\ (\therefore D = f)$$

$$= (1.769 \times 10^{-4} \times 0.1\ \text{mm})^2$$

$$\mathbf{= 3.129 \times 10^{-10}\ m^2}$$

and the intensity is given by

$$I = \frac{\text{Power } (P)}{\text{Area } (A)} = \frac{P}{A} = \frac{0.1\ \text{W}}{3.129 \times 10^{-10}\ \text{m}^2}$$

$$\mathbf{= 3.196 \times 10^8\ W/m^2}$$

EXAMPLE 9 For an ordinary source, the coherence time $\tau_c = 10^{-10}$ sec. Obtain the degree of non-monochromaticity for $\lambda_o = 5400$ Å.

SOLUTION Given $\tau_c = 10^{-10}$ sec

$$\Delta\nu = \frac{1}{\tau_c} = \frac{1}{10^{-10}} = 10^{10}\,\text{Hz}$$

For $\lambda_0 = 5400$ Å, $\nu_0 = \dfrac{c}{\lambda_0} = \dfrac{3.0\times10^{8}}{5400\times10^{-19}} = \dfrac{1}{18}\times10^{16}$

degree of non-monochromaticity

$$\frac{\Delta\nu}{\nu_0} = \frac{18\times10^{10}}{10^{16}} = 18\times10^{-6} = \mathbf{0.000018}$$

OBJECTIVE TYPE QUESTIONS

Q.1 LASER is a short form of
(a) Light Amplification Stimulated Emission Radiation
(b) Light Amplification by Stimulated Emission of Radiation
(c) Light Absorption by Stimulated Emission of Radiation
(d) Light Absorption by Spontaneous Emission of Radiation

Q.2 Mention the process under which an electron jumps from higher energy state to lower energy state by the influence of incident photon
(a) induced emission (b) spontaneous emission
(c) simple emission (d) none of these.

Q.3 Laser beam is
(a) highly monochromatic (b) highly coherent
(c) highly collimated (d) all of these.

Q.4 What is the life-time of electron in metastable state?
(a) 10^{-3} sec (b) 10^{-5} sec (c) 10^{-8} sec (d) 10^{-7} sec

Q.5 The number of atoms in the higher energy state is larger than lower energy state. This state is known as
(a) metastable state (b) ordinary state (c) excited state (c) none of these.

Q.6 In the population inversion
(a) the number of electrons in higher energy state is more than the ground state
(b) the number of electrons in lower energy state is more than higher energy state
(c) the number of electrons in higher and lower energy state are same
(d) none of them.

Q.7 The relations between Einstein's coefficient A and B is
(a) $\dfrac{8\pi h\nu^3}{c^3}$ (b) $\dfrac{8\pi^2 h^2\nu^3}{c^3}$ (c) $\left(\dfrac{2\pi h\nu}{c}\right)^3$ (d) $\dfrac{8\pi hc}{\lambda}$

Q.8 Laser beam is made of
(a) electrons (b) highly coherent photons
(c) very light and elastic particles (d) none of them

Q.9 In ruby laser which ions give rise to the laser action?
(a) Al_2O_3 (b) Al^{3+} (c) Cr^{3+} (d) none of them

Q.10 The output beam in ruby laser is
(a) continuous (b) discontinuous (c) both (a) & (b) (d) none of these

Q.11 Which one of the following laser have highest efficiency, ruby, He-Ne and semiconductor and carbon dioxide?
(a) ruby (b) semiconductor (c) He–Ne (d) carbon-dioxide

Q.12 The He–Ne laser produces the laser beam of wavelengths
(a) 6943Å (b) 6328Å (c) 6320Å (d) 6940Å.

Q.13 In He–Ne laser the ratio of the He to Ne is
(a) 10:1 (b) 1:10 (c) 100:1 (d) none of these.

Q.14 The method of population inversion to the laser action in He–Ne laser is:
(a) molecular collision (b) direction conversion
(c) electric discharge (d) electron impact.

Q.15 Ruby laser produces the laser beam of wavelength
(a) 6943 Å (b) 6328 Å (c) 6320 Å (d) 6940 Å.

Q.16 Characteristics of laser beam are
(a) highly directional (b) highly intense
(c) highly monochromatic (d) all of them.

Q.17 Holography was discovered by Dennis Gabor in
(a) 1948 (b) 1847 (c) 1748 (d) none of these.

Q.18 Holography records intensities and phases of light coming from an object on holographic plate has
(a) complete information of object (b) incomplete information of object
(c) no information of object (d) none of these.

Q.19 Holography produces the image
(a) real (b) virtual (c) both (a) & (b) (d) none of these.

Q.20 Which of the following statement is correct?
(a) Holography has been used to see the working condition of inner organs of the body in three dimension
(b) data storage
(c) in non-destructives testing of materials
(d) all of these.

Q.21 Information carrying capacity of hologram is
(a) large (b) small (c) zero (d) none of these.

PRACTICE PROBLEMS

Q.1 What do you mean by laser and its working principle, important requirements and applications?

Q.2 (a) Explain the term 'absorption', 'spontaneous' and 'stimulated' emission of radiation. Obtain a relation between transition probabilities of spontaneous and stimulated emission.
(b) What are Einstein's coefficient? Derive Einstein relation.

Q.3 Explain the construction and working principle of Ruby laser.

Q.4 What is the principle of laser?

Q.5 Discuss salient characteristics of laser beam.

Q.6 Describe various applications of laser?

Q.7 Explain the characteristics of laser beam. What are the necessary conditions for Lasing action?

Q.8 Explain the concept of directionality and monochromaticity as applied to lasers.

Q.9 Discuss Einstein's coefficients. Derive relation between them.

Q.10 Explain the terms: spontaneous and stimulated emission, population inversion, optical pumping.

Q.11 What is meant by population inversion and how is it achieved in practice.

Q.12 (i) What do you mean by spontaneous emission? Explain in detail various properties of Laser which makes it a superior source to study the phenomenon of interference.

(ii) Explain the working of Carbon-dioxide laser.

(iii) What is Nd YAG laser and explain its working principle?

Q.13 Define the following terms.

(a) Population inversion (b) Pumping (c) Active system

Q.14 What do you understand by solid state laser? Describe the principle, construction and working of Ruby Laser.

Q.15 Explain with the help of a neat diagram the working of Ruby laser.

Q.16 Explain the concept of coherence in lasers. What are necessary conditions for Lasing action? Give main components of a Ruby or He-Ne laser and principle of laser action.

Q.17 Explain the terms stimulated emission of radiation and optical pumping. Explain how lasers can be produced by He–Ne gas. How is it superior to a Ruby laser?

Q.18 Explain with the help of a neat diagram the principle and working of a He-Ne laser.

Q.19 What are the specialties of laser light? Give the description of He–Ne laser.

Q.20 Write a note on He–Ne laser.

Q.21 (a) Discuss with suitable diagrams the principle, construction and working of Helium–Neon laser.

(b) Describe the various applications of lasers.

Q.22 Discuss the essential requirements for producing laser action. Describe a He–Ne laser.

Q.23 What conditions must be fulfilled for a semiconductor laser? Explain.

Q.24 Give description of semiconductor laser and discuss the characteristics of laser beam.

Q.25 Describe the construction and working of a semiconductor laser with necessary diagrams.

Q.26 How does a semiconductor laser differ from other lasers? Explain main features of the semiconductor laser.

Q.27 Mention how does a semiconductor laser differ from Ruby laser.

Q.28 Write a note on the following.

(a) Laser (b) Solid state lasers (c) Semiconductor laser

Q.29 What is holography? What is the difference between holography and photography?

Q.30 What is the fundamental principle of a hologram? How is it produced and how is the image constructed from it?

Q.31 Give briefly the requirements for holography and mention the various properties of a hologram.

Fibre Optics

5

LEARNING OBJECTIVES

After reading this chapter you will be able to

- **LO1** Understand the concept of optical fibre
- **LO2** Know about types of optical fibres
- **LO3** Learn about acceptance angle, numerical aperture, skip distance and relative refractive index
- **LO4** Explain fibre optic communication
- **LO5** Illustrate optical fibre sensors, connectors, and couplers
- **LO6** Discuss the applications of optical fibre couplers

Introduction

In communication systems, there has been a frequent use of either the radiowaves or the microwaves in the form of carrier waves for sending the information. However, the advent of the laser in 1960 revolutionised the telecommunication and networking areas with an immediate appreciation of the potential benefits of sending information from one place to the other using light, as the laser is a coherent source of light waves. It is worth mentioning that at higher optical frequencies ($\sim 10^{15}$ Hz), one hundred thousand times more information can be carried compared to microwaves. However, the energy of light waves gets dissipated in open atmosphere. So it cannot travel long distances and hence a guiding channel is required to guide them just like a metal wire is required to guide electrical currents. This purpose is solved with the use of optical fibre. Optical fibre is a very thin glass or plastic conduit designed to guide light waves along the length of the fibre. As long as the refractive index of this fibre is greater than that of its surrounding medium, the light shall suffer a large number of total internal reflections and hence much of the light launched into one end will emerge from the other end due to small losses.

Fibre optics is a technology that uses glass, plastic, threads or fibres to transit data. A fibre optic cable consists of a bundle of glass threads (Fig. 5.1) which are protected by the cable's outer covering of treated paper, PVC or metal, called a jacket. Optical fibre has a number of advantages over the copper wire used to make connections electrically. For example, optical fibre, being made of glass or sometimes plastic, is protected from electromagnetic interference such as is caused by thunderstorms. A single optical fibre has its parts as core, cladding and sheath (protecting layer), as shown in Fig. 5.2. Core is thin glass centre of the fibre where the light travels. Cladding is outer optical material surrounding the core that reflects the light back into the core because cladding has lower refractive index. Sheath is plastic coating that protects the fibre from damage and moisture.

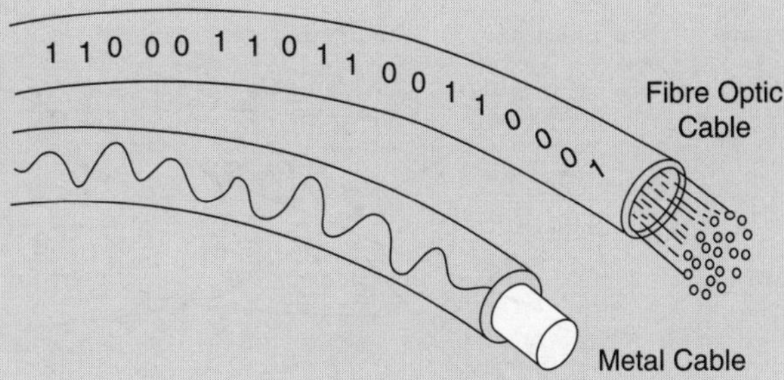

FIGURE 5.1

In order to understand the advantages of fibre optics, it is necessary to know about the bandwidth in general. Bandwidth is the difference between the upper and lower cutoff frequencies of a filter, a communication channel, or a signal spectrum. It is typically measured in Hertz. In the case of a lowpass filter or baseband signal, the bandwidth is equal to its upper cutoff frequency. In radio communications, bandwidth is the range of frequencies occupied by a modulated carrier wave. For example, an FM radio receiver's tuner spans a limited range of frequencies. In optics, it is the width of an individual spectral line or the entire spectral range.

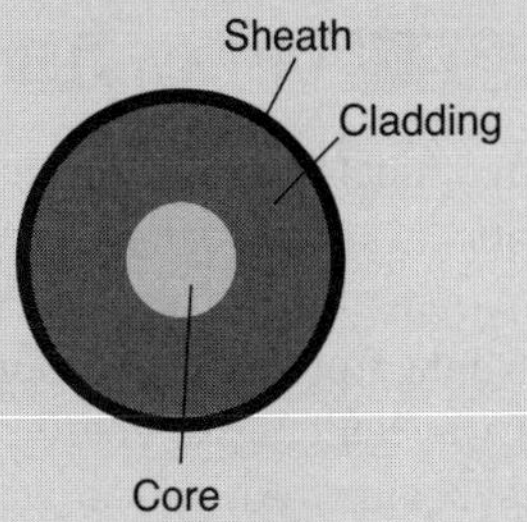

FIGURE 5.2

Fibre optics has many advantages compared with traditional metal communications lines, which are listed as follows:

(i) Fibre optic cables can carry more data as their bandwidth is greater than metal cables.
(ii) Fibre optic cables are less susceptible than metal cables to interference.
(iii) Fibre optic cables are much thinner and lighter than metal wires.
(iv) Through fibre optic cables the data can be transmitted digitally rather than analogically.
(v) Attenuation through fibre optic cables is very low in transmitting the data over a long distance, so there is so need of repeaters.

5.1 FUNDAMENTAL IDEAS ABOUT OPTICAL FIBRE — LO1

Optical fibres use light to carry digital signals and as mentioned earlier the base of this technology is the concept of total internal reflection. The digital signal that is carried by the light is reflected inside the optical cable and hence transfers the information. Main concepts of physics that are involved in optical fibres are refraction, refractive indices, critical angle and total internal reflection. In refraction, the light wave bends away from the normal when it propagates from a higher refractive index medium to a lower refractive index medium. The phenomenon of total internal reflection takes place when the angle of refraction becomes 90°. The incident angle at which the angle of refraction (transmitting) is equal to 90° is called *critical angle*. When a light wave propagating from a higher refractive index medium to a lower refractive index medium has a sufficiently large angle, i.e., greater than the critical angle, the light gets reflected back into the same medium. For a particular case of an optical fibre whose core is made of glass which is bounded by a plastic cladding, the critical angle is 82°. Therefore, the light when hits the plastic cladding at an angle more than 82° would be reflected back in the same medium, i.e., back to the glass core. This is shown in Fig. 5.3.

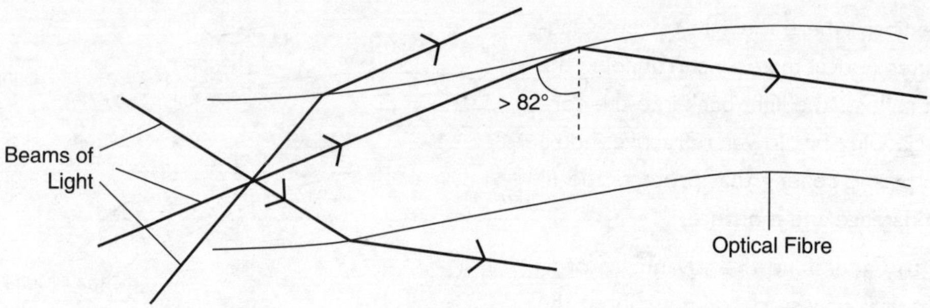

Figure 5.3

5.2 OPTICAL FIBRES AS A DIELECTRIC WAVEGUIDE LO2

An optical fibre is a dielectric waveguide with a very high bandwidth. It guides electromagnetic waves in an optical spectrum, the same way as microwaves are guided by rectangular or cylindrical metallic waveguides. An optical fibre confines the propagating waves inside it by utilizing the property of total internal reflection of light from a dielectric interface (i.e., the interface between two dielectric materials), whereas the waves in waveguides are confined within these structures by the reflection of the waves from the walls of the waveguides. An optical waveguide has a circular cross section, which is made in such a way that the outer dielectric (cladding) has lower refractive index than that of the inner one (core). Due to different refractive indices, the phenomenon of total internal reflection takes place in the optical fibre.

The transmission of light in optical fibres over large distances is possible with minimum loss of data. These fibres occupy less space and are light in weight in comparison with the waveguides or transmission lines. Moreover, the optical fibre has greater tensile strength. However, optical fibres are costly and also, they need more protection than waveguides or transmission lines. The other disadvantage is that attenuation of signals takes place in optical fibres. The attenuation is mainly due to Rayleigh scattering (which is inversely proportional to the fourth power of wavelength), absorption due to impurities and radiation of light due to bending of the core.

5.3 TYPES OF OPTICAL FIBRES LO2

Optical fibres are categorised based on their transmission properties and the structure. These can be classified into two types, one of which is single mode fibre and the second one is multimode fibre. The core size is the basic structural difference in the optical fibres.

5.3.1 Single Mode Step Index Fibre

A single mode fibre is called single (mono) mode step index fibre because the refractive index of the fibre "steps" up as we move from the cladding to the core and this fibre allows single mode to propagate at a time due to very small diameter of its core (Fig. 5.4a). In this fibre, the refractive indices of the cladding and the core remain constant. The size of its core (diameter) is typically around 10 μm. Single mode fibres have a lower signal loss and a higher information capacity or bandwidth than multimode fibres (introduced later) as the signal loss depends on the operational wavelength. These fibres are capable of transferring higher

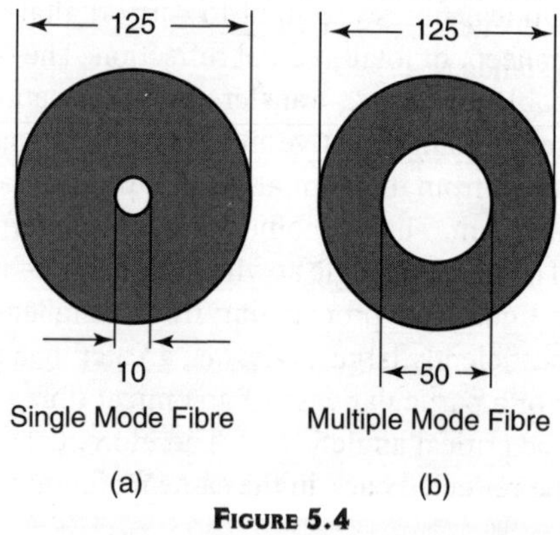

Figure 5.4

amount of data due to low fibre dispersion. In these fibres, the wavelength can increase or decrease the losses caused by fibre bending. In general, single mode fibres are considered to be low loss fibres, which increase system bandwidth and length. So these fibres are most useful for large bandwidth applications. Since these fibres are more resistant to attenuation, they can also be used in significantly longer cable runs.

5.3.2 Multimode Fibres

As the name implies multimode fibres allow more than one mode to propagate. Over 100 modes can propagate through multimode fibres at a time. Multimode fibre is sometimes abbreviated as MMF. The size of its core is typically around 50 μm (Fig. 5.4b). The multimode fibre is of two types, namely step index and graded index fibres.

5.3.2.1 Multimode Step Index Fibres

Multimode step index fibre is shown in Fig. 5.5 along with the refractive indices of its core and cladding. In this type of optical fibre, the number of propagating modes depends on the ratio of core diameter and the wavelength. This ratio is inversely proportional to the numerical aperture (abbreviated as NA and defined later). Typically the core diameter is 50 μm to 100 μm and NA varies from 0.20 to 0.29, respectively. Multimode fibre is used in short lengths, such as those used in Local Area Networks (LANs) and Storage Area Networks (SANs). Because the multimode optical fibre has higher NA and the large core size, fibre connections and launching of light has become very easy. Multimode fibres permit the use of light emitting diodes (LEDs). In such fibres, core-to-core alignment is less critical during fibre splicing. However, due to several modes the effect of dispersion gets increased, i.e., the modes arrive at the fibre end at slightly different times and so spreading of pulses takes place. This dispersion of the modes affects the system bandwidth. Therefore, the core diameter, NA, and index profile properties of multimode fibres are optimized to maximize the system bandwidth.

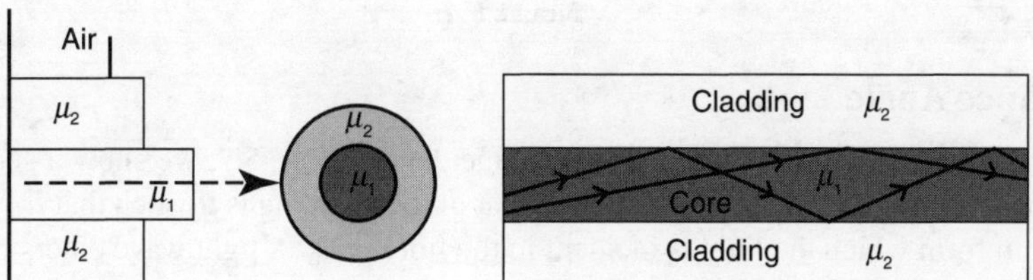

FIGURE 5.5

5.3.2.2 Multimode Graded Index Fibres

In a multimode graded index optical fibre, the refractive index of the core decreases with increasing radial distance from the fibre axis, which is the imaginary central axis running along the length of the fibre (Fig. 5.6). The value of the refractive index is highest at the centre of the core and decreases to a value at the edge of the core that equals the refractive index of the cladding. Therefore, the light waves in the outer zones of the core

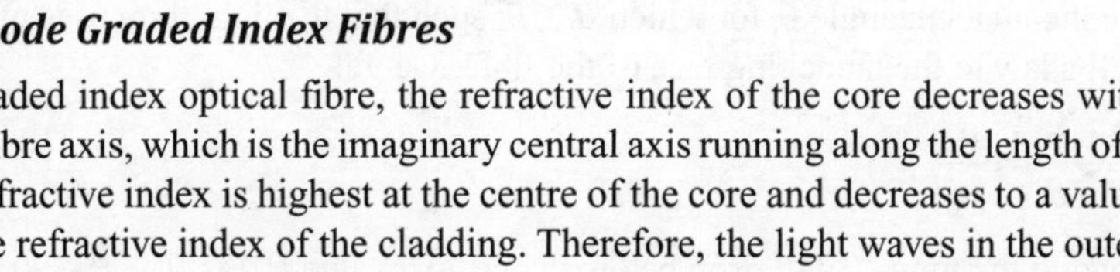

FIGURE 5.6

travel faster than those in the centre of the core. Thus the dispersion of the modes is compensated by this type of fibre design. Under this situation, the light waves follow sinusoidal paths along the fibre. In such fibres, the most common profile of the refractive index is very nearly parabolic that results in continual refocusing of the rays in the core, and minimizing modal dispersion. Standard graded index fibres typically have a core diameter of 50 μm or 62.5 μm and a cladding diameter of 125 μm. It is typically used for transmitting the information to the distances of a couple of kilometers. The advantage of the graded index fibre in comparison with multimode step index fibre is the considerable decrease in modal dispersion.

5.4 ACCEPTANCE ANGLE AND NUMERICAL APERTURE LO3

In order to propagate or transmit the light wave through the optical fibre, it is necessary to launch the light at angles that fall within certain range. The maximum limit of this angle is decided by the acceptance angle.

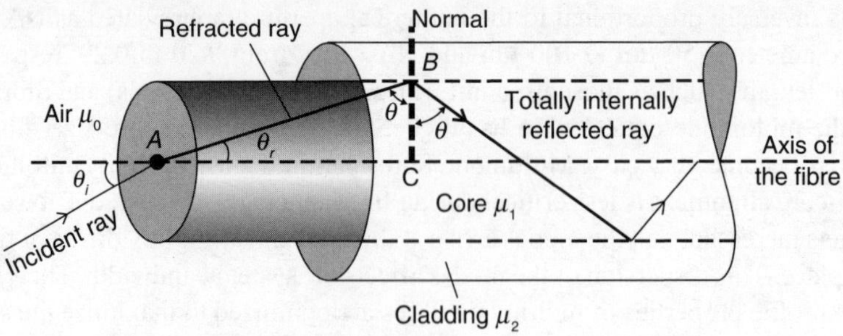

FIGURE 5.7

5.4.1 Acceptance Angle

Let us consider an optical fibre into which the light is incident. In Fig. 5.7, we show a section of cylindrical optical fibre. The refractive index of the core is μ_1 and that of the cladding is μ_2 such that $\mu_1 > \mu_2$. The refractive index of the medium form which the light is incident in the fibre is μ_0. A light wave enters the fibre at an angle θ_i with the axis of the fibre. This wave gets refracted at an angle θ_r and strikes core-cladding interface at an angle θ. If θ is greater than the critical angle θ_c, the wave undergoes total internal reflection at the interface, since $\mu_1 > \mu_2$. As long as the angle θ is greater than θ_c, the light will stay within the core of the fibre.

Let us now compute the incident angle θ_i for which $\theta_i \geq \theta_c$ such that the light refocuses within the core of the fibre. Applying Snell's law to the launching face of the fibre, we get

$$\frac{\sin\theta_i}{\sin\theta_r} = \frac{\mu_1}{\mu_0} \tag{i}$$

If θ_i is increased beyond the limit, θ will drop below the critical value θ_c (as $\theta_r + \theta = 90°$, in ΔABC) and the ray escapes from the side walls of the fibre. The largest value of θ_i occurs when $\theta = \theta_c$. This value of θ_i we represent by θ_i max. From the ΔABC, it is seen that

$$\sin\theta_r = \sin(90° - \theta) = \cos\theta, \text{ (as } \theta_r + \theta = 90°) \tag{ii}$$

From Eqs. (i) and (ii), we get

$$\frac{\sin\theta_i}{\sin\theta_r} = \frac{\mu_1}{\mu_0} \text{ or } \sin\theta_i = \frac{\mu_1}{\mu_0}\cos\theta$$

when $\theta = \theta_c$, $\quad \sin\theta_i = \dfrac{\mu_1}{\mu_0}\cos\theta_c$ (iii)

At critical angle, $\sin\theta_c = \dfrac{\mu_2}{\mu_1}$ (as $\theta = 90°$)

$$\therefore \quad \cos\theta_c = \sqrt{1-\sin^2\theta_c}$$

$$= \sqrt{1-\frac{\mu_2^2}{\mu_1^2}}$$

or $$\cos\theta_c = \sqrt{\frac{\mu_1^2-\mu_2^2}{\mu_1^2}} \quad \text{(iv)}$$

By putting the value of $\cos\theta_c$ from Eq. (iv) into Eq. (iii), we get

$$\sin\theta_{i\,\max} = \frac{\sqrt{\mu_1^2-\mu_2^2}}{\mu_0} \quad \text{(v)}$$

If the incident wave of light is launched from air medium (for which $\mu_0 = 1$), then putting $\theta_{i\,\max} = \theta_0$, Eq. (v) may be simplified to

$$\sin\theta_0 = \sqrt{\mu_1^2-\mu_2^2} \text{ or } \theta_0 = \sin^{-1}\left(\sqrt{\mu_1^2-\mu_2^2}\right) \quad \text{(vi)}$$

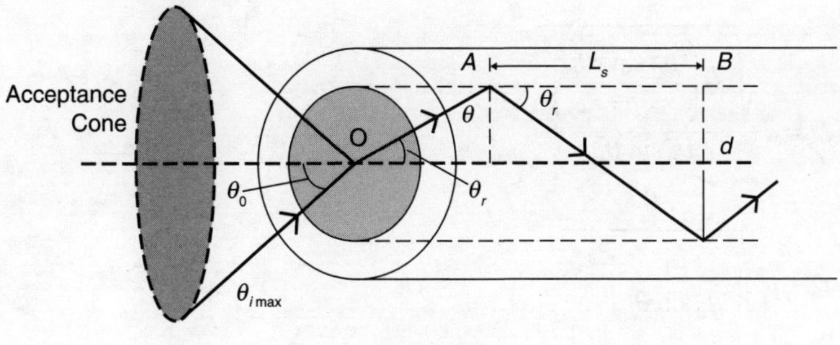

Figure 5.8

The angle θ_0 is called the acceptance angle of the fibre, which may be defined as the maximum angle that a light wave can have relative to the axis of the fibre for its propagation through the fibre. The light wave contained within the cone having a full angle $2\theta_0$ are accepted and transmitted along the fibre. Therefore, the cone associated with the angle $2\theta_0$ is called the *acceptance cone* (Fig. 5.8). The light incident at an angle beyond θ_0 refracts through the cladding. As at every internal reflection the light will be lost being incident at an angle less than the critical angle, the corresponding optical energy is lost. It is also obvious that the acceptance angle would be larger if the diameter of the cone is larger.

5.4.2 Numerical Aperture

Numerical aperture (NA) is the most important parameter of an optical fibre. It is a measure of how much light can be collected by an optical system such as an optical fibre or a microscope lens. Based on the refractive indices of core and cladding, we can measure the values of NA. It is defined as the sine of the acceptance angle if the end faces of the fibre are exposed to a medium for which $\mu_0 = 1$ (air). Otherwise, the numerical aperture is defined as $NA = \mu_0 \sin \theta_0$

For $\mu_0 = 1$, $\quad NA = \sin \theta_0 = \sqrt{\mu_1^2 - \mu_2^2}$

This relation shows that the light gathering ability of an optical fibre increases with its numerical aperture. Since the maximum value of $\sin \theta_0$ can be 1 only, the value of NA cannot exceed 1. It means the largest value of NA is unity. When $\theta_0 \approx 90°$, the fibre totally reflects all the light entering its face. Fibres with a wide variety or numerical apertures running from about 0.2 up to 1.0 and including 1.0 may commercially be obtained.

5.4.3 Skip Distance

It is well known that the light propagates in the optical fibre based on the principle of total internal reflection. The light ray gets reflected from the walls of the fibre. The distance between the two successive reflections of a ray of light propagating in the fibre is called the skip distance L_s. In Fig. 5.8, the distance AB is the skip distance, given by

$$L_s = d \cot \theta_r,$$

where d is the diameter of the core of the fibre and θ_r is the angle of refraction in the core. We can write the above relation in terms of incidence angle θ_i and the refractive indices μ_1 and μ_0 by using Snell's law as $\frac{\sin \theta_i}{\sin \theta_r} = \frac{\mu_1}{\mu_0}$. This gives

$$\sin \theta_r = \frac{\mu_0}{\mu_1} \sin \theta_i \quad \text{or} \quad \cos \theta_r = \sqrt{1 - \frac{\mu_0^2}{\mu_1^2} \sin^2 \theta_i}$$

Hence

$$L_s = d \frac{\sqrt{1 - \left(\frac{\mu_0^2 \sin^2 \theta_i}{\mu_1^2} \right)^2}}{\frac{\mu_0 \sin \theta_i}{\mu_1}}$$

or

$$L_s = d \sqrt{\left(\frac{\mu_1}{\mu_0 \sin \theta_i} \right)^2 - 1}$$

It is clear that the inverse of the skip distance L_s, i.e., $1/L_s$ will give the total number of reflections made by the light ray in a given length of the fibre. For example, in a fibre of length L, the number of reflections N_r would be

$$N_r = \frac{L}{d \sqrt{\left(\frac{\mu_1}{\mu_0 \sin \theta_i} \right)^2 - 1}}$$

For example, in the case $\mu_1 = 1.60$, $\mu_0 = 1$, $\theta_i = 30°$ and $d = 0.05$ mm, we get the skip distance as 0.152 mm. Therefore, in 1 m of fibre there will be 6580 reflections.

5.4.4 Relative Refractive Index

It has been established that the refractive indices of the core and the cladding of the fibre are different. The difference of these two indices gives a measure of the relative refractive index difference. In this light, it can be obtained that

$$\mu_1^2 - \mu_2^2 = (\mu_1 + \mu_2)(\mu_1 - \mu_2) = \frac{(\mu_1 + \mu_2)}{2}\frac{(\mu_1 - \mu_2)}{\mu_1}2\mu_1$$

Now $\frac{(\mu_1 + \mu_2)}{2}$ is very nearly equal to μ_1 in view of $\mu_1 >> \mu_2$. Further, we put $\frac{(\mu_1 - \mu_2)}{\mu_1} = \Delta\mu_r$ in the above equation and obtain $\mu_1^2 - \mu_2^2 = 2\mu_1^2\Delta\mu_r$. In term of this relation the numerical aperture NA can be written as

$$NA = \sqrt{\mu_1^2 - \mu_2^2} = \mu_1\sqrt{2\Delta\mu_r}$$

Here $\Delta\mu_r$ is called the relative refractive index difference or *fractional refractive index*.

5.5 FIBRE OPTICS COMMUNICATION LO4

A general communication system was oftenly used before the development of optical fibre communication. This system employs its essential components as modulator or transmitter, transmission medium and the demodulator or receiver, as shown in Fig. 5.9a. However, optical fibres have replaced copper coaxial cables due to their various advantages as they are very light weight, thin conduit cables which provide greater communications capacity with lower loss.

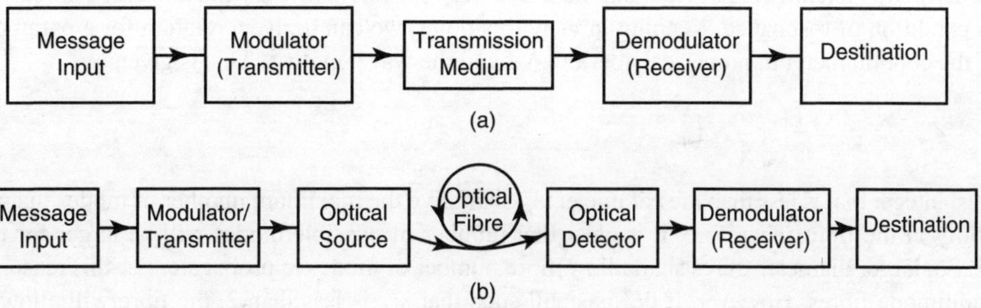

FIGURE 5.9

A fibre optic communication system from signal source to signal output is shown in Fig. 5.9b. Here, the information that is to be transmitted is first converted into an optical signal from an electrical signal. Then the optical signal is converted to an electrical signal after transmission by an optical fibre. Independent of the original nature of the signal, a fibre provides the choice of format of transmission as analog or digital because these two formats are convertible into one another. So the signal in analog or digital form is impressed onto the carrier wave by using a modulator. The carrier wave is generated from the carrier source which may be either light emitting diode (LED) or laser diode (LD). This carrier wave is modulated using various techniques viz., frequency modulation, amplitude modulation and digital modulation. The carrier source output into the optical fibre is represented by a single pulse. When a pulse is passed through a fibre, then it is attenuated and distorted due to several mechanism for example by intermodal distortion. Therefore, repeaters and regenerators are used to amplify the light signal at several positions of the fibre. And after that the light signal is coupled into a detector that may be a semiconductor device or most commonly a PIN diode at the end of a fibre. This changes the optical signal back into an electrical signal. The response of a detector should be well matched with the optical frequency of the signal received. The output of the detector then passes through the signal processor, which is used to capture the original electrical signal from the carrier by using the process of filtering, amplification and an analog to digital conversion. The signal output is finally communicated by the cathode ray tube (if it is video signal), by loudspeaker (if it is audio signal) or by computer input (if it is digital signal).

5.5.1 High Bit Rate Optical Fibre Communication

A high bit rate signal carried on copper wire transmission line is generally needed to be amplified every 300 m. However, high bit rate signals when carried on optical fibres need such amplification only every 100 km or so. As discussed earlier, a detector changes the optical signal back into an electrical signal, the light signal is coupled to the detector changes the optical signal back into an electrical signal, the light signal is coupled to the detector at the remote end of the fibre. This is done effectively when the response of the detector is well matched with the optical frequency of the signal received. Then a signal processor handles the detector output. The function of the signal processor is to recapture the original electrical signal from the carrier. This process involves filteration and amplification and a digital to analog conversion.

5.5.2 Allowed Modes and Normalised Frequency

It appears from the theory of acceptance cone that every ray shall propagate successfully once it enters the fibre within its acceptance cone. However, this is not the case always and only certain ray directions or modes are allowed to propagate successfully. Actually any ray represents plane waves that move up and down in the fibre. Evidently such waves overlap and interfere with one another. Only those waves will sustain which satisfy a condition of resonance. Keeping in view this point, we can derive a relation for a parameter m_m in terms of the core diameter d, numerical aperture NA and the wavelength λ. This is given by

$$m_m = \frac{1}{2}\left(\pi \frac{d}{\lambda} NA\right)^2$$

The largest integer that is less than the parameter m_m shall give the maximum number of modes that propagate successfully in the fibre. Therefore, it is clear that number of possible modes will be larger for the higher ratio d/λ. So, larger diameter fibres shall allow more number of modes to propagate. For this reason, they are called multimode fibres. However, if d/λ is small such that m_m is less than 2, the fibre will allow only one mode. So this type of fibre is called *single mode fibre* or *monomode fibre*. The condition $m_m < 2$ for a single mode fibre can be achieved if

$$\frac{d}{\lambda} < \frac{2}{\pi\,(NA)}$$

The above condition related to the diameter guarantees the performance of single mode fibre. However, a more careful analysis reveals that the single mode performance can be achieved even if

$$\frac{d}{\lambda} < \frac{2.4}{\pi\,(NA)}$$

As is evident, the parameter m_m decides the number of possible modes. Since this parameter depends on core diameter d and the numerical aperture NA, the number of allowed modes would be different for fibres of different core diameters. The word "number" intuitively adds a concept of normalised frequency, given by

$$V_n = \pi \frac{d}{\lambda} NA = \frac{\pi d}{\lambda}\sqrt{\mu_1^2 - \mu_2^2}$$

A careful look indicates that the normalised frequency is nothing but the factor carried by the parenthesis of the parameter m_m. Therefore, in terms of normalised frequency v_n, the parameter m_m is written as

$$m_m = \frac{v_n^2}{2}$$

5.5.3 Attenuation

When light travels along the fibre, there is a loss of optical power, which is called attenuation. Signal attenuation is defined as the ratio of optical input power (P_i) to the output power (P_0). Optical input power is the power transmitted into the fibre from an optical source. Optical output power is the power received at the fibre end. The following relation defines the signal attenuation or absorption coefficient in terms of length L of the fibre.

$$\alpha = \frac{10}{L}\log_{10}\frac{P_i}{P_0}$$

So signal attenuation is a log relationship. Length L of the fibre is expressed in kilometers. In view of this, the unit of attenuation is decibles/kilometre i.e., dB/km. The causes of attenuation in an optical fibre are absorption, scattering and bending losses. Each mechanism of loss is influenced by the properties of fibre material and fibre structure. However, loss is also present at fibre connections. Absorption losses over a length L of fibre can be described by the usual exponential law for light intensity (or irradiance) I

$$I = I_0 e^{-\alpha L}$$

where I_0 is the initial intensity or the irradiance of the light.

The attenuation profile for a single mode cable is depicted in Fig. 5.10, which shows that the amount of attenuation is also wavelength dependent. In the figure, two absorption peaks at 1.0μm and 1.4 μm are observed which are respectively due to the peculiarities of the single mode fibre and the traces of water remaining in the fibre as an impurity. The wavelengths 1.31 μm and 1.55 μm are the two standard single mode wavelengths that are commonly used due to this water absorption peaks. However, now the wavelength 1.55 μm are used in view of the need to extend the distance between repeaters.

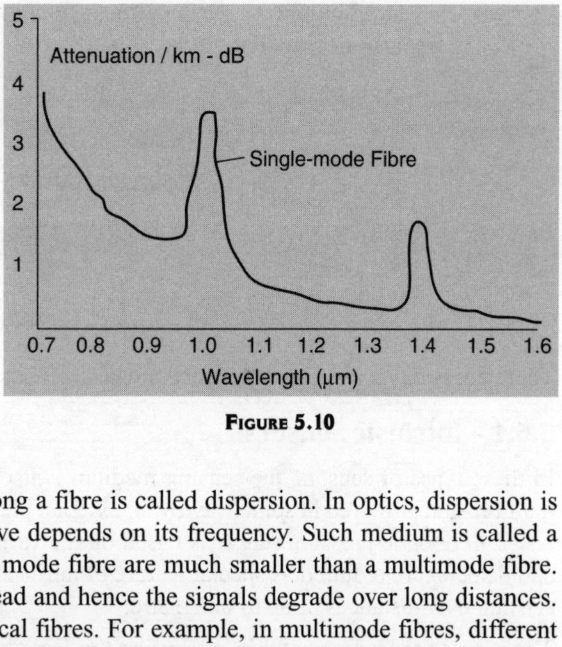

FIGURE 5.10

5.4.4 Pulse Dispersion in Optical Fibre

The spreading of pulses of light as they propagate along a fibre is called dispersion. In optics, dispersion is the phenomenon in which the phase velocity of a wave depends on its frequency. Such medium is called a *dispersive medium*. The dispersive effects in a single mode fibre are much smaller than a multimode fibre. Due to dispersion, optical pulses in optical fibres spread and hence the signals degrade over long distances. There are several factors that cause dispersion in optical fibres. For example, in multimode fibres, different axial speeds of different transverse modes cause intermodal dispersion that limits the performance of the fibre. In single mode fibres, though intermodal dispersion is eliminated, chromatic dispersion occurs because of the slight variation in the index of the glass with the wavelength of the light. Dispersion limits the bandwidth of the fibre because the spreading optical pulses limit the rate that pulses can follow one another on the fibre and still remain distinguishable at the receiver.

5.6 OPTICAL FIBRE SENSORS LO5

Optical fibre sensors are fibre based devices that are used for sensing some typical quantities like temperature or mechanical strain. These sensors are also sometimes used for several vibrations, pressure, acceleration, or concentrations of chemical species. The general principle of operation of fibre optic sensors is that when a light beam is sent through an optical fibre, then its parameters either in the fibre or in one or several fibre Bragg gratings experience subtle change. Then the light reaches a detector arrangement that measures these changes (Fig. 5.11). The light beam may be changed in five of its optical properties viz. intensity, phase, polarisation, wavelength and spectral distribution.

Optical fibre sensors have a number of advantages over other types of sensors.

(i) They consist of electrically insulating material, which makes possible their use in high voltage environments.

(ii) Since there is no risk of electrical sparks, even in the case of defects, these can be safely used in explosive environments.

(iii) They are immune to electromagnetic interference (EMI).

(iv) Their materials can be chemically inactive.

(v) They can operate over a broad range of temperature.

(vi) They have multiplexing capabilities, i.e., multiple sensors in a single fibre can be interrogated with a single optical source.

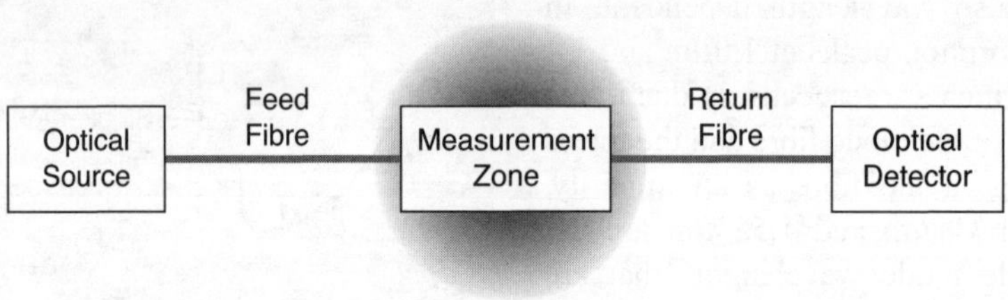

FIGURE 5.11

There are two types of sensors named intrinsic sensor and extrinsic sensor, which are dicussed below.

5.6.1 Intrinsic Sensors

In these types of sensors, the sensing medium is itself a fibre. It means the propagating light never leaves the fibre and is altered in some way by an external phenomenon. The simplest type of sensor called *intensity based fibre* optic pressure sensor is based on the variation of intensity, as in this case only a simple source and detector are required. A special feature of intrinsic fibre optic sensors is that they can provide distributed sensing over distances of up to one metre.

This type of sensor is useful in measuring the force being exerted between the two objects A and B, shown in Fig. 5.12. The fibre will become slightly deformed when the pressure is increased and it experiences increased microbending losses which results in a decrease in the light intensity received at the detector. A decrease in the pressure relieves stress on the fibre and hence there is an increase in transmitted light detected.

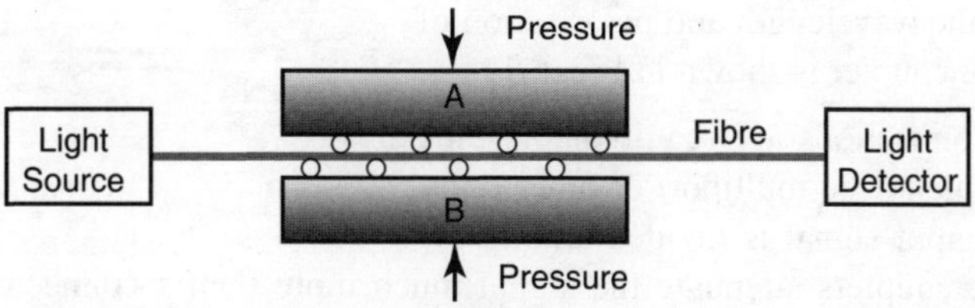

Figure 5.12

5.6.2 Extrinsic Sensors

In extrinsic sensors, the delivery of light and its collection is done by the fibre. Thus the propagating light leaves the fibre, is altered in some way, and is collected by the same or another fibre. These sensors are used to measure vibration, rotation, displacement, velocity, acceleration, torque, and twisting. A major benefit of these sensors is their ability to reach places which are otherwise inaccessible. For example, the temperature inside aircraft jet engines is measured by using a fibre that transmits radiation into a radiation pyrometer located outside the engine. The same way, extrinsic sensors can also be used to measure the internal temperature of electrical transformers, where the extreme electromagnetic fields present make other measurement techniques impossible.

An example of an intensity based extrinsic sensor is shown in Fig. 5.13, which detects any increase or decrease in the length/between the two fibres. The amount of light launched into the return fibre will decrease as the distance between the two fibres is increased. However, if the length is decreased the light intensity collected by the receiver will increase. This way these fibre optic sensors are capable of determining small shifts between objects.

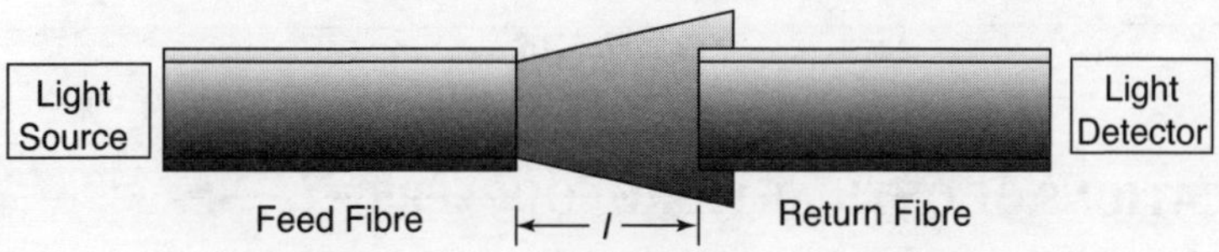

Figure 5.13

5.7 OPTICAL FIBRE CONNECTOR LO5

The end of an optical fibre is terminated by using an optical fibre connector, which enables quicker connection and disconnection than splicing. The coupling and alignment of the cones of fibres is done with the help of connector so that light can pass. Now a days a variety of optical fibre connectors is available. In general, the connectors are differentiated by their dimensions and methods of mechanical coupling. For example, some organisations standardize one kind of connector, depending on what equipment they commonly use, or per type of fibre. Now-a-days, small form factor connectors (for example, LC) and multifibre connectors (for example, MTP) are replacing the traditional connectors (for example, SC) in view of their datacom and telecom applications.

5.8 OPTICAL FIBRE COUPLERS LO5

Optical fibre couplers are fibre devices that are used for coupling light from one or several input fibres to one or several output fibres. Optical fibre couplers can distribute the optical signal (power) from one fibre among two or more fibres. So the light from an input fibre can appear at one or more outputs, with the distribution of power depending on the wavelength and polarisation of light. A two-by-two fibre coupler is shown in Fig. 5.14.

FIGURE 5.14

Simple point to point connections are required for some fibre optic data links, which need multiport or other types of connections. As the input signal is divided among the output ports, fibre optic couplers attenuate the signal much more than a connector or splice. Fibre optic couplers can be either active or inactive devices. An inactive coupler redistributes the optical signal without optical-to-electrical conversion, whereas the active couplers are electronic devices that split or combine the signal electrically and use fibre optic detectors and sources for input and output.

A basic fibre optic coupler is shown in Fig. 5.15 with N_1 input ports and N_2 output ports, which range from 1 to 64. The number of input ports and output ports vary depending on the intended application for the coupler. Types of fibre optic couplers include optical splitters, optical combiners, X couplers, star couplers, and tree couplers. Fibre couplers are usually directional couplers, which means that essentially no optical power sent into some input port can go back into one of the input ports.

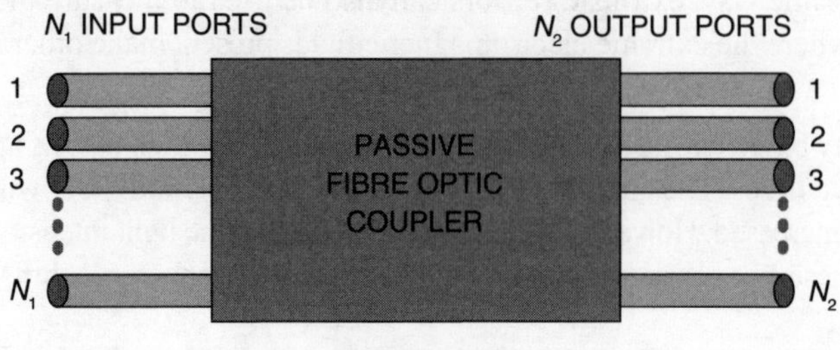

FIGURE 5.15

5.9 APPLICATIONS OF OPTICAL FIBRE COUPLERS LO6

Below we mention some typical applications of fibre couplers.

(i) Fibre couplers can be used in fibre interferometers.

(ii) In a cable TV system, the powerful signal from one transmitter is sent into a fibre splitter, which distributes the power signal over a large number of output fibres for different customers.

(iii) In fibre ring lasers, there is no resonator ends where light could be injected so a dichroic fibre coupler can be used to inject pump light within resonator. Then another fibre coupler is used as the output coupler.

(iv) For combining the radiation of several laser dioide in high power fibre lasers, we generally use multimode fibre couplers.

SUMMARY

The essence of the topics covered in this chapter is produced below.

- In communication systems, radiowaves or microwaves have been extensively used as the carrier waves for transmitting information. However, invention of laser led to the discovery of optical fibres which can carry hundred thousand times more information.

- Advantages of optical fibres over the traditional metal communication lines were discussed in view of their greater bandwidth, less susceptibility to interference, light weight, smaller thickness, and fast transmission of data.
- Based on transmission properties and the structure, we can categorize optical fibres as single mode fibre or multimode fibre. Typical diameter of core of the single mode fibre is 10 μm and that of multimode fibre ranges from 50 μm to 100 μm.
- Since the refractive index steps up when we move towards core side from the cladding side, these fibres are referred to as step index fibres.
- In order to compensate the mode dispersion, fibre is designed such that the refractive index of core and cladding match at their common boundary. Such type of fibre is called multimode graded index fibre where most commonly parabolic profile of the refractive index is used.
- It is not necessary that all the incident light rays shall transmit through the fibre. In this context, acceptance angle is an important parameter. The rays that fall within the acceptance cone are accepted for the transmission.
- Numerical aperture (NA) is the most important parameter of an optical fibre, which tells us how much light can be collected by an optical fibre.
- It is well known that the propagation of the light is based on its total internal reflection. The light gets reflected from the walls of the fibre. The distance between the two successive reflections of a light ray propagating in the fibre is called the skip distance. Inverse of this distance gives the total number of reflections made by the light in the fibre of a given length.
- Propagation mechanism of the information in optical fibres was discussed in detail along with a difference of components used in the general communication system.
- The acceptance cone only accepts the rays for their transmission in the fibre. It is not that the every ray shall propagate successfully once it enters within the acceptance cone. Only certain ray directions or modes are allowed to propagate successfully. Since ray represents plane waves that move up and down in the fibre, such waves overlap and interfere with one another. Only those waves will sustain which satisfy a condition of resonance. Such waves or modes are called allowed modes. Therefore, the concept of allowed modes and the normalised frequency were given and in support some theoretical relations were talked about.
- When light travels along the fibre, there is a loss of optical power. This is called attenuation. Signal attenuation is defined as the ratio of optical input power (P_i) to the optical output power (P_0) and is given by

$$\alpha = \frac{10}{L}\log_{10}\frac{P_i}{P_0}$$

- The unit of attenuation is decibels/kilometre, i.e., dB/km.
- In addition to the loss of power of the signal (pulse of light) that propagates in the fibre, there is spreading of pulses of light. This is called dispersion. So the dispersion was talked about in the case of optical fibre.
- The wonderful application of the optical fibres is in fibre optic sensors, which are fibre based devices that are used for sensing some typical quantities like temperature or mechanical strain. These sensors

are also sometimes used for sensing vibrations, pressure, acceleration, or concentrations of chemical species.

- The types of optical fibre sensors, namely intrinsic sensors and extrinsic sensors, were discussed. In the intrinsic sensors, the sensing medium is itself a fibre. So the propagating light never leaves the fibre and is altered by an external phenomenon. On the other hand, in the extrinsic sensors, the delivery of light and its collection is done by the fibre. Thus the propagating light leaves the fibre, is altered in some way, and is collected by the same or another fibre. The extrinsic sensors are used to measure vibration, rotation, displacement, velocity, acceleration, torque, and twisting.
- Another application of fibres is in optical fibre connectors and couplers. Optical fibre couplers are fibre devices that are used for coupling light from one or several input fibres to one or several output fibres. Optical fibre couplers can distribute the optical signal (power) from one fibre among two or more fibres. The fibre couplers also have applications in fibre interferometers, cable TV system, fibre ring lasers, etc.

Solved Examples

Example 1 The refractive indices for core and cladding for a step index fibre are 1.52 and 1.41 respectively. Calculate (i) critical angle (ii) numerical aperture and (iii) the maximum incidence angle.

Solution Given $\mu_{core} = \mu_1 = 1.52$, $\mu_{clad} = \mu_2 = 1.41$

Critical angle $(\theta_c) = \sin^{-1}\left(\frac{\mu_2}{\mu_1}\right)$

Numerical aperture (NA) = $\sqrt{(\mu_1^2 - \mu_2^2)}$ and

Maximum incidence angle $(\theta_0) = \sin^{-1}[\sqrt{(\mu_1^2 - \mu_2^2)}]$

(i) $$\theta_c = \sin^{-1}\left[\frac{1.41}{1.52}\right] = 68.06°$$

or $$\boldsymbol{\theta_c = 68.1°}$$

(ii) $$NA = \sqrt{(\mu_1^2 - \mu_2^2)} = \sqrt{(1.52)^2 - (1.41)^2} = 0.5677$$
$$= \mathbf{0.568}$$

(iii) $$\theta_0 = \sin^{-1}[\sqrt{(\mu_1^2 - \mu_2^2)}] = \sin^{-1}\sqrt{(1.52)^2 - (1.41)^2}$$
$$= \mathbf{\sin^{-1}[0.568]}$$
$$\theta_0 = 34.59° = \mathbf{34.6°}$$

Example 2 Find out the numerical aperture and acceptance angle of an optical fibre, if the refractive indices for core and cladding are 1.6 and 1.5, respectively.

Solution Given $\mu_{core} = \mu_1 = 1.6$, $\mu_{clad} = \mu_2 = 1.5$

Numerical aperture $(NA) = \sqrt{(\mu_1^2 - \mu_2^2)}$

Acceptance angle $(\theta_0) = \sin^{-1}[\sqrt{(\mu_1^2 - \mu_2^2)}]$

$$NA = \sqrt{(1.6)^2 - (1.5)^2} = \sqrt{0.31} = 0.556$$

or $$NA = \mathbf{0.556}$$

$$\theta_0 = \sin^{-1}(0.556) = \mathbf{33.78°}$$

EXAMPLE 3 A light ray enters from air to a fibre. The refractive index of air is 1.0. The fibre has refractive index of core is equal to 1.5 and that of cladding is 1.48. Find the critical angle, the fractional refractive index, the acceptance angle and numerical aperture.

SOLUTION Given $\mu_{air} = \mu_0 = 1.0$, $\mu_{core} = \mu_1 = 1.5$, $\mu_{clad} = \mu_2 = 1.48$.

Critical angle $(\theta_c) = \sin^{-1}\left(\frac{\mu_2}{\mu_1}\right)$

Fractional refractive index $(\Delta\mu_r) = \frac{\mu_1 - \mu_2}{\mu_1}$

Acceptance angle $(\theta_0) = \sin^{-1}[\sqrt{(\mu_1^2 - \mu_2^2)}]$

Numercial aperture $(NA) = \sqrt{(\mu_1^2 - \mu_2^2)}$

$$\theta_c = \sin^{-1}\left(\frac{\mu_2}{\mu_1}\right) = \sin^{-1}\left(\frac{1.48}{1.50}\right) = 80.63°$$

or $$\theta_c = \mathbf{80.63°}$$

$$\Delta\mu_r = \frac{\mu_1 - \mu_2}{\mu_1} = \frac{1.50 - 1.48}{1.48} = 0.0133$$

or $$\Delta\mu_r = \mathbf{1.33\%\ of\ light.}$$

$$NA = \sqrt{(\mu_1^2 - \mu_2^2)} = \sqrt{(1.50)^2 - (1.48)^2} = \mathbf{0.244}$$

$$\theta_0 = \sin^{-1}[\sqrt{(\mu_1^2 - \mu_2^2)}] = \sin^{-1}[0.244]$$

$\therefore$ $$\theta_0 = \mathbf{14.13°}$$

EXAMPLE 4 Calculate the numerical aperture and acceptance angle of optical fibre of refractive indices for core and cladding as 1.62 and 1.52, respectively.

SOLUTION Given $\mu_{core} = \mu_1 = 1.62$ and $\mu_{clad} = \mu_2 = 1.52$

Numerical aperture $(NA) = \sqrt{(\mu_1^2 - \mu_2^2)}$ and

Acceptance angle $(\theta_0) = \sin^{-1}[\sqrt{(\mu_1^2 - \mu_2^2)}]$

$$NA = \sqrt{(1.62)^2 - (1.52)^2} = [\sqrt{0.314}]$$

$$NA = \mathbf{0.56}$$

$$\theta_0 = \sin^{-1}(NA) = \sin^{-1}(0.56) = \mathbf{34.06°}$$

$$\theta_0 = \mathbf{34.1°}$$

EXAMPLE 5 Calculate the refractive indices of the core and cladding material of a fibre from the following data: $NA = 0.22$, $\Delta\mu_r = 0.012$, where NA is numerical aperture,

$$\Delta\mu_r = \frac{\mu_{core} - \mu_{clad}}{\mu_{core}}$$

μ_{core} and μ_{clad} have usual meanings.

Solution Given: $NA = 0.22$, $\Delta\mu_r = 0.012$, $\mu_{core} = \mu_1$ and $\mu_{clad} = \mu_2$

Formula used are

$$\Delta\mu_r = \frac{\mu_1 - \mu_2}{\mu_1} \text{ and } NA = \sqrt{\mu_1^2 - \mu_2^2}$$

$$\Delta\mu_r = 0.012 = \frac{\mu_1 - \mu_2}{\mu_1} = 1 - \frac{\mu_2}{\mu_1}$$

or $\mu_2 = 0.988\,\mu_1$

$$NA = 0.22 = \sqrt{\mu_1^2 - (0.988\mu_1)^2}$$

or $0.0484 = \mu_1^2\,[0.023856]$

or $\mu_1 = \mathbf{1.424} = \mu_{core}$

and $\mu_2 = 0.988\,\mu_1 = \mathbf{1.41}$

$\mu_2 = \mathbf{1.41} = \mu_{clad}$

Example 6 The refractive indices for core and cladding for a step index fibre of diameter 0.064 mm are 1.53 and 1.39, respectively. Calculate (i) numerical aperture of the fibre (ii) acceptance angle (iii) number of reflections in 90 cm of fibre for a ray at the maximum incidence angle and for one at half this angle.

Solution Given $d = 0.064$ mm, $\mu_{core} = \mu_1 = 1.53$ and $\mu_{clad} = \mu_2 = 1.39$

Numerical aperture $(NA) = \sqrt{(\mu_1^2 - \mu_2^2)}$

Acceptance angle $(\theta_0) = \sin^{-1}[\sqrt{(\mu_1^2 - \mu_2^2)}]$

Number of reflections $(N_r) = \dfrac{L}{d\sqrt{\left[\dfrac{\mu_1}{\mu_0 \sin\theta_i}\right]^2 - 1}}$

$$NA = \sqrt{(1.53)^2 - (1.39)^2} = 0.639$$

$NA = \mathbf{0.64}$

$\theta_0 = \sin^{-1}(NA) = \sin^{-1}(0.64) = 39.79°$

$\theta_0 = \mathbf{39.8°}$

N_r at $\theta_i = \theta_0$, then

$$N_r = \frac{90}{0.0064\sqrt{\left(\dfrac{1.53}{1 \times 0.64}\right)^2 - 1}}$$

$= \mathbf{6476}$

N_r at $\theta_i = \dfrac{\theta_0}{2}$, then

$$N_r = \frac{90}{0.0064\sqrt{\left(\dfrac{1.53}{1 \times \sin 19.9°}\right)^2 - 1}} = \frac{90}{0.0064\sqrt{\left(\dfrac{1.53}{0.34}\right)^2 - 1}}$$

$= 3205.14$

$N_r = \mathbf{3205}$

EXAMPLE 7 A graded index fibre has a core diameter of 0.05 mm and numerical aperture of 0.22 at a wavelength of 8500 Å. What are the normalised frequency (v_n) and number of modes guided in the core?

SOLUTION Given $d = 0.05$ mm, $NA = 0.22$, $\lambda = 0.00085$ mm

Normalised frequency $(v_n) = \dfrac{\pi d}{\lambda} NA$ and

Maximum number of modes guided or propagated $(m_m) = \dfrac{1}{2}\left[\dfrac{\pi d}{\lambda} NA\right]^2$

$$v_n = \frac{3.14 \times 0.05 \times 10^{-3} \times 0.22}{0.85 \times 10^{-6}}$$

$$= 40.63$$

$$v_n = \mathbf{40.63}$$

and $$m_m = \frac{1}{2}(v_n)^2$$

$$= 825.398$$

$$m_m = \mathbf{825}$$

EXAMPLE 8 The refractive indices of core and cladding of a fibre are 1.465 and 1.460, respectively, and the light of wavelength 1.25 μm is used. What should be the diameter of core for a single mode propagation? If the core diameter is given as 50 μm, how many modes can propagate through the fibre?

SOLUTION Given $\mu_{core} = \mu_1 = 1.465$, $\mu_{clad} = \mu_2 = 1.460$ and $\lambda = 1.25 \times 10^{-6}$ m, $d = ?$

For single mode propagation, $d < \dfrac{2.4 \times \lambda}{\pi NA}$,

Number of modes propagated $(m_m) = \dfrac{1}{2}\left[\dfrac{\pi d}{\lambda} NA\right]^2$

Numerical aperture $(NA) = \sqrt{(\mu_1^2 - \mu_2^2)}$

So, $$NA = \sqrt{(1.465)^2 - (1.460)^2} = 0.121$$

$$d < \frac{2.4 \times 1.25 \times 10^{-6}}{3.14 \times 0.121} = 7.896 \times 10^{-6}\,\text{m}$$

$$\mathbf{d < 7.9\ \mu m}$$

$$m_m = \frac{1}{2}\left[\frac{3.14 \times 50 \times 10^{-6} \times 0.121}{1.25 \times 10^{-6}}\right]^2$$

$$= \frac{1}{2}(15.197)^2$$

$$= \frac{230.94}{2}$$

$$= 115.47$$

∴ Number of modes

$$= \mathbf{115}$$

EXAMPLE 9 How many modes can propagate in a step-index fibre with a core diameter as 40 μm, if the refractive indices of its core and cladding are 1.461 and 1.456, respectively, and the light of wavelength is 8500 Å?

SOLUTION Given $\mu_{core} = \mu_1 = 1.461$, $\mu_{clad} = \mu_2 = 1.456$, $\lambda = 0.85 \times 10^{-6}$ m and $d = 4.0 \times 10^{-5}$ m.

Maximum mode propagated $(m_m) = \frac{1}{2}\left[\frac{\pi d}{\lambda} NA\right]^2$

and numerical aperture $(NA) = \sqrt{(\mu_1^2 - \mu_2^2)}$

So,
$$NA = \sqrt{(1.461)^2 - (1.456)^2}$$
$$NA = 0.121$$
$$m_m = \frac{1}{2}\left[\frac{3.14 \times 4.0 \times 10^{-5}}{0.85 \times 10^{-6}} \times 0.121\right]^2 = 159.83$$
$$m_m = \mathbf{159}$$

EXAMPLE 10 Consider a slab waveguide made of Al Ga As having refractive indices for core and cladding as 3.6 and 3.55, respectively. Find how many modes can propagate in this waveguide if (i) $d = 5\lambda$ and (ii) $d = 50\lambda$?

SOLUTION Given $\mu_{core} = \mu_1 = 3.6$, $\mu_{clad} = \mu_2 = 3.55$

Number of modes propagated $(m_m) = \frac{1}{2}\left[\frac{\pi d}{2} NA\right]^2$

and numerical aperture $(NA) = \sqrt{(\mu_1^2 - \mu_2^2)}$

$$NA = \sqrt{(3.6)^2 - (3.55)^2} = 0.5979$$

(i) $d = 5\lambda$, then

$$m_m = \frac{1}{2}\left[\frac{3.14 \times 5\lambda}{\lambda} \times .5979\right]^2 = 44.06$$
$$m_m = \mathbf{44}$$

(ii) $d = 50\lambda$, then

$$m_m = \frac{1}{2}\left[\frac{3.14 \times 50\lambda}{\lambda} \times 0.5979\right]^2 = 4405.81$$
$$m_m = \mathbf{4405}$$

EXAMPLE 11 Find out the maximum core diameter of an optical fibre whose core and cladding have refractive indices as 1.460 and 1.457, respectively, and which supports only one mode at 1.25×10^{-6} m wavelength.

SOLUTION Given: $\mu_{core} = \mu_1 = 1.460$, $\mu_{clad} = \mu_2 = 1.457$ and $\lambda = 1.25 \times 10^{-6}$ m.

Diameter of core $(d) < \frac{2.4\lambda}{\pi\, NA}$ and numerical aperture $(NA) = \sqrt{(\mu_1^2 - \mu_2^2)}$

So
$$NA = \sqrt{(1.46)^2 - (1.457)^2} = 0.0935$$

$\therefore$
$$d < \frac{2.4 \times 1.25 \times 10^{-6}}{3.14 \times 0.0935}$$
$$d < 10.22\ \mu\text{m}$$

$\therefore$ Maximum core diameter = **10.22 μm**

EXAMPLE 12 A signal of power 5 μW exists just inside the entrance of 0.1 km long fibre. Calculate the absorption coefficient of the fibre if the power inside the fibre be 1 μW.

SOLUTION Given $L = 0.1$ km, $P_i = 5 \times 10^{-6}$ W and $P_0 = 1 \times 10^{-6}$ W.

$$\text{Absoprtion coefficient } (\alpha) = \frac{10}{L}\log\left(\frac{P_i}{P_0}\right)$$

or $$\alpha = \left(\frac{10}{0.1}\right)\log_{10}\left(\frac{5\times10^{-6}}{1.0\times10^{-6}}\right) = 69.89 \text{ dB/km}$$

or $$\alpha = \mathbf{70\ dB/km}$$

EXAMPLE 13 An optical fibre cable 3.0 km long is made up of three 1.0 km length spliced together. The losses due to each length and splice are respectively 5 dB and 1.0 dB. What would be out put power if the input power is 5 mW?

SOLUTION Given $\alpha = 18/3 = 6$ dB/km, $P_i = 5$ mW.

$\because$ $$\alpha = \left(\frac{10}{L}\right)\log_{10}\left(\frac{P_i}{P_0}\right)$$

$$\frac{\alpha L}{10} = \frac{1}{2.303}\ln\left(\frac{P_i}{P_0}\right)$$

$$\ln\left(\frac{P_i}{P_0}\right) = \frac{6\times3\times2.303}{10} = 4.1454$$

$$\frac{P_i}{P_0} = e^{4.1454}$$

or $$P_0 = \frac{P_i}{e^{4.1454}} = \frac{5\times10^{-3}}{63.143}$$

or $$P_0 = 0.079 \times 10^{-3} \text{ W}$$

or $$P_0 = \mathbf{0.080\ mW}$$

EXAMPLE 14 A step-index fibre has a core index of refraction of $n_1 = 1.425$. The cut-off angle for light entering the fibre from air is found to be 8.50°. (a) Calculate the numerical aperture of the fibre. (b) Find the index of refraction of the cladding of this fibre (c) What would be the new numerical aperture and cut-off angle if the fibre were submersed in water?

SOLUTION

(a) The index of refraction for air $n_0 = n_{air} = 1.0003$.
The numerical aperture is found from the formula

$$NA = n_0 \sin\theta_{0max} = (1.0003)\sin(8.50°) = \mathbf{0.1479}$$

(b) The index of refraction of the cladding can be found from the numerical aperture using the formula

$$n_1^2 - n_2^2 = NA^2$$

This gives $n_2^2 = n_2^1 - NA^2 = (1.425)^2 - (0.1479)^2 = 2.0088$

$$n_2 = \mathbf{1.417}$$

(c) The index of refraction for water $n_0 = n_{water} = 1.33$. Since the numerical aperture is a property of the fibre and only depends upon n_1 and n_2, it will not change when the medium outside the fibre changes.
The cut-off angle will change in case the numerical aperture is to be kept unaffected by a change in n_0. It means $NA = \mathbf{0.1479}$.
Using $\sin\theta_{0max} = NA/n_0$, we get

$$\theta_{0max} = \sin^{-1}(NA/n_0) = \sin^{-1}(0.1479/1.33) = \sin^{-1}(0.1112) = \mathbf{6.38°}.$$

OBJECTIVE TYPE QUESTIONS

Q.1 Optical fibre communication uses carrier wave as
(a) radiowave (b) laser wave (c) ordinary light (d) microwaves.

Q.2 Optical fibre communication is based on the phenomenon of
(a) refraction (b) total internal reflection
(c) polarisation (d) diffraction.

Q.3 The loss in intensity of light in optical fibre is due to
(a) reflection (b) absorption (c) scattering (d) all of these.

Q.4 In single mode fibre the diameter of core is nearly equal to
(a) 10 μm (b) 100 μm (c) 50 μm (d) 125 μm.

Q.5 The inner most part of the optical fibre in known as
(a) core (b) cladding (c) sheath (d) optical fibre axis.

Q.6 The refractive indices of core (μ_1) and cladding (μ_2) of an optical fibre satisfy the relation
(a) $\mu_1 > \mu_2$ (b) $\mu_1 < \mu_2$ (c) $\mu_1 = \mu_2$ (d) none of them.

Q.7 By increasing the refractive index of core, the number of modes of propagation in an optical fibre cable
(a) remains unchange (b) increases
(c) decreases (d) none of these.

Q.8 In graded index optical fibre the refractive index of core is
(a) non-uniform (b) increase towards the axis of core
(c) same at core-cladding interface (d) all of these.

Q.9 In multimode step index fibre, the core diameter is of the order of
(a) 10 to 20 μm (b) 20 to 30 μm (c) 300 to 400 μm (d) 50 to 200 μm.

Q.10 The acceptance angle in terms of refractive index of core (μ_1) and cladding (μ_2), when the end face of an optical fibre is exposed by the air is equal to
(a) $\cos^{-1}(\mu_1^2 - \mu_2^2)$ (b) $\sin^{-1}(\mu_1^2 - \mu_2^2)$ (c) $\sin^{-1}\sqrt{(\mu_1^2 - \mu_2^2)}$ (d) $\sin^{-1}\sqrt{(\mu_2^2 - \mu_1^2)}$.

Q.11 If μ_1 be the refractive index of core, μ_2 that of cladding and μ_0 of the medium meeting end face of fibre, the value of numerical aperture (*NA*) can exceed 1 when
(a) $\mu_0 = 1$ (b) $\mu_0 > 1$ (c) $\mu_2 < \mu_1$ () $\mu_2 > \mu_1$.

Q.12 In intrinsic optical fibre sensor, the sensing medium is
(a) fibre (b) laser light (c) light detector (d) none of these.

SHORT-ANSWER QUESTIONS

Q.1 What is fibre optics?

Q.2 What do you mean by numerical aperture?

Q.3 What do you understand by core and cladding?

Q.4 Give name of various types of fibres.

Q.5 Why optical fibre communications are so important?

Q.6 Can more than one signal be propagated in single mode fibre?

Practice Problems

General Questions

Q.1 What is an optical fibre? Define and explain the terms

(a) Acceptance angle
(b) Acceptance cone
(c) Numerical aperture
(d) Relative refractive index difference
(e) Propagating modes and
(f) Normalised frequency.

Q.2 What are single mode, multimode and graded index fibres? Also explain in detail the difference in structures of single mode step index and multimode graded index fibre.

Q.3 Discuss the physical significance of numerical aperture. How does it depend on refractive indices of core and cladding?

Q.4 Explain the allowed modes in an optical fibre. How are they related to normalized frequency?

Q.5 Discuss the propagation mechanisms of light waves in optical fibre.

Q.6 Describe schematically the basic elements of optical fibre communication system.

Q.7 Explain why does fraction of power of a signal get lost due to bending of fibre.

Q.8 Discuss the attenuation and dispersion of signals in optical fibre.

Q.9 What do you understand by optical fibre sensors. How many types of optical fibre sensors are commonly used?

Q.10 How are optical fibre connectors and couplers needed in communication?

Q.11 What are the advantages of using optical fibre communication systems?

Q.12 Enumerate some applications of optical fibre communication system.

Q.13 Write a note on

(a) Fibre optics
(b) Application of optical fibre
(c) Numerical aperture and its physical significance
(d) Optical fibre sensors and couplers.

Electron Optics

6

LEARNING OBJECTIVES

After reading this chapter you will be able to

LO1 Understand specific charge of an electron and Thomson's method

LO2 Learn about motion of an electron in uniform and magnetic fields

LO3 Illustrate electrostatic and magnetostatic focusing

LO4 Explain Scanning Electron Microscope (SEM), its principle, components and applications

LO5 Discuss working of Scanning Tunneling Microscope (STM)

Introduction

The branch of physics which is concerned with beam of electrons and their deflection by means of electric and magnetic fields is referred to as electron optics. Electron optics is also concerned with interference of beam of electrons when they cross each other and their deflection when they pass through the spacing in its submicroscopic structure. Electron optics is related to the wave properties of electrons which can be treated based on quantum theory.

6.1 SPECIFIC CHARGE OF AN ELECTRON LO1

It is very difficult to measure the elementary charge of an electron. However, it is easier to find its specific charge which is the ratio between the charge and mass of the electron. In order to find the specific charge of the electron, we analyse the deflection of an electron in a constant magnetic field. When a beam of electrons enters perpendicular to the direction of magnetic field, electrons move in a circle, which is called its trajectory. This is due to the force exerted by magnetic field. This force is called Lorentz force and it acts perpendicular both to the magnetic field and the direction of beam propagation. If the mass, charge and velocity of electron are m, q, v, respectively and it moves in the circle of radius r, then the Lorentz force in that simple case is

$$F = qvB$$

This force compensates the centripetal force of the circular path. So we get

$$\frac{mv^2}{r} = qvB$$

or

$$\frac{q}{m} = \frac{v}{Br}$$

Therefore, we can determine q/m if we are able to measure v, B and r.

6.2 DETERMINATION OF SPECIFIC CHARGE OF AN ELECTRON: THOMSON'S METHOD

LO1

The idea to measure the specific charge of electron was given by *J. J. Thomson* in 1897. The basic concept of its working is related to the case when a beam of electrons is passed through the perpendicular electric and magnetic fields. The beam of electrons will remain undeflected if the forces on the electrons of the beam due to electric and magnetic fields are equal but opposite in direction.

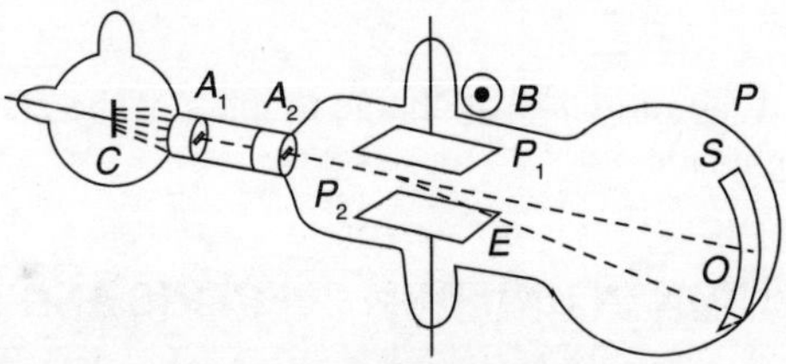

FIGURE 6.1

The apparatus used by *Thomson* is shown in Fig. 6.1. This apparatus consist of a highly evacuated gas container P. The electrons from the hot cathode C are accelerated and a beam of electrons is formed by a potential difference V between the anodes A_1 and A_2. An electric field is applied perpendicular to the path of electron beam by using two metal plates P_1 and P_2. Now a magnetic field is also applied perpendicular to the plane of the paper (pointed out of the plane of the paper) on the beam of electrons at the same place where electric field is acting. So after passing through the perpendicular electric and magnetic fields, the electron beam strikes the screen S at point O. The screen is coated with a material that glows at the point of impact.

In the absence of electric and magnetic fields the beam remains undeflected and strikes the screen at point O. However, in the presence of electric and magnetic fields the beam will deflect. Now, we adjust the strength of $\vec{E}$ and $\vec{B}$ fields so that the beam of electrons meets the screen S at the same point O i.e., at undeflected position. So in this case, the forces due to electric and magnetic fields balance each other.

Consider the electron of mass m and charge q which is moving with velocity v when it comes out through the anode. The force on the electron due to electric field $\vec{E}$ is

$$F_e = qE$$

The force on the electron due to the magnetic field $\vec{B}$ is

$$F_B = qvB$$

As discussed, these forces balance each other when we obtain the undeflected position.

$$qE = qvB$$

$$v = \frac{E}{B}$$

The accelerating potential V determines the speed v of the electrons. Since, the potential energy of an electron at the cathode appears as a gain in its kinetic energy at the anode when the electron beam is accelerated from cathode to anode. Hence, we get

$$\frac{1}{2}mv^2 = qV$$

$$\frac{q}{m} = \frac{1}{2}\frac{v^2}{V}$$

Substituting the value of speed v in the above relation we get

$$\frac{q}{m} = \frac{1}{2V}\left(\frac{E}{B}\right)^2$$

so $$\frac{q}{m} = \frac{E^2}{2VB^2}$$

It is clear from the above relation that the ratio q/m of charge to mass of the electron can be determined by measuring all the quantities on the right side.

6.3 MOTION OF AN ELECTRON IN UNIFORM ELECTRIC AND MAGNETIC FIELDS

LO2

As we know the electron is a charged particle. So it experiences a force when an electric field is applied. Once it gains energy, i.e it acquires some velocity, the Lorentz force can significantly change its direction of motion. This is based on the fact that a magnetic force alone cannot work. Under the situation of combined application of electric and magnetic fields, the electron attains different trajectories. Below we discuss some cases and calculate the trajectories of the electron. These calculations finally contribute to the electric and magnetic focusing of the electron.

6.3.1 Uniform Magnetic Field ($\vec{E}$ = 0)

Let an electron move in a uniform magnetic field, which is constant in time also. The direction of the magnetic field is parallel to the z-axis i.e., $\vec{B} = B\hat{z}$. As we know that the kinetic energy of the electron is not changed by the application of the magnetic field but is changed by an electric field. Consequently, the speed of the electron remains unchanged in the constant and uniform magnetic field. In this case, the equation of motion is

$$\vec{F} = m\vec{a} = m\frac{d\vec{v}}{dt} = q(\vec{v} \times \vec{B})$$

or $$m\frac{d\vec{v}}{dt} = -e(\vec{v} \times \vec{B}) \qquad \text{(i)}$$

In the above equation we have used $q = -e$. In Cartesian coordinate system, the components of the above equation can be written as

$$\frac{dv_x}{dt} = -\frac{eB}{m}v_y \qquad \text{(iia)}$$

$$\frac{dv_y}{dt} = \frac{eB}{m} v_x \tag{iib}$$

$$\frac{dv_z}{dt} = 0 \tag{iic}$$

From Eq. (iic), we obtain

$$v_z = v_{z0} \tag{iii}$$

This shows that the component of the velocity, which is parallel to the direction of the magnetic field, i.e., in z direction, is equal to a constant value v_{z0}. It means that electron does not accelerate in z direction. Differentiating Eqs. (iia) and (iib) w.r.t. t, we obtain the following relations

$$\frac{d^2v_x}{dt^2} = -\frac{eB}{m}\frac{dv_y}{dt} \quad \text{and} \quad \frac{d^2v_y}{dt^2} = \frac{eB}{m}\frac{dv_x}{dt}$$

$$\frac{d^2v_x}{dt^2} = -\left(\frac{eB}{m}\right)^2 v_x \quad \text{and} \quad \frac{d^2v_y}{dt^2} = -\left(\frac{eB}{m}\right)^2 v_y \tag{iv}$$

Above equations describe a simple harmonic oscillator at frequency ω_c called cyclotron frequency or *gyromagnetic frequency*. This frequency is defined as

$$\omega_c = \frac{eB}{m} \tag{v}$$

Note that ω_c is a non-negative quantity or remains positive always.

Now, multiplying Eq. (iib) with i and then adding to Eq. (iia), we get a single equation that will govern the motion of the electron in the x and y direction. This equation is produced below.

$$\frac{dv_x}{dt} + i\frac{dv_y}{dt} = \frac{eB}{m}(-v_y + iv_x),$$

$$\frac{d(v_x + iv_y)}{dt} = i\frac{eB}{m}(v_x + iv_y),$$

or

$$\frac{dV}{dt} = i\frac{eB}{m}V \tag{vi}$$

Here $\quad V = v_x + iv_y$

Again Eq. (vi) can be written as

$$\frac{dV}{dt} = i\omega_c V \tag{vii}$$

The solution of Eq. (vii) is given as:

$$V = v_{\perp 0}e^{i(\omega_c t + \phi_0)} \tag{viii}$$

Where $v_{\perp 0}$ and ϕ_0 are constants, that depend upon the initial conditions of the electron motion. Now solution (viii) can be written as

$$v_x + iv_y = v_{\perp 0}\{\cos(\omega_c t + \phi_0) + i \sin(\omega_c t + \phi_0)\}$$

From the above Eq. x and y components of the velocity are obtained as

$$v_x = v_{\perp 0} \cos(\omega_c t + \phi_0) \tag{ix}$$

$$v_y = v_{\perp 0} \sin(\omega_c t + \phi_0) \tag{x}$$

and z-component of the velocity from Eq. (iii) is written as

$$v_z = v_{z0}$$

After squaring and adding Eqs (ix) and (x), we get, $\sqrt{v_x^2 + v_y^2} = v_{\perp 0}$, which is the component of velocity perpendicular to the magnetic field. Thus,

$$v^2 = v_x^2 + v_y^2 + v_z^2 = v_{\perp 0}^2 + v_{z0}^2 = \text{Constant} \tag{xi}$$

Eq. (xi) shows that the speed of the electron is unchanged in uniform magnetic field, where $\vec{E} = 0$.

Now we want to analyse the trajectory of the electron, for which the coordinate x, y and z need to be obtained. For this, we integrate Eqs (iii), (ix) and (x) under the limits x: $x_0 \to x$, y: $y_0 \to y$, $z_0 \to z$ and $t : 0 \to t$ and for getting

$$x = x_0 + \frac{v_{\perp 0}}{\omega_c}\{\sin(\omega_c t + \phi_0) - \sin\phi_0\} \tag{xii}$$

$$y = y_0 - \frac{v_{\perp 0}}{\omega_c}\{\cos(\omega_c t + \phi_0) - \cos\phi_0\} \tag{xiii}$$

and

$$z = z_0 + v_{z0} t \tag{xiv}$$

So $r = (x_0, y_0, z_0)$ represents the initial position of the particle. On squaring and adding Eqs (xii) and (xiii), we get the relation

$$(x - \tilde{x}_0)^2 + (y - \tilde{y}_0)^2 = \left(\frac{v_{\perp 0}}{\omega_c}\right)^2 \tag{xv}$$

This represents the equation of a circle with centre $(\tilde{x}_0, \tilde{y}_0)$ and radius $\frac{v_{\perp 0}}{\omega_c}$. This radius is called *Larmour radius* $\mathbf{r}_L$. The coordinates $\tilde{x}_0$ and $\tilde{y}_0$ are defined as

$$\tilde{x}_0 = x_0 - \frac{v_{\perp 0}}{\omega_c}\sin\phi_0 \quad \text{and} \quad \tilde{y}_0 = y_0 + \frac{v_{\perp 0}}{\omega_c}\cos\phi_0 \tag{xvi}$$

From Eq. (xv), we observe that the motion of the electron lies in xy plane, which is perpendicular to the magnetic field direction. The trajectory of the electron is a circle with center $(\tilde{x}_0, \tilde{y}_0)$ and radius $\frac{v_{\perp 0}}{\omega_c}$. The direction of gyration of the electron is such that the magnetic field generated by the electron is always opposite to the externally applied field. It can be seen from Eqs (xii) and (xiii) that the electron is moving in the counterclockwise direction along the circumference of the circle with a uniform angular velocity ω_c, as shown in Fig. 6.2.

From Eq. (xiv), we observe that the z coordinate of the electron, which is parallel to the direction of the magnetic field, is increasing uniformly with time. Hence, the trajectory of the electron is a helix with axis parallel to the magnetic field direction (z-axis) and passing through $(\tilde{x}_0, \tilde{y}_0)$. The radius of the helix is equal to the Larmour radius and the pitch is equal to $2\pi v_{z0}/\omega_c$; pitch is the distance traveled by the electron in completing one revolution. Hence, it is clear that the parameters of the helix depend on the initial velocity and position of the electron. If we consider another case when the perpendicular component of the velocity is zero i.e., $v_{\perp 0} = 0$, the trajectory of the electron motion is a straight line along the magnetic field direction. Also if $v_{z0} = 0$, i.e., z component of velocity which is parallel to the direction of the magnetic field is zero, the trajectory of the electron motion is a circle with centre $(\tilde{x}_0, \tilde{y}_0)$ that is called *guiding centre*. Generally, guiding centre moves in the direction of the magnetic field with a constant velocity v_{z0} when the electron motion is helical. Thus, we can say that the helical motion may be separated in two ways. The first one is the uniform motion along the magnetic field and the other one is the circular motion around the field lines.

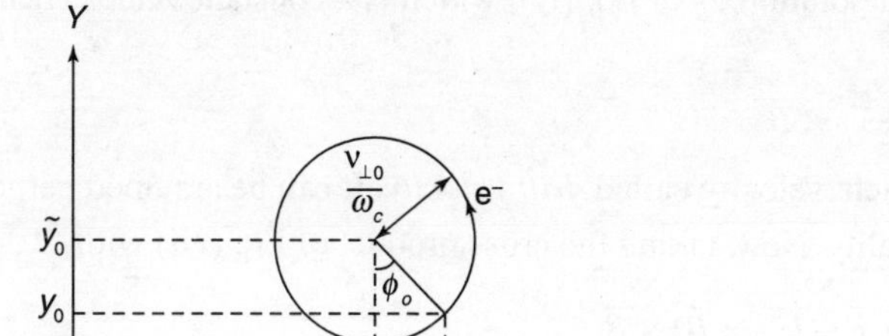

FIGURE 6.2

6.3.2 Constant and Uniform Electric and Magnetic Fields

In the presence of the electric field, the motion of an electron will be the sum of two motions i.e., the circular Larmour gyration due to the magnetic field in addition with a drift of the guiding centre due to the electric field. For convenience, let us separate the electron velocity and electric field in two parts, which are as follows

$$\vec{E} = \hat{z}E_z = \vec{E}_\perp \quad \text{and} \quad \vec{v} = \hat{z}v_z + \vec{v}_\perp \tag{i}$$

Here $\vec{E}_\perp$ and $\vec{v}_\perp$ are the components which are perpendicular to the magnetic field. These are called perpendicular components. In the presence of both electric and magnetic fields the equation of motion for a moving electron becomes

$$m\frac{d\vec{v}}{dt} = -e(\vec{E} + \vec{v} \times \vec{B}) \tag{ii}$$

We separate out the parallel and perpendicular components of Eq. (ii).

$$m\frac{dv_z}{dt} = -eE_z \tag{iii}$$

and

$$m\frac{d\vec{v}_\perp}{dt} = -e(\vec{E}_\perp + \vec{v}_\perp \times \vec{B}) \tag{iv}$$

From the integration of Eq. (iii), we get

$$v_z = -\frac{e}{m}E_z t + v_{z0} \tag{v}$$

Again integrating Eq. (v), we get the *z*-coordinate,

$$z = -\frac{e}{2m}E_z t^2 + v_{z0}t + z_0 \tag{vi}$$

Here v_{z0} and z_0 are the initial values of v_z and z, respectively.

Let us consider a particular solution v_E of Eq. (iv), which is a constant value. Then, by substituting v_E to be constant in (iv), we get

$$0 = \vec{E}_\perp + \vec{v}_E \times \vec{B} \tag{vii}$$

Note that $\vec{v}_E$ is the cross-field velocity called *drift velocity*. It can be assumed perpendicular to the magnetic field without loss of generality. Now, taking the cross product of Eq. (vii) with $\vec{B}$, we find that

$$0 = \vec{E}_\perp \times \vec{B} + (\vec{v}_E \times \vec{B}) \times \vec{B}$$

$$0 = \vec{E}_\perp \times \vec{B} + \vec{B}(\vec{v}_E \cdot \vec{B}) - \vec{v}_E(\vec{B} \cdot \vec{B})$$

$$0 = \vec{E}_\perp \times \vec{B} - \vec{v}_E B^2$$

$$\vec{v}_E = \left(\frac{\vec{E}_\perp \times \vec{B}}{B^2}\right) \tag{viii}$$

where $\vec{v}_E \cdot \vec{B} = 0$, as $\vec{v}_E$ is perpendicular to $\vec{B}$. We can see from Eq. (viii) that the $\vec{E} \times \vec{B}$ drift is independent of the charge (–*e*), mass (*m*) and velocity (*v*) of the electron. The above result verifies that $\vec{v}_E$ is a particular solution of the Eq. (iv). The neglect of term $\frac{d\vec{v}_\perp}{dt}$ is also justified, as this term gives only the circular motion at ω_c. So, the general solution of Eq. (iv) is looked in the form

$$\vec{v}_\perp = \vec{v}_E + \vec{v}_l \tag{ix}$$

Here $\vec{v}_l$ is due to the applied magnetic field. Substituting this relation, Eq. (iv) takes the form or

$$m\frac{d\vec{v}_l}{dt} = -e(\vec{E}_\perp + \vec{v}_E \times \vec{B} + \vec{v}_l \times \vec{B}) \quad \because \vec{v}_E \text{ is constant.}$$

or

$$m\frac{d\vec{v}_l}{dt} = -e(\vec{v}_l \times \vec{B}) \tag{x}$$

Here $\vec{v}_l$ corresponds to the orbital motion of the electron due to the application of magnetic field alone. It is called *gyration velocity*. Thus, we can say that in the presence of both electric and magnetic fields, the velocity perpendicular to the magnetic field consists of a constant drift velocity. It is also perpendicular to the electric field direction superimposed on an orbital motion that is same as in the constant and uniform magnetic field alone.

In order to discuss the trajectory of the electron, it is assumed that $E_z = 0$, $v_{z0} = 0$ and $z_0 = 0$. So in this case, electron motion is in the *XY* plane. Further, it is taken that the electric field is directed in the *y*-direction, i.e., $\vec{E} = \hat{y}E_y$. Since drift velocity (a constant velocity) satisfies the relation,

$$0 = \vec{E}_\perp + \vec{v}_E \times \vec{B}$$

the sum of this drift velocity with the velocity which is obtained when only magnetic field is applied, i.e., the gyrating motion, will satisfy the equation of motion given by Eq. (ii) .

By using the x and y components of velocity from the previous case (when $\vec{E} = 0$) and Eq. (viii) and (ix), we get the relations

$$v_x = \frac{E_y}{B} + v_{\perp 0} \cos(\omega_c t + \phi_0) \tag{xi}$$

$$v_y = v_{\perp 0} \sin(\omega_c t + \phi_0) \tag{xii}$$

We can obtain the initial values of $v_{\perp 0}$ and ϕ_0 in terms of initial velocities v_{x0} and v_{y0}.

$$v_{\perp 0} = \left[\left(v_{x0} - \frac{E_y}{B}\right)^2 + v_{y0}^2\right]^{1/2} \quad \text{and} \quad \tan\phi_0 = \frac{v_{y0}}{v_{x0} - E_y/B} \tag{xiii}$$

For a particular case in which electron is initially at rest ($v_{x0} = v_{y0} = 0$) Eq. (xiii) results

$$v_{\perp 0} = -\frac{E_y}{B} \quad \text{and} \quad \phi_0 = 0 \tag{xiv}$$

We can obtain x and y coordinates by integrating Eqs. (xi) and (xii) and then substituting the value of $v_{\perp 0}$ and ϕ_0. The calculated coordinates are obtained as follows

$$x = \frac{E_y}{B} t + \frac{v_{\perp 0}}{\omega_c}[\sin(\omega_c t + \phi_0) - \sin\phi_0] \tag{xv}$$

$$y = -\frac{v_{\perp 0}}{\omega_c}[\cos(\omega_c t + \phi_0) - \cos\phi_0] \tag{xvi}$$

The trajectory defined by the above two equations is a *cycloid* and the coordinates of the origin are considered to coincide with the initial position of the particle. It is possible to get different trajectories by applying different initial conditions. Some of the trajectories with different initial conditions are discussed below.

6.3.2.1 When $v_{x0} = v_{y0} = 0$

In this case, initially the electron is at rest. Therefore, the force $\vec{v} \times \vec{B}$ corresponding to magnetic field does not act on the electron. However, at this time the electric field that is perpendicular to the magnetic field is acting. So, the electric force is directed towards the negative y-direction because of the electric field direction along the positive y-axis. So, the electron gets accelerated in the negative y-direction with the action of the electric force. As the electron acquires some velocity, the $\vec{v} \times \vec{B}$ force will start acting upon it. This will modify the trajectory of the electron and shall force it to move in the positive x-direction. So the electron moves in the positive X and negative y-direction, i.e., in XY plane.

From Eqs (xi) and (xiv), we obtain the relation

$$\frac{v_x B}{E_y} = 1 - \cos\omega_c t$$

or

$$\tilde{v}_x = 1 - \cos\omega_c t \quad \text{where} \quad \frac{v_x B}{E_y} = \tilde{v}_x$$

Similarly, Eq. (xii) can be written as

$$\tilde{v}_y = -\sin \omega_c t, \quad \text{where} \quad \frac{v_y B}{E_y} = \tilde{v}_y$$

Since E_y/B has the dimensions of velocity, it follows that $\tilde{v}_x$ and $\tilde{v}_y$ are dimensionless quantities. From Eqs. (xv) and (xiv), we obtain that x coordinate as follows

$$x = \frac{E_y}{B}t - \frac{E_y}{B\omega_c}\sin \omega_c t$$

or

$$\frac{xB\omega_c}{E_y} = \omega_c t - \sin \omega_c t$$

or

$$\tilde{x} = \omega_c t - \sin \omega_c t \quad \text{where} \quad \frac{xB\omega_c}{E_y} = \tilde{x}$$

Similarly Eq. (xvi) can be written as

$$\tilde{y} = \cos \omega_c t - 1 \quad \text{where} \quad \frac{yB\omega_c}{E_y} = \tilde{y}$$

Since $E_y/\omega_c B$ has the dimensions of length, it follows that $\tilde{x}$ and $\tilde{y}$ are dimensionless quantities. Finally, we summarise these results as

$$\tilde{v}_x = 1 - \cos \omega_c t \qquad \tilde{v}_y = -\sin \omega_c t \tag{xvii}$$

$$\tilde{x} = \omega_c t - \sin \omega_c t \qquad \tilde{y} = \cos \omega_c t - 1 \tag{xviii}$$

The above relations show that the motion of the electron is a cycloid motion between $t = 0$ and $t = 2\pi/\omega_c$. Initially, the Lorentz force does not act on the electron as it starts from rest. The electric field forces the electron in the negative y direction. As the velocity of the electron increases the Lorentz force tends to curve the trajectory in the positive x direction. It means the electron moves in the positive X and negative y directions. So, v_y becomes more and more negative, reaches a minimum, and finally becomes zero at $t = 2\pi/\omega_c$ after getting less and less negative. During the time $0 \le t \le \pi/\omega_c$, the velocity component v_x is increased to reach a maximum value. For $t \ge \pi/\omega_c$, the Lorentz force continues to curve the electron in anticlockwise direction. Since v_y is positive, the electron starts moving in the positive y direction. Finally v_y becomes more and more positive and reaches a maximum. Then it starts decreasing and finally zero at t = $2\pi/\omega_c$. During $\pi/\omega_c \le t \le 2\pi/\omega_c$, v_x gets decreased and finally reaches zero at $t = 2\pi/\omega_c$. So at $t = 2\pi/\omega_c$, the electron is again at rest. This cycloid repeats itself every $2\pi/\omega_c$ seconds. The trajectory given by Eq. (xviii) is shown in Fig. 6.3a.

6.3.2.2 When $v_{x0} = E_y/B, v_{y0} = 0$

In this case, we consider that the electron is moving with some initial velocity that is equivalent to drift velocity (E_y/B). The perpendicular component of velocity $v_{\perp 0}$ is zero in this case (from Eq. xiii).

With initial conditions, Eqs (xi), (xii), (xv) and (xvi) are defined as follows:

$$\tilde{v}_x = 1 \qquad \tilde{v}_y = 0 \tag{xix}$$

$$\tilde{x} = \omega_c t \qquad \tilde{y} = 0 \tag{xx}$$

The above relations state that the electron moves uniformly along the x-direction with the initial velocity, which is equal to the drift velocity. In the present case, the force due to the electric field is exactly cancelled by the force due to the magnetic force. So, here we can say that the total force acting on the electron is zero. Consequently, the electron initial velocity is maintained continuously. The electron trajectory from this case is shown in Fig. 6.3b.

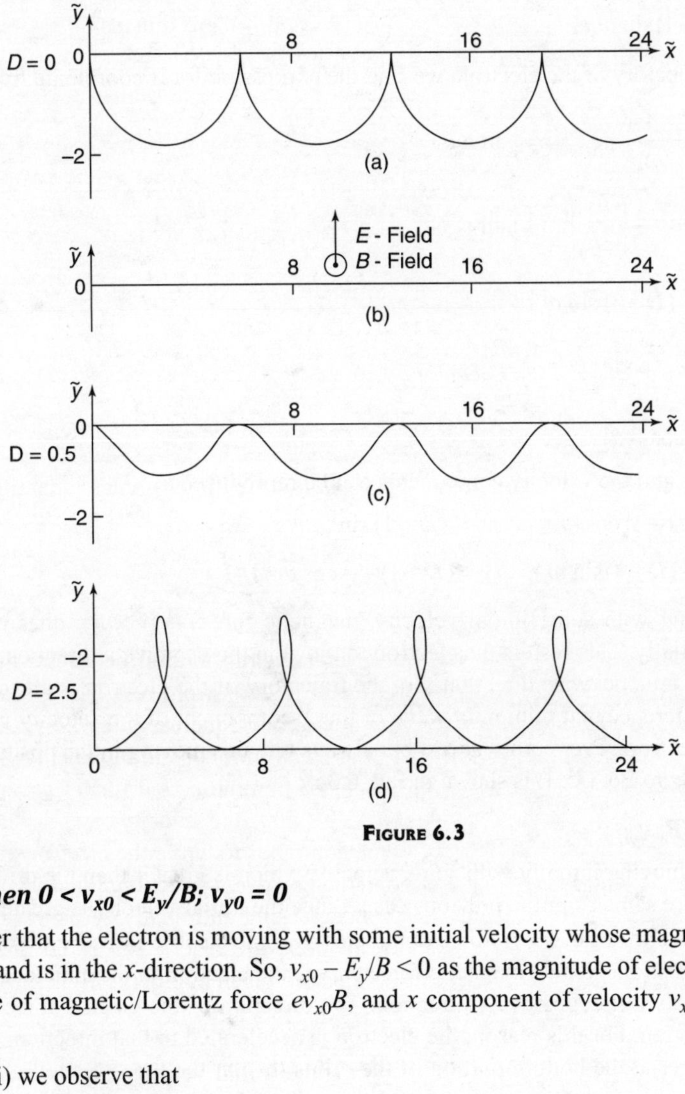

FIGURE 6.3

6.3.2.3 When $0 < v_{x0} < E_y/B$; $v_{y0} = 0$

Let us consider that the electron is moving with some initial velocity whose magnitude lies between zero and drift velocity and is in the x-direction. So, $v_{x0} - E_y/B < 0$ as the magnitude of electric force eE_y is greater than the magnitude of magnetic/Lorentz force $ev_{x0}B$, and x component of velocity v_x attains the minimum value v_{x0}.

From Eq. (xiii) we observe that

$$v_{\perp 0} = v_{x0} - E_y/B$$

Using this, Eq. (xi) becomes

$$v_x = \frac{E_y}{B} + \left(v_{x0} - \frac{E_y}{B}\right)\cos \omega_c t$$

or,

$$\frac{v_x B}{E_y} = 1 + \left(\frac{v_{x0}B}{E_y} - 1\right)\cos \omega_c t$$

or, $\tilde{v}_x = 1 + (D-1)\cos\omega_c t$

where $D = \dfrac{v_{x0}B}{E_y}$ and $\tilde{v}_x = \dfrac{v_x B}{E_y}$

Similarly, $\tilde{v}_y = (D-1)\sin\omega_c t$

In view of obtaining the trajectory of the electron, we find the expression for X coordinate from Eq. (xv), as

$$x = \frac{E_y}{B}t + \frac{1}{\omega_c}\left(v_{x0} - \frac{E_y}{B}\right)\sin\omega_c t$$

or,
$$\frac{xB\omega_c}{E_y} = \omega_c t + \left(\frac{Bv_{x0}}{E_y} - 1\right)\sin\omega_c t$$

or,
$$\tilde{x} = \omega_c t + (D-1)\sin\omega_c t$$

where $\tilde{x} = \dfrac{xB\omega_c}{E_y}$ and $D = \dfrac{v_{x0}B}{E_y}$

Similarly, $\tilde{y} = (D-1)(1-\cos\omega_c t)$

So, the position coordinates and the velocity components can be reproduced as

$$\tilde{v}_x = 1 + (D-1)\cos\omega_c t \qquad \tilde{v}_y = (D-1)\sin\omega_c t \tag{xxi}$$

$$\tilde{x} = \omega_c t + (D-1)\sin\omega_c t \qquad \tilde{y} = (D-1)(1-\cos\omega_c t) \tag{xxii}$$

Since the electron is moving with some initial velocity, magnetic force $-e(\vec{v}\times\vec{B})$ does not completely cancel the electric force initially. This results the electron to move in the negative y-direction. The magnetic force turns the orbit in the anticlockwise direction. So, the trajectory of the electron is a cycloid, the same as in case 6.3.2.1 with a difference that minimum value of the X-component of the velocity v_x is a non-zero positive quantity. So, the electron never comes at rest but always remains moving in the positive x-direction. The trajectory corresponding to Eq. (xxii) is shown in Fig. 6.3c.

6.3.2.4 When $v_{x0} > Ey/B$; $v_{y0} = 0$

In this case, the electron is moving initially with some velocity, which is greater than the drift velocity. The velocities and coordinates are same as in the previous case. Since the initial velocity is greater than the drift velocity in the positive x-direction, the $\vec{v}\times\vec{B}$ force dominates over the electric force ($\because ev_{x0}B > eE_y$) that is directed toward negative direction. Hence, magnetic force curves the electron in the anticlockwise direction and the electron motion starts in the positive y-direction. The direction of the electric force is opposite to the direction of electron motion. For this reason, the electron is decelerated in that direction. Consequently, speed of the electron is larger at the bottom portion of the orbits than at the top. So, the Larmour radius is longer at the bottom and smaller at the top, which means the radius of curvature of the trajectory is smaller at the top and larger at the bottom. This difference in Larmour radii at the top and the bottom portions of the trajectory results in a drift in the positive x-direction. So, the guiding center moves perpendicular to electric and magnetic fields. The trajectory for this case is shown in Fig. 6.3d.

6.4 ELECTROSTATIC AND MAGNETOSTATIC FOCUSING LO3

As discussed in the previous section, the application of electrostatic field and magnetostatic field can control the electron motion. This observation has led to the concept of focusing of electron beam either by using the electrostatic field or the magnetostatic field.

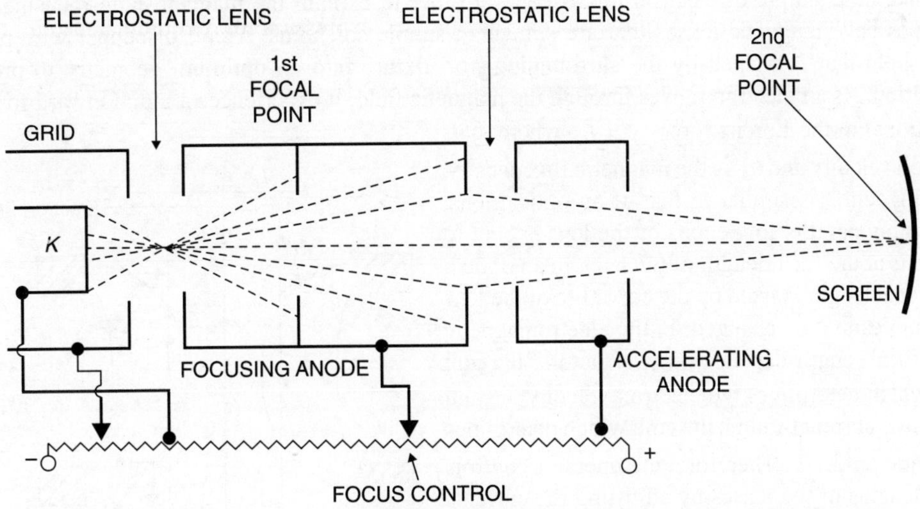

FIGURE 6.4

6.4.1 Electrostatic Focusing

The example of electrostatic focusing is an electron gun. In the electron gun, the electrons are focused by an electrostatic field. Electrostatic lenses are formed when negative and positive fields are near to each other. The electron gun is formed with several parts. For example, a heater and a cathode are used to generate electrons, a control grid is used to control electron flow, and also two anodes are used. The main purpose of the first anode is to focus the electrons into a narrow beam on the screen. Hence, it is called *focusing anode*. The second anode accelerates the electrons as they pass through it. So this is called *accelerating anode*. The control grid is a hollow metal tube placed over the cathode having a small opening in the center of a plate at the end opposite to the cathode. It can control the number of electrons that are emitted because it is near the cathode. This is based on the fact that the negative voltage of the grid can be varied either to control electron flow or to stop it completely. The anodes consist of two cylinders that contain plates with small holes in their centers. The cathode is indirectly heated, so it emits a cloud of electrons. The control grid is maintained at a negative potential with respect to the cathode to keep the electrons bunched together. A high positive potential on the anodes pulls electrons through the hole in the grid. Now, two electrostatic fields that exist between the control grid and first anode and between the first and second anodes focus the electron beam. The motion of the electrons through the electron gun (by dashed lines) and the relative voltage relationships on the electron gun elements is shown in Fig. 6.4.

The cathode (K) is at a fixed positive voltage with respect to the ground. The grid is at a variable negative voltage with respect to the cathode. A fixed positive voltage of several thousand volts is connected to the accelerating anode. The potential of the focusing anode is less positive than the potential of the accelerating anode. The electrostatic field areas are often referred to as lenses because the fields bend electron streams in the same manner as the optical lenses do with light rays. The first electrostatic lens causes the electrons to cross at the first focal point within the field. The second lens bends the spreading streams and returns them to a new second focal point at the screen.

6.4.2 Magnetostatic Focusing

We can take an example of electron microscope in order to explain the magnetostatic focusing. Electron microscopes have magnetic lenses that are similar to simple solenoids. A coil of copper wire produces a magnetic field that is shaped by the surrounding iron fixture into an optimum geometry to produce the lensing action. As an electron moves through the magnetic field, it experiences a radial inward force, which is proportional to the Lorenz force, $\vec{v} \times \vec{B}$, where $\vec{v}$ is the electron velocity and $\vec{B}$ is the magnetic flux density. The lensing action is similar to that of an optical lens, in which a ray parallel to the axis of the lens is bent to the lens axis at the focal length of the lens. In an optical lens, the focal length is fixed by the curvature of the lens surfaces and cannot be changed. In the electromagnetic lens, the focal length depends on two factors: the gun voltage, which determines the electron velocity $\vec{v}$, and the amount of current through the coil, which determines the magnetic field, $\vec{B}$. Therefore, the operator controls the focal lengths of the lenses by adjusting the currents supplied to them. An increase in current increases the radial force experienced by the beam and thus reduces the focal length. A typical magnetic lens is shown in Fig. 6.5. The focal length f of such a lens is given by

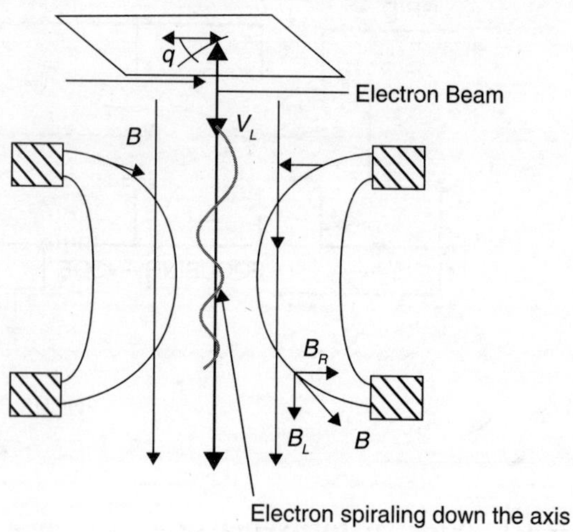

FIGURE 6.5

$$f = \frac{C_1 V}{N^2 I^2},$$

where V is the accelerating voltage, N is the number of turns in the coil, I is the current in the coil and C_1 is a constant.

For achieving good lens characteristics, it is essential to have constant accelerating voltage and constant lens current. Like the optical lenses, these lenses also suffer from defects, namely spherical aberration, chromatic aberration and astigmatism. Spherical aberration is caused by lens field acting inhomogeneously on the off axis rays, i.e., inability of a lens to focus all of a parallel incident beam to a point. This is reduced by using stop down lens. Chromatic aberration is caused by variation in the electron energy and thus the electrons are not monochromatic. So the electrons with different energies have different wavelengths and focus at different points. This demands for constant accelerating voltage V. Also source of electrons needs to be coherent, i.e., it should be of narrow range of energies. Astigmatism is caused by asymmetry in lens geometry. So, additional coils that introduce astigmatism in a controlled way are used to correct this effect.

6.5 SCANNING ELECTRON MICROSCOPE (SEM) LO4

Scanning Electron Microscope (SEM) is a microscope that uses electrons rather than light to form an image. These microscopes were developed due to the limitations of Light Microscopes. The first SEM was made available commercially around 1965. Its late development was due to the electronics involved in "scanning" the beam of electrons across the sample. SEM is one of the most heavily used instruments in academic/lab research areas and industry due to the combination of higher magnification, larger depth of field, greater resolution, and compositional and crystallographic information.

6.5.1 SEM Principle

The SEM uses a focused beam of high energy electrons to generate a variety of signals at the surface of solid specimens (Fig. 6.6). With the help of signals emitted as a result of electron specimen interaction we gather information about the sample including external morphology (texture), chemical composition, and crystalline structure and orientation of materials making up the sample.

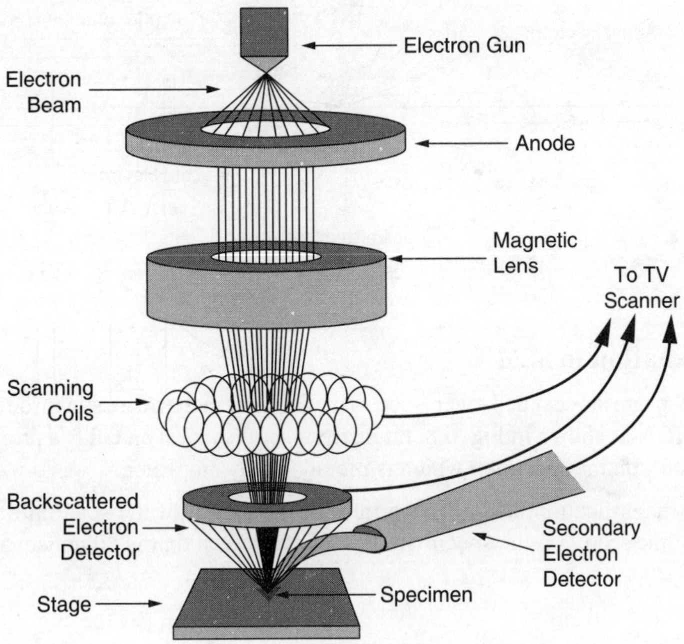

Figure 6.6

In most applications, data are collected over a selected area of the surface of the sample, and a 2D image is generated that displays spatial variations in these properties. The SEM is also capable of performing analyses of selected point locations on the sample. Depending on the incident energy of the electron beam, a variety of electrons (auger, secondary and back scattered), *X*-rays (characteristic and Bremsstrahlung), light (cathodoluminescence) and heat (phonons) are emitted (Fig. 6.7). Several of these interactions are used for imaging, semi-quantitative analysis and/or quantitation analysis.

6.5.2 SEM Components

All the SEMs consist of a column, a specimen chamber, detectors and viewing system. A column is used to generate a beam of electrons. The electron beam interacts with the sample in a specimen chamber. Detectors are used to monitor the different signals that result from the electron beam and sample interaction. A viewing system is used to build an image from the detector signal. Essential components of all SEMs are electron source (gun), electron lenses, sample stage, detectors for all signals of interest, display/data output devices and infrastructure requirements like power supply, vacuum system, cooling system, etc.

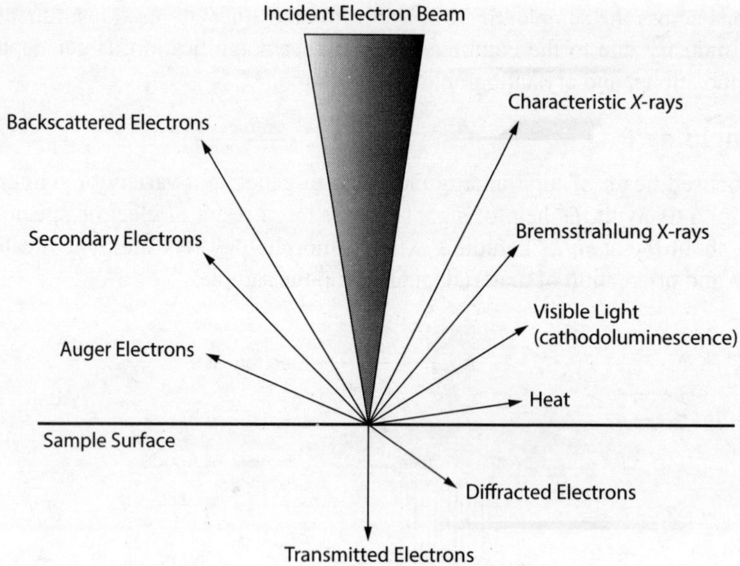

FIGURE 6.7

6.5.3 Image Formation in SEM

The primary electron beam is scanned over specimen in raster pattern in synchronisation with beam in Cathode Ray Tube (CRT), as shown in Fig. 6.8. Intensity at spot (say P_1) on CRT is proportional to the signal detected from the spot P_1 on the specimen which is modulated by amplifier.

Magnification The magnification is simply the ratio of the length of the scan on the Cathode Ray Tube (CRT) to the length of the scan l on the specimen. For a CRT screen that is 20 cm square, the magnification is given by

$$M = \frac{C}{l} = \frac{20\text{ cm}}{l}$$

So this is clear that the increased magnification M can be achieved by decreasing l.

Image Contrast If S_A and S_B represent signal generated from two points of scanned area (as shown in Fig. 6.8), then the image contrast is defined as

$$C = \frac{S_A - S_B}{S_A} = \frac{\Delta S}{S_A}$$

In order to detect small objects of poor contrast, it is necessary to use a high beam and a slow scan speed. This improves the signal to noise ratio, i.e., S/N ratio. Secondary electrons give topographical contrast and back scattered electrons give atomic number contrast (Fig. 6.8).

Resolution The maximum resolution that can be achieved using a microscope means the smallest interval distinguishable between two adjacent points. Any magnification exceeding such maximum would not make sense since further information cannot be provided. The maximum resolution mainly depends on the wavelength of the radiation selected for the image. Beams entering the lens and aperture system of the microscope produces overlapping diffraction patterns for each object point. The distance r_m between

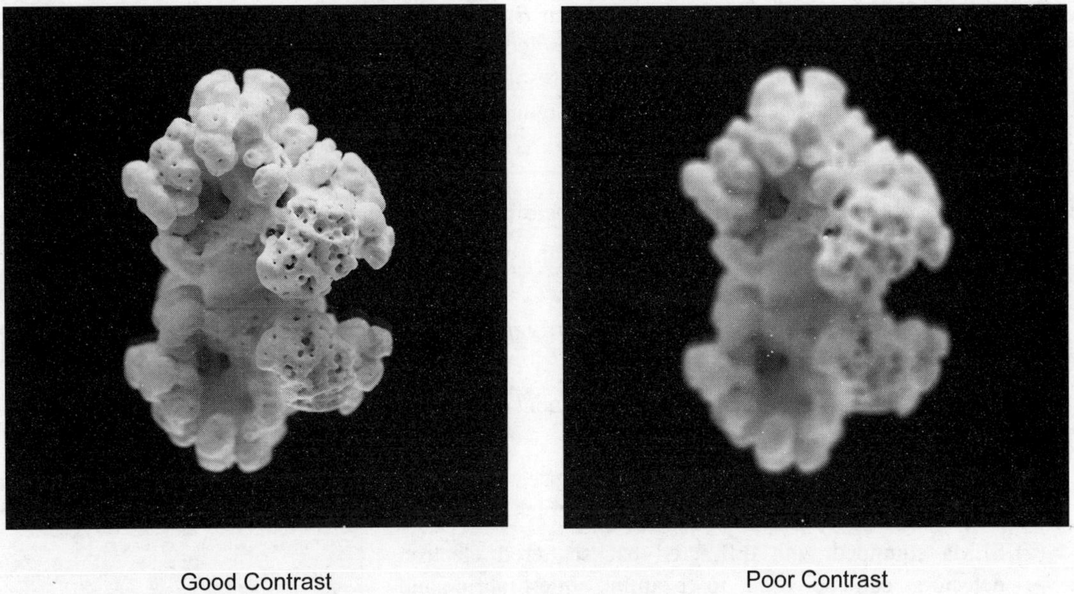

Figure 6.8

two diffraction maxima must exceed full width half maximum (FWHM), otherwise the diffraction maxima cannot be seen clearly (separate). This is clear from Fig. 6.9, which is prepared based on Rayleigh criterion. According to this, the distinction is possible when the maximum of the zero order coincides with the first minimum of the second diffraction pattern. The distance between the two first minima, i.e., r_1 is inversely proportional to the diameter of the aperture. Diffraction patterns are dependent on the wavelength λ, the index of refraction of the surrounding medium μ, and the angle θ formed by the optical axis and the edge beam, which can only just pass through the aperture. In view of this r_m is given as

$$r_m = \frac{r_1}{2} = \frac{0.61\lambda}{\mu \sin \theta}$$

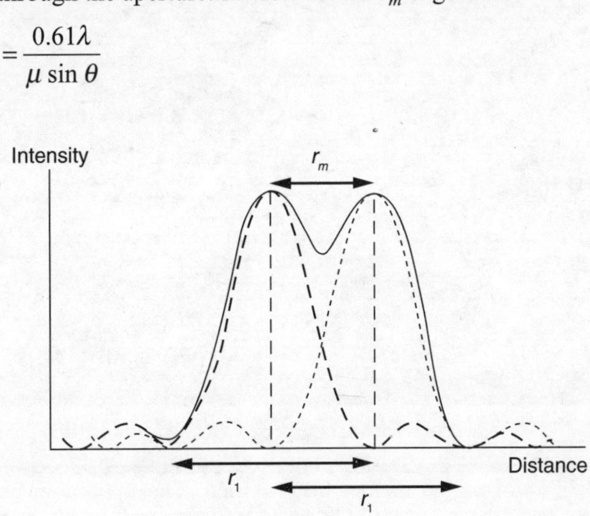

Figure 6.9

As discussed in previous chapter, the product μ sin θ is referred to as numerical aperture. In SEM, the resolution is limited to few nm, which is due to the electron probe size that in turn depends on the quality of the objective lens and the electron gun. Ultimate resolution obtainable in an SEM image is limited by the minimum probe size that can generate an adequate signal at the sample.

6.5.4 Applications of SEM

The SEM has got applications in various fields, some of which are mentioned below.

(i) The SEM is routinely used to generate high resolution images of shapes of objects and to show spatial variations in chemical compositions.

(ii) This instrument is also widely used to identify phases based on qualitative chemical analysis and/or crystalline structure.

(iii) This instrument is used in precise measurement of very small features and objects down to 50 nm in size.

(iv) Backscattered electron images can be used for rapid discrimination of phases in multiphase samples.

(v) SEMs equipped with diffracted backscattered electron detectors can be used to examine microfabric and crystallographic orientation in many materials.

(vi) Figures 6.10 and 6.11 show SEM photomicrographs of MESFET ohmic contacts and gate contact with poor surface morphology, respectively. Poor ohmic metal surface morphology leads to poor contact resistance of source drain pads (Fig. 6.10), whereas underdeveloped gate pattern leads to poor metal semiconductor interface (Fig. 6.11).

FIGURE 6.10

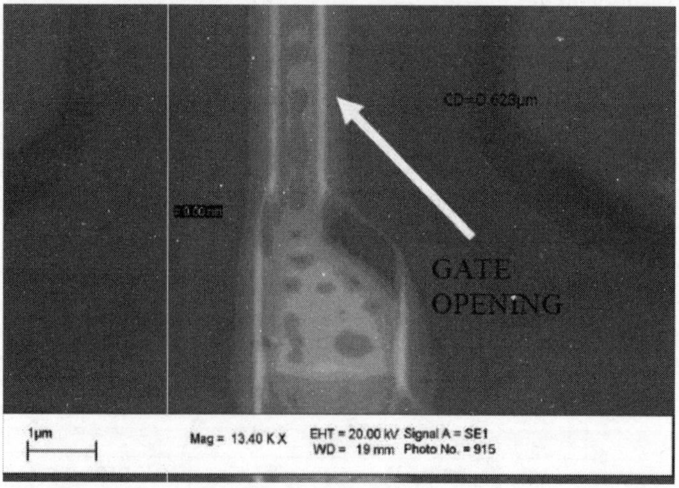

FIGURE 6.11

6.6 SCANNING TUNNELING MICROSCOPE (STM) LO4

Gerd Binnig and *Heinrich Rohrer* invented Scanning Tunneling Microscope (STM) in early 1980s. It is a microscopical technique that allows the investigation of electrically conducting surfaces down to atomic scale. The basic principle of STM is the tunneling of electrons between the sharp metallic tip of a probe and the surface of a sample under study. By etching or tearing a thin metal wire we can obtain these tips with only one atom at the tip. The tip is mechanically connected to the scanner, an *XYZ* positioning device realised by means of piezoelectric materials (Fig. 6.12). A small current called tunneling current flows if the tip is in contact with the sample which is positively and negatively biased. This small tunneling current is then amplified and measured. The flow of tunneling current is very sensitive to the distance between the tip and the sample. With the help of the tunneling current the feedback electronic keeps the distance between tip and sample constant. If the tunneling current exceeds its preset value, the distance between tip and sample is increased and if it falls below this value, the feedback decreases the distance. For keeping the flow of current constant, the height of the tip is continually adjusted when tip is swept over the surface. By recording the height fluctuations of the tip accurately, a map of the "bumps" on the surface is obtained. The tip is scanned line by line above the sample surface following the topography of the sample.

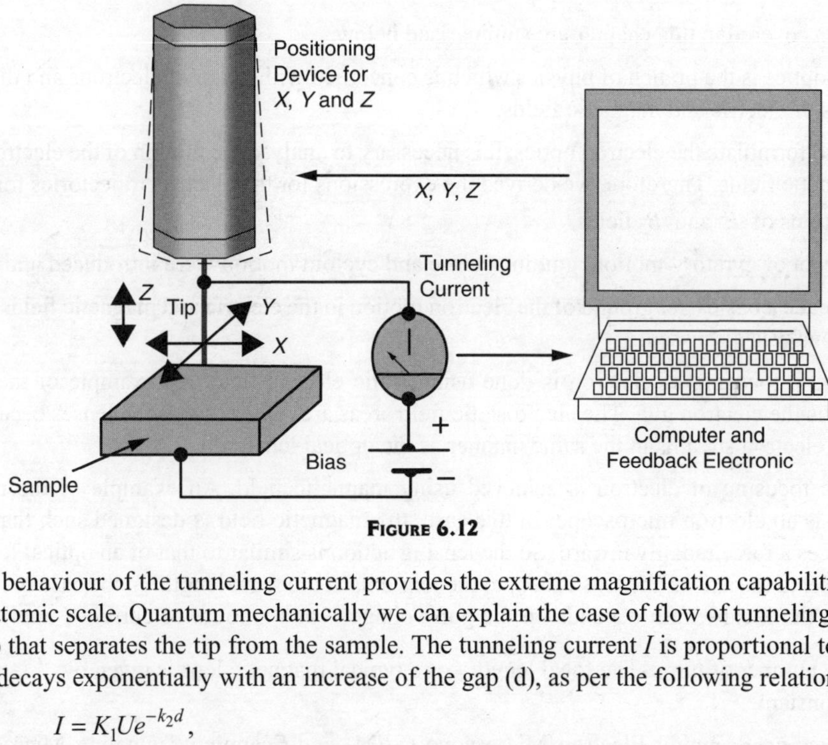

FIGURE 6.12

The physical behaviour of the tunneling current provides the extreme magnification capabilities of the STM down to the atomic scale. Quantum mechanically we can explain the case of flow of tunneling current across the small gap that separates the tip from the sample. The tunneling current *I* is proportional to the tunneling bias *U* but it decays exponentially with an increase of the gap (d), as per the following relation

$$I = K_1 U e^{-k_2 d},$$

where K_1 and k_2 are constants. The variation of *I* with the gap *d* is shown in Fig. 6.13. It is clear from the figure that a very small change in the tip sample separation induces a large change in tunneling current (please see δd and corresponding δI). So, the tip separation is controlled very exactly and tunneling current is carried by the outermost atom of the tip. A feedback loop constantly monitors the tunneling current and makes adjustments to the tip to maintain a constant tunneling current. These adjustments are recorded by the

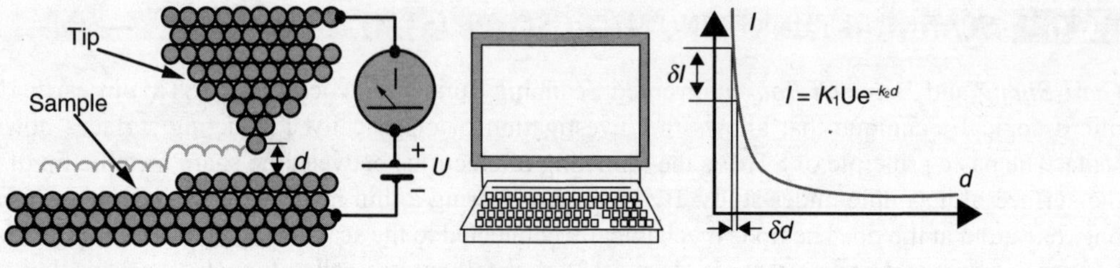

FIGURE 6.13

computer and presented as an image in the STM software. Such a setup is called a "constant current" image. In addition, for very flat surfaces, the feedback loop can be turned off and only the current is displayed. This is called a "constant height" image.

SUMMARY

The main topics covered in this chapter are summarised below.

- Electron optics is the branch of physics which is concerned with beam of electrons and their deflection by means of electric and magnetic fields.
- In order to formulate the electron optics, it is necessary to analyse the motion of the electrons in electric and magnetic fields. Therefore, we derived the expressions for the electron trajectories for the different combinations of $\vec{E}$ and $\vec{B}$ fields.
- The concept of gyratory motion, guiding centre and cycloid motion were introduced and explained.
- Having made a basic background of the electron motion in the electric and magnetic fields, the focusing of electrons was discussed.
- Electrostatic focusing of electron is done using static electric field. An example of the electrostatic focusing is the electron gun. The electrostatic field areas are referred to as the lenses because the fields bend the electron streams in the same manner as the optical lenses do.
- Magnetic focusing of electron is achieved using magnetic field. An example of the magnetostatic focusing is an electron microscope. In this case, the magnetic field is designed such that the electron experiences a force radially inward. So the lensing action is similar to that of an optical lens in which a ray parallel to the axis of the lens is bent to the lens axis at the focal length. In the case of magnetostatic focusing, the focal length depends on the gun voltage V and amount of current I applied to the coil having N number of turns. The focal length f of a typical magnetic lens is given by $f = \dfrac{C_1 V}{N^2 I^2}$, where C_1 is a constant.
- New topics on Scanning Electron Microscope (SEM) and Scanning Tunneling Microscope (STM) were discussed in detail. SEM uses electrons rather than the light to form an image. This microscope was developed due to the limitations of light microscopes.

SOLVED EXAMPLES

EXAMPLE 1 The voltage across the electrodes of a cathode ray gun is 500 eV. Calculate

(i) the energy gained by electron.

(ii) the speed of the electron.

(iii) the momentum of electron

Given mass of electron = 9×10^{-31} kg.

SOLUTION Given $V = 500$ eV, $m = 9 \times 10^{-31}$ kg and $e = 1.6 \times 10^{-19}$ C.

So,

(i) The energy gained by electron = eV

$$= 1.6 \times 10^{-19} \times 500 = \mathbf{8 \times 10^{-17}\ J}$$

(ii) When an electron is accelerated under a potential difference V then electron acquires a speed v. So

$$\text{K.E gained by electron} = \frac{1}{2}mv^2 = eV$$

or
$$v = \sqrt{\frac{2\,eV}{m}} = \sqrt{\frac{2 \times 8 \times 10^{-17}}{9 \times 10^{-31}}} = 1.33 \times 10^7\ \text{m/sec}$$

(iii) Momentum of the electron = m

$$= 9 \times 10^{-31} \times 1.33 \times 10^{7}$$

$$= \mathbf{12 \times 10^{-24}\ kg/sec}$$

EXAMPLE 2 What is the momentum of acceleration of an electron of speed 2.5×10^6 m/sec in a magnetic field of 2.0 G? Given that e/m = 1.76×10^{11} C/kg?

SOLUTION Given $v = 2.5 \times 10^6$ m/sec, $B = 2.0$ G $= 2.0 \times 10^{-4}$ T

and $e/m = 1.76 \times 10^{11}$ C/kg.

Force on electron due to magnetic field provides the required centripetal force. Therefore,

$$\frac{mv^2}{r} = Bev$$

$$\text{Centripetal acceleration} = \frac{v^2}{r}$$

$$\frac{v^2}{r} = \frac{Bev}{m}$$

$$= 2.0 \times 10^{-4} \times (1.76 \times 10^{11}) \times 2.5 \times 10^{6}$$

$$= \mathbf{8.80 \times 10^{13}\ m/sec^2}$$

EXAMPLE 3 In a Thomson's set up for determining e/m, the same high tension d.c. supply provides potential to the anode of accelerating column, as also to the positive deflecting plate in the region of crossed fields. If the supply voltage is doubled, by what factor should the magnetic field is increased to keep the electron beam undeflected?

SOLUTION *e*/*m* of electrons in a Thomson's experiment is given by

$$\frac{e}{m}=\frac{E^2}{2VB^2} \quad \text{where} \quad E=\frac{V}{d}$$

$$\therefore \qquad \frac{e}{m}=\frac{V^2}{2V\,B^2d^2}=\frac{V}{2B^2d^2}$$

This is also the condition for undeflected beam.

Since *d* is fixed from the above expression we get

$$B^2 \propto V \quad \text{or} \quad B \propto \sqrt{V}$$

So, when *V* is doubled ($V' = 2V$) then B' will be given by

$$B' \propto \sqrt{2V} = \sqrt{2}\mathbf{B}$$

It means that if *V* is doubled, *B* should be increased to $\sqrt{2}B$ to keep the electron beam undeflected.

EXAMPLE 4 (a) A monoenergetic electron beam with electron speed of 5.20×10^6 m/sec is subject to a magnetic field of 1.30×10^{-4} T, normal to the beam velocity. What is the radius of the circle traced by the beam?

Given e/m for an electron as 1.76×10^{11} C/kg.

(b) Is the formula you employed in (a) valid for calculating radius of the 210 MeV electron beam? If not, in what way is it modified?

SOLUTION Given $v = 5.20 \times 10^6$ m/sec,

$$B = 1.3 \times 10^{-4}\ \text{T}$$

$$e/m = 1.76 \times 10^{11}\ \text{C/kg}$$

Force exerted by the magnetic field on the electron

$$F = e|\bar{v} \times \bar{B}| = evB \sin\theta = evB \qquad (\because \theta = 90°)$$

Since, the normal magnetic field provides the centripetal force

$$evB = \frac{mv^2}{r}$$

$$r = \frac{mv}{qB} = \frac{v}{(e/m)B}$$

$$r = \frac{5.20 \times 10^6}{1.76 \times 10^{11} \times 1.30 \times 10^{-4}}$$

$$= 0.227\ \text{m} = \mathbf{22.7\ cm}$$

(b) Energy of the electron beam = 20 MeV

$$= 20 \times 1.6 \times 10^{-19} \times 10^6\ \text{J}$$

$$= 20 \times 10^{-13} \times 1.6 = \frac{1}{2}mv^2$$

$$\therefore \qquad v = \left(\frac{2 \times 20 \times 10^{-13} \times 1.6}{9 \times 10^{-31}}\right)^{\frac{1}{2}}$$

$$v = \mathbf{2.67 \times 10^9\ m/sec}$$

which is greater than the velocity of light.

So, the formula $r = \frac{mv}{qB}$ is not valid for calculating the radius of the path of 20 *MeV* electron beam because electron with such a high energy has velocity in the relativistic domain (comparable with the velocity of light). Since the mass varies at such speed, we use relativistic formula as follows

$$r = \frac{mv}{qB} = \frac{m_o}{\sqrt{1 - v^2/c^2}}\left(\frac{v}{qB}\right)$$

EXAMPLE 5 In a Thomson set up for determination of *e/m*, electrons accelerated by 2.5 kV enter the region of crossed electric and magnetic fields of strengths 3.6×10^4 *V/m* and 1.2×10^{-3} T respectively and go through undeflected. Determine the e/m of an electron.

SOLUTION Given $V = 2.5$ kV, $E = 3.6 \times 10^4$ V/m, and $B = 1.2 \times 10^{-3}$ T.

Energy gained by the electron = eV

$$eV = \frac{1}{2}mv^2$$

or

$$\frac{e}{m} = \frac{v^2}{2V}$$

Since electrons go through undeflected from the region of crossed electric and magnetic field, we have

$$eE = Bev \Rightarrow v = \frac{E}{B}$$

$$\frac{e}{m} = \frac{(E/B)^2}{2V} = \frac{[3.6 \times 10^4/1.2 \times 10^{-3}]^2}{2 \times 2.5 \times 10^3} = \frac{9 \times 10^{14}}{5 \times 10^3}$$

$$= \mathbf{1.8 \times 10^{11}\ C/kg}$$

EXAMPLE 6 An electron moves in the earth's magnetic field of 5×10^{-5} T with the energy of 10 keV. Find the Larmour radius of the electron neglecting its velocity component parallel to the magnetic field.

SOLUTION

$$E = \frac{1}{2}mv_{\perp 0}^2 \Rightarrow v_{\perp 0} = \sqrt{2E/m}$$

Formula used $$r_L = \frac{v_{\perp 0}}{\omega_c} = \frac{mv_{\perp 0}}{eB}$$

Now, $$v_{\perp 0} = \sqrt{\frac{2 \times 10 \times 10^3 \times 1.6 \times 10^{-19}}{9.1 \times 10^{-31}}} = 5.93 \times 10^7 \text{ m/s}$$

∴ $$r_L = \frac{9.1 \times 10^{-31} \times 5.93 \times 10^7}{1.6 \times 10^{-19} \times 5 \times 10^{-5}} = \mathbf{6.75\ m}$$

EXAMPLE 7 A solar wind proton is streaming with velocity $v_{\perp 0} = 3 \times 10^5$ m/sec in the magnetic field of 5×10^{-9} T. Compute the Larmour radius. Given that mass of proton = $m_p = 1.67 \times 10^{-27}$ kg.

SOLUTION Given $B = 5 \times 10^9$ T, $m_p = 1.67 \times 10^{-27}$ kg and $v_{\perp 0} = 3 \times 10^5$ m/sec.

Larmour radius

$$r_L = \frac{m_p v_{\perp 0}}{eB} = \frac{1.67 \times 10^{-27} \times 3 \times 10^5}{1.6 \times 10^{-19} \times 5 \times 10^{-9}}$$

$$= \mathbf{6.26 \times 10^5\ m}$$

EXAMPLE 8 Calculate the area traced by the trajectory of a 1 keV He^+ ion in the solar atmosphere near a sunspot, where $B = 5 \times 10^{-2}$ T.

SOLUTION Energy $E = 1$ keV $= 1 \times 10^3 \times 1.6 \times 10^{-19}$ J

$$E = \frac{1}{2} m v_{\perp 0}^2 \Rightarrow v_{\perp 0} = \sqrt{\frac{2E}{m_{He^+}}}$$

$$\Rightarrow \qquad v_{\perp 0} = \sqrt{\frac{2 \times 10^3 \times 1.6 \times 10^{-19}}{4 \times 1.67 \times 10^{-27}}} = 2.19 \times 10^5 \text{ m/sec.}$$

Here we have used the mass of He^+ as 4 m_p.

Now He^+ shall have gyratory motion, whose Larmour radius $r_L = \dfrac{v_{\perp 0} m_{He^+}}{eB}$

$$\Rightarrow \qquad r_L = \frac{2.19 \times 10^5 \times 4 \times 1.67 \times 10^{-27}}{1.6 \times 10^{-19} \times 5 \times 10^{-2}} = 0.183 \text{ m}$$

Since the trajectory of He^+ is simply the circular motion, the area traced by this

$$= \pi r_L^2$$

$$= 3.14 \times 0.183^2 = \mathbf{0.105\ m^2}$$

EXAMPLE 9 An electron is moving in uniform electric and magnetic fields which are perpendicular to each other. Find the drift of the guiding centre if the magnitudes of the electric field and magnetic field be 100 V/m and 10^{-3} T, respectively.

SOLUTION Under the actions of uniform electric and magnetic fields, the trajectory of the electrons is slanted helix with changing pitch. The drift of the guiding centre is

$$\vec{v_B} = (\vec{E} \times \vec{B}) / B^2$$

$$(v_B) = \frac{E}{B}$$

$$= \frac{100}{10^{-3}} = \mathbf{1 \times 10^5\ m/sec}$$

EXAMPLE 10 An ion engine has a 1 T magnetic field, and a collection of H^+ (behaving collectively) is to be shot out at an $\vec{E} \times \vec{B}$ velocity of 1×10^6 m/s. How much internal electric field must be present in the collection of ions (H^+)?

SOLUTION $\vec{E} \times \vec{B}$ drift is the drift of guiding centre; given by $\vec{v_E} = \dfrac{\vec{E} \times \vec{B}}{B^2}$

$$v_E = \frac{E}{B} \text{ or } E = Bv_E$$

or $$E = 1 \times 1 \times 10^6 = \mathbf{10^6\ V/m}$$

EXAMPLE 11 Magnification related studies were conducted using Scanning Electron Microscope. By keeping length of the scan on the Cathode Ray Tube fixed, the length of the scan on the specimen was halved. What will be the ratio of new magnification to the old magnification?

SOLUTION Magnification M is defined as

$$M = \frac{\text{Length of scan on CRT (C)}}{\text{Length of scan on specimen}}$$

$$M_1 = \frac{C}{L_1}$$

Similarly, $M_2 = \frac{C}{L_2} = \frac{C}{L_1/2} = \frac{2C}{L_1}$

or $$\frac{M_2}{M_1} = \frac{2C}{L_1} \times \frac{L_1}{C} = \mathbf{2}$$

EXAMPLE 12 Calculate the ratio of new focal length to the old focal length, if the number of turns in the coil used to form magnetic lens in magnetic focusing is increased by 10% and other parameters are kept fixed.

SOLUTION New no. of turns = $N + \frac{N}{10} = \frac{11N}{10}$

Focal length of magnetic lens

$$f = \frac{C_1 V}{N^2 I^2}$$

$$\Rightarrow \quad f_1 = \frac{C_1 V}{N^2 I^2} \quad \text{and} \quad f_2 = \frac{C_1 V}{\left(\frac{11}{10}\right)^2 N^2 I^2}$$

$$\Rightarrow \quad \frac{f_2}{f_1} = \left(\frac{10}{11}\right)^2 = \mathbf{0.826}$$

It means the new focal length is reduced by increasing the number of turns of the coil.

OBJECTIVE TYPE QUESTIONS

Q.1 Specific charge of an electron is equal to
(a) -1.6×10^{-19} C (b) 9.1×10^{-31} kg (c) 1.76×10^{12} C/kg (d) 1.6×10^{-19}C.

Q.2 In Thomson's method for the determination of specific charge of an electron, the $\vec{E}$ and $\vec{B}$ fields are applied such that
(a) electric force is double of magnetic force
(b) electron gets affected only by Lorentz force
(c) electric force and Lorentz force are in balance
(d) electron follows a helical path.

Q.3 The trajectory of an electron under the influence of a uniform magnetic field, when it is injected in the perpendicular direction to the magnetic field is
(a) circular (b) helix (c) linear (d) none of these.

Q.4 Larmour radius r_L of the electron motion in uniform and constant electric and magnetic fields depends on
(a) mass and charge of the electron only (b) magnetic field only
(c) perpendicular velocity only (d) all of these.

Q.5 Electron optics is the branch of physics which is concerned with
(a) beam of electrons and their deflections by $\vec{E}$ field
(b) beam of electrons and their deflections by $\vec{E}$ and $\vec{B}$ fields
(c) the radiation when electron jumps from the excited state to ground state
(d) none of the above.

Q.6 Electrostatic focusing is achieved when
(a) constant electric field is applied to focus the electron beam
(b) electric field and magnetic field both are applied in the same direction
(c) two opposite electric fields and a non-uniform magnetic field are applied
(d) none of the above.

Q.7 An example of magnetostatic focusing is
(a) electron microscope (b) electron gun
(c) anode (d) cathode.

Q.8 Lenses in electrostatic focusing are related to
(a) the electrostatic field areas (b) the magnetostatic field areas
(c) plane glass plates (d) convex lens used to focus the light.

Q.9 Magnetic lens formed for magnetostatic focusing suffer from defects like
(a) spherical aberration (b) chromatic aberration
(c) astigmatism (d) none of the above.

Q.10 In STM, the tunneling current decays
(a) linearly with the increase of gap between tip and sample
(b) exponentially with the increase of gap between tip and sample
(c) quadratically with the gap between tip and sample
(d) none of the above.

PRACTICE PROBLEMS

Q.1 Define electron optics and discuss the physics behind the coupling of the two words 'electron' and 'optics'.

Q.2 Discuss J.J. Thomson's method for the determination of specific charge of an electron.

Q.3 Derive the expression for cyclotron frequency of an electron under the action of a uniform magnetic field. Discuss its trajectory based on proper mathematical expressions.

Q.4 Prove that the motion of an election in uniform $\vec{E}$ and $\vec{B}$ fields is the sum of simple gyratory motion and the motion of guiding centre when both $\vec{E}$ and $\vec{B}$ fields are perpendicular to each other.

Q.5 Discuss electrostatic focusing of electron beam. Give an example of this focusing.

Q.6 How do you achieve magnetostatic focusing with the help of magnetic field. Discuss the similarities between magnetic lens and the optical lens.

Q.7 Discuss the principle of SEM and the image formation.

Q.8 Discuss in short the applications of the SEM.

Q.9 Discuss the principle of STM and the behavior of tunneling current I with the gap between the tip and the sample.

Waves and Oscillations

7

LEARNING OBJECTIVES

After reading this chapter you will be able to

LO 1 Differentiate between translational and oscillatory motion

LO 2 Learn about Simple Harmonic Motion (SHM) and its differential equation

LO 3 Know about simple pendulum, mass string system, and damped harmonic oscillator

LO 4 Explain attenuation coefficients of a vibrating system

LO 5 Discuss forced vibration

LO 6 Understand resonance

Introduction

The motion of things can be broadly classified into two classes. It is according to whether the thing that is moving stays near one place or travels from one place to another. The examples of things that stay near one place are an oscillating pendulum, a vibrating violin string, electron vibration in atoms, etc. The examples of things that travel from one place to another are a sliding hockey puck, a pulse traveling down a long stretched rope plucked at one end, ocean waves rolling towards the beach, electron beam of a television tube, etc. The motion of physical bodies may be classified mainly into two categories, namely translational motion and vibrational or oscillatory motion.

7.1 TRANSLATIONAL MOTION — LO1

If the position of a body varies linearly with time, for example, when a ball rolls on the ground or train moves on a straight track, them such motions are called translational motions.

7.2 VIBRATIONAL OR OSCILLATORY MOTION — LO1

A motion of a body that repeats itself after regular intervals of times and when the body moves back and forth over the same path is called vibrational or oscillatory motion. For example, the oscillations of the arms of a walking person, the bob of the pendulum clock, beating of heart, etc. are in vibrational motion.

7.3 SIMPLE HARMONIC MOTION (SHM) LO2

If the acceleration of a particle in a periodic motion is always directly proportional to its displacement from its equilibrium position and is always directed towards equilibrium position, then the motion of the particle is said to be Simple Harmonic Motion (SHM).

7.3.1 Type of SHM

Simple harmonic motion can be broadly classified into two classes, namely linear simple harmonic motion and angular simple harmonic motion.

Linear Simple Harmonic Motion

The motion is said to be linear simple harmonic motion, if the displacement of a particle executing SHM is linear. The examples are the motion of simple pendulum, the motion of a point mass tied with a spring, etc.

Angular Simple Harmonic Motion

The motion is said to be angular simple harmonic motion, if the displacement of a particle executing SHM is angular. The examples of angular SHM are torsional oscillations and oscillations of a compound pendulum.

7.3.2 Essential Condition of SHM

If f be the linear acceleration and x the displacement from the equilibrium position, then essential condition for linear SHM is

$$f \propto -x$$

If α be the angular acceleration and θ the angular displacement from the equilibrium position, then the condition of angular SHM is

$$\alpha \propto -\theta$$

Time Period: The smallest time interval during which the oscillation repeats itself is called the *time period.* It is denoted by T and its unit is seconds.

Frequency: The number of oscillations that a body completes in one second is called the *frequency* of periodic motion. It is the reciprocal of the time period T and is given by

$$n = \frac{1}{T}$$

unit of frequency is per second or Hertz, represented by Hz.

Amplitude: The amplitude of an SHM is the maximum displacement of the body from its mean position.

Phase: It is the physical quantity that expresses the instantaneous position and direction of motion of an oscillating system. The quantity $(\omega t + \delta)$ of the sine function is called the *phase* of the motion, if the motion is represented by $y = A \sin(\omega t + \delta)$, together with ω as the angular frequency.

7.4 DIFFERENTIAL EQUATION OF SHM AND ITS SOLUTION LO2

Let us consider a particle of mass m executing an SHM along a straight line with x as its displacement from the mean position at any time t. Then from the basic condition of the SHM, the restoring force F will be proportional to the displacement and will be directed opposite to this. Therefore,

$$F \propto -x \quad \text{or} \quad F = -kx \tag{i}$$

where k is a proportionality constant and is known as *force constant.*

If $f = \dfrac{d^2x}{dt^2}$ be the acceleration at any instant of time, then

$$m\frac{d^2x}{dt^2} = -kx$$

$$\frac{d^2x}{dt^2} + \frac{k}{m}x = 0 \tag{ii}$$

Putting $\dfrac{k}{m} = \omega^2$, we have

$$\frac{d^2x}{dt^2} + \omega^2 x = 0 \tag{iii}$$

The above Eqs. (ii) and (iii) are known as the *differential equations* of the SHM

In order to get the solution of Eq. (iii), we assume x as

$$x = Ce^{\alpha t} \tag{iv}$$

where C and α are constants. Then

$$\frac{dx}{dt} = C\alpha e^{\alpha t}$$

and

$$\frac{d^2x}{dt^2} = C\alpha^2 e^{\alpha t} \tag{v}$$

By using Eqs. (iv) and (v), Eq. (iii) becomes

$$C\alpha^2 e^{\alpha t} + \omega^2 Ce^{\alpha t} = 0$$

$$Ce^{\alpha t}(\alpha^2 + \omega^2) = 0$$

$Ce^{\alpha t} \neq 0$, as it is the displacement x

$$\alpha^2 + \omega^2 = 0$$

or

$$\alpha = \pm i\omega$$

where i is the imaginary number, given by $i = \sqrt{-1}$.

Since α has two values, possible solutions of Eq. (iii) are

$$x = Ce^{i\omega t} \text{ and } x = C^{-i\omega t},$$

the linear combination of which will give the general solution Eq. (iii) as

$$x = C_1 re^{i\omega t} + C_2 e^{-i\omega t} \tag{vi}$$

where C_1 and C_2 are constants.

Equation (vi) can also be written as

$$x = C_1(\cos \omega t + i \sin \omega t) + C_2(\cos \omega t - i \sin \omega t)$$

$$= (C_1 + C_2) \cos \omega t + (iC_1 - iC_2) \sin \omega t \tag{vii}$$

Let $C_1 + C_2 = A \sin \delta$ and $iC_1 - iC_2 = A \cos \delta$

Then Eq. (vii) gives

$$x = A \sin \delta \cos \omega t + A \cos \delta \sin \omega t$$

$$x = A\sin (\omega t + \delta) \tag{viii}$$

This is the desired solution of Eq. (iii), which gives the displacement of the particle executing SHM at any instant of time. Here A is the maximum displacement of the particle

In Eq. (viii), we replaced t by $\left[t + \frac{2\pi}{\omega}\right]$, then we have

$$x = A \sin \left[\omega\left(t + \frac{2\pi}{\omega}\right) + \delta\right]$$

$$= A \sin (\omega t + 2\pi + \delta)$$

or $$x = A \sin (\omega t + \delta)$$

Here, it may be noted that this is the same as Eq. (viii) and shows that the motion is repeated after an interval of $\frac{2\pi}{\omega}$. So this interval will be the time period of the SHM, given by $T = \frac{2\pi}{\omega}$

As we know that

$$\omega^2 = \frac{k}{m} \Rightarrow \omega = \sqrt{\frac{k}{m}}$$

$\therefore$ $$T = \frac{2\pi}{\omega} = 2\pi\sqrt{\frac{m}{k}} \tag{ix}$$

Frequency:

$$n = \frac{1}{T} = \frac{\omega}{2\pi} = \frac{1}{2\pi}\sqrt{\frac{k}{m}} \tag{x}$$

Phase: The quantity $(\omega t + \delta)$ is known as the phase of the vibrating particle. If $t = 0$ then $\omega t + \delta = \delta$, so that initial phase will be δ. If a particle starts motion from its mean position them δ will be zero but if it starts motion from the extreme position then δ will be $\pi/2$.

Velocity and acceleration: The particle executing SHM is a *harmonic oscilliator.*

We can find its velocity from the expression of its displacement, given below

$$x = A \sin (\omega t + \delta)$$

Differentiating it w.r.t. time, we get

$$v = \frac{dx}{dt} = A\omega \cos (\omega t + \delta) \tag{xi}$$

$$= A\omega\sqrt{1 - \sin^2(\omega t + \delta)} = \omega\sqrt{A^2 - A^2 \sin^2(\omega t + \delta)}$$

or $$v = \omega\sqrt{A^2 - x^2}$$

This is the expression for velocity of the particle at any displacement x. The maximum velocity is obtained by putting $x = 0$.

$$\therefore \qquad v_{max} = \omega A$$

Since $x = 0$ corresponds to its mean position, the particle has maximum velocity when it is at the mean position. At the maximum displacement, i.e., at the extreme position of the particle the velocity is obtained as zero. This extreme position is

$$x = A$$

Differentiating Eq. (xi) w.r.t. time t, we get

$$f = \frac{dv}{dt} = -A\omega^2 \sin(\omega t + \delta)$$

or
$$f = -\omega^2 x \tag{xiii}$$

The above equation gives acceleration of the oscillating particle at any displacement. This equation is the standard equation of SHM

This is clear from Eq. (xiii) that for the maximum acceleration

$$x = A \text{ (the extreme position)}$$

$\therefore$ Maximum acceleration,

$$f_{max} = \omega^2 A \qquad \text{at the extreme position}$$

Minimum acceleration is obtained by putting $x = 0$

$$f_{min} = 0 \quad \text{at the mean position}$$

7.4.1 Total Energy of a Harmonic Oscillator

The total energy of a simple harmonic oscillator at any point would be the sum of kinetic energy (K.E.) and potential energy (P.E.), as it executes vibrations under the action of a free restoring force.

The potential energy of a system is given by the amount of work done to move the system from the position 0 to x by applying a force. Therefore,

$$\text{P.E.} = \int \vec{F} \cdot \vec{dx} \int_0^x kx\, dx = \frac{1}{2}kx^2 \tag{xiv}$$

Here, we have not put the negative sign, as it simply tells us whether the work is done by the system or on the system. Putting the values of k and x, we get

$$\text{P.E.} = \frac{1}{2}m\omega^2 A^2 \sin^2(\omega t + \delta) \tag{xv}$$

The kinetic energy of the harmonic oscillator is

$$\text{K.E.} = \frac{1}{2}mv^2 = \frac{1}{2}m\left[\frac{dx}{dt}\right]^2$$

$$= \frac{1}{2}m\omega^2 A^2 \cos^2(\omega t + \delta) \tag{xvi}$$

The total energy E of the oscillator at any displacement can be obtained as

$$E = \text{P.E.} + \text{K.E.}$$

$$= \frac{1}{2} m\omega^2 A^2 \sin^2(\omega t + \delta) + \frac{1}{2} m\omega^2 A^2 \cos^2(\omega t + \delta)$$

$$E = \frac{1}{2} m\omega^2 A^2 \tag{xvii}$$

Since the total energy E is constant, it is obvious that the maximum possible value of K.E. or P.E. would be the same as $\frac{1}{2} m\omega^2 A^2$. It is interesting to learn how the kinetic and potential energies of the harmonic oscillator vary with time, if the oscillations start at $t = 0$ from its extreme position. This is shown in Fig. 7.1.

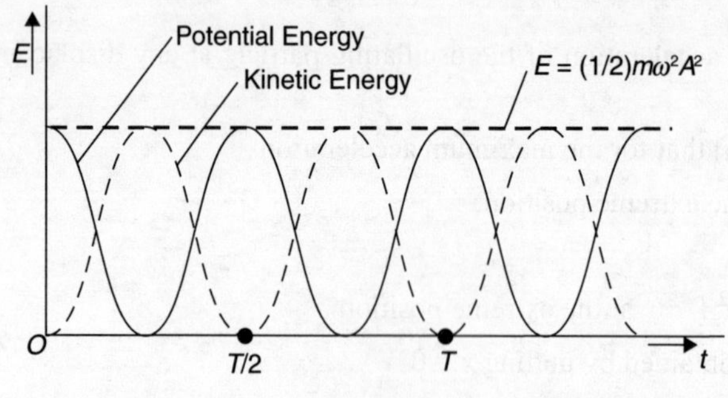

FIGURE 7.1

The average potential energy of the simple harmonic oscillator for one complete cycle is

$$\langle\text{P.E.}\rangle = \frac{1}{T}\int_0^T \frac{1}{2} kx^2 dt$$

$$= \frac{\int_0^T \frac{1}{2} kA^2 \sin^2(\omega t + \delta)\, dt}{T}$$

$$= \frac{\frac{1}{2}\int_0^T m\omega^2 A^2 \sin^2(\omega t + \delta)\, dt}{T}$$

$$= \frac{1}{2} m\omega^2 A^2 \frac{\int_0^T \sin^2(\omega t + \delta)\, dt}{T}$$

$$\langle\text{P.E.}\rangle = \frac{1}{4} m\omega^2 A^2 \qquad \left[\text{as } \frac{\int_0^T \sin^2(\omega t + \delta)\, dt}{T} = \frac{1}{2}\right]$$

Also, the average kinetic energy for one complete cycle is

$$\langle K.E.\rangle = \frac{1}{T}\int_0^T \frac{1}{2}m\left(\frac{dx}{dt}\right)^2 dt$$

$$= \frac{\int_0^T \frac{1}{2}mA^2\omega^2\cos^2(\omega t+\delta)\,dt}{T}$$

$$= \frac{1}{2}m\omega^2A^2\,\frac{\int_0^T \cos^2(\omega t+\delta)\,dt}{T}$$

$$\langle K.E.\rangle = \frac{1}{4}m\omega^2A^2$$

$$\Rightarrow \qquad \langle K.E.\rangle = \langle P.E.\rangle = \frac{1}{4}m\omega^2A^2$$

From the above calculations, it is clear that the average kinetic energy is equal to the average potential energy for a harmonic oscillator over a complete cycle

Now the total average energy over a complete cycle is

$$\langle E\rangle = \langle K.E.\rangle + \langle P.E.\rangle = \frac{1}{4}m\omega^2A^2 + \frac{1}{4}m\omega^2A^2$$

$$= \frac{1}{2}m\omega^2A^2$$

which is equal to the total energy of the harmonic oscillator.

7.5 SIMPLE PENDULUM LO3

An ideal simple pendulum is formed when a heavy point mass is suspended, by a weightless, inextensible and perfectly flexible string from a rigid support. However, these requirements cannot be fulfilled in practice. Therefore, a simple pendulum consists of small heavy metallic sphere suspended by a long fine string from a rigid support.

The arrangement of a simple pendulum is shown in Fig. 7.2, where pendulum AO is in equilibrium position. When the mass of the pendulum is displaced from its equilibrium position and released, it starts oscillating around its equilibrium position. The motion of the point mass is said to be SHM Suppose at any instant the particle is at point P at a distance x from its equilibrium position and makes an small angle α. The length of the string is l.

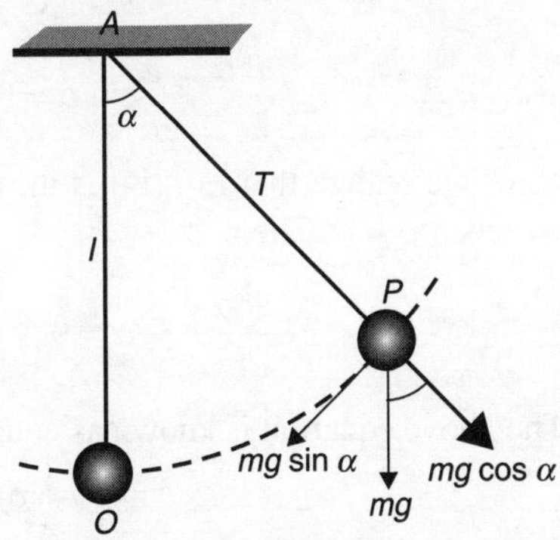

FIGURE 7.2

At point P, the particle is under the action of various forces: (i) the weight mg of the bob acting vertically downward (ii) tension T in string along PA. The weight mg has two components (a) radial component $mg\cos\alpha$, (b) transverse component $mg\sin\alpha$.

It is clear from the above arrangement that the tension in the string is opposed by the radial component mg cos α. Therefore, the force $T - mg \cos \alpha$ provides centripetal force for circular arc and the tangential component $mg \sin \alpha$ tends to bring the bob back to its initial position. Thus $mg \sin \alpha$ is often known as restoring force and therefore

$$F = -mg \sin \alpha \tag{i}$$

The negative sign indicates that the acceleration and the displacement are oppositely directed. If $\frac{d^2x}{dt^2}$ be the acceleration at any time t in the direction of increasing x, then the force

$$F = m\frac{d^2x}{dt^2} \tag{ii}$$

Therefore,

$$m\frac{d^2x}{dt^2} = -mg \sin \alpha$$

or
$$\frac{d^2x}{dt^2} = -g \sin \alpha \tag{iii}$$

For small angle α, the distance x (arc) can be written in terms of l and α, as

$$x = l\alpha$$

which on differentiation gives

$$\frac{d^2x}{dt^2} = l\frac{d^2\alpha}{dt^2}$$

or
$$l\frac{d^2\alpha}{dt^2} = -g \sin \alpha$$

or
$$\frac{d^2\alpha}{dt^2} = -\frac{g}{l}\sin \alpha$$

or
$$\frac{d^2\alpha}{dt^2} + \frac{g}{l}\sin \alpha = 0 \tag{iv}$$

Now, we realize that Eq. (iv) is the equation of motion of the pendulum. For a small deflection α, we can write $\sin \alpha \approx \alpha$. Then

$$\frac{d^2\alpha}{dt^2} + \frac{g}{l}\alpha = 0$$

The above equation is known as equation of motion of simple pendulum, whose solution can be written as

$$\alpha = \alpha_0 \sin(\omega t + \delta)$$

where, $\omega = \left(\frac{g}{l}\right)^{1/2}$ and δ is the initial phase.

Therefore, the time period is given by

$$T = \frac{2\pi}{\omega} = 2\pi\sqrt{\frac{l}{g}}$$

From the above expression, it is clear that the time period of a simple pendulum is independent of mass and shape of the bob.

7.6 MASS-STRING SYSTEM LO3

When a spring is compressed or stretched by applying a force on it, then the resulting compression does not have a simple linear relationship with the applied force. The relationship is linear only for small displacements. Here, the elastic force F produced in the spring is given by

$$F = -kx$$

together with x as the change in length of the spring and k as the spring constant. Now we shall discuss two simple cases.

7.6.1 Horizontal Oscillations

Here we assume a massless spring, one end of which is connected to a mass m and the other end is connected to a fixed point, as shown in Fig. 7.3

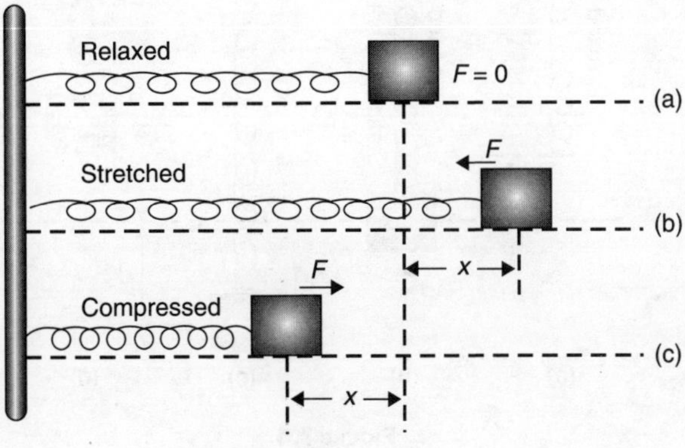

FIGURE 7.3

The mass m is free to move on a frictionless horizontal surface. The static equilibrium position is shown in Fig. 7.3(a) as relaxed and no force is acting on it. When the mass m is pulled to the right [Fig. 7.3(b)], through a small distance x, then the restoring force exerted by the spring is directed towards the left and is given by

$$F = -kx$$

where the negative sign indicate that the force and displacement are oppositely directed. Here, the mass starts moving with linear acceleration $\frac{d^2x}{dt^2}$. Then, we have

$$F = m\frac{d^2x}{dt^2}$$

or $$-kx = m\frac{d^2x}{dt^2} \quad \text{[By Newton's second law]}$$

or $$\frac{d^2x}{dt^2} + \frac{k}{m}x = 0, \text{ i.e., } \frac{d^2x}{dt^2} + \omega^2 x = 0$$

This is the equation of SHM, having time period

$$T = \frac{2\pi}{\omega} = 2\pi\sqrt{\frac{m}{k}}$$

7.6.2 Vertical Oscillations

Let us consider a perfectly elastic and massless spring of length L hanging freely from a support, as shown in Fig. 7.4(a). When a mass m is attached to its lower end, it is stretched through a distance x' by the force mg.

$$F = -kx'$$

where k is the force constant of the spring. The another force mg (weight) is acting downward on the spring. Since in this situation no net force acts on the body of mass m, i.e., $mg = -kx'$. Now we pull down the mass m through a small distance y' from the equilibrium position and release, then its starts oscillating, as shown in Fig. 7.4(c), (d). $F = -ky'$ is the restoring force and is oppositely directed to the displacement.

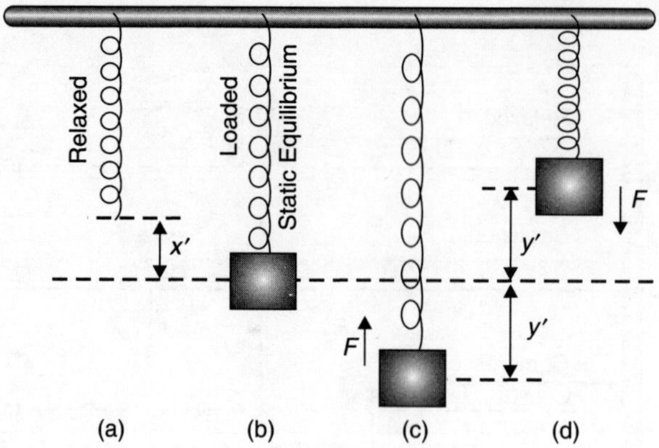

Figure 7.4

By applying Newton's second law, we get

$$F = m\frac{d^2y'}{dt^2} = -ky'$$

or $$\frac{d^2y'}{dt^2} = -\frac{k}{m}y' = -\omega^2 y'$$

where $\omega^2 = \frac{k}{m}$. This is the equation of motion of mass m executing SHM with time period

$$T = \frac{2\pi}{\omega} = 2\pi\sqrt{\frac{m}{k}}$$

By using this formula, we can calculate the time period of mass-spring system.

7.7 DAMPED HARMONIC OSCILLATOR

LO3

In the previous section free oscillations were discussed in some oscillating systems. In such systems, no frictional force or resistance is offered. Therefore, the body will keep on vibrating indefinitely and such vibrations are called *free vibrations*. But in real situation, there is always some resistance offered to the oscillating system. In real sense if a body set into vibrations will have its amplitude continuously decreasing due to fractional resistance and so the vibrations will die after some time. The motion is said to be damped by the friction and is called as *damped vibrations*. For example, when a pendulum is displaced from its equilibrium position, it oscillates with a decreasing amplitude and finally comes to the rest.

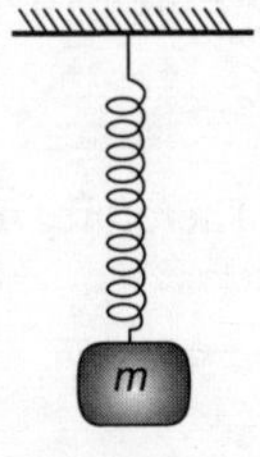

FIGURE 7.5

Suppose a system has a body of mass m attached to a spring whose force constant is k (Fig. 7.5). Again consider x' to be the displacement of the body from the equilibrium state at any instant of time. Then $\frac{dx'}{dt}$ will be the instantaneous velocity.

Here two types of forces are acting on the body: a restoring force ($F_r = -kx'$) which is proportional to the displacement x' and acts in the opposite direction to the displacement and a damping force $-q'\frac{dx'}{dt}$ which is proportional to the velocity and is oppositely directed to the motion. The total force acting on the body is thus given by

$$F = -kx' - q'\frac{dx'}{dt}$$

or $$m\frac{d^2x'}{dt^2} = -kx' - q'\frac{dx'}{dt}$$

or $$\frac{d^2x'}{dt^2} + \frac{q'}{m}\frac{dx'}{dt} + \frac{k}{m}x' = 0 \qquad \text{(i)}$$

If we put $\frac{q'}{m} = 2s$ and $\frac{k}{m} = \omega^2$, then above equation takes the form

$$\frac{d^2x'}{dt^2} + 2s\frac{dx'}{dt} + \omega^2 x' = 0 \qquad \text{(ii)}$$

which is the equation of second degree. We assume its solution as

$$x' = Ae^{\alpha t} \qquad \text{(iii)}$$

where A and α are arbitrary constants.

Differentiating Eq. (iii) w.r.t. t, we have

$$\frac{dx'}{dt} = A\alpha e^{\alpha t} \text{ and } \frac{d^2x'}{dt^2} = A\alpha^2 e^{\alpha t}$$

By putting these values in Eq. (ii), we have

$$A\alpha^2 e^{\alpha t} + 2sA\alpha e^{\alpha t} + \omega^2 Ae^{\alpha t} = 0$$

or $$Ae^{\alpha t}[\alpha^2 + 2s\alpha + \omega^2] = 0$$

$\because$ $$Ae^{\alpha t} \neq 0$$

$\therefore \qquad \alpha^2 + 2s\alpha + \omega^2 = 0$

This gives,

$$\alpha = -s \pm \sqrt{s^2 - \omega^2}$$

Therefore, α has two roots

$$\alpha_1 = -s + \sqrt{s^2 - \omega^2}$$

$$\alpha_2 = -s - \sqrt{s^2 - \omega^2}$$

Therefore, the general solution of Eq. (ii) is given by

$$x' = A_1 e^{(-s+\sqrt{s^2-\omega^2})t} + A_2 e^{(-s-\sqrt{s^2-\omega^2})t} \qquad \text{(iv)}$$

where A_1 and A_2 are arbitrary constants. The actual solution depends upon whether $s^2 > \omega^2$, $s^2 = \omega^2$ and $s^2 < \omega^2$. Now we will discuss all these cases.

Case-A: $s^2 > \omega^2$

The term $\sqrt{s^2 - \omega^2}$ is real and less than s. Therefore both exponents i.e., $[-s + \sqrt{s^2 - \omega^2}]$ and $[-s - \sqrt{s^2 - \omega^2}]$ in Eq. (iv) are negative. Due to this reason the displacement x' continuously decreases exponentially to zero without performing any oscillation [Fig. 7.6(a)]. This motion is known as *overdamped.*

Case-B: $s^2 = \omega^2$

For this case, Eq. (iv) does not satisfy differential Eq. (ii).Suppose that $\sqrt{s^2 - \omega^2}$ is not exactly zero but it is equal to very small quantity β. Now, Eq. (iv) gives

$$x' = A_1 \exp(-s + \beta)t + A_2 \exp(-s - \beta)\, t$$

$$x' = e^{-st}[A_1 e^{\beta t} + A_2 e^{-\beta t}]$$

$$= e^{-st}\,[A_1(1 + \beta t + \ldots) + A_2(1 - \beta t + \ldots)]$$

Here we have neglected the terms containing β^2, β^3, β^4 is very small

$$\therefore \qquad x' = e^{-st}[(A_1 + A_2) + \beta t(A_1 - A_2) + \ldots]$$

$$x' = e^{-st}[P' + Q't]$$

where $P' = (A_1 + A_2)$ and $Q' = \beta(A_1 - A_2)$ (v)

Equation (v) represents a possible solution of this form. From this equation, it is clear that as t increases the term $(P' + Q't)$ increases but the term e^{-st} gets decreased. Because of this fact the displacement x' first increases due to the term $(P' + Q't)$ but it decreases because of the exponential term e^{-st} and finally it approaches to zero as t increases. When we compare cases A and B it is noticed that the exponent is $-st$ in case B while it as more than $-st$ in case of A. Therefore, in case A the particle acquires its position of equilibrium rapidly than in case B[Fig. 7.6(b)]. This type of motion is called *critical damped motion.* Examples are voltmeter, ammeter, etc. In these instruments, pointer moves to the correct position and comes to the rest without any oscillations.

Case-C: $s^2 < \omega^2$

The term $\sqrt{s^2 - \omega^2}$ is imaginary, which can be written as

$$\sqrt{s^2 - \omega^2} = i\sqrt{\omega^2 - s^2} = i\beta'$$

where $\beta' = \sqrt{\omega^2 - s^2}$ and $i\ \sqrt{-1}$

Now, Eq. (iv) becomes

$$x' = A_1 e^{(-s + i\beta')t} + A_2 e^{(-s - i\beta')t}$$

$$x' = e^{-st}[A_1 e^{i\beta' t} + A_2 e^{-i\beta' t}]$$

$$= e^{-st}[A_1 \cos \beta' t + i \sin \beta' t) + A_2(\cos \beta' t - i \sin \beta' t)$$

$$= e^{-st}[(A_1 + A_2) \cos \beta' t + i[(A_1 - A_2) \sin \beta' t]$$

$$= e^{-st}[(A \sin \delta \cos \beta' t + A \cos \delta \sin \beta' t]$$

where $A \sin \delta = A_1 + A_2$ and $A \cos \delta = i(A_1 - A_2)$

$$\therefore \quad x' = e^{-st} A \sin (\beta' t + \delta)$$

Putting the value of β' in the above questions, we get

$$x' = Ae^{-st} \sin [\sqrt{(\omega^2 - s^2)}t + \delta] \tag{vi}$$

The above equation shows the oscillatory motion and represents the damped harmonic oscillator. The oscillations are not simple harmonic because the amplitude (Ae^{-st}) is not constant but decreases with time (t). However, the decay of amplitude depends upon the damping factor s. The motion is known as *under damped motion* [Fig. 7.6(c)]. The motion of a pendulum in air, the motion of the coil of ballistic galvanometer, etc. are the example of this case.

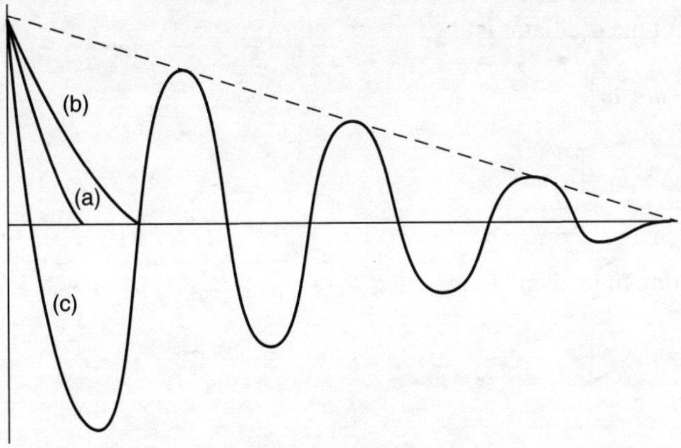

FIGURE 7.6

7.8 ATTENUATION COEFFICIENTS OF A VIBRATING SYSTEM LO4

By the definition, the energy of an oscillator is proportional to the square of its amplitude. In damped oscillator, the amplitude decays exponentially with time as e^{-st}. It means that the energy will decay as $(e^{-st})^2$, i.e., e^{-2st}. So the decay rate of energy depends upon s. The following three characteristics namely logarithmic decrement, relaxation time and quality factor may give the attenuation of a vibrating system.

7.8.1 Logarithmic Decrements

The rate at which the amplitude dies away is measured by logarithmic decrement. The amplitude of damped harmonic oscillator is given by a factor Ae^{-st}. Therefore, at time $t = 0$ the amplitude will be maximum (i.e., $A = A_0$). If $A_1, A_2, A_3, \ldots$ be the amplitude at time $t = T, 2T, 3T, \ldots$ respectively where T is the time period of oscillations, then.

$$A_1 = Ae^{-sT}, A_2 = Ae^{-s(2T)}, A_3 = Ae^{-s(3T)} \text{ and so on.}$$

This yields

$$\frac{A_0}{A_1} = \frac{A_1}{A_2} = \frac{A_2}{A_3} = \ldots = e^{sT} = e^{\lambda} \qquad \text{(where } sT = \lambda\text{)}$$

Here, λ is called *logarithmic decrement.*

Now, by taking the natural logarithmic, we have

$$\ln\frac{A_0}{A_1} = \ln\frac{A_1}{A_2} = \ln\frac{A_2}{A_3} = \lambda$$

Hence, logarithmic decrement is the natural logarithm of ratio between two successive maximum amplitudes, which are separated by one period.

7.8.2 Relaxation Time

It is the time taken by damped harmonic oscillator for decaying total mechanical energy by the factor $\frac{1}{e}$ of its initial value.

The mechanical energy of the oscillator is

$$E = \frac{1}{2}mA^2\omega^2e^{-2st} \qquad \text{(i)}$$

At $\quad t = 0, E = E_0 = \frac{1}{2}mA^2\omega_0^2$

$\therefore \quad$ Total energy, $E = E_0e^{-2st}$ (ii)

Suppose τ be the relaxation time, then at time $t = \tau$,

$$E = \frac{E_0}{e} \qquad \text{(By definition)}$$

By using Eq. (ii), we get

$$\frac{E_0}{e} = E_0e^{-2st}$$

or $\quad e^l = e^{2s\tau}$

or $\quad 1 = 2s\tau$

or $$\tau = \frac{1}{2s} \qquad \text{(iii)}$$

Therefore, the dissipated energy in terms of relaxation time is written as

$$E = E_0 e^{-t/\tau} \tag{iv}$$

7.8.3 Quality Factor

It is defined as 2π times the ratio of energy stored in the system to the energy lost per cycle. This factor of a damped oscillator shows the quality of oscillator so far as damping is concerned.

$$Q = 2\pi \frac{E}{P_d T} \tag{v}$$

where P_d is the power dissipation and T is the periodic time. Then,

$$Q = 2\pi \frac{E}{(E/\tau)T} = \frac{2\pi\tau}{T} \qquad \left[\because P_d = \frac{E}{\tau}\right]$$

$$Q = \omega\tau \qquad \left[\because \omega = \frac{2\pi}{T}\right] \tag{vi}$$

From the above equation, it is clear that the value of relaxation time τ will be higher (or damping will be lower) for higher value of Q.

For the force constant k and the mass m of the vibrating system

$$\omega = \sqrt{\frac{k}{m}} \text{ and } \tau = \frac{1}{2s} \quad \text{[from Eq. (iii)]}$$

$$\therefore \qquad Q = \frac{1}{2s}\sqrt{\frac{k}{m}}$$

Since lower values of s lead to lower damping, it is clear that for low damping, the quality factor would be higher.

7.9 FORCED VIBRATIONS LO5

We have discussed earlier the vibrations in which the body vibrates with its own frequency without being placed to any external force. The different situation arises if the body is placed to an external force while it is vibrating. These vibrations are known as *forced vibrations.* For example, if a bob of simple pendulum is held in hand and then given number of swings by the hand. In this case, the pendulum vibrates due to external force and not due to its natural frequency. So forced vibrations can also be defined as the vibrations in which the body vibrates with frequency other than its natural frequency, which is due to some external periodic force. The tuning fork is also one of the examples of forced vibrations.

Theory of Forced Vibrations

Suppose a particle of mass m is connected to a spring. When it is displaced from its mean position, the oscillations are started and the particle experiences different kinds of forces viz, a restoring force ($-kx$), a damping force $\left(-q'\frac{dx}{dt}\right)$ and the external periodic force ($F_0 \sin \omega t$). The total force acting on the particle is, therefore

$$F = F_0 \sin \omega t - q'\frac{dx}{dt} - kx \qquad \text{(i)}$$

By Newton's second law of motion

$$F = m\frac{d^2x}{dt^2}$$

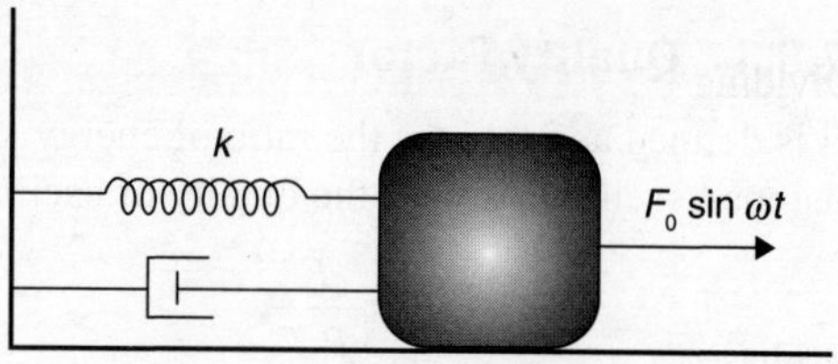

FIGURE 7.7

Hence,

$$m\frac{d^2x}{dt^2} = F_0 \sin \omega t - q'\frac{dx}{dt} - kx$$

or

$$\frac{d^2x}{dt^2} + \frac{q'}{m}\frac{dx}{dt} + \frac{k}{m}x = \frac{F_0 \sin \omega t}{m} \qquad \text{(ii)}$$

Eq (ii) is the differential equation of motion of the particle.

Substitute $\frac{q'}{m} = 2s$, $\frac{k}{m} = \omega_0^2$ and $\frac{F_0}{m} = f$, then Eq, (ii) becomes

$$\frac{d^2x}{dt^2} + 2s\frac{dx}{dt} + \omega_0^2 x = f \sin \omega t \qquad \text{(iii)}$$

In the steady state, the solution of the above equation should be

$$x = A \sin (\omega t - \delta) \qquad \text{(iv)}$$

where A is the amplitude of vibrations in the steady state. By differentiating Eq. (iv) twice w.r.t. t, we have

$$\frac{dx}{dt} = \omega A \cos (\omega t - \delta)$$

and

$$\frac{d^2x}{dt^2} = -\omega^2 A \sin (\omega t - \delta)$$

By substituting the values of x, $\frac{dx}{dt}$ and $\frac{d^2x}{dt^2}$ in Eq. (iii), we have

$$-\omega^2 A \sin (\omega t - \delta) + 2s\,\omega A \cos (\omega t - \delta) + \omega_0^2 A \sin (\omega t - \delta)$$
$$= f \sin \{(\omega t - \delta) + \delta\}$$

or

$$A(\omega_0^2 - \omega^2) \sin (\omega t - \delta) + 2s\omega A \cos (\omega t - \delta)$$
$$= f \sin (\omega t - \delta) \cos \delta + f \cos (\omega t - \delta) \sin \delta \qquad \text{(v)}$$

If Eq. (v) holds good for all values of t, then the coefficients of $\sin (\omega t - \delta)$ and $\cos (\omega t - \delta)$ must be equal on both the sides, then

$$A(\omega_0^2 - \omega^2) = f \cos \delta \qquad \text{(vi)}$$

and

$$2s\,\omega A = f \sin \delta \qquad \text{(vii)}$$

By squaring and adding Eqs. (vi) and (vii), we have

$$A^2(\omega_0^2 - \omega^2)^2 + 4s^2\,\omega^2 A^2 = f^2$$

or $$A = \frac{f}{\sqrt{(\omega_0^2 - \omega^2)^2 + 4s^2\omega^2}} \quad \text{(viii)}$$

Dividing Eq. (vii) by Eq. (vi), then we get

$$\tan\delta = \frac{2s\omega}{\omega_0^2 - \omega^2}$$

or phase, $$\delta = \tan^{-1}\left[\frac{2s\omega}{\omega_0^2 - \omega^2}\right] \quad \text{(ix)}$$

From Eq. (viii) and (ix), it is clear that the amplitude and phase of the forced oscillations depend upon $(\omega_0^2 - \omega^2)$, i.e., these depend upon the driving frequency (ω) and the natural frequency (ω_0) of the oscillator. The amplitude and the phase are explained as below.

Case - A: Very low driving frequency, i.e., $\omega << \omega_0$. In this case, the amplitude of the vibrations

$$A = \frac{f}{\sqrt{(\omega_0^2 - \omega^2)^2 + 4s^2\omega^2}} \approx \frac{f}{\omega_0^2}$$

$$= \frac{F_0}{m\omega_0^2} \qquad \left[\because f = \frac{F_0}{m}\right]$$

or $$A = \frac{F_0}{k} \qquad \left[\because \omega_0^2 = \frac{k}{m}\right]$$

Hence, the amplitude depends on the force constant of the spring and the magnitude of the applied force.

Phase, $$\delta = \tan^{-1}\left[\frac{2s\omega}{\omega_0^2 - \omega^2}\right] = \tan^{-1}\left[\frac{2s\omega}{\omega_0^2}\right]$$

Since $\omega_0^2 >> \omega$, $2s\omega/\omega_0^2 \to 0$ and $\delta \to 0$ or ≈ 0. Therefore, under this situation the driving force and the displacement are in phase.

Case - B: Same driving and natural frequencies, i.e., $\omega = \omega_0$. This frequency is called resonant frequency. Under this situation, the amplitude of vibrations

$$A = \frac{f}{\sqrt{(\omega_0^2 - \omega^2)^2 + 4s^2\omega^2}}$$

$$= \frac{f}{2s\omega_0} = \frac{F_0/m}{(q'/m)\omega_0} = \frac{F_0}{q'\omega_0}$$

Hence, the amplitude of vibrations depends upon the damping and applied force. Now

Phase, $$\delta = \tan^{-1}\left[\frac{2s\omega}{(\omega_0^2 - \omega^2)}\right] = \tan^{-1}\left[\frac{2s\omega}{0}\right]$$

$$= \tan^{-1}[\infty] = \frac{\pi}{2}$$

Thus, the displacement lags behind the force by a phase of $\frac{\pi}{2}$, as $x = A \sin(\omega t - \delta)$ and the applied force if $F_0 \sin \omega t$.

Case - C: Very large driving frequency, i.e., $\omega >> \omega_0$. Here, the amplitude of vibrations

$$A = \frac{f}{\sqrt{(\omega_0^2 - \omega^2)^2 + 4s^2\omega^2}}$$

Since ω is very large $\omega^4 >> 4\omega^2 s^2$

$$A = \frac{f}{\omega^2} = \frac{F_0}{m\omega^2}$$

Phase,
$$\delta = \tan^{-1}\left[\frac{2s\omega}{(\omega_0^2 - \omega^2)}\right] = \tan^{-1}\left[\frac{2s\omega}{-\omega^2}\right]$$

$$\approx \tan^{-1}\left[\frac{2s}{-\omega}\right] = \tan^{-1}[-0] = \pi \qquad \text{[Since } \omega \text{ is very large, } 1/\omega = 0]$$

Therefore, under the situation $\omega >> \omega_0$, the displacement lags behind the force by a phase of π.

7.10 RESONANCE LO6

So far we have discussed various types of oscillations. It is clear that if a vibrating system consists of a number of oscillators coupled together, the resultant motion would be complicated. However, if we choose the starting conditions correctly, it is possible to cause the system vibrate in such a way that every part has the same frequency. Such simple vibrations are known as normal oscillations or *normal modes* and the associated frequencies are called the normal frequencies or the *natural frequencies.* So this is evident that all mechanical structures, for example, buildings, airplanes, bridges, etc. have one or more natural frequencies. Now if such a structure is subject to a driving frequency, which is equal to one of the natural frequencies, the resulting oscillations will have large amplitude that can have disastrous consequences. Shattering a wine glass with a sound wave that matches one of the natural frequencies of the glass is one demonstration of this phenomenon of resonance. Another outcome of this effect is the collapse of roadways and bridges in earthquakes.

In order to make clear the phenomenon of resonance, we take an example of forced oscillations that occur at the frequency of the external force (driving frequency ω) and not at the natural frequency (ω_0) of the vibrating system. As has been seen, however, the amplitude of oscillations depends on the relationship between the driving frequency of the applied force and the natural frequency. A succession of small impulses, if applied at the proper frequency, can develop an oscillation of large amplitude. You can take an example of pushing a friend on a swing. You would have noticed that by applying pushes precisely at the same time in each cycle, you can cause your friend to move in an increasingly large arc.

In Fig.7.8, we show the variation of amplitude of the forced vibrations with the driving frequency ω for three cases of large damping, medium damping and small damping. Medium damping corresponds to twice the damping force and large damping corresponds to four times the damping force. It is clear from the figure that the amplitude decreases if the frequency ω is far from the condition $\omega = \omega_0$, called condition of

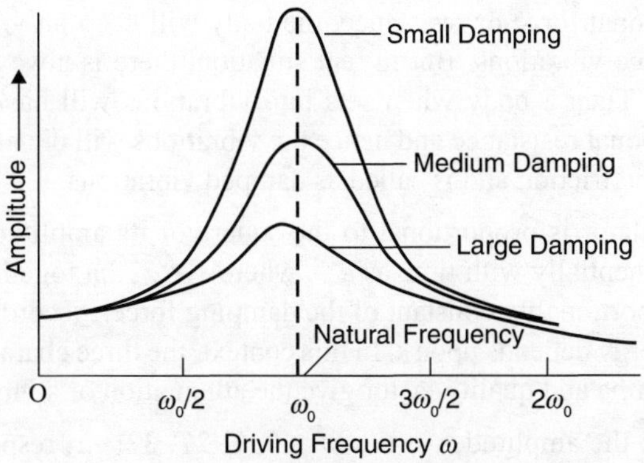

FIGURE 7.8

resonance. Moreover, when the damping is small, the amplitude of the forced oscillations increases rapidly as ω approaches ω_0. The amplitude reaches its maximum when $\omega = \omega_0$. For medium damping also, the amplitude gets increased but it does not increase so rapidly near the resonance ($\omega = \omega_0$). However, for the largest damping the resonant frequency is displaced slightly from the natural frequency.

SUMMARY

The topics covered in this chapter are summarised below.

- The motion of physical bodies is broadly classified into two categories, namely translational motion and vibrational or oscillatory motion. If the position of a body varies linearly with time, then such motions are called translational motions. The examples of translational motion are a ball that rolls on the ground and a train that moves on a straight track. A motion of a body that repeats itself after regular intervals of times and when the body moves back and forth over the same path is called vibrational or oscillatory motion. The example of vibrational motion are the oscillations of the arms of a walking person, the bob of the pendulum clock, beating of heart, etc.
- If the acceleration of a particle in a periodic motion is always directly proportional to its displacement from its equilibrium position and is always directed towards equilibrium position, then the motion of the particle is said to be Simple Harmonic Motion (SHM).
- For f as the linear acceleration of the particle and x as its displacement from the equilibrium position, the essential condition for linear SHM is $f \propto -x$. However, if α be the angular acceleration and θ be the angular displacement from the equilibrium position, then the condition for angular SHM is $\alpha \propto -\theta$.
- For the displacement x and the angular frequency ω ($= \sqrt{k/m}$, where k is the force constant and m is the mass of the particle), $\frac{d^2x}{dt^2} + \omega^2 x = 0$ represents the differential equation of the SHM
- The solution $x = A \sin(\omega t + \delta)$ gives the displacement of the particle executing SHM at any instant of time t. Here A represents the maximum displacement of the particle, which is called the amplitude of oscillations. The velocity of the particle is given by $v = \omega\sqrt{(A^2 - x^2)}$ and the acceleration is $f = -\omega^2 x$. The energy of a harmonic oscillator is given by $E = \frac{1}{2} m\omega^2 A^2$.

- When there is no frictional force or resistance, the body will keep on vibration indefinitely and such vibrations are called free vibrations. But in real situation, there is always some resistance offered to the oscillating system. Then a body when sets into vibrations will have its amplitude continuously decreasing due to fractional resistance and hence the vibrations will die after some time. The motion is said to be damped by the friction and is called as damped vibrations.
- The energy of an oscillator is proportional to the square of its amplitude. In damped oscillator, the amplitude decays exponentially with time as e^{-st}, where $s = q'/2m$ together with m as the mass of the body and q' as the proportionality constant of the damping force. Accordingly the energy also decays. So the decay rate of energy depends upon s. In this context, the three characteristics namely logarithmic decrement, relaxation time and quality factor give the attenuation of a vibrating system.
- If $A_0, A_1, A_2, A_3, \ldots$ be the amplitudes at time $t = 0, T, 2T, 3T, \ldots$, respectively, where T is the time period of oscillations, then logarithmic decrement is defined as $\lambda = \ln\frac{A_0}{A_1} = \ln\frac{A_1}{A_2} = \ln\frac{A_2}{A_3}$.
- The relation time is the time taken by damped harmonic oscillator for decaying total mechanical energy the factor of 1/e of its initial value. It is given by $\tau = \frac{1}{2s}$ where $s = q'/2m$ together with m as the mass of the body and q' as the proportionality constant of the damping force.
- Quality factor Q of the oscillator is defined as 2π times the ratio of energy stored in the system to the energy lost per cycle. This factor shows the quality of the oscillator and is given by $Q = \omega\tau = \frac{1}{2s}\sqrt{\frac{k}{m}}$. The higher value of Q means lower damping of the oscillator.
- All mechanical structures, for example, buildings, airplanes, bridges, etc. have one or more natural frequencies. If such a structure is subject to a driving frequency (say ω), which is equal to one of the natural frequencies (say ω_0), the resulting oscillations will have large amplitude that can have disastrous consequences. Shattering a wine glass with a sound wave that matches one of the natural frequencies of the glass is one demonstration of this phenomenon of resonance. Another outcome of this effect is the collapse of roadways and bridges in earthquakes. The condition (for example, for forced oscillation) where the driving frequency ω and the natural frequency ω_0 of the vibrating system match with each other is known as resonance. At the resonance, the amplitude of oscillations reaches its maximum.

SOLVED EXAMPLES

EXAMPLE 1 The total energy particle executing a SHM of period 2π seconds in 10.24×10^{-4} Joule. The displacement of a particle at $\pi/4$ second is $0.08\sqrt{2}m$. Find the amplitude and mass of the particle.

SOLUTION Given $E = 10.24 \times 10^{-4}$ J, $T = 2\pi$ sec and $x = 0.08\sqrt{2}m$ at $t = \pi/4$ sec.

In SHM, the displacement of a particle is given by

$$x = A \sin \omega t = A \sin\left(\frac{2\pi}{T}t\right)$$

$$0.08\sqrt{2} = A \sin\frac{2\pi}{2\pi}\frac{\pi}{4} = A\sin\frac{\pi}{4} = \frac{A}{\sqrt{2}}$$

or $\quad$ $\mathbf{A = 0.16\ m}$

Total energy is given by $E = \frac{2m\pi^2 A^2}{T^2}$ or $m = \frac{ET^2}{2\pi^2 A^2}$

or $$m = \frac{10.24 \times 10^{-4} \times (2\pi)^2}{2\pi^2 \times (0.16)^2} = \frac{20.48 \times 10^{-4}}{0.0256}$$

$$= 0.08 \text{ kg}$$

$$m = \mathbf{80\ g}$$

EXAMPLE 2 A particle executes SHM of period 10 sec. and amplitude 5 cm. Calcualte the maximum amplitude of velocity

SOLUTION Given displacement amplitude (A) = 0.05 m and T = 10 s.

Formula used for maximum amplitude of velocity $= A\omega = 0.05 \times \frac{2\pi}{T}$

$$= \frac{0.05 \times 2 \times 3.14}{10} = \mathbf{0.0314\ m/s}$$

EXAMPLE 3 Calculate the force constant and time period, if the potential of a harmonic oscillator of mass 2 kg in its resting position is 5.0 J and total energy is 9.0 J, when the amplitude is 1.0 m.

SOLUTION Given E = 9.0 J, U = 5.0 J, K.E = $E - U$ = 4.0 J and A = 1.0 m.

The kinetic energy of maximum displacement will be

$$\text{K.E.} = \frac{1}{2}kA^2 \quad \text{or} \quad 4.0 \text{ J} = \frac{1}{2}k(1.0)^2$$

$$k = 8.0 \text{ J/m}$$

$$T = 2\pi\sqrt{\frac{m}{k}} = 2\pi\sqrt{\frac{2}{8}} = \pi = \mathbf{3.14\ s}$$

EXAMPLE 4 A particle is executing SHM of amplitude 0.06 m and a period of 6 s. Find out the time taken by it in moving from one end of its path to a position 0.03 m from the equilibrium position on the same side.

SOLUTION Given A = 0.06 m, T = 6.0 s and x = 0.03 m

Displacement of a particle executing simple harmonic motion

$$x = A \sin(\omega t + \delta) \tag{i}$$

At $t = 0$, particle is at one end so that at $t = 0$, $x = A$, then

By using Eq. (i)

$$A = A \sin(0 + \delta) \text{ or } \sin\delta = 1 = \sin \pi/2$$

So, $$\delta = \pi/2$$

Putting this value of δ in Eq. (i), we get

$$x = A \sin\left(\omega t + \frac{\pi}{2}\right) = A \cos \omega t = A \cos \frac{2\pi}{T} t$$

or $$0.03 = 0.06 \cos \frac{2\pi}{6.0} t$$

or $$\cos\left(\frac{\pi}{3}t\right) = \frac{1}{2} = \cos\left(\frac{\pi}{3}\right)$$

or $$\frac{\pi}{3}t = \frac{\pi}{3}$$

or $$\mathbf{t = 1.0\ s}$$

EXAMPLE 5 Find the maximum velocity and acceleration of a particle executing SHM of period 10π second and amplitude 5×10^{-2} m.

SOLUTION Given $T = 10\pi$ sec and $A = 5 \times 10^{-2}$ m.

The equation of simple harmonic motion is $x = A \sin(\omega t + \delta)$

$$v = \frac{dx}{dt} = A\,\omega \cos(\omega t + \delta)$$

v will be maximum for $\cos(\omega t + \delta) = 1$

$$V_{\max} = A\omega = 5 \times 10^{-2} \times \frac{2\pi}{T} = 5 \times 10^{-2} \times \frac{2\pi}{10\pi}$$

$$= 1.0 \times 10^{-2} \text{ m/s}$$

$$\text{Acceleration } (f) = \frac{d^2x}{dt^2} = -A\omega^2 \sin(\omega t + \delta)$$

f will be maximum for $\sin(\omega t + \delta) = 1$, then

$$f = A\omega^2 = 5.0 \times 10^{-2} \times \left(\frac{2\pi}{T}\right)^2$$

$$= 5.0 \times 10^{-2} \times \left(\frac{2\pi}{10\pi}\right)^2$$

$$= \mathbf{2.0 \times 10^{-3}\ m/sec^2}$$

EXAMPLE 6 Calculate the maximum velocity of a particle that executes SHM of amplitude 0.06 m with time period of 10 π s.

SOLUTIOIN Given $A = 0.06$ m and $T = 10\pi$ s.

The equation of SHM is $x = A \sin(\omega t + \delta)$

$$v = \frac{dx}{dt} = A\omega \cos(\omega t + \delta)$$

v will b maximum if $\cos(\omega t + \delta) = 1$

$$v_{\max} = A\omega = 6.0 \times 10^{-2} \times \frac{2\pi}{10\pi}$$

$$= \mathbf{1.2 \times 10^{-2}\ m/s}$$

EXAMPLE 7 A mass of 1.0 kg is attached to a spring of stiffness constant 16 N/m. Find the natural frequency

SOLUTION Given $k = 16$ N/m and $m = 1.0$ kg.

Fornula used for natural frequency $n = \frac{1}{2\pi}\sqrt{\frac{k}{m}} = \frac{1}{2\pi}\sqrt{\frac{16}{1}} = \frac{2}{\pi}$

$$\mathbf{n = 0.64\ Hz}$$

EXAMPLE 8 A simple pendulum of one meter length is hanging at one end. Considering the oscillations to be of small displacement, find the period of oscillation if the mass of pendulum is 2.0 kg.

SOLUTION Given $l = 1.0$ m and $m = 2.0$ kg

Time period $T = 2\pi\sqrt{\frac{l}{g}} = 2\pi\sqrt{\frac{1.0}{9.8}} = 2 \times 3.14 \times \sqrt{\frac{1.0}{9.8}}$

$$= \mathbf{2.0\ s}$$

EXAMPLE 9 A particle of mass 100 gm is placed in a field of potential $U = 5x^2 + 10$ ergs/gm. Find the frequency.

SOLUTION Given $U = 5x^2 + 10$ ergs/g and $m = 100$ g

$$F = -\frac{dU}{dx} = -10x$$

$$F = m\frac{d^2x}{dt^2} = -10x$$

or $$\frac{d^2x}{dt^2} = -\frac{10}{m}x \qquad \text{(i)}$$

Now $$\frac{d^2x}{dt^2} = -\omega^2 x \qquad \text{(ii)}$$

Comparing Eqs. (i) and (ii),

$$\omega^2 = \frac{10}{m} \text{ or } \omega = \sqrt{\frac{10}{m}}$$

$$\frac{2\pi}{T} = \sqrt{\frac{10}{m}} \text{ or } n = \frac{1}{2\pi}\sqrt{\frac{10}{m}} = \frac{1}{2 \times 3.14}\sqrt{\frac{10}{100}}$$

$\boldsymbol{n = 0.05}$ **Hz**

EXAMPLE 10 A lift is ascending at acceleration of 3 m/s^2. What is the period of oscillation of simple pendulum of length one meter suspended in the lift?

SOLUTION Given $f = 3$ m/s^2 and $l = 1.0$ m

The lift is ascending with an acceleration 3 m/s^2 and acceleration due to gravity $g = 9.8$ m/s^2. Hence, total acceleration is $9.8 + 3 = 12.8$ m/s^2

Time period $$T = 2\pi\sqrt{\frac{l}{g'}} = 2\pi\sqrt{\frac{1.0}{12.8}}$$

$$= 1.755 \text{ s}$$

$\boldsymbol{T = 1.76}$ **s**

EXAMPLE 11 A mass of 6 kg stretches a spring 0.3 m from its equilibrium position. The mass is removed and another body of mass 1.0 kg is hanged from the spring. What would be the period of motion if the spring is now stretched and released?

SOLUTION

$$F = kx,\ k = \frac{F}{x} = \frac{mg}{x} = \frac{6 \times 9.8}{0.3}$$

$$k = 196 \text{ N/m}$$

$$T = 2\pi\sqrt{\frac{m}{k}} = 2 \times 3.14\sqrt{\frac{1.0}{196}}$$

$$= \mathbf{0.45\ s}$$

EXAMPLE 12 A spring when compressed by 10 cm develops a restoring force of 10 N. A body of mass 4 kg is attached to it. Calculate the compression of the spring due to the weight of the body and calculate the period of oscillation.

SOLUTION Here $F = kx$ or $k = \frac{F}{x} = \frac{10}{0.1} = 100$ N/m

The force applied $F' = mg = 4 \times 9.8 = 39.2$ N

$$x' = \frac{F'}{k} = \frac{39.2}{100} = 0.392 \text{ m}$$

Time period $T = 2\pi\sqrt{\frac{m}{k}} = 2 \times 3.14\sqrt{\frac{4}{100}}$

$$\mathbf{T = 1.26\ s}$$

EXAMPLE 13 The relaxation time for damped harmonic oscillator is 50 s. Determine the time in which the amplitude and energy of oscillator falls to $1/e$ times of its initial value.

SOLUTION The amplitude of dampled harmonic oscillator at time t is given by

$$A(t) = A_0 e^{-st}$$

Relaxation time $\tau = \frac{1}{2s}$

given $\tau = 50$ s

Now $\tau = \frac{1}{2s} = \frac{1}{2 \times 50} = \frac{1}{100}$ per s

A_0 is the amplitude at $t = 0$ and at time t the amplitude will be A_0/e. Hence

$$\frac{A_0}{e} = A_0 e^{-st} \Rightarrow \frac{1}{e} = e^{-st} \Rightarrow -1 = -st$$

or $t = \frac{1}{s} = \mathbf{100\ s}$

EXAMPLE 14 Considering quality factor of sonometer wire of frequency 260 Hz as 2000, calculate the time in which the amplitude decreases to $1/e^2$ of its initial value.

SOLUTION The quality factor is given by

$$Q = \omega\tau$$

Here $Q = 2000$ and $\omega = 2\pi n = 2 \times 3.14 \times 260$ rad/s

Relaxation time $\tau = \frac{Q}{\omega} = \frac{2000}{2 \times 260 \times 3.14}$

$$= 1.225 \text{ s}$$

The formula for amplitude of damped oscillator at time t is

$$A(t) = A_0 e^{-st}$$

Given $A(t) = \frac{A_0}{e^2}$

$\therefore$ $\frac{A_0}{e^2} = \frac{A_0}{e^{st}}$

or $t = \frac{2}{s} = 2\tau$

$$= 2 \times 1.225 = \mathbf{2.450\ s}$$

OBJECTIVE TYPE QUESTIONS

Q1. Which one of the following is not necessary for SHM?
(a) elasticity (b) inertia (c) restoring force (d) gravity

Q.2 The velocity of a particle making SHM is maximum at
(a) equilibrium position (b) position of $1/3^{rd}$ of amplitude
(c) extreme position (d) mid way (between zero and maximum)

Q.3 For a particle making SHM the phase difference between displacement and velocity is
(a) 0 (b) π (c) 2π (d) $\pi/2$

Q.4 Total energy of the particle executing SHM is same at
(a) Equilibrium position (b) midway (between zero and maximum)
(c) extreme position (d) all of them

Q.5 The potential energy of a particle executing SHM is maximum at
(a) equilibrium position (b) extreme position
(c) mid way (between zero and maximum) (d) position of $1/3^{rd}$ of amplitude

Q.6 The kinetic and potential energies of a particle executing SHM are same at displacement (from mean position)
(a) A (b) $A/2$ (c) $A/\sqrt{2}$ (d) $A/\sqrt{3}$

Q.7 The periodic motion which is not oscillatory, is
(a) simple pendulum (b) compound pendulum
(c) acoustic harmonic oscillator (d) motion of earth around sun

Q.8 The amplitude of a simple harmonic oscillator is doubled. The new time period would be
(a) same (b) double (c) half (d) none of these

Q.9 The unit of spring constant in SI system of units is
(a) Nm^2 (b) Nm^{-1} (c) Nm^{-2} (d) Nm

Q.10 In which of the following oscillations the amplitude varies with time
(a) damped oscillator (b) forced oscillator
(c) undamped oscillator (d) none of these

SHORT-ANSWER QUESTIONS

Q.1 What is periodic motion?

Q.2 What do you mean by simple harmonic motion?

Q.3 Write the expression for the average energy of simple harmonic oscillator.

Q.4 State and explain the equation for simple harmonic motion.

Q.5 What is the time period for simple pendulum executing SHM and how it is modified for an harmonic motion?

Q.6 What are the dimensions of forced constant of vibrating spring?

Q.7 Name the periodic motion which is not oscillatory.

Q.8 Are all the periodic motions simple harmonic? Is the reverse true? Explain.

Q.9 The amplitude of a simple harmonic oscillator is doubled. How does this effect the time period, total energy and maximum velocity of the oscillator?

Q.10 What is damping? On what factors the damping depends?

Q.11 What is the effect of damping on the natural frequency of an oscillator?

Q.13 What do you understand by 'quality factor'?

PRACTICE PROBLEMS

General Questions

Q.1 Differentiate simple harmonic motion and oscillatory motion. Define simple harmonic motion.

Q.2 Derive a general differential equation of motion for a simple harmonic oscillator and obtain its solution.

Q.3 Simple harmonic motion is called sinusoidal or co-sinusoidal. Justify.

Q.4 Discuss the characteristics of simple harmonic oscillations. What is quality factor and how do you define it?

Q.5 Derive an expression for the total energy of a harmonic oscillator and show that it is constant and proportional to the square of the amplitude.

Q.6 When the displacement is one half of the maximum amplitude, what fraction of the total energy is kinetic and what fraction is potential in simple harmonic motion?

Q.7 Derive a relation between restoring force of a spring and potential energy.

Q.8 Show that for a particle executing SHM the average values of kinetic and potential energies are the same and each is equal to half of the total energy.

Q.9 Define damped harmonic oscillations. Solve the differential equation and discuss the case of oscillatory motion. What is the quality factor and how do you define it?

Q.10 Discuss the theory of forced harmonic oscillations. How does sharpness of resonance depend on damping?

Q.11 What are damped vibrations? Establish a differential equation, of motion for a damped harmonic oscillator and obtain an expression of displacement. Discuss the case of heavy damping.

Q.12 Define damped harmonic oscillations. Write the differential equation for a damped harmonic oscillator. Solve the differential equation and discuss special cases of oscillatory motion.

Q.13 Write down the equation of damped simple harmonic oscillator. Find the expression for displacement and discuss when we get oscillatory damped simple harmonic motion.

Q.14 Discuss the methods (logarithmic decrement, relaxation time and quality factor) for quantitative measurement of damping effect in a damped simple harmonic oscillator.

Q.15 Explain free vibrations, damped vibrations, forced vibrations and resonance, giving one example of each.

Q.16 Write note on

(i) harmonic oscillator (ii) Forced oscillations

Simple Harmonic Motion and Sound Waves

8

Learning Objectives

After reading this chapter you will be able to

LO 1 Understand superposition of two simple Harmonic Motions (SHMs)

LO 2 Know about sound wave, its velocity, and sound displacement

LO 3 Learn basics of standing waves, node, and anti-node along with detailed description of their formation in air columns

LO 4 Understand Doppler effect, supersonic waves and shock waves along with derivation of sound speed and intensity of sound, and the level of sound intensity

LO 5 Know about interference of sound waves in time (beats), and relation between displacement and pressure amplitude

LO 6 Learn Lissajous figures and endoscopy together with its kinds

Introduction

Any periodic or oscillatory motion where the restoring force is proportional to the displacement and acts opposite to the displacement is called a Simple Harmonic Motion (SHM). A simple example is the weight attached to one end of a spring, the other end being tied to a rigid support such as a wall. If the mass is displaced from the mean (equilibrium) position, the spring exerts a restoring force, according to the Hooke's law, $F = -kx$, where k is the spring constant of the spring and x is the displacement from the mean position. According to the superposition principle, the net resultant at a given time and space, when two or more SHMs combine, is the sum of the resultants by each of the SHMs. In physics, a standing wave, also known as a stationary wave, is a wave in a medium in which each point on the axis of the wave has a constant amplitude.

A vibration that propagates as a typically audible mechanical wave of pressure and displacement through a transmission medium such as air or water is called sound. The change in frequency or the wavelength of a sound wave when the observer moves relative to the source, is termed as the Doppler effect. A common example is the change in the pitch of the sound when the source moves towards or away from the observer and vice-versa. A beat is an interference pattern between two sounds of slightly different frequencies. The beat frequency is equal to the difference between the interfering frequencies.

8.1 SUPERPOSITION OF TWO SHMS LO1

For this, we consider the superposition of two Simple Harmonic Motions (SHMs), which produce a displacement of the particle along the same line. We assume the case when both the motions have the same frequency. The displacement of the particle produced by each simple harmonic motion is given by

$$y_1 = a_1 \cos \omega t \tag{i}$$

$$y_2 = a_2 \cos (\omega t + \delta) \tag{ii}$$

Here, δ is the phase difference between y_1 and y_2.

The resulting displacement of the particle can be calculated by the linear combination

$$y = y_1 + y_2 = a_1 \cos \omega t + a_2 \cos (\omega t + \delta) \tag{iii}$$

If $\delta = 0$, i.e., the two motions are in phase, then the resultant motion is

$$y = a_1 \cos \omega t + a_2 \cos \omega t = (a_1 + a_2) \cos \omega t \tag{iv}$$

This relation shows that the resultant motion is also an SHM with the same angular frequency ω. However, the motion has an amplitude equal to the sum of the amplitudes of the two motions, i.e., $a = a_1 + a_2$.

8.2 SOUND WAVE AND ITS VELOCITY LO1

Let us look at the mechanism of sound-wave generation in air. We exert a pressure on the air molecules while we speak because of which compression of these molecules takes place (i.e., they come closer to each other). Then these molecules try to come back to their original position but cross the equilibrium position, due to which rarefaction is produced. In the next instance, these rarefactions are converted into compressions. Hence, the longitudinal wave in air is generated where the molecules execute SHMs. This is the way sound waves propagate in air. In general, we can say that the sound wave is the propagation of pressure variations in elastic media. Both the longitudinal sound waves and transverse sound waves, however, occur in solid elastic media. In a three-dimensional homogeneous (uniform) medium, the sound from an ideal point source propagates in the form of spherical waves.

The sound velocity depends on the properties of the medium, but for the large amplitudes, this also depends on the amplitude. The sound velocity in gases (say, C_{sg}) depends on the adiabatic coefficient $\gamma \left(\equiv \dfrac{C_P}{C_V} \right)$, density ρ and the pressure p or the temperature T of the gas. This is given by

$$C_{sg} = \sqrt{\frac{\gamma p}{\rho}} = \sqrt{\gamma R_s T} \tag{i}$$

Here, R_s is the specific gas constant in J/K kg. In most gases, the sound velocity is in the range of 200–1300 m/s.

The sound velocity in liquids, (say, C_{sl}) depends on the compression modulus K (measured in N/m^2) and the density ρ of the liquid. Specifically, it is given by

$$C_{sl} = \sqrt{\frac{K}{\rho}} \tag{ii}$$

In most liquids, the sound velocity is in the range of 1100–2000 m/s, and in water it is 1480 m/s at 20°C temperature.

The sound velocity in solids (say, C_{ss}) depends on the elasticity modulus E (measured in N/m^2) and the density ρ of the solid. Specifically, it is given by

$$C_{ss} = \sqrt{\frac{E}{\rho}} \quad \text{(iii)}$$

As mentioned earlier, sound waves in solids may be longitudinal waves or transverse waves. The sound velocity in most solids is in the range of 1200–6000 m/s, and in iron, it is 5000 m/s.

8.3 SOUND DISPLACEMENT LO1

So far, we have understood that the particles of the medium execute SHM in order to generate the sound wave, which means they vibrate about their equilibrium position. *Sound displacement* is defined as the displacement of the vibrating particles of the medium from their rest positions. This can be represented by

$$y(x, t) = y_0 \sin(\omega t - kx) \quad \text{(i)}$$

where ω is the angular frequency of oscillations ($\omega = 2\pi f$) and k is the wave number ($k = 2\pi/\lambda$, λ is the wavelength). If we differentiate y w.r.t. time t, we get the velocity of vibrating particles, which is also known as *sound-particle velocity*.
This is obtained as

$$v = \frac{dy}{dt} = y_0 \omega \cos(\omega t - kx)$$

or $$v(x, t) = y_0\, \omega \cos(\omega t - kx) \quad \text{(ii)}$$

This equation shows that there is a phase difference of $\pi/2$ between y and v. Moreover, the velocity v is different for different sound particles.

8.4 STANDING WAVES LO1

Standing waves are produced by superposition of two waves of equal frequency, amplitude and phase, while they propagate in opposite directions. It means the wave vectors $\vec{k}$ of these waves should be anti-parallel. If we consider the superposing waves to propagate in x and $-x$ directions, these can be represented by

$$y_1(x, t) = a \cos(kx - \omega t) \quad \text{(i)}$$

$$y_2(x, t) = a \cos(-kx - \omega t) \quad \text{(ii)}$$

The resultant wave achieved after the superposition is obtained as

$$y(x, t) = y_1 + y_2 = -2a \cos \omega t \cos kx \quad \text{(iii)}$$

The above equation represents the standing wave, whose displacement $y(x, t)$ is shown in Figure 8.1 as a function of x for different instants of time t. Here, it is seen that the minima and maxima of the standing wave are fixed in space, unlike the other propagating or travelling waves. The notation for a space-fixed minimum of a standing wave is termed *node*, while that for a space-fixed maximum is known as *anti-node*.

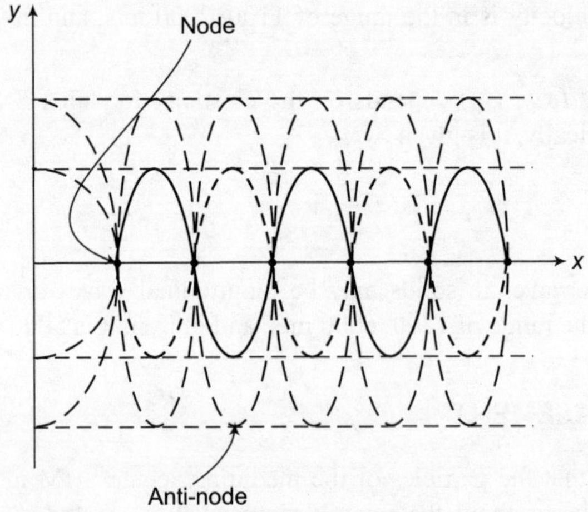

FIGURE 8.1

8.5 STANDING WAVES IN AIR COLUMNS

LO1

Standing waves can be set up in a tube of air, for example an organ pipe, by the superposition of longitudinal sound waves travelling in opposite directions. Here, the incident wave can be reflected from one end of the pipe. The phase relationship between the incident and reflected waves depends on whether that end is closed or open.

In a pipe closed at one end, a displacement node will always be found at the closed end because the wall at this end does not allow longitudinal motion of the air molecules. It also means that the reflected sound wave is 180° out of phase with the incident wave at a closed end of the pipe. The closed end of such an air column corresponds to a pressure anti-node (a point of maximum pressure variation), since the pressure wave is 90° out of phase with the displacement wave. On the other hand, a pressure node or displacement anti-node appears approximately at the open end of the air column. At the open end, pressure variation does not occur as this end of the air column is open to the atmosphere, where the pressure must remain constant at the atmospheric pressure.

Standing waves arise in the air column (length L) in a pipe with its one end closed, if the wavelength λ_n satisfies the following condition, which is also known as *resonance condition*.

$$\lambda_n = \frac{4L}{2n+1} \tag{i}$$

The frequencies corresponding to λ_n are referred to as *natural frequencies* f_n. For the sound waves (velocity C_s), these are given as

$$f_n = \frac{C_s}{\lambda_n} = \frac{C_s(2n+1)}{4L} \tag{ii}$$

Fundamental vibration corresponds to the standing wave with $n = 0$ (n is a non-negative integer). It means the fundamental wavelength λ_0 and the fundamental frequency f_0 are obtained as

$$\lambda_0 = 4L, f_0 = \frac{C_s}{4L} \tag{iii}$$

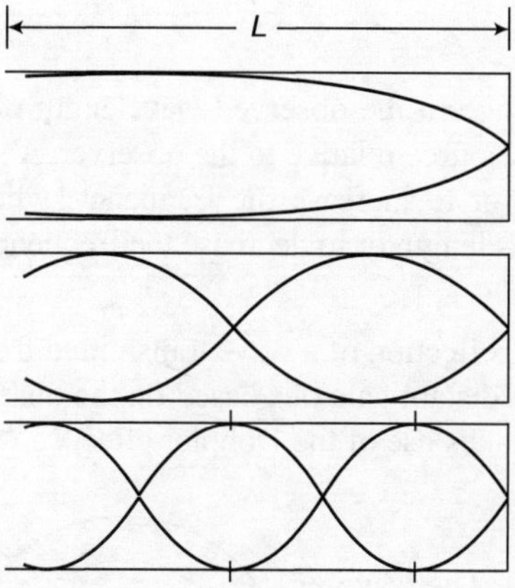

$$\lambda_0 = 4L,\ f_0 = \frac{C_s}{4L} \quad 1^{\text{st}} \text{ harmonic}$$

$$\lambda_1 = \frac{4L}{3},\ f_1 = \frac{3C_s}{4L} = 3f_0 \quad 3^{\text{rd}} \text{ harmonic}$$

$$\lambda_2 = \frac{4L}{5},\ f_2 = \frac{5C_s}{4L} = 5f_0 \quad 5^{\text{th}} \text{ harmonic}$$

FIGURE 8.2(a)

A standing wave with a node number n different from zero ($n > 0$) is called the *harmonic*. It is clear that the natural frequencies or the places of nodes and anti-nodes depend on the length of the pipe, L. The standing waves in a pipe closed at one end are shown in Figure 8.2(a). Here, it can be seen that the natural frequencies of oscillation form a harmonic series that includes only odd integral multiplies of the fundamental frequencies.

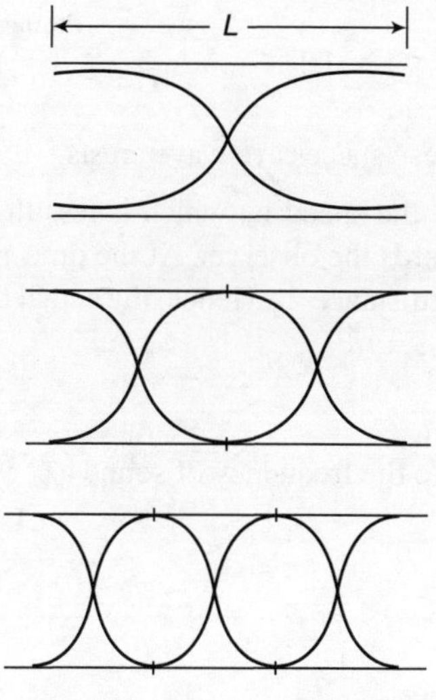

$$\lambda_0 = 2L,\ f_0 = \frac{C_s}{2L} \quad 1^{\text{st}} \text{ harmonic}$$

$$\lambda_1 = L,\ f_1 = \frac{C_s}{L} = 2f_0 \quad 2^{\text{nd}} \text{ harmonic}$$

$$\lambda_2 = \frac{2L}{3},\ f_2 = \frac{3C_s}{2L} = 3f_0 \quad 3^{\text{rd}} \text{ harmonic}$$

FIGURE 8.2(b)

In an open pipe (two ends free), the resonance condition reads

$$\lambda_n = \frac{2L}{n+1} \quad \text{or} \quad f_n = \frac{C_s(n+1)}{2L} \tag{iv}$$

It means the fundamental variation has wavelength λ_0 and frequency f_0, given below.

$$\lambda_0 = 2L,\ f_0 = \frac{C_s}{2L} \tag{v}$$

The first three normal modes of oscillation of a pipe open at both ends are shown in Figure 8.2(b).

8.6 DOPPLER EFFECT

LO1

Doppler effect is the phenomenon that gives rise to a change in the observed wavelength or frequency of sound or radiation, which results from the movement of its source relative to the observer. A source moving towards the observer appears to decrease the wavelength or to increase the frequency, whereas a source moving away from the observer appears to increase the wavelength or to decrease the frequency. The change in wavelength is known as the *Doppler shift*.

The Doppler effect is widely used to measure velocities by reflection of a wave transmitted from the moving object. This effect is also used to measure the velocities of distant galaxies based on the observed red shift. Not only this, ultrasound for blood movement in arteries makes use of the Doppler effect.

8.6.1 Moving Source

Consider a source of sound wave to be at rest (Figure 8.3). The wave crests corresponding to the emitted sound wave can be represented by circles whose centre is at the position of the source. If the frequency of sound is f_0, then these crests are generated at the frequency f_0 only, and the separation between successive crests will be the wavelength λ_0 of the sound. In view of C_s as the sound speed, f_0 and λ_0 are related to each other as follows:

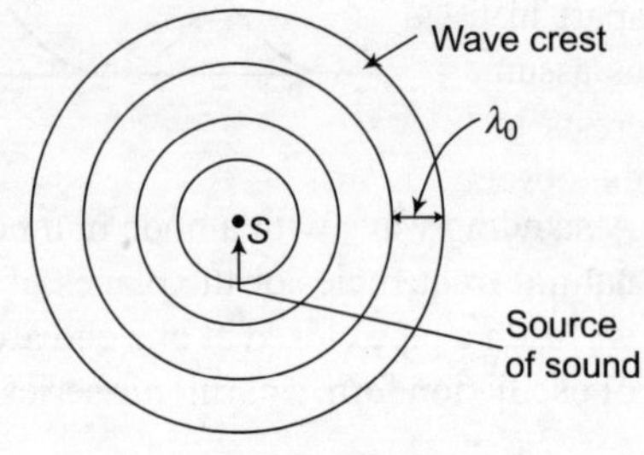

FIGURE 8.3

$$\lambda_0 = \frac{C_s}{f_0} = C_s \tau_0 \qquad \text{(i)}$$

Here, τ_0 is the time interval at which the observer receives these successive wave crests.

Now, consider the source S to move towards the observer at the speed v_s, which is less than C_s, i.e., $v_s << C_s$. Then in time τ_0, the source will cover a distance $v_s\tau_0$ towards the observer. At the time τ_0, the previously emitted crest will itself have moved towards the observer by a distance λ_0. Hence, the actual distance between the successive crests emitted towards the observer will be

$$\lambda' = \lambda_0 - v_s\,\tau_0 \qquad \text{(ii)}$$

Corresponding to the wavelength λ', the observer will observe the frequency of sound as

$$f' = \frac{C_s}{\lambda'} = \frac{C_s}{\lambda_0 - v_s\tau_0} = \frac{C_s/\lambda_0}{(1 - v_s\tau_0/\lambda_0)}$$

or
$$f' = \frac{f_0}{(1 - v_s/C_s)} \qquad \text{(iii)}$$

For the common case of $\frac{v_s}{C_s} << 1$, Eq. (iii) can be approximated as

$$f' = f_0\left(1 + \frac{v_s}{C_s}\right) \qquad \text{(iv)}$$

The above observation is picturized in Figure 8.4.

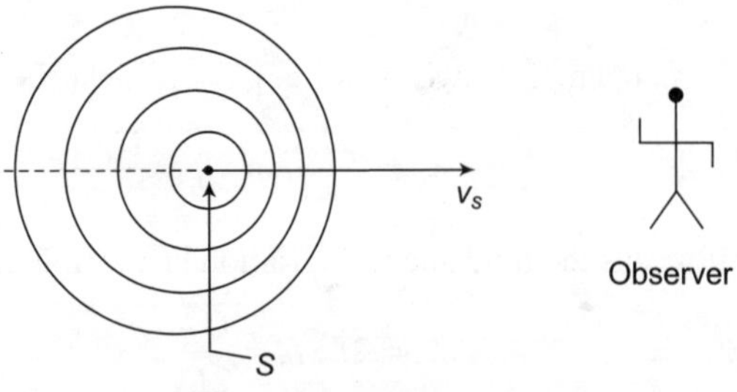

FIGURE 8.4

Based on the above argument, the frequency observed by an observer in the case of the source moving away from this would be

$$f' = \frac{f_0}{(1 + v_s/C_s)} \tag{v}$$

8.6.2 Stationary Source and Moving Observer

We can also discuss the case when the source of the sound wave remains at rest but the observer moves towards it at the speed v_{obs}. It is better to recall here that the sound-wave crests are λ_0 apart in the air and moving at the speed C_s. Let us assume that the observer meets the successive crests in time τ (different from τ_0). In this time, the observer moves a distance $v_{obs}\tau$ and the crests also move towards the observer by a distance $C_s\tau$ (Figure 8.5).

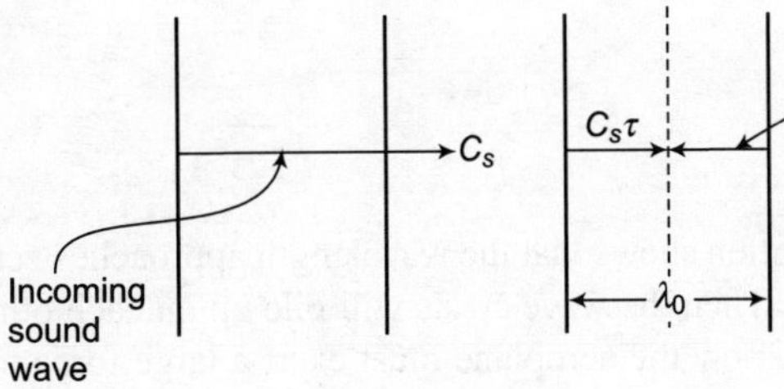

FIGURE 8.5

From Figure 8.5, it is clear that

$$\tau = \frac{\lambda_0}{C_s + v_{obs}} \tag{vi}$$

This gives the frequency of sound observed by the observer as

$$f = \frac{1}{\tau} = \frac{C_s + v_{obs}}{\lambda_0}$$

or $$f = \frac{C_s}{\lambda_0}\left(1 + \frac{v_{obs}}{C_s}\right)$$

$$= f'\left(1 + \frac{v_{obs}}{C_s}\right) \tag{vii}$$

Finally, we can also discuss the case when both the source and observer move towards each other, keeping in view Eqs (vii) and (iii), the observed frequency can be obtained as

$$f = f_0\frac{(1 + v_{obs}/C_s)}{(1 - v_s/C_s)}$$

or $$f = f_0\frac{(C_s + v_{obs})}{(C_s - v_s)} \tag{viii}$$

It means the frequency observed in the case of a moving source and observer towards each other is larger than the frequency observed when either source or observer moves alone and approach the other.

8.7 SUPERSONIC AND SHOCK WAVES LO1

In the case when the speed of a source is equal to the speed of sound, the wavefronts produced by the source cannot escape it. Hence, the waves will pile up and form a large amplitude *sound barrier*. This barrier makes persistent flight at this speed not only difficult but also risky. We will explain this phenomenon below.

Let us represent the speed of source, for example, an aeroplane, as v relative to the air. If the speed v is less than the speed of sound C_s, the waves in front of the aeroplane will be crowded together with a wavelength given by

$$\lambda = \frac{C_s - v}{f} \tag{i}$$

This equation shows that the wavelength approaches zero as the speed of the aeroplane approaches the speed of sound. Then the wave crests will pile up on each other, as shown in Figure 8.6(a) According to Newton's third law, now the aeroplane must exert a large force to compress the air in front of it and the air exerts an equal and opposite large force on the aeroplane. It means there is a large enhancement in aerodynamic drag or the air resistance, as the aeroplane approaches the speed of sound. This phenomenon is known as the *sound barrier*.

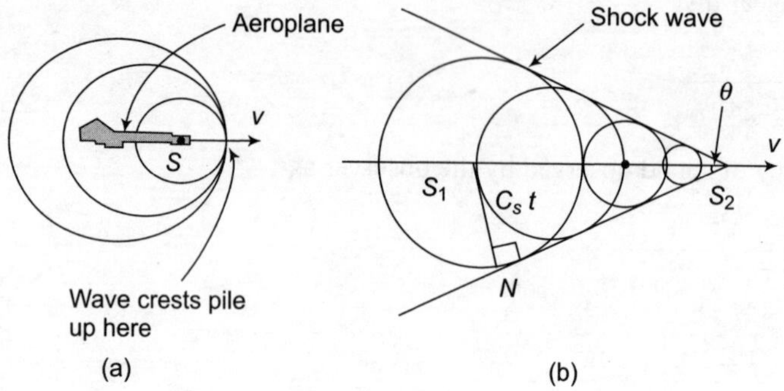

FIGURE 8.6

On the other hand, when the speed v of the source is greater than the speed C_s of the sound, the source of the sound, the aeroplane, is called *supersonic*. The aeroplane during its motion produces sound by displacing the surrounding air, and a series of wave crests is emitted from the front side (nose) of the aeroplane. Each of the wave crests spreads out in a circle centred at the position of the aeroplane. After a time t, the crest emitted from the initial position of the aeroplane (say, S_1) spreads to a circle of radius $C_s t$, whereas the aeroplane moves a greater distance vt (say, to the position S_2) in view of $v > C_s$. Under this situation, it can be seen that the circular crests interfere constructively at the points along the line which makes an angle θ with the direction of the velocity of the aeroplane. This leads to a very large amplitude wave crest along this line. This large amplitude crest is known as a *shock wave*. It is also evident that a shock wave forms a cone around the direction of motion of the source.

We can calculate the angle θ from Figure 8.6(b), where the right-angled triangle ($\Delta S_1 N S_2$) shows that

$$\sin\theta = \frac{C_s t}{vt} = \frac{C_s}{v} \tag{ii}$$

In view of $v > C_s$, the ratio v/C_s is called the *Mach number*, which is greater than unity for all supersonic speeds. Until the source such as a supersonic jet aeroplane or a rifle bullet moves with constant velocity, the angle θ remains a constant, and the shock-wave cone moves along with the source only. Here, it will be worth mentioning that the sonic boom we hear after a supersonic aeroplane has passed by is the arrival of this shock wave only. Unlike ordinary sound waves, the speed of a shock wave depends on its amplitude. The shock wave speed is always greater than the sound speed in the fluid, and it decreases as the amplitude of the wave reduces. This also means that the shock wave will die and reduce to an ordinary sound wave when the speed of the shock wave equals the normal speed of the sound.

8.8 DERIVATION OF SOUND SPEED

LO1

Let us take the example of sound waves in air or neutral gas. We represent the mass density by ρ, pressure by p and the velocity of air molecules by $\vec{v}$. In order to derive the sound speed, we can use the concept of conservation of mass (continuity equation) and the equation of motion (Navier–Stokes equation). Considering adiabatic compression, we can write the pressure in terms of density ρ, temperature T and the ratio γ of specific heats at constant pressure C_p and at constant volume C_v.

$$\frac{\nabla p}{p} = \gamma \frac{\nabla \rho}{\rho} \tag{i}$$

With this, the equation of motion reads

$$\rho \frac{\partial \vec{v}}{\partial t} = -\vec{\nabla} p = -\gamma p \frac{\vec{\nabla} \rho}{\rho} \tag{ii}$$

The continuity equation is written as

$$\frac{\partial \rho}{\partial t} + \vec{\nabla} \cdot (\rho \vec{v}) = 0 \tag{iii}$$

For the one-dimensional case $\left(\vec{\nabla} = \hat{x}\frac{\partial}{\partial x}\right)$, the linearized form of Eqs (ii) and (iii) can be found by taking $\rho = \rho_0 + \rho_1$ and $v = v_1$. Here, the quantity ρ_0 is the density in equilibrium and ρ_1 is due to oscillations or generation of wave. Similarly, v_1 is the velocity of air molecules during oscillations. Hence, we get

$$\rho_0 \frac{\partial v_1}{\partial t} = -\frac{\gamma p_0}{\rho_0} \frac{\partial \rho_1}{\partial x} \tag{iv}$$

$$\frac{\partial \rho_1}{\partial t} + \frac{\partial}{\partial x}(\rho_0 v_1) = 0 \tag{v}$$

For the wave of frequency ω and wave number k, the oscillating quantities can be taken to have dependence as

$$\rho_1 = \rho_1 \, e^{i(kx - \omega t)} \tag{vi}$$

$$v_1 = v_1 \, e^{i(kx - \omega t)} \tag{vii}$$

These equations show that $\frac{\partial}{\partial x}$ can be replaced with ik and $\frac{\partial}{\partial t}$ with $-i\omega$. So Eqs (iv) and (v) read

$$-i\omega\rho_0 v_1 = -\frac{\gamma p_0}{\rho_0} ik\rho_1 \tag{viii}$$

$$-i\omega\,\rho_1 + \rho_0\, ikv_1 = 0 \tag{ix}$$

We solve Eqs (viii) and (ix) for v_1 and obtain

$$\left(\omega^2 - k^2\frac{\gamma p_0}{\rho_0}\right)v_1 = 0 \tag{x}$$

Now, for the occurrence of wave, $v_1 \neq 0$.

Hence, $\omega^2 - k^2\dfrac{\gamma p_0}{\rho_0} = 0$

$$\text{or} \qquad \frac{\omega}{k} = \sqrt{\frac{\gamma p_0}{\rho_0}} \tag{xi}$$

The ratio $\dfrac{\omega}{k}$ is the phase velocity of the wave, which we call *sound speed* (say, C_s). In terms of temperature T of the gas or air, this can be written as

$$\frac{\omega}{k} = C_s = \sqrt{\frac{\gamma KTn_0}{n_0 M}} \quad (n_0 \text{ is the number density})$$

$$\Rightarrow \qquad C_s = \sqrt{\frac{\gamma KT}{M}} \tag{xii}$$

Here, M is the mass of each air molecule.

8.9 INTENSITY OF SOUND AND ITS DERIVATION LO1

Intensity of sound at a point in a progressive wave is defined as the sound energy per unit area per unit time perpendicular to the direction of propagation of the wave. It is measured in W/m^2 in the SI system of units.

We consider a plane progressive simple harmonic wave travelling along the positive x-direction with velocity $v(= \omega/k)$. The displacement y at a time t can be represented as

$$y = a\sin(\omega t - kx) \tag{i}$$

From this, we find the velocity of particle by differentiating it w.r.t. time

$$\frac{dy}{dt} = \omega a\cos(\omega t - kx) \tag{ii}$$

In order to calculate the energy or the intensity of sound, we consider a medium of density ρ. Taking unit area of medium having thickness dx perpendicular to the direction of propagation of wave, we find the kinetic energy as

$$dK = \frac{1}{2}\rho\, dx\left(\frac{dy}{dt}\right)^2 \tag{iii}$$

Putting the value of $\frac{dy}{dt}$ from Eq. (ii) and using $\omega = 2\pi f$, where f is the linear frequency of the wave, Eq. (iii) reads

$$dK = \frac{1}{2}\rho a^2 (2\pi f)^2 \cos^2 (\omega t - kx)\, dx$$

$$= 2\pi^2 a^2 \rho f^2 \cos^2 (\omega t - kx)\, dx \qquad \text{(iv)}$$

This shall give the total energy of the wave as

$$dE = dK_{max} \text{ (when potential energy is zero)}$$

$$= 2\pi^2 a^2 \rho f^2\, dx \qquad \text{(v)}$$

dx can be written in terms of the velocity v as $dx = vdt$.

Hence,

$$dE = 2\pi^2 a^2 f^2 \rho\, v\, dt \qquad \text{(vi)}$$

The integration gives

$$E = 2\pi^2 a^2 f^2 \rho\, vt \qquad \text{(vii)}$$

The energy flow per unit time is obtained from Eq. (vii) as $2\pi^2 a^2 f^2 \rho v$, which is nothing but the intensity of the sound wave. Hence,

$$I = 2\pi^2 a^2 f^2 \rho\, v \qquad \text{(viii)}$$

The other forms of the formula of sound intensity I are

$$I = 2\pi^2 \rho f^2\, v\, A^2_{max} \text{ (in terms of displacement)}$$

$$I = \frac{\rho v \Delta v^2_{max}}{2} \quad \left(\text{in terms of velocity, where } \Delta v_{max} = 2\pi f\, A_{max}\right)$$

$$I = \frac{\Delta p^2_{max}}{2\rho v} \quad \left(\text{in terms of pressure, where } \Delta p_{max} = 2\pi \rho f v\, A_{max}\right)$$

8.10 SOUND-INTENSITY LEVEL

LO1

The level of sound intensity (say I_L) can be defined in terms of decibels (dB) and neper (Np), as follows:

$$I_L = 10 \log_{10}\left(\frac{I}{I_0}\right) \quad \text{in dB} \qquad \text{(i)}$$

$$\text{or} \quad I_L = \log_{10}\left(\frac{I}{I_0}\right) \quad \text{in B (Bel)}$$

$$I_L = \frac{1}{2}\log_e\left(\frac{I}{I_0}\right) \quad \text{in Np} \qquad \text{(ii)}$$

Here, I is the sound intensity, I_0 is the reference intensity and B is the unit bel (1 B = 10 dB).
Since the sound intensity I is directly proportional to square of the pressure p, we have

$$\frac{I}{I_0} = \frac{p^2}{p_0^2} \quad (p_0 \text{ is the reference pressure})$$

With the help of this, I_L can be defined in terms of p as follows:

$$I_L = 10 \log_{10}\left(\frac{p^2}{p_0^2}\right)$$

$$= 20 \log_{10}\left(\frac{p}{p_0}\right) \quad \text{(iii)}$$

8.11 INTERFERENCE OF SOUND WAVES IN TIME: BEATS

LO1

A *beat* is an interference between two sound waves having slightly different frequencies. So beating is obtained when two sound waves of different frequencies superimpose with each other. Under this situation, alternating constructive and destructive interference cause the sound to be alternatively loud and soft. This phenomenon is called *beating* or *producing beats*.
The *beat frequency* is given by the difference of frequencies of respective waves, i.e.,

$$f_{\text{beat}} = |f_1 - f_2| = \frac{1}{T_{\text{beat}}} \quad \text{(i)}$$

Let us consider two sound waves having different frequencies (ω_1 and ω_2), which superimpose. Then, the displacement perceived will be given by

$$h(t) = \sin \omega_1 t + \sin \omega_2 t$$

$$= 2\sin\left(\frac{\omega_1 + \omega_2}{2}\right)t \cdot \cos\left(\frac{\omega_1 - \omega_2}{2}\right)t \quad \text{(ii)}$$

If we represent $\dfrac{\omega_1 + \omega_2}{2} = \omega_0$ as the average frequency and $\dfrac{\omega_1 - \omega_2}{2} = \Delta\omega$ as the difference in frequencies, then

$$h(t) = 2 \sin (\omega_0 t). \cos (\Delta\omega t) \quad \text{(iii)}$$

Clearly, the time period

$$T_{\text{beat}} = \frac{2\pi}{\Delta\omega} = \frac{2\pi}{\dfrac{\omega_1 - \omega_2}{2}} = \frac{1}{f_1 - f_2} \quad \text{(iv)}$$

Hence, the beat frequency

$$f_{\text{beat}} = \frac{1}{T_{\text{beat}}} = f_1 - f_2 \quad \text{(v)}$$

If $f_2 > f_1$, then $f_{\text{beat}} = f_2 - f_1$. This is the reason the modulus of the difference is taken in Eq. (i).

8.12 RELATION BETWEEN DISPLACEMENT AND PRESSURE AMPLITUDE LO1

Consider a plane monochromatic wave of sound travelling with a phase velocity v_0 along the length of the tube filled with a gas with density ρ_0. We will find the relationship between amplitude of the molecular displacement and that of pressure oscillations.

Proof

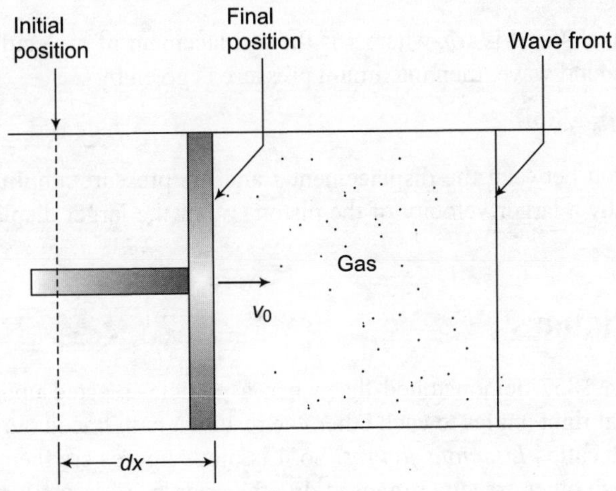

FIGURE 8.7

Consider a long tube of cross-sectional area A, in which a movable piston is fitted easily in the left side. The tube is open at another end with ambient surrounding. Firstly, the piston is at rest and then we apply a force F on the piston so that it moves towards the right side. This way we produce compression so that the wavefront moves dx distance towards right during the small interval dt. If the wavefront moves with velocity v_0, then

$$dx = v_0 dt \tag{i}$$

Due to the moment or displacement of the piston, the gas molecules also move with lower velocity v_m. Since the distance moved by the piston is very small, we can consider that the molecules under that volume have the same speed as that of the wavefront. So the total mass moved by the gas is obtained as

$$dm = \rho_0 A\, dx \tag{ii}$$

Here, ρ_0 is the mass density of the gas.

Linear momentum is given by

$$dP = dm \cdot v_m$$

$$= \rho_0 A\, dx\, v_m \tag{iii}$$

Putting the value of dx from Eq. (i), we get

$$dP = \rho_0 A\, v_m\, v_0\, dt \tag{iv}$$

From Eq. (iv), we can obtain the force as follows:

$$F = \frac{dp}{dt}$$

and the pressure $P = \frac{1}{A}\frac{dp}{dt}$

$$= \rho_0 v_0 v_m \quad \text{(v)}$$

If the maximum longitudinal speed is $s\omega$, where s is the displacement of molecules of the gas and ω is the angular frequency of the sound wave, then maximum pressure is given by

$$P_{\max} = \rho_0 v_0 s\omega \quad \text{(vi)}$$

This is the required relation between the displacement s and the pressure amplitude P or $P_{\max}$. Clearly, a larger pressure is exerted by a larger velocity of the piston (v_0) or the larger displacement of the molecules (s) of the sound.

8.13 LISSAJOUS FIGURES

LO1

Jules Antoine Lissajous in 1857 demonstrated that when a particle is acted upon simultaneously by two simple harmonic motions at right angles to each other, the resultant path traced out by the particle is a curve. The curves so obtained are called *Lissajous figures*. So, Lissajous figures are the patterns formed when two vibrations transverse to each other are superimposed. In other words, a Lissajous figure is a parametric plot of a harmonic system.

Let us consider that each vibration is a simple harmonic motion represented by a sinusoidal wave. Their amplitudes are a and b and frequencies are ω_1 and ω_2; ϕ_1 and ϕ_2 are their phases and t is the time. We plot an input $x(t)$ and output $y(t)$ of a linear time-independent system.

$$x = a \sin (\omega_1 t + \phi_1), y = b \sin (\omega_2 t + \phi_2) \quad \text{(i)}$$

After time t, the point P of the coordinates (x, y) will trace a curve. The equation of the curve can be found by eliminating t in Eq. (i). For example, when both the frequencies are same and there is no phase difference, i.e., $\omega_1 = \omega_2 = \omega$ and $\phi_1 = \phi_2 = 0°$, then we have

$$x = a \sin \omega t, y = b \sin \omega t$$

$$\Rightarrow \quad \frac{x}{a} = \frac{y}{b} \text{ and thus } y = \frac{b}{a}x \quad \text{(ii)}$$

This represents an equation of a straight line. It means when two sine waves are of equal frequency and are in phase, then we get a diagonal line to the right, as shown in Figure 8.8.

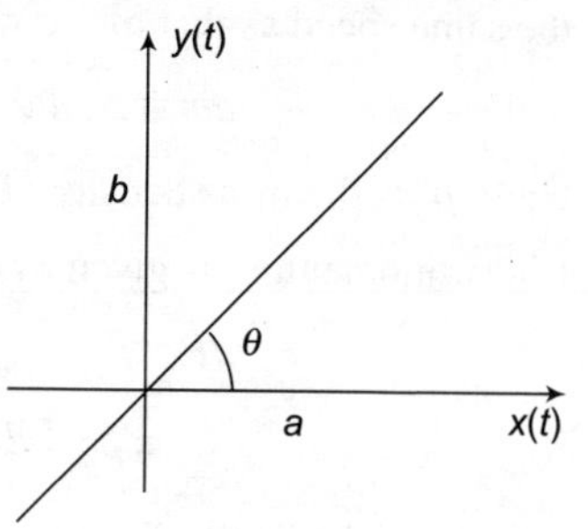

FIGURE 8.8

If the frequencies are same and the two vibrations are out of phase, i.e., $\omega_1 = \omega_2 = \omega$ and $\phi_1 - \phi_2 = \pi$, then Eq. (i) gives

$$y = -\frac{b}{a}x \quad \text{(iii)}$$

This is also an equation of a straight line. It means when two sine waves are of equal frequencies and 180° out of phase, we get a diagonal line to the left, as shown in Figure 8.9.

Now, we consider the case of $\pi/2$ phase difference, i.e., $\omega_1 = \omega_2 = \omega$ and $\phi_1 - \phi_2 = \pi/2$. Under this case, Eq. (i) gives

$$x = a \sin \omega t, \quad y = b \sin (\omega t + \pi/2) = b \cos \omega t \tag{iv}$$

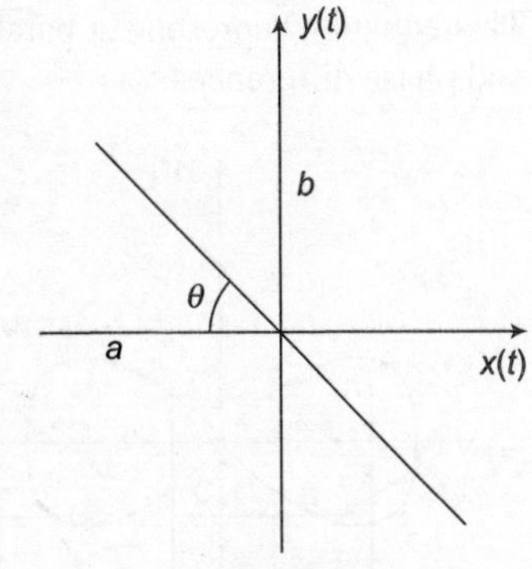

FIGURE 8.9

Equation (iv) yields

$$\frac{x^2}{a^2} = \sin^2 \omega t \text{ and } \frac{y^2}{b^2} = \cos^2 \omega t$$

which on adding gives

$$\frac{x^2}{a^2} + \frac{y^2}{b^2} = 1 \tag{v}$$

The above equation represents an ellipse as shown in Figure 8.10.

If $a = b$, then, Eq. (v) represents a circle $x^2 + y^2 = a^2$, as shown in Figure 8.11.

It means the circular path (Lissajous figure) is obtained when two sine waves are 90° out of phase superimposing each other.

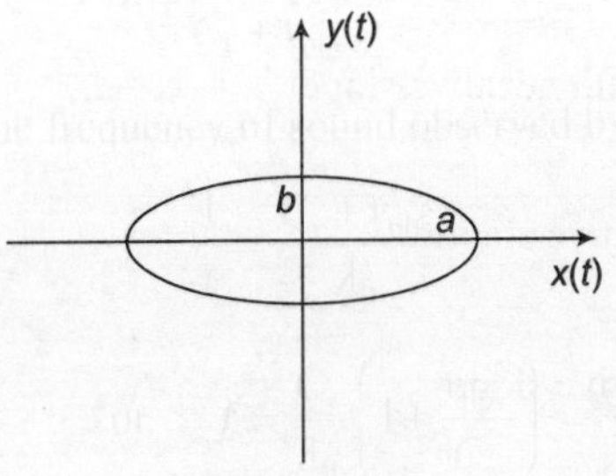

FIGURE 8.10

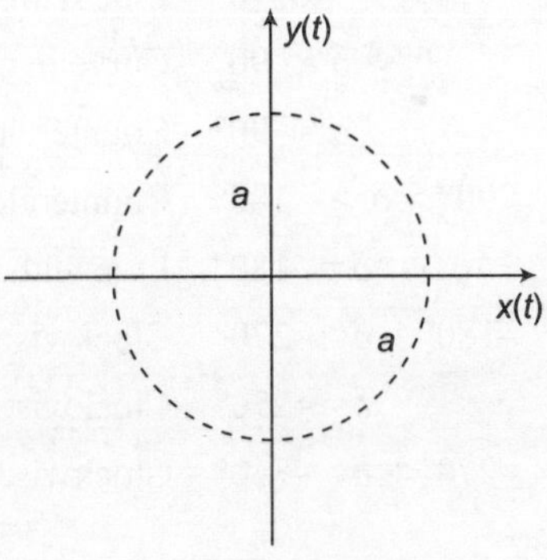

FIGURE 8.11

Finally, it can be seen that if the phase difference varies continuously, then the ellipse will slowly change its orientation and shape.

This is shown in Figure 8.12.

The curve becomes more complex when we consider the case of unequal frequencies. If we consider $\omega_2 = 2\omega_1$ and $\phi_1 - \phi_2 = \pi/2$, then Eq. (i) reads

$$\begin{aligned} x = a \sin \omega t, \; y &= b \sin (2\omega t + \pi/2) \\ &= b \cos 2\,\omega t \\ &= b[1 - 2 \sin^2 \omega t] \end{aligned} \tag{vi}$$

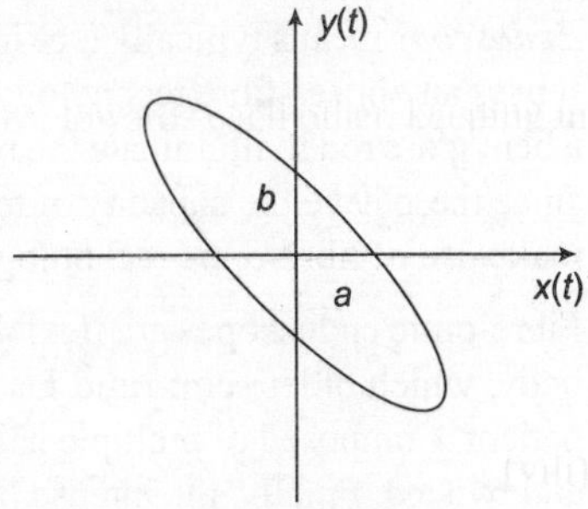

FIGURE 8.12

From Eq. (vi), we get

$$y = b\left[1 - 2\left(\frac{x}{a}\right)^2\right]$$

This equation represents a parabola. Following curves can also be obtained based on different frequencies and phase differences.

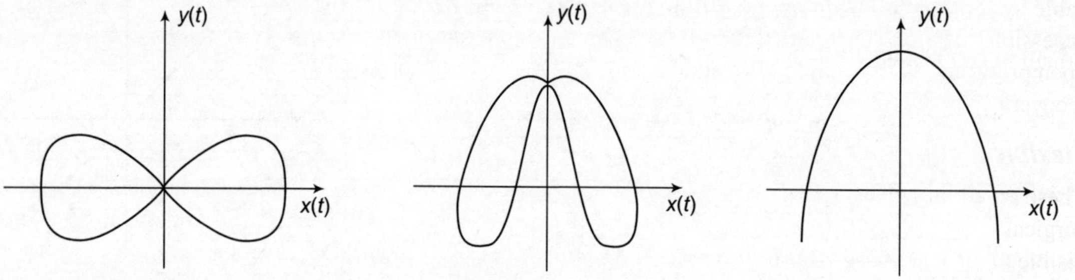

FIGURE 8.13

If we know both the angle of the major axis of a Lissajous curve and the direction of the curve's rotation, then we can determine the quadrant of phase shift $\delta \cdot (\equiv \phi_1 - \phi_2)$.

Taking $\phi_1 - \phi_2 = \delta$ for the case of $\omega_1 = \omega_2 = \omega$, we can summarize the above results as below:

$\delta = 0°$ Line with positive slope

$0° > \delta > -90°$ Curve in counterclockwise direction with positive slope

$\delta = -90°$ Counterclockwise circle

$-90° > \delta > -180°$ Counterclockwise curve with negative slope

$\delta = -180°$ Line with negative slope

$-180° > \delta > -270°$ Clockwise curve with negative slope

$\delta = -270°$ Clockwise circle

$-270° > \delta > -360°$ Clockwise curve with positive slope

8.14 ENDOSCOPY LO1

Endoscopy means typically looking inside the human body for medical reasons. This is done by an *endoscope*. It uses an optical-fibre technology that dates back to the 1880s when Dr. Roth and Prof. Reuss of Vienna used a bent glass rod to illuminate the body cavity. Based on endoscopy, many medicines have been revolutionized, since these have the capacity of reaching inside the human body in the least harmful way. All the endoscopes make use of fibre-optic technology to illuminate their images.

Fibre-optic endoscopes are flexible and highly maneuverable instruments that allow access to channels in the body, which older semi-rigid instruments cannot access at all or can access only at great discomfort to the patient. Composed of multiple hairlike glass rods bundled together, these instruments can be more easily bent and twisted. Finally, the intense light enables the endoscopist to see around.

8.14.1 Types of Endoscopes

There are two major types of endoscopes, namely, rigid endoscopes and flexible endoscopes. The details of these are given below.

Rigid Endoscopes

Also known as a *laparoscopes*, these are basically medical periscopes. A laparoscope is a long fibre-optic cable system that allows viewing of the affected area by inserting the cable from a more distant but more accessible location. There are a number of advantages to the patients having laparoscopic surgery over the open procedure. For example, pain and haemorrhaging are reduced due to smaller incisions. On top of it, the recovery times are also shorter.

Flexible Endoscopes

These are based on either fibre optics or LCD. The use of flexible endoscopes is common in both medical and surgical specialities. These endoscopes provide the unique ability to reach cavities and viscera, which are not visible to the naked eye. These allow for minimally invasive investigation of symptoms, diagnosis pathology and application of directed therapies. Advances in the imaging systems, newer endoscopes with 'self-drive' capabilities and enhancement of targeted therapeutics are future applications of flexible endoscopy.

SUMMARY

- The principle of superposition states that when two waves of the same kind meet at a point in space, the resultant displacement at that point is the vector sum of the displacements that the two waves would separately produce at that point. Superposing of two or more coherent waves to produce regions of maxima and minima in space results in the production of interference pattern. Constructive interference occurs when two or more waves arrive at the screen in phase (phase difference 0 or 360 degree) with each other, so that the resultant wave amplitude is the sum of the amplitude of the individual waves. Destructive interference occurs when the two or more waves arrive π out of phase with each other.
- Simple harmonic motion is any motion where a restoring force is applied that is proportional to the displacement and in the opposite direction of that displacement. In other words this also means that the acceleration is proportional to displacement but they are in opposite directions.
- Standing waves or stationary waves are produced when two different waves of same kind having same frequency, amplitude, and phase, superimpose when travelling in opposite direction. It results in the formation of regions of maxima and minima. The region of minimum or zero displacement is called a node, while the region of maximum displacement is called an antinode.
- Standing waves arise in a number of situations, for example an air column. The one end of the pipe has to be closed for waves to occur. The wavelength of the wave must satisfy a condition called the resonance condition, which is a function of 'n'. The frequency corresponding to different wavelengths are called *natural frequency.* Standing wave corresponding to $n = 0$ corresponds to *Fundamental vibrations.*
- Doppler Effect is the phenomena which results in the change in the frequency of the sound waves whenever there is relative motion between the source and the observer. The corresponding change in the frequency or wavelength is called *Doppler Shift.* Doppler shift is used to measure the velocities of distant galaxies based on their *recessional red shift.* This has led to the observation that the universe is expanding.
- In Doppler Effect, it is not the motion of the individual source or the observer that matters, but the relative motion between the source and the observer.

- When the speed of a source equals the speed of sound ($v = c$) the wave fronts cannot escape the source. The resulting pile of waves forms a large amplitude "sound barrier" that makes sustained flight at this speed difficult and risky. When the speed of a source exceeds the speed of sound ($v > c$) the wave fronts lag behind the source in a cone-shaped region with the source at the vertex. The edge of the cone forms a supersonic wave front with an unusually large amplitude called a "shock wave". When a shock wave reaches an observer a "sonic boom" is heard.
- A beat is an interference pattern between two sounds of slightly different frequencies. The beat frequency is equal to the difference between the interfering frequencies.
- Lissajous figure, also called *Bowditch Curve*, are the patterns produced by the intersection of two sinusoidal curves or waves which are at right angles to each other. First studied by the American mathematician Nathaniel Bowditch in 1815, the curves were investigated independently by the French mathematician Jules–Antoine Lissajous in 1857–58. So basically these are the patterns which are formed when two sinusoidal waves interfere at right angle to each other.
- Lissajous figures can also be used as an experimental setup in laboratory to find the frequency of an unknown source when frequency of one of the sources is known.
- Endoscopy means looking inside the human body for medical reasons which is done by endoscope. Fibre optic endoscopes are flexible and highly maneuverable instruments which allow access to channels in the body, which older semi rigid instruments cannot access at all or can access only at great discomfort to the patient.
- Two types of endoscopes are generally used. These are named as rigid endoscope and flexible endoscope. The rigid endoscope are also known as laparoscopes.

SOLVED EXAMPLES

EXAMPLE 1 A source of sound is travelling east at 10 m/s toward you. You are travelling at 2 m/s east. It is 20°C. When the source is not moving, it emits a sound of 3000 Hz frequency. What frequency do you hear? Sound in air at 20°C travels at 343 m/s.

SOLUTION Given $u_s = 10$ m/s, $u_{obs} = -2$ m/s

$$f_0 = 3000 \text{ Hz}, \quad v = 343 \text{ m/s}$$

$$f' = f_0\left(\frac{1 + u_{obs}/v}{1 - u_s/v}\right)$$

$$= 3000\left(\frac{1 - 2/343}{1 - 10/343}\right) = \mathbf{3072\ Hz}$$

EXAMPLE 2 Suppose a train is approaching you while you are standing on the platform at the station. As the train approaches the station, it slows down but the engineer is sounding the hooter at a constant frequency of 400 Hz. Describe the pitch of the hooter and the changes in pitch of the hooter that you hear as the train approaches you. Take the speed of sound in air as 340 m/s.

SOLUTION The frequency of the sound gradually increases as the train moves towards you. The pitch increases and you shall hear a higher pitched sound.

EXAMPLE 3 Passengers on a train hear its whistle at a frequency of 750 Hz. Ram is standing next to the train tracks. If the train moves directly towards him at a speed of 30 m/s. what frequency does he hear? Take the speed of sound in air as 340 m/s.

SOLUTION

$$f_L = ((v + v_L)/(v + v_S)) \times f_S = (340 + 0)/(340 - 30)) \times (750) = \mathbf{822.58\ Hz}$$

EXAMPLE 4 A small aircraft is taxiing directly away from you down a runway. The noise of its engine, as the pilot hears, has a frequency 1.20 times the frequency that you hear. What will be the speed of the plane? Take the speed of sound in air as 340 m/s.

SOLUTION The velocity of listener (you) is 0 and the source is moving away from you at an unknown velocity. This velocity must be positive. We also know that

$$f_S = 1.20\, f_L$$
$$f_L = (v + v_L)/(v + v_S) \times f_S = (340 + 0)/(340 + v_S) \times (1.20\, f_L)$$
$$340 + v_S = (340)(1.20)$$

This gives the velocity of the source v_S = **68 m/s**

EXAMPLE 5 Suresh is in his car moving at the speed of 0.50c towards Sheetal who is sitting in her stationary car. It is getting dark and Sheetal does not have her headlights on, so Suresh flashes his brights at Sheetal. If the frequency of the light which Suresh emits from his headlights is 4400 Hz, at what frequency does Sheetal hear the sound?

SOLUTION Since the source is moving at speed v_S, the appropriate formula is $f^1 = f/(1 \pm v_S/v)$. Here f^1 is the changed frequency and f is the initial frequency. The changed frequency will be faster because Suresh's car is coming towards Sheetal. So the bottom of the fraction should be less than one. It means we should use the minus sign rather than the plus sign. Therefore, the formula yields

$$f^1 = (4400)/(1 - (.50c)/(c)) = \mathbf{8800\ Hz}$$

EXAMPLE 6 Deepika is walking down the streets of downtown New Delhi and comes to an intersection where the walk signal is blinking indicating to stop walking. Deepika thinks she is smarter than the signal and tries to make a last minute run to the other side of the street. She realises that she is not going to make it when a speeding truck coming towards her at the speed of 25 m/s is honking at the frequency of 6040 Hz. With what frequency is the wave reaching Deepika right before she gets struck by the truck, if the speed of sound is 343 m/s?

SOLUTION Formula to be used is

$$f^1 = f/(1 - v_S/v)$$

This gives

$$f^1 = 6040/(1 - 25/343) = \mathbf{6514.86\ Hz}$$

EXAMPLE 7 As a train pulls out of the station going 60 m/s, it blasts its horn. What would be the frequency heard by the passengers in train if the passengers still at the station are hearing 380 Hz?

SOLUTION Since the frequency of the sound heard by the passengers at the station must be lower, the formula should be

$$f^1 = f/(1 + v_S/v)$$

This gives

$$380 = f/(1 + 60/343) \text{ or } f = \mathbf{446\ Hz}$$

EXAMPLE 8 Mukesh is on a motorcycle speeding down the highway at 45 m/s until he sees a traffic jam ahead. The honking made by the stopped cars is 780 Hz, what frequency does Mukesh hear the sound at?

SOLUTION Since the frequency Mukesh hears will be higher than what the light is actually emitting, the appropriate formula is

$$f^1 = f(1 + v_S/v)$$

This gives

$$f^1 = 780(1 + 45/343) = \mathbf{882\ Hz.}$$

EXAMPLE 9 The speed of sound waves in air is found to be 340 m/s. Determine the fundamental frequency (first harmonic) of an open-end air column that has a length of 67.5 cm.

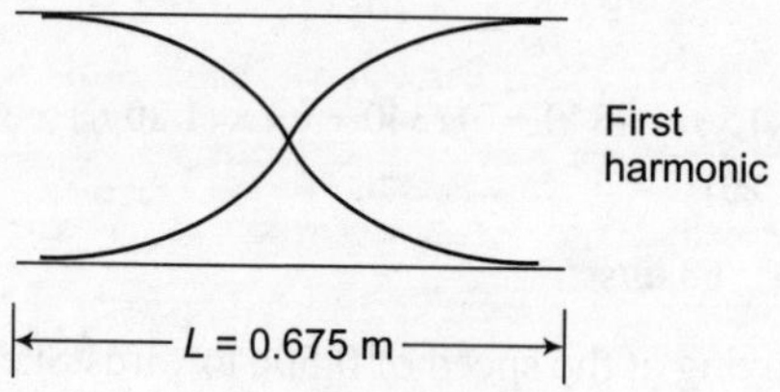

FIGURE 8.14

SOLUTION Given, $v = 340$ m/s, $L = 0.675$ m

For the first harmonic, wavelength is twice the length.

$$\lambda_1 = 2L = 2 \times 0.675 \text{ m} = 1.35 \text{ m}$$

$$f_1 = \frac{v}{\lambda_1} = \frac{v}{2L} = \frac{340 \text{ m/s}}{1.35 \text{ m}} = \mathbf{252\ Hz}$$

EXAMPLE 10 A wave of 1000 Hz frequency travels in air of 1.2 kg m^{-3} density at 340 m/s. If the wave has 10 μ Wm^{-2} intensity, find the displacement and pressure amplitudes.

SOLUTION

$$I = \frac{1}{2}(\rho v)(A\omega)^2 \qquad v \rightarrow \text{wave speed}$$

$$\Rightarrow \quad A = \sqrt{\frac{2I}{\rho v \omega^2}}$$

$$= \sqrt{\frac{2 \times 10^{-6}}{1.2 \times 340 \times (2\pi \times 1000)^2}}$$

$$= 11 \text{ nm}$$

$$p_0 = \rho v A \omega$$

$$= 1.2 \times 340 \times 11 \times 10^{-3} \times 2\pi \times 1000$$

$$= \mathbf{28\ mPa}$$

EXAMPLE 11 Assuming $\rho = 1.29$ kg/m^3 for the density of air and $v = 331$ m/s for the speed of sound, find the pressure amplitude corresponding to the threshold of hearing intensity of 10^{-12} W/m^3.

Solution

$$I = \frac{1}{2} P_{max}^2 / \rho_0 v$$

$$\Rightarrow \quad P_{max} = \sqrt{2I \rho_0 v}$$

$$= \sqrt{2 \times 10^{-12} \times 1.29 \times 331}$$

$$= \mathbf{2.92 \times 10^{-5}\ N/m^2}$$

Example 12 For ordinary conservation, the intensity level is given as 60 dB. What is the intensity of the wave?

Solution

$$I_L = 10 \log \frac{I}{I_0}$$

$$60 = 10 \log \frac{I}{10^{-12}}$$

$$\log I + \log 10^{12} = 6$$

$$\log I = -6$$

$$\therefore \quad I = 10^{-6}\ \text{W/m}^2 = \mathbf{1\ \mu W/m^2}$$

Example 13 A small source of sound radiates energy uniformly at a rate of 4 W. Calculate the intensity level at a point 25 cm from the source if there is no absorption.

Solution

$$I = \frac{\text{Power}}{4\pi r^2} = \frac{4}{4\pi \times 25^2} = 5.093 \times 10^{-4}\ \text{W/m}^2$$

$$I_L = 10 \log \frac{I}{I_0}$$

$$= 10 \log \frac{5.093 \times 10^{-4}}{10^{-12}}$$

$$= 10 \log (5.093 \times 10^8)$$

$$= 10[\log 5.093 + 8]$$

$$= \mathbf{87\ dB}$$

Example 14 The maximum pressure variation that the ear can tolerate is about 29 N/m^2. Find the corresponding maximum displacement for a sound wave in air having a frequency of 2000 Hz. Assume the density of air as 1.22 kg/m^3 and the speed of sound as 331 m/s.

Solution

$$A = \frac{P_{max}}{k \rho_0 v^2} = \frac{P_{max}}{2\pi \rho_0 f v}$$

$$k = 2\pi/\lambda \quad \text{and} \quad v = f\lambda$$

$$A = \frac{29}{2 \times 3.14 \times 1.22 \times 331 \times 2000} = \mathbf{5.7 \times 10^{-6}\ m}$$

Example 15 If two sound waves, one in air and the other in water, have equal pressure amplitude, what is the ratio of intensities of waves? Assume that the density of air is 1.293 kg/m^3, and the speeds of sound in air and water are 330 and 1450 m/s respectively.

SOLUTION $$I = \frac{P_{max}^2}{2\rho_0 v}$$

$$P_{max}\text{ (air)} = P_{max}\text{ (water)}$$

$$\therefore \quad \frac{I_{Water}}{I_{Air}} = \frac{\rho_A v_A}{\rho_W v_W}$$

$$= \frac{1.293 \times 330}{1000 \times 1450} = \mathbf{2.94 \times 10^{-4}}$$

EXAMPLE 16 The pressure in a progressive sound wave is given by the equation $P = 2.4 \sin \pi(x - 330\ t)$, where x is in metres, t in seconds and P in N/m^2. Find (a) pressure amplitude, (b) frequency, (c) wavelength, and (d) speed of wave.

SOLUTION $$P = 2.4 \sin \pi(x - 330\ t)$$

$$= 2.4 \sin 2\pi\left(\frac{1}{2}x - 165\ t\right)$$

$$P = P_{max} \sin 2\pi\left(\frac{x}{\lambda} - ft\right)$$

On comparing, we get

Pressure amplitude = 2.4 N/m^2
Frequency = 165 Hz
Wavelength = 2.0 m
Speed of wave $v = f\lambda = 165 \times 2 = 330$ m/s

OBJECTIVE TYPE QUESTIONS

Q.1 Damping of periodic motion always results in
(a) decreasing amplitude (b) energy gain in vibration
(c) resonance (d) energy loss to friction
(e) none of these

Q.2. If the frequency of a 0.75 m simple pendulum is 1.5 Hz, the angular frequency on a corresponding reference circle is (in rad/s)
(a) 1.5π (b) 0.33π (c) 3π (d) 0.5π (e) 2π

Q.3 If you are given the force constant of a spring as 100 N/m, the mass on the spring as 1 kg, and the amplitude as 0.04 m, then the period in seconds of the SHM is
(a) 20π (b) $\pi/5$ (c) 5π (d) $5/\pi$ (e) $\pi/20$

Q.4 If two vibrations of the same frequency are superimposed on a system with equal amplitudes, the maximum resultant amplitude shall be
(a) zero (b) $\sqrt{(2A)}$ (c) $A/\sqrt{2}$ (d) 2A (e) A

Q.5 Speed of sound waves in a fluid
(a) depends directly on the density of the medium
(b) varies as square of bulk modulus of the medium
(c) is inversely proportional to the square root of density
(d) depends directly on the square root of the bulk modulus of the medium

Q.6 A transverse harmonic wave on a string is described by $y(x, t) = 3.0 \sin (36t + 0.018x + \pi/4)$ where x and y are in cm and t is in s. The positive direction of x is from left to right. Then
(a) the wave is travelling from right to left
(b) the speed of the wave is 20 m/s
(c) the frequency of the wave is 5.7 Hz
(d) the least distance between two successive crests in the wave is 2.5 cm

Q.7 An object is vibrating at its natural frequency. Repeated and periodic vibrations of the same natural frequency impinge upon the vibrating object and the amplitude of its vibrations are observed to increase. This phenomenon is known as
(a) beats (b) resonance (c) interference (d) overtone

Q.8 Standing waves are produced in a wire by vibrating one end at a frequency of 100 Hz. The distance between the 2nd and the 5th nodes is 60.0 cm. The wavelength of the original traveling wave in cm is
(a) 50.0 (b) 40.0 (c) 30.0 (d) 20.0

Q.9 Which phenomena can be applied to estimate the velocity of star with respect to earth
(a) Doppler effect
(b) interference of waves
(c) beats phenomena
(d) all of these

Q.10 Doppler Effect applies to
(a) sound waves only
(b) light waves only
(c) both sound and light waves
(d) neither sound wave nor light waves

Q.11 The Lissajous patterns help in the measurement of
(a) Phase difference between two sine waves
(b) Frequency of one waveform if the frequency of other waveform is known
(c) both (a) and (b)
(d) none of these

Q.12 If the two input waveforms of equal amplitude and 90 degree phase difference is applied to the CRO, then the Lissajous patterns obtained will be
(a) straight line tilted at 45 degree with respect to x-axis
(b) circle
(c) ellipse
(d) vertical straight line

Q.13 A body executing SHM has a velocity of 2.0 cm/s when its displacement is 7.0 cm and a velocity of 7.0 cm/s. What is the square of the amplitude of oscillation when its displacement is 2.0 cm?
(a) 26.0 cm^2 (b) 53.0 cm^2 (c) 79.0 cm^2 (d) 106.0 cm^2

Q.14 A body executing linear SHM has a velocity of 3.0 cm/sec when its displacement is 8.0 cm, and a velocity of 8.0 cm/sec when its displacement is 3.0 cm. If the oscillator mass is 5.0 kg, find the approximate total energy of the oscillator?
(a) 4.5 mJ (b) 9.1 mJ (c) 13.7 mJ (d) 18.2 mJ

Q.15 Two block, each of mass $m = 2.0$ kg, are connected by a spring of force constant $R = 3.0$ N/m and placed on a horizontal frictionless surface, as shown in the following diagram. If an equal force of $F = 2.0$ N is applied to each block in the direction of arrow, what is the approximate time-period of the system when the force is removed.
(a) 1.2 sec (b) 2.4 sec (c) 3.6 sec (d) 4.8 sec

Q.16 In SHM of a simple pendulum, component of weight directed towards mean position is
(a) mg cos θ (b) mg sin θ (c) 0 (d) mg tan θ

Q.17 Which of the following quantities are always positive in a SHM?
(a) $\vec{F}, \vec{a}$ (b) $\vec{V}, \vec{r}$ (c) $\vec{a}, \vec{r}$ (d) $\vec{F}, \vec{r}$

Q.18 A small block oscillates back and forth on a smooth concave surface of radius *R*. The time period of small oscillation is
(a) $T = 2\pi\sqrt{R/g}$ (b) $T = 2\pi\sqrt{\frac{2R}{g}}$ (c) $T = 2\pi\sqrt{R/2g}$ (d) None of these

Q.19 When two mutually perpendicular simple Harmonic motions of same frequency, amplitude and phase are superimposed,
(a) Resulting motion is Uniform circular motion.
(b) Resulting motion is a linear SHM along a straight line inclined equally to the straight lines of motion of component ones.
(c) Resulting motion is an elliptical motion, symmetrical about the lines of motion of the component.
(d) The two SHM's will cancel each other.

Q.20 A simple pendulum has some time period *T*. What will be the percentage change in its time period if its amplitude is decreased by 5%?
(a) 6% (b) 3% (c) 1.5% (d) 0%

PRACTICE PROBLEMS

General Questions

Q.1 On what factors the velocity of the sound depends?
Q.2 Do you agree that sound waves are mechanical waves?
Q.3 How are stationary waves formed?
Q.4 Define simple harmonic motion (SHM). Give two examples of SHM. Why SHM is important in the study of waves and oscillations?
Q.5 Name three parameters of any SHM. Explain the meaning of each of them.
Q.6 Derive an expression for the energy of a harmonic oscillator of mass *m*, amplitude *A*, and frequency *v*. Find out the displacement at which energy is half kinetic and half potential.
Q.7 All simple harmonic motions are periodic but all the periodic motions are not simple harmonic. Explain this observation.
Q.8 Briefly explain how a simple harmonic motion can be represented by a rotating vector.

What is a shock waves?

Q.9 How are sound waves different from shock waves?
Q.10 Define the intensity of the sound?
Q.11 What do you understand by beats? Discuss their theory.

UNSOLVED QUESTIONS

Q.1 The force constant of a spring is 10 N/m. Find the period of a 100-g mass on the end of this spring.
Ans. 0.63 s

Q.2 Find the maximum velocity of the mass in problem 1 if the amplitude of oscillation is 2.0 cm.
Ans. 20 cm/s

Q.3 Find the velocity of the mass in Problem 2 when it is 1 cm from its equilibrium position.
Ans. 17 cm/s

Q.4 A mass on the end of a spring is released from a point 2 cm from its equilibrium position. The frequency of oscillation is 4 Hz. Write the equation for the position of the mass as a function of time.
Ans. 0.02 cos ($8\pi t$) m

Q.5 The period of a simple pendulum is 2.00 sec, find the length of the pendulum. **Ans. 0.993 m**

Q.6 Find the maximum energy stored in the spring of Problem 1 when it is compressed 2 cm from its equilibrium position. **Ans. 0.02 J**

Q.7 The speed of sound waves in air is found to be 340 m/s. Determine the fundamental frequency (1st harmonic) of an open-end air column that has a length of 67.5 cm. **Ans. 252 Hz**

Sound Waves and Acoustics of Buildings

9

Learning Objectives

After reading this chapter you will be able to

- **L01** Know about audible, ultrasonic, and infrasonic waves
- **L02** Learn about Production of ultrasonic waves, magnetostriction method, Piezoelectric method, ultrasonic transducer, Galton Whistle method
- **L03** Understand the absorption, dispersion, and detection of ultrasonic waves
- **L04** Analyse the applications of ultrasonic waves
- **L05** Explain the types of acoustics
- **L06** Discuss on acoustics of buildings
- **L07** Evaluate the factors affecting the architectural acoustics

Introduction

A vibration refers to the oscillating motion of any medium and sound is a vibration in an elastic medium. These vibrations transmitting through a solid, liquid, or gas, are composed of frequencies within the range of hearing and are of a level sufficiently strong to be heard. In the case of human hearing it is the vibrations in air that simulate our hearing organs and give a sensation of sound. When sound enters a new medium, it is reflected, transmitted, or absorbed. This scientific study of the propagation, absorption, and reflection of sound waves is called *acoustics*.

Acoustics is the interdisciplinary science that deals with the study of sound, ultrasound and infrasound (all mechanical waves in gases, liquids, and solids). In a broad sense, acoustics may be defined as generation, transmission and reception of energy in the form of vibration waves in matter.

9.1 AUDIBLE, ULTRASONIC AND INFRASONIC WAVES LO1

The simplest form of sound waves is sinusoidal waves of definite frequency, wavelength and amplitude. The frequency range of waves from 20 Hz to 20,000 Hz are said to be audible waves for which range human ears are sensitive but the waves of frequency above the audible range are called ultrasonic waves and below the audible range are known as infrasonic waves.

Ultrasonics is the study and application of the energy of sound waves vibrating at frequencies greater than 20,000 Hz, i.e., beyond the range of human hearing. The application of sound energy in the audible range is limited almost entirely to communications, since increasing the pressure, or intensity, of sound waves increases loudness and therefore causes discomfort to human beings. Ultrasonic waves, however, being inaudible, have little or no effect on the ear even at high intensities. They are produced, commonly, by a transducer containing a piezoelectric substance, e.g. a quartz-crystal oscillator that converts high-frequency electric current into vibrating ultrasonic waves.

Sound waves, particularly in the atmosphere, whose frequencies are below the audible range, i.e., lower than about 20 Hz are called infrasonic waves. Earthquake and seismic waves are elastic waves which occur at infrasonic frequencies in the Earth's crust and in the oceans and seas. The physical laws of propagation in the atmosphere are essentially the same as for audible sound. The local speed of infrasound in air at ambient temperatures near 20°C is about 340 m/s, the same as for audible sound.

9.2 PRODUCTION OF ULTRASONIC WAVES LO2

In most applications, ultrasonic waves are generated by applying an electric current to a special kind of crystal known as a piezoelectric crystal. The crystal converts electrical energy into mechanical energy, which, in turn, causes the crystal to vibrate at a high frequency. In another technique, a magnetic field is applied to a crystal, causing it to emit ultrasonic waves. Although the bulk attention is given to more popular types of transducers which are based on magnetostriction and piezoelectric effect, there are other means of generating ultrasonic waves some of which exhibit great promise.

9.2.1 Magnetostriction Method

Before discussing this method for the generation of ultrasonic waves, we shall talk about the magnetostriction effect.

9.2.1.1 Magnetostriction Effect

When a rod of ferromagnetic material such as iron, nickel or cobalt is placed in a magnetic field keeping its length parallel to the direction of magnetic field, the rod experiences a small change in its length. This effect is termed as magnetostriction effect. The change in length of the rod depends on the intensity of the applied magnetic field and nature of the ferromagnetic material. However, the change in the length is independent of the direction of the field. Since the change is not so great in the other dimensions of the rod, the rod is generally put with its length parallel to the direction of the magnetic field. The cause of change in material's dimensions can be understood as follows. Actually ferromagnetic materials have a structure that is divided into domains, each of which is a region of uniform magnetic polarisation. Under the application of an external magnetic field, the boundaries between the domains shift and the domains rotate. These two effects lead to a change in the dimensions of the materials.

9.2.1.2 Principle Involved

The general principle involved in producing ultrasonic waves is to cause ferromagnetic materials to vibrate very rapidly. These vibrations cause surrounding air to vibrate with the same frequency, which spreads out in the form of ultrasonic waves.

When the rod is placed inside a magnetic coil carrying alternating current, it suffers a change in length for each half of the alternating current. It means the rod vibrates at a frequency twice that of the frequency of the alternating current. Usually the amplitudes of vibrations are small, but these can be enhanced by achieving the resonance condition, i.e., by matching the frequency of the alternating current with the natural frequency of the material of the rod.

9.2.1.3 *Construction and Working*

In Fig. 9.1, a rod (ferromagnetic material) with its ends *A* and *B* is winded by the coils L_1 and L_2. The coil L_2 is connected to the collector of the transistor whereas the coil L_1 is connected to its base. In view of an *LC* circuit, we can adjust the frequency of the oscillatory circuit $\left(\frac{1}{2\pi\sqrt{L_2C}}\right)$ by adjusting the value of the capacitor *C*. The current flowing in the circuit can be determined by the milliammeter connected across the coil L_2. A necessary biasing, i.e., the emitter as forward biased and the collector as reverse biased for the *NPN* transistor, is achieved by the battery (current) connected between the emitter and the collector of the transistor. The alternating current passing through the coil L_2 causes a corresponding change in the magnetization of the rod and hence the rod starts vibrating due to the magnetostriction effect.

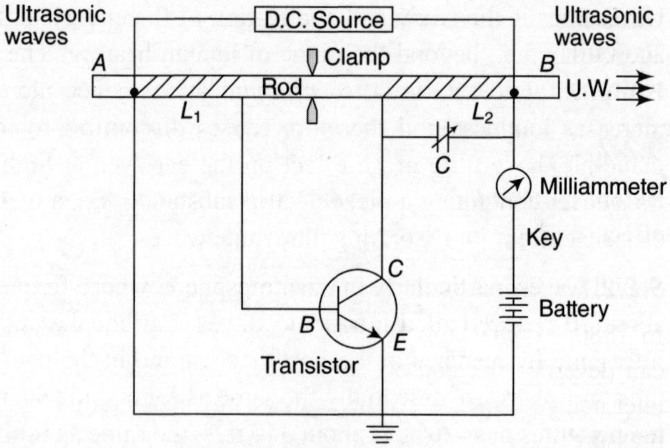

FIGURE 9.1

In the above situation, an emf is also induced in the coil which is called as converse magnetostriction effect. Due to this effect an emf is induced in the coil L_1. This induced emf is fed to the base of the transistor, which acts as a feed back continuously. This way the current is built up in the transistor and the vibrations of the rod are maintained for the generation of ultrasonic waves. When the frequency of the oscillatory circuit matches with the natural frequency of the vibrating rod, the resonance occurs. At the resonance, the rod vibrates longitudinally with larger amplitude and produces ultrasonic waves of high frequency along both the ends of the rod.

If the Young's modulus of the material of the rod is *Y*, its density is ρ and the length of the rod is *l*, then the frequency of vibrations of the rod is given by $\frac{1}{2l}\sqrt{\frac{Y}{\rho}}$. When this frequency matches with the frequency of the oscillatory circuit $\frac{1}{2\pi\sqrt{L_2C}}$, the resonance occurs. Based on this we get

$$\frac{1}{2\pi\sqrt{L_2C}} = \frac{1}{2l}\sqrt{\frac{Y}{\rho}}.$$

9.2.1.4 *Advantages and Limitations*

There are several advantages of magnetostriction method of generating ultrasonic waves. For example, magnetostrictive or ferromagnetic materials are easily available and inexpensive. The design of the oscillatory circuit is simple and it involves low cost of the materials. Moreover, at low ultrasonic frequencies, the large power output can be produced without any risk of damage of the oscillatory circuit.

On the other hand, this method has some limitations also. For example, through this method we cannot achieve ultrasonic waves of frequencies larger than about 3 MHz. The frequency of oscillations also depends on the temperature and degree of magnetization. So hearing effect may change the frequency of ultrasonic waves and there will be losses of energy due to hysteresis and eddy current. Finally the condition of resonance shows that we need to reduce the length of the rod in order to produce higher frequency ultrasonic waves, which is not practically feasible.

9.2.2 Piezoelectric Method

LO2

Before discussing this method for the generation of ultrasonic waves, we shall talk about the piezoelectric effect.

9.2.2.1 Piezoelectric Effect

In 1880, the brothers *Pierre* and *Jacques Curie* discovered in an experiment the fact that certain crystals can develop an electric charge when a mechanical pressure or tension is applied. This phenomenon was later named as Piezoelectric effect. They showed that there was a direct proportion between the mechanical pressure and the resultant charge, and sign of the charge changed when pressure changed to tension or vice versa. This may take the form of a separation of electric charge across the crystal lattice. If the material is not short-circuited, the applied charge induces a voltage across the material. The crystals which acquire a charge when compressed, twisted or distorted are said to be piezoelectric. The effect is present in many crystals but it is useful in *Quartz* and *6*.

In addition, certain ceramics and biological matter such as DNA, bone and various proteins show the piezoelectric effect or the piezoelectricity in response to applied mechanical stress. The word piezoelectricity, which means electricity resulting from pressure, was derived from the Greek words *piezō* or *piezein* that means to squeeze or press and *ēlektron* that means amber (an ancient source of electric charge). Piezoelectric effect finds useful applications such as the production and detection of sound, generation of high voltages, electronic frequency generation, microbalances, and ultrasonic focusing of optical assemblies.

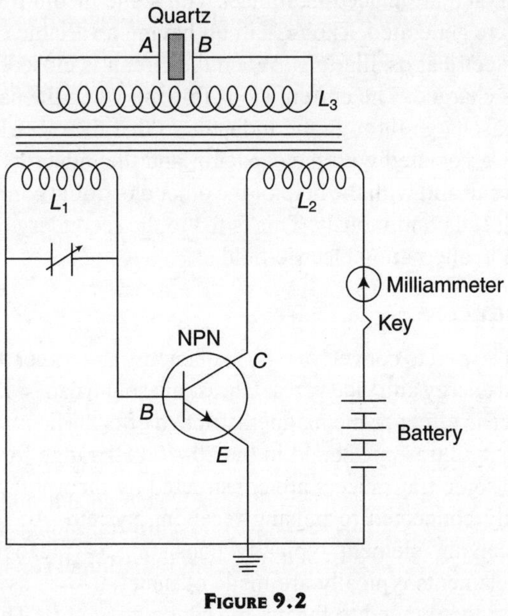

FIGURE 9.2

9.2.2.2 Principle Involved

When a slab of a piezoelectric crystal such as quartz is placed between two metal plates and resonant mechanical vibrations are produced in the crystal due to the linear expansion and contraction, elastic waves are propagated in the metallic plates which generate ultrasonic waves. An efficient generation of ultrasonic waves takes place when the crystal oscillates at the maximum amplitude. This happens when the frequency of the oscillatory circuit matches with the natural frequency of one of the modes of vibrations of the crystal. The frequency of the generated ultrasonic waves depends on the Young's modulus and the density of the piezoelectric material.

9.2.2.3 Construction and Working

Piezoelectric generator that works on the piezoelectric effect is used for generating ultrasonic waves of high frequency of about 50 MHz. For this a slice of quartz crystal is placed between two metal plates A and B in order to form a parallel plate capacitor having the quartz crystal as a dielectric medium. Quartz is preferred because it possesses rare physical and chemical properties. The metal plates are connected to the terminals of a coil which is inductively coupled to the oscillating circuit, as shown in Figure 9.2. Due to this electrical circuit, an alternating potential difference is developed across the plates of the capacitor because of which a tensile pressure appears on the crystal. This produces alternate contraction and expansion of the crystal and the opposite charges are generated on the faces of the crystal lying towards A and B. Through piezoelectric effect the crystal produces sound waves and when the frequency of electrical oscillations is in the ultrasonic range then ultrasonic waves are generated.

As shown in Fig. 9.2, the variable capacitor C is adjusted in order to match the frequency of the oscillatory circuit with the natural frequency of one of the modes of vibrations of the crystal. This way we are able to produce resonant mechanical vibrations in the crystal due to the linear expansion and contraction. If one or both the faces of the crystal are placed in contact with some medium in which elastic waves can be propagated, ultrasonic waves are generated. The LC circuit having a variable capacitor C and an inductor L_2 decides the frequency of the electrical oscillations. When the circuit is closed, the current flows through the LC circuit and the capacitor is charged. The current stops flowing when the capacitor is fully charged. After that the capacitor is made to discharge through the inductor so that the electric energy is stored in the form of electric and magnetic fields associated with the capacitor and the inductor, respectively. This way we get electrical oscillations in the circuit and with the help of the other electronic components including a transistor, electrical oscillations are produced continuously. This is fed to the secondary circuit and the crystal vibrates, as it is continuously subjected to alternating electric field.

9.2.3 Ultrasonic Transducer LO2

A transducer is a device which is used to convert one form of energy to another. Ultrasonic transducers convert electrical energy to mechanical energy and vice versa. Ultrasonic sound can be produced by transducers which operate either by the piezoelectric effect or the magnetostrictive effect. The magnetostrictive transducers can be used to produce high intensity ultrasonic sound in the 20–40 kHz range for ultrasonic cleaning and other mechanical applications. Ultrasonic transducers are constructed by incorporating one or more piezoelectric vibrators which are electrically connected to pulsing-receiving system. An ultrasonic transducer includes an ultrasonic transmitting/receiving element typically consisting of piezoelectric element connected to electrodes. The piezoelectric elements typically are made of material such as lead zirconate titanate (PZT), with a plurality of elements being arranged to form a transducer assembly. The transducer assembly is then further assembled into a housing possibly including control electronics, in the form of electronic circuit boards, the combination of which forms an ultrasonic probe.

The active element is the heart of the transducer as it converts the electrical energy to acoustic energy, and vice versa. The active element is basically a piece of polarised material (i.e., some parts of the molecule are positively charged, while other parts of the molecule are negatively charged) with electrodes attached to two of its opposite faces. When an electric field is applied across the material, the polarised molecules will align themselves with the electric field, resulting in induced dipoles within the molecular or crystal structure of the material. This alignment of molecules will cause the material to change dimensions. This phenomenon is known as *electrostriction.* In addition, a permanently-polarised material such as quartz (SiO_2) or barium titanate ($BaTiO_3$) will produce an electric field when the material changes dimensions as a result of an imposed mechanical force.

The thickness of the active element is determined by the desired frequency of the transducer. A thin wafer element vibrates with a wavelength that is twice its thickness. Therefore, piezoelectric crystals are cut to a thickness that is half the desired radiated wavelength. The higher the frequency of the transducer, the thinner is the active element. The primary reason that high frequency contact transducers are not produced is because the element is very thin and too fragile.

9.2.3.1 *Uses of Ultrasonic Transducers*

Ultrasonic transducers are useful for various applications. Ultrasonic testing equipment is used in a variety of applications such as for measuring flow, determining flaws, measuring thickness, and gauging corrosion. Ultrasonic diagnostic imaging systems are in widespread use for performing ultrasonic imaging and measurements of the human body through the use of probes which are used to view the internal structure of a body by creating a scan plane.

9.2.4 Galton Whistle Method

LO2

Galton's whistle was invented in 1876 by Francis Galton and is mentioned in his book named *Inquiries into Human Faculty and its Development.* In this book, he described experiments to test the range of frequencies that could be heard by various animals. This whistle most commonly known as dog whistle or silent whistle is a type of whistle that emits sound in the ultrasonic range, which people cannot hear but some other animals can, including dogs and domestic cats.

Galton and subsequent researchers used these whistles to create increasingly higher frequency tones to test research subjects as well as animal abilities to hear different tones, Galton was able to determine that the normal upper limit of human hearing was about 18 kHz. He also noted that the ability to hear higher frequencies declined with age.

Below we discuss the principle involved and the construction and working of the Galton's whistle with regard to the production of ultrasonic waves.

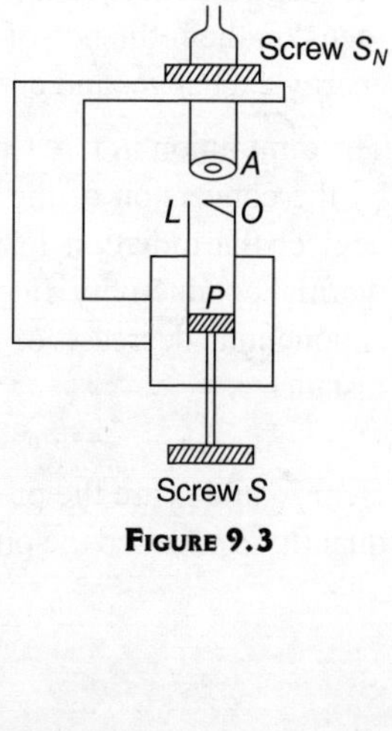

FIGURE 9.3

9.2.4.1 *Principle Involved*

Galton whistle works on the principle of organ pipe, where the distance of annular nozzle from the edge of a pipe and the pressure of air blast are suitably adjusted in order to set the pipe into resonant vibrations at the ultrasonic frequency with the help of the length and the diameter of the pipe.

9.2.4.2 *Construction and Working*

As shown in Fig. 9.3, Galton whistle consists of a closed end air column whose length can be adjusted with the help of a movable piston *P*. A screw *S* is connected to this piston which can move the piston to the desired position. The open end of the pipe O is fitted with a lip L, and the gap between the ends O and A can be adjusted with the help of another screw S_N which can move the pipe A up or down. A nozzle N is fitted on the top through which an air blast is blow towards lip L. When the blast of air strikes against the lip L, the column of air in the pipe is set into vibration. The resonant position is achieved in order to produce the ultrasonic waves by adjusting the length of the air column in O. Clearly the resonance frequency depends on the size of the pipe, i.e., its length and diameter.

The wavelength λ of the sound wave depends on the length l of the air column in O and the end correction x. This is given by

$$\lambda = 4(L + x)$$

From this we can calculate the frequency of the sound or ultrasonic wave as

$$f = \frac{V}{\lambda} = \frac{V}{4(l + x)}$$

Here V is the velocity of the waves produced by Galton's whistle. This whistle can produce ultrasonic waves of low frequencies up to 100 kHz and interestingly the micrometer screw S can be calibrated to give directly this frequency.

9.3 ABSORPTION AND DISPERSION OF ULTRASONIC WAVES LO3

When an ultrasonic wave passes through a medium, a part of its energy is converted into heat due to the alternative compression and rarefaction taken place in the wave phenomenon and hence its intensity goes down. The compressions produce the heat that increases the temperature of the medium whereas the rarefactions reduce the temperature, leading to the absorption of these waves in the medium and the wave is said to be attenuated. Two main mechanisms namely absorption and scattering (dispersion) are responsible for the ultrasound attenuation. Different mechanisms such as thermal conductance effects, chemical effects, viscous effects and nonlinearity are responsible for the absorption phenomenon. The phenomena responsible for the ultrasound absorption in biological tissues have not been so far completely understood. In liquids the viscous forces between neighbouring particles moving with different velocities are the major sources of the wave absorption, whereas viscoelastic forces are the main contributors to the wave absorption in homogeneous solids. For example, viscous losses may explain well the sound wave absorption in water where attenuation varies with the square of the frequency. However, this model of viscosity does not explain the experimental measurements of the absorption of the ultrasonic waves in soft biological tissues and bone in the diagnostic frequency range. We can say that the absorption and dispersion of ultrasonic waves generally focus on the influence of ultrasonics on molecular processes in liquids and gases, including hydrodynamics, energy exchange, and chemical reactions.

The diminution in intensity of the amplitude of a planar longitudinal wave passing through a liquid is caused by the conversion of the organised collective motion of the sonic pulse into random thermal motion, if we neglect the radiation losses. The total attenuation comprises contributions arising from viscous loss and thermal conductivity. In general, overall attenuation of the ultrasonic waves is characterised by the following exponential decrease of the pressure amplitude p and of the intensity amplitude I with the propagating distance z.

$$p = p_0 e^{-\alpha z} \quad \text{and} \quad I = I_0 e^{2-\alpha z}$$

Here p_0 and I_0 are the pressure and intensity at $z = 0$, i.e., when the wave starts penetrating the medium. The quantity α is called the pressure frequency-dependent attenuation coefficient, which is expressed in cm^{-1}. The factor of 2 in the exponential term of the intensity equation results from the transformation of the pressure into intensity, as the intensity is proportional to the square of the pressure. The commonly used units for α in biomedical ultrasonics are dB (decibel).

The dispersion of the ultrasonic wave is referred to the change in its velocity with frequency. In viscous liquids such as glycerine and castor oil the change in velocity with frequency or dispersion cannot be observed in the frequency regime or ultrasonic waves. However, the dispersion of these waves has been observed indirectly by determining the change in wavelength of the waves.

9.4 DETECTION OF ULTRASONIC WAVES LO3

There are several methods of detecting the ultrasonic waves, which include Kundt's tube method, sensitive method, and piezoelectric detection method.

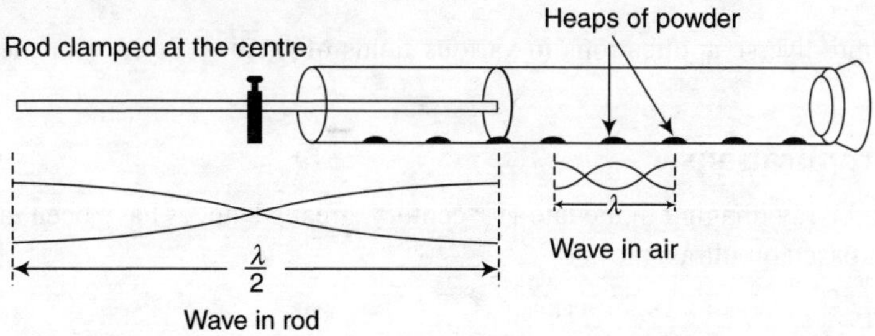

FIGURE 9.4

9.4.1 Kundt's Tube Method

Kundt's tube has been very efficiently used for the detection of ordinary sound waves and the similar method can be employed for detecting the ultrasonic waves. This tube is a long glass tube supported horizontally with an air column in it. A horizontal rod is clamped at the center of the tube, as shown in Fig. 9.4. This tube contains lycopodium powder scattered in it. When the ultrasonic waves are passed through the tube, the lycopodium powder collects in the form of heaps which are found to be situated at the nodal points whereas the powder is found to be blown off at the antinodal points. The average distance between two adjacent heaps gives rise to the value of half wavelength from which the wavelength of the waves can be calculated. In view of the use of the powder, the method is suitable for the detection of the ultrasonic waves of appreciable wavelengths and it cannot be employed if the wavelength of the waves is very small, i.e., less than few millimeters. However, in the case of liquid medium, powered coke is used in place lycopodium powder to detect the position of nodes and the wavelength of the waves.

If the average distance between the adjacent heaps is d and the frequency of the ultrasonic waves is f, then the velocity of the wave is given by $V = 2fd$.

9.4.2 Sensitive Flame Method

This method works on the basis of interaction of wave and a sensitive flame, where the change in pressure is noticed. In this method of detection of ultrasonic waves, a narrow sensitive flame is moved along the medium and the change in its intensity is noticed. At the positions of antinodes, the flame is found to be steady (stationary), while the flame is found to flicker at the positions of nodes due to a change taken place in the pressure. The positions of the nodes and antinodes are found out in the medium, and the average distance between the two adjacent nodes gives rise to the value half wavelength. If the value of the frequency of ultrasonic wave is known, the velocity of the wave passing through the medium can be calculated using the same formula as used in the method of Kundt's tube.

9.4.3 Piezoelectric Detector

Piezoelectric effect, which is being used in the production of ultrasonic waves based on quartz crystal, can also be used to detect the ultrasonic waves. The underlying principle is as follows. If ultrasonic waves comprising of compressions and rarefactions are allowed to fall upon a quartz crystal, a certain potential difference is developed across the faces of the crystal and varying electric charges are produced. These small charges after amplification by an electronic circuit are used to detect the ultrasonic waves.

9.5 APPLICATIONS OF ULTRASONIC WAVES LO4

Ultrasonics have found diverse applications in various fields of medical science or medicine, industry and communication.

9.5.1 Medical Applications

After the discovery of *X*-ray imaging in the late 19th century, great advances have been made to diagnosis and treatment equipment based on ultrasonics.

9.5.1.1 Diagnosis

Scanning of internal organs, vessels and tissues of patient's body based on ultrasonic waves is called ultrasonography. This makes use of high frequency sound waves to produce the images of internal organs and structures for the medical examination and it is possibly the best of all ultrasonic medical applications. The ultrasonic scans are less costly, quicker and easier to use than MRI (magnetic resonance imaging) and CT (computerized tomography) scans. Hence, these are frequently used to monitor and diagnose the condition of organs such as kidneys, liver or gallbladder. In order to diagnose and follow up heart conditions, doctors make efficient use of EVG (echocardiograms) or ultrasonic scans of the heart of the patient.

9.5.1.2 Surgery

The technology based on ultrasound is increasingly being used in surgery. Here ultrasonic surgical instruments convert an ultrasonic signal into a mechanical vibration by using a transducer. A waveguide is then used to amplify and propagate the vibration to a desired position. The ultrasonic surgical instruments are highly useful in diverse medical procedures, as these can cut bone and other tissue. At the same time reduce bleeding by coagulating tissue. Finally this reduces the average length of surgery and damage to tissue, resulting in fewer complications only.

9.5.1.3 Non-invasive Therapeutic Applications

Ultrasound energy can be used as non- or minimally invasive high intensity focused ultrasound (HIFU) or high intensity therapeutic ultrasound (HITU). By applying ultrasound energy to heat and destroy diseased tissues, these methods can be used to remove body tissue while treating the cancers and other conditions. Ultrasound imaging systems locate and target liver, kidney or gallbladder stones. These are smashed into pieces by ultrasound pulses and are finally evacuated naturally through urination. Other treatments using ultrasound technology include bone healing and physiotherapy for inflammation caused by joint injuries. Drug delivery is also done based on HIFU/HITU to treat tumours, especially in the brain where it may be difficult to achieve. Cosmetic applications, such as non-invasive liposuction and for a number of therapies to improve skin tone, scars and sun based damage also make use of ultrasound technology.

9.5.1.4 Dental Care

Another application of ultrasonics is in dental care as descalers to remove plaque. Ultrasonic descalers have a tip that vibrates at high frequency to break down the bacterial matter to which plaque and calculus stick. The ultrasonic waves have been found quite useful for painless dental cutting. This technology enables a smoother and less painful experience.

9.5.1.5 Hygiene Safely

All medical and dental equipment must be absolutely clean before use, otherwise the introduction of pathogenic microbes can lead to infection. It is very important to clean, disinfect and sterlize all multiple use instruments and devices after their use on a patient or surgery. In this direction, ultrasonic cleaning uses a special wash solution to reach and effectively remove organic waste from difficult-to-clean areas, such as equipment or devices with joints and crevices.

9.5.2 Industrial Applications

Industrial Applications of ultrasonics include ultrasonic machining, welding, cleaning, etc.

9.5.2.1 Machining

Ultrasonic machining is a vibratory process which is now in common use for the mechanical treatment of hard and brittle solids such as glasses, ceramics, precious stones, semiconductors and hard alloys. A glass rod oscillating with ultrasonic frequency can be used to bore holes in steel and other hard metals.

9.5.2.2 Welding

With regard to the application of ultrasonics for welding it is believed that practically all metals and plastics can be welded ultrasonic waves of suitable energy. Here the ultrasonic energy converts into heat at the contact area as a result of friction arising between the surfaces. As the temperature of surfaces' layers exceeds the crystallization point, both the layers melt and make a bond together to form a strong joint. Since this process induces negligible stress at the spot of welding, this is quite attractive that the structure of materials remains unchanged.

9.5.2.3 Cleaning

Towards the cleaning applications of ultrasound waves, it is worth mentioning that these waves with frequencies 20 kHz to 40 kHz are used for cleaning of jewellery, optical parts, surgical instruments, industrial parts etc. They are used for cleaning clothes and parts of watches. Printing industry used ultrasonic as a method of cleaning complicated and problematic parts has been available for many years with in a wide range of industries. The main advantages are that components of the most complicated shapes can be cleaned efficiently, speedily and comprehensively. Here ultrasonic millions of tiny bubbles within the fluid which act on the surface of the component behave as a brush in many ways. The scrubbing action of this brush can be made as vigorous or gentle as per the requirement.

9.5.2.4 Structural Composition and Analysis

Ultrasonic waves are used for producing alloys of uniform composition. Further, these waves are employed to detect cracks or flaws in metal structure.

9.5.3 Applications in Food Technology

By tuning frequency, ultrasound can be utilized in food technology. Since ultrasound techniques are relatively cheap, simple and energy saving, these have became an emerging technology for probing and modifying food products. Low power (high frequency) ultrasound is used for monitoring the composition and physiochemical properties of food components and products during processing and storage. However, high power (low frequency) ultrasound induces mechanical, physical and chemical (biochemical) changes through cavitation, which supports many food processing operations such as extraction, freezing, drying, emulsification and inactivation of pathogenic bacteria on food contact surfaces.

Using ultrasound, full reproducible food processes can now be completed in seconds or minutes. This can be done with high reproducibility, reducing the processing cost, simplifying manipulation and giving higher purity of the final product. This also eliminates post-treatment of waste water and consumes only a fraction of the time and energy normally needed for conventional processes.

9.5.4 Applications in Communications

Ultrasonic waves can be produced in the form of beams in the desired direction. These can travel long distances in water before being absorbed. This makes them suitable for the submarine applications. Submarine ultrasonic transmitters have been developed for detecting the presence of iceberg or submarines. These are used for signaling from ship to ship, especially in submerged submarines and also in determination of the depth of sea, position of a ship and submarine. The ship is equipped with the source and receiver of a particular frequency at its bottom. The source is used to transmit the short ultrasound pulses and the reflected pulses are received by the receiver for the detection. Actually the time interval (t) between sending and receiving the pulses is measured, which gives rise to the depth of the ocean as

$$d = \frac{Vt}{2}$$

Here V is the velocity of the ultrasonic waves.

9.5.5 Detection of Velocity of Sound in Liquid

The stationary wave method is applied to find the velocity of ultrasonic waves in liquid and gases. The velocity of ultrasonic waves in these medium can be calculated from the relation $V = f\lambda_{\text{ult}}$, where f is frequency and λ_{ult} is the wavelength of the ultrasonic waves in the medium. This method is more suitable for finding the velocity of ultrasonics in liquid and gases that are available in small quantity.

In order to find the velocity of sound in a liquid, say kerosene oil, a quartz crystal is placed between two metal plates and the plates are connected to an audio frequency oscillator. The assembly of the crystal and the plates is kept inside the liquid cell. The crystal is made to vibrate in resonance with the oscillator by adjusting the frequency to produce the ultrasonic waves. The reflections of the wave from the sides of the liquid cell form a standing wave pattern with nodes and antinodes at regular intervals. This leads to a particular distribution of the liquid density with maximum density at the nodes and minimum density at the antinodes. Accordingly the refractive index of the liquid is varied and it works as the diffraction grating, known as acoustic grating.

The acoustic grating is mounted on a prism table of spectrometer and a parallel beam of sodium light from collimator is allowed to fall normally on the grating. The diffracted light is found to form a diffraction pattern which is viewed through a telescope. The diffraction pattern consists of central maximum and principal maxima on either sides. The positions of the principal maxima satisfy the following relation

$$\lambda_{\text{ult}} \sin \theta_n = n\lambda$$

Here λ_{ult} is the wavelength of sound in the liquid, λ is the wavelength of incident sodium light (monochromatic) and θ_n is the angle of nth other diffraction. We can find the wavelength λ_{ult} of the wave in the liquid. If f be the frequency of vibrations of the crysttal, then the velocity of the ultrasonic wave in the liquid can be obtained using the relation $V = f\lambda_{ult}$.

9.6 TYPES OF ACOUSTICS LO5

There are various types of the acoustics. These are discussed below.

9.6.1 Physical Acoustics

Physical acoustics encompasses propagation and absorption of sound at all frequencies in air and other gases, liquids, semi-solids and solids. It deals with airborne, audible sound, infrasound and ultrasound. Physical acoustics includes both linear processes such as the propagation of sound from traffic, and nonlinear processes such as the shock waves that are generated by planes flying faster than the speed of sound.

9.6.2 Engineering Acoustics

Engineering acoustics deals with the development of devices to generate (e.g., loudspeakers), record (e.g., microphones) and analyse (e.g., frequency analysers) sound of all kinds. The field of sound production, recording and reproduction, with all its attendant electronics and measuring instruments, is an important part of engineering acoustics.

9.6.3 Architectural Acoustics

Architectural acoustics is concerned with sound in buildings. One aspect of this field is the control of sound within rooms to maximise the acceptability of music or intelligibility of speech. This branch of architectural acoustics deals with sound in lecture theatres, concert halls, meeting rooms and classrooms.

9.6.4 Musical Acoustics

Musical acoustics considers the workings of traditional, experimental and electronic musical instruments. The interaction of musicians, instruments, listeners and performance spaces means that many branches of acoustics influence work in this field.

9.6.5 Psychological Acoustics

Psychological acoustics studies the brain's signal-processing function, which takes nerve impulses from the ear and interprets them. Physiological acoustics deals with models and theories of the operation of the ear and its anatomy. One practical application of this field is the study of the elements important to achieve a stereophonic effect. Another is the determination of those factors that make one sound unpleasant or annoying and another reverse. There is no direct correlation between loudness and annoyance.

9.6.6 Bioacoustics

Bioacoustics studies all aspects of acoustic behaviour in animals and biological media in general. This field includes topics such as sound production by animals, bio-sonar, sound reception by animals, effects of noise on animals and medical diagnostics using acoustics, especially ultrasonics.

9.7 ACOUSTICS OF BUILDINGS LO6

The branch of the science which deals with the planning of a building or a hall with a view to provide best audible sound to the audience is called acoustics of building or architectural acoustics. *WC Sabine* in 1911, first of all scientifically tackled the problem of satisfactory speech and music in a hall.

9.7.1 Reverberation

When a sound is produced in a building, it lasts too long after its production. It reaches to a listener a number of times. Once it reaches directly from the source and subsequently after reflection from the walls, windows, ceiling and floor of the hall. The listener, therefore, receives series of sounds of diminishing intensity (since part of energy is lost at each reflection); the sound becomes muddy, garbled. The most important factor in the design of an auditorium is reverberation. Reverberation is nothing but the prolonged reflection of sound from the walls, floor and ceiling of a room. It is also defined as the persistence of audible sound after the source has stopped to emit sound. The duration for which the sound persists is called *reverberation time*. The time of reverberation is also defined as the time taken for the sound to fall below the minimum audibility measured from the instant when the source stops sounding. Sabine, using an organ pipe of frequency 512 Hz found that its sound becomes inaudible when its intensity fall to one millionth of its intensity just before stopping the organ pipe. Hence, *Sabine* defines the standard revereberation time as the time taken by sound to fall to one millionth of its intensity just before the source is cut off. Sabine found that the time of reverberation depends upon the size of the hall, loudness of the sound and upon the kind of the music or sound for which hall is to be used. For a sound of frequency 512 Hz, the best time of reverberation was found to be 1 to 1.5 sec and 1.5 to 2 sec for halls of 50,000 and 40,000 cubic feet, respectively.

Based on the range of values of revereberation time for specific purposes, we can determine a relationship between room volume and internal surface area. This assumes the use of standard auditorium construction materials.

9.7.2 Basic Requirement for Acoustically Good Halls

Before 1900, the architects and building engineers had no consideration about the acoustical properties of rooms and halls etc. Sometimes, a building was found to be unsatisfactory for the purpose for which it was built. According to Sabine the following essential features are required for the good acoustics.

(i) The sound heard must be sufficiently loud in every part of the hall and no echoes should be present.

(ii) The total quality of the speech and music must be unchanged, i.e., the relative intensities of the several components of a complex sound must be maintained.

(iii) For the sake of clarity, the successive syllables spoken must be clear and distinct, i.e., there must be no confusion due to overlapping of syllables.

(iv) The reverberation should be quite proper, i.e., neither too large nor too small. The reverberation time should be 1 to 2 sec for music and 0.5 to 1 sec for speech.

(v) There should be no concentration of sound in any part of the hall.

(vi) The boundaries should be sufficiently sound proof to exclude extraneous noise.

(vii) There should be no Echelon effect.

(viii) There should be resonance within the building.

9.7.3 Transmission of Sound and Transmission Loss

When sound is produced in a hall, it proceeds outward in spherical waves and strikes the boundaries of the hall. The sound waves undergo reflection, absorption and transmission in varying amounts. The amounts of these three processes depend upon the frequency of sound and the characteristics of the wall of the room (i.e., thickness of walls, weight, material, nature of surface). The transmission of sound means the sound energy transmitted through the walls.

The loss of sound energy across a wall or a barrier is defined as transmission loss. So when the sound is transmitted from the source to adjoining room or area, then there is a reduction in the sound intensity. This is known as *transmission loss*. Thus, the transmission loss is numerically equivalent to loss in the intensity of sound. This is expressed in decibels (dB).

(i) If 50 dB and 10 dB are the sound levels measured on either side of a wall, then transmission loss of the wall would be, 50 dB – 10 dB = 40 dB.

(ii) Larger the transmission loss, greater is the sound illusion.

(iii) Transmission loss varies with material used for the construction.

(iv) The methods used for the construction also affect the value of transmission loss.

(v) Transmission loss varies with frequency of sound.

9.7.4 Acoustic Environments and Sound Fields

There are two basic environments in which one makes measurements of sound and noise: outdoors and indoors. An outdoor acoustic environment may be quite often referred to as a free field. A sound field is said to be a free field if it is uniform, free from boundaries, and undisturbed by other sources of sound. Anechoic chambers and well-above-the-ground outdoors are free fields. Sound radiated by a source in a free field propagates away from the source and is never reflected back. Sound spreads three-dimensionally from the source such that the intensity falls off as $1/r^2$ (6 dB decrease from source doubles).

The indoors acoustic environment introduces boundaries which reflect sound. If the boundaries completely reflect all incident sound without any absorption then the resulting sound field is termed as diffuse or reverberant. In a diffuse sound field the time average of the mean square sound pressure is the same everywhere through the enclosure. The flow of energy is equally probable in all directions. If the boundaries absorb some of the incident sound and reflect the rest, then the sound field is called semi-reverberant. Energy flows in more than one direction. Semi-reverberant fields are the most widely encountered in the majority of architectural acoustic environments.

9.7.5 Indoor Acoustics

When a sound source is enclosed, the radiated sound energy is retained within the enclosure. If the boundaries are perfectly reflective then the sound energy inside the enclosure could theoretically grow until a pressure is reached that would be explosive. Fortunately, most realistic boundaries are at least partly absorbing (air also absorbs sound) and the kinds of sound sources usually encountered in a room (for example, human speech) are not extremely powerful. For example, the sound power produced by human speech is very small. Typical male and female speakers generate 34 μW and 18 μW, respectively, at a distance of 3.28 ft. So, common sound sources are not excessively powerful, the sound energy in the enclosure travels about the enclosure and slowly decays as it is absorbed by the boundaries and the medium.

9.7.6 Sabine's Formula for Reverberation Time

Consider a source of the sound in an enclosure. We shall assume that

(i) The rate of emission of energy from the source is constant.

(ii) The energy is transmitted equally in all directions and its distribution is uniform throughout the enclosure.

(iii) The dissipation of energy in the air is negligible and that it is confined only to the boundary walls.

(iv) The power of absorption of the absorbent material is independent of the intensity of the sound.

(v) The effect of superposition is negligible.

The reverberation time can be measured by producing white noise in the enclosure at a level 60 dB above background, switching it off, and measuring the time it takes to drop back to background. The drawbacks to this method are that it takes a high power source and the noise produced is unpleasantly loud. Measuring the time required for a smaller intensity drop and extrapolating to the 60 dB time is the usual procedure. This requires some calculation and a mathematical model for the decay curve of the sound in the enclosure.

If a sound source of intensity I_0 is switched off at time $t = 0$, then the sound starts to die away and will be 60 dB down (or $10^{-6}\ I_0$) after one reverberation time later, i.e., at $t = T$, where T is the revereberation time (sometimes called RT_{60} as a mnemonic for its defintion). To evaluate the scaling constant a, note that

$$e^{-at/T} = 10^{-6} \qquad \text{at } t = T \tag{i}$$

where a is called scaling constant.

Taking the natural logarithm of both sides, we get

$$-a = -6 \text{ In } (10) = -13.82$$

This evaluation will permit a calculation of the reverberation time T if any two intensities I_0 and I are measured along with the time interval t between the two measurements. From Eq. (i), the relationship for T is

$$T = \frac{13.28t}{\text{In } (I_0/I)} \tag{ii}$$

It is found that rather than the intensity, the voltage is displayed from the microphone and the intensity is proportional to the square of the voltage. In terms of the measured voltages, the reverberation time relationship is obtained as

$$T = \frac{13.28t}{\text{In } (V_0^2/V^2)} \tag{iii}$$

This is the relationship which is to be used in the experiment to determine the reverberation time of the room. It could then be compared with the approximate relationship

$$T = \frac{0.161\text{ V}}{aS} \tag{iv}$$

where V is room volume in cubic meters, S is surface area in square meters, and a is average absorption coefficient. The rough absorption coefficient obtained in this manner could help to evaluate the change, which might be made in the reverberation time by changing the surfaces (carpet, curtains, etc.).

Absorption coefficient for standard building materials is given by Sabine's formula

$$T = \frac{0.161\,V}{\Sigma aS}$$

where ΣaS is the total acoustic absorption in the enclosure.

9.7.7 Absorption Coefficient and Its Measurement

What are the essential parameters of a typical room necessary to determine its acoustical behaviour? First, an enclosed space that has an internal volume V. Second, it has a total boundary surface area S. Third, each of the individual surface areas has an absorption coefficient. The average absorption coefficient for all surfaces together is given by the individual surface areas and are the individual absorption coefficients of the individual surface areas. Since the boundaries of the room reflect incident sound energy, the sound signal received by a listener at some location in the room will consist of sound which arrives directly from the source, sound which arrives after reflecting from one surface, and sound which has undergone several reflections. The average distance between reflections in such a space is called the *mean free path* and is related to the dimensions of the room. This is because each time the sound interacts with a surface in the room it loses some of its energy due to absorption. The absorption coefficient of the material of a surface is a, then a fraction $(1 - a)$ of the sound energy incident upon the surface is returned to the side from which it came. The absorption coefficient of sound is essentially the dissipation of energy into heat, and in so far as it is affected by a bounding surface it is mainly due to one of two causes, porosity and flexural vibrations. The absorption coefficient of any material, as originally defined by Sabine, is the ratio of the sound absorbed by that material to that absorbed by an equivalent area of open window. Thus, a perfectly absorbent material would have an absorption coefficient of 1.

Measurement Two audio frequency sources of powers P_1 and P_2 are employed in the test chamber. The actual value of P_1 and P_2 need not be known, but their ratio must be known. Then

$$A = \frac{4V}{c}\frac{\log(P_1/P_2)}{T_1 - T_2}$$

where T_1 and T_2 are the respective times of decay to the threshold of audibility. From this relation mean coefficient of absorption can be obtained since $A - aS$, where S is the area of the absorbing surface is the measure of mean coefficient of absorption. The test chamber for the experiment exclude all extraneous and must have a long time of reverberation when empty.

Another reverberation method employed by Sabine was to measure the absorbing power of a room in terms of that of an open window. Let aS and $(aS + w)$ be the total absorption of the room with windows closed and open, respectively and t_0 and t_c are the time taken for the sound to decay to the threshold of audibility in the two cases, respectively. Then for same intensity of sound in both the cases,

$$\frac{w}{aS} = \left(\frac{t_0}{t_c} - 1\right)$$

and for same energy P of source

$$\log\left(1 + \frac{w}{aS}\right) = \frac{c}{4V}(aS(t_c - t_0) - wt_0)$$

From the above relation aS can be calculated in terms of w.

9.8 FACTORS AFFECTING THE ARCHITECTURAL ACOUSTICS LO7

By an acoustically good hall we mean that in which every syllable or musical note reaches an audible level of loudness at every point of the hall and then quickly dies away to make room for the next syllable or group of notes. The departure from this makes the hall defective acoustically. Following factors affect the architectural acoustics:

(i) **Reverberation**: The reverberation can be controlled by the following factors:

(a) By providing windows and ventilators which can be opened and closed in order to optimise the value of the time of reverberation optimum.

(b) By covering the floor with carpets.

(c) By heavy curtains with folds.

(d) By full capacity of audience.

(ii) **Adequate Loudness:** It can be reduced by using large sounding boards behind the speaker and facing the audience. Large polished wooden reflecting surfaces immediately above the speakers are also helpful. Low ceiling are also of great help in reflecting the sound energy towards the audience by providing additional sound energy with the help of equipment like loudspeakers.

(iii) **Focusing due to Walls and Ceiling**: For uniform distribution of sound energy in the hall, there should be curved surfaces. If such surfaces are present, they should be covered with absorbent material. Ceiling should be low. A paraboloidal reflected surface arranged with the speaker at the focus is also helpful in sending a uniform reflected beam of sound in the hall.

(iv) **Echoes**

(v) **Resonance**

(vi) **Echelon Effect**

(vii) **Extraneous Noise and Sound Insulation**: Generally, there are three types of noises, which are very troublesome. These are Air Borne Noise, Structure Borne Noise and Inside Noise.

Air Borne Noise:

The noises are transmitted through the air. Sound insulation for the reduction of air borne noise can be achieved by the following methods:

(a) By avoiding opening for pipes and ventilators.

(b) By allotting proper places for doors and windows.

(c) Using double doors and windows with separate frames and having insulating material between them.

(d) By making arrangements for perfectly shutting doors and windows.

(e) Using heavy glass in doors, windows and ventilators.

(f) By providing double wall construction, floating floor construction, suspended ceiling construction, box type construction, etc.

Structure Borne Noise:

The most common sources of this type of sound are footsteps, street traffic, hammering, drilling, operating machinery, moving of furniture, etc. Sound insulation for the reducing of structure borne noise is done by the following ways:

(a) By breaking the continuity by interposing layers of some acoustical insulator.

(b) By using double walls with air space between them.

(c) By using anti-vibrations mounts.

(d) By soft floor finishing (carpet, rubber, etc.)

(e) By insulating the machinery, as the mechanical equipments like refrigerators, lifts, fans etc., produce vibrations in the structure.

Inside Noise:

The following methods are used for sound insulation of inside noise.

(a) The machinery (like typewriters) should be placed on absorbent pads.

(b) Any engine inside the hall should be fitted on the floor with a layer of wood.

(c) The floor should be covered with carpet.

(d) The wall, floors and ceiling should be provided with sound absorbing materials.

(e) The sound absorbing materials should be mounted on the surface near the source of noise.

9.8.1 Sound Absorbing Materials

Special materials used to increase absorption of sound waves or to reduce the reflection of sound waves in a room or hall are known as sound absorbing material. The material should have the following requirements:

(a) It should have good resistance to fire.

(b) It should be efficient over a wide range of frequencies.

(c) It should have high sound absorbing efficiency.

(d) It should be cheap, easily available, easy to fix, good looking, light in weight and waterproof and should have economical maintenance and sufficient structural strength. The sound absorbing materials are broadly classified into the following four categories, namely porous absorbents, cavity resonators, composite types of absorbents, and resonant absorbents or panel absorbers.

Summary

The main topics covered in this chapter are summarised below.

- Scientific study of the propagation, absorption, and reflection of sound waves is called acoustics. Acoustics is the interdisciplinary science that deals with the study of sound, ultrasound and infrasound (all mechanical waves in gases, liquids, and solids). In a broad sense, acoustics may be defined as generation, transmission and reception of energy in the form of vibration waves in matter.
- Various types of acoustics, namely physical acoustics, engineering acoustics, architectural acoustics, musical acoustics, psychological acoustics, bioacoustics, were discussed.
- Description of audible waves, ultrasonic waves, and infrasonic waves were given.
- Certain crystals can develop an electric charge when a mechanical pressure or tension is applied. This phenomenon is named as Piezoelectric effect.

- A transducer is a device which is used to convert one form of energy to another. Ultrasonic transducers convert electrical energy to mechanical energy and vice versa. Ultrasonic sound can be produced by transducers which operate either by the piezoelectric effect or by the magnetostrictive effect. The magnetostrictive transducers can be used to produce high intensity ultrasonic sound in the 20–40 kHz range for ultrasonic cleaning and other mechanical applications.
- Principle of ultrasonic transducer was discussed.
- It was discussed how ultrasonic waves are produced. Their applications were talked about.
- Acoustics of buildings was discussed in detail. Reverberation was introduced and it was said that the reverberation is nothing but the prolonged reflection of sound from the walls, floor and ceiling of a room. It is also defined as the persistence of audible sound after the source has stopped to emit sound. The duration for which the sound persists is called reverberation time. The time of reverberation is also defined as the time taken for the sound to fall below the minimum audibility measured from the instant when the source stops sounding.
- Basic requirement for the acoustically good halls were discussed. These are the following.
  - (a) The sound heard must be sufficiently loud in every part of the hall and no echoes should be present.
  - (b) The total quality of the speech and music must be unchanged, i.e., the relative intensities of the several components of a complex sound must be maintained.
  - (c) For the sake of clarity, the successive syllables spoken must be clear and distinct, i.e., there must be no confusion due to overlapping of syllables.
  - (d) The reverberation should be quite proper, i.e., neither too large nor too small. The reverberation time should be 1 to 2 seconds for music and 0.5 to 1 second for speech.
  - (e) There should be no concentration of sound in any part of the hall.
  - (f) The boundaries should be sufficiently sound proof to exclude extraneous noise.
  - (g) There should be no Echelon effect.
  - (h) There should be resonance within the building.
- Transmission of sound and transmission loss were discussed in detail.
- Sabine's formula for reverberation time was derived and its theoretical as well as physical aspects were talked about.
- Finally, the absorption coefficient was introduced and methods were talked about for its measurement.
- Factors affecting the architectural acoustics were discussed in detail and the methods of its removal were talked about.

SOLVED EXAMPLES

EXAMPLE 1 The frequency limits of the range of human hearing ear is from about 20 Hz to 20 kHz. The speed of sound is about 34,500 cm/sec. What is the wavelength of the wave in cm?

SOLUTION The frequency range is given as 20 Hz to 20 kHz.

The speed of sound = 34500 cm/sec = 345 m/sec

We know that the frequency of sound wave is related to its wavelength by $v = f\lambda$

$$\therefore \qquad \lambda = \frac{345}{20}$$

So for the frequency of 20 Hz, wavelength $\lambda = \frac{345}{20}\text{m} = 17.25\text{ m} = 1725\text{ cm}$

For the frequency of 20 kHz, wavelength $\lambda = \frac{345}{20\times 10^3}\text{m} = 17.25\times 10^{-3}\text{ m} = \mathbf{1.725\ cm}$

Therefore, wavelength range of the sound wave is 1.725 cm to 1725 cm.

Example 2 Calculate the velocity of the sound in air in cm per sec at 100°C if the density of air at S.T.P. is 0.001293 g/cm^3, the density of the mercury at 0°C is 13.60 g/cm^3, the specific heat of air at constant pressure is 0.2417 and the specific heat of air at constant volume is 0.1715.

Solution The velocity of sound in air is given by

$$v = \sqrt{\frac{\gamma p}{\rho}} \text{ with usual notation.}$$

The quantity ρ of the air is 0.001293 g/cm^3
The pressure p is given by

$$p = h\rho g = 76\times 13.6\times 980 \text{ dynes/cm}^3$$

Now
$$\gamma = \frac{C_p}{C_v} = \frac{0.2417}{0.1715}$$

$$\therefore \qquad v = \sqrt{\frac{0.2417\times 76\times 13.6\times 980}{0.1715\times 0.001293}}$$

The velocity v is proportional to $\sqrt{T}$, where T is the temperature of the air. Thus, if v' be the velocity at 100°C and v at 0°C, then

$$\frac{v'}{v} = \sqrt{\frac{273+100}{273}} = \sqrt{\frac{373}{273}}$$

$$v' = \sqrt{\frac{373}{273}}v$$

$$= \sqrt{\frac{373\times 0.2417\times 76\times 13.6\times 980}{273\times 0.1715\times 0.001293}} = 38839.12 \text{ cm/sec}$$

$$= \mathbf{38839\ cm/sec}$$

Example 3 The wavelength of the gas emitted by a tuning fork of frequency 512 vibration/sec in air at 17°C is 66.5 cm. If the density of air at S.T.P. is 1.293 mg/cm^3, calculate the ratio of two principal specific heats of air. Assume that the density of mercury is 13.6 g/cm^3.

Solution Since $v = f\lambda$, the velocity of sound at 17°C is given by

$$v = 512\times 66.5 \text{ cm per sec}$$

Now,
$$v_0 = \sqrt{\frac{\gamma p}{\rho}}$$

Here $p = 76$ cm of mercury = 76 × 13.6× 980 dynes/cm^3. The density of air $\rho = \dfrac{1.293}{1000}$ g/cm^3. If v_0 be the velocity at 0°C and since the velocity is proportional to $\sqrt{T}$,

$$v_0 = \sqrt{\frac{273}{290}}v = \sqrt{\frac{273}{290}} \times 512 \times 66.5$$

Now, $$v_0 = \sqrt{\frac{\gamma p}{\rho}}$$

∴ $$\gamma = \frac{v_0^2 \rho}{p} = \frac{273 \times (512 \times 66.5)^2 \times 1.293}{290 \times 1000 \times 76 \times 13.6 \times 980} = \mathbf{1.39}$$

EXAMPLE 4 A hall of floors is 15 × 30 m^2 along with height of 6 m, in which 500 people occupy upholstered seat and the remainder sit on wooden chairs. Optimum reverberation time for orchestral music is 1.36 sec and absorption coefficient per person is 0.44.

(a) Calculate the coefficient of absorption to be provided by the walls, floor and ceiling when the hall is fully occupied.

(b) Calculate the reverberation time if only the half upholstered seats are occupied.

SOLUTION

(a) The optimum reverberation time is $T = 1.36$ sec

Using Sabine's formula equation of SI unit

$$T = 0.161\frac{V}{aS}$$

$$1.36 = \frac{0.161 \times (15 \times 30 \times 6)}{aS}$$

$$aS = 319 \text{ SI units}$$

Absorption due to audience = 500 × 0.44

= 220 SI units

Therefore, the absorption provided by the walls, floor and ceiling is

319 – 220 = **99 SI unit**

(b) When the hall is only half filled the absorption will also be provided by vacant seats in addition to the absortion by the audience.

250 × 0.44 = 110 SI unit

The absorption by vacant wooden seats = 250 × 0.02 = 5 SI unit

So the total absorption of the hall = 99 + 110 + 5 = **214 SI unit**

Here the reverberation time, given by Sabine's formula, is now

$$T = \frac{0.161 \times (15 \times 30 \times 6)}{214} = \frac{0.161 \times (15 \times 30 \times 6)}{214}$$

$$= \mathbf{2.03 \text{ sec}}$$

EXAMPLE 5 Calculate the total absorption coefficient of cinema hall, whose volume is 8000 m^3 and reverberation time required is 1.8 sec.

SOLUTION

The reverberation time is given by

$$T = \frac{0.161V}{aS} = \frac{0.161V}{\text{Total absorption in hall}}$$

$\therefore$ $$\text{Total absorption in hall} = \frac{0.161V}{T} = \frac{0.161 \times 8000}{1.8}$$

$$= \mathbf{715.55\ O.W.U.}$$

EXAMPLE 6 Find out reverberation time of empty hall of volume 1700 m^3 having a seating capacity for 150 persons with following data

Surface	*Area*	*Coefficient of absorption in O. W. U.*
Plastered wall	98 m^2	0.03
Plastered ceiling	144 m^2	0.04
Wooden door	15 m^2	0.06
Cushioned chairs	88 m^2	1.0

SOLUTION Given $V = 1700$ m^3

Based on the given data, the absorption by

Plastered wall = 98 × 0.03 = 2.94

Plastered ceiling = 144 × 0.04 = 5.76

Wooden door = 15 × 0.06 = 0.90

Cushioned chairs = 88 × 1.0 = 88.0

$\therefore$ Total absorption = 97.6

$$\text{Reverberation time } T = \frac{0.161V}{aS} = \frac{0.161 \times 1700}{97.6}$$

or **T = 2.80 sec**

EXAMPLE 7 Calculate the reverberation time for a hall of volume 1400 m^3, which has seating capacity of 110 persons with full capacity of audience and when audience are occupying only cushioned seats. Relevant data may be taken from Ex. 6.

SOLUTION We have total absorption in hall (from Ex. 6) = 97.6, $V = 1400$ m^3.

When the hall is with full capacity of 110 person, the absorption due to them

= 110 × 4.7 = 517

Now total absorption = 97.6 + 517 = 614.6

Reverberation time

$$T = \frac{0.161V}{aS} = \frac{0.161 \times 1400}{614.6}$$

$\mathbf{T = 0.367\ sec}$

EXAMPLE 8 The volume of a room is 980 m^3. The wall area of the room is 150 m^2, ceiling area is 95 m^2 and floor area is 90m^2. The average sound absorption coefficient (i) for wall is 0.03, (ii) for ceiling is 0.80 and (iii) for the floor is 0.06. Calculate the average sound absorption coefficient and the reverberation time.

SOLUTION The average sound absorption coefficient

$$a = \frac{a_1S_1 + a_2S_2 + a_3S_3}{S_1 + S_2 + S_3}$$

$$= \frac{0.03 \times 150 + 0.80 \times 95 + 0.06 \times 90}{150 + 95 + 90}$$

$$= 0.256$$

and total area $S = 150 + 95 + 90 = 335$

Now total absorption of the room

$$= aS$$

$$= 0.256 \times 335 = 85.76 \text{ metric Sabines}$$

Reverberation time

$$T = \frac{0.161V}{aS}$$

$$= \frac{0.161 \times 980}{85.76}$$

$$= \mathbf{1.84\ sec}$$

EXAMPLE 9 How much acoustic power enters the window of area 1.58 m^2, via the sound wave (standard intensity level = 10^{-16} W/cm^2). The window opens on a street where the street noise results in an intensity level at the window of 60 dB.

SOLUTION Given the intensity level at window = 60 dB

Area of the window = 1.58 m^2

Standard intensity level $I_0 = 10^{-16}$ W/cm^2 = 10^{-12} W/m^2.

We know that intensity level = 10 $\log_{10}(I/I_0)$ dB

$$\therefore \quad 60 = 10 \log_{10} (I/10^{-12}) \text{ dB}$$

$$I = 9.98 \times 10^{-7} \text{ W/m}^2$$

Acoustic power = intensity × area = $9.98 \times 10^{-7} \times 1.58 = 1.576 \times 10^{-6}$ W = $\mathbf{1.58 \times 10^{-6}}$ **W**

EXAMPLE 10 Find the frequency to which a piezoelectric oscillator circuit should be turned so that a piezoelectric crystal of 0.1 cm thickness vibrates in its fundamental mode to generate ultrasonic waves. Young's modulus and density of material of the crystal are 8×10^{10} Nm^{-2} and 2.654×10^3 kg m^{-3} respectively.

SOLUTION Given, thickness of the crystal $t = 1 \times 10^{-3}$ m, density (D) = 2.654×10^3 and $Y = 8 \times 10^{10}$ Nm^{-2}

From the relation, the fundamental frequency of piezoelectric oscillator

$$f = \frac{p}{2t}\sqrt{\frac{Y}{D}} = \frac{1}{2 \times 0.001}\sqrt{\frac{8 \times 10^{10}}{2.654 \times 10^3}}$$

$$= \mathbf{2.75 \times 10^6\ Hz}$$

EXAMPLE 11 Calculate the natural frequency of 30 mm of iron rod. The density of iron rod and Young's modulus are 7.25×10^3 kg/m^3 and 115×10^9 N/m^2 respectively. Can you use it in magnetostriction oscillator to produce ultrasonic waves?

SOLUTION Given, $l = 3 \times 10^{-2}$ m, $D = 7.25 \times 10^3$ kg/m^3 and $Y = 115 \times 10^9$ N/m^2

From the relation, the frequency of ultrasonic waves by magnetostriction oscillator is

$$f = \frac{1}{2l}\sqrt{\frac{Y}{D}} = \frac{1}{2\times 3\times 10^{-2}}\sqrt{\frac{115\times 10^{9}}{7.25\times 10^{3}}}$$

$$= 66.38\times 10^{3} \text{ Hz}$$

$$= \mathbf{66.38\ kHz}$$

EXAMPLE 12 Calculate the fundamental frequency of a quartz crystal of 3×10^{-3} m thickness. The density of the crystal is 2650 kg m^{-3} and Young's modulus is 7.9×10^{10} N/m^2.

SOLUTION Given, $l = 3 \times 10^{-3}$ m, $Y = 7.9 \times 10^{10}$ N/m^2 and $D = 2650$ kg/m^3

From the relation, the fundamental frequency of quartz crystal

$$f = \frac{1}{2l}\sqrt{\frac{Y}{D}} = \frac{1}{2\times 3\times 10^{-3}}\sqrt{\frac{7.9\times 10^{10}}{2650}}$$

$$= 9.1\times 10^{5} \text{ Hz}$$

$$= \mathbf{0.91\ MHz}$$

EXAMPLE 13 Calculate the natural frequency of iron of 0.03 m length, the density of iron is 7.23×10^3 kg/m^3 and Young's modulus 116×10^{10} N/m^2.

SOLUTION Given, $l = 0.03$ m, $D = 7.23 \times 10^3$ kg/m^3 and $Y = 116 \times 10^{10}$ N/m^2

Formula used is

$$f = \frac{1}{2l}\sqrt{\frac{Y}{D}} = \frac{1}{2\times 0.03}\sqrt{\frac{116\times 10^{10}}{7.23\times 10^{3}}}$$

$$f = 0.211\times 10^{6} \text{ Hz}$$

$$= \mathbf{0.21\ MHz}$$

EXAMPLE 14 An ultrasonic source of 0.67 MHz sends down a pulse towards sea bed which come back after 1 sec. Find out the depth of sea and the wavelength of pulse. The velocity of sound in sea water is 1690 m/sec.

SOLUTION Given, $f = 0.67 \times 10^6$ Hz, $t = 1$ sec and $v = 1690$ m/sec

By using the formula

$$2h = vt \quad \text{and} \quad v = f\lambda$$

where h is depth of the sea, we get

$$2\times h = 1690\times 1$$

$$h = 845 \text{ m}$$

and

$$\lambda = \frac{v}{f} = \frac{1690}{0.67\times 10^{6}} = \mathbf{0.00252\ m}$$

EXAMPLE 15 Calculate the capacitance to produce ultrasonic waves of 10^6 Hz with an inductance of 1 Henry.

SOLUTION Given, $f = 10^6$ Hz and $L = 1$ Henry

Formula used is

$$f = \frac{1}{2\pi\sqrt{LC}}$$

or

$$C = \frac{1}{4\pi^2 L f^2}$$

$$C = \frac{1}{4\times(3.14)^2\times 1\times(10^6)^2} = \frac{10^{-12}}{4\times(3.14)^2}$$

$$= \mathbf{0.0254\ pF}$$

Example 16 A quartz crystal of thickness 1 mm is vibrating at resonance. Calculate the fundamental frequency. Given Y for quartz = 7.9×10^{10} N/m^2 and ρ for quartz = 2650 kg/m^3.

Solution The fundamental frequency of the vibration is given by

$$f = \frac{1}{2l}\sqrt{\frac{Y}{e}}$$

$$f = \frac{1}{2\times 0.001}\sqrt{\frac{7.9\times 10^{10}}{2650}}$$

$$= 2.72998 \times 10^6 \text{ Hz}$$

The fundamental frequency of the quartz crystal = 2.730×10^6 Hz = **2.73 MHz**

Objective Type Questions

Q.1 If the period of a wave is decreased, then
(a) the amplitude of the wave decreases (b) the amplitude of the wave increases
(c) the frequency of the wave decreases (d) the frequency of the wave increases

Q.2 Which of the following frequency range is audible to the human ear?
(a) 50-100 Hz (b) 500-1000 Hz (c) 5000-10000 Hz (d) all of the above

Q.3 A property of sound which is most closely associated with the pitch of a musical note is
(a) amplitude (b) frequency (c) wave velocity (d) all of the above

Q.4 Two tones have the same amplitude. The statement which is true is
(a) the sounds must be equally loud to the ear
(b) the sounds must have the same pitch
(c) the sounds must have the same timbre
(d) none of the above must be true (though they may be true)

Q.5 When the amplitude of a sound pressure wave is doubled, the sound pressure level (in decibels)
(a) is doubled (b) is halved (c) decreases by 2 dB (d) increases by 6 dB

Q.6 "This note is higher in pitch than the note" is a statement about
(a) relative pitch (b) absolute pitch (c) binaural hearing (d) none of these

Q.7 Lying on the floor, you will exert pressure on the floor, compared to standing on the floor
(a) more (b) less (c) the same (d) not known

Q.8 A mass oscillates on a spring. As the mass passes through the point of equilibrium,
(a) kinetic energy is maximum and potential energy is maximum
(b) kinetic energy is minimum and potential energy is maximum

(c) kinetic energy is maximum and potential energy is minimum
(d) kinetic energy is minimum and potential energy is minimum

Q.9 The difference between the time each of your ears hears a sound can help you judge of the sound.
(a) direction (b) pitch (c) loudness (d) speed

Q.10 As the temperature of the air increases, the speed of sound in air
(a) increases (b) decreases (c) does not change (d) none of these

Q.11 Which of the following is the false statement?
(a) sound waves are longitudinal waves (b) sound can travel through a vacuum
(c) light travels much faster than sound (d) transverse waves in a guitar string are different from sound waves

Q.12 The Doppler shift explains
(a) why a sound grows quieter as we move away from the source
(b) why the siren on a police car changes pitch as it races past us
(c) the phenomenon of beats
(d) sound diffraction

Q.13 The two tones of 440 Hz and 444 Hz are played. The beat frequency is
(a) 440Hz (b) 444 Hz (c) 442 Hz (d) 4 Hz

Q.14 Which of the following will not affect the fundamental frequency of vibration of a string?
(a) changing the amplitude of vibration (b) changing the tension of the string
(c) changing the length of the string (d) changing the density of the string

Q.15 Compared to the velocity of a 400 Hz sound through air, the velocity of a 200 Hz sound through air is
(a) twice as great (b) one-half as great (c) the same (d) times larger

Q.16 In general, sound travels fastest through
(a) gases (b) liquids (c) solids (d) vacuum

Q.17 In a sound wave, at a place where there is a node in the air pressure wavy, the air molecule displacement
(a) has a node (b) has an antinode
(c) oscillates between node and antinode (d) none of the above

Q.18 When a sound wave passes from air into water, it changes direction. This phenomenon is known as
(a) refraction (b) diffraction (c) reflection (d) polarisation

Q.19 A sound pressure level of 110 decibels would be considered
(a) very loud (b) average for speaking
(c) very soft (d) nothing can be said

Q.20 In damped harmonic motion, the following quantity decreases.
(a) amplitude (b) frequency (c) period (d) velocity

Q.21 If the displacement of vibrating particles is perpendicular to the direction of propagation of the wave, the wave is said to be
(a) linear (b) transverse (c) longitudinal (d) standing

Q.22 There are two organ pipes of the same length. One has one end closed, while the other has both ends open. The one with a closed end will emit sound of
(a) a higher frequency (b) a lower frequency
(c) the same frequency (d) double amplitude

Q.23 When a stretched wire (fixed at both ends) is vibrating in the second harmonic, there is/are
(a) one node (b) two nodes (c) three nodes (d) infinite nodes

Q.24 Two pure tones cause resonance in different positions along the basilar of membrane. These tones have different
(a) amplitude (b) frequency (c) timbre (d) intensity

Q.25 Overtones have wavelengths, compared to the fundamental
(a) longer (b) shorter (c) the same (d) times larger

Q.26 In order to double the wavelength of a sound wave, you should only
(a) double its amplitude (b) double its frequency
(c) halve its amplitude (d) halve its frequency

Q.27 When two sine waves that are 180° out of phase are added together, the amplitude of the sum is
(a) always zero (b) always less than the amplitude of either wave
(c) equal to the amplitude of the smaller wave (d) always less than the amplitude of the larger wave
(e) always greater than the amplitude of the smaller wave

Q.28 A sound wave has sound intensity level SIL = 50 dB. Recall that SIL = 10 log ($[I/10^{-12}$ W/m^2)]). The intensity I of this wave, in W/m^2, is therefore
(a) 50 (b) 5 (c) 10^{-5} (d) 10^{-7} (e) 10^{-10}

Q.29 A sound wave with SIL = 50 dB is reflected by a cloth-covered wall that absorbs 75% of its intensity. The SIL of the reflected wave is
(a) 75 dB (b) 47 dB (c) 44 dB (d) 25 dB (e) 12.5 dB

Q.30 Light and sound are both waves; yet we can hear a car that is coming from behind the corner of a building before we can see the car. This is because
(a) sound travels faster than light (b) sound $\lambda_{sound} > \lambda_{light}$, sound diffracts more than light
(c) sound is not reflected by buildings (d) sound and light interfere, with sound winning out

Q.31 A moving locomotive is sounding its horn as it crosses a highway. There are people in all directions from the locomotive – in front, in back, to the right and left. Compared to the "true" pitch, as heard by the engineer, the horn's pitch heard by these people is
(a) higher (b) lower
(c) the same for all of the people (d) higher for some, true for others, and lower for yet others of the people

Q.32 The frequency of the note B4 is close to 500 Hz. The period of this vibration is
(a) 500 sec (b) 1 sec (c) 0.2 sec (d) 2 msec
(e) none of these

Q.33 A sine wave and a square wave cannot have the same
(a) loudness (b) wavelength (c) frequency (d) tone quality (e) pitch

Q.34 An electric bell is operating in a vacuum. We cannot hear the sound of the bell because
(a) air is needed to conduct the electric current to the bell
(b) the bell's metal cannot vibrate in vacuum
(c) there is no air to conduct the vibrations to our ears
(d) the vacuum jar absorbs the sound
(e) the noise of the pump is louder than the noise of the bell

Q.35 The wavelength of "shortwave" radio waves is smaller than that of standard broadcast (AM) radio waves. They both propagate at the same speed. This allows you to conclude that, compared to AM waves, the "shortwaves" have
(a) lower frequency (b) longer period (c) higher frequency (d) smaller amplitudes

Q.36 When a sound wave enters from air into a metal, in which the speed of sound is much larger than in air, it does not change its

(a) wavelength (b) frequency (c) speed (d) all of these (a-c) change

Q.37 Sound moves at 345 m/sec towards a rock wall, reflects, and returns (as an echo). The roundtrip takes 2 sec. How far away is the wall?

(a) 70 m (b) 170 m (c) 340 m (d) 345 m (e) 350 m

Q.38 Two identical sound sources differ in distance from the listener by $^1/_2$ wavelength. The result will be

(a) no sound at the listener
(b) constructive interference
(c) sound which is twice as loud as one source
(d) beats

Q.39 Which of the following crystals show piezoelectric effect?

(a) NaCl (b) Barium Titanate (c) Diamond (d) Quartz

Q.40 The frequency of vibration of the D.C. magnetized rod in the magnetostriction generator is

(a) Equal to the frequency of alternating current
(b) Twice the frequency of alternating current
(c) Half the frequency of alternating current
(d) None

TRUE OR FALSE

State whether True or False

Q.1 A steady tone played on a violin is an almost perfect sine wave.

Q.2 Different vowel sounds differ mainly in the relative frequency and amplitude of the first two formats.

Q.3 The threshold of hearing is at 0 phons for all frequencies.

Q.4 A standing wave remains constant, without any change in time whatever.

Q.5 The precedence effect enables us to hear the fundamental frequency of a complex wave, even when that frequency is absent in the Fourier spectrum.

Q.6 In the well-tempered scale, only the octaves are perfect intervals.

Q.7 Light is a longitudinal wave, whereas sound is transverse.

Q.8 A triangle wave contains higher-frequency Fourier components than a sine wave of the same periodicity.

Q.9 In a CD player the disk rotates at a constant linear velocity.

Q.10 In a dynamic loudspeaker the sound is produced by vibration of a permanent magnet.

Q.11 In an audio system, AM-FM tuner, tape recorder, and CD player each requires its own separate amplifier and loudspeaker.

Q.12 To make the acoustics of an auditorium more live, the wall, ceiling, and floor surfaces should be made as sound absorbent as possible.

Q.13 In white noise all frequencies are present, and all have the same intensity.

Q.14 Attack transients help determine the tone quality of a musical note.

Q.15 The sound quality of a violin is due only to the resonances of its strings; the violin body has no resonances of its own.

Q.16 Percussion instruments have only a single resonance.

Q.17 For two electrical devices connected to the same voltage, the one with the smaller resistance draws the smaller current.

Q.18 In order to avoid interference between different AM stations, each station uses a carrier wave that is different from that of any of the other stations.

Q.19 Magnetostriction oscillator can generate ultrasonic waves of single frequency.

Q.20 We can generate ultrasonic waves of frequency f, $2f$, ..., where f is the fundamental frequency.

Q.21 The magnitude of the Piezoelectric effect of a crystalline material does not depend on direction.

Q.22 The Piezoelectric effect of a material is something to do with the crystal symmetry.

Q.23 The Piezoelectric effect of a material depends on its crystal structure.

PRACTICE PROBLEMS

Q.1 What is piezoelectric effect? Describe the construction of a piezoelectric oscillator for the production of ultrasonic waves.

Q.2 Give the theoretical treatment of Sabine's law. Define the term 'period of revereberation'.

Q.3 Sketch a graph of pressure vs time for two sound waves that differ only in pitch. Sketch a graph of pressure vs time for two sound waves that differ only in timbre.

Q.4 A mass on a spring is found to oscillate naturally at a frequency of 0.5 Hz. This mass-spring system is then driven by an oscillator. Describe what happens as the frequency of the oscillator is varied from 0.2 Hz to 0.8 Hz.

Q.5 Sketch the first two normal modes of sound pressure in a tube open at one end, closed at the other end. If the fundamental mode has a frequency of 440 Hz, what is the frequency of the other mode? Is it harmonic?

Q.6 Ram is in a fire truck rushing toward the scene of a fire. Shyam is standing at the scene of the fire. There is no wind. Who hears a higher pitch for the fire truck's siren? Explain why.

Q.7 Explain what is meant by a restoring force? Why is it necessary for vibrations to occur?

Q.8 What is the wavelength of a 440 Hz sound in air, if the speed of sound in air is 340 m/s? Would the wavelength be longer or shorter if the sound were passing through water?

Q.9 A wave pulse travels down the length of a wave machine like the one in the front of the lecture hall. The pulse reflects from the end. Describe the difference you would notice between a wave machine with the end free to move and a wave machine with the end fixed.

Q.10 Define diffraction. How would you demonstrate diffraction of sound waves?

Q.11 One sound is made up of equal amplitudes of 110 Hz, 220 Hz, and 440 Hz pure tones. A second sound is made up of equal amplitudes of 110 Hz, 330 Hz, and 550 Hz pure tones. In what way(s) are these two sounds the same? In what way(s) are these two sounds different? What is the term given to this combining of pure tones to get a complex tone?

Q.12 Two pure tones are played, one at a constant frequency of 550 Hz, the other has a variable frequency. Describe all the phenomena you hear as the frequency of the second tone is varied gradually from 550 Hz to 1100 Hz.

Q.13 Why not ultrasonics be produced by passing high frequency alternating current through a loud speaker?

Dielectrics

10

LEARNING OBJECTIVES

After reading this chapter you will be able to

- **LO 1** Understand the concept of dielectric constant
- **LO 2** Know about types of dielectrics and polarisation of dielectrics
- **LO 3** Learn about types of polarisation
- **LO 4** Explain Gauss's law in the presence of dielectrics
- **LO 5** Discuss dielectric loss and the dependent factors
- **LO 6** Describe Lorentz field and validity of Clausius-Mosotti equation for non-polar dielectrics of cubic crystal structure

Introduction

A dielectric is an insulating material in which all the electrons are tightly bound to the nuclei of the atoms and there are no free electrons available for the conduction of current. Therefore, the electrical conductivity of a dielectric is very low. The conductivity of an ideal dielectric is zero. On the basis of band theory, the forbidden gap (E_g) is very large in dielectrics. Materials such as glass, polymers, mica, oil and paper are examples of dielectrics. They prevent flow of current through them. Therefore, they can be used for insulating purposes.

10.1 DIELECTRIC CONSTANT (LO1)

It is found experimentally that the capacitance of a capacitor is increased if the space between its plates is filled with a dielectric material. To understand this fact, Faraday took two identical capacitors, one was evacuated and the other was filled with dielectric material, as shown in Fig. 10.1.

Then these two capacitors were charged with a battery of same potential difference. He found that the charge on the capacitor filled with dielectric is larger than that of the other filled with air. If C_0 be the capacitance in

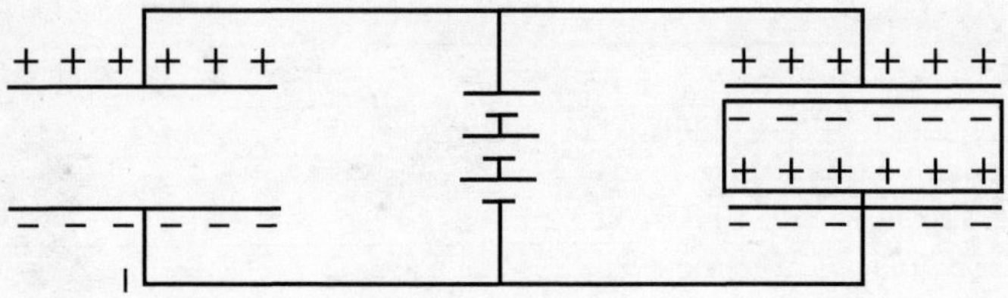

FIGURE 10.1

vacuum and C the capacitance when the space is filled with a dielectric material, then the dielectric constant of the material

$$K = \frac{C}{C_0}$$

Thus, the dielectric constant of a material is the ratio of the capacitance of a given capacitor completely filled with that material to the capacitance of the same capacitor in vacuum. In other words, the ratio of permittivity of medium to that of the vacuum is also known as *dielectric constant*, i.e.,

$$K = \frac{\varepsilon}{\varepsilon_0} = \varepsilon_r$$

This is also known as *relative permittivity* (ε_r). It is found to be independent of the shape and dimension of the capacitor.

10.2 TYPES OF DIELECTRICS LO2

A molecule is a neutral system in which the algebraic sum of all the charges is zero. Based on the dipole moment, the molecules of dielectrics are termed as non-polar and polar molecules. Accordingly these dielectrics are referred to as non-polar and polar dielectrics.

10.2.1 Non-polar Dielectrics

A 'non-polar' molecule is the one in which the centre of gravity of the positive (protons) and negative charges (electrons) coincide. So such molecule does not have any permanent dipole moment, as shown in Fig. 10.2a. Few common examples of non-polar molecules are oxygen (O_2), nitrogen (N_2) and hydrogen (H_2). As mentioned earlier, the dielectrics having non-polar molecules are known as non-polar dielectrics.

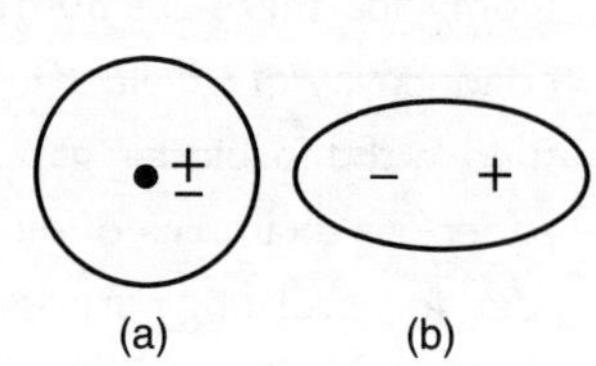

FIGURE 10.2

10.2.2 Polar Dielectrics

A polar molecule is the one in which the centre of gravity of the positive charges is separated by finite distance from that of the negative charges. Unbalanced electric charges, usually valence electrons, of such molecules result in a dipole moment and orientation. Therefore, these molecules possess permanent electric dipole (Fig. 10.2b). Few examples of polar molecules are N_2O, H_2O and HCl. The dielectrics having polar molecules are known as polar dielectrics.

10.3 POLARISATION OF DIELECTRICS

LO2

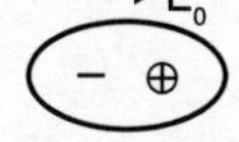

FIGURE 10.3

When an electric field is applied to a dielectric material; it exerts a force on each charged particle and pushes the positive charge in its own direction while the negative charge is displaced in opposite direction, as shown in Fig. 10.3. Consequently, the centres of positive and negative charges of each atom are displaced from their equilibrium positions. Such a molecule (or atom) is then called as *induced electric dipole* and this process is known as *dielectric polarisation.*

We consider a parallel plate capacitor which has vacuum initially between its plates. When it is charged with a battery, the electric field of strength E_0 is set up between the plates of the capacitor (Fig. 10.4a). If σ and $-\sigma$ are the surface charge densities of the two plates of the capacitor, then the electric field developed between the plates is given by

$$E_0 = \frac{\sigma}{\varepsilon_0} \tag{i}$$

If now a slab of dielectric material is placed between the two plates of the capacitor (Fig. 10.4b), then it becomes electrically polarised. Hence, its molecules become electric dipole oriented in the direction of the field. Because of this the centre of positive and negative charges gets displaced from each other. Therefore, in the interior of the dielectric as marked by dotted lines these charges cancel. However, the polarisation charges on the opposite faces of the dielectric slab are not cancelled. These charges produces their own electric field E_p, which opposes the external applied field E_0. Under this situation, the net electric field in the dielectric is given by

$$\vec{E} = \vec{E}_0 - \vec{E}_p \tag{ii}$$

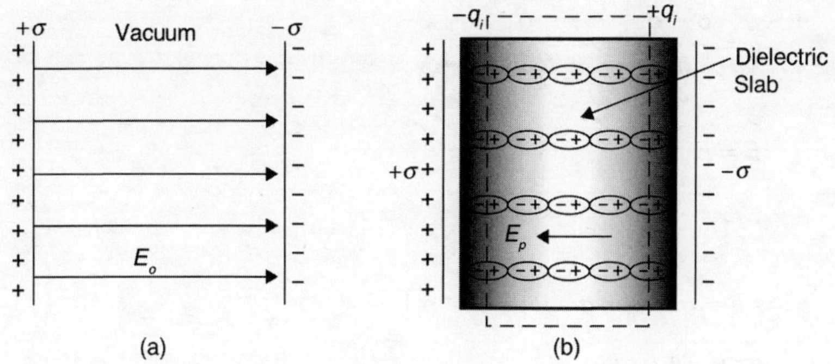

FIGURE 10. 4

10.3.1 Polarisation Density

The induced dipole moment developed per unit volume in a dielectric slab on placing it inside an electric field is known as polarisation density. It is denoted by a symbol P. If p is induced dipole moment of individual atom and N is the number of atoms in a unit volume, then polarisation density is

$$\vec{P} = N\vec{p} \tag{iii}$$

The induced dipole moment of an individual atom is found to be proportional to the applied electric field $\vec{E}$ and is given by

$$\vec{P} = \alpha\varepsilon_0\vec{E} \tag{iv}$$

where α is the proportionality constant and is also known as atomic polarisability. From Eqs. (iii) and (iv), we get

$$\vec{P} = N\alpha\varepsilon_0\vec{E} \tag{v}$$

Suppose S is the area of each plate of the capacitor and d is the separation between them (Fig. 10.5). Then the volume of the dielectric slab is Sd. Since $-q_i$ and $+q_i$ are the induced charges developed on the two faces of the dielectric slab, the total dipole moment of the slab will be equal to $q_i d$. From the definition of the polarisation density

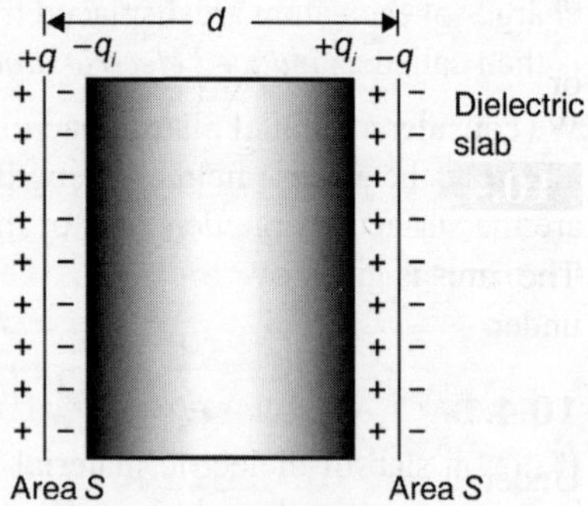

FIGURE 10.5

$$P = \frac{\text{Total dipole moment}}{\text{Volume of slab}}$$

$$= \frac{q_i d}{Sd} = \frac{q_i}{S} = \sigma_p$$

$$\therefore \qquad P = \sigma_p \tag{vi}$$

On placing the dielectric material between the two plates of the capacitor, the reduced value of the electric field may be evaluated as follows

$$E = \frac{\sigma - \sigma_p}{\varepsilon_0} = \frac{\sigma}{\varepsilon_0} - \frac{\sigma_p}{\varepsilon_0} \tag{vii}$$

or

$$E = E_0 - \frac{P}{\varepsilon_0} \tag{viii}$$

From Eq. (vii)

$$\varepsilon_0 E = \sigma - \sigma_p = \sigma - P \; [\because P = \sigma_p]$$

or

$$\sigma = \varepsilon_0 E + P \tag{ix}$$

The quantity $(\varepsilon_0 E + P)$ is of special significance and is known as the electric displacement vector $\vec{D}$ given by

$$\vec{D} = \varepsilon_0\vec{E} + \vec{P} \tag{x}$$

10.3.2 Relation between Dielectric Constant and Electric Susceptibility

The polarisation density of a dielectric is proportional to the effective value of electric field $\vec{E}$ and is given by

$$\vec{P} = \chi\varepsilon_0\vec{E} \tag{xi}$$

where χ is constant of proportionality and is known as susceptibility of dielectric material.

By using Eq. (viii), we get

$$E = E_0 - \frac{\chi \varepsilon_0 E}{\varepsilon_0} = E_0 - \chi E$$

or $$E_0 = E(1+\chi) \quad \text{or} \quad E_0/E = 1+\chi$$

or $$K = 1+\chi \qquad [\because \; K = E_0/E]$$

10.4 TYPES OF POLARISATION

LO3

The important types of polarisation are categorised as under.

10.4.1 Electronic Polarisation

Under the action of an external field, the electron clouds of atoms are displaced with respect to heavy fixed nuclei to a distance less than the dimensions of the atom (Fig. 10.6). This is called *electronic polarisation*, which does not depend on temperature. The electronic polarisation is represented as below

$$\vec{P}_e = N\alpha_e \vec{E} \qquad \text{(i)}$$

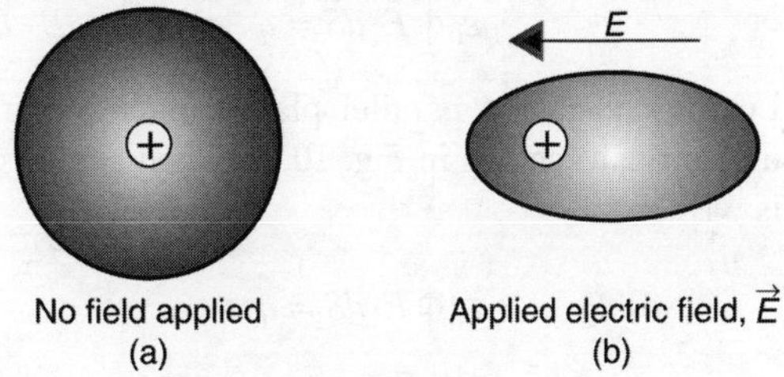

FIGURE 10.6

10.4.2 Ionic Polarisation

This type of polarisation occurs in ionic crystals, for example in sodium chloride crystal. In the presence of an external electric field, the positive and negative ions are displaced in opposite directions until ionic bonding forces stop the process (Fig. 10.7). This way, the dipoles get induced. The ionic polarisation does not depends upon temperature.

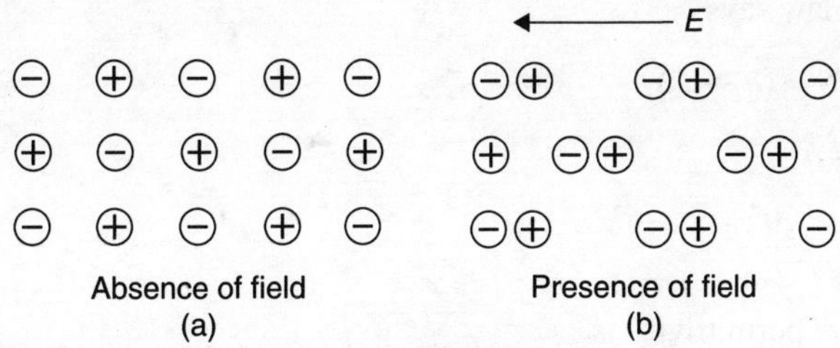

FIGURE 10.7

10.4.3 Orientation Polarisation

This types of polarisation is applicable in polar dielectrics. In the absence of an external electric field, the permanent dipoles are oriented randomly such that they cancel the effects of each other (Fig. 10.8a). When the electric field is applied, these dipoles tend to rotate and align in the direction of the applied filed (Fig. 10.8b). This is known as orientation polarisation, which depends upon temperature.

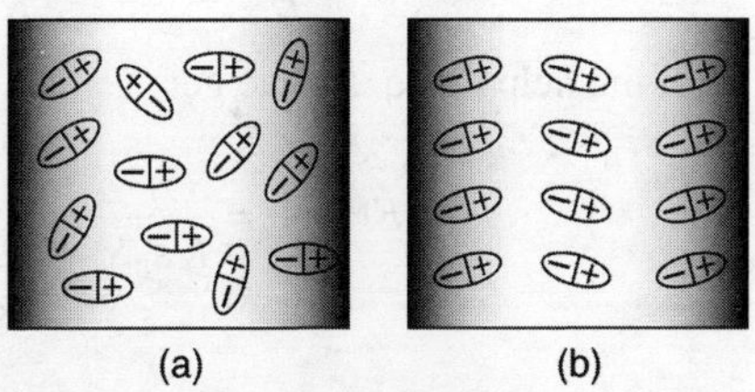

FIGURE 10.8

In view of all these polarisations, the total polarisation is the sum of the electronic, ionic and orientation polarisations. This is given by

$$\vec{P} = \vec{P}_e + \vec{P}_i + \vec{P}_o$$

10.5 GAUSS'S LAW IN DIELECTRICS LO4

Gauss's law states that "the surface integral of electric field vector over a closed surface is equal to $1/\varepsilon_0$ times the net charge enclosed by the surface", i.e.,

$$\oint \vec{E} \cdot d\vec{S} = \frac{1}{\varepsilon_0} q$$

or $$\varepsilon_0 \oint \vec{E} \cdot d\vec{S} = q \quad \text{or} \quad \oint \varepsilon_0 \vec{E} \cdot d\vec{S} = q$$

Let us consider a parallel plate capacitor without dielectric, as shown in Fig. 10.9a. Then Gauss's law is written as

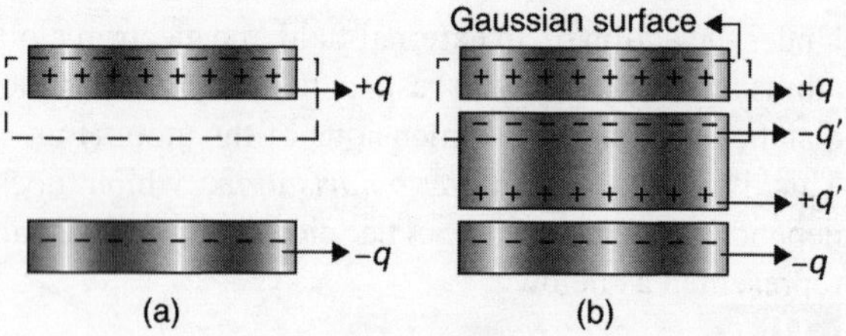

FIGURE 10.9

$$\varepsilon_0 \oint \vec{E} \cdot d\vec{S} = q$$

or $$\varepsilon_0 E_0 S = q$$

or $$E_0 = \frac{q}{\varepsilon_0 S} \tag{i}$$

where E_0 is the electric field between the plates.

In the presence of dielectric material between capacitor plates, it is clear from Fig. 10.9b that the total charge enclosed by Gaussian surface is $(q-q')$, where q' is the induced charge in the dielectric material due to polarisation. Then Gauss's law says

$$\varepsilon_0 \oint \vec{E} \cdot d\vec{S} = (q - q') \tag{ii}$$

or $$\varepsilon_0 E S = q - q'$$

or $$E = \frac{q}{\varepsilon_0 S} - \frac{q'}{\varepsilon_0 S} \tag{iii}$$

As we know that the relative permittivity is

$$K = \frac{E_0}{E} \quad \text{or} \quad E = \frac{E_0}{K} \tag{iv}$$

With the help of Eq. (i), we get

$$E = \frac{E_0}{K} = \frac{q}{K\varepsilon_0 S} \tag{v}$$

Using Eq. (v) in Eq. (iii) we obtain

$$\frac{q}{K\varepsilon_0 S} = \frac{q}{\varepsilon_0 S} - \frac{q'}{\varepsilon_0 S}$$

or $$\frac{q}{K} = q - q' \tag{vi}$$

The induced charge is therefore

$$q' = q\left(1 - \frac{1}{K}\right)$$

From the above relation it is clear that the induced charge q' is less than the free charge q.

By substituting the value of $(q - q') = \frac{q}{K}$ in Eq. (ii), we get for Gauss's law

$$\varepsilon_0 \oint \vec{E}\cdot d\vec{S} = \frac{q}{K} \quad \text{or} \quad \oint \vec{E}\cdot d\vec{S} = \frac{q}{\varepsilon_0 K} = \frac{q}{\varepsilon} \quad \text{or} \quad \oint \varepsilon \vec{E}\cdot d\vec{S} = q$$

i.e., $$\oint \vec{D}\cdot d\vec{S} = q$$

This is the Gauss's law in the presence of a dielectric.

10.6 DIELECTRIC LOSS LO5

When a dielectric material is placed in an alternating electric field (Fig. 10.10a), a part of the energy is wasted, which is known as *dielectric loss*. This is because of the fact that the reversing nature of the field causes the direction of the dipoles to reverse. The dielectric loss depends on the frequency and the mechanism by which the polarisation is produced in the material. An ideal dielectric does not absorb electrical energy. However, in a real dielectric, there is always a loss of some electrical energy.

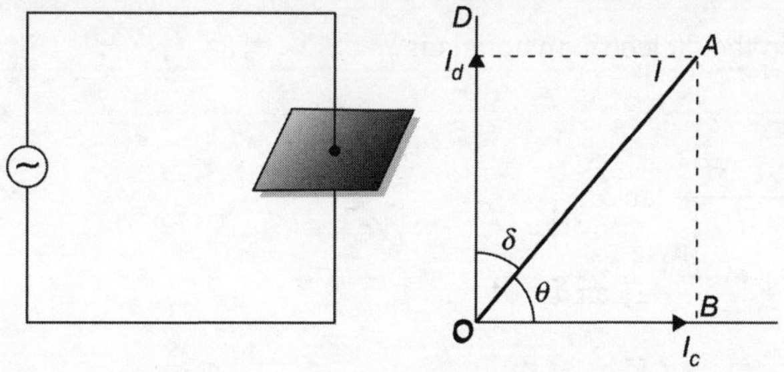

FIGURE 10.10

Consider a parallel plate capacitor of capacity C, whose plates having area S are separated by a distance d. The space between the plates of the capacitor is filled with dielectric material having permittivity ε. The sinusoidal voltage V of angular frequency ω is applied to the capacitor. Then the current through the capacitor is

$$I = \frac{Q}{t} + \frac{V}{R} = \frac{CV}{t} + \frac{V}{R}$$

Since ω is inversely proportional to the time period, the current I may be written as

$$I = \omega CV + \frac{V}{R}$$

or $$I = I_c + I_d \quad \text{(i)}$$

where I_c and I_d are conduction and displacement currents, respectively. From the above relation, it is clear that there are two kinds of currents that flow through the dielectric.

The Current I_c, I_d and I are plotted in Fig. 10.10b, from which it is clear that the resultant current $I = (I_c^2 + I_d^2)^{1/2}$ lags behind the displacement current by an angle δ. In an ideal dielectric $R = \infty$, which means $I_c = \dfrac{V}{R}$ becomes zero. In this situation the resultant current would be

$$I = I_d = \omega CV$$

Now $C = \dfrac{\varepsilon_0 S}{d}$ free space and $\dfrac{\varepsilon_0 \varepsilon_r S}{d}$ for the capacitor filled with dielectric material. Therefore,

$$I = I_d = \frac{\varepsilon_0 \varepsilon_r S \omega V}{d} \quad \text{(ii)}$$

The angle δ is known as *loss angle,* which can be calculated from Fig. 10.10b. In ΔOAD, $\tan\delta = \dfrac{I_c}{I_d}$

$$\text{so } I_c = I_d \tan\delta \quad \text{(iii)}$$

Eqs. (ii) and (iii) yield

$$I_c = \frac{(\omega \varepsilon_0 \varepsilon_r SV)\tan\delta}{d}$$

Thus, the real power loss in the dielectric materials is

$$P_l = VI_c$$

$$= \frac{\omega \varepsilon_0 \varepsilon_r SV^2}{d} \tan\delta$$

$$= \frac{\omega \varepsilon_0 \varepsilon_r (Sd)V^2}{d^2} \tan\delta$$

$$= 2\pi f \varepsilon_0 \varepsilon_r V \left(\frac{V}{d}\right)^2 \tan\delta$$

$$= (2\pi\varepsilon_0)\varepsilon_r f VE^2 \tan\delta$$

$$P_l = 5.54 \times 10^{-11} VE^2 \varepsilon_r f \tan\delta$$

The above expression shows that the power loss depends on the volume V of the dielectric, its dielectric constant ε_r, frequency f of the alternating field together with its amplitude E.

10.7 CLAUSIUS-MOSOTTI EQUATION LO6

The interaction between the atoms of gases can be neglected as the atoms are separated by sufficiently large distance. So atoms of gases feel the same field as applied to them. However, in solids and liquids, the atoms are closely surrounded (on all the sides) by other atoms which may be polarised under the action of an external field. Hence, the internal intensity of the electric field at a given point of the dielectric is generally not equal to the intensity of the applied field. The internal field is actually the electric field acting at the location of a given atom and is equal to the sum of the electric field created by neighbouring atoms and the applied field. In case of crystal possessing cubic symmetry, the internal field is given by

$$\vec{E}_i = \vec{E} + \vec{P}/3\varepsilon_0$$

This field is called *Lorentz field* and also some time referred to as local field. In linear and isotropic dielectric the molecular dipole moment $\vec{p}$ is directly proportional to the internal field E_i, that is

$$\vec{p} = \alpha \vec{E}_i$$

$$\vec{p} = \alpha(\vec{E} + \vec{P}/3\varepsilon_0) \quad \text{(i)}$$

As mentioned earlier, the proportionality constant α here is known as molecular polarisability. In terms of number of molecules per unit volume N, the polarisation density $\vec{P}$ is given by

$$\vec{P} = N\vec{p} = N\alpha(\vec{E} + \vec{P}/3\varepsilon_0) \quad \text{(ii)}$$

The above equation for $\vec{P}$ gives

$$\vec{P} = \frac{N\alpha}{1 - N\alpha/3\varepsilon_0}\vec{E} \quad \text{(iii)}$$

In terms of the electric susceptibility χ, the above equation is written as

$$\vec{P} = \varepsilon_0 \chi \vec{E} \quad \text{(iv)}$$

A comparison of Eqs. (iii) and (iv) yields

$$\frac{N\alpha}{1 - N\alpha/3\varepsilon_0} = \varepsilon_0 \chi \quad \text{(v)}$$

Along with the use of $K = 1 + \chi$ as $\varepsilon = \varepsilon_0 (1 + \chi)$ and $K = \varepsilon/\varepsilon_0$, Eq. (v) gives

$$\alpha = \frac{3\varepsilon_0(K - 1)}{N(K + 2)}$$

or

$$\frac{K - 1}{K + 2} = \frac{N\alpha}{3\varepsilon_0}$$

The above equation is known as *Clausius-Mosotti equation*. This equation is valid for non-polar dielectrics having cubic crystal structure. The Clausius-Mosotti equation is also known as *Lorentz-Lorentz* equation in view of its application in optics.

10.7.1 Physical Significance

Clausius–Mossotti equation or relation seems to hold best for gases but gives reasonably good results for many liquids and solids too. This is clear from the relation that it connects the relative permittivity K of a dielectric medium to the polarizability α of the atoms (or molecules) constituting the dielectric. Since the relative permittivity is a bulk or macroscopic property and the polarisability is a microscopic property of matter, the Clausius-Mosotti equation bridges the gap between a directly observable macroscopic property with a microscopic molecular property.

SUMMARY

The topics covered in this chapter are summarised below.

- A dielectric is an insulating material in which all the electrons are tightly bound to the nuclei of the atoms and there are no free electrons available for the conduction of current. Therefore, the electrical conductivity of a dielectric is very low. The forbidden gap (E_g) is very large in dielectrics. Materials such as glass, polymers, mica, oil and paper are a few examples of dielectrics.
- A non-polar molecule is the one in which the centre of gravity of the positive charge (protons) and negative charge (electrons) coincide. So such molecules do not have any permanent dipole moment. Nitrogen (N_2) and hydrogen (H_2) are the examples of non-polar molecules.
- A polar molecule is the one in which the centre of gravity of the positive charges is separated by finite distance from that of the negative charges. Unbalanced electric charges, usually valence electrons, of such molecules result in a dipole moment and orientation. Therefore, these molecules possess permanent electric dipole. Examples of polar molecules are N_2O, H_2O and HCl.
- An external electric field, when applied to a dielectric material, exerts a force on each charged particle and pushes the positive charge in its own direction while the negative charge is displaced in opposite direction. Consequently, the centres of positive and negative charges of each atom are displaced from their equilibrium positions. Such a molecule (or atom) is then called as induced electric dipole and this process is known as dielectric polarisation.
- The induced dipole moment developed per unit volume in a dielectric on placing it inside an electric field is known as polarisation density $\vec{P}$. If N be the number of atoms in a unit volume and α the atomic polarisability, then polarisation density is $\vec{P} = N\alpha\varepsilon_0\vec{E}$.
- Electric susceptibility χ and the dielectric constant K are related as $K = 1 + \chi$.
- Polarisation is of three types, namely electronic polarisation, ionic polarisation and orientation polarisation.
- Gauss's law states that the surface integral of electric field vector $\vec{E}$ over a closed surface is equal to $1/\varepsilon_0$ times the net charge enclosed by the surface, i.e., $\oint \vec{E}\cdot dS = \dfrac{q}{\varepsilon_0}$
- The energy stored in an electrostatic field $\vec{E}$ is $u = \dfrac{1}{2}\varepsilon_0 KE^2$ which takes the form $u = \dfrac{1}{2}\varepsilon_0 E^2$ in the free space.
- The internal intensity of the electric field at a given point of the dielectric is generally not equal to the intensity of the applied field. The internal field is actually the electric field acting at the location

of a given atom and is equal to the sum of the electric field created by neighbouring atoms and the applied field. In case of crystal possessing cubic symmetry, a relation $\frac{K-1}{K+2}=\frac{N\alpha}{3\varepsilon_0}$ exists between the dielectric constant K, atomic polarisability α and permittivity ε_0. This equation is known as Clausius-Mosotti equation, which is also known as Lorentz-Lorentz equation in view of its application in optics.

SOLVED EXAMPLES

EXAMPLE 1 Two parallel plates having equal and opposite charges are separated by a 2 cm thick slab that has dielectric constant 3. If the electric filed inside is 10^6 V/m. Calculate the polarisation and displacement vector.

SOLUTION Given $\vec{E}=10^6\text{ V/m}=10^6\text{ N/C}, K=3, \varepsilon_0=8.85\times10^{-12}\text{C}^2\text{N}^{-1}\text{m}^{-2}$

Formula used is $\vec{D}=\varepsilon_0\vec{E}+\vec{P}$

Also $\vec{D}=\varepsilon_0 K\vec{E}$

$$\text{or}\quad D=8.85\times10^{-12}\times3\times10^6$$
$$=2.655\times10^{-5}\text{C/m}^2$$
$$\vec{P}=\vec{D}-\varepsilon_0\vec{E}$$
$$=2.655\times10^{-5}-8.85\times10^{-12}\times10^6$$
$$=\mathbf{1.77\times10^{-5}\ C/m^2}$$

EXAMPLE 2 Two parallel plates have equal and opposite charges. When the space between them is evacuated the electric intensity is 3×10^5 V/m and when the space is filled with dielectric the electric intensity is 1.0×10^5 V/m. What is the included charge density on the surface of the dielectric?

SOLUTION Given $\vec{E}_0=3\times10^5$ V/m, and $\vec{E}=1\times10^5$ V/m

Formula used is

$$\vec{E}=\vec{E}_0-\frac{\vec{P}}{\varepsilon_0}\quad\text{or}\quad P=\sigma_p=\varepsilon_0(\vec{E}_0-\vec{E})\qquad[\because P=\sigma_p]$$
$$\sigma_p=8.85\times10^{-12}\,[3-1]\times10^5$$
$$\sigma_p=\mathbf{1.77\times10^{-6}\,C/m^2}$$

EXAMPLE 3 Two parallel plates of capacitor having equal and opposite charges are separated by 6.0 mm thick dielectric material of dielectric constant 2.8. If the electric field strength inside be 10^5 V/m, determine polarisation vector, displacement vector and energy density in the dielectric.

SOLUTION Given E = 10^5 V/m = 10^5 N/C and K = 2.8

$$\vec{P}=\varepsilon_0(K-1)\vec{E},\ \vec{D}=\varepsilon_0\,K\vec{E}\ \text{ and energy density}=\frac{1}{2}K\,\varepsilon_0E^2$$
$$P=8.85\times10^{-12}\times(2.8-1)\times10^5=1.593\times10^{-6}\,\text{C/m}^2$$
$$=1.6\times10^{-6}\,\text{C/m}^2$$
$$D=8.85\times10^{-12}\times2.8\times10^5=2.478\times10^{-6}\,\text{C/m}^2$$
$$=2.5\times10^{-6}\,\text{C/m}^2$$

$$\text{Energy density} = \frac{1}{2} \times 2.8 \times 8.85 \times 10^{-12} \times (10^5)^2$$

$$= 12.39 \times 10^{-2}\,\text{J/m}^3$$

$$= \mathbf{0.124\ J/m^3}$$

EXAMPLE 4 An isotropic material of relative permittivity ε_r is placed normal to a uniform external electric field with an electric displacement vector of magnitude $5 \times 10^{-4}\,\text{C/m}^2$. If the volume of the slab is $0.5\,\text{m}^3$ and magnitude of polarisation is $4 \times 10^{-4}\,\text{C/m}^2$, find the value of ε_r and total dipole moment of the slab.

SOLUTION Given $\vec{D} = 5 \times 10^{-4}\,\text{C/m}^2$, $\vec{P} = 4 \times 10^{-4}\,\text{C/m}^2$ and $V = 0.5\,\text{m}^3$ $\varepsilon_r = K = ?$

Formula used is

$$\because \quad \vec{D} = \varepsilon_0 \vec{E} + \vec{P}$$

$$\vec{E} = (\vec{D} - \vec{P})/\varepsilon_0$$

$$= \frac{(5-4) \times 10^{-4}}{8.85 \times 10^{-12}}$$

$$= 1.13 \times 10^7\ \text{V/m}$$

$$\text{or} \quad K = \varepsilon_r = \frac{D}{\varepsilon_0 E} = \frac{5 \times 10^{-4}}{10^{-4}} = 5$$

$$P = \frac{\text{Total dipole moment}}{\text{Volume}} = \frac{p}{V}$$

$$p = PV = 4 \times 10^{-4} \times 0.5 = \mathbf{2.0 \times 10^{-4}\ C\text{-}m}$$

EXAMPLE 5 Dielectric constant of a gas at N.T.P is 1.00074. Calculate dipole moment of each atom of the gas when it is held in an external field of $3 \times 10^4\,\text{V/m}$.

SOLUTION Given $\vec{E} = 3 \times 10^4\ \text{V/m} = 3 \times 10^4\ \text{N/C}$ and $K = \varepsilon_r = 1.00074$

Formula used is $K = 1 + \chi$

$$\text{or} \quad \chi = K - 1 = 1.00074 - 1 = 0.00074$$

and polarisation density is

$$P = \chi\, \varepsilon_0\, E = 0.74 \times 10^{-3} \times 8.85 \times 10^{-12} \times 3 \times 10^4$$

$$= 19.647 \times 10^{-11}\,\text{C/m}$$

No. of atoms of gas per cubic metre (N)

$$= \frac{6.06 \times 10^{23}}{22.4 \times 10^{-3}} = 2.7 \times 10^{25}$$

$$\text{Induced dipole moment of each atom } (p) = \frac{P}{N} = \frac{19.647 \times 10^{-11}}{2.7 \times 10^{25}}$$

$$\text{or} \quad p = \mathbf{7.27 \times 10^{-36}\ C\text{-}m}$$

EXAMPLE 6 Determine the electric susceptibility at 0°C for a gas whose dielectric constant at 0°C is 1.000041.

SOLUTION Given $K = 1.000041$ and $T = 0°\text{C}$

Formula used is

$$K = 1 + \chi$$

$$\text{or} \quad \chi = K - 1 = 1.000041 - 1 = 0.41 \times 10^{-4}$$

$$= \mathbf{4.1 \times 10^{-5}}$$

OBJECTIVE TYPE QUESTIONS

Q.1 Dimension of atomic polarisability in SI units is
(a) Cm^{-2} (b) $CV^{-1}m^2$ (c) CVm (d) none of them

Q.2 A non-polar molecule is the one in which the centre of gravity of positive and negative charges
(a) coincides (b) gets separated by 10^{-8} m
(c) gets separated by 1Å (d) none of these

Q.3 The net charge inside a dielectric before and after polarisation remains
(a) negative (b) positive (c) same (d) none of these

Q.4 In vacuum, electric susceptibility is
(a) less than 1 (b) greater than 1 (c) small but –ve (d) zero

Q.5 The electric susceptibility and dielectric constant are related as
(a) $\chi = K \sim 1$ (b) $\chi = 1 + K$ (c) $\chi = \varepsilon_0 K$ (d) $\chi = 1 + \varepsilon_0 K$

Q.6 Dimension of displacement vector in SI unit is
(a) C-m (b) C-m^{-1} (c) C-m^{-2} (d) C-m^2

Q.7 The relation between three electric vectors $\vec{E}$, $\vec{D}$ and $\vec{P}$ is
(a) $\vec{D} = \varepsilon_0 \vec{E} + \vec{P}$ (b) $\vec{D} = \varepsilon_0(\vec{E} + \vec{P})$ (c) $\vec{D} = \vec{E} + \varepsilon_0 \vec{P}$ (d) $\vec{D} = \frac{1}{\varepsilon_0}\vec{E} + \vec{P}$

Q.8 Polarisation density is
(a) dipole moment per unit area (b) surface charge density
(c) dipole moment per unit volume (d) (b) and (c)

Q.9 Which of the following relation reperesent the Clausius-Mosotti equation
(a) $\frac{K-1}{K+2} = \frac{N\alpha}{3\varepsilon_0}$ (b) $\frac{K+1}{K-2} = N\alpha$ (c) $\frac{K+2}{K-1} = \frac{N\alpha}{3\varepsilon_0}$ (d) None of these

Q.10 Which type of polarisation is depend on temperature
(a) Orientational (b) Electronic (c) Ionic (d) none of these

Q.11 Gauss's law in dielectrics is
(a) $\oint \vec{E} \cdot d\vec{S} = \frac{1}{\varepsilon_0} q$ (b) $\oint \vec{D} \cdot d\vec{S} = q$ (c) $\oint \vec{E} \cdot d\vec{S} = \varepsilon_0 q$ (d) both (a) and (b)

Q.12 Dimension of permittivity of free space in SI unit is
(a) Fm^{-1} (b) Cm^{-1} (c) Hm^{-1} (d) dimensionless

Q.13 Dimension of relative permittivity in SI unit is
(a) Fm^{-1} (b) Cm^{-1} (c) Hm^{-1} (d) dimensionless

Q.14 What changes in the capacitance of a capacitor occur if the dielectric material between the plates of the capacitor is replaced by air or vacuum?
(a) increases (b) decreases (c) remains same (d) none of these

Q.15 Dielectrics are the substances which are
(a) conductors (b) insulators (c) semi-conductor (d) none of these

Q.16 Which of the statement is true?
(a) Ionic polarisation decreases with increase of temperature
(b) Ionic polarisation does not depend on temperature
(c) Ionic polarisation increases with rise in temperature
(d) none of these

Q.17 Which of the following statement is true?
(a) Orientational polarisation decreases with rise in temperature
(b) Orientational polarisation increases with rise in temperature
(c) Orientational polarisation does not depends on temperature
(d) none of these

Short-Answer Questions

Q.1 What are polar and non-polar molecules?
Q.2 Discuss different polarisation mechanisms in dielectrics?
Q.3 What is a dielectric?
Q.4 How long does polarisation of non-polar molecules last?
Q.5 State and prove Gauss law in dielectrics.
Q.6 What are dielectric losses?
Q.7 Discuss the behaviour of a dielectric in a.c. field.
Q.8 Write the Clausius-Mossotti equation.
Q.9 Will an atom having spherically symmetric charge distribution be polar or non-polar? Explain.
Q.10 What do you understand by polarisation of dielectric and dielectric susceptibility? Find the relation between the two.
Q.11 Write note on
(i) Dielectrics (ii) Three electric vectors (iii) Dielectric losses.

Practice Problems

General Questions

Q.1 What is a dielectric substance? Give examples. Discuss the importance of dielectrics.
Q.2 What are polar and non-polar molecules? Discuss the effect of electric field on polar dielectrics. What is meant by polarisation of dielectric?
Q.3 Discuss different types of polarisations in dielectrics.
Q.4 What happens when a non-polar molecule is placed in an electric field? Define atomic dipole moment and atomic polarisability. What are their dimensions? Give their S.I. units.
Q.5 What is atomic polarisability? Find a relation between dipole moment and atomic polarisability or show that $\vec{p} = \alpha \vec{E}_0$.

Q.6 Show that the electric field inside a polarised dielectric due to induced polarisation charge is $\vec{E} = -\frac{\vec{P}}{\varepsilon_0}$ where $\vec{P}$ is the polarisation density vector.

Q.7 Explain the terms dielectric polarisation, susceptibility, permittivity and dielectric coefficient. Derive their inter-relation equation.

Q.8 Define and explain the three electric vectors $\vec{P}, \vec{E}$ and $\vec{D}$. Why electric field inside a dielectric decreases due to polarisation? Show that $\vec{D} = \varepsilon_0\vec{E} + \vec{P}$. Also give their units.

Q.9 Show that $\vec{D} = \varepsilon_0\vec{E} + \vec{P}$, where the symbols have their usual meanings.

Q.10 What are three electric vectors in dielectrics? Name and find relation between them.

Q.11 What do you understand by polarisation of dielectric and dielectric susceptibility? Find the relation between the two.

Q.12 Explain the phenomenon of polarisation of dielectric medium and show that $K = 1 + \chi_e$. Here the symbols have their usual meanings.

Q.13 Define the terms dielectric constant K and electric susceptibility χ_e. Prove the relation $K = 1 + \chi_e$.

Q.14 Find the relation between induced charge and free charge when a dielectric material of dielectric constant K is placed between the plates of a parallel plate capacitor.

Q.15 Prove that induced charge varies with the dielectric as $K = \left[1 - \frac{\sigma_p}{\sigma_{\text{free}}}\right]^{-1}$, where σ_p and σ_{free} are the induced and free surface charge densities, respectively. Hence show that for a metal $K = \infty$.

Q.16 The electric field between the plates of a parallel plate capacitor is $\vec{E}_0$ without dielectric. But if dielectric of relative permittivity ε_r is introduced between the plates what will the electric field be?

Q.17 What is the effect of temperature on the dielectric constant of a substance containing molecules of permanent dipole moment?

Q.18 Derive a relation between electric susceptibility and atomic polarisability on the basis of microscopic description of matters at atomic level.

Q.19 Derive Clausius-Mosotti relation for non-polar dielectrics.

Q.20 Discuss the effect of introducing a dielectric between the plates of a capacitor. Show that the capacitance of a charged capacitor when a dielectric material of dielectric constant K is introduced between the plates is given by $K\,\varepsilon_0 \frac{A}{d}$.

Q.21 Explain why the introduction of a dielectric slab between the plates of a capacitor changes its capacitance?

Q.22 State and prove Gauss's law in dielectrics.

Q.23 Derive an expression for Gauss's law in the presence of dielectric. Prove that divergence of displacement vector is equal to density of free charge or $\vec{\nabla} \cdot \vec{D} = \rho_{\text{free}}$. Also discuss integral form of Gauss's law.

Q.24 Using Gauss's law in dielectric medium, show that $\vec{\nabla} \cdot \vec{D} = \rho_{\text{free}}$, where symbols have their usual meanings.

Q.25 Explain the mechanism contributing to dielectric polarisation. Discuss the behaviour of a dielectric in an alternating field.

Q.26 Show that the electrostatic energy per unit volume in a dielectric is $\frac{1}{2}\vec{D} \cdot \vec{E}$, where symbols have their usual meanings.

Q.27 Deduce an expression for energy stored in dielectric in electrostatic field.

Electromagnetism

11

LEARNING OBJECTIVES

After reading this chapter you will be able to

- **LO 1** Understand charge density, del operator, gradient, divergence and curl
- **LO 2** Explain fundamental theorem of calculus and for gradient
- **LO 3** Know about Gauss's theorem and its functional correlations
- **LO 4** Discuss Stokes theorem and its use in calculating electric field and electric potential
- **LO 5** Know how to derive Poisson's and Laplace's equations
- **LO 6** Know about capacitor and their important configurations
- **LO 7** Understand the concepts of magnetic flux density, magnetic field strength and Ampere's circuital law
- **LO 8** Explain electrostatic boundary conditions
- **LO 9** Discuss scalar and vector potential, continuity equation, maxwell's equation in differential and integral form
- **LO 10** Evaluate significance of Maxwell's equation, Maxwell's displacement current and correction in Ampere's Law
- **LO 11** Learn about electromagnetic wave propagation, Transverse nature EMW, Maxwell's equation in Isotropic dielectric medium/conducting medium, electromagnetic energy density
- **LO 12** Elaborate on Poynting vector and Poynting theorem, waveguide, coaxial cables

Introduction

Ordinary matter is made up of atoms which have positively charged nuclei and negatively charged electrons surrounding them. Charge is quantized in terms of the electronic charge –*e*. One electronic charge e is equal to 1.602×10^{-19} Coulombs. One Coulomb of charge is the charge which would flow through a 220 W light bulb (220 Vac) in one second. Do you know two charges of one Coulomb each separated by one meter would repel each other with a force of about a million tons! The separation of charges produces electric field, whereas the motion of charges generates current and hence the magnetic field. When these fields are time varying they are coupled with each other through the Maxwell's equations.

With the help of the Maxwell's equations, we can derive wave equation, based on which the propagation of electromagnetic waves can be investigated in different media.

11.1 CHARGE DENSITY

LO1

If a charge is distributed continuously in a medium, it can be expressed in terms of a physical quantity known as charge density. There are three types of charge densities, namely linear charge density, surface charge density and volume charge density.

11.1.1 Linear Charge Density (λ)

If a charge is distributed continuously on a linear conductor, the charge on its unit length is called the *linear charge density*. It is generally represented by λ and is measured in the units of C/m. If l be the total length of a conductor and λ the linear charge density, then the total charge on the conductor will be

$$q = \int_0^l \lambda dl$$

For a uniform charge distribution, λ is constant, and is given as $\lambda = q/l$. Here q is the total charge, given by $q = \lambda l$.

11.1.2 Surface Charge Density (σ)

If the charge is distributed over a surface, the charge on the unit area of the conductor is called the *surface charge density* (σ) and its unit is C/m^2.

If the surface charge density at a point of the conductor is σ, then the charge contained in a small element of area dS will be $\sigma\, d$S. Therefore, the total charge on the surface of the conductor,

$$q = \int_S \sigma dS$$

If the charge distribution is uniform, then the value of σ will be constant, and is given by

$$\sigma = q/S$$

where S is total area of the surface and q is the total charge, given by $q = \sigma S$.

11.1.3 Volume Charge Density (ρ)

If the charge is distributed in the volume of a conductor, the charge contained in a unit volume of the conductor is called the *volume charge density* (ρ) and its unit is C/m^3.

If the volume charge density at a point of the volume of a conductor is ρ, then the charge contained in a small element of the volume dV will be $\rho\, d$V. Therefore, the total charge contained in the conductor,

$$q = \int_V \rho dV$$

For the uniform distribution of the charge, ρ will be constant, and is given by

$$\rho = q/V$$

or

$$q = \rho V$$

where q is the total charge and V is the total volume, which is equal to $\frac{4}{3}\pi r^3$ for a spherical shape of the conductor of radius r.

11.2 DEL OPERATOR

LO1

The del operator is the differential operator, which is represented by $\vec{\nabla}$ and is given by $\vec{\nabla} = \hat{i}\frac{\partial}{\partial x} + \hat{j}\frac{\partial}{\partial y} + \hat{k}\frac{\partial}{\partial z}$.

It is not a vector in itself, but when it operates on a scalar function it provides the resultant as a vector. For example, when $\vec{\nabla}$ is operated on a scalar function $F(x, y, z)$, we get

$$\vec{\nabla}F = \hat{i}\frac{\partial F}{\partial x} + \hat{j}\frac{\partial F}{\partial y} + \hat{k}\frac{\partial F}{\partial z}$$

$\vec{\nabla}F$ does not mean a multiplication of $\vec{\nabla}$ with F rather it is an instruction to differentiate. Here we should say that $\vec{\nabla}$ is a vector operator that acts upon F.

The del operator $\vec{\nabla}$ can act in different ways. For example, when it acts on a scalar function F, the resultant $\vec{\nabla}F$ is called the gradient of a scalar function F. When it acts on a vector function $\vec{A}$ via the dot product, the resultant is $\vec{\nabla}\cdot\vec{A}$, which is called the divergence of a vector $\vec{A}$. When it acts on a vector function $\vec{A}$ via the cross product, the resultant is $\vec{\nabla}\times\vec{A}$, which is called the curl of a vector $\vec{A}$. Finally, the Laplacian of a scalar function F is written as $\nabla^2 F$ together with $\nabla^2 = \frac{\partial^2}{\partial x^2} + \frac{\partial^2}{\partial y^2} + \frac{\partial^2}{\partial z^2}$.

The del operator can be written in spherical polar coordinate system (Fig. 11.1a) as

$$\vec{\nabla} = \hat{r}\frac{\partial}{\partial r} + \hat{\theta}\frac{1}{r}\frac{\partial}{\partial \theta} + \hat{\phi}\frac{1}{r\sin\theta}\frac{\partial}{\partial \phi}$$

In cylindrical coordinate system (Fig. 11.1b), it is given by

$$\vec{\nabla} = \hat{s}\frac{\partial}{\partial s} + \hat{\phi}\frac{1}{s}\frac{\partial}{\partial \phi} + \hat{z}\frac{\partial}{\partial z}$$

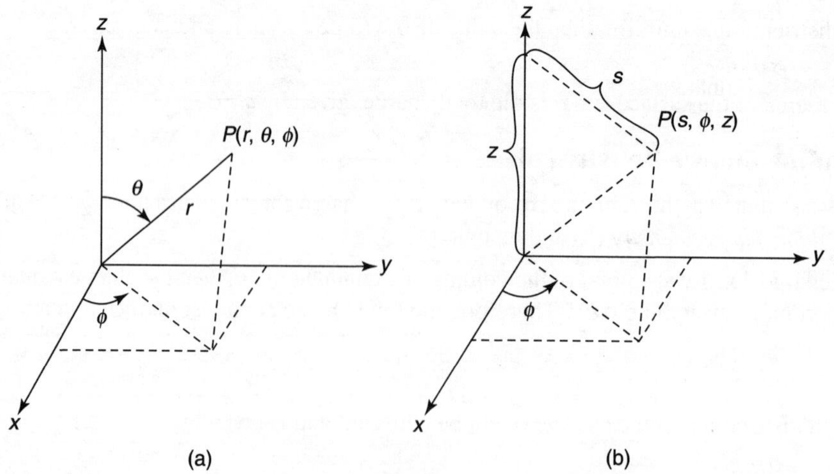

FIGURE 11.1

11.3 GRADIENT

LO1

If we think of the derivative of a function of one variable we notice that it simply tells us how fast the function varies if we move a small distance. It means the gradient is the rate of change of a quantity with distance. For example, temperature gradient in a metal bar is the rate of change of temperature along the bar. However, for a function of three variables the situation is more complicated, as it will depend on what direction we choose to move. For a function $F(x, y, z)$ of three variables, we obtain from a theorem on partial derivative

$$dF = \frac{\partial F}{\partial x}dx + \frac{\partial F}{\partial y}dy + \frac{\partial F}{\partial z}dz$$

Here dF is a measure of changes in F that occurs when we alter all three variables by small amounts dx, dy and dz.

The above expression for dF can be written in terms of a dot product of vectors as

$$dF = \vec{\nabla}F \cdot d\vec{l}$$

where $\vec{\nabla}F = \hat{i}\frac{\partial F}{\partial x} + \hat{j}\frac{\partial F}{\partial y} + \hat{k}\frac{\partial F}{\partial z}$ is nothing but gradient of F. Clearly the gradient is a vector quantity, i.e., it has both magnitude and direction. The meaning of gradient becomes clearer when we write

$$dF = \vec{\nabla}F \cdot d\vec{l}$$

or

$$dF = |\vec{\nabla}F||d\vec{l}|\cos\alpha$$

where α is the angle between $\vec{\nabla}F$ and $d\vec{l}$. Now we fix the distance dl, i.e., magnitude $|d\vec{l}|$, and look around in various directions for the maximum change in F. Clearly the maximum change in F takes place in the direction $\alpha = 0$. It means dF is largest when we move in the direction of $\vec{\nabla}F$. This can also be said that the gradient $\vec{\nabla}F$ points in the direction of maximum increase of the function F. Hence, the gradient of a scalar field F is a vector quantity which represents both the magnitude and the direction of the maximum space rate of increase of F.

The gradient of F in Cylindrical coordinate system is written as

$$\vec{\nabla}F = \frac{\partial F}{\partial s}\hat{s} + \frac{1}{s}\frac{\partial F}{\partial \phi}\hat{\phi} + \frac{\partial F}{\partial z}\hat{z}$$

In spherical polar coordinate system, the gradient of F is written as

$$\vec{\nabla}F = \frac{\partial F}{\partial r}\hat{r} + \frac{1}{r}\frac{\partial F}{\partial \theta}\hat{\theta} + \frac{1}{r\sin\theta}\frac{\partial F}{\partial \phi}\hat{\phi}$$

11.4 DIVERGENCE

LO1

As mentioned earlier, the divergence of a vector field $\vec{A}$ is represented by $\vec{\nabla} \cdot \vec{A}$. So it is given by

$$\vec{\nabla} \cdot \vec{A} = \frac{\partial A_x}{\partial x} + \frac{\partial A_y}{\partial y} + \frac{\partial A_z}{\partial z}$$

Clearly the divergence of a vector field is a scalar. Also, the divergence of a scalar cannot be obtained, as the dot product of $\vec{\nabla}$ with a scalar is not possible.

In order to make clear the meaning of the divergence, we consider the net flux $\oint \vec{A} \cdot d\vec{S}$ of a vector field $\vec{A}$ from a closed surface S. The divergence of $\vec{A}$ is defined as the net outward flux per unit volume over a closed surface. Actually the divergence of $\vec{A}$ at a given point is a measure of how much the vector $\vec{A}$ spreads out, i.e., diverges, from that point. Fig. 11.2a shows that the divergence of a vector field at point O is positive, as the vector spreads out. However, in Fig. 11.2b, the vector converges and hence the divergence at O is negative. In Fig. 11.2c, we can notice that the divergence of a vector field is zero, as the magnitude of vectors remains the same.

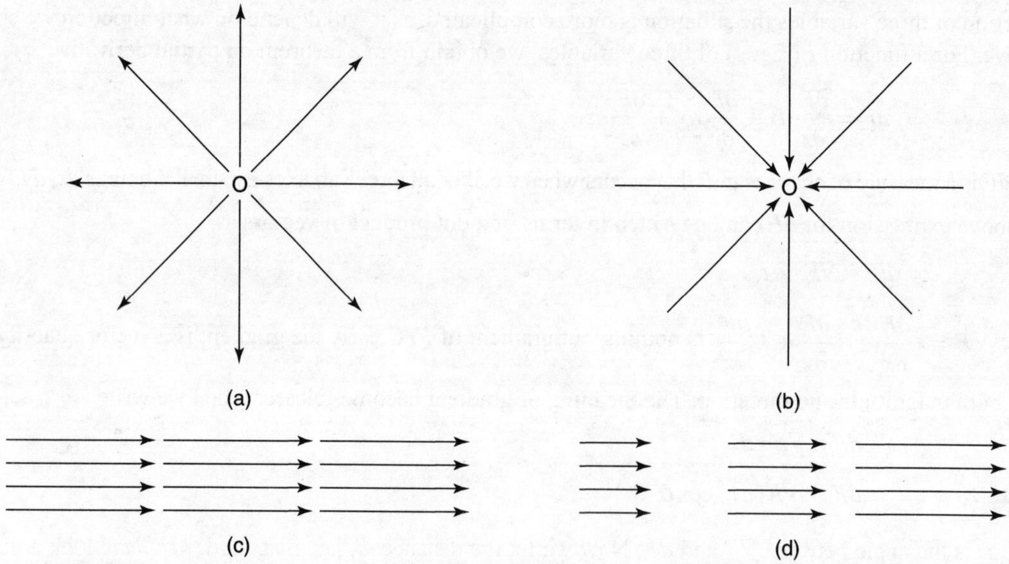

FIGURE 11.2

By now the difference between vector field and vector would have been clear to you. We call $\vec{A}$ as vector field because its values at different points are different, which are vectors with different magnitudes. For example, in Fig. 11.2d the magnitudes of the vectors get increased as we move towards right. Hence, the divergence of such a vector field is not zero but positive.

The divergence of a vector $\vec{A}$ in cylindrical coordinate system can be written as

$$\vec{\nabla}\cdot\vec{A} = \frac{1}{s}\frac{\partial(sA_s)}{\partial s} + \frac{1}{s}\frac{\partial A_\phi}{\partial \phi} + \frac{\partial A_z}{\partial z}$$

In spherical polar coordinate system, it is written as

$$\vec{\nabla}\cdot\vec{A} = \frac{1}{r^2}\frac{\partial}{\partial r}(r^2 A_r) + \frac{1}{r\sin\theta}\frac{\partial}{\partial\theta}(\sin\theta\, A_\theta) + \frac{1}{r\sin\theta}\frac{\partial A_\phi}{\partial\phi}$$

Finally, we mention that the vector field $\vec{A}$ is said to be solenoidal or divergenceless if $\vec{\nabla}\cdot\vec{A} = 0$.

11.5 CURL

LO1

As mentioned earlier, the curl of vector field $\vec{A}$ is represented by $\vec{\nabla}\times\vec{A}$. So it is given by

$$\vec{\nabla}\times\vec{A} = \hat{i}\left(\frac{\partial A_z}{\partial y} - \frac{\partial A_y}{\partial z}\right) + \hat{j}\left(\frac{\partial A_x}{\partial z} - \frac{\partial A_z}{\partial x}\right) + \hat{k}\left(\frac{\partial A_y}{\partial x} - \frac{\partial A_x}{\partial y}\right)$$

Clearly the curl is a vector quantity, i.e., it has both direction and magnitude. Also, it is evident that we cannot have the curl of a scalar as the cross product of $\vec{\nabla}$ with a scalar is meaningless.

In order to make clear the meaning of curl, we consider the circulation of a vector field $\vec{A}$ around a closed path, i.e., $\oint \vec{A}\cdot d\vec{l}$. It is evident that the curl of $\vec{A}$ is a rotational vector. Its magnitude would be the maximum circulation of $\vec{A}$ per unit area. Its direction is the normal direction of the area when the area is oriented so as to make the circulation maximum. Actually curl of $\vec{A}$ at some point O is a measure of how much the vector $\vec{A}$ curls around the point O. So, you may notice that the curl of vector field $\vec{A}$ in Fig. 11.2 is zero. However, in Fig. 11.3, the curl is finite and is pointing in the x-direction as per right hand rule.

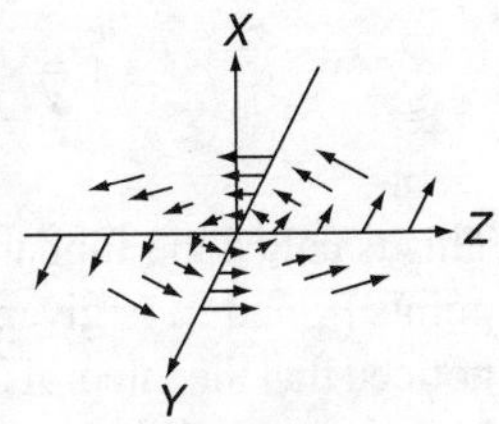

Figure 11.3

The curl of a vector $\vec{A}$ in cylindrical coordinate system can be written as

$$\vec{\nabla}\times\vec{A}=\left(\frac{1}{s}\frac{\partial A_z}{\partial\phi}-\frac{\partial A_\phi}{\partial z}\right)\hat{s}+\left(\frac{\partial A_s}{\partial z}-\frac{\partial A_z}{\partial s}\right)\hat{\phi}+\frac{1}{s}\left(\frac{\partial}{\partial s}(sA_\phi)-\frac{\partial A_s}{\partial\varphi}\right)\hat{z}$$

In spherical polar coordinate system, it is written as

$$\vec{\nabla}\times\vec{A}=\frac{1}{r\sin\theta}\left[\frac{\partial}{\partial\theta}(\sin\theta A_\phi)-\frac{\partial A_\theta}{\partial\phi}\right]\hat{r}+\frac{1}{r}\left(\frac{1}{\sin\theta}\frac{\partial A_r}{\partial\phi}-\frac{\partial}{\partial r}(rA_\phi)\right)\hat{\theta}+\frac{1}{r}\left[\frac{\partial}{\partial r}(rA_\theta)-\frac{\partial A_r}{\partial\theta}\right]\hat{\phi}$$

Finally, we mention that the vector field $\vec{A}$ is said to be irrotational or potential if $\vec{\nabla}\times\vec{A}=0$.

11.6 FUNDAMENTAL THEOREM OF CALCULUS LO2

If g(x) be a function of one variable, then dg would represent infinitesimal change in g (x) when we move from x to $x + dx$. This change in g(x) is given by $dg=\left(\frac{dg}{dx}\right)dx$. If we move from point a_1 to point a_2, then the total change in g(x) can be obtained by using the fundamental theorem of calculus. This theorem states that

$$\int_{a_1}^{a_2}\left(\frac{dg}{dx}\right)dx=g(a_2)-g(a_1)$$

It means the total change in the function can be obtained by simply subtracting the values of the function at the points a_2 and a_1. For the movement from point a_1 to point a_2, these points can be treated as the end points. In view of this, the fundamental theorem says that the integral of a derivative $\left(\frac{dg}{dx}\right)$ over an interval $a_1\to a_2$ is given by the value of the function at the end points. The end points represent the boundaries.

Since we have three types of derivatives, namely gradient, divergence and curl in vector calculus, there are three fundamental theorems for these derivatives. The fundamental theorem for divergence is also known as Gauss's theorem or Green's theorem. Similarly, the fundamental theorem for curl is also known as Stokes' theorem.

11.7 FUNDAMENTAL THEOREM FOR GRADIENT LO2

We know that the gradient is defined only for a scalar function. So we consider a scalar function of three variables $F(x, y, z)$. As discussed earlier, the change in this function is defined as $dF=\vec{\nabla}F\cdot d\vec{l}$. If we move from point a_1 to point a_2 then the total change in F can be calculated using fundamental theorem of calculus. Therefore, the total change in F in moving from a_1 to a_2 is given by

$$\int_{a_1}^{a_2} \vec{\nabla} F \cdot d\vec{l} = F(a_2) - F(a_1)$$

This is called the fundamental theorem for gradient, according to which the integral of a derivative over end points a_1 and a_2 is given by the value of the function at the boundaries, i.e., points a_1 and a_2. Here it can be noticed that the integral is line integral, the derivative is the gradient and the boundaries are the points a_1 and a_2. Moreover, it can be seen that the integral or the total change in function F is independent of path taken from a_1 to a_2. Also, if we take both the end points same, i.e., we evaluate the close integral, then the total change in the function F comes out to be zero (as the beginning and end points are identical).

11.8 GAUSS'S OR GREEN'S THEOREM

LO3

According to this theorem,

$$\int_v (\vec{\nabla} \cdot \vec{F}) dV = \oint_s \vec{F} \cdot d\vec{S}$$

Here dV is the volume element and $d\vec{S}$ is the surface element. In the same manner as we stated fundamental theorem for gradient, the Green's theorem states that the integral of a derivative (here the divergence) over a region (here the volume) is equal to the value of the function at the boundary (here the surface). Since the boundary of a volume is always a closed surface, the R.H.S. is the integral over closed surface.

Evidently this theorem converts the volume integral into the surface integral. Therefore, this theorem is very useful in the situations where it is difficult to calculate the volume integral.

The net outward electric flux through any closed surface drawn in an electric field is equal to $\frac{1}{\varepsilon_0}$ times the total charge enclosed within the surface. It is expressed as

$$\phi = \oint_s \vec{E} \cdot d\vec{S} = \frac{1}{\varepsilon_0} \sum q = \frac{1}{\varepsilon_0} Q$$

where Q is the sum of all the charges present within the surface. The charge outside of the surface is not counted, as the lines entering and leaving the surface due to this charge are the same in number. Therefore the flux ϕ due to the charge q sitting outside the surface is expressed as

$$\phi = \oint_s \vec{E} \cdot d\vec{S} = 0$$

11.8.1 Differential Form of Gauss's Theorem

When a charge is distributed over a volume such that ρ is the volume charge density, then the charge enclosed by the surface enclosing the volume is given by

$$q = \int_v \rho dV \quad \text{(i)}$$

Substituting this expression of q in $\oint_s \vec{E}\cdot d\vec{S} = \frac{q}{\varepsilon_0}$, we get

$$\varepsilon_0 \oint_s \vec{E}\cdot d\vec{S} = \int_v \rho dV \tag{ii}$$

According to Gauss's divergence theorem, we can convert the surface integral into the volume integral as

$$\oint_s \vec{E}\cdot d\vec{S} = \int_v \text{div}\,\vec{E}\,dV \tag{iii}$$

Therefore, from Eqs (ii) and (iii), we have

$$\varepsilon_0 \int_s \text{div}\,\vec{E}dV = \int_v \rho\,dV$$

Since the above equality is true for every volume, the integrands of left and right sides should be equal, i.e.,

$$\varepsilon_0 \text{div}\vec{E} = \rho$$

$$\vec{\nabla}\cdot(\varepsilon_0\vec{E}) = \rho \tag{iv}$$

$$\vec{\nabla}\cdot\vec{D} = \rho \tag{v}$$

where, $\vec{D}$ is the electric flux density, given by $\vec{D} = \varepsilon_0\vec{E}$. Eq. (v) is the differential form of Gauss's theorem.

11.9 STOKES' THEOREM

LO4

Stokes' theorem states that the integral of the curl of a vector function over a patch of surface is equal to the value of the function at the perimeter of the patch. So here the derivative is the curl, region is the surface and the boundary is the perimeter of the patch of the surface. Therefore,

$$\int_s (\vec{\nabla}\times\vec{F})\cdot d\vec{S} = \oint_c \vec{F}\cdot d\vec{l}$$

Clearly, this theorem converts the surface integral into the line integral. Here the L.H.S. is the surface integral whereas the R.H.S. is the closed line integral. So a point of confusion is that which way we should go around, *i.e.,* clockwise or anticlockwise when we integrate the line integral. Moreover, we should know about the direction of the surface element $d\vec{S}$. For example, for a closed surface $d\vec{S}$ points outwards normal but for an open surface which way is out? In order to overcome this confusion, we apply right hand rule. So if our fingers point in the direction of line integral, then the thumb gives the direction of $d\vec{S}$.

Based on the statement of Stokes' theorem, we can make some more observations. For example, $\int(\vec{\nabla}\times\vec{F})\cdot d\vec{S}$ does not depend on the shape of the surface rather it depends on the boundary line. Also for a closed surface, $\int_s(\vec{\nabla}\times\vec{F})\cdot d\vec{S} = 0$ as the boundary line shrinks down to a point.

11.10 ELECTRIC FIELD AND ELECTRIC POTENTIAL

LO4

Electric field is the region around a charge in which another charge experiences a force. The electrostatic field $\vec{E}$ is a special kind of field whose curl is always zero, i.e., $\vec{\nabla}\times\vec{E} = 0$. Since any vector whose curl is zero is equal to the gradient of some scalar quantity, we make use of this property to introduce the scalar quantity as the electric potential V. In order to find this, we use the Stokes' theorem $\int(\vec{\nabla}\times\vec{E})\cdot d\vec{S}=\oint\vec{E}\cdot d\vec{l}$. This gives

$$\oint\vec{E}\cdot d\vec{l}=0 \tag{i}$$

It means the line integral of $\vec{E}$ from point a_1 to point a_2 will be the same for all the paths between these points. Hence, the line integral of $\vec{E}$ is independent of path. Since changing the path would not alter the value of integral, we can define a function, say V, such that

$$V=-\int_a^r\vec{E}\cdot d\vec{l} \tag{ii}$$

The differential form of the above equation is written as

$$\vec{E}=-\vec{\nabla}V \tag{iii}$$

Here V is called the electric potential. Actually all the potentials are relative and there is no absolute zero potential. However, convention is that the potential is zero at infinite distance from the charge. In view of this, the lower limit a in Eq. (ii) is called as a standard reference point where V is zero. The upper limit is nothing but the point where V is to be calculated. So V depends only on the point $\vec{r}$.

11.10.1 Superposition Principle

According to the original principle of superposition of electrodynamics, the total force $\vec{F}$ on a charge q (test charge) is equal to the vector sum of the forces due to all the source charges (considering them individually). It means

$$\vec{F}=\vec{F}_1+\vec{F}_2+\vec{F}_3+\ldots \tag{iv}$$

Since $\vec{F}=q\vec{E}$, from Eq.(iv) we find the following for the electric field $\vec{E}$

$$\vec{E}=\vec{E}_1+\vec{E}_2+\vec{E}_3+\ldots \tag{v}$$

If we write a for the common reference point, the above equation can be written as

$$-\int_a^r\vec{E}\cdot d\vec{l}=-\int_a^r\vec{E}_1\cdot d\vec{l}-\int_a^r\vec{E}_2\cdot d\vec{l}-\int_a^r\vec{E}_3\cdot d\vec{l}-\ldots \tag{vi}$$

or
$$\vec{\nabla}V=\vec{\nabla}V_1+\vec{\nabla}V_2+\vec{\nabla}V_3+\ldots \tag{vii}$$

Now it is clear from Eq. (vii) that

$$V=V_1+V_2+V_3+\ldots \tag{viii}$$

The above equation reveals that the potential V at a given point $\vec{r}$ is the sum of the potentials due to all the charges. It means the electric potential also satisfies the principle of superposition and the sum is simply an ordinary sum. However, from Eq. (v) it is clear that in case of the electric field $\vec{E}$ this sum is the vector sum.

11.11 POISSON'S AND LAPLACE'S EQUATIONS LO5

For deriving these equations, we start with the following Gauss's law for a linear medium

$$\vec{\nabla}\cdot\vec{D}=\rho$$

or
$$\vec{\nabla}\cdot\varepsilon\vec{E}=\rho \tag{i}$$

Here e is basically the free charge density (volume) and $\vec{D}$ is the electric displacement (talked in detail later).

Since $\vec{E} = -\vec{\nabla}V$, the above equation for a homogeneous medium (where ε is constant) can be written as

$$\nabla^2 V = -\rho/\varepsilon \quad \text{(ii)}$$

This equation is called as Poisson's equation. For a charge free region, i.e., where $\rho = 0$, the Poisson's equation takes the form

$$\nabla^2 V = 0. \quad \text{(iii)}$$

This equation is known as Laplace's equation. This equation is much useful in solving electrostatic problems where a set of conductors are maintained at different potentials; for example, capacitors and vacuum tube diodes.

Using the expressions for Laplacian operator ∇^2 in Cartesian, cylindrical and spherical coordinate systems, we can write Laplace's Eq. (iii) in these coordinate systems, respectively, as

$$\frac{\partial^2 V}{\partial x^2} + \frac{\partial^2 V}{\partial y^2} + \frac{\partial^2 V}{\partial z^2} = 0$$

$$\frac{1}{s}\frac{\partial}{\partial s}\left(s\frac{\partial V}{\partial s}\right) + \frac{1}{s^2}\frac{\partial^2 V}{\partial \phi^2} + \frac{\partial^2 V}{\partial z^2} = 0$$

$$\frac{1}{r^2}\frac{\partial}{\partial r}\left(r^2\frac{\partial V}{\partial r}\right) + \frac{1}{r^2 \sin\theta}\frac{\partial}{\partial \theta}\left(\sin\theta\frac{\partial V}{\partial \theta}\right) + \frac{1}{r^2 \sin^2\theta}\frac{\partial^2 V}{\partial \phi^2} = 0$$

11.12 CAPACITOR

LO6

In order to construct a capacitor, we need at least two conductors that carry equal but opposite charges. If it is so, then flux lines leaving one conductor terminate at the surface of the other conductor. These conductors are also known as the plates of the capacitor, which may be separated by a dielectric or simply by free space. The ratio of the magnitude of the charge on one of the plates (say q) and the potential difference (say V) between the plates is known as the capacitance of the capacitor (represented by C). So

$$C = \frac{q}{V}$$

Some important capacitor configurations with two conductors are parallel plate capacitor, coaxial capacitor and spherical capacitor. These are discussed below.

11.12.1 Parallel Plate Capacitor

This type of capacitor has two plane conductors, which are parallel to each other. If the area of each plate be S and they are separated by a distance d, then the capacitance of this capacitor is given as

$$C = \varepsilon\frac{S}{d}$$

Here ε is the permittivity of the dielectric, which is filled between the plates. If a potential difference V is applied between the plates then the energy stored in the capacitor would be

$$U = \frac{1}{2}CV^2 = \frac{q^2}{2C}$$

11.12.2 Coaxial Capacitor

Coaxial capacitor is simply a coaxial cable. This is also referred to as coaxial cylindrical capacitor. If L be the length of the coaxial conductors, and the radii of inner and outer conductors be r_{in} and r_{out}, then the capacitance of the capacitor is obtained as

$$C = \frac{2\pi\varepsilon L}{\ln \frac{r_{out}}{r_{in}}}$$

Here also ε is the permittivity of the dielectric filling the space between the two conductors.

11.12.3 Spherical Capacitor

As name suggests, in this case the two conductors are in the form of spheres and these are concentric. Let the radius of inner sphere be r_{in} and of the outer sphere be r_{out}. Also these spheres are separated by a dielectric medium of permittivity ε. Then the capacitance of this type of capacitor is obtained as

$$C = \frac{4\pi\varepsilon}{\frac{1}{r_{in}} - \frac{1}{r_{out}}}$$

11.13 MAGNETIC FLUX DENSITY ($\vec{B}$) LO7

When a magnetic material is placed in an external magnetic field, it gets magnetised. The magnetism thus produced in the material is known as induced magnetism and this phenomenon is referred to as magnetic induction. The magnetic lines of force inside such magnetised materials are called magnetic lines of induction. The number of magnetic lines of induction crossing unit area at right angles to the flux is called the magnetic flux density $\vec{B}$. Its unit is the Tesla which is equal to 1 Wb/m^2.

11.14 MAGNETIC FIELD STRENGTH ($\vec{H}$) LO7

As mentioned earlier, a magnetic material becomes magnetised when placed in a magnetic field. The actual magnetic field inside the material is the sum of external field and the field due to its magnetisation.

$$\vec{H} = \frac{\vec{B}}{\mu_0} - \vec{M} \quad \text{or} \quad \vec{B} = \mu_0(\vec{H} + \vec{M})$$

Magnetic field strength at a point in a magnetic field is the magnitude of the force experienced by a unit pole situated at that point. The SI unit, corresponding to force of 1 Newton, is the A/m. The CGS unit, corresponding to a force of 1 dyne is the Oersted which is equal to 79.6 A/m.

11.15 AMPERE'S CIRCUITAL LAW LO7

Ampere's circuital law in magnetostatics is analogous to the Gauss's law used in electrostatics. This law says that the line integral of magnetic field $\vec{B}$ around any closed loop is equal to μ_0 times the net current I flowing through the area enclosed by the loop i.e.,

$$\oint \vec{B}\cdot d\vec{l} = \mu_0 I$$

Here, μ_0 is the permeability of the free space.

Proof: Consider a long straight conductor carrying a current I. By Biot-Savart law, the magnitude of the magnetic field at a point O, at a distance r from the conductor, is given by

$$B = \frac{\mu_0}{4\pi}\frac{2I}{r} \qquad \text{(i)}$$

Let us draw a circle with a radius r taking C as centre around the current carrying conductor Fig. 11.4. $\vec{B}$ will be the same in magnitude at all points on this circle. Again we consider a circle element of length dl at the point O. From the figure it is clear that $d\vec{l}$ and $\vec{B}$ are in the same direction.

$$\therefore \quad \oint \vec{B}\cdot d\vec{l} = \oint B dl \cos\theta$$

$$= B\oint dl \qquad [\because \theta = 0°]$$

$$= \frac{\mu_0}{4\pi}\frac{2I}{r}2\pi r = \mu_0 I$$

$$\oint \vec{B}\cdot d\vec{l} = \mu_0 I$$

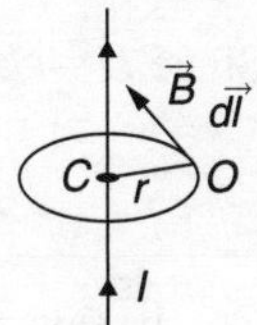

FIGURE 11.4

This is the required Ampere's circuital law. This law can be written in terms of volume current density $\vec{J}$, if we apply Stokes' theorem $\oint \vec{B}\cdot d\vec{l} = \int(\vec{\nabla}\times\vec{B})\cdot d\vec{S}$.

Since $I = \int \vec{J}\cdot d\vec{S}$, we get $\vec{\nabla}\times\vec{B} = \mu_0\vec{J}$ from $\oint \vec{B}\cdot d\vec{l} = \mu_0 I$. This is the another form of Ampere's law.

11.16 ELECTROSTATIC BOUNDARY CONDITIONS LO8

Consider the situation where the electric field exists in a region, which has two different media with permittivities as ε_1 (in region 1) and ε_2 (in region 2). Then the conditions, which should be satisfied by the field at the interface separating the two media or at the common boundary of these media, are called boundary conditions. It is seen that when we cross a boundary surface charge σ, the electric field does not remain continuous and it always undergoes a discontinuity.

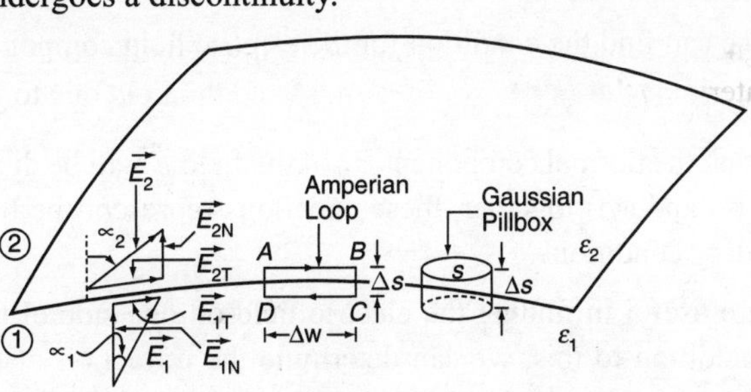

FIGURE 11.5

We can calculate the amount by which the electric field $\vec{E}$ changes at such a boundary, shown in Fig. 11.5. $\vec{E}_1$ is the field in the region 1 and $\vec{E}_2$ is the field in the region 2. These fields can be decomposed into two components, out of which one is tangential to the boundary (say $\vec{E}_T$) and the other is perpendicular to the boundary (say $\vec{E}_N$). So we can write

$$\vec{E}_1 = \vec{E}_{1N} + \vec{E}_{1T} \quad \text{and} \quad \vec{E}_2 = \vec{E}_{2N} + \vec{E}_{2T}$$

Now we can apply Ampere's and Gauss's laws for calculating the amount of discontinuity. For example, Ampere's law $\oint \vec{E}\cdot d\vec{l} = 0$ for the closed path *ABCDA*, whose length is Δw and the width is Δs, follows

$$0 = E_{2T}\Delta w - E_{2N}\frac{\Delta s}{2} - E_{1N}\frac{\Delta s}{2} - E_{1T}\Delta w + E_{1N}\frac{\Delta s}{2} + E_{2N}\frac{\Delta s}{2} \quad \text{(i)}$$

In the limit $\Delta s \to 0$, i.e., when the width of the loop is small so that we are well close to the boundary, the above equation says

$$E_{2T}\Delta w = E_{1T}\Delta w$$

or

$$E_{2T} = E_{1T} \quad \text{(ii)}$$

Since $\vec{D} = \varepsilon\vec{E}$, the above equation reads for the displacement vector $\vec{D}_T$

$$\frac{D_{2T}}{\varepsilon_2} = \frac{D_{1T}}{\varepsilon_1} \quad \text{(iii)}$$

Eq. (ii) says that the tangential component of the electric field remains continuous across the boundary, as its values just below and just above the boundary are equal. It means the electric field component $\vec{E}_T$ undergoes no change on the boundary. However, you can see from Eq. (iii) that the field component $\vec{D}_T$ undergoes some change across the boundary. It means $\vec{D}_T$ is discontinuous across the boundary.

In order to check the continuity of the normal component $\vec{E}_N$ of the field $\vec{E}$, we select Gaussian pillbox of area $\vec{S}$ (upper and lower surfaces) and the thickness Δs. Now we apply Gauss's law $\oint \vec{D}\cdot d\vec{S} = q$ and obtain the following under the limit of $\Delta s \to 0$

$$D_{2N}S - D_{1N}S = \sigma S$$

or

$$D_{2N} - D_{1N} = \sigma \quad \text{(iv)}$$

Here σ is the free charge density placed at the boundary. It is clear from Eq. (iv) that the normal component of $\vec{D}$ is discontinuous and this discontinuity amounts to the free charge density σ. If there is no free charge, the normal component of the field $\vec{D}$ will be continuous at the boundary, as in the absence of σ Eq. (iv) follows

$$D_{2N} - D_{1N} = 0 \quad \text{(v)}$$

From the above equation, we can find the condition for the electric field component $\vec{E}_N$ as

$$\varepsilon_2 E_{2N} - \varepsilon_1 E_{1N} = 0 \quad \text{(vi)}$$

The above equation shows that the normal component $\vec{E}_N$ of the field $\vec{E}$ will be discontinuous at the boundary. If we write Eqs (ii), (iii), (iv) and (vi) together, these equations represent the boundary conditions, named dielectric – dielectric boundary conditions.

The boundary conditions are useful in finding the electric field on one side of the boundary if the field on the other side is given. In addition to this, we can determine the refraction of the electric field across the boundary. If the field $\vec{E}_1$ (or $\vec{D}_1$) in the region 1 makes an angle α_1 and the field $\vec{E}_2$ (or $\vec{D}_2$) in the region 2 makes an angle α_2 with the normal to the boundary, then from Eq. (ii) we get

$$E_1 \sin\alpha_1 = E_2 \sin\alpha_2 \quad \text{(vii)}$$

Similarly, Eq. (vi) yields

$$\varepsilon_1 E_1 \cos\alpha_1 = \varepsilon_2 E_2 \cos\alpha_2 \quad \text{(viii)}$$

From Eqs (vii) and (viii), we finally get

$$\frac{\tan\alpha_1}{\varepsilon_1} = \frac{\tan\alpha_2}{\varepsilon_2}$$

or $$\frac{\tan\alpha_1}{\varepsilon_{r_1}} = \frac{\tan\alpha_2}{\varepsilon_{r_2}} \qquad [\because \varepsilon_1 = \varepsilon_0\varepsilon_{r1},\ \varepsilon_2 = \varepsilon_0\varepsilon_{r2}] \qquad \text{(ix)}$$

Here ε_{r1} is the dielectric constant of medium 1 and ε_{r2} is the dielectric constant of medium 2. The above equation is known as law of refraction of the electric field at the interface which is free of charge. This is also clear from Eq. (ix) that in general a boundary between two dielectrics produces bending of the flux lines. This is attributed to unequal polarisation charges accumulated on the sides of the boundary.

Now we can discuss some special cases of the above boundary conditions. For example, if the dielectric in the medium 2 is replaced with a conductor whose conductivity is infinite. In this case, it is obtained that the field does not exist within the conductor, i.e., $\vec{E} = 0$. Since $\vec{E} = -\vec{\nabla}V$, a potential difference between any two points in the conductor cannot exist. In other words, we say that a conductor is an equipotential body. Under this situation, only the normal component of the electric field survives, i.e., $D_T = \varepsilon_0\varepsilon_r E_T = 0$, $D_N = \varepsilon_0\varepsilon_r E_N = \sigma$. Similarly, when the dielectric in medium 1 is replaced with a conductor and the dielectric in medium 2 is replaced with free space ($\varepsilon_r = 1$), then the boundary conditions take the form $D_T = \varepsilon_0 E_T = 0$, $D_N = \varepsilon_0 E_N = \sigma$.

11.17 SCALAR AND VECTOR POTENTIALS LO9

As mentioned earlier, the zero curl of electrostatic field $\vec{E}$, i.e., $\vec{\nabla}\times\vec{E} = 0$, introduces a scalar potential V such that $\vec{E} = -\vec{\nabla}V$. When we analyze $\vec{\nabla}\cdot\vec{B} = 0$, we find that the field $\vec{B}$ can be written as a curl of another vector quantity (say $\vec{A}$), i.e.,

$$\vec{B} = \vec{\nabla}\times\vec{A} \qquad \text{(i)}$$

Since a field can be completely determined if we know its divergence as well as its curl, the divergence of $\vec{A}$ remains to be explored. In this context, with the use of Eq. (i), Ampere's law reads

$$\vec{\nabla}\times\vec{B} = \vec{\nabla}\times(\vec{\nabla}\times\vec{A}) = \vec{\nabla}(\vec{\nabla}\cdot\vec{A}) - \nabla^2\vec{A} = \mu_0\vec{J} \qquad \text{(ii)}$$

It is clear that Eq.(ii) will resemble Poisson's equation, if $\vec{\nabla}\cdot\vec{A} = 0$. This condition is known as Coulomb gauge. With the application of this condition, the Ampere's law simply yields

$$\nabla^2\vec{A} = -\mu_0\vec{J} \qquad \text{(iii)}$$

The solution of the above equation can be obtained if the current density $\vec{J}$ vanishes at infinity. Then the solution comes out to be

$$\vec{A}(\vec{r}) = \frac{\mu_0}{4\pi}\int\frac{\vec{J}}{r}\,dX \qquad \text{(iv)}$$

Here dX is the volume element and the vector $\vec{A}$ is called magnetic vector potential. Like the electric scalar potential V, the magnetic vector potential $\vec{A}$ cannot be uniquely defined as we can add to it another vector whose curl is zero. This addition does not change the field $\vec{B}$. On the other side, it is a point of observation that we cannot introduce a magnetic scalar potential U such that $\vec{B} = -\vec{\nabla}U$. The reason is that it is incompatible with Ampere's law, since the curl of a gradient is always zero.

11.18 CONTINUITY EQUATION LO9

The continuity equation says that the total current flowing out of some volume must be equal to the rate of decrease of the charge within that volume, if the charge is neither being created nor lost. Since the charge is flowing, we consider that the charge density ρ is a function of time. The transportation of the charge constitutes the current, i.e.,

$$I = \frac{dq}{dt} = \frac{d}{dt}\int_v \rho dV \tag{i}$$

Here, it is assumed that the current is extended in space of volume V closed by a surface S. The net amount of charge which crosses a unit area normal to the directed surface in unit time is defined as the *current density* $\vec{J}$. This current density $\vec{J}$ is related to the total current I flowing through the surface S as

$$I = \oint_s \vec{J} \cdot d\vec{S} \tag{ii}$$

Here the integral is over closed surface, as the surface bounding the volume is closed surface. From Eqs (i) and (ii), we have

$$\oint_s \vec{J} \cdot d\vec{S} = -\frac{dq}{dt} = \frac{-d}{dt}\int_v \rho dV \tag{iii}$$

The minus sign above is needed in view of decreasing charge ρ in the volume V. So

$$\oint_s \vec{J} \cdot d\vec{S} = -\int_v \frac{\partial \rho}{\partial t} dV \tag{iv}$$

From Gauss's divergence theorem, we have

$$\oint_s \vec{J} \cdot d\vec{S} = \int_v (\text{div}\,\vec{J}) dV$$

or

$$\int_v (\text{div}\,\vec{J})\, dV = -\int_v \frac{\partial \rho}{\partial t} dV \tag{v}$$

Since the Eq. (v) holds good for any arbitrary volume, we can put the integrands to be equal. Then

$$\text{div}\,\vec{J} + \frac{\partial \rho}{\partial t} = 0 \tag{vi}$$

This is the continuity equation.

In case of stationary currents, i.e., when the charge density at any point within the region remains constant, but the charges are moving.

$$\frac{\partial \rho}{\partial t} = 0, \tag{vii}$$

so that $\text{div}\,\vec{J} = 0$ or $\vec{\nabla} \cdot \vec{J} = 0$

which expresses the fact that there is no net outward flux of current density $\vec{J}$. Here the situation is the same as shown in Fig. 11.2c for zero divergence.

11.19 MAXWELL'S EQUATIONS: DIFFERENTIAL FORM LO10

When the charges are in motion, the electric and magnetic fields are associated with this motion which will have variations in both the space and the time. These electric and magnetic fields are inter related. This phenomenon is called *electromagnetism* which is summarised by the set of equations, known as Maxwell's equations. The Maxwell's equations are nothing but are the representation of the basic laws of electromagnetism.

First Maxwell's equation is the Gauss's law of electrostatics, i.e., $\vec{\nabla} \cdot \vec{E} = \rho/\varepsilon_0$. Second Maxwell's equation is the Gauss's law of magnetostatics, i.e., $\vec{\nabla} \cdot \vec{B} = 0$. Faraday's law of electromagnetic induction, i.e., $\vec{\nabla} \times \vec{E} = \frac{-\partial \vec{B}}{\partial t}$, is called Maxwell's third equation. Fourth Maxwell's equation is the modified Ampere's circuital law, i.e., $\vec{\nabla} \times \vec{B} = \mu_0 \vec{J} + \mu_0 \varepsilon_0 \frac{\partial \vec{E}}{\partial t}$.

Now we look for the convenient form of Maxwell's equations while we are working with materials (having permittivity ε and permeability μ) that are subject to electric and magnetic polarizations. Electric polarization $\vec{P}$ provides bound charges with volume density e_b, given by

$$\rho_b = -\vec{\nabla} \cdot \vec{P} \tag{i}$$

Similarly, a magnetic polarization or magnetization $\vec{M}$ results in a volume bound current density, given by

$$\vec{J}_b = \vec{\nabla} \times \vec{M} \tag{ii}$$

Equations (i) and (ii) represent the static case of uniforms polarization $\vec{P}$ and uniform magnetization $\vec{M}$. However, any change in polarization $\vec{P}$ gives rise to the polarization current density, given by

$$\vec{J}_P = \frac{\partial \vec{P}}{\partial t} \tag{iii}$$

Since $\vec{J}_P$ satisfies the continuity equation, it is evident that $\vec{J}_P$ is essential to account for the conservation of bound charge. On the other hand, a changing magnetization $\vec{M}$ does not lead to any analogous accumulation of charge or current. We do not have direct control on the bound charge and current. Hence, it is advisable to reformulate Maxwell's equations such that these make explicit reference only to those sources which we control directly. These are the free charges (ρ_f) and currents ($\vec{J}_f$). The total volume charge density can be written as

$$\begin{aligned} \rho &= \rho_f + \rho_b \\ &= \rho_f - \vec{\nabla} \cdot \vec{P}, \end{aligned} \tag{iv}$$

whereas the total volume current density is written as

$$\vec{J} = \vec{J}_f + \vec{J}_b + \vec{J}_P = \vec{J}_f + \vec{\nabla} \times \vec{M} + \frac{\partial \vec{P}}{\partial t} \tag{v}$$

In view of total charge density ρ, Gauss's law reads

$$\vec{\nabla} \cdot \vec{E} = \frac{1}{\varepsilon_0}(\rho_f - \vec{\nabla} \cdot \vec{P})$$

$$\vec{\nabla} \cdot (\varepsilon_0 \vec{E} + \vec{P}) = \rho_f$$

Recall that $\varepsilon_0 \vec{E} + \vec{P} = \vec{D}$. Hence, the first Maxwell's equation in materials takes the form

$$\vec{\nabla} \cdot \vec{D} = \rho_f \tag{vi}$$

Similarly, the modified Ampere's circuital law yields

$$\vec{\nabla} \times \vec{B} = \mu_0\left(\vec{J}_f + \vec{\nabla} \times \vec{M} + \frac{\partial \vec{P}}{\partial t}\right) + \mu_0\varepsilon_0\frac{\partial \vec{E}}{\partial t}$$

$$\vec{\nabla} \times \left(\frac{\vec{B}}{\mu_0} - \vec{M}\right) = \vec{J}_f + \frac{\partial}{\partial t}(\varepsilon_0\vec{E} + \vec{P})$$

$$\vec{\nabla} \times \vec{H} = \vec{J}_f + \frac{\partial \vec{D}}{\partial t} \qquad \left(\because \vec{H} = \frac{\vec{B}}{\mu_0} - \vec{M}\right) \tag{vii}$$

The four Maxwell's equations in terms of free charges (density ρ_f) and currents are written as

$$\vec{\nabla} \cdot \vec{D} = \rho_f \qquad \text{First equation}$$

$$\vec{\nabla} \cdot \vec{B} = 0 \qquad \text{Second equation}$$

$$\vec{\nabla} \times \vec{E} = -\frac{\partial \vec{B}}{\partial t} \qquad \text{Third equation}$$

$$\vec{\nabla} \times \vec{H} = \vec{J}_f + \frac{\partial \vec{D}}{\partial t} \qquad \text{Fourth equation}$$

However, for the sake of simplicity we shall write ρ for ρ_f and $\vec{J}$ for $\vec{J}_f$ in the derivation of Maxwell's equations, unless specified.

In free space, dielectric medium or conducting medium, the first and fourth Maxwell's equations assume different forms. For example, in free space and dielectric medium, free charge $\rho_f = 0$ and free current density $\vec{J}_f = 0$. Hence, the

First equation in free space yields $\vec{\nabla} \cdot \vec{D} = 0$

$$\Rightarrow \quad \vec{\nabla} \cdot (\varepsilon_0\vec{E}) = 0 \qquad (\because \vec{P} = 0 \text{ in free space})$$

or $\quad \vec{\nabla} \cdot \vec{E} = 0$

However, the first equation in dielectric medium gives

$$\vec{\nabla} \cdot \vec{D} = 0 \;\Rightarrow\; \varepsilon\vec{\nabla} \cdot \vec{E} = 0 \qquad (\because \vec{D} = \varepsilon\vec{E})$$

or $\quad \vec{\nabla} \cdot \vec{E} = 0$

In conducting medium, any free charge resides on its surface, i.e., $\rho_f = 0$ in the medium. Hence, the first equation again gives $\vec{\nabla} \cdot \vec{E} = 0$. This can also be understand as follows.
The continuity equations for free charges reads

$$\frac{\partial \rho_f}{\partial t} + \vec{\nabla} \cdot \vec{J}_f = 0$$

or

$$\frac{\partial \rho_f}{\partial t} = -\vec{\nabla} \cdot \vec{J}_f \tag{viii}$$

Since we want to see what happens when a free charge is given to a conductor, we find Eq. (viii) in terms of ρ_f by using Ohm's law $\vec{J}_f = \sigma\vec{E}$ (σ is conductivity) and Gauss's law $\vec{\nabla} \cdot \vec{D} = \rho_f$. Hence

$$\frac{\partial \rho_f}{\partial t} = -\vec{\nabla}\cdot(\sigma\vec{E})$$

$$\Rightarrow \quad \frac{\partial \rho_f}{\partial t} = -\frac{\sigma}{\varepsilon}\vec{\nabla}\cdot(\varepsilon\vec{E})$$

$$\Rightarrow \quad \frac{\partial \rho_f}{\partial t} = -\frac{\sigma}{\varepsilon}\rho_f \tag{ix}$$

Eq. (ix) is written as $\frac{1}{\rho_f}\frac{\partial \rho_f}{\partial t} = -\frac{\sigma}{\varepsilon}$ and is integrated to get

$$\rho_f(t) = \rho_f(0)e^{-\frac{\sigma}{\varepsilon}t} \tag{x}$$

Here $\rho_f(0)$ is the initial charge given to the conductor. For good conductors $\sigma \approx \infty$; means $\rho_f \to 0$ very quickly. This proves that the charge will flow out to the edges of conductor within very less time. This characteristic time is given by

$$\tau = \frac{\varepsilon}{\sigma} \tag{xi}$$

So this is clear that the Maxwell's first equation reads $\vec{\nabla}\cdot\vec{E} = 0$ in free space, dielectric and conducting medium.

Now we discuss different forms of Maxwell's fourth equation $\vec{\nabla}\times\vec{H} = \vec{J}_f + \frac{\partial \vec{D}}{\partial t}$. In free space $\vec{J}_f = 0$ and $\vec{D} = \varepsilon_0\vec{E}$ ($\because\ \vec{P} = 0$), Hence,

$$\vec{\nabla}\times\vec{H} = \varepsilon_0\frac{\partial \vec{E}}{\partial t}$$

or
$$\mu_0\vec{\nabla}\times\vec{H} = \mu_0\varepsilon_0\frac{\partial \vec{E}}{\partial t}$$

or
$$\nabla\times\vec{B} = \mu_0\varepsilon_0\frac{\partial \vec{E}}{\partial t} \tag{xii}$$

In dielectric medium $\vec{J}_f = 0$ and $\vec{D} = \varepsilon\vec{E}$. Hence

$$\vec{\nabla}\times\vec{H} = \varepsilon\frac{\partial \vec{E}}{\partial t}$$

or
$$\vec{\nabla}\times(\mu\vec{H}) = \mu\varepsilon\frac{\partial \vec{E}}{\partial t}$$

or
$$\vec{\nabla}\times\vec{B} = \mu\varepsilon\frac{\partial \vec{E}}{\partial t} \tag{xiii}$$

In conducting medium, $\vec{J}_f = \sigma\vec{E}$ and $\vec{D} = \varepsilon\vec{E}$.

Hence,

$$\vec{\nabla}\times\vec{H} = \sigma\vec{E} + \varepsilon\frac{\partial \vec{E}}{\partial t} \tag{xiv}$$

In view of the above discussion, this is clear that second and third Maxwell's equations remain unchanged in all types of the media.

11.19.1 Derivation of Maxwell's First Equation

Let us consider a surface S bounding a volume V in a dielectric medium, which is kept in the $\vec{E}$ field. The application of external field $\vec{E}$ polarises the dielectric medium and charges are induced, called bound charges or charges due to polarisation. The total charge density at a point in a small volume element dV would then be $(\rho + \rho_p)$, where ρ_p is the polarisation charge density (the same as $\rho_{\hat{b}}$), given by $\rho_P = -\text{div}\,\vec{P}$ and ρ is the free charge density at that point in the small volume element dV.

Thus, the total charge density at that point will be $\rho - (\text{div}\vec{P})$. Then Gauss's theorem can be expressed as

$$\oint_s \vec{E}\cdot d\vec{S} = \int_v (\text{div}\,\vec{E})dV = \frac{1}{\varepsilon_0}\int_v (\rho - \text{div}\,\vec{P})\,dV$$

or $$\varepsilon_0\int_v (\text{div}\,\vec{E})dV = \int_v (\rho - \text{div}\,\vec{P})\,dV$$

or $$\int_v \text{div}(\varepsilon_0\vec{E} + \vec{P})\,dV = \int_v \rho dV$$

The quantity $(\varepsilon_0\vec{E} + \vec{P})$ is denoted by a quantity $\vec{D}$, called the electric displacement. Therefore,

$$\int_v (\text{div}\,\vec{D})dV = \int_v \rho dV$$

Since this equation is true for all the arbitrary volumes, the integrands in this equation must be equal, i.e.,

$$\text{div}\vec{D} = \rho$$

or $$\vec{\nabla}\cdot\vec{D} = \rho$$

This is the Maxwell's first equation.

When the medium is isotropic, the three vectors $\vec{D}$, $\vec{E}$ and $\vec{P}$ are in the same direction and for small field $\vec{E}$, $\vec{D}$ is proportional to $\vec{E}$, i.e.,

$$\vec{D} = \varepsilon\vec{E}$$

where ε is called the permittivity of the dielectric medium. The ratio $\varepsilon/\varepsilon_0$ is called the dielectric constant of the medium.

11.19.2 Derivation of Maxwell's Second Equation

Since the magnetic lines of force are either closed or go off to infinity, the number of magnetic lines of force entering any arbitrary surface is exactly the same as leaving it. It means the flux of magnetic induction $\vec{B}$ across any closed surface is always zero, i.e.,

$$\oint_s \vec{B}\cdot d\vec{S} = 0$$

Transforming the surface integral to volume integral using Gauss's divergence theorem, we have

$$\oint_s \vec{B}\cdot d\vec{S} = \int_v (\text{div}\,\vec{B})dV = 0$$

The integrand in the above equation should vanish for the surface boundary as the volume is arbitrary. Therefore

$$\text{div}\,\vec{B} = 0 \quad \text{or} \quad \vec{\nabla}\cdot\vec{B} = 0$$

This is the Maxwell's second equation.

11.19.3 Derivation of Maxwell's Third Equation

According to Faraday's law, the emf induced in a closed loop is given by

$$E_{\text{emf}} = -\frac{\partial \phi}{\partial t} = -\int_s \frac{\partial \vec{B}}{\partial t}\cdot d\vec{S} = -\frac{\partial}{\partial t}\int_s \vec{B}\cdot d\vec{S}$$

Here the flux $\phi = \int_s \vec{B}\cdot d\vec{S}$ where S is any surface having the loop as boundary. The emf (E_{emf}) can also be found by calculating the work done in carrying a unit charge completely around the loop. Thus,

$$E_{\text{emf}} = \oint_c \vec{E}\cdot d\vec{l}$$

Here $\vec{E}$ is the intensity of the field associated with the induced emf. On equating the above two equations, we get

$$\oint_c \vec{E}\cdot d\vec{l} = -\int_s \frac{\partial \vec{B}}{\partial t}\cdot d\vec{S}$$

According to Stokes' theorem, the line integral can be transformed into surface integral with the help of $\oint_c \vec{E}\cdot d\vec{l} = \int_s (\vec{\nabla}\times\vec{E})\cdot d\vec{S}$. Therefore

$$\int_s (\vec{\nabla}\times\vec{E})\cdot d\vec{S} = -\int_s \frac{\partial \vec{B}}{\partial t}\cdot d\vec{S}$$

This equation must be true for any surface whether small or large in the field. So the two vectors in the integrands must be equal at every point, i.e.,

$$\vec{\nabla}\times\vec{E} = -\frac{\partial \vec{B}}{\partial t}$$

This is the Maxwell's third equation.

11.19.4 Derivation of Maxwell's Fourth Equation

According to the Ampere's law, the work done in carrying a unit magnetic pole once around a closed arbitrary path linked with the current is expressed by

$$\oint_c \vec{H}\cdot d\vec{l} = I$$

or

$$\oint_c \vec{H}\cdot d\vec{l} = \int_c \vec{J}\cdot d\vec{S}$$

As per Stokes' theorem,

$$\oint_c \vec{H}\cdot d\vec{l} = \int_s (\vec{\nabla}\times\vec{H})\cdot d\vec{S}$$

Therefore,

$$\int_s (\vec{\nabla}\times\vec{H})\cdot d\vec{S} = \int_s \vec{J}\cdot d\vec{S}$$

This gives

$$\vec{\nabla}\times\vec{H} = \vec{J}$$

The above relation is derived on the basis of Ampere's law, which holds good only for the steady current. However, for the changing electric fields, the current density should be modified. The difficulty with the above equation is that, if we take divergence of this equation, then

$$\text{div}(\vec{\nabla} \times \vec{H}) = \text{div}\,\vec{J}$$

$$\Rightarrow \quad 0 = \text{div}\,\vec{J} \qquad \text{[Since divergence of a curl = 0]}$$

$$\Rightarrow \quad \text{div}\,\vec{J} = 0$$

which conflicts with the continuity equation, as

$$\text{div}\,\vec{J} = -\frac{\partial \rho}{\partial t}$$

Therefore, Maxwell realised that the definition of the current density is incomplete and suggested to add another density $\vec{J}'$. Therefore

$$\text{curl}\,\vec{H} = \vec{J} + \vec{J}'$$

Now, taking divergence of the above equation, we get

$$\text{div}\,(\text{curl}\,\vec{H}) = \text{div}\,\vec{J} + \text{div}\,\vec{J}'$$

or

$$0 = \text{div}\,\vec{J} + \text{div}\,\vec{J}'$$

$$\text{div}\,\vec{J}' = -\,\text{div}\,\vec{J} = \frac{\partial \rho}{\partial t}$$

using continuity equation

Since,

$$\rho = \vec{\nabla} \cdot \vec{D}$$

$$\text{div}\,\vec{J}' = \frac{\partial}{\partial t}(\vec{\nabla} \cdot \vec{D})$$

$$\vec{\nabla} \cdot \vec{J}' = \vec{\nabla} \cdot \frac{\partial \vec{D}}{\partial t}$$

Hence $\vec{J}' = \dfrac{\partial \vec{D}}{\partial t}$

Therefore, the Maxwell's fourth equation can be written as

$$\vec{\nabla} \times \vec{H} = \vec{J} + \frac{\partial \vec{D}}{\partial t}$$

The last term of R.H.S. of this equation is called Maxwell's correction and is known as *displacement current density*. The above equation is called modified Ampere's law for unsteady or changing current which is responsible for the electromagnetic fields.

11.20 MAXWELL'S EQUATIONS: INTEGRAL FORM LO10

There are situations where the integral form of Maxwell's equations is useful. Therefore, now we derive these equations in integral form.

11.20.1 Maxwell's First Equation

Differential form of the Maxwell's first equation is

$$\vec{\nabla}\cdot\vec{D}=\rho \tag{i}$$

On integrating Eq. (i) over a volume V, we have

$$\int_v (\vec{\nabla}\cdot\vec{D})dV=\int_v \rho dV$$

Using Gauss's divergence theorem, the above equation reads

$$\oint_s \vec{D}\cdot d\vec{S}=\int_v \rho dV=q$$

or

$$\oint_s \vec{D}\cdot d\vec{S}=q$$

Here q is the net charge contained in the volume V and S is the surface bounding the volume V. This integral form of the Maxwell's first equation says that the total electric displacement through the surface S enclosing a volume V is equal to the total charge contained within this volume.

This statement can also be put in the following form: The total outward flux corresponding to the displacement vector $\vec{D}$ through a closed surface $\vec{S}$ is equal to the total charge q within the volume V enclosed by the surface $\vec{S}$.

11.20.2 Maxwell's Second Equation

Differential form of the Maxwell's second equation is

$$\vec{\nabla}\cdot\vec{B}=0 \tag{ii}$$

Exactly in a manner adopted above, we can show that

$$\oint_s \vec{B}\cdot d\vec{S}=0$$

which signifies that the total outward flux of magnetic induction $\vec{B}$ through any closed surface $\vec{S}$ is equal to zero.

11.20.3 Maxwell's Third Equation

Differential form of the Maxwell's third equation is

$$\vec{\nabla}\times\vec{E}=-\frac{\partial\vec{B}}{\partial t} \tag{iii}$$

On integrating Eq. (iii) over a surface $\vec{S}$ bounded by a closed path, we have

$$\int_s (\vec{\nabla}\times\vec{E})\cdot d\vec{S}=\int_s \frac{\partial\vec{B}}{\partial t}\cdot d\vec{S}$$

Converting surface integral into line integral by Stokes' theorem, we get

$$\oint_c \vec{E}\cdot d\vec{l}=-\frac{\partial}{\partial t}\int_s \vec{B}\cdot d\vec{S}$$

which signifies that the electromotive force around a closed path is equal to the time derivative of the magnetic flux through any closed surface bounded by that path.

11.20.4 Maxwell's Fourth Equation

Differential form of the Maxwell's fourth equation is

$$\vec{\nabla} \times \vec{H} = \vec{J} + \frac{\partial \vec{D}}{\partial t} \tag{iv}$$

Exactly, in a manner adopted above, we can have this equation in the following form

$$\oint_c \vec{H} \cdot d\vec{l} = \int_s \left(\vec{J} + \frac{\partial \vec{D}}{\partial t} \right) \cdot d\vec{S}$$

The above equation signifies that the magnetomotive force around a closed path is a measure of the conduction current plus the time derivative of the electric flux through any surface bounded by that path.

11.21 SIGNIFICANCE OF MAXWELL'S EQUATIONS LO10

Maxwell's equations represent concisely the fundamentals of electricity and magnetism. From them one can develop most of the working relationships in the field.

Maxwell's first equation represents the Gauss's law for electricity which says that the electric flux out of any closed surface is proportional to the total charge enclosed within the surface. The integral form of this equation finds applications in calculating electric fields around charge objects. It is consistent with Coulomb's law when applied to the electric field of a point charge. The area integral of the electric field gives a measure of the net charge enclosed. However, the divergence of the electric field gives a measure of the density of sources.

As mentioned, the area integral of a vector field determines the net source of the field (function). The integral form $\oint_s \vec{B} \cdot d\vec{S} = 0$ of the Maxwell's second equation says that the net magnetic flux out of any closed surface is zero. This is because the magnetic flux directed inward toward the south pole, of a magnetic dipole kept in any closed surface, will be equal to the flux outward the north pole. Therefore, the net flux is zero for dipole sources. If we imagine a magnetic monopole source, the area integral $\oint_s \vec{B} \cdot d\vec{S}$ would have some finite value. Because of this and since the divergence of a vector field is proportional to the density of point source, this form of the Gauss's law for magnetic field simply says that there are no magnetic monopoles.

The Maxwell's third equation when written in the integral form states that the line integral of the electric field around a closed loop is equal to the negative of the rate of change of the magnetic flux through the area enclosed by the loop. The line integral basically is the generated voltage or emf in the loop. Therefore, the physical interpretation of Maxwell's third equation is that the changing magnetic field induces electric field.

For static electric field $\vec{E}$, the second term of the R.H.S. of the Maxwell's fourth equation vanishes and then the integral form of this equation says that the line integral of the magnetic field around a closed loop is proportional to the electric current flowing through the loop. This form of the Maxwell's equation is useful for calculating the magnetic field for simple geometries. However, this equation more specifically reveals that the changing electric field induces magnetic field. This seems complimentary to the meaning of the Maxwell's third equation. Therefore, they together yield the formulation of electromagnetic fields or electromagnetic waves, where both electric and magnetic fields propagate together and the change in one field induces the other field.

11.22 MAXWELL'S DISPLACEMENT CURRENT AND CORRECTION IN AMPERE'S LAW

LO10

When a current flows in a conductor, magnetic field is produced around it. A relation between the conduction current (I) and the magnetic field vector ($\vec{B}$) was given by Ampere, according to which the line integral of the field $\vec{B}$ along a closed curve in the magnetic field is equal to μ_0 times the current I flowing in the conductor.

$$\oint_c \vec{B}\cdot d\vec{l} = \mu_0 I = \mu_0 \int_s \vec{J}\cdot d\vec{S} \tag{i}$$

or $$\int_s (\text{curl}\,\vec{B})\cdot d\vec{S} = \mu_0 \int_s \vec{J}\cdot d\vec{S} \qquad \text{(Using Stokes' theorem)}$$

or $$\int_s (\text{curl}\,\vec{B} - \mu_0\vec{J})\cdot d\vec{S} = 0$$

For arbitrary surfaces, the above integral is true, so we get

$$\text{curl}\,\vec{B} - \mu_0\vec{J} = 0$$

or $$\text{curl}\,\vec{B} = \mu_0\vec{J} \tag{ii}$$

or $$\text{div}(\text{curl}\,\vec{B}) = \text{div}(\mu_0\vec{J})$$

or $$0 = \mu_0\text{div}(\vec{J}) \qquad [\because \text{ divergence of curl of a vector field is always zero}]$$

or $$\text{div}\,\vec{J} = 0 \tag{iii}$$

Using continuity equation $\text{div}\,\vec{J} + \frac{\partial\rho}{\partial t} = 0$ or $\text{div}\,\vec{J} = -\frac{\partial\rho}{\partial t}$, we get $\text{div}\,\vec{J} = 0$ only if $\frac{\partial\rho}{\partial t} = 0$, i.e., for static charge density.

Thus, the Ampere's law given by Eq. (i) is valid only for the steady state conditions, i.e., for the fields that do not vary with time. This law is not valid for the time varying fields, such as charging and discharging of a capacitor, where $\frac{\partial\rho}{\partial t} \neq 0$.

Ampere's law was modified by Maxwell for the time varying fields. His concept was based on the Faraday's law of electromagnetic induction, according to which a changing magnetic field produces an electric field. On the basis of the fact that the magnetic field around a conductor is produced by the current flowing in it, Maxwell hypothesized that changing electric field should also induce a magnetic field. A changing electric field is equivalent to a current called displacement current (I_d) which flows as long as the electric field is changing. The displacement current produces the magnetic field the same way as the conduction current (I).

Thus, the total magnetic field ($\vec{B}$) will be the sum of the two terms (a) due to current $I(\vec{B}_1)$ and (b) due to current $I_d(\vec{B}_2)$, i.e.,

$$\vec{B} = \vec{B}_1 + \vec{B}_2$$

$\therefore$ $$\text{curl}\,\vec{B} = \text{curl}\,\vec{B}_1 + \text{curl}\,\vec{B}_2$$

or $$\text{curl}\,\vec{B} = \mu_0\vec{J} + \mu_0\vec{J}_d$$

or $$\text{curl}\,\vec{B} = \mu_0(\vec{J} + \vec{J}_d) \tag{iv}$$

where $\vec{J}_d$ is the displacement current density. Also, in the integral form

$$\oint_c \vec{B}\cdot d\vec{l} = \mu_0 I + \mu_0 I_d$$

or
$$\oint_c \vec{B}\cdot d\vec{l} = \mu_0 (I + I_d) \tag{v}$$

In analogy to the Faraday's law of induction $\oint_c \vec{E}\cdot d\vec{l} = -\frac{d\phi_B}{dt}$, I_d should correspond to $\varepsilon_0 \frac{d\phi_E}{dt}$. With this Eq. (v) can be written as

$$\oint_c \vec{B}\cdot d\vec{l} = \mu_0 \left(I + \varepsilon_0 \frac{d\phi_E}{dt} \right) \tag{vi}$$

Thus,
$$I_d = \varepsilon_0 \frac{d\phi_E}{dt} = \varepsilon_0 S \frac{d\vec{E}}{dt} = S\frac{d\vec{D}}{dt} = S\vec{J}_d \tag{vii}$$

where $\vec{D}$ is the electric displacement vector and S is the area.

Value of J_d can also be determined by taking divergence of Eq. (iv), i.e.,

$$\text{div curl } \vec{B} = \text{div } (\mu_0 \vec{J} + \mu_0 \vec{J}_d) = 0 \qquad [\because \text{div curl } \vec{B} = 0]$$

or
$$\text{div } \vec{J}_d = -\text{div } \vec{J} = \frac{\partial \rho}{\partial t} \qquad \left[\because \text{div } \vec{J} + \frac{\partial \rho}{\partial t} = 0\right]$$

or
$$\text{div } \vec{J}_d = \frac{\partial}{\partial t}(\text{div } \vec{D}) = \text{div}\left(\frac{\partial \vec{D}}{\partial t}\right) \qquad [\because \text{div } \vec{D} = \rho]$$

∴
$$\vec{J}_d = \frac{\partial \vec{D}}{\partial t} \tag{viii}$$

Therefore, the modified form of Ampere's law is

$$\text{curl } \vec{B} = \mu_0 \left(\vec{J} + \frac{\partial \vec{D}}{\partial t} \right) \tag{ix}$$

or
$$\text{curl } \vec{H} = \left(\vec{J} + \frac{\partial \vec{D}}{\partial t} \right) \qquad (\because \vec{B} = \mu_0 \vec{H})$$

or
$$\text{curl } \vec{H} = \vec{J} + \varepsilon_0 \frac{\partial \vec{E}}{\partial t} \tag{x}$$

The Eqs (v), (ix) or (x) represent the Maxwell's fourth equation in different form which is nothing but the modified form of Ampere's law.

11.23 ELETROMAGNETIC (EM) WAVE PROPAGATION IN FREE SPACE LO11

Maxwell's equations for free space are given as follows

$$\text{div } \vec{E} = 0 \tag{i}$$

$$\text{div } \vec{H} = 0 \tag{ii}$$

$$\text{curl}\,\vec{E} = -\mu_0 \frac{\partial \vec{H}}{\partial t} \tag{iii}$$

$$\text{curl}\,\vec{H} = \varepsilon_0 \frac{\partial \vec{E}}{\partial t} \tag{iv}$$

Taking curl of Eq. (iii), we get $\vec{\nabla} \times (\vec{\nabla} \times \vec{E}) = -\mu_0 \frac{\partial}{\partial t}(\vec{\nabla} \times \vec{H})$ or $\vec{\nabla}(\vec{\nabla} \cdot \vec{E}) - \nabla^2 \vec{E} = -\mu_0 \frac{\partial}{\partial t}\left[\varepsilon_0 \frac{\partial \vec{E}}{\partial t}\right]$

Here we have used Eq. (iv) for the value of $\vec{\nabla} \times \vec{H}$. Now from Eq. (i) $\vec{\nabla} \cdot \vec{E} = 0$. Hence

$$\nabla^2 \vec{E} - \mu_0 \varepsilon_0 \frac{\partial^2 \vec{E}}{\partial t^2} = 0$$

This is the wave equation governing the field $\vec{E}$. In view of the dimensions of $(\mu_0 \varepsilon_0)^{-1/2}$ as of velocity (say, v), we can write this as

$$\nabla^2 \vec{E} - \frac{1}{v^2}\frac{\partial^2 \vec{E}}{\partial t^2} = 0 \tag{v}$$

Similarly the curl of Eq. (iv) gives rise to the wave equation for the field $\vec{H}$ as

$$\nabla^2 \vec{H} - \mu_0 \varepsilon_0 \frac{\partial^2 \vec{H}}{\partial t^2} = 0$$

or
$$\nabla^2 \vec{H} - \frac{1}{v^2}\frac{\partial^2 \vec{H}}{\partial t^2} = 0 \tag{vi}$$

The plane wave solutions of Eqs. (v) and (vi) may be written as

$$\vec{E}(\vec{r}, t) = \vec{E}_0 e^{i(\vec{k}\cdot\vec{r} - \omega t)} \quad \text{and} \quad \vec{H}(\vec{r}, t) = \vec{H}_0 e^{i(\vec{k}\cdot\vec{r} - \omega t)}$$

where ω is the angular frequency of the variation of the fields $\vec{E}$ and $\vec{H}$ and $\vec{k}$ is the wave vector which tells the direction of propagation of the fields or wave. The ratio ω/k gives the phase velocity of the wave.

Now, $\vec{\nabla} = \left(\frac{\partial}{\partial x}\hat{i} + \frac{\partial}{\partial y}\hat{j} + \frac{\partial}{\partial z}\hat{k}\right)$, $\vec{E}_0 = (E_{0x}\hat{i} + E_{0y}\hat{j} + E_{0z}\hat{k})$

$$\vec{k} = (k_x\hat{i} + k_y\hat{j} + k_z\hat{k}), \vec{r} = (x\hat{i} + y\hat{j} + z\hat{k})$$

$\Rightarrow$
$$\vec{k}\cdot\vec{r} = (k_x x + k_y y + k_z z)$$

$\therefore$
$$\text{curl}\,\vec{E} = \vec{\nabla} \times [\vec{E}_0 e^{i(\vec{k}\cdot\vec{r} - \omega t)}]$$

$$= \vec{\nabla} \times \{(\vec{E}_0 e^{i\vec{k}\cdot\vec{r}})e^{-i\omega t}\}$$

$$= \vec{\nabla} \times \{[(E_{0x}\hat{i} + E_{0y}\hat{j} + E_{0z}\hat{k})e^{i(k_x x + k_y y + k_z z)}]e^{-i\omega t}\}$$

Here note that i appeared in exponential term is such that $i = \sqrt{-1}$, whereas $\hat{i}$ is the unit vector along the x-direction. When we solve the above equation with the help of expansion of curl, we obtain

$$\text{curl } \vec{E} = i\{\hat{i}[E_{0z}k_y - E_{0y}k_z] + \hat{j}[E_{0x}k_z - E_{0z}k_x] + \hat{k}[E_{0y}k_x - E_{0x}k_y]\}e^{i(k_x x + k_y y + k_z z)}e^{-i\omega t}$$

$$= i[\vec{k} \times \vec{E}_0]e^{i(\vec{k}\cdot\vec{r} - \omega t)}$$

or $$\text{curl } \vec{E} = i[\vec{k} \times \vec{E}] \quad \text{(vii)}$$

Here we have used

$$\vec{E}(\vec{r}, t) = \vec{E}_0 e^{i(\vec{k}\cdot\vec{r} - \omega t)}$$

Using Eqs. (iii) and (vii), we obtain

$$-\mu_0 \frac{\partial \vec{H}}{\partial t} = i[\vec{k} \times \vec{E}]$$

L.H.S. of the above equation is written as

$$-\mu_0 \frac{\partial}{\partial t}[\vec{H}_0 e^{i(\vec{k}\cdot\vec{r} - \omega t)}] = i\omega\mu_0 \vec{H}$$

Hence $$\vec{k} \times \vec{E} = \omega\mu_0 \vec{H} \quad \text{(viii)}$$

Similarly, it can be shown using Eq. (iv) that

$$\vec{k} \times \vec{H} = -\omega\varepsilon_0 \vec{E} \quad \text{(ix)}$$

From Eq. (viii) it is obvious that the magnetic field vector $\vec{H}$ is perpendicular to both the propagation vector $\vec{k}$ and the electric field vector $\vec{E}$ and according to Eq. (ix) $\vec{E}$ is perpendicular to both $\vec{k}$ and $\vec{H}$. Therefore, it may be concluded that the electric and magnetic vectors are normal to each other as well as to the direction of propagation of the wave or $\vec{E}$, $\vec{H}$ and direction of wave propagation $\vec{k}$ form a set of orthogonal vectors. Further, we can prove that the electromagnetic field or wave travels at the speed of light c in free space. For this, the cross product of $\vec{k}$ with Eq. (viii) gives

$$\vec{k} \times (\vec{k} \times \vec{E}) = \omega\mu_0(\vec{k} \times \vec{H})$$

$$\vec{k}(\vec{k} \cdot \vec{E}) - k^2\vec{E} = \omega\mu_0[-\omega\varepsilon_0\vec{E}] \quad \text{[Putting the value of } \vec{k} \times \vec{H} \text{ from Eq. (ix)]}$$

Since $\vec{k}$ and $\vec{E}$ are perpendicular to each other, $\vec{k} \cdot \vec{E} = 0$ and the above equation reads

$$k^2\vec{E} - \omega^2\mu_0\varepsilon_0\vec{E} = 0$$

$$(k^2 - \omega^2\mu_0\varepsilon_0)\vec{E} = 0$$

This relation between ω and k is known as dispersion relation.

Since $\vec{E}$ cannot be zero for the wave, $k^2 - \omega^2\mu_0\varepsilon_0 = 0$

$$\Rightarrow \quad \frac{\omega}{k} = \frac{1}{\sqrt{\mu_0\varepsilon_0}} = 3 \times 10^8 \text{ m/sec} = c, \text{ the speed of light.}$$

Therefore, the phase velocity $\frac{\omega}{k}$ of the electromagnetic wave is equal to the speed of light c in free space or vacuum.

11.24 TRANSVERSE NATURE OF ELECTROMAGNETIC WAVES LO11

Transverse nature of electromagnetic waves can be proved with the help of Maxwell's equations div $\vec{E} = 0$ and div $\vec{H} = 0$ for free space. Using different relations as discussed in the previous section div $\vec{E}$ and div $\vec{H}$ can be calculated as follows.

$$\vec{\nabla}\cdot\vec{E} = \left(\frac{\partial}{\partial x}\hat{i} + \frac{\partial}{\partial y}\hat{j} + \frac{\partial}{\partial z}\hat{k}\right)\cdot \vec{E}_0 e^{i(\vec{k}\cdot\vec{r}-\omega t)}$$

$$= \left(\frac{\partial}{\partial x}\hat{i} + \frac{\partial}{\partial y}\hat{j} + \frac{\partial}{\partial z}\hat{k}\right)\cdot[(E_{0x}\hat{i} + E_{0y}\hat{j} + E_{0z}\hat{k})e^{i(k_x x+k_y y+k_z z-\omega t)}]$$

$$= \frac{\partial}{\partial x}[E_{0x}e^{i(k_x x+k_y y+k_z z-\omega t)}] + \frac{\partial}{\partial y}[E_{0y}e^{i(k_x x+k_y y+k_z z-\omega t)}] + \frac{\partial}{\partial z}[E_{0z}e^{i(k_x x+k_y y+k_z z-\omega t)}]$$

$$= (E_{0x}ik_x + E_{0y}ik_y + E_{0z}ik_z)e^{i(k_x x+k_y y+k_z z-\omega t)}$$

$$= i(k_x E_{0x} + k_y E_{0y} + k_z E_{0z})e^{i(\vec{k}\cdot\vec{r}-\omega t)}$$

$$= i(\vec{k}\cdot\vec{E}_0)e^{i(\vec{k}\cdot\vec{r}-\omega t)}$$

$$= i\vec{k}\cdot\vec{E} \qquad (\because \vec{E}_0 e^{i(\vec{k}\cdot\vec{r}-\omega t)} = \vec{E})$$

$\therefore$ For free space, $\vec{\nabla}\cdot\vec{E} = 0$

$$i\vec{k}\cdot\vec{E} = 0 \quad \text{or} \quad \vec{k}\cdot\vec{E} = 0$$

It means the wave vector $\vec{k}$ is perpendicular to $\vec{E}$.

Similarly, from $\vec{\nabla}\cdot\vec{H} = 0$ it can be shown that $\vec{k}\cdot\vec{H} = 0$.

Hence, the wave vector $\vec{k}$ is perpendicular to $\vec{H}$. Therefore, the relations $\vec{k}\cdot\vec{E} = 0$ and $\vec{k}\cdot\vec{H} = 0$ indicate that the electromagnetic field vectors $\vec{E}$ and $\vec{H}$ (or $\vec{B}$, as $\vec{B} = \mu_0\vec{H}$) both are perpendicular to the direction of propagation vector $\vec{k}$. It means that the electromagnetic waves are transverse in nature.

11.25 MAXWELL'S EQUATIONS IN ISOTROPIC DIELECTRIC MEDIUM: EM WAVE PROPOGATION LO11

In an isotropic dielectric medium, the conduction or free current density $\vec{J}$ and volume charge density ρ are zero. Further, the displacement vector $\vec{D}$ and the magnetic field $\vec{B}$ are defined as $\vec{D} = \varepsilon\vec{E}$ and $\vec{B} = \mu\vec{H}$. In fact $\vec{D} = \varepsilon_0\vec{E} + \vec{P} = \varepsilon_0\vec{E} + \varepsilon_0\chi_e\vec{E} \equiv \varepsilon\vec{E}$ together with $\varepsilon = \varepsilon_0(1 + \chi_e)$ and $\vec{B} = \mu_0\vec{H} + \mu_0\vec{M} = \mu_0\vec{H} + \mu_0\chi_m\vec{H} \equiv \mu\vec{H}$ together with $\mu = \mu_0(1 + \chi_m)$ for the isotropic linear dielectric (polarizable and magnetic) medium. Here, the vectors $\vec{P}$ and $\vec{M}$ give respectively the measure of polarization and magnetization of the medium. However, for the dielectric medium, it would be sufficient to remember that ε_0 and μ_0 of free space have been simply replaced with ε and μ. Hence, for dielectric medium

$$J = 0 \quad (\text{or } \sigma = 0, f = 0, D = \varepsilon E \text{ and } \vec{B} = \mu\vec{H})$$

where ε and μ, which are respectively the absolute permittivity and permeability of the medium. Under this situation, we can express the Maxwell's equation as

$$\vec{\nabla}\cdot\vec{E} = 0 \tag{i}$$

$$\vec{\nabla} \cdot \vec{H} = 0 \tag{ii}$$

$$\vec{\nabla} \times \vec{E} = -\mu \frac{\partial \vec{H}}{\partial t} \tag{iii}$$

$$\vec{\nabla} \times \vec{H} = \varepsilon \frac{\partial \vec{E}}{\partial t} \tag{iv}$$

Taking curl of Eq. (iii), we get

$$\vec{\nabla} \times (\vec{\nabla} \times \vec{E}) = \vec{\nabla} \times \left[-\mu \frac{\partial \vec{H}}{\partial t} \right]$$

or
$$\vec{\nabla}(\vec{\nabla} \cdot \vec{E}) - \nabla^2 \vec{E} = -\mu \frac{\partial}{\partial t}(\vec{\nabla} \times \vec{H})$$

or
$$0 - \nabla^2 E = -\mu\varepsilon \frac{\partial^2 \vec{E}}{\partial t} \qquad \text{[Using Eqs (i) and (iv)]}$$

$$\nabla^2 \vec{E} = \mu\varepsilon \frac{\partial^2 \vec{E}}{\partial t^2} \tag{v}$$

Similarly, taking curl of Eq. (iv) and using Eqs. (ii) and (iii), we get

$$\nabla^2 \vec{H} = \mu\varepsilon \frac{\partial^2 \vec{H}}{\partial t^2} \tag{vi}$$

As discussed earlier, $\frac{1}{\sqrt{\mu\varepsilon}}$ gives the phase velocity of the wave in the medium. If we represent this as v, we obtain from Eqs. (v) and (vi)

$$\nabla^2 \vec{E} - \frac{1}{v^2}\frac{\partial^2 \vec{E}}{\partial t^2} = 0$$

and
$$\nabla^2 \vec{H} - \frac{1}{v^2}\frac{\partial^2 \vec{H}}{\partial t^2} = 0$$

Eqs. (v) and (vi) are the wave equations in an isotropic linear dielectric medium.

Now, $v = \frac{1}{\sqrt{\mu\varepsilon}} = \frac{1}{\sqrt{\mu_0\mu_r\varepsilon_0\varepsilon_r}}$ $\qquad (\because \mu = \mu_0\mu_r \text{ and } \varepsilon = \varepsilon_0\varepsilon_r)$

or
$$v = \frac{c}{\sqrt{\mu_r\varepsilon_r}} \qquad \left[\because \; c = \frac{1}{\sqrt{\mu_0\varepsilon_0}} \right] \tag{vii}$$

Eq. (vii) shows that the propagation velocity of an electromagnetic wave in a dielectric medium is less than that in free space.

Also, refractive index $= \frac{c}{v} = \sqrt{\mu_r\varepsilon_r}$ (viii)

For non-magnetic dielectric medium $\mu_r \approx 1$. Hence, refractive index $= \sqrt{\varepsilon_r}$ or Refractive index $= \sqrt{\text{Relative permittivity}}$

11.26 MAXWELL'S EQUATIONS IN CONDUCTING MEDIUM: EM WAVE PROPAGATION AND SKIN DEPTH

LO11

We consider a linear and isotropic conducting medium whose permeability is μ, permittivity is ε and the conductivity is σ. Under this situation, we can write the Maxwell's equation as

$$\vec{\nabla}\cdot\vec{E}=0 \tag{i}$$

$$\vec{\nabla}\cdot\vec{H}=0 \tag{ii}$$

$$\vec{\nabla}\times\vec{E}=-\frac{\partial\vec{B}}{\partial t}=-\mu\frac{\partial\vec{H}}{\partial t} \tag{iii}$$

$$\vec{\nabla}\times\vec{H}=\vec{J}+\varepsilon\frac{\partial\vec{E}}{\partial t}$$

or $$\vec{\nabla}\times\vec{H}=\sigma\vec{E}+\varepsilon\frac{\partial\vec{E}}{\partial t} \qquad [\because \vec{J}=\sigma\vec{E}] \tag{iv}$$

Taking curl of Eq. (iii), we have

$$\vec{\nabla}\times(\vec{\nabla}\times\vec{E})=\vec{\nabla}\times\left[-\mu\frac{\partial\vec{H}}{\partial t}\right]$$

$$=-\mu\left[\vec{\nabla}\times\frac{\partial\vec{H}}{\partial t}\right]$$

$$=-\mu\frac{\partial}{\partial t}[\vec{\nabla}\times\vec{H}]$$

$$=-\mu\frac{\partial}{\partial t}\left[\sigma\vec{E}+\varepsilon\frac{\partial\vec{E}}{\partial t}\right] \qquad \text{[using Eq. (iv)]}$$

or $$\vec{\nabla}\times(\vec{\nabla}\times\vec{E})=-\mu\sigma\frac{\partial\vec{E}}{\partial t}-\mu\varepsilon\frac{\partial^2\vec{E}}{\partial t^2}$$

Also, $\vec{\nabla}\times(\vec{\nabla}\times\vec{E})=\vec{\nabla}(\vec{\nabla}\cdot\vec{E})-(\vec{\nabla}\cdot\vec{\nabla})\vec{E}=-\nabla^2\vec{E}$ $\qquad [\therefore \vec{\nabla}\cdot\vec{E}=0$ from Eq. (i)]

$\therefore$ $$-\nabla^2\vec{E}=-\mu\sigma\frac{\partial\vec{E}}{\partial t}-\mu\varepsilon\frac{\partial^2\vec{E}}{\partial t^2}$$

or $$\nabla^2\vec{E}=+\mu\sigma\frac{\partial\vec{E}}{\partial t}+\mu\varepsilon\frac{\partial^2\vec{E}}{\partial t^2} \tag{v}$$

Eq. (v) is the electromagnetic wave equation for the electric field $\vec{E}$ in a conducting medium.

In case of non-conducting medium $\sigma = 0$. Hence, from Eq. (v)

$$\nabla^2\vec{E}=\mu\varepsilon\frac{\partial^2\vec{E}}{\partial t^2} \tag{vi}$$

Eqs (v) and (vi) show that the term $\mu\sigma\frac{\partial\vec{E}}{\partial t}$ is the dissipative term which allows the current to flow through the medium due to the appearance of conductivity σ.

Now, by taking curl of Eq. (iv), we obtain

$$\vec{\nabla} \times (\vec{\nabla} \times \vec{H}) = \vec{\nabla} \times \left[\sigma \vec{E} + \varepsilon \frac{\partial \vec{E}}{\partial t} \right]$$

$$= \sigma(\vec{\nabla} \times \vec{E}) + \varepsilon \frac{\partial}{\partial t}(\vec{\nabla} \times \vec{E})$$

$$= \sigma\left(-\mu \frac{\partial \vec{H}}{\partial t}\right) + \varepsilon \frac{\partial}{\partial t}\left(-\mu \frac{\partial \vec{H}}{\partial t}\right) \quad \text{[Using Eq. (iii)]}$$

$$= -\mu\sigma \frac{\partial \vec{H}}{\partial t} - \varepsilon\mu \frac{\partial^2 \vec{H}}{\partial t^2}$$

$\because$
$$\vec{\nabla} \times (\vec{\nabla} \times \vec{H}) = \vec{\nabla}(\vec{\nabla} \cdot \vec{H}) - (\vec{\nabla} \cdot \vec{\nabla})\vec{H}$$

$$= 0 - \nabla^2 \vec{H} \quad [\therefore \vec{\nabla} \cdot \vec{H} = 0 \text{ from Eq. (ii)}]$$

$\therefore$
$$-\nabla^2 \vec{H} = -\mu\sigma \frac{\partial \vec{H}}{\partial t} - \mu\varepsilon \frac{\partial^2 \vec{H}}{\partial t^2}$$

or
$$\nabla^2 \vec{H} = \mu\sigma \frac{\partial \vec{H}}{\partial t} + \mu\varepsilon \frac{\partial^2 \vec{H}}{\partial t^2} \quad \text{(vii)}$$

Equations (vii) represents the electromagnetic wave equation for magnetic field ($\vec{H}$) in conducting medium. In case of non-conducting medium $\sigma = 0$, the wave equation takes the form

$$\nabla^2 \vec{H} = \mu\varepsilon \frac{\partial^2 \vec{H}}{\partial t^2} \quad \text{(viii)}$$

Equations (vii) show that the term $\mu\sigma \frac{\partial \vec{H}}{\partial t}$ is the dissipative term which allows the current to flow through the conducting medium.

11.26.1 Solution of Wave Equation

Equations (v) and (vii) are called inhomogeneous wave equation due to the presence of dissipative term. These equations in one-dimension (along z-axis) are written as

$$\frac{\partial^2 \vec{E}}{\partial z^2} = \mu\sigma \frac{\partial \vec{E}}{\partial t} + \mu\varepsilon \frac{\partial^2 \vec{E}}{\partial t^2} \quad \text{(ix)}$$

$$\frac{\partial^2 \vec{H}}{\partial z^2} = \mu\sigma \frac{\partial \vec{H}}{\partial t} + \mu\varepsilon \frac{\partial^2 \vec{H}}{\partial t^2} \quad \text{(x)}$$

We assume the following plane wave solutions to the above equations

$$\vec{E}(z, t) = \vec{E}_0 \, e^{i(kz - \omega t)} \quad \text{(xi)}$$

and
$$\vec{H}(z, t) = \vec{H}_0 \, e^{i(kz - \omega t)} \quad \text{(xii)}$$

The use of Eq. (xi) in Eq. (ix) and Eq. (xii) in Eq. (x) leads

$$k^2 = \mu\varepsilon\omega^2 + i\mu\sigma\omega \quad \text{(xiii)}$$

This relation is called dispersion relation that governs the electromagnetic wave propagation in a conducting medium. Equation (xiii) suggests that the quantity k, i.e., wave number, will be a complex quantity. So, we assume

$$k = k_r + ik_i \quad \text{(xiv)}$$

With this the fields $\vec{E}$ and $\vec{H}$ become

$$\vec{E}(z,t) = \vec{E}_0 \bar{e}^{k_i z} \cdot e^{i(k_r z - \omega t)} \tag{xv}$$

and
$$\vec{H}(z,t) = \vec{H}_0 \bar{e}^{k_i z}\, e^{i(k_r z - \omega t)} \tag{xvi}$$

11.26.2 Skin Depth

The expressions (xv) and (xvi) follow that the amplitude of the electric field $\vec{E}$ is $E_0 \bar{e}^{k_i z}$ and that of the magnetic field $\vec{H}$ is $H_0 \bar{e}^{k_i z}$. Hence the amplitude of electromagnetic wave will decrease exponentially as it propagates through the conductor. This is called the attenuation of the wave and the distance through which the amplitude is reduced by a factor of $1/e$ is called the skin depth or penetration depth δ. At $z = \delta$, the amplitude is E_0/e. Hence

$$E_0 \bar{e}^{k_i \delta} = E_0/e \tag{i}$$

This gives the skin depth as

$$\delta = \frac{1}{k_i} \tag{ii}$$

Equation (ii) shows that the imaginary part of the wave number k is the measure of the skin depth. However, the real part k_r of k determines the wave propagation characteristics in the following manner.

Wavelength $\lambda = 2\pi/k_r$ (iii)

Phase velocity $v = \omega/k_r$ (iv)

Refractive index $n = \dfrac{c}{v} = \dfrac{ck_r}{\omega}$ (v)

By putting $k = k_r + i\,k_i$ in Eq. (xiii) we obtain

$$k_r = \omega\sqrt{\frac{\mu\varepsilon}{2}\left[\sqrt{1+\left(\frac{\sigma}{\omega\varepsilon}\right)^2}+1\right]} \tag{vi}$$

and
$$k_i = \omega\sqrt{\frac{\mu\varepsilon}{2}\left[\sqrt{1+\left(\frac{\sigma}{\omega\varepsilon}\right)^2}-1\right]} \tag{vii}$$

For good conductors, $\sigma >> \omega\varepsilon$. This condition when put in Eqs. (vi) and (vii) gives

$$k_r = k_i = \omega\sqrt{\frac{\mu\varepsilon}{2}\frac{\sigma}{\omega\varepsilon}}$$

$$\Rightarrow \quad k_r = k_i = \sqrt{\frac{\omega\mu\sigma}{2}} = \sqrt{\pi f \mu\sigma} \tag{viii}$$

Hence, the skin depth is given by

$$\delta = \frac{1}{\sqrt{\pi f \mu\sigma}} \tag{ix}$$

Since δ is inversely proportional to f, which is the frequency of electromagnetic wave, high frequency waves are found to penetrate less into the conductor. Also, the penetration will be less in the medium having high conductivity σ. Ideally an electromagnetic wave will not penetrate into a perfect conductor as $\sigma = \infty$.

11.26.3 Phase Relationship of $\vec{E}$ and $\vec{B}$ Fields

In view of the imaginary wave number k_i, we can also make another observation with regard to the phase difference between $\vec{E}$ and $\vec{H}$ vectors. If we take the direction of $\vec{E}$ field along the x-axis, then $\vec{\nabla} \times \vec{E} = -\mu \frac{\partial \vec{H}}{\partial t}$ gives

$$\vec{H}(z,t) = \hat{j}\frac{kE_0}{\omega\mu} e^{-k_i z} e^{i(k_r z - \omega t)} \quad \text{(x)}$$

Since k is the complex quantity, it can be represented as $k = k' e^{i\theta_k}$. Here $k' = \sqrt{k_r^2 + k_i^2}$ and $\theta_k = \tan^{-1}\left(\frac{k_i'}{k_r}\right)$. Then the expression of $\vec{H}$ becomes

$$\vec{H}(z,t) = \hat{j}\frac{k'E_0}{\omega\mu} e^{-k_i z} e^{i(k_r z - \omega t + \theta_k)} \quad \text{(xi)}$$

A comparison of Eq. (xi) with $\vec{E}(z,t) = \hat{i}E_0 e^{-k_i z} e^{i(k_r z - \omega t)}$ reveals that the electric field and magnetic field vectors do not remain in phase when electromagnetic wave propagates in a conducing medium. This is in contrast to the cases of vacuum and dielectrics.

11.27 ELECTROMAGNETIC ENERGY DENSITY — LO11

It can be proved that the work done in assembling a static charge distribution (number n) against the Coulomb repulsion of like charges is

$$W_E = \frac{1}{2}\sum_{j=1}^{n} q_j V(\vec{r}_j) \quad \text{(i)}$$

For a volume charge density ρ, this equation takes the form

$$W_E = \frac{1}{2}\int \rho V dX \quad \text{(ii)}$$

Here dX is the volume element. Now the above equation can be written in terms of the resulting electric field $\vec{E}$ if we apply Gauss's law $\varepsilon_0 \vec{\nabla} \cdot \vec{E} = \rho$ and mention the potential V in terms of $\vec{E}$. This yields the following relation where the integration is over all the space containing the whole charge distribution.

$$W_E = \frac{\varepsilon_0}{2}\int E^2 dX \quad \text{(iii)}$$

The same way we can derive an expression for the work done on a unit charge against the back emf in one trip around the circuit, as follows

$$W_B = \frac{1}{2\mu_0}\int B^2 dX \quad \text{(iv)}$$

Here B is the resulting magnetic field. Eqs. (iii) and (iv) suggest that the total energy stored in electromagnetic field would be

$$W_{EM} = \frac{1}{2}\int\left(\varepsilon_0 E^2 + \frac{B^2}{\mu_0}\right) dX \quad \text{(v)}$$

Therefore, the electromagnetic energy density can be obtained as

$$U_{EM} = \frac{1}{2}\left(\varepsilon_0 E^2 + \frac{B^2}{\mu_0}\right) \tag{vi}$$

For a monochromatic plane electromagnetic wave, $B = \frac{E}{c}$, where $c \left(\equiv \frac{1}{\sqrt{\mu_0 \varepsilon_0}}\right)$ is the speed of light. Hence, it can be seen that the contribution of magnetic field $\vec{B}$ to the energy density is the same as that of the electric field $\vec{E}$.

11.28 POYNTING VECTOR AND POYNTING THEOREM LO12

The electromagnetic waves carry energy when they propagate and there is an energy density associated with both the electric and magnetic fields.

The amount of energy flowing through unit area, perpendicular to the direction of energy propagation per unit time, i.e., the rate of energy transport per unit area, is called the *poynting vector*. It is also termed as instantaneous energy flux density and is represented by $\vec{S}$ (or $\vec{P}$, sometimes). Mathematically it is defined

$$\vec{S} = \vec{E} \times \vec{H}$$

where $\vec{E}$ and $\vec{H}$ represent the instantaneous values of the electric and magnetic field vectors. This is clear that the rate of energy transport $\vec{S}$ is perpendicular to both $\vec{E}$ and $\vec{H}$ and is in the direction of propagation of the wave, as $\vec{E} \times \vec{H}$ is in the direction of $\vec{k}$. Since the poynting vector represents the rate of energy transport per unit area, its units are W/m^2.

Derivation: We can calculate the energy density carried by electromagnetic waves with the help of Maxwell's equations given below.

$$\text{div}\, \vec{D} = 0 \tag{i}$$

$$\text{div}\, \vec{B} = 0 \tag{ii}$$

$$\text{curl}\, \vec{E} = \frac{-\partial \vec{B}}{\partial t} \tag{iii}$$

$$\text{curl}\, \vec{H} = \vec{J} + \frac{\partial \vec{D}}{\partial t} \tag{iv}$$

Take scalar (dot) product of Eq. (iii) and Eq. (iv) with $\vec{H}$ and $\vec{E}$ respectively, i.e.,

$$\vec{H} \cdot \text{curl}\, \vec{E} = -\vec{H} \cdot \frac{\partial \vec{B}}{\partial t} \tag{v}$$

and

$$\vec{E} \cdot \text{curl}\, \vec{H} = \vec{E} \cdot \vec{J} + \vec{E} \cdot \frac{\partial \vec{D}}{\partial t} \tag{vi}$$

Subtract Eq. (vi) from Eq. (v), i.e.,

$$\vec{H} \cdot \text{curl}\, \vec{E} - \vec{E} \cdot \text{curl}\, \vec{H} = -\vec{H} \cdot \frac{\partial \vec{B}}{\partial t} - \vec{E} \cdot \vec{J} - \vec{E} \cdot \frac{\partial \vec{D}}{\partial t}$$

$$= -\left(\vec{H} \cdot \frac{\partial \vec{B}}{\partial t} + \vec{E} \cdot \frac{\partial \vec{D}}{\partial t}\right) - \vec{E} \cdot \vec{J}$$

or
$$\text{div}(\vec{E}\times\vec{H}) = -\left(\vec{H}\cdot\frac{\partial\vec{B}}{\partial t} + \vec{E}\cdot\frac{\partial\vec{D}}{\partial t}\right) - \vec{E}\cdot\vec{J} \tag{vii}$$

$$[\because \text{div}(\vec{A}\times\vec{B}) = \vec{B}\cdot\text{curl}\,\vec{A} - \vec{A}\cdot\text{curl}\,\vec{B}]$$

Using the relations $\vec{B} = \mu\vec{H}$ and $\vec{D} = \varepsilon\vec{E}$, we can get

$$\vec{E}\cdot\frac{\partial\vec{D}}{\partial t} = \vec{E}\cdot\frac{\partial}{\partial t}(\varepsilon\vec{E}) = \frac{1}{2}\varepsilon\frac{\partial}{\partial t}(E)^2 = \frac{\partial}{\partial t}\left(\frac{1}{2}\vec{E}\cdot\vec{D}\right) \qquad [\because E^2 = \vec{E}\cdot\vec{E}]$$

$$\vec{H}\cdot\frac{\partial\vec{B}}{\partial t} = \vec{H}\cdot\frac{\partial}{\partial t}(\mu\vec{H}) = \frac{1}{2}\mu\frac{\partial}{\partial t}(H)^2 = \frac{\partial}{\partial t}\left(\frac{1}{2}\vec{H}\cdot\vec{B}\right) \qquad [\because H^2 = \vec{H}\cdot\vec{H}]$$

Now Eq. (vii) can be written as

$$\text{div}(\vec{E}\times\vec{H}) = \frac{\partial}{\partial t}\left[\frac{1}{2}(\vec{H}\cdot\vec{B} + \vec{E}\cdot\vec{D})\right] - \vec{E}\cdot\vec{J}$$

or
$$\vec{E}\cdot\vec{J} = \frac{\partial}{\partial t}\left[\frac{1}{2}(\vec{H}\cdot\vec{B} + \vec{E}\cdot\vec{D})\right] - \text{div}(\vec{E}\times\vec{H}) \tag{viii}$$

Integrating Eq. (viii) over a volume V enclosed by a surface S, we get

$$\int_v \vec{E}\cdot\vec{J}dV = -\int_v\left[\frac{\partial}{\partial t}\left\{\frac{1}{2}(\vec{H}\cdot\vec{B} + \vec{E}\cdot\vec{D})\right\}\right]dV - \int_v \text{div}(\vec{E}\times\vec{H})dV$$

or
$$\int_v \vec{E}\cdot\vec{J}dV = -\frac{\partial}{\partial t}\int_v\left[\left(\frac{1}{2}\mu H^2 + \frac{1}{2}\varepsilon E^2\right)\right]dV - \oint_s(\vec{E}\times\vec{H})\cdot d\vec{S} \tag{ix}$$

$$[\because \ \vec{B} = \mu\vec{H}, \vec{D} = \varepsilon\vec{E} \ \text{ and } \ \int_v \text{div}(\vec{E}\times\vec{H})dV = \oint_s(\vec{E}\times\vec{H})\cdot d\vec{S}]$$

Eq. (ix) can also be written as

$$\int_v \vec{E}\cdot\vec{J}\,dV = -\frac{\partial}{\partial t}\int_v\left[\frac{1}{2}\mu H^2 + \frac{1}{2}\varepsilon E^2\right]dV - \oint_s(\vec{E}\times\vec{H})\cdot d\vec{S} \tag{x}$$

Interpretation

(a) $\int_v(\vec{E}\cdot\vec{J})dV$: This term represents the rate of energy transferred into the electromagnetic field through the motion of charges in the volume V, i.e., the total power dissipated in a volume V.

(b) $\frac{\partial}{\partial t}\int_v\left[\frac{1}{2}\mu H^2 + \frac{1}{2}\varepsilon E^2\right]dV$: The terms $\frac{1}{2}\mu H^2$ and $\frac{1}{2}\varepsilon E^2$ represent the energy stored in electric and magnetic fields respectively and their sum will be equal to the total energy stored in electromagnetic field. Therefore, this total expression represents the rate of decrease of energy stored in volume V due to electric and magnetic fields.

(c) $\int_s(\vec{E}\times\vec{H})\cdot d\vec{S}$: This term represents the amount of electromagnetic energy crossing the closed surface per second or the rate of flow of outward energy through the surface S enclosing volume V. The vector $(\vec{E}\times\vec{H})$ is known as the *poynting vector* $\vec{S}$ or $\vec{S} = (\vec{E}\times\vec{H})$.

Equation (x) is also known as *poynting theorem* or *work-energy theorem*, according to which the power transferred into the electromagnetic field is equal to the sum of the time rate of change of electromagnetic energy within a certain volume and the time rate of the energy flowing out through the boundary surface. This is also called as the *energy conservation* law in electromagnetism.

11.29 WAVE PROPAGATION IN BOUNDED SYSTEM: WAVEGUIDE

LO12

The first waveguide was proposed by *Thomson* in 1893 and was experimentally verified by *Lodge* in 1894. A structure that can guide waves like electromagnetic waves, light waves or sound waves is called *waveguide*. For each type of the wave there are different types of waveguides. For example, depending on the frequency of electromagnetic wave the waveguide can be constructed from either conductive or dielectric material. Such electromagnetic waveguides are especially useful in the microwave and optical frequency ranges. The waveguides used at optical frequencies (optical waveguides) are typically dielectric waveguides. In such waveguides, a dielectric material with high permittivity (i.e., high index of refraction) is surrounded by a material with lower permittivity. This type of structure guides optical waves by the process of total internal reflection. The most common optical waveguide is optical fibre. On the other hand, a structure that guides sound waves is called an *acoustic waveguide*. A duct for sound propagation also behaves as a transmission line. The duct contains some medium like air which supports the propagation of sound waves.

11.29.1 Electromagnetic Waveguides

The original and the most common meaning of waveguide is a hollow metal pipe used for guiding the waves. The electromagnetic waves in such waveguides may be imagined as waves travelling down the guide in a zig zag path as these waves are repeatedly reflected between opposite walls of the guide (for example, rectangular waveguide). The first mathematical analysis of the propagating modes (waves) within a hollow metal cylinder was performed by *Rayleigh* in 1897.

To function properly, a waveguide must have a certain minimum diameter relative to the wavelength of the signal. If the waveguide is too narrow or the frequency is too low (the wavelength is too long), the electromagnetic field cannot propagate. There is a minimum frequency, known as *cutoff frequency* for the propagation of the wave, i.e., a wave can propagate only if its frequency is larger than the cutoff frequency. The cutoff frequency is decided by the dimensions of the waveguide.

11.29.2 Modes in Waveguides

In order to analyse the mode (wave) propagation in the waveguide, we solve the Maxwell's equations along with appropriate bounding conditions determined by the properties of the materials and their interfaces. These equations admit multiple solutions, or modes, which are origin functions of the equation system. The propagation of the waveguide modes depends on the operating wavelength and polarization, and shape and size of the waveguide. The longitudinal mode can be realised in a cavity (closed end waveguide), the longitudinal mode is particularly standing wave pattern formed by the waves confined in the cavity. However, a number of transverse modes can be excited in the waveguide, which are classified below.

(1) ***Transverse Electric (TE) Modes:*** These modes do not have electric field in the direction of propagation. So electric field vector is in transverse direction.

(2) ***Transverse Magnetic (TM) Modes:*** These modes have no magnetic field in the direction of propagation. So magnetic field vector is in transverse direction.

(3) ***Transverse Electromagnetic (TEM) Modes:*** These modes have no electric and magnetic fields in the direction of mode propagation. In hollow waveguides, TEM modes are not possible because as per Maxwell's equation the electric field then must have zero divergence, zero curl and be zero at the boundaries. This will result in a zero field or $\nabla^2\vec{E} = 0$. However, TEM modes can propagate in a coaxial cable.

(4) ***Hybrid Modes:*** These modes have both electric and magnetic field components in the direction of propagation. The mode for which the cutoff frequency is the minimum is called the *fundamental mode*. For example, Transverse Electric TE_{10} mode is the fundamental mode for rectangular waveguide whereas TE_{11} mode is the fundamental mode for circular waveguide.

11.30 COAXIAL CABLE LO12

A transmission line is the material medium that forms a path from one place to another for directing the transmission of energy like electromagnetic waves, acoustic waves or electric power. Open wire transmission lines have the property that the electromagnetic wave propagating down the line extends into the space surrounding the parallel wires. The transmission lines cannot be bent, twisted or otherwise shaped without changing their characteristic impedance. They also cannot be attached to anything conductive, as the extended fields will induce currents in the nearby conductors. This will cause unwanted radiation and detuning of the line. However, this problem is solved by coaxial lines that confine the electromagnetic wave to the area inside the cable, as the coaxial cable consists of a round conducting wire surrounded by an insulating spaces (called the dielectric), surrounded by a cylindrical conducting sheath that is usually surrounded by a final insulating layer (called jacket). Here the transmission of energy occurs totally through the dielectric inside the cable between the conductors. Coaxial lines (cables) can therefore be bent and moderately twisted without negative effects. Also, they can be strapped to conductive supports without inducing unwanted currents in them. Coaxial cable is used as a high-frequency transmission line to carry a high frequency or broadband signals. Coaxial cables may be flexible or rigid depending on the type of sheath. Rigid types have a solid sheath, whereas flexible types have an interwoven sheath usually of thin copper wire. The inner insulator has a significant effect on the cable's properties such as its characteristic impedance and its attenuation. The characteristic impedance in ohms is calculated from the ratio of the inner diameter d and outer diameter D of the dielectric and the dielectric constant ε_r as below

$$Z_0 = \frac{138}{\sqrt{\varepsilon_r}} \log\left(\frac{D}{d}\right)$$

The most common impedances that are widely used are 50 or 52 Ω for industrial and commercial radio frequency applications, and 75 Ω for domestic television and radio, although other impedances are available for specified applications.

11.30.1 Signal Propagation in Coaxial Cables

In radio-frequency applications up to a few GHz (10^9Hz) the wave propagates only in the transverse electromagnetic (TEM) mode which means the electric and magnetic fields are both perpendicular to the direction of propagation. However, above a certain cutoff frequency, transverse electric (*TE*) and/or transverse magnetic (*TM*) modes can also propagate, as they do in a waveguide. It is usually undesirable to transmit signals above the cutoff frequency, since it may cause multiple modes with different phase velocities to propagate, interfering with each other. This cutoff frequency is roughly inversely proportional to the outer diameter.

11.30.2 Uses of Coaxial Cables

Short length coaxial cables are commonly used to connect home video equipment and in measurement electronics. They are used to be common for implementing computer networks (particularly ethernet) but twisted pair cables have replaced them in most applications except in the growing consumer cable modem market for broadband internet access.

Long distance coaxial cable is used to connect radio networks and television networks, though this has largely been superseded by other more high-tech methods which use fibre optics, satellite etc. Micro coaxial cables are used in a range of consumer devices, military, equipment and also in ultrasound scanning equipment.

SUMMARY

The topics covered in this chapter are summarised below.

- The separation of charges produces electric field, whereas the motion of charges generates current and hence the magnetic field. When these fields are time varying they are coupled with each other through the Maxwell's equations. With the help of the Maxwell's equations, we can derive wave equation, based on which the propagation of electromagnetic waves can be investigated in different media.
- If a charge is distributed continuously in a medium, it can be expressed in terms of a physical quantity known as charge density. There are three types of charge densities, namely linear charge density λ, surface charge density σ and volume charge density ρ.
- Gauss's law of electricity states that the net outward electric flux through any closed surface drawn in an electric field is equal to $1/\varepsilon_0$ times the total charge enclosed within the surface. Here it is clear that the charge located outside the surface does not contribute to the electric flux and hence the electric field.
- The del operator is the differential operator, which is represented by $\vec{\nabla}$ and is given by $\vec{\nabla} = \hat{i}\frac{\partial}{\partial x} + \hat{j}\frac{\partial}{\partial y} + \hat{k}\frac{\partial}{\partial z}$. It is not a vector in itself, but when it operates on a scalar function it provides the resultant as a vector. $\vec{\nabla}F$ does not mean a multiplication of $\vec{\nabla}$ with F rather it is an instruction to differentiate. Here we should say that $\vec{\nabla}$ is a vector operator that acts upon F.
- The gradient of a scalar field F is a vector quantity which represents both the magnitude and the direction of the maximum space rate of increase of F. The gradient of a vector field cannot be realised.
- The divergence of a vector field $\vec{A}$ is defined as the net outward flux per unit volume over a closed surface. The gradient at a given point is a measure of how much the vector $\vec{A}$ spreads out, i.e., diverges, from that point.
- The curl of a vector field $\vec{A}$ is a rotational vector. Its magnitude would be the maximum circulation of $\vec{A}$ per unit area. Its direction is the normal direction of the area when the area is oriented so as to make the circulation maximum. The curl of $\vec{A}$ at some point O is a measure of how much the vector $\vec{A}$ curls around the point O.
- For a function $g(x)$, the fundamental theorem of calculus states that $\int_{a_1}^{a_2}\left(\frac{dg}{dx}\right)dx = g(a_2) - g(a_1)$ It means the total change in the function $g(x)$ can be obtained by simply subtracting the values of the function at the points a_2 and a_1.

- For a scalar function F, $\int_{a_1}^{a_2} \vec{\nabla}F \cdot d\vec{l} = F(a_2) - F(a_1)$. This is called the fundamental theorem for gradient, according to which the integral of a derivative over end points a_1 and a_2 is given by the value of the function at the boundaries, i.e., the points a_1 and a_2.
- According to Gauss's or Green's theorem, $\int_V (\vec{\nabla} \cdot \vec{F}) dV = \oint_S \vec{F} \cdot d\vec{S}$. Here dV is the volume element and $d\vec{S}$ is the surface element. This theorem states that the integral of a derivative (here the divergence) over a region (here the volume) is equal to the value of the function at the boundary (here the surface). Since the boundary of a volume is always a closed surface, the R.H.S. is the integral over closed surface. Evidently this theorem converts the volume integral into the surface integral.
- Stokes' theorem states that the integral of the curl of a vector function over a patch of surface is equal to the value of the function at the perimeter of the patch. So here the derivative is the curl, region is the surface and the boundary is the perimeter of the patch of the surface. Therefore,

$$\int_S (\vec{\nabla} \times \vec{F}) \cdot d\vec{S} = \oint_C \vec{F} \cdot d\vec{l}$$

Clearly, this theorem converts the surface integral into the line integral.
- If we represent the charge density by ρ and the current density by $\vec{J}$, then the continuity equation is written as $\vec{\nabla} \cdot \vec{J} + \frac{\partial \rho}{\partial t} = 0$. In this equation, the second term is the time rate of change of charge density, i.e., the current. The first term represents the divergence of $\vec{J}$, i.e., the measure of spreading of the current through a surface. This spreading is balanced by the time rate of change of charge density. Therefore, this equation just tells us the conservation of charges.
- When the charges are in motion, the electric and magnetic fields are associated with this motion, which will have variations in both the space and time. These electric and magnetic fields are interrelated. This phenomenon is called electromagnetism, which is summarised by the set of equations, known as Maxwell's equations. The Maxwell's equations are nothing but are the representation of the basic laws of electromagnetism. If the electric field is represented by $\vec{E}$, magnetic field by $\vec{B}$, current density by $\vec{J}$, then the Maxwell's equations are , $\vec{\nabla} \cdot \vec{D} = \rho$, $\vec{\nabla} \cdot \vec{B} = 0$, $\vec{\nabla} \times \vec{E} = -\frac{\partial \vec{B}}{\partial t}$ and $\vec{\nabla} \times \vec{H} = \vec{J} + \frac{\partial \vec{D}}{\partial t}$. Here $\vec{B} = \mu \vec{H}$ and $\vec{D} = \varepsilon \vec{E}$ for the linear media.
- The Maxwell's first equation $\vec{\nabla} \cdot \vec{D} = \rho$ represents the Gauss's Law for electricity along with ρ as the free charge density.
- The area integral of a vector field determines the net source of the field (function). Since the Maxwell's second equation $\vec{\nabla} \cdot \vec{B} = 0$ can be written as $\oint \vec{B} \cdot d\vec{S} = 0$, it says that the net magnetic flux out of any closed surface is zero. This is because the magnetic flux directed inward toward the south pole, of a magnetic dipole kept in any closed surface, will be equal to the flux outward the north pole. Therefore, the net flux is zero for dipole sources. For a magnetic monopole source, the value of the area integral $\oint \vec{B} \cdot d\vec{S}$ would be finite. Since the divergence of a vector field is proportional to the density of point source, the Maxwell's second equation simply says that there are no magnetic monopoles.

- The Maxwell's third equation $\vec{\nabla} \times \vec{E} = -\frac{\partial \vec{B}}{\partial t}$ when written in the integral form states that the line integral of the electric field around a closed loop is equal to the negative rate of change of the magnetic flux through the area enclosed by the loop. The line integral basically is the generated voltage or emf in the loop. Therefore, the physical interpretation of Maxwell's third equation is that the changing magnetic field induces electric field.
- For static electric field $\vec{E}$, i.e., when $\vec{E}$ does not change with time , the second term of the R.H.S. of the Maxwell's fourth equation $\vec{\nabla} \times \vec{H} = \vec{J} + \frac{\partial \vec{D}}{\partial t}$ vanishes and then the integral form of this equation says that the line integral of the magnetic field around a closed loop is proportional to the electric current flowing through the loop. This form of the Maxwell's equation is useful for calculating the magnetic field for simple geometries. However, this equation more specifically reveals that the changing electric field induces magnetic field as $\frac{\partial \vec{D}}{\partial t}$ is nothing but the rate of change of electric field. This is complimentary to the meaning of the Maxwell's third equation. Therefore, they together form the electromagnetic fields or electromagnetic waves, where both electric and magnetic fields propagate together and the change in one field induces the other field.
- The vector $\vec{D}(= \varepsilon\vec{E})$ is called the displacement vector and the term $\frac{\partial \vec{D}}{\partial t}$ is called the displacement current. The displacement current is postulated in a dielectric when electric stress or potential gradient is varied. It is different from a normal or conduction current, as it is not accompanied by the motion of current carriers in the dielectrics. The concept of introducing this term in the Ampere's law was given by Maxwell for the completion of his electromagnetic equations, as the Ampere's law is valid only for the steady state or static fields. In order to make it consistent with the time varying fields, i.e., also for satisfying the continuity equation, this additional term is required.
- Electromagnetic fields, i.e., the electric field and magnetic field are transverse in nature that always remain perpendicular to each other and also in the transverse direction to the direction of wave propagation.
- An electromagnetic wave of frequency ω and wave number k satisfies the dispersion relation $k^2 - \omega^2\mu_0\varepsilon_0 = 0$ in free space or vacuum. The wave propagates at the phase velocity $v_p = \frac{\omega}{k} = c$, the speed of light, in this medium.
- An electromagnetic wave of frequency ω and wave number k satisfies the dispersion relation $k^2 - \omega^2\mu\varepsilon = 0$ in a dielectric medium whose permittivity is ε and permeability is μ. Its phase velocity is $v_p = \frac{\omega}{k} = \frac{1}{\sqrt{\mu\varepsilon}} = \frac{c}{\sqrt{\mu_r\varepsilon_r}}$. It means this wave propagates at slower speed in dielectric medium compared with the vacuum or free space.
- For non-magnetic dielectric medium, $\mu_r = 1$ and its refractive index = $\sqrt{\text{relative permittivity}}$.
- The wave equation for an electromagnetic wave in a conducting medium is modified due to the term $\mu\sigma\frac{\partial \vec{E}}{\partial t}$ or $\mu\sigma\frac{\partial \vec{H}}{\partial t}$. This term is called dissipative term which allows the current to flow through the medium due to the appearance of conductivity σ.

- The wave number k in the case of the propagation of an electromagnetic wave in a conducting medium does not remain a real quantity rather becomes a complex quantity due to the development of its imaginary part. The imaginary wave number leads to the attenuation of the wave in the medium, i.e., exponential decay of the amplitude/field of the wave when it enters the conductors.
- The distance through which the amplitude of the wave is reduced by a factor of $1/e$ is called the skin depth or penetration depth δ, given by $\delta = 1/\sqrt{\pi f \mu \sigma}$, where f is the frequency of the wave, μ is the permeability and σ is the conductivity of the conducting medium.
- Unlike the cases of vacuum/free space and dielectric medium, the electric field and magnetic field vectors do not remain in phase when electromagnetic wave propagates in a conducting medium. This happens due to the presence of imaginary part of the wave number.
- The real part k_r of the wave number k determines the wave propagation characteristics when an electromagnetic wave propagates in a conducting medium.

  For example,

  Phase velocity $v_p = \omega/k_r$

  Wavelength $\lambda = 2\pi/k_r$

  Refractive index $n = \dfrac{c}{v_p} = \dfrac{ck_r}{\omega}$
- The electromagnetic waves carry energy when they propagate and there is an energy density associated with both the electric and magnetic fields. The amount of energy flowing through unit area, perpendicular to the direction of energy propagation per unit time, i.e., the rate of energy transport per unit area, is called the poynting vector. It is also known as instantaneous energy flux density and is represented by $\vec{S} = \vec{E} \times \vec{H}$.
- There are structures that can guide waves like electromagnetic waves, light waves or sound waves. These structures are called waveguides. For each type of the wave there are different types of waveguides. For example, depending on the frequency of electromagnetic wave the waveguide can be constructed from either conductive or dielectric material. Such electromagnetic waveguides are especially useful in the microwave and optical frequency ranges. The waveguides used at optical frequencies (optical waveguides) are typically dielectric waveguides.
- The waveguides support different types of wave propagation, which are called modes. In order to analyze the mode propagation, we solve the Maxwell's equations along with appropriate boundary conditions determined by the properties of the materials and their interfaces. These equations admit multiple solutions, or modes, which are origin functions of the equation system. The mode propagation in the waveguides depends on the operating wavelength and polarization, and shape and size of the waveguide.
- A number of transverse modes can be excited in the waveguide. For example, transverse electric (*TE*) modes which do not have electric field in the direction of propagation, transverse magnetic (*TM*) modes which have no magnetic field in the direction of propagation, transverse electromagnetic (*TEM*) modes which have no electric and magnetic fields in the direction of mode propagation. In hollow waveguides, *TEM* modes are not possible. However, *TEM* modes can propagate in a coaxial cable. Also hybrid modes can be excited which have both electric and magnetic field components in the direction of propagation.
- The mode for which the cutoff frequency is the minimum is called the fundamental mode. For example, TE_{10} mode is the fundamental mode for rectangular waveguide whereas TE_{11} mode is the fundamental mode for circular waveguide.

- A transmission line forms a path from one place to another for directing the transmission of energy like electromagnetic waves, acoustic waves or electric power. The transmission lines cannot be bent, twisted or otherwise shaped without changing their characteristic impedance. They cannot be attached to anything conductive, as the extended fields will induce currents in the nearby conductors. This will cause unwanted radiation and detuning of the line. However, coaxial lines that confine the electromagnetic wave to the area inside the cable solve this problem. Here the transmission of energy occurs totally through the dielectric inside the cable between the conductors. Coaxial lines (cables) can therefore be bent and moderately twisted without negative effects. Also, they can be strapped to conductive supports without inducing unwanted currents in them.
- In radio frequency applications up to a few GHz (10^9Hz) the wave propagates only in the form of transverse electromagnetic (*TEM*) mode. However, above a certain cutoff frequency, transverse electric (*TE*) and/or transverse magnetic (*TM*) modes can also propagate, as they do in a waveguide. It is usually undesirable to transmit signals above the cutoff frequency, since it may cause multiple modes with different phase velocities to propagate, interfering with each other.

SOLVED EXAMPLES

EXAMPLE 1 Find the value of $\vec{\nabla} r^n$ where $\vec{r} = x\hat{i} + y\hat{j} + z\hat{k}$.

SOLUTION
$$\vec{\nabla} r^n = \left(\hat{i}\frac{\partial}{\partial x} + \hat{j}\frac{\partial}{\partial y} + \hat{k}\frac{\partial}{\partial z}\right)(x^2 + y^2 + z^2)^{n/2}$$

where $r = (x^2 + y^2 + z^2)^{1/2}$

$$\vec{\nabla} r^n = \left\{\frac{n}{2}(x^2 + y^2 + z^2)^{\frac{n}{2}-1} 2x\right\}\hat{i} + \left\{\frac{n}{2}(x^2 + y^2 + z^2)^{\frac{n}{2}-1} 2y\right\}\hat{j} + \left\{\frac{n}{2}(x^2 + y^2 + z^2)\frac{n}{2} - 1\right\}\hat{k}$$

$$= n(x^2 + y^2 + z^2)^{\left(\frac{n}{2}-1\right)}(\hat{i}x + \hat{i}y + \hat{k}z)$$

$$= \boldsymbol{n r^{n-2} \vec{r}}$$

EXAMPLE 2 Prove that $\vec{\nabla}\left[\frac{1}{r^n}\right] = -\frac{n}{r^{n+2}}\vec{r}$, where $r = \sqrt{x^2 + y^2 + z^2}$.

SOLUTION
$$\vec{\nabla}\left(\frac{1}{r^n}\right) = \left(\hat{i}\frac{\partial}{\partial x} + \hat{j}\frac{\partial}{\partial y} + \hat{k}\frac{\partial}{\partial z}\right) r^{-n}$$

$$= \left(\hat{i}\frac{\partial}{\partial x} + \hat{j}\frac{\partial}{\partial y} + \hat{k}\frac{\partial}{\partial z}\right)(x^2 + y^2 + z^2)^{\frac{-n}{2}}$$

$$= \left\{\frac{-n}{2}(x^2 + y^2 + z^2)^{\frac{-n}{2}-1} 2x\right\}\hat{i} + \left\{\frac{-n}{2}(x^2 + y^2 + z^2)^{\frac{-n}{2}-1} 2y\right\}\hat{j} + \left\{\frac{-n}{2}(x^2 + y^2 + z^2)^{\frac{n}{2}-1}\right\}\hat{k}$$

$$= -n(x^2 + y^2 + z^2)^{-\left(\frac{n+2}{2}\right)}(\hat{i}x + \hat{j}y + \hat{k}z)$$

$$= \boldsymbol{\frac{-n}{r^{n+2}}\vec{r}}$$

Hence proved.

Example 3 If $\phi = x^{3/2} + y^{3/2} + z^{3/2}$, find $\vec{\nabla}\phi$.

Solution $\vec{\nabla}\phi = \left(\hat{i}\frac{\partial}{\partial x} + \hat{j}\frac{\partial}{\partial y} + \hat{k}\frac{\partial}{\partial z}\right)(x^{3/2} + y^{3/2} + z^{3/2})$

$$= \frac{3}{2}(\hat{i}x^{1/2} + \hat{j}y^{1/2} + \hat{k}z^{1/2})$$

Example 4 If $f(x, y, z) = 3x^2y - yz^2$, find grad ϕ at point (1, 2, –1).

Solution grad $\phi = \vec{\nabla}\phi = \left(\hat{i}\frac{\partial}{\partial x} + \hat{j}\frac{\partial}{\partial y} + \hat{k}\frac{\partial}{\partial z}\right)(3x^2y - yz^2)$

$$= \hat{i}(6xy) + \hat{j}(3x^2 - z^2) + \hat{k}(-2yz)$$

Value of grad ϕ at point (1, 2, –1) $= 12\hat{i} + 2\hat{j} + 4\hat{k}$

Example 5 If ϕ is a scalar field and $\vec{A}$ is vector field find the value of div $(\phi\vec{A})$.

Solution div $(\phi\vec{A}) = \vec{\nabla}\cdot(\phi\vec{A})$

$$= \left(\hat{i}\frac{\partial}{\partial x} + \hat{j}\frac{\partial}{\partial y} + \hat{k}\frac{\partial}{\partial x}\right)\cdot(\phi A_x\hat{i} + \phi A_y\hat{j} + \phi A_z\hat{k})$$

$$= \frac{\partial}{\partial x}(\phi A_x) + \frac{\partial}{\partial y}(\phi A_y) + \frac{\partial}{\partial z}(\phi A_z)$$

$$= \frac{\partial\phi}{\partial x}A_x + \phi\frac{\partial A_x}{\partial x} + \frac{\partial\phi}{\partial y}A_y + \phi\frac{\partial A_y}{\partial y} + \frac{\partial\phi}{\partial z}A_z + \phi\frac{\partial A_z}{\partial z}$$

$$= \left(\frac{\partial\phi}{\partial x}A_x + \frac{\partial\phi}{\partial y}A_y + \frac{\partial\phi}{\partial z}A_z\right) + \phi\left(\frac{\partial A_x}{\partial x} + \frac{\partial A_y}{\partial y} + \frac{\partial A_z}{\partial z}\right)$$

$$= \left(\frac{\partial\phi}{\partial x}\hat{i} + \frac{\partial\phi}{\partial y}\hat{j} + \frac{\partial\phi}{\partial z}\hat{k}\right)\cdot(\hat{i}A_x + \hat{j}A_y + \hat{k}A_z) + \phi\left(\frac{\partial A_x}{\partial x} + \frac{\partial A_y}{\partial y} + \frac{\partial A_z}{\partial z}\right)$$

$$\mathbf{div}(\phi\vec{A}) = \vec{\nabla}\phi\cdot\vec{A} + \phi(\vec{\nabla}\cdot A)$$

Example 6 Prove that $\vec{\nabla}\cdot(\vec{A} + \vec{B}) = \vec{\nabla}\cdot\vec{A} + \vec{\nabla}\cdot\vec{B}$ where A and B are differentiable vector functions.

Solution $\vec{A} = \hat{i}A_x + \hat{j}A_y + \hat{k}A_z$, $\vec{B} = \hat{i}B_x + \hat{j}B_y + \hat{k}B_z$

and $$(\vec{A} + \vec{B}) = \hat{i}(A_x + B_x) + \hat{j}(A_y + B_y) + \hat{k}(A_z + B_z) \quad \text{(i)}$$

Taking divergence on both sides of Eq. (i), we have

$$\vec{\nabla}\cdot(\vec{A} + \vec{B}) = \left(\hat{i}\frac{\partial}{\partial x} + \hat{j}\frac{\partial}{\partial y} + \hat{k}\frac{\partial}{\partial z}\right)\cdot[(A_x + B_x)\hat{i} + (A_y + B_y)\hat{j} + (A_z + B_z)\hat{k}]$$

$$= \frac{\partial}{\partial x}(A_x + B_x) + \frac{\partial}{\partial y}(A_y + B_y) + \frac{\partial}{\partial z}(A_z + B_z)$$

$$= \left(\frac{\partial A_x}{\partial x} + \frac{\partial A_y}{\partial y} + \frac{\partial A_z}{\partial z}\right) + \left(\frac{\partial B_x}{\partial x} + \frac{\partial B_y}{\partial y} + \frac{\partial B_z}{\partial z}\right)$$

$$\vec{\nabla}\cdot(\vec{\boldsymbol{A}} + \vec{\boldsymbol{B}}) = \vec{\nabla}\cdot\vec{\boldsymbol{A}} + \vec{\nabla}\cdot\vec{\boldsymbol{B}}$$

EXAMPLE 7 Prove that $\vec{A} = 3y^2z^2\hat{i} + 3x^2z^2\hat{j} + 3x^2y^2\hat{k}$ is a solenoidal vector.

SOLUTION $\vec{A} = 3y^2z^2\hat{i} + 3x^2z^2\hat{j} + 3x^2y^2\hat{k}$

$$\vec{\nabla}\cdot\vec{A} = \left(\hat{i}\frac{\partial}{\partial x} + \hat{j}\frac{\partial}{\partial y} + \hat{k}\frac{\partial}{\partial z}\right)\cdot(3y^2z^2\hat{i} + 3x^2z^2\hat{j} + 3x^2y^2\hat{k})$$

$$= \frac{\partial}{\partial x}(3y^2z^2) + \frac{\partial}{\partial y}(3x^2z^2) + \frac{\partial}{\partial z}(3x^2y^2)$$

$$= \mathbf{0}$$

As the divergence of a vector field $\vec{A}$ is zero, the vector $\vec{A}$ is solenoidal.

EXAMPLE 8 Find the constant a, the the vector $\vec{A} = (x + 3y)\hat{i} + (2y + 3z)\hat{j} + (x + az)\hat{k}$ is a solenoidal vector.

SOLUTION $\vec{A} = (x + 3y)\hat{i} + (2y + 3z)\hat{j} + (x + az)\hat{k}$

$$\text{div } A = \vec{\nabla}\cdot\vec{A} = \left(\hat{i}\frac{\partial}{\partial x} + \hat{j}\frac{\partial}{\partial y} + \hat{k}\frac{\partial}{\partial z}\right)\cdot[(x + 3y)\hat{i} + (2y + 3z)\hat{j} + (x + az)\hat{k}] = 0$$

or $$\frac{\partial}{\partial x}(x + 3y) + \frac{\partial}{\partial y}(2y + 3z) + \frac{\partial}{\partial z}(x + az) = 0$$

$$1 + 2 + a = 0$$

or $$a = \mathbf{-3}$$

EXAMPLE 9 Calculate the value of $\vec{\nabla}\cdot(r^3\vec{r})$ where $\vec{r} = (\hat{i}x + \hat{j}y + \hat{k}z)$.

SOLUTION Given $\vec{r} = (\hat{i}x + \hat{j}y + \hat{k}z)$

$$\vec{\nabla}\cdot(r^3\vec{r}) = \left(\hat{i}\frac{\partial}{\partial x} + \hat{j}\frac{\partial}{\partial y} + \frac{\hat{k}\partial}{\partial z}\right)\cdot[(x^2 + y^2 + z^2)^{3/2}(\hat{i}x + \hat{j}y + \hat{k}z)]$$

$$= \frac{\partial}{\partial x}[x(x^2 + y^2 + z^2)^{3/2}] + \frac{\partial}{\partial y}[y(x^2 + y^2 + z^2)^{3/2}] + \frac{\partial}{\partial z}[z(x^2 + y^2 + z^2)^{3/2}]$$

Now $$\frac{\partial}{\partial x}[x(x^2 + y^2 + z^2)^{3/2}] = (x^2 + y^2 + z^2)^{3/2} + x\frac{3}{2}2x[x^2 + y^2 + z^2]^{1/2}$$

$$= r^3 + 3x^2r$$

Similarly, $$\frac{\partial}{\partial y}[y(x^2 + y^2 + z^2)^{3/2}] = r^3 + 3y^2r$$

and $$\frac{\partial}{\partial z}[z(x^2 + y^2 + z^2)^{3/2}] = r^3 + 3z^2r$$

so $$\vec{\nabla}(r^3\vec{r}) = 3r^3 + 3(x^2 + y^2 + z^2)r$$

$$= 3r^3 + 3r^2\, r = \mathbf{6r^3}$$

EXAMPLE 10 Show that the vector field $\vec{A} = \frac{-2z^2y}{x^3}\hat{i} + \frac{z^2}{x^2}\hat{j} + \frac{2yz}{x^2}\hat{k}$ is irrotational.

SOLUTION Given $\vec{A} = \frac{-2z^2y}{x^3}\hat{i} + \frac{z^2}{x^2}\hat{j} + \frac{2yz}{x^2}\hat{k}$

If a vector field $\vec{A}$ is irrotational, then

Curl $\vec{A} = \vec{\nabla} \times \vec{A} = 0$

$$\text{Now } \vec{\nabla} \times \vec{A} = \begin{vmatrix} \hat{i} & \hat{j} & \hat{k} \\ \frac{\partial}{\partial x} & \frac{\partial}{\partial y} & \frac{\partial}{\partial z} \\ \frac{-2z^2 y}{x^3} & \frac{z^2}{x^2} & \frac{2yz}{x^2} \end{vmatrix}$$

$$= \hat{i}\left[\frac{\partial}{\partial y}\left(\frac{2yz}{x^2}\right) - \frac{\partial}{\partial z}\left(\frac{z^2}{x^2}\right)\right] + \hat{j}\left[\frac{\partial}{\partial z}\left(\frac{-2z^2 y}{x^3}\right) - \frac{\partial}{\partial x}\left(\frac{2yz}{x^2}\right)\right] + \hat{k}\left[\frac{\partial}{\partial x}\left(\frac{z^2}{x^2}\right) - \frac{\partial}{\partial y}\left(\frac{-2z^2 y}{x^3}\right)\right]$$

$$= \hat{i}\left[\frac{2z}{x^2} - \frac{2z}{x^2}\right] + \hat{j}\left[\frac{-4yz}{x^3} + \frac{4yz}{x^3}\right] + \hat{k}\left[\frac{-2z^2}{x^3} + \frac{2z^2}{x^3}\right]$$

$$\vec{\nabla} \times \vec{A} = \mathbf{0}$$

Hence the curl of a vector field $\vec{A}$ is zero. So the vector field $\vec{A}$ is zero, So the vector field $\vec{A}$ is irrotational.

Example 11 Consider a vector field $\vec{A} = x^2\hat{i} + y^2\hat{j} + z^2\hat{k}$

(i) Is the field solenoidal?

(ii) Is the field irrotational?

Solution Given $\vec{A} = x^2\hat{i} + y^2\hat{j} + z^2\hat{k}$ (i)

Now $\vec{A} = \hat{i}A_x + \hat{j}A_y + \hat{k}A_z$ (ii)

By using Eqs (i) and (ii), We have

$A_x = x^2, A_y = y^2$ and $A_z = z^2$

$$\text{Now } \vec{\nabla} \cdot \vec{A} = \left(\hat{i}\frac{\partial}{\partial x} + \hat{j}\frac{\partial}{\partial y} + \hat{k}\frac{\partial}{\partial z}\right).(\hat{i}A_x + \hat{j}A_y + \hat{k}A_z)$$

$$= \frac{\partial A_x}{\partial x} + \frac{\partial A_y}{\partial y} + \frac{\partial A_z}{\partial z}$$

$$= \frac{\partial x^2}{\partial x} + \frac{\partial y^2}{\partial y} + \frac{\partial z^2}{\partial z} = 2x + 2y + 2z$$

$$\vec{\nabla} \cdot \vec{A} \neq \mathbf{0}$$

From the above, it is clear that divergence of vector field $\vec{A}$ is not equal to zero. Hence, this field is not solenoidal.

(ii) $$\vec{\nabla} \cdot \vec{A} = \begin{vmatrix} \hat{i} & \hat{j} & \hat{k} \\ \frac{\partial}{\partial x} & \frac{\partial}{\partial y} & \frac{\partial}{\partial z} \\ x^2 & y^2 & z^2 \end{vmatrix} = \hat{i}\left[\frac{\partial}{\partial y}(z^2) - \frac{\partial}{\partial z}(y^2)\right] + \hat{j}\left[\frac{\partial}{\partial z}(x^2) - \frac{\partial}{\partial x}(z^2)\right] + \hat{k}\left[\frac{\partial}{\partial x}(y^2) - \frac{\partial}{\partial y}(x^2)\right]$$

$$\therefore \quad \vec{\nabla} \cdot \vec{A} = \mathbf{0}$$

Since curl of the vector field $\vec{A}$ is zero, the field is irrotational.

Example 12 A vector field is given by $\vec{A} = yz\hat{i} + xz\hat{j} + xy\hat{k}$. Show that it is both irrotational and solenoidal.

Solution Given $\vec{A} = yz\hat{i} + xz\hat{j} + xy\hat{k}$

Comparing it with $\vec{A} = A_x\hat{i} + A_y\hat{j} + A_z\hat{k}$

So $A_x = yz, A_y = xz$ and $A_z = xy$

Divergence of vector field $\vec{A}$

$$\vec{\nabla}\cdot\vec{A}=\left(\hat{i}\frac{\partial}{\partial x}+\hat{j}\frac{\partial}{\partial y}+\hat{k}\frac{\partial}{\partial z}\right)\cdot(\hat{i}A_x+\hat{j}A_y+\hat{k}A_z)$$

$$=\frac{\partial A_x}{\partial x}+\frac{\partial A_y}{\partial y}+\frac{\partial A_z}{\partial z}$$

$$=\frac{\partial(yz)}{\partial x}+\frac{\partial(xz)}{\partial y}+\frac{\partial(xy)}{\partial z}$$

$$\therefore \qquad \vec{\nabla}\cdot\vec{A}=\mathbf{0}$$

Hence vector field $\vec{A}$ is solenoidal.

$$\text{curl }\vec{A}=\vec{\nabla}\times\vec{A}=\begin{vmatrix}\hat{i} & \hat{j} & \hat{k}\\ \frac{\partial}{\partial x} & \frac{\partial}{\partial y} & \frac{\partial}{\partial z}\\ yz & xz & xy\end{vmatrix}$$

$$=\hat{i}\left(\frac{\partial xy}{\partial y}-\frac{\partial xz}{\partial z}\right)+\hat{j}\left(\frac{\partial yz}{\partial z}-\frac{\partial xy}{\partial x}\right)+\hat{k}\left(\frac{\partial xz}{\partial x}-\frac{\partial yz}{\partial y}\right)$$

$$=\hat{i}(x-x)+\hat{j}(y-y)+\hat{k}(z-z)$$

$$\therefore \qquad \vec{\nabla}\times\vec{A}=\mathbf{0}$$

Hence vector field $\vec{A}$ is irrotational.

EXAMPLE 13 Given $\vec{A}=x^2y\hat{i}+(x-y)\hat{k}$. Find (i) $\vec{\nabla}\cdot\vec{A}$ and (ii) $\vec{\nabla}\times\vec{A}$.

SOLUTION Given $\vec{A}=x^2y\hat{i}+(x-y)\hat{k}$

$$\vec{A}=A_x\hat{i}+A_y\hat{j}+A_z\hat{k}$$

$$A_x=x^2y,\ A_y=0 \text{ and } A_z=(x-y)$$

(i) $$\vec{\nabla}\cdot\vec{A}=\left(\hat{i}\frac{\partial}{\partial x}+\hat{j}\frac{\partial}{\partial y}+\hat{k}\frac{\partial}{\partial z}\right)\cdot(\hat{i}A_x+\hat{j}A_y+\hat{k}A_z)$$

$$=\frac{\partial A_x}{\partial x}+\frac{\partial A_y}{\partial y}+\frac{\partial A_z}{\partial z}=\frac{\partial(x^2y)}{\partial x}+\frac{\partial(0)}{\partial y}+\frac{\partial(x-y)}{\partial z}$$

$$\vec{\nabla}\cdot\vec{A}=\mathbf{2xy}$$

(ii) $$\text{curl }\vec{A}=\vec{\nabla}\times\vec{A}=\begin{vmatrix}\hat{i} & \hat{j} & \hat{k}\\ \frac{\partial}{\partial x} & \frac{\partial}{\partial y} & \frac{\partial}{\partial z}\\ x^2y & 0 & (x-y)\end{vmatrix}$$

$$=\hat{i}\left(\frac{\partial(x-y)}{\partial y}-\frac{\partial(0)}{\partial z}\right)+\hat{j}\left(\frac{\partial x^2y}{\partial z}-\frac{\partial(x-y)}{\partial x}\right)+\hat{k}\left(\frac{\partial(0)}{\partial x}-\frac{\partial x^2y}{\partial z}\right)$$

$$=\hat{i}(-1)+\hat{j}(-1)-x^2\hat{k}$$

$$\vec{\nabla}\times\vec{A}=-\hat{\boldsymbol{i}}-\hat{\boldsymbol{j}}-\boldsymbol{x^2}\hat{\boldsymbol{k}}$$

Example 14 Check whether the electrostatic field represented by $\vec{E} = axy^2(y\hat{i} + x\hat{j})$ is conservative or not?

Solution If $\vec{\nabla} \times \vec{E}$ is zero, then the electrostatic field is conservative.

Given $\vec{E} = axy^3\hat{i} + ax^2y^2\hat{j}$

$$\vec{E} = \hat{i}E_x + \hat{j}E_y + \hat{k}E_z$$

So $E_x = axy^3$, $E_y = ax^2y^2$ and $E_z = 0$

$$\text{Then, } \vec{\nabla} \times \vec{E} = \begin{vmatrix} \hat{i} & \hat{j} & \hat{k} \\ \frac{\partial}{\partial x} & \frac{\partial}{\partial y} & \frac{\partial}{\partial z} \\ axy^3 & ax^2y^2 & 0 \end{vmatrix} \neq 0$$

$$= \hat{i}\left(\frac{\partial(0)}{\partial y} - \frac{\partial}{\partial z}(ax^2y^2)\right) + j\left(\frac{\partial}{\partial z}^{axy^3} - \frac{\partial}{\partial x}(0)\right) + k\left(\frac{\partial}{\partial x}^{(ax^2y^2)} - \frac{\partial}{\partial y}^{(axy^3)}\right)$$

$$= 2axy^2 - 3axy^2 \neq 0$$

Hence $\vec{E}$ is nonconservative field.

Example 15 If 2000 flux through lines enter a given volume of space and 4000 lines diverge from it, find the total charge within the volume.

Solution Given $\phi_1 = 2000\,\text{Vm}$ and $\phi_2 = 4000$ Vm.

According to Gauss's theorem,

$$\phi = \frac{q}{\varepsilon_0} \tag{i}$$

Net flux emerging out of the surface, i.e.,

$$\phi = \phi_2 - \phi_1 = 4000 - 2000 = 2000 \text{ Vm}$$

By using Eq. (i), we get

$$q = \varepsilon_0\phi = 8.85 \times 10^{-12} \times 2000$$
$$= \mathbf{1.77 \times 10^{-8}\ C}$$

Example 16 Find the total charge enclosed by a closed surface if number of lines entering is 20,000 and emerging out is 45000.

Solution Given $\phi_1 = 20{,}000\,\text{Vm}$ and $\phi_2 = 45{,}000\,\text{Vm}$.

$$\phi = \phi_2 - \phi_1 = \text{net flux emerging out the surface}$$
$$\phi = 45000 - 20000$$
$$= 25{,}000 \text{ Vm}$$

According to Gauss's theorem

$$\phi = \frac{q}{\varepsilon_0} \text{ or } q = \varepsilon_0\phi$$

or
$$q = 8.85 \times 10^{-12} \times 25{,}000$$
$$= \mathbf{22.125 \times 10^{-8}\ C}$$

Example 17 A point charge of 13.5 Micro Coulomb is enclosed at the centre of the cube of side 6.0 cm. Find the electric flux (i) through the whole volume and (ii) through one face of the cube.

Solution Given $q = 13.5\,\mu\text{C} = 13.5\times10^{-6}\text{C}$ and $a = 6.0\,\text{cm}$.

(i) According to Gauss's theorem, the total flux through the whole volume

$$\phi = \frac{q}{\varepsilon_0}$$

$$= \frac{13.5 \times 10^{-6}}{8.85 \times 10^{-12}}$$

$$= 1.525 \times 10^{6} \text{ Nm}^2/\text{C}$$

Since a cube has 6 faces of equal area, the flux through one face of the cube would be

$$= \frac{1}{6}\frac{q}{\varepsilon_0} = \frac{1.525}{6} \times 10^{6} \text{ Nm}^2/\text{C}$$

$$= \mathbf{2.54 \times 10^{5}\ Nm^2/C}$$

EXAMPLE 18 A point charge of 11 Coulomb is located at the centre of a cube of side 5.0 cm. Calculate the electric flux through each surface.

SOLUTION Given $q = 11$ C and $a = 5.0$ cm

As a cube has six faces of equal area, so the flux through each surface of the cube is

$$= \frac{1}{6}\frac{q}{\varepsilon_0} = \frac{11}{6 \times 8.85 \times 10^{-12}}$$

$$= \mathbf{2.07 \times 10^{11}\ Nm^2/C}$$

EXAMPLE 19 A hollow metallic sphere of radius 0.1 m has 10^{-8} Coulomb of charge uniformly spread over it. Determine the electric field intensity (i) on the surface of the sphere (ii) at point 7 cm away from the centre and (iii) at point 0.5 m away from the centre.

SOLUTION Given radius of the hollow sphere (R) = 0.1 m and charge on it $q = 10^{-8}$ C.

Formula used for electric field intensity

$$E = \frac{1}{4\pi\varepsilon_0}\frac{q}{r^2}$$

(i) Intensity on the surface of the sphere ($r = R$) is

$$E = \frac{1}{4\pi\varepsilon_0}\frac{q}{R^2}$$

$$= \frac{1}{4 \times 3.14 \times 8.85 \times 10^{-12}} \times \frac{10^{-8}}{(0.1)^2}$$

$$= 9 \times 10^{9} \times 10^{-6}$$

$$= \mathbf{9 \times 10^{3}\ N/C}$$

(ii) Intensity at distance 7.0 cm away from the centre. This point lies inside the sphere so that inside the sphere electric field will be zero, i.e.,

$$E = 0$$

(iii) Intensity at 0.5m away from the centre

$$E = 9 \times 10^{9} \times \frac{10^{-8}}{(0.5)^2}$$

$$E = \mathbf{0.36 \times 10^{3} N/C}$$

EXAMPLE 20 If the charge on a proton is 1.6×10^{-19} Coulomb, find the magnitude of the electric field at a distance of 1 Å from the proton.

SOLUTION Given $q_p = 1.6 \times 10^{-19}$ C and $r = 10^{-10}$ m.

$$E = 9\times10^9 \times \frac{q}{r^2} = 9\times10^9 \times \frac{1.6\times10^{-19}}{(10^{-10})^2}$$

$$= \mathbf{1.44\times10^{11}\ V/m}$$

EXAMPLE 21 Determine the energy gained by an α-particle when it is accelerated through a potential of 1000 volts.

SOLUTION Given $q_\alpha = 2 \times 1.6 \times 10^{-19}$ C and $V = 1000$ V.

The energy gained by α-particle is $= qV$

$$= 3.2 \times 10^{-19} \times 1000 = \mathbf{3.2 \times 10^{-16}\ J}$$

EXAMPLE 22 If the charge on a proton is 1.6×10^{-19} C, find

(i) the electrostatic potential and potential energy at a distance of 1.0 Å from the proton.

(ii) the potential difference between two points 1 Å and 0.2 Å from the proton.

SOLUTION (i) $q = 1.6 \times 10^{-19}$ and $r = 1.0 \times 10^{-10}$ m.

Potential $$V = \frac{1}{4\pi\varepsilon_0}\frac{q}{r}$$

$$= 9\times10^9 \times \frac{1.6\times10^{-19}}{10^{-10}}$$

$$= 14.4\ \text{V}$$

Potential energy $$= -\frac{1}{4\pi\varepsilon_0}\frac{q^2}{r}$$

$$= -9\times10^9 \frac{(1.6\times10^{-19})^2}{10^{-10}}$$

$$= -23.04\times10^{-19}\ \text{J}$$

$$= \mathbf{-14.4\ eV}$$

(ii) Potential difference $= V_1 - V_2 = \frac{1}{4\pi\varepsilon_0}\left(\frac{q}{r_1} - \frac{q}{r_2}\right)$

$$= 9\times10^9\times1.6\times10^{-19}\left[\frac{1}{0.2\times10^{-10}} - \frac{1}{10^{-10}}\right]$$

$$= \mathbf{57.6\ V}$$

EXAMPLE 23 Consider a point charge 15×10^{-6} C. What is the radius of the equipotential surface having potential 30V?

SOLUTION Given potential = 30 V and $q = 1.5 \times 10^{-6}$ C.

Potential $= \frac{1}{4\pi\varepsilon_0}\frac{q}{r}$

$$30 = 9\times10^9\times\frac{1.5\times10^{-6}}{r}$$

or $$r = 9\times10^9\times\frac{1.5\times10^{-16}}{30} = \mathbf{450\ m}$$

EXAMPLE 24 Calculate the value of poynting vector at the surface of the sun if the power radiated by the sun is 3.8×10^{26} W and its radius is 7×10^8 m.

SOLUTION Given $P = 3.8 \times 10^{26}$ W and $r = 7 \times 10^8$ m.

Formula used is

$$S_{av} \times 4\pi r^2 = P$$

where S_{av} is the average poynting vector at surface of the sun

$$S_{av} = \frac{P}{4\pi r^2} = \frac{3.8\times10^{26}}{4\times3.14\times(7\times10^8)^2}$$

$$= \mathbf{6.174\times10^7\ W/m^2}$$

EXAMPLE 25 Calculate the radiation pressure at the surface of the earth and sun assuming that solar constant has a value of 2 cal/cm^2 min at the surface of the earth and the radius of the sun is 7×10^8 m and the average distance between earth and sun is 1.5×10^{11} m.

SOLUTION Given

$$S_E = \frac{2\text{ cal}}{\text{cm}^2\text{ min}} = \frac{2\times4.2}{10^{-4}\times60}$$

$$= 1.4\times10^3\,\frac{\text{J}}{\text{m}^2\text{ sec}}$$

$$[P_{\text{rad}}] = \frac{S_E}{c} = \frac{1.4\times10^3}{3\times10^8} = 4.67\times10^{-6}\ \text{N/m}^2$$

Further as

$$S_s r_s^2 = S_E r_{\text{ES}}^2$$

$$S_S = S_E\left(\frac{r_{ES}}{r_s}\right)^2 = 1.4\times10^3\times\left(\frac{1.4\times10^{11}}{7\times10^8}\right)^2$$

$$= 5.6\times10^7\ \text{W/m}^2$$

$\therefore$ $$[P_{\text{rad}}]_S = \frac{S_s}{c} = \frac{5.6\times10^7}{3\times10^8}$$

$$= \mathbf{0.187\ N/m^2}$$

EXAMPLE 26 Derive Coulomb's law of electrostatics with the help of Maxwell's first equation.

SOLUTION From the Maxwell's first equation

$$\text{div}\,\vec{E} = \frac{\rho}{\varepsilon_0} \quad \text{or} \quad \vec{\nabla}\cdot\vec{E} = \frac{\rho}{\varepsilon_0}.$$

Integrating it over a volume V enclosed by a surface S in a medium, we get

$$\int_V \vec{\nabla}\cdot\vec{E}\,dV = \int_V \left(\frac{\rho}{\varepsilon_0}\right) dV$$

But from Gauss's divergence theorem,

$$\int_V \vec{\nabla}\cdot\vec{E}\,dV = \oint_s \vec{E}\cdot\vec{ds}$$

$$\therefore \qquad \oint_s \vec{E}\cdot\vec{ds} = \int_V \left(\frac{\rho}{\varepsilon_0}\right) dV = \frac{1}{\varepsilon_0}\int_V \rho\,dV$$

or

$$\oint_s \vec{E}\cdot\vec{ds} = \frac{1}{\varepsilon_0}\times q = \frac{q}{\varepsilon_0}$$

If the electric field around the charge is symmetrical, then

$$E\times 4\pi r^2 = \frac{q}{\varepsilon_0}$$

or

$$E = \frac{1}{4\pi\varepsilon_0}\times\frac{q}{r^2}$$

Hence, the force on a test charge q_0 in the electric field E, is

$$\boldsymbol{F = q_0 E = \frac{1}{4\pi\varepsilon_0}\frac{qq_0}{r^2}}$$

This is the Coulomb's law.

EXAMPLE 27 A plane electromagnetic wave propagating along the x-direction has a wavelength 5.0 mm. The electric field is in the y-direction and its maximum magnitude is 38 V/m. Find the time and space varying equations for the electric and magnetic fields.

SOLUTION The equations of electric and magnetic fields of a plane electromagnetic wave are given by

$$E = E_0 \sin\left[\frac{2\pi}{\lambda}(ct-x)\right] \quad \text{and} \quad H = H_0 \sin\left[\frac{2\pi}{\lambda}(ct-x)\right]$$

Given $E_0 = 38$ V/m and $\lambda = 5.0$ mm $= 5\times10^{-3}$ m.

Hence,

$$E = 38\sin\left[\frac{2\pi}{5\times10^{-3}}(ct-x)\right]$$

or

$$\boldsymbol{E = 38\sin\left[\frac{2\pi\times10^3}{5}(ct-x)\right]} \qquad \text{(i)}$$

The magnitude of the magnetic field is given by

$$H_0 = \frac{E_0}{c} = \frac{38}{3\times10^8} = 1.27\times10^{-7}\ \text{Wb/m}^2$$

$$\therefore \qquad \boldsymbol{H = 1.27\times10^{-7}\sin\left[\frac{2\pi\times10^3}{5}(ct-x)\right]} \qquad \text{(ii)}$$

The electric field is along y-axis and the magnetic field along z-axis.

EXAMPLE 28 If the earth receives 2 cal min^{-1} cm^{-2} solar energy, what would be the amplitudes of electric and magnetic fields of radiation.

SOLUTION Here, solar energy which the earth receives is 2 cal min^{-1} cm^{-2}

$$\therefore \quad \frac{E_0}{H_0} = \sqrt{\frac{\mu_0}{\varepsilon_0}} = \sqrt{4\pi \times 4\pi \times 9 \times 10^9 \times 10^{-7}} = \sqrt{9 \times 4\pi \times 4\pi \times 10^2}$$

$$= 4\pi \times 3 \times 10 = 120\pi$$

$$= 120 \times 3.14$$

$$= 376.8 \approx 377$$

Poynting vector,

$$\vec{P} = \vec{E} \times \vec{H}$$

$$= EH \text{ J m}^{-2} \text{ sec}^{-1}$$

$$= \frac{2 \times 4.2}{60 \times 10^{-4}} \text{Jm}^{-2}\text{s}^{-1} = 1400 \text{ Jm}^{-2} \text{ sec}^{-1}$$

$$\therefore \quad EH = 1400$$

$$\frac{E_0}{H_0} = 377 = \frac{E}{H}$$

$$\therefore \quad E^2 = 1400 \times 377$$

or

$$E = 10 \times \sqrt{14 \times 377} = 726.5 \text{ V/m}$$

$$H = \frac{E}{377} = \frac{726.5}{377} = 1.927$$

Amplitudes of electric and magnetic field vectors are

$$E_0 = E\sqrt{2} = \mathbf{1024.3\ V/m}$$

$$H_0 = H\sqrt{2} = \mathbf{2.717\ A/m}$$

EXAMPLE 29 If the magnitude of $\vec{H}$ in a plane wave is 1 A/m, find the magnitude of $\vec{E}$ for plane wave in free space.

SOLUTION We know that, $\frac{H_0}{E_0} = \sqrt{\frac{\varepsilon_0}{\mu_0}}$ or $E_0 = \left(\sqrt{\frac{\mu_0}{\varepsilon_0}}\right) H_0$

Here $H_0 = 1$ A/m, $\mu_0 = 4\pi \times 10^{-7}$ Wb/A-m

and $e_0 = 8.85 \times 10^{-12}$ C/Nm^2

$$\therefore \quad E_0 = 1 \times \sqrt{\frac{4\pi \times 10^{-7}}{8.85 \times 10^{-12}}} = \mathbf{376.72\ V/m}$$

EXAMPLE 30 Show that equation of continuity div $\vec{J} + \frac{\partial \rho}{\partial t} = 0$ is contained in Maxwell's equations.

SOLUTION According to Maxwell's fourth equation,

$$\text{curl } \vec{H} = \vec{J} + \frac{\partial \vec{D}}{\partial t}$$

Taking divergence of both sides, we get

$$\text{div (curl } \vec{H}) = \text{div}\left(\vec{J} + \frac{\partial \vec{D}}{\partial t}\right)$$

But, div (curl $\vec{H}$) = 0

$$\therefore \quad \text{div}\left(\vec{J} + \frac{\partial \vec{D}}{\partial t}\right) = 0 \quad \text{or} \quad \text{div}\,\vec{J} + \text{div}\left(\frac{\partial \vec{D}}{\partial t}\right) = 0$$

or $$\text{div}\,\vec{J} + \frac{\partial}{\partial t}(\text{div}\,\vec{D}) = 0$$ [∴ space and time operations are interchangeable]

From Maxwell's first equation,

$$\text{div}\,\vec{D} = \rho,$$

where ρ is the surface charge density.

$$\therefore \quad \textbf{div}\,\vec{J} + \frac{\partial \rho}{\partial t} = \mathbf{0}$$

EXAMPLE 31 Considering that all the energy from a 1000 W lamp is radiated uniformly, calculate the average value of the intensity of electric field of radiation at a distance of 2 m from the lamp.

SOLUTION As the total power (P) is radiated uniformly, energy flux per unit area per second at a distance r from the lamp is given by

$$S = \frac{P_0}{4\pi r^2} = \frac{1000}{4\pi(2)^2}\,\text{W/m}^2$$

From the poynting theorem, $|\vec{S}| = |\vec{E} \times \vec{H}| = EH \sin 90°$

∵ E and H are perpendiculars to each other,

$$\therefore \quad EH = \frac{1000}{16\pi} \tag{i}$$

But $$\frac{E}{H} = \sqrt{\frac{\mu_0}{\varepsilon_0}} = 376.72 \text{ ohm} \tag{ii}$$

Multiplying Eqs. (i) and (ii), we get

$$EH \times \frac{E}{H} = \frac{376.72 \times 1000}{16\pi} \Rightarrow E = \left(\frac{376.72 \times 1000}{16 \times 3.14}\right)^{1/2} = \mathbf{86.59\ V/m}$$

EXAMPLE 32 If the relative permittivity of distilled water be 81. Calculate refractive index and velocity of light in it.

SOLUTION We know that, $\mu' = \sqrt{\dfrac{\mu\varepsilon}{\mu_0\varepsilon_0}}$ and $v = \dfrac{c}{\mu'}$

Here $\varepsilon = 81\varepsilon_0$ and for distilled water $\mu \approx \mu_0$

$$\therefore \quad \mu' = \sqrt{81} = 9 \quad \text{and} \quad v = \frac{3 \times 10^8}{9} = \mathbf{3.33 \times 10^7\ m/sec}$$

Example 33 Consider an infinite conducting sheet in the *xy*-plane with a time dependent current density *kti*, where *k* is constant. The vector potential at (*x*, *y*, *z*) is given by $A = \dfrac{\mu_0 k}{4c}(ct - z)^2$. Find the magnetic field *B*.

Solution

$$B = \vec{\nabla} \times \vec{A}$$

$$= \begin{vmatrix} \hat{i} & \hat{j} & \hat{k} \\ \dfrac{\partial}{\partial x} & \dfrac{\partial}{\partial y} & \dfrac{\partial}{\partial z} \\ \dfrac{\mu_0 k}{4c}(ct - z)^2 & 0 & 0 \end{vmatrix}$$

$$= \frac{\mu_0 k}{4c}\frac{\partial}{\partial z}(ct - z)^2\,\hat{j}$$

$$= \boldsymbol{\frac{\mu_0 k}{2c}(z - ct)\hat{j}}$$

Example 34 A magnetic field $\vec{B} = B_0(\hat{i} + 2\hat{j} - 4\hat{k})$ exists at a point. If a test charge moving with a velocity $\vec{v} = v_0(3\hat{i} - \hat{j} + 2\hat{k})$ experiences no force at a certain point, what will be the electric field at that point in SI units?

Solution Lorentz force is given by

$$\vec{F} = q\vec{E} + q(\vec{v} \times \vec{B})$$

$$= 0 \quad \text{(as per question)}$$

$$\therefore \quad \vec{E} = -(\vec{v} \times \vec{B})$$

$$\vec{E} = v_0 B_0 \begin{vmatrix} \hat{i} & \hat{j} & \hat{k} \\ 1 & 2 & -4 \\ 3 & -1 & 2 \end{vmatrix}$$

$$\vec{E} = -v_0 B_0[\hat{i}(0) - \hat{j}(14) + \hat{k}(-7)]$$

$$\boldsymbol{\vec{E} = 7v_0 B_0(2\hat{j} + \hat{k})\,\text{V/m}}$$

Example 35 Find the electric and magnetic fields $\vec{E}(z, t)$ and $\vec{B}(z, t)$, respectively, corresponding to the scalar potential $\phi(z, t) = 0$ and vector potential $A = zt\,\hat{i}$

Solution

$$\vec{B} = \vec{\nabla} \times \vec{A} = \begin{vmatrix} \hat{i} & \hat{j} & \hat{k} \\ \dfrac{\partial}{\partial x} & \dfrac{\partial}{\partial y} & \dfrac{\partial}{\partial z} \\ tz & 0 & 0 \end{vmatrix}$$

$$\vec{B} = \frac{\partial}{\partial z}(tz) = t$$

$$\vec{E} = \phi - \frac{\partial A}{\partial t} = 0 - \frac{\partial}{\partial t}(zt) = -z\hat{i}$$

$$\boldsymbol{\vec{E} = -z\hat{i}\ \text{V/m}}$$

Example 36 A plane polarized electromagnetic wave in free space at time $t = 0$ is given by $\vec{E}(x, z) = 10\hat{j}\ \exp[i(6x + 8y)]$. Find the corresponding magnetic field.

Solution From expression of $\vec{E}, \vec{k} = 6\hat{i} + 8\hat{j}$

Maxwell's equation $\vec{\nabla} \times \vec{E} = -\dfrac{\partial \vec{B}}{\partial t}$ or $\vec{\nabla} \times \vec{E} = -\mu_0 \dfrac{\partial \vec{H}}{\partial t}$ gives

$$i\vec{k} \times \vec{E} = \mu_0 i\omega \vec{H} \Rightarrow \vec{H} = \frac{k}{\mu_0 \omega}\hat{k} \times \vec{E}$$

or
$$\vec{H} = \frac{1}{\mu_0 c}\hat{k} \times \vec{E} \qquad (\because \omega/k = c \text{ in free space})$$

$$= \frac{10}{\mu_0 c}\left[\frac{(6\hat{i} + 8\hat{j})}{10} \times \hat{j}\right]\exp[i(6x + 8y)]$$

$$= \frac{\mathbf{6}}{\boldsymbol{\mu_0 c}}\mathbf{exp}(\boldsymbol{i}(\mathbf{6}\boldsymbol{x} + \mathbf{8}\boldsymbol{y}))\boldsymbol{\hat{k}}\ \mathbf{A/m}$$

Example 37 A plane electromagnetic wave has magnetic field

$$\vec{B}(x, y, z, t) = B_0 \sin\left[(x + y)\frac{k}{\sqrt{2}} + \omega t\right]\hat{k}$$

where k is wave number and $\hat{i}$, $\hat{j}$ and $\hat{k}$ are cartesian unit vectors in x, y and z directions respectively. Find the electric field $\vec{E}(x, y, z, t)$ and average poynting vector.

Solution Using $\vec{k} \times \vec{H} = -\varepsilon_0 \omega \vec{E}$

$$\vec{E} = -\frac{1}{\varepsilon_0 \omega}\vec{k} \times \vec{H}$$

$$= -\frac{B_0 k}{\mu_0 \varepsilon_0 \omega}\sin\left[(x + y)\frac{k}{\sqrt{2}} + \omega t\right]\left[\frac{(\hat{i} + \hat{j})}{\sqrt{2}} \times \hat{k}\right]$$

$$= -\frac{B_0 k}{\mu_0 \varepsilon_0 \omega}\sin\left[(x + y)\frac{k}{\sqrt{2}} + \omega t\right]\left(\frac{-\hat{j} + \hat{i}}{\sqrt{2}}\right)$$

$$\boldsymbol{\vec{E}} = -\frac{\boldsymbol{B_0 k}}{\boldsymbol{\mu_0 \varepsilon_0 \omega}}\mathbf{sin}\left[(\boldsymbol{x} + \boldsymbol{y})\frac{\boldsymbol{k}}{\sqrt{\mathbf{2}}} + \boldsymbol{\omega t}\right]\left(\frac{\boldsymbol{\hat{i}} - \boldsymbol{\hat{j}}}{\sqrt{\mathbf{2}}}\right)\mathbf{units}$$

Average poynting vector is given by

$$<P> = \vec{E} \times \vec{H} = \vec{E} \times \frac{\vec{B}}{\mu_0}$$

$$= \frac{-B_0^2 k}{\mu_0^2 \varepsilon_0 \omega}\sin^2\left[(x + y)\frac{k}{\sqrt{2}} + \omega t\right]\left(\frac{\hat{i} - \hat{j}}{\sqrt{2}}\right) \times \hat{k}$$

$$<P> = \frac{-B_0^2 k}{2\mu_0^2 \varepsilon_0 \omega}\left(\frac{-\hat{j} - \hat{i}}{\sqrt{2}}\right)$$

$$= -\frac{\boldsymbol{B_0^2 c^2 k}}{\boldsymbol{2\mu_0 \omega}}\left(\frac{\boldsymbol{\hat{i}} + \boldsymbol{\hat{j}}}{\mathbf{2}}\right)\mathbf{units}$$

EXAMPLE 38 In a non-conducting medium characterized by $\varepsilon = \varepsilon_0$, $\mu = \mu_0$ and conductivity $\sigma = 0$, the electric field (in Vm^{-1}) is given by $\vec{F} = 20 \sin[10^8 t - kz]\,\hat{j}$. Find the magnetic field, H (in Am^{-1}).

SOLUTION For non-conducting medium,

$$\vec{H} = \frac{1}{\mu\omega}(\vec{k} \times \vec{E})$$

$$\vec{H} = \frac{20k}{\mu 10^8}(\hat{k} \times \hat{j})\sin[10^8 t - kz]$$

$$\boldsymbol{\vec{H} = -\frac{20k}{10^8 \mu}\sin[10^8 t - kz]\hat{i}\ \text{A/m}}$$

EXAMPLE 39 Consider two concentric conducting *spherical shells* with inner and outer radii a, b, c and d, as shown. Both the shells are given Q amount of positive charges. Find electric field in different regions.

SOLUTION Conducting shells are given, means the supplied charge Q will be distributed over their outer surfaces (see the figure). Moreover, a charge — Q will be induced on the inner surface of the outer shell due to the positive charge Q on the inner shell. Due to induction, a charge $+Q$ will be induced on the outer surface of the outer shell. It means the total charge on its outer surface will be as $2Q$.

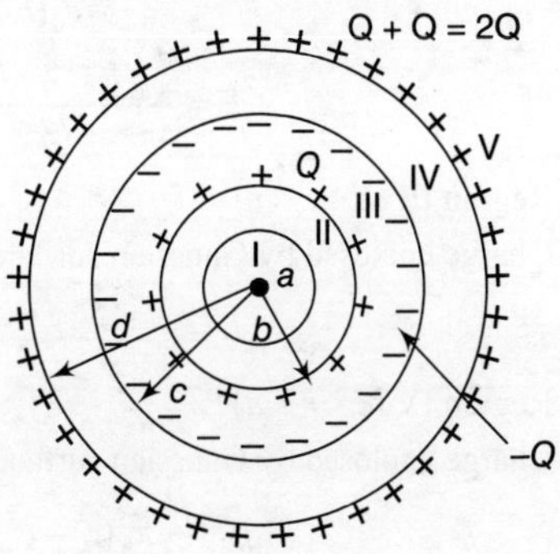

Field in region I ($r < a$)

There is no charge enclosed by the Gaussian surface within this region.

Hence, Gauss's law reads $\oint \vec{E} \cdot d\vec{S} = \frac{q_{enc}}{\varepsilon_0} = 0$

$\Rightarrow$ $\qquad \vec{E} = 0 \quad \text{for} \quad r < a$

Field in region II ($b < r < a$)

In this region also, the charge enclosed by the Gaussian surface would be zero.

$\Rightarrow$ $\qquad \vec{E} = 0 \quad \text{for} \quad a < r < b.$

Field in region III ($b < r < c$)

Here enclosed charged will be as Q.

$\therefore$ $\qquad$ Gauss's law reads $\oint \vec{E} \cdot d\vec{S} = Q/\varepsilon_0$

or $\qquad E 4\pi r^2 = Q/\varepsilon_0 \Rightarrow \vec{E} = \frac{Q}{4\pi\varepsilon_0 r^2}\hat{r}$

Field in region IV ($c < r < d$)

In this region, the enclosed charge $= -Q + Q = 0$

$\Rightarrow$ The field $\vec{E} = 0$.

Field in region V ($r > d$)

The charge enclosed by Gaussian surface in this region will be $2Q - Q + Q = 2Q$

$\therefore$ Gauss's law reads $\oint \vec{E} \cdot d\vec{S} = \frac{2Q}{\varepsilon_0}$

$\Rightarrow$ $\qquad \vec{E} = \frac{2Q}{4\pi\varepsilon_0 r^2}\hat{r} = \frac{Q}{2\pi\varepsilon_0 r^2}\hat{r}$

EXAMPLE 40 Consider two concentric uniformly charged spherical shells with inner and outer radii a, b, c and d, as shown. Both the shells carry equal amount of positive charge Q. Find electric field in different regions.

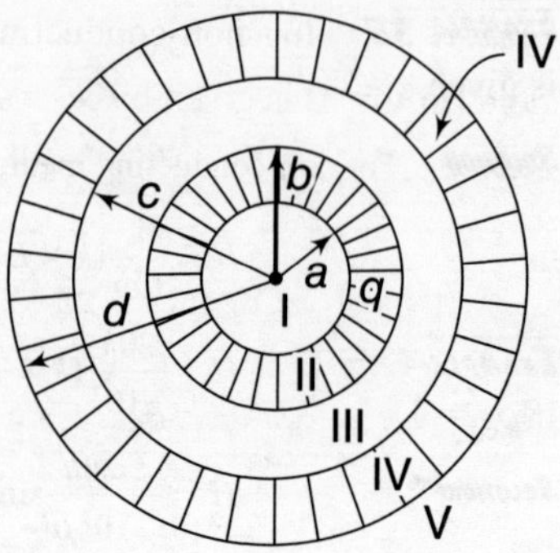

Solution Charge density of inner spherical shell

$$\rho_1 = \frac{Q}{\frac{4}{3}\pi(b^3 - a^3)}$$

Charge density of outer spherical shell $\rho_2 = \dfrac{Q}{\frac{4}{3}\pi(d^3 - c^3)}$

Region I ($r < a$)

Charge enclosed by Gaussian surface for $r < a$ will be zero.

Hence, Gauss's law $\oint \vec{E} \cdot d\vec{S} = 0 \Rightarrow \vec{E} = 0$.

Region II ($a < r < b$)

In this region, the charge enclosed by Gaussian surface will be $\rho_1 \cdot \frac{4}{3}\pi\ (r^3 - a^3)$

Hence, Gauss's Law reads

$$\oint \vec{E} \cdot d\vec{S} = \frac{Q}{\varepsilon_0 \frac{4}{3}\pi(b^3 - a^3)} \cdot \frac{4}{3}\pi(r^3 - a^3)$$

$$\Rightarrow \qquad E \cdot 4\pi r^2 = \frac{Q(r^3 - a^3)}{\varepsilon_0(b^3 - a^3)}$$

or $$\vec{E} = \frac{Q(r^3 - a^3)}{4\pi\varepsilon_0 r^2(b^3 - a^3)}\hat{r}$$

Region III ($b < r < c$)

Charge enclosed by Gaussian surface will be Q.

$$\Rightarrow \qquad \vec{E} = \frac{Q}{4\pi\varepsilon_0 r^2}\hat{r}$$

Region IV ($c < r < d$)

Charge enclosed by Gaussian surface will be

$$\rho_2 \cdot \frac{4}{3}\pi\ (r^3 - c^3)$$

∴ Gauss's law reads

$$\oint \vec{E} \cdot d\vec{S} = \frac{Q}{\varepsilon_0 \frac{4}{3}\pi(d^3 - c^3)} \cdot \frac{4}{3}\pi(r^3 - c^3)$$

or $$E 4\pi r^2 = \frac{Q(r^3 - c^3)}{\varepsilon_0(d^3 - c^3)}$$

or $$\vec{E} = \frac{Q(r^3 - c^3)}{4\pi\varepsilon_0 r^2(d^3 - c^3)}\hat{r}$$

Region V ($r > d$)

Charge enclosed by Gaussian surface will be

$$Q + Q = 2Q$$

Hence, $$\oint \vec{E} \cdot d\vec{S} = \frac{2Q}{\varepsilon_0}$$

or $$\vec{E} = \frac{2Q}{4\pi\varepsilon_0 r^2}\hat{r}$$

$$= \frac{Q}{2\pi\varepsilon_0 r^2}\hat{r}$$

EXAMPLE 41 The electric field of a uniform plane wave propagating in a dielectric, non-conducting medium is given by $\vec{E} = \hat{x}\,10\cos(6\pi \times 10^7 t - 0.4\pi z)$ V/m. Find phase velocity of wave.

SOLUTION For the plane wave whose electric field is given by $\vec{E} = \vec{E}_0 \cos(\omega t - kz)$, the phase velocity is $v_p = \omega/k$. Hence

$$v_p = \frac{\omega}{k} = \frac{6\pi \times 10^7}{0.4 \times 3.14} = \mathbf{1.5 \times 10^8\ m/s}$$

EXAMPLE 42 Determine penetration depth by which an electromagnetic wave enters into copper, if $\rho_{cu} = 1.69 \times 10^{-8}$ μm and frequency = 10^4 MHz.

SOLUTION

$$\delta = \sqrt{\frac{2}{\mu\sigma\omega}} = \sqrt{\frac{\rho}{\pi f \mu}}$$

$$= \sqrt{\frac{1.69 \times 10^{-8} \times 10^{-6}}{3.14 \times 10^{10} \times 4\pi \times 10^{-7}}}$$

$$= 0.0654 \times 10^{-8}\ \text{m} = \mathbf{0.654\ nm}$$

EXAMPLE 43 In free space, $H = 0.1\cos(2 \times 10^8 t - kx)\,\hat{j}$ H/m. Calculate k, λ and T.

SOLUTION

$$k = \frac{\omega}{v}$$

For free space, $v = c$

$$k = \frac{\omega}{c} = \frac{2 \times 10^8}{3 \times 10^8} = 0.667\ \text{rad/m}$$

$$k = \frac{2\pi}{\lambda},\ \lambda = \frac{2\pi}{k} = \frac{2 \times 3.14}{0.667} = 9.42\ \text{m}$$

T is the period, $T = \dfrac{2\pi}{\omega} = \dfrac{2 \times 3.14}{2 \times 10^8} = \mathbf{3.14 \times 10^{-8}\ s}$

EXAMPLE 44 In a lossless medium for which $Z = 60\pi$, $\mu_r = 1$, and $\vec{H} = -0.1\cos(\omega t - z)\,\hat{i} + 0.5\sin(\omega t - z)$ A/m. Calculate ε_r, ω and E.

SOLUTION

$$Z = \sqrt{\frac{\mu}{\varepsilon}} = \sqrt{\frac{\mu_0}{\varepsilon_0}}\sqrt{\frac{\mu_r}{\varepsilon_r}} = \frac{120\pi}{\sqrt{\varepsilon_r}}$$

$$\Rightarrow \qquad \sqrt{\varepsilon_r} = \frac{120\pi}{60\pi} = 2 \Rightarrow \varepsilon_r = 4$$

$$\beta = \omega\sqrt{\mu\varepsilon} = \omega\sqrt{\mu_0\varepsilon_0}\sqrt{\mu_r\varepsilon_r} = \frac{\omega}{c}\sqrt{4} = \frac{2\omega}{c}$$

$$\omega = \frac{\beta c}{2} = \frac{3 \times 10^8}{2} = 1.5 \times 10^8\ \text{rad/s} \quad (\because \beta = 1 \text{ due to } \beta z = z)$$

$$\vec{\nabla} \times \vec{H} = \varepsilon\frac{\partial \vec{E}}{\partial t} \Rightarrow \vec{E} = \frac{1}{\varepsilon}\int \vec{\nabla} \times \vec{H}\,dt$$

$$\vec{\nabla}\times\vec{H}=\begin{vmatrix}\hat{i} & \hat{j} & \hat{k}\\ \partial/\partial x & \partial/\partial y & \partial/\partial z\\ H_x & H_y & 0\end{vmatrix}=-\frac{\partial H_y}{\partial z}\hat{i}+\frac{\partial H_x}{\partial z}\times\hat{j}$$

$$=0.5\cos(\omega t-z)\hat{i}-0.1\sin(\omega t-z)\hat{j}$$

Hence
$$\vec{E}=\frac{1}{\varepsilon}\int[+0.5\cos(\omega t-z)\hat{i}dt-0.1\sin(\omega t-z)\hat{j}dt]$$

$$=\frac{0.5}{\varepsilon\omega}\sin(\omega t-z)\hat{i}+\frac{0.1}{\varepsilon\omega}\cos(\omega t-z)\hat{j}$$

or
$$\vec{E}=\frac{0.5}{\varepsilon_0\varepsilon_r\omega}\sin(\omega t-z)\hat{i}+\frac{0.1}{\varepsilon_0\varepsilon_r\omega}\cos(\omega t-z)\hat{j}$$

$$=\frac{0.5}{8.85\times10^{-12}\times4\times1.5\times10^{8}}\sin(\omega t-z)\hat{i}+\frac{0.1}{8.85\times10^{-12}\times4\times1.5\times10^{8}}\cos(\omega t-z)\hat{j}$$

$\Rightarrow$
$$\boldsymbol{\vec{E}=94.16\sin(\omega t-z)\hat{i}+18.83\cos(\omega t-z)\hat{j}}$$

Example 45 If $\varepsilon_r = 1$, $\mu_r = 20$ and $\sigma = 3$ mhos/m for a medium and the electric field of an electromagnetic wave is

$$\vec{E}=2e^{-\alpha z}\sin(10^8 t-\beta z)\hat{j}\ \text{V/m}$$

Find α, β and $\vec{H}$.

Solution As
$$\frac{\sigma}{\varepsilon\omega}=\frac{3}{8.85\times10^{-12}\times10^{8}}$$
$$=3389 >> 1.$$

the medium is good conductor at frequency of operation.

$\Rightarrow$
$$\alpha=\beta=\sqrt{\frac{\mu\omega\sigma}{2}}$$

$$=\left[\frac{4\pi\times10^{-7}\times20\times10^{8}\times3}{2}\right]^{1/2}$$

$$\beta=61.4\ \text{rad/m}$$

$$\eta=\sqrt{\frac{\mu\omega}{\sigma}}=\sqrt{\frac{4\pi\times10^{-7}\times20\times10^{8}}{3}}$$

$$=\sqrt{\frac{800\pi}{3}}\ \Omega$$

$$H=H_0e^{-\alpha z}\sin\left(\omega t-\beta z-\frac{\pi}{4}\right)$$

$$H_0=\frac{E_0}{\eta}=2\sqrt{\frac{3}{800\pi}}=69.1\times10^{-3}\ \text{A/m}$$

$$\vec{H}=69.1\times e^{-61.4z}\sin\left(10^8 t-61.4z-\frac{\pi}{4}\right)(\vec{k}\times\hat{j})$$

$$\vec{H}=-69.1e^{-61.4z}\sin\left(10^8 t-61.4z-\frac{\pi}{4}\right)\hat{i}\ \text{mA/m}$$

Example 46 In non-magnetic medium

$$\vec{E} = 4 \sin(2\pi \times 10^7 t - 0.8x)\,\hat{k} \text{ V/m}$$

Find power crossing 100 cm^2 of plane

$$2x + y = 5$$

Find impedance (η), and ε_r.

Solution

$$\beta = \omega\sqrt{\mu\varepsilon} = \omega\sqrt{\mu_0\varepsilon_0\varepsilon_r} = \frac{\omega\sqrt{\varepsilon_r}}{c}$$

$$\sqrt{\varepsilon_r} = \frac{\beta c}{\omega} = \frac{0.8 \times 3 \times 10^8}{2\pi \times 10^7} = \frac{12}{\pi}$$

$$\varepsilon_r = 14.59$$

$$\eta = \sqrt{\frac{\mu}{\varepsilon}} = \sqrt{\frac{\mu_0}{\varepsilon_0\varepsilon_r}} = \frac{120\pi}{\sqrt{\varepsilon_r}} = \frac{120\pi \cdot \pi}{12}$$

$$= 10\pi^2\,\Omega$$

$$\vec{P} = \vec{E} \times \vec{H} = \frac{E_0^2}{\eta}\sin^2(\omega t - \beta x)\hat{i}$$

$$P_{\text{av}} = \frac{1}{t}\int_n^T \vec{P}dt = \frac{E_0^2}{2\eta}\hat{i}$$

$$= \frac{16}{2 \times 10\pi^2}\hat{i}$$

On plane $2x + y = 5$, the normal is $\hat{n} = \dfrac{2\hat{i} + \hat{j}}{5}$

$$\therefore \quad \vec{P}_{\text{avg}} = \int \vec{P}_{\text{av}} \cdot d\vec{S} = P_a \cdot s_{an}$$

$$= (81 \times 10^{-3}\hat{a}_x) \cdot \left(\frac{100 \times 10^{-4}}{\sqrt{5}}(2\hat{i} + \hat{j})\right)$$

$$= \mathbf{724.5\ \mu W}$$

Example 47 Consider two concentric spherical conducting shells (negligible thickness) having equal charge q. The radius of inner shell is "a" and that of the outer shell is "b". Calculate the electric field in three regions

(a) $r < a$

(b) $a < r < b$

(c) $r > b$

Solution In Figure, dark black circles represent the spherical shells, which carry equal charge q, and grey color circles represent the Gaussian surfaces in all the three regions.

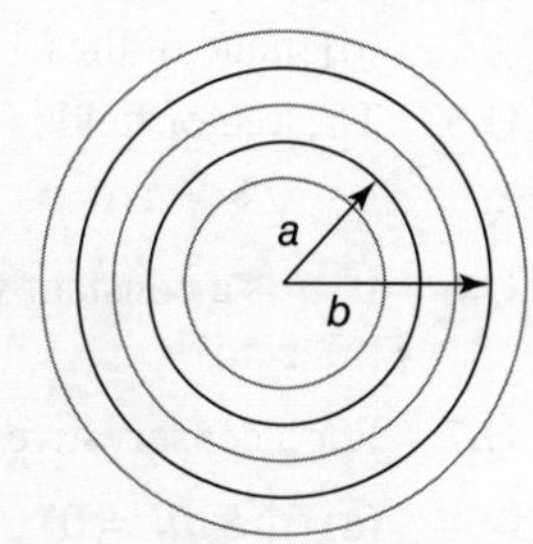

(i) For interior region ($r < a$)

Charge enclosed by the Gaussian surface drawn in this region is zero. Using Gauss's law for spherical symmetry

$$\oint \vec{E} \bullet d\vec{A} = \frac{q_{\text{enc}}}{\varepsilon_0}$$

Since the field E would be the same on the Gaussian surface

$$E \cdot 4\pi r^2 = \frac{0}{\varepsilon_0}$$

$$\mathbf{E = 0}$$

(ii) For middle region ($a < r < b$)

Charge enclosed by the Gaussian surface drawn in this region is q. Using Gauss's law, we get

$$E \cdot 4\pi r^2 = \frac{q}{\varepsilon_0}$$

$$\vec{E} = \frac{q}{4\pi\varepsilon_0 r^2}\hat{r}$$

(iii) For outer region ($r > b$)

Charge enclosed by the Gaussian surface drawn in this region is $2q$. Using Gauss's law, we get

$$\oint \vec{E} \bullet d\vec{A} = \frac{q_{\text{enc}}}{\varepsilon_0}$$

$$E \cdot 4\pi r^2 = \frac{2q}{\varepsilon_0}$$

$$\vec{E} = \frac{q}{2\pi\varepsilon_0 r^2}\hat{r}$$

OBJECTIVE TYPE QUESTIONS

Q.1 The direction of $\vec{\nabla}\phi$ is always
(a) parallel to the surface (b) perpendicular to the surface
(c) depends upon the shape of surface (d) none of these

Q.2 The vector field $\vec{A}$ is solenoidal if
(a) $\vec{\nabla} \times \vec{A} = 0$ (b) $\vec{A} = 0$ (c) $\vec{\nabla} \cdot \vec{A} = 0$ (d) none of these

Q.3 A vector field $\vec{A}$ is conservative if
(a) $\vec{\nabla} \times \vec{A} \neq 0$ (b) $\vec{\nabla} \cdot \vec{A} = 0$ (c) $\vec{A} = \vec{\nabla}\phi$ (d) none of these

Q.4 The vector function $\vec{\nabla} \cdot \vec{A}$ represents
(a) the total flux over any arbitrary closed surface
(b) the inward flux density at the point (x,y,z)
(c) the outward flux density at the point (x,y,z)
(d) none of these

Q.5 The vector field is irrotational if
(a) $\vec{\nabla} \times \vec{A} = 0$ (b) $\vec{\nabla} \cdot \vec{A} = 1$ (c) $\vec{\nabla} \cdot \vec{A} = 0$ (d) $\vec{\nabla} \times \vec{A} = 1$

Q.6 If $\vec{a}$ is a constant vector then $\vec{\nabla} \times (\vec{a} \times \vec{r})$ is equal to
(a) zero (b) $2\vec{a}$ (c) $\vec{a}/2$ (d) $\vec{a}$

Q.7 For a conservative field $\vec{E}$
(a) $\oint \vec{E} \cdot d\vec{l} = 0$ (b) $\oint \vec{E} \cdot \vec{ds} = 0$ (c) $\vec{\nabla} \cdot \vec{E} \neq 0$ (d) $\vec{\nabla} \times \vec{E} \neq 0$

Q.8 The divergence of curl of a vector is always
(a) 1 (b) $\pi/2$ (c) 1/2 (d) zero

Q.9 The unit of ε_0 in SI system of units is
(a) Nm^2/C^2 (b) C/S (c) C^2/Nm^2 (d) N/C

Q.10 The unit of electric flux in SI system of units is
(a) Weber (b) Gauss (c) Nm^2/C (d) N/C

Q.11 A charge of 15 μC is placed at one corner of a cube. The electric flux will be
(a) zero through each face
(b) same through each face
(c) zero through three faces meeting at the location
(d) none of these.

Q.12 The variation of electric field intensity E with distance r from the centre of hollow spherical shell of radius R is

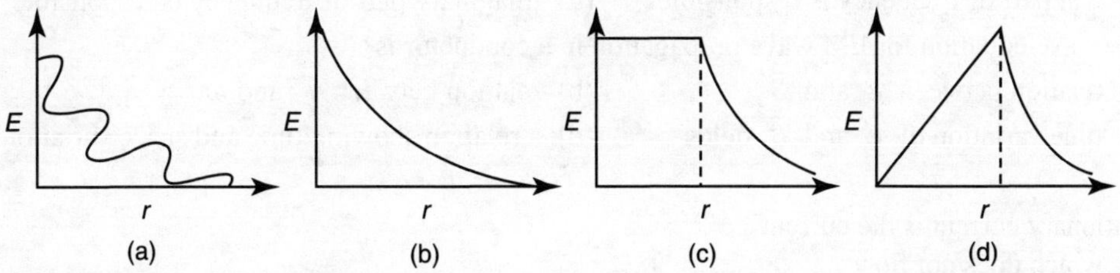

Q.13 The electric field intensity E inside a uniformly charged sphere varies with distance r of the observation point as
(a) $E \propto r$ (b) $E \propto \frac{1}{r}$ (c) $E \propto r^2$ (d) $E \propto \frac{1}{r^2}$

Q.14 The electric field between two oppositely charged plates having equal charge density σ is given by
(a) σ/ε_0 (b) $\sigma/2\varepsilon_0$ (c) zero (d) $2\sigma/\varepsilon_0$

Q.15 Which of the following is zero
(a) grad div (b) div grad (c) curl grad (d) curl curl

Q.16 The relation between electric field and potential is
(a) $\vec{\nabla} \cdot \vec{E} = V$ (b) $E = -\nabla^2 V$ (c) $\vec{E} = -\vec{\nabla} V$ (d) $E = \nabla^2 V$

Q.17 The work done in displacing a charge 2C through 0.5 m on an equipotential surface is
(a) zero (b) 4 J (c) 1 J (d) none of these

Q.18 A point charge q is located at the origin. The amount of work done in bringing a unit positive charge from infinity to the origin is
(a) zero (b) finite (c) infinite (d) none of these

Q.19 Which of the following equations tells us about the non-existence of the magnetic monopole?
(a) curl $\vec{E} = -\frac{\partial \vec{B}}{\partial t}$ (b) div $\vec{B}=0$ (c) div $\vec{D}=\rho$ (d) curl $\vec{H} = \vec{J} + \frac{\partial \vec{B}}{\partial t}$

Q.20 Displacement current appears because of
(a) time varying electric field (b) time varying magnetic field
(c) negative charge only (d) positive charge only

Q.21 Skin depth is the distance
(a) travelled by EM wave in conductor
(b) by which electric field amplitude of EM wave falls to $\frac{1}{e}$ times of its initial value
(c) by which wavelength of EM wave falls to $\frac{1}{e}$ times of its initial value
(d) by which wavelength of EM wave falls to $\frac{1}{2}$ times of its initial value

Q.22 EM wave attenuation is
(a) sudden increase in wave amplitude (b) very slow increase in wave velocity
(c) exponential decay of wave amplitude (d) exponential increase of wave velocity

Q.23 Dispersion relation of an EM wave is
(a) $v = f\lambda$ (b) $v = \omega/k$
(c) $f = \omega/2\pi$ (d) relation between ω and k

Q.24 For wave attenuation
(a) real part of wave number is responsible (b) imaginary part of wave number is responsible
(c) real part of frequency is responsible (d) imaginary part of frequency is responsible

Q.25 The wave equation for EM wave propagation in a conductor is
(a) relation between $\vec{E}$ and $\vec{B}$ (b) relation between $\vec{E}$ and ω
(c) time variation of $\vec{E}$ and $\vec{B}$ fields (d) a relation showing time and space variation of either $\vec{E}$ or $\vec{B}$

Q.26 Stationary current is the current
(a) which does not flow
(b) produced by holes
(c) where charge density remains constant during motion of charges
(d) where charge density vanishes

Q.27 In EM wave
(a) electrons produce magnetic field only
(b) electron produce electric field only
(c) time variation of electric field produces magnetic field and vice-versa
(d) time variation of electric field guides the wave

Q.28 Following expression represents the wave motion
(a) $E = E_0 \sin \omega t$ (b) $E = E_0 \sin (\omega t - kx)$
(c) $E= E_0 \cos kx$ (d) $E = E_0 \sin \cos \omega t$

SHORT-ANSWER QUESTIONS

Q.1 Discuss conservative versus nonconservative fields by giving examples.
Q.2 How do you find the magnetic flux crossing an infinitesimal surface?
Q.3 Provide physical interpretations for the closed surface integrals of any two vectors.
Q.4 What are different ways in which an emf is induced around a loop?
Q.5 State Ampere's circuital law.
Q.6 What is displacement current?
Q.7 State Gauss's law for electric field.
Q.8 State Gauss's law for magnetic field.

Q.9 What is the physical interpretation of Gauss' law for magnetic field?
Q.10 State the law of conservation of charge.
Q.11 How is Gauss's law dependent on Ampere's circuital law?
Q.12 What is Gaussian surface?
Q.13 What is the electric flux through a closed surface surrounding a dipole?
Q.14 Can we apply Gauss's law to calculate the electric field due to electric dipole? Explain.
Q.15 A Gaussian surface encloses no net charge. Does it mean $E = 0$ on its surface.
Q.16 What do you understand by pointing vector?
Q.17 Write Maxwell's equations.
Q.18 State Faraday's laws of electromagnetic induction.
Q.19 What do you mean by a waveguide?
Q.20 Write an expression for characteristic impedance of a co-axial cable.

PRACTICE PROBLEMS

General Questions

Q.1 Describe gradient of a scalar field in Cartesian coordinates. Explain its physical significance.
Q.2 The gradient of a scalar field is a vector. Hence explain how can you produce a vector from a scalar field.
Q.3 Give the physical interpretation of grad V.
Q.4 Define divergence of a vector field. What is its physical meaning? Give two examples.
Q.5 Divergence of a vector field is a scalar quantity. Hence explain how you can produce a scalar field from a vector field.
Q.6 Derive an expression for divergence of a vector field in Cartesian coordinates from first principle.
Q.7 What do you mean by a solenoidal vector field? Give one example. What is the meaning of $\vec{\nabla} \cdot \vec{E} \neq 0$?
Q.8 State and prove Gauss's divergence theorem.
Q.9 Prove that the volume integral of the divergence of a vector field $\vec{A}$ taken over any volume is equal to the surface integral of $\vec{A}$ over the closed surface surrounding the volume.
Q.10 Define curl of a vector field and give its physical significance. Show that curl of a vector field is a vector quantity.
Q.11 Calculate the value of the curl of a vector in terms of Cartesian coordinates.
Q.12 What is an irrotational field? Give one example.
Q.13 Prove that the curl of linear velocity of the particles of a rigid body rotating about an axis passing through it is twice the angular velocity.
Q.14 If $\vec{\omega} \times \vec{r} = \vec{V}$, prove that $\vec{\omega} = \frac{1}{2}$ curl $\vec{V}$, where $\vec{\omega}$ is a constant vector.
Q.15 Show by actual computation that curl gradient of a scalar function is always zero or curl of grad $\phi = 0$.
Q.16 Show that the curl of a uniform electric field is zero.
Q.17 Show that a vector field whose curl is everywhere zero can be expressed as the gradient of another suitable scalar field. What is this type of field called?

Q.18 Prove that div curl $\vec{A} = 0$.

Q.19 If a vector $\vec{B}$ is curl of another vector $\vec{A}$, then prove that the divergence of such vector is zero.

Q.20 Show that a vector field whose divergence is everywhere zero can be expressed as curl of some other suitable vector field.

Q.21 State and prove Stokes' theorem. Discuss its importance.

Q.22 What is a conservative field? Show that a conservative field is the gradient of a scalar field and curl of such a field is zero.

Q.23 Show that electric field is conservative and curl $\vec{E} = 0$.

Q.24 What is the difference between a conservative and non-conservative field? Give one example of each.

Q.25 What do you understand by the term charge density?

Q.26 What is line charge density? Derive an expression for the electric field due to an infinitely long uniformly charged straight wire using Coulomb's law.

Q.27 A thin non-conducting rod of length l carries a positive charge distributed uniformly over its length. If the linear charge density is λ, find the intensity of the electric field at a point at a distance a from the near end of the rod and on its axis.

Q.28 Two parallel infinite wires have uniform line charge densities λ_1 and λ_2 separated by a distance x. Calculate the electric force per unit length on one wire as a result of the other.

Q.29 Derive an expression for electric field at a point situated on the axis of a uniformly charged ring.

Q.30 Define surface charge density and volume charge density. State the relation between electric intensity and charge density.

Q.31 Is volume charge density invariant (under Lorentz transformations)?

Q.32 Find the electric field due to a circular charged disc at a point on a line perpendicular to the disc and passing through its centre. Hence calculate electric field due to an infinitely large plane conducting sheet of charge.

Q.33 Calculate the electric field strength due to a uniform charged circular sheet on the axis.

Q.34 Explain the meaning of the term electric flux. What are its dimensions and S.I. units.

Q.35 State and prove Gauss's theorem in electrostatics. Prove that total flux over a surface due to a charge lying outside is zero.

Q.36 State Gauss's theorem. Derive the differential form of this Gauss's theorem.

Q.37 Write the law for a volume distribution of charge.

Q.38 Apply Gauss's theorem to calculate the electric field due to a uniformly charged solid cylinder.

Q.39 Prove that the electric field at a point inside a uniformly charged cylinder of infinite length is proportional to the distance of the point from the axis.

Q.40 Apply Gauss's theorem to find the electric field strength $\vec{E}$ near a plane non-conducting thin sheet of charge of infinite extent. Hence show that the field is independent of the distance of the observation point from the sheet.

Q.41 Apply Gauss's law to calculate

(i) The electric field at any point due to two parallel sheets of charge.

(ii) Calculate the intensity of the electric field at a point between oppositely charged parallel plates.

Q.42 Using Gauss's theorem calculate the electric field due to a uniform spherical shell of charge at a point (i) Outside the shell and (ii) inside the shell. Hence show that for points lying external to it a uniformly charged spherical shell behaves as if the entire charge were concentrated at its centre and for point lying inside it the electric field is zero.

Q.43 Using Gauss's theorem calculate the electric field due to a uniformly charged non-conducting solid sphere at a point

(i) Outside the sphere

(ii) On the surface of the sphere, and

(iii) Inside the sphere.

Q.44 State and prove Gauss's law or Gauss's theorem. Express it in differential form and show that $\vec{\nabla}\cdot\vec{E} = \rho/\varepsilon_0$

Q.45 Show that Coulomb's law can be deduced from Gauss's law and considerations of symmetry.

Q.46 Coulomb's law is a special case of Gauss's law. Explain.

Q.47 Prove that the electric field on the surface of a conductor is $\dfrac{\sigma}{\varepsilon_0}$ where σ is the surface charge density. Hence find the electric field near a charged conducting sheet with the same surface density of charge.

Q.48 State and prove Ampere's circuital law of magnetic field. Deduce Ampere's law in the form $\oint \vec{B}\cdot d\vec{l} = \mu_0 I$, where the symbols have their usual meaning.

Q.49 Show that the line integral of the magnetic field over a closed path is independent of the shape of the path.

Q.50 Using Ampere's law obtain an expression for the magnetic field due to a current carrying straight conductor of infinite length.

Q.51 Using Ampere's law calculate the magnetic field at point inside a long current carrying solenoid.

Q.52 Explain the concept of Maxwell's displacement current and show how it led to the modification of Ampere's law.

Q.53 What is equation of continuity? Explain it. How could Maxwell correct and present Ampere's law in its generalized form?

Q.54 What are Maxwell's equations? Derive Maxwell's equations (differential form). Discuss integral form of above equations. What are the significance of these equations to electricity and magnetism?

Q.55 Obtain the electromagnetic wave equations, using Maxwell's equation, in an isotropic dielectric medium and show that the speed of wave is less than its speed in vacuum.

Q.56 Obtain Maxwell's equations and deduce an expression for the velocity of propagation of a plane electromagnetic wave in a medium of dielectric constant ε and relative permeability μ.

Q.57 Define poynting vector. Derive an expression for it and explain its physical significance for electromagnetic wave in free space.

Q.58 Derive the electromagnetic wave equation from Maxwell field equations. Consider plane wave solutions of this equation and prove that the energy density associated with such a wave in a stationary homogeneous non-conducting medium propagates with the same speed with which the field vectors do.

Q.59 A plane monochromatic electromagnetic wave propagates in a conducting medium. Show that attenuation is equal to phase vector.

Q.60 Discuss the propagation of plane monochromatic electromagnetic waves in conducting media. Derive the dispersion equation and thus obtain: (i) phase velocity (ii) refractive index (iii) skin depth.

Q.61 Show that inside the conducting medium the wave is damped and obtain an expression for the skin depth δ.

Q.62 What is a wave guide? Describe the propagation of electromagnetic wave along a hollow wave guide of uniform cross section.

Q.63 Give a brief note on coaxial cables with special reference to characteristic impedance.

Q.64 Write note on

(i) Displacement current

(ii) Poynting vector

Q.65 What do you understand by dispersion relation of an electromagnetic wave?

Q.66 Using dispersion relation, find the wavelength and phase velocity of an electromagnetic wave in a dielectric medium having $\varepsilon = 4\varepsilon_0$ and $\mu = \mu_0$.

Q.67 What do you understand by wave attenuation? Find an expression for the skin depth of an electromagnetic wave in a conductor.

Q.68 Discuss in detail the electromagnetic wave propagation in a conducting medium. What are the roles of real and imaginary parts of the wave number of the wave?

Q.69 Find the time taken by a charge Q placed in the interior of a copper to drop to 36.8 percent of its initial value. Take $\sigma = 5.8 \times 10^7$ mhos/m and $\varepsilon = \varepsilon_0$.

Q.70 Plot a graph between the skin depth in copper and frequency of an electromagnetic wave when $\sigma = 5.8 \times 10^7$ mhos/m, and $\mu = \mu_0$.

Q.71 Prove that electric and magnetic field vectors of an electromagnetic wave do not remain in phase when it propagates in a conductor. Also show that $\theta_H - \theta_E = \theta_K$, where θ_H is the phase of magnetic field $\vec{H}$, θ_E is the phase of electric field $\vec{E}$ and θ_K is the phase of wave vector $\vec{k}$.

Theory of Relativity

12

LEARNING OBJECTIVES

After reading this chapter you will be able to

- **LO1** Understand the inertial and non-inertial frames of reference
- **LO2** Learn about Galilean transformation and Michelson-Morley experiment
- **LO3** Know postulates of special theory of relativity and Lorentz transformation
- **LO4** Explain length contraction, time dilation
- **LO5** Discuss addition of velocities
- **LO6** Evaluate variation of mass with velocity and Einstein's mass energy relation

Introduction

Before the beginning of the 20th century, main branches of physics, namely mechanics and electromagnetism, had developed independently. It was a firm opinion of the physicists that these two have no strong relation with each other. However, early in the 20th century they started facing many new and basic problems. For example, Newton's second law of motion did not give correct results when applied to the objects moving with high speed comparable to the speed of light. Moreover, they noticed that for two observers, which are in relative motion, the same set of transformation equations cannot be used to transform the laws of mechanics and electromagnetism from the frame of reference of one observer to the frame of reference of the other observer. However, the introduction of special theory of relativity by *Einstein* in 1905 led to the solution of these and other difficulties. The special theory of relativity deals with the objects or frames of references, which are moving with uniform velocity relative to each other. For dealing with accelerated frames of references, Einstein developed the general theory of relativity in 1915.

12.1 FRAME OF REFERENCE LO1

It is well known that the motion of a body is better understood if it is described with respect to some well defined coordinate system. This coordinate system is called *frame of reference*. Thus, a system of coordinate axes which defines the position of a particle in two-or three-dimensional space is called *frame of reference*.

It is divided into two categories:

(i) Inertial i.e., unaccelerated system

(ii) Non-inertial i.e., accelerated system

12.1.1 Inertial Frame of Reference

Newton's first law of motion says that a system at rest will remain at rest or a system in uniform motion will remain in uniform motion till an external force is applied on it. The systems in which the law of inertia holds good are called inertial systems or inertial frames of reference. Actually the earth is not an inertial frame of reference because of its orbital and rotational motions. However, for most of the purposes, the earth may be regarded as an inertial frame of reference.

12.1.2 Non-inertial Frame of Reference

A non-inertial system is the one in which Newton's first law of motion does not hold good. A frame of reference which is in accelerated motion with respect to an inertial frame is called non-inertial frame of reference.

12.2 GALILEAN TRANSFORMATION LO2

There are the transformations which can transform the coordinates of a particle from one inertial system to another. Consider two inertial systems F and F', where F' is moving with uniform velocity v relative to F along the +ve direction of x-axis (Fig. 12.1). We further consider that the origin of both the systems coincide at time $t = t' = 0$. Let an event happen at point P whose coordinates are (x, y, z, t) and (x', y', z', t') with respect to the frames of references F, and F' respectively. It can easily be seen that these coordinates are related, as

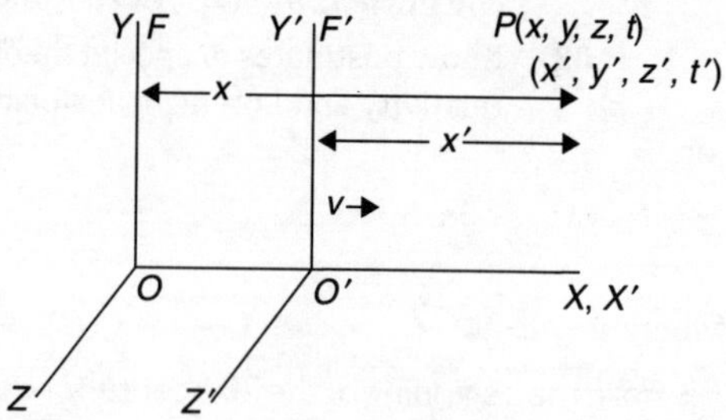

FIGURE 12.1

$$\left.\begin{aligned} x' &= x - vt \\ y' &= y \\ z' &= z \\ t' &= t \end{aligned}\right\} \qquad \text{(i)}$$

Eq. (i) is known as Galilean transformation for position.

Now, consider that frame F' is moving along any direction with velocity v relative to F such that

$$\vec{v} = \vec{v}_x\hat{i} + \vec{v}_y\hat{j} + \vec{v}_z\hat{k}$$

where v_x, v_y and v_z are the components of v along X, Y and Z-axes, respectively, as shown in Fig. 12.2. Suppose the origins of the two systems F and F' coincide at $t = t' = 0$. Let (x, y, z, t) and (x', y', z', t') are the coordinates of the event happening at point P. At the time of event the frame F' is separated from frame F by a distance v_xt, v_yt and v_zt along X, Y and Z-axes respectively. Then we have,

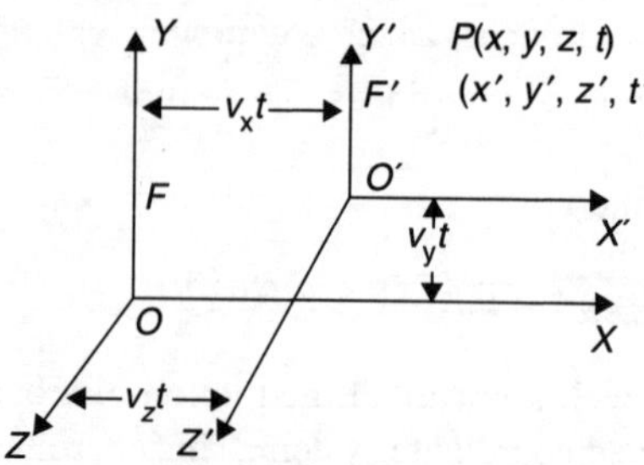

FIGURE 12.2

$$\left.\begin{aligned} x' &= x - v_x t \\ y' &= y - v_y t \\ z' &= z - v_z t \\ t' &= t \end{aligned}\right\} \quad \text{(ii)}$$

Eq. (ii) is also known as Galilean transformation for position.

Galilean velocity transformation of the particle can be obtained by differentiating Eq. (ii), with respect to time. By using $\frac{d}{dt} = \frac{d}{dt'}$ and v_x, v_y and v_z to be constant, we obtain

$$\left.\begin{aligned} \frac{dx'}{dt'} &= \frac{dx}{dt} - v_x \\ \frac{dy'}{dt'} &= \frac{dy}{dt} - v_y \\ \frac{dz'}{dt'} &= \frac{dz}{dt} - v_z \end{aligned}\right\} \quad \text{(iii)}$$

or

$$\left.\begin{aligned} u'_x &= u_x - v_x \\ u'_y &= u_y - v_y \\ u'_z &= u_z - v_z \end{aligned}\right\} \quad \text{(iv)}$$

where u_x, u_y and u_z are the velocities of the particle observed by an observer O in system F and u'_x, u'_y and u'_z are the velocities of the particle observed by O' in system F' along X, Y and Z-axes, respectively. From Eq. (iv), we have,

$$u'_x\hat{i} + u'_y\hat{j} + u'_z\hat{k} = u_x\hat{i} + u_y\hat{j} + u_z\hat{k} - v_x\hat{i} - v_y\hat{j} - v_z\hat{k}$$

or

$$\vec{u}' = \vec{u} - \vec{v} \quad \text{(v)}$$

where $\hat{i}$, $\hat{j}$, $\hat{k}$ are unit vectors along X, Y and Z-axes, respectively. Eq. (v) represents the Galilean transformation of velocity of particle.

Similarly, Galilean acceleration transformation of the particle can be represented by the following equations by knowing the fact that the acceleration of a particle is the time derivative of its velocity.

$$a_x = \frac{du_x}{dt}, a_y = \frac{du_y}{dt} \quad \text{and} \quad a_z = \frac{du_z}{dt}$$

To find the Galilean acceleration transformations, we differentiate the velocity transformation and use the fact that $t' = t$ and v_x, v_y and v_z are the constants. This yields

$$a_x' = a_x, a_y' = a_y \quad \text{and} \quad a_z' = a_z$$

12.3 MICHELSON-MORLEY EXPERIMENT LO2

In 19th century, scientists had assumed that a hypothetical medium called luminiferrous ether is required for the propagation of the light. It was considered that the ether exists uniformly in the space and it is at rest relative to the earth and other planets. The basic purpose of Michelson-Morley experiment was to confirm the existence of stationary ether. The existence of stationary ether (an absolute frame of reference) can be confirmed if we can measure the absolute velocity of earth with respect to the stationary ether.

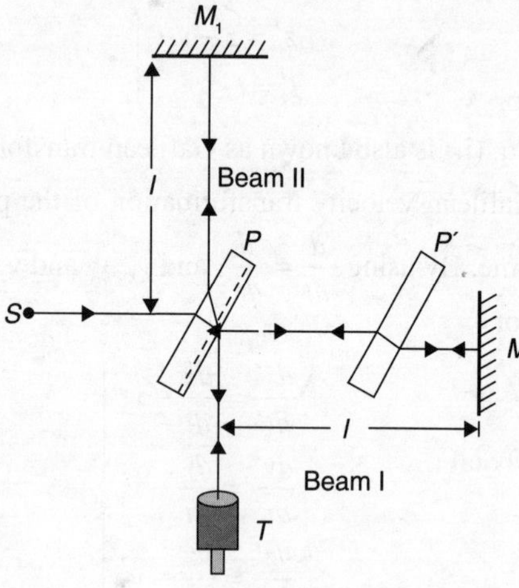

FIGURE 12.3

Figure 12.3 shows an arrangement of Michelson-Morley experiment. Its main components are mirror M_1 and M_2, glass plates P and P' of the same size, a source of light S and a telescope T. The light beam moves to the mirror M_1 and the other moves to mirror M_2. These two beams are reflected back by these mirrors and again are recombined at plate P. Finally, they enter the telescope and produce interference. In the context of interference, this experiment has been discussed in detail in Chapter 1.

In this experiment, the mirrors M_1 and M_2 are set such that $PM_1 = PM_2 = l$. According to Galilean transformation, the velocity of light in a frame moving with constant velocity v relative to stationary either from P to M_2 is $(c - v)$ while from M_2 to P is $(c + v)$. If t_1 be the time taken by the transmitted beam from P to M_2 and back, then

$$t_1 = \frac{l}{c-v} + \frac{l}{c+v} \qquad \text{(i)}$$

or
$$t_1 = \frac{2lc}{c^2 - v^2} = \frac{2l}{c}\left[\frac{1}{1 - v^2/c^2}\right]$$

or
$$t_1 = \frac{2l}{c}\left[1 - \frac{v^2}{c^2}\right]^{-1}$$

By using Binomial expansion theorem for the case of $v/c << 1$, we get

$$t_1 = \frac{2l}{c}\left[1 + \frac{v^2}{c^2}\right] \qquad \text{[Neglecting higher order terms]} \qquad \text{(ii)}$$

The distance travelled by the light beam in time t_1 is given by

$$x_1 = t_1 c = 2l\left[1 + \frac{v^2}{c^2}\right] \qquad \text{(iii)}$$

If t_2' be the time taken by beam-II from P to M_1 and in the same time distance travelled by this beam is ct_2'. In this time t_2', the mirror M_1 shifts to M_1' and travels a horizontal distance vt_2'.

With the help of Fig. 12.4, we get $PO = l$, $PO' = ct_2'$ and $OO' = vt_2'$.

$$(PO')^2 = (PO)^2 + (OO')^2$$

$$(ct_2')^2 = (l)^2 + (vt_2')^2$$

or $$(c^2 - v^2)t_2'^2 = l^2$$

$$t_2' = \left(\frac{l^2}{c^2 - v^2}\right)^{1/2}$$

or $$t_2' = \frac{1}{c} \times \frac{1}{\sqrt{1 - v^2/c^2}} \qquad \text{(iv)}$$

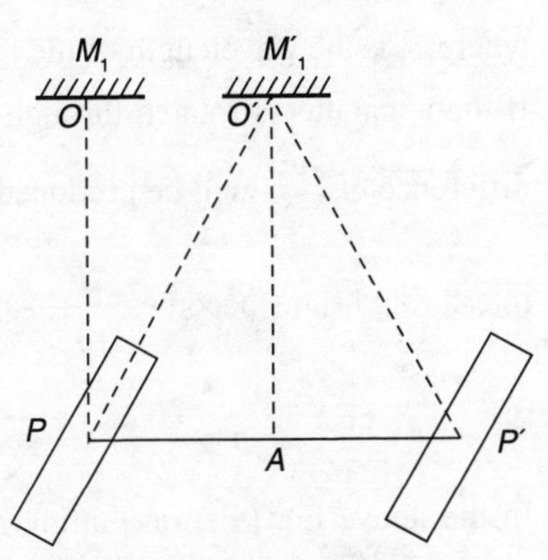

FIGURE 12.4

Total time, $$t_2 = 2t_2' = \frac{2l}{c} \times \frac{1}{\sqrt{1 - v^2/c^2}}$$

or $$t_2 = \frac{2l}{c}\left(1 - \frac{v^2}{c^2}\right)^{-1/2}$$

By using Binomial expansion theorem, we obtain

$$t_2 = \frac{2l}{c}\left(1 + \frac{v^2}{2c^2}\right) \qquad \text{[Neglecting the higher order terms]} \qquad \text{(v)}$$

The distance travelled by the beam in time t_2 is given by

$$x_2 = ct_2$$

$$x_2 = 2l\left(1 + \frac{v^2}{2c^2}\right) \qquad \text{(vi)}$$

With the help of Eqs. (iii) and (vi), we get the path difference

$$\Delta x = x_1 - x_2 = 2l\left(1 + \frac{v^2}{c^2}\right) - 2l\left(1 + \frac{v^2}{2c^2}\right)$$

$$= 2l\left[1 + \frac{v^2}{c^2} - 1 - \frac{v^2}{2c^2}\right]$$

$\therefore$ $$\Delta x = \frac{lv^2}{c^2}$$

Because of the introduction of this path difference of the two beams, the interference pattern would be shifted as,

$$n = \frac{lv^2}{\lambda c^2} \qquad [\because \text{ path difference} = n\lambda \text{ for constructive interference}]$$

where, λ is the wavelength of the light used.

If the apparatus is rotated through 90°, the reflected and transmitted beams get interchanged and the path difference of $\frac{lv^2}{c^2}$ will be produced in the opposite direction. This way the total path difference between the interfering beams becomes $\frac{2lv^2}{c^2}$ and the interference pattern would be shifted as,

$$n = \frac{2lv^2}{\lambda c^2} \quad \text{(vii)}$$

In the above Eq. (vii) on putting $l = 11$ m, velocity of earth $v = 3 \times 10^4$ m/sec, $c = 3 \times 10^8$ m/sec and $\lambda = 5.5 \times 10^{-7}$ m, the expected fringe shift comes out to be

$$n = \frac{2lv^2}{\lambda c^2} = \frac{2 \times 11 \times (3 \times 10^4)^2}{5.5 \times 10^{-7} \times (3 \times 10^8)^2}$$

or $$n = 0.4$$

This displacement of fringe width could have been observed easily since the apparatus used in this experiment was capable of observing a fringe shift as small as 0.01. However, experimentally no significant fringe shift could be observed. This experiment was repeated at different places on the earth, at different times of the day and different seasons of the year. However, no fringe shift was observed in any case. This negative result observed by the experiments suggests that the medium or space in which light propagates is not moving relative to earth.

12.3.1 Explanation of Negative Results of Michelson-Morley Experiment

Three explanations were given in order to explain the negative results of Michelson-Morley experiment.

12.3.1.1 Constancy of Speed of Light

If we accept that the measured speed of light is the same for all directions in every inertial frame, then this could be one of the basis to explain the negative results of Michelson-Morley experiment. If it is so, then the speeds $c + v$ and $c - v$ will simply be equal to c. Hence, the times t_1 and t_2 will be equal to $2l/c$. Thus $\Delta x = x_2 - x_1$ will be zero. So according to the principle of constancy of speed of light, there should not be any shift of fringes. Thus, the principle of constancy of speed of light successfully explains the negative result of Michelson-Morley experiment. However, under this situation, there would be no experimental evidence to indicate the existence of absolute ether. This is because all the inertial frames would be equivalent for the propagation of light as it does not depend on the motion of the observer.

12.3.1.2 Ether Drag Hypothesis

In order to explain the negative results of Michelson-Morley experiment, we consider that the ether is carried by the bodies moving through it or in simple words ether is dragged by the bodies moving through it. As per this hypothesis, ether is also carried by the earth with its own velocity. Hence, there would be no relative motion between the ether and the earth. Therefore, no shift of fringes is obtained. However, this is contrary to stellar aberrations and Fizeau convection coefficient, which are the experimentally observed phenomena.

12.3.1.3 Lorentz FitzGerald Contraction Hypothesis

As per this hypothesis, all bodies moving with velocity v are contracted in the direction of motion by a factor $\sqrt{1-v^2/c^2}$. So if L_0 be the length of a body at rest with respect to ether and L be its length when the body is in motion with velocity v with respect to ether, then $L = L_0\sqrt{1-v^2/c^2}$. With this $t_1 = \frac{2L}{c}(1+v^2/c^2) = \frac{2L_0\sqrt{(1-v^2/c^2)}}{c}(1+v^2/c^2) = \frac{2L_0(1-v^2/2c^2)}{c}(1+v^2/c^2)$. If we neglect the higher powers of v/c in view of $v < c$, then the time $t_1 = \frac{2L_0}{c}(1+v^2/2c^2) = t_2$. It means the time taken or path traversed by the reflected and transmitted beams is the same. So no shift of fringes is observed. However, this concept could not be applied for explaining the negative results of Michelson-Morley experiment when the two arms of the interferometer are not equal.

In view of the above explanations, it is clear that there is no acceptable experimental basis for the idea of ether or absolute frame of reference. However, Einstein in 1905 gave a revolutionary idea that the motion through ether is not a meaningful concept rather the motion relative to a frame of reference has physical significance. Here the frame of reference could be the earth's surface, the sun, the centre of galaxy, etc; though in every case we must specify it. This idea finally developed the theory of relativity.

12.4 POSTULATES OF SPECIAL THEORY OF RELATIVITY LO3

We simply list the two postulates of the special theory of relativity, as

(1) All the fundamental laws of physics retain the same form in all the inertial frames of reference.

(2) The velocity of light in free space is constant and is independent of the relative motion of the source and the observer.

12.5 LORENTZ TRANSFORMATION LO3

The invariance of speed of light in all inertial frames implies that Galilean transformation equations are not suitable. Therefore, we have to introduce new transformation equations which are consistent with the new concept of the invariance of speed of light in free space. These transformation equations were derived by Lorentz and are known as *Lorentz transformation equations.*

Let us consider two inertial frames F and F', as shown in Fig. 12.5. Again consider two observers O and O' situated at the origin in the frame F and F', respectively. Two coordinate systems coincide initially at $t = t' = 0$. Let a pulse of light be generated at time $t = 0$ from the origin which spreads out in the space and at the same time the frame F' starts moving with constant velocity v along +ve direction of x-axis relative to the frame F. This pulse reaches at point P, whose coordinates of position and time are (x, y, z, t) and (x', y', z', t') measured by the observer O and O', respectively.

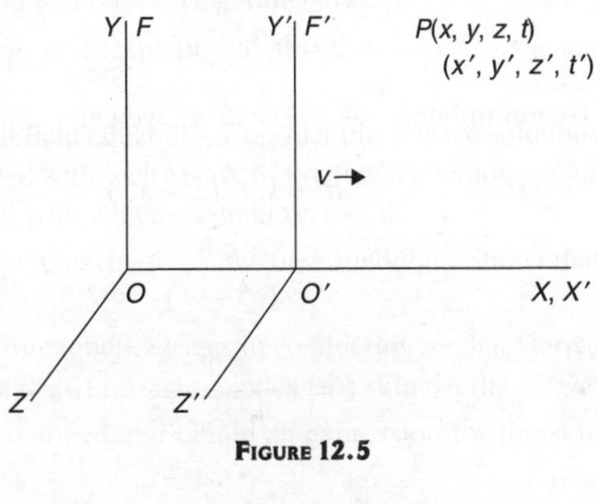

FIGURE 12.5

Therefore, the transformation equations of x and x' can be written as,

$$x' = k(x - vt) \tag{i}$$

where, k is a constant of proportionality and is independent of x and t.

The inverse relation can be written as,

$$x = k(x' + vt') \tag{ii}$$

Keeping in mind that the time t and t' are not equal, we put the value of x' from Eq. (i) in Eq. (ii) in order to get

$$x = k[k(x - vt) + vt']$$

or $$\frac{x}{k} = kx - kvt + vt'$$

or $$t' = \frac{x}{kv} - \frac{kx}{v} + kt$$

$\therefore$ $$t' = kt - \frac{kx}{v}\left(1 - \frac{1}{k^2}\right)$$

$\therefore$ $$t' = kt - \frac{kx}{v}\left(1 - \frac{1}{k^2}\right) \tag{iii}$$

Now according to second postulate of special theory of relativity speed of light c remains constant. So the velocity of pulse of light which spreads out from the common origin observed by observers O and O' should be the same. Therefore,

$$\left.\begin{aligned} x &= ct \\ x' &= ct' \end{aligned}\right\} \tag{iv}$$

By putting the values of x and x' from Eq. (iv) in Eqs. (i) and (ii), we have,

$$ct' = k(x - vt) = k(ct - vt)$$

or $$ct' = kt(c - v) \tag{v}$$

and $$ct = k(ct' + vt')$$

or $$ct = kt'(c + v) \tag{vi}$$

By multiplying Eqs. (v) and (vi), we get

$$c^2tt' = k^2tt'(c^2 - v^2)$$

$$k^2 = \frac{c^2}{(c^2 - v^2)}$$

or $$k = \pm\frac{1}{\sqrt{1 - v^2/c^2}} \tag{vii}$$

or
$$\frac{1}{k^2} = 1 - \frac{v^2}{c^2}$$

or
$$1 - \frac{1}{k^2} = \frac{v^2}{c^2} \tag{viii}$$

Using Eqs. (i), (iii), (vii) and (viii), we have

$$x' = \frac{x - vt}{\sqrt{1 - v^2/c^2}} \tag{ix}$$

$$t' = kt - \frac{kx}{v}\left(\frac{v^2}{c^2}\right)$$

$$= kt - \frac{kxv}{c^2} = k\left(t - \frac{xv}{c^2}\right)$$

or
$$t' = \frac{\left(t - \frac{xv}{c^2}\right)}{\sqrt{1 - v^2/c^2}} \tag{x}$$

$$y' = y \text{ and } z' = z \tag{xi}$$

Hence, the transformation equations become

$$x' = \frac{x - vt}{\sqrt{1 - v^2/c^2}}, \; y' = y, \; z' = z \quad \text{and} \quad t' = \frac{\left(t - \frac{xv}{c^2}\right)}{\sqrt{1 - v^2/c^2}}$$

Imagine if the frame F is moving with velocity v along the –ve direction of x-axis relative to frame F', then we get transformation equations of the form

$$x = \frac{x' + vt'}{\sqrt{1 - v^2/c^2}}, \; y = y', \; z = z' \quad \text{and} \quad t = \frac{\left(t' + \frac{vx'}{c^2}\right)}{\sqrt{1 - v^2/c^2}}$$

These equations are known as *inverse Lorentz transformation equations.*

If the speed of moving frame is much smaller than the velocity of light c, (i.e., $v \ll c$) then the Lorentz transformation equations reduce to Galilean transformation equations.

12.6 LENGTH CONTRACTION LO4

In classical mechanics, the length of an object is independent of the velocity of the moving observer relative to the object. However, on the basis of the theory of relativity, the length of an object depends upon the velocity of the observer with respect to the object.

Let us consider two inertial systems F and F'. The system F' is moving with velocity v relative to the system F along x-axis, as shown in Fig. 12.6. Let a rod AB be at rest in moving system F' relative to the observer O' and L_0 be the length of the rod in this frame measured by observer O' at any instant. This length L_0 measured from the system in which the rod is at rest is called proper length. So L_0 will be given by

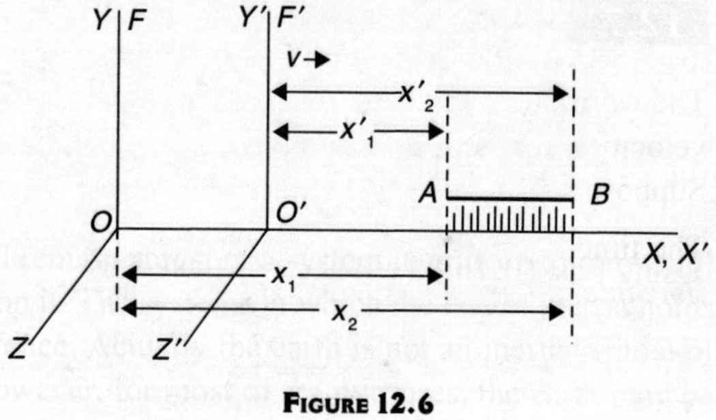

FIGURE 12.6

$$L_0 = x_2' - x_1' \quad \text{(i)}$$

where x_1' and x_2' are the coordinates of two ends of the rod at any instant. At the same time, the length of this rod (say L) measured by an observer O in the stationary frame F is given by

$$L = x_2 - x_1 \quad \text{(ii)}$$

where, x_1 and x_2 are the abscissae of the ends of the rod in the frame F.

Since it would be appropriate (accurate in crude manner) to measure the length by the observer O' in the frame F', we use Lorentz transformation equations for getting the length L_0. For this, we have

$$x_1' = \frac{x_1 - vt}{\sqrt{1 - v^2/c^2}} \quad \text{(iii)}$$

$$x_2' = \frac{x_2 - vt}{\sqrt{1 - v^2/c^2}} \quad \text{(iv)}$$

Substracting Eq. (iii) from Eq. (iv) we have,

$$x_2' - x_1' = \frac{x_2 - x_1}{\sqrt{1 - v^2/c^2}}$$

or

$$L_0 = \frac{L}{\sqrt{1 - v^2/c^2}} \quad \text{(v)}$$

$$L = L_0\sqrt{1 - \frac{v^2}{c^2}} \quad \text{(vi)}$$

From Eq. (vi), we see that $L < L_0$. Thus, the length of the rod is reduced in the ratio $\sqrt{1 - v^2/c^2} : 1$ as measured by the observer moving with velocity v with respect to the rod.

12.6.1 Physical Insight

Length contraction is also called Lorentz contraction or Lorentz-FitzGerald contraction. The contraction takes place only in the direction parallel to the direction in which the observed body travels. For example, in the present case the length contraction takes place in the x-direction only. This effect is negligible at everyday speeds for standard objects and can be ignored for all regular purposes. However, the effect becomes dominant as the magnitude of the velocity approaches the speed of light. So the length contraction is the phenomenon which is usually noticeable at a substantial fraction of the speed of light.

12.7 TIME DILATION

LO4

The word dilation means to lengthen. Consider two coordinate systems F and F' such that F' is moving with velocity v along the x-axis relative to F. Imagine a gun placed at the fixed position (x', y', z') in the frame F'. Suppose it fires two shots at time intervals t_1' and t_2' measured by observer O' in the frame F'.

The time interval $(t_2' - t_1')$ of two shots measured by the clock at rest in moving frame F' is called *proper time interval* and is given by

$$t_2' - t_1' = t_0 \tag{i}$$

As the motion is relative, we may assume that F is moving with velocity $-v$ along the $+x$-axis relative to F'. In the frame F, the observer O, which is at rest, observes these two shots at different times t_1 and t_2. The time interval appears to him is given by

$$t_2 - t_1 = t \tag{ii}$$

Since it would be appropriate (accurate in crude manner) to measure the time by the observer O in the frame F, we use inverse Lorentz transformation equations for getting the time difference t. For this, we have

$$t_1 = \frac{t_1' + vx'/c^2}{\sqrt{1 - v^2/c^2}} \tag{iii}$$

$$t_2 = \frac{t_2' + vx'/c^2}{\sqrt{1 - v^2/c^2}} \tag{iv}$$

By using Eqs. (iii) and (iv), we get,

$$t = t_2 - t_1$$

or

$$t = \frac{t_2' - t_1'}{\sqrt{1 - v^2/c^2}}$$

$$t = \frac{t_0}{\sqrt{1 - v^2/c^2}} \tag{v}$$

Eq. (v) shows that $t > t_0$, i.e., the time interval appears to be lengthened by a factor $\dfrac{1}{\sqrt{1 - v^2/c^2}}$ which is observed by the observer O in frame F. This is known as *time dilation*.

12.7.1 Physical Insight

The time dilation is actually a difference of elapsed time between two events, when measured by the observers that are moving relative to each other, i.e., they have relative motion; the same effect is observed when the observers are definitely situated from a gravitational mass (or masses). The two clocks with the two observers may be measured to tick at different rates, which arises neither from technical aspects of the clocks nor from the propagation time of signals. This takes place due to the nature of space-time, which is such that the time measured along different trajectories is affected by differences in either velocity or gravity, as each of these affects the time in different ways.

12.8 ADDITION OF VELOCITIES

LO5

The classical laws of addition of velocities need to be modified at very high velocities. Consider two frames of references F and F' such that the frame F' is moving with a velocity v relative to F along x-axis. Suppose a particle is moving relative to both the systems F and F'. Let u and u' be the velocities of the particle measured in frames F and F', respectively. Then the velocity components are given as

$$\left.\begin{aligned} u_x &= \frac{dx}{dt}, \quad u_y = \frac{dy}{dt} \quad \text{and} \quad u_z = \frac{dz}{dt}, \\ u'_x &= \frac{dx'}{dt'}, \quad u'_y = \frac{dy'}{dt'} \quad \text{and} \quad u'_z = \frac{dz'}{dt'}, \end{aligned}\right\} \tag{i}$$

From inverse Lorentz transformations, we have

$$x = \frac{x' + vt'}{\sqrt{1 - v^2/c^2}}, \; y = y', \; z = z' \quad \text{and} \quad t = \frac{t' + vx'/c^2}{\sqrt{1 - v^2/c^2}} \tag{ii}$$

By differentiating these equations, we get

$$dx = \frac{dx' + vdt'}{\sqrt{1 - v^2/c^2}}, \; dy = dy', \; dz = dz' \quad \text{and} \quad dt = \frac{dt' + \dfrac{vdx'}{c^2}}{\sqrt{1 - v^2/c^2}} \tag{iii}$$

From Eqs. (i) and (iii), we have

$$u_x = \frac{dx}{dt} = \frac{dx' + vdt'}{dt' + \dfrac{vdx'}{c^2}} = \frac{\dfrac{dx'}{dt'} + v}{1 + \dfrac{v}{c^2}\dfrac{dx'}{dt'}} = \frac{u'_x + v}{1 + \dfrac{v}{c^2}u'_x}$$

$$u_x = \frac{u'_x + v}{1 + \dfrac{vu'_x}{c^2}} \tag{iv}$$

Similarly,

$$u_y = \frac{dy}{dt} = \frac{dy'\sqrt{1 - v^2/c^2}}{dt' + \dfrac{vdx'}{c^2}} = \frac{\dfrac{dy'}{dt'}\sqrt{1 - v^2/c^2}}{1 + \dfrac{v}{c^2}\dfrac{dx'}{dt'}} \quad [\because \; dy = dy']$$

$$u_y = \frac{u'_y\sqrt{1 - v^2/c^2}}{1 + \dfrac{vu'_x}{c^2}} \tag{v}$$

Similarly,

$$u_z = \frac{u'_z\sqrt{1-v^2/c^2}}{1+\dfrac{vu'_x}{c^2}} \tag{vi}$$

Eqs. (iv), (v) and (vi) represent the relativistic laws of addition of velocities whereas in classical mechanics u_x is simply represented by

$$u_x = u_x' + v$$

If $u_x' = c$, i.e., if the light is emitted in the moving frame F' along its direction of motion relative to F, then

$$u_x = \frac{u'_x+v}{1+\dfrac{vu'_x}{c^2}} = \frac{c+v}{1+\dfrac{vc}{c^2}} = \frac{c(c+v)}{(c+v)} = c$$

Thus, from the above expression it is clear that the speed of light is the same in all inertial frames.

If u_x' and v are smaller as compared to c, then $\dfrac{vu'_x}{c^2}$ can be neglected as compared to unity and u_x becomes $u_x = u_x' + v$, the law of addition of velocity which is similar to the one in classical mechanics.

12.9 VARIATION OF MASS WITH VELOCITY LO6

In classical mechanics, mass of a particle is considered to be a constant quantity and independent of its velocity. However, in relativistic mechanics, like length and time, the mass also depends on its velocity. Consider two frames of references F and F' such that F' is moving with a constant velocity v relative to F in the positive direction of X, as shown in Fig. 12.7. Suppose, two similar elastic balls B_1 and B_2 each having the same mass m approach each other in the frame F' with equal speeds (u and $-u$) and collide with each other in such a way that they coalesce into one body. By applying the law of conservation of linear momentum, we have

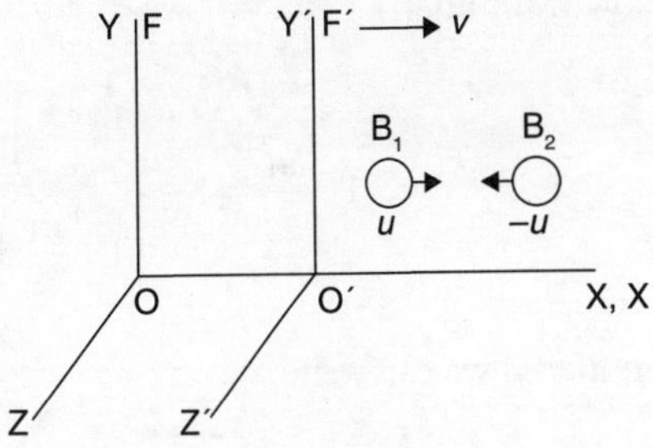

FIGURE 12.7

Momentum of ball B_1 + momentum of ball B_2 = momentum of coalesced body

$$(mu) + (-mu) = 0$$

Thus, the coalesced body must be at rest in frame F'. Now, let us consider the collision with regard to the frame of reference F, where u_1 and u_2 are the velocities of the balls. Then, according to Lorentz velocity transformations

$$u_1 = \frac{u+v}{1+uv/c^2} \tag{i}$$

$$u_2 = \frac{-u+v}{1-uv/c^2} \tag{ii}$$

After an inelastic collision, the coalesced body moves with the velocity of frame F', (as it remains at rest in F'). Thus, v is the observed velocity in frame F. Let mass of the ball B_1 moving with velocity u_1 is m_1 and that of ball B_2 moving with velocity u_2 is m_2 in the frame of reference F. By applying conservation of linear momentum, we have

$$m_1u_1 + m_2u_2 = (m_1 + m_2)v \tag{iii}$$

On substituting u_1 and u_2 from Eqs. (i) and (ii) into Eq. (iii), we have

$$m_1\left[\frac{u+v}{1+uv/c^2}\right] + m_2\left[\frac{-u+v}{1-uv/c^2}\right] = (m_1 + m_2)v$$

$$m_1\left[\frac{u+v}{1+uv/c^2}\right] - m_1v = m_2v - m_2\left[\frac{-u+v}{1-uv/c^2}\right]$$

$$m_1\left[\frac{u+v}{1+uv/c^2} - v\right] = m_2\left[v - \frac{-u+v}{1-uv/c^2}\right]$$

$$m_1\left[\frac{u(1-v^2/c^2)}{1+uv/c^2}\right] = m_2\left[\frac{u(1-v^2/c^2)}{1-uv/c^2}\right]$$

or

$$\frac{m_1}{m_2} = \frac{1+uv/c^2}{1-uv/c^2} \tag{iv}$$

Now from Eq. (i),

$$1 - \frac{u_1^2}{c^2} = 1 - \frac{\left(\frac{u+v}{c}\right)^2}{\left(1+\frac{uv}{c^2}\right)} = \frac{(1-u^2/c^2)(1-v^2/c^2)}{(1+uv/c^2)^2} \tag{v}$$

Similarly, we can write

$$1 - \frac{u_2^2}{c^2} = \frac{(1-u^2/c^2)(1-v^2/c^2)}{(1-uv/c^2)^2} \tag{vi}$$

On dividing Eq. (vi) by Eq. (v) we have

$$\frac{1-u_2^2/c^2}{1-u_1^2/c^2} = \frac{(1+uv/c^2)^2}{(1-uv/c^2)^2}$$

$$\frac{\sqrt{1-u_2^2/c^2}}{\sqrt{1-u_1^2/c^2}} = \frac{1+uv/c^2}{1-uv/c^2} \tag{vii}$$

Thus, from Eqs.(iv) and (vii), we have,

$$\frac{m_1}{m_2} = \frac{\sqrt{1 - u_2^2/c^2}}{\sqrt{1 - u_1^2/c^2}}$$

$$m_1\left[\sqrt{1 - u_1^2/c^2}\right] = m_2\left[\sqrt{1 - u_2^2/c^2}\right] \quad \text{(viii)}$$

From Eq. (viii), it is clear that the left hand side and right hand side are independent of one another. This result may be correct only if each is a constant. Therefore,

$$m_1\left[\sqrt{1 - u_1^2/c^2}\right] = m_2\left[\sqrt{1 - u_2^2/c^2}\right] = m_0$$

where, m_0 is the rest mass of the body.

Thus, $$m_1 = \frac{m_0}{\sqrt{1 - u_1^2/c^2}} \quad \text{(ix)}$$

and $$m_2 = \frac{m_0}{\sqrt{1 - u_2^2/c^2}} \quad \text{(x)}$$

In view of Eqs. (ix) and (x), we conclude that if m_0 be the rest mass of the body then its mass m when it moves at speed v will appear as

$$m = \frac{m_0}{\sqrt{1 - v^2/c^2}} \quad \text{(xi)}$$

This is the relativistic formula for the variation of mass with velocity.

If we substitute $v = c$ in Eq. (xi), then m becomes ∞, which means an object travelling with the velocity of light would acquire infinite mass. Thus, no material particle can have a velocity equal to or greater than the velocity of light. The variation of mass m with the velocity v/c is graphically shown in Fig. 12.8.

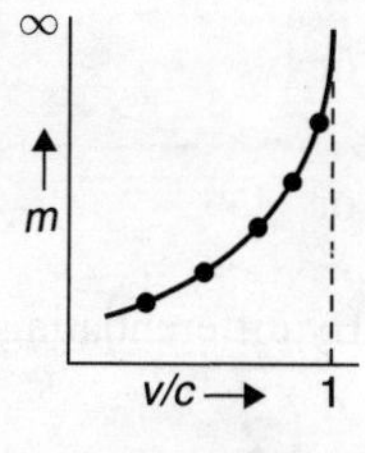

FIGURE 12.8

12.10 EINSTEIN'S MASS ENERGY RELATION LO6

In classical mechanics, the mass of a moving particle is independent of its velocity. Contrary to this aspect, by Einstein's special theory of relativity, the mass of a moving body depends upon its velocity and is given by

$$m = \frac{m_0}{\sqrt{1 - v^2/c^2}} \quad \text{(i)}$$

where v is the velocity of the body, m_0 is its rest mass and c is the velocity of light. The increase in energy of the particle by the applications of force may be defined in terms of work done which is the product of the force and the displacement. According to Newton's second law of motion, the rate of change of momentum of the particle is equal to the force applied on it. Thus

$$F = \frac{d(mv)}{dt} \tag{ii}$$

If the particle is displaced a distance dx by the application of force F, the work done Fdx is stored as kinetic energy (E_K) in the body. Then

$$dW = dE_K = Fdx \tag{iii}$$

By using Eqs. (ii) and (iii), we get

$$dE_K = \frac{d(mv)}{dt}dx$$

or $$dE_K = \frac{dx}{dt}d(mv) \tag{iv}$$

or $$dE_K = v[vdm + mdv] \quad \text{(as } dx/dt{=}v\text{)}$$

$$= v^2 dm + mvdv \tag{v}$$

But $$m = \frac{m_0}{\sqrt{1 - v^2/c^2}}$$

or $$m^2 = \frac{m_0^2 c^2}{c^2 - v^2}$$

or $$m^2c^2 - m^2v^2 = m_0^2c^2 \tag{vi}$$

By differentiating Eq. (vi), we have,

$$2mdmc^2 - 2mdmv^2 - 2vdvm^2 = 0 \quad \text{(as } m_0 \text{ and } c \text{ are constants)}$$

or $$dmc^2 = v^2dm + mvdv \tag{vii}$$

By using Eqs. (v) and (vii), we get

$$dE_K = dmc^2 \tag{viii}$$

Thus, from Eq. (viii) it is clear that the change in kinetic energy is directly proportional to the change in mass. If the body is at rest, its velocity will be zero and hence the change in kinetic energy will be zero. Therefore, its mass will be m_0. If the body moves with velocity v, then its mass becomes m and its kinetic energy becomes E_K. Therefore, by integrating equation (viii), we get

$$\int_0^{E_K} dE_K = c^2 \int_{m_0}^{m} dm$$

or $$E_K = c^2[m - m_0] = mc^2 - m_0c^2$$

or $$E = mc^2 = E_K + m_0c^2 \tag{ix}$$

From Eq. (ix), it is clear that $E = mc^2$ is the total energy. It is the sum of kinetic and rest mass energy. That is,

$$E = mc^2 \tag{x}$$

This relation is called *Einstein's mass energy relation.*

SUMMARY

The topics covered in this chapter are summarized below.

- Newton's second law of motion did not give correct results when applied to the objects moving with high speed comparable to the speed of light. Moreover, for two observers, which are in relative motion, the same set of transformation equations cannot be used to transform the laws of mechanics and electromagnetism from the frame of reference of one observer to the frame of reference of the other observer. The efforts of finding solutions to such problems finally led to the development of special theory of relativity by Einstein in 1905.
- The systems in which the law of inertia holds good are called inertial systems or inertial frames of reference. However, a non-inertial system is the one in which Newton's first law of motion does not hold good. A frame of reference which is in accelerated motion with respect to an inertial frame is called non-inertial frame of reference.
- If the coordinates of a point P are (x, y, z, t) in a frame of reference F and (x', y', z', t') in another frame of reference F' which is moving with uniform velocity v relative to F, then $x' = x - v_xt$, $y' = y - v_yt$, $z' = z - v_zt$ and $t' = t$ are known as Galilean transformation of position. If u_x, u_y, u_z be the velocity components of the particle in the frame of reference F and u'_x, u'_y, u'_z be the velocity components in the frame of reference F', then $u'_x = u_x - v_x$, $u'_y = u_y - v_y$, $u_z = u_z - v_z$ represent Galilean transformation of velocity of particle. Galilean transformation of acceleration of particle is $a'_x = a_x$, $a'_y = a_y$, $a'_z = a_z$.
- The existence of stationary ether (absolute frame of reference) can be confirmed if we can measure the absolute velocity of earth with respect to this ether. For this, Michelson-Morley experiment was conducted at different places on the earth, at different times of the day and in different seasons of the year. However, no shift of fringes was obtained. Thus the motion of the earth through ether could not be experimentally detected.
- Three separate explanations were given for the negative results of Michelson-Morley experiment. These explanations were based on ether drag hypothesis, Lorentz Fitzgerald contraction hypothesis and the constancy of speed of light hypothesis.
- Invariance of speed of light in all inertial frames implies that Galilean transformation equations are not suitable. Therefore, another transformation, named Lorentz transformation was introduced. The Lorentz transformation equations are consistent with the new concept of the invariance of speed of

light in free space. The transformation equations $x' = \frac{x - vt}{\sqrt{1 - v^2/c^2}}$, $y' = y$, $z' = z$, $t' = \frac{t - vx/c^2}{\sqrt{1 - v^2/c^2}}$ are called Lorentz transformation equations, where (x, y, z, t) are the coordinates in a frame of reference F and (x', y', z', t') in another frame of reference F', which is moving with uniform velocity v relative to F.

- In classical mechanics, length of an object is independent of velocity of the moving observer relative to the object. However, as per theory of relativity the lengths measured in a frame F, which is at rest, and in frame F', which is moving with velocity v relative to F, are not the same. However, $L = L_0\sqrt{1 - v^2/c^2}$, where L_0 is the length when measured in the system F' in which the object is kept and L is the length when measured in the frame F which is at rest.
- As per theory of relativity the time intervals measured in a frame F, which is at rest, and in frame F', which is moving with velocity v relative to F, are not the same. However, $t = t_0/\sqrt{1 - v^2/c^2}$, where t_0 is the time interval when measured in the system F' in which the clock is kept and t is the time interval when measured in the frame F which is at rest.
- Classical laws of addition of velocities are modified at very high velocities. The addition of velocities in relativistic mechanics reads $u_x = \frac{u'_x + v}{1 + vu'_x/c^2}$, $u_y = \frac{u'_y\sqrt{1 - v^2/c^2}}{1 + vu'_x/c^2}$, $u_z = \frac{u'_z\sqrt{1 - v^2/c^2}}{1 + vu'_x/c^2}$
- In relativistic mechanics, like length and time, the mass also depends on its velocity. If the mass of the body is m_0 when it is at rest, then its mass m when it moves with velocity v becomes $m = m_0/\sqrt{1 - v^2/c^2}$.
- In classical physics, the force F acting on a body is defined as the time rate of change of momentum. However, in relativistic mechanics, it is the time rate of change of relativistic momentum. In view of this, the kinetic energy of a particle of mass m_0 which acquires velocity v when a force F acts on it through a distance x in time t, is given by $E_K = (\gamma - 1)m_0c^2$, where $\gamma = 1/\sqrt{1 - v^2/c^2}$. This equation states that the increase in kinetic energy of a particle is due to an increase in its mass. The total energy of a particle $E = mc^2$ is equal to the sum of rest mass energy (m_0c^2) and the kinetic energy of the particle. $E = mc^2$ is called the Einstein mass energy relation. This relation simply represents the total energy of the particle in relativistic mechanics.
- The separate laws for the conservation of mass and the conservation of energy are replaced in the theory of relativity by a single law called the conservation of mass energy or the law of conservation of total relativistic energy. According to this law, the total relativistic energy is invariant under Lorentz transformation. It means for an isolated system the total relativistic energy is the same as the energy observed from any inertial system.
- A simple and very useful relation between relativistic momentum p, rest mass energy m_0c^2 and the total energy E is given as $E^2 = p^2c^2 + m_0^{\,2}c^4$.

SOLVED EXAMPLES

EXAMPLE 1 Show that if the variation of mass with velocity is taken into account, the kinetic energy of a particle of rest mass m_0 and moving with velocity v is given by

$$K = m_0c^2\left[\left(1-\frac{v^2}{c^2}\right)^{-1/2} - 1\right]$$

SOLUTION We know,

$$K = (m - m_0)c^2 \quad \text{and} \quad m = \frac{m_0}{\sqrt{1-v^2/c^2}}$$

$$\therefore \quad K = \left(\frac{m_0}{\sqrt{1-v^2/c^2}} - m_0\right)c^2$$

$$= m_0\left(\frac{1}{\sqrt{1-v^2/c^2}} - 1\right)c^2$$

$$= \mathbf{m_0c^2[(1-v^2/c^2)^{-1/2} - 1]}$$

EXAMPLE 2 Show that the relativistic form of Newton's second law, when $\vec{F}$ is parallel to $\vec{v}$, is $\vec{F} = m_0\dfrac{d\vec{v}}{dt}\left(1-\dfrac{v^2}{c^2}\right)^{-3/2}$.

SOLUTION We know that

$$m = \frac{m_0}{\sqrt{1-v^2/c^2}}, \; p = mv \quad \text{and} \quad F = \frac{dp}{dt} = \frac{d}{dt}(mv)$$

$$F = \frac{d}{dt}\left[\frac{m_0}{\sqrt{1-v^2/c^2}}v\right] = m_0\frac{d}{dt}\left[\frac{v}{(1-v^2/c^2)^{1/2}}\right]$$

$$= m_0\left[\frac{1}{(1-v^2/c^2)^{1/2}}\frac{dv}{dt} + v\frac{(-1/2)(-2v/c^2)}{(1-v^2/c^2)^{3/2}}\frac{dv}{dt}\right]$$

$$= m_0\frac{dv}{dt}\left[\frac{(1-v^2/c^2)}{(1-v^2/c^2)^{3/2}} + \frac{(v^2/c^2)}{(1-v^2/c^2)^{3/2}}\right]$$

$$= m_0\frac{dv}{dt}\frac{1}{(1-v^2/c^2)^{3/2}}\left[\left(1-\frac{v^2}{c^2}\right) + \frac{v^2}{c^2}\right]$$

$$= m_0\frac{dv}{dt}\left(1-\frac{v^2}{c^2}\right)^{-3/2}$$

$$\therefore \quad \boldsymbol{\vec{F} = m_0\frac{d\vec{v}}{dt}\left(1-\frac{v^2}{c^2}\right)^{-3/2}}$$

EXAMPLE 3 Derive energy-momentum relation or prove the relation

$$E^2 - c^2p^2 = m_0{}^2c^4.$$

SOLUTION Momentum-energy relation: According to mass-energy relation, we have, energy $E = mc^2$ and momentum $p = mv$

$$E^2 - c^2p^2 = (mc^2)^2 - c^2(mv)^2$$

$$E^2 - c^2p^2 = m^2c^4 - c^2m^2v^2$$

But according to mass-velocity relation,

$$m = \frac{m_0}{\sqrt{1 - v^2/c^2}}$$

$$\therefore \quad E^2 - c^2p^2 = \left[\frac{m_0}{\sqrt{1 - v^2/c^2)}}\right]^2 c^4 - \left[\frac{m_0}{\sqrt{1 - v^2/c^2}}\right]^2 c^2v^2$$

$$= \frac{m_0^2c^4}{(1 - v^2/c^2)} - \frac{m_0^2c^2v^2}{(1 - v^2/c^2)}$$

$$= m_0^2c^2\left[\frac{c^2 - v^2}{1 - v^2/c^2}\right]$$

$$= m_0^2c^4\left[\frac{1 - v^2/c^2}{1 - v^2/c^2}\right]$$

$$E^2 - c^2p^2 = m_0^2c^4$$

$$\mathbf{E = c^2p^2 + m_0^2c^4}$$

which is the required relation for energy and momentum.

EXAMPLE 4 Show that $x^2 + y^2 + z^2 - c^2t^2 = x'^2 + y'^2 + z'^2 - c^2t'^2$ or $x^2 + y^2 + z^2 - c^2t^2$ is invariant under Lorentz transformation.

SOLUTION According to Lorentz transformation, we have

$$x' = \frac{x - vt}{\sqrt{1 - v^2/c^2}}$$

$$y' = y$$

$$z' = z$$

and

$$t' = \frac{t - vx/c^2}{\sqrt{1 - v^2/c^2}}$$

$$\therefore \quad x'^2 + y'^2 + z'^2 - c^2t'^2 = \left[\frac{x - vt}{\sqrt{1 - v^2/c^2}}\right]^2 + y^2 + z^2 - c^2\left(\frac{t - vx/c^2}{\sqrt{1 - v^2/c^2}}\right)^2$$

$$= \frac{(x - vt)^2 - c^2(t - vx/c^2)^2}{1 - v^2/c^2} + y^2 + z^2$$

$$= \frac{x^2 + v^2t^2 - 2xvt - c^2t^2 - v^2x^2/c^2 + 2xvt}{1 - v^2/c^2} + y^2 + z^2$$

$$= \frac{x^2(1 - v^2/c^2) - c^2t^2(1 - v^2/c^2)}{1 - v^2/c^2} + y^2 + z^2$$

$$= \frac{(x^2 - c^2t^2)(1 - v^2/c^2)}{1 - v^2/c^2} + y^2 + z^2$$

$$= x^2 - c^2t^2 + y^2 + z^2$$

$$= \mathbf{x^2 + y^2 + z^2 - c^2t^2}$$

This shows that the quantity $x^2 + y^2 + z^2 - c^2t^2$ is same in both frames of references. So, this is invariant under Lorentz transformation.

Example 5 An event occurs at $x = 100\,\text{m}$, $y = 10\,\text{m}$, $z = 5\,\text{m}$ and $t = 1\times10^{-4}$ second in a reference frame F. Calculate the coordinates of the event in a reference frame F' which is moving with velocity 2.7×10^8 m/sec with respect to the frame F along the common XX' axis using (a) Galilean transformation and (b) Lorentz transformation.

Solution (a) According to Galilean transformation, we have

$$x' = x - vt,\ y' = y,\ z' = z \quad \text{and} \quad t' = t$$

Substituting the given values, we get

$$x' = 100 - 2.7 \times 10^8 \times 1 \times 10^{-4} = -26900 \text{ m}$$

$$y' = y = 10 \text{ m}$$

$$z' = z = 5 \text{ m}$$

and

$$t' = t = 1 \times 10^{-4} \text{ sec}$$

Thus, the coordinates in frame F' are

$$\mathbf{x' = 26900 \text{ m}, y' = 10 \text{ m}, z' = 5 \text{ m and } t = 10^{-4} \text{ sec.}}$$

(b) According to Lorentz transformation, we have

$$x' = \frac{x - vt}{\sqrt{1 - v^2/c^2}},\ t' = \frac{t - vx/c^2}{\sqrt{1 - v^2/c^2}}$$

Substituting the given values, we get

$$x' = \frac{100 - 2.7 \times 10^8 \times 1 \times 10^{-4}}{\sqrt{1 - \{(2.7 \times 10^8)/(3 \times 10^8)\}^2}} = -61720 \text{ m}$$

$$y' = y = 10 \text{ m},$$

$$z' = z = 5 \text{ m}$$

$$t' = \frac{1 \times 10^{-4} - \{(2.7 \times 10^8 \times 100)/(3 \times 10^8)^2\}}{\sqrt{1 - \{(2.7 \times 10^8)/(3 \times 10^8)\}}}$$

$$t' = 2.288 \times 10^{-4} \text{ sec}$$

Thus, the coordinates in F' are

$$x' = -61720 \text{ m}, y' = 10 \text{ m}, z' = 5 \text{ m and } t = 2.288 \times 10^{-4} \text{ sec}$$

Example 6 At what speed must a particle move for its mass to be four times its rest mass?

Solution Given $m = 4m_o$. $v = ?$

Formula used is

$$m = \frac{m_0}{\sqrt{1 - v^2/c^2}} \quad \text{or} \quad 4m_0 = \frac{m_0}{\sqrt{1 - v^2/c^2}}$$

$$1 - \frac{v^2}{c^2} = \frac{1}{16} \quad \text{or} \quad \frac{v^2}{c^2} = 1 - \frac{1}{16} = \frac{15}{16}$$

or $$\mathbf{v = 2.9 \times 10^8\ m/sec}$$

Example 7 With what velocity a particle should move so that its mass appears to increase by 20% of its rest mass?

Solution Given $m = m_0 + \frac{20}{100} m = 1.2 m_0$

Formula used is

$$m = \frac{m_0}{\sqrt{1 - v^2/c^2}} \quad \text{or} \quad 1.2m_0 = \frac{m_0}{\sqrt{1 - v^2/c^2}}$$

$$1 - \frac{v^2}{c^2} = \frac{1}{1.44} \quad \text{or} \quad \frac{v^2}{c^2} = 1 - \frac{1}{1.44} = 0.3055$$

$$\mathbf{v = 0.553\ c}$$

Example 8 Show that the momentum of a particle of rest mass m_0 and kinetic energy K_E is given by the expression.

$$p = \sqrt{\frac{K_E^2}{c^2} + 2m_0 K_E}$$

Solution We know that

$$m = \frac{m_0}{\sqrt{1 - v^2/c^2}}$$

$\therefore$ $$m^2(1 - v^2/c^2) = m_0^2$$

or $$\frac{v^2}{c^2} = \frac{m^2 - m_0^2}{m^2}$$

or
$$\begin{aligned} m^2v^2c^2 &= [(m - m_0)c^2][(m + m_0)c^2] \\ &= [(m - m_0)c^2][(m - m_0)c^2 + 2m_0c^2] \\ &= K_E[K_E + 2m_0c^2] \\ &= K_E^2 + 2m_0c^2K_E \end{aligned}$$

or $$m^2v^2 = \frac{K_E^2}{c^2} + 2m_0K_E$$

$\therefore$ $$\boldsymbol{p = mv = \sqrt{\frac{K_E^2}{c^2} + 2m_0K_E}}$$

EXAMPLE 9 Calculate the mass and speed of 2 MeV electron.

SOLUTION The relativistic kinetic energy of the electron is

$$\text{KE} = mc^2 - m_0c^2$$

$$= \frac{m_0c^2}{\sqrt{1 - v^2/c^2}} - m_0c^2$$

$$= m_0c^2\left[\frac{1}{\sqrt{1 - (v^2/c^2)}} - 1\right]$$

If an electron acquires this energy by a potential V volts, then

$$m_0c^2\left[\frac{1}{\sqrt{1 - (v^2/c^2)}} - 1\right] = eV$$

$$\frac{1}{\sqrt{1 - (v^2/c^2)}} = 1 + \frac{eV}{m_0c^2}$$

$$= 1 + \frac{1.6 \times 10^{-19} \times 2 \times 10^6}{9.1 \times 10^{-31} \times (3 \times 10^8)^2}$$

$$= 1 + \frac{3.2 \times 10^{-13}}{8.19 \times 10^{-14}}$$

$$= 1 + 3.90 = 4.90$$

$$\sqrt{1 - \frac{v^2}{c^2}} = \frac{1}{4.90} = 0.204$$

$$1 - \frac{v^2}{c^2} = 0.0416$$

This gives

$$\boldsymbol{v = 0.98\,c}$$

$$\text{Mass, } m = \frac{m_0}{\sqrt{1 - v^2/c^2}}$$

$$m = \frac{m_0}{\sqrt{1 - \left(\frac{0.98c}{c}\right)^2}} = \frac{m_0}{\sqrt{1 - 0.96}}$$

$$m = \frac{m_0}{0.2} = 5m_0 = 5 \times 9.1 \times 10^{-31}\,\text{kg}$$

$$\boldsymbol{m = 45.5 \times 10^{-31}\,\textbf{kg}}$$

Example 10 Depict that no signal can travel faster than the velocity of light.

Solution

$$u_x = \frac{u'_x + v}{1 + \frac{vu'_x}{c^2}}$$

If $u'_x = c$ and $v = c$, where c = speed of light

$$\therefore \qquad u_x = \frac{c + c}{1 + \frac{c \cdot c}{c^2}} = \frac{2c}{2} = c$$

Thus, addition of any velocity to the velocity of light simply reproduces the velocity of light. Hence, it can be concluded that no signal can travel faster than the velocity of light.

Example 11 At what velocity will the mass of a body is 2.25 times its rest mass?

Solution The mass of a body moving with velocity v is given by

$$m = \frac{m_0}{\sqrt{1 - v^2/c^2}}$$

Given $\qquad m = 2.25\, m_0$

$$2.25 m_0 = \frac{m_0}{\sqrt{1 - v^2/c^2}}$$

$$1 - \frac{v^2}{c^2} = \frac{1}{(2.25)^2} = \frac{1}{5.0625}$$

$$1 - \frac{1}{5.0625} = \frac{v^2}{c^2}$$

$$\frac{v^2}{c^2} = 0.8024$$

or $\qquad \mathbf{v = 2.68 \times 10^8\ m/sec}$

Example 12 If the kinetic energy of a body is double of its rest mass energy, calculate its velocity.

Solution $\text{KE} = (m - m_0)\, c^2$

Given that

$$\text{KE} = 2 m_0 c^2$$

$$2 m_0 c^2 = mc^2 - m_0 c^2$$

$$3 m_0 c^2 = mc^2$$

$$3 m_0 = \frac{m_0}{\sqrt{1 - v^2/c^2}}$$

$$\sqrt{1-v^2/c^2} = 1/3$$

or $$1 - v^2/c^2 = 1/9 \quad \text{or} \quad 1-(1/9) = v^2/c^2$$

$$8/9 = v^2/c^2 \quad \text{or} \quad v = \sqrt{\frac{8}{9}}\, c$$

$\therefore$ $$v = \frac{\mathbf{2\sqrt{2}}}{\mathbf{3}} c = \mathbf{0.94\, c}$$

Example 13 The mass of a moving electron is 11 times its rest mass. Calculate its kinetic energy and momentum.

Solution $m = 11 m_0$

$$\begin{aligned} KE = E - E_0 &= mc^2 - m_0c^2 \\ &= 11\, m_0c^2 - m_0c^2 \\ &= 10\, m_0c^2 \\ &= 10 \times 9.1 \times 10^{-31} \times (3 \times 10^8)^2 \text{ J} \\ &= 9.1 \times 9 \times 10^{-14} \text{ J} \\ &= 8.2 \times 10^{-13} \text{ J} \\ &= \frac{8.2 \times 10^{-13}}{1.6 \times 10^{-19}} \text{eV} \\ &= \mathbf{5.1\, MeV} \end{aligned}$$

Also,

$$m = \frac{m_0}{\sqrt{1-v^2/c^2}}$$

$$11\, m_0 = \frac{m_0}{\sqrt{1-v^2/c^2}}$$

$$1 - v^2/c^2 = \frac{1}{121} \quad \text{or} \quad v^2/c^2 = 1 - \frac{1}{121}$$

$$v = 2.98 \times 10^8 \text{ m/sec.}$$

$\therefore$ $$\begin{aligned} p &= mv \\ &= 11 \times 9.1 \times 10^{-31} \times 2.98 \times 10^8 \\ &= 298.298 \times 10^{-23} \\ &= \mathbf{2.98 \times 10^{-21}\ N\text{-}sec} \end{aligned}$$

Example 14 How fast must an electron move in order to have its mass equal the rest mass of the proton?

Solution $$m = \frac{m_0}{\sqrt{1-v^2/c^2}}$$

Given: $m = 1.67 \times 10^{-27}$ kg (rest mass of proton)

$$m_0 = 9.1 \times 10^{-31} \text{ kg (rest mass of electron)}$$

$$1.67 \times 10^{-27} \text{ kg} = \frac{9.1 \times 10^{-31} \text{ kg}}{\sqrt{1 - v^2/c^2}}$$

$$\sqrt{1 - v^2/c^2} = 5.45 \times 10^{-4}$$

$$\mathbf{v = 2.999 \times 10^8 \ m/s}$$

EXAMPLE 15 Find the velocity of a 0.1 MeV electron according to classical and relativistic mechanics.

SOLUTION Classical mechanics gives

$$\text{KE} = (1/2)mv^2$$

$$0.1 \times 10^6 \times 1.6 \times 10^{-19} = (1/2) \times 9.1 \times 10^{-31} v^2$$

$$0.0351 \times 10^{18} = v^2$$

$$v = 1.87 \times 10^8 \text{ m/sec}$$

Relativistic mechanics gives

$$\text{KE} = (m - m_0)c^2$$

$$\text{KE} = \left(\frac{m_0}{\sqrt{1 - v^2/c^2}} - m_0\right)c^2$$

$$\frac{\text{KE}}{m_0 c^2} = \frac{1}{\sqrt{1 - v^2/c^2}} - 1$$

$$1 + \frac{\text{KE}}{m_0 c^2} = \frac{1}{\sqrt{1 - v^2/c^2}}$$

$$1 + \frac{0.1 \times 10^6 \times 1.6 \times 10^{-19}}{9.1 \times 10^{-31} \times (3 \times 10^8)^2} = \frac{1}{\sqrt{1 - v^2/c^2}}$$

$$1 + 0.195 = \frac{1}{\sqrt{1 - v^2/c^2}}$$

$$1.195 = \frac{1}{\sqrt{1 - v^2/c^2}}$$

$$1 - (v^2/c^2) = 0.7$$

This gives $v = 0.54c = 0.54 \times 3 \times 10^8 =$ **1.64×10^8 m/sec**

EXAMPLE 16 Prove that (1/2) mv^2, where $m = \dfrac{m_0}{\sqrt{1 - v^2/c^2}}$ does not equal to the kinetic energy of a particle moving at relativistic velocity.

SOLUTION Relativistic kinetic energy is given by

$$KE = (m - m_0)c^2 = \left\{ \frac{m_0}{\sqrt{1 - v^2/c^2}} - m_0 \right\} c^2$$

$$= \left\{ \frac{1}{\sqrt{1 - v^2 c^2}} - 1 \right\} m_0 c^2$$

If $v << c$, then $\left(1 - \frac{v^2}{c^2}\right)^{-1/2} = 1 + \frac{v^2}{2c^2} + \frac{3}{8}\frac{v^4}{c^4} + \cdots = 1 + \frac{v^2}{2c^2}$

Since $\frac{v^2}{c^2}$ is very small, the term $\frac{v^4}{c^4}$ and higher order terms are neglected.

$$\therefore \quad KE = \left\{ 1 + \frac{v^2}{2c^2} - 1 \right\} m_0 c^2$$

$$KE = (1/2)\, m_0 v^2$$

As $\qquad \boldsymbol{m \neq m_0}$

Hence, $(1/2)\, mv^2$ does not equal to the KE of a particle moving at relativistic velocity.

EXAMPLE 17 Kinetic energy of a particle is (i) 3 times (ii) equal to its rest mass energy. What is its velocity?

SOLUTION Formula used is $E_K = (m - m_0)c^2$

$$m_0 c^2 = \text{rest mass energy}$$

(i) Given, Kinetic energy = 3 × Rest mass energy

$$(m - m_0)c^2 = 3 \times m_0 c^2$$

or
$$m = 4m_0$$

and
$$m = \frac{m_0}{\sqrt{1 - v^2/c^2}} \quad \text{or} \quad 4m_0 = \frac{m_0}{\sqrt{1 - v^2/c^2}}$$

$$1 - \frac{v^2}{c} = \left(\frac{1}{4}\right)^2 \quad \text{or} \quad \frac{v^2}{c^2} = 1 - \frac{1}{16} = \frac{15}{16}$$

or
$$v = \sqrt{\frac{15}{16}}\, c$$

or
$$v = 0.968c = \mathbf{2.9 \times 10^8\ m/sec}$$

(ii) Kinetic energy = Rest mass energy

$$(m - m_0)c^2 = m_0 c^2$$

$$m = 2m_0$$

and
$$m = \frac{m_0}{\sqrt{1 - v^2/c^2}} \quad \text{or} \quad 2m_0 = \frac{m_0}{\sqrt{1 - v^2/c^2}}$$

$$\frac{v^2}{c^2} = 1 - \frac{1}{4} = \frac{3}{4} \quad \text{or} \quad v = \frac{\sqrt{3}}{2} c = 0.866c$$

$$= \mathbf{2.6 \times 10^8\ m/sec}$$

EXAMPLE 18 Show that the circle $x^2 + y^2 = a^2$ in frame F appears to be an ellipse in frame F' which is moving with velocity v relative to F.

SOLUTION The equation of circle in a stationary frame is $x^2 + y^2 = a^2$

$$x = x'\sqrt{1 - v^2/c^2} \quad \text{and} \quad y = y'$$

Substituting these values in the equation of the circle,

$$x'^2\left(1 - \frac{v^2}{c^2}\right) + y'^2 = a^2$$

$$\frac{x'^2}{a^2}\left(1 - \frac{v^2}{c^2}\right) + \frac{y'^2}{a^2} = 1.$$

$$\text{Suppose } b^2 = \frac{a^2}{1 - (v^2/c^2)}$$

$$\therefore \quad \frac{x'^2}{b^2} + \frac{y'^2}{a^2} = 1 \quad \text{(This is the equation of an ellipse)}$$

EXAMPLE 19 Calculate the mass m and speed v of an electron having kinetic energy 1.5 MeV. [Take rest mass of electron $m_0 = 9.11 \times 10^{-31}$ kg and velocity of light $c = 3 \times 10^8$ m/sec].

SOLUTION The relativistic kinetic energy $K = (m - m_0)c^2$

$$K = 1.5 \text{ MeV} = 1.5 \times 10^6 \times 1.6 \times 10^{-19} \text{ J}$$

$$1.5 \times 10^6 \times 1.6 \times 10^{-19} = (m - 9.11 \times 10^{-31})(3 \times 10^8)^2$$

$$m - 9.11 \times 10^{-31} = \frac{1.5 \times 10^6 \times 1.6 \times 10^{-19}}{(3 \times 10^8)^2}$$

$$m = 3.58 \times 10^{-30} \text{ kg}$$

$$m = \frac{m_0}{\sqrt{1 - v^2/c^2}}$$

or

$$1 - (v^2/c^2) = (m_0/m)^2$$

$$v = c\sqrt{1 - (m_0/m)^2} = 3 \times 10^8 \times \sqrt{1 - \left(\frac{9.11 \times 10^{-31}}{3.58 \times 10^{-30}}\right)^2}$$

$$\mathbf{v = 2.9 \times 10^8 \text{ m/sec}}$$

EXAMPLE 20 What is the length of a metre stick moving parallel to its length when its mass is (3/2) times of its rest mass?

SOLUTION We know that

$$L = L_0\sqrt{1 - (v^2/c^2)} \quad \text{and} \quad m = \frac{m_0}{\sqrt{1 - (v^2/c^2)}}$$

$$\therefore \quad L = L_0(m_0/m)$$

$$= L_0[1/m/m_0] = L_0[1/(3/2)] \quad [\because \ m/m_0 = 3/2]$$

or

$$L = (2/3)L_0 = 0.67\ L_0$$

For $L_0 = 1$ m, $L = 0.67 \times 1 = \mathbf{0.67\ m}$

Example 21 A circular lamina moves with its plane parallel to the x-y plane of a reference frame S at rest. Assuming its motion to be along the axis of x (or y), calculate the velocity at which surface area would appear to be reduced to half to an observer in frame S.

Solution For an observer in frame S at rest, circular lamina, when in motion along the axis of x (or y), will appear to be an ellipse. If diameter of the circle is D_0, its value (say D_x) during motion will be

or $$D_x = D_0\sqrt{1 - v^2/c^2}$$

Area of elliptical lamina

$$(A_e) = \pi(a)(b) = \pi\left(\frac{D_0}{2}\right)\left(\frac{D_0}{2}\sqrt{1 - v^2/c^2}\right)$$

$$= \pi\left(\frac{D_0}{2}\right)^2\sqrt{1 - \frac{v^2}{c^2}}$$

Area of circular lamina

$$(A_c) = \pi\left(\frac{D_0}{2}\right)^2 = \frac{\pi D_0^2}{4}$$

Given, $$A_e = A_c/2$$

$$\therefore \quad \frac{\pi D_0^2}{4}\sqrt{1 - \frac{v^2}{c^2}} = \left(\frac{\pi D_0^2}{4}\right)\left(\frac{1}{2}\right)$$

$$\left(1 - \frac{v^2}{c^2}\right) = \frac{1}{4} \quad \text{or} \quad \frac{v^2}{c^2} = 1 - \frac{1}{4} = \frac{3}{4}$$

$$v = \frac{\sqrt{3}}{2}c = \mathbf{2.6 \times 10^8\ m/sec}$$

Example 22 At what speed should a clock be moved so that it may appear to lose 1 minute in each hour?

Solution The clock loses 1 minute in 1 hour, means clock must record 59 minutes for each 1 hour. So that,
Proper time $t_0 = 59$ min, apparent time $t = 60$ min.
According to Lorentz transformation, time dilation is given by

$$t = \frac{t_0}{\sqrt{1 - v^2/c^2}} \tag{i}$$

Substituting the values in Eq. (i), we have

$$60 = \frac{59}{\sqrt{1 - v^2/c^2}}$$

$$(1 - v^2/c^2) = (59/60)^2$$

$$v^2/c^2 = 1 - (59/60)^2$$

$$v^2 = [1 - (59/60)^2](3 \times 10^8)^2$$

$$v = \mathbf{5.45 \times 10^7\ m/sec}$$

Example 23 The proper life of π^+-mesons is 2.5×10^{-8} s. If a beam of these mesons of velocity 0.8 c is produced, compute the distance the beam can travel before the flux of the meson beam is reduced to $1/e^2$ times the initial flux.

Solution

$$t = \frac{t_0}{\sqrt{1-v^2/c^2}} = \frac{2.5\times10^{-8}}{\sqrt{1-(0.8c/c)^2}} = \frac{2.5\times10^{-8}}{0.6}$$

$$= 4.16\times10^{-8}\,\text{s.}$$

If N_0 is the initial flux and N is the flux after time t', we have

$$N = N_0 e^{-t'/\tau}, \text{ where } \tau \text{ is mean life time}$$

$$\therefore \quad N = (1/e^2)N_0$$

$$\therefore \quad N_0/e^2 = N_0 e^{-t'/\tau}$$

$$e^{t'/\tau} = e^2$$

$$t'/\tau = 2$$

$$t' = 2\tau = 2t$$

The distance travelled by the beam before the flux is reduced to $1/e^2$ times the initial flux $= 2t\times0.8c = 2\times4.16\times10^{-8} \times0.8\times3\times10^8 =$ **19.96 m**

Example 24 A space ship moving away from the earth with velocity 0.6 c fires a rocket whose velocity relative to the spaceship is $0.7c$ (i) away from the earth (ii) towards the earth. What will be the velocity of the rocket, as observed from the earth in two cases?

Solution Formula used for relativistic velocity

$$u = \frac{u'+v}{1+\dfrac{vu'}{c^2}}$$

where, u' is the velocity of rocket relative to space ship and v is the velocity of space ship relative to the earth. The velocity away from the earth is taken as +ve and towards the earth as –ve.
Given $u' = 0.7c$ and $v = 0.6c$.

(i) Rocket fired away from the earth, then

$$u = \frac{0.7c+0.6c}{1+0.7\times0.6} = \frac{1.3c}{1.42} = 0.915c$$

$$= \mathbf{0.92c}$$

(ii) Rocket fired towards the earth, then

$$u = \frac{-u+v}{uv} = \frac{(-0.7+0.6)c}{(1-0.7\times0.6)} = -0.17c$$

$$= \mathbf{0.17c}$$

Example 25 A 1.0 m long rod is moving along its length with a velocity $0.6c$. Calculate its length as it appears to an observer on the earth.

Solution Given $L_0 = 1.0\ m$ and $v = 0.6c$.

Formula used is,

$$L = L_0\sqrt{1-\frac{v^2}{c^2}}$$

or

$$L = 1\sqrt{1-\left(\frac{0.6c}{c}\right)^2} = \sqrt{1-(0.6)^2}$$

$$= \sqrt{0.64} = \mathbf{0.8\ m}$$

EXAMPLE 26 A rod has a length of 2.0 m. Find its length when it is carried in a rocket with a speed of 2.7×10^8 m/sec.

SOLUTION Given $v = 2.7 \times 10^8$ m/sec and $L_0 = 2.0$ m.

Formula used is

$$L = L_0\sqrt{1-\frac{v^2}{c^2}}$$

$$= 2.0\sqrt{1-\left(\frac{2.7\times10^8}{3\times10^8}\right)^2}$$

$$= 2.0\sqrt{1-\left(\frac{2.7}{3}\right)^2}$$

$$= 2.0\sqrt{1-0.81}$$

$$= 2.0\sqrt{0.19}$$

$$= \mathbf{0.872\ m}$$

EXAMPLE 27 Calculate the percentage contraction of a rod moving with a velocity 0.8 times the velocity of light in a direction at 60° to its own length.

SOLUTION Given $v = 0.8c$.

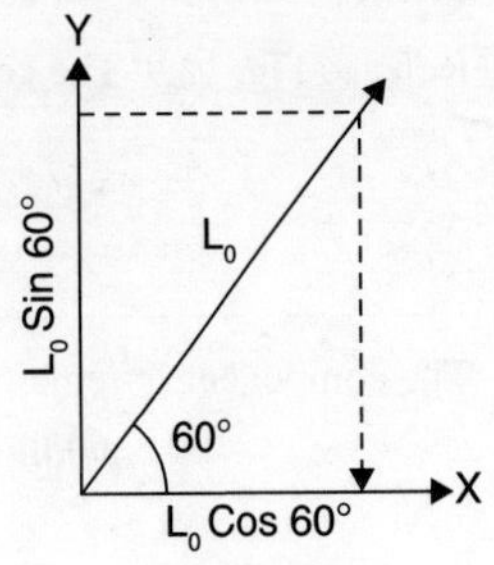

FIGURE 12.9

In Fig. 12.9, component of length along the direction of motion $L_x = L_0 \cos 60° = \dfrac{L_0}{2}$

and perpendicular to the direction of motion $L_y = L_0 \sin 60° = \dfrac{\sqrt{3}}{2}L_0$.

The relativistic contraction occurs only along the direction of motion i.e.,

$$L'_x = L_x\sqrt{1-\frac{v^2}{c^2}} = \frac{L_0}{2}\sqrt{1-(0.8)^2} = 0.3L_0$$

and the component L_y along y-direction remains unchanged.

$$L'_y = L_y = \frac{\sqrt{3}}{2}L_0 = 0.87\,L_0$$

The length of the rod in moving frame, i.e.,

$$L' = \sqrt{L_x'^2 + L_y'^2} = \sqrt{(0.3\,L_0)^2 + (0.87\,L_0)^2}$$

$$= 0.196\,L_0$$

Percentage Length contraction

$$= \frac{L_0 - 0.916\, L_0}{L_0} \times 100$$

$$= \mathbf{8.4\%}$$

EXAMPLE 28 The length of a rod is found to be half of its length when at rest. What is the speed of rod relative to the observer?

SOLUTION Given $L = L_0/2$, $v = ?$

Formula used is

$$L = L_0\sqrt{1 - \frac{v^2}{c^2}}$$

or $$\frac{L_0}{2} = L_0\sqrt{1 - \frac{v^2}{c^2}}$$

or $$1 - \frac{v^2}{c^2} = \frac{1}{4} \quad \text{or} \quad \frac{v^2}{c^2} = 1 - \frac{1}{4} = \frac{3}{4}$$

or $$\frac{v}{c} = \frac{\sqrt{3}}{2} \quad \text{or} \quad v = \frac{\sqrt{3}}{2}c$$

or $$v = \mathbf{0.866\, c}$$

EXAMPLE 29 Calculate the length and orientation of rod of length 5m in a frame of reference which is moving with a velocity $0.6c$ in a direction making an angle of 30° with the rod.

SOLUTION Given $v = 0.6c$ and $L_0 = 5$m.

Refer to Fig. 12.9. The component of the length of the rod along x-direction will be

$$L'_x = L_0 \cos 30° = 5\frac{\sqrt{3}}{2}$$

$$= 4.33$$

The component L_y remains unchanged, i.e.,

$$L'_y = L_y = L_0 \sin 30° = \frac{L_0}{2} = 2.5$$

The length of the rod in a moving frame (L'), i.e.,

$$L' = \sqrt{L'^2_x + L'^2_y} = \sqrt{(4.33)^2 + (2.5)^2} = 4.99 \text{ m}$$

The orientation of the rod is given by angle θ

$$\tan\theta = \frac{L'_y}{L'_x} = \frac{2.5}{3.464} = 0.57$$

$$= \mathbf{30.00°}$$

EXAMPLE 30 Half-life of a particle at rest is 17.8 nanosecond. What will be the half-life when its speed is $0.8c$?

Solution Given $T_0 = 17.8 \times 10^{-9}$ second and $v = 0.8c$

$$T = \frac{T_0}{\sqrt{1-\frac{v^2}{c^2}}} = \frac{17.8\times10^{-9}}{\sqrt{1-\left(\frac{0.8c}{c}\right)^2}} = \frac{17.8\times10^{-9}}{\sqrt{1-0.64}}$$

$$= \frac{17.8\times10^{-9}}{0.6} = \mathbf{29.67\ n\ sec}$$

Example 31 A clock keeps correct time on earth. It is put on the space ship moving uniformly with a speed of 1×10^8 m/sec. How many hours does it appear to lose per day?

Solution Given the time observed by the observer moving with the clock as 24 hrs, i.e., $T = 24$ hrs and the time observed by the observer on the earth = T_0

Formula used is

$$T = \frac{T_0}{\sqrt{1-v^2/c^2}} = \frac{T_0}{\sqrt{1-\left(\frac{1\times10^8}{3\times10^8}\right)^2}} = \frac{T_0}{\sqrt{\frac{8}{9}}} \Rightarrow T_0 = 24\times\frac{2\sqrt{2}}{3}$$

or $$T_0 = 8\times2\sqrt{2} = 16\sqrt{2}$$

$$= 22.63 \text{ sec}$$

Time lost per day = 24 – 22.63 = **1 hr 22 min 12 sec**

Example 32 With what velocity should a rocket move so that every year spent on it corresponds to 4 years on earth?

Solution Given $T_0 = 1$ year (proper time on the rocket) and

$T = 4$ year (relativistic time, corresponding time on the earth)

Formula used is

$$T = \frac{T_0}{\sqrt{1-v^2/c^2}} \quad \text{or} \quad v = \frac{1}{\sqrt{1-v^2/c^2}}$$

or $$1-\frac{v^2}{c^2} = \frac{1}{16} \quad \text{or} \quad \frac{v^2}{c^2} = 1-\frac{1}{16} = \frac{15}{16}$$

or $$v = \sqrt{\frac{15}{16}}\times c = 0.968c$$

$$v = \mathbf{0.97c}$$

Example 33 Determine the time (as measured by a clock at rest on the rocket) taken by a rocket to reach a distant star and return to earth with a constant velocity v equal to $\sqrt{0.9999}\,c$, if the distance to the star is 4 light years.

Solution Given distance of the star from the earth = 4 light year, T_0 = time observed by observer in the rocket.

The time taken by the rocket to go to star from the earth and back with speed $\sqrt{0.9999}\,c$ is

$$T = \frac{2 \times 4c}{\sqrt{0.9999c}} \text{ year} = \frac{8}{\sqrt{0.9999}} = 8.0004 \text{ years}$$

Now $$T = \frac{T_0}{\sqrt{1 - v^2/c^2}} \quad \text{or} \quad T_0 = T\sqrt{1 - v^2/c^2}$$

$$= 8.0004\sqrt{1 - 0.9999} \text{ year}$$

∴ $$\mathbf{T_0 = 0.08 \text{ year}}$$

Example 34 In the laboratory, the life-time of particle moving with speed 2.8×10^8m/sec is found to be 2×10^{-7} sec. Calculate the proper life-time of the particle.

Solution Given $T = 2 \times 10^{-7}$ sec and $v = 2.8 \times 10^8$ m/sec.

Formula used is

$$T = \frac{T_0}{\sqrt{1 - v^2/c^2}} \quad \text{or} \quad T_0 = T\sqrt{1 - v^2/c^2} = 2 \times 10^{-7}\sqrt{1 - \left(\frac{2.8}{3}\right)^2}$$

$$= \mathbf{7.18 \times 10^{-8} \text{ sec}}$$

Example 35 Two electron beams travel along the same straight line but in opposite directions with velocities $v = 0.9c$ relative to the laboratory frame. Find the relative velocity of electrons according to Newtonian mechanics. What will be the velocity measured by an observer moving with one of the electron beams?

Solution According to Newtonian mechanics, the relative velocity between electron beams will be

$$u' = u - v = 0.9c - (-0.9c) = 1.8c$$

The velocity measured by an observer moving with one of the electron beam

$$u' = \frac{u - v}{1 - \frac{uv}{c^2}} = \frac{0.9c - (-0.9c)}{1 + 0.9 \times 0.9} = \frac{1.8c}{1.81}$$

$$= \mathbf{0.994\, c}$$

From the above, it is clear that the relative velocity according to Newtonian mechanics is found to be greater than the velocity of light that is not possible.

Example 36 Two photons approach each other, what is their relative velocity?

Solution Let velocity of each photon be c.

Formula used is

$$u = \frac{u' + v}{1 + \frac{u'v}{c^2}} = \frac{c + c}{1 + \frac{c^2}{c^2}} = c$$

i.e., the relative velocity of photons approaching each other is equal to the velocity of light.

Example 37 A proton has a total relativistic energy as 900 MeV. If the rest mass of the proton is 1.6×10^{-27} kg, find its speed and kinetic energy.

Solution Given

$$E = 900 \text{ MeV} = 900 \times 10^6 \times 1.6 \times 10^{-19} \text{ J}$$

$$= 1.44 \times 10^{-10} \text{ J}$$

Formula used is, $E = mc^2$

$$\therefore \quad \text{Relativistic mass} = m = \frac{E}{c^2} = \frac{1.44 \times 10^{-10}}{(3 \times 10^8)^2} = \mathbf{1.6 \times 10^{-27}\ kg}$$

From the above it is clear that

Rest mass of proton = Relativistic mass of proton.

So that proton is at rest and hence its speed and kinetic energy will be zero.

EXAMPLE 38 Dynamite liberates 5.4×10^6 J/kg when it explodes. What fraction of total energy content is this?

SOLUTION Total energy in one kg content $= m_0 c^2$

$$= 1 \times (3 \times 10^8)^2 = 9 \times 10^{16}\ \text{J}$$

$$\text{Dynamite liberates energy (per kg)} = 5.4 \times 10^6\ \text{J}$$

$$\text{Fraction of total energy} = \frac{5.4 \times 10^6}{9 \times 10^{16}} = \mathbf{0.6 \times 10^{-10}}$$

EXAMPLE 39 Calculate the speed of the electron which has kinetic energy as 1.02 MeV. Given rest mass energy of the electron = 0.51 MeV.

SOLUTION E = Kinetic energy + rest mass energy

$$= E_K + m_0 c^2$$

Given,

$$E_K = 1.02\ \text{MeV} = 2 \times 0.51 = 2m_0 c^2$$

$$= 2 \times \text{rest mass energy}$$

$$E = mc^2 = \frac{m_0 c^2}{\sqrt{1 - v^2/c^2}}$$

$$\therefore \quad E_K + m_0 c^2 = \frac{m_0 c^2}{\sqrt{1 - v^2/c^2}}$$

$$2m_0 c^2 + m_0 c^2 = \frac{m_0 c^2}{\sqrt{1 - v^2/c^2}}$$

$$1 - \frac{v^2}{c^2} = \frac{1}{9} \quad \text{or} \quad \frac{v^2}{c^2} = \frac{8}{9}$$

$$\text{or} \quad v = \frac{2\sqrt{2}}{3} c = 0.943c$$

$$= \mathbf{2.83 \times 10^8\ m/sec}$$

EXAMPLE 40 The earth receives 1400 W/m^2 of solar energy. The distance between the earth and the sun is 1.5×10^{11} m. Estimate the rate of decrease of the mass of the sun.

SOLUTION Solar energy received by the earth = 1400 W/m^2

= 1400 J/m^2 sec

Distance of the earth from the sun $R = 1.5 \times 10^{11}$ m

Total energy liberated by the sun per second

$$= 4\pi R^2 \times 1400 = 4 \times 3.14 \times (1.5 \times 10^{11})^2 \times 1400$$

$$= 3.96 \times 10^{26}\ \text{J/sec}$$

Energy at the sun is generated as per Einstein mass-energy relation, i.e.,

$$E = mc^2 = \mathbf{3.96 \times 10^{26}} \quad \text{or} \quad m = \mathbf{4.4 \times 10^9\ kg/sec}$$

EXAMPLE 41 Calculate the expected fringe-shift in a Michelson-Morley experiment if the distance from each path is 2 meters and light has wavelength 6000 Å. Given, $v = 3 \times 10^4$ m/sec and $c = 3 \times 10^8$ m/scc.

SOLUTION Given, $l = 2$ m, $\lambda = 6.0 \times 10^{-7}$ m, $v = 3 \times 10^4$ m/sec and $c = 3 \times 10^8$ m/sec.

The relation for fringe shift in Michelson-Morley Experiment is given by

$$n = \frac{2lv^2}{\lambda c^2} = \frac{2 \times 2 \times (3 \times 10^4)^2}{6 \times 10^{-7} \times (3 \times 10^8)^2} = \frac{4 \times 10^8}{6 \times 10^{16} \times 10}$$

$$= \mathbf{0.067}$$

EXAMPLE 42 A clock is moving with a speed of 0.95c relative to an observer stationed on the earth. If the speed is increased by 5% by what % does time dilation increases?

SOLUTION Given, $v = 0.95c$ and $v' = \left(v + \frac{5}{100}v\right) = 1.05\ V$

$$= 1.05 \times 0.95c$$

The relation of time dilation is given by

$$\Delta t' = \frac{\Delta t}{\sqrt{1 - \frac{v^2}{c^2}}} \quad \text{or} \quad \frac{\Delta t'}{\Delta t} = \frac{1}{\sqrt{1 - \frac{v^2}{c^2}}}$$

$$\frac{\Delta t'}{\Delta t} = \frac{1}{\sqrt{1 - \left(\frac{0.95c}{c}\right)^2}} = \frac{1}{\sqrt{1 - (0.95)^2}} = 3.2$$

Now $\quad v' = 1.05\ v = 1.05 = 0.95\ c = 0.9975$ c

Then $\quad \left(\frac{\Delta t'}{\Delta t}\right)^1 = \frac{1}{\sqrt{1 - (0.9975)^2}} = 14.14$

$$\text{\% time dilation increases} = \frac{14.14 - 3.2}{14.14} \times 100 = \mathbf{77.3\%}$$

EXAMPLE 43 A beam of particle of half-life 2×10^{-8} sec travels in the laboratory with speed 0.96 c. How much distance does the beam travel before the number of particle is reduced to half-times of the initial value.

SOLUTION The time interval in the particles own frame of reference in which the flux reduces to half of its initial flux is the proper half-time ($\Delta t'$), given by formula

$$\Delta t' = \frac{\Delta t}{\sqrt{1 - \frac{v^2}{c^2}}} = \frac{2 \times 10^{-8}}{\sqrt{1 - (0.96)^2}} = 7.1 \times 10^{-8}\ \text{sec}$$

The distance travelled by the beam in this time in the laboratory frame

$$= 0.96c \times 7.1 \times 10^{-8} = 0.96 \times 3 \times 10^8 \times 7.1 \times 10^{-8}$$

$$= \mathbf{20.45\ m}$$

Example 44 At what sped a body must move so as to have its mass double.

Solution Given, $m = 2m_0$

Formula used is

$$m = \frac{m_0}{1 - \frac{v^2}{c^2}}$$

$$2m_0 = \frac{m_0}{\sqrt{1 - \frac{v^2}{c^2}}}$$

or
$$\frac{v^2}{c^2} = 1 - \frac{1}{4} = \frac{3}{4}$$

$$v = \frac{\sqrt{3}}{2}c = \frac{\sqrt{3}}{2} \times 3 \times 10^8 \text{ m/sec}$$

$$= \mathbf{2.6 \times 10^8 \text{ m/sec}}$$

Example 45 A muon decays with a mean life time of 22×10^{-6} seconds measured in a frame of reference in which it is at rest. If the muon velocity is 0.99c with respect to the laboratory, what is its mean life as observed from laboratory frame?

Solution Given, $\Delta t' = 22 \times 10^{-6}$, $v = 0.99c$

Formula used is

$$\Delta t = \frac{\Delta t'}{\sqrt{1 - \frac{v^2}{c^2}}} = \frac{22 \times 10^{-6}}{\sqrt{1 - (0.99)^2}}$$

$$= \mathbf{1.57 \times 10^{-4} \text{ sec}}$$

Example 46 A stationary body explodes into two fragments of rest mass 1 kg that move apart at speed of 0.6c relative to original body. Find the mass of original body.

Solution Given rest mass of two fragments $(m_0) = 1$ kg and velocity of each fragment i.e., $v_1 = 0.6c$ and $v_2 = -0.6c$

Using the relation, $m = \frac{m_0}{\sqrt{1 - \frac{v^2}{c^2}}}$

For fragment first, $m_1 = \frac{m_0}{\sqrt{1 - \frac{v_1^2}{c^2}}} = \frac{1}{\sqrt{1 - \frac{0.36c^2}{c^2}}} = \frac{1}{0.64}$

$= 1.25$ kg

For fragment second, $m_2 = \frac{m_0}{\sqrt{1 - \frac{v_2^2}{c^2}}} = \frac{1}{\sqrt{1 - \frac{0.36c^2}{c^2}}} = \frac{1}{0.64}$

$= 1.25$ kg

By the law of conservation of mass, the mass of original body will be

$$M = m_1 + m_2 = (1.25 + 1.25) \text{ kg} = \mathbf{2.5 \text{ kg}}$$

EXAMPLE 47 What is the speed of particle whose KE is equal to its rest mass energy?

SOLUTION Given, Rest mass energy = m_0c^2 and

Relavistic KE = $(m - m_0)c^2$

According to the problem, $(m - m_0)c^2 = m_0c^2$

$$mc^2 = 2m_0c^2$$

$$m = 2m_0$$

$$\frac{m_0}{\sqrt{1-\frac{v^2}{c^2}}} = 2m_0 \quad \text{or} \quad \sqrt{1-\frac{v^2}{c^2}} = \frac{1}{2}$$

$$v = \frac{\sqrt{3}}{2}c \quad \text{or} \quad \mathbf{v = 2.6 \times 10^8\ m/sec.}$$

EXAMPLE 48 Find the energy equivalent to a mass of 5.0 mg.

SOLUTION Given, Rest mass m_0 = 5.0 mg = 5.0×10^{-6} kg

As rest mass energy $E = m_0c^2$

$$E = m_0c^2 = 5 \times 10^{-6} \times (3 \times 10^8)^2$$

$$= \mathbf{45 \times 10^{10}\ Joules.}$$

EXAMPLE 49 Show that $E^2/c^2 - p^2$ is invariant.

SOLUTION Consider two frames of reference F and F'. The frame F' is moving with velocity v along x-axis with respect to F. Let p be the momentum and E be the energy of a particle in reference frame F and p' and E' in reference frame F'

$$\therefore \quad p^2 = p_x^2 + p_y^2 + p_z^2 \tag{i}$$

$$p'^2 = p_x'^2 + p_y'^2 + p_z'^2$$

and
$$k = \frac{1}{\sqrt{\left(1-\frac{v^2}{c^2}\right)}} \tag{ii}$$

$$p_x' = k\left[p_x - \frac{V_E}{c^2}\right]$$

$$\therefore \quad p_y' = p_y \text{ and } p_z' = p_z$$

and
$$E' = k[E - vp_x]$$

$$\frac{E'^2}{c^2} - p'^2 = \frac{k^2}{c^2}(E - vp_x)^2 - (p_x'^2 + p_y'^2 + p_z'^2)$$

$$= \frac{k^2}{c^2}[E - vp_x]^2 - k^2\left[p_x - \frac{vE}{c^2}\right]^2 - p_y^2 - p_z^2$$

$$= \frac{k^2}{c^2}[E^2 + v^2p_x'^2 - 2vEp_x - p_x^2c^2 - \frac{v^2E^2}{c^2} + 2vEp_x] - p_y^2 - p_z^2$$

$$= \left[\frac{k^2}{c^2}\left\{E^2 - \frac{v^2E^2}{c^2}\right\} + \left\{-p_x^2c^2 + v^2p_x^2\right\} - p_y^2 - p_z^2\right]$$

$$= \left[\frac{k^2}{c^2}\left\{E^2\left(1-\frac{v^2}{c^2}\right) - p_x^2c^2\left(1-\frac{v^2}{c^2}\right) - p_y^2 - p_z^2\right\}\right]$$

Put $$1-\frac{v^2}{c^2}=\frac{1}{k^2}$$

$$\frac{E'^2}{c^2}-p'^2=\frac{k^2}{c^2}\left[\frac{E^2}{k^2}-\frac{px^2c^2}{k^2}\right]-p_y^2-p_z^2$$

$$=\frac{E^2}{c^2}-(p_x^2+p_y^2+p_z^2)$$

$$\frac{E'^2}{c^2}-p'^2=\frac{E^2}{c^2}-p^2$$

Hence, $\frac{E^2}{c^2}-p^2$ **is invariant.**

EXAMPLE 50 A relativistic electron (m_0 = 0.511 MeV/c^2 and a photon (m_0 = 0) both have momenta of 2.0 MeV/c. Find the total energy of each.

SOLUTION Rest mass of electron m_0 = 0.511 MeV/c^2 and $p(m_0)_{\text{photon}} = 0$, $p_{\text{electron}} = p_{\text{photon}}$ = 2.0 MeV/c

The momentum and energy relation for electron

$$E^2 = e^2p^2 + m_0^2c^4$$

$$= c^2\cdot\frac{(2.0)^2}{c^2}+\frac{(0.511)^2}{c^4}\cdot c^4$$

$$= (2.0)^2 + (0.511)^2 = 4.2611$$

$$E = 2.0642 \text{ MeV}$$

The total energy for photon

$$E = cp = c\cdot 2.0\frac{\text{MeV}}{c} = \mathbf{2.0\ MeV}$$

EXAMPLE 51 Show from Lorentz transformation that two events simultaneous ($t_1 = t_2$) at different positions ($x_1 \neq x_2$) is a reference frame S are not, in general simultaneous in another reference.

SOLUTION Consider a frame F' moving relative to a frame F with a velocity v along x-axis. Let two event occur simultaneously ($t_1 = t_2$) at different positions x_1 and x_2 ($x_1 \neq x_2$) in frame F and the corresponding times of occurrence in frame F' and t_1' and t_2'. According to Lorentz transformations,

$$x_1'=\frac{x_1-vt_1}{\sqrt{1-\left(\frac{v^2}{c^2}\right)}} \quad\text{and}\quad x_2'=\frac{x_2-vt_2}{\sqrt{1-\left(\frac{v^2}{c^2}\right)}}$$

$$t_1'=\frac{t_1-\frac{x_1v}{c^2}}{\sqrt{1-\left(\frac{v^2}{c^2}\right)}} \quad\text{and}\quad t_2'=\frac{t_2-\frac{x_2v}{c_2}}{\sqrt{1-\left(\frac{v^2}{c^2}\right)}}$$

$\therefore$ $$x_2'-x_1'=\frac{(x_2-x_1)-v(t_2-t_1)}{\sqrt{1-\left(\frac{v^2}{c^2}\right)}}$$

For $t_1 = t_2$

$$x_2' - x_1' = \frac{x_2 - x_1}{\sqrt{1 - \left(\frac{v^2}{c^2}\right)}}$$

Also
$$t_2' - t_1' = \frac{(t_2 - t_1) - \frac{v}{c^2}(x_2 - x_1)}{\sqrt{1 - \left(\frac{v^2}{c^2}\right)}}$$

For $t_1 = t_2$,
$$t_2' - t' = -\frac{v}{c^2}\frac{(x_2' - x_1')}{\sqrt{1 - \left(\frac{v^2}{c^2}\right)}}$$

or
$$t_2' - t_1' = -\frac{v}{c^2}(x_2' - x_1')$$

Since $x_1' \neq x_2'$ hence $t_1' \neq t_2'$. This shows that the two events which are simultaneous ($t_1 = t_2$) at positions ($x_1 \neq x_2$) in frame F are not simultaneous in reference frame F'. The negative sign in above relations shows that the events occur at x_2'. First and then at x_1' in frame F'.

EXAMPLE 52 A rod 1.0 m long is moving along its length with velocity 0.6c. Calculate the length as it appears to an observer on the surface of earth.

SOLUTION Let a rod be at rest in moving frame F' relative to observer o' and L_0 be the lengths of the rod in this frame i.e., $L_0 = 1.0$ m.

$$L_o = \frac{L}{\sqrt{1 - \left(\frac{v^2}{c^2}\right)}} \quad \text{or} \quad L = L_o\sqrt{\left(1 - \frac{v^2}{c^2}\right)}$$

$$L = 100 \text{ cm} \times \sqrt{1 - \left(\frac{0.6c}{c}\right)^2} = 100\sqrt{1 - 0.36}$$

$$= 100\sqrt{0.64} = \mathbf{80\ cm}$$

EXAMPLE 53 In an inertial F_1 a red light and a blue light are separated by a distance $\Delta x = 2.45$ km, with the red light at the longer value of x. The blue light flashes and 5.36 μs later the red light flashes. Frame F' is moving in the direction of increasing x with speed of $v = 0.855$c. What is the distance between the two flashes and the time between them as measured in F'?

SOLUTION Given: $\Delta x = x_2 - x_1 = 2.45$ km, $\Delta t = t_2 - t_1 = 5.33$ μs, $v = 0.855$c

Let (x_1, t_1) and (x_2, t_2) represent the position and time of blue and red light in frame F and (x_1', t_1') and (x_2', t_2') are the corresponding values in frame F'

∴
$$x_1' = \frac{x_1 - vt_1}{\sqrt{1 - \frac{v^2}{c^2}}}, \quad x_2' = \frac{x_2 - vt_2}{\sqrt{1 - \frac{v^2}{c^2}}}$$

$$t_1' = \frac{t_1 - vx_1/c^2}{\sqrt{1 - \frac{v^2}{c^2}}}, \quad t_2' = \frac{t_2 - vx_2/c^2}{\sqrt{1 - \frac{v^2}{c^2}}}$$

Let x' and t' represent the distance between the two flashes and the time between them, respectively, as measured in F'.

$$\therefore \qquad x' = x_2' - x_1' = \frac{(x_2 - x_1) - v(t_2 - t_1)}{\sqrt{\left(1 - \frac{v^2}{c^2}\right)}} = \frac{2.54 \times 10^3 - 0.855 \times 3 \times 10^8 \times 5.35 \times 10^{-6}}{\sqrt{1 - (0.855)^2}} = 2.08 \text{ km}$$

and
$$t' = t_2' - t_1' = \frac{(t_2 - t_1) - v/c^2 (x_2 - x_1)}{\sqrt{\left(1 - \frac{v^2}{c^2}\right)}}$$

$$= \frac{5.35 \times 10^{-6} - \frac{0.855}{(3 \times 10^8)^2} \times 2.45 \times 10^3}{\sqrt{1 - (0.855)^2}} = \mathbf{-3.15\ \mu s}$$

This result shows that when measurements are made from F', the red flash comes before the blue flash in F'.

EXAMPLE 54 A particle of rest mass m_0 moves with speed $\frac{c}{\sqrt{2}}$. What are mass, momentum, total energy and kinetic energy.

SOLUTION Given: m_0 = rest mass, $v = \frac{c}{\sqrt{2}}$

By the relation $m = \frac{m_0}{\sqrt{\left(1 - \frac{v^2}{c^2}\right)}} = \frac{m_0}{\sqrt{1 - \frac{\left(\frac{c}{\sqrt{2}}\right)^2}{c^2}}}$

or
$$m = \frac{m_0}{\sqrt{\left(1 - \frac{1}{2}\right)}} = \sqrt{2}\, m_0$$

$$m = \sqrt{2}\, m_0$$

Momentum $p = mv = \sqrt{2}\, m_0 \times \frac{c}{2} = m_0 c$

or
$$p = m_0 c$$

Total energy $E = mc^2 = \sqrt{2}\, m_0 c^2$

$$E = \sqrt{2}\, m_0 c^2$$

and kinetic energy (KE) = $(m - m_0)c^2 = (\sqrt{2} - 1)m_0 c^2$

or
$$\text{KE} = (1.414 - 1)m_0 c^2 = 0.414\, m_0 c^2$$

$$\text{KE} = \mathbf{0.414\, m_0 c^2}$$

EXAMPLE 55 How fast must have an electron move in order to have its mass equal the rest mass of the proton?

SOLUTION Given rest mass of proton (m) = 1.67×10^{-27} kg

$$m_0 = 9.1 \times 10^{-31} \text{ kg.}$$

By using the relation

$$m = \frac{m_0}{\sqrt{1-\left(\frac{v^2}{c^2}\right)}}$$

$$1.67 \times 10^{-27} = \frac{9.1 \times 10^{-31}}{\sqrt{\left(1 - \frac{v^2}{(3 \times 10^8)}\right)}}$$

$$\sqrt{1 - \frac{v^2}{(3 \times 10^8)^2}} = 5.45 \times 10^{-4}$$

$$\frac{v^2}{(3 \times 10^8)^2} = 29.69 \times 10^{-8}$$

$$\mathbf{v = 2.999 \times 10^8 \ m/sec.}$$

Example 56 Calculate the velocity that one atomic mass unit will have if its kinetic energy is equal to twice the rest mass energy.

Solution Given kinetic energy = $2m_0c^2$

As E = kinetic energy + rest mass energy

$$mc^2 = KE + m_0c^2 \quad [\because E = mc^2]$$

$$mc^2 = 2m_0c^2 + m_0c^2 = 3m_0c^2$$

or

$$\frac{m}{m_0} = 3 \quad \text{or} \quad m = 3m_0$$

As

$$m = \frac{m_0}{\sqrt{\left[1-\left(\frac{v^2}{c^2}\right)\right]}} \quad \text{or} \quad 3m_0 = \frac{m_0}{\sqrt{\left(1-\frac{v^2}{c^2}\right)}}$$

$$\left(1 - \frac{v^2}{c^2}\right) = \frac{1}{9}$$

$$\frac{v^2}{c^2} = 1 - \frac{1}{9} = \frac{8}{9}$$

$$v = \sqrt{\left(\frac{8}{9}\right)} \cdot c$$

$$v = 0.94 \times 3 \times 10^8$$

$$\mathbf{v = 2.82 \times 10^8 \ m/sec}$$

Example 57 A nucleus of mass m emits a gamma ray photon of frequency v. Show that the decrease in internal energy of nucleus is not hv, but $hv\left[1+\left(\frac{hv}{2mc^2}\right)\right]$.

Solution Given; frequency of gamma rays photon = v

The momentum of photon of frequency v is

$$p = \frac{hv}{c}$$

The nucleus of mass m recoils back with a momentum $\frac{hv}{c}$ after emitting out a γ-ray photon. The energy used in recoil is

$$E = \frac{1}{2}mv^2 = \frac{p^2}{2m} = \frac{\left(\frac{hv}{c}\right)^2}{2m} = \frac{(hv)^2}{2mc}$$

Hence, the total decrease in internal energy of nucleus is given by

$$\mathbf{\frac{(hv)^2}{2mc^2} + hv = hv\left[1 + \left(\frac{hv}{2mc^2}\right)\right]}$$

Hence proved.

Example 58 What is the total energy of a 2.5 MeV electron?

Solution KE = 2.5 MeV

The total energy of a electron = KE + rest mass energy

$$E = 2.5 \text{ MeV} + \frac{m_0c^2}{1.6\times10^{-19}} = 2.5 \text{ MeV} + \frac{9.1\times10^{-31}\times(3\times10^8)^2}{1.6\times10^{-19}}\text{eV}$$

$$= 2.5 \text{ MeV} + 0.512\times10^6 \text{ eV} = 2.5 \text{ MeV} + 0.512 \text{ MeV}$$

$$= \mathbf{3.012 \text{ MeV}}$$

Example 59 Having the same momentum, which will move faster an electron or a photon?

Solution Given: $p_e = p_p$

or $m_e v_e = m_p v_p$

$m_e >> m_p$, then

$$v_p = \frac{m_e}{m_p} v_e$$

As $m_e >> m_p$, then $\frac{m_e}{m_p} >> 1$, so

$$\mathbf{v_p >> v_e}$$

Photon will travel faster than electron.

Example 60 Find the amount of work to be done to increase the speed of an electron from 0.6c to 0.8c. Take rest energy of electron = 0.5 MeV.

Solution Given, $m_0c^2 = 0.5\times10^6$ eV

K = kinetic energy, $E = K + m_0c^2$

$$K = mc^2 - m_0c^2 = c^2\left[\frac{m_0}{\sqrt{1-\frac{v^2}{c^2}}} - m_0\right]$$

$$K = m_0c^2\left[\frac{1}{\sqrt{1-\frac{v^2}{c^2}}} - 1\right]$$

$$K_1 = m_0c^2\left[\frac{1}{\sqrt{1-\left(\frac{0.6c}{c}\right)^2}} - 1\right] = m_0c^2\left[\frac{1}{0.8} - 1\right]$$

$$= 0.25\times m_0c^2 = 0.25\times0.5\times10^6 \text{ eV}$$

$$= 1.25\times10^5 \text{ eV}$$

Similarly

$$K_2 = m_0c^2\left[\frac{1}{0.6}-1\right] = 0.67 \times 0.5 \times 10^6$$
$$= 3.35 \times 10^5 \text{ eV}$$

The amount of work done to be done $= K_2 - K_1$

$$= 2.1 \times 10^5 \text{ eV}$$
$$= 2.1 \times 10^5 \times 1.6 \times 10^{-19} \text{ J}$$
$$= \mathbf{3.36 \times 10^{-14} \text{ Joule}}$$

EXAMPLE 61 What is the length of a meter stick moving parallel to its length when its mass is $\frac{3}{2}$ times of its rest mass?

SOLUTION Given: $m = \frac{3}{2}m_0$, $L_0 = 1.0$ meter

We know that

$$L = L_0\sqrt{\left(1-\frac{v^2}{c^2}\right)}$$

and

$$m = \frac{m_0}{\sqrt{\left(1-\frac{v_2}{c_2}\right)}}$$

$\Rightarrow$

$$\frac{m_0}{m} = \sqrt{\left(1-\frac{v^2}{c^2}\right)}$$

Or

$$L = L_o\left(\frac{m_0}{m}\right) = \left(\frac{2}{3}\right)L_o \qquad \left[\because \frac{m}{m_0} = \frac{3}{2}\right]$$

$$L = 0.67 \times 1.0 = 0.67 \text{ meter}$$

$$\mathbf{L = 0.67 \text{ meter}}$$

EXAMPLE 62 How fast would a rocket ship have to go relative to an observer for its length to be contracted to 99 per cent of its length at rest?

SOLUTION Given $l = \frac{99}{100}l_o$

Formula used $l = l_o\sqrt{\left(1-\frac{v^2}{c^2}\right)}$

$$\frac{99}{100}l_o = l_o\sqrt{\left(1-\frac{v^2}{c^2}\right)} \quad \text{or} \quad 1-\frac{v^2}{c^2} = \left(\frac{99}{100}\right)^2$$

$$v^2 = 0.0199\text{c} = 0.0199 \times (3 \times 10^8)^2$$

$$v = \sqrt{(0.0199)}\ 3 \times 10^8$$

$$\mathbf{v = 42.3 \times 10^6 \text{ m/sec}}$$

Example 63 A muon decays with a mean life time of 22×10^{-6} seconds measured in a frame of reference in which it is at rest. If the muon velocity is 0.99c with respect to the laboratory, what is its mean life as observed from laboratory frame?

Solution Given: Proper mean life-time $t_0 = 22 \times 10^{-6}$ sec, $v = 0.99c$

$\therefore$ Apparent mean life-time

$$t = \frac{t_0}{\sqrt{1-\frac{v^2}{c^2}}} = \frac{22\times10^{-6}}{\sqrt{\left[1-\frac{(0.99c)^2}{c^2}\right]}}$$

$$= \frac{22\times10^{-6}}{[1-(0.99)^2]} = 1.57\times10^{-5}\ \text{sec}$$

$$\mathbf{t = 1.57 \times 10^{-5}\ sec}$$

Example 64 At what speed should a rocket be move so that it may appear to lose 1 minute in each hour.

Solution Let the clock loses 1 minute in 1 hour, means clock must record 59 min for each 1 hour. So that

Proper time $t_0 = 59$ min., Apparent time $t = 60$ min. According to Lorentz transformation,

$$t = \frac{t_0}{\sqrt{\left(1-\frac{v^2}{c^2}\right)}}$$

Substituting the values in above equation, we have

$$60 = \frac{59}{\sqrt{1-\frac{v^2}{c^2}}} \quad \text{or} \quad \left(\frac{59}{60}\right)^2 = \left(1-\frac{v^2}{c^2}\right)$$

or
$$\frac{v^2}{c^2} = 1-\left(\frac{59}{60}\right)^2 \quad \text{or} \quad v^2 = \left[1-\left(\frac{59}{60}\right)^2\right]\times 3\times10^8$$

$$\mathbf{v = 5.45 \times 10^7\ m/sec}$$

Example 65 An electron has an initial speed of 1.4×10^8 m/sec. How much additional energy must be imported to it for its speed to double?

Solution Given: $v = 1.4 \times 10^8$ m/sec

Let rest mass of electron $m_0 = 9.1 \times 10^{-31}$ kg

The mass of electron at speed 1.4×10^8 m/sec

$$m = \frac{m_0}{\sqrt{1-\frac{v^2}{c^2}}}$$

or
$$m = \frac{9.1\times10^{-31}}{\sqrt{1-\left(\frac{1.4\times10^8}{3\times10^8}\right)^2}}$$

$$= 1.029\times10^{-30}\ \text{kg}$$

Then corresponding energy

$$E_1 = mc^2 = 1.029 \times 10^{-30} \times (3 \times 10^8)^2$$
$$= 9.261 \times 10^{-14} \text{ Joules}$$

If $v_1 = 2v = 2 \times 1.4 \times 10^8 = 2.8 \times 10^8$ m/sec

$$m' = \frac{9.1\times10^{-31}}{\sqrt{1-\left(\frac{2.8\times10^8}{3.0\times10^8}\right)^2}} = \frac{9.1\times10^{-31}}{0.359}$$
$$= 2.5348\times10^{-30}\text{kg}$$
$$E_2 = mc^2 = 2.5348 \times 10^{-30} \times (3 \times 10^8)^2$$
$$= 22.813 \times 10^{-14} \text{ Joules}$$

Hence, additional energy $\Delta E = E_2 - E_1$

$$\Delta E = 22.813 \times 10^{-14} - 9.26 \times 10^{-14}$$
$$= 12.542 \times 10^{-14} \text{ Joules}$$
$$\mathbf{\Delta E = 1.254 \times 10^{-13} \text{ Joules}}$$

Example 66 At what speed does a clock move if it runs at a rate which is one-third the rate of a clock at rest?

Solution Let a clock moves with velocity v w.r.t. another similar clock which is at rest. Let the time interval observed by the clock at rest is t and the time interval observed by the clock moving with velocity v is t_0.

The time interval measured by an observer that is stationary w.r.t. to the moving clock is proper time. From concept of time interval

$$t = \gamma t_0 \text{ where } \gamma = \frac{1}{\sqrt{1-\frac{v^2}{c^2}}}$$

Given that $t = 3\ t_0$,

$$3t_0 = \gamma t_0$$
$$\gamma = 3$$

Velocity of the moving clock is

$$\frac{1}{\sqrt{1-\frac{v^2}{c^2}}} = 3$$

$$\mathbf{v = 0.9428c}$$

Example 67 At what speed does a meter stick move if its length is observed to shrink to 0.6 m?

Solution From the length contraction, when the observer and the object have relative motion, length appears to be shortened as per the following relation

$$L = \frac{L_0}{\gamma} \text{ where } v = \frac{1}{\sqrt{1-v^2/c^2}}$$

Given that

$$L = 0.6 \text{ m}; L_0 = 1.0 \text{ m}$$

This implies

$$\gamma = 1.666$$

or
$$\frac{1}{\sqrt{1-\frac{v^2}{c^2}}} = 1.666$$

$$v = 0.7998c$$

The speed of a meter stick is **0.7998*c***.

Example 68 The average lifetime of a π meson in its own frame of reference is 26.0 ns. If the π meson moves with speed 0.9c with respect to the Earth,

(a) What is its lifetime as measured by an observer at rest on Earth?
(b) What is the average distance it travels before decaying as measured by an observer at rest on Earth?

Solution

(a) The life time of π meson in its frame is 26.0 ns. This is the proper time observation as there is no relative motion between observer and the object (say t_0). The velocity of the π meson w.r.t. Earth is 0.9c.

Let the observed life time of π meson by an observer at rest on Earth be t. Then

$$t = \gamma t_0 \text{ where } \gamma = \frac{1}{\sqrt{1-\frac{v^2}{c^2}}}$$

$$t = \frac{26.0}{\sqrt{1-\frac{(0.9c)^2}{c^2}}}$$

$$t = \frac{26.0}{\sqrt{1-0.81}}\text{ns}$$

$$t = \frac{26.0}{\sqrt{0.19}}\text{ns}$$

$$t = 2.294 \times 26.0 \text{ ns}$$

$$\mathbf{t = 59.64 \text{ ns}}$$

(b) Average distance it travels before decaying as measured by an observer at rest on Earth is given as L = lifetime × velocity

$$L = 59.64 \times 10^{-9} \times 0.9 \times 3 \times 10^8$$

$$\mathbf{L = 16.10 \text{ m}}$$

Example 69 Electrons in projection television sets are accelerated through a potential difference of 60 kV.

(a) Calculate the speed of the electrons using the relativistic form of kinetic energy assuming the electrons start from rest.
(b) Calculate the speed of the electrons using the classical form of kinetic energy.
(c) Is the difference in speed significant in the design of this set?

Solution

(a) Total energy of a relativistic particles is given as E = KE + rest mass energy

$$KE = E - m_0c^2$$

$$\text{KE} = \gamma m_0c^2 - m_0c^2 = (\gamma - 1)m_0c^2$$

Rest mass energy of electron is m_0c^2

$$= 9.1 \times 10^{-31} \times 3 \times 10^8 \times 3 \times 10^8 = 81.9 \times 10^{-15}\text{ J}$$

$$= 0.511\text{ MeV} = 511\text{ keV}$$

Kinetic energy, KE = $(\gamma - 1)m_0c^2$

$60 = (\gamma - 1) \times 511$ [Potential difference of 60 kV ⇒ kinetic energy = 60 keV]

or $$\gamma = 1 + \frac{60}{511} = \frac{571}{511} = 1.1174$$

$$\gamma = \frac{1}{\sqrt{1 - \frac{v^2}{c^2}}} = 1.1174$$

or $$v_{rel} = 1.3376 \times 10^8 = \mathbf{0.446c}$$

(b) From classical point of view kinetic energy = $\frac{1}{2}mv^2$

$$60 \times 1000 \times 1.6 \times 10^{-19} = \frac{1}{2} \times 9.1 \times 10^{-31} \times v^2$$

$$v^2_{classical} = \frac{60 \times 1000 \times 1.6 \times 10^{-19} \times 2}{9.1 \times 10^{-31}} = 21098.90 \times 10^{12} = 2.1098 \times 10^{16}$$

$$v_{classical} = \sqrt{2.1098 \times 10^{16}} = 1.4525 \times 10^8 \frac{m}{s} = \mathbf{0.484c}$$

(c) The difference between the speeds of electrons accelerated through the potential of 60 kV is not significant in case of TV as the distance travelled is of the order of 50 cm and the time lag thus produced is in of the order of few nano-second. This can be neglected in present case.

Example 70 Two powerless rockets are heading towards each other on a collision course. As measured by a stationary observer at Earth, rocket A has speed 0.800c, rocket B has speed 0.600c, both rockets are 50.0 m in length, and they are initially 2.52 Tm apart.

(a) What are their respective proper lengths?
(b) What is the length of each rocket as observed by a stationary observer in the other rocket?
(c) According to observer at earth, how long before the rockets collide?
(d) According to Rocket A, how long before they collide?
(e) According to Rocket B, how long before they collide?
(f) If the crew are able to evacuate their rockets safety within 50 min (their own time), will they be able to do so before the collision?

Solution

(a) Given velocity of rocket A w.r.t. Earth = 0.8c

Length of rocket A w.r.t. Earth = 50 m

Velocity of rocket B w.r.t. Earth = 0.6 c

Length of rocket B w.r.t. Earth = 50 m

The distance between two rockets w.r.t. Earth = 2.52 Tm = 2.52×10^{12} m.

The proper length of rocket A

$$L = \frac{L_0}{\gamma}$$

$$L_0 = \gamma L = \frac{1}{\sqrt{1 - \frac{v^2}{c^2}}} \times L = \frac{1}{\sqrt{1 - \frac{(.8c)^2}{c^2}}} \times 50\,\text{m}$$

$$L_0 = \frac{1}{0.6} \times 50\,\text{m} = \mathbf{83.33\ m}$$

Similarly, the proper length of rocket B

$$L_0 = \gamma L = \frac{1}{\sqrt{1 - \frac{v^2}{c^2}}} \times L = \frac{1}{\sqrt{1 - \frac{(.6c)^2}{c^2}}} \times 50\ \text{m}$$

$$L_0 = \frac{1}{0.8} \times 50\ \text{m} = \mathbf{62.50\ m}$$

(b) To find the length of each rocket w.r.t. other rocket, we need to know the relative speed of each rocket w.r.t. each other.

Thus, from relative speed formula, the speed of rocket A w.r.t. rocket B is

$$u'_x = \frac{u_x - v}{1 - \frac{u_x v}{c^2}}$$

or $$u'_x = \frac{0.8c - (-.6c)}{1 - \frac{.8c(-.6c)}{c^2}}$$

or $$u'_x = \frac{1.4c}{1.48} \approx \mathbf{0.946}c$$

Similarly the velocity of Rocket B as measured by a stationary observer in Rocket A is **0:946 c**.

Then

The length of rocket A w.r.t. rocket B is $L = \dfrac{L_0}{\gamma}$

or $$L = \frac{83.33}{3.083} = \mathbf{27.02\ m}$$

The length of rocket B w.r.t. rocket A is

$$L = \frac{62.5}{3.083} = \mathbf{20.27\ m}$$

(c) From Earth, the time taken in collision of two rockets is

$$0.8c \times \Delta t + 0.6c \times \Delta t = 2.52 \times 10^{12}$$

$$\Delta t = \frac{2.52 \times 10^{12}}{1.4 \times 3 \times 10^8} = 6000\ \text{s} = \mathbf{100\ min.}$$

(d) The time taken before collision, according to rocket A

$$t = \frac{1}{\sqrt{1-\frac{v^2}{c^2}}} t_0$$

$$100 = \frac{1}{\sqrt{1-\frac{(.8c)^2}{c^2}}} t_0$$

$$t_0 = 100 \times \sqrt{1-\frac{(.8c)^2}{c^2}} = 100 \times 0.6 = \mathbf{60\ min.}$$

(e) The time taken before collision, according to rocket B

$$t = \frac{1}{\sqrt{1-\frac{v^2}{c^2}}} t_0$$

$$100 = \frac{1}{\sqrt{1-\frac{(.6c)^2}{c^2}}} t_0$$

$$t_0 = 100 \times \sqrt{1-\frac{(.6c)^2}{c^2}} = 100 \times 0.8 = \mathbf{80\ min.}$$

(f) All the crew members can evacuate their rockets before the collision as they get sufficient time for the same.

EXAMPLE 71 Superfast muons ($v = .998c$) can be produced by the collision of cosmic radiation with atoms high in the atomosphere. Slow-moving muons in the laboratory fame have a lifetime as 2.2 μs. Experiments show that a large number of muons do reach the sea surface. Explain this phenomenon with time dilation.

SOLUTION The life time of muons is 2.2 μs in the laboratory frame i.e., this is the observation with no relative motion between muons and observer. From time dilations, the life time of muons is different and increased from an observer (Earth) with respect to which muons are moving.

Thus

$$t = \gamma t_0 \text{ where } \gamma = \frac{1}{\sqrt{1-\frac{v^2}{c^2}}}$$

Putting $t_0 = 2.2$ μs and $v = 0.998c$

$$t = \frac{1}{\sqrt{1-\frac{v^2}{c^2}}} t_0 = \frac{1}{\sqrt{1-\frac{(.998c)^2}{c^2}}} 2.2\ \mu\text{s} \qquad (\because d = ct)$$

$$t = 15.81 \times 2.2\ \mu\text{s} = \mathbf{34.78\ \mu s}$$

The distance covered by high speed muons during this life time calculated in Earth frame of reference is 10.41 km. This results that a muon which is produced at height 10 km from sea level can travel the distance to reach at sea level with speed .998c and life time 34.78 μs.

EXAMPLE 72 The period of a pendulum is measured to be 5 s in the reference frame of the pendulum. What is the period when measured by an observer moving at a speed of 0.90*c* relative to the pendulum? What if we increase the speed of the observer by 10%? Does the dilated time interval increase by 10% or more?

SOLUTION The time period of the pendulum gets changed when the reference frame is changed from stationary frame to moving frame. Thus

$$t = \gamma t_0 \text{ where } \gamma = \frac{1}{\sqrt{1-\dfrac{v^2}{c^2}}}$$

Here t_0 = 3s and v = 0.95c then

$$t = \frac{1}{\sqrt{1-\dfrac{(.9c)^2}{c^2}}} 3 = 2.294 \times 5\text{s} = \mathbf{11.47\ s}$$

If the speed of observer is increased by 10%, then speed becomes

$$v = .9c \times 1.10 = \mathbf{0.99c}$$

With 10% increased velocity, the value of gamma factor is modified from 2.294 to 7.0888 which in turn, increases the dilated time period of pendulum by over 200%.

EXAMPLE 73 A spacecraft is measured to be 150.0 m long and 30.0 m in diameter while at rest relative to an observer. If this spacecraft now flies by the observer with a speed of 0.95*c*, what length and diameter does the observer measure?

SOLUTION From length contraction, the length observations made from moving frame are shortened by factor γ. Here,

Actual length L_0 = 150.0 m
Actual diameter d_0 = 30.0 m
Let observations made by observer while spacecraft is moving with velocity 0.95*c* be *L* and *d*.
Then

$$L = \frac{L_0}{\gamma} = L_0\sqrt{1-\frac{v^2}{c^2}} = 120\sqrt{1-\frac{(.95c)^2}{c^2}} = 120\sqrt{1-.9025} = 120\sqrt{0.0975}$$

$$L = 150 \times .3122 \text{ m}$$

$$L = \mathbf{46.84\ m}$$

The length of the space craft will appear to be 46.84 m if space craft is moving with velocity 0.95*c* relative to observer.

However, diameter will be appear to be of 30 m as there is no motion of spacecraft in the axis of diameter.

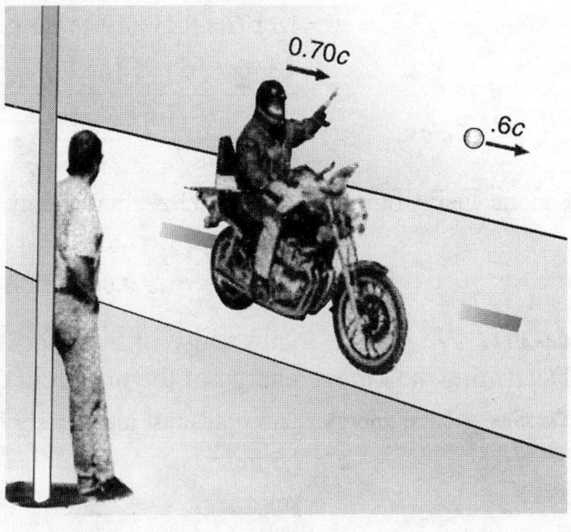

EXAMPLE 74 Imagine a motorcycle moving with a speed 0.70*c* past a stationary observer, as shown in adjoining figure. If the rider tosses a ball in the forward direction with a speed of 0.60*c* relative to himself, what is the speed of the ball relative to the stationary observer?

SOLUTION The speed of the motorcycle relative to the stationary observer is $v = 0.7c$. The speed of the ball in the frame of reference of the motorcyclist is $u_x' = 0.6c$. Therefore, the speed u_x of the ball relative to the stationary observer is

$$u_x = \frac{u'_x + v}{1 + \frac{u'_x v}{c^2}}$$

$$u_x = \frac{.6c + .7c}{1 + \frac{.6c + .7c}{c^2}} = \frac{1.3c}{1.42} = .9154c$$

The speed of the ball relative to the stationary observer is **0.9154c**.

EXAMPLE 75 An electron, which has a mass of 9.11×10^{-31} kg, moves with a speed of 0.850*c*. Find its relativistic momentum and compare this value with the momentum calculated from the classical expression.

SOLUTION Relativistic momentum of the electron is given as $p_{\text{rel}} = mv = \gamma m_0 v$

Classical momentum of the electron is given as $p_{\text{classical}} = m_0 v$

$$p_{\text{rel}} = \gamma m_0 v = \frac{1}{\sqrt{1 - \frac{v^2}{c^2}}} m_0 v$$

$$p_{\text{rel}} = \frac{1}{\sqrt{1 - \frac{(.85c)^2}{c^2}}} 9.11 \times 10^{-31} \times .85 \times 3 \times 10^8$$

$$p_{\text{rel}} = 44.0988 \times 10^{-23} = \mathbf{4.049 \times 10^{-22}\ kg.m/s}$$

$$p_{\text{classical}} = 9.11 \times 10^{-31} \times .85 \times 3 \times 10^8 = 23.23 \times 10^{-23} = \mathbf{2.323 \times 10^{-22}\ kg.m/s}$$

Relativistic momentum is approximately 89% greater than the classical momentum.

EXAMPLE 76 An electron in a television picture tube typically moves with a speed $u = 0.450c$. Find its total energy and kinetic energy in electron volts.

SOLUTION Total energy of a relativistic particle $E = \gamma m_0 c^2 = \frac{1}{\sqrt{1 - \frac{v^2}{c^2}}} m_0 c^2$

$$E = \frac{1}{\sqrt{1 - \frac{(.45c)^2}{c^2}}} \times 9.11 \times 10^{-31} \times 3 \times 3 \times 10^8 \times 10^8\ \text{J}$$

$$E = 1.1197 \times 9.11 \times 10^{-31} \times 3 \times 3 \times 10^8 \times 10^8\ \text{J}$$

$$E = \frac{1.1197 \times 9.11 \times 10^{-31} \times 3 \times 3 \times 10^8 \times 10^8}{1.6 \times 10^{-19}}\ \text{eV}$$

$$E \approx 0.574\ \text{MeV}$$

Kinetic energy of relativistic particle = total energy – rest mass energy

= (.574 – .511) MeV ($\because$ rest mass energy = $m_0 c^2$)

= **0.063 MeV**

EXAMPLE 77 If the total energy of a proton is 2.5 times its rest energy, what is the speed of the proton? Determine the kinetic energy of the proton in electron volts. What is the proton's momentum?

SOLUTION Total energy = 2.5 times rest mass energy

$E = 2.5\ m_0 c^2$;

$E = \gamma m_0 c^2$;

From above relations, $\gamma = 2.5 = \dfrac{1}{\sqrt{1-\dfrac{v^2}{c^2}}}$

$$v = .9165c = .9165 \times 3 \times 10^8 = \mathbf{2.747 \times 10^8\ m/s}$$

Kinetic energy of proton = $(\gamma - 1) \times$ rest mass energy = $(\gamma - 1)\, m_0 c^2$

$$KE = (2.5 - 1) \times 1.67 \times 10^{-27} \times 3 \times 3 \times 10^8 \times 10^8 = 22.54 \times 10^{-11}$$
$$= 2.254 \times 10^{-10}\ \text{J}$$

$$KE = \frac{2.254 \times 10^{-10}}{1.6 \times 10^{-19}}\ \text{eV} = 1.40875 \times 10^9\ \text{eV} = \mathbf{1408.75\ MeV}$$

For relativistic momentum,

Using expression for total energy $E^2 = p^2c^2 + m_0{}^2c^4$; $E = 2.5m_0c^2$

$$p^2c^2 = E^2 - m_0{}^2c^4$$
$$p^2c^2 = 6.25m_0{}^2c^4 - m_0{}^2c^4 = 5.25m_0{}^2c^4$$
$$pc = \sqrt{5.25}\, m_0c^2;\ m_0c^2 = 938\ \text{MeV}$$
$$p = \sqrt{5.25}\,\frac{938}{c}\,\text{MeV}$$
$$p = \mathbf{2149.2\,\frac{MeV}{c}}$$

Example 78 A crew watches a movie that is 2 hours long in a spacecraft that is moving at high speed through space. Will an Earthbound observer, who is watching the movie through a powerful telescope, measure the duration of the movie to be (a) longer than, (b) shorter than, or (c) equal to 2 hours?

Solution The two events are the beginning and the end of the movie, both of which take place at rest with respect to the space craft crew. Thus, the crew measures the proper time interval of 2 h. Any observer in motion with respect to the spacecraft, which includes the observer on Earth, will measure a longer time interval due to time dilation.

Example 79 Suppose astronauts are paid according to the amount of time they spend travelling in space. After a long voyage travelling at a speed approaching c, would a crew rather be paid according to (a) an Earth-based clock, (b) their spacecraft's clock, or (c) either clock?

Solution (a) If their on-duty time is based on clocks that remain on the Earth, they will have larger paycheques. A shorter time interval will have passed for the astronauts in their frame of reference than for their employer back on the Earth.

Objective Type Questions

Q.1 The aim of Michelson-Morley experiment was to
(a) prove existence of ether (b) measure speed of light
(c) measure speed of earth relative to ether (d) test the isotropy of space

Q.2 The postulates of special theory of relativity are applicable to
(a) accelerated frames (b) inertial frames
(c) stationary frames (d) none of these

Q.3 Frame F is stationary and the frame F' is moving along the positive x-direction. A rod placed along y-axis if observed from F frame appears
(a) contracted (b) elongated
(c) unchanged (d) nothing can be said

Q.4 A rocket is moving with a velocity $0.70c$. Velocity of the light with respect to the rocket is
(a) $0.7c$ (b) c (c) $1.4c$ (d) $0.35c$

Q.5 The relative velocity of two photons when they approach each other will be
(a) less than c (b) 0 (c) more than c (d) c

Q.6 The energy produced by one kg of mass, which is fully converted into energy, will be equal to
(a) 3×10^{10} J (b) 9×10^{16} J (c) 10^{18} J (d) 1 J

Q.7 A body of mass *m* falls through *h* meters. The decrease in its mass is equivalent to
(a) mgh/c^2 (b) mgh (c) $mghc^2$ (d) mgh/c

Q.8 At what velocity the kinetic energy of a body is equal to its rest mass energy
(a) $\sqrt{3}c/2$ (b) $c/2$ (c) $c/3$ (d) $\sqrt{2}c$

Q.9 Relativistic transformations were suggested by
(a) Newton (b) Einstein (c) Huygens (d) Lorentz

Q.10 The apparent length of a meter rod moving parallel to its length with velocity $0.6c$ will be
(a) 0.8 m (b) 0.6 m (c) 1 m (d) 1.2 m

Q.11 When a body of rest mass 1 kg moves with velocity of light, its mass becomes
(a) 0 (b) ∞ (c) 2 kg (d) 100 kg

Q.12 Einsten's famous mass energy relation is
(a) $E = m_0c^2$ (b) $E = mc^2$ (c) $E = \frac{1}{2} m_0 c^2$ (d) none of these

Q.13 A rod of length L_0 is kept in a frame F' which is moving with velocity of light in the direction of length. The observed length of rod from a stationary frame of reference (earth) would be
(a) ∞ (b) 0 (c) $10L_0$ (d) $3 \times 10^8 L_0$

Q.14 The negative result of Michelson-Morley experiment was that
(a) it could not measure speed of light (b) it could not prove the existence of ether
(c) it could not show the shifting of fringes (d) it could not prove the electromagnetic nature of light waves

SHORT-ANSWER QUESTIONS

Q.1 What do you understand by frame of reference?
Q.2 What are inertial frames of reference?
Q.3 Is Earth an inertial frame? If not why?
Q.4 What are Galilean transformations?
Q.5 Write Lorentz transformation equation.
Q.6 Write inverse Lorentz transformation equations.
Q.7 What are time dilation and length contraction?
Q.8 What are postulates of special theory of relativity?
Q.9 Explain the variation of mass with velocity?
Q.10 What do you understand by mass-energy equivalence relation?
Q.11 What is the rest mass of photon?

Practice Problems

General Questions

Q.1 Distinguish between inertial and non-inertial frames of references. Give one example of each. Is earth an inertial frame? Give reasons.

Q.2 What is Newtonian principle of relativity? Discuss with examples. Why should laws of nature be the same in all inertial frames of reference?

Q.3 What are Galilean transformations? Derive Galilean transformation equations for two inertial frames. State and prove Galilean invariance.

Q.4 Prove that Newton's law of motion are invariant under Galilean transformations?

Q.5 What are the quantities which are invariant under Galilean transformations?

Q.6 Show that a frame of reference having a uniform translatory motion (or moving with constant velocity) relative to an inertial frame is also inertial.

Q.7 Show that the laws of conservation of momentum and energy are invariant to Galilean transformations.

Q.8 What was the objective of conducting the Michelson-Morley experiment? Describe the experiment. How is the negative result of the experiment interpreted?

Q.9 What do you conclude from Michelson-Morley experiment? If ether does not exist in what medium does light travel? What vibrates in light waves?

Q.10 (a) What efforts were made to explain the null results of Michelson-Morley experiment on the basis of ether hypothesis?

(b) Draw the ray diagram in ether frame after 90° rotation of the apparatus.

Q.11 (a) Why the apparatus of Michelson-Morley experiment was rotated through 90°?

(b) Why did Michelson and Morley repeat the experiment during day and night and during all seasons of the year?

Q.12 State and explain the fundamental (basic) postulates of special theory of relativity and derive Lorentz space time trasformation equations on their basis.

Q.13 Derive Lorentz transformation equations for space and time coordinates and show that these equations become the Galilean equations at very low speeds.

Q.14 Show by means of Lorentz transformation equations that

$$x'^2 - c^2t'^2 = x^2 - c^2t^2$$

Q.15 Derive Lorentz transformation equations and using them prove that moving clock appears to go slow.

Q.16 (a) On the basis of Lorentz transformations derive an expression for length contraction.

(b) Define proper length.

(c) A circle and a square are moving along x-axis. How will they appear to stationary observer?

Q.17 Apply Lorentz transformation to derive expression for length contraction and time dilation.

Q.18 What do you mean by length contraction at relativistic speed? Deduce the necessary expression for it?

Q.19 (a) What do you understand by time dilation? Establish a relation between proper and improper interval of time.

(b) Give an example to show that time dilation is a real effect.

Q.20 What do you understand by time dilation? On the basis of Lorentz transformations discuss the variation of time with velocity according to the special theory of relativity. Explain why does a moving clock appear to run slow. Explain the terms, 'proper time' and 'improper time'. Show that when $v << c$ Lorentz transformations for time reduce to Galilean transformations.

Q.21 Deduce an expression for variation of mass with velocity and depict it graphically. Also prove that no material particle can have a velocity equal or greater than the velocity of light (c).

Q.22 Obtain the relativistic formula for the addition of velocities and also show that the velocity of light is an absolute constant independent of the frame of reference.

Q.23 (a) Starting from Lorentz transformation equations for space and time co-ordinates derive equations for relativistic addition of velocities. Hence, prove that no material particle can move with a velocity greater than that of light.

(b) Show that the law agrees with velocity addition formula for non-relativistic velocities.

Q.24 Starting with Einstein's velocity addition formula show that it is in conformity with principle of constancy of speed of light.

Q.25 (a) Derive the formula for relativistic variation of mass with velocity, i.e., $m = \dfrac{m_0}{\sqrt{1 - v^2/c^2}}$

(b) Hence prove that it is not possible for a material particle to have a velocity equal to or greater than the velocity of light.

Q.26 Obtain Einstein's mass energy relation and discuss it. Give some evidence showing its validity.

Q.27 Establish mathematically Einstein's mass energy relationship. Explain physical significance of this relation. Mention nuclear phenomena supporting this relation.

Q.28 Write notes on the following

(i) Michelson-Morely experiment and its results
(ii) Variation of mass with velocity
(iii) Lorentz-FitzGerald contraction
(iv) Time dilation
(v) Mass-energy equivalence

UNSOLVED QUESTIONS

Q.1 A space ship is 50 metre long on the ground, when it is in flight its length appears to be 49 metres to an observer on the ground. Find the speed of the space ship. **[Ans: 0.6 c]**

Q.2 Calculate the percentage contraction of a rod moving with a velocity 0.8 times the velocity of light in a direction inclined at 45° to its own length. **[Ans: 17.5 %]**

Q.3 The length of a rod is 100 m. If the length of this rod is measured by the observer moving parallel to its length is 51 m, find the speed of the observer. **[Ans: 0.86 c]**

Q.4 A burst of 10^4 π^+ mesons travels in a circular path of radius 20 m at a speed $v = 0.99c$. The proper mean life of π^+ meason is 2.5×10^{-8} s.

(i) How many mesons would be left in a burst that had remained at rest at the origin for the same period of time?

(ii) How many mesons survive when the burst returns to the point of origin?

[Ans: (i) $N_0\pi^+$ meson would survive, (ii) 920]

Q.5 Calculate the velocity at which electron mass is $\sqrt{3}$ times the rest mass. **[Ans: 2.45 · 10^8 m/sec]**

Q.6 What should be the speed of electron so that its relativistic mass is twice its rest mass? **[Ans: 0.87 c]**

Q.7 Kinetic energy of a particle is twice its rest mass energy. What is its velocity? **[Ans: 0.943 c]**

Q.8 Calculate the velocity of 1.0 MeV electron. **[Ans: 2.82 · 10^8 m/sec]**

Q.9 If one gram of a substance is fully converted into energy in one second, how many calories of heat will be produced and how much power will be generated? **[Ans: 9 · 10^{13} J, 9 · 10^7 MW]**

Applied Nuclear Physics

13

Learning Objectives

After reading this chapter you will be able to

LO 1 Understand charge, mass, and size of nucleus, angular momentum, magnetic, electric, and statistical properties, parity

LO 2 Know about charge independence and meson theory

LO 3 Learn about binding energy of nucleus

LO 4 Discuss nuclear stability

LO 5 Explain nuclear shell model and its theory and applications and nuclear liquid drop model

LO 6 Learn about radioactivity together with laws of radioactive disintegration alpha decay, beta decay, gamma decay and nuclear radiation detectors

LO 7 Know about discovery of neutron

LO 8 Understand nuclear reactions along with disintegration energy and threshold energy

LO 9 Learn about nuclear fission, nuclear fusion and controlled fusion together with plasma, ignition temperature, Lawson criterion and methods of fusion,

LO 10 Learn about particle accelerators including linear accelerator, cyclotron, betatron and plasma-based accelerators

Introduction

For many chemists, the atomic nucleus is nothing but a point charge, which carries most of the mass of the atom. However, physicists' perspective is different and it has been the field of research for them to investigate how the protons and neutrons of the nucleus play important roles in the history and structure of the universe. Actually, rapid progress in nuclear physics began after the discovery of the neutron in 1932. The discovery of the neutron solved a known puzzle related to the spin of the nitrogen-14 nucleus, which was experimentally measured as 1 basic unit of angular momentum, but at that time physicists could not find any way to arrange 21 particles (14 protons and 7 electrons of ${}^{14}{}_{7}N$) so as to give a spin of 1. However, the presence of the neutron as an uncharged particle in the nucleus with spin ½ solved this problem. Moreover, the concept of neutron was used to explain spin differences in many different nuclides (nuclide is an atomic nucleus as characterised by its atomic number, its mass number and nuclear energy state). On the other hand, neutrons play an important role in achieving energy through nuclear reactions. For example, nuclear fission is initiated by a slow neutron and the resulting reactions produce 2.4 neutrons on an average. This way, a chain reaction is generated that allows a self-sustaining mode of operation. However, neutrons are not effective initiators for light element fusion reactors; the reason being their consumption during reaction rather than their production in fusion reactors. Since no chain reaction takes place, applied fuel is needed to be added continually to keep the operation continuous. Now researchers have been investigating different means to achieve fusion, for example by using laser field, accelerated particles, etc. Application of plasma has been a very attractive field and hence we discuss in this chapter new topics such as plasma, Lawson criterion, fusion by inertial confinement, magnetic confinement, laser fusion, etc. As mentioned particle acceleration also contributes to this field, various types of accelerators, namely linear accelerator, cyclotron, betatron and plasma-based accelerators have been discussed. Other topics on radioactive disintegration of nuclear, nucleus radiation and radiation detectors also have been covered.

13.1 BASIC PROPERTIES OF NUCLEUS — LO1

According to Rutherford model, most of the mass of the atom is concentrated in a small spherical volume called the nucleus, and the electrons are distributed around the nucleus. The radius of the nucleus is about 10^{-13} cm and the spherical volume where electrons are distributed carries radius of about 10^{-8} cm. The protons and neutrons are the only constituents of the nucleus, which are called nucleons.

13.1.1 Charge, Mass and Size

A nucleus consists of Z number of protons each with charge $+e$, and N number of neutrons, which are neutral. The total number of nucleons is called mass number A, i.e., $A = Z + N$. The charge of the nucleus is +Ze.

The rest mass of the proton, m_p, is 1.00727663 amu, whereas the mass of the neutron, m_n, is 1.0086654 amu. It is seen that the mass of the nucleus is not exactly equal to $Zm_p + Nm_n$, but it is less than this value. The energy equivalent to this mass difference is known as binding energy, which is responsible for holding neutrons and protons together in the nucleus.

The scattering of α-particles by the nucleus has been an important source of information with regard to the nuclei, and in each case this showed departures from Coulomb scattering. Based on these experiments and some theoretical models, it was deduced that the size of any atom is almost constant, but the size of any nucleus depends on its mass number A. Clearly the size of the nucleus having large number of nucleons should be large. Assuming a nucleus to be in the form of a sphere, its radius R is given by

$$R = r_0 A^{1/3}$$

The value of r_0 is found to depend on the type of experiment. In general, it is between 1.2×10^{-13} cm and 1.48×10^{-13} cm.

13.1.2 Angular Momentum

Each nucleon in the nucleus is assumed to have both orbital and spinning motions just like an electron has in an atom. It means each nucleon has both orbital and spin angular momenta. The magnitude of the spin angular momentum is $\frac{\hbar}{2}\left(\hbar = \frac{h}{2\pi}\right)$. Its orientation in space can be described by only two states: the spin axis is either parallel or antiparallel to any given direction (say z-axis). So the component of spin along z-axis is either $\hbar/2$ or $-\hbar/2$. In view of this, the total angular momentum i of each nucleon is

$$i = l \pm s,$$

where l is the orbital angular momentum and s is the spin angular momentum. For nuclei having more than one nucleon, l is replaced with L and s is replaced with S, which represent the corresponding total momentum of all the nucleons. Hence, the total angular momentum of the nucleus is given by

$$I = L \pm S$$

I is actually a vector, whose magnitude is the maximum possible component in any given direction (*z*-axis), and is represented by italic *I*. The value of *I* is an integral multiple of $\hbar$ for the nuclei with even mass numbers, and it is an odd half-integral multiple of $\hbar$ for the nuclei with odd mass numbers. In particular, even-even nuclei (nuclei with both *Z* and *N* even) carry zero value of *I*. This is also called that even-even nuclei have zero spin, where nuclear spin refers to the total nuclear quantum number. The name nuclear spin which is frequently used for total angular momentum of the nucleus is actually misleading. This incorrect usage was introduced before the problem of the internal structure of nuclei had attained its present importance. Since then it has been continuing.

13.1.3 Magnetic Property

Magnetic property of a nucleus is associated with the nuclear magnetic moment. The motion of the nucleons inside the nucleus should give rise to the nuclear magnetic moments like the electrons' motion in the atom provides the magnetic moment. This is true always unless the total angular momentum of the nucleus or the nuclear spin is zero. If we assume spherically symmetric charge distribution of the nucleus, the nucleus will give rise to a magnetic dipole moment only. The nuclear magnetic moments μ_I are measured in terms of nuclear magnetons μ_N, given by $\mu_N = e\hbar/2\, m_H$. Here m_H is the mass of hydrogen atom, which is equal to mass of proton m_p. Hence $\mu_N = \dfrac{e\hbar}{2m_p} = \dfrac{\mu_B}{1836}$. μ_B is known as Bohr magnetons, which is defined as the magnetic moment associated with an atomic electron in orbital motion with an angular momentum of $1\hbar$. This is given by $\mu_B = \dfrac{e\hbar}{2m_e} = 0.927 \times 10^{-23}$ J/Wb/m^2. The value of nuclear magnetons μ_N is thus obtained as 5.05×10^{-27} J/wb/m^2.

The measured values of μ_I are between $-3\mu_N$ and $+10\mu_N$. When the magnetic moment of the nucleus is in the opposite direction to the direction of nuclear spin, μ_I carries negative values. The positive valve of μ_I means the directions of the magnetic moment of the nucleus is the same as that of the nuclear spin. The magnetic moment of a proton is $+2.79276\ \mu_N$, whereas that of neutron is $-1.191315\ \mu_N$. This indicates that the proton and neutron have a non-uniform charge distribution, which is also very complex.

The magnetic moment of a nucleus can also be represented in terms of nuclear gyromagnetic ratio γ_I and nuclear g-factor g_I, as $\mu_I = \gamma_I \hbar I = g_I \mu_N I$.

Since $I = 0$ for nuclei containing even numbers of protons and neutrons, the even-even nuclei have no magnetic moment.

13.1.4 Electric Property

Electric property of a nucleus is associated with the electric quadrupole moment, which is highly important in connection with the shape of the nucleus. The electric quadrupole moment is a measure of the deviation of the nucleus from its spherical symmetry. Under the situation of a deviation, the nucleus can be imagined to be an ellipsoid of revolution with its diameter 2*b* along the axis of symmetry and diameter 2*a* along the axis perpendicular to this. The quadrupole moment *Q* of the nucleus, when its electric charge is uniformly distributed throughout the ellipsoid, is given by

$$Q = \frac{2}{5} Z(b^2 - a^2)$$

Clearly *Q* is zero for the nuclei having spherical symmetry ($a = b$) and whose charge is uniformly distributed. As the formula suggests, the magnitude of electric quadrupole moment depends on the magnitude of nuclear charge *Z*, size of the nucleus (magnitudes of *b* and *a*) and the extent of deviation (difference in *b* and *a*) from spherical symmetry. The sign of Q may be positive or negative, depending on nature of Z and the values of b and a. The Q value for a nucleus (deuterium) with one proton and one neutron is $+0.00274 \times 10^{-24}$ cm^2, whereas an isotope of lutecium has a Q value of 7×10^{-24} cm^2 since it has 176 nucleons.

13.1.5 Statistical Property

The concept of statistics is related to the behaviour of large number of particles. In general, the properties of assemblies of electrons, protons, neutrons, photons and atomic nuclei cannot be described based on classical statistics rather these follow the quantum statistics, i.e., Bose-Einstein statistics and Fermi-Dirac statistics. In this connection, wave function is an important quantity that describes the particular system. A nucleon is described by a function of its three space coordinates and the value of its spin. The wave function is said to be anti-symmetric if it changes sign when three spatial and one spin coordinates of two identical particles are interchanged, otherwise it is symmetric. Fermi-Dirac statistics apply to the system of particles which are governed by anti-symmetric wave function. It also follows that the Pauli exclusion principle applies to particles obeying the Fermi-Dirac statistics. The electrons, protons and neutrons obey this statistics as is done by neuclei of odd mass number. On the other hand, all the neuclei with even mass number obey the Bose-Einstein statistics. This statistics apply to the systems of particles which are governed by symmetric wave function. Since nuclei with odd mass number have total angular momenta as odd half-integral multiples of $\hbar$ and nuclei with even mass number have momenta as integral multiple of $\hbar$, there is a direct correlation between the total angular momentum of a nucleus and its statistics.

13.1.6 Parity

The wave function of a nucleus, to a good approximation, may be expressed as a product of two functions: one of the space coordinates and the other depending only on the spin orientation. If the spatial part of its wave function remains unchanged when the space coordinates (x, y, z) are replaced by $(-x, -y, -z)$, the motion of the nucleus is said to have even parity. If the spatial part of the wave function changes sign on such transformation of coordinates, the motion of the nucleus is said to have odd parity.

The parity of a nucleus in a given state is related to the orbital angular momentum L. If the value of L is even, the parity is even and if the value of L is odd, the parity is odd.

13.2 NUCLEAR FORCES LO2

The forces between nucleons, i.e., between proton and neutron, neutron and neutron, and neutron and proton, are referred to as nuclear forces. These forces must be attractive; if it is not so, there would be no stable nuclei. In view of Coulombic repulsion, these forces should be much stronger than Coulomb forces between protons. Again if it is not so, there would be no stable nuclei. Study of structures of atoms and molecules reveals that the nuclear forces do not contribute to the formation or the properties of atoms and molecules. In view of this, the nuclear forces must be short-range. These forces extend from the centre to the surface of the nucleus or only a little beyond its surface. Figure 13.1 shows the cases when a proton

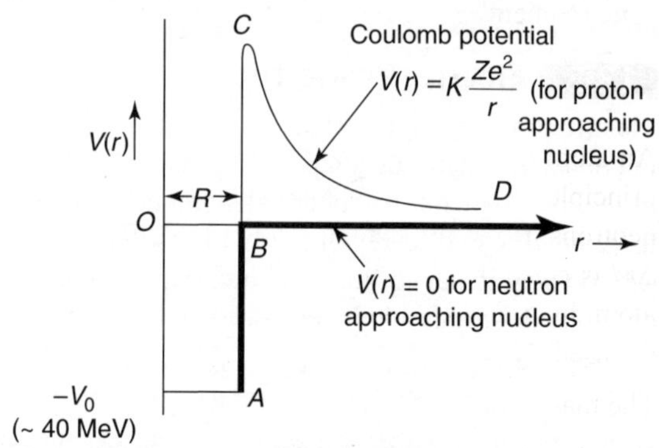

FIGURE 13.1

or a neutron is brought from an infinite distance toward a nucleus at O. The radius of the nucleus is R. When the proton (thin line) is brought toward the nucleus, (case $r > R$), it experiences the potential $V(r)$ varying as $1/r$ and the Coulomb force of repulsion varying as $1/r^2$. However, for this case ($r > R$), the neutron does not feel any force and hence $V(r) = 0$ (thick line). For $r < R$, the proton as well as the neutron starts feeling the attractive force of the nucleus. Since this attractive nuclear force is much stronger than Coulomb repulsive force, it is represented by a negative potential $-V_0$($\simeq$ typically 40 MeV). The exact form of the potential inside nucleus, i.e., between 0 and R is still not known.

13.2.1 Charge Independence

The charge independence of nuclear forces implies that the force between two protons (p-p), the force between two neutrons (n-n) and the force between a neutron and a proton (n-p) are almost equal. The forces are said to be charge symmetric, if only $p\text{-}p \simeq n\text{-}n$. From the observation that $N \simeq Z \simeq A/2$ for the light and medium weight stable nuclei for which Coulomb repulsion can be neglected, it is deduced that neutrons and protons have a tendency to go in pairs. Examples of extraordinary stable nuclei are $^4_2$He, $^8_4$Be, $^{12}_6$C, $^{16}_8$O, etc. Experimental evidences indicate that the nuclear forces show saturation.

13.2.2 Meson Theory

In 1935, Japanese physicist Yukawa established a fact theoretically, which was suggested by Heisenberg in 1932 that nuclear forces result from the constant exchange of massive particles between two nucleons. Since this phenomenon is like the exchange of photons resulting in Coulomb force between two charged particles, the massive particles taking part in nuclear forces were also called heavy quanta. According to Yukawa, because of short range nature of nuclear forces, a nucleon is surrounded by a cloud of virtual massive particles which are constantly emitted and absorbed by the nucleon; the same way as the electrical charge is surrounded by a cloud of virtual photons. When a nucleon is brought near to another nucleon, a particle emitted by one may be absorbed by the other or vice-versa. This way there is a constant transfer of momentum from one nucleon to the other and hence a force is exerted between them.

The massive particle or the heavy quanta involved in nuclear exchange was given the name meson (π) which can exist in three different forms. Its neutral form is called neutral pi meson or pion (π^o), its negative form is called negative pion (π^-) and its positive form is called positive pion (π^+). The exchange of neutral or charged pions results in the interactions between the nucleons. In this process, neutron is converted into proton and the proton is converted into neutron.

13.3 BINDING ENERGY OF NUCLEUS

LO3

An atom denoted by $^A_Z X$ is formed by bringing together Z protons, Z electrons and $(A - Z)$ neutrons. In principle, the mass of the atom should be equal to the sum of masses of Z protons, Z electrons and $(A - Z)$ neutrons. However, the mass $M(A, Z)$ of the atom is less than this sum in the free state. This decrease in mass ΔM is converted into energy $\Delta E = \Delta Mc^2$. This much energy is released in the process of formation of the atom. In order to break the atom into its constituents, we will have to supply energy equivalent to this energy released. This energy is called the binding energy BE.

The mass defect ΔM can be written as $\Delta M = Zm_p + Zm_e + (A - Z)\, m_n - M(A, Z)$

Here m_p, m_e and m_n are the masses of the proton, electron and neutron, respectively. The mass $m_p + m_e$ can be written as m_H, i.e., mass of the hydrogen atom, if we neglect the small binding energy of the hydrogen atom.

Hence, the mass defect ΔM becomes

$$\Delta M = Zm_H + (A - Z)\, m_n - M(A, Z)$$
$$= 1.0081437\, Z + 1.0089830\, (A - Z) - M(A, Z)$$

Since one amu corresponds to 931.145 MeV, the binding energy of the nucleus is written as

$$BE \text{ (in MeV)} = 931.145\, [1.0081437Z + 1.0089830\, (A - Z) - M(A, Z)]$$

From this we obtain average BE per nucleon as $BE/A = \dfrac{931.145}{A}\, [1.0081437Z + 1.0089830\, (A - Z) - M(A, Z)]$

A graph of the binding energy per nucleon as a function of the mass numer is called binding energy curve, which is shown in Fig. 13.2. With the exception of ${}_2^4\text{He}$, ${}_6^{12}\text{C}$ and ${}_8^{16}\text{O}$, the value of binding energy per nucleon for almost all the nuclei lie on or close to the binding energy curve. From the curve, we notice some of the outstanding features, as

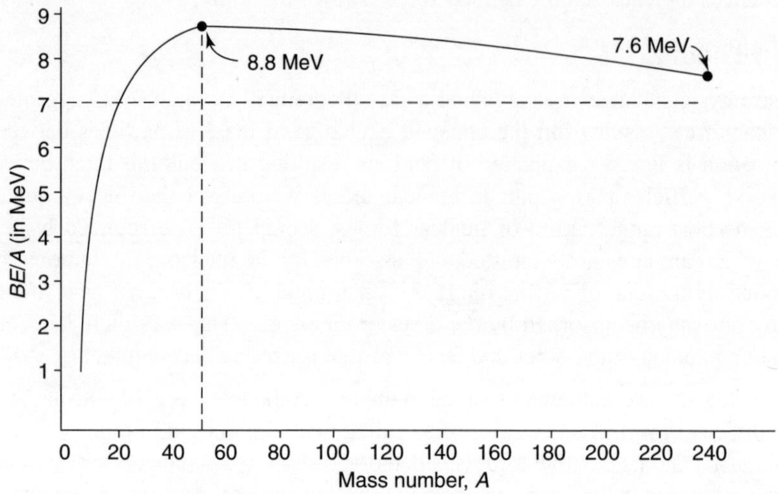

Figure 13.2

(i) BE/A curve attains a flat maximum around $A = 50$ with the value ~8.8 MeV.

(ii) BE/A is low for nuclei with low A, but it increases very rapidly with increasing A (say upto $A \sim 20$).

(iii) Average BE/A in the region $A \sim 20$ and $A \sim 160$ is about 8.5 MeV and does not show much variation

(iv) BE/A decreases slowly and continuously for $A>140$ and it reaches a value of 7.6 MeV at $A = 238$ for ${}^{238}\text{U}$.

13.3.1 Explanation

Consider a nucleus as the drop of liquid in which protons and neutrons take the place of the molecules in the drop. Two cases of nuclei are shown in Fig. 13.3 with small number of nucleons (light nuclei) and sufficiently large number of nucleons (medium or heavy nuclei). This is

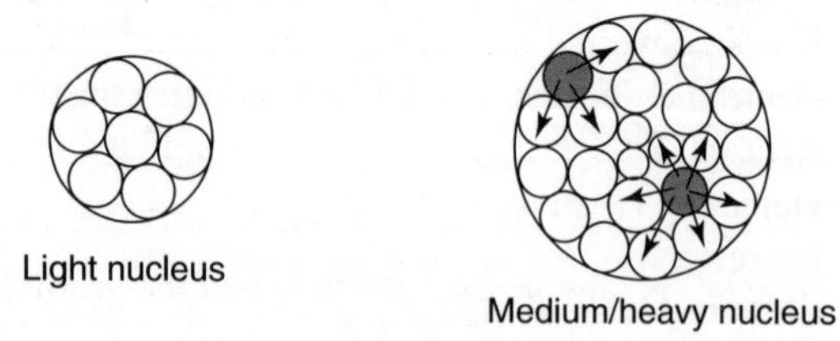

Figure 13.3

clear from the figure that the nucleons, which are deep inside the nucleus, are attracted from all sides by neighbouring nucleons. However, the nucleons setting on the surface are attracted from one side only or by very less number of nucleons. This way the binding energy for the nucleons at the surface of the nucleus is smaller than the binding energy of the nucleons inside the nucleus. For nuclei having smaller number of A, there is a large fraction of nucleons at the surface. For this reason, BE/A is small for small A. On the other side, BE/A for nuclei with large A is due to the Coulomb repulsive force between protons which reduces the binding energy.

13.4 NUCLEAR STABILITY LO4

In nuclei with smaller number of nucleons, Coulomb repulsion can be neglected in view of strong nuclear forces. However, Coulomb repulsion becomes important for the nuclei having large number of nucleons. Since this repulsion has the effect of making nuclei unstable, heavy nuclei tend to be made up of more number of neutrons than protons. If we plot number of neutrons N versus number of protons Z for all known nuclei, we notice a great deviation from the line $N/Z = 1$ in the case of heavy nuclei. The deviation reaches $N/Z = 1.6$ for $A = 238$. The curve $N/Z = 1$ is called the stability curve. The isotopes on both sides of the stability curve are radioactive, which decay in such a way that the final product lies on the stability curve and is now stable.

The total binding energy of a nucleus depends not only on the ratio N/Z but also on whether these numbers of neutrons and protons are odd or even. This is called odd-even effect. All the stable nuclei can be classified into four groups, namely even-even, even-odd, odd-even and odd-odd, based on number of protons and neutrons, respectively. Even-even nuclei having even number of mass number A are the most abundant, i.e., these nuclei are most stable, whereas odd-odd nuclei are very few. The stability of odd-even and even-odd lies between the two extremes.

Data collected for stable nuclei suggests that nucleons tend to form neutron-proton pairs. This is called pairing of nucleons, according to which nuclei that satisfy the condition $A/2 = Z$ or $A = 2Z$ are more strongly bound together and any deviation from $A = 2Z$ should decrease the binding energy.

13.5 NUCLEAR MODELS LO5

In the absence of a detailed theory of nuclear structure, attempts were made to correlate nuclear data in terms of various models. Several models were proposed. Each of them was based on a set of simplifying assumptions and hence, was useful in a limited way only. The nuclear models include the shell or independent particle model, liquid drop model, collective nuclear model and the optical model for nuclear reactions. The shell model and the liquid drop models are the most important and useful models of nuclear structure.

13.6 NUCLEAR SHELL MODEL

If you look at the structure of the atom, quantum mechanics correctly predicted the closure or filling of electronic shells. For example, the electronic shells for the elements He(Z = 2), Ne(Z = 10), Ar(Z = 18), Kr(Z = 36) and Xe(Z = 54) are completely filled or closed. Moreover, the ionization potentials for these elements are found to be higher than those of their neighbouring elements. The numbers 2, 10, 18, 36 and 54 are referred to as the atomic magic numbers of the periodic table. On the other hand, detailed investigations of the curve of BE/A versus A and other properties of nuclei reveal that many nuclear properties vary periodically in a sense similar to that of the periodic system of the elements. It was understood that the periodicities in the properties of the nuclei might be due to a nuclear shell structure similar to the atomic shell structure. Most of the nuclear properties show discontinuities near certain even values of Z or N. Experiments show that stable nuclei result when either Z or N is equal to one of the numbers 2, 8, 20, 50, 82 and 126. These number are called nuclear magic numbers, which have been interpreted as forming closed shells of neutrons or protons in analogy with the filling of electron shells in atoms. The proton and neutron shells appear to be independent of each other.

Following are some of the evidences for the nucleus having a shell structure and hence for the existence of nuclear magic numbers.

(i) As the elements corresponding to atomic magic numbers were found to be chemically inactive, the nuclei for which N or Z corresponded to nuclear magic numbers are found to be more stable than their neighbours.

(ii) Nuclei with the above nuclear magic numbers have many more isotopes than their neighbours. For example, Sn with $Z = 50$ has 10 stable isotopes whereas In with $Z = 49$ and *Sb* with $Z = 51$ each have only 2 isotopes.

(iii) When BE/A obtained from nuclear disintegration data and mass spectrographic measurements is plotted against A, the binding energy curve is found to have several kinks or breaks corresponding to sudden increase in the value of BE/A. They have been found to occur for the nuclei corresponding to the nuclear magic numbers.

(iv) Nuclei with magic numbers should have very low cross section, as the closed shells mean that there is no more vacancy. The plot of neutron-absorption cross section versus number of neutrons (N) indicated this to be true.

The above discussion indicates that neutrons and protons within the nucleus are arranged into shells within the nucleus like electrons in atoms. Each shell is limited to a certain maximum number of neutrons or protons. The resulting configuration is particularly stable and has an unusually low energy when a shell is filled or closed.

13.6.1.1 Theory

In the case of the structure of the atom we knew the form of Coulomb potential (proportional to $1/r$) and arrived at the system of electronic shells and magic numbers theoretically by solving the Schrödinger equation. However, in the case of the structure of the nucleus we do not yet know the exact form of the nuclear potential or the nuclear force, though investigations show that nuclear force are strongly attractive and they extend over a very short range from the centre of the nucleus. In several theories it is assumed that each nucleon moves in its orbit within the nucleus and its orbit is determined by a potential energy function $V(r)$ which represent the average effect of all interactions with other nucleons and is the same for each nucleon. Since each nucleon is regarded as an independent particle which is governed by the potential $V(r)$, nuclear shell model is also called the independent particle model. The potential energy $V(r)$ is analogous to the Coulomb energy but this potential $V(r)$ describing the nuclear attractions is quite different from the Coulomb potential. It has a form between the square well potential $V = -V_0$ and the so called oscillator potential $V(r) = -V_0 + \frac{kr^2}{2}$, where r is the distance between the nucleon and the centre of force and k is a constant. The solution of the Schrödinger equation with the square well potential and the oscillator potential does not give all the magic numbers. However, the combination of these two potentials, i.e., when $V(r) = -V_0\left(1 - \frac{r^2}{k^2}\right)$ with R as the nucleus radius, gives rise to all the magic numbers except magic number 28.

13.6.1.2 Applications

This model has been applied successfully to a variety of nuclear problems. For example, based on this it could be possible to predict the total angular momenta of nuclei and that too in good agreement with the experiments. So it is possible to assign values of the total angular momenta to nuclei for which the same has not been measured, viz. β-radioactive nuclei. Based on the shell model a correlation between the distribution of isomers and the magic numbers has also been found. Groupings or iselands of isomers are found just below the magic numbers 50, 82 and 126, and there is a break at each of these numbers. It means isomerism disappears when a shell is filled and does not appear again until the next shell is about half full. The shell model when used to predict the total angular momenta of low-lying excited levels established that the conditions for isomerism should exist below the magic numbers 50, 82 and 126 but not immediately above them. This model also predicted correctly when isomerism should appear in the unfilled shells. Finally, the experimental data on magnetic moments and electric quadrupole moments have also been interpreted in terms of this model. For example, the quadrupole moment is zero or small at the proton numbers 2, 8, 20, 50 and 82.

13.7 NUCLEAR LIQUID DROP MODEL — LO5

The nature of variation of binding energy per nucleon curve was explained based on the concept of liquid drop model of the nucleus. The nucleus was considered to be analogous to a drop of incompressible fluid of very high density of about 10^{17} kg/m^3. Saturation and short range properties of the nuclear force are analogous to the properties of the forces which hold a liquid drop together. In this connection, mass and binding energy of the nucleus become important for which a formula would be very useful. Such a formula is called semi-empirical mass formula or semi-empirical binding energy formula. The main contribution to this formula comes from various terms, namely volume energy, surface energy, symmetry effect, Coulomb energy and odd-even effect of the nucleus.

Since the volume of the nucleus is proportional to the mass number A, the volume energy is given by $E_V = a_V$Å where a_V is a constant. The Coulomb energy between the protons tends to lower the binding energy. So its effect should be a term with minus sign. This is given by

$$E_C = -4a_C \frac{Z(Z-1)}{A^{1/3}}, \text{ where } a_C \text{ is a constant.}$$

This has already been seen that the binding energy is reduced because the nucleus has a surface. Since the radius of nucleus will be proportional to $A^{2/3}$, the surface energy is represented by $E_S = -a_S A^{2/3}$, where a_s is a constant.

For a given value of A, there is a particular value of Z for the most stable nuclide. The Coulomb effect is small in the light nuclides, where the condition $Z = A/2$ or $A = 2Z$ is applicable for the stable or symmetric (spherical) nuclides. It means the departure from the symmetry or the term corresponding to the symmetry effect would be proportional to some power of neutron excess $(A - 2Z)$.

A detailed study of the symmetry effect shows that the term corresponding to this effect is given by

$$E_r = -a_r \frac{(A-2Z)^2}{A}, \text{ where } a_r \text{ is a constant.}$$

Finally odd-even effect can be represented by a term E_δ whose value depends on Z and N. For nuclides containing even Z and even N, $E_\delta = +\frac{\delta}{2A}$ whereas for the nuclides with odd Z and odd N, $E_\delta = -\frac{\delta}{2A}$.

By combinining all the terms we obtain the following formula for the binding energy BE

$$BE = a_V A - a_S A^{2/3} - 4a_C \frac{Z(Z-1)}{A^{1/3}} - a_r \frac{(A-2Z)^2}{A} + E_\delta$$

The values of the constants a_V, a_S, a_C and a_r are determined by a combination of theoretical calculations and adjustments to fit experimental values of the masses or binding energies. These are given by

$$a_V = 14 \text{ MeV}, a_S = 13.1 \text{ MeV}, a_C = 0.146 \text{ MeV and } a_r = 19.4 \text{ MeV}.$$

With these the semiempirical binding energy formula is obtained as

$$BE \text{ (in MeV)} = 14\text{A} - 13.1A^{2/3} - 0.584 \frac{Z(Z-1)}{A}$$

$$-19.4 \frac{(A-2Z)^2}{A} + E_\delta$$

Here $E_\delta = 135/A$ for even A, even Z and $E_\delta = -135/A$ for even A, odd Z.
For odd A, even Z and odd A, odd Z, $E_\delta = 0$.

13.8 RADIOACTIVITY LO6

Radioactivity is the disintegration of certain natural heavy elements such as radium, actinium, uranium and thorium. The radioactive disintegration is a spontaneous process. This disintegration is accompanied by the emission of α-rays (positively charged helium nuclei), β-rays (fast electrons) and γ-rays (short X-rays). The ultimate end product of the radioactive disintegration process is an isotope of lead. The radioactivity is of two types, namely, artificial radioactivity and induced radioactivity.

13.8.1 Artificial Radioactivity

Artificial radioactivity is the radiation obtained from isotopes after high energy bombardment in an accelerator (discussed later) by α-particles, protons and other light nuclei, or by neutrons in a nuclear reactor.

13.8.2 Induced Radioactivity

Induced radioactivity is the radioactivity induced in non-radioactive elements by neutrons in a reactor, or protons or deuterons in a cyclotron or linear accelerator. *X*-rays or γ-rays do not induce radioactivity unless their energies are exceptionally high.

13.8.3 Laws of Radioactive Disintegration

In the process of radioactive disintegration, any nucleus disintegrates by emitting a particle or a γ-ray or by capturing an electron from the atomic shell. This way the original nucleus gets changed during the process of radioactive decay. As mentioned, the radioactive disintegration is a spontaneous process. Depending upon the type of the nucleus, the overall activity is a prolonged process varying from a few seconds to millions of years. As disintegrations occur in a random manner, we can say that radioactive decay is a statistical process.

13.8.3.1 Disintegration or Decay Constant

Let us assume a constant probability λ per unit time for the decay of each nucleus of a given element. Then statistically, the probability of decay of each nucleus in time dt is λdt. If there are N undecayed nuclei at time t, the number dN that will decay in time interval between t and $t + dt$ is given by

$$dN = -\lambda dtN \quad \text{(i)}$$

where the negative sign shows that N decreases with increasing t. The probability constant λ is called the disintegration constant or decay constant. From Eq. (i) we have

$$\frac{dN}{N} = -\lambda dt$$

The above equation can be integrated under the limits of N and t. So we assume that initially there are N_0 radioactive atoms at time $t = 0$. With this, we get

$$\int_{N_0}^{N} \frac{dN}{N} = -\int_{0}^{t} \lambda dt$$

or $$\ln\left(\frac{N}{N_0}\right) = -\lambda t$$

or $$N = N_0 e^{-\lambda t} \quad \text{(ii)}$$

where N is the number of atoms present at time t.

13.8.3.2 Activity

The activity A is nothing but the number of disintegrations per second of a sample. By knowing A, we can detect the presence of a radioactive sample not by the radioactive atoms present but the radiation emitted by these atoms when they disintegrate. So, the activity A is obtained from Eq. (ii) as

$$A = \left|\frac{dN}{dt}\right| = \lambda N_0 e^{-\lambda t} = \lambda N \quad \text{(iii)}$$

13.8.3.3 Half-Life Time

The half-life time, $T_{1/2}$, of any sample is defined as the time interval in which the number of undecayed atoms decreases by half. Clearly in the half-life time, the activity drops to half, i.e., to $A_0/2$. If N_0 be the number of radioactive atoms at $t = 0$, then after $t = T_{1/2}$, N_0 becomes $N_0/2$. So, from Eq. (ii) we obtain

$$\frac{N_0}{2} = N_0 e^{-\lambda T_{1/2}}$$

or $$T_{1/2} = \frac{\ln 2}{\lambda} = \frac{0.693}{\lambda} \quad \text{(iv)}$$

It is clear from this equation that the unit of disintegration constant or decay constant is 1/sec as the units of $T_{1/2}$ is sec.

13.8.3.4 Relation between Half-Life Time and Mean Life Time

Equation (ii) due to the exponential nature of decay states that a given radioactive sample takes an infinite time to disintegrate completely. Individual radioactive atoms may have life times between zero and infinity. Hence, it is meaningful to talk about the average or mean life time τ, which is different from the half-life time of a radionuclide. The mean life of a radioactive nuclide is defined as

$$d\tau = \frac{t_1 dN_1 + t_2 dN_2 + t_3 dN_3 + \cdots}{dN_1 + dN_2 + dN_3 + \cdots} \quad \text{(v)}$$

The above equation states that dN_1 nuclei have life time t_1, dN_2 have life time t_2, etc. If we consider dN to be small, Eq. (v) can be written in the integral form by noting that $dN_1 + dN_2 + dN_3 + \ldots\ldots = N_0$. So

$$\tau = \frac{\int_0^{N_0} t dN}{\int_0^{N_0} dN} = \frac{\int_0^{N_0} t dN}{N_0} \tag{vi}$$

From Eq. (ii), we get

$$dN = -N_0 \lambda e^{-\lambda t} dt$$

The limit of N from 0 to N_0 corresponds to the limit of time t as ∞ to 0 [from Eq. (ii)]. Hence

$$\tau = \frac{-\int_\infty^0 \lambda t N_0 e^{-\lambda t} dt}{N_0} = \int_0^\infty \lambda t e^{-\lambda t} dt = \frac{1}{\lambda}$$

or

$$\tau = \frac{1}{\lambda} \tag{vii}$$

Therefore, the mean life time of an element is the reciprocal of its decay probability per unit time. Therefore, using Eq. (iv), we get

$$\tau = \frac{T_{1/2}}{0.693} = 1.44\, T_{1/2}$$

This is the relation between mean life time and half-life time of a radioactive nuclide. Clearly the mean life time is larger than the half-life time.

13.8.4 Alpha Decay

If a nucleus contains 210 or more nucleons, i.e., when nuclei are so large that the short-range nuclear forces that hold them together are barely able to counterbalance the mutual repulsion of their protons, then the process of alpha decay takes place in order to reduce the repulsion, or in other words, to increase the stability of such nuclei by reducing their size.

Let us consider a parent nucleus X which disintegrates into daughter nucleus Y and an alpha particle in alpha decay. Then

$${}^{A}_{Z}X \rightarrow {}^{A-4}_{Z-2}Y + {}^{4}_{2}\text{He} \tag{i}$$

In this process, ${}^{4}_{2}\text{He}$ is called α particle. It is clear that with the emission of one alpha particle from the nucleus, the atomic number is decreased by 2 and atomic weight is decreased by 4. Hence, because of the different value of Z, the chemical nature of the daughter nucleus is different from the parent nucleus. The condition for alpha decay can be obtained by applying the principle of conservation of energy and linear momentum. Let M_p, m_d and m_α be the rest masses of the parent nucleus, daughter nucleus and alpha particles, respectively. Since initially the parent nucleus remains at rest before decay, its linear momentum is zero. Therefore, the directions of daughter nucleus and alpha particle should be just opposite to conserve momentum. Let T_i be the total energy before decay and T_f be the total energy after decay. According to the law of conservation of energy,

or $T_i = T_f$

$$M_p c^2 = m_d c^2 + U_d + m_\alpha c^2 + U_\alpha \tag{ii}$$

where U_d and U_α are the nuclear energies of the daughter and alpha particle, given by $U_d = \frac{1}{2}m_d v_d^2$ and $U_\alpha = \frac{1}{2}m_\alpha v_\alpha^2$.

Now Eq. (ii) can be written as

$$U_d + U_\alpha = (M_p - m_d - m_\alpha)c^2. \tag{iii}$$

This equation represents total disintegration energy or Q-value and its value must be positive for spontaneous emission. Hence, the condition for spontaneous alpha decay is that the rest mass of the parent nucleus must be greater than the sum of the masses of the daughter and alpha particle.

In order to calculate the kinetic energy of the alpha particle, we use the conservation laws of linear momentum and energy. Conservation of momentum yields

$$m_d v_d = m_\alpha v_\alpha \tag{iv}$$

as the parent nucleus is at rest initially. Also,

$$Q = U_d + U_\alpha = \frac{1}{2}m_d v_d^2 + \frac{1}{2}m_\alpha v_\alpha^2 \tag{v}$$

By substituting the value of v_d from (iv) we get

$$Q = \frac{1}{2}m_d\left(\frac{m_\alpha v_\alpha}{m_d}\right)^2 + \frac{1}{2}m_\alpha v_\alpha^2,$$

or

$$Q = U_\alpha\left(\frac{m_\alpha}{m_d} + 1\right),$$

or

$$U_\alpha = \frac{Q}{1 + (m_\alpha / m_d)}.$$

From Eq. (i) $m_\alpha/m_d \cong 4/A - 4$. Therefore, the kinetic energy of the α particle can be written as

$$U_\alpha \cong \frac{A-4}{A}|Q|$$

From the above expression, it is clear that the alpha particle carries most of the disintegration energy, as A is very large. The decay of nuclei by alpha emission cannot be explained classically but quantum mechanically we can explain it. It is presumed that the parent nucleus before decay consists of the daughter nucleus and an alpha particle. In the language of quantum mechanics, an alpha particle exists in one of the discrete energy states of the daughter nucleus (say T_0) to a spherical region and the potential barrier created by the daughter nucleus restricts its motion. Classically, an alpha particle does not have enough energy to climb the barrier but quantum mechanically the wave associated with the alpha particle has some probability to penetrate the barrier. This effect is called quantum tunneling. The probability of finding the alpha particle one side of the barrier is much less than to find it in the other side of the barrier, and inside the barrier the probability decreases exponentially. This explains why alpha-particle emitters have long half lives for low T_0.

13.8.5 Beta Decay

Beta decay is a radioactive decay in which a beta particle that may be either an electron or a positron is emitted. Since an electron or a positron cannot exist in the nucleus, we assume that this particle is created at the time when the nucleus disintegrates. Electron emission takes place when a neutron is simultaneously converted into a proton. Actually, three mechanisms are involved in beta decay. These are

1. ***β^- decay (Electron or β^- emission)***: In this mechanism, a nucleus decays by emission of an electron. It is known as β^- decay. Since β^- is also called negatron, it is also sometimes referred to as negatron emission.
2. ***β^+ decay (or Positron emission)***: In this mechanism, a nucleus decays by emission of a positron. It is known as β^+ decay.
3. ***Electron capture***: In this mechanism, a nucleus decays by capturing an extra nuclear atomic electron. The electron disappears because its mass is converted into energy.

The condition for spontaneous decay is obtained by using the principle of conservation of energy. By using this, we can show whether a given unstable beta nucleus will decay by β^- emission, β^+ emission, or electron capture. If we represent X and Y as the parent and daughter nuclei respectively, then the processes, which we have discussed above, can be given as

$$\beta^- \text{ decay:} \qquad {}^A_Z X \rightarrow {}^{\;\;A}_{Z+1}Y + {}^{\;\;0}_{-1}e \tag{i}$$

$$\beta^+ \text{ decay:} \qquad {}^A_Z X \rightarrow {}^{\;\;A}_{Z-1}Y + {}^{\;\;0}_{+1}e \tag{ii}$$

$$\text{Electron capture} \qquad {}^A_Z X + {}^{\;\;0}_{-1}e \rightarrow {}^{\;\;A}_{Z-1}Y \tag{iii}$$

The above equations are correct if we take the kinetic energy of the emitted beta particle as maximum. Let M_p, m_d and m_e be the masses of X, Y and electron, respectively, and U_d and U_e^{max} be the kinetic energies of the daughter nucleus and the electron. Since the kinetic energy of X is zero, i.e., $U_p = 0$ the principle of conservation of energy yields

$$M_p c^2 = m_d c^2 + U_d + m_e c^2 + U_e^{max} \tag{iv}$$

The disintegration energy Q of this decay is given as

$$Q = U_d + U_e^{max} = (M_p - m_d - m_e)c^2 \tag{v}$$

We can replace the nuclear masses M_p and m_d by the atomic masses $M(Z)$ and $M(Z+1)$ by making use of relations

$$\begin{aligned} M(Z) &= M_p + Zm_e \\ M(Z+1) &= m_d + (Z+1)m_e \end{aligned} \tag{vi}$$

In view of these equations, Eq. (v) can be written as

$$Q = [M(Z) - M(Z+1)]c^2 \tag{vii}$$

As for spontaneous decay, Q must be positive, the above equation states that β^- decay will occur only if the atomic mass of the parent atom is greater than that of the daughter atom.

We can find the following Q-value for β^+ decay by applying principle of conservation of energy to Eq. (ii)

$$Q = [M(Z) - M(Z-1) - 2m_e]c^2 \tag{viii}$$

Similarly, for electron capture

$$Q = [M(Z) - M(Z-1)]c^2 \tag{ix}$$

The mass number A of the nucleus does not change, i.e., $\Delta A = 0$ in all the above mentioned processes, namely β^- emission, β^+ emission and electron capture. So we can say that beta decay is an isobaric transformation. In order to conserve charge in β^- decay, a neutron is simultaneously converted into proton, when an electron is emitted.

In the process of beta decay, if there are only two particles named daughter nucleus and beta particle (β^- or β^+), this situation is similar to a two-body problem. The linear momentum is, however, not conserved in beta decay because the observed directions of certain nuclides and electrons are almost never exactly opposite. Since the daughter nucleus is very large compared to that of the beta particle, the beta particle should take away all the disintegration energy. However, experimentally it is not observed, but is found that almost all the emitted beta particles have energies less than the maximum energy or end-point energy, i.e., between 0 and $U_e^{\max}$. It means the energy conservation does not take place in the process of beta decay. Moreover, the angular momentum is also not conserved in beta decay because spin is not conserved in the nuclear reaction. This can be explained with the help of spin of particles.

Let there be A nucleons before decay; the number remains same after decay also. Since the spin of each nucleon is ½, in beta decay the spin is not conserved. For example,

$$n \rightarrow p + e^-$$

The spin of the neutron is ½, proton is ½ and also the spin of electron is ½. So from the above relation, we observe that the spin is not conserved and hence angular momentum is also not conserved.

All the above discrepancies would be removed if an uncharged particle of small or zero rest mass and spin ½ is emitted in beta decay. In 1930, Pauli postulated the existence of a neutrino. The neutrino carries off energy equal to the difference between $U_e^{\max}$ and the actual kinetic energy. The momentum of a neutrino also exactly balances those of the electron and the recoiling daughter nucleus. Actually in beta decay, two types of neutrinos are involved. The first one is called neutrino (ν) and the other one is called antineutrino ($\overline{\nu}$).

β^-decay: $$n \rightarrow p + \beta^- + \overline{\nu} \qquad \text{(x)}$$

β^+decay: $$p \rightarrow n + \beta^+ + \nu \qquad \text{(xi)}$$

Electron capture: $$p + {}_{-1}^{0}e \rightarrow n + \nu \qquad \text{(xii)}$$

Based on the above equations, we can conclude that to conserve energy, linear momentum and angular momentum, the particle neutrino must have zero rest mass, zero charge but a spin of ½.

13.8.6 Gamma Decay

By emitting alpha, beta or other particles, the nucleus disintegrates and is usually left in the excited state. If the excited nucleus does not have sufficient energy to emit another particle, like an excited atom, it returns to its ground state by emitting photons. The energy of these photons is equal to the energy differences between the various initial and final energy levels up to several MeV. These emitted photons from nuclei are called gamma rays. A nucleus in the higher energy state T_i makes a transition to lower excited energy state (or ground state) T_f, then the excess energy is governed by the equation

$$\Delta T = T_i - T_f$$

This excess energy is emitted either by gamma-ray emission or by internal conversion. The process of internal conversion is more frequent in a heavy excited nuclei. In this process, an excited nucleus is returned into its ground state by giving up its excitation energy to one of the orbital electrons, which is very close to it. The internal conversion is an alternative of gamma ray emission. Both these processes are caused by electromagnetic interaction. In the gamma ray emission, the gamma ray spectrum consists of discrete energy levels. This indicates that the nucleus has discrete energy levels. The relationship between decay schemes and energy levels is shown in Fig. 13.4. This figure shows the beta decay of ${}^{27}_{12}Mg$ to ${}^{27}_{13}Al$. The half-life of the decay is 9.5 min and it may take place in either of the two excited states of ${}^{27}_{13}Al$. The resulting excited nucleus ${}^{27}_{13}Al^*$ then undergoes one or two gamma decays to reach the ground state.

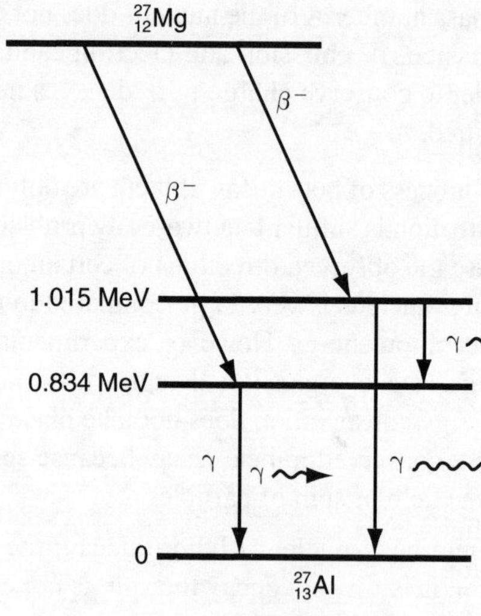

FIGURE 13.4

Let us suppose that an electron is in the *K*-shell. Then the nucleus de-excites by giving its excitation energy to that nearest *K*-shell electron. Such electrons are thereby knocked out of the atoms entirely. These electrons are called conversion electrons and the kinetic energy of the conversion electron is given by

$$U_e = \Delta T - U_B$$

where ΔT is the energy with which gamma ray would have been emitted and U_B is the atomic binding energy. It means the emitted electron has an energy equal to the loss in nuclear kinetic energy minus the binding energy of the electron. This process is similar to the photoelectric effect in which a nuclear photon is absorbed by an atomic electron.

13.8.7 Nuclear-Radiation Detector

The detection of nuclear radiations depends upon their interaction with matter and especially on the excitation and ionisation processes. Nuclear–radiation detector is a device in which the presence of radiation induces physical change that is observable. In such studies, it is essential to know the exact number of incident radiations and their energy at the detector. Different types of detectors have been used for detecting these radiations. The most common detectors are discussed below in detail.

13.8.7.1 Ionisation Chamber

The ionisation chamber works on the principle that a charged particle ionises the gas molecule when it is passed through the gas. In the ionisation process, the number of ion pairs are formed and they give the valuable information about the nature of the radiation, i.e., whether it is an α or β-particle, and energy of the incident particle.

The ionisation chamber consists of a cylindrical chamber fitted with a pair of electrodes, which are kept at some distance from each other, and a high potential difference is applied between them (Figure 13.5.) The chamber is filled with a gas like air or argon at normal pressure. One end of

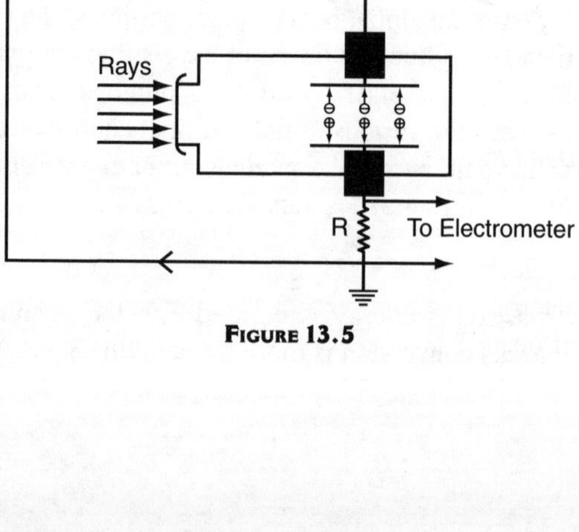

FIGURE 13.5

the chamber is slated with a window, which is made of mica or nylon of thickness of about 0.002 mm and is coated with graphite to make it conducting.

When a particle (or radiation) passes through the chamber, the gas in the chamber gets ionised and ions are collected by opposite electrodes. They give rise to current to the external circuit, which is measured by a meter. This ionisation current is proportional to number of pairs of ions and hence to the number of radiations entering the chamber. The ionisation chamber is much less sensitive to β-particles in comparison with α-particles. It is also too insensitive to γ-rays because they do not produce enough ionisation. With the help of this chamber we cannot count individual particles, but only the average effect of these particles can be determined.

13.8.7.2 Geiger–Mueller Counter

Geiger–Mueller or GM counter is most efficient, accurate and useful device, which is used for detecting individual particles such as α-, β-, γ- and *X*-rays. It consists of a metallic cylindrical tube fitted with an axial fine tungsten wire (Fig. 13.6). One end window of this tube is sealed by thin mica sheet through which radiation can enter the tube. The whole arrangement is enclosed within a thin glass chamber. This tube is filled with a gaseous mixture of about 90% argon and 10% ethyl alcohol at a pressure of 10 cm of Hg. The potential of the order of 1000 volts is applied between anode and cathode. The value of this applied voltage is adjusted to be somewhat below the breakdown potential of the gaseous mixture. When radiation enters the GM tube, some of the argon atoms get ionised and produce number of ion-pairs and then electrons are moved to the anode. Due to the shape of electrodes, the electrostatic field is radial and it acts strongly near the anode. When electrons move towards the anode, they collide with gas molecules and produce further ionisation. In this process, the multiplication of ions continues and as a result, avalanche of electrons is obtained. If the exciting potential is sufficiently high, the secondary ionisation takes place and further another avalanche of electrons is obtained. Thus, within no time almost entire volume of the gas in the tube is ionised and it leads to an amplification as high as 10^8. In ionisation process, the total number of ions produced does not depend upon radiation entered.

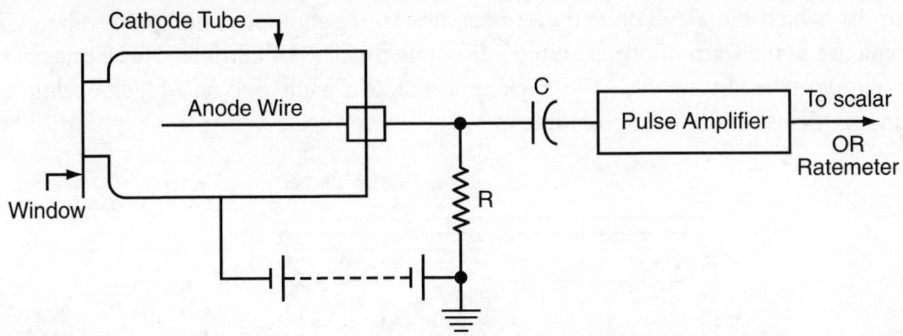

FIGURE 13.6

Dead time The time during which the counter is incapable of responding to ions is known as *dead time*. The reason of incapability of counter can be explained as follows. The electrons are collected rapidly on the anode because of their light mass and leave behind a space charge of slow-moving positive ions. After the collection of electrons, the space charge of positive ions becomes large enough and it is sufficient to cancel the applied electric field and then further ionisation is stopped. The counter remains ionisation dead till space charge of positive ions is collected by the cathode. After the removal of space charge, the counter again becomes sensitive for further pulse recording. Actually, an electronic circuit is used to quench the discharge and pass on an impulse to record the event.

13.8.7.3 Scintillation Counter

It is a very sensitive device used for detection and measurement of high energy nuclear radiations, viz, α-, β-, and γ-rays. These radiations are detected by means of fluorescence which they produce in certain materials called *scintillants*. The selection of scintillant depends upon the radiation to be detected. For α-, β-, and γ-particles ZnS, naphthalene and NaI crystal, respectively, are used as scintillants.

The scintillation counter consists of scintillation chamber, photo-multiplier tube and electronic counter. The scintillation chamber is made of an aluminium casing in which suitable scintillation crystal is placed. When the radiation enters the crystal, it produces a tiny flash of light. This tiny flash of light is made to fall on a transparent photosensitive layer of a photomultiplier tube where it ejects photoelectrons. These photoelectrons are accelerated by the successive dynodes which are kept at progressively higher voltage. The photoelectrons are pulled to the dynode 1 where a number of secondary electrons are emitted for each primary photoelectron. These electrons are pulled to the dynode 2 where the electrons are further multiplied by the secondary emission. Ultimately, the number of electrons is so much increased in successive stages that a measurable current pulse is obtained. These pulses are sent to an electronic system where they are counted.

This counter has many advantages over a GM counter. The efficiency of this counter for counting γ-rays is comparatively much higher. The time of flight of the electrons through this tube is so small that it can count about 10^6 particles per second.

13.8.7.4 Wilson's Cloud Chamber: Visual Detector

Wilson in 1911 designed a cloud chamber, as shown in Fig. 13.7 to detect the radioactive particles and to measure their energy. The dust-free air mixed with saturated vapour of water and any other liquid like alcohol or ether is filled in a chamber fitted with a piston. α- or β-particles enter the chamber such that each particle produces ions by ionising the air molecules on its path. Suddenly the piston is dropped down in order to reduce the temperature by which the air is cooled and becomes supersaturated with water vapour. This way, the vapour will condense in the form of drops on the ions along the path of particle. These tracks are illuminated with light and photographed by camera. The track appears like a white line on a black background. Different types of ionising particles produce different tracks.

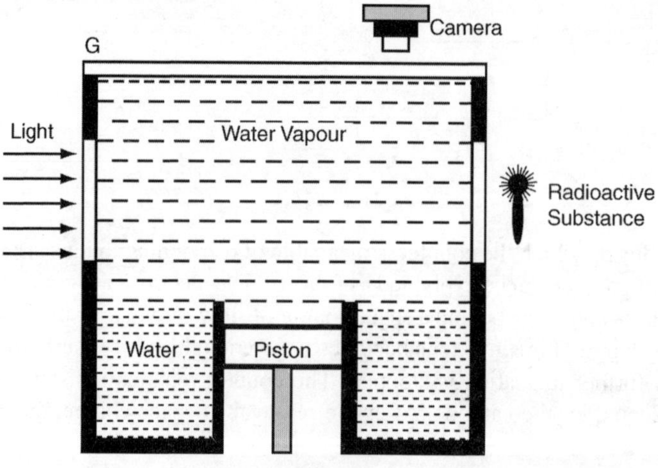

FIGURE 13.7

The track of α-particles is thick, short and continuous. On the other hand, the β-particles track is thin, dotted and long (Fig. 13.8). Thus, these particles can be distinguished on the basis of tracks.

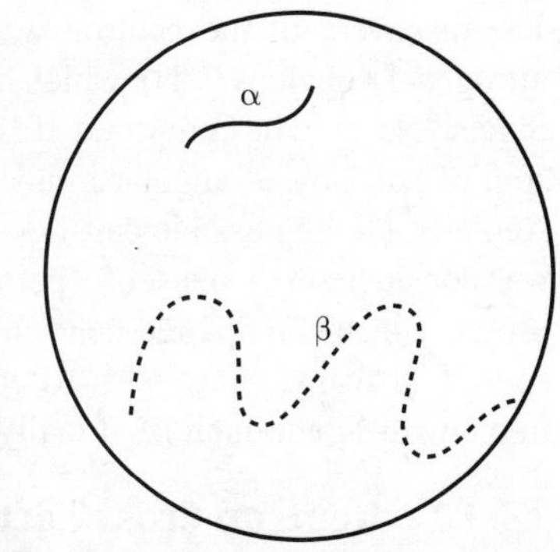

Figure 13.8

13.8.7.5 Bubble Chamber: High Energy Detector

The cloud chamber is not suitable to detect highly energetic particles. To overcome this problem, Glaser in 1952 invented the bubble chamber, which is almost inverse of cloud chamber. A bubble chamber mainly consists of a heavy-walled pyrex bulb filled with a low-boiling-point liquid like liquid-propane or liquid-hydrogen. This liquid is compressed by passing air through a pressure-regulating device and the proper temperature around the pyrex bulb is maintained with the help of thermostat-controlled oil-bath.

It is well-known that the boiling point of the liquid can be raised by increasing pressure on its surface. If the pressure on its surface is increased (like in a pressure cooker), the boiling does not start till a higher temperature is reached. If the pressure is suddenly released then liquid becomes superheated. This superheated state can be maintained for a few seconds. If an ionising particle passes through the liquid immediately after the pressure is released, it leaves a trail of ions behind it. These ions, left in the track of a particle, act as condensation centres and form vapour bubbles. This track of bubbles can be immediately illuminated and photographed.

13.8.7.6 Semiconductor Detector

This device, shown in Fig. 13.9, is used as a particle detector. It consists of a *p-n* junction which is connected in reverse bias. The purpose of applying reverse bias is to increase the thickness of the depletion layer. This depletion region has no carriers of either sign. When an ionising particle enters the depletion region, the number of electron-hole pairs are produced. Under the influence of applied reverse bias, the electrons and holes are swept rapidly to the +ve and –ve electrodes. Thus, it produces a current pulse across the resistor R which is amplified and then counted.

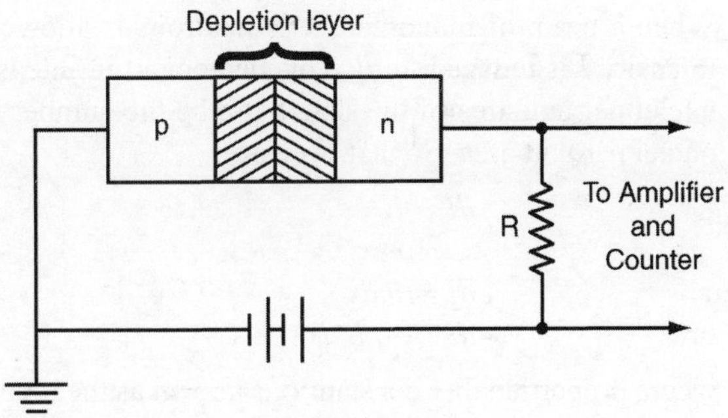

Figure 13.9

13.9 DISCOVERY OF NEUTRON — LO7

It was 1931 when Bothe and Becker in Germany observed that if very energetic α-particles (emitted from Po) fell on certain light elements (Be, B, Li), an unusually penetrating radiation was produced. At first, it was assumed to be γ-radiation, but the experimental results were very difficult to be interpreted based on this. In 1932, Curie and Joliot in Paris showed that if this unknown radiation fell on paraffin or any other hydrogen-containing compound, it ejects protons of very high energy. In the same year of 1932, based on a series of experiments, the physicist James Chadwick showed that the new radiation consisted of uncharged particles of approximately mass of the proton. These uncharged particles were called *neutrons*.

The discovery of the neutron was quite helpful in explaining a known puzzle involving the spin of the nitrogen-14 nucleus (^{14}N), which had been experimentally measured to be 1 basic unit of angular momentum. Since $^{14}_{7}N$ would be composed of 14 protons and 7 electrons and both protons and electrons carried an intrinsic spin of half unit of angular momentum, there was no way to arrange these 21 particles to give a spin of 1. Actually, all the possible pairings could give a net spin of half. However, after the discovery of neutron, $^{14}_{7}N$ was considered to consist of 3 pairs of protons and neutrons together with an additional unpaired neutron and proton. Since an unpaired neutron and proton each contributed a spin of half in the same direction, the total spin of $^{14}_{7}N$ came out to be 1. After this, the concept of nuclear neutrons was used to explain spin differences in many different nuclides. Finally, the neutron was accepted as a basic structural unit of atomic nuclei.

13.9.1 Neutron Cross-Section

The probability that a bombarding particle will interact in a certain way with a target particle is represented in terms of nuclear reaction cross-section. Each target presents a certain area, called its cross-section, to the incident particle. The incident particle interacts with the target if it is directed to this area.

Neutron absorption cross-section is the cross-section for a nuclear reaction which is initiated by neutrons. For many materials, this rises to a large value at particular neutron energies due to resonance effects. For example, a thin sheet of Cd forms an almost impenetrable barrier to thermal neutrons. The same way we can define neutron scattering cross-section. Neutron cross-section is the measure of probability of scattering and absorption of neutron when it approaches a nucleus. The incident neutron interacts with the nucleus if it is directed to this area. The neutron cross-section is denoted by the symbol σ and depends on the energy of the incident neutron and size and nature of the target. This is expressed in barns (1 barn = 10^{-28} m^2).

When a beam of mono-energetic neutrons is allowed to fall on a target slab having thickness dx, then its intensity I is reduced to dI. This decrement in intensity depends upon incident intensity (I) and number of nuclei per unit area of the slab. Let n be the number of nuclei per unit volume. Then ndx will be number of nuclei per unit area. So that, we have,

$$dI \propto I$$

$$\propto ndx$$

$\therefore$ $$dI \propto Indx$$

or $$dI = -\sigma\, Indx \tag{i}$$

where proportionality constant σ is known as the total cross-section of neutrons and the quantity σn is known as *attenuation coefficient*. The negative sign is appearing as the intensity gets decreased with distance. Eq. (i) can be written as

$$\frac{dI}{I} = -\sigma ndx, \tag{ii}$$

Which on integration gives

$$\int_{I_0}^{I} \frac{dI}{I} = -\int_0^x \sigma ndx$$

or $$[\ln I]_{I_0}^{I} = -\sigma n[x]_0^x \quad \Rightarrow \quad \ln I - \ln I_0 = -\sigma nx$$

or $$\ln_e\left(\frac{I}{I_0}\right) = -\sigma nx \tag{iii}$$

or $$\frac{I}{I_0} = e^{-\sigma nx} \tag{iv}$$

$$I = I_0 e^{-\sigma nx}$$

This expression shows that the intensity of a transmitted beam gets decreased exponentially with increasing thickness of the slab. On the other hand, the total cross-section σ is calculated from Eq. (iii) as

$$\sigma = -\frac{1}{nx}\ln\left(\frac{I}{I_0}\right)$$

$$\sigma = \frac{1}{nx}\ln\left(\frac{I_0}{I}\right) \tag{v}$$

13.10 NUCLEAR REACTIONS: CONSERVATION LAWS

LO8

If a target material is bombarded by fast-moving particles such as protons, neutrons, electrons, deuterons or alpha particles, then the target nuclei after the bombardment are usually different from what they were before. The target nuclei may change their mass number or atomic number or both due to interaction with an incident particle. This is called a transmutation and the reaction is called *nuclear reaction.*

The collision of a bombarding particle of rest mass m_b and kinetic energy U_b with a target containing nuclei of rest mass M_T and kinetic energy U_T results in the *nuclear reaction.*

$$U_b + N_T \rightarrow N_R + y \tag{i}$$

In this interaction a recoil nucleus N_R of rest mass M_R with kinetic energy U_R and a light particle y of rest mass m_y with kinetic energy U_y are emitted, as shown in Fig. 13.10. As per the law of conservation of energy, total initial energy T_i that is the sum of the rest mass energies and kinetic energies must be equal to the total final energy, T_f. That is

$$T_i = T_f$$

or $$m_bc^2 + U_b + M_Tc^2 + U_T = M_Rc^2 + U_R + m_yc^2 + U_y \tag{ii}$$

or $$[(U_R + U_y) - (U_T - U_b)] = [(M_T + m_b)c^2 - (M_R + m_y)c^2] \tag{iii}$$

or $$U_f - U_i = M_ic^2 - M_fc^2 \tag{iv}$$

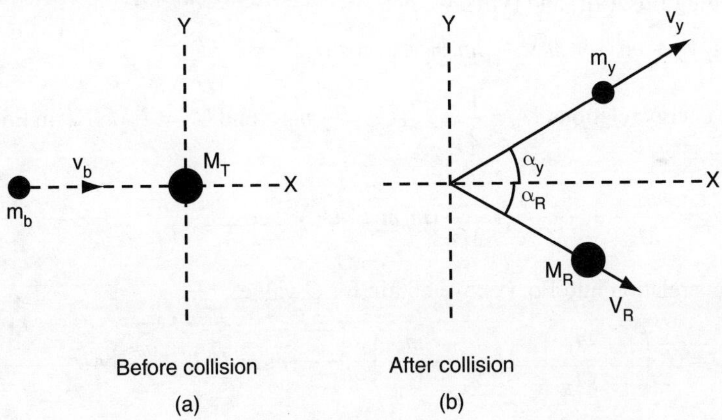

FIGURE 13.10

where $U_i = U_T + U_b$ and $U_f = U_R + U_y$ together with U_i and U_f as the initial and final kinetic energies and M_i and M_f as the total initial and final rest masses, given by $M_i = M_T + m_b$ and $M_f = M_R + m_y$. The above equation states that the net increase in kinetic energy is equal to the net decrease in rest mass energy that is equal to the disintegration energy or Q-value. Hence Q-value is defined as

$$Q = U_f - U_i = M_i c^2 - M_f c^2$$

13.10.1 Exoergic and Endoergic Reactions

If the Q-value is positive, then the nuclear reaction is called exoergic reaction and in this case $U_f > U_i$ or $M_i c^2 > M_f c^2$. If Q-value is a negative quantity, then the nuclear reaction is an endoergic reaction and in this case $U_f < U_i$ or $M_i c^2 < M_f c^2$. If initially the target nucleus is at rest, then its kinetic energy $U_T = 0$. Hence, Q-value of a reaction is given by

$$Q = (U_R + U_y) - U_b \tag{v}$$

$$= [(M_T + m_b) - (M_R + m_y)]c^2 \tag{vi}$$

13.10.2 Disintegration Energy

We can find the disintegration energy by knowing the values of U_b, U_y and U_R. The kinetic energy U_R of the recoiling nucleus N_R is very small. It is difficult to measure it accurately because the mass of this nucleus is very large in comparison with the light particle y. Therefore, by using the law of conservation of linear momentum we can calculate the Q-value.

Let the recoiling nucleus N_R be emitted with a velocity V_R at an angle α_R and the particle y is emitted with the velocity v_y at an angle α_y, as shown in Fig. 13.7. Now, from the law of conservation of momentum we get the equations

$$M_R V_R \cos \alpha_R + m_y v_y \cos \alpha_y = m_b v_b$$

or
$$M_R V_R \cos \alpha_R = m_b v_b - m_y v_y \cos \alpha_y \tag{vii}$$

and
$$m_y v_y \sin \alpha_y - M_R V_R \sin \alpha_R = 0$$

or
$$M_R V_R \sin \alpha_R = m_y v_y \sin \alpha_y \tag{viii}$$

By squaring and adding Eqs. (vii) and (viii), we get

$$M_R^2 V_R^2 = m_b^2 v_b^2 + m_y^2 v_y^2 - 2 m_b m_y v_b v_y \cos \alpha_y \tag{ix}$$

By using the kinetic energy relations $U_b = \frac{1}{2} m_b v_b^2$, $U_y = \frac{1}{2} m_y v_y^2$ and $U_R = \frac{1}{2} M_R V_R^2$ in Eq. (ix), the expression for U_R becomes

$$U_R = \frac{m_b U_b}{M_R} + \frac{m_y U_y}{M_R} - \frac{2}{M_R}(m_b m_y U_b U_y)^{1/2} \cos \alpha_y \tag{x}$$

Now, substituting this relation into Eq. (v), we obtain for Q-value

$$Q = U_y \left(1 + \frac{m_y}{M_R}\right) - U_b \left(1 - \frac{m_b}{M_R}\right) - \frac{2}{M_R}(m_b m_y U_b U_y)^{1/2} \cos \alpha_y \tag{xi}$$

13.10.3 Threshold Energy

The above relation shows that the Q-value is independent of the mass of the target nucleus M_T and the kinetic energy of the recoiling nucleus, U_R. As mentioned earlier, Q is positive for an exoergic reaction and from Eq. (v) the final kinetic energy is the sum of Q and U_b. If $U_b = 0$, $U_R + U_y = Q > 0$. Hence, we can say that an exoergic reaction is energetically possible if the bombarding particle has zero kinetic energy. On the other hand, endoergic reaction is energetically possible only when $U_b > |Q|$. The minimum kinetic energy of the bombarding particle which is necessary to initiate the endoergic reaction is called *threshold energy*. We can calculate this energy by using the centre-of-mass coordinate system, in which the linear momentum is always zero before and after the reaction. If U'_b be the kinetic energy of the incident particle in this coordinate system, the endoergic reaction is energetically possible only if

$$U'_b \geq |Q| \tag{xii}$$

In terms of the reduced mass m_r of the incident particle and the target nucleus, given by $m_r = \dfrac{m_b M_T}{m_b + M_T}$, the above condition can be written as

$$\frac{1}{2}\frac{m_b M_T}{m_b + M_T} v_b^2 \geq |Q| \tag{xiii}$$

or

$$\frac{1}{2} m_b v_b^2 \geq \left(\frac{M_T + m_b}{M_T}\right)|Q| \tag{xiv}$$

Hence, the minimum energy required for endoergic reaction to take place, i.e., threshold energy, should be

$$(U_b)_{\min} = \left(\frac{M_T + m_b}{M_T}\right)|Q| = \left(1 + \frac{m_b}{M_T}\right)|Q|$$

This is clear from the above relation that the threshold energy is greater than the Q-value or the disintegration energy by a factor of $\left[1 + \dfrac{m_b}{M_T}\right]$.

13.11 NUCLEAR FISSION LO9

The phenomenon of breaking of a heavy nucleus into two or more light nuclei of almost equal masses together with the release of a huge amount of energy is known as *nuclear fission.* The process of nuclear fission was first discovered by the German scientists, Otto Hahn and Strassman, in the year 1939. In this process, when uranium nucleus ($^{235}_{92}U$) was bombarded with slow neutrons, this nucleus was found to split up into two radioactive nuclei which were identified as isotopes of barium ($^{144}_{56}Ba$) and krypton ($^{89}_{36}Kr$). It is given by the following nuclear reaction.

$$^{235}_{92}\text{U} + {}^1_0n \rightarrow \left[{}^{236}_{92}\text{U}\right] \rightarrow {}^{144}_{56}\text{Ba} + {}^{89}_{36}\text{Kr} + 3^1_0n + \text{energy}$$

It is not that barium and krypton are the only isotopes to be obtained by the fission of ^{235}U. Actually, this is a very complicated phenomenon and more than 100 isotopes of over 20 different elements have been obtained in it. All these elements fall in the middle 75 to 160 mass number region of the periodic table.

13.11.1 Theory: Liquid-Drop Model

The mechanism of nuclear fission was first explained by Bohr and Wheeler on the basis of liquid-drop model of the nucleus. According to this model, the nucleus is assumed to be similar to a liquid drop, which remains in equilibrium by a balance between the short-range, attractive forces between the nucleons and the repulsive electrostatic forces between the protons. This inter-nucleon force gives rise to surface tension forces to maintain a spherical shape of the nucleus. Thus, there is a similarity in the forces acting on the nucleus and liquid drop. When nucleus-drop captures slow or thermal neutron, oscillations set up within the drop. These oscillations tend to distort the spherical shape so that the drop becomes ellipsoid in shape (Fig. 13.11). The surface-tension forces try to make the drop return to its original spherical shape while the excitation energy tends to distort the shape still further. If the excitation energy and hence oscillations are sufficiently large, the drop attains the *dumbbell* shape (Fig. 13.11). The Coulombic repulsive forces then push the nucleus into two similar drops. Then each drop (bell) tries to attain the shape for which the potential energy is minimum, for example, spherical shape.

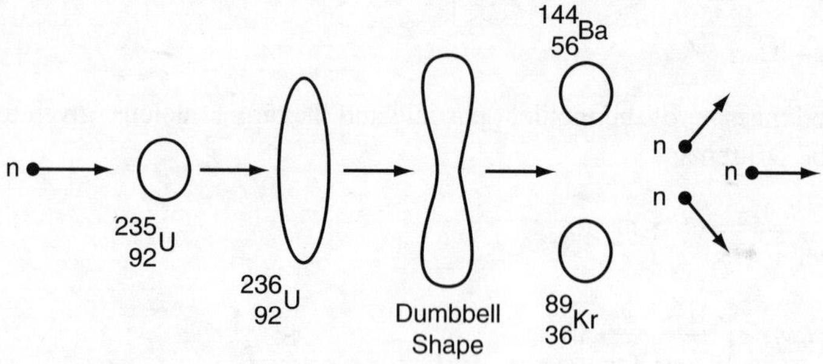

FIGURE 13.11

13.11.2 Nuclear Energy

In nuclear fission, a huge amount of energy is liberated, which is known as *nuclear energy*. An estimation of this energy can be made as follows. The mass of

$$^{235}_{92}U + ^{1}_{0}n = 234.99 \text{ amu} + 1.01 \text{ amu} = 236.0 \text{ amu}$$

Similarly, the mass of

$$\begin{aligned} ^{144}_{56}Ba + ^{89}_{36}Kr + 3\,^{1}_{0}n &= 143.87 \text{ amu} + 88.90 \text{ amu} + 3 \times 1.01 \text{ amu} \\ &= 235.80 \text{ amu} \end{aligned}$$

$\therefore$ Mass defect $\Delta m = 236.00 - 235.80 = 0.20$ amu

According to Einstein's mass-energy relation, $E = mc^2$, 1 amu mass is equivalent to 931 MeV energy. So energy released in each fission process = $0.20 \times 931 \approx 190$ MeV. This energy is millions of times more than what we get by any chemical reaction.

13.11.3 Chain Reaction

When uranium is bombarded by neutrons, each uranium nucleus is broken into two nearly equal fragments and a huge amount of energy is liberated and two or three fresh neutrons are emitted. If the conditions are favourable, these neutrons take part in the fission of other uranium nuclei in the same way. This leads to a chain of nuclear fissions which continues till the whole of uranium is fissioned within a fraction of time (Fig. 13.12). Thus the energy produced in nuclear fission goes on multiplying. This energy takes a tremendous magnitude very soon and is released as a violent explosion. Such a chain reaction is known as *uncontrolled* chain reaction. This happens in an atom bomb.

If by some means, the reaction is controlled in such a way that only one of the neutrons produced in each fission is able to cause further fission, then the fission process is slow and the energy is released steadily. This chain reaction is known as *controlled* chain reaction. This happens in nuclear reactors.

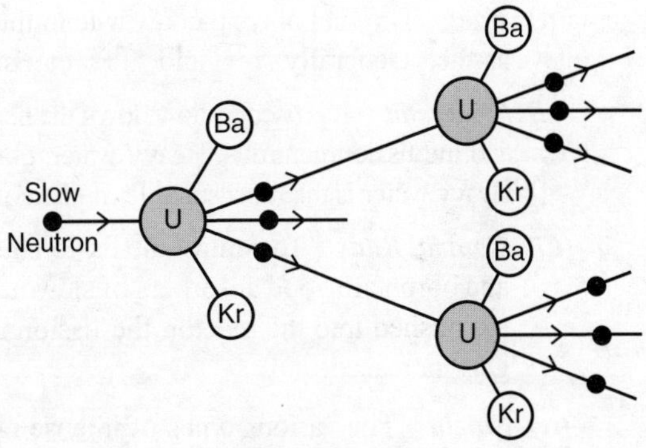

Figure 13.12

13.11.4 Critical Size of Nucleus

We consider a uranium in sphere shape of radius r, which has N_1 as the number of neutrons produced in a given time interval, N_2 as the number of neutrons lost in non-fission process and N_3 as the number of neutrons escaped through the surface in the same time interval. N_1 and N_2 will be proportional to the volume, whereas N_3 will be proportional to the surface area of the sphere, i.e.,

$$N_1 \propto \frac{4}{3}\pi r^3 = k_1 r^3$$

$$N_2 \propto \frac{4}{3}\pi r^3 = k_2 r^3$$

and $$N_3 \propto 4\pi r^2 = k_3 r^2$$

where k_1, k_2 and k_3 are proportionality constants. The chain reaction is possible only when the rate of emission of neutrons is greater than the total number of neutrons absorbed within the substance and going out of the substance, i.e.,

$$N_1 > N_2 + N_3$$

$$\Rightarrow \quad k_1 r^3 > k_2 r^3 + k_3 r^2$$

or $$(k_1 - k_2) r > k_3$$

or $$r > \frac{k_3}{k_1 - k_2} = k \text{ (say)}$$

where k is known as the critical size of the nucleus. Thus, in order to achieve a self-sustained chain reaction, the size of the sample must be greater than a critical value k. Below this critical value the chain reaction will stop.

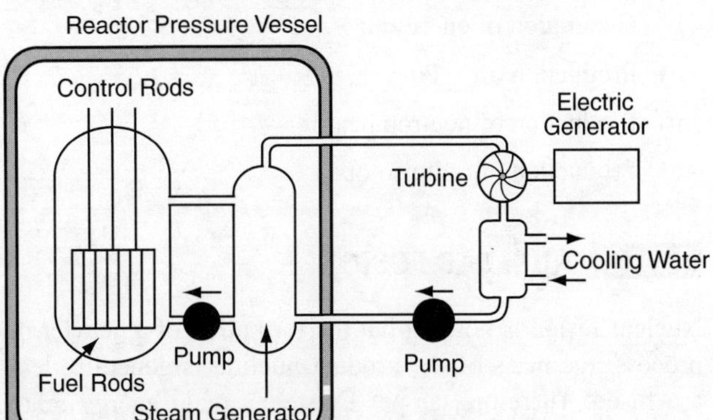

Figure 13.13

13.11.5 Nuclear Reactor

It is a device that produces a self-sustained and controlled chain reaction in a fissionable material. One type of nuclear reactor is shown in Fig. 13.13. A modern reactor has the following important parts.

(i) ***Fuel*** The fuel plays the key role in the operation of the reactor. The fissionable material is known as fuel. Generally, ^{235}U and ^{239}Pu can be used as fuel.

(ii) ***Moderator*** It is used to slow down the neutrons to thermal energies by elastic collisions between its nuclei and the fission neutrons. Heavy water, graphite or beryllium oxide are commonly used for this purpose. Heavy water is the most suitable moderator.

(iii) ***Control Rods*** To control the fission rate in the reactor, we use cadmium and boron rods. Cadmium and boron are good absorbers of slow neutrons. These rods are fixed in the reactor-walls. When they are pushed into the reactor, the fission rate decreases and when they are pulled out the fission rate gets increased.

(iv) ***Shield*** The various types of intense rays, like α-, β-, γ-rays in radioactivity are emitted from the reactor. These rays may be injurious to the health of people working near the reactor. For protection, the reactor is therefore surrounded by a concrete wall of about 2 meter thick and containing high protection elements like iron.

(v) ***Coolant*** The reactor generates heat energy due to the fission reaction which is removed by means of a cooling agent. For this purpose, air, water, carbon dioxide etc. are generally used as coolant. Coolant is circulated through the interior of a reactor by a pumping system.

(vi) ***Safety Device*** If the reactor begins to go too fast, a special set of control rods, known as shut-off rods drop inside automatically. They absorb all the neutrons so that the chain reaction stops immediately.

13.11.5.1 *Working of Nuclear Reactor*

To start the reactor, no external source is required. Even a single neutron is capable of starting fission, although few neutrons are always present there. The reactor is started by pulling out the control rods. Then the neutron strikes ^{235}U nucleus and fission it along with the emission of two or three fast neutrons. These neutrons are slowed down by moderator (graphite), after which they induce further fission of ^{235}U. The reaction once started is controlled with the help of control rods by moving them inside and outside.

13.11.5.2 *Applications of Nuclear Reactor*

The nuclear reactors are used mainly for the following purposes.

(i) Generation of energy

(ii) Production of ^{239}Pu

(iii) Production of neutron beam

(iv) Production of radioisotopes

13.12 NUCLEAR FUSION

Nuclear fusion is nothing but the formation of a heavier nuclide by the fusing of two light nuclides. In this process, the mass of the product nuclide is generally less than the sum of masses of the nuclides which are fused. Therefore, as per Einstein's mass energy relation $E = mc^2$, an enormous amount of energy is released, which is called nuclear energy. The first artificial fusion reaction was the hydrogen bomb which was tested in November 1952. Fusion reactions are thermonuclear reactions which occur at extremely high temperatures. For example, in order to fuse deuterium ($_1^2$H) and tritium ($_1^3$H), the force of repulsion (called Coulomb potential barrier) of these two positively charged particles must be overcome.

The following fusion reaction is possible for the fusion of two heavy hydrogen nuclides $_1^2$H.

$$_1^2\text{H} + _1^2\text{H} \rightarrow _1^3\text{H} + _1^1\text{H} + 4.0 \text{ MeV (energy)}$$

The nucleus of tritium ($_1^3$H) can again fuse with heavy hydrogen nucleus

$$_1^3\text{H} + _1^2\text{H} \rightarrow _2^4\text{H} + _0^1n + 17.6 \text{ MeV (energy)}$$

Thus, the combined form is

$$_1^2\text{H} + _1^2\text{H} + _1^2\text{H} \rightarrow _2^4\text{He} + _0^1n + _1^1\text{H} + 21.6 \text{ MeV (energy)}$$

From the above equation, it is clear that three deuterium nuclei fuse together to form a helium nucleus and liberate 21.60 MeV energy which is obtained in the form of kinetic energy of proton ($_1^1$H) and neutron ($_0^1n$).

The above reaction can also be possible in the following way:

$$_1^2\text{H} + _1^2\text{H} \rightarrow _2^3\text{H} + _0^1n + 3.30 \text{ MeV (energy)}$$

$$_2^3\text{H} + _1^2\text{H} \rightarrow _2^4\text{He} + _1^1\text{H} + 18.30 \text{ MeV (energy)}$$

13.13 CONTROLLED FUSION LO9

Nuclear fusion is the main energy source that powers the sun and stars. As discussed earlier, in fusion the nuclei of light elements (for example, hydrogen) fuse together to make heavier elements (for example, helium), resulting in an enormous amount of energy. All over the world, efforts have been made to achieve this fusion in a controlled manner for the utilisation of energy. The International Thermonuclear Experimental Reactor (ITER) is a new attempt that will use a *tokamak* concept. Tokamak is a doughnut-shaped vessel in which a strong, helical magnetic field guides the charged particles. Hence, a magnetic configuration is used to create and maintain the conditions for controlled fusion reactions on earth. In ITER, superconducting magnet coils around a toroidal vessel will confine and control a mixture of charged particles (called plasma) and induce an electrical current through it. In this configuration, fusion reactions will occur when the plasma is hot enough, dense enough and contained long enough (Lawson criterion, discussed later) for the atomic nuclei in the plasma to start fusing together. The idea behind controlled fusion is to use magnetic fields to confine a high temperature plasma of deuterium and tritium.

13.13.1 Plasma: The Fourth State of Matter

You are aware of three states of the matter. These are solid, liquid and gas. In solids, the atoms are packed very close to each other and are fixed at definite positions. These are connected with each other by the interatomic forces. When we supply energy to the atoms of the solids, they start oscillating about their equilibrium positions and as a result, the interatomic forces become weaker and the atoms are separated significantly. This way the solid takes the form of liquid. The liquid has a specific volume but does not have a specific shape. So it changes shape according to the shape of the container in which it is kept. Now if we again supply energy to the atoms, the interatomic forces become insignificant, the atoms get separated and start moving freely. It means the liquid has taken the form of gas. In gas, the atoms are not connected with each other and hence can move in any direction. The gas does not have specific shape and volume also. It takes the shape and volume of the container in which it is kept. If more energy is supplied to the gas species (atoms or molecules), the electrons from the outermost level of the atoms get easily detached and hence the atoms become ionised. As a result, we are left with the collection of ions, electrons and some neutrals (atoms that are not ionised). This collection of charged and neutral particles is referred to as plasma, which is sometimes called the fourth state of matter. This is because it is found in natural conditions; for example, the gases near the sun are always in ionised state that qualify for plasma.

The species of the plasma being charged are connected with each other by the electromagnetic forces. This can be explained as follows. Since the charges separated with each other give electric field, the plasma species produce electric field. However, the separation of charges of plasma is not fixed (as the species do not remain stationary; they keep on moving/oscillating). So, this electric field is a time-varying field, which will generate magnetic field according to the Maxwell's fourth equation $\vec{\nabla} \times \vec{H} = \vec{J} + \varepsilon_0 \frac{\partial \vec{E}}{\partial t}$. However, the motion of charges generates current and hence the magnetic field. In view of this, the plasma species produce a time-varying magnetic field, which will induce electric field according to the Maxwell's third equation $\vec{\nabla} \times \vec{E} = -\frac{\partial \vec{B}}{\partial t}$. Therefore, it can be said that the plasma species are connected with each other by the electromagnetic fields. Since the number of ions and electrons in the plasma is almost equal, the plasma as a whole is neutral. Since we cannot neglect the internal forces at the same time, the plasma is however quasineutral. Moreover, if we attempt to perturb a part of the plasma, the whole body of the plasma will get perturbed due to the connection of all the species with each other. This property is known as collective behaviour of the plasma. Therefore, an ionised gas can qualify for plasma state, if it is quasineutral and shows collective behaviour.

Another interesting property of the plasma is its ability to shield out the field which is applied on it. This happens when we insert the electrodes of a battery into the plasma. The positive (negative) electrode attracts the electrons (ions) whose number is decided by the charge carried by the electrode. Under this situation, an electron cloud is developed around the electrode which shields/cancels the external field. The thickness of this electron cloud is known as Debye length λ_{De}. Since electrons are light species compared with ions, the shielding is generally done by the electrons only. Clearly the field exists within the cloud or the Debye sphere (sphere with the radius λ_{De}). Now imagine if the Debye length is much less than the dimension (L) of the plasma, the bulk of the plasma will remain neutral. Therefore, the required condition for quasineutrality is $\lambda_{De} << L$. Moreover, if the number of electrons in the Debye sphere (say, N_{De}) is much larger than unity, i.e., $N_{De} >> 1$, the condition of collective behaviour will be fulfilled.

Any distance in the plasma system is measured in terms of Debye length λ_{De} and the time is measured in terms of inverse of plasma frequency f_{pe}. The plasma frequency is nothing but the natural frequency of the plasma, the same as all the materials have their natural frequencies. Actually, this is the frequency of oscillations made by the electrons about their equilibrium positions. The Debye length λ_{De} and the plasma frequency f_{pe} in SI system of units are given by

$$\lambda_{De} = \sqrt{\left(\frac{\varepsilon_0 k T_e}{n_0 e^2}\right)} \quad \text{and} \quad f_{pe} = \frac{1}{2\pi}\sqrt{\left(\frac{n_0 e^2}{\varepsilon_0 m_e}\right)}$$

Here k is the Boltzmann constant (= 1.38×10^{-23} J/K), n_0 is the plasma density, which is the common density of ions (n_i) and electrons (n_e), i.e., $n_0 = n_i = n_e$, T_e is the electron temperature, e is the electron charge and m_e is its mass.

In plasmas, generally we do not talk specifically about the temperature of the ions and electrons, but we focus on their energies. That is, the temperature is written in terms of energy. For example, 1 eV energy of the electron would be equal to its thermal energy kT_e (for a 2-D system). So

$$1 \text{ eV} = kT_e$$

or $$1.6 \times 10^{-19} \text{ (J)} = 1.38 \times 10^{-23} \times T_e \text{ (J/K)}$$

or $$T_e = 11{,}600 \text{ K}$$

It means 1 eV of energy is equivalent to 11,600 K temperature. In laboratory plasmas, generally the electron temperature varies from 1 eV to 5 eV. For a plasma with density $10^{18}/\text{m}^3$ and temperature 2 eV, the Debye length is of the order of μm and the plasma frequency is of the order of GHz (10^9 Hz).

13.13.2 Ignition Temperature

A hot plasma at thermonuclear temperature loses a considerable amount of its energy in the form of radiation. Therefore, it is required that the nuclear fusion produce more energy than is lost from the radiation of the plasma. This requirement determines the minimum temperature for a nuclear fusion reactor to be self-sustained. This temperature is called ignition temperature at which alpha-particle heating can sustain the fusion reaction. As the temperature is increased, the production energy as well as the radiation losses get increased. However, the fusion-energy production increases faster than the radiation loss.

13.13.3 Lawson Criterion

For obtaining a net yield of energy from a fusion reaction, it is required, that in addition, to providing a sufficiently high temperature to enable the particles to overcome the Coulomb barrier, this temperature must be maintained for a sufficient confinement time and with a sufficient ion density. The overall conditions that must be met for a yield of more energy than is required for heating of the plasma are generally stated in terms of the product of ion density (n_0) and confinement time (τ). This condition is called Lawson criterion. For deuterium-tritium (DT) fusion, it is

$$n_0 \tau \geq 10^{14} \text{ sec/cm}^3$$

However, for deuterium-deuterium (DD) fusion, the Lawson criterion reads

$$n_0 \tau \geq 10^{16} \text{ sec/cm}^3$$

13.13.4 Fusion by Inertial Confinement

Inertial confinement fusion (ICF) is an attempt for generating commercial energy using controlled thermonuclear explosions. In this scheme, heavy isotopes of hydrogen, called fuel, are heated to temperatures of around 10 keV. Actually, laser or particle beams are focused onto the surface of a capsule (diameter of few millimetres), containing a small quantity of fuel. Due to evaporation and ionisation of the outer layer of the material, a plasma crown is formed, which expands and, as in a rocket, generates an inward moving compression front which heats up the inner layers of the material. This way, the core of the fuel is compressed to as much as one thousand times its liquid density. Then ignition takes place when the core temperature reaches about one hundred million degrees. Thermonuclear combustion spreads rapidly through the compressed fuel and produces energy equivalent to several times the amount deposited on the capsule by the laser or particle beams. The period of time during which these thermonuclear reactions occur is limited by the inertia of the fuel itself. Because of this reason, such type of fusion is called fusion by inertial confinement or inertial confinement fusion.

There are the following four basic phases of an inertial fusion pellet implosion.

(i) Irradiation In this process, irradiation of the pellet surface and formation of plasma is achieved with the help of an intense energy beam such as laser. For achieving this, the laser is bombarded upon the pellet (Fig. 13.14a).

(ii) Compression The compression of the pellet and fuel is driven by rocketlike blowoff of the surface material (Fig. 13.14b).

(iii) Ignition The central fuel core is ignited to about 1000 to 10,000 times liquid deuterium-tritium density and a temperature of 108°C (Fig. 13.14c).

(iv) Burning While the compressed fuel is inertially confined, it is burnt to achieve the fusion (Fig. 13.14d).

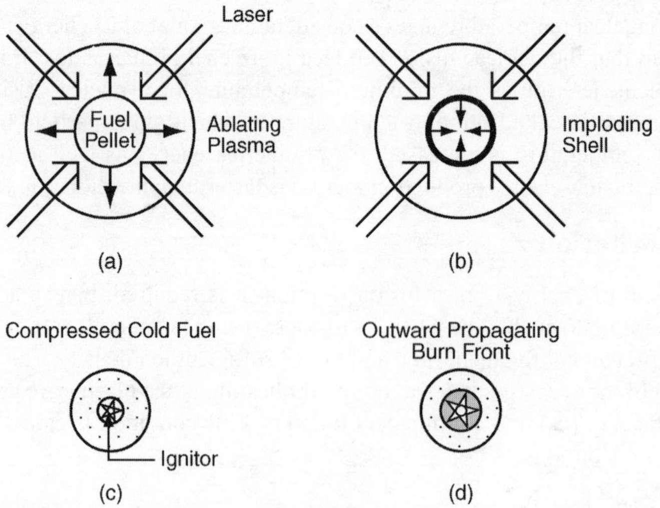

Figure 13.14

13.13.4.1 *Laser Fusion*

Laser fusion works on the concept of inertial confinement. If technically feasible, it would eliminate the problems of magnetic instabilities. There are following two ways to achieve laser fusion.

(i) Laser-Gas-Fusion In this mechanism, CO_2 laser beam is used to ionise and heat a long column of gaseous deuterium and tritium at the density of $n_0 = 10^{17}/\text{cm}^3$. Here, the light of the laser is absorbed by a process known as inverse bremstrahlung. This is because of the resistive damping of light wave due to electron ion collisions. However, this process is not sufficient as for the density $n_0 << n_c$, where n_c is the critical density, the absorption length is very large (in kilometres).

(ii) Laser-Pellet-Fusion In this mechanism, laser light is focused on to a small pellet of solid deuterium–tritium (DT), which has a number density $n_0 \approx 5 \times 10^{22}/\text{cm}^3$ and mass density $\rho = 0.2\,\text{g/cm}^3$. In this case since $n_0 >> n_c$, the radiation is reflected as soon as plasma density of $10^{21}/\text{cm}^3$ is formed on the pellet surface. This depends on anomalous absorption by parametric decay instability to ionise the rest of the pellet and heat it to a high temperature of 10 keV. As mentioned earlier, at this energy nuclei can penetrate the Coulomb potential barrier and hence the nuclear reactions can take place.

13.13.5 Magnetic Confinement

The basic problem in achieving controlled fusion is to generate plasma at very high temperatures and hold its species (particles) together long enough for a substantial number of fusion reactions to occur. Since the plasma is a mixture of charged particles, it can be controlled and influenced by external magnetic fields. Charged particles gyrate around the magnetic field lines and also move along these field lines, as shown in Fig. 13.15. These particles move in such a way that the direction of magnetic field generated by their motion is opposite to the direction of the applied field.

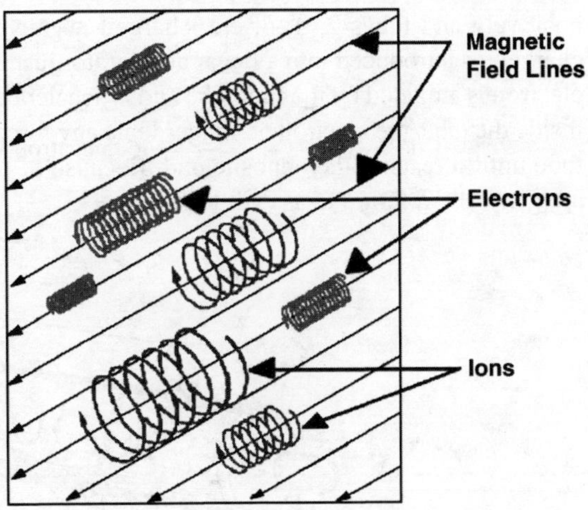

FIGURE 13.15

It is well established that the frequency at which a charged particle, of charge q and mass m, gyrates in the presence of magnetic field B_0 is given by

$$\omega_c = \frac{|q| B_0}{m}$$

If $v_\perp$ be the perpendicular component of its velocity with the direction of the magnetic field, the radius of the circle (in which the particle gyrates), called Larmour radius, can be given as

$$r_L = \frac{v_\perp}{\omega_c} = \frac{v_\perp m}{|q| B_0}$$

It means depending upon the strength and direction of the magnetic field, we can confine these charged particles. Thus, these particles are confined by the application of magnetic field. However, if these field lines are open–ended, theses particles are lost. In order to stop this loss of the particles at both ends, the magnetic field lines can be closed to a ring. This configuration, which can be established by arranging a set of coils in a ring, is called a torus.

13.14 PARTICLE ACCELERATORS

LO10

As discussed, in inertial confinement fusion, laser or particle beams are focused onto the surface of a pellet, containing a small quantity of fuel, in order to evaporate and ionise the outer layer of the material. So highly accelerated particles will contribute to achieve the controlled fusion. In this context, the acceleration of particles and hence the particle accelerators are desirable. A particle accelerator is a device where we use electric fields to propel electrically charged particles to high speeds and to contain them in well-defined beams. An ordinary cathode ray tube (CRT) television set is a simple form of an accelerator. Basically, there are two types of accelerators, namely, linear accelerators and circular accelerators.

13.14.1 Linear Accelerator

In a linear accelerator, abbreviated as LINAC, charged particles are accelerated in a straight line with a target of interest at one end. Linear accelerators operate on the general principle as used in a Van deGraaff generator except that now a particle is exposed to a series of electrical fields, each of which increases the velocity of the particle.

Typically, a linear accelerator consists of a few hundred or a few thousand cylindrical metal tubes which are arranged one in front of another (Fig. 13.16). These tubes are electrically charged such that each carries a charge opposite to that of the tube on either side of it. For example, tubes 1, 3, 5, etc. might be charged positively and tubes 2, 4, 6, etc. charged negatively. Now imagine that an electron, which is negatively charged, is introduced into a linear accelerator just in front of the first tube. Under the said configuration, the electron is attracted by the first tube and is accelerated toward it. Then the electron passes into that tube. Once inside the tube, the electron no longer feels any force of attraction or repulsion. So it merely drifts through the tube until it reaches the opposite end. Because of this behavior, the cylindrical tubes in a linear accelerator are generally referred to as drift tubes.

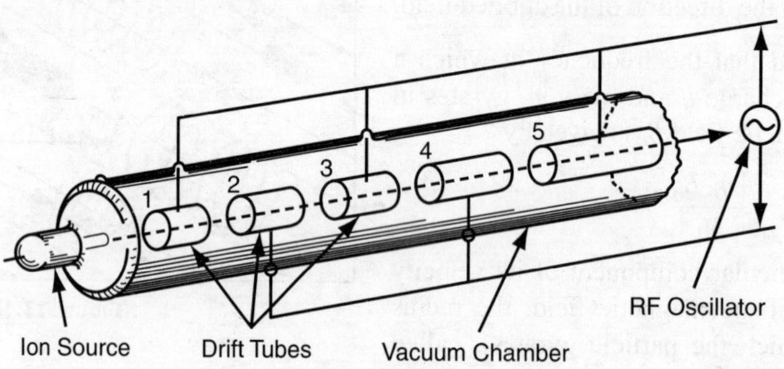

FIGURE 13.16

If the electron after leaving the first tube sees the next tube as positively charged, it will further accelerate. In view of this, the moment that the electron leaves the first drift tube, the charge on all drift tubes is reversed. So, tubes 1, 3, 5, etc., are now negatively charged and tubes 2, 4, 6, etc., are positively charged. Therefore, the electron exiting the first tube now finds itself repelled by the tube it has just left. At the same time, it feels attracted by the second tube. These forces of attraction and repulsion provide a kind of kick that accelerates the electron in a forward direction in steps through different tubes.

As the electron moves through the linear accelerator, the electric charge on all drift tubes reverses in a regular pattern. As mentioned earlier, the electron is repelled by the tube behind it and is attracted to the tube ahead of it. This way the electron gains energy at every step. As a result, the electron moves faster in each new tube it enters. Hence, it will cover a greater distance in the same amount of time. In order to make sure that the electron exits a tube at just the right moment, each tube is made slightly longer than the one before it.

Stanford Linear Accelerator, located at the Stanford Linear Accelerator Center (SLAC) in Stanford, California, is the largest LINAC in the world. This accelerator is 3 kilometres in length that holds 82,650 drift tubes along with the magnetic, electrical, and auxiliary equipment needed for the machine's operation. In this accelerator, the electrons have been found to be accelerated up to 32 GeV (32×10^9 eV).

13.14.2 Cyclotron

The cyclotron is a particle accelerator which is also known as Lawrence Cyclotron, as it was conceived by Lawrence in 1929. A cyclotron consists of two large dipole magnets designed to produce a semicircular region of uniform magnetic field, pointing uniformly downward. Because of their *D*-shape these are called *D*'s. The two *D*'s are placed back-to-back with their straight sides parallel but slightly separated, as shown in Fig. 13.17.

Now in order to produce an electric field across this gap, we apply an oscillating voltage. Particles, which are injected into the magnetic field region of a *D*, trace out a semicircular path until they reach the gap. However, as the particles pass across the gap they are accelerated by the applied electric field. After gaining energy, these particles follow a semicircular path in the next *D* with larger radius. Then they reach the gap again but with the opposite direction of motion. Hence, the electric field frequency must be such that the direction of the field is reversed by the time of arrival of the particles at the gap. This way the particles are accelerated by the field in the gap. After gaining energy, these particles enter the first *D* again. Because of larger speed they trace a larger arc and so they always take the same time to reach the gap. This way a constant frequency electric field oscillation continues to always accelerate the particles across the gap. However, there is a limitation on the energy that can be achieved in such a device. This depends on the size of the magnets that form the *D*'s and the strength of their magnetic fields.

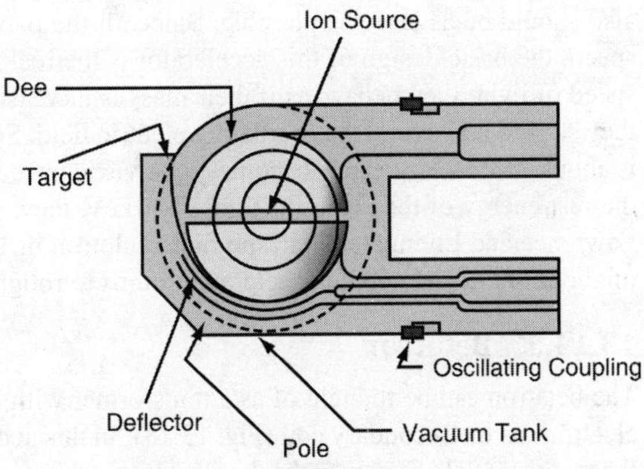

FIGURE 13.17

The cyclotron uses electric and magnetic fields and the whole accelerator remains in a uniform magnetic field. In this accelerator, charged particles moving in the field feel a force acting at 90° to their direction of motion. Hence, they move in circles. Here the Lorentz force due to the magnetic field provides the necessary centripetal force for the circular motion with radius R. It means for a charged particle of charge q and mass m circulating with velocity v,

$$qvB = \frac{mv^2}{R} \quad \text{or} \quad R = \left(\frac{m}{qB}\right)v$$

This equation says that the particles move in circles with radii proportional to their speeds.

Hence the time period of this motion

$$T = \frac{2\pi R}{v} = \frac{2\pi}{v}\left(\frac{m}{qB}\right)v = \frac{2\pi m}{qB}$$

Inverse of this time period gives the cyclotron frequency as

$$f = \frac{qB}{2\pi m}$$

Clearly, it is not the function of the velocity v. This expression also shows that similar particles of same charge and masses will all orbit at the same rate, regardless of their speed.

The particles' orbits inside two metal semicircles are called Dees. These are connected to opposite terminals of an ac supply which is set to the cyclotron frequency calculated above. The particles are accelerated in bunches by the electric field between the plates. Since the cyclotron frequency is same for all the particles, they all will be accelerated by the field. It is clear from $R = mv/q\text{B}$ that the accelerated particles will move in larger and larger circles until reaching a deflector at the edge of the machine which directs them to the target.

The maximum speed of the particles corresponding to maximum radius of circle would be $v_{max} = \frac{qBR_{max}}{m}$.

Therefore, maximum energy of the particles would be $\frac{1}{2}mv_{max}^2 = \frac{q^2B^2R_{max}^2}{2m}$. It is clear from this expression that for obtaining high energy, the strength of the magnetic field should be large and the radius of machine also should be as large as possible. Since all the particles must orbit at the same frequency, whatever their speed, the basic design of this accelerator is the real limiting factor. It is seen that as particles approach the speed of light, they behave as if their mass is increasing. Therefore, their frequency becomes less and hence they start to lag behind the oscillating electric field. So far, maximum energy gain achieved using a cyclotron is about 20 MeV. A particle becomes relativistic once its kinetic energy is comparable to its rest energy. Since the rest energy of the electron is only 500 keV, they cannot be accelerated to a useful energy in a cyclotron. Lawrence and Livingston built the first cyclotron in 1932, which was about 0.3 m across in a magnetic field of about 0.5 T. They could accelerate protons to roughly 1.2 MeV.

13.14.3 Betatron

The betatron can be thought of as a transformer with a ring of electrons as the secondary coil (Fig. 13.18). In this accelerator, the magnetic field used to make the electrons move in a circle is also the one used to accelerate them, as it is a rapidly alternating field. However, the magnet must be carefully designed so that the field strength at the orbit radius (i.e., B_{orbit}) is equal to half the average field strength $\bar{B}$ linking the orbit, i.e., $B_{\text{orbit}} = \frac{\bar{B}}{2}$.

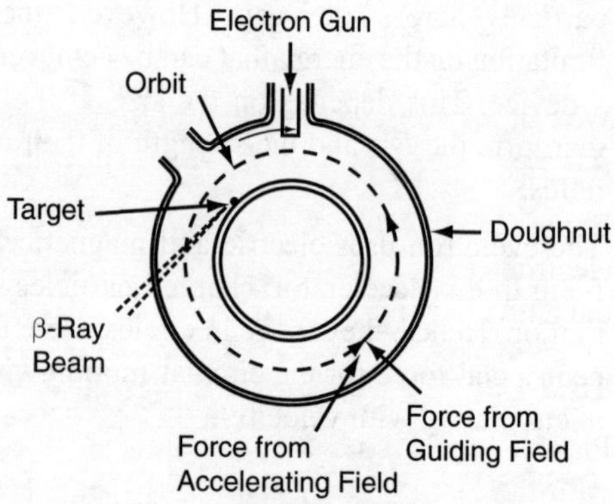

FIGURE 13.18

We can change the flux by increasing the magnetic field. Since the flux links the loop of electrons, an induced emf accelerates the electrons. As the electrons get faster they need a larger magnetic field to keep moving at a constant radius. This is provided by the increasing field. The field is changed by passing an alternating current through the primary coils. The particle acceleration takes place on the first quarter of the voltage sine wave's cycle. Although the last quarter of the cycle also has a changing field that would accelerate the electrons, it is in the wrong direction for them to move in the correct circle. In order to get effective acceleration, the target is bombarded with pulses of particles at the frequency of the ac supply.

The particles gain maximum energy when the magnetic field is at its strongest value. However, the formula used for the cyclotron will not be appropriate for betatrons because the electron will be relativistic. If the total energy is much greater than the rest energy, then $E = pc$ is a good approximation. As the centripetal force is again provided by the Lorentz force, we have

$$\frac{mv^2}{R} = qvB$$

Here, B is the strength of the magnetic field which is needed to keep the particle in the orbit. From the above relation, we obtain the maximum momentum as $p = qBR$. The maximum possible energy would be $E = pc = RqcB$, where c is the speed of light.

The formula for the electron's momentum can also be derived by using Faraday's law of electromagnetic induction. Since emf is equal to $N\frac{d\phi}{dt}$, we can write

$$\text{emf} = NA\frac{d\bar{B}}{dt}$$

For $N = 1$ and $A = \pi R^2$ together with emf as $\oint \vec{E} \cdot d\vec{l}$, we obtain the electric field E from the above relation as below.

$$\oint \vec{E}\cdot d\vec{l} = \pi R^2 \frac{d\bar{B}}{dt}$$

or
$$E\cdot 2\pi R = \pi R^2 \frac{d\bar{B}}{dt}$$

or
$$E = \frac{R}{2}\frac{d\bar{B}}{dt}$$

The magnitude of force on the electron will be given by $F = qE$. Now as per Newton's second law of motion, we obtain

$$F = \frac{dp}{dt} \quad \text{or} \quad \frac{qR}{2}\frac{d\bar{B}}{dt} = \frac{dp}{dt}$$

We can obtain the momentum of the particle if we integrate the above equation. This gives $p = \frac{qR\bar{B}}{2}$. Since $\bar{B} = 2B_{\text{orbit}}$, the momentum $p = qRB_{\text{orbit}}$.

Being compact in size, the betatrons are used in industry and medicine. Since cyclotrons cannot accelerate electrons to useful energies, they are not as useful as betatrons are. A 315 MeV betatron was built in 1949 in the University of Chicago.

13.14.4 Plasma-based Particle Accelerators

Plasma acceleration is a technique for accelerating charged particles like electrons, positrons, ions, etc., with the help of an electric field. This electric field is associated with an electron plasma wave, which is produced either using electron pulses or by very short laser pulses. If we use laser pulses to excite this wave, the technique is known as laser plasma acceleration. Through these techniques, we can achieve high-performance particle accelerators along with much smaller size than conventional accelerators, for example, RF linear accelerator (RF LINAC). This is because of the coherency, which is attained in such devices. These devices show accelerating gradients of several orders of magnitude higher than the one of current particle accelerators. For example, in an experimental laser plasma accelerator at Lawrence Berkeley National Laboratory, the electrons can be accelerated to 1 GeV over about a 3.3 cm distance. However, the SLAC conventional accelerator requires 64 m to reach the same energy. In another technique, called plasma wake field acceleration (at SLAC) an energy gain of 42 GeV was achieved over 85 cm. It is believed that the plasma acceleration technology could replace many of the traditional RF accelerators currently found in hospitals and research facilities.

Since the plasmas are already in ionised state, unlike the conventional LINACs, there is no problem of the electric field breakdown. The plasma can sustain the wave-breaking field $E_0 = cm_e\omega_{pe}/e$, where $\omega_{pe} = \sqrt{4\pi n_0 e^2/m_e}$ is the electron plasma frequency (in CGS system of units) and n_0 is the electron density. For example, when $n_0 = 10^{18}/\text{cm}^3$, the maximum attainable field $E_0 \approx 100$ GV/m. There are several variants of plasma-based accelerators. Among those most investigated particle accelerators are plasma wake field accelerators (PWFA), plasma beat wave accelerator (PBWA), laser wake field accelerator (LWFA) and self-modulated laser wake field accelerator (SMLWFA).

13.14.4.1 Plasma Wake Field Accelerator (PWFA)

In order to understand the concept of wake field acceleration, it is necessary to understand what wake field is. When a speed-boat travels in the water, it produces two types of waves, namely, bow waves and wake field waves. The bow waves are conical waves having a tip at the front end of the boat. These are produced as the velocity of the boat exceeds that of the water waves. The wake field waves are the waves set up at the back (or wake) of the boat. These waves travel with the phase velocity equal to the velocity of the boat.

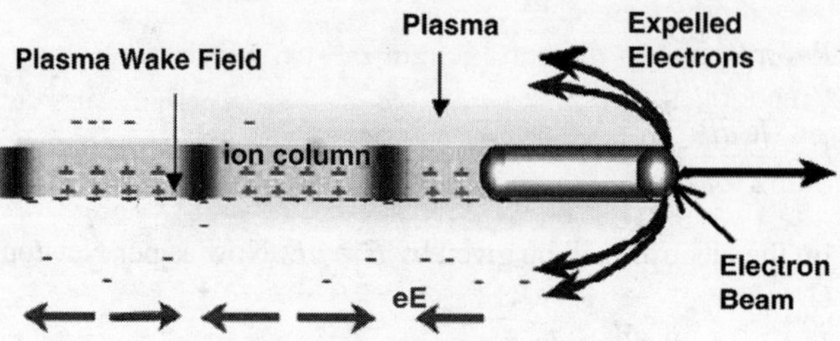

FIGURE 13.19

In a plasma wake field accelerator, the electron plasma wave is driven by one or more electron beams. Effectively, the plasma wake fields can be excited by a relativistic electron beam (Fig. 13.19). This can be achieved if the electron beam terminates in a time shorter than the plasma period ω_{pe}^{-1}. In such a scheme, the ratio of energy gain to the drive beam energy (called the transformer ratio R_t) is limited to $R_t \leq 2$ for a symmetric driving beam in the linear regime. However, it can be increased by using an asymmetric drive beam. The idea of enhancing the wake field amplitude was also introduced by researchers, where they proposed to use multiple electron drive bunches spaced at the plasma period.

13.14.4.2 Laser Beat Wave Accelerator (LBWA)

In the LBWA method, the plasma wave is exited by beating two optimal waves of significantly different frequencies. Two long pulse laser beams having same direction of polarisation, and frequencies ω_1 and ω_2 are used to resonantly excite a plasma wave (Fig. 13.20). When these pulses travel in a plasma of uniform density n_0(corresponding plasma frequency ω_{pe}), they will beat at $\Delta\omega = \omega_1 - \omega_2$ frequency. Here, the excitation of plasma wave is done by appropriately adjusting the laser frequencies and plasma density such that the resonance condition $\omega_1 - \omega_2 = \omega_{pe}$ is satisfied. Since the beat wave moves with laser pulse and plasma wave also moves with a phase velocity equal to the group velocity of a laser wave, a properly placed bunch of electrons with a velocity slightly lesser than the laser group velocity will get accelerated by the mechanism of transfer of energy from wave to particle. However, in such mechanism, there is a problem of phase detuning between the accelerated electrons and the plasma wave. This problem can be overcome if we make the use of a transverse magnetic field.

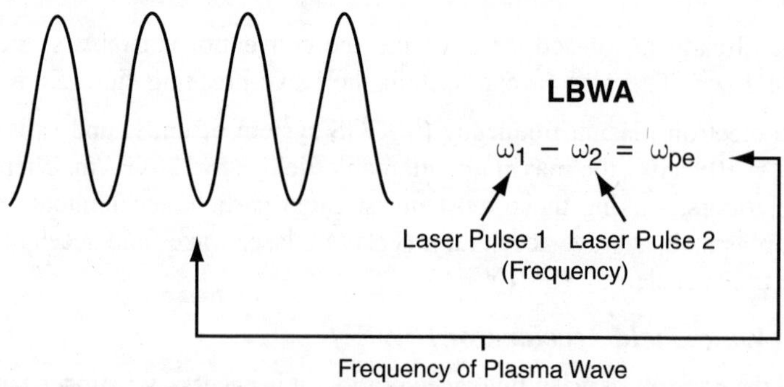

FIGURE 13.20

13.14.4.3 Laser Wake Field Accelerator (LWFA)

In this technique, we use a short (time duration $\tau \le 10^{-12}$ sec: 1 ps) laser pulse of very high intensity ($\ge 10^{18}$ W/cm^2). As the laser pulse being an electromagnetic radiation propagates with the speed of light *c*, its time duration τ will correspond to the pulse length $L = c\tau$, which is nothing but the distance traveled by the pulse in time τ. As mentioned earlier, the wake is the electron plasma wave excited behind the pulse and this wave propagates with the group velocity of the laser pulse. The group velocity of the pulse is almost equal to the speed of light *c*. So, the wavelength λ_p of this wake wave would be c/f_{pe}, where f_{pe} is the frequency of the wave, equal to the electron plasma frequency. Under this situation, if the laser pulse length L matches the plasma wavelength c/f_{pe}, then high amplitude wake field will be produced due to resonance (Fig. 13.21). Therefore, a correctly placed traveling bunch of electrons can be accelerated by the longitudinal field of the plasma wave or the wake field, which is the field corresponding to the plasma wave.

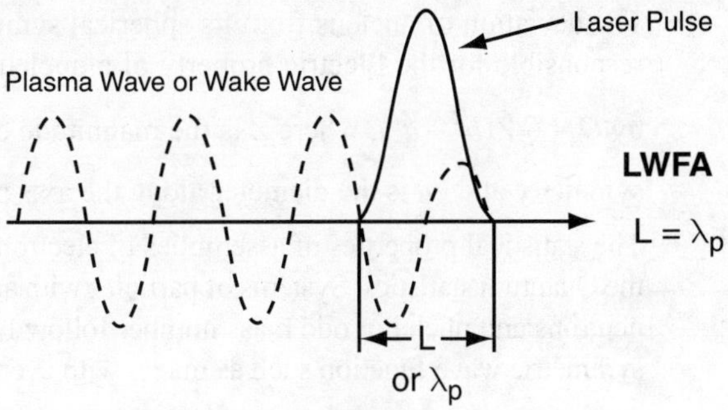

FIGURE 13.21

13.14.4.4 Self-Phase Modulation LWFA

In this scheme, the electron plasma wave is excited resonantly by the modulation of the laser pulse envelope. The self-modulated LWFA uses a single short ($\le$ 1 ps) ultrahigh intensity ($\ge 10^{18}$ W/cm^2) laser pulse, as in the case of LWFA. The self-modulated LWFA, however, operates at higher plasma densities than the standard LWFA. As we know the plasma frequency f_{pe} gets increased for the higher density and hence the plasma wavelength λ_p becomes smaller. So the condition $L > \lambda_p$, which says that the laser pulse length is long compared to the plasma wavelength, is satisfied easily if the plasma density is higher. Moreover, the laser power P is taken somewhat larger than the critical power P_c. Under this situation, in this high density regime, the laser pulse undergoes a self modulation instability which causes the pulse to become axially modulated at the plasma period (Fig. 13.22). Hence, the plasma wave is generated resonantly by the modulated pulse.

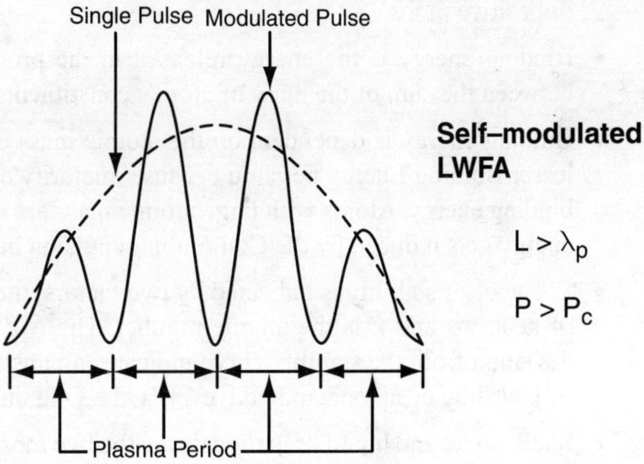

FIGURE 13.22

SUMMARY

The topics covered in this chapter are summarised below.

- The radius of the nucleus of an atom depends on the number of nucleons, *A*. $R = r_0 A^{1/3}$, where $1.2 \times 10^{-3} \le r_0 \le 1.48 \times 10^{-13}$ cm.

- The total angular momentum of the nucleons is given by $I = L \pm S$; where I is a vector with magnitude equal to maximum possible component of angular momentum in any direction, S is the total spin angular momentum of all nucleons and L is the total orbital angular momentum.
- The nuclear magnetic moment μI is measured in terms of nuclear magneton μn, where value of $\mu_N = 5.05 \times 10^{-27}$ J/$\omega b/m^2$. $-3\mu_N \leq \mu_I \leq +10\ \mu_N$. $\mu_I = 0$ for even-even- nuclei.
- The deviation of nucleus from its spherical symmetry results in electric quadrupole moment which is responsible for the Electric property of a nucleus. The quadrupole moment Q of the nucleus is given by $Q = \frac{2}{5}Z(b^2 - a^2)$, where Z is the magnitude of nuclear charge; $2b$ is the diameter along the axis of symmetry and $2a$ is the diameter along the axis perpendicular to it.
- The statistical properties of assemblies of electrons, protons, neutrons, photons and atomic nuclei follow the Quantum statistics. Systems of particles with anti-symmetric wave function such as electrons, protons, neutrons and nuclei of odd mass number follow Fermi-Dirac statistics whereas systems of particles with symmetric wave function such as nuclei with even mass number obey the Bose-Einstein statistics.
- The parity of a nucleus in a given state is related to the orbital angular momentum L. It is even for even value of L and odd for odd value of L.
- Nuclear forces are of short range and are attractive in nature.
- Nuclear forces are charge independent. Extraordinarily stable nuclei are charge symmetric.
- The existence of mesons was predicted by Japanese physicist Yukawa. The mesons are massive particles being constantly exchanged between two nucleons. Messons (π) can be neutral pions, negative pions or positive pions.
- Binding energy is the energy released in the process of formation of the atom. It is the difference between the sum of the mass of atom's constituents and the mass of the atom.
- Binding energy is dependent on the atomic mass of atom. Atoms with every small atomic mass have lesser Binding Energy per amu because majority of nucleons in this case are on surface, having lesser binding energy. Atoms with large atomic mass also have less binding energy per amu since the binding energy gets reduced by the Coulombic repulsion between protons.
- The nuclear stability is indicated by two factors; the N/Z ratio and the odd–even effect. N is the number of neutrons and Z is the atomic number. The $N/Z = 1$ curve is called the stability curve and greater deviation from the stability curve indicates an unstable nucleus. Even–even nuclei are the most stable, followed by even–odd and odd–even. Odd–Odd nuclei are the least stable.
- Shell model and liquid drop models are the two most important models of nuclear structure. The nuclear shell model considers the fact that there is a periodicity in the nuclear properties in terms of atomic mass. There is a periodic increase in BE/A in case of nuclei with either Z or N equal to 2, 8, 20, 50, 82 and 126, called the magic numbers. The stability corresponding to the magic numbers is explained by the formation of closed shells of protons or neutrons. The proton and neutron shells appear to be independent of each other.
- In accordance with the shell theory of nuclear structure, the potential $V(r)$ of nucleons is given by $-V_o\left(1 - \frac{r^2}{k^2}\right)$; where V_o is the maximum potential between nucleons; r is the distance between the nucleon and the centre of force and k is a constant.

- The shell model of nuclear structure has enabled the prediction of total angular momenta of nuclei, occurrence of isomerism and zero quadrupole moment at proton number 2, 8, 20, 50, and 82.
- According to Liquid Drop Model, the nucleus is considered analogous to a drop of incompressible, high density liquid. Taking into account the volume energy, surface energy, symmetry effect, Coulomb energy and the odd–even effect of the nucleus; the Binding energy for a nucleus is given by—

$$BE = 14A - 13.1A^{2/3} - \frac{0.584Z(Z-1)}{A} - \frac{19.4(A-2Z)^2}{A} + E_\delta \text{ MeV}$$

$E_\delta = \frac{135}{A}$ for even–even nuclei, $-\frac{135}{A}$ for odd–odd nuclei and O for even–odd and odd–even nuclei.

- The discovery of neutron in 1932 solved a known puzzle related to the spin of the nitrogen-14 nucleus, which was experimentally measured as 1 basic unit of angular momentum, but at that time, physicists could not find any way to arrange 21 particles of $^{14}_{7}N$ so as to give a spin of 1. However, the presence of neutron as an uncharged particle in the nucleus with spin ½ solved this problem. Another importance of neutron is in nuclear fission as a slow neutron initiates the fusion and the resulting reactions produce on an average 2.4 neutrons. This way, a chain reaction is generated that allows a self-sustaining mode of operation.
- Radioactivity is the disintegration of certain natural heavy elements, which is accompanied by the emission of α-rays (positively charged helium nuclei), β-rays (fast electrons) and γ-rays (short X-rays). The ultimate end product of the radioactive disintegration process is an isotope of lead. The radioactivity is of two types, namely, artificial radioactivity and induced radioactivity.
- If λ be the disintegration constant or decay constant, N_0 be the initial number of nuclides at time $t = 0$ and N be undecayed nuclei at time t, then the decay takes place as per the relation $N = N_0 e^{-\lambda t}$.
- The activity A is the number of disintegrations per second of a sample. It is given by $A = \left|\frac{dN}{dt}\right| = \lambda N_0 e^{-\lambda t} = \lambda N$.
- The half-life time, $T_{1/2}$, of any sample is defined as the time interval in which the number of undecayed atoms decreases by half. It is given by $T_{1/2} = \frac{0.693}{\lambda}$.
- The mean life time τ of a nuclide is the reciprocal of its decay probability per unit time. The mean life time and half-life time $T_{1/2}$ are related to each other as $\tau = \frac{T_{1/2}}{0.693} = 1.44\, T_{1/2}$.
- If a nucleus contains 210 or more nucleons, i.e., when nuclei are so large that the short-range nuclear forces that hold them together are barely able to counterbalance the mutual repulsion of their protons, the nuclei decay by the process of alpha ($^{4}_{2}He$) decay in order to increase their stability. The α decay takes place as per $^{A}_{Z}X \rightarrow {}^{A-4}_{Z-2}Y + {}^{4}_{2}He$ relation, if a parent nucleus X disintegrates into daughter nucleus Y.
- Beta decay is a radioactive decay in which a beta particle that may be either an electron or a positron is emitted. Actually, three mechanisms are involved in beta decay. These are β^- decay (electron or β^- emission, or negatron emission), β^+ decay (or positron emission) or electron capture, in which a nucleus decays by capturing an extra nuclear atomic electron and the electron is disappeared because its mass is converted into energy.

- By emitting alpha, beta or other particles, the nucleus disintegrates and is usually left in the excited state. If the excited nucleus does not have sufficient energy to emit another particle, like an excited atom, it returns to its ground state by emitting photons. These emitted photons from nuclei are called gamma rays.
- The detection of nuclear radiations depends upon their interaction with matter and especially on the excitation and ionisation processes. A nuclear-radiation detector is a device in which presence of radiation induces physical change that is observable. These detectors include ionisation chamber, Geiger–Mueller counter, scintillation counter, Wilson's cloud chamber, bubble chamber and semiconductor detector.
- Neutron absorption cross-section is the cross-section for a nuclear reaction which is initiated by neutrons. The same way we can define neutron scattering cross-section. Neutron cross-section is the measure of probability of scattering and absorption of neutron when it approaches a nucleus. This is expressed in barns (1 barn = 10^{-28} m^2).
- If a target material is bombarded by fast-moving particles such as protons, neutrons, electrons, deuterons or alpha particles, then the target nuclei after the bombardment are usually different from what they were before. The target nuclei may change their mass number or atomic number or both due to its interaction with an incident particle. This is called a transmutation and the reaction is called nuclear reaction.

  The nuclear reactions are of two types: exoergic reaction and endoergic reaction. If the disintegration energy or the Q-value is positive, then the nuclear reaction is called exoergic reaction. If Q-value is a negative quantity, then the nuclear reaction is an endoergic reaction. The minimum kinetic energy of the bombarding particle which is necessary to initiate the endoergic reaction is called threshold energy.
- The phenomenon of breaking of a heavy nucleus into two or more light nuclei of almost equal masses together with the release of huge amount of energy is known as nuclear fission. The released energy in this process is called nuclear energy. In order to achieve a self-sustained chain reaction, the size of the sample must be greater than a critical value, which is called the critical size of the nucleus.
- Nuclear fission is achieved in a nuclear reactor, which produces a self-sustained and controlled chain reaction in a fissionable material. The important parts of the reactor are fuel, moderator, control rods, shield, coolant and safety device.
- Nuclear fusion is the formation of a heavier nuclide by the fusing of two light nuclides. The first artificial fusion reaction was the hydrogen bomb which was tested in November 1952. Fusion reactions are thermonuclear reactions which occur at extremely high temperatures. For example, in order to fuse deuterium (^2_1H) and tritium (^3_1H), the force of repulsion (called Coulomb potential barrier) of these two positively charged particles must be overcome.
- Nuclear fusion is the main energy source that powers the sun and stars. All over the world, efforts have been made to achieve this fusion in controlled manner for the utilisation of energy. This is called controlled fusion. International Thermonuclear Experimental Reactor (ITER) is a new attempt that will use the concept of a tokamak, which is a doughnut-shaped vessel in which a strong, helical magnetic field guides the charged particles. The idea behind controlled fusion is to use magnetic fields to confine a high temperature plasma of deuterium and tritium.
- Plasma is a collection of charged and neutral particles. The charged particles are nothing but ions and electrons. These ions and electrons are almost in equal numbers in the plasma and the plasma as a

whole is neutral. Since we cannot neglect the internal forces at the same time, the plasma is however quasineutral. Moreover, if we attempt to perturb a part of the plasma, the whole body of the plasma will get perturbed due to the connection of all the species with each other. This property is known as collective behaviour of the plasma. Therefore, an ionised gas can qualify for plasma state, if it is quasineutral and it shows collective behaviour.

- A hot plasma at thermonuclear temperature loses a considerable amount of its energy in the form of radiation. In order to produce more nuclear fusion energy than is lost from the radiation, there is the requirement of a minimum temperature for a nuclear reactor to be self-sustained. This temperature is called ignition temperature at which alpha particle heating can sustain the fusion reaction.
- The overall conditions that must be met for a yield of more energy from the nuclear reactor than is required for heating of the plasma are generally stated in terms of the product of ion density (n_0) and confinement time (τ). This condition is called Lawson criterion. For deuterium–tritium (DT) fusion, it is $n_0\tau \geq 10^{14}$ sec/cm^3. However, for deuterium–deuterium (DD) fusion, the Lawson criterion reads $n_0\tau \geq 10^{16}$ sec/cm^3.
- In inertial confinement fusion (ICF), heavy isotopes of hydrogen, called fuel, are heated to temperatures of around 10 keV. Laser or particle beams are focused onto the surface of a capsule containing a small quantity of fuel. Due to evaporation and ionisation of the outer layer of the material, a plasma crown is formed, which expands and generates an inward moving compression front which heats up the inner layers of the material. So the core of the fuel is compressed to as much as one thousand times its liquid density. Then ignition takes place when the core temperature reaches about one hundred million degrees. Thermonuclear combustion spreads rapidly through the compressed fuel and produces energy equivalent to several times the amount deposited on the capsule by the laser or particle beams.
- Laser fusion works on the concept of inertial confinement. If technically feasible, it would eliminate the problems of magnetic instabilities. There are two ways to achieve laser fusion, namely, laser-gas-fusion and laser-pellet-fusion.
- The basic problem in achieving controlled fusion is to generate plasma at very high temperatures and hold its species together long enough for a substantial number of fusion reactions to occur. Since the plasma is a mixture of charged particles, it can be controlled and influenced by external magnetic fields. Charged particles gyrate around the magnetic field lines and also move along these field lines; so by applying the magnetic field with a proper configuration, we can confine the plasma. This is called magnetic confinement.
- Highly accelerated particles contribute to achieve the controlled fusion, as the laser or particle beams are focused onto the surface of a pellet, containing a small quantity of fuel, in order to evaporate and ionise the outer layer of the material. Therefore, the acceleration of particles is a topic of interest. A particle accelerator is a device where we use electric fields to propel electrically charged particles to high speeds and to contain them in well-defined beams. An ordinary cathode ray tube (CRT) television set is a simple form of accelerator. Various types of accelerators, namely, linear accelerator, cyclotron, betatron, and plasma-based accelerators were discussed. Plasma-based accelerators include plasma wake field accelerator, laser beat wave accelerator, laser wake field accelerator and self-phase modulation laser wake field accelerator.

SOLVED EXAMPLES

EXAMPLE 1 In an absorption experiment with 1.14 MeV γ-radiation from ^{65}Zn it is found that 20 cm of aluminium reduces the beam intensity to 3%. Calculate the half-value thickness and mass absorption coefficient of Al for this radiation. Density of aluminium = 2700 kg/m^3.

SOLUTION Given $I/I_0 = 3\% = \frac{3}{100}$, $x = 20$ cm = 0.20 m and $\rho = 2700\,\text{kg/m}^3$

The formula used is

$$I = I_0\, e^{-\mu x} \tag{i}$$

where μ is linear absorption coefficient and x is thickness of aluminium.

$$\frac{I}{I_0} = e^{-\mu x} \quad \text{or} \quad \ln\frac{I}{I_0} = -\mu x$$

or

$$\mu x = \ln\frac{I_0}{I} \tag{ii}$$

$$x = 0.2\text{ m} = \frac{2}{10}\text{m} = \frac{1}{5}\text{m}$$

Putting the value of x in Eq. (ii), we get

$$\frac{\mu}{5} = \ln\left(\frac{100}{3}\right) = \ln 100 - \ln 3$$

$$= 4.60517 - 1.0986$$

$$\frac{\mu}{5} = 3.5065577$$

$$\mu = 17.5328\,\text{m}^{-1}$$

Mass absorption coefficient $= \frac{\mu}{\rho} = \frac{17.5328}{2700}\,\text{m}^2/\text{kg}$

or $= 0.00649363\,\text{m}^2/\text{kg}$

Half-value thickness $(x_{1/2}) = \frac{0.693}{\mu}$

or $x_{1/2} = \frac{0.693}{17.5328} = \mathbf{0.0395\ m}$

EXAMPLE 2 In an absorption experiment with 1.1 MeV γ-radiation from ^{65}Zn it is found that 25 cm of Al reduces the beam intensity to 2%. Calculate the half-thickness, and the mass attenuation coefficient of Al for this radiation. Density of Al = 2700 kg/m^3.

SOLUTION Given $I/I_0 = 2\% = \frac{2}{100} = \frac{1}{50}$, $x = 0.25\text{ m} = \frac{1}{4}\text{m}$

Formula used is $I = I_0 e^{-\mu x}$

or

$$\mu x = \ln\left(\frac{I_0}{I}\right)$$

$$\frac{\mu}{4} = \ln(I_0/I) \quad \text{or} \quad \mu = 4\ln(50)$$

$$\mu = 4 \ln 50 = 4 \times 3.912 \text{ m}^{-1}$$
$$= 15.648 \text{ m}^{-1}$$

Mass attenuation (or absorption) coefficient

$$= \frac{\mu}{\rho} = \frac{15.648}{2700}$$
$$= 5.8 \times 10^{-3} \text{ m}^2/\text{kg}$$

and half-value thickness $(x_{1/2}) = \frac{0.693}{\mu} = \frac{0.693}{15.648} = \mathbf{0.0443\ m}$

EXAMPLE 3 Half-life of $^{23}_{11}$Na is 15 hours. How long does it take for 93.75 % of a sample of this isotope to decay?

SOLUTION Given Half-life of $^{23}_{11}$Na, i.e., $T_{1/2}$ = 15 hours.

The radioactive constant $\lambda = \frac{0.693}{T_{1/2}}$

or
$$\lambda = \frac{0.693}{15} \text{ hr}^{-1}$$
$$= 0.0462 \text{ hr}^{-1}$$

Now, $\frac{N}{N_0} = e^{-\lambda t}$

Here N_0 is the number of atoms that existed in beginning and N is the number of atoms left behind after time t. Then

$$\frac{N}{N_0} = \frac{6.25}{100} = e^{-\lambda t}$$

or
$$\frac{1}{16} = e^{-\lambda t} \quad \text{or} \quad \ln\left(\frac{1}{16}\right) = -\lambda t$$

or
$$t = \frac{\ln(16)}{\lambda} = \frac{2.7726}{0.04621}$$
$$= \mathbf{60\ hrs}$$

EXAMPLE 4 Half-life of a radioactive element is 4 years. After what time will the element present in specimen reduce to 1/64 of its original mass?

SOLUTION Given $T_{1/2}$ = 4 yrs and N/N_0 = 1/64

Decay constant $(\lambda) = \frac{0.693}{T_{1/2}} = \frac{0.693}{4} = 0.17327$ per yr

and
$$\frac{N}{N_0} = \frac{1}{64} = e^{-\lambda t} \quad \text{or} \quad \lambda t = \ln 64$$
$$t = \frac{\ln 64}{0.17327} = 24.002326$$
$$= \mathbf{24\ yrs}$$

EXAMPLE 5 The half-life of a radioactive substance is 15 years. Calculate the period in which 2.5% of the initial quantity will be left over.

SOLUTION Given $T_{1/2} = 15$ yrs and $N/N_0 = \frac{2.5}{100} = \frac{1}{40}$

$$\text{Decay constant } \lambda = \frac{0.693}{T_{1/2}} = \frac{0.693}{15} = 0.0462 \text{ per year}$$

and
$$\frac{N}{N_0} = \frac{1}{40} = e^{-\lambda t} \quad \text{or} \quad \lambda t = \ln(40)$$

$$t = \frac{\ln 40}{\lambda} = \frac{3.689}{0.0462}$$
$$= \mathbf{79.85\ yrs}$$

EXAMPLE 6 How long does it take for 60% of a sample of radon to decay? $T_{1/2}$ for radon is 3.8 days.

SOLUTION Given $T_{1/2} = 3.8$ days. If 60% radon decays, it means 40% of it is left behind.

$$\text{Decay constant } (\lambda) = \frac{0.693}{3.8}$$
$$= 0.18237$$
$$= 0.1824\,\text{d}^{-1}$$

Now $N = N_0 e^{-\lambda t}$

or
$$40 = 100 \times e^{-\lambda t} \quad \text{or} \quad \lambda t = \ln\frac{100}{40} = \ln 2.5$$

or
$$t = \frac{\ln 2.5}{\lambda} = \frac{0.9163}{0.1824}$$
$$t = \mathbf{5.024\ d}$$

EXAMPLE 7 Calculate the half-life time and mean life time of the radioactive substance whose decay constant is 4.28×10^{-4} per year.

SOLUTION Given $\lambda = 4.28 \times 10^{-4}$ per year.

$$\text{Half-life time } T_{1/2} = \frac{0.693}{\lambda}$$

or
$$T_{1/2} = \frac{0.693}{4.28 \times 10^{-4}}$$
$$= 1619.16 \text{ yrs}$$

$$\text{Now mean life time } \tau = \frac{1}{\lambda}$$

or
$$\tau = \frac{1}{4.28 \times 10^{-4}} = \mathbf{2336.45\ yrs}$$

EXAMPLE 8 Find the half-life of a radioactive material if its activity drops to 1/64 of initial activity in 30 years.

SOLUTION Given $t = 30$ years and $A = \frac{1}{64} A_0$.

Activity $A = A_0 e^{-\lambda t}$

$$\frac{A_0}{64} = A_0 e^{-\lambda t} \quad \text{or} \quad \frac{1}{64} = e^{-\lambda t}$$

or $$\lambda t = \ln 64$$

$$\lambda = \frac{\ln 64}{t} = \frac{4.1589}{30}$$

$$\lambda = 0.1386 \text{ yrs}^{-1}$$

Half-life time $$T_{1/2} = \frac{0.693}{\lambda} = \frac{0.693}{0.1386}$$
$$= 4.999 \text{ yrs}$$
$$= \mathbf{5.0\ yrs}$$

EXAMPLE 9 What is the decay constant of a nucleus whose half-life is 2.1 min?

SOLUTION Given $T_{1/2} = 2.1$ min.

Decay constant

$$\lambda = \frac{0.693}{T_{1/2}} = \frac{0.693}{2.1} \text{ per min}$$
$$= \mathbf{0.33\ min^{-1}}$$

EXAMPLE 10 Calculate the decay constant for ^{198}Au whose half-life is 2.7 days. If at some time, a sample contains 10^{-6}gm of ^{198}Au, what would be its activity? Calculate decays occurring per second after 8 days.

SOLUTION Given $T_{1/2} = 2.7$ d, $m = 10^{-6}$ g and $t = 8$ d.

Decay constant

$$\lambda = \frac{0.693}{T_{1/2}} = \frac{0.693}{2.7 \times 24 \times 60 \times 60}$$
$$= 2.971 \times 10^{-6} \text{ sec}^{-1} \qquad \text{(i)}$$

The number of nuclei in sample N can be calculated as

$$N = \frac{\text{mass} \times \text{Avogadro number}}{\text{molar mass}}$$
$$= \frac{(10^{-6}\ g)(6.02 \times 10^{23} \text{ atoms/mol})}{(198 \text{ g/mol})}$$
$$= 3.04 \times 10^{15} \text{ atoms}$$

But putting the value of λ and N in $A_0 = \lambda N$ we get the following for activity at $t = 0$

$$A_0 = 2.971 \times 10^{-6} \times 3.04 \times 10^{15}$$
$$= 9.032 \times 10^{9} \text{ number of atoms disintegrated per second}$$

Now Activity $A = A_0 e^{-\lambda t}$

$$A = (9.032 \times 10^{9}) \times e^{-2.054}$$
$$= 9.032 \times 10^{9} \times 0.12822$$
$$A = \mathbf{1.1581 \times 10^{9}\ decays/sec}$$

Example 11 What fraction of a sample is left after 3 half-lives?

Solution Let N_0 be the original number of nuclei and $N_0/2$ be the number of nuclei remains after half-lives.

$$\therefore \quad \text{The fraction after half-lives } = \frac{N_0/2}{N_0} = \frac{1}{2}$$

$$\text{So fraction after 3 half-lives } = \frac{N_0/8}{N_0} = \frac{1}{8}$$

$$= \mathbf{0.125}$$

Example 12 Ten milligrams of a radioactive substance of life period 2 years is kept for four years. How much of the substance remained unchanged?

Solution The substance remains unchanged after 4 years would be

$$N = \frac{N_0}{4} = \frac{10 \text{ mg}}{4} = \mathbf{2.5\ mg}$$

Example 13 One gram of radium is reduced by 2.1 mg in five years by α-decay. Calculate decay constant, half-lives of sample, and average life.

Solution Given $N_0 = 1.0$ g and $N = 1 - 0.0021$ g $= 0.9979$ g.

$$\text{Now } \frac{N}{N_0} = e^{-\lambda t} \quad \text{or} \quad \lambda t = \ln\frac{N_0}{N}$$

$$\text{or decay constant } \lambda = \frac{\ln\left(\frac{1.0}{0.9979}\right)}{t} = \frac{\ln(1.0021)}{5} = \frac{2.1022 \times 10^{-3}}{5}$$

$$= 3.996 \times 10^{-4} \text{ per year}$$

$$= 0.0004 \text{ per year}$$

$$\text{Half-life time } T_{1/2} = \frac{0.693}{0.0004} = 1732.5 \text{ yrs}$$

$$\text{Average life } \tau = \frac{1}{\lambda} = \frac{1}{4 \times 10^{-4}}$$

$$= \mathbf{2500\ yrs}$$

Example 14 The activity of certain radio nuclide decreases to 15% of its original value in 10 days. Find its half-life.

Solution Given $N_0 = 100$, $N = 15$ and $t = 10$ days

$$\text{Now } N = N_0 e^{-\lambda t} \quad \text{or} \quad \frac{N}{N_0} = e^{-\lambda t}$$

$$\lambda t = \ln\left(\frac{N_0}{N}\right) = \ln\left(\frac{100}{15}\right)$$

$$\lambda = \frac{1.897}{10} = 0.1897\, d^{-1}$$

$$\text{Half-life } T_{1/2} = \frac{0.693}{\lambda} = \frac{0.693}{0.1897}$$

$$= \mathbf{3.65\ d}$$

EXAMPLE 15 What fraction of a radioactive isotope remains after 50 years, if its half-life is 13.3 years?

SOLUTION Given $t = 50$ years and $T_{1/2} = 12.3$ years.

Fraction of a radioactive isotope

$$\frac{N}{N_0} = e^{-\lambda t} \tag{i}$$

where decay constant

$$\lambda = \frac{0.693}{T_{1/2}} = \frac{0.693}{12.3 \text{ years}}$$

$$= 0.05634 \text{ yr}^{-1}$$

and $$\lambda t = 2.817.$$

Now putting the value of λt in Eq. (i), we get

$$\frac{N}{N_0} = e^{-2.817}$$

$$= 0.0598 = \mathbf{0.06}$$

EXAMPLE 16 Calculate the mass of ^{214}Pb having radioactivity of 1 Curie. Half-life of ^{214}Pb is equal to 26.8 minutes.

SOLUTION Given $T_{1/2} = 26.8$ min $= 26.8 \times 60$ sec, and 1.0 Curie $= 3.7 \times 10^{10}$ disintegrations/sec.

Let m gram mass of ^{214}Pb has an activity of 1 Curie, then the number of atoms in m gram of ^{214}Pb.

$$N = \frac{6.023 \times 10^{23} \times m}{214}$$

Disintegration constant $$\lambda = \frac{0.693}{T_{1/2}} = \frac{0.693}{1608 \text{ sec}}$$

$$= 4.31 \times 10^{-4} \text{ sec}^{-1}$$

and activity

$$A = N\lambda$$

So $$A = \frac{6.023 \times 10^{23} \times m \times 4.31 \times 10^{-4}}{214}$$

$$3.7 \times 10^{10} = \frac{6.023 \times 10^{23} \times m \times 4.31 \times 10^{-4}}{214}$$

or $$m = \mathbf{3.05 \times 10^{-8}\ g}$$

EXAMPLE 17 Calculate the weight in grams of ^{214}Pb from the half-life of 26.8 minutes when its activity is 10^6.

SOLUTION As done in Ex. 16, $N = \dfrac{6.025 \times 10^{23} \times m}{214}$,

decay constant $\lambda = \dfrac{0.693}{26.8 \times 60 \text{ sec}} = 4.31 \times 10^{-4} \text{ sec}^{-1}$

and activity $A = N\lambda = \dfrac{6.023 \times 10^{23} \times m \times 4.31 \times 10^{-4}}{214}$

or $$10^6 = \frac{6.023\times10^{23}\times4.31\times10^{-4}\times m}{214}$$

or $$m = \mathbf{8.24\times10^{-13}\ g}$$

EXAMPLE 18 One gram of ^{226}Ra has an activity of one Curie. Calculate the mean life and half-life of radium.

SOLUTION Given activity $A = 1$ Curie $= 3.7\times10^{10}$ disintegrations per second,

and $$N = \frac{6.023\times10^{23}\times1}{226}$$

Activity $A = N\lambda$

or $$\lambda = \frac{A}{N} = \frac{3.7\times10^{10}\times226}{6.023\times10^{23}}$$

$$\lambda = 1.38\times10^{-11}\ \text{sec}^{-1}$$

$$\text{Mean life } = \frac{1}{\lambda} = \frac{1}{1.38\times10^{-11}}$$

$$= 7.25\times10^{10}\ \text{sec}$$

and half-life $$T_{1/2} = \frac{0.693}{\lambda} = \frac{0.693}{1.38\times10^{-11}}$$

$$= \mathbf{5.02\times10^{10}\ sec}$$

EXAMPLE 19 Calculate the activity of 0.1 mg sample of ^{90}Sr at $t = 9$ sec when half-life period of ^{90}Sr is 28 years.

SOLUTION Given $T_{1/2} = 28$ yrs $= 28\times365\times24\times60\times60$ sec
$= 8.83\times10^8$ sec

Decay constant

$$\lambda = \frac{0.693}{T_{1/2}} = \frac{0.693}{8.83\times10^8}$$

$$= 7.85\times10^{-10}\ \text{sec}^{-1}$$

Number of disintegration per second $= N_0 - N$

or $$N_0 - N_0e^{-\lambda t} = N_0 - N_0[1 - \lambda t + \ldots]$$

$$= N_0[\lambda t] = N_0\lambda t$$

Now $$N_0 = \frac{6.023\times10^{23}\times10^{-4}}{90} = 6.69\times10^{17} \qquad \left[\because\ N_0 = \frac{\text{Avogadro numbers}\times\text{mass}}{\text{Molar mass}}\right]$$

So number of disintegration per sec $= N_0\lambda t$

$$= 6.69\times10^{17}\times7.85\times10^{-10}\times9$$

$$= \mathbf{5.25\times10^{8}}$$

EXAMPLE 20 The half-life of radium (226) is 1600 years and that of radon (222) is 3.8 days. Calculate the mass of radon that will be in equilibrium with one g of radium.

SOLUTION Half-life of radium $(T_{1/2}) = T_1 = 1600$ yrs and half-life of radon $(T_{1/2}) = T_2 = 3.8$ days.

Let N_1 be the number of atoms in one g of radium, i.e.,

$$N_1 = \frac{N}{226} \qquad \text{(i)}$$

and suppose m is the mass of radon-222 which is in equilibrium with 1.0 g of radium. Then the number of atoms in m gram radon, i.e.,

$$N_2 = \frac{mN}{222} \qquad \text{(ii)}$$

Again consider λ_1 and λ_2 as the radioactive decay constants for radium and radon, respectively, and T_1 and T_2 are corresponding half-life periods. In equilibrium

$$N_1\lambda_1 = N_2\lambda_2 \quad \text{or} \quad N_1\frac{0.693}{T_1} = N_2\frac{0.693}{T_2} \qquad \text{(iii)}$$

By using Eqs. (i) and (ii) in Eq. (iii), we get

$$\frac{N}{226}\frac{1}{T_1} = \frac{Nm}{222}\frac{1}{T_2}$$

or

$$\frac{1}{226}\frac{1}{365 \times 1600} = \frac{m}{222}\frac{1}{3.8}$$

or

$$m = \mathbf{6.39 \times 10^{-6}\ g}$$

EXAMPLE 21 How much energy would a γ-ray photon have if it is to split an α-particle into a tritium ${}^3_1H$ and proton ${}^1_1H$? Given masses of ${}^4_2He$, ${}^3_1H$ and ${}^1_1H$ as 4.002603 a.m.u., 3.016056 a.m.u. and 1.007276 a.m.u. respectively.

SOLUTION Given $m_\alpha = 4.002603$ a.m.u., $m_t = 3.016056$ a.m.u. and $m_p = 1.007276$ a.m.u.

According to the problem, the reaction may be

$${}^4_2He + \gamma = {}^3_1H + {}^1_1H \qquad \text{(i)}$$

By putting the values of masses of various constituents in Eq. (i), we get

$$4.002603 + \gamma = 3.016056 + 1.007276$$

The mass of γ-ray photon = 4.023332 – 4.002603

= 0.020729 a.m.u.

Equivalent energy of γ-ray photon is = 0.020729 × 931 MeV

= **19.298 MeV**

EXAMPLE 22 A tritium gas target (${}^3_1H$) is bombarded with a beam of protons (${}^1_1H$) of kinetic energy 3MeV. Determine Q value of the following reaction and specify the type of reaction.

$${}^1_1H + {}^3_1H \rightarrow {}^3_2He + {}^1_0n + Q$$

Given $m({}^1_1H) = 1.007276$ a.m.u.; $m({}^3_1H) = 3.016056$ a.m.u.; $m({}^1_0n) = 1.008665$ a.m.u.; $m({}^3_2He) = 3.016036$ a.m.u.

SOLUTION The given reaction is

$${}^1_1H + {}^3_1H \rightarrow {}^3_2He + {}^1_0n + Q \qquad \text{(i)}$$

By putting the values of masses of various constituents in Eq. (i), we get

$$1.007276 + 3.016056 = 3.016036 + 1.008665 + Q$$

$$
\begin{aligned}
Q &= 4.023332 - 4.024701 \\
&= -0.001369 \text{ a.m.u.} \\
&= -0.001369 \times 931.5 \text{ MeV} \\
&= \mathbf{-1.2745\ MeV}
\end{aligned}
$$

The negative sign in Q value indicates that 1.2745 MeV is required for the reaction to occur. Hence, this reaction is endoergic reaction.

EXAMPLE 23 Assuming that 200 MeV of energy is released per fission. Calculate the energy released in Joules and also the heat produced by complete disintegration of 10 mg of $^{235}_{92}$U.

SOLUTION The energy released per fission of $^{235}_{92}$U atoms

$$
\begin{aligned}
&= 200 \text{ MeV} = 200 \times 10^6 \times 1.6 \times 10^{-19} \text{ J} \\
&= 3.2 \times 10^{-11} \text{ J}
\end{aligned}
$$

The number of atoms in one gram of $^{235}_{92}$U is given by Avogadro's number, i.e., 6.023×10^{23}.

∴ The energy released by one gram atom of the $^{235}_{92}$U

$$= \frac{3.2 \times 10^{-11} \times 6.03 \times 10^{23}}{235}$$

Hence the energy released by fission of 10×10^{-3} g of $^{235}_{92}$U

$$
\begin{aligned}
&= \frac{3.2 \times 10^{-11} \times 6.03 \times 10^{23} \times 10^{-2}}{235} \\
&= 8.21 \times 10^8 \text{ J}
\end{aligned}
$$

$$\text{Heat produced} = \frac{8.21 \times 10^8}{4.186} \text{ calories} = \mathbf{1.961 \times 10^8\ calories}$$

EXAMPLE 24 Considering the average energy released per fission as 200 MeV, determine the energy released by fission of 1.0 kg of ^{235}U. Given Avogadro number as 6.03×10^{26} per kg atom.

SOLUTION The energy per fission of ^{235}U atom is

$$
\begin{aligned}
&= 200 \text{ MeV} = 200 \times 10^6 \times 1.6 \times 10^{-19} \text{ J} \\
&= 3.2 \times 10^{-11} \text{ J}
\end{aligned}
$$

$$\text{The energy released by the fission of 1.0 kg of } ^{235}\text{U is} = \frac{3.2 \times 10^{-11} \times 6.03 \times 10^{26}}{235}$$

$$= \mathbf{8.21 \times 10^{13}\ J}$$

EXAMPLE 25 On an average, 1 GW electric power is required to enlighten a city. If a nuclear reactor of efficiency 30% is used for the same purpose with ^{235}U as a nuclear fuel, what amount of fuel would be required per day. Consider the energy released per fission of ^{235}U as 3.2×10^{-11} J.

SOLUTION Given energy required per second = 1 GW

$$= 1.0 \times 10^9 \text{ J},$$

Energy released per fission = 200 MeV

$$
\begin{aligned}
&= 200 \times 10^6 \times 1.6 \times 10^{-19} \text{ J} \\
&= 3.2 \times 10^{-11} \text{ J}
\end{aligned}
$$

Efficiency of reactor is 30%, so the actual energy released per fission $= 3.2 \times 10^{-11} \times \frac{30}{100}$

$= 9.5 \times 10^{-12}$ J

So, the number of fission required per second $= \frac{1.0 \times 10^{9}}{9.6 \times 10^{-12}} = 1.04167 \times 10^{20}$

Number of atoms required undergoing fission per day $= 1.04167 \times 10^{20} \times 24 \times 60 \times 60$

$= 9.0 \times 10^{24}$

Number of atoms in one kg of ^{235}U $= \frac{6.03 \times 10^{26}}{235}$

$= 2.5659574 \times 10^{24}$

Amount of fuel required per day for operation of reactor with 30% efficiency (weight of ^{235}U for 9.0×10^{24} atoms)

$$= \frac{9.0 \times 10^{24}}{2.5659574 \times 10^{24}} = 3.51 \text{ kg}$$

$= \mathbf{3.51\ kg}$

Example 26 In a nuclear reactor, the fission of ^{235}U atom yields 200 MeV. If energy of 3.7 kg uranium is consumed in a day, find the power output of the reactor assuming that the reactor is 20% efficient.

Solution The number of atoms in 1.0 kg of ^{235}U

$$= \frac{6.03 \times 10^{26}}{235}$$

$= 2.5659574 \times 10^{24}$

and the total number of atoms in 3.7 kg of ^{235}U

$= 3.7 \times 2.5659574 \times 10^{24}$

$= 9.494 \times 10^{24}$

Energy released per fission = 200 MeV $= 200 \times 10^{6}$ eV

$= 2.0 \times 10^{8} \times 1.6 \times 10^{-19}$ J

$= 3.2 \times 10^{-11}$ J

The efficiency of the reactor is 20%, so the net energy released per fission $= 3.2 \times 10^{-11} \times \frac{20}{100}$

$= 6.4 \times 10^{-12}$ J

The net energy released due to consumption of 3.7 kg of U^{235} per day $= 9.494 \times 10^{24} \times 6.4 \times 10^{-12}$ J/day

$$= \frac{60.7616 \times 10^{12}}{24 \times 60 \times 60} \text{ J/s}$$

$= 7.033 \times 10^{8}$ W

$= \mathbf{0.703\ GW}$

Example 27 In a reaction, the energy is produced by the fusion of the three helium nuclei to form a $^{12}_{6}$C nucleus. How much energy is produced by each reaction? Consider the mass of helium atom, electron and $^{12}_{6}$C as 4.00260 a.m.u., 0.00055 a.m.u. and 12.0000 a.m.u., respectively.

Solution Mass of 3 helium nuclei = 3×4.00260 a.m.u.

$= 12.00780$ a.m.u.

Mass of $^{12}_{6}C$ atom = 12.0000 a.m.u.

Mass defect (Δm) = $12.00780 - 12.0000$

$= 0.00780$ a.m.u.

Energy produced = 0.00780×931 MeV

$=$ **7.2618 MeV**

Example 28 In an industry, the energy is produced using the fusion reaction $^{2}_{1}H + ^{2}_{1}H \rightarrow ^{4}_{2}H$ + energy. If the efficiency of fusion reactor is 33%, calculate how much deuterium will be consumed per day for production of 50 MW energy. Consider mass of $^{2}_{1}H$ and $^{4}_{2}He$ as 2.01478 and 4.00388 a.m.u. respectively.

Solution According to the fusion reaction, the mass difference

$$= 2.01478 + 2.01478 - 4.00388 = 0.02568 \text{ a.m.u.}$$

∴ Equivalent energy (i.e., energy produce per fission)

$$= 0.02568 \times 931 = 23.908 \text{ MeV}$$
$$= 23.91 \text{ MeV}$$

Efficiency of fusion reactor is 33% $= \dfrac{33}{100}$

i.e., $$\frac{\text{Energy output}}{\text{Energy input}} = \frac{33}{100}$$

∴ Energy output $\dfrac{33}{100} \times 23.91 = 7.89 \text{ MeV} = 1.262 \times 10^{-12}$ J

The actual energy output per deuterium atom

$$= \frac{1.262 \times 10^{-12}}{2} = 0.631 \times 10^{-12} \text{ J}$$

The number of deuterium atoms required per sec for production of 50 MW energy

$$= \frac{50 \times 10^{6} \text{ J/sec}}{0.631 \times 10^{-12} \text{ J}} = 7.924 \times 10^{19} \text{ atoms/sec}$$

Mass of one deuterium atom = 2.01478 a.m.u.

$$= \frac{2.01478}{6.03 \times 10^{26}} \text{ kg}$$
$$= 3.3413 \times 10^{-27} \text{ kg}$$

The equivalent mass of deuterium atoms consumed in production of energy per second

$$= 7.924 \times 10^{19} \times 3.3413 \times 10^{-27}$$
$$= 2.65 \times 10^{-7} \text{ kg}$$

The net amount of deuterium consumed per day

$$= 2.65 \times 10^{-7} \times 24 \times 60 \times 60$$
$$= 0.02287 \text{ kg}$$

$=$ **0.023 kg**

EXAMPLE 29 A cyclotron with Dee's of diameter 1.8 m has a magnetic field of 0.8 Tesla. Calculate the energy to which the doubly ionised helium ion He^{++} can be accelerated. Also calculate the number of revolutions the particle makes in attaining this energy. Mass of He^{++} is 6.68×10^{-27} kg.

SOLUTION Given $B = 0.8\ T$ and mass of α-particle $= 6.68 \times 10^{-27}$ kg, charge on α-particle $(q_\alpha) = 2 \times 1.6 \times 10^{-19}$ C and $r = 0.9$ m.

Maximum energy attained $E = \dfrac{B^2 q_\alpha^2 r^2}{2m}$

$$= \frac{(0.8)^2 \times (3.2 \times 10^{-19})^2 \times (0.9)^2}{2 \times 6.68 \times 10^{-27}}$$

$$= 0.39734 \times 10^{-11}\ \text{J}$$

$$= \frac{0.39734 \times 10^{-11}}{1.6 \times 10^{-19}} = 24.83 \times 10^6\ \text{eV}$$

$$= 24.83\ \text{MeV}$$

Frequency can be obtained by using the relation

$$f = \frac{Bq_\alpha}{2\pi m} = \frac{0.8 \times 3.2 \times 10^{-19}}{2 \times 3.14 \times 6.68 \times 10^{-27}}$$

$$= 0.061 \times 10^8$$

$$= 6.1 \times 10^6\ \text{Hz}$$

Hence number of complete revolution performed by helium ion in obtaining the above energy

$$= \frac{f}{2} = \frac{6.1}{2} \times 10^6$$

$$= \mathbf{3.05 \times 10^6\ per\ sec}$$

EXAMPLE 30 A cyclotron has an oscillator frequency of 12×10^6 Hz and Dee radius of 21 inches. What is the value of magnetic induction needed to accelerate deuteron in it?

SOLUTION Given $f = 12 \times 10^6$ Hz, $r = 21$ inch $= 0.53$ m,

$q_d = 1.6 \times 10^{-19}$ C and $m_d = 2m_p = 3.34 \times 10^{-27}$ kg

Formula used is

$$B = \frac{2\pi f m_d}{q_d} = \frac{2 \times 3.14 \times 12 \times 10^6 \times 3.34 \times 10^{-27}}{1.6 \times 10^{-19}}$$

$$= \mathbf{1.573\ T}$$

EXAMPLE 31 Deuteron in a cyclotron describes a circle of radius 0.32 m just before emerging out of the Dee's. The frequency of the applied e.m.f. is 10 MHz. Find the flux density of the magnetic field and the velocity of the deuterons emerging out of the cyclotron. Mass of deuteron is 3.32×10^{-27} kg and charge 1.6×10^{-19} C.

SOLUTION Given $m_d = 3.32 \times 10^{-27}$ kg, $q_d = 1.6 \times 10^{-19}$ C,

$f = 10 \times 10^6$ Hz and $r = 0.32$ m.

Formula used is

$$B = \frac{2\pi m_d f}{q_d} = \frac{2 \times 3.14 \times 3.32 \times 10^{-27} \times 10 \times 10^6}{1.6 \times 10^{-19}}$$

$$= 1.303\ \text{T}$$

Velocity of deuteron is

$$\frac{mv^2}{r} = qvB \quad \text{or} \quad v = \frac{qBr}{m_d}$$

$$v = \frac{1.6\times10^{-19}\times1.303\times0.32}{3.32\times10^{-27}} = 2.009\times10^7 \text{ m/sec}$$

$$= \mathbf{2.01\times10^7\ m/sec}$$

Example 32 A betatron working on an operating frequency of 60 Hz has a stable orbit of 1.6 m diameter. Find the energy gained per turn and also the final energy if the magnetic field at the orbit is 0.5 Tesla.

Solution Given $r = d/2 = 0.8$ m, $f = 60$ Hz and $B = 0.5$ T.

Average energy gained by the electron per turn is $4e\omega r^2B$. This can be proved as follows.

Let us consider that magnetic flux in the betatron is given by the relation

$$\phi = \phi_0 \sin \omega t$$

The increasing magnetic flux is obtained during the quarter cycle for a given direction in which the current in the electromagnet increases from zero to maximum value.

$$\therefore \qquad \text{Time of acceleration} = \frac{T}{4} = \frac{2\pi}{4\omega} = \frac{\pi}{2\omega}$$

Where T is the time period of the changing magnetic flux and w is the corresponding angular frequency.

Energy gained by the electron per turn $= eE$

$$= e\frac{d\phi}{dt} = e\frac{d}{dt}(\phi_0 \sin \omega t)$$

$$= e\phi_0 \frac{d}{dt}(\sin \omega t)$$

As this energy is gained in a time $\frac{T}{4} = \frac{\pi}{2\omega}$

Average value of energy per turn

$$= \frac{e\phi_0}{\pi/2\omega} \int_0^{\pi/2\omega} \frac{d}{dt'}(\sin \omega t)dt'$$

$$= \frac{2e\omega\phi_0}{\pi}$$

To maintain a stable orbit of constant radius, tangential force on the electron must be zero. From this condition, we get

$$\phi_0 = 2\pi r^2 B$$

$$\therefore \qquad \text{Average energy per turn} = \frac{2e\omega}{\pi} \times 2\pi r^2 B$$

$$= 4e\omega r^2 B \text{ Joule}$$

$$= \frac{4\times1.6\times10^{-19}\times2\times3.14\times60\times(0.8)^2\times0.5}{1.6\times10^{-19}} \text{ eV}$$

$$= 482.3 \text{ eV}$$

The total (or final) energy

= number of revolution done × average energy per revolution

$$= \frac{c}{4\omega r} \times 4e\omega r^2 B$$

$$= cerB$$

$$= 3 \times 10^8 \times 1.6 \times 10^{-19} \times 0.8 \times 0.5 \text{ J}$$

$$= 1.92 \times 10^{-11} \text{ J}$$

$$= \mathbf{120 \ MeV}$$

EXAMPLE 33 In a 70 MeV betatron synchrotron, the radius of the stable electron orbit is 28 cm. Find the value of magnetic field B at the orbit for the given energy.

SOLUTION Given E = 70×10^6 eV = $70 \times 1.6 \times 10^{-13}$ J and $r = 0.28$ m.

Formula used is $B = \dfrac{E}{cer} = \dfrac{70 \times 1.6 \times 10^{-13}}{3 \times 10^8 \times 1.6 \times 10^{-19} \times 0.28}$

$$= \mathbf{0.83 \ T.}$$

EXAMPLE 34 A sample of uranium emitting α-particles of energy 4.18 MeV is placed near an ionisation chamber. Assuming that 12 particles per second enter the chamber, calculate the current produced, if an ion pair requires energy of 40 eV. Charge on the electron $e = 1.6 \times 10^{-19}$ C.

SOLUTION Given $E_\alpha = 4.18$ MeV = 4.18×10^6 eV.

Energy required to produce an ion pair = 40 eV

Number of α-particles entering the chamber per second = 12

$\therefore$ Energy required = $12 \times 4.18 \times 10^6$ eV

= 50.16×10^6 eV

Therefore, number of ion pair produced per second

$$(n) = \frac{\text{Total energy supplied to the system}}{\text{Energy required to produce one ion pair}}$$

$$n = \frac{50.16 \times 10^6 \text{ eV}}{40 \text{ eV}}$$

$$n = 1254 \times 10^3$$

Current (i) = time rate of collection of charge = $1254 \times 10^3 \times 1.6 \times 10^{-19}$ A

$$= \mathbf{2.0 \times 10^{-13} \ A}$$

EXAMPLE 35 A GM counter collects 10^8 electrons per discharge when the counting rate is 500 counts per minutes. What will be the average current in the circuit?

SOLUTION Number of discharge per second $\dfrac{500}{60} = 8.333$

The average current in the circuit (i)

$$= 8.333 \times 10^8 \times 1.6 \times 10^{-19}$$

$$= \mathbf{1.33 \times 10^{-10} \ A}$$

EXAMPLE 36 Neglecting parallel component of velocity, calculate cyclotron frequency and Larmour radius for a 10 keV electron in the earth's magnetic field of 5×10^{-5} Tesla.

SOLUTION Given $B = 5 \times 10^{-5}$ T and $E = 10$ keV = 10^4 eV

$$\text{Cyclotron frequency } (\omega_c) = \frac{qB}{m}$$

$$= \frac{1.6 \times 10^{-19} \times 5 \times 10^{-5}}{9.1 \times 10^{-31}}$$

$$= 0.879 \times 10^7 \text{ sec}^{-1}$$

$$E = \frac{1}{2} m v_\perp^2 \quad \text{or} \quad v_\perp = \left[\frac{2E}{m}\right]^{1/2} = \left[\frac{2 \times 10^4 \times 1.6 \times 10^{-19}}{9.1 \times 10^{-31}}\right]^{1/2}$$

$$= 5.93 \times 10^7 \text{ m/s}$$

$$\text{and Larmour radius } (r_L) = \frac{v_\perp}{\omega_c} = \frac{5.93 \times 10^7}{0.879 \times 10^7}$$

$$= \mathbf{6.746\ m}$$

EXAMPLE 37 What would be the cyclotron frequency of a solar wind proton streaming under the effect of magnetic field B = 5 × 10^{-9} Tesla. If the proton streams with velocity 3 × 10^5 m/s, what would be the Larmour radius. Neglect the parallel component of velocity.

SOLUTION Given $B = 5 \times 10^9$ T and $v_\perp = 3 \times 10^5$ m/sec

$$\text{Larmour radius}(r_L) = \frac{m v_\perp}{eB} = \frac{1.67 \times 10^{-27} \times 3 \times 10^5}{1.6 \times 10^{-19} \times 5 \times 10^{-9}}$$

$$= \mathbf{6.26 \times 10^5\ m}$$

EXAMPLE 38 A He$^+$ ion of energy 1 keV is gyrating in a circle of Larmour radius of 0.188 m under the effect of external magnetic field. Calculate the magnetic field B by neglecting the parallel component of velocity.

SOLUTION Given $E = 103$ eV and $r_L = 0.188$ m.

$$\text{Energy } E = \frac{1}{2} m v_\perp^2 \quad \text{or} \quad v_\perp = \left[\frac{2E}{m}\right]^{1/2}$$

$$v_\perp = \left[\frac{2 \times 10^3 \times 1.6 \times 10^{-19}}{4 \times 1.67 \times 10^{-27}}\right]^{1/2}$$

$$= 2.19 \times 10^5 \text{ m/sec}$$

and

$$B = \frac{m v_\perp}{e r_L} = \frac{4 \times 1.67 \times 10^{-27} \times 2.19 \times 10^5}{1.6 \times 10^{-19} \times 0.183}$$

$$= \mathbf{4.996 \times 10^{-2}\ T}$$

EXAMPLE 39 Calculate the Larmour radius for a 3.5 MeV He^{++} ash particle in an 8 T DT fusion reactor by neglecting the parallel component of velocity.

SOLUTION Given E = 3.5 × 10^6 eV.

$$\text{Energy } E = \frac{1}{2} m v_\perp^2 \quad \text{or} \quad v_\perp = \left[\frac{2E}{m}\right]^{1/2}$$

$$= \left[\frac{2 \times 3.5 \times 10^6 \times 1.6 \times 10^{-19}}{4 \times 1.67 \times 10^{-27}}\right]^{1/2}$$

$$V_\perp = 1.29 \times 10^7 \text{ m/s}$$

and Larmour radius

$$r_L = \frac{mv_\perp}{qB} = \frac{4\times 1.67\times 10^{-27}\times 1.29\times 10^{7}}{2\times 1.6\times 10^{-19}\times 8}$$

$$= \mathbf{3.36\times 10^{-2}\ m}$$

EXAMPLE 40 Calculate the Debye length (λ_{De}) and plasma frequency f_{Pe} for the plasma of earth's ionosphere having electron density of 10^{12} per m^3 and thermal energy KT_e as 0.1 eV.

SOLUTION Given $n = 10^{12}$ per m^3 and $KT_e = 0.1$ eV.

$$\text{Debye length } \lambda_{De} = \left[\frac{\varepsilon_0 KT_e}{ne^2}\right]^{1/2} = \left[\frac{8.85\times 10^{-12}\times 0.1\times 1.6\times 10^{-19}}{10^{12}\times 1.6\times 10^{-19}\times 1.6\times 10^{-19}}\right]^{1/2}$$

$$= 2.35\times 10^{-3}\ \text{m}$$

$$\omega_{Pe} = \left[\frac{ne^2}{m\varepsilon_0}\right]^{1/2} = \left[\frac{10^{12}\times(1.6\times 10^{-19})^2}{9.1\times 10^{-31}\times 8.85\times 10^{-12}}\right]^{1/2}$$

$$\omega_{Pe} = 0.5638\times 10^{8}\ \text{rad/sec}$$

So the plasma frequency $(f_{Pe}) = \dfrac{\omega_{Pe}}{2\pi}$

or $$(f_{Pe}) = \frac{0.5638\times 10^{8}}{2\times 3.14}$$

$$= 0.08977707\times 10^{8} = \mathbf{8.98\ MHz}$$

EXAMPLE 41 Calculate the plasma frequency and Debye length for a glow discharge of density 10^{16} per m^3 and thermal energy 2 eV.

SOLUTION Given density $n = 10^{16}$ per m^3 and $KT_e = 2\ \text{eV} = 2\times 1.6\times 10^{-19}$ J

$$\text{Debye length } \lambda_{De} = \left[\frac{\varepsilon_0 KT_e}{ne^2}\right]^{1/2}$$

$$= \left(\frac{8.85\times 10^{-12}\times 2\times 1.6\times 10^{-19}}{10^{16}\times(1.6\times 10^{-19})^2}\right)^{1/2} = [11.0625\times 10^{-9}]^{1/2}$$

$$= 1.0518\times 10^{-4}\ \text{m}$$

Angular plasma frequency

$$(\omega_{Pe}) = \left(\frac{ne^2}{\varepsilon_0 m}\right)^{1/2}$$

$$= \left(\frac{10^{16}\times(1.6\times 10^{-19})^2}{8.85\times 10^{-12}\times 9.1\times 10^{-31}}\right)^{1/2}$$

$$= 5.637\times 10^{9}\ \text{rad/sec}$$

So plasma frequency

$$f_{Pe} = \frac{\omega_{Pe}}{2\pi}$$

$$= \frac{5.637\times 10^{9}}{2\times 3.14}$$

$$= \mathbf{8.977\times 10^{8}\ Hz}$$

OBJECTIVE TYPE QUESTIONS

Q.1 The radius of a nucleus depends on its mass number A and it is proportional to
(a) A (b) $A^{1/3}$ (c) $A^{2/3}$ (d) A^3

Q.2 The magnitude of the spin angular momentum of a nucleon in the nucleus is
(a) $\hbar$ (b) $\hbar/2$ (c) 0 (d) $\hbar^2$

Q.3 The total angular momentum of a nucleus having odd mass number A is
(a) 0 (b) $\hbar$
(c) odd half-integral multiple of $\hbar$ (d) integral multiple of $\hbar$

Q.4 The total angular momentum of a nucleus having even mass number A is
(a) 0 (b) $\hbar$
(c) odd half-integral multiple of $\hbar$ (d) integral multiple of $\hbar$

Q.5 Bohr magnetons μ_B is given by
(a) $\mu_B = \dfrac{e\hbar}{2m_e}$ (b) $\mu_B = \dfrac{e\hbar}{2m_p}$ (c) $\mu_B = \dfrac{eh}{2m_e}$ (d) $\mu_B = \dfrac{eh}{2m_p}$

Q.6 The magnetic moment of a proton is
(a) 0 (b) positive (c) negative (d) undefined

Q.7 Electric quadrupole moment of a nucleus is
(a) always zero (b) a measure of deviation of the nucleus from its spherical symmetry
(c) simply the charge (d) undefined quantity

Q.8 Nuclear forces are
(a) short-range forces (b) long-range forces
(c) always repulsive (d) Coulomb forces between protons

Q.9 Yukawa gave a theory for nuclear forces based on the exchange of
(a) electrons between protons and neutrons (b) mesons between protons and neutrons
(c) electromagnetic photons (d) X-rays between protons and neutrons

Q.10 Binding energy curve is the curve between
(a) BE/A and A (b) BE and A
(c) BE/A and charge (d) BE/A and Coulomb energy

Q.11 Binding energy of light nuclei is small because
(a) more nucleons reside at the nucleus surface
(b) mass of the nucleus is small
(c) charge of the nucleus is small
(d) Coulomb force supercedes nuclear forces

Q.12 Nuclides lying on the stability curve are
(a) more stable (b) more unstable
(c) having large number of neutrons (d) having large number of protons

Q.13 As per nuclear shell model
(a) neutrons and protons move in the same orbit within the nucleus
(b) neutrons start moving in electronic shell lying near the nucleus
(c) protons are converted into neutrons and vice-versa
(d) neutrons and protons move in their separate orbits within the nucleus

Q.14 Nuclear magic numbers represent
(a) total mass number of the nucleus
(b) difference in number of neutrons and protons
(c) the atomic magic numbers
(d) even number of Z or N so that nuclear shells are completely filled.

Q.15 Nuclei with nuclear magic numbers
(a) have always one proton and one neutron
(b) do not have any isotope
(c) have more number of isotopes than their neighbours
(d) have equal number of isotopes as their neighbours

Q.16 As per nuclear shell model, each shell
(a) is limited to a certain maximum number of neutrons or protons
(b) has equal number of neutrons and protons
(c) is half filled
(d) is two-third filled

Q.17 Nuclear shell model gives results which are in good agreement with experiments, when the potential energy function $V(r)$ is
(a) independent of r
(b) rectangular
(c) oscillatory
(d) a combination of square well potential and the oscillator potential

Q.18 Nuclear drop model assumes
(a) drops of protons and neutrons
(b) the shape of nucleus as hollow sphere
(c) protons to remain on the surface of nucleus
(d) nucleus as a liquid drop containing neutrons and protons as its molecules

Q.19 In nuclear drop model
(a) drops of protons are converted into drops of neutrons
(b) surface energy is neglected
(c) some properties of nuclear forces are analogous to the properties of forces that hold a liquid drop together
(d) protons and neutrons move freely

Q.20 A Curie is a standard unit of radioactivity. Its value is
(a) 10^9 disintegrations/sec
(b) 2.7×10^{10} disintegrations/s
(c) 3.7×10^{10} disintegrations/sec
(d) none of these

Q.21 Radioactivity is the phenomenon associated with
(a) electron emission from atom
(b) fission of nuclei
(c) transformation of nuclei
(d) none of these

Q.22 The half-life period and radioactive decay constant are related as
(a) $T_{1/2} = \dfrac{1}{\lambda}$
(b) $T_{1/2} = \dfrac{\ln 2}{\lambda}$
(c) $T_{1/2} = \dfrac{\lambda}{0.6931}$
(d) none of these

Q.23 The half-life of ${}^{24}_{11}\text{Na}$ is 15 hours. How long does it take for 93.75% of a sample of this isotope to decay?
(a) 35 hr
(b) 60 hr
(c) 15 hr
(d) 90 hr

Q.24 The activity of a certain radionuclide decreases to 15% of its original value in 10 days. Its half-life would be
(a) 10 days (b) 5 days (c) 2.65 days (d) 3.65 days

Q.25 β-decay corresponds to
(a) an electron detached from atom's outermost orbit
(b) the emission of proton from the nucleus
(c) electromagnetic wave pulse
(d) an electron emitted by the nucleus

Q.26 The end product of uranium series $A = (U_n + 2)$ is
(a) ^{206}Pb (b) ^{208}Pb (c) ^{207}Pb (d) None of these

Q.27 The method(s) for determining the age of a sample is (are)
(a) cranium dating (b) carbon dating
(c) both (a) and (b) (d) none of these

Q.28 The radiation obtained from radioactive substance are
(a) α-rays (b) β-rays (c) γ-rays (d) all of these

Q.29 Factors on which the range of α-particle depends are
(a) the initial energy of the particle (b) the ionisation potential of the gas
(c) both (a) and (b) (d) none of these

Q.30 Geiger Nuttal rule gives the range of
(a) α-particle (b) β-particle (c) β^+-particle (d) γ-rays

Q.31 Parity is not conserved in
(a) α-decay (b) γ-decay (c) β-decay (d) none of these

Q.32 A long-lived excited nucleus is called
(a) isotone (b) isotope (c) isomer (d) none of these

Q.33 The time during which pulses are recorded but are of smaller duration in a GM Counter is called
(a) recovery time (b) resolving time (c) dead time (d) none of these

Q.34 The time during which pulses are not recorded in a GM Counter is called
(a) dead time (b) recovery time (c) resolving time (d) none of these

Q.35 Which scintillator is used for detection of α-particle in a scintillator counter?
(a) NaI (b) Zinc sulphide (c) Anthracene (d) None of these

Q.36 The radiation detector/detectors based on image formation is/are
(a) bubble chamber (b) Wilson's cloud chamber
(c) nuclear emulsion charged particle detector (d) all of these

Q.37 If σ be the microscopic cross-section and n be the number of nuclei per unit volume, then the microscopic cross-section is the product
(a) σn (b) σndx (c) $nd\sigma dx$ (d) σnx

Q.38 The energy of the fast neutron is of the order of
(a) 1.0 MeV (b) above 1.2 MeV upto 10 MeV
(c) 1.0 eV (d) none of these

Q.39 Which particle cannot be accelerated by cyclotron?
(a) neutron (b) proton (c) deutron (d) α-particle

Q.40 In a nuclear fission reactor
(a) control rods are used to slow down the fast neutrons
(b) moderator is used to slow down the fast neutron
(c) coolant is used to slow down the fast neutrons
(d) none of these

Q.41 Plasma is
(a) state between third and fourth states of matter
(b) every ionised gas
(c) neutral gas with a few electrons
(d) ionised gas with ions and electrons in almost equal numbers

Q.42 In laboratory plasma, the electron temperature is usually
(a) 1 eV to 5 eV (b) 1 keV to 3 keV (c) 0 K (d) 1 MeV to 3 MeV

Q.43 If λ_{De} and L be the Debye length and dimension of plasma, respectively, then the condition for quasineutrality of plasma is
(a) $\lambda_{De} \simeq L$ (b) $\lambda_{De} << L$ (c) $\lambda_{De} >> L$ (d) $\lambda_{De} = L/2$

Q.44 For the purpose of particle acceleration, a linear accelerator uses
(a) a static electric field
(b) static magnetic field
(c) combination of electric and magnetic fields
(d) laser radiation

Q.45 A wave breaking field $\bar{E}$ in the plasma, whose frequency is ω_{Pe}, is almost equal to
(a) $cm_e\ \omega_{Pe}/e$ (b) $cem_e\ \omega_{Pe}$ (c) $c^2m_e\ \omega_{Pe}/e$ (d) $10\ cm_e\ \omega_{Pe}/e$

Q.46 In plasma wake field accelerator, the plasma wave is generated by
(a) electron beam (b) laser pulse
(c) static electric field (d) static magnetic field

Q.47 In a laser wake field accelerator a short laser pulse of pulse length L is used in a plasma of frequency ω_{pe}/e and excites the wake wave of wavelength λ_p under the condition
(a) $L = \lambda_p$ (b) $L = \lambda_{p/2}$ (c) $L = \lambda_{p/3}$ (d) $c/L = \omega_{pe}/2\pi$

Q.48 In a self-phase modulation laser wake filed accelerator, the density of plasma should be such that the laser pulse length L and wake length λ_p satisfy the relation
(a) $L = \lambda_p$ (b) $L = \lambda_{p/2}$ (c) $L > \lambda_p$ (d) $L = \lambda_{p/3}$

Q.49 In a laser beat wave accelerator, where two laser pulses of frequencies ω_1 and ω_2 are used in a plasma of frequency ω_{pe}, the following condition should be satisfied
(a) $\omega_1 + \omega_2 = \omega_{pe}$ (b) $\omega_1 - \omega_2 = \omega_{pe}$ (c) $\omega_1 + \omega_{pe} = \omega_2$ (d) $\omega_1 + 2\omega_2 = \omega_{pe}$

Q.50 The binding energy of a nucleus is equivalent to the
(a) mass of protons (b) mass of nucleus
(c) mass defect of nucleus (d) none of these

Q.51 The fusion process is only possible at high temperature because
(a) nuclei disintegrate
(b) molecule disintegrate
(c) the nuclei get sufficient energy to overcome the Columbian attractive forces
(d) none of these

Q.52 Which of the following is not a particle accelerator?
(a) Betatron (b) Cyclotron
(c) Scintillation counter (d) none of these

Q.53 Which of the following nuclear reaction is known as exothermic when its Q value is
(a) negative (b) zero (c) positive (d) none of these

Q.54 Which of the following nuclear reaction is known as endothermic when its Q value is
(a) negative (b) positive (c) zero (d) none of these

Q.55 Most of the suitable material is used as moderator in a nuclear reactor
(a) cadmium (b) heavy water (c) slow neutrons (d) none of these

SHORT-ANSWER QUESTIONS

Q.1 Define radioactivity.
Q.2 Describe properties of α, β and γ-rays.
Q.3 What is decay constant? How is it related to the decay probability per nucleus per second?
Q.4 What types of radiations are emitted in radioactive disintegration?
Q.5 Compare the properties of α, γ and γ-rays.
Q.6 What is mean life of radioactive isotope?
Q.7 What do you mean by half-life?
Q.8 What is half-life of radium, uranium and polonium?
Q.9 What do you mean by activity of a radioactive substance?
Q.10 What is basic mechanism of detection of radiation? Explain briefly.
Q.11 Which is the most sensitive and accurate method used for the deflection of nuclear radiation?
Q.12 Distinguish between ionisation chamber and GM counter.
Q.13 Briefly explain solid-state detector and radiation detectors.
Q.14 Define neutron cross-section. What are its units?
Q.15 What is fission?
Q.16 What do you understand by liquid-drop model?
Q.17 How can a fission chain reaction be controlled?
Q.18 What is nuclear fusion?
Q.19 What do you understand by ignition temperature?
Q.20 What is Lawson criterion?
Q.21 Differentiate between inertial and magnetic confinements?
Q.22 What are charged particle accelerators?
Q.23 Discuss plasma wake field accelerator in short.
Q.24 Discuss laser accelerator in short.
Q.25 Write short-note on
(a) Radioactivity
(b) Nuclear fission and its applications
(c) Nuclear fusion
(d) Neutron cross-section
(e) Cloud chamber
(f) Bubble chamber

(g) Radiation detectors
(h) GM counter
(i) Scintillation counter
(j) Solid state detectors
(k) Nuclear reactors
(l) Reactor criticality
(m) Lawrence cyclotron
(n) Betatron

Practice Problems

General Questions

Q.1 Discuss basic properties of a nucleus in detail.

Q.2 Write a note on angular momentum of a nucleus.

Q.3 Discuss magnetic and electric properties of a nucleus.

Q.4 What do you understand by parity of a nucleus.

Q.5 Discuss charge independence property of nuclear forces.

Q.6 Discuss the meson theory of nuclear forces.

Q.7 Write down the correlation between binding energy and stability of nuclei.

Q.8 Write down the facts of nuclear shell model.

Q.9 Discuss theory of nuclear shell model.

Q.10 What are the applications of nuclear shell model?

Q.11 Discuss nuclear magic numbers and their significance.

Q.12 Brief out the nuclear liquid drop model.

Q.13 What are various terms that contribute to the calculation of binding energy of nucleus?

Q.14 Discuss the volumes and surfaces energies used in the nuclear drop model.

Q.15 Write down the facts of nuclear liquid drop model and the semiempirical binding energy formula.

Q.16 Discuss in brief the way out for determining various constants appeared in nuclear liquid drop model.

Q.17 What is natural radioactivity? Explain what is radioactive disintegration. State the laws of radioactive decay and deduce them from first principles using probability concepts.

Q.18 What is mean life of a radioactive isotope? Show that the mean life is the time for nuclei to decay to $1/e$ times their original number.

Q.19 Define radioactive constant and half-life period. Prove that the radioactive constant of a substance is the reciprocal of the time after which the number of atoms of the substance falls to $1/e$ of its original value.

Q.20 Define mean life of a radioactive nuclide. Derive a relation between mean life time and radioactive constant.

Q.21 Define half-life and radioactive nuclide. Derive a relation between half-life and radioactive constant.

Q.22 What is the difference between half-life and mean life in radioactivity?

Q.23 What is the cause of radioactivity? Give various types of radioactive decays and discuss the process involved in all these decays?

Q.24 What are α-particles? How will you show experimentally that α-particle is an ionised helium atom?

Q.25 State the conditions for α-decay and explain why in α-decay of a radioactive nuclide the kinetic energy of the emitted α-particle is little less than the disintegration energy?

Q.26 What is β-decay? Is β-spectrum discrete or continuous?

Q.27 Show that the law of conservation energy and momentum are not obeyed in β-decay. Show how neutrino hypothesis explains this discrepancy and accounts for continuous β-ray spectrum.

Q.28 Discuss gas ionisation based radiation detectors, suggest how one can detect thermal neutrons?

Q.29 What are radiation detectors based upon ionisation phenomenon. Explain the phenomenon and discuss in detail any one of the ionisation based detectors. State briefly the applications also.

Q.30 Give three properties of nuclear radiation used in detection instruments. What are gas filled ionisation based nuclear detectors?

Q.31 Discuss the principle and technical details of the working of ionisation method for detecting radiation.

Q.32 Describe the construction, principle and working of an ionisation chamber. Explain the difference between ionisation chamber and GM counter. Why is an ionisation chamber less sensitive to β-particle?

Q.33 Describe the construction, working and principle of GM counter.

Q.34 Discuss in detail the construction and working of GM counter. What is the dead time and recovery time? What do you mean by quenching of a GM counter? What is its necessity? How is it achieved internally?

Q.35 Describe briefly operation and testing of a GM counter. What is plateau as applied to a GM counter? Give its applications. What are its limitations?

Q.36 Discuss the working and principle of a scintillation counter.

Q.37 Explain the principle, construction and operation of a scintillation radiation counter. Draw the block diagram to show the main components. Briefly describe a few types of scintillators used. In what way a scintillation counter is superior to a GM counter?

Q.38 Describe construction and working of a scintillation counter.

Q.39 Describe the construction and working of a cloud chamber.

Q.40 Describe Wilson's cloud chamber. How does a cloud chamber make visible the path of an incoming particle? What is its main advantage.

Q.41 Discuss the working principle of a cloud chamber. How is it different from a bubble chamber? Explain.

Q.42 What is a bubble chamber? Describe the process by which ionising particles make a visible track in a bubble chamber. Give its advantages and drawbacks.

Q.43 Discuss the construction and working of a semiconductor radiation counter. Give a few advantages of semiconductor detector.

Q.44 Define and explain the term nuclear reaction cross-section. What are its units? If a beam of N_0 particle is incident on a slab of thickness x of material, how many particles will emerge out of unit slab. Given that the slab contains n atoms per unit volume and σ is the cross-section of the reaction.

Q.45 What do you understand by nuclear fission? Explain the release of energy during nuclear fission.

Q.46 Give the main assumptions of liquid drop model of the nucleus. Justify the name liquid-drop model.

Q.47 Obtain the expression for the binding energy of a nucleus based on liquid-drop model.

Q.48 How can the fission chain reaction be controlled?

Q.49 Define chain reaction. What is nuclear reactor? State its principle. Explain the function of moderator in uranium fed nuclear reactors.

Q.50 What is nuclear fission? Explain why the fission of nucleus like ^{235}U, ^{233}U with slow neutrons is the only fission reaction that is used for producing energy?

Q.51 Distinguish between fission and fusion. Describe the principle of construction and working of a nuclear reactor.

Q.52 Describe the phenomenon of nuclear fission. Explain nuclear fission on the basis of liquid-drop model.

Q.53 Explaining the use of absorbers and methods of enrichment of ^{235}U. Give the construction, working and applications of nuclear reactor.

Q.54 Explain the term thermonuclear energy or nuclear fusion. Discuss its importance in universe. Where do the sun and other stars get their energy from?

Q.55 Describe a nuclear reactor. How does it work?

Q.56 What are similarities and dissimilarities between nuclear fission and fusion?

Q.57 Explain the terms: neutron cross-section, reactor criticality and shielding.

Q.58 Explain carefully the principle of linear accelerator. Deduce the expression for the energy of the particle and length of cylinders in terms of the constants of the apparatus.

Q.59 (a) What is the difference between linear and circular accelerator?
(b) Which accelerator makes use of electromagnetic radiations for accelerating particle?

Q.60 Describe the principle, construction and working of a cyclotron. Derive expression for the maximum kinetic energy achieved by a particle of mass m in terms of the applied magnetic field and Dee radius. Also state the relation in terms of the frequency of the applied electric field. Discuss its limitations.

Q.61 Can a cyclotron be used to accelerate electrons? If not why?

Q.62 What is a betatron? Derive the betatron condition for successful acceleration of electrons. Briefly describe its principle, construction and function of alternating magnetic field in it.

Q.63 What do you understand by plasma? Explain its quasineutrality and collective behaviour.

Q.64 What is plasma frequency? How does it depend on plasma density? Is it same for both the constituents of the plasma?

Q.65 What is Debye length? Why do you need Debye length to be much smaller than the dimension of the plasma?

Q.66 What are plasma-based particle accelerators? Name any three of them.

Q.67 Discuss plasma wake field accelerator. How is it different from laser wake field accelerator?

Q.68 Explain plasma beat wave accelerator. What are its merits and demerits compared with laser wake field accelerator?

Q.69 What do you understand by self-modulated laser wake field accelerator? Why do you need a dense plasma for the successful operation of this accelerator?

Unsolved Questions

Q.1 The linear absorption coefficient μ of lead for 1 MeV gamma rays is 0.74 cm calculate (a) half-thickness of lead for these γ-rays, and (b) thickness of lead required to reduce the intensity of γ-rays to $\frac{1}{1000}$ of its original value. **[Ans: (a) 0.94 cm (b) 9.32 cm]**

Q.2 One mg of radioactive material with half-life of 1600 years is kept for 2000 years. Calculate the mass which would have decayed by this time. **[Ans: 0.50 mg]**

Q.3 The half-life of a radioactive substance is 2.5 days. Calculate the percentage of original material left after 7.5 days. **[Ans: 12.5%]**

Q.4 One gram radioactive radium-226 decays with a half-life of 1620 years. Calculate decay constant and mean life? **[Ans: 1.36×10^{-11} per sec, 2337.3 yrs]**

Q.5 Calculate the activity of 1 mg radium - 226 which has a half-life of 1620 years. **[Ans: 0.98 milli Curie]**

Q.6 The half-life of ^{238}U against α-decay is 4.5×10^9 yrs. Find the activity of 1.0 kg of ^{238}U. **[Ans: 0.334 milli Curie]**

Q.7 When a nucleus of ^7Li is bombarded with a proton, two α-particles are formed. Calculate the kinetic energy of the α-particle assuming negligible energy of the bombarding proton. **[Ans: 8.67 MeV]**

Q.8 A reactor is producing energy at the rate of 1500 kW. How many atoms of ^{235}U undergo fission per second? How many kg of ^{235}U would be used in 1000 hours of operation assuming that on an average energy of 200 MeV is released per fission? **[Ans: 65.86×10^{-3} kg]**

Q.9 A cyclotron has a magnetic field of 10^4 Gauss and a radius of 80 cm. Calculate the frequency of the alternating electric field that must be applied and to what energy deutrons can be accelerated? Mass of deuteron = 2 a.m.u. **[Ans: 15.4 MeV]**

Q.10 A cyclotron oscillator frequency 1 MHz is used to accelerate protons. If the radius of the Dees be 60 cm, what would be the magnetic field in Tesla? **[Ans: 6.56 T]**

Q.11 A GM counter with dead time of 300 μs records 16000 counts per minute. What is the dead time loss in counting rate. **[Ans: 5×10^{-6} min]**

Crystal Structure

14

Learning Objectives

After reading this chapter you will be able to

LO 1 Understand crystalline, amorphous solids, primitive lattice and Wigner-Seitz primitive cell, and types of crystals

LO 2 Know about translation vectors, lattice planes, and significance and representation of Miller indices

LO 3 Illustrate structures of NaCl, CsCl, and diamond, coordination number of simple cubic lattice, bcc lattice, fcc lattice

LO 4 Learn about interplanar spacing and nearest neighbour distance and atomic radius

LO 5 Discuss packing fraction for sc, bcc, fcc, diamond, hcp, interatomic attractive/repulsive forces

LO 6 Explain ionic bond, covalent bond, metallic bond, van der Waals bond, hydrogen bond, crystal structure analysis i.e., Bragg's law and spectrometer, Laue method, powder method

LO 7 Evaluate vacancies, concentration of Schottky defects and Frenkel defects, compositional and electronic defect

Introduction

A crystal structure is a unique arrangement of atoms. It consists of a set of atoms which are identical in composition, arrangement and orientation, called basis and a lattice. Bases are located upon the points of a lattice, which is an array of points repeating periodically in three dimensions. The points can be thought of as forming identical tiny boxes, called *unit cells*, that fill the space of the lattice. The lengths of the edges of a unit cell and the angles between them are called the *lattice parameters*. A crystal structure and symmetry play an important role in determining many of its properties, like electronic band structure and optical properties.

It is clear that a crystal structure is formed by the addition of a basis of atoms to every lattice point. Mathematically, it can be represented as

$$\text{Crystal structure} = \text{Lattice} + \text{Basis}$$

14.1 TYPES OF SOLIDS LO1

The solids are broadly classified into two groups, namely, crystalline solids and amorphous solids.

14.1.1 Crystalline Solids

Crystalline solids are arranged in fixed geometric patterns or lattices. Ice, methanol and sodium chloride are a few examples of crystalline solids. They have orderly arranged units and are practically incompressible. Crystalline solids also show a definite melting point and so they pass rather sharply from solid to liquid state. There are various crystalline forms which are divided into seven crystal system or shapes. They are cubic, tetragonal, hexagonal, orthorhombic, monoclinic, trigonal and triclinic. The units that constitute these systems can be atoms, molecules or ions. Ionic and atomic crystals are hard and breakable with high melting points.

14.1.2 Amorphous Solids

A rigid material whose structure lacks crystalline periodicity is called an amorphous solid. It means the pattern of its constituent atoms or molecules does not repeat periodically in three dimensions. Even amorphous materials have some short range order at the atomic length scale due to the nature of chemical bonding. They are considered supercooled liquids in which the molecules are arranged in a random manner somewhat as in the liquid state. Glass and plastic are the examples of amorphous solids. Unlike crystalline solids, amorphous solids do not have definite melting points.

14.2 UNIT CELL LO1

The smallest portion of a space lattice which can generate the complete crystal by repeating its own dimensions in various directions is called a *unit cell*. A unit cell is defined by the length of its edges and by the angles between them, as shown in Fig. 14.1.

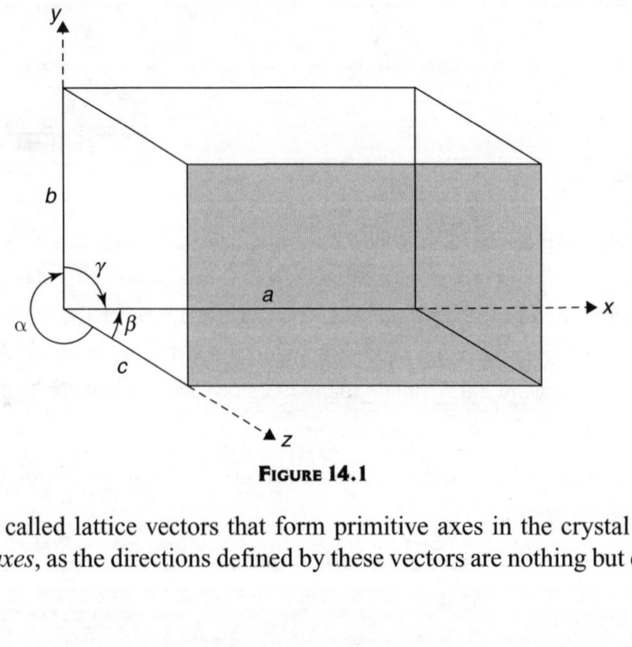

FIGURE 14.1

Vectors $\vec{a}, \vec{b}$ and $\vec{c}$ are called lattice vectors that form primitive axes in the crystal structure. We also call them *crystallographic axes*, as the directions defined by these vectors are nothing but crystal axes. These vectors are used in translation vector and hence are called *fundamental translation vectors*. The magnitudes of vectors $\vec{a}, \vec{b}$ and $\vec{c}$ are called *lattice constants* that specify the distances of the bases along the crystal axes.

14.2.1 Primitive Lattice Cell

The *parallelepiped* defined by primitive axes $\vec{a}, \vec{b}, \vec{c}$ is called a *primitive cell*. A primitive cell is a type of unit cell which fills all the space under the action of suitable crystal translation operation. A primitive cell is a minimum volume unit cell, as shown in Fig. 14.2. There is a density of one lattice point per primitive cell. The volume of a primitive cell is defined by primitive axes $\vec{a}, \vec{b}, \vec{c}$ as $|(\vec{a} \times \vec{b}) \cdot \vec{c}|$.

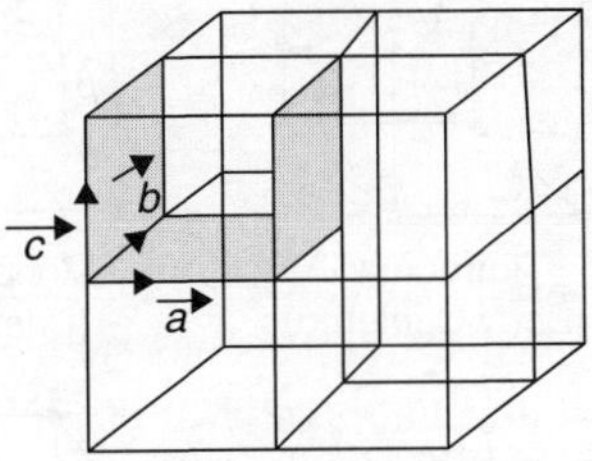

FIGURE 14.2

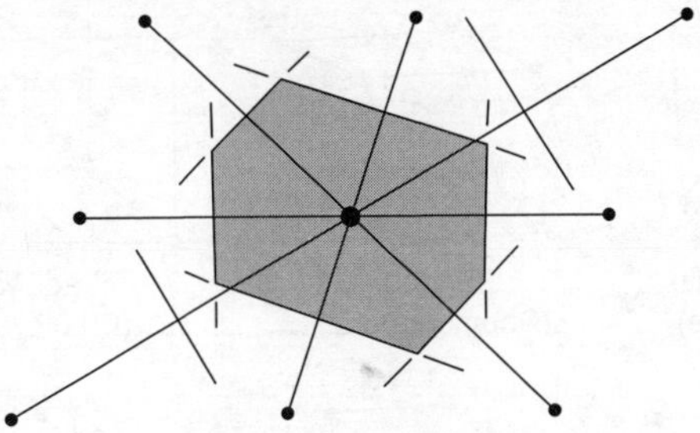

FIGURE 14.3

14.2.2 Wigner–Seitz Primitive cell

The smallest volume enclosed by the normal lines drawn from midpoints of the lines which connect lattice point to all nearby lattice points is called a Wigner–Seitz primitive cell. It is shown in Fig. 14.3.

14.3 TYPES OF CRYSTALS LO1

All crystals are classified into seven crystal systems on the basis of the shape of the unit cells. Bravais in 1948 explained that there are fourteen different types of crystal lattices under the seven crystal systems. These seven types of crystal systems are tabulated in Table 14.1, and shown in Fig. 14.4.

TABLE 14.1

S.No.	*Name of System*	*Relative Between*		*Number of Possible Lattices*	*Examples*
		Primitives	*Angles*		
1.	Cubic	$a = b = c$	$\alpha = \beta = \gamma = 90°$	3(P, F, I)	Nacl
2.	Trigonal	$a = b = c$	$\alpha = \beta = \gamma \neq 90°$	1(R)	$CaSO_4$
3.	Tetragonal	$a = b \neq c$	$\alpha = \beta = \gamma = 90°$	2(P, I)	$NiSO_4$
4.	Hexagonal	$a = b \neq c$	$\alpha = \beta = 90°, \gamma = 120°$	1(P)	Quartz
5.	Orthorhombic	$a \neq b \neq c$	$\alpha = \beta = \gamma = 90°$	4(P, C, F, I)	KNO_3
6.	Monoclinic	$a \neq b \neq c$	$\alpha = \gamma = 90°, \neq \beta$	2(P, C)	$FeSO_4$
7.	Triclinic	$a \neq b \neq c$	$\alpha \neq \beta \neq \gamma \neq 90°$	1(P)	$CuSO_4$

P = Primitive, C = Base centered, I = Body Centered, F = Face centered, R = Rhombohedral

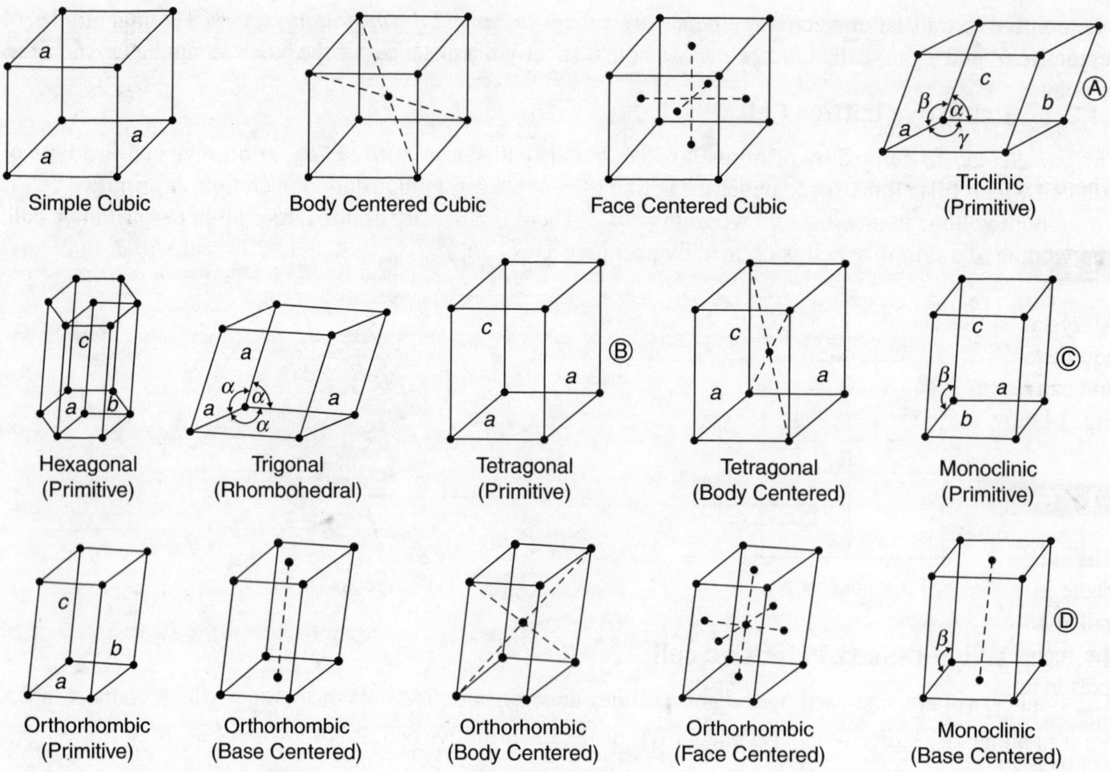

FIGURE 14.4

14.3.1 Cubic System

In a cubic system, there are three types of lattices, namely simple cubic, body centered cubic and face centered cubic. In addition to these structures, other structures are also depicted in Fig. 14.4.

(a) *Simple Cubic:* It contains lattice points at all eight corners of the unit cell. It is represented by sc.

(b) *Body Centered Cubic:* It contains one additional lattice point at the centre of the body of the unit cell including at all eight corners. It is represented by bcc.

(c) *Face Centered Cubic:* It contains lattice points at the centre of each face as well as at all eight corners. It is represented by fcc.

14.4 TRANSLATION VECTORS LO2

We take any lattice point O as an origin in a plane lattice shown in Fig. 14.5. Any other point in the two-dimensional lattice can be obtained by repeatedly translating the vectors $\vec{a}$ and $\vec{b}$. These vectors are known as basis vectors. Based on these basis vectors, we obtain the plane lattice by their repeated translation. The position vector of any other lattice point, i.e., translation vector, can be represented as

$$\vec{T} = n_1\vec{a} + n_2\vec{b}$$

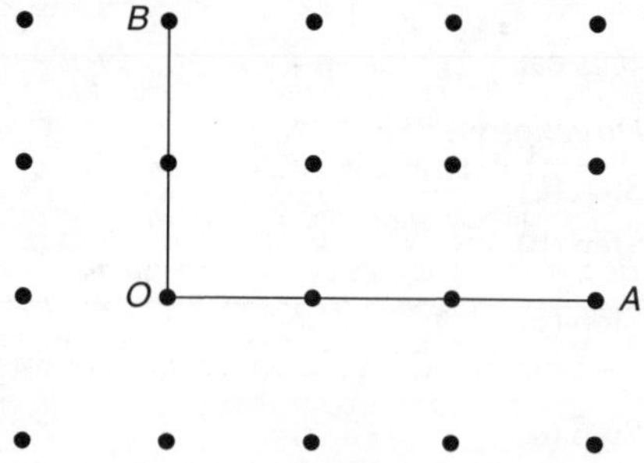

FIGURE 14.5

where n_1 and n_2 are the integers which represent the number of lattice points along the directions *OA* and *OB*, respectively, and $\vec{a}$ and $\vec{b}$ are the primitives. In case of three-dimensional crystal structures, the arrangement of points can be represented as

$$\vec{T} = n_1\vec{a} + n_2\vec{b} + n_3\vec{c}$$

where $\vec{a}, \vec{b}$ and $\vec{c}$ are the primitives along *X, Y* and *Z* axes, respectively.

14.5 LATTICE PLANES LO2

A crystal lattice is made of a large number of parallel equidistant planes (Fig. 14.6) known as lattice planes and can be chosen in a number of ways, as shown in Fig. 14.6a, b, c and d.

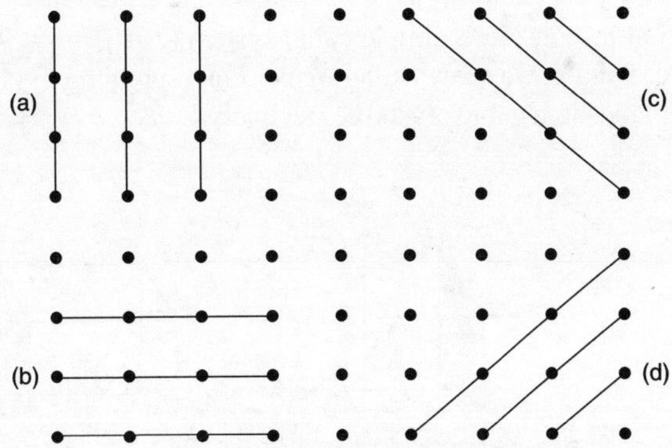

Figure 14.6

14.6 MILLER INDICES LO2

The integers which determine the orientation of a crystal plane in relation to the three crystallographic axes are called *Miller indices*. In order to find the Miller indices, the reciprocals of the intercepts of the plane on the axes in terms of lattice constants are reduced to the smallest integers in ratio. Miller indices are also called crystal indices.

Let us assume that *a, b* and *c* are the magnitudes of fundamental translation vectors along the three axes, respectively. Again we consider that *ABC* represents the plane whose Miller indices are to be obtained. Let *OA, OB* and *OC* are the intercepts made by this plane along the three axes (Fig. 14.7).

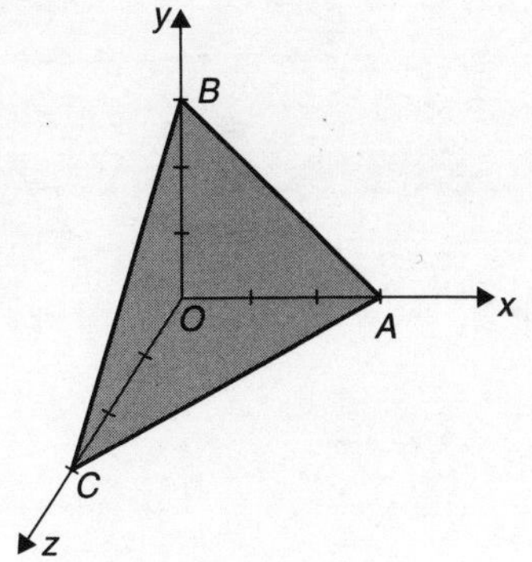

Figure 14.7

In order to find the Miller indices, we do the following.

(a) We find the intercepts along the three axes.
(b) We express these intercepts as multiple of lattice parameters.
(c) We divide these intercepts by lattice parameters.
(d) We take the reciprocal of these.
(e) We clear these fractions by taking the LCM of the denominators and multiplying it. This gives the required Miller indices.

This can be understood in a better way in the following steps corresponding to the above steps a to e.

Directions	x	y	z				
Step (a)	*OA*	*OB*	*OC*				
Step (b)	pa	qb	rc				
Step (c)	$\frac{pa}{a}$	$\frac{qb}{b}$	$\frac{rc}{c}$	or	p	q	r
Step (d)	$\frac{1}{p}$	$\frac{1}{q}$	$\frac{1}{r}$				
Step (e)	$(h$	k	$l)$				

14.6.1 Significance of Miller Indices

Same Miller indices are assigned to all parallel planes, for example, planes with coordinates x, y, z; $2x$, $2y$, $2z$; $-x$, $-y$, $-z$; etc., are represented by the same Miller indices. The corresponding index for a plane, which is parallel to one coordinate axis, is zero. The corresponding index is negative for a plane that cuts an axis on the negative side of the origin. The negative index is represented by putting a bar above the index.

14.6.2 Representation of Miller Indices

Let us consider a simple cubic system ($\alpha = \beta = \gamma = 90°$ and a = b = c). For this system in Fig. 14.8a, b, c, d, e, f and g, we show Miller planes corresponding to $(\bar{1}\ 0\ 0)$, (1 0 0), (0 1 0), (0 0 1), (1 1 0), (1 0 1) and (1 1 1) planes in a cubic crystal respectively.

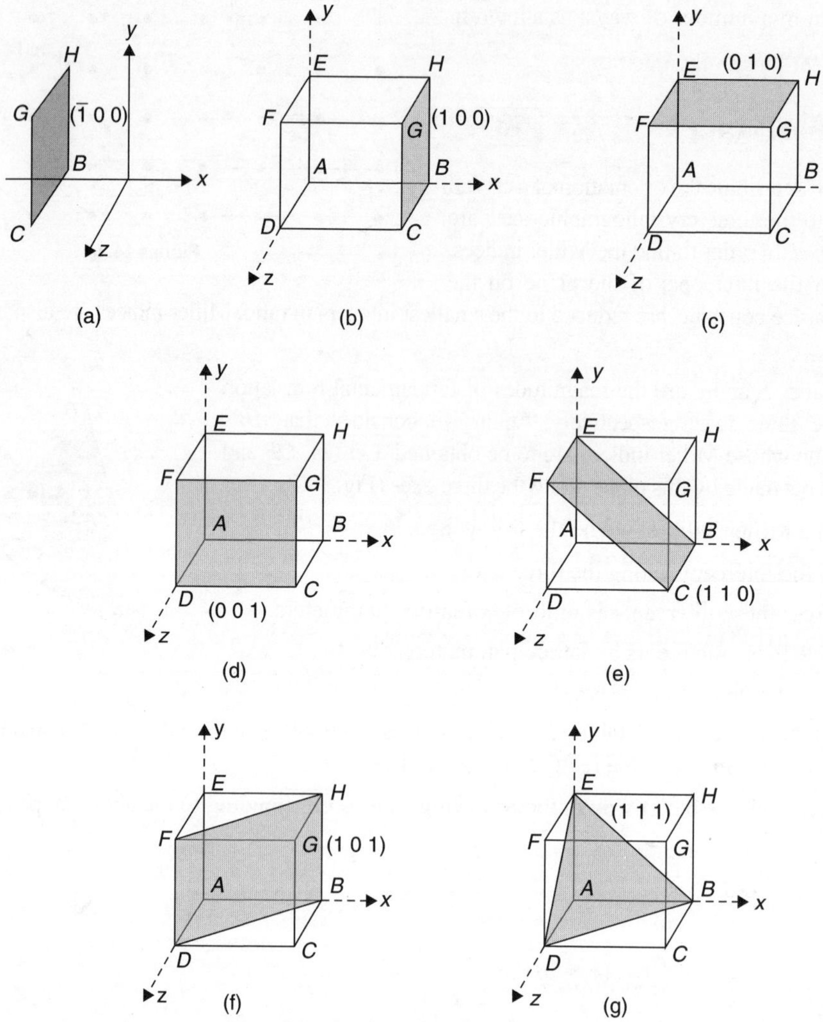

Figure 14.8

14.7 SIMPLE CRYSTAL STRUCTURES

LO3

Below we discuss simple crystal structures, for example, the structures of sodium chloride, cesium chloride and diamond.

14.7.1 Structure of NaCl

The sodium chloride structure is shown in Fig. 14.9. It consists of two face centered cubic sublattices, one of Na ion having its origin at the point (0, 0, 0) and the other of the Cl ion having its origin midway along a edges of the cube say at the point $\left(\frac{1}{2},\frac{1}{2},\frac{1}{2}\right)$. The space lattice is therefore truly fcc, with a basic of one Na ion and one Cl ion separated by one half the body diagonal of a unit cube. There are four $Na^+ - Cl^-$ ion pairs in each unit cube, with different ions in the positions.

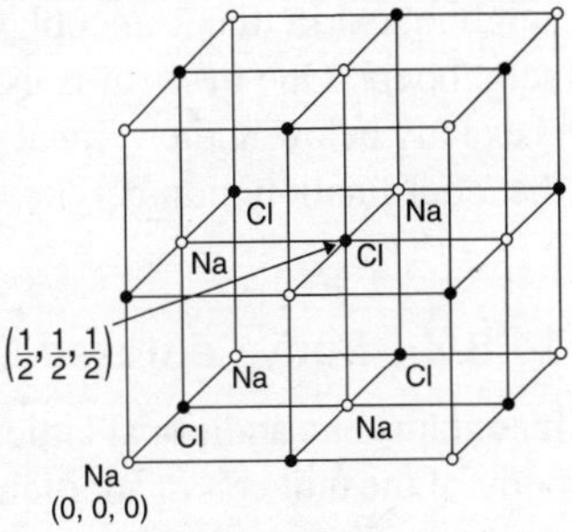

FIGURE 14.9

The coordinates of Na ions are defined as

$$(0,0,0),\left(\frac{1}{2},\frac{1}{2},0\right),\left(\frac{1}{2},0,\frac{1}{2}\right),\left(0,\frac{1}{2},\frac{1}{2}\right)$$

The coordinates of Cl ions

$$\left(\frac{1}{2},\frac{1}{2},\frac{1}{2}\right),\left(0,0,\frac{1}{2}\right),\left(0,\frac{1}{2},0\right),\left(\frac{1}{2},0,0\right)$$

14.7.2 Cesium Chloride (CsCl) Structure

In a CsCl crystal, we have a body centered cubic lattice, in which the Cs atom is placed at the origin (0, 0, 0), i.e., at the body center point and Cl atoms are placed at the corners of the bcc lattice whose coordinates are given by $\left(\frac{1}{2},\frac{1}{2},\frac{1}{2}\right)$. Structure of cesium chloride is shown in Fig. 14.10.

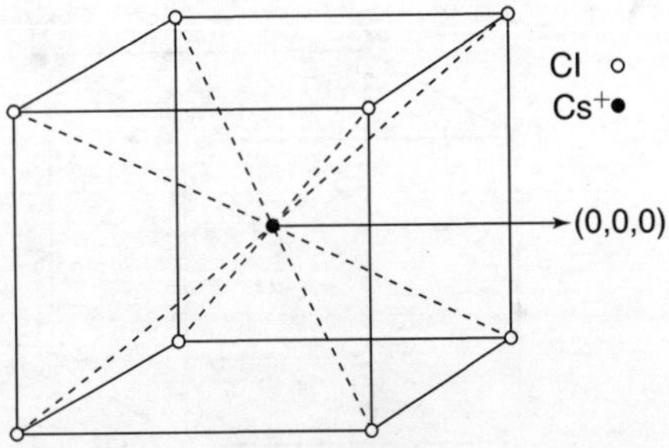

FIGURE 14.10

14.7.3 Diamond Structure

The space lattice of diamond is fcc. In diamond structure we have two fcc lattices placed at (0, 0, 0) and $\left(\frac{1}{4},\frac{1}{4},\frac{1}{4}\right)$ which superimpose each other. In diamond structure we have two carbon atoms placed at (0, 0, 0) and $\left(\frac{1}{4},\frac{1}{4},\frac{1}{4}\right)$. The diamond crystal structure is shown in Fig. 14.11.

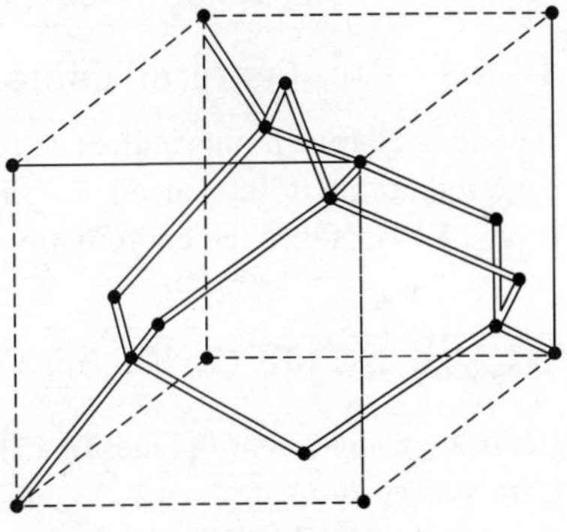

FIGURE 14.11

14.8 COORDINATION NUMBER

LO3

In a crystal, every atom is surrounded by the other atoms. The number of nearest neighbours to the given atom in crystal lattice is known as coordination number.

14.8.1 Simple Cubic Lattice

This simple cubic lattice contains lattice points at all eight corners of the unit cell. Any corner atom has four neighbours. In the same plane as well as other two neighbours. One of them is exactly above and the other is exactly below in the vertical plane (Fig. 14.12). Hence, the coordination number in a simple cubic lattice is 6 (= 4+2).

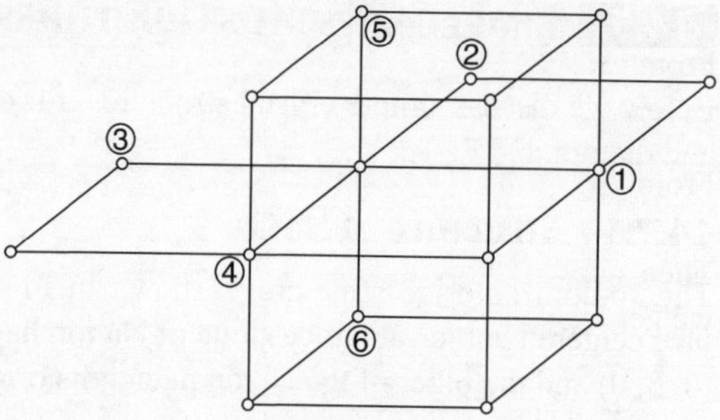

FIGURE 14.12

14.8.2 Body Centered Cubic Lattice

It contains one additional lattice point at the center of the body of the unit cell in addition to those at eight corners. In this lattice, every centered atom is surrounded by eight equidistant neighbours (Fig. 14.13). Hence, coordination number in a body centered cubic lattice is 8.

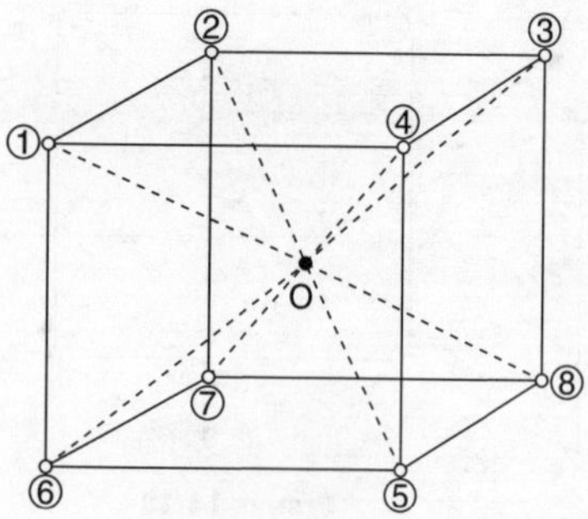

FIGURE 14.13

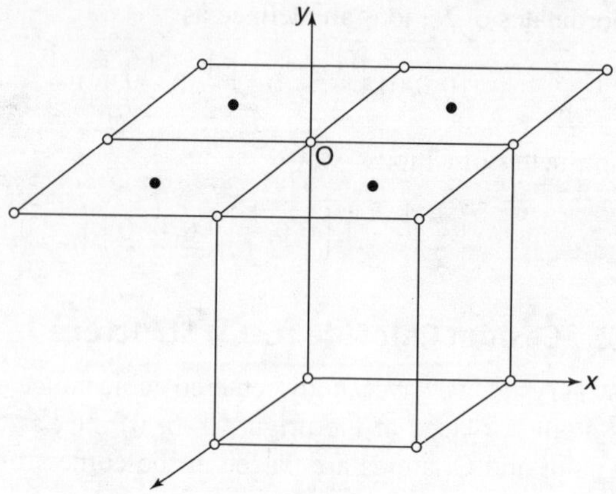

FIGURE 14.14

14.8.3 Face Centered Cubic Lattice

It contains lattice points at the centre of each face as well as at all eight corners. Any corner atom has 4 face centered atoms of the surrounding in its own plane and 4 face centered atoms above and below this plane (Fig. 14.14). Hence the coordination number for this case is 12 (= 4 × 3).

14.9 INTERPLANAR SPACING LO4

Let us consider a set of planes ($h\ k\ l$) in a unit cell. This plane PQR makes intercepts a/h, b/k, c/l on the three Cartesian co-ordinate axes X, Y and Z respectively, as shown in Fig. 14.15. Consider O as the origin at a lattice point. Let the plane PQR ($h\ k\ l$) be parallel to the plane passing through the origin. Draw a perpendicular ON from O (origin) to the plane. Thus ON = d, is the distance between adjacent planes or interplanar spacing between parallel planes. The normal ON makes angles α, β and γ with the x, y and z axes, respectively. From Fig. 14.15 we obtain $OP = a/h$, $OQ = b/k$ and $OR = c/l$.

Now from ΔONP, $\dfrac{ON}{OP} = \cos\alpha = \dfrac{d}{a/h} = \dfrac{dh}{a}$

From $\Delta ONQ, \dfrac{ON}{OQ} = \cos\beta = \dfrac{d}{b/k} = \dfrac{dk}{b}$

From $\Delta ONR, \dfrac{ON}{OR} = \cos\gamma = \dfrac{d}{c/l} = \dfrac{dl}{c}$

Then according to the law of direction cosines, we get

$$\cos^2\alpha + \cos^2\beta + \cos^2\gamma = 1$$

$$d^2\left[\frac{h^2}{a^2}+\frac{k^2}{b^2}+\frac{l^2}{c^2}\right]=1$$

or
$$d = \frac{1}{\sqrt{\left[\dfrac{h^2}{a^2}+\dfrac{k^2}{b^2}+\dfrac{l^2}{c^2}\right]}}$$

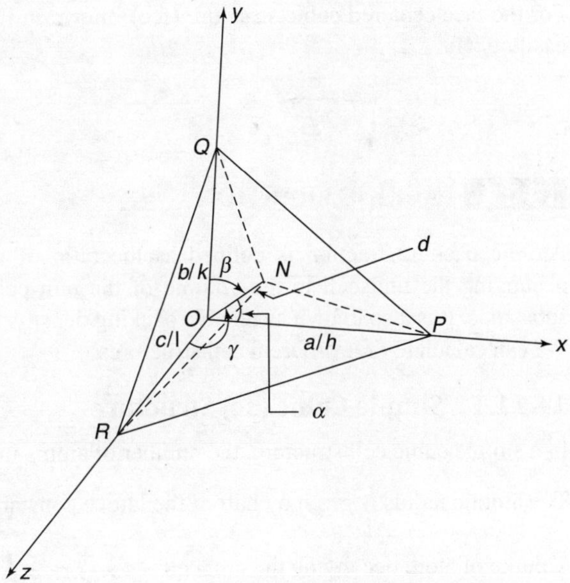

FIGURE 14.15

For cubic crystal $a = b = c$, we get

$$d = \frac{a}{\sqrt{[h^2+k^2+l^2]}}$$

For tetragonal crystal $a = b \neq c$, we get

$$d = \frac{1}{\sqrt{\dfrac{h^2+k^2}{a^2}+\dfrac{l^2}{c^2}}}$$

14.10 NEAREST NEIGHBOUR DISTANCE AND ATOMIC RADIUS

LO4

The distance between the centers of two neighbouring atoms is called nearest neighbour distance. For a closely packed crystal, this distance would be $2r$ for an atom of radius r. The distance r is called atomic radius, which can be represented in terms of edge of cube a for certain unit cell structures. For example, for simple cubic (sc) structure the diameter of an atom would be equal to the cube edge. That is $2r = a$ or $r = a/2$. However, in case of a body centered cubic (bcc) structure, shown in Fig. 14.16, the atomic arrangement is such that

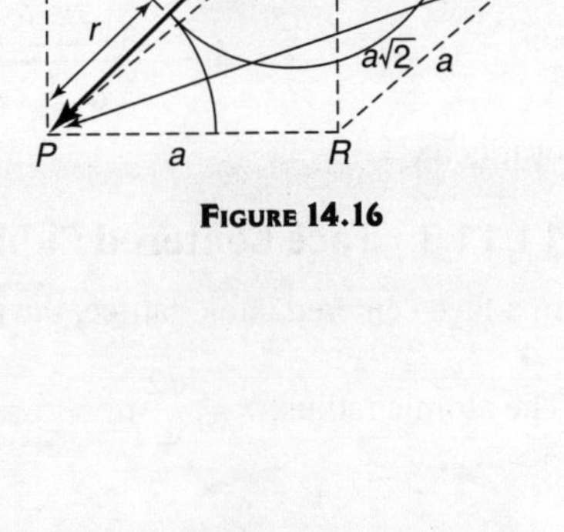

FIGURE 14.16

$$PQ = r + 2r + r$$

Since $\quad PR = RS = a, \quad PS = a\sqrt{2}$

Therefore, $\quad PQ = \sqrt{PS^2 + SQ^2}$

or $\quad (4r)^2 = 2a^2 + a^2$

or $\quad r = \dfrac{a\sqrt{3}}{4}$

For the face centered cubic structure (fcc), shown in Fig 14.17, one can easily obtain

$$r = \frac{a}{2\sqrt{2}}.$$

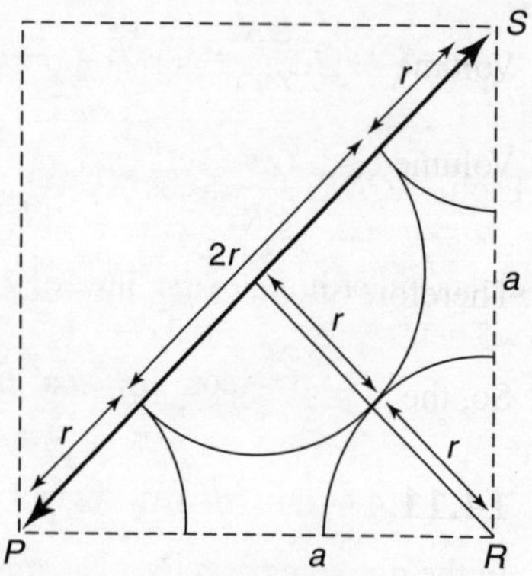

FIGURE 14.17

14.11 PACKING FRACTION

LO5

Atomic packing fraction is defined as the ratio of volume of atoms occupying the unit cell to the volume of the unit cell relating to that substance. It is also known as relative packing density. It is denoted by *f*. We can calculate *f* for different cell structures.

14.11.1 Simple Cubic (sc) Structure

In a simple cubic cell structure, the number of atoms in unit cell = 1.

The atomic radius is given by half of the lattice constant, i.e., $r = a/2$.

Volume of atom occupying the unit cell $= 1 \times \frac{4}{3}\pi r^3 = \frac{4}{3}\pi\left(\frac{a}{2}\right)^2$

Volume of the unit cell $= a^3$

Therefore, atomic packing fraction $f = \dfrac{\frac{4}{3}\pi\left(\frac{a}{2}\right)^3}{a^3} = \frac{\pi}{6} = 0.52$

It also means that the packing fraction is 52%.

14.11.2 Body Centered Cubic (bcc) Structure

In a body centered cubic lattice, the number of atoms per unit cell = 2

The atomic radius, $r = \frac{\sqrt{3}}{4}a$

Volume of atoms occupying the unit cell $= 2 \times \frac{4}{3}\pi r^3 = 2 \times \frac{4}{3}\pi\left(\frac{\sqrt{3}}{4}a\right)^3$

Volume of the unit cell $= a^3$

Therefore, the atomic packing fraction

$$f = \frac{2 \times \frac{4}{3}\left(\frac{\sqrt{3}}{4}a\right)^3}{a^3} = \frac{\sqrt{3}}{8}\pi = 0.68$$

or it is 68%.

14.11.3 Face Centered Cubic (fcc) Structure

In a face centered cubic lattice, the number of atoms per unit cell = 4

The atomic radius, $r = \frac{\sqrt{2}}{4}a$

Volume of atoms occupying the unit cell $= 4\times\frac{4}{3}\pi r^2 = 4\times\frac{4}{3}\pi\left(\frac{\sqrt{2}}{4}a\right)^3$

Volume of the unit cell = a^3

Therefore, atomic packing fraction $f = \frac{4\times\frac{4}{3}\pi\left(\frac{\sqrt{2}}{4}a\right)^3}{a^3} = \frac{\pi}{3\sqrt{2}} = 0.74$

So, the atomic packing fraction is 74%.

14.11.4 Diamond Structure

In the diamond structure, the number of atoms per unit cell = 8

The atomic radius, $r = \frac{\sqrt{3}}{8}a$

Volume of atoms occupying the unit cell $= 8\times\frac{4}{3}\pi r^3 = 8\times\frac{4}{3}\pi\left(\frac{\sqrt{3}}{8}a\right)^3$

Volume of the unit cell = a^3

Therefore, atomic packing fraction $f = \frac{8\times\frac{4}{3}\pi\left(\frac{\sqrt{3}}{8}a\right)^3}{a^3} = \frac{\sqrt{3}}{16}\pi = 0.34$

or the packing fraction is 34%.

14.11.5 Hexagonal Closed Packed (hcp) Structure

In the hexagonal cell, the number of atoms per unit cell = 6

The atomic radius, $r = \frac{a}{2}$

Volume of atoms occupying the unit cell $= 6\times\frac{4}{3}\pi r^3 = 6\times\frac{4}{3}\pi\left(\frac{a}{2}\right)^3$

Volume of the hexagonal unit cell $= \frac{3\sqrt{3}}{2}a^2c$ where $c = a\sqrt{\frac{8}{3}}$

$$\therefore \quad \text{Volume} = \frac{3\sqrt{3}}{2}a^2\sqrt{\frac{8}{3}}a = 3\sqrt{2}a^3$$

Therefore, atomic packing fraction $f = \frac{6\times\frac{4}{3}\pi\left(\frac{a}{2}\right)^3}{3\sqrt{2}a^3} = \frac{\pi}{3\sqrt{2}} = 0.74$

or the packing fraction is 74%.

14.12 POTENTIAL ENERGY CURVE AND NATURE OF INTERATOMIC FORCES LO5

The interatomic forces between atoms of the solids are electrostatic in nature which can either be attractive or repulsive.

(a) Interatomic Attractive Forces

These forces act between the atoms to hold them together. These forces are long range and act between the atoms at a finite distance. These forces become zero for infinite interatomic distances.

(b) Interatomic Repulsive Forces

A repulsive force also acts between the atoms to prevent the merging with each other and keep them at a finite distance. These forces are short-range and act only when the atoms come so close with each other that their electron cloud overlap with one another.

To understand interatomic forces, we consider a pair of atoms A and B, which are separated by a distance r at any time. Then, the potential energy due to force of attraction is given by

$$U_{\text{attr}} = -\frac{\alpha}{r^n} \tag{i}$$

and the potential energy due to force of repulsion is given by

$$U_{\text{rep}} = +\frac{\beta}{r^m} \tag{ii}$$

The total potential energy of the system is therefore given by

$$U = U_{\text{attr}} + U_{\text{rep}} \tag{iii}$$

$$U = -\frac{\alpha}{r^n} + \frac{\beta}{r^m}$$

where α, β, n and m are the characteristic constants of the molecule. From the above equation it is clear that if $r = \infty$, then $U = 0$.

The force acting between these two atoms can be written as

$$F = -\frac{dU}{dr} = -\frac{d}{dr}\left[-\frac{\alpha}{r^n} + \frac{\beta}{r^m}\right]$$

or

$$F = \frac{-n\alpha}{r^{n+1}} + \frac{m\beta}{r^{m+1}} \tag{iv}$$

For the fixed value of α, β, n and m, it is clear from this equation that the force $F = 0$ at a particular distance $r = r_0$. The potential energy and interatomic force curves are shown in Fig. 14.18.

When the distance r between the two atoms is very large, then it is clear from the Figs 13.18(a) and (b) that no force acts between the atoms and hence the total potential energy is zero. If the atoms approach each other and come close together nearly equal to the atomic diameter then repulsive force also begins to act and attractive force also gets very large. If the atoms come even more closer to each other then the repulsive force increases faster and the net force between atoms becomes repulsive in nature.

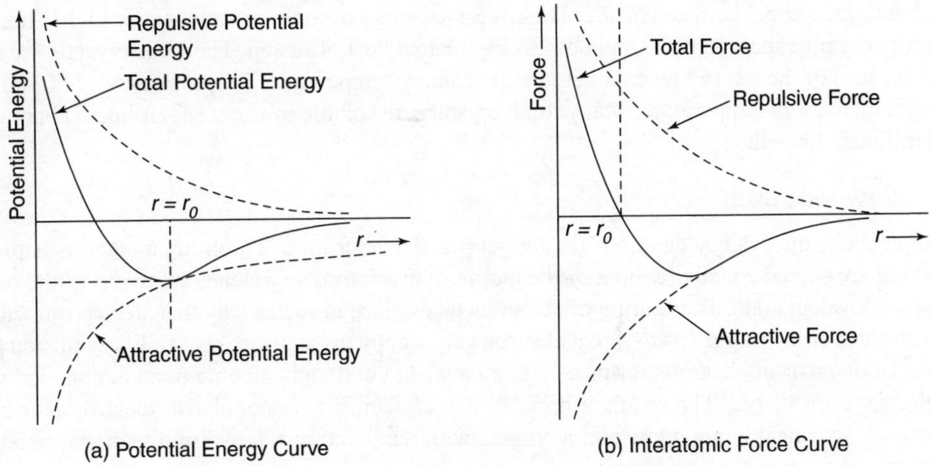

FIGURE 14.18

At a particular distance ($r = r_0$), the attractive and repulsive forces attain equal values and therefore no net force acts between the atoms. This position is known as equilibrium position in which the molecule becomes most stable because of the minimum potential energy.

14.13 DIFFERENT TYPES OF BONDING FORCES

LO6

The constituent particles of crystals have different types of charge distribution which provides different types of binding forces. The binding forces in most cases are electrostatic in nature but the distribution of electrons in various atoms are qualitatively different in different crystals. These binding forces are of different types, for example, ionic bond, covalent bond, metallic bond, molecular bond (or van der Waals bonds) and hydrogen bonds.

14.13.1 Ionic Bond

The ionic bond is formed due to transfer of one or more electrons from one type of atoms that lose electrons readily to the other type that have affinity for electrons. Due to the transfer of electrons, these atoms become positive and negative ions. The two types of atoms, which are involved in ionic bonding, are of different types. The arrangement of ions formed in the ionic bonding is such that the Coulomb attraction between ions of opposite charges is stronger than the Coulomb repulsion between ions of the same charges. Thus, the ionic bond results from the electrostatic interaction of oppositely charged ions. In this situation, both the ions attract each other and form ionic bond, which is clearly shown in Fig. 14.19. The examples of ionic crystals are NaCl, CsCl, KBr, KOH, etc. The ionic bond is non-directional in nature.

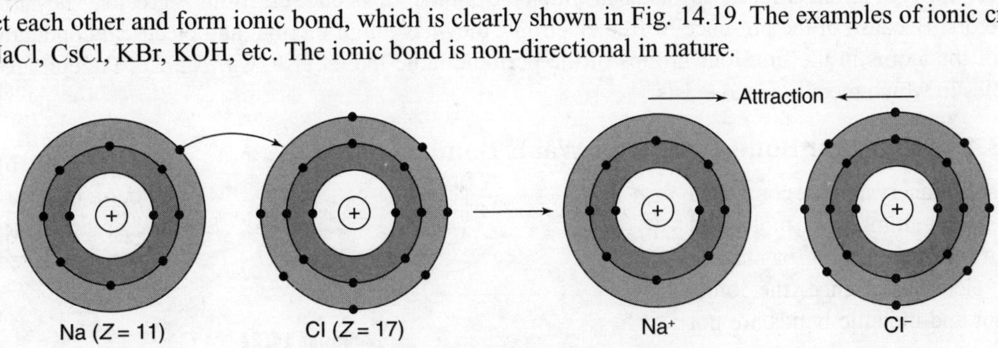

FIGURE 14.19

The ionic bonds are strong, hard and brittle. The cohesive energy of ionic crystals is very high. Thus the ionic crystals have very high melting point and possess high latent heat of fusion. The ionic crystals are insulators in general because of their very low conductivity at ordinary temperature. The conductivity of ionic crystals increase with increase in temperature. Many ionic crystals are soluble in water (polar liquid) but not soluble in nonpolar liquids like ether.

14.13.2 Covalent Bond

In covalent crystals, the valency electrons do not get transferred from one atom to another as happens in an ionic bond but are shared equally by both of the atoms. This sharing of valence electrons of the constituent atoms forms a covalent bond. The sharing of electrons takes place in such a way that an electron with spin up pairs with an electron with spin down, if that electron can occupy the states as per Pauli's exclusion principle. The sharing of one pair of electrons forms a singlet covalent bond and a double bond is obtained when two pairs of electrons are shared. The covalent bond is also known as a homopolar or electron pair bond. The conductivity of a covalent bond is low and increases with temperature. These are directional bonds and are very hard because the bond is very strong, for example in diamond.

The simplest example of covalent bonding is the H_2 molecule, as shown in Fig. 14.20. As the hydrogen atoms come close to each other, each of the two electrons are attracted by both the nuclei. In case of an oxygen molecule, two oxygen atoms share two pairs of electrons, thus forming a covalent double bond. A triple bond is formed by the sharing of three pairs of electrons in a nitrogen molecule. Covalent bonds are also formed between atoms of different elements like HCl, H_2O, NH_3, etc.

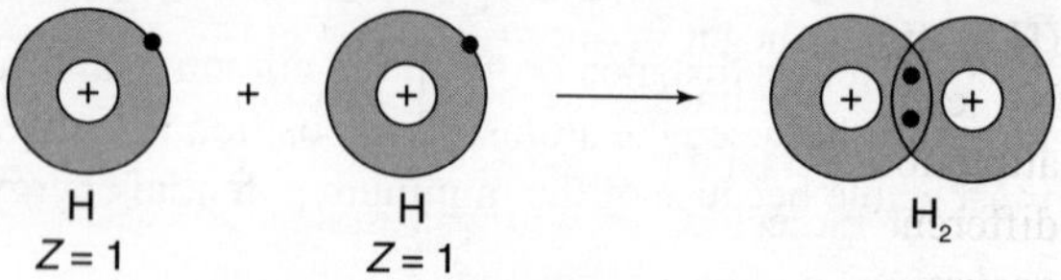

FIGURE 14.20

14.13.3 Metallic Bond

Metallic crystals are commonly known as metals. In the atoms of metals, the electrons in the outermost orbits are loosely bound as the ionisation energy is low in the case of metals. These electrons are free to move around among all the atoms and are called free electrons or conduction electrons. This way the metals have residual positive ions. The electrostatic attraction between these positive ions and negative electron gas is responsible for holding the solid together. This type of bonding is called metallic bonding.

The metallic bond is electrostatic in nature, though partially, and do not exert directional influence. The metallic bond is weaker in nature than the covalent bond because of the fewer electrons bonding the nuclei. However, it can be stronger for those metals in which the number of valence electrons is greater. They are good conductors. Because of the presence of free electrons, they have high thermal and electrical conductivities. Most of the atoms in the first four groups of the periodic table like Li, Na, Cu, Ag, Zn, Fe, etc. are good examples in which metallic bond exists.

14.13.4 Molecular Bond or van der Waals Bond

Many solids or crystals are composed of neutral atoms or molecules without any transferring or sharing of the electrons. Hence, the ionic, covalent and metallic bonds are not

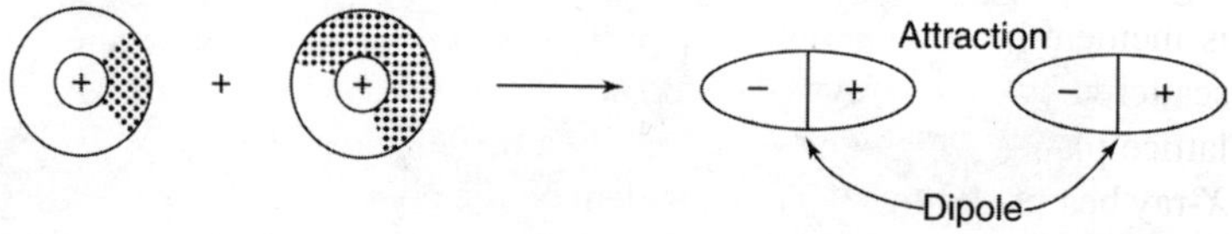

FIGURE 14.21

possible between these atoms. These substances remain bound by much weaker short-range attractive forces, which are called van der Waals forces. These forces are weaker than the atomic bonding forces. The van der Waals bonds are usually found in inert gases in which outermost electron orbits are completely filled, i.e., there are no valence electrons and hence they are incapable of forming any bond. These bonds are formed due to electrostatic attraction between oscillating or permanent dipoles (Fig. 14.21). As we know that the dipoles are formed due to the asymmetrical charge distribution around atoms, these dipoles are called oscillating dipoles.

In the case of inert gases, there is a very small attraction between the atoms due to closed outer shells. These gases condense when the temperature is reduced and hence a weak interatomic attraction is developed due to van der Waals forces. The van der Waals forces are non-directional in nature and a little energy is required to break the bonds because these are much weaker than ionic and covalent bonds. These types of bonds are found in inert gases like solid argon and in many organic symmetrical molecules like methane (CH_4).

14.13.5 Hydrogen Bond

In certain crystals, a positive hydrogen ion (H^+) attracts negative ions such as F^-, O^-, N^-, etc. Sometimes, due to electrostatic attraction, attachment between atoms in different molecules or within a molecule occurs in addition to bonds, which holds atoms together to form molecules. A hydrogen bond is formed when a hydrogen atom makes such an attachment or association with an electronegative atom like oxygen, nitrogen, fluorine, etc.

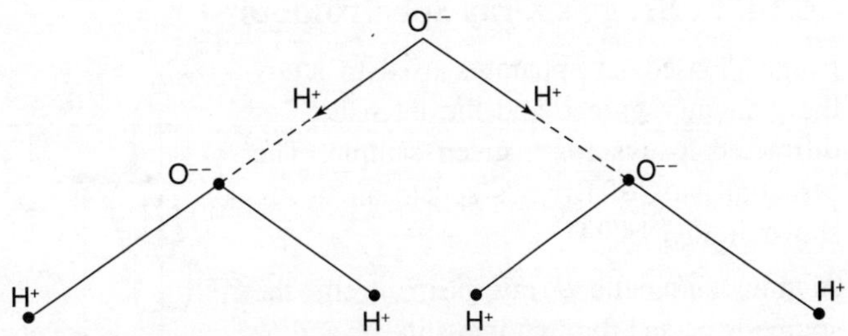

FIGURE 14.22

The hydrogen bond is found in H_2O, HF and in many organic molecules, particularly proteins and DNA molecules. In the water molecule (H_2O), the hydrogen and oxygen atoms are held together by covalent bonds. The positive dipole end i.e., hydrogen, can strongly attract the negative dipole end of water molecule. This bonding of the water molecule is shown in Fig. 14.22, where dashed lines represent the hydrogen bond.

14.14 CRYSTAL STRUCTURE ANALYSIS — LO6

As we know that the X-rays can penetrate solids. Since the wavelength of the X-rays (~1 Å) is of the order of interplanar spacing, these rays get strongly diffracted from different crystal planes. By analysing the different X-rays, we can gather information about the internal atomic arrangement of a crystal.

14.14.1 Bragg's Law: Diffraction of X-rays

The diffraction of *X*-rays from a single crystal is shown in Fig. 14.23. When a narrow beam of *X*-rays of wavelength λ is incident on the parallel planes at an angle θ, the *X*-rays are scattered by the atoms of the crystals. Consider a set of parallel lattice planes which are separated by a distance d. Further, two *X*-ray beams AO and BO' are incident on the crystal, as shown in the figure. The incident *X*-rays get diffracted at an angle θ at the

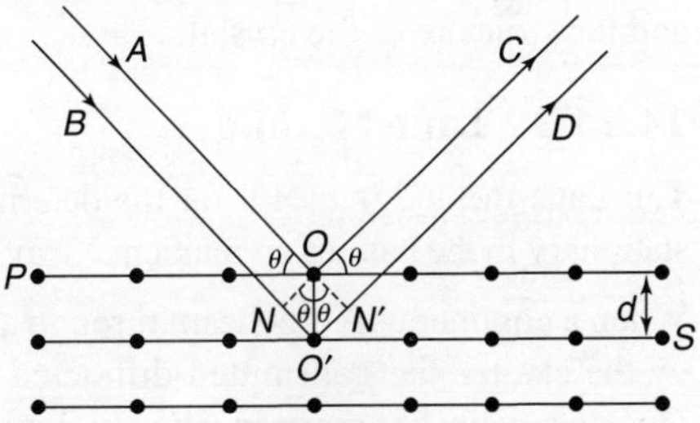

FIGURE 14.23

point O and O' of the crystal planes. The path difference between these two beams can be obtained by drawing perpendiculars ON on BO' and ON' on $O'D$, as

$$NO' + O'N' = d \sin \theta + d \sin \theta = 2d \sin \theta$$

For constructive interference (or for maximum intensity), we must have

$$2d \sin \theta = n\lambda \tag{i}$$

where n is an integer. The above relation is known as Bragg's law of diffraction. It is useful in calculating the distance d between crystal lattice planes once we know the wavelength λ of the X-rays and measure the angle of diffraction θ.

It is clear from Bragg's condition (i) that every X-ray will not get diffracted by the atoms of a crystal, only those will be diffracted whose wavelength λ and the angle θ exactly match this condition. The standard methods of X-ray diffraction used in the analysis of crystal structure are designed to achieve this. Bragg's X-ray spectrometer, Laue method and Powder method are such methods, which are discussed below.

14.14.2 Bragg's X-ray Spectrometer

Bragg devised an apparatus used to study the glancing angle θ and the intensities of diffracted X-rays for a given sample. The essential parts of Bragg's spectrometer are shown in Fig. 14.24.

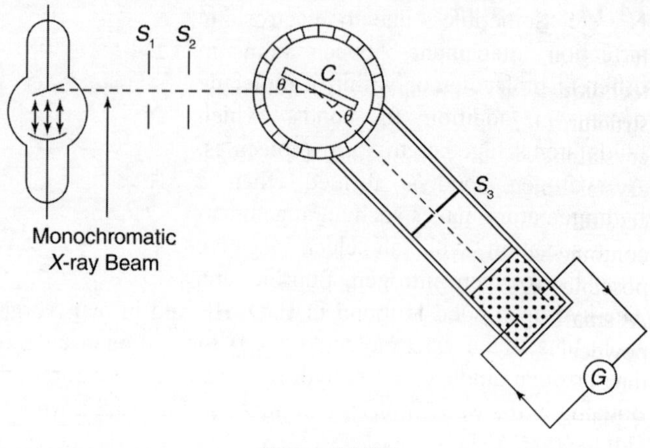

FIGURE 14.24

A monochromatic X-ray beam from the source is passed through two slits, S_1 and S_2 which collimate it into a fine narrow beam. This X-ray beam is then allowed to fall on the crystal sample C mounted at the centre of a turn table. A movable arm is also attached to the turn table for detecting the reflected X-ray beam from the crystal C. This turn table is capable of rotation about a vertical axis and the angle of rotation can be measured on the circular scale. The rates of rotation of the turn table and the detector (ionisation chamber) arm are such that the ionisation chamber always receives the diffracted beam. When the turn table rotates through an angle θ, the ionisation chamber arm automatically rotates through an angle 2θ with the direction of incident ray. This way the measurements of different diffracted X-ray beams' intensities and angles are recorded. Then using Bragg's condition, we get the interplanar spacing and the structure of the crystal.

14.14.3 Laue Method

The Laue method is useful for the determination of crystal structure. In this method, a single crystal is held stationary in the path of an incident X-ray beam, as shown in Fig. 14.25.

When a continuous X-ray beam through a pinhole is allowed to fall on the crystal, then this beam is diffracted by the crystal, the transmitted-diffracted and the reflected-diffracted beams are received by the films P and Q, respectively as shown in (Fig. 14.25a).

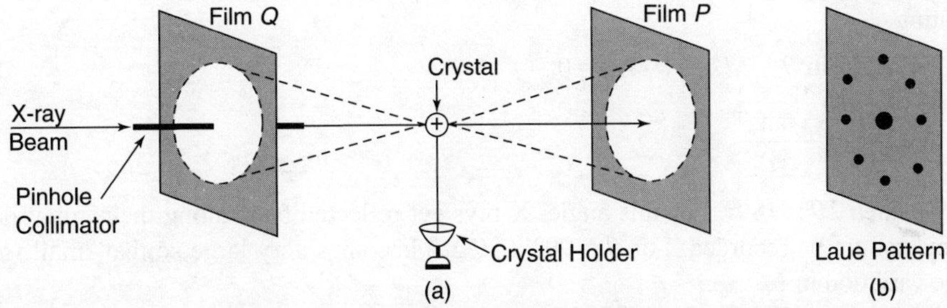

FIGURE 14.25

The transmitted-diffracted beams form a series of spots, which is the characteristic of crystal structure and is called Laue pattern (Fig. 14.25b). Each spot in the Laue pattern corresponds to the interference maxima for a set of crystal planes satisfying the Bragg's condition ($2d \sin\theta = n\lambda$) for a particular wavelength selected from the beam of incident light. By studying the position and intensities of these Laue spots, the crystal structure can be determined.

14.14.4 Powder Method

It is a standard and straight forward technique for analysing the crystal structure. In this technique, we use the crystal in powder form instead of single crystal so that its tiny crystals (i.e., crystallities) are randomly (i.e., almost continuously) oriented and make all possible angles with the incident beam. A small specimen of the crystalline powder is taken in a small capillary tube (P) of nondiffracting material and is placed in the path of fine monochromatic beam of X-ray (Fig 14.26). Thus all possible diffraction planes will be available for the Bragg diffraction ($2d \sin\theta = n\lambda$) to take place.

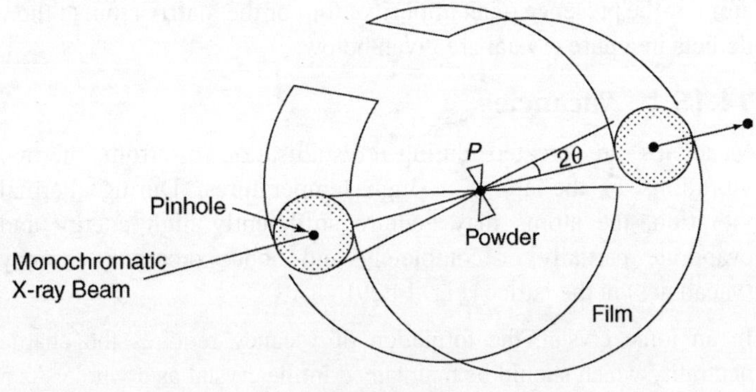

FIGURE 14.26

All these diffracted rays will lie on a conical surface having its apex at P and semivertical angle 2θ. The diffracted X-ray is recorded by the photographic film placed around the crystal and we get the arc of the circle on the photographic film, as shown in Fig. 14.27.

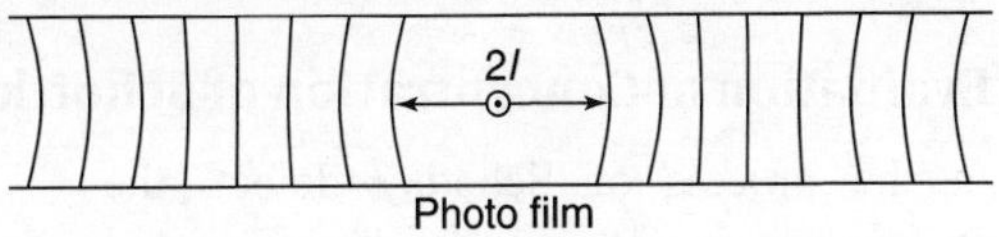

FIGURE 14.27

If $2l$ is the distance of the two arcs of the same circle on the photographic film and R is the radius of cylindrical camera (Fig. 14.28), then

$$2\theta = \frac{l}{R} \quad \text{or} \quad \theta = \frac{l}{2R}$$

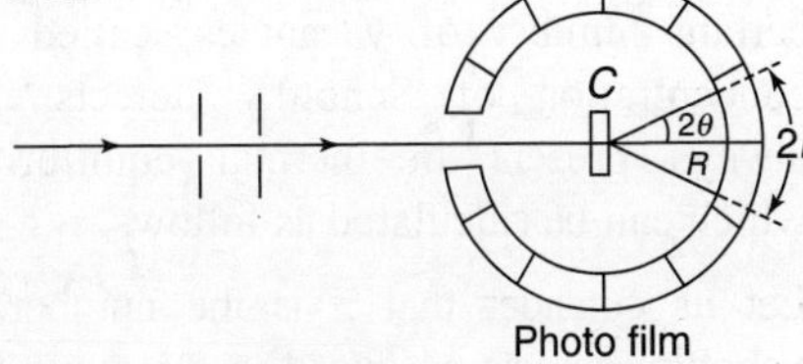

FIGURE 14.28

This value of θ will given the spacing between the planes with the help of Bragg's relation,

$$2d \sin\theta = n\lambda$$

By differentiating

$$2\Delta d \sin\theta + 2d\cos\theta\cdot\Delta\theta = 0$$

or
$$\frac{\Delta\theta}{\Delta d} = -\frac{\tan\theta}{d}$$

If θ tends to 90° then 2θ = 180°. For this angle, X-rays get reflected back along their initial path and such reflected beams cannot be recorded. For θ = 90°, ($\Delta\theta/\Delta d$) becomes very large so that small variations in d produces large variation in θ.

14.15 POINT DEFECTS IN SOLIDS

LO7

Point defect is a discontinuity in a crystal lattice. It consists of either a missing atom or an ion that creates a vacancy in the lattice (often known as Schottky defect). If an extra atom or ion exists between two normal lattice points it is said to create an interstitial position and if the missing atom or ion shifts to an interstitial position, then the vacancy is called Frenkel defect. Point defect occurs because of the absence of a matrix atom or the presence of an impurity atom or the matrix atom at the wrong place. The most common type point defects in a pure crystal are given below.

14.15.1 Vacancies

Vacancies are created during crystallisation or from thermal vibrations of the atoms at high temperatures. During thermal vibration, the atoms may acquire sufficiently high energy and evaporate partially or completely and hence create a vacancy (vacancies) in the lattice (Fig. 14.29).

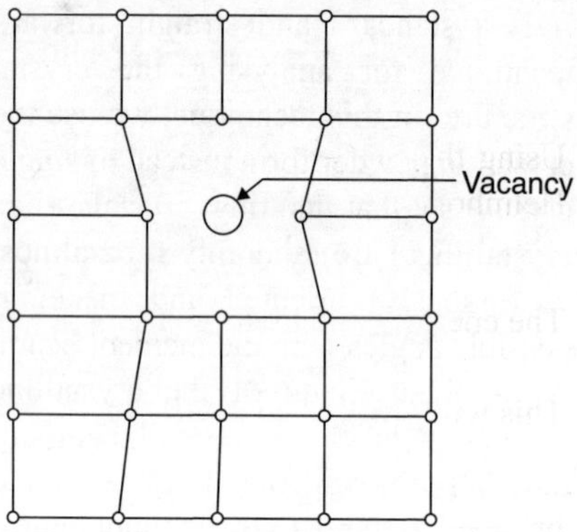

FIGURE 14.29

In an ionic crystal, the formation of vacancy requires the charge neutrality which should be maintained in the crystal as a whole. As a result, a pair of vacancies causes to missing of one cation and one anion from the structure. Such a pair of vacant sites is called a *Schottky defect* (Fig. 14.30). If a cation goes into an interstitial position, then the interstitially pair is known as *Frenkel Defect* (Fig. 14.30).

Evaluation of Concentration of Schottky Defects

In the process of Schottky defect, the absorbed energy is more than compensated by the resultant disorder in the lattice. Due to this excess energy the disorder increases that causes an enhancement in entropy. A certain number of vacancies, called the concentration of Schottky defects, are always present in thermal equilibrium which can be calculated as follows.

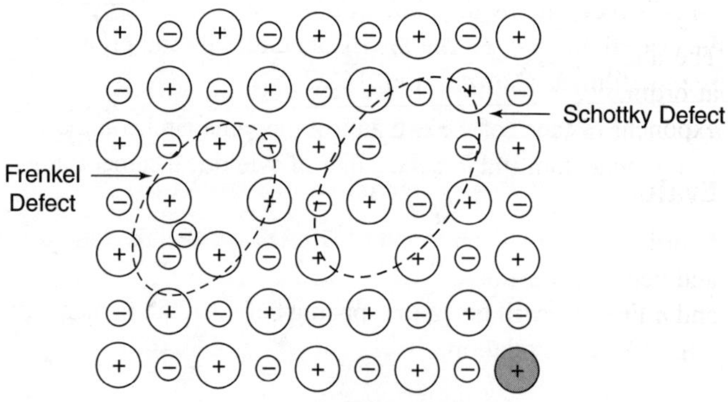

FIGURE 14.30

Let us consider that n is the number of Schottky defects produced in removing na^+ cations and na^- anions. Further, N is taken as the total number of cation-anion pairs. If U be the average energy, which is required to produce a Schottky defect, then nU would be the increase in energy associated with the generation of n vacancies. The number of different ways in which a cation or an anion can be removed is given by $\frac{N!}{(N-n)!n!}$. Since there are n cation and n anions, the total number of different ways in which n Schottky defects can be produced will be

$$W = \left[\frac{N!}{(N-n)!n!}\right]\left[\frac{N!}{(N-n)!n!}\right]$$

The increase in entropy is given by

$$S = k \ln W$$

$$= 2k \ln\left[\frac{N!}{(N-n)!n!}\right]$$

Here k is the Boltzmann constant. This increase in entropy produces a change in the Helmholtz free energy F, which can be obtained as

$$F = \text{Increase in energy} - \text{Temperature} \times \text{Increase in entropy}$$

$$= nU - 2kT \ln\left[\frac{N!}{(N-n)!n!}\right]$$

Using the Sterling's approximation $\ln y! = y \ln y - y$, we get the following expression for the change in Helmholtz free energy F

$$F = nU - 2kT[N \ln N - (N-n)\ln(N-n) - n \ln n]$$

The energy should be a constant at equilibrium.

This will give $\left.\frac{\partial F}{\partial n}\right|_T = 0$ or $U = 2kT \ln\left[\frac{(N-n)}{n}\right]$

or $\frac{N-n}{n} = \exp\left(\frac{U}{2kT}\right)$

For smaller number of Schottky defects, i.e., when $n << N$, we have $N - n \approx N$.

This gives, $n = N \exp\left(-\frac{U}{2kT}\right)$

The above expression gives the number of Schottky defects in binary ionic crystals like MgO and NaCl at ordinary temperature. It is clear from the expression that the fraction n/N of Schottky defects increases exponentially with the temperature.

Evaluation of Concentration of Frenkel Defects

In order to calculate the concentration of Frenkel defects, we consider a pure crystal that consists of positively and negatively charged ions in equal proportion. Further, we consider that it contains total N number of ions and n Frenkel defects. The number of different ways in which n Frenkel defects will be produced is given as

$$W = \frac{N!}{(N-n)!n!} \cdot \frac{N_i!}{(N_i-n)!n!} \quad \text{(i)}$$

where N_i is the number of interstitial sites. The Helmholtz free energy F of the crystal is given by the relation

$$F = U - ST \tag{ii}$$

where S is the increase in entropy and T is the temperature.

If E_i be the energy required to produced a vacancy, then U can be expressed as

$$U = nE_i \tag{iii}$$

and the associated increase in entropy is given by the Boltzmann relation

$$S = k \ln W = k \ln\left[\frac{N!}{(N-n)!n!} \cdot \frac{N_i!}{(N_i - n)!n!}\right] \tag{iv}$$

With the help of Eqs. (iii) and (iv), Eq. (ii) becomes

$$F = nE_i - kT \ln\left[\frac{N!}{(N-n)!n!}\frac{N_i}{(N_i - n)!n!}\right] \tag{v}$$

Sterling's approximation [$\ln y! = y! \ln y - y$] yields

$$\ln\left[\frac{N!}{(N-n)!n!}\frac{N_i!}{(N_i - n)!n!}\right] = N \ln N + N_i \ln N_i - \{(N-n)\ln(N-n)$$

$$+ (N_i - n)\ln(N_i - n) + 2n \ln n)$$

$$= nE_i - kT[N \ln N + N_i \ln N_i - (N-n)\ln(N-n)$$
$$- (N_i - n)\ln(N_i - \text{n}) - 2n \ln n] \tag{vi}$$

Differentiating Eq. (vi) w.r.t. n, we get

$$\left[\frac{\partial F}{\partial n}\right]_T = E_i - kT \ln\frac{(N-n)(N_i - n)}{n^2} \tag{vii}$$

Free energy remains constant, when the equilibrium position is attained at a given temperature T. It means

$$\left[\frac{\partial F}{\partial n}\right]_T = 0$$

$\therefore$ $$E_i - kT \ln\frac{(N-n)(N_i - n)}{n^2} = 0$$

or $$\ln\frac{(N-n)(N_i - n)}{n^2} = \frac{E_i}{kT}$$

For smaller number of Frenkel defects, i.e., when $N >> n$ and $N_i >> n$ the above relation reads

$$\ln\frac{NN_i}{n^2} = \frac{E_i}{kT}$$

or $$2 \ln n = \ln(NN_i) - \frac{E_i}{kT}$$

or $$\ln n = \frac{1}{2}\ln(NN_i) - \frac{E_i}{2kT}$$

or $$n = (NN_i)^{1/2} e^{-E_i/2kT}$$

From the above relation the concentration of Frenkel defects at a temperature T can be calculated.

14.15.2 Interstitial

As defined earlier, in interstitial defect an atom or an ion moves from its proper position to a place between regular lattice sites, as shown in Fig. 14.31. The interstitial is either due to a normal atom of the crystal or of a foreign atom.

14.15.3 Compositional Defect

The compositional defect occurs because of the replacement of a host atom by a foreign atom. The foreign atom remains at the regular lattice site, as shown in Fig 14.32.

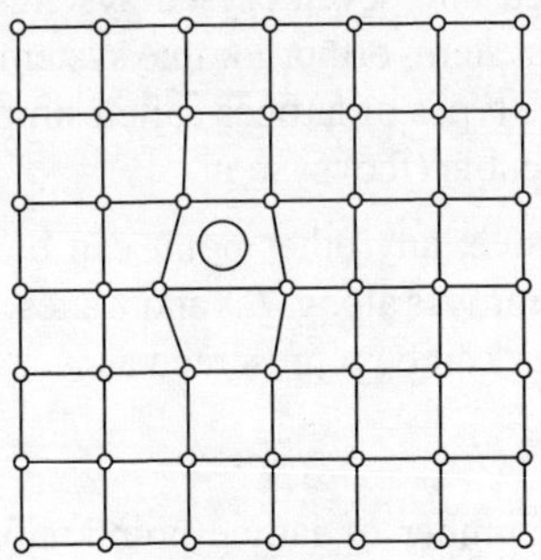

FIGURE 14.31

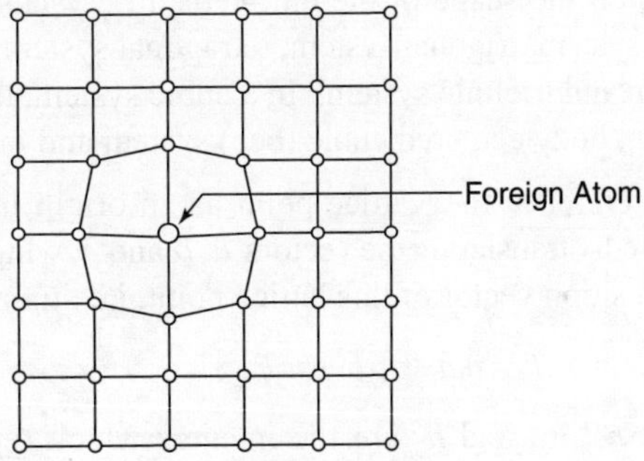

FIGURE 14.32

14.15.4 Electronic Defect

At absolute zero in a purely covalent crystal (e.g. *Si*), the electrons are tightly bound to the core and all are said to be in the valence band. Above absolute zero, some of the electrons are likely to occupy higher energy state depending upon the temperature. So in the crystal of pure silicon, some of the electrons from the covalent bonds get thermally released and become free to move, as shown in Fig. 14.33. This way the deficiency of electron creates a hole. Then the electrons and holes give rise to electronic imperfections.

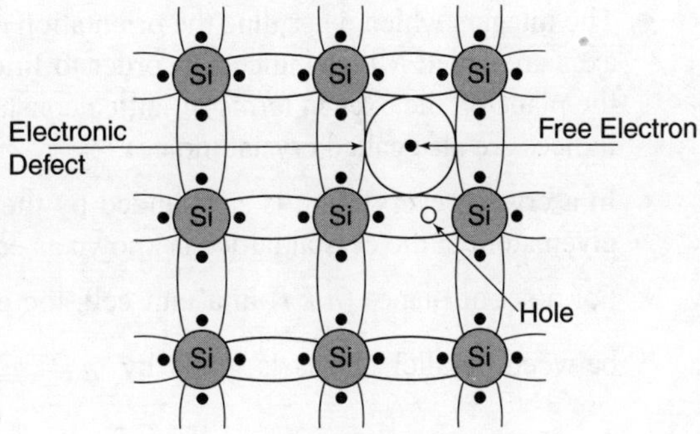

FIGURE 14.33

SUMMARY

The topics covered in this chapter are summarised below.

- A unique arrangement of atoms is called a crystal structure. It consists of a set of atoms which are identical in composition, arrangement and orientation called basis and a lattice. Bases are located upon the points of a lattice, which is an array of points repeating periodically in three dimensions. A crystal structure and symmetry play an important role in determining many of its properties such as electronic band structure and optical properties.

- The solids are broadly classified into two groups, namely crystalline solids and amorphous solids. Crystalline solids are arranged in fixed geometric patterns or lattices. Ice, methanol and sodium chloride are a few examples of crystalline solids. A rigid material whose structure lacks crystalline periodicity is called an amorphous solid. It means the pattern of its constituent atoms or molecules does not repeat periodically in three dimensions in amorphous solids.
- A crystal structure can be obtained by translation of a unit cell in three dimensions. A unit cell is a smallest pattern of a space lattice, which can generate the complete crystal by repeating its own dimensions in various directions in amorphous solids.
- Based on the shape of the unit cells all crystals are classified into seven crystal systems. These are cubic system, trigonal system, tetragonal system, hexagonal system, orthorhombic system, monoclinic system and triclinic system. In a cubic system, there are three types of lattices called simple cubic (sc) system, body centered cubic (bcc) system and face centered cubic (fcc) system.
- With reference to a lattice point as an origin in a plane lattice, any other point can be obtained by repeatedly translating the vectors $\vec{a}, \vec{b}$ and $\vec{c}$, which are the primitives along *X, Y* and *Z* axes, respectively. The position vector of this lattice point, i.e., translation vector, can be represented as

  $$\vec{T} = n_1\vec{a} + n_2\vec{b} + n_3\vec{c}$$

  where n_1, n_2, and n_3 are the integers which represent the number of lattice points along the three directions.
- A crystal lattice is made of large number of parallel equidistant planes known as lattice planes.
- The integers which determine the orientation of a crystal plane in relation to the three crystallographic axes are called Miller indices. In order to find the Miller indices, the reciprocals of the intercepts of the plane on the axes in terms of lattice constants are reduced to the smallest integers in ratio. Miller indices are also called crystal indices.
- In a crystal, every atom is surrounded by the other atoms. The number of nearest neighbours to the given atom in the crystal lattice is known as coordination number.
- For a set of planes $(h\ k\ l)$ in a unit cell, the distance between adjacent planes or interplanar spacing between parallel planes is given by $d = \dfrac{1}{\sqrt{\left[\dfrac{h^2}{a^2} + \dfrac{k^2}{b^2} + \dfrac{l^2}{c^2}\right]}}$. Here *a*, *b* and *c* are the fundamental translational vectors along the three axes.
- The distance between the centers of two neighbouring atoms is called the nearest neighbour distance. For a closely packed crystal, this distance is $2r$ for an atom of radius r. The distance r is called atomic radius, which is generally represented in terms of edge of cube a for certain unit cell structures.
- The ratio of volume of atoms occupying the unit cell to the volume of the unit cell relating to that substance is called atomic packing fraction. It is also known as relative packing density. It is denoted as f. The atomic packing fraction f for the simple cubic structure is 52%. For a body centered cubic (*bcc*) structure f = 68% and for a face centered cubic (*fcc*) and hexagonal closed packed structures f = 74%. The atomic packing fraction f for the diamond structure is 34%.
- The interatomic forces between atoms of the solids are electrostatic in nature which can either be attractive or repulsive. It is obtained that at a particular distance, the attractive and repulsive forces attain equal values and therefore no force acts between the atoms.

- ✦ The constituent particles of crystals have different types of charge distribution which provides different types of binding forces. The binding forces in most cases are electrostatic in nature but the distribution of electrons in various atoms are qualitatively different in different crystals. These binding forces are of different types, for example, ionic bond, covalent bond, metallic bond, molecular bond (or van der Waals bonds) and hydrogen bonds. Accordingly, the crystals are referred to as the ionic crystal, covalent crystal, metallic crystal, molecular crystal and hydrogen bonded crystal.
- ✦ Since the X-rays can penetrate solids and their wavelength (1 Å) is of the order of interplanar spacing, these rays can get strongly diffracted from different crystal planes. An analysis of the diffracted X-rays can provide the information about the structure of the crystal. The standard methods of X-ray diffraction include Bragg's X-ray spectrometer, Laue method, rotating crystal method and powder method.
- ✦ Point defect is a discontinuity in a crystal lattice. It consists of either a missing atom or an ion that creates a vacancy in the lattice (often known as Schottky defect). If an extra atom or ion exists between two normal lattice points, it is said to create an interstitial position and if the missing atom or ion shifts to an interstitial position, then the vacancy is called Frenkel defect. Point defect occurs because of the absence of a matrix atom or the presence of an impurity atom at the matrix atom in the wrong place.
- ✦ The number of Schottky defects in binary ionic crystals like MgO and NaCl at ordinary temperature is given by $n = N\exp\left(-\frac{U}{2kT}\right)$ where N is the total number of cation-anion pairs and U is the average energy required to produce the Schottky defects.
- ✦ The number of Frenkel defects in crystals at ordinary temperature is given by $n = (NN_i)^{1/2}\exp\left(-\frac{E_i}{2kT}\right)$ where N is the number of ions, N_i is the number of interstitial sites and E_i is the energy required to produce the vacancy.

SOLVED EXAMPLES

EXAMPLE 1 A plane cuts intercepts $2a$, $3b$ and c along the crystallographic axes in a crystal. Determine the Miller indices of plane.

SOLUTION Intercepts are $2a$, $3b$ and c.

Then from the law of rational indices, we have

$$2a:3b:c = \frac{a}{h}:\frac{b}{k}:\frac{c}{l}$$

or

$$\frac{1}{h}:\frac{1}{k}:\frac{1}{l} = 2:3:1$$

or

$$h:k:l = \frac{1}{2}:\frac{1}{3}:1 = 3:2:6$$

Therefore, the Miller indices of the plane are **(3 2 6)**.

EXAMPLE 2 In a triclinic crystal, a lattice plane makes intercepts at a length a, $2b$ and $-3c/2$. Find the Miller indices of the plane.

SOLUTION Intercepts are a, $2b$ and $-3c/2$.

$$\therefore \qquad a:2b:\frac{-3c}{2}=\frac{a}{h}:\frac{b}{k}:\frac{c}{l}$$

or
$$\frac{1}{h}:\frac{1}{k}:\frac{1}{l}=1:2:\frac{-3}{2}$$

or
$$h:k:l=1:\frac{1}{2}:\frac{-2}{3}=6:3:-4$$

Therefore, the Miller indices of the given plane are **(6 3 $\bar{4}$)**.

Example 3 Deduce the Miller indices for planes in each of the following sets which intercept $\vec{a}, \vec{b}$ and $\vec{c}$ at (i) $3a$, $3b$, $2c$ (ii) a, $2b$, ∞ (iii) a, $b/2$, c

Solution (i) Intercepts are $3a$, $3b$, $2c$.

Then,
$$3a:3b:2c=\frac{a}{h}:\frac{b}{k}:\frac{c}{l}$$

or
$$3:3:2=\frac{1}{h}:\frac{1}{k}:\frac{1}{l}$$

or
$$h:k:l=\frac{1}{3}:\frac{1}{3}:\frac{1}{2}$$

or
$$h:k:l=2:2:3$$

Therefore, the Miller indices are **(2 2 3)**.

(ii) Intercepts are a, $2b$, ∞.

Then,
$$a:2b:\infty=\frac{a}{h}:\frac{b}{k}:\frac{c}{l}$$

or
$$\frac{1}{h}:\frac{1}{k}:\frac{1}{l}=1:2:\infty$$

or
$$h:k:l=1:\frac{1}{2}:\frac{1}{\infty}=2:1:0$$

Therefore, the Millers indices are **(2 1 0)**.

(iii) Intercepts are a, $b/2$, c.

Then,
$$a:\frac{b}{2}:c=\frac{a}{h}:\frac{b}{k}:\frac{c}{l}$$

or
$$1:\frac{1}{2}:1=\frac{1}{h}:\frac{1}{k}:\frac{1}{l}$$

or
$$h:k:l=1:2:1$$

Therefore, the Miller indices are **(1 2 1)**.

Example 4 Calculate the spacing between (1 0 0) and (1 1 1) planes of a cubic system of lattice parameter a.

Solution Spacing between the planes of a cubic system of lattice parameter a.

$$d_{hkl}=\frac{a}{\sqrt{h^2+k^2+l^2}}$$

For plane (1 0 0), $d_{100} = \dfrac{a}{\sqrt{1^2+0^2+0^2}} = a$

and For plane (1 1 1), $d_{111} = \dfrac{a}{\sqrt{1^2+1^2+1^2}} = \dfrac{a}{\sqrt{3}}$

Example 5 Deduce the Miller indices of a set of parallel planes which make intercepts in the ratio of a: $2b$ on the x and y axis and are parallel to z-axis, $\vec{a}, \vec{b}, \vec{c}$ being primitive vectors of lattice. Also calculate the interplanar distance d of the plane taking the lattice to be cubic with $a = b = c = 5$Å.

Solution The parallel planes are parallel to z axis. It means that their intercepts on the z-axis are infinite. Thus, the intercepts are a, $2b$ and ∞. And, lattice constant $a = b = c = 5$Å.

Then,

$$a:2b:\infty c = \frac{a}{h}:\frac{b}{k}:\frac{c}{l}$$

or

$$\frac{1}{h}:\frac{1}{k}:\frac{1}{l} = 1:2:\infty$$

or

$$h:k:l = 1:\frac{1}{2}:\frac{1}{\infty} = 2:1:0$$

Therefore, the Miller indices are (2 1 0).

∴ Interplanar distance

$$d = \frac{a}{\sqrt{h^2+k^2+l^2}} = \frac{5\times10^{-10}}{\sqrt{2^2+1^2+0^2}}$$

$$= \frac{5\times10^{-10}}{\sqrt{5}} = \sqrt{5}\text{Å}$$

Example 6 Determine the Miller indices of plane parallel to the z axis and cut intercepts of 2 and 2/3 along x and y axes, respectively.

Solution Intercepts are $2a$, $\dfrac{2b}{3}$, ∞.

$$2a:\frac{2b}{3}:\infty c = \frac{a}{h}:\frac{b}{k}:\frac{c}{l}$$

or

$$\frac{1}{h}:\frac{1}{k}:\frac{1}{l} = 2:\frac{2}{3}:\infty$$

or

$$h:k:l = \frac{1}{2}:\frac{3}{2}:\frac{1}{\infty} = 1:3:0$$

Therefore, the Miller indices are **(1 3 0)**.

Example 7 Calculate the interplanar spacing for (2 3 1) plane of an fcc structure whose atomic radius is 0.175 nm.

Solution Given plane = (2 3 1) and atomic radius (r) = 0.175 nm.

Atomic radius (r) of *fcc* structure

$$= \frac{\sqrt{2}}{4}a$$

and interplanar spacing $d = \dfrac{a}{\sqrt{h^2 + k^2 + l^2}}$

$$\because \qquad r = \frac{\sqrt{2}a}{4} \Rightarrow a = \frac{4r}{\sqrt{2}} = \frac{4 \times 0.175 \times 10^{-9}}{\sqrt{2}}$$

$$\therefore \qquad d_{231} = \frac{a}{\sqrt{h^2 + k^2 + l^2}} = \frac{4 \times 0.175 \times 10^{-9}}{\sqrt{2} \times \sqrt{2^2 + 3^2 + 1}}$$

$$= \frac{4 \times 0.175 \times 10^{-9}}{\sqrt{2} \times \sqrt{14}} = \mathbf{0.132 \times 10^{-9}\ m}$$

Example 8 In a simple cubic crystal (i) find the ratio of intercepts of three axes by (1 2 3) plane and (ii) find the ratio of spacings of (1 1 0) and (1 1 1) planes.

Solution (i) Given $(h\ k\ l)$ of the plane as (1 2 3). Intercepts on the axes of a simple cubic crystal are given as a/h, a/k, a/l.

∴ The ratio of intercepts are

$$\frac{a}{1} : \frac{a}{2} : \frac{a}{3} = 1 : \frac{1}{2} : \frac{1}{3}$$

(ii) The spacings d of plane (h, k, l) in a simple cubic crystal of side a

$$= \frac{a}{\sqrt{h^2 + k^2 + l^2}}$$

∴ For plane (1 1 0), $d_{110} = \dfrac{a}{\sqrt{1^2 + 1^2 + 0}} = \dfrac{a}{\sqrt{2}}$

and for plane (1 1 1), $d_{111} = \dfrac{a}{\sqrt{1^2 + 1^2 + 1^2}} = \dfrac{a}{\sqrt{3}}$

Therefore the ratio of spacing between these two plane is $d_{110}/d_{111} = \sqrt{3/2} = \mathbf{1.225}$

Example 9 Calculate the distance between two atoms of basis of the diamond structure if the lattice constant of the structure is 5Å.

Solution Given lattice constant a = 5Å.

The distance between two atoms is equivalent to the nearest neighbor distance.

For diamond structure, nearest neighbour distance = $\dfrac{\sqrt{3}}{4}a$

∴ Distance between two atoms $= \dfrac{\sqrt{3}}{4} \times 5\text{Å} = \dfrac{1.732 \times 5}{4} = \mathbf{2.17Å}$

Example 10 What is the number of atoms in the primitive cell of diamond. Calculate the length of a primitive translation vector if the cube edge a = 3.56 Å.

Solution Diamond is a fcc lattice with two carbon atoms in a primitive cell. So, the number of atoms is 8.

Given the cube edge a = 3.56 Å.

∴ Primitive translation vector $= \dfrac{a}{\sqrt{2}} = \dfrac{3.56}{1.41} = \mathbf{2.52Å}$

EXAMPLE 11 Determine the number of atoms per unit cell of lead which has an fcc structure. Atomic weight of Pb = 207.2, density of Pb = 11.36×10^3 kg m^{-3}, and a = 3.2 Å and Avogadro's number = 6.023×10^{26}/kg mole.

SOLUTION Given atomic weight of Pb (M) = 207.2, density of Pb = 11.36×10^3 kg m^{-3} and a = 3.2 Å

Avogadro's number N = 6.023×10^{26} /kg mole.

Number of atoms $n = \dfrac{a^3 \rho N}{M}$

$$= \frac{(3.2\times10^{-10})^3 \times 11.36\times10^3 \times 6.023\times10^{26}}{207.2}$$

$$= \mathbf{1.082 \approx 1}$$

EXAMPLE 12 Calculate the lattice constant 'a' of a substance having fcc lattice, molecular weight 60.2 and density 6250 kg/m^3. (N=6.02×10^{26}/kg mole)

SOLUTION Given molecular weight M = 60.2, density ρ = 6250 kg/m^3, and $N = 6.02 \times 10^{26}$/kg mole.

For fcc lattice n = 4

Lattice constant $a = \left(\dfrac{4M}{N\rho}\right)^{1/3}$

$$\therefore \quad a = \left[\frac{4\times60.2}{6250\times6.02\times10^{26}}\right]^{1/3}$$

$$= \left[\frac{240.8}{37265\times10^{26}}\right]^{1/3}$$

$$= \mathbf{4\ Å}$$

EXAMPLE 13 In NaCl crystal, the spacing between the successive (1 0 0) plane is 2.82 Å. X-ray incident on the surface of the crystal is found to give rise to first order Bragg reflection at glancing angle 8.8°. Calculate the wavelength of X-ray.

SOLUTION Given d = 2.82 Å, θ = 8.8° and n = 1.

Formula used is $2d \sin\theta = n\lambda$.

$$2 \times 2.82 \times 10^{-10} \times \sin 8.8 = \lambda$$

$$\Rightarrow \quad \lambda = 5.64\times10^{-10} \times 0.153$$

$$= \mathbf{0.863\ Å}$$

EXAMPLE 14 The first-order diffraction is found to occur at a glancing angle of 9°. Calculate the wavelength of X-ray and the glancing angle for second order diffraction if the spacing between the adjacent plane is 2.51 Å.

SOLUTION Given n =1, θ = 9° and d = 2.51 Å = 2.51×10^{-10} m.

Formula used, is $2d \sin\theta = n\lambda$.

Therefore, $\lambda = 2 \times (2.51 \times 10^{-10}) \times \sin 9° = 0.7853$ Å.

For n = 2

$$\theta = \sin^{-1}\left(\frac{2\lambda}{2d}\right) = \sin^{-1}\left(\frac{\lambda}{d}\right)$$

$$= \sin^{-1}\left(\frac{0.7853\times10^{-10}}{2.51\times10^{-10}}\right)$$

$$= \mathbf{18.2°}$$

EXAMPLE 15 X-rays of wavelength 1.5 Å make a glancing angle 60º in the first-order when diffracted from NaCl crystal. Find the lattice constant of NaCl.

SOLUTION Given $\lambda = 1.5$ Å, $\theta = 60º$ and $n = 1$.

Formula used is $2d \sin \theta = n\lambda$.

$$\therefore \quad d = \left(\frac{n\lambda}{2 \sin \theta}\right) = \frac{1 \times 1.5 \times 10^{-10}}{2 \sin 60°} = \frac{1.5 \times 10^{-10} \times 2}{2 \times \sqrt{3}}$$

$$= \mathbf{0.87Å}$$

EXAMPLE 16 X-ray of wavelength 1.4 Å is found to be Bragg reflected from the (1 1 1) plane of an *fcc* structure. If the lattice parameter of the crystal is 5 Å. Find the angle at which the X-ray is incident on the (1 1 1) plane of the crystal.

SOLUTION Given $\lambda = 1.4$ Å, lattice parameter of *fcc* structure (a) = 5 Å and the plane of the fcc structure = (1 1 1).

Interplanar spacing $d_{hkl} = \dfrac{a}{\sqrt{h^2 + k^2 + l^2}}$

and $\quad 2d \sin \theta = n\lambda$

$$\therefore \quad d_{111} = \frac{a}{\sqrt{3}} = \frac{5 \times 10^{-10}}{\sqrt{3}} = 2.887 \times 10^{-10}\, m$$

and so, $$\theta_{111} = \sin^{-1}\left(\frac{n\lambda}{2d_{111}}\right) = \sin^{-1}\left[\frac{1 \times (1.4 \times 10^{-10})}{2 \times 2.887 \times 10^{-10}}\right]$$

$$= \mathbf{14°}$$

EXAMPLE 17 Calculate the glancing angle on the cube face (1 0 0) of a rock salt crystal (a = 2.184 Å) corresponding to second order reflection of X-rays of wavelength 0.710 Å.

SOLUTION Given $d = a = 2.814$ Å $= 2.814 \times 10^{-10}$ m for cube face (1 0 0), $n = 2$ for second order diffraction and $\lambda = 0.710 \times 10^{-10}$ m.

$$2d \sin \theta = n\lambda$$

$$\theta = \sin^{-1}\left[\frac{n\lambda}{2d}\right]$$

$$= \sin^{-1}\left[\frac{2 \times 0.710 \times 10^{-10}}{2 \times 2.814 \times 10^{-10}}\right]$$

$$= \mathbf{14.6°}$$

EXAMPLE 18 From the following data calculate the wavelength of neutron beam and its speed. Spacing between successive planes is 3.84 Å, glancing angle 30° and the order of Bragg reflection = 1.

SOLUTION Given $d = 3.84$ Å $= 3.84 \times 10^{-10}$ m, $\theta = 30°$ and $n = 1$.

Formula used are $2d \sin \theta = n\lambda$ and $\lambda = \dfrac{h}{mv}$.

Thus, $2 \times 3.84 \times 10^{-10} \times \sin 30° = \lambda$

or $$\lambda = 2 \times 3.84 \times 10^{-10} \times \frac{1}{2} = 3.84Å$$

$$\because \quad \lambda = \frac{h}{mv} \quad \text{[according to deBroglie relation]}$$

$$\therefore \qquad \lambda = \frac{h}{m\lambda} = \frac{6.62 \times 10^{-34}}{1.67 \times 10^{-27} \times 3.84 \times 10^{-10}}$$

$$\boldsymbol{\lambda = 1.03 \times 10^{3}\ \mathrm{m}}$$

EXAMPLE 19 Electrons accelerated from the state of rest by 120 V are reflected from an fcc crystal. The reflection maximum is observed at 22°. Determine the lattice parameter if the Bragg reflection occurs from the (1 1 1) plane.

SOLUTION Given $V = 120$ V, $\theta = 22°$ and $n = 1$.

Formula used are $\lambda = \dfrac{h}{(2\mathrm{meV})^{1/2}} \qquad \left[\because \frac{1}{2}mv^2 = \mathrm{eV}\right]$

and $\qquad 2d \sin\theta = n\lambda$

$$\therefore \qquad \lambda = \frac{h}{\sqrt{2meV}}$$

$$= \frac{6.6 \times 10^{-34}}{\sqrt{2 \times 9.1 \times 10^{-31} \times 1.6 \times 10^{-19} \times 120}} = 1.12 \times 10^{-10}\ \mathrm{m}$$

so, $$d_{111} = \frac{n\lambda}{2\sin\theta} = \frac{1 \times 1.12 \times 10}{2 \times \sin 22°} = 1.4949\text{Å}$$

and $$d_{111} = \frac{a}{\sqrt{3}}$$

$$\therefore \qquad a = d_{111} \times \sqrt{3} = \sqrt{3} \times 1.4949\text{Å} = \mathbf{2.589Å}$$

EXAMPLE 20 A monochromatic beam of *X*-rays of wavelength 1.24 Å is reflected by cubic crystal of KCl. Determine the interplanar distances for (1 0 0), (1 1 0) and (1 1 1) planes. Given density of KCl = 1980 kg/m^3 and molecular weight $M = 74.5$. Avogadro's number $N = 6.023 \times 10^{26}$/kg mole.

SOLUTION Given $M = 74.5$, $\rho = 1980$ kg/m^3 and $N = 6.023 \times 10^{26}$/kg mole.

Formulas used are

$$a^3 = \frac{nM}{N\rho}$$

and $$d_{hkl} = \frac{a}{\sqrt{h^2 + k^2 + l^2}} \text{ for cubic crystal.}$$

$$\therefore \qquad a^3 = \frac{nM}{N\rho} = \frac{4 \times 74.5}{6.023 \times 10^{26} \times 1.98 \times 10^{3}} = 24.99 \times 10^{-29}$$

$$a = [249.9 \times 10^{-30}]^{1/3} = 6.30 \times 10^{-10} = 6.3\text{Å}$$

$$d_{100} = \frac{a}{1^2 + 0^2 + 0} = \frac{a}{1} = \frac{6.3\text{Å}}{1} = 6.3\text{Å}$$

$$d_{110} = \frac{a}{\sqrt{1^2 + 1^2 + 0}} = \frac{a}{\sqrt{2}} = 4.38 \times 10^{-10}\ \mathrm{m} = 4.38\text{Å}$$

$$d_{111} = \frac{a}{\sqrt{1^2 + 1^2 + 1^2}} = \frac{a}{\sqrt{3}} = \frac{6.3 \times 10^{-10}}{\sqrt{3}} = 3.64 \times 10^{-10}\ \mathrm{m} = \mathbf{3.64Å}$$

Example 21 Determine the potential energy of K^+ and Cl^- ion when they are separated by a distance of 0.15 nm.

Solution Given separation distance $r_0 = 0.15$ nm $= 0.15 \times 10^{-9}$ m.

The potential energy of the ions $(V) = \dfrac{-e^2}{4\pi\varepsilon_0 r_0}\ J = \dfrac{-e}{4\pi\varepsilon_0 r_0}$ eV

$$\therefore \quad V = \frac{1.6\times10^{-19}}{4\times3.14\times8.85\times10^{-12}\times0.15\times10^{-9}}$$

$$= 0.0959\times10^{2}\text{ eV}$$

$$= 0.096\times10^{2}\text{ eV} = \mathbf{-9.6\ eV}$$

Example 22 From the following data determine the cohesive energy of NaCl. The equilibrium separation $r_0 = 0.32$ nm, $\alpha = 1.748$, $n = 9$, ionisation energy = 4 eV and electron affinity = –2.16 eV.

Solution Given $r_0 = 0.32$ nm $= 0.32 \times 10^{-9}$ m, $\alpha = 1.748$, $n = 9$, Ionisation energy = 4 eV and electron affinity = –2.16 eV.

$$V(r_0) = \frac{-\alpha e^2}{4\pi\varepsilon r_0}\left(1-\frac{1}{n}\right)\text{ Joule or } V(r_0) = \frac{-\alpha e}{4\pi\varepsilon_0 r_0}\left(1-\frac{1}{n}\right)\text{ electron volt}$$

$$= \frac{-1.748\times(1.6\times10^{-19})^2}{4\times3.14\times8.85\times10^{-12}\times0.32\times10^{-9}}\left(1-\frac{1}{9}\right)$$

$$= \mathbf{0.0698\ eV}$$

Example 23 Find the ratio of number of Schottky defects to the total number of cation–anion pairs for a binary ionic crystal of NaCl of the average energy required to produce a Schottky defect be 2.02 eV at room temperature. Given Boltzman constant $k = 1.38 \times 10^{-23}$ J/K.

Solution The number of Schottky defects is given by $n = N \exp\left(\dfrac{-U}{2kT}\right)$ where N is the total number of cation–anion pairs.

Room temperature $T = 27°C = 300$ K.

Hence
$$\frac{n}{N} = \exp\left(\frac{-2.02\times1.6\times10^{-19}}{2\times1.38\times10^{-23}\times300}\right)$$

$$= \mathbf{1.12\times10^{-17}}$$

Objective Type Questions

Q.1 Which one of the following Bravais lattices is not found in cubic crystals?
(a) simple cubic (b) body centered cubic
(c) face centered cubic (d) base centered cubic

Q.2 In hexagonal crystal, the angles between the axes are
(a) $\alpha = \beta = \gamma = 90°$ (b) $\alpha \neq \beta \neq \gamma \neq 90°$
(c) $\alpha = \beta = 90°, \gamma = 120°$ (d) $\alpha = \gamma = 90° \neq \beta$

Q.3 The coordination number of bcc lattice is
(a) 6 (b) 8 (c) 12 (d) 3

Q.4 The number of atoms present in the fcc lattice is
(a) 1 (b) 2 (c) 3 (d) 4

Q.5 The packing fraction of diamond crystal structure is
(a) 74% (b) 68% (c) 52% (d) 34%

Q.6 The number of ions pair per unit cell present in NaCl is
(a) 5 (b) 2 (c) 4 (d) 1

Q.7 X-rays can be deflected by
(a) magnetic fields (b) electric fields
(c) large–size obstacles (d) none of these

Q.8 The Miller indices of a set of parallel planes which make equal intercepts on the three axes are
(a) (1 2 1) (b) (1 1 1) (c) (1 0 0) (d) (1 0 1)

Q.9 In a simple cubic lattice the ratio $d_{100} : d_{110} : d_{111}$ is
(a) $\sqrt{6}:\sqrt{3}:\sqrt{2}$ (b) $\sqrt{3}:\sqrt{6}:1$ (c) $6:3:\sqrt{2}$ (d) 6:3:1

Q.10 Hexagonal closed pack structure is
(a) simple cubic structure (b) body centered cubic structure
(c) face centered cubic structure (d) base centered cubic structure

Q.11 Which one of the following crystals is an example of monoclinic-
(a) NaCl (b) Mg (c) $CaSO_4$ (d) $CuSO_4.5H_2O$

Q.12 Which one of the following crystals is the example of rhombohedral (or trigonal)-
(a) CsCl (b) As (c) Fe_3C (d) $K_2Cr_2O_7$

Q.13 The atomic radius for fcc lattice is
(a) $\frac{a}{2}$ (b) $\frac{a}{2\sqrt{2}}$ (c) $\frac{\sqrt{3}}{4}a$ (d) $\frac{a}{\sqrt{2}}$

Q.14 The packing fraction has maximum value for
(a) fcc structure (b) sc structure (c) hpc structure (d) bcc structure

Q.15 NaCl crystal has
(a) fcc structure (b) hpc structure (c) sc structure (d) bcc structure

Q.16 Electrons are transferred from one type of atom to the other type in
(a) covalent crystals (b) ionic crystals (c) metallic crystals (d) molecular crystals

Q.17 van der Waals bond is formed due to
(a) the sharing of valence electrons
(b) attraction between positive ion cores and electrons
(c) electrostatic attraction between oscillating or permanent dipoles
(d) electrostatic force of attraction between ions

Q.18 Which one of the following bonds has high electrical conductivity?
(a) ionic bond (b) covalent bond (c) metallic bond (d) H-bond

Q.19 The energy required to separate the atoms of a molecule is called
(a) ionisation energy (b) Coulomb energy
(c) cohesive energy (d) dissociation energy

Q.20 If n be the number of atoms in a unit cell of the cubic system, N and M be the Avogadro's number and atomic weight, respectively, and 3 be the density of the element, then the lattice constant a is given by
(a) $\left[\frac{nM}{N\rho}\right]^{1/3}$ (b) $\left[\frac{nM}{N\rho}\right]^{-1/3}$ (c) $\left[\frac{M\rho}{nN}\right]^{1/3}$ (d) $\frac{nM}{N\rho}$

Q.21 Which of the following shape of unit cells are correct for most of the crystals
(a) parallelopiped (b) elliptical (c) spherical (d) none of these

Q.22 A cubic system is completely represented by
(a) $a = b = c$
$\alpha = \beta = \gamma = 90°$
(b) $a = b \neq c$
$\alpha = \beta = \gamma = 90°$
(c) $a = b = c$
$\alpha = \beta = \gamma \neq 90°$
(b) none of these

Q.23 The Bravais lattice of CsCl structure is
(a) body centered cubic (b) face centered cubic
(c) simple cubic (d) none of these

Q.24 The arrangement of atoms in a crystal is known as
(a) crystal structure (b) lattice
(c) Bragg's lattice (d) none of these

Q.25 Covalent bond is formed
(a) by emission of electron from the atom
(b) by transferring of electrons from one to another atom
(c) by sharing of electrons
(d) none of these

Q.26 The potential energy between atoms in equilibrium is
(a) minimum (b) maximum (c) both (a) & (b) (d) none of these

Q.27 The resultant force between the atoms in equilibrium is
(a) large (b) zero (c) attractive (d) none of these

Q.28 Inter molecular bonds are
(a) H-bonds (b) dipole bonds (c) dispersion bonds (d) all of these

Q.29 If the Miller indices of a plane is (1 0 0), then
(a) the plane is perpendicular to x-axis (b) the plane parallel to x-axis
(c) the plane is perpendicular to y-axis (d) none of these

SHORT-ANSWER QUESTIONS

Q.1 What is a crystal?
Q.2 Why are most solids crystalline in nature?
Q.3 Define crystal lattice, plane lattice and space lattice.
Q.4 What is crystal structure?
Q.5 Define a primitive unit cell. Can a unit cell be primitive?
Q.6 Define mathematically a lattice. What conditions an ideal lattice must follow?
Q.7 Define translation vector, unit cell, coordination number and packing efficiency.
Q.8 Write coordination number of simple cubic, body centered cubic and face centered cubic lattices.
Q.9 What is coordination number of diamond crystal? Also state the number of next nearest neighbour.
Q.10 What is coordination number of hcp structure?
Q.11 How many lattice points are there in the cubic unit cell of bcc structure.

Q.12 Explain diamond structure. Calculate its packing fraction. Give examples of any two materials having this structure.

Q.13 Why *X*-rays are used for crystal structure analysis?

Q.14 Why γ-rays are not used to study crystal structure?

Q.15 What is Bragg's law?

Q.16 What is Bragg's equation?

Q.17 How does Bragg reflection differ from ordinary reflection?

Q.18 Explain the term bonding.

Q.19 What are different kinds of bonding?

Q.20 What is ionic crystal?

Q.21 Name various point defects in solids.

PRACTICE PROBLEMS

General Questions

Q.1 (a) Distinguish between a crystal and an amorphous solid.
(b) Give three main differences between crystalline and amorphous solids.

Q.2 (a) What is crystal structure? State the relation between crystal structure, lattice and basis.
(b) Define a primitive cell. Distinguish between a primitive unit cell and non-primitive unit cell with the help of diagram. Can a unit cell be primitive?

Q.3 What is the concept of Miller indices? Derive the formula for the distance between two adjacent planes of a simple cubic lattice.

Q.4 What is Bravais lattice? Explain different types of Bravais lattices in three dimensions.

Q.5 Draw the diagrams of the following structures: NaCl and CsCl. Give at least two examples of each structure.

Q.6 Explain the crystal structure of sodium chloride (NaCl). Draw a sketch of sodium chloride lattice and write down the coordinates of the atoms in the unit cell. What is the number of sodium ions in unit cell of NaCl?

Q.7 Explain the Crystal structure of diamond. In diamond crystal, what is the number of nearest neighbours, the number of atoms per unit cell and packing fraction? Show that it has comparatively loose packing.

Q.8 (a) Explain the concept of Miller indices. How are they calculated? How the orientation of a plane is specified by Miller indices? Define Miller indices of a direction. State their important features.
(b) Why the reciprocals of intercepts of the plane are taken to find Miller indices?

Q.9 Draw the planes (1 0 0), (0 1 0), (0 0 1), (1 1 0), (1 0 1), (0 1 1), (2 0 0), ($\bar{2}$ 0 0), ($\bar{1}$ 0 0), (2 0 1), (1 1 1), and (1 1 2) in a simple cubic unit cell.

Q.10 Derive the expression for the interplanar spacing between two parallel planes with Miller indices (*h k l*) and show that for a simple cubic lattice of lattice constant a

$$d_{hkl} = \frac{a}{\sqrt{h^2 + k^2 + l^2}}$$

Q.11 Derive Bragg's law of crystal diffraction $2d \sin \theta = n\lambda$ and give its significance. Discuss briefly the method of crystal structure determination.

Q.12 Is there any interdependence of coordination number and packing efficiency. Illustrate by giving examples.

Q.13 Discuss briefly the experimental method for crystal structure determination by *X*-ray diffraction.

Q.14 Name the standard experimental methods of *X*-ray diffraction.

Q.15 Describe in detail Laue method and also describe the usefulness of this method.

Q.16 Explain with necessary theory the powder method for *X*-ray analysis.

Q.17 Describe in detail the powder method and its usefulness.

Q.18 What are point defects in solids?

Q.19 What are different types of point defects? Explain.

Q.20 What are Schottky and Frenkel defects. Derive the necessary relation to show that Schottky defects in ionic crystal depend on the temperature.

Q.21 Show that the number of Frenkel defects in equilibrium at a given temperature is proportional to $(NN_i)^{1/2}$, where N and N_i are number of atoms and interstitial atoms respectively.

Q.22 Name various types of bonds in solids and given one example of each.

Q.23 Explain any four types of bondings in solids.

Q.24 Write short note on bonding in solids.

UNSOLVED QUESTIONS

Q.1 Find the Miller indices for planes in each of the following sets which intercept $\vec{a}, \vec{b}$, and $\vec{c}$, at (i) $3a$, $3b$, $2c$ and (ii) a, $2b$, ∞. **[Ans: (2,2,3) and (2,1,0)]**

Q.2 Lattice constant of a cubic lattice is a. Calculate the spacing between (i) (011), (ii) (101) and, (iii) (110) planes. **[Ans: (i) $a/\sqrt{2}$ (ii) $a/\sqrt{2}$ (iii) $a/\sqrt{2}$]**

Q.3 For a cubic lattice calculate the distance of (123) and (234) planes from a plane passing through the origin. **[Ans: $a/\sqrt{14}$ and $a/\sqrt{29}$]**

Q.4 Calculate the glancing angle at which *X*-rays with λ =1.549 Å will be reflected in first and second orders from a crystal with interplanar distance 4.225 Å. **[Ans: θ_1 = 10°31′ and θ_2 = 21°21′]**

Q.5 Using 2.02 as the value of lattice constant, calculate the wavelength of X-ray in second order, if angle of diffraction θ = 26°. **[Ans: 1.24 Å]**

Q.6 A crystal is mounted on an *X*-ray spectrometer. *X*-rays are incident at the glancing angle for three reflections are 5°28′, 12°1′ and 18°12′. Show that these are successive orders of reflections from the same crystal plane. Also find the spacing. [Given λ for *X*-rays used as 0.586 Å]. **[Ans: 2.817 Å, 2.817 Å and 2.817 Å]**

Q.7 A certain crystal reflects monochromatic *X*-rays strongly when Bragg glancing angle (first order) is 15°. What are the glancing angles for second and third order spectrum. **[Ans: 31.17°, 50.93°]**

Development of Quantum Mechanics

15

Learning Objectives

After reading this chapter you will be able to

LO 1 Learn about blackbody radiation and Planck's quantum hypothesis

LO 2 Understand the concept of quantum theory

LO 3 Know about wave particle duality and photoelectric effect and its theoretical applications

LO 4 Discuss the de Brogile waves and its demonstration by Davisson-Germer experiment

LO 5 Explain Compton effect and its verification

LO 6 Evaluate phase and group velocities and their interrelationship

Introduction

Newton's laws describe the motion of particles in classical mechanics and Maxwell's equations describe the electromagnetic fields in classical electromagnetism. The classical mechanics correctly explains the motion of celestial bodies like planets, stars, macroscopic and microscopic terrestrial bodies moving with non-relativistic speeds. However, classical theory does not hold in the region of atomic dimensions, i.e., it cannot explain the non-relativistic motion of electrons, protons etc. Classical theory could not explain the stability of atoms, spectral distribution of blackbody radiation, the origin of discrete spectra of atoms, etc. Also, classical mechanics could not explain a large number of observed phenomena like photoelectric effect, Compton effect, Raman effect, etc. So, the insufficiency of classical mechanics led to the development of quantum mechanics. Quantum mechanics is the description of motion and interaction of particles at the small scales where the discrete nature of the physical world becomes important. The quantum mechanics for the atomic system led to the explanation of discrete energy levels as well as the postulation of different quantum numbers. *Niels Bohr* had a large influence on the development of quantum mechanics through his so-called Copenhagen Interpretation, a philosophical construct that was formulated to provide a fundamental framework for understanding the implicit assumptions, limitations, and applicability of the theory of quantum mechanics.

The development of quantum mechanics took place in two stages. The first stage began with *Max Planck's hypothesis* according to which the radiation is emitted or absorbed by matter in discrete packets or quanta of energy. This energy is equal to $h\nu$, where h is Planck's constant and ν is the frequency of radiation. This hypothesis led to a theory which was not completely satisfactory being a mixture of classical and non-classical concepts. The second stage of quantum mechanics began in 1925 along with two points of views. For example, *matrix mechanics* was introduced by *Heisenberg*, in which only observed quantities like frequencies and intensities of spectral lines are taken into account and unobserved quantities like positions, velocities, etc. in electronic orbits are omitted. Another form of quantum mechanics is called *wave mechanics*, whose theory was developed by *Schroedinger* in 1926. In this mechanics, concepts of classical wave theory and deBroglie's wave particle relationship are combined with each other. With the application of quantum mechanics, several problems of atomic physics have been solved. However, this mechanics also has certain limitations. Therefore, a more complete theory of particles called quantum field theory has been accepted since 1947. In order to understand the development of wave mechanics, we begin with the blackbody radiation.

15.1 BLACKBODY RADIATION: SPECTRAL DISTRIBUTION LO1

A body that completely absorbs radiations of all wavelengths incident on it is referred to as a blackbody. When such a body is heated, it emits radiations which we call as blackbody radiations. A cavity made out of a hollow container of any material (iron or copper) with a narrow opening and painted with lampblack in the inside portion gives a close approximation to a perfectly blackbody. When any radiation falls on this hole, it enters the cavity, gets reflected by the wall of the cavity and eventually gets absorbed. Now if we heat the container at various temperatures, it will emit radiations of all the frequencies (or wavelengths). So, the emitted radiation from a blackbody is a continuous spectrum, i.e., it contains radiation of all the frequencies.

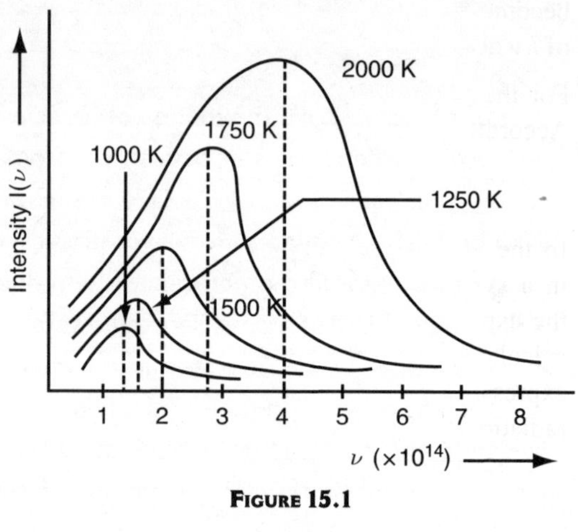

FIGURE 15.1

Let the intensity of emitted radiation be $I(\nu)\,d\nu$ between the frequencies ν and $\nu+d\nu$. The experimental measurements of intensity $I(\nu)$ with different ν is shown in Fig. 15.1 for different values of temperature T. These plots show that

(1) The distribution of frequencies is a function of temperature of the blackbody.
(2) With the increase in temperature, the total amount of emitted radiation $I(\nu)\,d\nu$ increases.
(3) The position of the maximum peak shifts toward higher frequencies with increasing equilibrium temperature.

The classical electromagnetic theory or wave theory together with classical thermodynamics does not explain the characteristics of the blackbody spectrum. However, Planck's hypothesis can explain these characteristics together with the use of classical thermodynamics.

Classical wave theory says that the electromagnetic radiation inside the cavity of the blackbody at an equilibrium temperature T forms the standing waves and the number of standing waves (possible modes) that can fit in the cavity depends on the wavelength. The number of possible modes in the cavity is large if the wavelength is small. However, for large wavelengths the number of possible modes is small. According to Rayleigh and Jeans, this increase in the number of modes is proportional to $1/\lambda^2$ or ν^2 and also each of the standing waves must be assigned an average kinetic energy kT, where k is the Boltzmann constant. This leads to the following Rayleigh-Jeans law (details discussed later)

$$I(\nu)d\nu = \frac{8\pi\nu^2}{c^3}kTd\nu$$

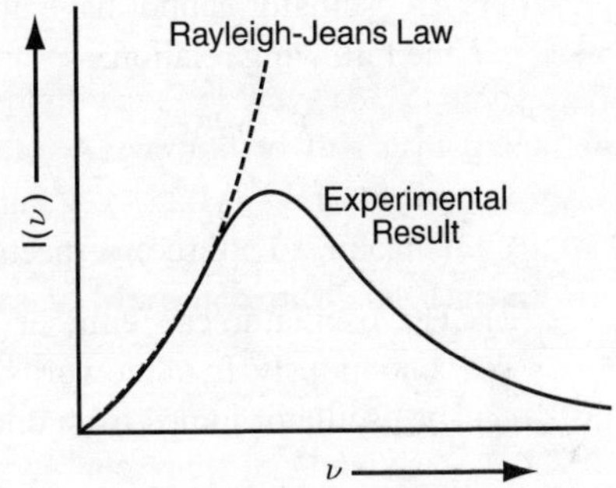

FIGURE 15.2

This relation shows that $I(\nu)$ is proportional to the square of ν. The corresponding plot is shown in Fig. 15.2. It is clear from the figure that the experimental data does not agree with the theory; the agreement is good only for smaller values of ν. The disagreement at high frequencies, i.e., in the UV region, is called ultraviolet catastrophe. Thus, the spectral distribution of a blackbody could not be explained on the basis of classical theory. This difficulty was resolved by Planck in 1900, when he stated that by assuming electromagnetic radiation to be emitted or absorbed in bundles of size $h\nu$, one could correctly predict the spectrum of blackbody radiation. As mentioned earlier, this bundle of energy is called a quantum. The quanta of high frequencies have high energies and those of low frequencies have low energies. Thus, the atoms and molecules in the cavity will emit radiation only if they have energy in the excess of $h\nu$. For low frequencies ν, there will be a large number of atoms and molecules that might have this excess energy. Since the bundles become quite bigger for higher frequencies ν, the number of atoms or molecules having energies in the excess of $h\nu$ decreases. It means for large ν, the intensity $I(\nu)$ does not increase rather decreases.

For the explanation of blackbody radiation, Planck made a use of the Maxwell-Boltzmann distribution. According to this distribution, the number of molecules N_n with energy E is given by

$$N_n = N_0 e^{-E/kT}$$

In the above expression, N_0 refers to the number of molecules in a system in equilibrium at temperature T. Planck combined the expression of N_n with his quantum hypothesis $E = nh\nu$ and calculated the mean energy. Finally, he arrived at the following expression for the distribution of the maximum intensity of radiation in the spectrum of blackbody.

$$I(\nu) = \frac{8\pi h\nu^3}{c^3}\frac{1}{e^{(h\nu/kT)} - 1}$$

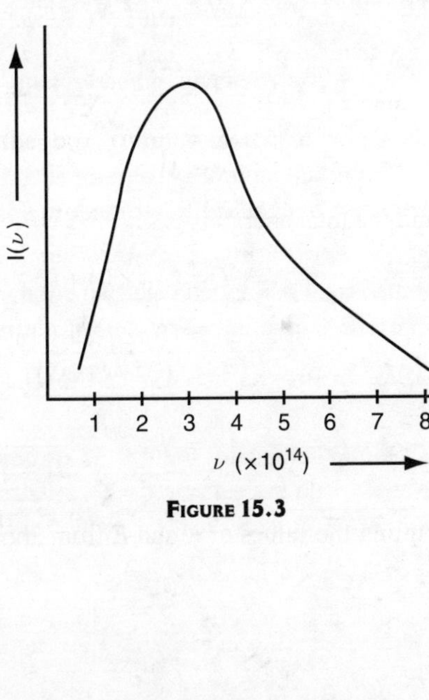

FIGURE 15.3

This expression is referred to as Planck's radiation law. This theoretical formula fits very well with the experimental data for the entire wavelength, as shown in Fig. 15.3. Thus, Planck's quantum theory was able to interpret fully different characteristics of blackbody radiation which classical theory could not.

15.2 PLANCK'S QUANTUM HYPOTHESIS LO1

Max Planck in 1900 introduced the quantum theory of radiation to explain the distribution of energy in the spectrum of *blackbody* radiation. He assumed that the atoms of the walls of a blackbody behave as oscillators and each has a characteristic frequency of oscillation. He made the following two revolutionary assumptions about the atomic oscillator.

(1) An oscillator cannot have any arbitrary value of energy but can have only discrete energies as per the following relation

$$E = nh\nu$$

where $n = 0,1,2,\ldots$; ν and h are known as frequency of oscillation and Planck's constant ($=6.62 \times 10^{-34}$ J sec), respectively. This relation shows that the energy of the oscillation is quantised.

(2) The oscillator can emit or absorb energy only in the form of packets of energy ($h\nu$) but not continuously. In other words, we can say that the emission or absorption of energy occurs only when the oscillator jumps from one energy state to another along with the energy difference given by

$$\Delta E = \Delta n h\nu$$

or $$E_2 - E_1 = (n_2 - n_1)h\nu$$

15.2.1 Average Energy of Planck's Oscillators

If N be the total number of oscillators and E as the total energy of these oscillators, then the average energy will be given by the relation

$$\bar{E} = \frac{E}{N} \tag{i}$$

Now consider $N_0, N_1, N_2, \ldots N_n$, as the number of oscillators having energy values 0, $h\nu$, $2h\nu$, …, $nh\nu$, respectively. Then by Maxwell's distribution formula $N_n = N_0 e^{-nh\nu/kT}$, we have

$$N = N_0 + N_1 + N_2 + \ldots.$$
$$= N_0(1 + e^{-h\nu/kT} + e^{-2h\nu/kT} + \ldots.)$$
$$= \frac{N_0}{(1 - e^{-h\nu/kT})} \tag{ii}$$

and the total energy is

$$E = (N_0 \times 0) + (N_1 \times h\nu) + (N_2 \times 2h\nu) + \ldots.$$
$$= (N_0 \times 0) + (N_0 e^{-h\nu/kT} \times h\nu) + (N_0 e^{-2h\nu/kT} \times 2h\nu) + \ldots.$$
$$= N_0 e^{-h\nu/kT} h\nu[1 + 2e^{-h\nu/kT} + 3e^{-2h\nu/kT} + \ldots]$$
$$= N_0 e^{-h\nu/kT} \frac{h\nu}{(1 - e^{-h\nu/kT})^2} \tag{iii}$$

Putting the values of N and E from above relation in Eq. (i), we get

$$\overline{E} = \frac{E}{N} = \frac{h\nu e^{-h\nu/kT}}{1 - e^{-h\nu/kT}} = \frac{h\nu}{e^{h\nu/kT} - 1}$$

$$\overline{E} = \frac{\boldsymbol{h\nu}}{\boldsymbol{e^{h\nu/kT} - 1}} \tag{iv}$$

This is the expression for average energy of a Planck's oscillator.

15.2.2 Planck's Radiation Formula

The energy density of radiation (u_ν) in the frequency range ν and $\nu + d\nu$ depending upon the average energy of an oscillator is given by

$$u_\nu d\nu = \frac{8\pi\nu^2}{c^3} d\nu \times \overline{E} \tag{v}$$

Putting the value of $\overline{E}$ from Eq. (iv) in Eq. (v) gives

$$u_\nu d\nu = \frac{8\pi\nu^2}{c^3} \frac{h\nu}{e^{h\nu/kT} - 1} d\nu$$

or $$\boldsymbol{u_\nu d\nu = \frac{8\pi h}{c^3} \frac{\nu^3 d\nu}{e^{h\nu/kT} - 1}} \tag{vi}$$

The above relation is known as Planck's radiation formula in terms of frequency. This law can also be expressed in terms of wavelength λ of the radiation. Since $\nu = \frac{c}{\lambda}$ for electromagnetic radiation, $d\nu = -\frac{c}{\lambda^2} d\lambda$.

Further, we know that the frequency is reciprocal of wavelength or in other words an increase in frequency corresponds to a decrease in wavelength. Therefore

$$u_\lambda d\lambda = -u_\nu d\nu$$

or $$u_\lambda d\lambda = -\frac{8\pi h}{c^3} \frac{\left(\frac{c}{\lambda}\right)^3 \left(-\frac{c}{\lambda^2} d\lambda\right)}{e^{hc/\lambda kT} - 1}$$

$$\boldsymbol{u_\lambda d\lambda = \frac{8\pi hc}{\lambda^5} \frac{1}{e^{hc/\lambda kT} - 1} d\lambda} \tag{vii}$$

The above relation is known as Planck's formula in terms of wavelength.

15.2.3 Wien's Law and Rayleigh-Jeans Law

With the help of Planck's radiation formula Wien's law and Rayleigh-Jeans law can be derived. When the wavelength λ and temperature T are very small, then $e^{hc/\lambda kT} >> 1$. Therefore, 1 can be neglected in the denominator of Eq. (vii).

Thus

$$u_\lambda d\lambda = \frac{8\pi hc}{\lambda^5} e^{-hc/\lambda kT} d\lambda$$

By substituting $8\pi hc = A$ and $\frac{hc}{k} = B$, we get

$$u_\lambda d\lambda = \frac{A}{\lambda^5} e^{-B/\lambda T} d\lambda \tag{viii}$$

This is known as *Wien's law*, which is valid at low temperature T and small wavelength λ.

For high temperature T and large wavelength λ, $e^{hc/\lambda kT}$ can be approximated to $1 + \frac{hc}{\lambda kT}$. Then we have from Eq. (vii)

$$u_\lambda d\lambda = \frac{8\pi hc}{\lambda^5\left(1 + \frac{hc}{\lambda kT} - 1\right)} d\lambda$$

$$u_\lambda d\lambda = \frac{8\pi kT}{\lambda^4} d\lambda \tag{ix}$$

This is known as *Rayleigh-Jeans Law.*

15.3 SIMPLE CONCEPT OF QUANTUM THEORY — LO2

As discussed earlier, Planck's hypothesis says that the radiation does not emit in continuous fashion rather it gets emitted in discrete packets of energy equal to $h\nu$. These packets are referred to as quanta or photons. Therefore, it can be said that the exchange of energy between the radiation and the matter takes place in discrete set of values. In view of the application of quantum theory, it is necessary to be aware of the photon.

15.3.1 Photon: Mass, Energy and Momentum

Photon is an elementary particle that is massless and has no charge. It is a bundle of energy or packet of energy emitted by a source of radiation. It moves with velocity of light. It can carry energy and momentum. We know that the mass m of the particle moving with v, comparable with the velocity of light c, is given by as per the special theory of relativity.

$$m = \frac{m_0}{\sqrt{1 - v^2/c^2}} \tag{i}$$

where m is the relativistic mass of the particle and m_0 is its rest mass. Since the photon is moving with the velocity of the light, we substitute $v = c$ in Eq. (i). With this the moving mass m of the photon becomes $m = \infty$, which is not possible. So, if the photon moves with the velocity of light then the zero in the numerator balances the zero in the denominator, i.e., $m = 0/0$; this is an indeterminate quantity. It means if we take rest mass of the photon to be zero, this value should not particularly disturb us due to the fact that the photons are never at rest and always keep moving with the velocity of light.

The energy of a photon is given below as

$$E = h\nu \tag{ii}$$

If m is the moving mass of the photon, then according to special theory of relativity, the following relation gives the energy

$$E = mc^2 \tag{iii}$$

so $mc^2 = h\nu$

$$\text{or} \quad m = \frac{h\nu}{c^2} \tag{iv}$$

Now the energy relation

$$E^2 = p^2c^2 + m_0^2 c^4 \tag{v}$$

Since $m_0 = 0$, $E = pc$ and the momentum of the photon is given by

$$p = \frac{E}{c} = \frac{mc^2}{c} = mc \tag{vi}$$

$$\text{or} \quad p = \frac{E}{c} = \frac{h\nu}{c} \tag{vii}$$

Thus, if a photon of frequency ν is to be treated as a particle, then the characteristics of the photon are given as

$$m_0 = 0, E = h\nu, m = h\nu/c^2 \text{ and } p = h\nu/c \tag{viii}$$

These characteristics of the photons are useful in the discussion of Compton effect, which establishes the photon hypothesis.

15.4 WAVE PARTICLE DUALITY LO3

The phenomena of interference, diffraction and polarisation can be explained on the basis of wave theory of light. However, the wave nature of light fails to explain the phenomena of Compton effect, photoelectric effect, the continuous *X*-ray spectrum and the blackbody radiation. In the light of these facts, physicists assumed the particle nature of electromagnetic radiation (light). These entire phenomena can be explained on the basis of quantum hypothesis, according to which electromagnetic radiation is propagated in small packets or bundles. These packets are called photons. It means that light or electromagnetic radiation exhibits wave and particle properties both. Hence, light or electromagnetic radiation has dual nature, i.e., it behaves like a particle as well as a wave. This dual characteristic property of radiation is called dual nature of light or wave particle duality.

15.5 PHOTOELECTRIC EFFECT LO3

The photoelectric effect refers to the emission or ejection of electrons from the surface of a metal (generally) in response to incident light. Energy contained within the incident light is absorbed by the electrons within the metal, gaining sufficient energy to be 'knocked' out of, i.e., emitted from the surface of the metal. According to the classical Maxwell wave theory of light, the more intense incident light should eject the electrons from the metal with their greater energy. It means the average energy carried by an ejected (photoelectric) electron should increase with the intensity of the incident light. In fact, *Lenard* found that this was not so. Rather, he observed the energies of the emitted electrons to be independent of the intensity of the incident radiation. In 1905, *Einstein* resolved this paradox successfully by proposing that the incident light consists of individual quanta, called photons, that interact with the electrons in the metal like discrete particles, rather than as continuous waves. He adopted the Planck's quantum hypothesis and applied it to the electromagnetic radiation. For a given frequency (ν) or colour (λ) of the incident radiation, each photon carries the energy $E = h\nu$. According to Einstein's model, increase in the intensity of the light corresponds to the enhancement in the number of incident photons per unit time (flux), while the energy of each photon remains the same as long as the frequency of the radiation is kept constant. It means, increasing the intensity of the incident radiation would cause greater numbers of electrons to be ejected and each electron would carry the same average energy because each incident photon carries the same energy. Likewise, in Einstein's model, increasing the frequency ν rather than the intensity of the incident radiation would increase the average energy of the emitted electrons. Both of these predictions were confirmed experimentally. It is interesting to note that the rate of increase of the energy of the ejected electrons with increasing frequency enables us to determine the value of Planck's constant *h*, as the frequency can be measured.

15.5.1 Theoretical Explanation

In photoemission, one quantum is absorbed by one electron. If the electron is some distance into material of the cathode, some energy will be lost as it moves towards the surface. There will always be some electrostatic cost as the electron leaves the surface. This is known as the work function ϕ_0. The electrons those are very close to the surface will be the most energetic, and they will leave the cathode with kinetic energy given by

$$E_K = h\nu - \phi_0$$

$$\text{or} \quad E_K = h\nu - h\nu_0$$

$$\text{where} \quad h\nu_0 = \phi_0$$

Therefore, it is clear that there is a minimum light frequency called *threshold frequency* ν_0 for a given metal for which the quantum of energy is equal to the work function. Light below that frequency, no matter how bright, will not cause photoemission.

15.5.2 Experiment

An experimental arrangement to the photoelectric effect is shown in Fig. 15.4. It consists of a vacuum tube *A*, which contains a metallic plate *B* and a charge collecting plate *C*. When light is incident on the plate *B* through the quartz window, electrons are ejected from the metallic surface. The collector is kept at positive potential *V* with respect to the metallic plate, which is at zero potential. So, due to this positive potential the collector *C* collects these ejected electrons. Therefore, a current i_e is produced, which can be measured by the galvanometer *G*. We can increase the current i_e by increasing the potential *V* until i_e reaches a constant value, i.e., it approaches a saturation.

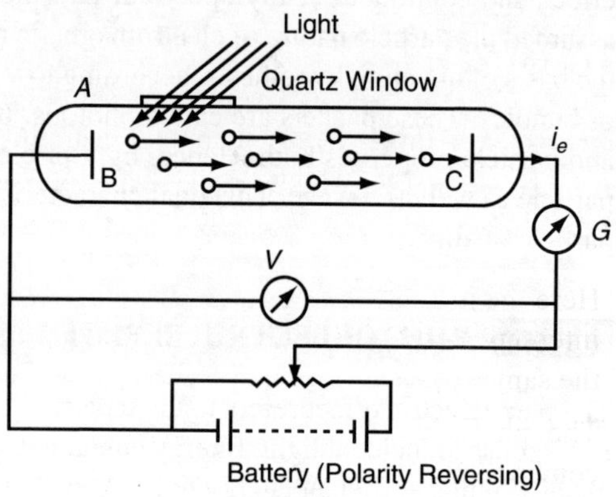

FIGURE 15.4

By using the reversing switch, we apply the negative potential to the collecting plate *C*. Under this situation, the electrons are repelled by *C* and only those electrons whose energy is greater than the potential energy *eV* will be able to reach the collector *C*. So we get some current in the galvanometer *G*. The applied potential for which the current i_e becomes zero, i.e., $i_e = 0$, is called stopping potential V_0. The relation between the maximum kinetic energy of the electrons E_K and stopping potential V_0 is given as below.

$$E_k = \frac{1}{2}mv_{\max}^2 = e|V_0|$$

We can obtain the following results by performing detailed experiment under various conditions.

(1) The photoelectric current i_e increases with the increasing intensity I of the incident radiation, if the frequency is kept constant.
(2) There is no time lag between illumination of the metal surface and the emission of electrons.
(3) If the frequency of the incident radiation is greater than the threshold frequency ν_0 (certain minimum frequency), only then the emission of electrons takes place.
(4) The maximum kinetic energy E_K of the photoelectrons is independent of the intensity I of the incident light. This is shown in Fig. 15.5 in which we observe that the stopping potential is same for the light of three different intensities having same frequency.
(5) The maximum kinetic energy of the photoelectrons depends on the frequency of the incident radiation. From Fig. 15.6, we observe that at different frequencies, stopping potential is also different but the saturation current remains the same.

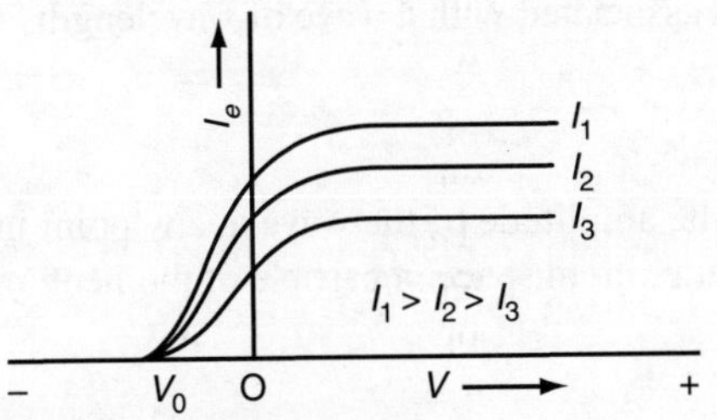

FIGURE 15.5

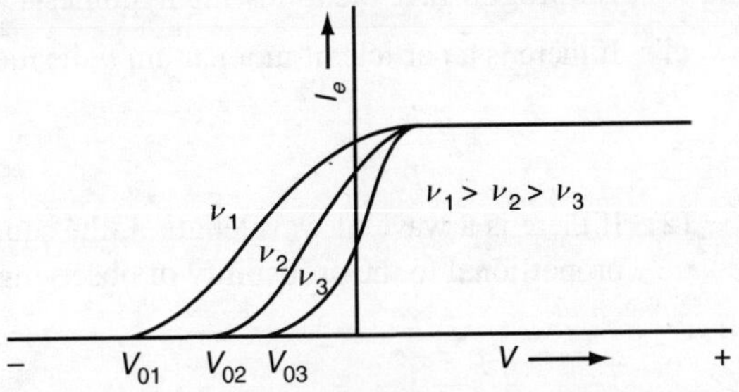

FIGURE 15.6

(6) There is a linear relation between maximum kinetic energy and frequency. We show this in Fig. 15.7 for three different metals cesium, potassium and tungsten, which satisfy the following relation

$$E_K = a_1\nu + a_2$$

Here a_1 is the slope of the straight line and a_2 is the intercept. From the figure it is clear that though a_1 remains the same for all surfaces, a_2 is different for different metals.

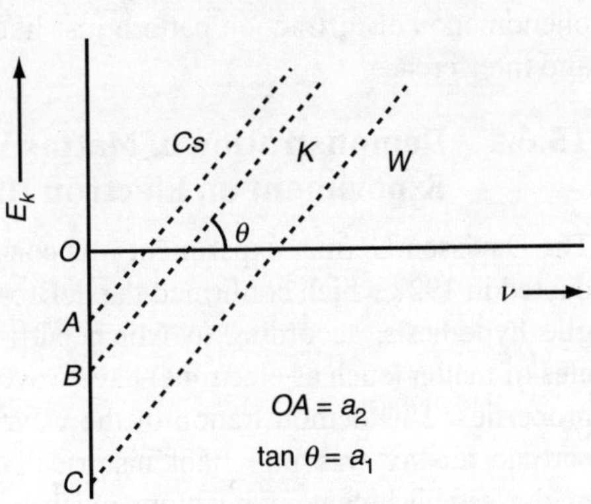

FIGURE 15.7

The photoelectric effect is perhaps the most direct and convincing evidence of the existence of photons and the *'corpuscular'* nature of light and electromagnetic radiation. That is, it provides undeniable evidence of the quantisation of the electromagnetic field and the limitations of the classical field equations of Maxwell. *Albert Einstein* received the Nobel Prize in Physics in 1921 for explaining the photoelectric effect and for his contributions to the theoretical physics.

15.6 de BROGLIE WAVES: MATTER WAVES

In 1924, Louis deBroglie proposed in his doctoral dissertation that there was a fundamental relation between waves and particles. Therefore, the energy of the photon according to special theory of relativity is given by

$$E = h\nu \tag{i}$$

and momentum p is

$$p = \frac{h\nu}{c} = \frac{h}{\lambda} \tag{ii}$$

Here, it can be noted that E and p are the characteristics of the partricles, and ν and λ are the characteristics of the waves. From above relations, we see that these sets of quantities are related to each other by the Planck's constant h. deBroglie also suggested that the dual nature of electromagnetic radiation may be extended to material particles such as electrons, protons, neutrons etc. It means that a moving particle, whatever its nature be, has wave properties associated with it. The waves associated with these particles are known as matter waves or deBroglie waves. The difference between the electromagnetic radiation and elementary particles is that in the case of photons, $m_0 = 0$ and $v = c$ but in the case of material particles $m_0 \neq 0$ and $v < c$. deBroglie gave the following hypothesis which is applicable to all matters, radiation and particles.

(1) If there is a particle of momentum p, its motion is associated with a wave of wavelength

$$\lambda = \frac{h}{p} \tag{iii}$$

(2) If there is a wave of wavelength λ, the square of the amplitude of the wave at any point in space is proportional to the probability of observing, at that point in space, a particle of momentum

$$p = \frac{h}{\lambda} \tag{iv}$$

The dual nature of matter can be proved if we could show that a beam of particles also exhibits the phenomenon of diffraction pattern just like the electromagnetic waves show the phenomena of diffraction and interference.

15.6.1 Demonstration of Matter Waves: Davisson-Germer Experiment on Electron Diffraction

The Davisson-Germer experiment was conducted in 1927 which confirmed the deBroglie hypothesis, according to which particles of matter (such as electrons) have wave properties. This demonstration of the wave particle duality was important historically in the establishment of quantum mechanics and of the Schröedinger equation. The experimental setup is shown in Fig. 15.8(a). Here the electrons from a heated filament or electron gun were accelerated by a voltage V and allowed to fall on the surface of nickel target. *Davisson* and *Germer* measured the intensity of the scattered electrons as a function of the angle ϕ and plotted it in the form of polar diagram. Fig. 15.8(b) shows the results from the accelerating voltage of 54 V. For this case, there is an intense scattering or a pronounced

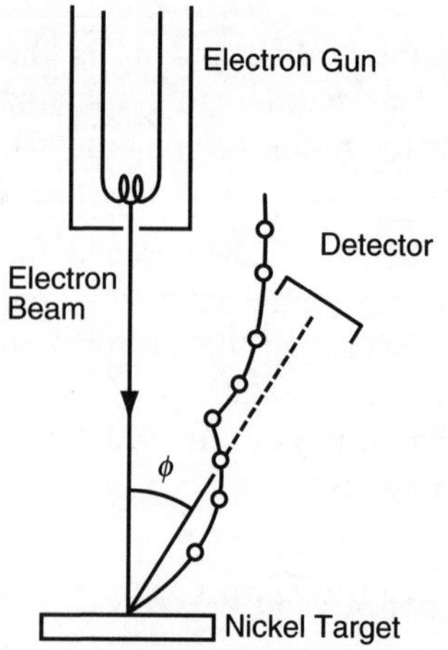

FIGURE 15.8(a)

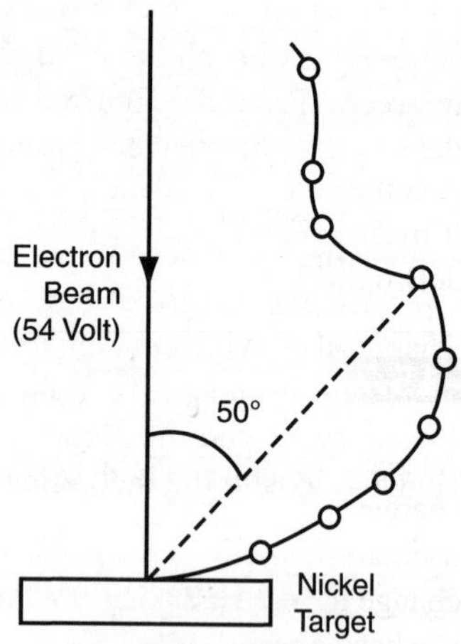

FIGURE 15.8(b)

peak at an angle of $\theta = 50^o$. Such deflection can be explained by assuming that the electron beam has a wave associated with it. This situation is similar to the Bragg's diffraction. So the waves associated with the electron beam were satisfying Bragg's law, which caused a diffraction peak. In order to prove this, consider Fig. 15.9 that shows atomic planes in Ni crystal. Here $\theta = 50^o$, $\phi = (180 - 50)/2 = 65^o$, $d = 0.91$ Å. Hence, for $n = 1$, Bragg's law $n\lambda = 2d \sin\theta$ gives

$$\lambda = 2d \sin\theta = 2 \times 0.91 \times \sin 65^o = 1.65 \text{ Å} \qquad (i)$$

Since Bragg's law basically talks about the diffraction of *X* rays, this experiment enables us to treat the electrons as waves and the wavelength associated with the electrons should be 1.65 Å, if they are scattered at $\theta = 50°$.

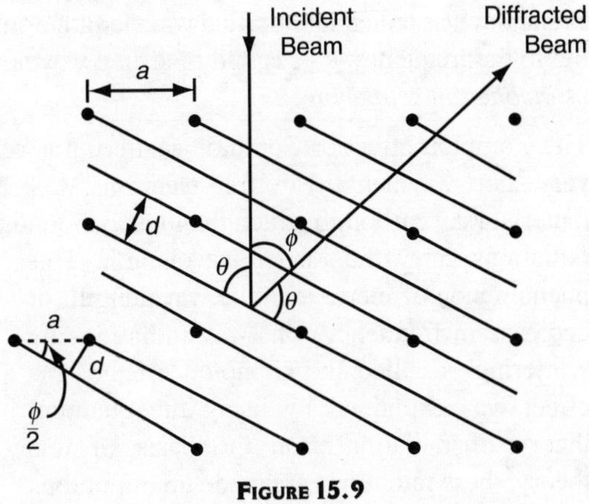

FIGURE 15.9

Now, we apply deBroglie's hypothesis. Since the electron of mass m gains the velocity v when it gets accelerated through a potential difference of V, we write the following relation for the energy for the nonrelativistic motion of the electron

$$\frac{1}{2}mv^2 = eV$$

So the deBroglie wavelength associated with the electron is given by

$$\lambda = \frac{h}{p} = \frac{h}{mv} = \frac{h}{\sqrt{2meV}}$$

or $$\lambda = \sqrt{\frac{150}{V}} \text{ Å}$$

Therefore, deBroglie wavelength associated with the electron that is accelerated by 54 V is given as

$$\lambda = \sqrt{\frac{150}{V}} \text{ Å} = \sqrt{\frac{150}{54}} \text{ Å} = 1.67 \text{ Å} \qquad \text{(ii)}$$

A comparison of Eq. (i) with Eq. (ii) shows that the value of the wavelength λ is the same in both the cases. It means there is a wave called deBroglie wave associated with the electrons. Therefore, this confirms the deBroglie hypothesis.

15.7 COMPTON EFFECT: COMPTON SCATTERING LO5

As per classical electromagnetic theory, when an electromagnetic radiation (frequency ν) is incident on free charges (say, electrons), the free charges absorb this radiation and start oscillating at frequency ν. Then these oscillating charges radiate electromagnetic waves of the same frequency ν. This type of scattering where the change in frequency or wavelength does not take place is called *coherent scattering*. This coherent scattering has been observed with the radiation in visible range and also at longer wavelengths. However, this predication of classical theory fails in the case of scattering of radiation of very short wavelengths like *X*-rays. Here the scattered *X*-rays are found to consist of two frequencies: ν and ν'. The wavelength λ corresponding to the frequency ν is called unmodified wavelength or unmodified radiation, whereas the wavelength λ' corresponding to the frequency ν' is called modified wavelength or modified radiation. This type of scattering is known as *incoherent scattering*.

The Compton effect or Compton scattering is related to the scattering of *X*-rays (electromagnetic waves of very short wavelength) by free electrons. A.H. Compton found that when *X*-rays are scattered by a solid material (say carbon in which the loosely bound electrons are assumed to be almost free) the scattered *X*-ray radiations carry the longer wavelength. This phenomenon of increase in the wavelength (or decrease in frequency) of *X*-ray radiations by scattering is called the Compton effect. This effect was explained by using the quantum theory of radiations. On the basis of this theory, these radiations are made up of photons of energy $h\nu$. These photons in the incident *X*-rays collide with the free electrons of the target (Fig. 15.10). If the collision is elastic, then the energy or wavelength of the scattered photons remains the same. If the collision is inelastic, then the incident photon transfers some of the energy to the electron. Thus, as a result, the energy of the scattered *X*-ray photon decreases (or wavelength increases).

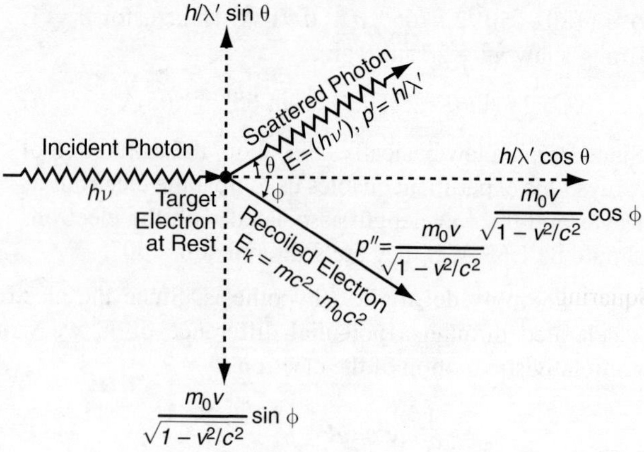

FIGURE 15.10

Let us assume,

λ = wavelength of incident *X*-rays

λ' = wavelength of scattered *X*-rays

$\Delta\lambda = \lambda' - \lambda$ = Compton shift

Energy of incident *X*-ray photon = $h\nu = hc/\lambda$

Momentum of incident *X*-ray photon = h/λ

Energy of the scattered *X*-ray photon $h\nu' = hc/\lambda'$

Momentum of the scattered *X*-ray photon = h/λ'

Kinetic energy of the recoiled electron $= (m - m_0)c^2 = \left[\dfrac{m_0}{\sqrt{1 - v^2/c^2}} - m_0\right]c^2$

Momentum of the recoiled electron $= mv = \dfrac{m_0 v}{\sqrt{1 - v^2/c^2}}$

m = moving mass or relativistic mass of the electron and m_0 is the rest mass.

According to the law of conservation of energy

Energy of the incident photon = Energy of the scattered photon + Energy of the recoiled electron.

$$E = E' + E_K$$

$$\frac{hc}{\lambda} = \frac{hc}{\lambda'} + (m - m_0)c^2$$

$$\text{or} \quad \frac{h}{\lambda} - \frac{h}{\lambda'} = \frac{m_0 c}{\sqrt{1 - v^2/c^2}} - m_0 c$$

$$\text{or} \quad \frac{h}{\lambda} - \frac{h}{\lambda'} + m_0 c = \frac{m_0 c}{\sqrt{1 - v^2/c^2}} \qquad \text{(i)}$$

According to the law of conservation of momentum,

$$\frac{h}{\lambda'} \sin\theta = \frac{m_0 v}{\sqrt{1 - v^2/c^2}} \sin\phi \qquad \text{(ii)}$$

$$\text{and} \quad \frac{h}{\lambda} = \frac{h}{\lambda'} \cos\theta + \frac{m_0 v}{\sqrt{1 - v^2/c^2}} \cos\phi$$

$$\text{or} \quad \frac{h}{\lambda} - \frac{h}{\lambda'} \cos\theta = \frac{m_0 v}{\sqrt{1 - v^2/c^2}} \cos\phi \qquad \text{(iii)}$$

Squaring and adding Eqs. (ii) and (iii), we get

$$\frac{h^2}{\lambda'^2} + \frac{h^2}{\lambda^2} - \frac{2h^2}{\lambda\lambda'} \cos\theta = \frac{m_0^2 v^2}{1 - v^2/c^2}$$

$$\text{or} \quad \frac{h^2}{\lambda'^2} + \frac{h^2}{\lambda^2} - \frac{2h^2}{\lambda\lambda'} \cos\theta = \frac{m_0^2 v^2 c^2}{c^2 - v^2} \qquad \text{(iv)}$$

On squaring Eq. (i), we get

$$\frac{h^2}{\lambda^2} + \frac{h^2}{\lambda'^2} + m_0^2 c^2 - \frac{2h^2}{\lambda\lambda'} - \frac{2hm_0 c}{\lambda'} + \frac{2hm_0 c}{\lambda} = \frac{m_0^2 c^2}{1 - v^2/c^2} = \frac{m_0^2 c^4}{c^2 - v^2}$$

$$\text{or} \quad \frac{h^2}{\lambda^2} + \frac{h^2}{\lambda'^2} - \frac{2h^2}{\lambda\lambda'} + 2hm_0 c\left(\frac{1}{\lambda} - \frac{1}{\lambda'}\right) = \frac{m_0^2 c^4}{c^2 - v^2} - m_0^2 c^2$$

$$\text{or} \quad \frac{h^2}{\lambda^2} + \frac{h^2}{\lambda'^2} - \frac{2h^2}{\lambda\lambda'} + 2hm_0 c\left(\frac{\lambda' - \lambda}{\lambda\lambda'}\right) = \frac{m_0^2 c^4 - m_0^2 c^4 + m_0^2 v^2 c^2}{c^2 - v^2}$$

$$\frac{h^2}{\lambda^2} + \frac{h^2}{\lambda'^2} - \frac{2h^2}{\lambda\lambda'} + 2hm_0 c\left(\frac{\lambda' - \lambda}{\lambda\lambda'}\right) = \frac{m_0^2 v^2 c^2}{c^2 - v^2} \qquad \text{(v)}$$

On comparing Eqs. (iv) and (v), we get

$$\frac{h^2}{\lambda^2} + \frac{h^2}{\lambda'^2} - \frac{2h^2}{\lambda\lambda'} + 2hm_0 c\left(\frac{\lambda' - \lambda}{\lambda\lambda'}\right) = \frac{h^2}{\lambda^2} + \frac{h^2}{\lambda'^2} - \frac{2h^2}{\lambda\lambda'} \cos\theta$$

$$\text{or} \quad 2hm_0 c\left(\frac{\lambda' - \lambda}{\lambda\lambda'}\right) = \frac{2h^2}{\lambda\lambda'}(1 - \cos\theta)$$

$$\text{or} \quad (\lambda' - \lambda) = \Delta\lambda = \frac{h}{m_0 c}(1 - \cos\theta)$$

$$\text{or} \quad \Delta\lambda = \frac{h}{m_0 c}(1 - \cos\theta) \qquad \text{(vi)}$$

where, h = Planck's constant. m_0 = rest mass of electron, c = velocity of light and θ = angle of scattering of the photon. This is just to emphasise that the RHS contains the angle of scattering of photon not of the electron.

15.7.1 Verification of Compton Effect

A beam of monochromatic X-ray of known wavelength is made incident on a graphite target. The intensity distribution with wavelength of monochromatic X-ray scattered at different angles is measured by Bragg's X-ray spectrometer. The intensity distribution with wavelength for different angles is shown in Fig. 15.11. It may be noted that diffraction patterns have two diffraction peaks-one corresponding to modified radiation and the other corresponding to unmodified radiation. The difference between two peaks on the wavelength axis provides the Compton shift. It can be concluded from the diffraction patterns that greater scattering angle yields greater Compton shift. For example, $\Delta\lambda = \lambda' - \lambda = h/m_0c\ (1 - \cos\theta)$ at $\theta = 90°$ gives $\Delta\lambda = 0.024$ Å.

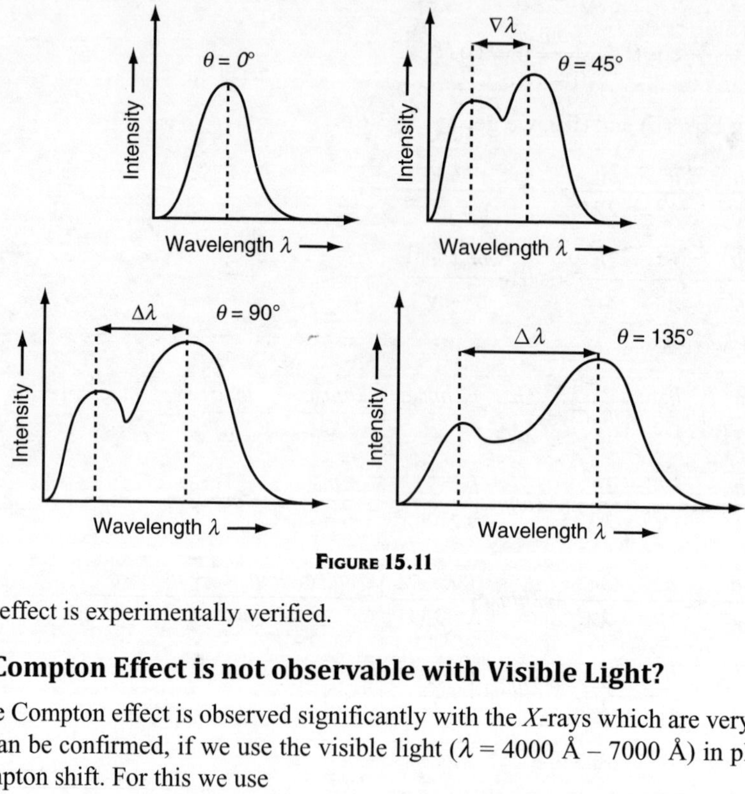

FIGURE 15.11

Hence, Compton effect is experimentally verified.

15.7.2 Why Compton Effect is not observable with Visible Light?

As mentioned, the Compton effect is observed significantly with the X-rays which are very short wavelength radiations. This can be confirmed, if we use the visible light (λ = 4000 Å – 7000 Å) in place of X-rays and calculate the Compton shift. For this we use

$$\Delta\lambda = \frac{h}{m_0 c}(1 - \cos\theta) = 0.0242(1 - \cos\theta)\text{Å}$$

For maximum shift, $\theta = 180°$. Hence

$$(\Delta\lambda)_{max} = 0.0484 \text{ Å}$$

The percentage Compton shift for λ = 4000 Å

$$= \frac{(\Delta\lambda)_{max}}{\lambda} \times 100 \simeq 0.001\%$$

Similarly, the percentage Compton shift for larger wavelength of visible light, i.e., for $\lambda = 7000$ Å, would be 0.0007%. So, you can see that the Compton shift for the case of visible light is not significant. For this reason, the X-rays are appropriate for realising the Compton effect or Compton scattering.

15.8 PHASE AND GROUP VELOCITIES: de BROGLIE WAVES LO6

Phase Velocity

Waves have already been discussed in Chapter 1. However, here we will discuss phase and group velocities in the context of deBroglie waves. We can write the deBroglie wave travelling along the $+x$ direction as

$$y = a \sin(\omega t - kx) \tag{i}$$

where a is the amplitude, ω $(=2\pi\nu)$ is the angular frequency and k $(=2\pi/\lambda)$ is the propagation constant of the wave. By the definition, the ratio of angular frequency ω to the propagation constant k is the phase (or wave) velocity. If we represent the phase velocity by u, then

$$u = \frac{\omega}{k}$$

$(\omega t - kx)$ is called the phase of the wave motion. It means the particle of the constant phase travels such that $(\omega t - kx)$ = constant.

$$\text{or} \quad \frac{d}{dt}(\omega t - kx) = 0$$

$$\omega - k\frac{dx}{dt} = 0$$

$$\text{or} \quad \boldsymbol{\frac{dx}{dt} = u = \frac{\omega}{k}} \tag{ii}$$

where $u = \dfrac{dx}{dt}$ is the phase (or wave) velocity. Thus the wave velocity is the velocity of planes of constant phase which advances through the medium. We can write the phase velocity $u = \nu\lambda$ and for an electromagnetic wave $E = h\nu$, or $\nu = E/h$

According to deBroglie $\lambda = \dfrac{h}{p} = \dfrac{h}{mv}$

$$u = \nu\lambda = \frac{E}{h} \times \frac{h}{mv} = \frac{mc^2}{mv} = \frac{c^2}{v}$$

$$u = \frac{c^2}{v} \tag{iii}$$

Since $c >> v$, Eq. (iii) implies that the phase velocity of deBroglie wave associated with the particle moving with velocity v is greater than c, the velocity of light.

Group Velocity

As we have seen, the phase velocity of a wave associated with a particle comes out to be greater than the velocity of light. This difficulty can be overcome by assuming that each moving particle is associated with a group of waves or a wave packet rather than a single wave. In this context, deBroglie waves are represented by a wave packet and hence we have *group velocity* associated with them. In order to realize the concept of group velocity, we consider the combination of two waves, resultant of which is shown in Fig. 15.12. The two waves are represented by the following relations

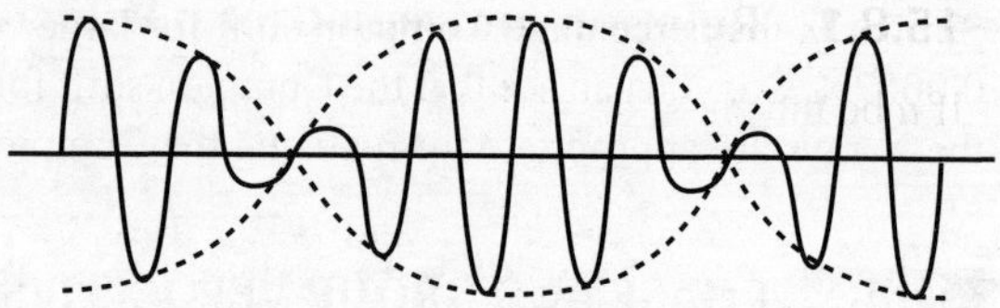

FIGURE 15.12

$$y_1 = a \sin (\omega_1 t - k_1 x) \quad \text{(i)}$$

and $$y_2 = a \sin (\omega_2 t - k_2 x) \quad \text{(ii)}$$

Their superposition gives

$$y = y_1 + y_2 = a\,[\sin (\omega_1 t - k_1 x) + \sin (\omega_2 t - k_2 x)]$$

or $$y = 2a \sin\left[\frac{(\omega_1+\omega_2)t}{2} - \frac{(k_1+k_2)x}{2}\right] \cos\left[\frac{(\omega_1-\omega_2)t}{2} - \frac{(k_1-k_2)x}{2}\right]$$

$$\therefore \quad y = 2a \cos\left[\frac{(\omega_1-\omega_2)t}{2} - \frac{(k_1-k_2)x}{2}\right] \sin (\omega t - kx) \quad \text{(iii)}$$

where $\omega = \dfrac{\omega_1+\omega_2}{2},\ k = \dfrac{k_1+k_2}{2}$

Eq. (iii) can be re-written as

$$\mathbf{y = 2a \cos\left[\frac{(\Delta\omega)t}{2} - \frac{(\Delta k)x}{2}\right] \sin(\omega t - kx)} \quad \text{(iv)}$$

where $\Delta\omega = \omega_1 - \omega_2$ and $\Delta k = k_1 - k_2$.

The resultant wave Eq. (iv) has two parts.

(i) A wave of frequency ω, propagation constant k and the velocity u, given by

$$\boldsymbol{u = \frac{\omega}{k} = \frac{2\pi\nu}{2\pi/\lambda} = \nu\lambda}$$

which is the phase velocity or wave velocity.

(ii) Another wave of frequency $\Delta\omega/2$, propagation constant $\Delta k/2$ and the velocity $G = \dfrac{\Delta\omega}{\Delta k}$. This velocity is the velocity of envelope of the group of waves, i.e., it is the velocity of the wave packet (shown by dotted lines) and is known as group velocity.

For the waves having small difference in their frequencies and wave numbers, we can write

$$G = \frac{\Delta\omega}{\Delta k} = \frac{\partial\omega}{\partial k} = \frac{\partial(2\pi\nu)}{\partial(2\pi/\lambda)} = \frac{\partial\nu}{\partial(1/\lambda)} = -\lambda^2\frac{\partial\nu}{\partial\lambda}$$

$$G = -\frac{\lambda^2}{2\pi}\frac{\partial\omega}{\partial\lambda}$$

This is the expression for the group velocity.

15.8.1 Relation between Group Velocity and Phase Velocity

If u be the phase (wave) velocity, then the group velocity can be written as

$$G = \frac{d\omega}{dk} = \frac{d}{dk}(uk) \quad \left[u = \frac{\omega}{k}\right]$$

or $$G = u + k\frac{du}{dk}$$

But $$k = \frac{2\pi}{\lambda} \Rightarrow dk = -\frac{2\pi}{\lambda^2}d\lambda \text{ and } \frac{k}{dk} = -\frac{\lambda}{d\lambda}$$

Therefore, the group velocity is given by

$$G = u + \left(-\frac{\lambda}{d\lambda}\right)du$$

or $$G = u - \lambda\frac{du}{d\lambda}$$

This relation shows that the group velocity G is less than the phase velocity u in a dispersive medium where u is a function of k or λ. However, in a non-dispersive medium, the velocity u is independent of k, i.e., the wave of all wavelength travel with the same speed, i.e., $\frac{du}{d\lambda} = 0$. Then $\boldsymbol{G = u}$. This is true for electromagnetic waves in vacuum and the elastic waves in homogenous medium.

15.8.2 Relation between Group Velocity and Particle Velocity

Consider a material particle of rest mass m_0. Let its mass be m when it moves with velocity v. Then its total energy E is given by

$$E = mc^2 = \frac{m_o c^2}{\sqrt{1 - v^2/c^2}}$$

Its momentum is given by

$$p = mv = \frac{m_o v}{\sqrt{1 - v^2/c^2}}$$

The frequency of the associated deBroglie wave is given by

$$\nu = \frac{E}{h} = \frac{m_o c^2}{\sqrt{1 - v^2/c^2}}\frac{1}{h}$$

$$\omega = 2\pi\nu = \frac{2\pi m_0 c^2}{h\sqrt{1 - v^2/c^2}}$$

or $$\frac{d\omega}{dv} = \frac{2\pi m_0 c^2}{h(1 - v^2/c^2)^{3/2}}\left[-\frac{1}{2} \times \left(-\frac{2v}{c^2}\right)\right]$$

$$\text{or} \quad \frac{d\omega}{dv} = \frac{2\pi m_0 v}{h(1 - v^2/c^2)^{3/2}}$$

The wavelength of the associated deBroglie wave is given by

$$\lambda = \frac{h}{p} = \frac{h(1 - v^2/c^2)^{1/2}}{m_0 v}$$

Hence propagation constant

$$k = \frac{2\pi}{\lambda} = \frac{2\pi m_0 v}{h(1 - v^2/c^2)^{1/2}}$$

$$\text{or} \quad \frac{dk}{dv} = \frac{2\pi m_0}{h} \frac{\left(1 - \frac{v^2}{c^2}\right)^{1/2} - v\frac{1}{2}\left(1 - \frac{v^2}{c^2}\right)^{-1/2}\left(-\frac{2v}{c^2}\right)}{\left(1 - \frac{v^2}{c^2}\right)}$$

$$\text{or} \quad \frac{dk}{dv} = \frac{2\pi m_0}{h}\left[\left(1 - \frac{v^2}{c^2}\right)^{-1/2} + \frac{v^2}{c^2}\left(1 - \frac{v^2}{c^2}\right)^{-3/2}\right]$$

$$= \frac{2\pi m_0}{h}\left(1 - \frac{v^2}{c^2}\right)^{-3/2}\left[1 - \frac{v^2}{c^2} + \frac{v^2}{c^2}\right]$$

$$= \frac{2\pi m_0}{h}\left(1 - \frac{v^2}{c^2}\right)^{-3/2}$$

$$\frac{dk}{dv} = \frac{2\pi m_0}{h(1 - v^2/c^2)^{3/2}}$$

Since the group velocity

$$G = \frac{d\omega}{dk} = \frac{d\omega/dv}{dk/dv}$$

$$G = \frac{\dfrac{2\pi m_0 v}{h(1 - v^2/c^2)^{3/2}}}{\dfrac{2\pi m_0}{h(1 - v^2/c^2)^{3/2}}} = v$$

$\mathbf{G} = \mathbf{v}$ = the particle velocity

Thus the wave group associated with the moving material particle travels with the same velocity as the particle. It proves that a material particle in motion is equivalent to group of waves or a wave packet.

SUMMARY

The topics covered in this chapter are summarised below.

- Insufficiency of classical mechanics in explaining phenomena like photoelectric effect, Compton effect, Raman effect, etc. led to the development of quantum mechanics.
- Quantum mechanics developed in two stages. The first stage began with Max Planck's hypothesis according to which the radiation is emitted or absorbed by matter in discrete packets or quanta of energy. This hypothesis led to a theory which was not completely satisfactory being a mixture of classical and nonclassical concepts. The second stage led to two types of mechanics, namely matrix mechanics and wave mechanics.
- Blackbody radiation shows that i) the distribution of frequencies is a function of temperature of the blackbody, ii) with the increase in temperature, the total amount of emitted radiation increases, and iii) the position of the maximum peak shifts toward higher frequencies with increasing equilibrium temperature.
- Planck explained blackbody radiation by using the Maxwell-Boltzmann distribution, i.e., $N_n = N_0 e^{-E/kT}$. He obtained the following expression (called Planck's radiation law) for the distribution of the maximum intensity of radiation in the spectrum of blackbody

$$I(\nu) = \frac{8\pi h\nu^3}{c^3}\frac{1}{e^{(k\nu/kT)} - 1}$$

- As per Planck's quantum hypothesis, the atoms of the wall of a blackbody behave as oscillators and each has a characteristic frequency of oscillation. Then average energy of these Planck's oscillators were calculated and finally Planck's radiation formula was derived.
- Wien's law is deduced from Planck's radiation formula under the condition when the wavelength λ and temperature T are very small. However, Planck's radiation formula under the condition of high temperature T and large wavelength λ takes the form of Rayleigh-Jeans law.
- Photon is an elementary particle that is massless and has no charge. It is a bundle of energy or packet of energy emitted by a source of radiation. It moves with velocity of light. It can carry energy and momentum. If a photon of frequency ν is to be treated as a particle, then the characteristics of the photon are given as

$$m_0 = 0, \quad E = h\nu, \quad m = h\nu/c^2 \quad \text{and} \quad p = h\nu/c.$$

- Light or electromagnetic radiation exhibits wave and particle properties both, it means it has dual nature, i.e., it behaves like a particle as well as a wave. The dual characteristic property of radiation is called wave particle duality.
- deBroglie suggested that the dual nature of electromagnetic radiation may be extended to material particles such as electrons, protons, neutrons etc. It means that a moving particle, whatever its nature be, has wave properties associated with it. The waves associated with these particles are known as matter waves or deBroglie waves. The Davisson-Germer experiment on electron diffraction demonstrated these waves.
- If there is a particle of momentum p, its motion is associated with a wave of wavelength $\lambda = \frac{h}{p}$.

- If there is a wave of wavelength λ, the square of the amplitude of the wave at any point in space is proportional to the probability of observing, at that point in space, a particle of momentum $p = \dfrac{h}{\lambda}$.
- The photoelectric effect refers to the emission or ejection of electrons from the surface of a metal (generally) in response to incident light. Energy contained within the incident light is absorbed by the electrons within the metal, gaining sufficient energy to be 'knocked' out of, i.e., emitted from, the surface of the metal. The photoelectric effect is perhaps the most direct and convincing evidence of the existence of photons and the 'corpuscular' nature of light and electromagnetic radiation. That is, it provides undeniable evidence of the quantisation of the electromagnetic field and the limitations of the classical field equations of Maxwell.
- When an electromagnetic radiation (frequency ν) is incident on free charges (say, electrons), the free charges absorb this radiation and start oscillating at frequency ν. Then these oscillating charges radiate electromagnetic waves of the same frequency ν. This type of scattering where the change in frequency or wavelength does not take place is called coherent scattering. This coherent scattering has been observed with the radiation in visible range and also at longer wavelengths.
- In the case of scattering of radiation of very short wavelengths like *X*-rays, the scattered rays are found to consist of two frequencies: ν and ν_1. The wavelength λ corresponding to the frequency ν is called unmodified wavelength or unmodified radiation, whereas the wavelength λ_1 corresponding to the frequency ν_1 is called modified wavelength or modified radiation. This type of scattering is known as incoherent scattering.
- Compton effect is not significantly observable with visible light, as the Compton shift is extremely small.

Solved Examples

Example 1 Calculate the frequency and wavelength of a photon whose energy is 75 eV.

Solution Given energy $E = 75 \text{ eV} = 75 \times 1.6 \times 10^{-19}$ J.

Formula used is

$$E = h\nu = \frac{hc}{\lambda}$$

$$\text{Frequency}(\nu) = \frac{E}{h} = \frac{75 \times 1.6 \times 10^{-19}}{6.62 \times 10^{-34}}$$

$$= 18.13 \times 10^{15} \text{ Hz}$$

and wavelength $$\lambda = \frac{c}{\nu} = \frac{3 \times 10^8}{18.13 \times 10^{15}}$$

$$= 1.655 \times 10^{-8} \text{ m}$$

$$= 165.5 \times 10^{-10} \text{ m}$$

or $\lambda = \mathbf{165.5 \text{ Å}}$

Example 2 Find the number of quanta of energy emitted per second if a radio station operates at a frequency of 98 MHz and radiates power of 2×10^5 W.

Solution Given $\nu = 98 \times 10^6$ cycles/sec and Power (P) = 2×10^5 W = 2×10^5 J/sec.

Energy of each quanta is

$$E = h\nu$$

$$\therefore E = 6.62 \times 10^{-34} \times 98 \times 10^6$$

$$= 6.4876 \times 10^{-26} \text{ J/quanta}$$

$$= 6.5 \times 10^{-26} \text{ J/quanta}$$

Number of quanta emitted per second

$$= \frac{\text{Power}}{\text{quantum energy}}$$

$$\frac{2 \times 10^5 \text{ (J/sec)}}{6.5 \times 10^{-26} \text{ (J/quanta)}}$$

$$= \mathbf{3.08 \times 10^{30}} \textbf{ quanta/sec}$$

Example 3 A certain spectral line has wavelength 4000 Å. Calculate the energy of the photon.

Solution Given $\lambda = 4.0 \times 10^{-7}$ m.

Formula used is

$$E_k = h\nu = \frac{hc}{\lambda}$$

$$= \frac{6.62 \times 10^{-34} \times 3 \times 10^8}{4 \times 10^{-7} \text{ m}}$$

$$= \mathbf{4.965 \quad 10^{-19} \text{ J}}$$

Example 4 Calculate the number of photons of green light of wavelength 5000 Å require to make one erg of energy.

Solution Given $\lambda = 5 \times 10^{-7}$ m.

Formula used is

$$E = \frac{hc}{\lambda} = \frac{6.62 \times 10^{-34} \times 3 \times 10^8}{5 \times 10^{-7}}$$

$$= 3.972 \times 10^{-19} \text{ J}$$

$$= 3.972 \times 10^{-12} \text{ erg}$$

Number of photons of green light emitted (per energy)

$$= \frac{1.0}{3.972 \times 10^{-12}}$$

$$= \mathbf{252 \times 10^9}$$

Example 5 Calculate the wavelength of a photon of energy 5×10^{-19} J.

Solution Given $E = 5 \times 10^{-19}$ J

Formula used is

$$E = \frac{hc}{\lambda}$$

$$\text{or} \quad \lambda = \frac{hc}{E} = \frac{6.62 \times 10^{-34} \times 3 \times 10^{8}}{5 \times 10^{-19}}$$

$$= 3.972 \times 10^{-7} \text{ m}$$

$$= \mathbf{4000\ Å}$$

Example 6 Calculate the energy of an electron of wavelength 4.35×10^{-7} m.

Solution Given wavelength (λ) = 4.35×10^{-7} m.

Formula used is

$$E = h\nu = \frac{hc}{\lambda}.$$

$$= \frac{6.62 \times 10^{-34} \times 3 \times 10^{8}}{4.35 \times 10^{-7}}$$

$$= \mathbf{4.566 \times 10^{-19}\ J}$$

Example 7 How many watts of power at the threshold is received by the eye, if it receives 120 photons per second of the visible light of wavelength = 5600 Å.

Solution Given $\lambda = 5.6 \times 10^{-7}$ m and number of photons = 120.

Energy of a photon $E = h\nu = \frac{hc}{\lambda}$

$$\text{or} \quad E = \frac{6.62 \times 10^{-34} \times 3 \times 10^{8}}{5.6 \times 10^{-7}} = 3.55 \times 10^{-19} \text{ J}$$

The energy received by the eye per second = 3.5464×120 J/sec

Because $3.55 \times 120 = 426$

3.5464×120 J/sec = **425.57 W** = $\mathbf{4.2557 \times 10^{-17}\ W}$

Example 8 How many photons of yellow light of wavelength 5500 Å constitute 1.5 J of energy.

Solution Given $\lambda = 5.5 \times 10^{-7}$ m and energy of n photons = 1.5 J

Formula used is $E = h\nu = \frac{hc}{\lambda}$

Energy of a photon of yellow light, i.e.,

$$E = \frac{hc}{\lambda} = \frac{6.62 \times 10^{-34} \times 3 \times 10^{8}}{5.6 \times 10^{-7}}$$

$$= 3.55 \times 10^{-19} \text{ J}$$

Given

$n \times$ energy of one photon = 1.5 J

$$\text{or} \quad n = \frac{1.5}{3.55 \times 10^{-19}}$$

$$= \mathbf{4.2 \times 10^{18}}$$

EXAMPLE 9 Calculate the work function, stopping potential and maximum velocity of photoelectrons for a light of wavelength 4350 Å when it incidents on sodium surface. Consider the threshold wavelength of photoelectrons to be 5420 Å.

SOLUTION Given $\lambda_0 = 5.42 \times 10^{-7}$ m and $\lambda = 4.35 \times 10^{-7}$ m.

Formulae used are

$$\phi_0 = \frac{hc}{\lambda_0} = h\nu_0$$

$$\frac{1}{2}mv_{\max}^2 = hc\left[\frac{1}{\lambda} - \frac{1}{\lambda_0}\right] \quad \text{and}$$

$$eV = h\nu - h\nu_0 = hc\left[\frac{1}{\lambda} - \frac{1}{\lambda_0}\right]$$

$$\text{or} \quad eV = \frac{1}{2}mv_{\max}^2 = (E_k)_{\max}$$

$$\phi_0 = \frac{hc}{\lambda_0} = \frac{6.62 \times 10^{-34} \times 3 \times 10^8}{5.42 \times 10^{-7}} = 3.664 \times 10^{-19}\ \text{J}$$

$$\frac{1}{2}mv_{\max}^2 = hc\left[\frac{1}{\lambda} - \frac{1}{\lambda_0}\right]$$

$$v_{\max}^2 = \frac{2hc}{m}\left[\frac{\lambda_0 - \lambda}{\lambda_0 \lambda}\right]$$

$$= \frac{2 \times 6.62 \times 10^{-34} \times 3 \times 10^8}{9.1 \times 10^{-31}}\left[\frac{(5.42 - 4.35) \times 10^{-7}}{5.42 \times 4.35 \times 10^{-14}}\right] = 0.1981 \times 10^{12}$$

$$\therefore \quad v_{\max} = 0.445 \times 10^6\ \text{m/sec}$$

$$= 4.45 \times 10^5\ \text{m/sec}$$

$$\text{eV} = \frac{1}{2}mv_{\max}^2$$

The stopping potential

$$V = \frac{mv_{\max}^2}{2e} = \frac{9.1 \times 10^{-31} \times (4.45 \times 10^5)^2}{2 \times 1.6 \times 10^{-19}}$$

$$\mathbf{= 0.56\ volts}$$

EXAMPLE 10 The threshold frequency for photoelectric emission in copper is 1.1×10^{15} Hz. Find the maximum energy in *eV* when light of frequency 1.2×10^{15} Hz is directed on the copper surface.

SOLUTION Given $\nu_0 = 1.1 \times 10^{15}$ Hz and $\nu = 1.2 \times 10^{15}$ Hz.

Formula used is

$$\frac{1}{2}mv_{\max}^2 = h\nu - h\nu_0 = h(\nu - \nu_0)$$

$$= 6.62 \times 10^{-34}[1.2 - 1.1] \times 10^{15}$$

$$= 0.662 \times 10^{-19}\ \text{J}$$

$$\mathbf{= 0.414 eV}$$

EXAMPLE 11 Calculate the work function in electron volts of a metal, given that photoelectric threshold is (i) 6200 Å (ii) 5000 Å.

SOLUTION Given (i) $\lambda_0 = 6.2 \times 10^{-7}$ m (ii) $\lambda_0 = 5.0 \times 10^{-7}$ m.

$$\phi_0 = h\nu_0 = \frac{hc}{\lambda_0}$$

$$\text{(i)} \quad \phi_0 = \frac{6.62 \times 10^{-34} \times 3 \times 10^8}{6.2 \times 10^{-7} \times 1.6 \times 10^{-19}} \text{eV}$$

$$= \mathbf{2.0\ eV}$$

$$\text{(ii)} \quad \lambda_0 = 5.0 \times 10^{-7} \text{ m}$$

$$\phi_0 = \frac{6.62 \times 10^{-34} \times 3 \times 10^8}{5 \times 10^{-7} \times 1.6 \times 10^{-19}}$$

$$= 2.483 \text{ eV}$$

$$= \mathbf{2.48\ eV}$$

EXAMPLE 12 Find out the maximum energy of the photoelectron, work function and threshold frequency when a light of wavelength 3132 Å is incident on a surface of cesium and the stopping potential for the photo electron is 1.98 volt.

SOLUTION Given $V = 1.98$ volts and $\lambda = 3.132 \times 10^{-7}$ m.

Formulae used are

$$E_k = \frac{1}{2} m v_{\text{max}}^2 = \text{eV}_0, \quad V_0 = \text{stopping potential}$$

$$\text{and } E_k = h(\nu - \nu_0) = hc\left(\frac{1}{\lambda} - \frac{1}{\lambda_0}\right)$$

Then maximum energy of the photoelectron (E_{max})

$$= eV_0 = 1.6 \times 10^{-19} \times 1.98 \text{ J}$$

$$E_k = 3.168 \times 10^{-19} \text{ J}$$

$$E_k = hc\left(\frac{1}{\lambda} - \frac{1}{\lambda_0}\right)$$

$$3.168 \times 10^{-19} = 6.62 \times 10^{-34} \times 3 \times 10^8 \left[\frac{1}{3.132 \times 10^{-7}} - \frac{1}{\lambda_0}\right]$$

$$\frac{1}{6.2689 \times 10^{-7}} = \frac{1}{3.132 \times 10^{-7}} - \frac{1}{\lambda_0}$$

$$\text{or } \frac{1}{\lambda_0} = 3.193 \times 10^6 - 1.595 \times 10^6 = 1.598 \times 10^6$$

$$\lambda_0 = \frac{1}{1.598 \times 10^6} = 6258 \text{ Å}$$

$$\text{Work function } (\phi_0) = \frac{hc}{\lambda_0} = \frac{6.62 \times 10^{-34} \times 3 \times 10^8}{6.258 \times 10^{-7}}$$

$$= \mathbf{3.174 \times 10^{-19}\ J}$$

EXAMPLE 13 Is it possible to liberate an electron from a metal surface having work function 4.8 eV with an incident radiation of wavelength (i) 5000 Å *and* (ii) 2000 Å.

SOLUTION Given $\phi_0 = 4.8$ eV.

Formula used is $E_k = \dfrac{hc}{\lambda}$.

$$\text{(i) Energy } (E_k) = \frac{hc}{\lambda} = \frac{6.62 \times 10^{-34} \times 3 \times 10^8}{5 \times 10^{-7}} \text{J}$$

$$= \frac{6.62 \times 10^{-34} \times 3 \times 10^8}{5 \times 10^{-7} \times 1.6 \times 10^{-19}} \text{eV}$$

$$= \mathbf{2.48\ eV}$$

From the above it is clear that the energy corresponding to wavelength 5000 Å is found to be less than the work function i.e., 4.8 eV. So it will not be able to liberate an electron.

$$\text{(ii)} \quad E_k = \frac{hc}{\lambda} = \frac{6.62 \times 10^{-34} \times 3 \times 10^8}{2.0 \times 10^{-7}} = 9.93 \times 10^{-19} \text{ J}$$

$$\text{or } E_k = \frac{9.93 \times 10^{-19}}{1.6 \times 10^{-19}} \text{ eV} = 6.206 \text{ eV}$$

$$E_k = \mathbf{6.21\ eV}$$

As the energy corresponding to wavelength 2000 Å is greater than the work function. So it is sufficient to liberate electrons.

EXAMPLE 14 Find the maximum energy of the photoelectron, the work function and threshold frequency, if the potassium surface is illuminated by a light of wavelength 5893 Å. The stopping potential for the emitted electron is 0.36 V.

SOLUTION Given stopping potential $V_0 = 0.36$ V and $\lambda = 5893$ Å.

Formula used is

$$E_k = \text{eV} = h\nu - \phi_0$$

$$E_k = \text{eV} = 0.36 \text{ eV}$$

Work function

$$(\phi_0) = h\nu - \text{eV} = \frac{hc}{\lambda} - \text{eV}$$

$$= \frac{6.62 \times 10^{-34} \times 3 \times 10^8}{5.893 \times 10^{-7} \times 1.6 \times 10^{-19}} - 0.36 \text{ eV}$$

$$= 2.11 - 0.36 = 1.75 \text{ eV}$$

Thus the work function is 1.75 eV.

Threshold frequency

$$\nu_0 = \frac{\phi}{h}$$

$$= \frac{1.75 \times 1.6 \times 10^{-19}}{6.62 \times 10^{-34}}$$

$$= \mathbf{4.23 \times 10^{14}\ cycles/sec}$$

EXAMPLE 15 Find the maximum kinetic energy of the emitted electrons and the stopping potential if the light of wavelength 5890 Å is incident on the surface for which threshold frequency is 7320 Å.

SOLUTION Given $\lambda = 5.89 \times 10^{-7}$ m and $\lambda_0 = 7.32 \times 10^{-7}$ m.

Formula used is

$$E_k = h\nu - h\nu_0 = hc\left(\frac{1}{\lambda} - \frac{1}{\lambda_0}\right)$$

$$= \frac{6.62 \times 10^{-34} \times 3 \times 10^8}{10^{-7}}\left[\frac{1}{5.89} - \frac{1}{7.32}\right]$$

$$= 19.86 \times 10^{-19}\left[\frac{7.32 - 5.89}{5.89 \times 7.32}\right]$$

$$= \mathbf{6.587 \times 10^{-20}\ J}$$

$$eV = E_k \quad \text{or} \quad V = \frac{E_k}{e}$$

$$\text{Stopping potential } (V) = \frac{6.587 \times 10^{-20}}{1.6 \times 10^{-19}}$$

$$= \mathbf{0.412\ V}$$

EXAMPLE 16 The threshold wavelength for photoelectric emission in Tungsten is 2300 Å. What wavelength of light must be used in order for electrons with a maximum energy of 1.5 eV to be ejected?

SOLUTION Given $\lambda_0 = 2.3 \times 10^{-7}$ m and $E_k = 1.5$ eV.

Formula used is

$$E = h\nu - h\nu_0 = hc\left(\frac{1}{\lambda} - \frac{1}{\lambda_0}\right)$$

$$\frac{1}{\lambda} - \frac{1}{\lambda_0} = \frac{E}{hc}$$

$$\text{or} \quad \frac{1}{\lambda} = \frac{1.5 \times 1.6 \times 10^{-19}}{6.62 \times 10^{-34} \times 3 \times 10^8} + \frac{1}{2.3 \times 10^{-7}}$$

$$= 1.2085 \times 10^6 + 4.3478 \times 10^6$$

$$\frac{1}{\lambda} = 5.556 \times 10^6$$

$$\text{or} \quad \lambda = 1.7998 \times 10^{-7}\ \text{m}$$

$$\lambda = \mathbf{1799.8\ Å}$$

EXAMPLE 17 The work function of Tungsten is 4.53 eV. If ultraviolet light of wavelength 1500 Å is incident on the surface, does it cause photoelectron emission? If so, what is the kinetic energy of the emitted electron?

SOLUTION Given work function $\phi_0 = 4.53$ eV and $\lambda = 1.5 \times 10^{-7}$ m.

Formula used is $E_k = \dfrac{hc}{\lambda}$

Energy corresponding to incident photon of wavelength 1.5×10^{-7} m

$$E_k = \frac{hc}{\lambda} = \frac{6.62 \times 10^{-34} \times 3 \times 10^8}{1.5 \times 10^{-7}} \text{ J}$$

$$= 13.24 \times 10^{-19} \text{ J}$$

$$E_k = 8.28 \text{ eV}$$

The kinetic energy of the electron

$$E_k = \frac{1}{2} m v_{max}^2 = h\nu - h\nu_0 = h\nu - \phi_0$$

$$= 8.28 - 4.53$$

$$= \mathbf{3.75 \text{ eV}}$$

EXAMPLE 18 The work function of sodium metal is 2.3 eV. What is the longest wavelength of light that cause photoelectric emission from sodium?

SOLUTION Given $\phi_0 = 2.3 \text{ eV} = 2.3 \times 1.6 \times 10^{-19}$ J.

$$\phi_0 = h\nu_0 = \frac{hc}{\lambda_0}$$

Longest wavelength = Threshold wavelength

$$\lambda_0 = \frac{hc}{\phi_0} = \frac{6.62 \times 10^{-34} \times 3 \times 10^8}{2.3 \times 1.6 \times 10^{-19}}$$

$$= \mathbf{5396.74 \text{ Å}}$$

EXAMPLE 19 Evaluate the threshold wavelength of photoelectric material whose work function is 2.0 eV.

SOLUTION Given $\phi_0 = 2.0 \text{ eV} = 2 \times 1.6 \times 10^{-19}$ J.

Formula used is

$$\lambda = \frac{hc}{\phi_0}$$

$$\text{or } \lambda = \frac{6.62 \times 10^{-34} \times 3 \times 10^8}{2.0 \times 1.6 \times 10^{-19}}$$

$$= \mathbf{6206 \text{ Å}}$$

EXAMPLE 20 Calculate the threshold wavelength and the wavelength of incident electromagnetic radiation so that the photoelectrons emitted from potassium have a maximum kinetic energy of 4eV. Take the work function of potassium as 2.2 eV.

SOLUTION Given $E_{max} = 4.0 \times 1.6 \times 10^{-19}$ J and $\phi_0 = 2.2 \times 1.6 \times 10^{-19}$ J.

Formulae used are

$$\phi_0 = h\nu_0 = \frac{hc}{\lambda_0} \quad \text{and} \quad E_k = h\nu - \phi_0 = h\nu - h\nu_0$$

$$\lambda_0 = \frac{hc}{\phi_0} = \frac{6.62\times10^{-34}\times3\times10^8}{2.2\times1.6\times10^{-19}}$$

$$\lambda_0 = 5642\ \text{Å (Threshold wavelength)}$$

$$E_k = 4\times1.6\times10^{-19} = hc\left[\frac{1}{\lambda} - \frac{1}{\lambda_0}\right]$$

$$\text{or}\quad \frac{1}{\lambda} = \frac{4\times1.6\times10^{-19}}{6.62\times10^{-34}\times3\times10^8} + \frac{1}{\lambda_0}$$

$$= 3.223\times10^6 + 1.772\times10^6$$

$$\frac{1}{\lambda} = 4.995\times10^6$$

$$\lambda = \frac{1}{4.995\times10^6}$$

$$= \mathbf{2002\ Å}$$

EXAMPLE 21 Ultraviolet light of wavelength 350 nm and intensity 1.0 watt/m^2 is directed at a potassium surface. (i) Find the maximum kinetic energy of photoelectron (ii) 0.5% of incident photons produce photoelectrons, how many photoelectrons are emitted per second if the surface of potassium is 1.0 cm^2. Work function of potassium is 2.1 eV.

SOLUTION Given $\lambda = 3.5\times10^{-7}$ m and $\phi_0 = 2.1$ eV.

(i) Formula used is

$$E_k = \frac{1}{2}mv_{max}^2 = h\nu - \phi_0.$$

$$= \frac{hc}{\lambda} - \phi_0 = \frac{6.62\times10^{-34}\times3\times10^8}{3.5\times10^{-7}\times1.6\times10^{-19}} - 2.1\ \text{eV}$$

$$E_k = (3.546 - 2.1)\ \text{eV} = 1.45\ \text{eV}$$

$$= 2.3136\times10^{-19}\ \text{J}$$

$$= \mathbf{2.314\times10^{-19}\ J}$$

(ii) Energy incident per second on 1.0 cm^2 surface of potassium = 10^{-4} Joule

The energy which produces photoelectron per second = 0.5%.

Effective energy which will be used to produce photoelectrons $= \frac{0.5}{100}\times10^{-4}\ \text{J} = 5\times10^{-7}\ \text{J}$

Minimum energy required to eject one electron from the surface

$$= 2.314\times10^{-19}\ \text{J}$$

So the number of electrons emitted per second from 1.0 cm^2 area of the surface of potassium will be $= \frac{5\times10^{-7}}{2.314\times10^{-19}}$.

$$= \mathbf{2.16\times10^{12}}$$

EXAMPLE 22 Calculate the value of Planck's constant from the following data, assuming that the electronic charge e has value of 1.6×10^{-19} Coulomb. A surface when irradiated with light of wavelength 5896 Å emits electrons for which the stopping potential is 0.12 volts. When the same surface is irradiated with light of wavelength 2830 Å, it emits electrons for which the stopping potential is 2.2 volts.

SOLUTION If the radiation of wavelength is incident on the surface of the metal having work function ϕ_0 and stopping potential V_0 for the emitted electrons, then ϕ_0 and V_0 satisfy the following relation.

$$eV_0 = \frac{hc}{\lambda} - Ve\,\phi_0 \tag{i}$$

(i) Given $\lambda = 5.896 \times 10^{-7}$ m and $V_0 = 0.12$ volts

$$\frac{hc}{\lambda} = eV_0 + \phi_0$$

$$\frac{h \times 3 \times 10^8}{5.896 \times 10^{-7}} = 1.6 \times 10^{-19} \times 0.12 + \phi_0 \tag{ii}$$

(ii) Given $\lambda = 2.83 \times 10^{-7}$ m and $V_0 = 2.2$ volts, then

$$\frac{h \times 3 \times 10^8}{2.83 \times 10^{-7}} = 1.6 \times 10^{-19} \times 2.2 + \phi_0 \tag{iii}$$

On subtracting Eq. (ii) from Eq. (iii), we get

$$h\left[\frac{3 \times 10^8}{2.83 \times 10^{-7}} - \frac{3 \times 10^8}{5.896 \times 10^{-7}}\right] = \left[1.6 \times 10^{-19} \times 2.2 - 1.6 \times 10^{-19} \times 0.12\right]$$

$$h \times \frac{3 \times 10^{15}[5.896 - 2.83]}{2.83 \times 5.896} = 1.6 \times 10^{-19} \times 2.08$$

$$h = \mathbf{6.04 \times 10^{-34}\ Jsec}$$

EXAMPLE 23 Calculate Compton shift if *X*-rays of wavelength 1.0 Å are scattered from a carbon block. The scattered radiation is viewed at 90° to the incident beam.

SOLUTION Given $\lambda = 1.0$ Å $= 10^{-10}$m and $\phi = 90°$.

Formula used is

$$\Delta\lambda = \frac{h}{m_0 c}(1 - \cos\phi)$$

$$= \frac{6.62 \times 10^{-34}}{9.1 \times 10^{-31} \times 3 \times 10^8}(1 - \cos 90°)$$

$$= 0.242 \times 10^{-11}\ \text{m}$$

$$= 0.024 \times 10^{-10}\ \text{m}$$

$$= \mathbf{0.0242\ Å}$$

EXAMPLE 24 An *X*-ray photon is found to have doubled its wavelength on being scattered by 90^0. Find the energy and wavelength of incident photon.

SOLUTION Given $\phi = 90°$.

Formula used is

$$\Delta\lambda = \frac{h}{m_0 c}(1 - \cos\phi) \tag{i}$$

$$= \frac{6.62 \times 10^{-34}}{9.1 \times 10^{-31} \times 3 \times 10^8}(1 - \cos 90°)$$

$$= \mathbf{0.242 \times 10^{-11}\ m = 0.024\ Å}$$

As $\Delta\lambda = \lambda' - \lambda$, where λ is the wavelength of incident photon and λ' is the wavelength of scattered photon, then

$$\lambda' = \lambda + \Delta\lambda \tag{ii}$$

Given $$\lambda' = 2\lambda \tag{iii}$$

From Eqs. (ii) and (iii), we get

$$2\lambda = \lambda + \Delta\lambda$$

or $$\lambda = \Delta\lambda = 0.0242 \times 10^{-10}\text{ m} = 0.0242\text{ Å}$$

Energy of the incident photon $(E) = h\nu = \dfrac{hc}{\lambda}$

$$= \frac{6.62 \times 10^{-34} \times 3 \times 10^{8}}{0.0242 \times 10^{-10}} = \mathbf{0.513\ MeV}$$

Example 25 Calculate the wavelength of incident *X*-ray photon which produces recoil electron of energy 4.0 KeV in Compton effect. The electron recoils in the direction incident photon and photon is scattered at an angle of 180°.

Solution $\phi = 180^{\circ}$ and energy of the recoiled electron = 4000 eV.

Let λ be the wavelength of incident *X*-ray photon and λ' be the scattered photon, then according to the law of conservation of energy.

$$\frac{hc}{\lambda} - \frac{hc}{\lambda'} = \text{Kinetic energy of the recoiled electron}$$

$$= \frac{1}{2}mv^2 = 4 \times 10^3\text{ eV} = 4 \times 10^3 \times 1.6 \times 10^{-19}\text{ J}$$

or $$\frac{hc}{\lambda} - \frac{hc}{\lambda'} = 6.4 \times 10^{-16}\text{ J} \tag{i}$$

According to the principal of conservation of linear momentum in the direction incident photon

$$\frac{h}{\lambda} = \frac{h}{\lambda'}\cos\phi + mv\cos\theta$$

$$= \frac{h}{\lambda'}\cos 180^{\circ} + mv\cos 0^{\circ} = -\frac{h}{\lambda'} + mv \tag{ii}$$

$$\frac{h}{\lambda} + \frac{h}{\lambda'} = mv \tag{iii}$$

Momentum ($p = mv$) can be calculated as

$$\frac{1}{2}mv^2 = 4.0k\text{ eV} = 4 \times 10^3 \times 1.6 \times 10^{-19}\text{ J}$$

$$= 6.4 \times 10^{-16}\text{ J}$$

$$mv^2\frac{m}{m} = 2m \times 6.4 \times 10^{-16}\,J = 2 \times 9.1 \times 10^{-31} \times 6.4 \times 10^{-16}$$

$$= 1.1648 \times 10^{-45} = 11.648 \times 10^{-46}$$

$$mv = 34.13 \times 10^{-24}\text{ kgm sec}^{-1} \tag{iv}$$

By using Eqs. (iii) and (iv) then, we get

$$\frac{h}{\lambda}+\frac{h}{\lambda'}=34.13\times10^{-24}$$

Multiplying by velocity of light

$$\frac{hc}{\lambda}+\frac{hc}{\lambda'}=102.4\times10^{-16} \quad \text{(v)}$$

By adding Eq. (v) with Eq. (i), we get

$$2\frac{hc}{\lambda}=(102.4+6.4)\times10^{-16}$$

$$=108.79\times10^{-16}$$

$$\lambda=\frac{2hc}{108.79\times10^{-16}}=0.365\times10^{-10}\text{ m}$$

$$\lambda=\mathbf{0.365\ Å}$$

Example 26 *X*-rays with $\lambda = 1$ Å are scattered from a carbon block. The scattered radiation is viewed at 90^0 to the incident beam.

1. What is Compton shift $\Delta\lambda$?
2. What kinetic energy is imparted to the recoil electron?

Solution Given $\lambda = 1\times10^{-10}$ m.

Formula used for Compton shift is

$$\Delta\lambda=\frac{h}{m_0c}(1-\cos\phi)$$

$$\phi=90°$$

$$\Delta\lambda=\frac{6.62\times10^{-34}}{9.1\times10^{-31}\times3\times10^{8}}(1-\cos 90°)$$

$$=2.425\times10^{-12}\text{ m}$$

Let λ be the wavelength of incident *X*-ray photon and λ' be the scattered photon, then according to the law of conservation of energy

$$\frac{hc}{\lambda}=\frac{hc}{\lambda'}+E_k=\frac{hc}{\lambda+\Delta\lambda}+E_k$$

$$E_k=\frac{hc}{\lambda}-\frac{hc}{\lambda+\Delta\lambda}=\frac{hc\Delta\lambda}{\lambda(\lambda+\Delta\lambda)}$$

$$=\frac{6.62\times10^{-34}\times3\times10^{8}\times2.425\times10^{-12}}{1\times10^{-10}\times(1+0.02425)\times10^{-10}}$$

$$=47.02\times10^{-18}\text{ J}$$

$$=\mathbf{294\ eV}$$

Example 27 *X*-ray of wavelength 0.144 Å are scattered from a carbon target. Find maximum shift in wavelength and maximum energy of recoil electron.

SOLUTION Given $\lambda = 0.144 \times 10^{-10}$ m.

Formula used for Compton shift

$$\Delta\lambda = \frac{h}{m_0 c}(1 - \cos\phi)$$

The Compton shift will be maximum if $\phi = 180^o$

$$[\Delta\lambda]_{\max} = \frac{6.62\times10^{-34}}{9.1\times10^{-31}\times3\times10^{8}}(1-\cos 180°)$$

$$= \frac{2\times6.62\times10^{-34}}{9.1\times10^{-31}\times3\times10^{8}}$$

$$= 0.485\times10^{-11}\,\text{m}$$

$$= 0.0485\ \text{Å}$$

The kinetic energy of the recoil electron is given by the relation

$$E_k = \frac{hc}{\lambda} - \frac{hc}{\lambda'} = hc\left(\frac{1}{\lambda} - \frac{1}{\lambda'}\right)$$

$\lambda' = \lambda + \Delta\lambda$, then

$$E_k = hc\left[\frac{1}{\lambda} - \frac{1}{\lambda+\Delta\lambda}\right]$$

$$= \frac{6.62\times10^{-34}\times3\times10^{8}}{10^{-10}}\left[\frac{1}{0.144} - \frac{1}{0.144+0.0485}\right]$$

$$= 34.75\times10^{-16}\,\text{J}$$

$$= \mathbf{21.72\ keV}$$

EXAMPLE 28 *X*- rays of wavelength 0.2 Å are scattered from a target. Calculate the wavelength of *X*-ray scattered through 45°. Also find the maximum kinetic energy of the recoil electron.

SOLUTION Given $\lambda = 0.2$ Å $= 0.2 \times 10^{-10}$ m and $\phi = 45^o$.

$$\Delta\lambda = \frac{h}{mc}(1-\cos 45°) = \frac{6.62\times10^{-34}}{9.1\times10^{-31}\times3\times10^{8}}[1-0.7071] = 0.0071\ \text{Å}$$

Therefore, wavelength of scattered *X*-rays

$$\lambda' = \lambda + \Delta\lambda = 0.2 + 0.0071 = 0.2071\ \text{Å}$$

$$E_k = hc\left(\frac{1}{\lambda} - \frac{1}{\lambda'}\right)$$

Thus kinetic energy is maximum if λ' is maximum. The maximum value of λ' can be obtained by the relation $\lambda' = \lambda + \Delta\lambda_m$. Maximum value of $\Delta\lambda$ is obtained at $\phi = 180^o$.

$$\Delta\lambda_m = \frac{h}{m_0 c}(1-\cos\phi) = \frac{6.62\times10^{-34}}{9.1\times10^{-31}\times3\times10^{8}}(1-\cos 180°)$$

$$= \frac{2\times6.62\times10^{-34}}{9.1\times10^{-31}\times3\times10^{8}} = 0.0485\ \text{Å}$$

$$\lambda' = 0.2 + 0.0485$$

$$= 0.2485\ \text{Å}$$

Hence, maximum kinetic energy i.e.,

$$E_k = \frac{6.62\times10^{-34}\times3\times10^8}{10^{-10}}\left[\frac{1}{0.2}-\frac{1}{0.2485}\right]$$

$$= \mathbf{19.38\times10^{-16}\ J}$$

EXAMPLE 29 Calculate the deBroglie wavelength associated with the automobile of mass 2×10^3 kg which is moving with a speed 96 km/hr.

SOLUTION Given $m = 2\times10^3\ \text{kg}, v = \dfrac{96\times10^3}{60\times60}\ \text{m/sec} = 26.67\ \text{m/sec}.$

deBroglie wavelength is given as

$$\lambda = \frac{h}{mv} = \frac{6.62\times10^{-34}}{2\times10^3\times26.67}$$

$$= 0.124\times10^{-37}\ \text{m}$$

$$= \mathbf{1.24\times10^{-38}\ m}$$

EXAMPLE 30 A particle of charge q and mass m is accelerated through a potential difference V. Find its deBroglie wavelength. Calculate the wavelength (λ), if the particle is an electron and $V = 50$ volts.

SOLUTION When a particle of charge q and mass m is accelerated through a potential V, then deBroglie wavelength is given by

$$\lambda = \frac{h}{mv} \qquad \text{(i)}$$

and $E_k = \frac{1}{2}mv^2 = qV$ or $m^2v^2 = 2mqV$

$$\text{or} \quad mv = \sqrt{2mqV} \qquad \text{(ii)}$$

By using Eqs. (i) and (ii), we obtain

$$\lambda = \frac{h}{\sqrt{2mqV}}$$

Given $q = 1.6\times10^{-19}$ C and $V = 50$ volts, then

$$\lambda = \frac{6.62\times10^{-34}}{\sqrt{2\times9.1\times10^{-31}\times1.6\times10^{-19}\times50}}$$

$$= \mathbf{1.74\ Å}$$

EXAMPLE 31 Calculate the wavelength of thermal neutrons at 27°C, given mass of neutron = 1.67×10^{-27} kg, Planck's constant $h = 6.6\times10^{-34}$ J sec and Boltzmann's constant $k = 1.37\times10^{-23}\ \text{JK}^{-1}$.

SOLUTION Given T = 27°C = 27 + 273 = 300K, $m = 1.67\times10^{-27}$kg, $h = 6.6\times10^{-34}$ Jsec and $k = 1.376\times10^{-23}\text{JK}^{-1}$.

deBroglie wavelength is given by

$$\lambda = \frac{h}{mv} \qquad \text{(i)}$$

$$E_t = \frac{1}{2}mv^2 = \frac{3}{2}kT \quad \text{or} \quad (mv)^2 = 3mkT$$

$$\text{or} \quad mv = \sqrt{3mkT} \qquad \text{(ii)}$$

$$\text{Then,}\quad \lambda = \frac{h}{\sqrt{3mkT}} = \frac{6.6\times10^{-34}}{\sqrt{3\times1.67\times10^{-27}\times1.376\times10^{-23}\times300}}$$

$$= 1.452\times10^{-10}\ \text{m}$$

$$\text{or}\quad \boldsymbol{\lambda = 1.452\ \text{Å}}$$

Example 32 A proton is moving with a speed 2×10^8 m/sec. Find the wavelength of matter wave associated with it.

Solution Given $v = 2 \times 10^8$ m/sec.

Formula used for deBroglie wavelength is

$$\lambda = \frac{6.62\times10^{-34}}{1.67\times10^{-27}\times2\times10^{8}}$$

$$= \mathbf{1.98\times10^{-15}\ m}$$

Example 33 The deBroglie wavelength associated with an electron is 0.1 Å. Find the potential difference by which the electron is accelerated.

Solution Given $\lambda = 0.1 \times 10^{-10}$ m.

deBroglie wavelength in terms of potential difference is given by

$$\lambda = \frac{h}{\sqrt{2mqV}}$$

$$\text{or}\quad 2mqV = \frac{h^2}{\lambda^2}$$

$$V = \frac{h^2}{2mq\lambda^2} = \frac{(6.62\times10^{-34})^2}{2\times9.1\times10^{-31}\times1.6\times10^{-19}\times(10^{-11})^2}$$

$$= \mathbf{15.05\ kV}$$

Example 34 Calculate the deBroglie wavelength of an α-particle accelerated through a potential difference of 200 volts.

Solution Given $V = 200$ volts, $q = q_\alpha = 2e = 3.2 \times 10^{-19}$ C and $m = m_\alpha = 4m_p$.

deBroglie wavelength in terms of potential difference

$$\lambda = \frac{h}{\sqrt{2m_\alpha qV}} = \frac{6.62\times10^{-34}}{\sqrt{2\times4\times1.67\times10^{-27}\times2\times1.6\times10^{-19}\times200}}$$

$$= \frac{6.62\times10^{-34}}{92.468\times10^{-23}} = 0.07159\times10^{-11}$$

$$\boldsymbol{\lambda = 7.16\times10^{-13}\ \text{m}}$$

Example 35 Calculate the deBroglie wavelength of an average Helium atom in furnace of 400 K. Given $k = 1.38 \times 10^{-23}$ J/K

Solution Given $T = 400$ K, $k = 1.38 \times 10^{-23}$ J/K and mass of Helium atom $= 4m_p = 4 \times 1.67 \times 10^{-27}$ kg.

deBroglie wavelength in terms of temperature i.e.,

$$\lambda = \frac{h}{\sqrt{3mkT}} = \frac{6.62 \times 10^{-34}}{\sqrt{3 \times 4 \times 1.67 \times 10^{-27} \times 1.38 \times 10^{-23} \times 400}}$$

$$= \frac{6.62 \times 10^{-34}}{105.176 \times 10^{-25}} = 0.6294 \text{ Å}$$

$$\lambda = \mathbf{0.6294 \text{ Å}}$$

EXAMPLE 36 Calculate the deBroglie wavelength associated with a neutron moving with a velocity of 2000 m/sec.

SOLUTION Given $v = 2000$ m/sec and $m = 1.67 \times 10^{-27}$ kg.

deBroglie wavelength

$$\lambda = \frac{h}{mv} = \frac{6.62 \times 10^{-34}}{1.67 \times 10^{-27} \times 2000}$$

$$= 1.98 \times 10^{-10} \text{ m}$$

$$= \mathbf{1.98 \text{ Å}}$$

EXAMPLE 37 Calculate the energy in eV corresponding to a wavelength of 1.0 Å for electron and neutron. Given $h = 6.6 \times 10^{-34}$ J sec, mass of electron $= 9.1 \times 10^{-31}$ kg and mass of the neutron $= 1.7 \times 10^{-27}$ kg.

SOLUTION Formula used is

$$\lambda = \frac{h}{mv} \text{ or } \frac{h}{\lambda m}$$

$$\text{or} \quad v = \frac{6.6 \times 10^{-34}}{1.0 \times 10^{-10} \times 1.7 \times 10^{-27}}$$

$$= \mathbf{3.88 \times 10^3 \text{ m/sec}}$$

If the velocity is much less than the velocity of light, it can be considered as non-relativistic case and hence deBroglie wavelength can be obtained by the relation.

$$\lambda = \frac{h}{\sqrt{2mE}} \quad \text{or} \quad \lambda^2 = \frac{h^2}{2mE}$$

For Electron

$$E = \frac{h^2}{2m\lambda^2} = \frac{(6.62 \times 10^{-34})^2}{2 \times 9.1 \times 10^{-31} \times (10^{-10})^2}$$

$$= \frac{43.8244 \times 10^{-68}}{18.2 \times 10^{-51}} = 2.41 \times 10^{-17} \text{ J}$$

$$= 1.51 \times 100 = 151 \text{ eV}$$

$$E = \mathbf{151 \text{ eV}}$$

For neutron

$$E = \frac{(6.62 \times 10^{-34})^2}{2 \times 1.7 \times 10^{-27} \times (10^{-10})^2} = \frac{43.8244 \times 10^{-68}}{3.4 \times 10^{-47}}$$

$$= 12.89 \times 10^{-21} \text{ J}$$

$$= \mathbf{0.081 \text{ eV}}$$

Example 38 Calculate deBroglie wavelength of an electron whose kinetic energy is (i) 500 eV, (ii) 50 eV and (iii) 1.0 eV.

Solution Formula used is

$$\lambda = \frac{h}{\sqrt{2mE}}$$

(i) $E = 500\text{ eV} = 500 \times 1.6 \times 10^{-19} = 8.0 \times 10^{-17}$ J

$$\lambda = \frac{6.62 \times 10^{-34}}{\sqrt{2 \times 9.1 \times 10^{-31} \times 8.0 \times 10^{-17}}} = 5.486 \times 10^{-11}\text{ m}$$

$$= \mathbf{0.5486\ \text{Å}}$$

(ii) $E = 500\text{ eV} = 50 \times 1.6 \times 10^{-19} = 8.0 \times 10^{-18}$

$$\lambda = \frac{6.62 \times 10^{-34}}{\sqrt{2 \times 9.1 \times 10^{-31} \times 8 \times 10^{-18}}}$$

$$\lambda = 1.735 \times 10^{-10}\text{ m}$$

or $\lambda = \mathbf{1.735\ \text{Å}}$

(iii) $E = 1.0\text{ eV} = 1.6 \times 10^{-19}$ J

$$\lambda = \frac{6.62 \times 10^{-34}}{\sqrt{2 \times 9.1 \times 10^{-31} \times 1.6 \times 10^{-19}}}$$

$$\lambda = \mathbf{12.267\ \text{Å}}$$

Example 39 Calculate the ratio of deBroglie wavelengths associated with the neutrons with kinetic energies of 1.0 eV and 510 eV.

Solution Formula used is

$$\lambda = \frac{h}{\sqrt{2mE}}$$

For $E = 1.0\text{ eV} = 1.6 \times 10^{-19}$ J and $m_n = 1.7 \times 10^{-27}$ kg

$$\lambda_1 = \frac{6.62 \times 10^{-34}}{\sqrt{2 \times 1.7 \times 10^{-27} \times 1.6 \times 10^{-19}}}$$

$$= 2.838 \times 10^{-11}$$

$$\lambda_1 = \mathbf{0.284\ \text{Å}}$$

For $E = 510\text{ eV} = 510 \times 1.6 \times 10^{-19} = 816 \times 10^{-19}$ J

$$\lambda_2 = \frac{6.62 \times 10^{-34}}{\sqrt{2 \times 1.7 \times 10^{-27} \times 816 \times 10^{-19}}}$$

$$\lambda_2 = 0.01257\ \text{Å}$$

$$= \mathbf{0.0126\ \text{Å}}$$

and ratio of deBroglie wavelength is

$$\frac{\lambda_1}{\lambda_2} = \frac{0.284}{0.0126} = \mathbf{22.54 : 1}$$

EXAMPLE 40 Calculate the ratio of deBroglie waves associated with a proton and an electron each having the kinetic energy as 20 M eV [$m_p = 1.67 \times 10^{-27}$ kg and $m_e = 9.1 \times 10^{-31}$ kg].

SOLUTION Given energy of each proton and electron is $20 \times 10^6 \times 10^{-19}$ J $= 3.2 \times 10^{-12}$ J.

Formula used is

$$\lambda = \frac{h}{\sqrt{3mE}}$$

For proton

$$\lambda_p = \frac{6.62 \times 10^{-34}}{\sqrt{2 \times 1.67 \times 10^{-27} \times 3.2 \times 10^{-12}}}$$

$$= 6.4 \times 10^{-15} \text{ m}$$

For electron

$$\lambda_e = \frac{6.62 \times 10^{-34}}{\sqrt{2 \times 9.1 \times 10^{-31} \times 3.2 \times 10^{-12}}}$$

$$= 2.74 \times 10^{-13} \text{ m}$$

The ratio of λ_p to λ_e is

$$\boldsymbol{\lambda_p : \lambda_e = 1{:}43}$$

EXAMPLE 41 Calculate the deBroglie wavelength of 1.0 M eV proton. Do we require relativistic calculation?

SOLUTION Given Energy $E = 1.0 \times 10^6 \times 1.6 \times 10^{-19}$ J $= 1.6 \times 10^{-13}$ J

Formula used for velocity of Proton

$$E = \frac{1}{2}mv^2 \quad \text{or} \quad v^2 = \frac{2E}{m}$$

$$\text{or} \qquad v = \sqrt{\frac{2E}{m}} = \sqrt{\frac{2 \times 1.6 \times 10^{-13}}{1.67 \times 10^{-27}}}$$

$$= \mathbf{1.38 \times 10^7 \text{ m/sec}}$$

From the above result it is clear that the velocity of proton is nearly one twentieth of the velocity of light. So the relativistic calculations are not required.

EXAMPLE 42 Calculate the deBroglie wavelength associated with a proton moving with a velocity equal to 1/20th of velocity of light.

SOLUTION Given $v = \frac{c}{20} = \frac{3 \times 10^8}{20} = 1.5 \times 10^7$ m/sec and $m = 1.67 \times 10^{-27}$ kg

Formula used is

$$\lambda = \frac{h}{p} = \frac{h}{mv} = \frac{6.62 \times 10^{-34}}{1.67 \times 10^{-27} \times 1.5 \times 10^7}$$

$$= \mathbf{2.643 \times 10^{-14} \text{ m}}$$

EXAMPLE 43 Calculate the kinetic energy of a proton and an electron so that the deBroglie wavelengths associated with them is the same and equal to 5000 Å.

SOLUTION Given wavelength of proton and electron = 5.0×10^{-7} m.
Formula used in

$$\lambda = \frac{h}{\sqrt{2mE}} \quad \text{or} \quad E = \frac{h^2}{2m\lambda^2}$$

For proton $m = m_p = 1.67 \times 10^{-27}$ kg and $\lambda = 5.0 \times 10^{-7}$ m

$$E = \frac{(6.62 \times 10^{-34})^2}{2 \times 1.67 \times 10^{-27} \times (5.0 \times 10^{-7})^2}$$

$$= \frac{43.8244 \times 10^{-68}}{83.5 \times 10^{-41}} = 0.5248 \times 10^{-27} \text{ J}$$

$$= \mathbf{5.248 \times 10^{-28} \ J}$$

For electron $m = m_e = 9.1 \times 10^{-31}$

$$E = \frac{(6.62 \times 10^{-34})^2}{2 \times 9.1 \times 10^{-31} \times (5 \times 10^{-7})^2} = \frac{43.8244 \times 10^{-68}}{4.55 \times 10^{-43}}$$

$$E = \mathbf{9.63 \times 10^{-25} \ J}$$

EXAMPLE 44 Find deBroglie wavelength of an electron in the first Bohr's orbit of hydrogen atom.

SOLUTION Energy of an electron in the first Bohr's orbit of hydrogen atom can be obtained by using the relation $E_n = \frac{-13.6}{n^2}$

$$E_1 = \frac{-13.6}{1^2} = -13.6 \text{ eV}$$

$$E_1 = -13.6 \times 1.6 \times 10^{-19} \text{ J} = -2.176 \times 10^{-18} \text{ J}$$

Magnitude of energy = 2.176×10^{-18} J

$$\text{Wavelength } \lambda = \frac{h}{\sqrt{2mE}} = \frac{6.62 \times 10^{-34}}{\sqrt{2 \times 9.1 \times 10^{-31} \times 21.76 \times 10^{-19}}}$$

$$= 3.3 \times 10^{-10} \text{ m}$$

$$= \mathbf{3.3 \ Å}$$

EXAMPLE 45 Calculate the ratio of deBroglie wavelengths of a hydrogen atom and helium atom at room temperature, when they move with thermal velocities. Given mass of hydrogen atom $m_H = 1.67 \times 10^{-27}$ kg and mass of helium atom $m_{He} = 4 \times m_p = 4 \times 1.67 \times 10^{-27}$ kg at room temperature $T = 27°C = 300$ K and Boltzmann's constant $k = 1.376 \times 10^{-23}$ J/K.

SOLUTION deBroglie wavelength can be calculated by the relation

$$\lambda = \frac{h}{\sqrt{3mkT}}$$

For Hydrogen atom

$$\lambda = \frac{6.62 \times 10^{-34}}{\sqrt{3 \times 1.67 \times 10^{-27} \times 1.376 \times 10^{-23} \times 300}}$$

$$= 1.456 \times 10^{-10} \text{ m}$$

$$\lambda = \mathbf{1.456 \ Å}$$

For Helium atom

$$\lambda_{He} = \frac{6.62\times10^{-34}}{\sqrt{3\times4\times1.67\times10^{-27}\times1.376\times10^{-23}\times300}}$$

$$= 0.728\times10^{-10}\text{ m}$$

$$\lambda = \mathbf{0.728\ Å}$$

The ratio of wavelengths i.e.,

$$\frac{\lambda_H}{\lambda_{He}} = \frac{1.456}{0.728} = \frac{2}{1}$$

$$\boldsymbol{\lambda_H : \lambda_{He} = 2:1}$$

EXAMPLE 46 A proton and a deuteron have the same kinetic energy. Which has a longer wavelength?

SOLUTION m_p=mass of proton, m_d=$2m_p$ and v_p and v_d are the velocities of proton and deuteron.

Kinetic energy of proton is given by

$$E_p = \frac{1}{2}m_p v_p^2$$

and kinetic energy of deuteron is

$$E_d = \frac{1}{2}m_d v_d^2 = \frac{1}{2}(2m_p)v_d^2$$

$$E_d = m_p v_d^2$$

But $E_p = E_d$, then

$$m_p v_d^2 = \frac{1}{2}m_p v_p^2$$

or $$v_d = \frac{v_p}{\sqrt{2}}$$

deBroglie wavelength corresponding to moving proton and deuteron are

$$\lambda_p = \frac{h}{m_p v_p} \text{ and}$$

$$\lambda_d = \frac{h}{m_d v_d} = \frac{h}{2m_p v_p/\sqrt{2}} = \frac{h}{\sqrt{2}m_p v_p}$$

$$\frac{\lambda_d}{\lambda_p} = \frac{h}{\sqrt{2}m_p v_p}\times\frac{m_p v_p}{h} = \frac{1}{\sqrt{2}}$$

$$\boldsymbol{\lambda_p = \sqrt{2}\lambda_d}$$

i.e., proton has a longer wavelength.

EXAMPLE 47 Find the phase and group velocities of an electron whose deBroglie wavelength is 1.2 Å.

SOLUTION Formula used is

$$\lambda = \frac{h}{mv} \tag{i}$$

v_g= Group velocity = Particle velocity = v

$$v = \frac{h}{m\lambda} = \frac{6.62 \times 10^{-34}}{9.1 \times 10^{-31} \times 1.2 \times 10^{-10}}$$

$$v = 6.06 \times 10^6 \text{ m/sec} = \text{group velocity} = v_g$$

or $v_g = 6.06 \times 10^6$ m/sec

Phase velocity $v_p = \dfrac{\omega}{k}$ (ii)

Energy $E = h\nu$

or $E = \dfrac{h}{2\pi} 2\pi\nu = \hbar\omega$ (iii)

and momentum $p = \dfrac{h}{\lambda}$

or $p = \dfrac{h}{2\pi}\dfrac{2\pi}{\lambda} = \hbar k$ (iv)

or $v_p = \dfrac{\omega}{k} = \dfrac{\hbar\omega}{\hbar k} = \dfrac{E}{p}$

and $E = \dfrac{1}{2}mv^2$ *and* $p = mv$ (v)

or $E = \dfrac{m^2v^2}{2m} = \dfrac{p^2}{2m}$ (vi)

$v_p = \dfrac{E}{p}$ From Eq. (v)

$$v_p = \frac{p^2/2m}{p} = \frac{p}{2m} = \frac{h/\lambda}{2m} = \frac{h}{2m\lambda}$$

$$= \frac{6.62 \times 10^{-34}}{2 \times 9.1 \times 10^{-31} \times 1.2 \times 10^{-10}}$$

$$v_p = \mathbf{3.03 \times 10^6 \text{ m/sec}}$$

From the above result it is clear that the phase velocity is just half of group velocity.

Example 48 Calculate the deBroglie wavelength of

(a) a particle accelerated by a potential difference of 30,000 V and

(b) an electron moving with a velocity of 0.01c, where *c* is the speed of light.

Solution Given $V = 30{,}000$ Volts, $e = 1.6 \times 10^{-19}$ Coulomb, $m_e = 9.1 \times 10^{-31}$ kg, $h = 6.63 \times 10^{-34}$ J.sec

and $E = eV = 1.6 \times 10^{-19} \times 30{,}000$

$= 4.8 \times 10^{-15}$ Joules

(a) Formula used is

$$\lambda = \frac{h}{mv} = \frac{h}{\sqrt{2mE}} \qquad \left[E = \frac{1}{2}mv^2\right]$$

$$\lambda = \frac{6.63 \times 10^{-34}}{\sqrt{2 \times 9.1 \times 10^{-31} \times 4.8 \times 10^{-15}}} = 7.09 \times 10^{-12} \text{m}$$

(b) Formula for deBrogile wavelength is given by

$$\lambda = \frac{h}{mv} = \frac{6.63\times10^{-34}}{9.1\times10^{-31}\times0.01\,c} = \frac{6.63\times10^{-34}}{9.1\times10^{-31}\times0.01\times3\times10^{8}}$$

$$= \mathbf{2.43\ Å}$$

EXAMPLE 49 Calculate the deBreoglie wavelength of virus particle of mass 1×10^{-15} kg moving at a speed of 2×10^{-3} m/sec.

SOLUTION Given, $v = 2\times10^{-3}$ m/sec; $m = 1\times10^{-15}$ kg

$$\lambda = \frac{h}{mv} = \frac{6.63\times10^{-34}}{1\times10^{-15}\times2\times10^{-3}} = \mathbf{3.315\times10^{-16}\,m}$$

OBJECTIVE TYPE QUESTIONS

Q.1 Which of the following phenomena cannot be explained by the classical theory?
(a) Photoelectric effect (b) Compton effect
(c) Raman effect (d) All of these

Q.2 Which of the following phenomena show the particle nature of light?
(a) Photoelectric effect (b) Raman effect
(c) Compton effect (d) All of these

Q.3 Wien's law is deduced from Planck's radiation formula under the condition of
(a) very small wavelength and temperature (b) large wavelength and high temperature
(c) small wavelength and high temperature (d) large wavelength and low temperature

Q.4 Rayleigh–Jeans law is deduced from the Planck's radiation formula under the condition of
(a) large wavelength and high temperature (b) small wavelength and low temperature
(c) small wavelength and high temperature (d) large wavelength and low temperature

Q.5 Which of the following characteristic(s) photon has(have)?
(a) $m_0 = 0$ (b) $E = hv$ (c) $m = \frac{hv}{c^2}$ (d) All of these

Q.6 Which of the following relation can be used to determine deBroglie wavelength associated with a particle of mass m and having energy E
(a) $\frac{h}{\sqrt{2mqV}}$ (b) $\frac{h}{\sqrt{3mkT}}$ (c) $\frac{h}{\sqrt{2mE}}$ (d) All of these

Q.7 The phase velocity of deBroglie wave associated with an electron is given by
(a) $\frac{E}{p}$ (b) hv (c) $\frac{hc}{\lambda}$ (d) $\hbar k$

Q.8 Electron behaves like a wave as it
(a) can be deflected by an electric field (b) can be deflected by a magnetic field
(c) they ionise a gas (d) can be diffracted by a crystal

Q.9 A material particle is in thermal equilibrium at temperature T. The wavelength of deBroglie wave associated with it is

(a) $\frac{h}{\sqrt{2kT}}$ (b) $\frac{h}{\sqrt{8\pi^2 mkT}}$ (c) $\frac{h}{\sqrt{2mkT}}$ (d) $\frac{h}{\sqrt{4\pi^2 mkT}}$

Q.10 Photoelectric effect involves only
(a) free-electron (b) bound electron
(c) both (a) and (b) (d) none of these

Q.11 The deBroglie hypothesis is concerned with
(a) wave nature of radiations (b) wave nature of all material particles
(c) wave nature of electrons only (d) wave nature of α-particles only

Q.12 A proton and a deuteron have the same kinetic energy. The relation between the wavelengths of waves associated with them is
(a) $\lambda_p > \lambda_d$ (b) $\lambda_d = \lambda_p$ (c) $\lambda_p = \lambda_d$ (d) none of these

Q.13 The group velocity of matter waves is
(a) equal to the particle velocity (b) greater than the particle velocity
(c) less than the particle velocity (d) same as phase velocity

Q.14 The ratio of deBroglie wavelengths of a hydrogen atom and helium atom at room temperature, when they move with thermal velocities, is
(a) 1:2 (b) 2:1 (c) 3:1 (d) 1:3

Q.15 The existence of matter wave is experimentally proved by
(a) Raman (b) Davisson and Germer
(c) deBroglie (d) none of these

Q.16 Dual character of matter was proposed by
(a) Davisson and Germer (b) deBroglie
(c) Planck (d) none of these

Q.17 Quantum theory successfully explains the phenomena of
(a) photoelectric and compton effects (b) interference, diffraction and polarisation
(c) black body radiations (d) all of these

Q.18 Matter waves
(a) show diffraction (b) show interference
(c) polarisation (d) none of these

Q.19 Matter waves are similar in nature to
(a) cathode rays (b) electromagnetic waves
(c) *X*-rays (d) both (a) & (b)

Q.20 Tick the correct target material in Davisson Germer experiment
(a) coper (b) nickel (c) silver (d) none of these

SHORT-ANSWER QUESTIONS

Q.1 What do you understand by blackbody radiation?
Q.2 What is Planck's quantum hypothesis?
Q.3 What do you mean by duality of matter?
Q.4 What is photoelectric effect?
Q.5 Explain Compton effect.

Q.6 What do you understand by wave velocity and group velocity in the context of deBroglie waves?

Q.7 What is the difference between phase velocity and group velocity in the context of deBroglie waves?

Q.8 Discuss the shortcomings of the classical physics and also explain how did quantum mechanics develop?

Q.9 What is Planck's constant? Discuss its importance.

Q.10 What are the limitations of old quantum theory?

PRACTICE PROBLEMS

General Questions

Q.1 What are the shortcomings of old quantum theory?

Q.2 Discuss the failures of classical physics and how does quantum mechanics overcome these failures?

Q.3 What is Planck's quantum hypothesis to explain the observed spectrum of a black body?

Q.4 Explain briefly quantum theory of radiation. What is a photon? State its properties. Express the linear momentum of a photon in terms of wave vector $|\vec{k}|$ and energy of photon in terms of angular velocity ω.

Q.5 Discuss and derive Planck's radiation formula. Explain Wien's law and Rayleigh–Jeans law as the special cases of it.

Q.6 What is photoelectric effect? Draw a labelled diagram of the apparatus you will use to demonstrate photoelectric effect. Write down its important results and show them graphically.

Q.7 (a) State the laws of photoelectric emission.

(b) In what way classical electromagnetic theory of light fails to explain the basic facts of photoelectricity.

Q.8 Derive Einstein's photoelectric equation. How does it explain the laws of photoelectric emission? Why all the photoelectrons do not have the same energy?

Q.9 What is meant by work function of a material? Show how you will measure it experimentally?

Q.10 Draw a curve showing stopping potential against frequency of a photosensitive material. How do you determine the following with the help of the curve. (a) Threshold frequency (b) Work function and (c) Planck's constant.

Q.11 Explain the concept of wave particle dualism. What led deBroglie to suggest that matter has wave characteristic?

Q.12 State the deBroglie hypothesis of matter waves. Derive an expression for deBroglie wavelength of matter particle in terms of kinetic energy and temperature.

Q.13 Derive a formula expressing deBroglie wavelength of an electron in terms of potential difference (V) in volts through which it is accelerated.

Q.14 Why are wave properties of particles normally observed only when we study very small particles?

Q.15 Why can't we observe deBroglie wavelength with a fast moving cricket ball?

Q.16 Describe with necessary theory the Davisson and Germer experiment for establishing wave nature of the electron?

Q.17 What is the effect of increasing the electron energy on the scattering angle in a Davisson and Germer experiment?

Q.18 What is Compton effect? Derive an expression for Compton shift and wavelength of scattered photon. Explain why Compton shift is not observed with visible light?

Q.19 What is Compton wavelength? Determine its value. Distinguish between Compton shift and Compton wavelength. What are the factors on which Compton shift depends?

Q.20 What is the difference between phase velocity and group velocity?

Q.21 Prove that the wave group associated with a moving particle travels with the same velocity as that of the particle?

Q.22 Show that group velocity and wave velocity are the same in a non-dispersive medium?

Q.23 Show that the group velocity $G = -\lambda^2 \dfrac{dv}{d\lambda}$, where the symbols have their usual meanings?

Q.24 Explain group velocity and phase velocity. Derive the expression for group velocity with which a wave group travels?

Q.25 What is the difference between phase and group velocities. Show that the deBroglie group velocity associated with the wave packet is equal to the velocity of the particle?

Q.26 Distinguish between phase velocity and group velocity. Show that of a non-relativistic free particle phase velocity is half of the group velocity?

UNSOLVED QUESTIONS

Q.1 Find out the wavelength associated with photon of energy 10^{-19} J and also find the energy in eV.
[Ans: 19800 Å & 0.63 eV]

Q.2 It source is operating at a frequency of 10^8 Hz and radiates a power 10^4 J/sec, what would be the number of quanta of energy emitted in one second. **[Ans: 1.51 × 10^{30}]**

Q.3 A 10 kilowatt transmitter operates at a frequency of 880 kHz. How many photons per second are emitted? **[Ans: 1.716 × 10^{31}]**

Q.4 How many photons of red light of wavelength 7800 Å constitute 2.0 J of energy? **[Ans: 7.85 ×10^{18}]**

Q.5 Calculate the work function in electron volts of a metal, when photoelectric threshold wavelength is 6800 Å. **[Ans: 1.83 eV]**

Q.6 The threshold wavelength for photoelectric emission in tungsten is 230 nm. What wavelength of incident light must be used in order to eject electrons with a maximum velocity of 5×10^5 m/sec
[Ans: 203 nm]

Q.7 Light of wavelength 2000 Å falls on a photosensitive material having work function 4.2 eV. What is the kinetic energy of the fastest and slowest photo-electron? Also calculate the stopping potential.
[Ans: 2.0 eV, Zero, 2.0 V]

Q.8 When *X*-rays of energy 0.1 MeV strike a target, they are scattered at an angle of 30°, Compute the energy of *X*-rays scattered and the energy of recoiled electron. **[Ans: 97.44 keV, 2.56 keV]**

Q.9 Compute the deBroglie wavelength of an electron whose kinetic energy is 50 eV. (Given $h = 6.62 \times 10^{-34}$ Jsec, $m = 9.1 \times 10^{-31}$ kg and eV = 1.6×10^{-19}J.) **[Ans: 1.73 Å]**

Q.10 Each of a photon and an electron has an energy of 1 keV. Calculate their corresponding wavelengths.
[Ans: 12.4 Å, 0.39 Å]

Q.11 Find the energy of neutron having deBroglie wavelength 10^{-14} m. Given rest mass of neutron as 1.6×10^{-27} kg. **[Ans: 8.5 MeV]**

Quantum Mechanics

16

LEARNING OBJECTIVES

After reading this chapter you will be able to

- **LO 1** Understand and learn the Heisenberg uncertainty principle and its applications
- **LO 2** Describe how to obtain time independent/dependent Schrödinger equation
- **LO 3** Know about operators associated with measurable parameters
- **LO 4** Explain applications of Schrödinger equation
- **LO 5** Discuss quantum statistics

Introduction

The wave like and particle behaviour of electrons and photons have been discussed in the previous chapter. However, all the subatomic particles like protons, neutrons, etc. show their dual nature, i.e., sometimes they behave as particle and sometimes as wave. Various types of explanation to understand this wave particle duality led to the development of quantum mechanics. Quantum mechanics deals with the behaviour and characteristics of matter, in the subatomic level, and energy. With the development of quantum theory, queries like stability of electron orbits and blackbody radiation could be explained scientifically.

Basics of quantum theory were developed by Planck, Einstein, Schrödinger and Heisenberg. As discussed earlier, *Planck* in 1900 established that all forms of matter emit or absorb energy in units, called quanta. Prior to this theory, it was assumed that energy existed only in the form of electromagnetic waves. In 1905, *Einstein* stated that not only energy but also radiation is quantifiable. He came to the conclusion that the energy (E) of light depends on its frequency (ν) as per the relation $E = h\nu$. *Schrödinger* discovered the wave equation and contributed to the development of quantum mechanics. In 1927 *Heisenberg* proposed the uncertainty principle according to which it is impossible to measure the precise values of momentum and position of a subatomic particle. This way the modern quantum theory was developed in the early 20th century. As we have already seen, quantum physics mainly deals with waves and the subatomic particles of matter. For this reason quantum theory is also referred to as quantum wave mechanics.

16.1 HEISENBERG UNCERTAINTY PRINCIPLE

LO1

Heisenberg uncertainty principle is perhaps the best known result of the wave particle duality, i.e., the concept of waves or wave packet associated with a moving particle. According to Heisenberg uncertainty principle it is impossible to determine simultaneously the exact position and momentum (or velocity) of a small moving particle like electron.

As discussed earlier, the quantity $|\psi(x, t)|^2\Delta x$ represents the probability that the particle is within the region between x and $x + \Delta x$. It means there is an uncertainty in the location of the position of the particle and Δx is a measure of the uncertainty. The uncertainty in the position would be less if Δx is smaller, i.e., if the wave packet is very narrow. The narrow wave packet means the range of wavelength $\Delta\lambda$ between λ and $\lambda + \Delta\lambda$ is smaller or the range of wave numbers Δk between k and $k + \Delta k$ is larger. So Δx is inversely proportional to Δk, i.e.,

$$\Delta x \propto \frac{1}{\Delta k}$$

We may approximate this as $\Delta x \Delta k \approx 1$. Taking $\hbar = \frac{h}{2\pi}$, we get $p = \frac{h}{\lambda} = \frac{h}{2\pi}\frac{2\pi}{\lambda} = \hbar k$, $\Delta k = \frac{\Delta p}{\hbar}$. Therefore

$$\Delta x \Delta p \approx \hbar$$

The above relation represents the lowest limit of accuracy. Therefore, we can write more generally.

$$\Delta x \Delta p \geq \hbar$$

The principle of uncertainty can also be represented in terms of energy E and time t. Since $\frac{\Delta p}{\Delta t} = \Delta F$, we can write

$$\Delta p = \Delta F \cdot \Delta t$$

Putting this value of Δp in the expression $\Delta x \Delta p \geq \hbar$ we obtain

$$\Delta x \times (\Delta F \times \Delta t) \geq \hbar$$

or

$$[\Delta F \times \Delta x]\ \Delta t \geq \hbar$$

$$\Delta E\ \Delta t \geq \hbar$$

The principle of uncertainty can also be expressed in terms of angular momentum and angle. Suppose we have a particle at a particular angular position θ and its angular momentum is L_θ. Then the limits in the uncertainties $\Delta\theta$ and ΔL_θ are given by the relation $\Delta\theta \Delta L_\theta \geq \hbar$.

16.1.1 Mathematical Proof

Heisenberg's uncertainty principle can be proved on the basis of deBroglie's wave concept that a material particle in motion is equivalent to a group of waves or wave packet, the group velocity G being equal to the particle velocity v. Consider a simple case of wave packet which is formed by the superposition of two simple harmonic plane waves of equal amplitudes a and having nearly equal frequencies ω_1 and ω_2. The two waves can be represented by the equations.

$$y_1 = a \sin(\omega_1 t - k_1 x)$$

$$y_2 = a \sin(\omega_2 t - k_2 x)$$

where k_1 and k_2 are their propagation constants and $\frac{\omega_1}{k_1}$ and $\frac{\omega_2}{k_2}$ are their respective phase velocities. The resultant wave due to superposition of these wave is given by

$$y = y_1 + y_2$$

$$y = 2a \sin(\omega t - kx) \cos\left[\frac{\Delta\omega}{2}t - \frac{\Delta k}{2}x\right] \tag{i}$$

where $\omega = (\omega_1 + \omega_2)/2$, $k = (k_1 + k_2)/2$, $\Delta\omega = \omega_1 - \omega_2$ and $\Delta k = k_1 - k_2$.

The resultant wave is shown in Fig. 16.1. The envelope (loop) of this wave travels with the group velocity G, given by

$$G = \frac{\Delta\omega}{\Delta k} = \frac{\omega_1 - \omega_2}{k_1 - k_2}$$

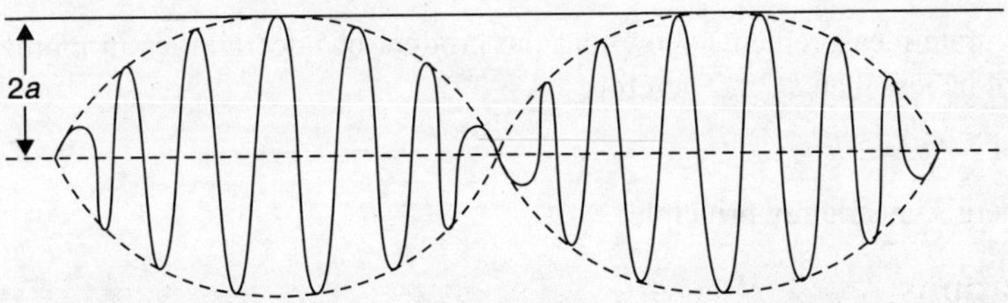

FIGURE 16.1

Since the group velocity of deBroglie wave group associated with the moving particle is equal to the particle velocity, the loop so formed is equivalent to the position of the particle. Then the particle may be anywhere within the loop. Now the condition of the formation of node from Eq. (i) is given by

$$\cos\left[\frac{\Delta\omega}{2}t - \frac{\Delta k}{2}x\right] = 0$$

or

$$\frac{\Delta\omega}{2}t - \frac{\Delta k}{2}x = \frac{\pi}{2}, \frac{3\pi}{2}, \ldots, \frac{(2n+1)\pi}{2} \tag{ii}$$

where $n = 0, 1, 2, \ldots.$

If x_1 and x_2 be the values of positions of two consecutive nodes, then from above equation by putting n and $(n+1)$, we get

$$\frac{\Delta\omega}{2}t - \frac{\Delta k}{2}x_1 = \frac{(2n+1)\pi}{2} \tag{iii}$$

and

$$\frac{\Delta\omega}{2}t - \frac{\Delta k}{2}x_2 = \frac{(2n+3)\pi}{2} \tag{iv}$$

From Eqs. (iii) and (iv), we have

$$\frac{\Delta k}{2}(x_1 - x_2) = \pi$$

or $$\frac{\Delta k}{2}\Delta x = \pi \tag{v}$$

or $$\Delta x = \frac{2\pi}{\Delta k}$$

but $$k = \frac{2\pi}{\lambda} = \frac{2\pi}{h/p} = \frac{2\pi p}{h}$$

$$\Delta k = \frac{2\pi}{h}\Delta p$$

where Δp is the error (uncertainty) in the measurement of momentum p. Therefore, from Eq. (v)

$$\Delta x = \frac{2\pi h}{2\pi \Delta p} = \frac{h}{\Delta p}$$

or $$\Delta p \Delta x = h$$

However, more accurate measurements show that the product of uncertainties in momentum (Δp) and the position (Δx) cannot be less than $h/2\pi$. Therefore

or $\Delta p \Delta x \geq \hbar$

This is the Heisenberg's uncertainty principle.

16.1.2 Applications

Some important applications of uncertainty principle are discussed below.

16.1.2.1 Non-Existence of Electron in the Nucleus

The radius of the nucleus of an atom is of the order of 10^{-14} m. If an electron is confined within the nucleus, the uncertainty in its position must not be greater than 10^{-14} m. According to uncertainty principle for the lowest limit of accuracy

$$\Delta x \Delta p = \frac{h}{2\pi} \tag{i}$$

where Δx is uncertainty in the position and Δp is the uncertainty in the momentum.

From Eq. (i),

$$\Delta p = \frac{h}{2\pi \Delta x} = \frac{6.625\times 10^{-34}}{2\times 3.14\times 2\times 10^{-14}} \quad \text{(as } \Delta x = \text{diameter of nucleus)}$$

$$\Delta p = 5.275\times 10^{-21}\ \text{kg m/sec}$$

This is the uncertainty in momentum of the electron. It means the momentum of the electron would not be less than Δp, rather it could be comparable to Δp. Thus

$$p = 5.275\cdot 10^{-21}\ \text{kg m/sec}$$

The kinetic energy of the electron can be obtained in terms of momentum as

$$T = \frac{1}{2}mv^2 = \frac{p^2}{2m}$$

$$= \frac{(5.275 \times 10^{-21})^2}{2 \times 9.1 \times 10^{-31}} \text{ J}$$

$$= \frac{(5.275 \times 10^{-21})^2}{2 \times 9.1 \times 10^{-31} \times 1.6 \times 10^{-19}} \text{ eV}$$

$$= 95.55 \times 10^6 \text{ eV}$$

$$\approx 96 \text{ MeV}$$

From the above result, it is clear that the electrons inside the nucleus may exist only when it possesses the energy of the order of 96 MeV. However, the maximum possible kinetic energy of an electron emitted by radioactive nuclei has been found about 4 MeV. Hence, it is concluded that the electron cannot reside inside the nucleus.

16.1.2.2 *Radius of Bohr's First Orbit*

If Δx and Δp be the uncertainties in determining the position and momentum of the electron in the first orbit, then from the uncertainty principle

$$\Delta x \Delta p \approx \hbar$$

or

$$\Delta p \approx \frac{\hbar}{\Delta x} \tag{i}$$

The uncertainty in kinetic energy (K.E.) of electron may be written as

$$\Delta T = \frac{(\Delta p)^2}{2\,m} \qquad \left[\text{K.E.} = T = \frac{p^2}{2m}\right] \tag{ii}$$

From Eqs. (i) and (ii), we have

$$\Delta T = \frac{1}{2m}\left[\frac{\hbar}{\Delta x}\right]^2$$

and the uncertainty in the potential energy of the same electron is given by

$$\Delta V = \frac{1}{4\pi\varepsilon_0}\frac{(Ze)(-e)}{\Delta x} \qquad \left[\because\ V = \frac{1}{4\pi\varepsilon_0}\frac{(Ze)(-e)}{x}\right]$$

The uncertainty in the total energy of electron together with Ze as the nucleus charge

$$\Delta E = \Delta T + \Delta V$$

$$= \frac{\hbar^2}{2m(\Delta x)^2} - \frac{1}{4\pi\varepsilon_0}\frac{Ze^2}{\Delta x}$$

The condition for this uncertainty in the energy to be minimum is

$$\frac{d(\Delta E)}{d(\Delta x)} = 0$$

or

$$-\frac{\hbar^2}{m(\Delta x)^3} + \frac{1}{4\pi\varepsilon_0}\frac{Ze^2}{(\Delta x)^2} = 0$$

$$\Delta x = \frac{\hbar^2 (4\pi\varepsilon_0)}{mZe^2}$$

Hence, the radius of first orbit

$$r = \Delta x = \frac{\hbar^2 (4\pi\varepsilon_0)}{mZe^2}$$

or $$\boldsymbol{r = \frac{\varepsilon_0 h^2}{\pi m Z e^2}}$$

This is the radius of first Bohr's orbit.

16.1.2.3 Energy of a Particle in a Box or Infinite Potential Well

Let us consider a particle having mass m in infinite potential well of width L. The maximum uncertainty in the position of the particle may be

$$(\Delta x)_{\text{max}} = L$$

From the uncertainty principle

$$\Delta x \Delta p = \hbar$$

or $$\Delta p = \frac{\hbar}{\Delta x} = \frac{\hbar}{L}$$

Kinetic energy

$$T = \frac{p^2}{2m} = \frac{\hbar^2}{2mL^2}$$

$$T = \frac{\hbar^2}{2mL^2}$$

This is the minimum kinetic energy of the particle in an infinite potential well of width L.

16.1.2.4 Ground State Energy of Linear Harmonic Oscillator

The total energy E of a linear harmonic oscillator is the sum of its kinetic energy (K.E.) and potential energy (P.E.).

$$E = \text{K.E} + \text{P.E.}$$

$$E = \frac{p^2}{2m} + \frac{1}{2}kx^2 \qquad \text{(i)}$$

Let a particle of mass m executes a simple harmonic motion along x-axis. The maximum uncertainty in the determination of its position can be taken as Δx. From the uncertainty relation, the uncertainty in momentum is then given by

$$\Delta p = \frac{\hbar}{2\Delta x} \quad \left[\text{Taking } \Delta p \Delta x = \frac{\hbar}{2} \text{ for more accuracy}\right] \qquad \text{(ii)}$$

For maximum uncertainties $\Delta p \simeq p$ and $\Delta x \simeq x$

Hence, the total energy E of the oscillator becomes

$$E = \frac{(\Delta p)^2}{2m} + \frac{1}{2}k(\Delta x)^2$$

or $$E = \left(\frac{\hbar}{2}\right)^2 \cdot \frac{1}{2m(\Delta x)^2} + \frac{1}{2}k(\Delta x)^2$$

or $$E = \frac{\hbar^2}{8m(\Delta x)^2} + \frac{1}{2}k(\Delta x)^2 \qquad \text{(iii)}$$

For a minimum value of energy,

$$\frac{\partial E}{\partial(\Delta x)} = 0$$

Then we get

$$-\frac{\hbar^2}{4m(\Delta x)^3} + k(\Delta x) = 0$$

or $$(\Delta x)^4 = \frac{\hbar^2}{4mk}$$

or $$(\Delta x) = \left(\frac{\hbar^2}{4mk}\right)^{1/4} \quad \text{(iv)}$$

Substituting value of Δx in Eq. (iii) from Eq. (iv), we get

$$E_{\text{min}} = \frac{\hbar^2}{8m}\left(\frac{4mk}{\hbar^2}\right)^{1/2} + \frac{1}{2}k\left(\frac{\hbar^2}{4mk}\right)^{1/2}$$

$$E_{\text{min}} = \frac{\hbar}{2}\left(\frac{k}{m}\right)^{1/2}$$

But $\sqrt{\frac{k}{m}} = \omega =$ angular frequency. Therefore, the minimum energy of harmonic oscillator is expressed by the following relation

$$\mathbf{E_{min} = \frac{1}{2}\hbar\omega}$$

Here it would be worth mentioning that the energy comes out to be $E_{\text{min}} = \hbar\omega$ if we use $\Delta p \Delta x = \hbar$ (less accuracy).

16.2 WAVE FUNCTION AND ITS PHYSICAL SIGNIFICANCE

Waves in general are associated with quantities that vary periodically. In case of matter waves, the quantity that varies periodically is called *wave function*. The wave function, represented by ψ, associated with the matter waves has no direct physical significance. It is not an observable quantity. However, the value of the wave function is related to the probability of finding the particle at a given place at a given time. The square of the absolute magnitude of the wave function of a body evaluated at a particular time at a particular place is proportional to the probability of finding the particle at that place at that instant.

The wave functions are usually complex. The probability in such a case is taken as $\psi^*\psi$, i.e., the product of the wave function with its complex conjugate, ψ^* being the complex conjugate. Since the probability of finding a particle somewhere is finite, we have the total probability over all space equal to unity. That is

$$\int_{-\infty}^{\infty} \psi^* \psi \, dV = 1 \quad \text{(i)}$$

where $dV = dxdydz$.

Equation (i) is called the normalisation condition and a wave function that obeys this equation is said to be normalised. Further, ψ must be a single valued since the probability can have only one value at a particular place and time. Besides being normalisable, a further condition that ψ must obey is that it and its partial derivatives $\frac{\partial\psi}{\partial x}$, $\frac{\partial\psi}{\partial y}$ and $\frac{\partial\psi}{\partial z}$ be continuous everywhere.

The important characteristics of the wave function are as follows.

(i) ψ must be finite, continuous and single valued everywhere.

(ii) $\frac{\partial\psi}{\partial x}$, $\frac{\partial\psi}{\partial y}$ and $\frac{\partial\psi}{\partial z}$ must be finite, continuous and single valued.

(iii) ψ must be normalisable.

16.3 TIME INDEPENDENT SCHRÖDINGER EQUATION LO2

Consider a system of stationary waves associated with a moving particle. The waves are said to be stationary w.r.t the particle. If the position coordinates of the particle are (x, y, z) and ψ be the periodic displacement for the matter waves at any instant of time t, then we can represent the motion of the wave by a differential equation as follows.

$$\frac{\partial^2\psi}{\partial x^2} + \frac{\partial^2\psi}{\partial y^2} + \frac{\partial^2\psi}{\partial z^2} = \frac{1}{u^2}\frac{\partial^2\psi}{\partial t^2} \tag{i}$$

where u is the velocity of wave associated with the particle. The solution of Eq. (i) gives ψ as a periodic displacement in terms of time, i.e.,

$$\psi(x, y, z, t) = \psi_0(x, y, z)e^{-i\omega t} \tag{ii}$$

where ψ_0 is the amplitude of the particle wave at the point (x, y, z) which is independent of time (t). It is a function of (x, y, z). i.e., the position r and not of time t, Here.

$$r = x\hat{i} + y\hat{j} + z\hat{k} \tag{iii}$$

Eq. (ii) may be expressed as

$$\psi(r, t) = \psi_0(r)e^{-i\omega t} \tag{iv}$$

Differentiating Eq. (iv) twice with respect to t, we get

$$\frac{\partial^2\psi}{\partial t^2} = -\omega^2\psi_0(r)e^{-i\omega t}$$

or

$$\frac{\partial^2\psi}{\partial t^2} = -\omega^2\psi \tag{v}$$

Substituting the value of $\frac{\partial^2\psi}{\partial t^2}$ from this equation in Eq. (i), we get

$$\frac{\partial^2\psi}{\partial x^2} + \frac{\partial^2\psi}{\partial y^2} + \frac{\partial^2\psi}{\partial z^2} + \frac{\omega^2}{u^2}\psi = 0 \tag{vi}$$

where $\omega = 2\pi\nu = 2\pi(u/\lambda)$ [as $u = \lambda\nu$]

so that

$$\frac{\omega}{u} = \frac{2\pi}{\lambda} \tag{vii}$$

Also

$$\frac{\partial^2 \psi}{\partial x^2} + \frac{\partial^2 \psi}{\partial y^2} + \frac{\partial^2 \psi}{\partial z^2} = \nabla^2 \psi \tag{viii}$$

where ∇^2 is known as Laplacian operator. Using Eqs. (vi), (vii) and (viii), we have

$$\nabla^2 \psi + \frac{4\pi^2}{\lambda^2} \psi = 0 \tag{ix}$$

Also from the deBroglie wave concept

$$\lambda = \frac{h}{mv}$$

Using this relation in Eq. (ix) gives

$$\nabla^2 \psi + \frac{4\pi^2 m^2 v^2}{h^2} \psi = 0 \tag{x}$$

Here it can be noted that the velocity of particle v has been introduced in the wave equation.

If E and V are respectively the total energy and potential energy of the particle then its kinetic energy is given by

$$\frac{1}{2} mv^2 = E - V$$

$$m^2 v^2 = 2m(E - V) \tag{xi}$$

The use of Eq. (xi) in Eq. (x) gives rise to

$$\nabla^2 \psi + \frac{8\pi^2 m}{h^2} (E - V) \psi = 0$$

or

$$\nabla^2 \psi + \frac{2m}{\hbar^2} (E - V) \psi = 0 \tag{xii}$$

This is the time independent Schrödinger equation, where the quantity ψ is known as *wave function*.

For a freely moving or free particle $V = 0$. Therefore, Eq. (xii) becomes

$$\nabla^2 \psi + \frac{2mE}{\hbar^2} \psi = 0 \tag{xiii}$$

This is called time independent Schrödinger equation for a free particle.

16.4 TIME DEPENDENT SCHRÖDINGER EQUATION

LO3

In order to obtain a time dependent Schrödinger equation, we eliminate the total energy E from time independent Schrödinger equation. For this we differentiate Eq. (iv) w.r.t. t and obtain

$$\frac{\partial \psi}{\partial t} = -i\omega\psi_0(r)e^{-i\omega t}$$

$$= i(2\pi v)\psi_0(r)e^{-i\omega t}$$

$$= -2\pi v i\psi = -2\pi i\frac{E}{h}\psi = -\frac{iE}{h}\times\frac{i}{i}\psi$$

$$\Rightarrow \qquad \frac{\partial \psi}{\partial t} = \frac{E\psi}{i\hbar}$$

or $$E\psi = i\hbar\frac{\partial \psi}{\partial t} \qquad \text{(xiv)}$$

Substituting the value of Eψ from Eq. (xiv) in Eq. (xii), we have

$$\nabla^2\psi + \frac{2m}{\hbar^2}\left[i\hbar\frac{\partial \psi}{\partial t} - V\psi\right] = 0$$

or $$\nabla^2\psi = -\frac{2m}{\hbar^2}\left[i\hbar\frac{\partial \psi}{\partial t} - V\psi\right]$$

or $$\left(-\frac{\hbar^2}{2m}\nabla^2 + V\right)\psi = i\hbar\frac{\partial \psi}{\partial t} \qquad \text{(xv)}$$

This equation is known as *Schrödinger's time dependent wave equation*. The operator $\left(-\frac{\hbar^2}{2m}\nabla^2 + V\right)$ is called *Hamilitonian operator* and is represented by H. If we see the RHS of Eq. (xv) and keep in mind Eq. (xiv), we notice that the operator $i\hbar\frac{\partial}{\partial t}$ operating on ψ gives E. Hence, Schrödinger equation can be written in operator form, as below

$$H\psi = E\psi$$

16.5 OPERATORS LO3

In a physical system, there is a quantum mechanical operator that is associated with each measurable parameter. In quantum mechanics, we deal with waves (wave function) rather than discrete particles whose motion and dynamics can be described with the deterministic equations of Newtonian physics. Generally an operator is anything that is capable of doing something to a function. There is an operator corresponding to every observable quantity. However, the choice of operator is arbitrary in quantum mechanics. When an operator operates on a wave function it must give observable quantity times the wave function. It is a must condition for an operator.

If we consider an operator represented by A corresponding to the observable quantity a, then

$$A\psi = a\psi$$

Wave function that satisfies the above equation is called *eigen function* and corresponding observable quantity is called *eigen value* and the equation is called eigen value equation. Some of those operators are tabled below.

Classical Quantity	Quantum Mechanical Operator
Position $x, y, z,$	x, y, z
Momentum p	$-i\hbar\vec{\nabla}$
Momentum components p_x, p_y, p_z	$-i\hbar\frac{\partial}{\partial x}, -i\hbar\frac{\partial}{\partial y}, -i\hbar\frac{\partial}{\partial z}$
Energy E	$i\hbar\frac{\partial}{\partial t}$
Hamiltonian (Time independent)	$-\frac{\hbar^2}{2m}\frac{\partial^2}{\partial x^2}+V(r)$
Kinetic energy	$-\frac{\hbar^2}{2m}\frac{\partial^2}{\partial r^2}$

16.6 APPLICATIONS OF SCHRÖDINGER EQUATION LO4

In classical mechanics, based on Newton's second law of motion ($\vec{F} = m\vec{a}$) we make a mathematical prediction of the path, a given system will take following set of known initial conditions. The analogue of Newton's law is Schrödinger equation in quantum mechanics for a quantum system such as atoms, molecules and subatomic particles. The subatomic particles may be free, bound or localized. Schrödinger equation describes the time evolution of the system's wave function.

Schrödinger's equation is extremely useful for investigating various quantum mechanical problems. With the help of this equation and boundary conditions, the expression for the wave function is obtained. Then the probability of finding the particle is calculated by using the wave function. In the following subsections, we discuss different quantum mechanical problems, viz. particle in a box, one-dimensional harmonic oscillator, step potential and step barrier.

16.6.1 Particle in a Box (Infinite Potential Well)

The simplest quantum mechanical problem is that of a particle trapped in a box with infinitely hard walls. Infinitely hard walls means the particle does not loose energy when it collides with such walls, i.e., its total energy remains constant. A physical example of this problem could be a molecule which is strictly confined in a box.

Let us consider a particle restricted to move along the x-axis between $x = 0$ and $x = L$, by ideally reflecting, infinitely high walls of the infinite potential well, as shown in Fig. 16.2. Suppose that the potential energy V of the particle is zero inside the box, but rises to infinity outside, that is,

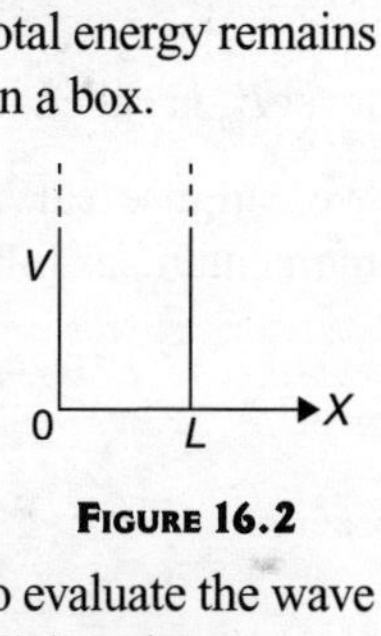

FIGURE 16.2

$$V = 0 \quad \text{for} \quad 0 \le x \le L$$
$$V = \infty \quad \text{for} \quad x < 0 \quad \text{and} \quad x > L$$

In such a case, the particle is said to be moving in an infinitely deep potential well. In order to evaluate the wave function ψ in the potential well, Schoredinger equation for the particle within the well ($V = 0$) is written as

$$\frac{\partial^2\psi}{\partial x^2}+\frac{8\pi^2 mE}{h^2}\psi = 0 \quad \text{(i)}$$

we put $\dfrac{8\pi^2 mE}{h^2} = k^2$ in the above equation for getting

$$\frac{\partial^2 \psi}{\partial x^2} + k^2 \psi = 0 \tag{ii}$$

The general solution of this differential equation is

$$\psi(x) = A \sin kx + B \cos kx \tag{iii}$$

where A and B are constants.

Applying the boundary condition $\psi(x) = 0$ at $x = 0$, which means the probability of finding particle at the wall $x = 0$ is zero, we obtain

$$A \sin(0) + B \cos(0) = 0 \quad \Rightarrow \quad B = 0$$

Again, we have $\psi(x) = 0$ at $x = L$, then

$$A \sin kL + B \cos kL = 0 \quad \Rightarrow \quad A \sin kL = 0$$

The above equation is satisfied when

$$kL = n\pi$$

or $$k = \frac{n\pi}{L} \quad \text{where } n = 1, 2, 3, \ldots$$

or $$k^2 = \frac{n^2\pi^2}{L^2} \tag{iv}$$

or $$\frac{8\pi^2 mE}{h^2} = \frac{n^2\pi^2}{L^2} \tag{v}$$

or in general we can write Eq. (v) as

$$E_n = \frac{n^2 h^2}{8mL^2} \quad \text{where } n = 1, 2, 3, \ldots$$

Thus, it can be concluded that in an infinite potential well the particle cannot have an arbitrary energy, but can take only certain discrete energy values corresponding to $n = 1, 2, 3, \ldots$. These are called the *eigen values* of the particle in the well and constitutes the energy levels of the system. The integer n corresponding to the energy level E_n is called its *quantum number*, as shown in Fig. 16.3.

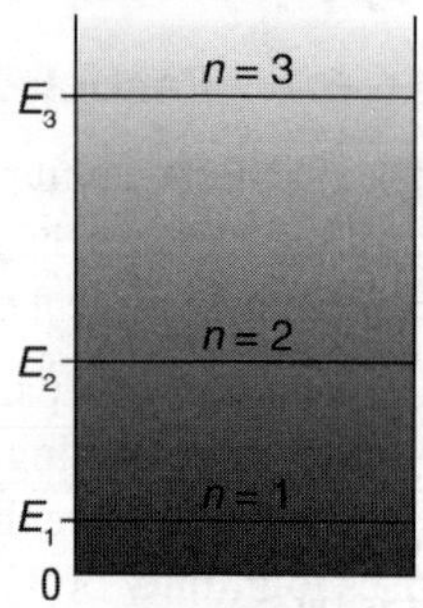

FIGURE 16.3

We can also calculate the momentum p of the particle or the eigen values of the momentum, as follows,

$$\text{Since } k = \frac{2\pi}{\lambda} = \frac{2\pi}{h/p} = \frac{p}{\hbar}$$

$$p = \hbar k = \frac{n\pi\hbar}{L}$$

The wave function (or eigen function) is given by Eq. (iii) along with the use of expression for k.

$$\psi_n(x) = A \sin\frac{n\pi x}{L}$$

To find the value of A, we use the normalisation condition.

$$\int_{-\infty}^{\infty} |\psi_n(x)|^2\, dx = 1$$

As mentioned earlier, the above expression simply says that the probability of finding the particle is 1. In the present case, the particle is within the box i.e., between $0 < x < L$. So the normalisation condition becomes

$$A^2 \int_0^L \sin^2 \frac{n\pi x}{L}\, dx = 1$$

$$A^2\left(\frac{L}{2}\right) = 1 \quad \text{or} \quad A = \sqrt{\frac{2}{L}}$$

The normalised eigen wave function of the particle is, therefore, given by

$$\psi_n(x) = \sqrt{\frac{2}{L}} \sin \frac{n\pi x}{L}$$

The first three eigen functions ψ_1, ψ_2, ψ_3 together with the probability densities $|\psi_1|^2$, $|\psi_2|^2$, $|\psi_3|^2$, are shown in Figs. 16.4(a) and (b), respectively.

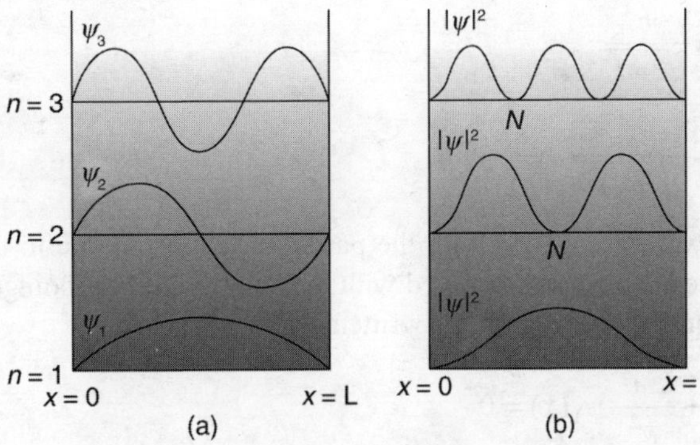

FIGURE 16.4

Classical mechanics predicts the same probability for the particle being anywhere in the well. Wave mechanics, on the other hand, predicts that the probability is different at different points and there are points (nodes) where the particle is never found. Further, at a particular point, the probability of finding the particle is different for different energy states. For example, a particle in the lowest energy state ($n = 1$) is more likely to be in the middle of the box, while in the next energy state ($n = 2$) it is never there since $|\psi_2|^2$ is zero there. It is $|\psi_n|^2$ which provides the probability of finding the particle within the potential well.

16.6.2 Finite Potential Step

A physical example of this quantum mechanical problem can be thought as the neutron which is trying to escape nucleus. The potential function of a potential step may be represented as

$$V(x) = 0 \qquad \text{for } x < 0 \text{ region I}$$

$$V(x) = V_0 \qquad \text{for } x > 0 \text{ region II} \qquad \text{(i)}$$

We consider that a particle of energy E is incident from left on the potential step of height V_0 as shown in Fig. 16.5. Further, we assume that the energy of the incident particle is greater than the step barrier height i.e., $E > V_0$. Since $E > V_0$, according to classical theory there should be no reflection at the boundary of the step potential barrier. However, quantum mechanically this is not true. It means that there will be some reflection from the boundary of the potential step.

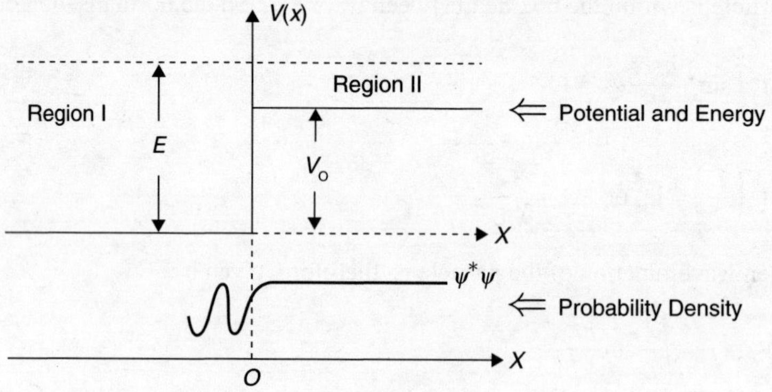

FIGURE 16.5

The wavelength of the particle suddenly changes from region I to region II and is given as follows

$$\lambda_1 = \frac{h}{p_1} = \frac{h}{\sqrt{2mE}} \tag{ii}$$

and
$$\lambda_2 = \frac{h}{p_2} = \frac{h}{\sqrt{2m(E - V_0)}} \tag{iii}$$

Hence, a small part of the wave associated with the particle is reflected due to this change in wavelength and the rest part is transmitted. This can be proved with the solution of Schrödinger wave equations for two regions. The Schrödinger equation for region I is written as

$$\frac{d^2\psi_1(x)}{dx^2} + \frac{2mE}{\hbar^2}\psi_1(x) = 0 \tag{iv}$$

Schrödinger equation for region II is written as

$$\frac{d^2\psi_2(x)}{dx^2} + \frac{2m(E - V_0)}{\hbar^2}\psi_2(x) = 0 \tag{v}$$

The solutions of Eqs. (iv) and (v) are written as

$$\psi_1(x) = A_1 e^{ik_1 x} + A_2 e^{-ik_1 x} \tag{vi}$$

$$\psi_2(x) = A_3 e^{ik_2 x} + A_4 e^{-ik_2 x} \tag{vii}$$

where $\psi_1(x)$ and $\psi_2(x)$ are the wave functions of region I and II and A_1, A_2, A_3 and A_4 are constants. k_1 and k_2 are defined as follows

$$k_1 = \frac{\sqrt{2mE}}{\hbar} \quad \text{and} \quad k_2 = \frac{\sqrt{2m(E - V_0)}}{\hbar}$$

The first term in Eq. (vi) represents the wave travelling in the positive x direction in the first region and second term represents the reflected part of the incident wave travelling in the negative x direction in region I. In Eq. (vii), first term represents the transmitted part of the incident particle wave travelling in the direction of positive x axis in region II. The second term of Eq. (vii) has no meaning, because the reflection of the particle cannot take place in region II. So, considering this, Eq. (vii) can be written as

$$\psi_2(x) = A_3 e^{ik_2 x} \tag{viii}$$

The boundary condition at $x = 0$ is defined as

$$\psi_1(0) = \psi_2(0), \tag{ix}$$

which means the wave function is continuous at the boundary. Also the derivative of ψ should be continuous at the boundary, i.e.,

$$\left.\frac{d\psi_1(x)}{dx}\right|_{x=0} = \left.\frac{d\psi_2(x)}{dx}\right|_{x=0} \tag{x}$$

Using these boundary conditions, we get

$$A_1 + A_2 = A_3 \tag{xi}$$

and $$ik_1(A_1 e^{ik_1 x} - A_2 e^{-ik_1 x})\big|_{x=0} = ik_2 A_3 e^{ik_2 x}\big|_{x=0}$$

or $$ik_1(A_1 - A_2) = ik_2 A_3 \tag{xii}$$

Solving Eq. (xi) and (xii), we get

$$\frac{A_2}{A_1} = \frac{k_1 - k_2}{k_1 + k_2} \tag{xiii}$$

and $$\frac{A_3}{A_1} = \frac{2k_1}{k_1 + k_2} \tag{xiv}$$

Since the reflection and transmission of the particle takes place, the problem can be investigated based on the reflection and transmission coefficients. Further, the coefficient A_1 is related to the wave function ψ_1, i.e., of the incident particle and A_3 is related to the wave function ψ_3, i.e., of the transmitted particle. It means the reflection and transmission coefficients can be defined as follows.

Reflection coefficient = Reflected intensity/Incident intensity
= (Reflected amplitude)^2/(Incident amplitude)^2 (xv)

Transmission coefficient = Transmitted intensity/Incident intensity
= (Transmitted amplitude)^2/(Incident amplitude)^2 (xvi)

The reflection coefficient is

$$R = \frac{|A_2|^2}{|A_1|^2} = \left(\frac{k_1 - k_2}{k_1 + k_2}\right)^2 \neq 0 \tag{xvii}$$

and transmission coefficient is

$$T = \frac{|A_3|^2}{|A_1|^2} = \left(\frac{2k_1}{k_1 + k_2}\right)^2 \neq 1 \tag{xviii}$$

From the above results we see that the reflection coefficient (R) is not zero and the transmission probability (T) is not unity in the quantum mechanical treatment of the particle behaviour in the finite potential step problem. However, classically the reflection coefficient should be zero and transmission coefficient should be equal to unity.

16.6.3 Finite Potential Barrier

A physical example of this quantum mechanical problem can be thought as the α particle which is trying to escape Coulomb barrier. For this case, the potential function is defined as

$$V(x) = 0 \quad \text{for } x < 0 \quad \text{Region I}$$
$$V(x) = V_0 \quad \text{for } 0 < x < a \quad \text{Region II}$$
$$V(x) = 0 \quad \text{for } x > a \quad \text{Region III}$$

The potential barrier is considered between $x = 0$ and $x = a$, as shown in Fig. 16.6. Here, we suppose that a particle incident on the barrier has energy E which is less than the barrier height V_0, i.e., $E < V_0$. Classically, when $E < V_0$, the particle can never penetrate the potential barrier and appear in region III. It means the particle is always reflected from the barrier. Therefore, the transmission coefficient is zero. However, quantum mechanically this is not true and there is some probability for a particle penetrate the barrier. It means a fraction of the particles incident from the left will cross the barrier and appear in region III.

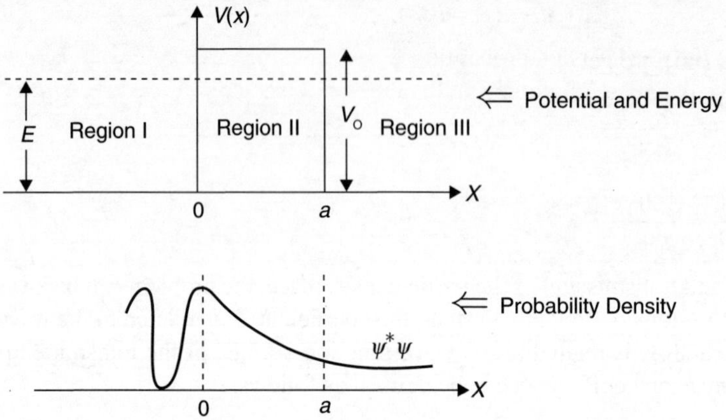

FIGURE 16.6

Schrödinger wave equations for the regions I and III are as follows

$$\frac{d^2\psi_1(x)}{dx^2} + \frac{2mE}{\hbar^2}\psi_1(x) = 0 \tag{ii}$$

and
$$\frac{d^2\psi_3(x)}{dx^2} + \frac{2mE}{\hbar^2}\psi_3(x) = 0 \tag{iii}$$

where $\psi_1(x)$ and $\psi_3(x)$ are the wave function of region I and III. The solutions of these equations are

$$\psi_1(x) = A_1e^{ik_1x} + A_2e^{-ik_1x} \tag{iv}$$

and
$$\psi_3(x) = A_3e^{ik_1x} + A_4e^{-ik_1x} \tag{v}$$

where $k_1 = \frac{\sqrt{2mE}}{\hbar} = \frac{p}{\hbar} = \frac{2\pi}{\lambda}$ and A_1, A_2, A_3 and A_4 are constants. The solution for ψ_1 is a combination of reflected and transmitted wave in region I. But in the region III, the reflected part of the wave is zero ($A_4 = 0$) and the transmitted wave is traveling in the positive x direction. So the solution in region III becomes

$$\psi_3(x) = A_3 e^{ik_1 x} \tag{vi}$$

Now, the Schrödinger equation for region II is written as

$$\frac{d^2\psi_2(x)}{dx^2} + \frac{2m(E - V_0)}{\hbar^2}\psi_2(x) = 0 \tag{vii}$$

But as we know that $E < V_0$ then it will be convenient to write this equation in the form

$$\frac{d^2\psi_2(x)}{dx^2} - \frac{2m(V_0 - E)}{\hbar^2}\psi_2(x) = 0 \tag{viii}$$

where ψ_2 is the wave function of region II. The solution of above equation is

$$\psi_2(x) = A_5 e^{-ik_2 x} + A_6 e^{ik_2 x} \tag{ix}$$

where $$k_2 = \frac{\sqrt{2m(V_0 - E)}}{\hbar}$$

In order to calculate the transmission probability T, we must apply the boundary conditions to wave function ψ_1, ψ_2 and ψ_3. These boundary conditions at the left hand (at $x = 0$) or at the right hand wall (at $x = a$) of the barrier are defined as

Boundary conditions at $x = 0$ are

$$\psi_1(0) = \psi_2(0) \tag{x}$$

and $$\frac{\partial\psi_1(0)}{\partial x} = \frac{\partial\psi_2(0)}{\partial x} \tag{xi}$$

At $x = a$ are

$$\psi_2(a) = \psi_3(a) \tag{xii}$$

and $$\frac{\partial\psi_2(a)}{\partial x} = \frac{\partial\psi_3(a)}{\partial x} \tag{xiii}$$

The above boundary conditions along with the use of wave functions ψ_1, ψ_2 and ψ_3 yield

$$A_1 + A_2 = A_5 + A_6 \tag{xiv}$$

$$ik_1A_1 - ik_1A_2 = -k_2A_5 + k_2A_6 \tag{xv}$$

$$A_5 e^{-k_2 a} + A_6 e^{k_2 a} = A_3 e^{ik_1 a} \tag{xvi}$$

$$-k_2 A_5 e^{-k_2 a} + k_2 A_6 e^{k_2 a} = ik_1 A_3 e^{ik_1 a} \tag{xvii}$$

Solving Eqs. (xiv) and (xv), we get

$$A_1 = \frac{(ik_1 - k_2)}{2ik_1}A_5 + \frac{(ik_1 + k_2)}{2ik_1}A_6 \tag{xviii}$$

and solving Eqs. (xvi) and (xvii), we get

$$A_6 = \frac{(k_2 + ik_1)e^{ik_1a}}{2k_2e^{k_2a}} A_3 \tag{xix}$$

and

$$A_5 = \frac{(k_2 - ik_1)e^{ik_1a}}{2k_2e^{-k_2a}} A_3 \tag{xx}$$

Substituting these values in Eq. (xviii), we get

$$\frac{A_1}{A_3} = \left[\frac{1}{2} + \frac{i}{4}\left(\frac{k_2}{k_1} - \frac{k_1}{k_2}\right)\right] e^{(ik_1 + k_2)a} + \left[\frac{1}{2} - \frac{i}{4}\left(\frac{k_2}{k_1} - \frac{k_1}{k_2}\right)\right] e^{(ik_1 - k_2)a} \tag{xxi}$$

As we assumed, the potential barrier to be much higher than the energy of the incident particles. In this condition $k_2/k_1 >> k_1/k_2$ and hence

$$\frac{k_2}{k_1} - \frac{k_1}{k_2} \approx \frac{k_2}{k_1} \tag{xxii}$$

Further we assume that the potential barrier wide enough so that for ψ_2 gets severely weakened between $x = 0$ and $x = a$. This means $k_2a >> 1$, i.e.,

$$e^{k_2a} >> e^{-k_2a} \tag{xxiii}$$

So Eq. (xxi) is approximated by

$$\frac{A_1}{A_3} = \left(\frac{1}{2} + \frac{ik_2}{4k_1}\right) e^{(ik_1 + k_2)a} \tag{xxiv}$$

The complex conjugate of this is written as

$$\left(\frac{A_1}{A_3}\right)^* = \left(\frac{1}{2} - \frac{ik_2}{4k_1}\right) e^{(-ik_1 + k_2)a} \tag{xxv}$$

On multiplying Eqs. (xxiv) and (xxv), we get

$$\text{So, } \frac{A_1A_1^*}{A_3A_3^*} = \left(\frac{1}{4} + \frac{k_2^2}{16k_1^2}\right) e^{2k_2a} \tag{xxvi}$$

Since the coefficient A_1 is related to the wave function ψ_1, i.e., of the incident particle and A_3 is related to the wavelength of ψ_3, of the transmitted particle, the transmission probability is equivalent to

$$T = \frac{A_3A_3^*}{A_1A_1^*} = \left(\frac{A_1A_1^*}{A_3A_3^*}\right)^{-1} = \left(\frac{16}{4 + (k_2/k_1)^2}\right) e^{-2k_2a} \tag{xxvii}$$

From the definitions of k_1 and k_2 we see that

$$\left(\frac{k_2}{k_1}\right)^2 = \frac{2m(V_0 - E)/\hbar^2}{2mE/\hbar^2} = \frac{V_0}{E} - 1 \tag{xxviii}$$

With this it can be seen that the quantity in the bracket varies slowly with E and V_0 than the variation of exponential term. So the approximated transmission probability is

$$T = e^{-k_2 a} \tag{xxix}$$

16.6.4 One-Dimensional Harmonic Oscillator

A physical example of this quantum mechanical problem can be thought as an atom of vibrating diatomic molecule. In general, a particle undergoing simple harmonic motion in one dimension is called one dimensional harmonic oscillator. The potential and total energy of such a system is shown in Fig. 16.7 where the probability density is also shown. In such a motion, the restoring force F is proportional to the particle's displacement x from the equilibrium position, i.e.,

$$F = -kx \tag{i}$$

where k is force constant. The potential energy V can be written as

$$V = \frac{1}{2}kx^2$$

Then, the Schrödinger's equation for the oscillator with $V = \frac{1}{2}kx^2$ is

$$\frac{d^2\psi}{dx^2} + \frac{2m}{\hbar^2}\left[E - \frac{1}{2}kx^2\right]\psi = 0$$

Putting $\hbar = \frac{h}{2\pi}$, $\frac{8\pi^2 mE}{h^2} = \alpha$ and $\left(\frac{4\pi^2 mk}{h^2}\right)^{1/2} = \beta$ in the above equation, we obtain

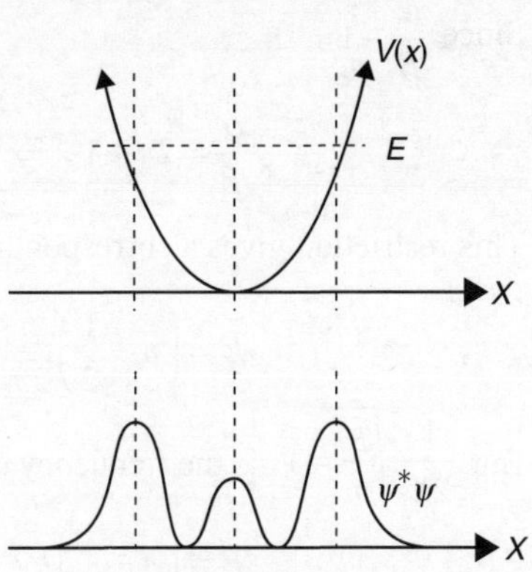

FIGURE 16.7

$$\frac{d^2\psi}{dx^2} + (\alpha - \beta^2 x^2)\psi = 0 \tag{ii}$$

Now we introduce a dimension less independent variable as $\xi = \sqrt{\beta}x$. Thus Eq. (ii) becomes.

$$\beta\frac{d^2\psi}{d\xi^2} + \left[\alpha - \beta^2\frac{\xi^2}{\beta}\right]\psi = 0$$

$$\frac{d^2\psi}{d\xi^2} + \left[\frac{\alpha}{\beta} - \xi^2\right]\psi = 0 \tag{iii}$$

The solution of this equation is

$$\psi = CUe^{-\xi^2/2} \tag{iv}$$

where U is a function of ξ. Then Eq. (iii) takes the form

$$\frac{d^2U}{d\xi^2} - 2\xi\frac{dU}{d\xi} + \left[\frac{\alpha}{\beta} - 1\right]U = 0$$

If we replace $\frac{\alpha}{\beta} - 1$ by $2n$, this equation becomes Hermite differential equation. Then function $U(\xi)$ may be replaced with Hermite polynomial H. So, we get

$$\frac{d^2H}{d\xi^2} - 2\xi\frac{dH}{d\xi} + 2nH = 0$$

Thus, the solution of Eq. (iii) is obtained by replacing U by Hermite polynomial H in Eq. (iv). Hence, we get

$$\psi = CHe^{-\xi^2/2}$$

In general, $\psi_n(\xi) = CH_n(\xi)e^{-\xi^2/2}$, where $n = 0, 1, 2,\ldots$

Eigen Values of Energy

Since $\frac{\alpha}{\beta} - 1 = 2n$

$$\Rightarrow \quad \frac{\alpha}{\beta} = 2n+1 \quad \Rightarrow \quad \alpha = (2n+1)\beta \quad \Rightarrow \quad \frac{8\pi^2 mE}{h^2} = (2n+1)\sqrt{\frac{4\pi^2 mk}{h^2}}$$

This restriction gives a corresponding restriction on E, i.e.,

$$E = \left(n + \frac{1}{2}\right)\frac{h}{2\pi}\sqrt{\frac{k}{m}}$$

But $\frac{1}{2\pi}\sqrt{\frac{k}{m}} = \nu$ is the frequency of oscillations. Hence, the energy can be written in terms of ν as

$$E = \left(n + \frac{1}{2}\right)h\nu$$

Thus, in general, the oscillator has finite, unambiguous and continuous solutions at values of E given by

$$E_n = \left(n + \frac{1}{2}\right)h\nu \qquad \text{(v)}$$

Following conclusions can be drawn from equation (v)

The particle executing simple harmonic motion can have only

(i) discrete energy levels that are equidistant and are separated by $h\nu$, as shown in Fig. 16.8

(ii) The energy levels are non-degenerate.

(iii) For $n = 0$, $E_0 = \frac{1}{2}h\nu$. It means the minimum energy is not zero.

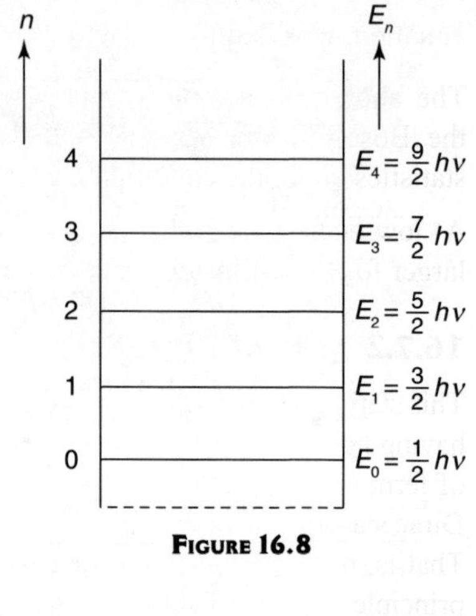

FIGURE 16.8

16.7 QUANTUM STATISTICS LO5

Classical statistics, i.e., Maxwell-Boltzmann statistics, successfully explained the energy and velocity distribution of molecules of an ideal gas but it failed to explain the energy distribution of electrons in metals, for example, electron gas and the energy distribution of photons in a photon gas. These phenomena can be explained on the basis of quantum statistics, where the particles of the system are considered to be indistinguishable contrary to the consideration of particles as distinguishable in classical statistics. If n_i particles are distributed in the g_i cell of the ith compartment in the phase space, then the number of particles per cell is defined as $n_i(E)/g_i(E)$. The factor $n_i(E)/g_i(E)$ is called *occupation index*. If $n_i(E)/g_i(E) \geq 0$ or 1 the particles are considered as indistinguishable which is the basic feature of the quantum statistics. If the indistinguishable particles have integral spin, we use the Bose-Einstein distribution function and if the particles have half-integral spins, then Fermi-Dirac distribution function is appropriate. The brief description of these statistics is given below.

16.7.1 Bose-Einstein Statistics

It is applicable to those systems which contain identical, indistinguishable particles of zero or integral spins. Such particles are called *bosons*. Examples of bosons are photons, phonons etc. Pauli exclusion principle does not apply to the bosons. Bose-Einstein distribution law is given by

$$n_i(E) = \frac{g_i(E)}{e^{\alpha+\beta E} - 1} \qquad \text{(i)}$$

where $\alpha = -E_F/kT$ and $\beta = 1/kT$. This law is also applicable in the case of photon gas for which $\beta = 0$ and $E = h\nu$. For the photon gas, the above equation reads

$$n_i(E) = \frac{g_i(E)}{e^{\beta E} - 1} \qquad \text{(ii)}$$

The plot of $n_i(E)/g_i(E)$ versus E is shown in Fig. 16.9 for two different temperatures with $T_2 > T_1$. If $E >> kT$, the exponential term in the above equation is very large and –1 may be dropped. It means

$$n_i(E) = g_i(E)e^{-E/kT}$$

The above relation represents Maxwell-Boltzmann statistics. So the Bose-Einstein statistics reduces to the Maxwell-Boltzmann statistics under the condition $E >> kT$.

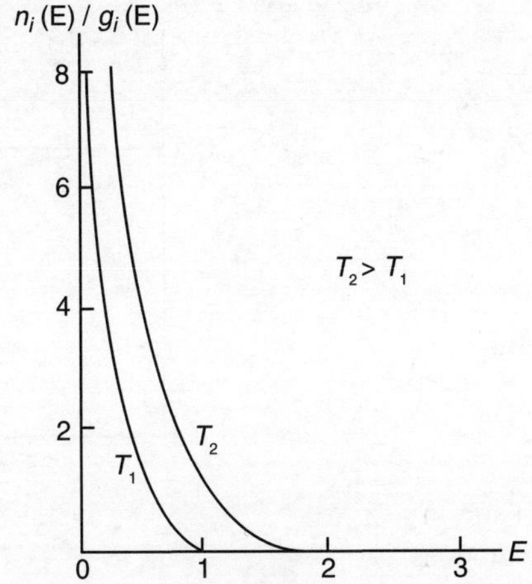

FIGURE 16.9

At low energy, i.e., when $E << kT$, $e^{\beta E}$ can be neglected as –1 predominates. This makes $n_i(E)/g_i(E)$ much larger for Bose-Einstein statistics than for Maxwell-Boltzmann statistics at low energies.

16.7.2 Fermi-Dirac Statistics

This statistics is applicable to systems, which consist of identical, independent and indistinguishable particles of having half-integral spins. The particles, which obey Fermi-Dirac statistics, are called *fermions*. The examples of fermions are electrons, protons, neutrons, etc. The fermion must obey Pauli exclusion principle. In Fermi-Dirac statistics, interchange of two particles of the system leaves the resultant system in an antisymmetric state. That is, the wave function of the system gets changed only with minus sign. As it obeys the Pauli exclusion principle, in Fermi-Dirac statistics, there can be only one particle in each state. Hence, the total number of particles must be less than or equal to the total number of states available. Under these considerations, fermions lead to the following distribution law, named Fermi-Dirac distribution law, given by

$$n_i(E) = \frac{g_i(E)}{e^{\alpha+\beta E} + 1} \qquad \text{(i)}$$

where $\alpha = -E_F/kT$ and $\beta = 1/kT$.

So, $$n_i(E) = \frac{g_i(E)}{e^{(E-E_F)/kT} + 1} \quad \text{(ii)}$$

$$\frac{n_i(E)}{g_i(E)} = \frac{1}{e^{(E-E_F)/kT} + 1} \quad \text{(iii)}$$

In the above equations, n_i is the number of particles in an energy state E, g_i is the statistical weight factor and E_F is the Fermi energy. Fermi energy is independent of temperature. The plots of $n_i(E)/g_i(E)$ versus E for different temperatures is shown in Fig. 16.10.

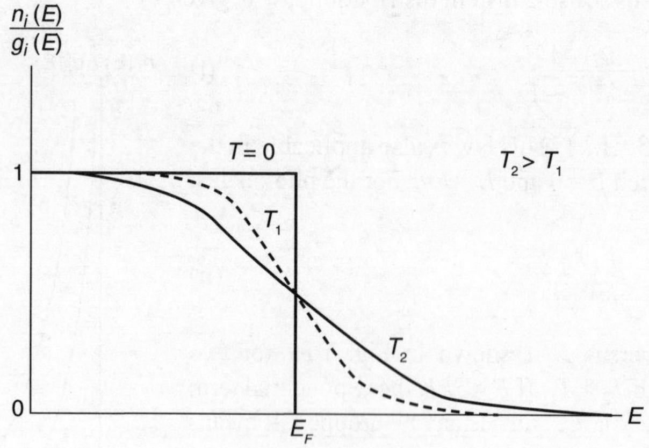

FIGURE 16.10

Now the different cases will be discussed.

(a) **At T = 0 K and when $E < E_F$, Eq. (iii)**

$$\frac{n_i(E)}{g_i(E)} = 1 \quad \text{(iv)}$$

It means that $n_i(E) = g_i(E)$, i.e., all the energy states will have one electron each.

(b) **At T = 0 K and $E > E_F$,**

$$\frac{n_i(E)}{g_i(E)} = 0 \quad \text{(v)}$$

It means that n_i = 0, i.e., all such energy states which have energies greater than Fermi energy are vacant. This clarifies that all states with energies up to E_F are filled while all states with energy greater than E_F are vacant. The plot between $n_i(E)/g_i(E)$ and E is shown in Fig. 16.11 for these conditions at T = 0 K.

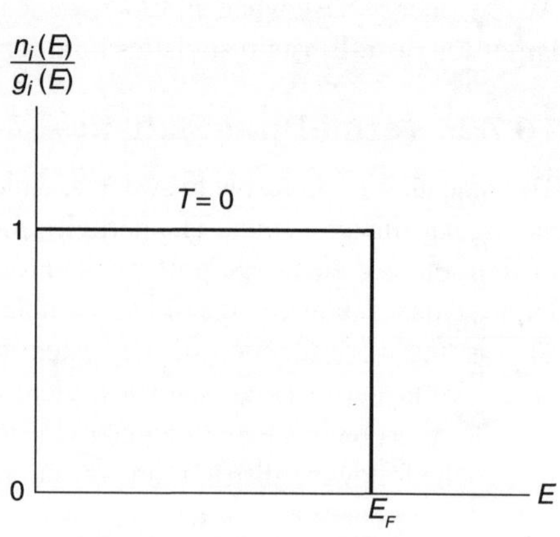

FIGURE 16.11

SUMMARY

The main topics covered in this chapter are summarised below.

- Quantum mechanics deals with the behaviour and characteristics of matter, in the subatomic level, and energy.
- Basics of quantum theory were particularly developed by Planck, Einstein, Schrödinger and Heisenberg.
- The best known result of the wave particle duality, i.e., the concept of waves or wave packet associated with a moving particle, is the Heisenberg uncertainty principle. According to this principle it is impossible to determine simultaneously the exact position and momentum (or velocity) of a small moving particle like electron. If Δx be the uncertainty in the position and Δp be the uncertainty in the momentum, then $\Delta x \Delta p \approx \hbar$ represents the lowest limit of accuracy. More generally, the Heisenberg uncertainty principle in mathematical form is written as $\Delta x \Delta p \approx \hbar$.
- Based on Heisenberg uncertainty principle (i) it was proved that the electron cannot reside inside the nucleus, (ii) radius of Bohr's first orbit was calculated, (iii) energy of a particle in a box was evaluated, and (iv) ground state energy of a linear harmonic oscillator was calculated.
- Waves in general are associated with quantities that vary periodically. In case of matter waves, the quantity that varies periodically is called wave function (ψ). The wave function ψ is generally a complex quantity and as such it has no physical meaning. However, the product $\psi\psi^*$, where ψ^* is its complex conjugate, represents the probability.
- Wave function ψ should be finite, continuous and single valued and so it is space derivative. Also, the wave function should be normalisable.
- Time independent Schrödinger equation is written as $\nabla^2\psi + \frac{2m}{\hbar^2}(E - V)\psi = 0$, where ψ is the wave function, m is the mass of the particle, E is the total energy of the particle and V is the potential.
- Time dependent Schrödinger equation is written as $H\psi = E\psi$, where $H = \frac{\hbar^2}{2m}\nabla^2 + V$ is Hamiltonian operator. E is the total energy of the particle.
- Particle in a box acquires only discrete energy values, the energy being given by $E_n = \frac{n^2h^2}{8mL^2}$, where m is the mass of the particle, L is the width of the box (infinite potential well) and n (=1, 2, 3,...) is called the eigen value or the quantum number. It is interesting to note that for $n = 1$, the particle is mostly found in the middle of the well, whereas the probability of finding the particle in middle for $n = 2$ is zero. It happens quantum mechanically.
- In the quantum mechanical problem of finite potential well, when the particle energy E is less than the barrier height V_0, we find that there is a probability that the particle would cross the barrier. Here the transmission probability was obtained as $T = e^{-k_2 a}$, where $k_2 = \frac{\sqrt{2m(V_0 - E)}}{\hbar}$.
- In quantum statistics, the particles of the system are considered to be indistinguishable contrary to the consideration of particles as distinguishable in classical statistics. If n_i particles are distributed in the g_i cell of the ith compartment in the phase space, then the number of particles per cell is defined

as $n_i(E)/g_i(E)$, which is called occupation index. If $n_i(E)/g_i(E) \geq 0$ or 1, the particles are considered as indistinguishable which is the basic feature of the quantum statistics.

- Quantum statistics has two branches, namely Bose-Einstein statistics and Fermi-Dirac statistics.
- Bose-Einstein statistics is applicable to those systems which contain identical, indistinguishable particles of zero or integral spins. Such particles are called bosons. Examples of bosons are photons, phonons etc. Pauli exclusion principle does not apply to the bosons.
- Fermi-Dirac statistics is applicable to systems, which consist of identical, independent and indistinguishable particles having half-integral spins. The particles, which obey Fermi-Dirac statistics, are called fermions. The examples of fermions are electrons, protons, neutrons, etc. The fermion must obey Pauli exclusion principle. In Fermi-Dirac statistics, interchange of two particles of the system leaves the resultant system in an antisymmetric state. That is, the wave function of the system gets changed only with minus sign.

SOLVED EXAMPLES

EXAMPLE 1 The position and momentum of a 1.0 keV electron are simultaneously measured. If the position is located within 1 Å, what is the percentage of uncertainty in momentum?

SOLUTION Given $\Delta x = 1.0 \times 10^{-10}$ m and $E = 1000 \times 1.6 \times 10^{-19}$ J $= 1.6 \times 10^{-16}$ J.

Heisenberg's uncertainty principle says

$$\Delta x \Delta p = \frac{\hbar}{2} \quad \text{and} \quad p = \sqrt{2mE}$$

$$p = \sqrt{2 \times 9.1 \times 10^{-31} \times 1.6 \times 10^{-16}}$$

$$= 1.71 \times 10^{-23} \text{ kg m/sec}$$

and

$$\Delta p = \frac{\hbar}{2\Delta x} = \frac{h}{2 \times 2\pi \times \Delta x} = \frac{6.62 \times 10^{-34}}{2 \times 2 \times 3.14 \times 1.0 \times 10^{-10}}$$

$$= 5.27 \times 10^{-25} \text{ km g/sec}$$

Percentage of uncertainty in momentum

$$= \frac{\Delta p}{p} \times 100 = \frac{5.27 \times 10^{-25}}{1.71 \times 10^{-23}} \times 100$$

$$= \mathbf{3.1\%}$$

EXAMPLE 2 The uncertainty in the location of a particle is equal to its deBroglie wavelength. Calculate the uncertainty in its velocity.

SOLUTION Given $\Delta x = \dfrac{h}{p}$.

Now

$$\Delta x \Delta p = \frac{\hbar}{2} = \frac{h}{4\pi}$$

or $$\Delta p = \Delta(mv) = \frac{h}{4\pi}\frac{1}{\Delta x} = \frac{h}{4\pi}\frac{p}{h} = \frac{mv}{4\pi}$$

$$m\Delta v = \frac{mv}{4\pi}$$

or $$\Delta v = \frac{v}{4\pi}$$

Example 3 The position and momentum of 0.5 keV electron are simultaneously determined. If its position is located within 0.2 nm, what is the percentage uncertainty in its momentum?

Solution Given $E = 0.5 \times 10^3 \times 1.6 \times 10^{-19} = 0.8 \times 10^{-16}$ J and $\Delta x = 0.2 \times 10^{-9}$ m.

Now

$$\Delta x \Delta p = \frac{\hbar}{2} \text{ and momentum } p = \sqrt{2mE}$$

so $$p = \sqrt{2 \times 9.1 \times 10^{-31} \times 0.8 \times 10^{-16}} = 12.06 \times 10^{-24}$$

or $$p = 1.21 \times 10^{-23} \text{ kg m/sec}$$

or $$\Delta p = \frac{\hbar}{2}\frac{1}{\Delta x} = \frac{h}{4\pi}\frac{1}{0.2 \times 10^{-9}}$$

$$\Delta p = \frac{6.62 \times 10^{-34}}{4 \times 3.14 \times 0.2 \times 10^{-9}} = 2.635 \times 10^{-25} \text{ kg m/sec}$$

∴ Percentage uncertainty in momentum

$$\frac{\Delta p}{p} \times 100 = \frac{2.635 \times 10^{-25}}{1.21 \times 10^{-23}} \times 100$$

$$= \frac{2.635 \times 10^{-23}}{1.21 \times 10^{-23}} = \mathbf{2.18\%}$$

Example 4 Wavelengths can be determined with accuracies of one part in 10^6. What is the uncertainty in the position of a 1 Å X-ray photon when its wavelength is simultaneously measured?

Solution Given $\lambda = 10^{-10}$ m.

By uncertainty principle,

$$\Delta x \Delta p = \frac{\hbar}{2} = \frac{h}{4\pi}$$

and $$\lambda = \frac{h}{p} \quad \text{or} \quad p\lambda = h \tag{i}$$

By differentiating

$$p\Delta\lambda + \lambda\Delta p = 0$$

or $$\Delta p = \frac{p\Delta\lambda}{\lambda} = \frac{h\Delta\lambda}{\lambda^2} \quad \left[\because\ p = \frac{h}{\lambda}\right] \tag{ii}$$

By using Eqs. (i) and (ii), we get

$$\Delta x \frac{h\Delta\lambda}{\lambda^2} = \frac{h}{4\pi}$$

or
$$\Delta x \Delta\lambda = \frac{\lambda^2}{4\pi} \tag{iii}$$

Wavelength can be measured with accuracy of one part in 10^6, it means the uncertainty in wavelength is

$$\frac{\Delta\lambda}{\lambda} = \frac{1}{10^6} = 10^{-6} \tag{iv}$$

By putting this value in Eq. (iii), then

$$\Delta x \frac{\Delta\lambda}{\lambda} = \frac{\lambda}{4\pi} \quad \text{or} \quad \Delta x \times 10^{-6} = \frac{\lambda}{4\pi}$$

or
$$\Delta x = \frac{10^6 \times \lambda}{4\pi} = \frac{10^6 \times 10^{-10}}{4 \times 3.14} = \mathbf{7.96\ \mu m}$$

EXAMPLE 5 Calculate the uncertainty in measurement of momentum of an electron if the uncertainty in locating it is 1 Å.

SOLUTION Given $\Delta x = 1.0 \times 10^{-10}$ m.

Formula used is

$$\Delta x \Delta p = \frac{\hbar}{2} = \frac{h}{4\pi}$$

$$\Delta p = \frac{h}{4\pi}\frac{1}{\Delta x} = \frac{6.62 \times 10^{-34}}{4 \times 3.14} \times \frac{1}{10^{-10}}$$

$$\Delta p = \mathbf{5.27 \times 10^{-25}\ kgm/sec}$$

EXAMPLE 6 An electron has a momentum 5.4×10^{-26} kg m/sec with an accuracy of 0.05%. Find the minimum uncertainty in the location of the electron.

SOLUTION Given $p = 5.4 \times 10^{-26}$ kg m/sec.

The uncertainty in the measurement of momentum

$$\Delta = \frac{5.4 \times 10^{-26} \times 0.05}{100}$$

$$= 2.7 \times 10^{-29} \text{ kg m/sec}$$

$$\Delta x \Delta p = \frac{\hbar}{2} = \frac{h}{4\pi}$$

$\therefore$
$$\Delta x = \frac{h}{4\pi}\frac{1}{\Delta p} = \frac{6.62 \times 10^{-34}}{4 \times 3.14} \times \frac{1}{2.7 \times 10^{-29}}$$

$$= 1.952 \times 10^{-6} \text{ m}$$

$$= \mathbf{1.952\ \mu m}$$

EXAMPLE 7 A hydrogen atom is 0.53 Å in radius. Use uncertainty principle to estimate the minimum energy an electron can have in this atom.

SOLUTION Given $\Delta x_{max} = 0.53$ Å.

Heisenberg's uncertainty principle

$$\Delta x \Delta p = \frac{\hbar}{2} = \frac{h}{4\pi}$$

$$(\Delta x)_{max} (\Delta p)_{min} = \frac{h}{4\pi} \qquad \text{(i)}$$

and $$(\text{K.E.})_{min} = \frac{p_{min}^2}{2m} = \frac{(\Delta p)_{min}^2}{2m} \qquad [\because \; p_{min} = \Delta p_{min}]$$

$$(\Delta p)_{min} = \frac{h}{4\pi} \frac{1}{\Delta x} = \frac{6.62 \times 10^{-34}}{4 \times 3.14} \frac{1}{0.53 \times 10^{-10}}$$

$$= 0.9945 \times 10^{-24}$$

$$= 9.945 \times 10^{-25} \text{ kg m/sec}$$

and $$(\text{K.E.})_{min} = \frac{(\Delta p)_{min}^2}{2m} = \frac{(9.945 \times 10^{-25})^2}{2 \times 9.1 \times 10^{-31}}$$

$$= \mathbf{5.434 \times 10^{-17}\ J}$$

EXAMPLE 8 The speed of an electron is measured to be 5.0×10^3 m/sec to an accuracy of 0.003%. Find the uncertainty in determining the position of this electron.

SOLUTION Given $v = 5.0 \times 10^3$ m/sec.

Formula used is

$$\Delta x \Delta p = \frac{\hbar}{2} = \frac{h}{4\pi}$$

$$\Delta v = v \times \frac{0.003}{100} = 5.0 \times 10^3 \times \frac{0.003}{100} = 0.15 \text{ m/sec}$$

and $$\Delta p = m\Delta v = 9.1 \times 10^{-31} \times 0.15 = 1.365 \times 10^{-31} \text{ kg m/sec}$$

$$\Delta x = \frac{6.62 \times 10^{-34}}{4 \times 3.14} \frac{1}{\Delta p} = \frac{6.62 \times 10^{-34}}{4 \times 3.14} \frac{1}{1.365 \times 10^{-31}}$$

$$= \mathbf{3.861 \times 10^{-4}\ m}$$

EXAMPLE 9 An electron has speed of 6.6×10^4 m/sec with an accuracy of 0.01%. Calculate the uncertainty in position of an electron. Given mass of an electron as 9.1×10^{-31} kg and Planck's constant h as 6.6×10^{-34} J sec.

SOLUTION Given $v = 6.6 \times 10^4$ m/sec and $\Delta v = 6.6 \times 10^4 \times \frac{0.01}{100}$ m/sec

$$= 6.6 \text{ m/sec.}$$

Formula used is

$$\Delta x \Delta p = \frac{\hbar}{2} = \frac{h}{4\pi} \quad \text{or} \quad \Delta x = \frac{h}{4\pi} \frac{1}{\Delta p}$$

$$\Delta p = m\Delta v = 9.1 \times 10^{-31} \times 6.6$$

or $$\Delta x = \frac{h}{4\pi} \frac{1}{\Delta p} = \frac{6.6 \times 10^{-34}}{4 \times 3.14 \times 9.1 \times 10^{-31} \times 6.6}$$

$$\Delta x = \mathbf{8.75 \times 10^{-6}\ m}$$

Example 10 Calculate the smallest possible uncertainty in the position of an electron moving with a velocity 3×10^7 m/sec.

Solution Given $v = 3 \times 10^7$ m/sec.

Formula used is

$$\Delta x \Delta p = \frac{\hbar}{2} = \frac{h}{4\pi}$$

$$\Delta p_{\min} \approx p = mv = \frac{m_0 v}{\sqrt{1 - v^2/c^2}}$$

$$\therefore \quad \Delta x = \frac{h}{4\pi}\frac{1}{\Delta p} = \frac{h}{4\pi}\left[\frac{\sqrt{1 - v^2/c^2}}{m_0 v}\right]$$

$$= \frac{6.62 \times 10^{-34}}{4 \times 3.14}\left[\frac{\sqrt{1 - \left(\frac{3 \times 10^7}{3 \times 10^8}\right)^2}}{9.1 \times 10^{-31} \times 3 \times 10^7}\right]$$

$$= \mathbf{1.92 \times 10^{-12}\ m}$$

Example 11 If an excited state of hydrogen atom has a life-time of 2.5×10^{-14} sec, what is the minimum error with which the energy of this state can be measured? Given $h = 6.62 \times 10^{-34}$ J sec.

Solution Given $\Delta t = 2.5 \times 10^{-14}$ sec.

Formula used is

$$\Delta E \Delta t = \frac{\hbar}{2} = \frac{h}{4\pi}$$

$$\Delta E = \frac{h}{4\pi}\frac{1}{\Delta t} = \frac{6.62 \times 10^{-34}}{4 \times 3.14} \times \frac{1}{2.5 \times 10^{-14}} = 0.211 \times 10^{-20}\ \text{J}$$

$$\Delta E = \mathbf{2.11 \times 10^{-21}\ J}$$

Example 12 An excited atom has an average life-time of 10^{-8} sec. During this time period it emits a photon and returns to the ground state. What is the minimum uncertainty in the frequency of this photon?

Solution Given $\Delta t = 10^{-8}$ sec.

Formula used is

$$\Delta E \Delta t = \frac{\hbar}{2} = \frac{h}{4\pi}$$

As $E = h\nu$ or $\Delta E = \Delta(h\nu) = h\Delta\nu$

or $$h\Delta\nu\Delta t = \frac{h}{4\pi} \quad \text{or} \quad \Delta\nu\Delta t = \frac{1}{4\pi}$$

or $$\Delta\nu = \frac{1}{4\pi}\frac{1}{\Delta t} = \frac{1}{4 \times 3.14} \times \frac{1}{10^{-8}}$$

$$\Delta\nu = \mathbf{7.96 \times 10^6\ sec}$$

Example 13 Compare the uncertainties in velocity of a proton and an electron contained in a 20 Å box.

SOLUTION Given $\Delta x = 2.0 \times 10^{-9}$ m.

Formula used is

$$\Delta p \Delta x = \frac{\hbar}{2} = \frac{h}{4\pi} \quad \text{or} \quad \Delta p = \frac{h}{4\pi \Delta x}$$

As uncertainty in momentum for electron and proton does not depend upon mass, we have

$$\Delta p = \Delta(mv) = m\Delta v \quad \text{or} \quad \Delta v = \frac{\Delta p}{m}$$

As $\Delta p_p = \Delta p_e$

$$\Delta v_p = \frac{\Delta p_p}{m_p}$$

and

$$\Delta v_e = \frac{\Delta p_e}{m_e}$$

$$\frac{\Delta v_p}{\Delta v_e} = \frac{\Delta p_p}{\Delta p_e}\frac{m_e}{m_p} = \frac{m_e}{m_p} \qquad [\because \ \Delta p_p = \Delta p_e]$$

$$= \frac{9.1 \times 10^{-31}}{1.67 \times 10^{-27}} = \mathbf{5.45 \times 10^{-4}}$$

EXAMPLE 14 Find the energy of an electron moving in one dimension in an infinitely high potential box of width 1.0 Å. Given $m = 9.1 \times 10^{-31}$ kg and $h = 6.62 \times 10^{-34}$ J sec.

SOLUTION Given $l = 1.0 \times 10^{-10}$ m, $m = 9.1 \times 10^{-31}$ kg and $h = 6.62 \times 10^{-34}$ J sec.

Formula used is

$$E_n = \frac{n^2 h^2}{8mL^2}$$

$$= \frac{n^2 (6.62 \times 10^{-34})^2}{8 \times 9.1 \times 10^{-31} \times (1.0 \times 10^{-10})^2}$$

$$= 0.602 \times 10^{-17} n^2 \text{ J}$$

for $n = 1$,

$$E_1 = 6.02 \times 10^{-18} \text{ J}$$

and for $n = 2$,

$$E_2 = 6.02 \times 10^{-18} \times 4 \text{ J}$$
$$= 2.408 \times 10^{-17} \text{ J}$$
$$= \mathbf{2.41 \times 10^{-17}} \text{ J}$$

EXAMPLE 15 Calculate the energy difference between the ground state and the first excited state for an electron in a box of length 1.0 Å.

SOLUTION Given $L = 1.0 \times 10^{-10}$ m.

Formula used is

$$E_n = \frac{n^2 h^2}{8mL^2}$$

Put $n = 1$ for ground state and $n = 2$ for first excited state

$$E_2 - E_1 = \frac{h^2}{8mL^2}[2^2 - 1^2] = \frac{(6.62 \times 10^{-34})^2 \times 3}{8 \times 9.1 \times 10^{-31} \times (1.0 \times 10^{-10})^2}$$

$$= \mathbf{1.81 \times 10^{-17}\ J}$$

EXAMPLE 16 Compute the energy of the lowest three levels for an electron in a square well of width 3 Å.

SOLUTION Given $L = 3 \times 10^{-10}$ m.

Formula used is

$$E_n = \frac{n^2 h^2}{8mL^2}$$

Put $n = 1, 2, 3$ for first three levels, then

$$E_1 = \frac{h^2}{8mL^2} = \frac{(6.62 \times 10^{-34})^2}{8 \times 9.1 \times 10^{-31} \times (3 \times 10^{-10})^2}$$

$$= 6.688 \times 10^{-19}\ \text{J}$$

$$= \mathbf{6.7 \times 10^{-19}\ J}$$

$$E_2 = 4E_1 = \mathbf{2.68 \times 10^{-18}\ J} \quad \text{and}$$

$$E_3 = 9E_1 = \mathbf{6.03 \times 10^{-18}\ J}$$

EXAMPLE 17 An electron is bound in one-dimensional potential box which has a width 2.5×10^{-10} m. Assuming the height of the box to be infinite, calculate the lowest two permitted energy values of the electron.

SOLUTION Given $L = 2.5 \times 10^{-10}$ m.

Formula used is

$$E_n = \frac{n^2 h^2}{8mL^2}$$

For lowest two permitted energy values of electrons, put $n = 1$ and 2. Then

for $n = 1$,

$$E_1 = \frac{h^2}{8mL^2} = \frac{(6.62 \times 10^{-34})^2}{8 \times 9.1 \times 10^{-31} \times (2.5 \times 10^{-10})^2}$$

$$= \mathbf{9.63 \times 10^{-19}\ J}$$

for $n = 2$,

and
$$E_2 = \frac{(2)^2 \times (6.62 \times 10^{-34})^2}{8 \times 9.1 \times 10^{-31} \times (2.5 \times 10^{-10})^2}$$

$$= \mathbf{3.853 \times 10^{-18}\ J}$$

EXAMPLE 18 Compute the lowest energy of a neutron confined to the nucleus which is considered as a box with a size of 10^{-14} m.

SOLUTION Given $L = 10^{-4}$ m, $h = 6.62 \times 10^{-34}$ J sec and $m = 1.67 \times 10^{-27}$ kg.

Formula used is

$$E_n = \frac{n^2 h^2}{8mL^2}$$

For lowest energy $n = 1$

$$E_1 = \frac{h^2}{8mL^2} = \frac{(6.62\times10^{-34})^2}{8\times1.67\times10^{-27}\times(10^{-14})^2}$$

$$E_1 = \mathbf{3.28\times10^{-13}\ J}$$

EXAMPLE 19 State the values of momentum and energy of *a* particle in one-dimensional box with impenetrable walls. Find their values for an electron in a box of length 1.0 Å for $n = 1$ and $n = 2$ energy states. Given $m = 9.1 \times 10^{-31}$ kg and $h = 6.63 \times 10^{-34}$ J sec.

SOLUTION Given $L = 1.0 \times 10^{-10}$ m, $n = 1$ and 2, $m = 9.1 \times 10^{-31}$ kg and $h = 6.63 \times 10^{-34}$ J sec.

The Formulae used are

$$p_n = \frac{nh}{2L} \tag{i}$$

$$E_n = \frac{p_n^2}{2m} = \frac{n^2h^2}{8mL^2} \tag{ii}$$

Momentum for $n = 1$ and 2 are

$$p_1 = \frac{1\times6.63\times10^{-34}}{2\times10^{-10}} = 3.315\times10^{-24}\ \text{kg m/sec}$$

and $$p_2 = \frac{2\times6.63\times10^{-34}}{2\times10^{-10}} = 6.63\times10^{-24}\ \text{kg m/sec}$$

Energy for $n = 1$ and 2 are

$$E_1 = \frac{n^2h^2}{8mL^2} = \frac{(1)^2(6.63\times10^{-34})^2}{8\times9.1\times10^{-31}\times(10^{-10})^2}$$

$$= \mathbf{6.04\times10^{-18}\ J}$$

For $n = 2$,

$$E_2 = \frac{n^2h^2}{8mL^2} = (2)^2E_1 = 2.416\times10^{-17}$$

$$= \mathbf{2.42\times10^{-17}\ J}$$

EXAMPLE 20 An electron is constrained to move in a one dimensional box of length 0.1 nm. Find the first three energy eigen values and the corresponding deBroglie wave lengths. Given $h = 6.63 \times 10^{-34}$ J sec.

SOLUTION Given $L = 1.0 \times 10^{-10}$ m.

Formulae used are

$$E_n = \frac{n^2h^2}{8mL^2} \quad \text{and} \quad p_n = \frac{nh}{2L}$$

for $n = 1$

$$E_1 = \frac{h^2}{8mL^2} = \frac{(6.63\times10^{-34})^2}{8\times(9.1\times10^{-31})\times(10^{-10})^2}$$

$$= 6.04\times10^{-18}\ \text{J}$$

Similarly, for $n = 2$, is equal to

$$E_2 = (2)^2 E_1 = 2.416 \times 10^{-17} \text{ J}$$

and for $n = 3$,

$$E_3 = (3)^2 E_1 = 5.36 \times 10^{-17} \text{ J}$$

As we know

$$\lambda_n = \frac{h}{p_n} \quad \text{and} \quad p_n = \frac{nh}{2L}$$

or
$$\lambda_n = \frac{2L}{n}$$

For $n = 1$

$$\lambda_1 = 2L = 2.0 \times 10^{-10} \text{ m} = 2 \text{ Å}$$

For $n = 2$,

$$\lambda_2 = \frac{2L}{2} = l = 1.0 \times 10^{-10} \text{m} = 1.0 \text{ Å}$$

For $n = 3$,

$$\lambda_3 = \frac{2L}{3} = 0.667 \times 10^{-10} \text{m} = \mathbf{0.667 \text{ Å}}$$

Example 21 The minimum energy possible for a particle entrapped in a one dimensional box is 3.2×10^{-18} J. What are the next three energies in eV the particle can have?

Solution Given $E_1 = 3.2 \times 10^{-18}$ J.

Formula used is

$$E_n = \frac{n^2 h^2}{8mL^2} \quad \text{or} \quad E_n \propto n^2 \tag{i}$$

$$\text{Now energy in } eV = \frac{32.0 \times 10^{-19}}{1.6 \times 10^{-19}}$$

$$E_1 = 20\ eV$$

Next three values of energy can be obtained by putting $n = 2$, 3 and 4.

$$E_2 = n^2 E_1 = (2)^2 E_1 = 4 \times 20 \text{ eV} = 80 \text{ eV}$$

$$E_3 = (3)^2 E_1 = 9 \times E_1 = 9 \times 20 \text{ eV} = 180 \text{ eV}$$

$$\text{and} \quad E_4 = (4)^2 E_1 = 16 \times 20 \text{ eV} = \mathbf{320 \text{ eV}}$$

Example 22 The energy of an electron constrained to move in a one dimensional box of length 4.0 Å is 9.664×10^{-17} J. Find out the order of excited state and the momentum of the electron in that state. Given $h = 6.63 \times 10^{-34}$ J sec.

Solution Given $E_n = 9.664 \times 10^{-17}$ J and $L = 4 \times 10^{-10}$ m

Formulae used are

$$E_n = \frac{n^2h^2}{8mL^2} \quad \text{and} \quad p_n = \frac{nh}{2L}$$

$$E_1 = \frac{h^2}{8mL^2} = \frac{(6.63 \times 10^{-34})^2}{8 \times 9.1 \times 10^{-31} \times (4 \times 10^{-10})^2}$$

$$= 3.774 \times 10^{-19} \text{ J}$$

$$E_n = \frac{n^2h^2}{8mL^2} = n^2E_1$$

or $$n^2 = \frac{E_n}{E_1} = \frac{966.4 \times 10^{-19} \text{ J}}{3.774 \times 10^{-19} \text{ J}}$$

or $$n = 16 \quad \text{(order of excited state)}$$

Momentum of electron for $n = 16$

$$p_n = \frac{nh}{2L} = \frac{16 \times 6.63 \times 10^{-34}}{2 \times 4 \times 10^{-10}}$$

$$= \mathbf{13.26 \times 10^{-24} \text{ kg m/sec}}$$

Example 23 Evaluate the first three energy levels of an electron enclosed in a box of width 10 Å. Compare it with those of glass marble of mass 1.0 gm, contained in a box of width 20 cm. Can these levels of the marble be measured experimentally?

Solution Given for an electron $n = 1$ and $L = 1.0 \times 10^{-9}$ m and for glass marble $n = 1$, $L = 0.2$ m and $m = 1.0 \times 10^{-3}$ kg.

Formula used is $E_n = \dfrac{n^2h^2}{8mL^2}$

For electron

$$E_1 = \frac{h^2}{8mL^2} = \frac{(6.62 \times 10^{-34})^2}{8 \times 9.1 \times 10^{-31} \times (1.0 \times 10^{-9})^2}$$

$$= \mathbf{6.02 \times 10^{-20} \text{ J}}$$

Similarly

$$E_2 = (2)^2E_1 = 4 \times E_1 = \mathbf{24.08 \times 10^{-20} \text{ J}}$$

and $$E_3 = (3)^2E_1 = 9 \times 6.02 \times 10^{-20} \text{ J}$$

and $$= \mathbf{54.18 \times 10^{-20} \text{ J}}$$

For glass marble

$$E_1 = \frac{(6.62 \times 10^{-34})^2}{8 \times 10^{-3} \times (0.2)^2}$$

$$= 1.3695 \times 10^{-63} \text{ J}$$

$$= \mathbf{1.37 \times 10^{-63} \text{ J}}$$

Similarly,

$$E_2 = (2)^2E_1 = \mathbf{5.48 \times 10^{-63} \text{ J}}$$

and $$E_3 = (3)^2E_1 = 9E_1 = \mathbf{12.33 \times 10^{-63} \text{ J}}$$

It is clear that the levels in case of marble are very small and are nearly zero. So it is not possible to measure them experimentally.

EXAMPLE 24 Find the smallest possible uncertainty in position of the electron moving the velocity 3×10^7 m/sec (Given, $h = 6.63 \times 10^{-34}$ J.sec, $m_0 = 9.1 \times 10^{-31}$ kg)

SOLUTION By using the formula,

$$\Delta X_{\min} \Delta p_{\max} = \frac{h}{2\pi}$$

$$\Delta p_{\max} = p = mv = \frac{m_0 v}{1 - \frac{v^2}{c}}$$

$$\Delta x_{\min} \times \frac{m_0 v}{\sqrt{1 - \frac{v^2}{c^2}}} = \frac{h}{2\pi}$$

$$\Delta x_{\min} = h \frac{\sqrt{1 - \frac{v^2}{c^2}}}{2\pi m_0 v} = \frac{6.63 \times 10^{-34} \sqrt{1 - \left(\frac{3 \times 10^7}{3 \times 10^8}\right)^2}}{2 \times 3.14 \times 9.1 \times 10^{-31} \times 3 \times 10^7}$$

$$= \mathbf{3.8 \times 10^{-12}\, m}$$

EXAMPLE 25 Show that the uncertainty in the location of the particle is equal to deBrogile wave length the uncertainty in its velocity is equal to the velocity.

SOLUTION Given, $\Delta x = A$

Formula used is

$$\Delta x \Delta p_x = h \qquad \text{or} \qquad x\Delta p = h$$

$$\Delta p_x = \frac{h}{x} = \frac{h}{\lambda} = p_x \quad [\because x = \lambda \text{ (given)}]$$

$$\Delta p_x = p_x$$

or $$m\Delta v_x = mv_x$$

or $$\Delta v_x = v_x \quad \textbf{Hence proved.}$$

EXAMPLE 26 An electron is confined to move between two right walls separated by 10^{-9} m. Find the deBrogile wavelengths respresenting the first three allowed energy state of the electron and the corresponding energies (electron mass is 9.1×10^{-31} kg and $h = 6.63 \times 10^{-34}$ J sec.

SOLUTION Given, $L = 10^{-9}$ m = 10 Å, $m_e = 9.1 \times 10^{-31}$ kg and $h = 6.63 \times 10^{-34}$ J·sec.

The electron moves forth and back between rigid walls will form a stationary wave-pattern with node at the walls, for this, the distance L between the wall must be a whole multiple of the debrogile half-wavelengths.

Thus, $$L = n\frac{\lambda}{2}, \text{ where } n = 1, 2, 3, \ldots$$

$$\lambda = \frac{2L}{n} = \frac{2 \times 10\text{ Å}}{n}$$

$$\lambda_1 = 20\text{ Å}, \lambda_2 = 10\text{ Å}, \lambda_3 = 6.7\text{ Å}$$

The corresponding energies are given by

$$E_n = \frac{n_L h^2}{8mL^2} = \frac{n^2 \times (6.63 \times 10^{-34})^2}{8 \times 9.1 \times 10^{-31} \times (10^{-9})^2}$$

$$= n^2 \times 6.04 \times 10^{-20} \text{ Joule}$$

$$= \frac{n^2 \times 6.04 \times 10^{-20}}{1.6 \times 10^{-19}} = 0.38\, n^2 eV$$

For $n = 1, 2, 3, \ldots$

$$\mathbf{E_1 = 0.38\ eV, E_2 = 1.52\ eV, E_3 = 3.45\ eV}$$

EXAMPLE 27 The wave function of a certain particle is $\psi = A\cos^2 x$ for $-\frac{\pi}{2} < x < \frac{\pi}{2}$

(i) Find the value of A.

(ii) Find the probability that the particle be found between $x = 0$ and $x = \frac{\pi}{4}$

SOLUTION

(i) Given, $\psi = A\cos^2 x$ for $-\frac{\pi}{2} < x < \frac{\pi}{2}$

By using the condition for normalisation

$$\int_{-\pi/2}^{\pi/2} \psi\psi^1 dx = 1 \quad \text{or} \quad 2A^2 \int_0^{\pi/2} \cos^4 x\, dx = 1$$

$$2A^2\left[\frac{\pi}{8} + \frac{\pi}{8}\right] = 1 \Rightarrow 2A^2 \times \frac{3\pi}{16} = 1$$

$$\mathbf{A = \sqrt{\frac{8}{3\pi}}}$$

(ii) Probability is given by

$$P = \int_0^{\pi/4} |\psi|^2\, dx = \int_0^{\pi/4} A^2 \cos^4 x\, dx$$

$$= A^2 \int_0^{\pi/4} \cos^4 x\, dx = \frac{8}{3\pi} \int_0^{\pi/4} \cos^4 x\, dx$$

$$= \frac{8}{3\pi}\left[\frac{3\pi + 8}{32}\right] = \frac{3\pi + 8}{12\pi} = \left[\frac{1}{4} + \frac{2}{37}\right]$$

$$P = 0.25 + 0.2123 = \mathbf{0.4623}$$

EXAMPLE 28 Normalise the wave function

$\psi(x) = 0$ outside the box of size l

$\psi(x) = \text{A} \sin kx$ for $0 < x < l$

where $$k = \frac{\pi}{l}$$

SOLUTION Given, $\psi(x) = A\sin kx$ and $k = \frac{\pi}{l}$

The condition for normalisation for the confinement 0 to l

$$\int_0^l |\psi|^2 \, dx = 1$$

For given problem, $\int_0^l A^2 \sin^2 kx = 1$

On solving the above eqs,

$$A = \sqrt{\frac{2}{l}}$$

Then, $$\psi(x) = \sqrt{\frac{2}{l}} \sin \frac{\pi x}{l}$$

EXAMPLE 29 Calculate the energy difference between the ground state and the first excited state for an electron in one-dimensional rigid box of length 10^{-8} cm. (Mass of electron is 9.1×10^{-31} kg and $h = 6.63 \times 10^{-34}$ J · sec).

SOLUTION Given, $m_e = 9.1 \times 10^{-31}$ kg, $h = 6.63 \times 10^{-34}$ J sec and $L = 10^{-8}$ cm.

The energy of particle of mass m in 1-D rigid box of side L is given by,

$$E_n = \frac{n^2 L^2}{8mL^2},\ n = 1, 2, 3, \ldots$$

$$= \frac{n^2 (6.63 \times 10^{-34})^2}{8 \times 9.1 \times 10^{-31} \times (10^{-10})^2} = 6 \times 10^{-18} \times n^2 \text{ Joule}$$

$$= \frac{6 \times 10^{-18}}{1.6 \times 10^{-19}} n^2 \text{eV}$$

For ground state, $n = 1$

$E_1 = 38$ eV

For first excited state, $n = 2$

$E_2 = 152$ eV

The energy difference $\Delta E = E_2 - E_1$

$= 152 - 38$

$= \mathbf{114\ eV}$

OBJECTIVE TYPE QUESTIONS

Q.1 Which of the following relations is correct for Heisenberg's uncertainty principle?

(a) $\Delta E \Delta t \geq \frac{\hbar}{2}$ (b) $\Delta x \Delta p \geq \frac{h}{4\pi}$ (c) $\Delta L \Delta \theta \geq \frac{\hbar}{2}$ (d) All of these

Q.2 Heisenberg uncertainty relation holds good for

(a) microscopic as well as macroscopic particles both

(b) only microscopic particles (c) only macroscopic particles

(d) none of these

Q.3 The energy of a particle in infinite potential well is

(a) proportional to n^2 (b) inversely proportional to n^2

(c) proportional to n (d) inversely proportional to n

Q.4 The momentum of a particle in infinite potential well of length l is

(a) proportional to l (b) inversely proportional to l
(c) proportional to l^2 (d) inversely proportional to l^2

Q.5 The momentum of a particle in infinite potential well is

(a) proportional to n (b) inversely proportional to n^2
(c) proportional to n^2 (d) inversely proportional to n

Q.6 Which one of the following operators is associated with energy

(a) $-\frac{\hbar^2}{2m}\nabla^2 + V$ (b) $-\frac{\hbar}{2m}\nabla^2$ (c) $\frac{\hbar^2}{i}\vec{\nabla}$ (d) $i\hbar\frac{\partial}{\partial t}$

Q.7 Which one of the following operators is associated with kinetic energy

(a) $-\frac{\hbar^2}{2m}\nabla^2$ (b) $-\frac{\hbar^2}{2m}\nabla^2 + V$ (c) $i\hbar\frac{\partial}{\partial x}$ (d) $-\frac{\hbar}{2m}\nabla^2$

Q.8 Which one of the following operators is associated with momentum

(a) $\frac{\hbar}{i}\vec{\nabla}$ (b) $-\frac{\hbar^2}{i}\nabla^2$ (c) $-\frac{\hbar^2}{i}\nabla^2$ (d) $-\frac{\hbar^2}{2m}\nabla^2$

Q.9 Which one of the following energy values of a particle in infinite potential well of length l is allowed

(a) $\frac{n^2\pi^2\hbar^2}{2ml}$ (b) $\frac{n^2\pi^2\hbar^2}{2ml^2}$ (c) $\frac{\pi^2\hbar^2}{2mn^2 l}$ (d) $\frac{n^2\hbar^2}{2ml^2}$

Q.10 Which one of the following is the radius of first Bohr's orbit

(a) $\frac{\varepsilon_0 h^2}{\pi m Z e^2}$ (b) $\frac{\varepsilon_0 \hbar^2}{\pi m Z^2 e}$ (c) $\frac{\varepsilon_0 \hbar}{\pi m Z e^2}$ (d) $\frac{\varepsilon_0 \hbar}{\pi^2 m^2 Z^2 e^2}$

Q.11 The entire information of a quantum system can be gathered with the help of

(a) position (b) eigen value
(c) momentum operator (d) wave function

Q.12 The expression $|\psi(x, t)|^2$ stands for

(a) normalisation (b) position
(c) time probability density (d) position probability density

Q.13 If ψ is normalised wave function, then the value of $\int_{-\infty}^{+\infty} \psi * \psi \, dV$ will be

(a) zero (b) 1 (c) ∞ (d) $-\infty$

Q.14 Which of the following relation is correct for Schrödinger's wave equation, moving also x-axis?

(a) $\frac{\partial^2\psi}{\partial x^2} + \frac{2m}{\hbar^2}(E - V)\psi = 0$ (b) $\frac{\partial^2\psi}{\partial x^2} - \frac{2m}{\hbar^2}(E - V)\psi = 0$

(c) $\frac{\partial^2\psi}{\partial x^2} + \frac{2m^2}{\hbar^2}(E - V)\psi = 0$ (d) none of these

Q.15 The wave function 'ψ' associated with matter waves has no direct physical significance. It

(a) is a complex quantity (b) is not an observable quantity
(c) both (a) and (b) (d) none of these

Q.16 The normalised eigen wave function of a particle in a box of length 'L' is
(a) $\sqrt{\frac{2}{L}} \sin \frac{n\pi x}{L}$ (b) $\sqrt{\frac{2}{L}} \sin \frac{n\pi x}{L}$ (c) $\frac{2}{L} \sin \frac{n\pi x}{L}$ (d) none of these

Q.17 The energy levels of a particle in a box are
(a) equally spaced (b) continuous
(c) not-equally spaced (d) none of these

Short-Answer Questions

Q.1 What is Heisenberg's uncertainty principle?
Q.2 What do you understand by wave function?
Q.3 What is normalised wave function?
Q.4 Write and explain Schrödinger's time dependent and time independent equations.
Q.5 What are the basic assumptions of Bose-Einstein statistics?
Q.6 Differentiate between classical and Quantum statistics.
Q.7 What do you understand by normalised and orthogonal wave functions?

Practice Problems

General Questions

Q.1 Starting from deBroglie's wave concept obtain Heisenberg's uncertainty principle.
Q.2 State Heisenberg's uncertainty principle and derive it from a hypothetical gamma ray microscope.
Q.3 Why is uncertainty principle important for microscopic particles but significant in practical life?
Q.4 Illustrate Heisenberg's uncertainty principle by diffraction of a beam of electrons by a narrow slit.

OR

Prove position momentum uncertainty principle using particle approach.
Q.5 By applying uncertainty principle explain non-existence of electrons in atomic nucleus.
Q.6 What other reasons show why electrons cannot exist inside the nucleus?
Q.7 Apply Hiesenberg's uncertainty principle to explain the following.
(a) Non-existence of electrons within the nucleus
(b) Existence of protons, neutrons and α-particles
(c) Existence of finite zero-point energy
(d) Binding energy of an electron in a hydrogen atom is of the order of 15 eV
Q.8 Explain the difference between quantum mechanics and classical mechanics.
Q.9 What do you understand by the wave function ψ of a moving particle?
Q.10 Give the physical significance of wave function. What does the square of wave function signify?
Q.11 What are the conditions and limitations that the wave function must obey?
Q.12 Starting from the wave equation and introducing energy and momentum of the particle obtain an expression for three dimensional Schrödinger's equation in time dependent form.

Q.13 Obtain three dimensional time independent Schrödinger's wave equation from time dependent Schrödinger's equation.

Q.14 Derive an expression for Schrödinger time independent and time dependent wave equations.

Q.15 Derive time dependent Schrödinger wave equation.

Q.16 Give the formulation of time dependent Schrödinger equation for a free particle. Discuss the interpretation of position, probability density and normalisation of wave function.

Q.17 Derive both time independent and time dependent Schrödinger equations for non-relativistic particle.

Q.18 Why should ψ and $\frac{d\psi}{dx}$ be continuous everywhere?

Q.19 What do you understand by orthogonal wave function? Explain orthogonality and orthonormality of wave functions.

Q.20 Obtain Schrödinger's wave equation for a particle in square well potential and discuss energy levels when the well is infinitely deep.

Q.21 Discuss quantum mechanically the problem of linear harmonic oscillator and obtain its eigen values. Also, write significance of zero point energy.

Q.22 Why are we not aware of quantisation in daily experience? Explain it.

UNSOLVED QUESTIONS

Q.1 An electron of mass 9.1×10^{-31} kg has a speed of 1.0 m/sec with an accuracy of 0.05%. Calculate the uncertainty with which the position of the electron can be located. **[Ans: 1.15×10^{-4} m]**

Q.2 The electron in hydrogen atom may be confined to a nucleus of radius 5×10^{-11} m. Find out the minimum uncertainty in the momentum of the electron and also find out the minimum kinetic energy of the electron. Given $m = 9.0 \times 10^{-31}$ kg and $h = 6.62 \times 10^{-34}$ J sec.

[Ans: 1.054×10^{-24} kg m/sec, 6.142 J]

Q.3 The speed of a bullet of mass 50 gm is measured to be 300 m/sec with an uncertainty of 0.01%. With what accuracy can we locate the position of the bullet if it is measured simultaneously with its speed.

[Ans: 3.5×10^{-32} m]

Q.4 Life time of a nucleus in the excited state is 10^{-12} sec. Calculate the probable uncertainty in energy and frequency of a γ-ray photon emitted by it. **[Ans: 1.054×10^{-22} J; 1.59×10^{11} Hz]**

Q.5 Compute the energy difference between the ground state and first excited state for an electron in a one-dimensional rigid box of length 10^{-8} cm. Given $m = 9.1 \times 10^{-31}$ kg and $h = 6.626 \times 10^{-34}$ J sec

[Ans: 114 eV]

Q.6 Calculate the value of lowest energy of an electron in one dimensional force free region of length 4 Å.

[Ans: 3.78×10^{-19} J]

Q.7 The lowest energy possible for a certain particle entrapped in a box is 40 eV. What are the next three higher energies the particle can have? **[Ans: 160 eV, 360 eV and 640 eV]**

Q.8 Find the energy levels of an electron in a box 1 nm wide. Mass of electron is 9.1×10^{-31} kg. Also find the energy levels of 10 gm marble in a box 10 cm wide.

[Ans: 6.02×10^{-20} J, 24.08×10^{-20} J and 54.18×10^{-20} J; and for marble 5.49×10^{-64} J, 21.96×10^{-64} J and 49.41×10^{-64} J]

Free Electron Theory

17

LEARNING OBJECTIVES

After reading this chapter you will be able to

- **LO 1** Understanding Lorentz-Drude theory
- **LO 2** Learn about application and limitation of Lorentz-Drude theory
- **LO 3** Explain quantum theory of free electrons
- **LO 4** Discuss thermionic emission, Richardson's equation

Introduction

The simplest metals are alkali metals, which include sodium (Na), potassium (K), lithium (Li) etc. The electronic configuration of Na atom is $1s^2$, $2s^2$, $2p^6$, $3s^1$. Thus, the valance electron is in the $3s$ state. This electron behaves as conduction electron in the metal. The remaining 10 electrons of Na^+ ion core fill the $1s$, $2s$ and $2p$ states, which contain 2, 2 and 6 electrons, respectively. The distribution of core electrons is the same as in the free ion in the metal. This way we can say that the metal crystal contains the positive ion with the free electrons. These free electrons behave like the molecules in a perfect gas and are called free electron gas.

According to free electron theory, a metal can be considered to consist of ion cores having the nucleus and electrons other than valence electrons. These valence electrons form an electron gas, surround the ion cores and are free to move anywhere within the metal. Thus, the valence electrons of the atom become conduction electrons. In the theory, the force between the conduction electrons and ion cores are neglected so that the total energy of the electron is all kinetic, i.e., the potential energy is taken to be zero. Hence, the motion of the electrons within the metal is free because there are no collisions, similar to the molecules of an ideal gas.

Long back, it was believed that many physical properties of metals including electrical and thermal conductivities can be understood by considering free electron model. Attempts have been made by Drude and Lorentz to explain quantitatively the conductivities of metals on the basis of free electron theory.

In this context it is necessary to understand first the main characteristics of metals, which are discussed below:

(i) Metals obey Ohm's law, i.e., in steady state, the current density $\vec{J}$ is proportional to the applied electric field strength $\vec{E}$. It means

$$\vec{J} \propto \vec{E} \quad \text{or} \quad \vec{J} = \sigma \vec{E}$$

where, σ is electrical conductivity.

(ii) Metals possess high electrical and thermal conductivities.

(iii) At low temperature, the resistivity ρ is proportional to the fifth power of absolute temperature, i.e.,

$$\rho \propto T^5, \quad \text{where} \quad \rho = \frac{1}{\sigma}$$

(iv) The resistance of metals increases with rise in temperature, i.e., they have positive temperature coefficient. The resistance of certain metals vanishes at absolute zero and they exhibit the phenomenon of superconductivity.

(v) For most of the metals, resistivity is inversely proportional to the pressure, i.e., $\rho \propto \frac{1}{P}$.

(vi) The ratio of thermal and electrical conductivities is directly proportional to the absolute temperature and this phenomenon is known as *Wiedemann–Franz Law*, i.e.,

$$\frac{K}{\sigma} \propto T \quad \text{or} \quad \frac{K}{\sigma T} = \text{constant}$$

17.1 LORENTZ-DRUDE THEORY: CLASSICAL FREE ELECTRON THEORY OF METALS — LO1

As mentioned earlier, the valence electrons in a metal form an electron gas which surrounds the ion cores and are free to move throughout the crystal lattice and behave as molecules in the perfect gas. These electrons are negatively charged. Therefore, the flow of electricity corresponds to their motion. But in the absence of externally applied field the electrons are moving in the random directions and collide with the residual ions frequently. Thus, the laws of classical theory of gases can be applied to free electron gas also. This way, the electrons can be assigned a mean free path λ, mean collision time τ and average velocity $\bar{v}$.

In 1900, *Drude* used the electron-gas model to explain electrical conductivity in metals theoretically. Later Drude theory was modified by Lorentz and named as Lorentz-Drude free-electron theory. Lorentz and Drude assumed that at ordinary temperature, free electrons move in metals randomly with an average speed of the order of 10^5 m/sec. During the random motion, these electrons collide with themselves and atoms or ions of lattice and have no practical contribution to electrical and thermal conductivities Hence, in the absence of external electric field, the contribution of electrons to current in metal will be zero. If an external electric field is applied to the metals, the electrons will be accelerated in the opposite direction to the applied electric field and produce a current. In thermal equilibrium, the free electrons are assumed to follow Maxwell–Boltzmann distribution.

17.2 APPLICATIONS OF LORENTZ-DRUDE THEORY LO2

The Lorentz–Drude theory can be applied to explain the properties of metals including their electrical and thermal conductivities.

17.2.1 Electrical Conductivity

We consider that there are n free electrons per cubic meter in the metal. If we apply an electric field to the metal, the electrons modify their random motion and move with an average drift velocity v_d in the opposite direction to that of the applied field. The magnitude of force experienced by the electron is given by

$$eE = ma \tag{i}$$

or $$a = \frac{eE}{m}$$

Thus, the electrons undergo an acceleration eE/m. However, the electron will not accelerate indefinitely but after a short period it collides with a +ve ion in the metal. At each collision its velocity is reduced to zero. So it is accelerated between two collisions only. If λ is the mean free path and t is the free time, then the time taken between two successive collisions

$$t = \frac{\lambda}{v} \tag{ii}$$

During this time the velocity will be $\frac{eE}{m}t$. Thus the velocity at the beginning of the path is zero and at its end is $\frac{eE}{m}t$. Hence, the average drift velocity v_d will be the mean of the two, i.e.,

$$v_d = \frac{1}{2}\frac{eE}{m}t \tag{iii}$$

The current density is then given by

$$J = nev_d = ne\left[\frac{eE}{2m}t\right]$$

or $$J = \frac{ne^2t}{2m}E \tag{iv}$$

Metals obey Ohm's law which states that in steady state the current density J is proportional to the electric field strength.

$$J \propto E \quad \text{or} \quad J = \sigma E \tag{v}$$

By using Eqs. (iv) and (v), we get

$$\sigma = \frac{ne^2t}{2m} \tag{vi}$$

By putting the value of t from Eq. (ii) in Eq. (vi) we get

$$\sigma = \frac{ne^2\lambda}{2mv} \tag{vii}$$

Since the energy

$$\frac{1}{2}mv^2 = \frac{3}{2}kT$$

$$\sigma = \frac{ne^2\lambda v}{6kT}$$

17.2.2 Thermal Conductivity

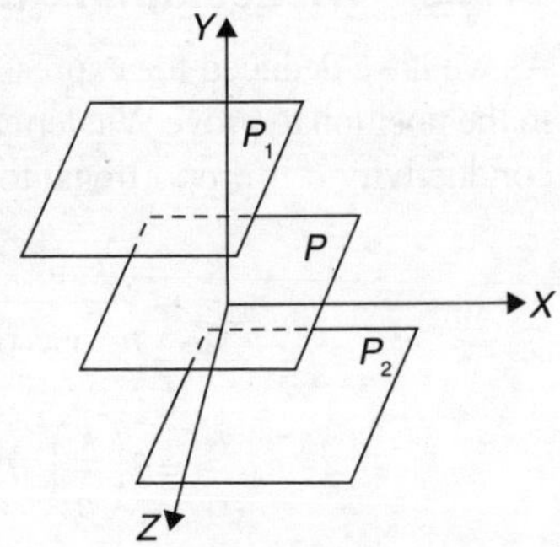

FIGURE 17.1

Free electrons also contribute to the conduction of heat energy in the metals. As we have already understood, free electrons behave similar to the molecules of a perfect gas. They possess greater kinetic energy at the hot end of the metal sheet than at the cold end. Suppose n number of electrons are moving randomly in all directions in the metal and their motion can be resolved along the three axes. Along any one particular direction only $\frac{n}{6}$ electrons will move, as there are six directions of their possible motions along three axes. We consider three planes of unit area such that the plane P_1 and P_2 are at the same distance of mean free path λ from the plane P. The temperatures of planes P_1 and P_2 are T_1 and T_2, respectively. If the temperature of plate P_1 is greater than that of plate P_2, the energy will transfer from plate P_1 to P_2 i.e., $\frac{nv}{6}$ electrons will transfer from P_1 to P_2. Since each electron has energy $\frac{1}{2}mv^2 = \frac{3}{2}kT_1$, ($k$ is the Boltzmann constant), the net energy transfer from P_1 to P_2 per unit area per second will be

$$\frac{nv}{6} \times \frac{3}{2}kT_1 \qquad \text{(i)}$$

and the same way, the net energy transfer from P_2 to P_1 will be

$$\frac{nv}{6} \times \frac{3}{2}kT_2 \qquad \text{(ii)}$$

Thus, the net transfer of energy from plate P_1 to plate P_2 through plate P per unit area per second will be

$$Q = \frac{nv}{6}\left[\frac{3kT_1}{2} - \frac{3kT_2}{2}\right]$$

$$= \frac{nvk}{4}[T_1 - T_2] \qquad \text{(iii)}$$

But $$Q = \frac{[T_1 - T_2]}{2\lambda}K \qquad \text{(iv)}$$

where K is thermal conductivity of metal.

By using Eqs. (iii) and (iv), we have

$$\frac{[T_1 - T_2]K}{2\lambda} = \frac{nvk}{4}[T_1 - T_2]$$

or
$$K = \frac{knv\lambda}{2} \qquad \text{(v)}$$

The above equation represents the expression for thermal conductivity of a metal.

17.2.3 Wiedemann-Franz Law

As we have deduced the expressions for thermal conductivity K and the electrical conductivity σ, now we are in the position to prove Wiedemann-Franz law which states that ratio of thermal conductivity K to the electrical conductivity σ is proportional to the absolute temperature. From the expressions of K and σ, we have

$$\frac{K}{\sigma} = \frac{kn v\lambda/2}{ne^2 v\lambda/6kT} \quad \left(\because \; \sigma = \frac{ne^2\lambda}{2mv} = \frac{ne^2\lambda v}{6kT}\right)$$

$$\frac{K}{\sigma} = 3\left(\frac{k}{e}\right)^2 T \qquad \text{(i)}$$

or
$$\frac{K}{\sigma T} = 3\left(\frac{k}{e}\right)^2 \qquad \text{(ii)}$$

Putting the values of Boltzmann constant k and the charge of electron, we find

$$\frac{K}{\sigma T} = 3\left(\frac{1.38 \times 10^{-23}}{1.6 \times 10^{-19}}\right)^2 = 2.23 \times 10^{-8}$$

Thus, $\frac{K}{\sigma T}$ has the same values at all the temperatures for all the metals or the ratio K/σ is directly proportional to absolute temperature. This is called Weidemann-Franz Law.

17.3 LIMITATIONS OF LORENTZ–DRUDE OR FREE ELECTRON THEORY LO2

The limitations of free electron theory are listed below.

(i) It does not explain why only some crystals are metallic.

(ii) It does not explain why the metals prefer only certain structures.

(iii) In real situations, the electrical conductivity depends on the temperature. The free electron theory does not explain the temperature variation of electrical conductivity.

(iv) The paramagnetism of metals is nearly independent of temperature. This result could also not be explained by this theory.

17.4 QUANTUM THEORY OF FREE ELECTRONS LO3

As we know that the metal contains a large number of conduction electrons which are not completely free (but partially), though they are not bound to any particular atomic system. The forces between conduction electrons and ion cores are neglected in the free electron approximation so that the electrons within the metal are treated as free. Further, the energy possessed by electron is kinetic, since the potential energy is taken to be zero. Consider an electron of mass m confined in a box of length L. Under this situation, the Schröedinger wave equation becomes

$$-\frac{\hbar^2}{2m}\nabla^2\psi = E\psi \tag{i}$$

The solution of the above equation is

$$\psi = \psi_0 \exp(i\,\vec{k}\cdot\vec{r}) \tag{ii}$$

where $\vec{k}$ is the wave vector with the magnitude $k = \frac{2\pi}{\lambda}$.

It can be shown from Eq. (ii) that

$$\frac{\partial^2\psi}{\partial x^2} = -k_x^2\psi;\quad \frac{\partial^2\psi}{\partial y^2} = -k_y^2\psi;\quad \frac{\partial^2\psi}{\partial z^2} = -k_z^2\psi$$

Then,

$$\nabla^2\psi = \frac{\partial^2\psi}{\partial x^2} + \frac{\partial^2\psi}{\partial y^2} + \frac{\partial^2\psi}{\partial z^2}$$

$$= -(k_x^2 + k_y^2 + k_z^2)\psi \tag{iii}$$

or

$$\nabla^2\psi = -k^2\psi \tag{iv}$$

By using Eqs. (i) and (iv), we have

$$-\frac{\hbar^2}{2m}\times(-k^2\psi) = E\psi$$

or

$$E(k) = \frac{\hbar^2 k^2}{2m} \tag{v}$$

Now we can write the total energy E in terms of momentum p with the help of following relations.

Since,

$$k = \frac{2\pi}{\lambda} \text{ and } \lambda = \frac{h}{p}$$

$$E = \frac{\hbar^2 k^2}{2m} = \frac{h^2}{4\pi^2}\frac{1}{2m}\frac{4\pi^2}{\lambda^2}$$

$$= \frac{h^2}{2m}\frac{1}{\lambda^2} = \frac{h^2}{2m}\frac{p^2}{h^2} = \frac{p^2}{2m}$$

or

$$E = \frac{p^2}{2m} \tag{vi}$$

Eq. (vi) represents the energy of a free particle (i.e., electron) and thus the energy is continuous. Here it may be mentioned that we have not considered the lattice periodicity and also assumed the constant potential inside the crystal to be zero. However, for cyclic boundary conditions, $k = \frac{2\pi n}{L}$, where L is the length of the cyclic chain (i.e., the solid). Therefore

$$E(n) = \frac{\hbar^2 k^2}{2m}$$

$$E(n) = \frac{n^2 h^2}{2mL^2} \quad \text{(vii)}$$

The first three lower energy state wavefunctions are represented in Fig. 17.2. The distribution of the available electrons among the various allowed energy levels and the evaluation of the related quantities can be understood better along with the treatment of the free electron gas in three-dimensional box of length L.

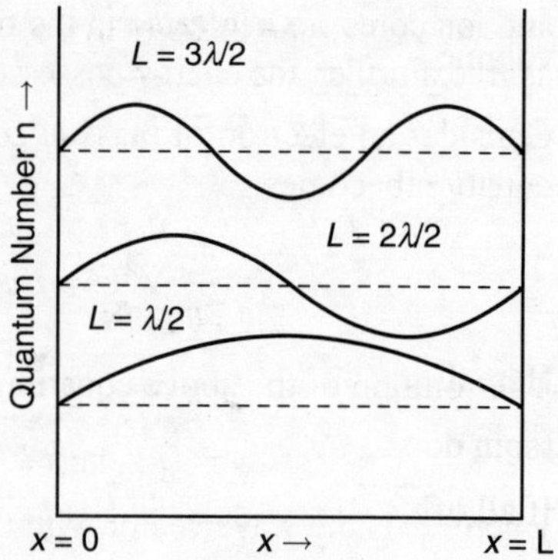

Figure 17.2

17.4.1 Fermi Energy

Consider that N free electrons are contained in a box at absolute temperature. At absolute zero all the energy levels below a certain level will be filled with electrons and the levels above this level will be empty. The energy level which divides the filled and empty levels is called '*Fermi level*' and the corresponding energy of that level is known as '*Fermi-energy*' E_F. In ground state of the system of N free electrons, the occupied states may be represented as a point inside a sphere in k-space as shown in Fig. 17.3. The k_x, k_y and k_z are the components of k_F along X, Y and Z axes, respectively. As per previous article, the energy of the electron is given by

$$E_k = \frac{\hbar^2 k^2}{2m} \quad \text{(i)}$$

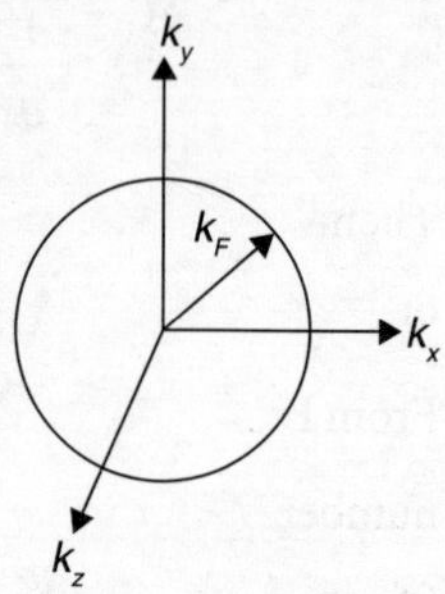

Figure 17.3

From the above relation it is clear that the energy increases as the square of distance from the origin of the k space coordinate system. All the electrons which lie on the same spherical shell of radius, k_F, have the same energy, which is called *Fermi Energy*. It is given by

$$E_F = \frac{\hbar^2}{2m} k_F^2 \quad \text{(ii)}$$

Since,
$$k = \frac{2\pi}{L} n$$

$$k_x = \frac{2\pi}{L} n_x$$

$$k_y = \frac{2\pi}{L} n_y$$

$$k_z = \frac{2\pi}{L} n_z$$

where n_x, n_y and n_z have the values 0, ±1, ±2, …. Therefore

$$k_x = 0, \pm\frac{2\pi}{L}, \pm\frac{4\pi}{L}, \pm\frac{6\pi}{L}, \ldots$$

k_y and k_z also have the same values. Suppose $\left(\frac{2\pi}{L}\right)^3$ is the volume of one shell in k-space (Fig. 17.4). Then in a sphere of

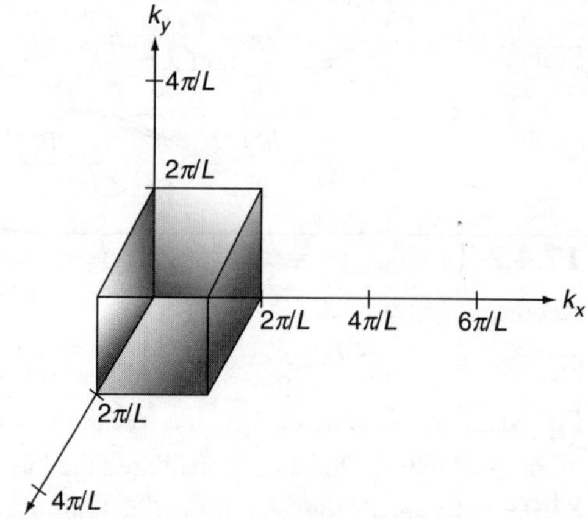

Figure 17.4

volume $\frac{4\pi}{3}k_F^3$, the total number of energy states (or shells) will be

$$N_F = 2\frac{\frac{4\pi}{3}k_F^3}{\left(\frac{2\pi}{L}\right)^3} \tag{iii}$$

Here the factor 2 represents the two allowed values of spin quantum number $m_s = +\frac{1}{2}$ (spin up) and $m_s = -\frac{1}{2}$ (spin down) for each allowed value of k.

If all of these energy shells are filled with electrons, then N will be equal to the number of electrons, i.e.,

$$2\frac{\frac{4\pi}{3}k_F^3}{\left(\frac{2\pi}{L}\right)^3} = N \quad \text{or} \quad N = \frac{V}{3\pi^2}k_F^3 \quad (\because\ V = L^3) \tag{iv}$$

or $$k_F^3 = \frac{3N\pi^2}{V} \quad \text{or} \quad k_F = \left(\frac{3N\pi^2}{V}\right)^{1/3} \tag{v}$$

From Eq. (v), it is clear that k_F depends upon electron concentration $\left(\frac{N}{V}\right)$ or in other wards k_F depends upon number of electrons per unit volume but it does not depend on the mass of electrons. Now the Fermi energy is

$$E_F = \frac{\hbar^2}{2m}k_F^2$$

The energy can be written as

$$E_F = \frac{1}{2}mv_F^2 \tag{vii}$$

where v_F is the velocity of electron in Fermi level, i.e., corresponding to Fermi energy. Then

$$\frac{1}{2}mv_F^2 = E_F = \frac{\hbar^2}{2m}\left[\frac{3N\pi^2}{V}\right]^{2/3}$$

$\therefore$ $$v_F = \frac{\hbar}{m}\left[\frac{3N\pi^2}{V}\right]^{1/3} \tag{viii}$$

17.4.2 Effect of Temperature on Fermi-Dirac Distribution

According to Fermi-Dirac distribution law, the most probable distribution is given by

$$n(E) = \frac{g(E)}{e^{(\alpha+\beta E)}+1} \tag{i}$$

where $\alpha = \frac{-E_F}{kT}$ and $\beta = \frac{1}{kT}$ together with k as Boltzmann constant. It is often convenient to introduce the Fermi-distribution function $f(E)$, which is defined as

$$f(E) = \frac{n(E)}{g(E)} = \frac{1}{e^{(\alpha + \beta E)} + 1}$$

With the values of α and β, this function can be written together with E_F as the Fermi energy

$$f(E) = \frac{1}{e^{(E - E_F)/kT} + 1}$$

At absolute zero ($T = 0$)

$$\frac{E - E_F}{kT} = -\infty, \quad \text{if } E < E_F$$

$$= +\infty \quad \text{if } E > E_F$$

and the Fermi distribution function

$$f(E) = \frac{1}{e^{-\infty} + 1} = 1 \quad \text{for } E < E_F$$

$$= \frac{1}{e^{+\infty} + 1} = 0 \quad \text{for } E > E_F$$

At any temperature T, and for E = E_F,

$$f(E) = \frac{1}{e^{(E - E_F)kT} + 1} = \frac{1}{e^0 + 1} = \frac{1}{1+1}$$

or $$f(E) = \frac{1}{2}$$

The energy level corresponding to $E = E_F$ is called Fermi level. Since at $E = E_F$, $f(E) = 1/2$, the Fermi level is defined as the energy level at which there is a 1/2 probability of finding an electron. It depends on the distribution of energy levels and the number of electrons available.

17.4.3 Density of States

It is defined as the number of energy states per unit energy range. It is denoted by the symbol $D(E)$. In other words, the 'density of states' for electrons in a band gives the number of orbitals (or states) in a certain energy range. Hence, the number of filled (i.e., density of electron state) states having the energy in the range E and $E + dE$ is

$$N(E)dE = E(E)f(E)dE \tag{i}$$

From the energy relation

$$E = \frac{\hbar^2 k^2}{2m} \tag{ii}$$

and $$N = \frac{V}{3\pi^2} k^3$$

we get $$E = \frac{\hbar^2}{2m}\left[\frac{3\pi^2 N}{V}\right]^{2/3} \tag{iii}$$

where E is total energy and N is the number of electrons. Therefore, from Eq. (iii), we have

$$\frac{3\pi^2 N}{V}=\left[\frac{2mE}{\hbar^2}\right]^{3/2}=\left[\frac{2mE}{h^2/4\pi^2}\right]^{3/2}$$

or $$N=\frac{8\pi V}{3h^3}(2mE)^{3/2} \qquad \text{(iv)}$$

By differentiating Eq. (iv) w.r.t. E, we get

$$\frac{dN}{dE}=\frac{8\pi V}{3h^3}(2m)^{3/2}\frac{3}{2}E^{1/2}=\frac{8\pi mV(2mE)^{1/2}}{h^3}$$

or $$\frac{dN}{dE}=\frac{V}{2\pi^2}\left[\frac{2m}{\hbar^2}\right]^{3/2}(E)^{1/2} \qquad \text{(v)}$$

The quantity $\left(\frac{dN}{dE}\right)$ is frequency referred to as the density of available state $D(E)$, which on multiplication with probability of occupation $f(E)$ gives density of occupied state $N(E)$, as shown in Fig. 17.5.

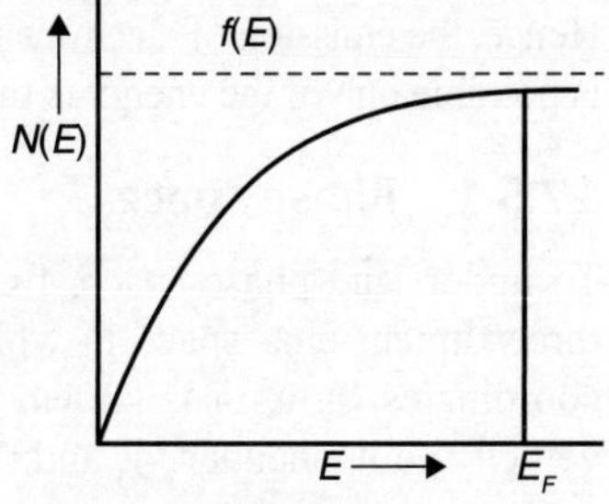

FIGURE 17.5

Thus, the number of electrons whose energies lie between E and $E + dE$ is given by

$$N(E)dE=\frac{dN}{dE}f(E)dE$$

$$N(E)dE=\frac{8\pi mV}{h^3}(2mE)^{1/2}\frac{dE}{e^{(E-E_F)/kT}+1}$$

17.4.4 Average Kinetic Energy of Free Electron Gas at 0 K

All the electrons have energy less than the Fermi energy E_F at 0 K, i.e., $E < E_F$. With this condition, the Fermi-Dirac distribution function becomes

$$f(E)=\frac{1}{e^{(E-E_F)/kT}+1}=1$$

If we consider that the average energy of an electron is $\overline{E}_e$. Since $f(E)= 1$, we can write

$$\overline{E}_e=\frac{1}{N}\int_0^{E_F}EN(E)dE=\frac{1}{N}\int_0^{E_F}E\frac{dN}{dE}f(E)dE$$

Substituting the value of $\frac{dN}{dE}$ from Eq. (v) and $f(E)$ as 1, in the above equation, we get

$$\overline{E}_e=\frac{1}{N}\left(\frac{8\pi mV}{h^3}\right)(2m)^{1/2}\int_0^{E_F}E^{3/2}dE$$

$$\overline{E}_e=\frac{1}{N}\left(\frac{8\pi mV}{h^3}\right)(2m)^{1/2}\frac{2}{5}E_F^{5/2}$$

Now the above relation with the value of N substituted from Eq. (iv) reads

$$\overline{E}_e=\frac{3}{5}E_F$$

17.5 THERMIONIC EMISSION

LO4

When a metal is heated, electrons are emitted from its surface. This phenomenon is known as thermionic emission. It is well known that the electrons in a metal are not completely free, but they are bound to the metal surface by some attractive forces. The term E_F is the Fermi energy, i.e., the energy of electron in metal at absolute zero, which is insufficient to escape out an electron from the metal surface (Fig. 17.6). According to free electron theory, W (potential energy) is the minimum energy which is required to emit an electron from the metal surface. Therefore, the minimum energy to be supplied to the electron for emission is $(W - E_F)$ which is known as *work function*. Hence, the emission of electrons from the surface of the metal is possible only if the energy is more than that of work function.

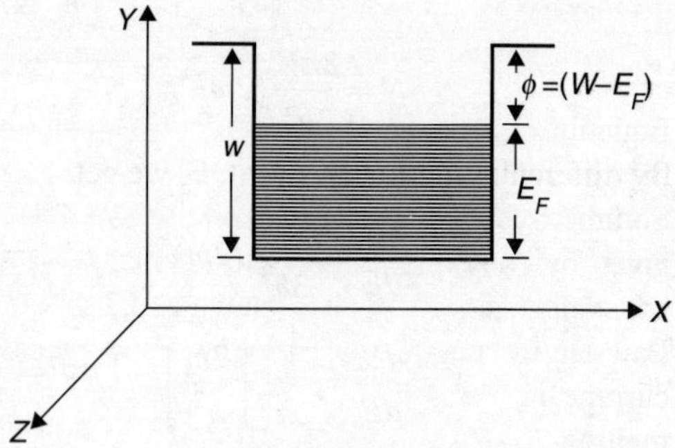

FIGURE 17.6

17.5.1 Phase space

To understand phase space, first of all we should know about position space and momentum space. The three dimensional space in which the location of a particle is completely specified by the three position coordinates (x, y, z) is known as *position space*. The instantaneous motion of a particle is described by velocity components v_x, v_y and v_z. However, for many purposes it is more convenient to use the corresponding momentum components p_x, p_y and p_z. The three dimensional space in which the momentum of a particle is determined by the three momentum coordinates is known as *momentum space*. The combination of position space and momentum space is known as *phase space*. The phase space is a six dimensional space (x, y, z, p_x, p_y, p_z). If dv and dp are the elements of volume enclosed by any particular cell in position space and momentum space, respectively, then

$$d\tau = dv\, dp$$

The elementary volume enclosed by this cell in the phase space is given by

$$\begin{aligned} d\tau &= dx\, dy\, dz\, dp_x\, dp_y\, dp_z \\ &= dx\, dp_x\, dy\, dp_y\, dz\, dp_z = h^3 \qquad (\because dx\, dp_x = dy\, dp_y = dz\, dp_z = h) \end{aligned}$$

$$\text{Total number of cells in phase} = \frac{\text{Total volume in phase space}}{\text{Volume of one cell}} = \frac{V\iiint dp_x\, dp_y \cdot dp_y}{h^3}$$

$$= \frac{\int 4\pi p^2 dp}{h^3} \qquad \text{(i)}$$

17.5.2 Richardson's Equation

Richardson's equation is well-known equation used for thermionic emission. This equation enables us to find the emission current density of electrons.

If we represent $\phi = W - E_F$, where W is the work function and E_F is the Fermi energy, then the emission current density J is given by

$$J = AT^2 e^{-\phi/kT} \qquad \text{(i)}$$

where A is a constant given by

$$A = \frac{4\pi mek^2}{h^3}, \text{ and } T \text{ is the temperature of the metal.}$$

Equation (i) is known as Richardson's equation. However, there have been various theoretical expressions for the constant A based on different physical assumptions. A significant work was done by Dushman, Fowler, Sommerfeld and Nordheim in addition to the remarkable work of Richardson. A modern theoretical treatment given by Modinos assumes the band theory of the emitting material. Then according to the form of the constant A, Eq. (i) is also known as Dushman's equation, Richardson-Dushman equation, and Richardson-Law-Dushman equation. The main understanding of this equation is that due to the exponential function the current increases rapidly with temperature when kT is less than ϕ. However, for essentially every material melting occurs well before $kT = \phi$.

SUMMARY

The topics covered in this chapter are summarised below.

- According to free electron theory, a metal can be considered to consist of ion cores having the nucleus and electrons other than valence electrons. These valence electrons form an electron gas, surround the ion cores and are free to move anywhere within the metal. Thus, the valence electrons of the atom become conduction electrons. In the theory, the force between the conduction electrons and ion cores are neglected so that the total energy of the electron is all kinetic, i.e., the potential energy is taken to be zero. Hence, the motion of the electrons within the metal is free because there are no collisions, similar to the molecules of an ideal gas.
- In the theory given by Lorentz-Drude, they assumed free electrons to move in metals randomly with an average speed. During this random motion the electrons were considered to collide with themselves and with atoms or ions of lattice. Further, it was assumed that these electrons have no practical contribution to the electrical and thermal conductivities. However, in the presence of external electric field, these electrons are accelerated and hence produce the current. In thermal equilibrium, these free electrons are assumed to follow the Maxwell-Boltzmann distribution.
- Based on Lorentz-Drude theory, the electrical and thermal conductivities of the metals were explained.
- Weidemann-Franz law states that the ratio of thermal conductivity K to the electrical conductivity of the metal is proportional to the absolute temperature.
- Limitations of free electron theory were talked about.
- In free electron approximation, the electrons within the metal are treated as free as the forces between conduction electrons and ion cores are neglected. However, the free electron was treated as confined in a box of length L as per quantum theory of free electrons. Using the Schröedinger equation, the energy of the electron was calculated and it was realised that the energy is not continuous rather it is quantized. Finally, the lower energy state wave functions were represented. However, the distribution of the available electrons among the various allowed energy levels and the evaluation of the related quantities can be understood better along with the treatment of the free electron gas in three-dimensional box.

- At absolute temperature, all the energy levels below a certain level are filled with electrons and the levels above this level are empty. The energy level which demarcates the filled and empty levels is called Fermi level. The energy corresponding to the Fermi level is referred to as Fermi energy.
- The concept of k space was given and the total number of energy sates (or shells) was calculated. Finally, the velocity of electrons in Fermi level, i.e., corresponding to Fermi energy, was obtained as $v_F = \frac{\hbar}{m}\left[\frac{3N\pi^2}{V}\right]^{1/3}$, where N is the number of electrons and V the volume of the box of length L.
- Effect of temperature on Fermi-Dirac distribution was discussed and then the concept of density of states $D(E)$ was given. The density of states is nothing but is the number of energy states per unit energy range. In other words, the density of sates for electrons in a band gives the number of orbitals (or states) in a certain energy range.
- Concept of phase space was introduced in order to explain the thermionic emission. The three-dimensional space in which the momentum of a particle is determined by the three momentum coordinates (p_x, p_y, p_z) is called the momentum space. The combination of position space (x,y,z) and the momentum space (p_x, p_y, p_z) is known as the phase space. So the phase space is represented by (x, y, z, p_x, p_y, p_z).
- The Richardsons's equation that enables us to find the emission current density of the electrons was mentioned. It is represented as $J = AT^2e^{-\phi/kT}$, where $\phi = W - E_F$ together with W as the work function and E_F as the Fermi energy. The constant $A = \left[\frac{4\pi mek^2}{h^3}\right]$ together with k as the Boltzmann constant and m and e as the mass and charge of the electron, respectively.

SOLVED EXAMPLES

EXAMPLE 1 Determine the average energy and speed of electron at its mean energy at 0 K, if the Fermi energy is 10 eV.

SOLUTION Given $E_F = 10$ eV

Average energy
$$\bar{E}_0 = \frac{3}{5}E_F$$
$$= \frac{3}{5}\times 10 \text{ eV}$$
$$= 6.0 \text{ eV}$$

and,
$$\frac{1}{2}mv^2 = \bar{E}_0 \quad \text{or} \quad v = \sqrt{\frac{2\bar{E}_0}{m}}$$
$$v = \sqrt{\frac{2\times 6.0\times 1.6\times 10^{-19}}{9.1\times 10^{-31}}}$$
$$\mathbf{v = 1.45\times 10^6\ m/sec}$$

Example 2 Fermi energy of a given substance is 7.9 eV. What is the average energy and speed of electron in this substance at 0 K?

Solution Given $E_F = 7.9$ eV

Average energy $\bar{E}_0 = \frac{3}{5}E_F$

$$\bar{E}_0 = \frac{3}{5} \times 7.9 \text{ eV}$$

$$= 4.74 \text{ eV}$$

and, $\frac{1}{2}mv^2 = \bar{E}_0$ or $v = \sqrt{\frac{2\bar{E}_0}{m}}$

or $$v = \sqrt{\frac{2 \times 4.74 \times 1.6 \times 10^{-19}}{9.1 \times 10^{-31}}} = \mathbf{1.29 \times 10^6\ m/sec}$$

Example 3 There are 2.5×10^{28} free electrons per cubic meter of sodium. Calculate the Fermi energy and Fermi velocity.

Solution Given $\frac{N}{V} = 2.5 \times 10^{28}$

Formula used is

$$E_F = \frac{1}{2}mv_F^2 = \frac{\hbar^2}{2m}\left(3\pi^2 \cdot \frac{N}{V}\right)^{2/3}$$

or $$E_F = \frac{h^2}{8\pi^2 m}\left(3\pi^2 \cdot \frac{N}{V}\right)^{2/3}$$

$$= \frac{(6.62 \times 10^{-34})^2}{8 \times (3.14)^2 \times 9.1 \times 10^{-31}}[3 \times (3.14)^2 \times 2.5 \times 10^{28}]^{2/3}$$

$$= 4.99 \times 10^{-19} \text{ J}$$

$$E_F = 5.0 \times 10^{-19} \text{ J}$$

or $$E_F = 3.12 \text{ eV}$$

and Fermi velocity $$v_F = \frac{h}{2\pi m}\left[3\pi^2 \cdot \frac{N}{V}\right]^{1/3}$$

$$= \frac{6.62 \times 10^{-34}}{2 \times 3.14 \times 9.1 \times 10^{-31}}(3 \times (3.14)^2 \times 2.5 \times 10^{28})^{1/3}$$

$$= \mathbf{1.05 \times 10^6\ m/sec}$$

Example 4 The density of copper is 8940 kg/m^3 and atomic energy weight is 63.55. Determine the Fermi energy of copper. Also obtain the average energy of free electrons of copper at 0 K.

Solution Given atomic weight = 63.55 kg and density of copper = 8940 kg/m^3.

Volume of 1 kg mole of copper, $V = \frac{63.55 \text{ kg}}{8940 \text{ kg/m}^2}$

Number of atoms per kg atom = 6.02×10^{26}

or $$\frac{N}{V} = \frac{6.02 \times 10^{26} \times 8940}{63.55}$$

Fermi energy $E_F = \frac{1}{2} m v_F^2 = \frac{\hbar^2}{2m}\left[3\pi^2 \frac{N}{V}\right]^{2/3}$

$$= \frac{h^2}{8\pi^2 m}\left[3\pi^2 \frac{N}{V}\right]^{2/3}$$

$$= \frac{(6.62\times10^{-34})^2}{8\times(3.14)^2\times9.1\times10^{-31}}\left[3\times(3.14)^2\times\frac{6.02\times10^{26}\times8940}{63.55}\right]^{2/3}$$

$$= 11.261\times10^{-19}\ \text{J}$$

$$E_F = 7.038\ \text{eV}$$

Average energy $\bar{E}_0 = \frac{3}{5} E_F = \frac{3}{5}\times7.038\ \text{eV}$

$$= \mathbf{4.22\ eV}$$

EXAMPLE 5 Consider silver in the metallic state with one free electron per atom. Calculate the Fermi energy. Given that density of silver is 10.5 g/cm^3 and atomic weight is 108.

SOLUTION Volume of 1 g mole of silver, $V = \frac{108\ \text{g}}{10.5\ \text{g/cm}^3}$ and number of atoms per g atom = 6.02 × 10^{23}.

$$\frac{N}{V} = \frac{6.2\times10^{23}\times10.5}{108}$$

$$= 5.85\times10^{22}\ \text{per cm}^3$$

$$= 5.85\times10^{28}\ \text{per cm}^3$$

Fermi energy $E_F = \frac{h^2}{8\pi^2 m}\left[3\pi^2\cdot\frac{N}{V}\right]^{2/3}$

$$= \frac{(6.62\times10^{-34})^2}{8\times(3.14)^2\times9.1\times10^{-31}}[3\times(3.14)^2\times5.85\times10^{28}]^{2/3}$$

$$= 8.799\times10^{-19}\ \text{J}$$

$$= 5.499\ \text{eV}$$

$$= \mathbf{5.5\ eV}$$

EXAMPLE 6 Aluminium metal crystallises in f.c.c. structure. If each atom contributes single electron as free electron and the lattice constant a is 4.0 Å, treating conduction electron as free electron Fermi gas, find (i) Fermi energy (E_F) and Fermi vector (k_F) and (ii) total kinetic energy of free electron gas per unit volume at 0 K.

SOLUTION In f.c.c. lattice number of electrons per unit cell will be (N) = 4 and volume of a unit cell is $a^3 = 64\times10^{-30}\,\text{m}^3$

and $\frac{N}{V} = \frac{4}{64\times10^{-30}}$

$$= 6.25\times10^{28}$$

Fermi energy $E_F = \dfrac{h^2}{8\pi^2 m}\left[3\pi^2 \cdot \dfrac{N}{V}\right]^{2/3}$

$$= \frac{(6.62\times10^{-34})^2}{8\times(3.14)^2\times9.1\times10^{-31}}[3\times(3.14)^2\times6.25\times10^{28}]^{2/3}$$

$$= 9.2\times10^{-19}\text{ J}$$

$$= 5.75\text{ eV}$$

Fermi vector $k_F = \left[3\pi^2 \cdot \dfrac{N}{V}\right]^{1/3} = [3\times(3.14)^2\times6.25\times10^{28}]^{1/3}$

$$= 1.23\times10^{10}\text{ per meter}$$

Total kinetic energy of free electrons per unit volume at 0 K = (Average energy per electron at 0 K) × (number of electrons per unit volume)

$$= \frac{3}{5}E_F \times \frac{N}{V} = \frac{3}{5}\times5.75\times6.25\times10^{28}\text{ eV}$$

$$= \mathbf{21.56\times10^{28}\text{ eV}}$$

EXAMPLE 7 Calculate the drift velocity of electrons in an aluminium wire of diameter 0.9 mm carrying current of 6 A. Assume that 4.5×10^{28} electrons/m^3 are available for conduction.

SOLUTION Given $I = 6$ A, $n = 4.5\times10^{28}$ electrons/m^3 and radius $r = \dfrac{d}{2} = \dfrac{0.9\times10^{-3}}{2} = 4.5\times10^{-4}$ m

Current density $J = \dfrac{I}{A} = \dfrac{6.0}{\pi\times(4.5\times10^{-4})^2}$

$$= \frac{6.0}{3.14\times(4.5\times10^{-4})^2}$$

$$= 9.44\times10^6\text{ A/m}^2$$

and drift velocity $v_d = \dfrac{J}{ne} = \dfrac{9.44\times10^6}{4.5\times10^{28}\times1.6\times10^{-19}}$

$$= \mathbf{1.311\times10^{-3}\text{ m/sec}}$$

EXAMPLE 8 The density of Cu is 8.92×10^3 kg/m^3 and its atomic weight is 63.5. Determine the current density if the current of 5.0 A is maintained in Cu wire of radius 0.7 mm. Assuming that only one electron of an atom takes part in conduction. Also calculate the drift velocity of electrons.

SOLUTION Given

Atomic weight = 63.5 kg,

Density of copper = 8.92×10^3 kg/m^3, $I = 5$ A

Radius = 0.7×10^{-3} m

Radio $\dfrac{N}{V} = \dfrac{6.02\times10^{28}\times8.92\times10^3}{63.5}$

$$= 8.456\times10^{30}\text{ electrons/m}^3$$

$$\text{Current density } J = \frac{I}{A} = \frac{I}{\pi r^2}$$

$$= \frac{5.0}{3.14 \times (0.7 \times 10^{-3})^2} = 3.25 \times 10^6 \text{ m/sec}$$

$$\text{Drift velocity } (v_d) = \frac{J}{ne} = \frac{3.25 \times 10^6}{8.456 \times 10^{30} \times 1.6 \times 10^{-19}}$$

$$= \mathbf{2.4 \times 10^{-6} \text{ m/sec}}$$

EXAMPLE 9 Using the data given, evaluate Fermi energy of the following alkali metals.

	Li	Na	K
Density (g/cm^3)ρ	0.534	0.971	0.86
Atomic Weight W_A	6.939	22.99	39.202

$h = 6.62 \times 10^{-34}$ J sec and $m = 9.1 \times 10^{-31}$ kg

SOLUTION Fermi energy $E_F = \dfrac{\hbar^2}{2m}\left[3\pi^2 \cdot \dfrac{N}{V}\right]^{2/3} = \dfrac{h^2}{8\pi^2 m}\left[3\pi^2 \dfrac{N}{V}\right]^{2/3}$

For Li $\quad \dfrac{N}{V} = \dfrac{N\rho}{W_A}$

$$\frac{N}{V} = \frac{6.023 \times 10^{26} \times 0.534 \times 10^3}{6.939} = 4.635 \times 10^{28} \text{ electrons/m}^3$$

$$E_F = \frac{(6.62 \times 10^{-34})^2}{8(3.14)^2 \times 9.1 \times 10^{-31}} \times [3 \times (3.14)^2 \times 4.635 \times 10^{28}]^{2/3}$$

$$= 7.535 \times 10^{-19} \text{ J}$$

$$= \mathbf{4.71 \text{ eV}}$$

Similarly for Na,

$$\frac{N}{V} = \frac{6.023 \times 10^{26} \times 0.971 \times 10^3}{22.99} = 2.58 \times 10^{28} \text{ electrons/m}^3$$

$$E_F = \frac{(6.62 \times 10^{-34})^2}{8 \times (3.14)^2 \times 9.1 \times 10^{-31}} \times [3 \times (3.14)^2 \times 2.58 \times 10^{28}]^{2/3}$$

$$E_F = 5.032 \times 10^{-19} \text{ J}$$

or

$$E_F = \mathbf{3.145 \text{ eV}}$$

and for K,

$$\frac{N}{V} = \frac{6.023 \times 10^{26} \times 0.86 \times 10^3}{39.202} = 1.321 \times 10^{28} \text{ electrons/m}^3$$

$$E_F = \frac{(6.62 \times 10^{-34})^2}{8 \times (3.14)^2 \times 9.1 \times 10^{-31}} \times [3 \times (3.14)^2 \times 1.321 \times 10^{28}]^{2/3}$$

$$E_F = 5.032 \times 10^{-19} \text{ J}$$

$$E_F = \mathbf{3.145 \text{ eV}}$$

Example 10 Calculate the energy difference between the ground state and first excited state for an electron in one-dimensional box of length 10^{-10} m, $V = 0$ for $0 \le x \le a$ and $V = \infty$ for $0 > x > a$.

Solution Under the given conditions, the energy for the nth level is

$$E_n = \frac{\hbar^2 k^2}{2m} = \frac{n^2 h^2}{8ma^2} \quad \text{[where } a \text{ is length of one-dimensional box]}$$

so,
$$E_1 = \frac{h^2}{8ma^2}(n=1) \quad \left[\text{and } k = \frac{n\pi}{a}\right]$$

and
$$E_2 = \frac{4h^2}{8ma^2}(n=2)$$

The energy difference between first excited and ground state

$$\Delta E = E_2 - E_1 = \frac{3h^2}{8ma^2} = \frac{3 \times (6.62 \times 10^{-34})^2}{8 \times 9.1 \times 10^{-31} \times (10^{-10})^2}$$

$$= 1.806 \times 10^{-17} \text{ J} = \mathbf{112.87\ eV}$$

Objective Type Questions

Q.1 With the increase in temperature, the resistance of a metal
(a) remains constant (b) increases
(c) decreases (d) becomes zero

Q.2 Metals
(a) obey Ohm's law
(b) have high electrical and thermal condítivities
(c) resistivity in proportional to temperature
(d) all of these

Q.3 Average kinetic energy $(\bar{E}_0)$ of a free electron gas at 0 K is
(a) $\frac{2}{5}E_F$ (b) $\frac{5}{3}E_F$ (c) $\frac{3}{5}E_F$ (d) E_F

Q.4 The density of states of electrons between the energy range E and $E+dE$ is proportional to
(a) $E^{1/2}$ (b) E^2 (c) E (d) $E^{3/2}$

Q.5 The phase space is a
(a) two dimensional space (b) one dimensional space
(c) three dimensional space (d) six dimensional space

Q.6 At low temperature, the resistivity of a metal is proportional to
(a) T^2 (b) T (c) T^5 (d) $T^{1/2}$

Q.7 Which one of the following relations is correct for current density
(a) $J = nev_d$ (b) $J = \frac{1}{nev_d}$ (c) $J = \frac{ne}{v_d}$ (d) $J = neAv_d$

Q.8 In the presence of applied field, the average distance travelled by an electron between two successive collisions is known as

(a) mean free path (b) drift velocity
(c) mobility of electron (d) none of these

Q.9 Which of the following is the correct form of Ohm's law

(a) $J = \sigma E$ (b) $J = \sigma / E$ (c) $J = \sigma E^2$ (d) none of these

Q.10 Which one of the following relations is correct for the conductivity of metals

(a) $\sigma = \dfrac{ne^3}{2mt}$ (b) $\sigma = \dfrac{2m}{ne^2 t}$ (c) $\sigma = \dfrac{ne^2 t}{2m}$ (d) $\sigma = \dfrac{n}{2me^2 t}$

Q.11 The value of Fermi-distribution function at absolute zero ($T = 0\,K$) is 1, i.e., $F(E)=1$, under the condition

(a) $E > E_F$ (b) $E < E_F$ (c) $E = E_F$ (d) $E >> E_F$

Q.12 At any temperature T and for $E = E_F$, the Fermi-distribution function becomes

(a) 0 (b) ∞ (c) 1 (d) $\dfrac{1}{2}$

Q.13 The free electron theory of metals was initiated by

(a) Pauli (b) Sommerfield
(c) Lorentz and Drude (d) Fermi-Dirac

Q.14 Which one of the following theory was developed by Lorentz and Drude

(a) quantum free electron theory (b) classical free electron theory
(c) zone theory (d) all of these

Q.15 Quantum theory of free electrons in metals explain:

(a) electrical conductivity and themionic emission
(b) specific heat & paramagnetism
(c) both (a) and (b)
(d) none of these

Short-Answer Questions

Q.1 What do you understand by free electron gas model of metals?

Q.2 What is the difference between electrical conductivity and thermal conductivity?

Q.3 Define and discuss Weidemann-Franz law.

Q.4 What are the limitations of free electron theory?

Q.5 Discuss the elements of classical free-electron theory. What were its limitations?

Q.6 What is the difference between classical and quantum theory of free electrons?

Q.7 Write expression for the Fermi energy at absolute zero.

Q.8 Write the expression for Fermi-Dirac distribution function.

Q.9 What are the drawbacks and successes of classical free electron theory?

Q.10 What is density of states? Discuss briefly.

Q.11 Write Richardson's thermionic equation and explain every term occurring in this.

PRACTICE PROBLEMS

General Questions

Q.1 What do you mean by free electron gas model of metals? Define free electron Fermi gas. Which properties of solids are explained by free electron gas theory?

Q.2 Discuss the successes and failures of the free electron theory.

Q.3 Obtain an expression for the electrical conductivity of a metal on the basis of free electron theory. Hence prove Ohm's law.

Q.4 (a) Obtain an expression for thermal conductivity of a metal on the basis of free electron theory.
(b) State Weidemann-Franz law.

Q.5 Derive an expression for Fermi energy and density of states of a system.

Q.6 What is free electron theory of metals? Derive an expression for conductivity of metals on the basis of Drude-Lorentz theory.

Q.7 Explain the quantum theory of free electrons in metals. Derive an expression for the Fermi-energy at absolute zero.

Q.8 Discuss quantum theory of free-electrons and explain the following (a) Fermi level, (b) Density of states (c) F-D distribution function.

Q.9 In terms of Fermi energy, calculate the kinetic energy at 0 K.

Q.10 Derive an expression for Fermi energy of free electrons. Discuss briefly the effect of temperature.

Q.11 Obtain an expression for energy levels in one dimensional free electron gas.

Q.12 What is Fermi gas? Does the Fermi energy of a metal depend upon the temperature?

Q.13 Considering the free electrons in a metal to form an electron gas obeying Fermi-Dirac statistics, obtain Richardson's equation for thermionic emission of electrons.

Q.14 State the difference between quantum and classical theories of free electron. Obtain Richardson-Dushman equation of thermionic equation.

Q.15 Discuss the phenomenon of thermionic emission in metals. Obtain Richardson-Dushman equation for the emission of current density.

Q.16 Derive the Richardson's thermionic emission equation.

Q.17 Show that the kinetic energy of a three-dimensional gas of M free electrons at 0 K is $\bar{E}_0 = \frac{3}{2} E_F$.

Q.18 Write a note on
(i) Fermi-Dirac distribution function.
(ii) Density of states in one-dimension.
(iii) Energy levels and wave function of free electrons in a box.

Band Theory of Solids and Photoconductivity

18

LEARNING OBJECTIVES

After reading this chapter you will be able to

- **LO 1** Understand Kronig-Penney model, energy E versus k diagram and one- and two-dimensional Brillouin zones
- **LO 2** Explain effective mass of an electron and deviation of electron behaviour in crystal lattice
- **LO 3** Differentiate between Insulators, n-type, p-type semiconductors and conductors (metals)
- **LO 4** Explain electron configuration in conduction band, hole concentration in valence band, Fermi level of intrinsic and extrinsic semiconductor
- **LO 5** Illustrate Hall effect, Hall voltage and Hall coefficient
- **LO 6** Learn simple model of photoconductor and gain factor
- **LO 7** Discuss effects of traps, applications of photoconductivity

Introduction

A solid contains an enormous number of atoms packed closely together. When N atoms of the solid are well separated, then these atoms lead to N-fold degenerate levels of the solid. As the atoms approach one another to form a solid, i.e., their separation reduces, a continuously increasing interaction occurs between them. This causes each of the levels to split into N distinct levels. It is the separation distance (say r) which specifies the amount of overlap that causes the splitting. Since a solid contains about 10^{23} atoms per mole, i.e., N is very large, the splitted energy levels become so numerous and close together that they form an almost continuous **energy band**.

The amount of splitting is different for different energy levels. For example, the lower energy levels are found to spread or split less than the higher levels. It means the lowest levels remain almost unsplit. The reason is that the electrons in lower levels are the ones which are in inner subshells of the atoms. So they are not significantly influenced by the presence of nearby atoms. Since the potential barriers between the atoms are for them relatively high and wide, these electrons are localised in particular atoms, even when r is small. However, the electrons in the higher levels are the valence electrons and are not localised at all for small r but they become part of the whole system. From the quantum point of view, the wave functions of the valence electrons overlap and the overlapping of their wave functions results in splitting or spreading of their energy levels.

The band formation of the higher energy levels of sodium, whose ground state atomic configuration is $1s^2\ 2s^2\ 2p^6\ 3s^1$, is shown in Fig. 18.1. In the figure, the dashed and vertical line indicates the observed interatomic separation in the solid sodium. It is clear from the figure that the bands overlap when the atomic separation decreases. This figure also shows that the allowed band corresponding to inner subshells, for example $2p$ in sodium, are extremely narrow and does not begin to split until the interatomic distance r becomes less than the value actually found in the crystal. As we move towards the higher energy states, the energy of the electrons become larger and also the region in which they can move becomes wider. Since they are also affected more by the nearby ions, it is seen that the bands become progressively wider for the outer occupied subshells and also for the unoccupied subshells of the atoms in its ground state. Therefore, with the increase of energy the successive allowed bands become wider and overlap each other in energy.

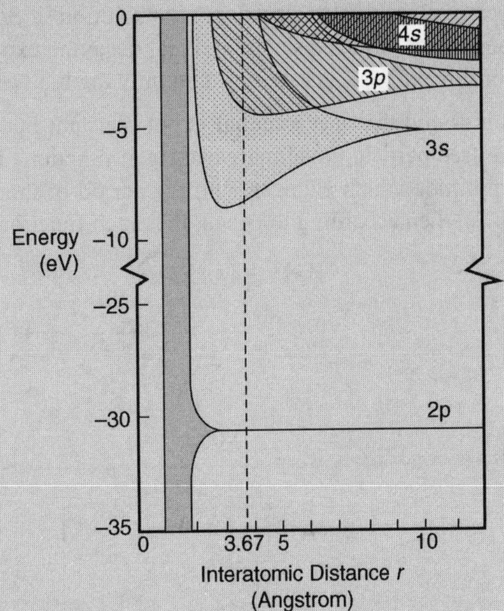

FIGURE 18.1

It is clear from the above discussion that the energy bands in a solid correspond to energy levels in an atom. Therefore, an electron in a solid can occupy only energy that falls within these energy bands. The overlapping of the bands depends on the structure of the solid. If the bands do not overlap, then the intervals between them represents energies which the electrons in the solid cannot occupy. These intervals are called *forbidden bands* or energy gaps. However, if the adjacent bands in the solid overlap, then the electrons possess a continuous distribution of allowed energies.

18.1 KRONIG-PENNEY MODEL

LO1

Free electron model ignores the effects those arise when the electrons interact with crystal lattice. However, now we consider this by making some general remarks about the effect of the periodic variation in the potential. Due to this periodicity in potential for an infinitely long lattice, the wave function does not remain sinusoidal travelling waves of constant amplitude but now they include the lattice periodicity in their amplitudes, and electrons may be scattered by the lattice. When the deBroglie wavelength of the electron corresponds to a periodicity in the spacing of the ions, the electron interacts strongly with the lattice. This situation is the same as an electromagnetic wave suffers Bragg's reflection, when the Bragg's condition is satisfied.

In order to find the allowed energies of electrons in solids, we consider the effect of formation of a solid when the individual constituent atoms are brought together. We solve the Schröedinger equation for periodic potential seen by an electron in a crystal lattice. We also consider that the periodic potential is a succession of rectangular wells and barriers. The solution of Schröedinger equation is a sinusoidal wave in certain energy ranges, i.e., allowed states, and real decaying exponential wave in the other ranges, i.e., the forbidden bands. For this purpose, here we present only qualitative approach.

It is found that the potential is not constant but varies periodically. The effect of periodicity is to change the free particle travelling wave eigen function. Therefore, the travelling wave eigen function has a varying amplitude which changes with the period of the lattice. If we consider that the space periodicity is a (Fig. 18.2), then according to Bloch, the eigen function for one-dimensional system has the form

$$\psi(x) = u_k(x)e^{ikx}$$

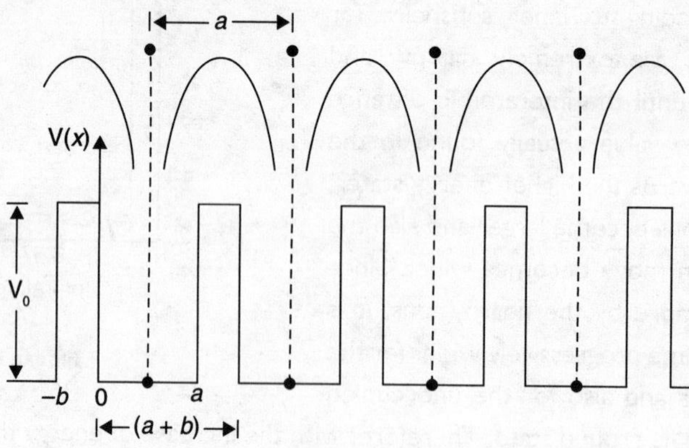

FIGURE 18.2

As is clear, this is different from the free travelling wave function $\psi(x) = Ae^{ikx}$ · $u_k(x)$ is the periodic function with the periodicity a of the periodic potential, i.e.,

$$u_k(x) = u_k(x + a)$$

In general,

$$u_k(x) = u_k(x + na)$$

where n is an integer. Hence, with the effect of periodicity, the complete wave function is

$$\psi(x, t) = u_k(x)e^{i(kx - \omega t)} \tag{i}$$

In the above equation, the exponential term indicates a wave of wavelength $\lambda = \dfrac{2\pi}{k}$ which travels along $+x$ direction if k is positive and it moves along $-x$ direction if the value of k is negative.

The exact form of the function $u_k(x)$ depends on the particular potential assumed and the value of k.

In 1930, *Kronig and Penney* proposed a one-dimensional model for the shape of rectangular potential wells and barriers having the lattice periodicity, as shown in Fig. 18.2. Each well represents an approximation to the potential produced by one ion. In the region such as $0 < x < a$, the potential energy is assumed to be zero while in the region $-b < x < 0$ or $a < x < (a + b)$, the potential energy is taken as V_0. The relevant Schröedinger equations for these two regions are

$$\frac{d^2\psi}{dx^2} + \left[\frac{2m}{\hbar^2}\right]E\psi = 0 \qquad \text{(ii) } [0 < x < a]$$

$$\frac{d^2\psi}{dx^2} + \left[\frac{2m}{\hbar^2}\right](E - V_0)\psi = 0 \qquad \text{(iii) } [-b < x < 0]$$

The electron of not too high energy is practically bound within one of the wells that are deep and widely spaced. So the lower energy eigen values are those of a single well. However, for the wells those are closer together the eigen function can penetrate the potential barriers more easily. Because of this, spreading of previously single energy level into a band of energy levels takes place. The band becomes wider with the decrease in the separation of the wells. Under the limit of zero barrier thickness, we obtain an infinitely wide single well in which all energies are allowed. So the present case is reduced to the free electron model. The comparison between the allowed energies of a single well and an array of wells (Kronig-Penney model) is shown in the Fig. 18.3. In this figure, we have assumed $b = a/16$ and the well strength as $2mV_0a^2/\hbar^2 = 121$. It is clear from the figure that each band corresponds to a single energy level of the single well. The forbidden bands appear even for energies $E > V_0$.

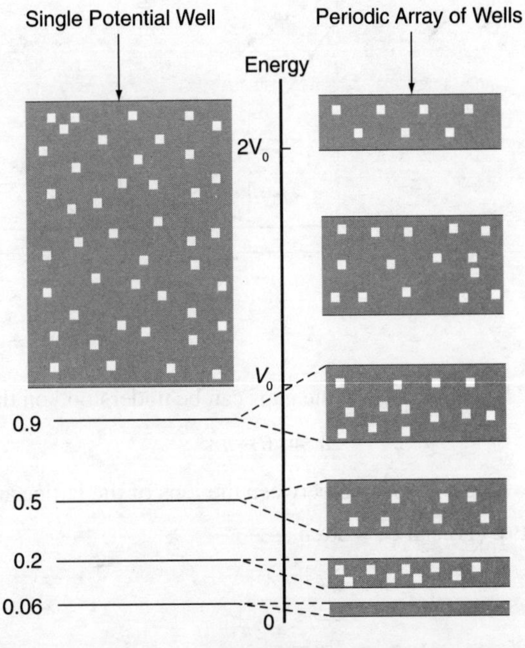

Figure 18.3

Here we will solve the Schröedinger wave equation for electron for Kronig-Penney potential under the condition that ψ and $\frac{d\psi}{dx}$ are continuous at the boundaries of the well. A complicated expression for the allowed energies in terms of k shows that gaps in energy are obtained at values such that

$$k = \pm\frac{\pi}{a}, \pm\frac{2\pi}{a}, \pm\frac{3\pi}{a}, \ldots \qquad \text{(iv)}$$

The solution of the Schröedinger wave equation for free-electrons results in the energy values given by

$$E = \frac{h^2k^2}{8\pi^2 m} = \frac{\hbar^2k^2}{2m} \qquad \text{(v)}$$

18.1.1 Energy *E* versus Wave Number *k* Diagram

From Eq. (v), it is clear that the relation between E and k is parabolic. The parabolic relation between E and k, valid in case of free electrons, is therefore, interrupted at different values of k, as shown in Fig. 18.4. It means the energies corresponding to the values of k given by Eq. (iv) are not permitted for electrons in the crystal. Thus, the energies of electrons are divided into forbidden and allowed bands (Fig. 18.4).

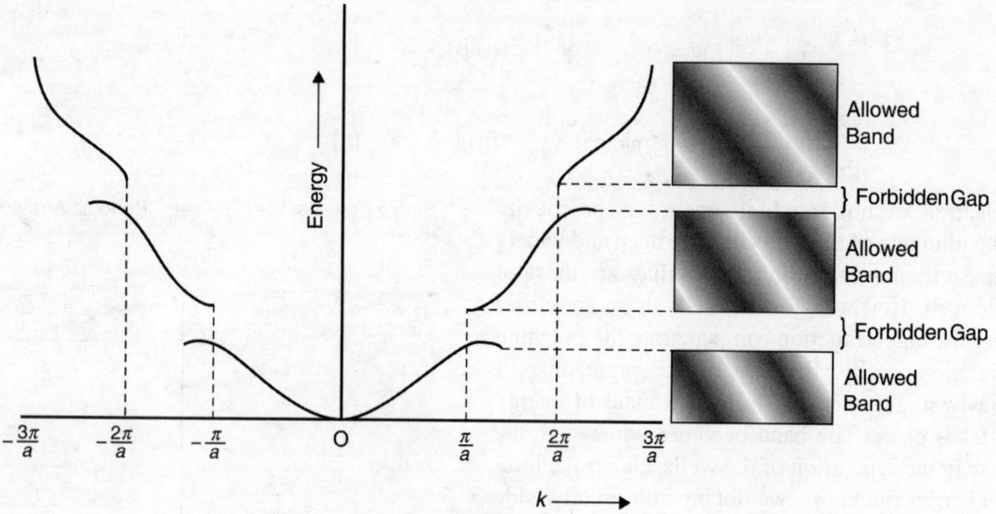

FIGURE 18.4

The occurrence of the gaps can be understood on the basis of Bragg's condition for the diffraction, given as

$$2a \sin \theta = n\lambda \qquad n = 1, 2, 3, \ldots \tag{vi}$$

where a is spacing between the ions of the lattice and θ is the angle of incidence.

Eq. (vi) can be written as

$$2a = n\lambda \qquad (\text{for } \theta = 90^\circ)$$

or $$2a = n\frac{2\pi}{k}$$

or $$k = \frac{n\pi}{a} \tag{vii}$$

or $$k = \pm\frac{\pi}{a}, \pm\frac{2\pi}{a}, \pm\frac{3\pi}{a}, \ldots$$

We have put ± signs because the incident wave can travel along +x-axis as well as along –x-axis. At all these values of k the gaps in energy occur, as shown in Fig. 18.4.

The waves corresponding to values of k satisfying the Bragg's condition are reflected and resulted in standing waves. On each subsequent Bragg reflection, the direction in which the wave is travelling is reversed again. The eigen function of incident and corresponding reflected waves for $k = \pm n\frac{\pi}{a}$ are therefore $e^{i(\pi/a)x}$ and $e^{-i(\pi/a)x}$. These two eigen functions can be combined in two different ways to give total eigen function

$$\psi_1 = e^{i(\pi/a)x} + e^{-i(\pi/a)x} = \cos(\pi/a)x$$

$$\psi_2 = e^{i(\pi/a)x} + e^{-i(\pi/a)x} = \sin(\pi/a)x$$

or $$\psi_1 \propto \cos(\pi/a)x \tag{viii}$$

and $$\psi_2 \propto \sin(\pi/a)x \tag{ix}$$

Hence, the two standing waves are obtained. The probability density curves for these two stationary waves, i.e., $|\psi_1|^2$ and $|\psi_2|^2$, are shown in Fig. 18.5. From this figure and Eq. (viii) it is clear that the value of $|\psi_1|^2$ is maximum at the positions of positive ions (i.e., $x = 0, \pm a, \pm 2a,...$). The value of $|\psi_2|^2$ is maximum in between the position of positive ions. From Fig. 18.2, it is evident that the potential energy of an electron is maximum between the ions and minimum at the positions of the ions. So an electron can have two different values of energies, i.e., E_1 and E_2 for $k = \frac{\pi}{a}$ corresponding to the two standing waves ψ_1 and ψ_2. Hence, no electron can have any energy between E_1 and E_2. This phenomenon creates a difference in energy ($E_1 \sim E_2$) which is known as *energy gap*.

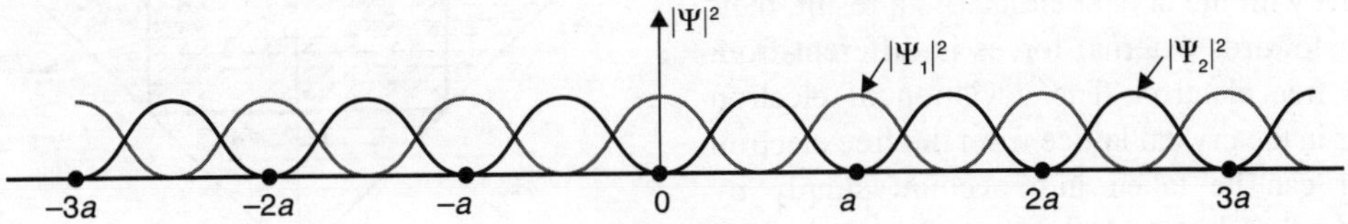

FIGURE 18.5

18.2 ONE-AND TWO-DIMENSIONAL BRILLOUIN ZONES LO1

In Kronig-Penney model, we have seen that the discontinuities in energy occurs when the wave number k satisfies the condition $k = n\pi/a$, where n takes the values $\pm 1, \pm 2, \pm 3,...$ etc. The graph between the total energy E and wave number k is shown in Fig. 18.4. It is clear from the figure that an electron has allowed energy values in the region between $-\pi/a$ to $+\pi/a$. This region is called the first Brillouin zone. As discussed earlier, there is a discontinuity of gap in the energy values after this allowed energy value. This gap is called forbidden gap or forbidden zone. Again there is another allowed energy zone, which is observed after this forbidden gap and is extended from $-\pi/a$ to $-2\pi/a$ and π/a to $2\pi/a$. This zone is called second Brillouin zone. Similarly, the other higher order Brillouin zones can be defined.

The first two Brillouin zones in one-dimensional case are shown in Fig. 18.6. We can extend the concept of the Brillouin zones to two-dimension by considering that the electron is moving in the two-dimensional square lattice. In this case, the wave number k has the two components, i.e., along the x-axis and y-axis. Let us represent them as k_x and k_y, respectively. The two-dimensional Brillouin zones are shown in Fig. 18.7. It can be seen that $k_x = k_y = \pm\pi/a$ limits the first Brillouin zone. In the figure, the first Brillouin zone is represented by a square passing through the points A, B, C and D. The second Brillouin zone for a two-dimensional lattice is represented by a square passing through E, F, G and H.

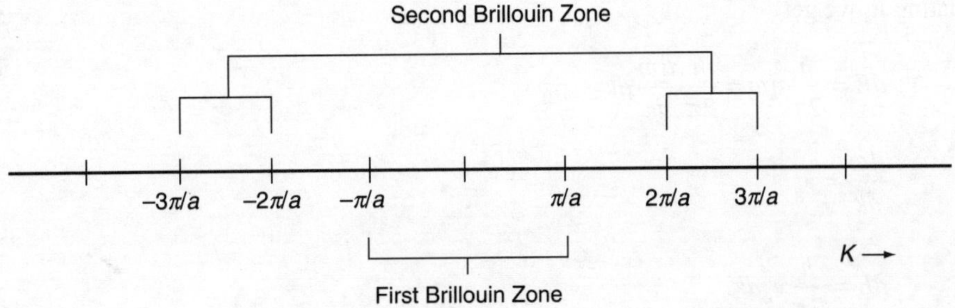

FIGURE 18.6

It is clear that the boundary for the second zone is given by $k_x=\pm 2\pi/a$ and $k_y=\pm 2\pi/a$. Thus the region between the squares *ABCD* and *EFGH* is called the second Brillouin zone.

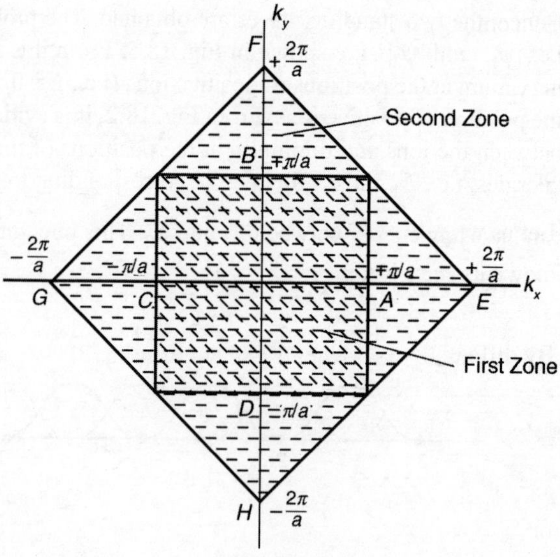

Figure 18.7

18.3 EFFECTIVE MASS OF AN ELECTRON LO2

The electrons in a crystal are not completely free but interact with the crystal lattice. As a result, their behaviour towards external forces is different from that of a free electron. The deviation of electron behaviour in the crystal lattice from the free electron behaviour can be taken into account simply by considering the electron to have an altered value of mass called the 'effective mass' m^* rather than its mass m^* in free space, which is different from the mass m of the electron in free space. The effective mass m^* depends on the nature of crystal lattice and varies with the direction of motion of the free electron in the lattice.

Suppose an electron is moving along the *x*-axis in a crystal in the presence of an external electric field E'. So it experiences a force eE'. If the electron gains velocity v over a distance dx in time dt under the action of this force, then

Work done $= dE = eE'\,dx = eE'\,v\,dt$ $\qquad \left[\because v = \dfrac{dx}{dt}\right]$

As we know that the velocity v of a particle (electron) is the same as the 'group velocity' $\left(v_g = \dfrac{d\omega}{dk}\right)$ of the de Broglie waves associated with the particle ($v = v_g$). Thus, we can write the work done in terms of v_g as

$$dE = eE'\,v_g\,dt \tag{i}$$

According to Einstein's de Broglie relation

$$E = h\nu = \frac{h}{2\pi}\omega \qquad [\because \omega = 2\pi\nu]$$

By differentiating it, we get

$$dE = \frac{h}{2\pi}d\omega = \frac{h}{2\pi}\frac{d\omega}{dk}dk \tag{ii}$$

$$\because \quad \frac{d\omega}{dk} = v_g$$

$$\because \quad dE = \frac{h}{2\pi}v_g\,dk \tag{iii}$$

By comparing Eq. (iii) with Eq. (i), we get

$$eE'dt = \frac{h}{2\pi} v_g dk$$

or $$\frac{dk}{dt} = \frac{2\pi}{h} eE' \quad \text{(iv)}$$

Let us write the group velocity v_g in terms of energy E from Eq. (ii)

$$v_g = \frac{d\omega}{dk} = \frac{2\pi}{h}\frac{dE}{dk} \quad \text{(v)}$$

By differentiating the above equation w.r.t. time, we get

$$\frac{dv_g}{dt} = \frac{2\pi}{h}\frac{d^2E}{dtdk} = \frac{2\pi}{h}\frac{d^2E}{(dk)^2}\frac{dk}{dt}$$

or $$\frac{dv_g}{dt} = \frac{2\pi}{h}\frac{d^2E}{(dk)^2}\frac{2\pi}{h} eE' \quad \text{(By using Eq. (iv))} \quad \text{(vi)}$$

Employing $v_g = v$ again, this can be written as

$$\frac{dv}{dt} = \left(\frac{4\pi^2}{h^2}\frac{d^2E}{dk^2}\right) eE' \quad \text{(vii)}$$

This equation connects the force eE' on the electron with the acceleration $\frac{dv}{dt}$ through the proportionality factor $\left(\frac{4\pi^2}{h^2}\frac{d^2E}{dk^2}\right)$.

Since $F = ma,\ a = F/m$ (viii)

A comparison of Eq. (viii) with Eq. (vii) yields

$$\frac{1}{m^*} = \frac{4\pi^2}{h^2}\frac{d^2E}{dk^2}$$

The quantity $\frac{1}{m^*}$ is the reciprocal of the effective mass of the electron in the crystal lattice.

18.4 DISTINCTION BETWEEN INSULATORS, SEMICONDUCTORS AND CONDUCTORS (METALS) LO3

The formation of bands in solids has already been discussed and it was shown that there is an energy gap, called forbidden band, representing energies which the electrons cannot occupy. Based on this energy gap and the conduction, the solids are classified into different categories named as insulators, semiconductors and conductors.

18.4.1 Insulators

For these types of solids, the band formation is like the one shown in Fig. 18.8a. In this case, the forbidden gap between the highest filled band (valence band) and the lowest empty band (conduction band) is very wide; it is about 3 eV to 6 eV. It is seen that a very few electrons from the filled band reach the empty band, even if we thermally excite them or apply an electric field to them. Moreover, Pauli exclusion principle restricts the electrons for moving about in the filled band. For this reason, a free electron current cannot be obtained and the solids of this type are poor conductors of electricity. This class of solid is known as insulators. Diamond, quartz, and most covalent and ionic solids like ZnO and AgCl are the examples of insulators.

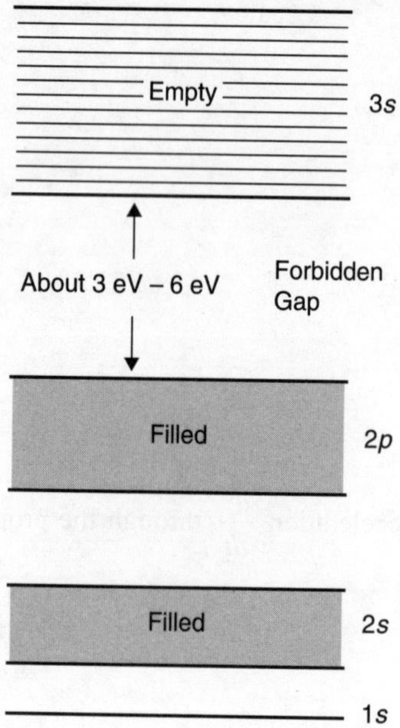

FIGURE 18.8a

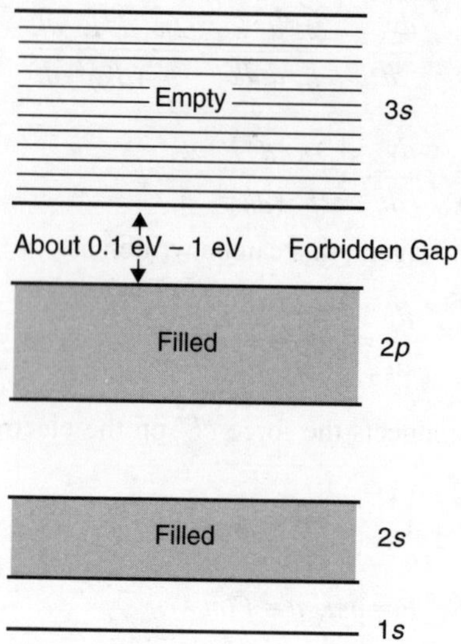

FIGURE 18.8b

18.4.2 Semiconductors

For these types of solids the band formation is like the one shown in Fig. 18.8b. In this case, the forbidden gap between the highest filled band (valence band) and the lowest empty band (conduction band) is very narrow; it is about 0.1 eV to 1 eV. Under this situation we can easily move the electrons from the highest filled band to the empty band. This can be achieved by thermal excitation or also by applying an electric field. For this reason, a free electron current can be obtained as a few electrons are available in the empty band. This class of solids is known as semiconductors. Silicon and germanium are the examples of semiconductors.

In semiconductors, there also exists another mechanism that causes the generation of electric current. Actually there are vacancies or the empty places left behind when the electron moves, which remain near the top of the uppermost filled band. These vacancies are called holes. The holes behave as positive electrons and can contribute to the generation of electric current. This is possible as the electron below the hole may gain enough energy to jump and occupy the hole due to the applied electric field. With such successive jumps of the electrons, the hole moves towards the lower energy state and contribute to the generation of electric current.

The semiconductors are mainly of two types, defined below.

18.4.2.1 *n-type Semiconductors*

The *n*-type semiconductors are the ones in which the electron conduction (negative) exceeds the hole conduction (positive). In such semiconductors, the donor impurity predominates. This can be understood if we introduce a small amount of phosphorous (*P*) or arsenic (*As*), i.e., an element of fifth group of the periodic table, as an impurity into a crystal of silicon (*Si*) or germanium (*Ge*). This addition of *P* or As means replacing an atom of *Si* or *Ge* at a lattice site by an atom of the impurity. Atoms of fifth group elements have five valence electrons whereas *Si* or *Ge* has four valence electrons. So four electrons of *P* or *As* form covalent bonds with the electrons of the atoms of *Si* or *Ge*. However, the fifth electron remains only very weakly bound to the *P* or *As* atom by electrostatic forces and this cannot be accommodated in the already filled original valence band. So, it occupies a discrete energy level which is just below the conduction band (with only a few tenths of an *eV*). Hence these extra electrons jump easily into the conduction band and contribute to the electric conductivity in addition to the electron hole pairs produced by thermal excitation of the pure semiconductor. This way the number of electrons sits more than holes to serve as charge carriers.

18.4.2.2 *p-type Semiconductors*

The *p*-type semiconductors are the one in which the hole conduction (positive) exceeds the electron conduction (negative). In such semiconductors, the acceptor impurity predominates. This can be understood if we introduce a small amount of *Al*, *Ga* or *In*, i.e., an element of third group of the periodic table, as an impurity into a crystal of silicon (*Si*) or germanium (*Ge*). Atoms of third group elements have three valence electrons whereas *Si* or *Ge* has four valence electrons. So three electrons of *Al* or *Ga* form covalent bonds with the electrons of the atoms of *Si* or *Ge*. However, the fourth available electron of the semiconductor lacks an electron with which it can form a bond. This is equivalent to as if a vacancy or hole has been created at the site of the impurity atom. Hence, the impurity atoms introduce vacant discrete energy levels very near the top of completely filled valence band of *Si* or *Ge*. So these extra holes move from an impurity atom. These holes behave as positive charge carriers and are available in excess. Since the crystals of this type have an excess of positive charge carriers, they are called positive semiconductors or p-type semiconductors.

18.4.3 Conductors or Metals

For these types of solids, the band formation is like the one shown in Fig. 18.8c. In this case, the valence band is either partially filled or the next allowed empty band overlaps with the filled band. In both the cases, there are unoccupied states for electrons in the uppermost band. So these electrons are available to generate the current. This class of solids is known as conductors. The conductors offer a low resistance to

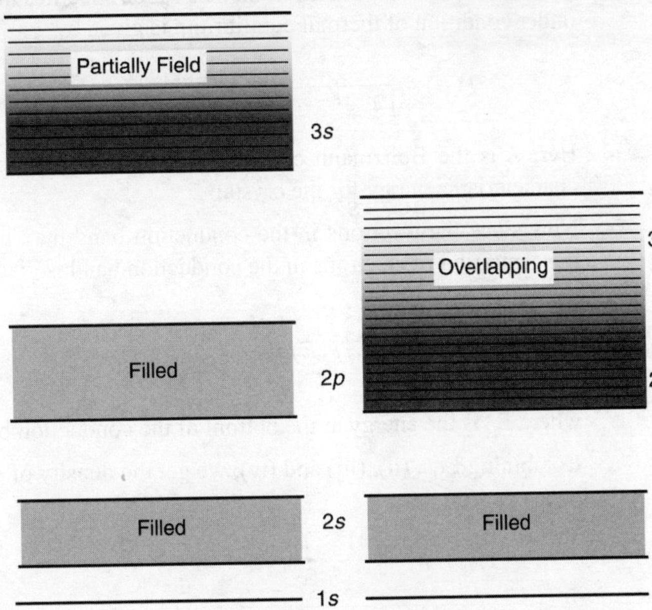

FIGURE 18.8c

the passage of an electric current. Silver, copper, iron, aluminium, etc. are the examples of conductors or metals.

18.5 INTRINSIC SEMICONDUCTOR

LO4

The semiconductors in which the transformation of electrons to the conduction band and the generation of holes in the valence band are achieved purely by thermal excitation are called intrinsic semiconductors. It means this effect is temperature dependent and produces equal numbers of electron and hole carriers. The electrons and holes are called intrinsic charge carriers and the resulting conductivity is known as intrinsic conductivity.

18.5.1 Electron Concentration in Conduction Band

For this we assume the following.

(i) Electrons in the conduction band behave as free particles with an effective mass *m*.

(ii) The number of conduction electrons per cubic meter whose energies lie between *E* and *E* + *dE* is given by

$$dn_C = N(E) f(E)\, dE \tag{i}$$

where *N(E)* is the density of states at the bottom of the conduction band and is given by as per quantum mechanics.

$$N(E) = \frac{4\pi}{h^3}(2m)^{3/2}(E - E_C)^{1/2} \tag{ii}$$

Here E_C is the energy at the bottom of the conduction band.

(iii) The Fermi-Dirac probability function which specifies the fraction of all states at energy *E* occupied under condition of thermal equilibrium is given by

$$f(E) = \frac{1}{1 + e^{(E - E_F)/kT}} \tag{iii}$$

Here *k* is the Boltzmann constant, *T* is the temperature in Kelvin and E_F is the Fermi level or characteristic energy for the crystal.

Now since the electrons in the conduction band may have energies lying between E_C and ∞, the concentration of electrons in the conduction band is given by

$$n_C = \int_{E_C}^{\infty} N(E) f(E) dE \tag{iv}$$

where E_C is the energy at the bottom of the conduction band.

Combining Eqs. (ii), (iii) and (iv), we get the density of electrons in the conduction band as

$$n_C = \frac{4\pi}{h^3}(2m)^{3/2} \int_{E_C}^{\infty} \frac{(E - E_C)^{1/2}}{1 + e^{(E-E_F)/kT}} dE \tag{v}$$

For $E \geq E_C$ and $E - E_F >> kT$, 1 in the denomination can be neglected and this equation reduces to

$$n_C = \frac{4\pi}{h^3}(2m)^{3/2} \int_{E_C}^{\infty} (E - E_C)^{1/2} e^{(E_F - E)/kT} dE$$

$$n_C = \frac{4\pi}{h^3}(2m)^{3/2} e^{(E_F - E_C)/kT} \int_{E_C}^{\infty} (E - E_C)^{1/2} e^{(E_C - E)/kT} dE$$

Substituting $(E - E_C) / kT = x$, so that $dE = kT\, dx$, we get

$$n_C = \frac{4\pi}{h^3}(2m)^{3/2} e^{(E_F - E_C)/kT} \int_0^{\infty} x^{1/2} (kT)^{1/2} e^{-x} kT\, dx$$

$$= \frac{4\pi}{h^3}(2m)^{3/2} (kT)^{3/2} e^{(E_F - E_C)/kT} \int_0^{\infty} x^{1/2} e^{-x} dx$$

$$= \frac{4\pi}{h^3}(2m)^{3/2} (kT)^{3/2} e^{(E_F - E_C)/kT} \frac{\sqrt{\pi}}{2} \qquad \left[\therefore \int_0^{\infty} x^{1/2} e^{-x} dx = \frac{\sqrt{\pi}}{2}\right]$$

$$n_C = 2\left[\frac{2\pi mkT}{h^2}\right]^{3/2} e^{(E_F - E_C)/kT} \tag{vi}$$

This relation gives the density or concentration of electrons in the conduction band of an intrinsic semiconductor. Please note that here m is the effective mass of the electron.

18.5.2 Hole Concentration in Valence Band

By a similar method as adopted above, we can get an expression for the concentration of holes in the valence band. Since a hole signifies a vacancy created by removal of an electron, i.e., an empty energy level, the Fermi function for a hole is $1 - f(E)$. Here $f(E)$ represents the probability that the level is occupied by an electron. Along with the expression of $f(E)$, the factor $1 - f(E)$ becomes

$$1 - f(E) = 1 - \frac{1}{1 + e^{(E - E_F)/kT}}$$

$$= 1 - [1 + e^{(E - E_F)/kT}]^{-1}$$

$$= 1 - [1 - e^{(E - E_F)/kT}] \quad \text{for} \quad (E - E_F)/kT << 1$$

$$= 1 - 1 + e^{(E - E_F)/kT}$$

$$1 - f(E) = e^{(E - E_F)/kT}$$

For the top of the valence band (the maximum energy), the density of states is given by

$$N(E) = \frac{4\pi}{h^3}(2m_h)^{3/2}(E_v - E)^{1/2}$$

Here m_h is the effective mass of holes near the top of the valence band, where the energy is E_V. With the above relation, the density of holes in the valence band is calculated as

$$n_h = \int_{-\infty}^{E_V} N(E)[1 - f(E)]dE$$

$$= \frac{4\pi}{h^3}(2m_h)^{3/2} \int_{-\infty}^{E_v} (E_v - E)^{1/2} e^{(E-E_F)/kT} dE$$

$$= \frac{4\pi}{h^3}(2m_h)^{3/2} e^{(E_V-E_F)/kT} \int_{-\infty}^{E_V} (E_V - E)^{1/2} e^{(E-E_V)/kT} dx$$

$$= \frac{4\pi}{h^3}(2m_h)^{3/2} e^{(E_V-E_F)/kT} \int_{+\infty}^{0} x^{1/2} (kT)^{1/2} e^{-x} kTdx$$

where we have substituted $\frac{E_V - E}{kT}$ so that $dE = - kT\, dx$. Now

$$n_h = \frac{4\pi}{h^3}(2m_h)^{2/3} (kT)^{3/2} e^{(E_V-E_F)/kT} \int_{0}^{\infty} x^{1/2} e^{-x} dx$$

$$n_h = 2\left[\frac{2\pi m_h kT}{h^2}\right]^{3/2} e^{(E_V-E_F)/kT} \tag{vii}$$

18.5.3 Intrinsic Concentration of Charge Carriers

Combining Eqs. (vi) and (vii), we get the following expression for the product of electron-hole concentration

$$n_C n_h = np = 4\left[\frac{2\pi kT}{h^2}\right]^3 (mm_h)^{3/2} e^{(E_V-E_C)/kT}$$

$$n_C n_h = AT^3 e^{-E_g/kT} \tag{viii}$$

where $E_g = E_C - E_V$ is the width of forbidden energy gap between conduction and valence bands and $A = \frac{32\pi^3 k^3}{h^6}(mm_h)^{3/2}$ is a constant. In most of the cases, n_C is written as n and n_h as p only.

Eq. (viii) shows that the product of holes and electron densities depends on the temperature T and the forbidden energy gap E_g, but is independent of the Fermi level E_F. Thus the product of electron and hole concentrations, for a given material, is constant at a given temperature. If an impurity is added to increase n, there will be a corresponding decrease in p such that the product np remains a constant. Since for an intrinsic semiconductor, $n = p = n_i$, we arrive at an important relationship, called the law of action

$$np = n_i^2 = AT^3 e^{-E_g/kT} \tag{ix}$$

where n_i is called the intrinsic density of either carrier. Equation (ix) is true for a semiconductor regardless of donor or acceptor concentrations.

18.5.4 Energy Band Diagram and Fermi Level

In an intrinsic semiconductor, the electrons and holes are always generated in pairs, i.e., $n = p = n_i$. Substituting the values of n and p from Eqs. (vi) and (vii) of the previous sections, we get

$$2\left[\frac{2\pi mkT}{h^2}\right]^{3/2} e^{(E_F - E_C)/kT} = 2\left[\frac{2\pi m_h kT}{h^2}\right]^{3/2} e^{(E_V - E_F)/kT}$$

or
$$e^{(2E_F - E_C - E_V)/kT} = \left(\frac{m_h}{m}\right)^{3/2}$$

Taking the logarithm of both sides

$$\frac{2E_F - E_C - E_V}{kT} = \frac{3}{2}\ln\left(\frac{m_h}{m}\right)$$

or
$$2E_F - E_C - E_V = \frac{3}{2}kT\ \ln\left(\frac{m_h}{m}\right)$$

or
$$E_F = \frac{E_C + E_V}{2} + \frac{3}{4}kT\ \ln\left(\frac{m_h}{m}\right)$$

If the effective mass of hole and a free electron is the same i.e., $m_h = m$, then

$$E_F = \frac{E_C + E_V}{2}$$

This shows that the Fermi level E_F lies exactly in the centre of the forbidden energy gap E_g as depicted in Fig. 18.9. The Fermi level can also be defined as the energy level at which there is a 0.5 probability of finding an electron. It depends on the distribution of energy levels and the number of electrons available.

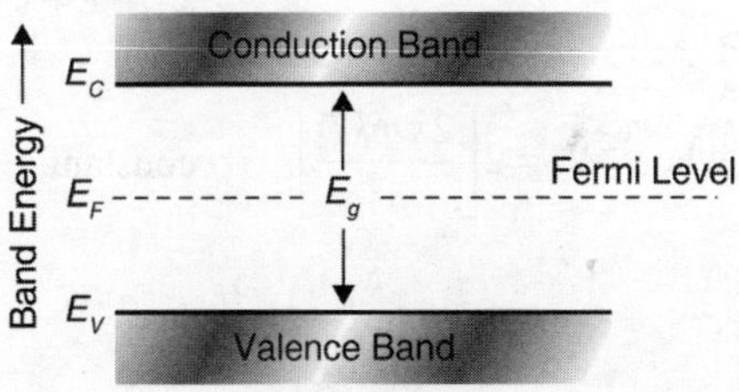

FIGURE 18.9

18.6 EXTRINSIC SEMICONDUCTOR LO4

The conductivity of an intrinsic semiconductor can be increased significantly by adding certain impurities to it. By doing so, we get impurity semiconductor which is also known as extrinsic semiconductor. In extrinsic semiconductors, current carriers are introduced by donor and acceptor impurities with locked energy levels near the top or bottom of the forbidden gap.

18.6.1 Energy Band Diagram and Fermi Level

(i) n-type Extrinsic Semiconductor

When a small amount of pentavalent impurity is added to the crystal, it creates extra electrons without adding any new holes. This impurity introduces new energy levels into the energy band picture. The location of these new levels is slightly below the bottom of the conduction band for intrinsic semiconductor. The width of the gap for germanium was previously stated as 0.72 eV. The energy required to move an electron from a donor impurity into the conduction band is of the order of 0.01 eV, and since at normal ambient temperature the thermal energy is considered to be about 0.02 eV, it is concluded that almost all the electrons are detached from the donor atoms and have conduction band energies. In the case of silicon doped with donor impurities, the energy required to move an electron from donor impurity into the conduction band is of the order of 0.05 eV. The energy band diagram for an n-type semiconductor is shown in Fig. 18.10. Here E_g represents the energy level corresponding to donor impurities.

In intrinsic semiconductor, Fermi level lies in the middle of the forbidden energy E_g indicating equal concentrations of free electrons and holes. When a donor type impurity is added to the crystal, then if we assume that all the donor atoms are ionised, the donor electrons will occupy the states near the bottom of the conduction band. Hence, it will be more difficult for the electrons from the valence band to cross the energy gap by thermal agitation. Consequently, the number of holes of the valence band is decreased. Since Fermi level is a measure of the probability of occupancy of the allowed energy states, E_F for n-type semiconductors must move closer to the conduction band, as shown in Fig. 18.10. At usual temperatures all the donor levels will be fully activated and the donor atoms will be ionised. It means the density of electrons in the conduction band will be approximately equal to the density of donor atoms, i.e., $n \approx N_d$ (N_d being the density of donor atoms). Then from Eq. (vi), we have

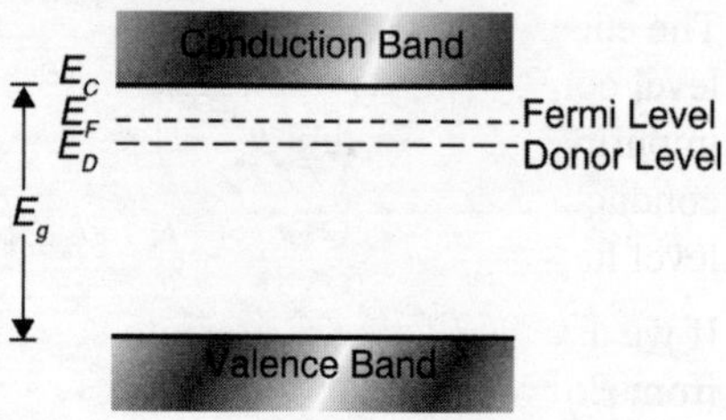

FIGURE 18.10

$$n = N_d = 2\left[\frac{2\pi mkT}{h^2}\right]^{3/2} e^{(E_F - E_C)/kT}$$

$$= N_c e^{(E_F - E_C)/kT}$$

where $N_c = 2\left[\dfrac{2\pi mkT}{h^2}\right]^{3/2}$ = constant

$$\therefore \qquad \frac{N_c}{N_d} = e^{-(E_F - E_C)/kT}$$

Taking the logarithm of both the sides

$$\ln\frac{N_c}{N_d} = -\frac{E_F - E_C}{kT}$$

or

$$E_F = E_C - kT\ln\frac{N_c}{N_d}$$

It shows that the Fermi level lies below the bottom of the conduction band, as shown in Fig. 18.10.

(ii) p-type Extrinsic Semiconductor

When an acceptor-type impurity is added, it also modifies the energy level diagram of semiconductor and makes the conduction easier. The presence of impurity creates new energy levels which are in the gap in the neighbourhood of the top of valence band of energies, as shown in Fig. 18.11. Ambient temperature results in ionisation of most acceptor atoms and thus an apparent movement of holes takes place from the acceptor levels to the valence band. The energies for holes are highest near the valence bond and decrease vertically upward in the energy level diagram. Alternatively, one may say that electrons are accepted by the acceptors and these electrons are supplied from the valence band, thus leaving a preponderance of holes in the valence band.

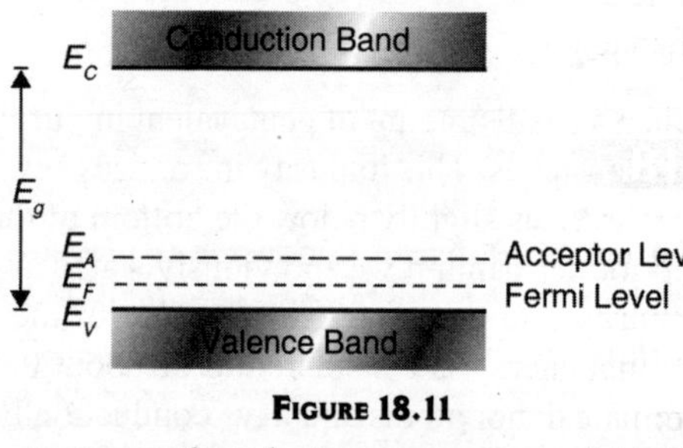

FIGURE 18.11

The energy band diagram for a p-type semiconductor is shown in Fig. 18.11 where E_A represents the energy level corresponding to the acceptor impurity. When an intrinsic semiconductor is doped with acceptor type impurity, the concentration of holes in the valence band is more than the concentration of electrons in the conduction band and the Fermi level shifts towards the valence band, as shown in Fig. 18.11. The acceptor level lies immediately above the Fermi level.

If we assume that there are only acceptor atoms present and that these are all ionised, we have $p = N_a$. Then from Eq. (vii), we get

$$p = N_a = 2\left[\frac{2\pi m_h kT}{h^2}\right]^{3/2} e^{(E_V - E_F)/kT}$$

$$= N_v e^{(E_V - E_F)/kT}$$

where $N_v = 2\left[\frac{2\pi m_h kT}{h^2}\right]^{3/2}$ = constant

$$\therefore \qquad \frac{N_v}{N_a} = e^{-(E_V - E_F)/kT}$$

Taking logarithm of both the sides,

$$\ln\frac{N_v}{N_a} = -\frac{E_V - E_F}{kT}$$

or

$$E_F = E_V + kT\ln\frac{N_V}{N_a}$$

It shows that the Fermi level lies above the top of valence band, as shown in Fig. 18.11.

18.6.2 Effect of Temperature

Let us see what happens if we increase the temperature of an n-type semiconductor. Since all the donors have already donated their free electrons at room temperature, the additional thermal energy will only increase the generation of electron-hole pairs. Thus the concentration of minority charge carriers increases. A temperature is ultimately reached when the number of covalent bonds broken is very large such that the number of holes and electrons is almost equal. The extrinsic semiconductor then behaves like an intrinsic semiconductor, although its conductivity is higher. This critical temperature is 85°C for germanium and 20°C for silicon. The same arrangement can be put forward for the p-type semiconductor. Thus with an increase in the temperature of an extrinsic (impurity) semiconductor, it behaves almost intrinsically.

18.7 HALL EFFECT

LO5

If a current carrying conductor is placed in a transverse magnetic field, a potential is developed in the conductor in the direction perpendicular to both the current and magnetic field. This phenomenon is known as *Hall effect*. It was discovered by *Hall* in 1879.

Let us consider a rectangular strip carrying current along x-axis (with electron flow towards x-axis) and magnetic field of strength B is applied along z-axis, as shown in Fig. 18.12. The force on electrons would be exerted due to the effect of magnetic field in negative y-direction. Therefore, the face MN collects a negative charge and the face PQ collects positive ions. This separation of charges sets up an electrostatic field inside

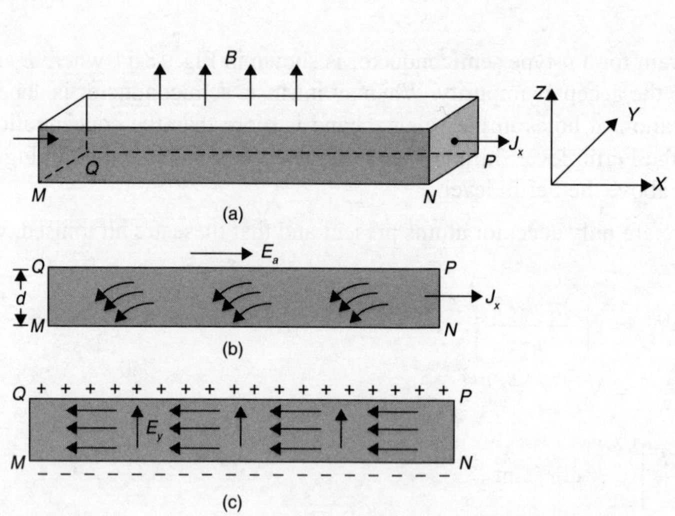

FIGURE 18.12

the conductor in y-direction. The accumulation of charge on the surfaces of the specimen continues until the force on moving charges due to electric field associated with the accumulated charge itself is large enough to cancel the force exerted by the magnetic field. Ultimately a steady state is reached in which the net force on the moving charges in y-direction vanishes and the electron can again move freely along the conductor. In stationary state the value of E is denoted by E_H and is called *Hall electric field*.

18.7.1 Hall Voltage and Hall Coefficient

We can calculate Hall voltage and Hall coefficient by using the following approach.

The force on the charge carriers $= q\vec{E}_H$

The force on charge carriers due to magnetic field $B = q(\vec{v}_d \times \vec{B})$

At the stationary state both these forces balance each other. Hence

$$q\vec{E}_H = q(\vec{v}_d \times \vec{B}) \quad \text{or} \quad \vec{E}_H = \vec{v}_d \times \vec{B} \tag{i}$$

Thus by measuring E_H and B, the drift velocity v_d can be determined both in magnitude and direction. If we talk about magnitude only, then Eq. (i) gives

$$E_H = v_d B \qquad [\because v_d \text{ and } B \text{ are at right angle}] \tag{ii}$$

The current density J in terms of the drift velocity v_d is written as

$$J = -nev_d \qquad [n = \text{number of charge carriers per unit volume}] \tag{iii}$$

From Eqs. (ii) and (iii), we get

$$E_H = -\frac{JB}{ne} \tag{iv}$$

The above relation shows that

$$E_H \propto B$$

If d be the width of the strip, then the transverse Hall electric field E_H can be related to the Hall potential difference V_H as

$$E_H = \frac{V_H}{d}$$

or
$$V_H = E_H d = -\frac{JBd}{ne} \tag{v}$$

V_H is also known as *Hall Voltage*.

The coefficient of proportionality $-\frac{1}{ne}$ is called *Hall coefficient* and is denoted by R_H. It is given below

$$R_H = -\frac{1}{ne} \tag{vi}$$

Hall coefficient is negative if the charge carriers are electrons and it will be positive if the charge carriers are holes.

From the above equation, we can develop a relation between the Hall electric field E_H and the electric field E_x, which causes the current to flow in the conductor. We can write the current density $J = \sigma E_x = \frac{1}{\rho} E_x$ where σ is the conductivity and ρ is the resistivity. Then from Eq. (iv)

$$E_H = -\frac{JB}{ne} = \frac{-E_x B/\rho}{ne}$$

$$= -\frac{E_x B}{en\rho}$$

$$\frac{E_H}{E_x} = \frac{-B}{ne\rho}$$

18.8 PHOTOCONDUCTIVITY LO6

Photoconductivity is an electrical phenomenon in which a material becomes more conductive due to the absorption of light radiation. When the energy of the incident radiation is higher than the energy gap E_g between conduction band and valence band, the electrons and holes raise the electrical conductivity. All the photoconductors are not the intrinsic type, but imperfections also play an important role. If impurities are present, electron and hole pairs are produced even if the energy of incident photon is below the threshold. The role of impurities is important to understand the experimental facts of photoconductivity. The presence of impurities produces discrete energy levels in the forbidden gap and they are often known as traps.

18.9 SIMPLE MODEL OF PHOTOCONDUCTOR LO6

We shall discuss a simple model of photoconductor in order to understand the photoconductivity. When light radiations fall on the crystal specimen, electron-hole pairs are produced throughout the volume of the crystal, as shown in Fig. 18.13.

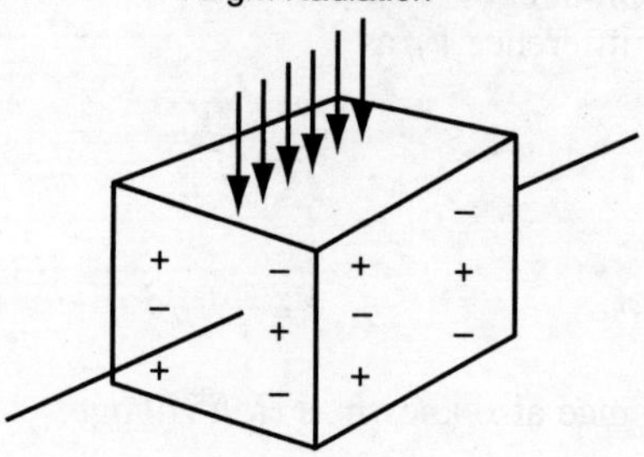

FIGURE 18.13

The recombination process occurs by direct annihilation of electrons with holes. We further suppose that as many electrons leave the crystal at one electrode the same number of electrons enters into the crystal from the opposite electrode. This model is highly hypothetical one. The production rate of electron-hole pairs is

$$\frac{dn}{dt} = L - An_e n_h = L - An^2 \qquad \text{(i)} \quad [n_e = n_h = n]$$

where L is the number of photons absorbed by the specimen, n_e and n_h are the number of electrons and holes, respectively, and A is known as *recombination coefficient*. The term $An_e n_h$ shows the recombination rate and it will be proportional to the rate of production of electron-hole pairs.

In the steady state,

$$\frac{dn}{dt} = 0$$

so that, Eq. (i) becomes

$$L - An_0^2 = 0$$

or

$$n_0 = \left(\frac{L}{A}\right)^{1/2} \qquad \text{(ii)}$$

Here n_0 is the number of electrons in the steady state.

The corresponding conductivity is given by

$$\sigma = n_0 e \mu_e = \left[\frac{L}{A}\right]^{1/2} e\mu_e \qquad \text{(iii)}$$

where, μ_e is the mobility of the electron.

Hence, the photocurrent density will be

$$J = \sigma E$$

$$J = \left[\frac{L}{A}\right]^{1/2} e\mu_e \left[\frac{V}{d}\right] \qquad \left[\therefore E = \frac{V}{d}\right]$$

or

$$J = \frac{n_0 e \mu_e V}{d} \qquad \text{(iv)}$$

where V is the voltage across the specimen, and d is the thickness of the specimen. If the light is switched off suddenly, then $L = 0$, by which Eq. (i), becomes

$$\frac{dn}{dt} = -An^2 \quad \text{or} \quad \frac{dn}{n^2} = -A\,dt$$

On integration, we get

$$\int \frac{dn}{n^2} = -A\int dt \quad \text{or} \quad \frac{n^{-2+1}}{-2+1} = -At - \text{constant}$$

or $$\frac{1}{n} = At + \frac{1}{n_0}(\text{constant}) \tag{v}$$

since at $t = 0$, $n = n_0$ (numbers of electrons at time $t = 0$). If light is turned off, i.e., the light falling on the specimen is stopped and generation of the electrons stops so that the electron concentration drops to $\frac{1}{2}n_0$ in time t_0, then

$$\frac{1}{n_0/2} = At_0 + \frac{1}{n_0}$$

or $$t_0 = \frac{1}{An_0} = \frac{1}{A(L/A)^{1/2}} = \frac{1}{(AL)^{1/2}} = (AL)^{-1/2}$$

or $$t_0 = \frac{1}{(AL)^{1/2}} \times \frac{L}{L} = \left(\frac{L}{A}\right)^{1/2}\frac{1}{L} = \frac{n_0}{L} \quad \text{or} \quad t_0 = \frac{1}{An_0} = (AL)^{-1/2} = \frac{n_0}{L} \tag{vi}$$

since $\sigma = n_0 e\mu_e$ or $n_0 = \dfrac{\sigma}{e\mu_e}$ Eq. (vi) becomes

$$t_0 = \frac{n_0}{L} = \frac{\sigma}{e\mu_e L} \tag{vii}$$

where time t_0 is known as the response time. From equation (vii) it is clear that the response time is directly proportional to the photoconductivity at a given light level L.

18.9.1 Sensitivity or Gain Factor

Sensitivity is the ratio of number of carriers crossing the specimen to the number of photons absorbed by the specimen. i.e.,

$$G = \frac{\text{Particle flux}}{\text{Ld}} = \frac{J_n}{\text{Ld}}$$

$$J_n = \frac{J}{e} = \frac{n_0\mu_e V}{d} = \left(\frac{L}{A}\right)^{1/2}\frac{\mu_e V}{d} \qquad \text{[using Eqs. (ii) and (iv)]}$$

The multiplication and division of Ld in the above expression leads to

$$J_n = \frac{\mu_e V}{(LA)^{1/2}d^2}(\text{Ld})$$

$\therefore$ $$G = \frac{J_n}{\text{Ld}} = \frac{\mu_e V}{(LA)^{1/2}d^2} \tag{viii}$$

The gain factor is also expressed in terms of transit time (T_d) of an electron between the electrodes and life time T_e of the electron before recombination. Then the gain factor is defined as

$$G = T_e/T_d \tag{ix}$$

A comparison of Eq. (ix) with Eq. (viii) yields

$$T_e = (LA)^{-1/2}$$

$$T_d = d^2/\mu_e V$$

18.10 EFFECT OF TRAPS

The trap is an energy level in the forbidden energy gap of specimen which is capable of capturing either an electron or a hole. The captured electron or hole may be re-emitted at any time and can further move to another trap (Fig. 18.14). There are two types of the traps which differ from each other according to their operation. One of the type helps recombining electron and holes for assisting to achieve thermal equilibrium. The other type of trap does not contribute directly in recombination process but they contribute indirectly. Let us consider a specimen of the crystal having N electron trap levels per unit volume at low temperature. In the present aspect, Eq. (i) is

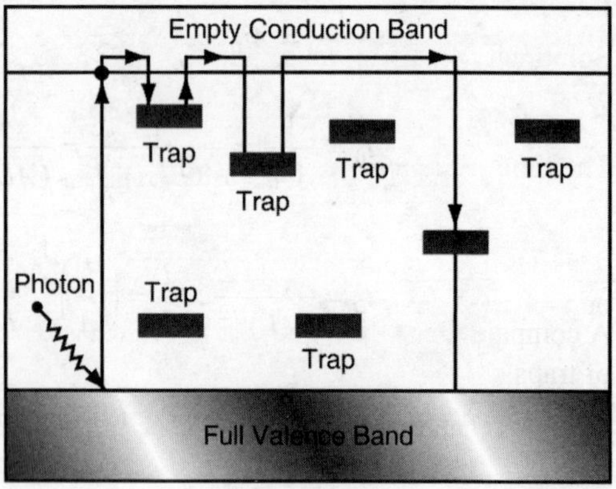

FIGURE 18.14

$$\frac{dn}{dt} = L - An(n+N) \tag{x}$$

where A is the recombination coefficient, which is the same for electron-hole recombination, and n is the electron concentration in the conduction band. Thermal ionisation of carriers from traps back into the conduction band contributes to the production rate through the term, say B_{n_1}. Hence Eq. (x) takes the form

$$\frac{dn}{dt} = L - An(n+N) + B_{n_1} \tag{xi}$$

In the steady state, we have

$$\frac{dn}{dt} = 0, n \to n_0 \quad \text{and} \quad B_{n_1} = 0. \text{ Therefore Eq. (xi) gives}$$

$$n_0(n_0 + N) = \left(\frac{L}{A}\right) \tag{xii}$$

18.10.1 Limiting Cases

Now there are two limiting cases to be discussed. At low current, the carrier concentration n_0 may be very much less than electron trap levels (N), i.e., $n_0 << N$.

Then from Eq. (xii), we have

$$n_0 = \frac{L}{NA} \tag{xiii}$$

If $n_0 >> N$, then from Eq. (xii) we get

$$n_0 = \left(\frac{L}{A}\right)^{1/2}$$

18.10.2 Response Time

The response time is related to the decay of carriers on switching off the light. This is the time in which the carrier concentration falls to $1/e$ time of its initial value n_0. This can be obtained by setting $L = 0$ in Eq. (x), which yields

$$\ln\frac{n+N}{n} - \ln\frac{n_0+N}{n_0} = NAt \qquad \text{(xiv)}$$

Solution of Eq. (xiv) under the limit of $N >> n_0$ is represented as

$$n = n_0 e^{-NAt} \qquad \text{(xv)}$$

The time for the photocurrent to fall to e^{-1} of its initial value can be obtained from Eq. (xv) as

$$t_0 = \frac{1}{NA} \qquad \text{(xvi)}$$

A comparison of this equation with Eq. (vi) obtained for the case of absence of traps yields that the presence of traps reduces the response time.

18.11 APPLICATIONS OF PHOTOCONDUCTIVITY

There are many applications of photoconductivity, some of which are listed below.

(i) This is used to measure the intensity of illumination.
(ii) This is used to control the street light.
(iii) This is used in exposure meters for cameras.
(iv) This is used for relay control.
(v) This is used in moving object counters.
(vi) This is used in voltage regulator or systems.
(vii) This is also used for burglar alarms.
(viii) The important application of photocell, a device based on photoconducting, is in the reproduction of sound in cinema, television and photography.
(ix) This is used in determination of opacity of solids and liquids.
(x) The temperature of celestial bodies is measured and their spectra are studied by using cells obtained based on the concept of photoconductivity.
(xi) This is used to control temperature in furnaces and chemical reactions.
(xii) Photoelectric cells are used in obtaining electrical energy from sunlight during space travel. Thus, the electric current is used to charge the battery.

SUMMARY

The topics covered in this chapter are summarised below.

- When N atoms of the solid are well–separated, then these atoms lead to N-fold degenerate levels of the solid. However, a continuously increasing interaction occurs between them when the atoms approach one another to form a solid. This causes each of the levels to split into N distinct levels. Since a solid contains about 10^{23} atoms per mole, i.e., N is very large, the splitted energy levels become so numerous and close together that they form an almost continuous energy band. From the quantum point of view, the wave functions of the valence electrons overlap and the overlapping of their wave functions results in splitting or spreading of their energy levels.
- Free electron model ignores the effects those arise when the electrons interact with crystal lattice. However, due to the effect of the periodic variation in the potential for an infinitely long lattice, the wave function does not remain sinusoidal travelling waves of constant amplitude but now they include the lattice periodicity in their amplitudes and electrons may be scattered by the lattice. In 1930, Kronig and Penney proposed a one-dimensional model by assuming the shape of rectangular potential wells and barriers along with the lattice periodicity. This is known as Kronig-Penney model.
- Kronig-Penney model tells about the discontinuities in energy when the wave number k satisfies the condition $k = n\pi/a$, where $n = \pm1, \pm2, \pm3,...$ etc. The E versus k diagram shows that an electron has allowed energy values in the region between $-\pi/a$ to $+\pi/a$. This region is called the first Brillouin zone. Again there is another allowed energy zone, which is observed after the forbidden gap and is extended from $-\pi/a$ to $-2\pi/a$ and π/a to $2\pi/a$. This zone is called second Brillouin zone.

  The concept of the Brillouin zones was extended to two-dimension by considering the wave number k to have the two components k_x and k_y, i.e., along the x-axis and y-axis, respectively. This way $k_x = k_y = \pm2\pi/a$ limits the first Brilluion zone. The second Brillouin zone for a two-dimensional lattice is represented by $k_x = \pm2\pi/a$ and $k_y = \pm2\pi/a$ excluding the space occupied by the first Brillouin zone.
- The concept of effective mass of an electron was introduced. This effective mass m^* is given by $\frac{1}{m^*} = \frac{4\pi^2}{h^2}\frac{d^2E}{dk^2}$, where E is the energy of the electron and k is the wave number.
- During the formation of bands in solids it is observed that there is an energy gap, called forbidden band. The forbidden gap represents energies which the electrons cannot occupy. Based on this energy gap and the conduction, the solids are classified into different categories named as insulators, semiconductors and conductors. A detailed description of these classes of the solids was given.
- The semiconductors in which the transformation of electrons to the conduction band and the generation of holes in the valence band are achieved purely by thermal excitation are called intrinsic semiconductors. During this effect equal numbers of electron and hole carriers are produced. The electrons and holes are called intrinsic charge carriers and the resulting conductivity is known as intrinsic conductivity. Here we obtained the electron concentration in the conduction band, hole concentration in valence band, concentration of charge carriers, the energy band diagram and the Fermi level.
- The conductivity of an intrinsic semiconductor can be increased significantly by adding certain impurities to it. By doing so, we get impurity semiconductor which is also known as extrinsic semiconductor. In extrinsic semiconductors, current carriers are introduced by donor and acceptor impurities with locked energy levels near the top or bottom of the forbidden gap.

The extrinsic semiconductors are of two types, namely n-type and p-type semiconductors. The concept of Fermi energy was discussed in terms of energy diagram for both these types of the semiconductors.

- The effect of temperature was discussed on both n-type and p-type semiconductors. Since all the donors donate their free electrons at room temperature, the additional thermal energy increases the generation of electron-hole pairs. Thus the concentration of minority charge carriers increases. A temperature is ultimately reached when the number of covalent bonds broken is very large such that the number of holes and electrons is almost equal. The extrinsic semiconductor then behaves like an intrinsic semiconductor, although its conductivity is higher. Thus with an increase in the temperature of an extrinsic semiconductor, it behaves almost intrinsically.
- If a current carrying conductor is placed in a transverse magnetic field, a potential is developed in the conductor in the direction perpendicular to both the current and magnetic field. This phenomenon is known as Hall effect.
- Under Hall effect, with the application of magnetic field B, the accumulation of charge on the surfaces of the specimen continues until the force on moving charges due to electric field associated with the accumulated charge itself is large enough to cancel the force exerted by the magnetic field. So a steady state condition is achieved. In this state, the value of electric field is called Hall electric field and the potential difference so developed is called Hall voltage. The Hall voltage V_H is given by $V_H = -\frac{JBd}{ne}$, where J is the current density, d is the width of the specimen (strip), n is the number of electrons per unit volume and e is the electronic charge. The coefficient $-\frac{1}{ne}$ is called the Hall coefficient R_H.
- Photoconductivity is an electrical phenomenon in which a material becomes more conductive due to the absorption of light radiation. When the energy of the incident radiation is higher than the energy gap E_g between conduction band and valence band, the electron–hole pairs are produced in the crystal. The electrons are in the conduction band and the holes are in the valence band of the crystal. These electron–hole pairs are the carriers of electrical conductivity.
- Gudden and Pohl discovered some basic experimental facts of the phenomenon of photoconductivity. These are listed as follows.

  (i) For a given material, the absorption of light and the excitation of photoconductivity by the light have a similar dependence on the wavelength of light.

  (ii) The region of photoresistivity gets extended to longer wavelengths in the presence of impurities.
- Simple model of photoconductor was discussed in detail by considering that the electron–hole pairs are produced throughout the volume of the crystal when the light radiation falls on the crystal. In this context, the response time t_0 was defined as the time in which the electron concentration is dropped to half of the number of electrons in the steady state, when the light falling on the specimen is stopped. The response time t_0 is given by $t_0 = \frac{\sigma}{e\mu_e L}$, where σ is the conductivity, μ_e is the mobility of the electrons and L is the number of photons absorbed by the crystal.
- We defined the sensitivity or the gain factor as the ratio of carriers crossing the specimen to the number of photons absorbed by the specimen. The gain factor is also expressed in terms of transit time T_d of an electron between the electrodes and the life time T_e of the electron before recombination. The gain factor G is equal to T_e/T_d.

- The presence of impurities produces discrete energy levels in the forbidden gap. They are known as traps. It means the trap is an energy level in the forbidden energy gap of specimen, which is capable of capturing either an electron or a hole. The captured electron or hole may be re-emitted at any time and can further move to another trap.

SOLVED EXAMPLES

EXAMPLE 1 Consider two-dimensional square lattice of side 3.0 Å. At what electron momentum values do the sides of first Brillouin zone appear? What is the energy of free electron with this momentum?

SOLUTION Given $a = 3.0 \times 10^{-10}$ m.

Formula used for momentum of electron

$$p = \hbar k$$

For first Brillouin zone $k = \pm\dfrac{\pi}{a}$, then

$$p = \frac{h}{2\pi}\frac{\pi}{a} = \frac{h}{2a} = \frac{6.62 \times 10^{-34}}{2 \times 3 \times 10^{-10}}$$

$$= 1.1 \times 10^{-24} \text{ kg m/sec}$$

$$\text{Energy } E = \frac{p^2}{2m} = \frac{(1.1 \times 10^{-24})^2}{2 \times 9.1 \times 10^{-31}}$$

$$= 6.648 \times 10^{-19} \text{ J}$$

$$= \mathbf{4.155 \text{ eV}}$$

EXAMPLE 2 Find the position of Fermi level E_F at room temperature (= 27°C) for germanium crystal having 5×10^{22} atoms/m^3.

SOLUTION Given $T = 27°\text{C} = 300$ K and $n_C = 5 \times 10^{22}$ per m^3

Formula used is

$$n_C = 2\left(\frac{2\pi mkT}{h^2}\right)^{3/2} e^{(E_F - E_C)/kT}$$

$$e^{(E_F - E_C)/kT} = \frac{n_C}{2\left(\dfrac{2\pi mkT}{h^2}\right)^{3/2}}$$

$$e^{(E_F - E_C)/kT} = \frac{5 \times 10^{22}}{2\left[\dfrac{2 \times 3.14 \times 9.1 \times 10^{-31} \times 1.381 \times 10^{-23} \times 300}{(6.62 \times 10^{-34})^2}\right]^{3/2}}$$

$$= \frac{5 \times 10^{22}}{25.115 \times 10^{24}}$$

$$e^{-(E_c-E_F)/kT} = 0.1991\times10^{-2}$$

$$e^{(E_C-E_F)/kT} = 502.296 \quad \text{or} \quad E_C - E_F = \ln 502.296$$

$$\frac{E_C-E_F}{kT} = \mathbf{6.2192} \quad \text{or} \quad E_C - E_F = \mathbf{0.161\ eV}$$

EXAMPLE 3 Consider the Fermi 0.3 eV below the conduction band at room temperature (=27°C) in an *n*-type semiconductor. If the temperature is raised to 57°C, what would be the new position of Fermi level?

SOLUTION Given $E_C - E_F = 0.3$ eV, $T_1 = 27°C = 300$ K and $T_2 = 57°C = 330$ K.

Formula used is

$$E_F = E_C - kT \ln\left(\frac{N_c}{N_d}\right)$$

$$E_C - E_F = kT \ln\left(\frac{N_c}{N_d}\right)$$

At temperature T_1, $E_C - E_F = kT_1 \ln\left(\frac{N_c}{N_d}\right)$

At temperature T_2, $E_C - E'_F = kT_2 \ln\left(\frac{N_c}{N_d}\right)$

$$\frac{E_C - E'_F}{E_C - E_F} = \frac{T_2}{T_1}$$

or

$$E_C - E'_F = \frac{T_2}{T_1}(E_C - E_F)$$

$$= \frac{330}{300}\times 0.3 \text{ eV}$$

$$= \mathbf{0.33\ eV}$$

EXAMPLE 4 For an intrinsic semiconductor having band gap $E_g = 0.7$ eV, calculate the density of holes and electrons at room temperature (= 27°C).

SOLUTION Given $E_g = 0.7$ eV.

In intrinsic semiconductor, the concentration of electrons and holes are same. So

$$n_e = n_h = 2\left[\frac{2\pi kTm}{h^2}\right]^{3/2} e^{(E_F-E_C)/kT}$$

The Fermi level lies exactly in the middle of condition and valence band.

i.e.

$$E_F = \frac{E_C + E_V}{2}$$

∴

$$E_F - E_C = \frac{E_C + E_V}{2} - E_C = -\frac{(E_C - E_V)}{2} = \frac{-E_g}{2}$$

$$\therefore \qquad n_e = n_h = 2\left[\frac{2\pi kTm}{h^2}\right]^{3/2} e^{-E_g/2kT}$$

$$= 2\times\left[\frac{2\times 3.14\times 1.38\times 10^{-23}\times 300\times 9.1\times 10^{-31}}{6.662\times 10^{-34}}\right]^{3/2} e^{-\left[\frac{0.7}{2\times 0.026}\right]}$$

$$= \mathbf{3.6\times 10^{19}\ per\ m^3}$$

Example 5 Assuming that there are 5×10^{28} atoms/m^3 in copper, find the Hall coefficient.

Solution Given $n = 5 \times 10^{28}$ atoms /m^3.

Formula used is

$$R_H = -\frac{1}{ne} = \frac{1}{5\times 10^{28}\times 1.6\times 10^{-19}}$$

$$= \mathbf{-\,0.125\times 10^{-9}\ m^3/C}$$

Example 6 Using free electron model, find the Hall coefficient of sodium assuming bcc structure for *Na* of cell side 4.28 Å.

Solution Given $a = 42.8 \times 10^{-10}$m.

Unit cell of sodium atom (Na) of volume a^3 has 2 atoms, i.e.,

$$n = 2\frac{1}{a^3} = \frac{2}{(4.28\times 10^{-10})^3} = 2.55\times 10^{28} \text{ per m}^3$$

Hall coefficient $R_H = -\dfrac{1}{ne} = \dfrac{-1}{2.551\times 10^{28}\times 1.6\times 10^{-19}}$

$$= \mathbf{-\,0.245\times 10^{-9}\ m^3/C}$$

Objective Type Questions

Q.1 The energy eigen value in a free electron model is given by

(a) $E = \dfrac{h^2}{8\pi^2 k^2 m}$ (b) $E = \dfrac{h^2 k^2}{2m}$ (c) $E = \dfrac{\hbar^2 k^2}{2m}$ (d) $E = 2m\hbar^2 k^2$

Q.2 The first Brillouin zone is defined between the region

(a) $k = 0$ to $\dfrac{\pi}{a}$ (b) $-\dfrac{2\pi}{a}$ to $\dfrac{\pi}{a}$

(c) $k = -\dfrac{\pi}{a}$ to $\dfrac{2\pi}{a}$ (d) $-\dfrac{\pi}{a}$ to $\dfrac{\pi}{a}$

Q.3 The second Brillouin zone is defined between the region

(a) $k = \dfrac{-\pi}{a}$ to $\dfrac{+\pi}{a}$ (b) $k = \dfrac{-2\pi}{a}$ to $\dfrac{-\pi}{a}$ and $\dfrac{\pi}{a}$ to $\dfrac{2\pi}{a}$

(c) $k = \dfrac{-\pi}{a}$ to $\dfrac{2\pi}{a}$ (d) $k = \dfrac{-2\pi}{a}$ to $\dfrac{2\pi}{a}$

Q.4 Bragg's conditions for diffraction is given by the relation

(a) $2 \sin \theta = n\lambda$ (b) $2a \sin \theta = \lambda$

(c) $2a \sin \theta = n\lambda$ (d) $a \sin \theta = n\lambda$

Q.5 Which one of the following relations is correct for effective mass of an electron

(a) $m^* = \frac{h^2}{4\pi^2}\frac{d^2E}{dk^2}$ (b) $m^* = \frac{h^2}{4\pi^2}\left[\frac{dk^2}{d^2E}\right]^{-1}$

(c) $m^* = h^2\left[\frac{d^2E}{dk^2}\right]^{-1}$ (d) $m^* = \frac{h^2}{4\pi^2}\left[\frac{d^2E}{dk^2}\right]^{-1}$

Q.6 When we increase the temperature of extrinsic semiconductor, after a certain temperature it behaves like

(a) an insulator (b) an intrinsic semiconductor

(c) a conductor (d) a superconductor

Q.7 Pure semiconductor behaves as an insulator at

(a) 273 K (b) –273°C (c) 373K (d) none of these

Q.8 The group velocity of the electron in one–dimensional lattice is defined as

(a) $v_g = \frac{dk}{d\omega}$ (b) $v_g = \frac{d\omega}{dk}$ (c) $v_g = \frac{\omega}{k}$ (d) $v_g = \omega k$

Q.9 Hall coefficient is given by the relation

(a) $R_H = -neJ$ (b) $R_H = \frac{J}{ne}$ (c) $R_H = -\frac{1}{ne}$ (d) $R_H = -\frac{-J}{ne}$

Q.10 Which type of semiconductor material has negative Hall coefficient

(a) *p*-type (b) *n*-type (c) intrinsic (d) none of these

Q.11 The electrical conductivity of pure semiconductor at 0K is

(a) zero (b) large (c) finite (d) none of these

Q.12 The electrical current in an intrinsic semiconductor is because of

(a) electrons in conduction band (b) holes in valance band

(c) both (a) and (b) (d) none of these

Q.13 The majority carriers in *n*-type semiconductors are

(a) electrons in conduction band (b) holes in valence band

(c) both (a) and (b) (d) none of these

Q.14 By doping fifth group elements in germanium crystal, which type of semiconductor material is obtained

(a) positive (b) negative (c) neutral (d) none of these

Q.15 In an *n*-type semiconductor the Fermi-level lies

(a) above the top of the valence band (b) below the bottom of the conduction band

(c) In the middle of the forbidden gap (d) near the conduction band

Q.16 On increase of temperature, the Fermi level shifts upwards in

(a) *p*-type semiconductor (b) *n*-type semiconductor

(c) intrinsic semiconductor (d) none of these

Q.17 The concentration of electrons in the conduction band of an intrinsic semiconductor is proportional to
(a) T (b) T^2 (c) $T^{3/2}$ (d) T^3

Q.18 The energy gap between the valence and conduction bands in a semiconductor is of the order of
(a) 26 eV (b) 1.0 eV (c) 7.0 eV (d) 0.001 eV

Q.19 Which of the following relation is correct for product of electron-hole concentration (np)
(a) $AT^3e^{-Eg/KT}$ (b) $AT^5e^{-Eg/KT}$ (c) $AT^7e^{-Eg/KT}$ (d) none of these

Q.20 The product of electron-hole concentration (np) changes with
(a) temperature (b) doping concentration
(c) pressure (d) none of these

Q.21 Which one of the following is correct for Kronig-Penney model
(a) real model (b) approximate model
(c) virtual model (d) none of these

Q.22 Which of the following theory proposes about the free electron inside periodic lattice
(a) zone theory (b) quantum theory of free electron
(c) classical theory of free electron (d) none of these

Q.23 In the absence of potential, per Kronig-Penney model
(a) forbidden energy regions are not there (b) forbidden energy regions are there
(c) all values of energy are allowed (d) none of these

Q.24 In the absence of potential barrier, the *E-k* curve is
(a) parabola with discontinuities (b) continuous parabola
(c) discontinuous energy levels (d) none of these

Q.25 The discontinuities occur in *E-k* curve at
(a) $k = \pm\frac{n\pi}{a^2}$ (b) $k = \pm\frac{n\pi}{a}$ (c) $k = \pm\frac{a}{n\pi}$ (d) none of these

Q.26 The effective mass of an electron may be
(a) positive (b) negative (c) infinity (d) all of these

Q.27 The Hall-effect is used in determining
(a) mobility of charged carriers (b) density of charged carriers in extrinsic semiconductor
(c) type of extrinsic semiconductor (d) all of these

Q.28 In the phenomenon of photoconductivity a material becomes
(a) more conductive (b) more resistive
(c) less conductive (d) none of these

Q.29 The phenomenon of decrease of resistivity of an insulator when light radiation falls on it, is called
(a) photoelectric effect (b) Frenkel effect
(c) Compton effect (d) photoconductivity

Q.30 The photoconductivity is mainly due to
(a) extrinsic excitations (b) intrinsic excitations
(c) both (a) and (b) (d) none of these

Q.31 The minimum energy which is required for intra excitation is
(a) greater than the forbidden gap E_g (b) less than the forbidden gap E_g
(c) equal to the forbidden gap E_g (d) none of these

Q.32 Which one of the following statements is not correct for photoconductive cells?
(a) they are made of a single photoresistive material
(b) they are also called photoresistors
(c) they have a forward biased *p–n* junction
(d) they have a high dark-to-light resistance ratio

Q.33 The photocurrent density of a photoconductor is given by the relation
(a) $J + \dfrac{d}{n_0 e\mu_e V}$ (b) $J \dfrac{n_0 e\mu_e}{dV}$ (c) $J = \dfrac{n_0 e\mu_e V}{d}$ (d) $J = n_0 e\mu_e V d$

Q.34 The presence of traps in a photoconductor has the following effect
(a) it reduces the response time (b) it increases the response time
(c) it has no effect on response time (d) it has no relevance with photoconductor

Q.35 Which one of the following relations is correct for response time of a photoconductor
(a) $t_0 = \sigma e\mu_e L$ (b) $t_\theta = \dfrac{\sigma}{e\mu_e L}$ (c) $t_\theta = \dfrac{1}{\sigma e\mu_e L}$ (d) $t_0 = \dfrac{\sigma\mu e}{eL}$

Q.36 In a photoconductive cell, the internal resistance changes with a change in
(a) frequency of light (b) intensity of light
(c) both of these (d) none of these

Q.37 Photoconductor is also known as
(a) photovoltaic cell (b) solar cell (c) photoresistor (d) photodiode

PRACTICE PROBLEMS

General Questions

Q.1 Explain how the atomic energy levels split into bands when a number of atoms are brought close together to form a crystal?

Q.2 Discuss Kronig-Penney model. Using the model show the energy spectrum of electron consisting of a number of allowed energy bands separated by forbidden bands.

Q.3 What is the effect of periodic potential on the energy of electrons in a metal? Explain it on the basis of Kronig-Penney model and explain the formation of energy bands.

Q.4 Discuss how the concept of bands was originated in solids. Give necessary theory. What is *E-K* diagram and what do you infer from them?

Q.5 What are Brillouin zones? Explain using *E-K* diagrams.

Q.6 Discuss the formation of Brillouin zones for (i) linear lattice (ii) two–dimensional lattice.

Q.7 Define (m^*) and prove that effective mass of an electron $m^* = \dfrac{\hbar^2}{d^2E/dk^2}$

Give the physical basis of effective mass and explain its physical significance.

Q.8 Write notes on following
(a) Intrinsic and extrinsic semiconductors
(b) Effective mass

Q.9 Explain origin of bands in solids.

Q.10 Write short note on band theory of solids.

Q.11 Discuss the band theory of solids and explain the formation of bands and concept of holes.

Q.12 Discuss the motion of electrons in a periodic field of a crystal and show that the effective mass of an electron in a crystal is inversely proportional to the second derivative of *E-K* curve. Under what conditions, the effective mass of an electron can become *+ve*, *–ve* and infinity?

Q.13 Based on band theory of solids, distinguish between conductors, semiconductors and insulators.

Q.14 Name three semiconductors along with the value of band gaps.

Q.15 Derive an expression for the carrier concentration in an extrinsic semiconductor. What would be the position of Fermi level? Explain.

Q.16 Write short notes on intrinsic and extrinsic semiconductors.

Q.17 Show that Hall coefficient is independent of the applied magnetic field and is inversely proportional to the current density and electronic charge.

Q.18 What is Hall effect? Give an experimental method of calculating concentration and type of charges in a given semiconductor.

Q.19 Write short note on Hall effect and its application.

Q.20 Explain how Hall effect helps in determining the sign of charge carrier in a material. Also prove that $\phi = \mu H$ where ϕ is the Hall angle, μ is mobility of charge carrier and H is the applied magnetic field.

Q.21 What is photoconductivity?

Q.22 State the principle of photoconductive cell. Describe its construction, working and uses. Show the illumination characteristics and spectral response.

Q.23 What are traps? Discuss the effect of traps on photoconductivity of a photoconductor.

Q.24 What are traps? Discuss a simple model to show the effect of traps on the photoconductivity.

Q.25 Why does the electrical conductivity increase when certain solids are exposed to light of suitable wavelengths? Suggest simple model of a photoconductor and explain the following (a) gain, (b) response time, (c) effect of traps.

Q.26 Explain with proper diagram the working of a simple model of photoconductor.

Q.27 Discuss sensitivity or gain factor in detail.

Magnetic Properties of Solids

19

Learning Objectives

After reading this chapter you will be able to

- **LO 1** Define magnetic susceptibility, relative permeability and magnetic moment of an electron
- **LO 2** Discuss classification of magnetic materials
- **LO 3** Compare properties of paramagnetic, diamagnetic and ferromagnetic materials
- **LO 4** Explain classical theory of diamagnetism and paramagnetism
- **LO 5** Know about classical theory of ferromagnetism
- **LO 6** Discuss hysteresis, energy loss due to hysteresis, and importance of hysteresis curve
- **LO 7** Learn about magnetic circuits
- **LO 5** Analyse magnetic materials and their applications

Introduction

All materials, i.e., metals, semiconductors and insulators, reveal the phenomenon of magnetism. Magnetic materials play an important role in modern technology as they are frequently used in industrial electronics, computer industry, etc. The traditional methods of information storage and retrieval are rapidly being replaced by magnetic storage. The magnetism of materials is mainly an outcome of the interactions of magnetic moments of their constituent atoms or molecules. Magnetic materials can be classified into three categories on the basis of their permeability or susceptibility. The magnetic materials for which susceptibility χ_m is negative (permeability $\mu_r \leq 1$) are said to be *diamagnetic materials*, whereas the materials with positive susceptibility (permeability $\mu_r \geq 1$) are said to be *paramagnetic materials*. If the susceptibility is much larger than zero and permeability $\mu_r >> 1$, then the magnetic materials are called *ferromagnetic materials*. Paramagnetic and diamagnetic materials have a linear relationship between $\vec{B}$ and $\vec{H}$. However, the relationship between $\vec{B}$ and $\vec{H}$ is nonlinear for ferromagnetic materials. A material is said to be nonmagnetic if susceptibility $\chi_m = 0$ (or $\mu_r = 1$), it is magnetic otherwise. Depending on the alignment of magnetic moments within the materials, these are further classified into five important groups, namely, diamagnetic, paramagnetic, ferromagnetic, anti-ferromagnetic and ferrimagnetic materials. Since diamagnetism, paramagnetism and antiferromagnetism are weak effects, the materials which exhibit these phenomena are known to be nonmagnetic. However, ferromagnetism and ferrrimagnetism are very strong effects. Therefore, in a large number of devices, these two magnetic phenomena are prominently utilised.

The magnetic materials are of two types, namely, soft materials and hard materials. Soft magnetic materials are used in ac applications, since they are easily magnetised and demagnetised. However, hard magnetic materials are used in producing permanent magnets, since they retain magnetism on a permanent basis. Due to such properties, these materials are significantly used in information storage devices. In order to realise the operating principles of different magnetic devices, it is essential to understand the magnetic phenomena. So at first we define various terms, viz., intensity of magnetisation, magnetic susceptibility, relative permeability, etc. Magnetic flux density $\vec{B}$ and magnetic field strength $\vec{H}$ have already been discussed in detail in Chapter 10.

Intensity of Magnetisation (I)

It is defined as the magnetic moment per unit volume of the magnetised substance

$$I = \frac{M}{V}$$

which for a substance of length $2l$ and cross-sectional area a becomes

$$I = \frac{m \times 2l}{a \times 2l} = \frac{m}{a}$$

Thus, it can also be defined as pole-strength per unit area of cross-section. The intensity of magnetisation is sometimes represented by M. In that case, another symbol is used for the magnetic moment.

Magnetic Susceptibility (χ_m)

It is the ratio of the magnetic moment per unit volume (I) to the magnetic field strength (H) of the magnetising field.

$$\chi_m = \frac{I}{H}$$

It is positive for a paramagnetic material and negative for a diamagnetic one.

Relative Permeability (μ_r)

It is the ratio of the magnetic permeability (μ) of the substance to the permeability of the free space (μ_0).

$$\mu_r = \frac{\mu}{\mu_0}$$

This can also be defined as the ratio of the magnetic flux density produced in the medium to that which would be produced in a vacuum by the same magnetising force.

Relation between Permeability (μ_r) and Magnetic Susceptibility (χ_m)

As discussed earlier, the magnetic flux density B can be written in terms of the magnetic field strength H and the intensity of magnetisation I as

$$B = \mu_0(H + I) \quad \text{(i)}$$

$$\chi_m = \frac{I}{H} \quad \text{(ii)}$$

Therefore,

$$B = \mu_0(H + I) = \mu_0 H\left(1 + \frac{I}{H}\right) = \mu_0 H(1 + \chi_m)$$

or $$\frac{B}{H} = \mu = \mu_0(1 + \chi_m)$$

$\therefore$ Relative permeability $\mu_r = \dfrac{\mu}{\mu_0} = 1 + \chi_m$

where μ_0 is the magnetic permeability of free space.

19.1 MAGNETIC MOMENT OF AN ELECTRON LO1

The magnetic properties of solids originate due to the motion of electrons. An atom is made up of a positively charged nucleus placed at the centre and negatively charged electrons which move around the nucleus. The orbital motion of each electron around the nucleus may be treated as a current loop. As a result, it sets up a magnetic field.

The current produced,

$$i = \frac{\text{charge}}{\text{time period}} = \frac{e}{T} \quad \text{(i)}$$

$$T = \frac{\text{distance}}{\text{velocity}} = \frac{2\pi r}{v} \quad \text{(ii)}$$

where v is the linear velocity of the electron in a circular orbit of radius r. Then,

$$i = \frac{e}{2\pi r/v} = \frac{ev}{2\pi r} \quad \text{(iii)}$$

Magnetic moment of a magnetic dipole is

$$M = iA = \frac{ev}{2\pi r}\pi r^2 = \frac{1}{2}evr \quad \text{(iv)}$$

According to Bohr, the magnetic moment can also be expressed as

$$mvr = \frac{nh}{2\pi}$$

or $$vr = \frac{nh}{2\pi m} \quad \text{(v)}$$

By putting the value of vr from Eq. (v) in Eq. (iv), we get

$$M = \frac{e}{2}\left(\frac{nh}{2\pi m}\right) = n\left[\frac{eh}{4\pi m}\right], \text{ where } n = 1, 2, 3, \ldots$$

The above relation gives the magnetic moment of an electron orbiting around a nucleus. The quantity $\frac{eh}{4\pi m}$ is called Bohr magneton, represented by μ_B.

19.2 CLASSIFICATION OF MAGNETIC MATERIALS LO2

Magnetic materials can be classified into five different categories, namely, diamagnetic materials, paramagnetic materials, ferromagentic materials, anti-ferromagnetic materials and ferrimagnetic materials. We discuss below all the materials in detail.

19.2.1 Diamagnetic Materials

On placing in an external magnetic field, the materials which acquire feeble magnetism in the direction opposite to that of the applied field are called *diamagnetic materials*. This property is found in the substances whose outermost orbit has an even number of electrons. Since the electrons have spins opposite to each other, the net magnetic moment of each atom is zero. The magnetism of diamagnetic materials is called *diamagnetism*. If these materials are brought close to the pole of a powerful electromagnet, they are repelled away from the magnet.

Examples of diamagnetic materials are bismuth, zinc, copper, silver, gold, lead, water, etc.

19.2.2 Paramagnetic Materials

On placing in an external magnetic field, the materials which acquire feeble magnetism in the direction of the applied field are called *paramagnetic materials,* and their magnetism is known as *paramagnetism*. This property is found in the substances whose outermost orbit has an odd number of electrons. The source of paramagnetism is the permanent magnetic moment possessed by the atoms of paramagnetic materials. If these substances are brought close to a pole of a powerful electromagnet, they get attracted towards the magnet.

Examples of paramagnetic materials are aluminium, odium, platinum, manganese, copper chloride, liquid oxygen, etc.

19.2.3 Ferromagnetic Materials

On placing in an external magnetic field, the materials which acquire strong magnetism in the direction of the applied field are called *ferromagnetic materials* and their magnetism is called *ferromagnetism*. This property is found in the substances which are generally like paramagnetic materials. These are strongly attracted by magnets.

Examples of ferromagnetic materials are iron, nickel, cobalt, magnetite (Fe_3O_4), etc.

19.2.4 Anti-ferromagnetic Materials

Anti-ferromagnetic substances are crystalline materials. In these materials, the dipole moments of the neighbouring dipoles are equal and opposite in orientation so that the net magnetisation vanishes. If they are placed in the magnetic field, they are feebly magnetised in the direction of the field. Such materials are called *anti-ferromagnetic materials* and their magnetism is called *anti-ferromagnetism*. Examples of anti-ferromagnetic materials are: MnO, FeO, CaO, NiO, MnO_4, MnS, etc. Susceptibility of these materials vary with temperature. It increases with increasing temperature and reaches a maximum at a particular temperature called the Neel temperature (T_N). Above this temperature, these materials behave like paramagnetic materials.

19.2.5 Ferrimagnetic Materials

If the spins of the atoms are such that there is a net magnetic moment in one direction, the materials are called *ferrimagnetic materials*. Examples of ferrimagnetic materials are ferrites which consist of mainly ferric oxide Fe_2O_3 combined with one or more oxides of divalent metals.

19.3 COMPARISON OF PROPERTIES OF PARAMAGNETIC, DIAMAGNETIC AND FERROMAGNETIC MATERIALS LO3

S.No.	*Paramagnetic Materials*	*Diamagnetic Materials*	*Ferromagnetic Materials*
1.	These materials show positive magnetic susceptibility (of the order of 10^{-6}).	These materials show negative susceptibility (of the order of 10^{-6}).	These materials show positive and high magnetic susceptibility ($\sim 10^6$).
2.	The relative permeability is slightly more than unity ($\mu_r > 1$).	μ_r is slightly less than unity ($\mu_r < 1$).	The μ_r for a ferromagnetic material is of the order of few thousands.
3.	The magnetic susceptibility is strongly dependent on temperature and varies inversely with temperature.	The magnetic susceptibility of diamagnetic materials is practically independent of temperature.	The magnetic susceptibility decreases with increase in temperature.
4.	When a bar of a paramagnetic material is suspended between the poles of a magnet, it stays parallel to the lines of force.	When a bar of these materials is suspended between the poles of a magnet, it stays parallel to the magnetic field.	When a bar of these materials is suspended between the poles of a magnet, it behaves like a paramagnetic material.
5.	If these materials are placed in a non-uniform field, they are attracted towards the stronger field.	If these materials are placed in a non-uniform field, they are attracted towards the weaker field.	These materials behave like paramagnetic substances, if placed in a non-uniform field.

19.4 CLASSICAL THEORY OF DIAMAGNETISM (LANGEVIN'S THEORY) LO4

The classical theory of diamagnetism was developed by the French Physicists Paul Langevin in 1905. We have discussed earlier that atoms of diamagnetic materials do not possess a permanent magnetic moment. When the material is placed in an external magnetic field, the orbital motion of electrons changes and hence atoms acquire an induced magnetic moment. The direction of the induced dipole will be opposite to that of the external magnetic field.

Suppose in the absence of an external magnetic field, an electron of mass m moves with linear velocity v_0 (or angular velocity ω_0) in a circular orbit of radius r (Fig. 19.1).

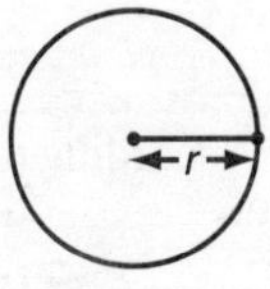

FIGURE 19.1

The magnetic moment equivalent to this motion is given by

$$M = iA \qquad \text{(i)}$$

By the definition of current,

$$i = \frac{e}{T} = \frac{e}{1/n} = ne, \text{ where } n \text{ is the frequency}$$

or $$i = \frac{\omega_0 e}{2\pi} \qquad \text{(ii)} \quad [\because \omega_0 = 2\pi n]$$

with the help of Eqs. (i) and (ii), we have

$$M = \frac{e\omega_0}{2\pi} A = \frac{e\omega_0}{2\pi} \pi r^2$$

or $$M = \frac{er^2\omega_0}{2} \qquad \text{(iii)}$$

Under the above condition (when $B=0$), the centripetal force on the electron is given by

$$F_c = \frac{mv_0^2}{r} = m\omega_0^2 r \qquad \text{(iv)}$$

Now, if an external magnetic field B is applied to the plane of the electron orbit normally Fig. 19.2, an additional magnetic force will act on the electron, given by

$$\vec{F}_m = -e(\vec{v} \times \vec{B}) \qquad \text{(v)}$$

$\Rightarrow$ $$F_m = -evB$$

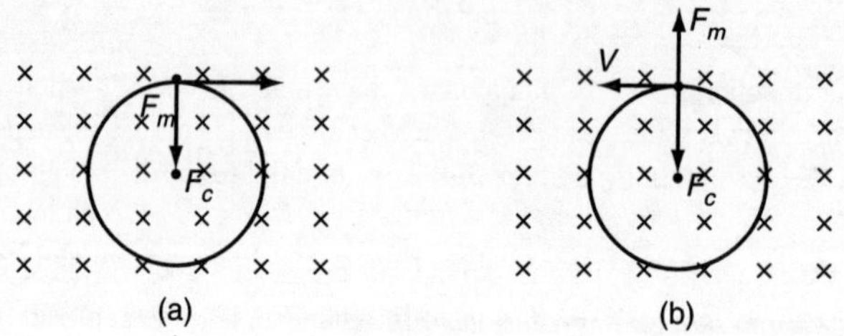

FIGURE 19.2

By Fleming's left hand rule, this magnetic force acts on the electron radially inward or outward if the electron moves in clockwise Fig. 19.2a or anti-clockwise Fig. 19.2b direction, respectively. Hence, the total force on the electron will be

$$\vec{F}_t = \vec{F}_c \pm \vec{F}_m$$

$$\vec{F}_t = m\omega_0^2 r \mp evB \qquad \text{(vi)}$$

Due to the magnetic force, the angular frequency changes from ω_0 to ω. This change in angular frequency can be calculated as under. By Faraday's law of induction

$$|\varepsilon| = -\frac{d\phi}{dt} \qquad \text{(vii)}$$

According to the definition of induced e.m.f.

$$\varepsilon = \oint \vec{E} \cdot d\vec{l} = E2\pi r$$

or $$E = \frac{\varepsilon}{2\pi r} \tag{viii}$$

where r is the radius of the orbit.

By Newton's second law of motion

$$F = m\frac{dv}{dt} = -eE \tag{ix}$$

By using Eqs. (viii) and (ix), we get

$$m\frac{dv}{dt} = -\frac{e}{2\pi r}\varepsilon \tag{x}$$

By using Eqs. (vii) and (x), we get

$$m\frac{dv}{dt} = -\frac{e}{2\pi r}\frac{d\phi}{dt}$$

or $$dv = -\frac{e}{2\pi rm}d\phi$$

or $$\Delta v = -\frac{e}{2\pi rm}\Delta\phi$$

As the external magnetic field changes from 0 to B, the corresponding flux changes from 0 to $\pi r^2 B$.

Then

$$\Delta v = -\frac{e}{2\pi rm}(\pi r^2 B - 0) = -\frac{e}{2\pi rm}\pi r^2 B$$

or $$\Delta v = -\frac{eBr}{2m} \tag{xi}$$

Therefore, the change in angular velocity will be

$$\Delta\omega = \frac{\Delta v}{r} = -\frac{eB}{2m} \tag{xii}$$

This change in angular frequency is also known as *Larmour frequency.*

On application of an external magnetic field, the angular frequency of the orbital electron gets changed which leads to a charge in magnetic moment. Therefore, in both the cases of electron rotating in clockwise or anti-clockwise directions, the magnetic moment of the orbital electron changes and hence it can be obtained as

$$\Delta M = \frac{er^2}{2}\Delta\omega \qquad \text{[Using Eq. (iii)]}$$

$$= \frac{er^2}{2}\left[-\frac{eB}{2m}\right]$$

$$\therefore \quad \Delta M = -\frac{e^2 B}{4m}r^2 \tag{xiii}$$

As seen above, the negative sign shows that the induced magnetic moment is always opposite to change in the magnetic field. In deriving the above equation, we have assumed that the orbit of the electron is normal to the applied field. But these orbits can have any orientation with the field. Since there are a number of randomly oriented electron's orbits in the atom that show the spherical symmetry, the total induced magnetic moment in the atom is given by

$$\Delta M = -\frac{Ze^2 B}{4m}\overline{r}^2$$

Here Z is the atomic number (i.e., the number of electrons in an atom) and $\vec{r}^2$ is the mean square radius of the electrons orbits. Let (x, y, z) be the coordinates of any point on the spherical orbit of radius r. Then

$$r^2 = x^2 + y^2 + z^2$$

Again consider $\overline{x}, \overline{y}, \overline{z}$ as the average values of components of radii for all the electrons along the three axes. Then

$$\vec{r}^2 = R^2 = \overline{x}^2 + \overline{y}^2 + \overline{z}^2$$

For the spherically symmetric charge distribution,

$$\overline{x}^2 = \overline{y}^2 = \overline{z}^2 = \frac{R^2}{3}$$

If the direction of the magnetic field is along the z-axis then considering the plane normal to the direction of magnetic field we get

$$\overline{r}^2 = \overline{x}^2 = \overline{y}^2 = \frac{2R^2}{3}$$

Therefore, the total induced magnetic moment in the atom will be

$$\Delta M = -\frac{Ze^2 B}{4m} \times \frac{2R^2}{3} \qquad \text{(xiv)}$$

$$\therefore \qquad \Delta M = -\frac{Ze^2 R^2 B}{6m}$$

If N is the number of atoms per unit volume of the substance, then the induced magnetic moment per unit volume is

$$I_{\text{induced}} = -\frac{NZe^2 R^2 B}{6m} \qquad \text{(xv)}$$

Since $B = \mu_0 H$,

$$I_{\text{induced}} = -\frac{NZe^2 R^2 \mu_0 H}{6m}$$

By definition, the magnetic susceptibility of a diamagnetic substance is

$$\chi_m = \frac{I_{\text{induced}}}{H} = -\frac{\mu_0 NZe^2 R^2}{6m} \qquad \text{(xvi)}$$

From the above equation, it is clear that the diamagnetic susceptibility is negative, which does not depend on the temperature and the intensity of the external magnetic field.

19.5 CLASSICAL THEORY OF PARAMAGNETISM (LANGEVIN'S THEORY) LO4

The classical theory of paramagnetism was developed by Langevin in 1905. In paramagnetic materials, each individual atom has a permanent magnetic moment. This is because of the fact that the magnetic moment of the individual electrons due to their orbital as well as spin motions do not cancel out. So an individual atom has a permanent dipole. This property is found in those substances whose outermost orbit has an odd number of electrons. In the absence of an external magnetic field, these atomic magnets are randomly oriented in the substance and the total magnetic moment of the substance is zero. When a magnetic field is applied, these atomic dipoles tend to line up in the direction of the external magnetic field but the temperature opposes this alignment.

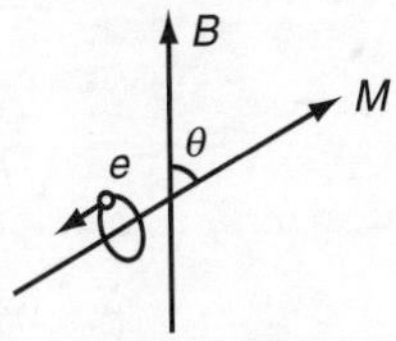

FIGURE 19.3

Let the paramagnetic substance contain N atoms per unit volume at a temperature $T(K)$. Each atom possesses a permanent magnetic moment M. The potential energy of an atomic magnetic dipole inclined at an angle θ with direction of external field, as shown in Fig. 19.3 is

$$U = -MB\cos\theta \tag{i}$$

According to Boltzmann's classical law, the number of atoms per unit volume in the energy ranging from U to $U + dU$ is given by

$$\begin{aligned} dN &= Ce^{-U/kT}\,dU \\ &= Ce^{MB\cos\theta/kT}\,MB\sin\theta\,d\theta \end{aligned} \tag{ii}$$

where k is the Boltzmann constant and C is the constant of proportionality depending on the atoms. Integrating Eq. (ii) for θ from 0 to π, we can find total number of atoms per unit volume of the substance as

$$\begin{aligned} N &= \int_0^\pi dN \\ &= C\int_0^\pi e^{MB\cos\theta/kT}\,MB\sin\theta\,d\theta \end{aligned} \tag{iii}$$

Thus,

$$C = \frac{N}{\int_0^\pi e^{MB\cos\theta/kT}\,MB\sin\theta\,d\theta} \tag{iv}$$

We know that M is the magnetic moment of each magnetic dipole that makes an angle θ with the direction of the external magnetic field. So, its component in the direction of external field will be $M\cos\theta$. Thus, resultant magnetic moment due to atoms along the external field will be $M\cos\theta\,dN$. Hence the total magnetic moment per unit volume of the substance (i.e., the intensity of magnetisation) is given by

$$I = \int_0^\pi M\cos\theta\,dN \tag{v}$$

By using Eqs. (ii) and (v), we get

$$I = \int_0^{\pi} M \cos\theta \, Ce^{MB\cos\theta/kT} MB \sin\theta \, d\theta$$

$$I = CM^2 B \int_0^{\pi} e^{MB\cos\theta/kT} \sin\theta \cos\theta \, d\theta \qquad \text{(vi)}$$

By putting the value of C in Eq. (vi) from Eq. (iv), we have

$$I = NM^2 B \frac{\int_0^{\pi} e^{MB\cos\theta/kT} \sin\theta \cos\theta \, d\theta}{\int_0^{\pi} e^{MB\cos\theta/kT} MB \sin\theta \, d\theta}$$

$$= NM \frac{\int_0^{\pi} e^{MB\cos\theta/kT} \sin\theta \cos\theta \, d\theta}{\int_0^{\pi} e^{MB\cos\theta/kT} \sin\theta \, d\theta} \qquad \text{(vii)}$$

By putting $\frac{MB}{kT} = x$, $\cos\theta = y$, so $-\sin\theta \, d\theta = dy$ in Eq. (vii), we have

$$I = NM \frac{-\int_{+1}^{-1} e^{xy} y \, dy}{-\int_{+1}^{-1} e^{xy} dy} = NM \frac{\int_{+1}^{-1} e^{xy} y \, dy}{\int_{+1}^{-1} e^{xy} dy} = NM \frac{\int_{-1}^{+1} e^{xy} y \, dy}{\int_{-1}^{+1} e^{xy} dy}$$

$$I = NM \frac{\left[\frac{ye^{xy}}{x} - \frac{e^{xy}}{x^2} \right]_{-1}^{+1}}{\left[\frac{e^{xy}}{x} \right]_{-1}^{+1}} = NM \left[\frac{(e^x + e^{-x})}{(e^x - e^{-x})} - \frac{1}{x} \right]$$

or

$$I = I_0 \left[\cot hx - \frac{1}{x} \right] \qquad \text{(viii)}$$

where $NM = I_0$ shows the saturation value of the intensity of magnetisation (I) when all the magnetic dipoles get aligned in the direction of the external magnetic field.

In Eq. (viii), the function $I/I_0 = \cot hx - 1/x$ is called the *Langevin function* and is represented by $L(x)$. The variation of $L(x)$ with x is shown in Fig. 19.4. From the figure, it clear that

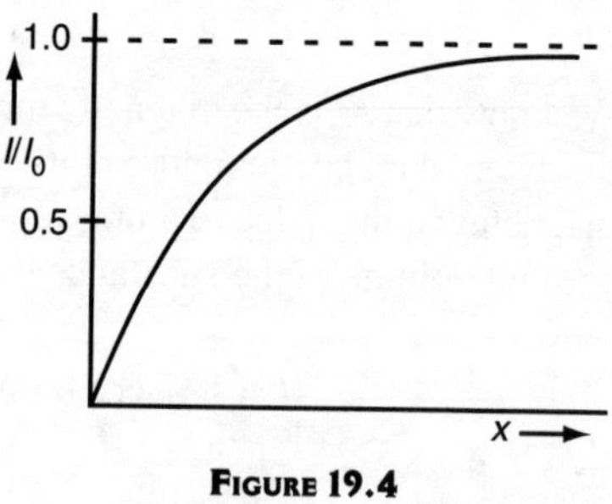

FIGURE 19.4

(i) If $x \left(= \frac{MB}{kT} \right)$ is large, i.e., temperature is very low, then

$$L(x) \approx \cot hx$$

or $\quad I = I_0$

Thus, for a very low temperature (or in the strong magnetic field) all the magnetic dipoles are aligned in the direction of external field and a saturation is obtained.

(ii) If $x\left(=\dfrac{MB}{kT}\right)$ is very small i.e., the temperature is very high or in other words, the external magnetic field is weak, then we have

$$I = I_0\left[\cot hx - \frac{1}{x}\right] = I_0\left[\frac{1}{x}\left(1+\frac{x^2}{3}\right)-\frac{1}{x}\right]$$

$$= I_0\frac{x}{3} = I_0\frac{MB}{3kT} = NM\frac{MB}{3kT}$$

$$I = \frac{NM^2}{3kT}B = \frac{NM^2}{3kT}\mu H \tag{ix}$$

$$\chi_m = \frac{I}{H} = \frac{\mu NM^2}{3kT} = \frac{\mu(NM)^2}{3NkT} \qquad [\because NM = I_0]$$

or

$$\chi_m = \frac{\mu I_0^2}{3NkT} \tag{x}$$

Eq. (x) is known as the *Curie law*, which can also be written as

$$\chi_m = \frac{C}{T}$$

where $C = \dfrac{\mu I_0^2}{3Nk}$ is known as *Curie constant.*

This equation shows that the magnetic susceptibility χ_m of a paramagnetic material depends on the temperature T and it varies inversely with T.

19.5.1 Failure of Langevin's Theory

This theory was not able to explain the complicated dependence of susceptibility on the temperature as shown by several paramagnetic substances. In view of this, Langevin's theory was further modified by Curie and Weiss.

19.5.2 Curie–Weiss Law

In 1907, Weiss had modified the Langevin's theory of paramagnetism. He assumed that in a paramagnetic substance an internal molecular magnetising field is generated because of mutual interaction between the atomic magnetic dipoles. If the molecular magnetising field (H_i) generated at any point because of a neighbouring atomic magnet is proportional to the intensity of magnetisation (I), then we have

$$H_i \propto I \quad \text{or} \quad H_i = \lambda I \tag{xi}$$

where λ is the molecular field coefficient and is independent of temperature. Hence, the effective magnetising field within the substance may be expressed as

$$H_{\text{effective}} = H + \lambda I \tag{xii}$$

With which Eq. (ix) becomes

$$I = \frac{NM^2 \mu H_{\text{effective}}}{3kT} = \frac{NM^2 \mu (H + \lambda I)}{3kT}$$

or $$I\left(1 - \frac{NM^2 \mu \lambda}{3kT}\right) = \frac{NM^2 \mu H}{3kT}$$

or $$I = \frac{NM^2 \mu H}{3kT - NM^2 \mu \lambda} \quad \text{(xiii)}$$

The magnetic susceptibility then becomes

$$\chi_m = \frac{I}{H} = \frac{NM^2 \mu}{3kT - NM^2 \mu \lambda} = \frac{NM^2 \mu}{3k\left[T - \dfrac{NM^2 \mu \lambda}{3k}\right]}$$

$$\chi_m = \frac{\dfrac{NM^2 \mu}{3k}}{T - \dfrac{NM^2 \mu \lambda}{3k}} = \frac{C}{T - \theta} \qquad \left[\because C = \frac{\mu I_0^2}{3Nk} = \frac{NM^2 \mu}{3k}\right] \quad \text{(xiv)}$$

where $$\theta = \frac{NM^2 \mu \lambda}{3k}$$

The relation (xiv) is called the Curie–Weiss law and the constant θ is known as *Curie temperature*. It is clear from the relation that if $T < \theta$, then the magnetic susceptibility of the paramagnetic substance becomes negative and it behaves like a diamagnetic substance. Hence, Curie–Weiss law is applicable only for temperatures $T > \theta$.

19.6 CLASSICAL THEORY OF FERROMAGNETISM

LO5

We know that each atom of the ferromagnetic materials such as iron, nickel, cobalt, etc. has a permanent magnetic moment like paramagnetic substances. The magnetic susceptibility of a ferromagnetic substance is a thousand times greater than that of a paramagnetic substance. In general, a specimen of a ferromagnetic substance contains a number of small regions called *domains*. These domains contain a large number of atoms, nearly 10^{17} to 10^{22}, and have the dimensions of about 10^{-6} cm^3 to 10^{-2} cm^3. Every domain is magnetically saturated and the direction of magnetisation in different domains is different.

If the external magnetic field is absent, all the magnetic domains in the substance are randomly oriented and their resultant magnetic moment in any direction will be zero. As per modern theory, it is assumed that in the absence of an external magnetic field, these domains form closed loops within the substance so that the net magnetic moment of the whole substance is zero, as shown in Fig. 19.5a.

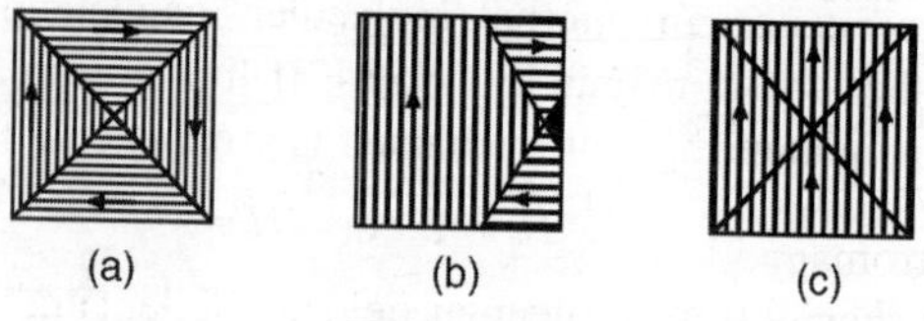

FIGURE 19.5

When a magnetic field is applied to the ferromagnetic substance, the substance becomes magnetised. The intensity of magnetisation of the substance depends upon the intensity of an external magnetic field.

When the substance is placed in a weak external magnetic field, the magnetisation produced is due to the displacement of boundaries of domains Fig. 19.5b and if the external magnetic field is strong, the magnetisation produced is mainly by the rotation of domains Fig. 19.5c. Fig. 19.6 represents the magnetisation curve for the ferromagnetic substance. In a very weak magnetic field, as represented in the part *OP* of the curve, the displacement of boundaries of domains is reversible and if we removed the external magnetic field, the boundaries of domains again come back to their original positions. If we increase the external magnetic field, as represented in the part *PQ* of the curve, the displacement of boundaries of domains is irreversible and the material immediately becomes magnetised. If we again increase the magnetic field, as represented in the part *QR* of the curve, the magnetisation of the substance is because of rotation of domains in the direction of magnetising field.

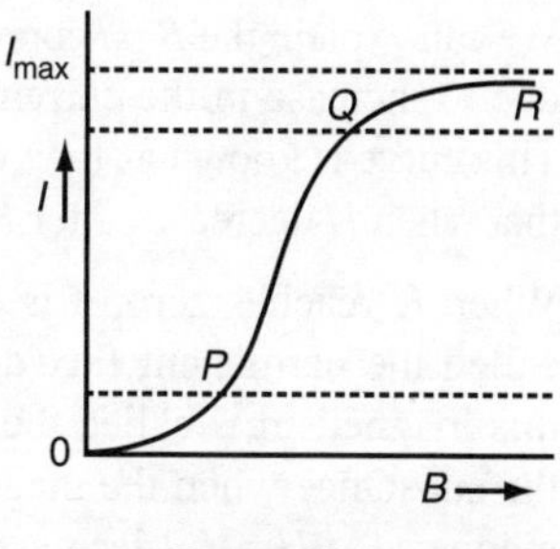

FIGURE 19.6

Thus, the net effective magnetic field of the ferromagnetic substance is given by

$$H_{\text{effective}} = H + H_i$$

where H_i is the magnetic field generated due to the mutual interaction between magnetic dipoles. By using Curie–Weiss law as represented in Eq. (xiv), the magnetic susceptibility of the ferromagnetic substance is

$$\chi_m = \frac{C}{T - \theta}$$

where *C* and θ are the Curie constant and the Curie temperature, respectively.

On the basis of the above relation, the following conclusions can be drawn.

(a) If $T = \theta$, the magnetic susceptibility will approach to infinity.

(b) If $T < \theta$, the magnetic susceptibility will be negative. In this condition, the Curie–Weiss law is not applicable because the ferromagnetic substance gets magnetised even in the absence of external magnetic field.

(c) If $T > \theta$, the magnetic susceptibility decreases with the increase in temperature. In this condition, the ferromagnetic properties disappear and the substance becomes paramagnetic.

19.7 HYSTERESIS: NONLINEAR RELATIONSHIP BETWEEN $\vec{B}$ AND $\vec{H}$ LO6

Ferromagnetic materials like iron and steel are used for screening (or shielding) that protect sensitive electrical devices from disturbances from strong magnetic fields. An example of an iron shield is the compass, which without shielding gives an erroneous reading due to the effect of external magnetic field. For perfect screening, it is required that the shield have infinite permeability ($\mu_r = \infty$).

Even though $\vec{B} = \mu_0(\vec{H} + \vec{I})$ holds good for all materials including ferromagnetic materials, the relationship between $\vec{B}$ and $\vec{H}$ depends on previous magnetisation of a ferromagnetic material, i.e., on its magnetic history. Instead of having a linear relationship between $\vec{B}$ and $\vec{H}$ (i.e., $\vec{B} = \mu\vec{H}$), it is only possible to represent the relationship by a magnetisation curve or a *B–H* curve, as shown in Fig. 19.7. At any point on this curve, μ is given by the ratio *B–H*, and not by the slope of the curve (*dB*/*dH*).

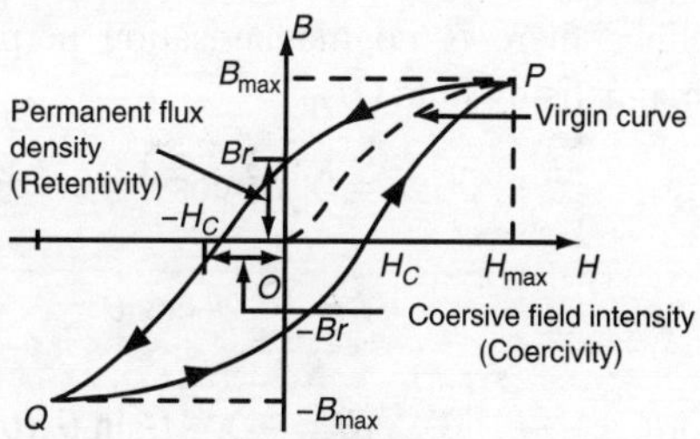

FIGURE 19.7

We can explain the B–H curve as follows. Initially, a ferromagnetic material is unmagnetised. As H is increased due to increase in the current from O to the maximum applied field intensity H_{max}, a curve OP is produced. This curve is known as the virgin or initial magnetisation curve. Now we move back and decrease H. It is seen that when H decreases after P, B does not follow the initial curve but lags behind H. This is called hysteresis.

When H reaches zero, it is obtained that $B \neq 0$, i.e., the material possesses some finite B. This finite B_r is called the permanent flux density or residual magnetism which depends on H_{max}. The power of retaining this magnetism is called the retentivity of the substance. It is a measure of the remaining magnetisation in the substance when the magnetising field is removed. The existence of B_r is the cause of having permanent magnets. At $H = H_c$ (decreased by reversing the current) $B = 0$. This value of H, i.e., H_c, is called the coercive field intensity or coercivity of the substance. It is a measure of the reverse magnetising field required to destroy the residual magnetism of the substance. Materials for which H_c is small are said to be magnetically hard. H_c also depends on H_{max}. Further increase in H to reach Q and in reverse direction to reach P gives a closed hysteresis loop. The shape of this loop varies from one material to another. Some ferrites have an almost rectangular hysteresis loop. These ferrites are used in digital computers as magnetic information storage devices. The area of the loop represents energy loss (hysteresis loss)/ unit volume during one cycle of the periodic magnetisation of the ferromagnetic material. This energy loss is in the form of heat. Therefore, it is desirable that materials used in electric generators, motors and transformers should have a tall but narrow hysteresis loop for minimal losses.

19.8 ENERGY LOSS DUE TO HYSTERESIS LO6

During the process of magnetisation, a loss of energy is always involved in aligning the domains in the direction of the applied magnetic field. When the direction of an external magnetic field is reversed, the absorbed energy is not completely recovered and rest energy in sample is lost in the form of heat. This loss of energy is called hysteresis loss.

19.8.1 Calculation of Hysteresis Loss

It can be proved that the energy lost per unit volume of the substance in a complete cycle of magnetisation is equal to the area of the hysteresis loop. We consider a unit volume of the ferromagnetic substance, which has N magnetic domains. Let M be the magnetic moment of each magnetic domain which makes an angle θ with the direction of the magnetising field H.

So, the total magnetic moment per unit volume in the direction of magnetising field

$$= \sum_N M \cos\theta$$

$$= \text{Intensity of magnetisation } (I) \tag{i}$$

Since there is no magnetisation perpendicular to H, the total magnetic moment perpendicular to the magnetising field (H),

$$= \sum_N M \cos\theta = 0 \tag{ii}$$

$$\therefore \quad I = \sum_N M \cos\theta$$

or

$$dI = -\sum_N M \sin\theta \, d\theta \tag{iii}$$

Here the negative sign shows that I decreases with increasing θ.

The work done in rotating the domain from this direction by an angle $d\theta$ towards the H is

$$dW = -\mu_0 \sum_N MH \sin\theta \, d\theta$$

$$= \mu_0 H dI \text{ [By using Eq. (iii)]}$$

The work done per unit volume of substance in the complete cycle is (iv)

$$W = \oint \mu_0 H \, dI$$

$$= \mu_0 \oint H \, dI$$

$$= \mu_0 \times (\text{area of } I\text{-}H \text{ loop}) \quad \text{(v)}$$

Hence, the work done per unit volume of the substance per cycle of magnetisation is equal to μ_0 times the area of I–H curve and this energy is lost in the form of heat.

19.8.1.1 Hysteresis Loss due to B–H curve

The magnetic flux density (B) in substance is due to the magnetising field (H) and the intensity of magnetisation I. They are related as

$$B = \mu_0(H + I) \quad \text{(vi)}$$

or $$dB = \mu_0(dH + dI)$$

or $$dI = \frac{1}{\mu_0} dB - dH \quad \text{(vii)}$$

From Eqs. (v) and (vii) we get

$$W = \oint H \, dB - \mu_0 \oint H \, dH \quad \text{(viii)}$$

The value of $\oint H \, dH$ will be zero, because the curve between H and H is a straight line and will not enclose any area, i.e.,

$$\oint H \, dH = 0$$

Then Eq. (viii), takes the form,

$$W = \oint H \, dB = \text{area of } B\text{–}H \text{ curve.}$$

19.9 IMPORTANCE OF HYSTERESIS CURVE LO6

The hysteresis curves (B–H loops or I–H loops) of a ferromagnetic material provide very useful information regarding the magnetic properties of a material. The size and shape of the hysteresis loop of ferromagnetic materials can be used to obtain information regarding retentivity, coercivity, susceptibility, permeability and loss in energy per cycle of magnetisation. On the basis of these properties of the magnetic materials, it is possible to choose a particular material for a particular purpose.

19.10 MAGNETIC CIRCUITS LO7

The concept of a magnetic circuit is based on solving some magnetic field problems using circuit approach. Transformers, motors, generators and relays are magnetic devices which may be considered as magnetic circuits. By exploiting an analogy between a magnetic circuit and an electric circuit, the analysis of such circuits is made simple. However, these two types of circuits have some differences. For example, magnetic flux does not flow unlike an electric circuit where the current I flows. Further, the permeability μ varies with the flux density $\vec{B}$ in a magnetic circuit, whereas the conductivity σ is independent of the current density $\vec{J}$ in an electric field. This is because ferromagnetic (nonlinear) materials are normally used in most practical magnetic devices.

19.11 FORCES ON MAGNETIC MATERIALS

Consider an electromagnet made of iron of constant relative permeability, whose coil has N turns and carries a current I, as shown in Fig. 19.8. Further, we take that the magnetic field in the air gap dl is the same as in iron ($B_{1n}=B_{2n}$).

Under the above situation, the attractive force between the two pieces of iron can be obtained as

$$F=-2\left(\frac{B^2S}{2\mu_0}\right)$$

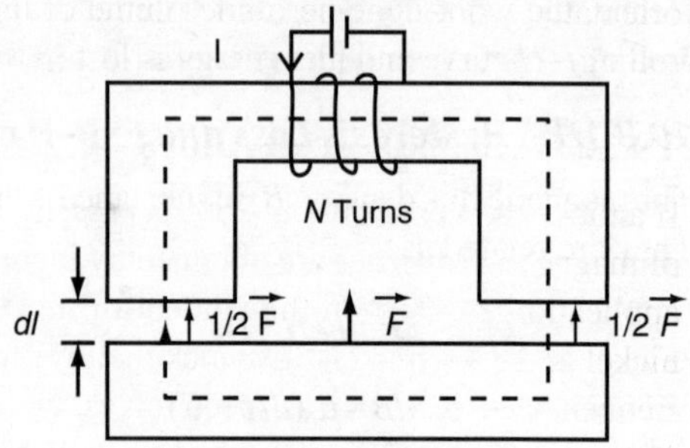

FIGURE 19.8

where S is the area of the coil. The force F is exerted on the lower piece and not on the current-carrying upper piece giving rise to the field. So the tractive force across a single gap would be

$$F=-\frac{B^2S}{2\mu_0}$$

where B is the magnetic field at the surface of the material. The tractive pressure is nothing but the energy density in the air, given by

$$P=\frac{B^2}{2\mu_0}=\frac{1}{2}BH$$

19.12 MAGNETIC MATERIALS AND THEIR APPLICATIONS LO8

Electrical devices like power transformers, motors, generators, electromagnets, etc., use soft magnetic materials. Electrical steels are used as core materials in them. For retaining magnetic fields of permanent magnets, hard magnetic materials are used in fabrications. Different types of magnetic materials are used for different applications.

19.12.1 Low-Carbon Steel

Pure iron, although has higher permeability, causes more eddy current losses due to its higher electrical conductivity. Low–carbon steel has relatively small permeability and higher resistivity. It is the lowest-grade core material.

19.12.2 Iron–Silicon Alloys

Adding of about 3–4% silicon to iron produces iron–silicon alloys with improved characteristics. Silicon increases the electrical resistivity of low-carbon steel and thus reduces the eddy current losses. It also increases the magnetic permeability and lowers hysteresis losses. It reduces the magnetorestriction and therefore reduces transformer noise. However, iron–silicon alloys are not useful for communication applications due to their low magnetic permeability at low fields, because in communication applications much higher permeabilities are required at low fields.

19.12.2.1 Grain Orientation

By using favourable grain orientation in the material, the hysteresis losses can be decreased and permeability can be substantially increased. The <100> direction is the easy direction in the case of iron crystals and spin moments in a virgin crystal are aligned along <100> directions. The <100> direction is parallel to the rolling direction when steel of iron alloys are manufactured by rolling and annealing. Thus, cold rolled grain orientated (CRGO) steel carries better magnetic properties in the same direction as that of the direction of rolling. Consequently, less material is required for cores.

19.12.3 Nickel–Iron Alloys

If a nickel content of about 25% is present, a pure nickel–iron alloy is practically nonmagnetic. Wide ranges of magnetic properties are obtained by increasing the nickel content. So, nickel–iron alloys are used for these applications. Based on the content of nickel, these alloys are divided into three groups: 36% nickel, 50% nickel and 77% nickel. 36% nickel alloys have high resistivity and low permeability and are used for high–frequency devices such as speed relays, wideband transformers and inductors. Having moderate permeability of about 25,000 and high saturation induction, the 50% nickel alloys are used where low loss and small size are required, such as in small motors, synchores, etc. The 79% nickel alloys have high permeability but lower saturation induction and are used in recording heads, pulse transformers, sensitive relays, etc.

19.12.4 Mumetal

Multicomponent nickel–iron alloys like permalloy, supermalloy, etc. have the highest permeabilities of the order of 10^5. Mumetal having a component of 77% nickel, 16% iron, 5% copper and 2% chromium can be rolled into thin sheets and is used to shield electronic equipment from stray magnetic fields.

19.12.5 Alnico Alloys

Alnico alloys containing Al, Ni, Co and Fe and minor constituents of Cu and Ti are used for making permanent magnets. These are characterised by a high energy product, a high remanent induction and a moderate coercivity. Besides being mechanically hard and breakable, magnetic properties of alnico alloys are highly stable against temperature variation, shock, etc. The properties are improved by heat treatments in alnico 2 or by cooling the alloy in magnetic field.

19.12.6 Other Alloys

Rare earth magnetic alloys like Sm–Co alloys are superior to alnico alloys in terms of magnetic properties and they have an energy product up to 2.4×10^5 J/m^3 and coercivities of about 3.2×10^6 A/m. They are used in medical devices such as thin motors in implantable pumps and valves.

Fe–Cr–Co alloys, which are similar to alnico alloys, are used in making permanent magnets for modern telephone receivers. Similarly, Nd–Fe–B magnetic alloys have a high energy product of the order of 3×10^5 J/m^3 and are used mainly in making light and compact electric motors.

19.12.7 Soft Ferrites

Due to high electrical properties of dielectrics combined with the magnetic properties of ferromagnetic materials, ferrites can be used for high frequency applications without eddy current losses. They also have high electrical resistance (10^5 to 10^{15} times the resistance of metallic ferromagnets). The soft ferrites are used for low signal, memory core, audiovisual and recording head applications. Major applications include deflection yoke cores, flyback transformers and convergence coils for television receivers. Mn–Zn ferrites are used for operations of up to 500 kHz whereas Ni–Zn ferrites are effective for the use for high frequency operation up to 100 MHz.

Mg–Mn ferrites, Mn–Cu ferrites and Li–Ni ferrites are used as memory or logic operation devices in computers, as switching devices, and in information storage. They are made in the form of tiny rings called cores, which are assembled into large matrix software containing cores at each junction. Microwave devices like modulators, couplers, circulators, phase shifters, matching devices are made using microwave ferrites, mainly manganese ferrite, nickel ferrite, cobalt ferrite, etc.

19.12.8 Hard Ferrites

Hard ferrites are also used in making permanent magnets. Barium ferrites (trade name Ferroxdure) are being replaced by strontium ferrites having superior magnetic properties. They find major applications in generator relays, loudspeakers, telephone ringers, toys, etc. Hard ferrite powders are often mixed with plastic materials to form flexible magnets for door closers and other holding devices.

19.12.9 Magnetic Storage

Magnetic materials find significant use in the storage of information. Credit cards are popularly used which also have magnetic strips. To store larger quantities of information at low cost, computers are usually backed up with magnetic disks.

The recording head consisting of a laminated electromagnet is made of permalloy or soft ferrite having 0.3 m wide air gap. Here the data written by the electrical signal generates a magnetic field across the gap within the coil. Finally, the stored information is read using the same head, and an alternating e.m.f is induced in the coil of the head by moving tap or disk in the read or playback mode. This e.m.f is amplified and fed to a suitable output device.

SUMMARY

The topics covered in the chapter are summarised below.

- Based on their permeability or susceptibility, the magnetic materials can be broadly classified into three categories, namely diamagnetic, paramagnetic and ferromagnetic materials. The magnetic materials for which susceptibility χ_m is negative (permeability $\mu_r \le 1$) are said to be diamagnetic materials, whereas the materials with positive susceptibility (permeability $\mu_r \ge 1$) are said to be paramagnetic materials. If the susceptibility is much larger than zero and permeability $\mu_r \gg 1$, then the magnetic materials are called ferromagnetic materials.
- Paramagnetic and diamagnetic materials have a linear relationship between $\vec{B}$ and $\vec{H}$. However, the relationship between $\vec{B}$ and $\vec{H}$ is nonlinear for ferromagnetic materials. A material is said to be nonmagnetic if susceptibility $\chi_m = 0$ (or $\mu_r = 1$), it is magnetic otherwise.

- The magnetic materials are of two types, namely soft materials and hard materials. Soft magnetic materials are used in ac applications, since they are easily magnetised and demagnetised. However, hard magnetic materials are used in producing permanent magnets, since they retain magnetism on a permanent basis. Due to such properties, these materials are significantly used in information storage devices.
- The magnetic properties of solids originate due to the motion of electrons. The magnetic moment M of an electron is given by $M = \frac{neh}{4\pi m}$, where $n = 1, 2, 3\ldots$, e is the electronic charge, m is the electron mass and h is the Planck's constant.
- When placed in an external magnetic field, the materials which acquire feeble magnetism in the opposite direction to that of the applied field are called diamagnetic materials. The substances whose outermost orbits have an even number of electrons show the property of diamagnetism. Bismuth, zinc, copper, silver, gold, lead, water, etc., are the examples of diamagnetic materials.
- When placed in an external magnetic field, the materials which acquire feeble magnetism in the direction of an applied field are called paramagnetic materials. The source of paramagnetism is the permanent magnetic moment possessed by the atoms of paramagnetic materials. Aluminium, odium, platinum, manganese, copper chloride, liquid oxygen, etc., are the examples of paramagnetic materials.
- When placed in an external magnetic field, the materials which acquire strong magnetism in the direction of an applied field are called ferromagnetic materials. This property is found in the substances which are generally like paramagnetic materials. These are strongly attracted by magnets. Iron, nickel, cobalt, magnetite (Fe_3O_4), etc., are the examples of ferromagnetic materials.
- Anti-ferromagnetic substances are crystalline materials, in which the dipole moments of the neighbouring dipoles are equal and opposite in the orientation so that the net magnetisation vanishes. If they are placed in the magnetic field, they are feebly magnetised in the direction of the field. The susceptibility of these materials varies with temperature. It increases with increasing temperature and reaches a maximum at a particular temperature called the Neel temperature (T_N). MnO, FeO, CaO, MnO_4, MnS, etc., are the examples of anti-ferromagnetic materials.
- If the spins of the atoms are such that there is a net magnetic moment in one direction, the materials are called ferrimagnetic materials. The examples of ferrimagnetic materials are ferrites which consist of mainly ferric oxide Fe_2O_3, combined with one or more oxides of divalent metals.
- The classical theory of diamagnetism was developed by the French Physicist Paul Langevin in 1905, according to which the magnetic susceptibility of a diamagnetic material is given by $\chi_m = -\frac{\mu_0 NZe^2 R^2}{6m}$. Here, N is the number of atoms per unit volume of the substance, R is the average value of radii for all the electrons along three axes, μ_0 is the permeability of free space and Z is the atomic number.
- According to Langevin's theory of paramagnetism, the magnetic susceptibility of a paramagnetic material is given by $\chi_m = \frac{\mu I_0^2}{3NkT}$. Here, I_0 is the saturation value of the intensity of magnetisation I when all the magnetic dipoles get aligned in the direction of an external magnetic field, k is the Boltzmann constant, T is the temperature and $\frac{\mu I_0^2}{3Nk}$ is called Curie constant. This equation shows

that the magnetic susceptibility of a paramagnetic material depends on the temperature T and it varies inversely with T.

- Langevin's theory was not able to explain the complicated dependence of susceptibility on the temperature, as shown by several paramagnetic substances. In view of this, Langevin's theory was further modified by Curie and Weiss. Weiss assumed that in a paramagnetic substance an internal molecular magnetising field is generated because of a mutual interaction between the atomic magnetic dipoles. Finally, the magnetic susceptibility of a paramagnetic material is given by $\chi_m = \dfrac{C}{T-\theta}$, where $\theta = \dfrac{NM^2\mu\lambda}{3k}$ is called Curie temperature together with M as the permanent magnetic moment of the atoms and λ as the molecular field coefficient. $C = \dfrac{NM^2\mu}{3k}$ is the Curie constant.
- In general, a specimen of a ferromagnetic substance contains a number of small regions called domains. According to the classical theory of ferromagnetism, every domain is magnetically saturated and the direction of magnetisation in different domains is different. In the absence of an external magnetic field, all the magnetic domains are randomly oriented and hence their resultant magnetic moment in any direction will be zero. As per modern theory, it is assumed that in the absence of an external magnetic field, these domains form closed loops within the substance so that the net magnetic moment of the whole substance is zero. When a magnetic field is applied to the ferromagnetic substance, the substance becomes magnetised. When the substance is placed in a weak external magnetic field, the magnetisation produced is due to the displacement of boundaries of domains and if the external magnetic field is strong, the magnetisation produced is mainly by the rotation of domains.
- Ferromagnetic materials like iron and steel are used for screening or shielding that protects sensitive electrical devices from disturbances from strong magnetic fields. For perfect screening, it is required that the shield has infinite permeability ($\mu_r = \infty$).
- Even though $\vec{B} = \mu_0(\vec{H} + \vec{I})$ holds good for all materials including ferromagnetics, the relationship between $\vec{B}$ and $\vec{H}$ depends on previous magnetisation of a ferromagnetic material, i.e., its magnetic history. Instead of having a linear relationship between $\vec{B}$ and $\vec{H}$ (i.e., $\vec{B} = \mu\vec{H}$), it is only possible to represent the relationship by a magnetisation curve or B–H curve. Hysteresis is defined as the lagging of intensity of magnetisation from the magnetising field.
- During the process of magnetisation, a loss of energy is always involved in aligning the domains in the direction of the applied magnetic field. When the direction of an external magnetic field is reversed, the absorbed energy is not completely recovered and the rest of the energy in the sample is lost in the form of heat. This loss of energy is called hysteresis loss. The energy lost per unit volume of the substance in a complete cycle of magnetisation is equal to the area of the hysteresis loop.
- The concept of a magnetic circuit is based on solving some magnetic field problems using the circuit approach. Transformers, motors, generators and relays are magnetic devices which may be considered as magnetic circuits. By exploiting an analogy between magnetic circuits and electric circuits the analysis of such circuits is made simple.
- Magnetic materials find diverse applications in various modern technologies. Different types of magnetic materials are used for different applications. Electrical devices like power transformers, motors, generators, electromagnets, etc. use soft magnetic materials. Electrical steels are used as core materials in them. For retaining magnetic field of permanent magnets, hard magnetic materials are used in fabrications.

- In view of the application of magnetic materials, different materials were discussed, viz., low-carbon steel, iron–silicon alloys, nickel–iron alloys, mumetal, alnico alloys, soft ferrites, hard ferrites, etc.

SOLVED EXAMPLES

EXAMPLE 1 In hydrogen atom, an electron revolves around a nucleus in an orbit of 0.53 Å radius. If the frequency of revolution of an electron is 6.6×10^{15} Hz, find the magnetic moment of the orbiting electron and calculate numerical value of Bohr magneton.

SOLUTION Given $r=0.53\times10^{-10}$ m and $n=6.6\times10^{15}$ Hz.

Magnetic Moment $M=iA$

$$i=\frac{e}{T}=\frac{e}{1/n}=en=1.6\times10^{-19}\times6.6\times10^{15}\ \text{A}$$

$$\text{Area}=\pi r^2=3.14\times(0.53\times10^{-10})^2$$

$$\therefore\ M=iA=1.6\times10^{-19}\times6.6\times10^{15}\times3.14\times(0.53\times10^{-10})^2$$

$$=9.314\times10^{-24}\ \text{Am}^2$$

Bohr magneton is the smallest value of the orbital magnetic moment of the electron. For $n = 1$, Bohr magneton

$$\mu_B=\frac{eh}{4\pi m}$$

$$=\frac{1.6\times10^{-19}\times6.6\times10^{-34}}{4\times3.14\times9.1\times10^{-31}}$$

$$=9.239\times10^{-24}$$

$$\mathbf{=9.24\times10^{-24}\ J/T}$$

EXAMPLE 2 Determine the magnetisation and flux density in silicon, if its magnetic susceptibility is -4.2×10^{-6} and the magnetic field in it is $1.19\times10^5\ \text{Am}^{-1}$. What would be the value of the relative permeability of the material?

SOLUTION Given $\chi=-4.2\times10^{-6}$ and $H=1.19\times10^5\ \text{Am}^{-1}$.

The formulae used are

Magnetisation $I=\chi H$

$$=-4.2\times10^{-6}\times1.19\times10^5\ \text{Am}^{-1}$$

$$=-0.4998\ \text{Am}^{-1}$$

$$\mathbf{=-0.50\ Am^{-1}}$$

Flux density $B=\mu_0(H+I)$

$$=4\pi\times10^{-7}\times(1.19\times10^5-0.50)$$

$$=0.1495\ \text{T}$$

$$\mathbf{=0.150\ T}$$

Relative permeability

$$B = \mu H = \mu_0 H\left(1 + \frac{I}{H}\right)$$

or
$$\mu_r = \frac{\mu}{\mu_0} = 1 + \frac{I}{H}$$

$$= 1 + \frac{-0.50}{1.19 \times 10^5}$$

$$= 1 - 0.42 \times 10^{-5}$$

$$\mathbf{= 0.999}$$

Example 3 Find the percentage increase in magnetic induction when the space within a current-carrying toroid is filled with magnesium. Given χ for magnesium as 1.2×10^{-5}.

Solution Given $\chi = 1.2 \times 10^{-5}$.

Magnetic flux density

$$B = \mu_0 H \quad \text{(i)}$$

When the free space is filled with magnesium, then

$$B' = \mu_r \mu_0 \text{H} \quad \text{(ii)}$$

and
$$\mu_r = 1 + \chi \quad \text{(iii)}$$

From Eqs. (ii) and (iii)

$$B' = (1 + \chi)B \quad \text{(iv)}$$

Hence, the percentage increase in magnetic induction

$$= \frac{B' - B}{B} \times 100 \quad \text{(v)}$$

By using Eqs. (i) and (iv), Eq. (v) becomes

$$= \frac{(1 + \chi)B - B}{B} \times 100 = \chi \times 100$$

$$= 1.2 \times 10^{-5} \times 100$$

$$= 1.2 \times 10^{-3}\%$$

$$\mathbf{= 0.0012\%}$$

Example 4 Determine the magnetisation and flux density of the diamagnetic material if its magnetic susceptibility is -0.4×10^{-5} and the magnetic field in it is 10^4Am^{-1}.

Solution Given $\chi = -0.4 \times 10^{-5}$ and $H = 10^4$ A m^{-1}.

Magnetisation

$$I = \chi H$$

$$= -0.4 \times 10^{-5} \times 10^4$$

$$\mathbf{= -0.04\ Am^{-1}}$$

Magnetic flux density

$$B = \mu_0(H + I)$$

$$= 4\pi \times 10^{-7} \times [10^4 - 0.04]$$

$$\mathbf{= 0.01256\ T}$$

Example 5 The magnetic susceptibility of aluminium is 2.3×10^{-5}. Find its permeability and relative permeability.

SOLUTION Given $\chi = 2.3 \times 10^{-5}$.

Permeability

$$\mu = \mu_0 \mu_r \text{ and}$$

Relative permeability

$$\mu_r = 1 + \chi$$

$$\therefore \quad \mu_r = 1 + 2.3 \times 10^{-5}$$

$$\mu_r = 1.000023$$

$$\mu = \mu_0 \mu_r = 4\pi \times 10^{-7} \times 1.000023$$

$$= \mathbf{12.56 \times 10^{-7}\ N/A^2}$$

EXAMPLE 6 The magnetic susceptibility of a medium is 940×10^{-4}. Calculate its absolute and relative permeability.

SOLUTION Given $\chi = 0.094$.

Absolute permeability

$$\mu = \mu_0 \mu_r$$

Relative permeability $\mu_r = 1 + \chi$

$$\text{so, } \mu_r = 1 + 0.094 = 1.094$$

$$\mu = \mu_0 \mu_r = 4\pi \times 10^{-7} \times 1.094$$

$$= \mathbf{13.74 \times 10^{-7}\ N/A^2}$$

EXAMPLE 7 The maximum value of the permeability of a material is 0.126 N/A^2. What is the relative permeability and magnetic susceptibility?

SOLUTION Given $\mu = 0.126 \text{N/A}^2$.

Relative permeability $\mu_r = \dfrac{\mu}{\mu_0}$ and susceptibility is $\chi = \mu_r - 1$

$$\therefore \quad \mu_r = \frac{\mu}{\mu_0} = \frac{0.126}{4\pi \times 10^{-7}}$$

$$= 10^5$$

$$\chi = \mu_r - 1 = 10^5 - 1$$

$$= \mathbf{99999}$$

EXAMPLE 8 Calculate the diamagnetic susceptibility of He assuming that the two electrons are contributing to its diamagnetism. Consider the mean radius of the atom as 0.6 Å and $N = 28 \times 10^{26}$ per m^3.

SOLUTION Given $N = 28 \times 10^{26}$ per m^3 and $R = 0.6 \times 10^{-10}$ m.

Susceptibility of diamagnetic material

$$\chi_{\text{dia}} = \frac{-\mu_0 Z e^2 N R^2}{6m}$$

$$= -\frac{(4\pi \times 10^{-7}) \times 2 \times (1.6 \times 10^{-19})^2 \times 28 \times 10^{26} \times (0.6 \times 10^{-10})^2}{6 \times 9.1 \times 10^{-31}}$$

$$= \mathbf{-11.872 \times 10^{-8}}$$

EXAMPLE 9 A magnetising field of 1000 A/m produces a magnetic flux of 2×10^{-5} Weber in a bar of iron of 0.2cm^2 cross-section. Calculate permeability and susceptibility of the bar.

SOLUTION Given $H=10^3$ A/m, $\phi=2\times10^{-5}$ Wb and $A=0.2\times10^{-4}$ m^2.

Magnetic flux density $B=\dfrac{\phi}{A}$

Permeability $\mu=\dfrac{B}{H}$

Susceptibility $\chi=\mu_r-1=\dfrac{\mu}{\mu_0}-1$

$$\therefore \quad B=\frac{\phi}{A}=\frac{2\times10^{-5}}{2\times10^{-5}}=1.0 \text{ Wb/m}^2$$

$$\mu=\frac{B}{H}=\frac{1}{10^3}=10^{-3} \text{ N/A}^2$$

and
$$\chi=\frac{\mu}{\mu_0}-1=\frac{10^{-3}}{4\pi\times10^{-7}}-1=\mathbf{795.18}$$

EXAMPLE 10 An iron rod of 1.0 m length and cross-section 4 sq cm is in the form of a closed ring. If the permeability of iron is 50×10^{-4} Hm^{-1}. Show that the number of ampere turns required to produce a magnetic flux of 4×10^{-4} Wb through the closed ring is 200.

SOLUTION Given $L=1.0$ m, $A=4\times10^{-4}$m^2, $\mu=50\times10^{-4}$ H/m and $\phi=4\times10^{-4}$ Wb.

Magnetic flux density $B=\dfrac{\phi}{A}$

$$=\frac{4\times10^{-4}}{4\times10^{-4}}=1.0 \text{ Wb/m}^2$$

Also $B=\mu NI$

$\therefore$ Ampere turn $NI=\dfrac{B}{\mu}$

$$=\frac{1.0}{50\times10^{-4}}=\mathbf{200\ A/m}$$

EXAMPLE 11 The mean length of an iron ring having 200 turns of wire upon it is 0.5 m and its cross-section is 4×10^{-4} m^2. What current through the winding should be sent to produce a flux of 4×10^{-4} Wb in the ring? Permeability of iron is 65×10^{-4} Wb/Am.

SOLUTION Given $\mu=6.5\times10^{-4}$ Wb/Am, $\phi=4\times10^{-4}$ Wb and $A=4\times10^{-4}$ m^2.

The formula used is $B=\dfrac{\phi}{A}=\dfrac{4\times10^{-4}}{4\times10^{-4}}=1.0$ Wb/m^2

Also
$$B=\mu NI$$

or
$$I=\frac{B}{\mu N}$$

where N is the number of turns per metre, i.e.,

$$N=\frac{200}{0.5}=400 \text{ turns/m}$$

Then, current $I = \dfrac{B}{\mu N} = \dfrac{1.0}{6.5\times10^{-4}\times400}$

$\mathbf{I = 3.85\ A}$

EXAMPLE 12 Assuming the susceptibility of a diamagnetic material as -5.6×10^{-6} and its structure as a body-centred cubic with a lattice constant 2.55 Å.

Calculate the radius of its atom, if only one electron per atom is contributing to diamagnetism.

SOLUTION Given $\chi = -5.6\times10^{-6}$ and $a = 2.55$ Å $= 2.55\times10^{-10}$m.

$$\chi = -\frac{\mu_0 Ze^2 NR^2}{6m}$$

or
$$R = \left[\frac{-\chi 6m}{\mu_0 Ze^2 N}\right]^{1/2}$$

where N is the number of atoms per unit volume

i.e.,
$$N = 2\frac{1}{V} = 2\times\frac{1}{a^3} = 2\times\frac{1}{(2.55\times10^{-10})^3}$$

where the factor 2 arises because the body-centred cubic has two electrons per unit cell.

or
$$N = 1.206\times10^{29} \text{ per m}^3$$

$\therefore$
$$|R| = \left[\frac{5.6\times10^{-6}\times6\times9.1\times10^{-31}}{4\pi\times10^{-7}\times1\times(1.6\times10^{-19})^2\times1.206\times10^{29}}\right]^{\frac{1}{2}}$$

$$\mathbf{= 0.89\ Å}$$

EXAMPLE 13 A paramagnetic substance contains 6.5×10^{25} atoms per m^3 and the magnetic moment of each atom is one Bohr magneton. Find the susceptibility at room temperature.

SOLUTION Given $N = 6.5\times10^{25}$ atoms/m^3.

T corresponding to room temperature $= 27 + 273 = 300$ K

Susceptibility
$$\chi = \frac{\mu_0 NM^2}{3kT} \qquad \text{(i)}$$

The magnetic moment of each atom

$$M = n\left(\frac{eh}{4\pi m}\right)$$

$$= 1\left(\frac{1.6\times10^{-19}\times6.6\times10^{-34}}{4\times3.14\times9.1\times10^{-31}}\right)$$

$$= 9.24\times10^{-24}\ \text{Am}^2$$

From Eq. (i), we get

$$\chi = \frac{(4\pi\times10^{-7})\times6.5\times10^{25}\times(9.24\times10^{-24})^2}{3\times1.38\times10^{-23}\times300}$$

$$= 5.612099\times10^{-7}$$

$$\mathbf{= 5.612\times10^{-7}}$$

EXAMPLE 14 The molecular weight and density of a paramagnetic substance are 168.5 and 4370 kg/m^3, respectively, at room temperature. Considering the contribution to paramagnetism as two Bohr magnetons per molecule, calculate its susceptibility and magnetisation produced in it in a field of 2×10^5 Am^{-1}.

SOLUTION Given molecular weight $M_0=168.5$, molecular density $f=4370$ kg/m^3, room temperature $(T)=27°C$ $=27+273=300$ K, $H=2\times10^5$ A/m, Bohr magnetons $(\mu_B)=9.24\times10^{-24}$ A m^2.

$$\text{Number of atoms per unit volume } N=\frac{fN_A}{M_0}=\frac{4370\times6.02\times10^{26}}{168.5}$$

$$=1.56\times10^{28}\text{ per m}^3$$

$$\text{Susceptibility } \chi=\mu_0 N\mu^2{}_B/3kT$$

$$=\frac{(4\pi\times10^{-7})\times(1.56\times10^{28})\times(2\times9.24\times10^{-24})^2}{3\times(1.38\times10^{-23})\times300}$$

In the above expression factor of 2 comes before μ_B, as the contribution is two Bohr magnetons per molecule

$$\therefore \qquad \chi_m=5.388\times10^{-4}$$

Now magnetisation $I=\chi H$

$$=5.388\times10^{-4}\times2\times10^5$$

$$=\mathbf{107.76\ A/\mathit{m}}$$

EXAMPLE 15 The hysteresis loop of a transformer has an area of 2500 ergs/cm^3. Calculate the loss of energy per hour at 50 Hz frequency. The density of iron is 7.5 g/cm^3 and weight is 10 kg.

SOLUTION Given $m=10$ kg$=10^4$ g and $d=7.5$ g/cm^3.

Area of hysteresis loop$=2500$ ergs/cm^3

The loss of energy per unit volume per hour

$$=50\times60\times60\times2500$$

$$=4.5\times10^8\text{ ergs/cm}^3$$

$$\text{Volume of iron } (V)=\frac{m}{d}=\frac{10^4}{7\cdot5}=1\cdot33\times10^3\text{ cm}^3$$

Hence, the total loss of energy per hour

$$=4.5\times10^8\times1.33\times10^3$$

$$=\mathbf{6.0\times10^{11}\ ergs}$$

EXAMPLE 16 A bar magnet has a coercivity of 5×10^3 A/m. It is desired to demagnetise it by inserting it inside a 10 cm long solenoid having 50 turns. What current should be sent through the solenoid?

SOLUTION Here, coercivity $H=5\times10^3$ A/m, $l=10$ cm and total twins$=50$,

Turns per meter $N=50\times10=500$ turns/m

Now, $\qquad H=Ni$

$$5\times10^3=500\times i$$

or $\qquad \mathbf{\mathit{i}=10\ A}$

EXAMPLE 17 An iron rod of 50 cm length and 4 sq cm cross-section area is in the form of a circular ring. If the permeability of iron is 65×10^{-4} H/m, compute the number of turns required to produce a flux of 4×10^{-5} Weber.

Solution Here, $A=4\ cm^2=4\times10^{-4}\ m^2$, $l=50\ cm=0.5\ m$,

$\mu=65\times10^{-4}$ H/m and $\phi=4\times10^{-5}$ Wb.

As $$H=\frac{N_i}{l} \text{ or } N_i=Hl$$

$$B=\frac{\phi}{A}=\frac{4\times10^{-5}}{4\times10^{-4}}=0.1\ \text{Wb/m}^2$$

$$\because \quad \mu=\frac{B}{H} \quad H=\frac{B}{\mu}=\frac{0.1}{65\times10^{-4}}=15.38\ \text{A-turns/m}$$

Number of turns $=N=Hl=15.38\times0.5=$ **7.69**

Example 18 A magnetising field of 600 Am^{-1} produces a magnetic flux of 2.4×10^{-5} Weber in an iron bar of 0.2 cm^2 cross-sectional area. Compute the permeability and susceptibility of the bar.

Solution Given $\phi=2.4\times10^{-5}$ Wb and $A=0.2\ cm^2=0.2\times10^{-4}m^2$. The magnetic flux density is given by

$$B=\frac{\phi}{A}=\frac{2.4\times10^{-5}}{0.2\times10^{-4}}=\mathbf{1.2\ wb/m^2}$$

The permeability is given by

$$\mu=\frac{B}{H}$$

or $$\mu=\frac{1.2}{600}=\mathbf{0.002\ N/A^2}$$

The susceptibility is given by

$$\chi=\frac{\mu}{\mu_0}-1$$

$$=\frac{0.002}{4\times3.14\times10^{-7}}-1=1592-1$$

$$=\mathbf{1591}$$

Example 19 The magnetic susceptibility of medium is 950×10^{-11}. Compute the permeability and relative permeability.

Solution Magnetic susceptibility $\chi=950\times10^{-11}$.

As $$\mu=\mu_0(1+\chi)$$

and permeability of free space $\mu_0=4\pi\times10^{-7}$ H/m

$$\therefore \quad \mu=4\pi\times10^{-7}\times(1+950\times10^{-11})$$

Hence, μ is slightly greater than μ_0

Now relative permeability $\mu_r=\dfrac{\mu}{\mu_0}=\mathbf{1+950\times10^{-11}}$

Example 20 Find the energy loss per hour in an iron core of a transformer, if the area of the *B–H* loop is 250 J/m^3 and the frequency of the alternating current is 50 Hz. The density of iron is $7.5\times10^3\ kg/m^3$ and the mass of the core is 100 kg.

Solution Area of *B–H* loop $=250\ J/m^3$ and frequency $f=50$ Hz, density $\rho=7.5\times10^3\ kg/m^3$ and mass $m=100$ kg.

Volume of core, $V=\dfrac{m}{\rho}=\dfrac{100}{7.5\times10^3}=13.3\times10^{-3}m^3$

Number of ac cycles in 1 hour $=(60\times60)\times50$

$=180000$

Hysteresis loss per cycle per unit volume = area of *B–H* loop = 250 J/m^3

$\therefore$ Hysteresis loss in the entire core in 1 hour

$=(250)\times13.3\times10^{-3}\times180000=\mathbf{59.85\times10^4\,J}$

EXAMPLE 21 In *B–H* loop, the maximum value of B_{max} is 1.375 Weber/m^2 and the area of the loop is 0.513 cm^2. If the value of 1 cm on the *x*-axis is 10 A/cm and the value of 1 cm on the *y*-axis is 1 Weber/m^2, calculate the hysteresis power loss when an alternating magnetic flux density of 1.375 Weber/m^2 intensity and 50-Hz frequency is applied on 10^{-3}m^3 volume of the specimen.

SOLUTION 1 cm on the *x*-axis $=10$ A/cm $=10\times100$ A/m

1 cm on the *y*-axis $=1$ Wb/m^2

$\therefore$ Area of the *B–H* loop $=0.513\text{ cm}^2=0.513\times(10\times100)\times1=513\text{ J/m}^3$

Hysteresis loss per cycle per m^3 = 513 J

But volume of specimen $=10^{-3}$m^3

and frequency = 50 Hz (= number of cycles per second)

$\therefore$ Hysteresis loss per second (or hysteresis power loss)

$=513\times10^{-3}\times50$

$=\mathbf{25.65\,W}$

OBJECTIVE TYPE QUESTIONS

Q.1 Which of the following statements is/are true about magnetic susceptibility χ?
(a) χ may be positive or negative
(b) χ for a paramagnetic material has values close to 1
(c) At a given temperature, the value of χ_m increases with increasing magnetic field
(d) For paramagnetic substances, χ is inversely proportional to the absolute temperature of the same

Q.2 The relative permeability of a medium is the permeability relative to that of
(a) vacuum (b) iron (c) water (d) none of these

Q.3 The dipole moment of current loop does not depend upon the
(a) current in the loop (b) shape of the loop
(c) area of the loop (d) number of turns in the loop

Q.4 The dimensions of magnetic susceptibility are
(a) Wb/*m* (b) amp/*m* (c) Wb/*m*^{2} (d) dimension less

Q.5 The magnetic susceptibility of a diamagnetic substance is
(a) positive (b) negative (c) zero (d) none of these

Q.6 The magnetic susceptibility is negative for a substance; which is:
(a) diamagnetic (b) ferromagnetic (c) paramagnetic (d) none of these

Q.7 Which of the following materials are feebly attracted by external magnetic fields?
(a) ferromagnetic material (b) ferrimagnetic material
(c) parramagnetic material (d) none of these

Q.8 The magnetic susceptibility of a diamagnetic substance
(a) does not depend on temperature (b) increases with increase in temperature
(c) decrease with decrease in temperature (d) none of these

Q.9 Which of the following substances are diamagnetic material?
(a) alluminium and platinum (b) bismuth and lead
(c) copper and gold (d) both (b) and (c)

Q.10 A current loop of magnetic moment $\vec{P}_m$ when placed in a non-uniform magnetic field experiences a force
(a) $\vec{\nabla}(\vec{P}_m \cdot \vec{B})$ (b) $(\vec{\nabla} \cdot \vec{B})$ (c) $\vec{\nabla} \cdot \vec{P}_m \vec{B}$ (d) $(\vec{P}_m \cdot \vec{\nabla})\vec{B}$

Q.11 The relation between magnetising field H and intensity of magnetisation is
(a) $I \propto H^{-1}$ (b) $I \propto H$ (c) $I \propto H^2$ (d) $I \propto H^{1/2}$

Q.12 In classical theory of diamagnetism, the term $eB/2\,\text{m}$ is known as
(a) Bohr magneton (b) Debye angular frequency
(c) Larmour angular frequency (d) none of these

Q.13 Numerical value of Bohr magneton is
(a) 9.24×10^{-24} J/T (b) 6.6×10^{-24} J/T (c) 6.6×10^{-14} J/T (d) none of these

Q.14 The value of susceptibility of a diamagnetic substance is about
(a) 10^5 (b) 10^{-6} (c) -10^{-7} (d) 10^7

Q.15 The magnetic moment of an atom is due to
(a) orbital motion of electrons only (b) spin motion of electron
(c) both orbital and spin motion (d) none of these

Q.16 The value of susceptibility of a ferromagnetic substance is of the order of
(a) 10^5 (b) 10^6 (c) 10^{-7} (d) -10^{-6}

Q.17 Which of the following substance/substances can have positive permeability and negative susceptibility?
(a) diamagnetic (b) ferromagnetic (c) paramagnetic (d) none of these

Q.18 Feritites show
(a) paramagnetic behaviour (b) ferrimagnetic behaviour
(c) ferromagnetic behaviour (d) none of these

Q.19 On application of magnetic field on a material, if it does not show any effect, then the material is a
(a) non-magnetic material (b) diamagnetic material
(c) anti-ferromagnetic material (d) none of these

Q.20 The boundary wall between domains is known as
(a) potential wall (b) Bloch wall (c) magnetic wall (d) none of these

Q.21 Which of the following statement is correct?
(a) Magnetic dipole moment per unit volume of material is known as magnetisation of substance
(b) In a ferromagnetic material, susceptibility is very small and positive
(c) The most suitable material for making permanent magnet is steel
(d) all of these

Q.22 Which of the following statement is true?
(a) Hysteresis is lagging of an effect behind the cause of effect
(b) In soft magnetic material, the hysterisis loss is less
(c) The area of hysteresis loop of a ferromagnetic gives the energy that is consumed during each cycle
(d) all of these

Short-Answer Questions

Q.1 What do you mean by magnetisation, permeability and susceptibility of a magnetic substance?
Q.2 What is the magnetic dipole moment associated with a loop-carrying current?
Q.3 Define atomic magnetic moment and discuss orbital diamagnetism.
Q.4 What is diamagnetism?
Q.5 Discuss ferromagnetism.
Q.6 What do you mean by ferromagnetic domain?
Q.7 How do you account for the magnetic properties of materials? Explain.
Q.8 Is it meaningful to say that an atom is ferromagnetic?
Q.9 Why is ferromagnetism found in solids only and not in fluids?
Q.10 What are the characteristics of diamagnetic, paramagnetic and ferromagnetic substances?
Q.11 What is Curie Point or Curie temperature? The magnetic behaviour of magnetic substances decreases with increasing temperature. Comment.
Q.12 Explain the temperature dependence of the behaviour of paramagnetic, diamagnetic and ferromagnetic substances.
Q.13 What does the area of a *B–H* loop represent?

Practice Problems

General Questions

Q.1 Explain magnetic flux density (B), intensity of magnetisation (M), magnetic flux density (H). How are they related to each other?

Q.2 Define magnetic susceptibility (χ) and relative magnetic permeability (μ_r) and establish a relation $\mu = \mu_0 (1 + \chi)$.

Q.3 Distinguish between dia, para and ferromagnetic materials. Derive an expression for magnetic susceptibility of a paramagnetic substance.

Q.4 Differentiate paramagnetic, diamagnetic and ferromagnetic substances by illustrating simple experiments.

Q.5 Derive an expression for diamagnetic susceptibility on the basis of Langevin's theory and show that it is independent of temperature.

Q.6 Discuss diamagnetic, paramagnetic, ferromagnetic, anti-ferromagnetic and ferrimagnetic substances citing one example of each.

Q.7 Prove that the change is the same whether the electron is orbiting around the nucleus in clockwise direction or anti-clockwise direction. Hence, discuss the diamagnetic behaviour of the substance according to Langevin's theory of diamagnetism.

Q.8 Based on Langevin's theory of diamagnetism, show that the diamagnetic susceptibility is negative and independent of temperature and field strength.

Q.9 Give Langevin's electronic theory of paramagnetism and hence prove that susceptibility (χ) of paramagnetic substance is inversely proportional to absolute temperature.

Q.10 Why are some substances diamagnetic while others paramagnetic? Explain.

Q.11 How do you classify a material as dia, para or ferromagnetic? Discuss the classical theory of paramagnetism.

Q.12 Explain the origin of atomic dipole moments and derive Langevin's equation for paramagnetic susceptibility.

Q.13 What are the physical basis of diamagnetism and paramagnetism of materials? Describe the Weiss's molecular theory of ferromagnetism and derive the Curie–Weiss Law.

Q.14 What are the distinguishing features of ferromagnetism? Give the theory of magnetic domains in ferromagnetic materials.

Q.15 What is ferromagnetism? Explain ferromagnetism on the basis of domain theory. Why does a piece of iron ordinarily not behave as a magnet?

Q.16 What do you understand by hysteresis remanence (retentivity) and coercivity? How do you determine the value of remanence and coercivity from a hysteresis loop?

Q.17 Show that the loss of energy due to hysteresis per unit volume of the material per cycle of magnetisation is given by (i) $\mu_0 \times$ area of I–H loop and (ii) area of B–H loop.

Q.18 What type of material should be used for making

(a) permanent magnets, and

(b) electromagnets?

Q.19 Explain the use of a hysteresis curve. What type of magnetic material is suitable for transformer cores, telephone diaphragm and chokes?

20

Superconductivity

LEARNING OBJECTIVES

After reading this chapter you will be able to

LO 1 Gain knowledge/Learn about electrical resistivity of solids and phonons

LO 2 Understand the properties and classification of superconductors

LO 3 Learn about effect of magnetic field and Isotope effect on superconductivity

LO 4 Know how London equations explain zero resistance and ideal diamagnetism of superconductors

LO 5 Discuss penetration depth of supercurrent and magnetic flux in superconductors

LO 6 Explain formation of Cooper pairs and its relation to Bose–Einstein condensation

LO 7 Understand basis of BCS theory and coherence length

LO 8 Analyse high temperature conductivity and applications of conductivity

Introduction

The phenomenon of *superconductivity* was first discovered by Kammerlingh Onnes in 1911. He found that electrical resistivity of some metals, alloys and compounds drops suddenly to zero when they are cooled below a certain temperature. This phenomenon is known as *superconductivity* and the materials that exhibit this behaviour are called as *superconductors*. However, all the materials cannot superconduct even at 0 K. The temperature at which a normal material turns into a superconducting state is called critical temperature T_c. Each superconducting material has its own critical temperature. Kammerlingh Onnes discovered that the electrical resistance of highly purified mercury dropped abruptly to zero at 4.15 K, as shown in Fig. 20.1.

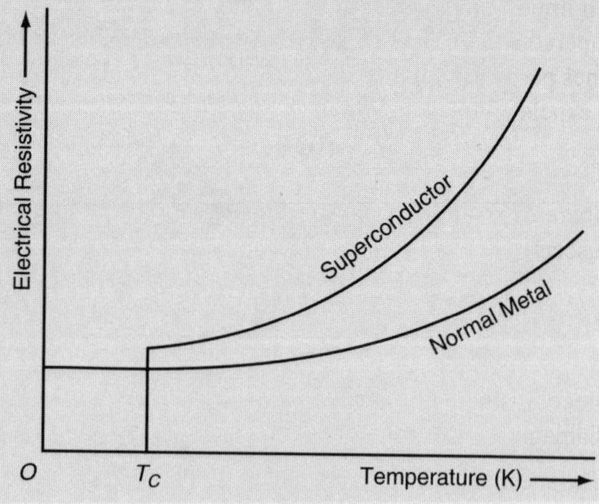

FIGURE 20.1

Generally good conductors like Au, Ag, Cu, Li, Na, K, etc. do not show superconductivity even at absolute zero.

The interaction of electrons with one another and with the lattice ions were averaged out by the free electron theory (model) approximation. This could be responsible for resistance to the flow of electrons under normal conditions. This independent particle model was unable to explain superconductivity. The clear understanding of the phenomenon of superconductivity requires the consideration of collective behavior of electrons and ions. This is called many body effects in solids.

20.1 ELECTRICAL RESISTIVITY OF SOLIDS AND PHONONS LO1

There are several factors that contribute to the electrical resistivity of a solid. For example, the deviations from a perfect lattice, which may be due to impurities or structural defects in crystal, can scatter the electrons. Moreover, the vibrations of lattice ions take place in normal modes. These vibrations constitute acoustic waves which travel through the solid. These waves are called phonons, which carry momentum. It is obvious that the number of phonons will increase if the temperature is raised. In the presence of phonons, now there is an interaction between the electrons and phonons. This interaction scatters conduction electrons and hence causes more resistance. Therefore, it is clear that the electrical resistance of a solid will decrease if we cool the solid.

20.2 PROPERTIES OF SUPERCONDUCTORS LO2

Electrical, magnetic and thermal properties are the main properties of the superconductors. These are described below.

20.2.1 Electrical Property

A superconductor is characterised by zero electrical resistivity. Once the current is started to flow, it will continue for years without any detectable decay (ideally) even if the applied voltage is removed.

20.2.2 Magnetic Property: Meissner Effect

An important property of the superconducting phase is the repulsion of magnetic flux lines from the bulk of superconductor. It is called *Meissner effect*. When a specimen is placed in a magnetic field, the magnetic flux lines pass through it (Fig. 20.2a). Now, if the temperature is decreased below the transition temperature (T_c), it expels all the magnetic flux lines from inside of the specimen (Fig.20.2b). Hence, we get

$$B = \mu_0(H + M) = 0 \quad \text{or} \quad M = -H$$

where M is the intensity of magnetisation due to applied magnetic field H. By the definition of magnetic susceptibility

$$\chi_m = \frac{M}{H} = -1$$

Since diamagnetic materials have negative magnetic susceptibility, the specimen becomes an 'ideal diamagnetic' in superconducting state.

If a specimen of superconductor is placed in a strong magnetic field, the specimen loses its property of superconductivity and becomes normal material as shown in Fig. 20.2.

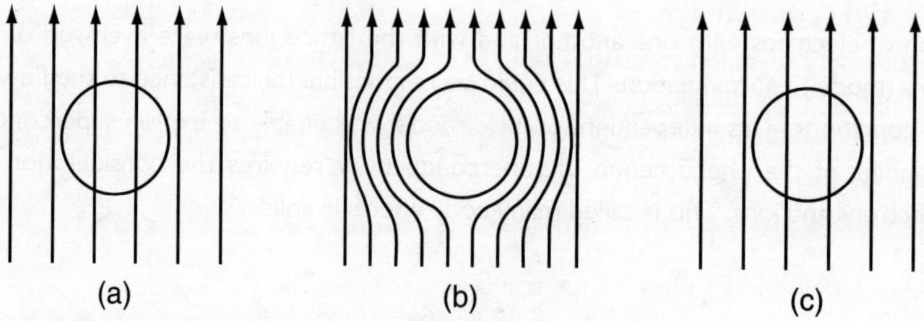

FIGURE 20.2

20.2.3 Thermal Properties

Thermal properties include the entropy, specific heat and thermal conductivity, which are discussed below.

20.2.3.1 Entropy

We know that the entropy is a measure of the disorder of a system. In all superconductors, the entropy decreases significantly on cooling below the critical temperature T_C. Therefore, the observed decrease in entropy between the normal state and superconducting state shows that the superconducting state is more ordered than the normal state. For aluminium, the change in entropy was observed to be small of the order of $10^{-14}\,k$ per atom, where k is the Boltzmann constant. The variation of entropy of aluminium in the normal and superconducting states with temperature is shown in Fig. 20.3.

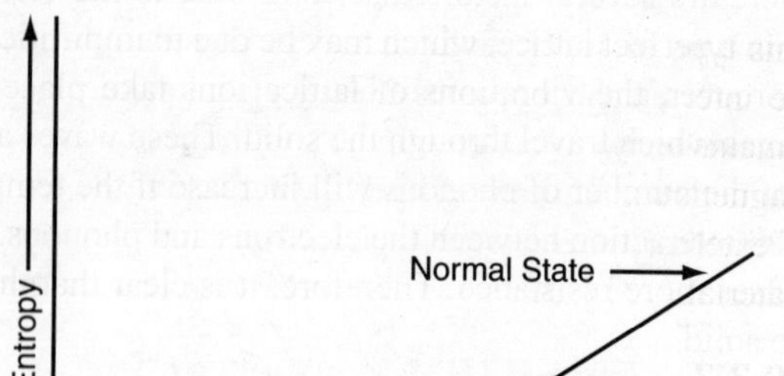

FIGURE 20.3

20.2.3.2 Specific Heat

The specific heat of normal metal is found to vary with temperature. The variation follows the following trend

$$C_n(T) = \gamma T + \beta T^3$$

In this relation, the first term is the specific heat of electrons in metal whereas the second term is due to the contribution of lattice vibrations at low temperature. On the other side, the specific heat of the superconductor shows a jump at T_c (Fig. 20.4). As we know that the superconductivity affects electrons mainly, so we may assume that the lattice vibration part remains unaffected and has the same value βT^3 in the normal and superconducting states.

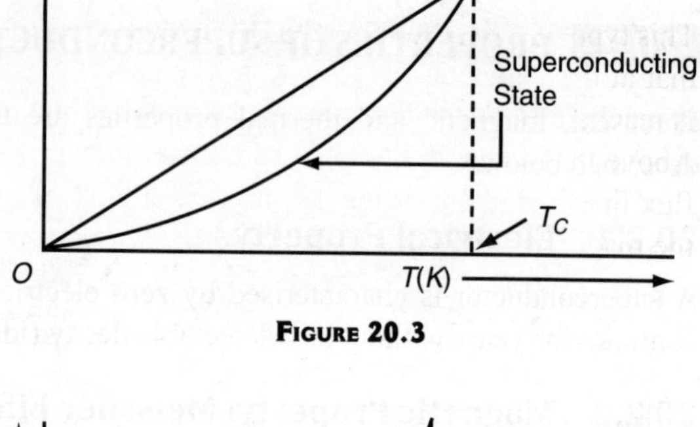

FIGURE 20.4

20.2.3.3 *Thermal Conductivity*

The thermal conductivity of superconductors undergoes a continuous change between the two phases. It is usually lower in the superconducting phase which shows that the electronic contribution goes down. This suggests that the superconducting electrons possibly play no role in the transfer of heat.

20.3 CLASSIFICATION OF SUPERCONDUCTORS

LO2

On the basis of magnetising behaviour, superconductors can be classified as type-I (or soft) and type-II (or hard) superconductors.

20.3.1 Type-I (Soft) Superconductor

This type of superconductor obeys complete Meissner effect. It expels all the magnetic field abruptly from the interior and becomes an 'ideal diamagnetic material'. Magnetisation produced in the superconductor remains in the direction opposite to the applied external magnetic field, as shown in Fig. 20.5a. At the critical magnetising field, the magnetisation decreases abruptly and the material becomes normal. For all the values of external magnetic field above the critical value, the magnetic flux lines penetrate completely inside the material.

20.3.2 Type-II (Hard) Superconductor

This type of superconductor loses magnetisation gradually rather than abruptly. From the Fig. 20.5b it is clear that at the magnetic field H_{C1}, the flux starts penetrating into the material until the upper critical field H_{C2} is reached. Between the two critical magnetic fields H_{C1} and H_{C2}, the material is said to be in a mixed state. Above the magnetic field H_{C2}, the material becomes normal conductor. Inspite of the fact that the magnetic flux lines penetrate inside the material in the mixed state, the electrical resistivity continues to be zero upto the magnetic field H_{C2}.

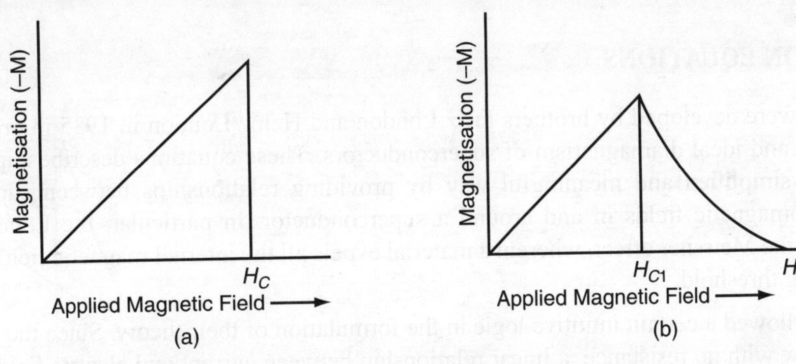

FIGURE 20.5

20.4 EFFECT OF MAGNETIC FIELD

LO3

In order to achieve the superconducting state in the metal, it is required that some combination of temperature and field strength should be less than a critical value. The superconductivity of a specimen can be destroyed by applying a critical magnetic field. So a superconductor becomes a normal conductor when it is placed in an intense magnetic field. Therefore, it retains its resistance. The critical magnetic field is a function of temperature also and it is defined as

$$H_c = H_0\left[1-\left(\frac{T}{T_c}\right)^2\right]$$

Here H_0 is the critical magnetic field at 0 *K*. We can see that at $T = 0$ K, $H_c = H_0$, and at $T = T_c$, $H_c = 0$. This relation between H_c and T shows that the critical magnetic field varies parabolically with the temperature. This curve demarcates the two states, i.e., it defines the boundary below which superconductivity is present and outside of which it behaves as a normal conductor.

20.5 ISOTOPE EFFECT LO3

It has been observed that the critical temperature T_c of superconductors varies with isotopic mass. Higher T_c is found in samples with lighter nuclei. In mercury, T_c varies form 4.85 K to 4.146 K as the average atomic mass M varies from 199.5 a.m.u. to 203.4 a.m.u. It is also found that the transition temperature changes smoothly when we mix different isotopes of the same element. The dependence of T_c on the atomic mass reveals that lattice vibrations and hence electron phonon interaction is deeply involved in the superconductivity. Based on experimental results it is found that

$$T_c \propto M^{-\alpha}$$

or $$T_c M^{\alpha} = \text{Constant}$$

Here, M is the atomic mass, T_c is the critical temperature and $\alpha = 0.49 \pm 0.01$. In view of this value of α it was thought that $\alpha = 0.5$ is valid for most of the materials. With this we get

$$T_c M^{1/2} = \text{Constant}$$

or $$T_{c_1} M_1^{1/2} = T_{c_2} M_2^{1/2}$$

20.6 LONDON EQUATIONS LO4

London equations were developed by brothers Fritz London and Heinz London in 1935 in order to explain the zero resistance and ideal diamagnetism of superconductors. These equations describe superconducting phenomenon in a simplified and meaningful way by providing relationships between superconducting current and electromagnetic fields in and around a superconductor. In particular, the London equations beautifully explain the Meissner effect, wherein a material expels all the internal magnetic fields as it crosses the superconducting threshold.

London brothers followed a certain intuitive logic in the formulation of their theory. Since the electrons in a superconductor flow with no resistance, a linear relationship between current and electric field (Ohm's law) may not be possible in a superconductor. In view of this, London imagined the electrons as if these were free electrons under the influence of a uniform external electric field. According to the Lorentz force law these electrons should encounter a uniform force and thus accelerate uniformly. Hence, the equation of motion for the superconducting electrons in the presence of applied electric field E is given by

$$m\frac{d\vec{v}}{dt} = \vec{F} = -e\vec{E} \tag{i}$$

The current density $\vec{J}$ can be expressed as

$$\vec{J} = -ne\vec{v} \tag{ii}$$

where n is the number of electrons per unit volume i.e., n is the number density of superconducting carriers. By differentiating Eq. (ii) w.r.t. time, we get

$$\frac{d\vec{J}}{dt} = -ne\frac{d\vec{v}}{dt} = -ne\left[-\frac{e\vec{E}}{m}\right]$$

$$\frac{d\vec{J}}{dt} = \frac{ne^2}{m}\vec{E} \tag{iii}$$

Equation (iii) is known as *first London equation*. According to London's theory, it was assumed that two types of the electrons, i.e., normal and superconducting electrons are present in the superconductors. The normal electrons don't respond to the electric field and only the superconducting electrons respond to the electric field. Now Maxwell's equation can be written as

$$\text{curl } \vec{E} = -\frac{d\vec{B}}{dt}$$

or $$\vec{\nabla} \times \vec{E} = -\mu_0 \frac{d\vec{H}}{dt} \tag{iv}$$

as $\vec{B} = \mu_0 \vec{H}$.

Taking curl of Eq. (iii), we get

$$\text{curl}\frac{d\vec{J}}{dt} = \frac{ne^2}{m}\text{curl } \vec{E}$$

or $$\vec{\nabla} \times \frac{d\vec{J}}{dt} = \frac{ne^2}{m}(\vec{\nabla} \times \vec{E}) \tag{v}$$

By using Eqs. (iv) and (v), we get

$$\vec{\nabla} \times \frac{d\vec{J}}{dt} = -\frac{\mu_0 ne^2}{m}\frac{d\vec{H}}{dt} \tag{vi}$$

By integrating Eq. (vi) w.r.t. time, we get

$$\text{curl } \vec{J} = -\frac{\mu_0 ne^2}{m}[\vec{H} - \vec{H}_0] \tag{vii}$$

where H_0 is a constant of integration. As we know that Meissner effect exhibits complete absence of magnetic field inside the superconductor. Therefore, $\vec{H}_0$ must be zero. Then

$$\text{curl } \vec{J} = -\frac{\mu_0 ne^2}{m}\vec{H}$$

or $$\text{curl } \vec{J} = -\frac{ne^2}{m}\vec{B} \tag{viii}$$

Equation (viii) is known as *second London equation* which explains Meissner effect as well.

20.7 PENETRATION DEPTH LO5

The second London equation (Eq. (viii)) explains the Meissner effect. This equation also predicts the penetration of supercurrent and magnetic flux in a superconductor, which will be shown below.

We take the curl of Eq. (viii)

$$\vec{\nabla} \times (\vec{\nabla} \times \vec{J}) = -\frac{\mu_0 ne^2}{m}\vec{\nabla} \times \vec{H} \quad (\because \vec{B} = \mu_0 \vec{H})$$

$$\vec{\nabla}(\vec{\nabla} \cdot \vec{J}) - \nabla^2 \vec{J} = -\frac{\mu_0 ne^2}{m}\vec{\nabla} \times \vec{H}$$

Since $\vec{\nabla} \cdot \vec{J} = 0$, we obtain

$$\nabla^2 \vec{J} = \frac{\mu_0 ne^2}{m}\vec{\nabla} \times \vec{H} \tag{ix}$$

The Maxwell's equation for direct current is

$$\vec{\nabla} \times \vec{H} = \vec{J}(x) \qquad \because \partial \vec{E}/\partial t = 0 \tag{x}$$

From Eqs. (ix) and (x), we find

$$\nabla^2 \vec{J} = \frac{\mu_0 ne^2}{m}\vec{J}$$

$$\nabla^2 \vec{J} = \frac{\vec{J}}{\lambda_L^2} \tag{xi}$$

where $$\lambda_L^2 = \frac{m}{\mu_0 e^2 n} \tag{xii}$$

The parameter λ_L has the dimensions of length and is called London penetration depth. So in terms of the penetration depth, Eq. (xi) can be written in one dimension as

$$\frac{d^2 \vec{J}}{dx^2} = \frac{\vec{J}}{\lambda_L^2}$$

Solution of this equation is as follows

$$J = A_1 e^{x/\lambda_L} + A_2 e^{-x/\lambda_L}$$

where A_1 and A_2 are constants. x represents the distance into the metal (superconductor) from the surface. The first term in the above solution gets increased when x is increased. However, this is contrary to the fact. Therefore, we neglect coefficient A_1 and write the above solution as

$$J = A_2 e^{-x/\lambda_L}$$

Now at $x = 0$, J has some finite value, which we consider as J_0, i.e., $J = J_0$. With this $J = J_0 e^{-x/\lambda_L}$.

or $$J = J_0 e^{-x/\lambda_L}$$

When $x = \lambda_L$, the above solutions reads $\frac{J}{J_0} = \frac{1}{e}$.

Thus, the penetration depth is the depth where the current drops to $1/e$ times of its value at the surface. It is a fundamental length that characterises a superconductor. Therefore, the London equations provide a characteristic length scale λ_L, over which external magnetic fields are exponentially suppressed. In order to understand the physical meaning of the London penetration depth, we consider a superconductor within free space where a constant magnetic field outside the superconductor is pointed parallel to the superconducting boundary plane in the z-direction. Then for the x-direction, which is perpendicular to the boundary, the solution inside the superconductor may be shown to be $B_z(x) = B_0 e^{-x/\lambda_L}$. This explains the exponential suppression of the external magnetic field in the superconductor.

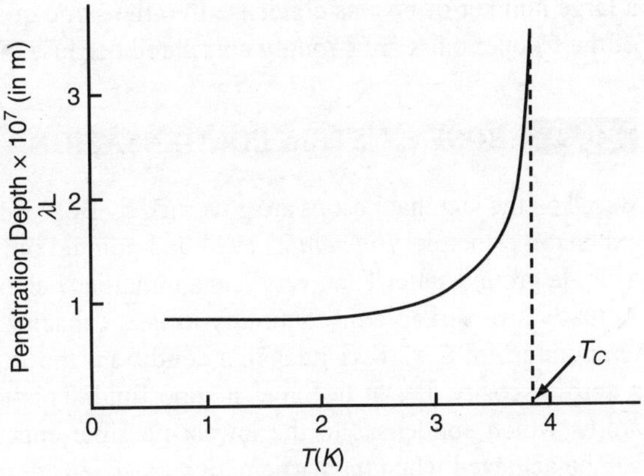

FIGURE 20.6

The penetration depth depends on temperature and it gets significantly increased as T approaches T_c (Fig. 20.6). The variation of penetration depth with temperature is according to the following relation

$$\frac{\lambda(T)}{\lambda(0)} = \left[1 - \left(\frac{T}{T_c}\right)^4\right]^{-1/2}$$

20.8 COOPER PAIRS

LO6

As mentioned earlier, the superconductors offer zero resistance. So, when an electron in a solid passes by adjacent ions in the lattice, it imparts a momentum to these ions due to Coulomb interaction. It means the ion starts vibrating and hence phonon gets excited. Now consider a second electron which is subsequently passing through the moving region of increased positive charge density. It will experience an attractive Coulomb interaction and can absorb all momentum of vibrating region. Under this situation, the vibrations of ions are stopped and hence the phonon is absorbed by this second electron. So in this interaction, the momentum which was imparted by the first electron is taken up by the second electron and hence these electrons undergo an interaction. This interaction would be an attractive interaction because exchange of momentum takes place via Coulomb attraction interaction; of course through phonon. According to BCS theory, under certain conditions this attractive interaction overcomes the force of repulsion between the two electrons. Therefore, the electrons are loosely bound together. This pair of electrons is called *Cooper pair*. According to BCS theory, the binding energy of Cooper pair at absolute zero is $\sim 3kT_c$. This binding energy gets reduced as the temperature is raised. The binding energy becomes zero when temperature equals the critical temperature T_c. Hence, a Cooper pair is not bound at temperature ~ T_c.

In view of the above discussion, it is clear that the Cooper pairs are formed due to the electron lattice interactions. The two electrons of a cooper pair have equal and opposite momenta. They also have opposite spins, i.e., one electron is spin up and another is spin down. Thus, the bound Cooper pair is a spin zero object and is a boson. The bound Cooper pairs overlap each other because the Cooper pair wavefunction is very large (few hundred nanometers in diameter). As we know that when identical bosons overlap each other, then a large number of bosons condense into the same quantum state (Bose Einstein condensation). The motions of the Cooper pairs are strongly correlated because all of the Cooper pairs are in the same quantum state.

20.9 BOSE EINSTEIN CONDENSATION LO6

Based on the fact that bosons are governed by Bose Einstein statistics and they are not constrained by the Pauli exclusion principle, *Einstein* in 1924 had pointed out that bosons could condense in unlimited numbers into a single ground state. Then very late anomalous behaviour of liquid helium was noticed at low temperatures. Actually a remarkable discontinuity in heat capacity of helium was observed when it was cooled to critical temperature of 2.17 K. Under this condition, the liquid density drops and a fraction of the liquid attains a zero viscosity, i.e., it becomes a superfluid. This superfluidity takes place due to the fraction of helium atoms which condenses to the lowest possible energy. This is called *Bose Einstein Condensation*, which can be achieved when the participating particles are identical. This condition of indistinguishability requires that the deBroglie wavelengths of the particles overlap significantly. This is possible only at extremely low temperature as the deBroglie wavelengths become longer. This also requires a high density of particles to narrow the gap between the particles. *Cornell* and *Wieman* together with *Ketterle* received the 2001 Nobel Prize for achieving Bose Einstein Condensation in dilute gases of alkali atoms.

20.10 BCS THEORY: QUALITATIVE EXPLANATION LO7

The basis of a quantum theory of superconductivity was led by the classic 1957 paper of Bardeen, Cooper and Schrieffer. This is now called the BCS theory. It was widely applicable, for example, from He^3 atoms in their condensed phase to type I and type II metallic superconductors. This theory involves the electron interaction through phonons as mediators. The formulation of BCS theory is based on two experimental facts viz. the isotope effect and the variation of specific heat of superconductors. The isotope effect indicated that the lattice vibrations play an important role in achieving the superconducting state. Further, T_c attains a value zero when atomic mass M approaches infinity. This implies that non-zero transition temperature T_c is a consequence of finite mass of ions. According to second observation, the jump in the value of specific heat at transition temperature indicates the presence of an energy gap in the energy specturm of the electron in the superconducting state. We discuss below the quantum theory of superconductors due to BCS.

When an electron approaches a positive ion core then it suffers attractive Coulomb interaction. So, the ion core sets in motion due to this attraction. Consequently the lattice gets distorted. It is obvious that this distortion will be greater if the mass of the positive ion core is small. Now suppose that another electron interacts with the distorted lattice. Due to this interaction, the energy of the second electron gets lowered. Thus we can say that the two electrons interact via the lattice distortion or the phonon field. So, lowering of energy for the electrons takes place which implies that the force between the two electrons is attractive. This type of interaction is called as electron-lattice-electron interaction. This interaction would be strongest when the two electrons have equal and opposite momenta and spins. The above interaction can be interpreted as the electron electron interaction through phonons as the mediator, because the oscillatory distortion of lattice is quantized in terms of phonons.

The superconductivity occurs when an attractive interaction (as mentioned above) between two electrons due to phonon exchange dominates the usual repulsive interaction. This is the fundamental postulate of the BCS theory. As discussed earlier also, the two electrons, which interact attractively, are called a Cooper pair. The energy of the pair of electrons in bound state is less than the energy of the pair in free state. This difference of energy is called binding of Cooper pairs. It means by applying this amount of energy we can break this pair. The pairing is complete at $T = 0$ K and completely broken at $T = T_c$. It was observed that the binding energy of the Cooper pair is maximum when electrons forming the pair have opposite momenta and spins.

In addition, there is a BCS wavefunction composed of particle pairs. When treated by *BCS* theory, it gives the familiar electronic superconductivity observed in metals and exhibits the energy gap. For the accomplishment of BCS wavefunction, we need that

(i) An interaction (attraction) between electrons can lead to a ground state separated from excited states by an energy gap. The thermal properties and most of the electromagnetic properties are consequences of this energy gap.

(ii) The magnetic flux through a superconducting ring is quantized and effective unit of charge is $2e$ rather than e.

20.11 COHERENCE LENGTH LO7

The concept of coherence is related to the idea that superconductivity is due to the mutual interaction and correlation of the behavior of electrons that extend over a considerable distance. The coherence length is the maximum distance up to which the states of pairs of electrons are correlated to produce superconductivity. It is represented by $\in_0$. In view of this, $\in_0^3$ represents a volume known as *coherence volume*. The properties of a superconductor depend on the correlation of electrons within this coherence volume. A large number of electrons in such a volume forms Cooper pairs and acts together in the superconductivity so that a transition is very sharp. The ratio of London penetration depth to the coherence length is given by

$$k_k = \frac{\lambda_L}{\varepsilon_0}$$

where k_k is a number known as Ginzburg-Landau parameter. This number demarcates the two types of superconductors. For type-I superconductors, $0 < k_k < 1/\sqrt{2}$ and for type-II superconductors, $k_k > 1/\sqrt{2}$. BCS theory says that the coherence length $\in_0$ is related to the energy gap according to

$$\varepsilon_0 = \frac{h v_F}{2\pi \cdot 2\Delta}$$

Here 2Δ is the energy gap and v_F is the Fermi velocity, given by $v_F = \sqrt{2E_F/m}$.

20.12 HIGH TEMPERATURE (HI-T_C) SUPERCONDUCTIVITY LO8

The high temperature superconductors represent a new class of materials, those bear extraordinary super conducting and magnetic properties. These types of superconductors have great potential for wide ranging technological applications. The need to keep the superconducting circuits at Liquid helium temperatures has revealed that all such applications are expensive and specialized. Bendroz and Muller discovered a class of superconductors with higher critical temperatures. The high temperature superconductors have transition temperature above 40 K or may be even 90 K. Liquid nitrogen is better coolant with temperature (T_c > 77K) than helium because of its larger heat capacity. It is inexpensive also. Since it has a temperature of 77 K, the new materials can be maintained in their superconducting state relatively easily and cheaply. There are at least four different crystal structures that result in high temperature superconductivity, all composed of layers of copper-oxygen structures sandwiched around layers of other elements. These are as follows

(i) Yttrium barium copper oxide ($YBa_2Cu_3O_7$) with 92 K as T_c.

(ii) Bismuth strontium calcium copper oxide {$(BiPb)_2Sr_2Ca_2Cu_3O_x$} with 105 K as T_c.

(iii) Thallium barium calcium copper oxide ($TlBa_2Ca_2Cu_3O_y$) with 115 K as T_c.

(iv) Mercury barium calcium copper oxide ($HgBa_2Ca_2Cu_3O_y$) with 135 K as T_c.

In view of their diverse applications, there has been a tremendous effort to find still more of them with even high T_c. The hi-T_c materials are all type-II superconductors and have extremely high upper critical fields. However, new high temperature superconductors have a drawback as they are very brittle and do not carry enough current. Moreover, in the new hi-T_c superconductors, the formation of electrons pairs is not caused by the same electron lattice interaction as in the BCS theory. So far it not well understood that what does cause the formation of pairs. Intense research is being conducted in this direction.

20.13 APPLICATION OF SUPERCONDUCTIVITY LO8

The attractive property of zero resistance (absence of Joule heating) and Meissner effect make superconductors very useful for many applications. Some of the applications are listed below.

(i) Superconducting cables can be used to transmit electric power over long distances without power losses.

(ii) Superconductors can be used to check unwanted magnetic flux by using their diamagnetic property.

(iii) By using superconductor, very fast and accurate computers can be developed.

(iv) Superconductors are used to detect brain wave activities.

(v) Superconductors are also used in harnessing the various forms of nuclear energies.

(vi) Most successful application is to use superconducting wire to carry the large currents in high field electromagnets. Such superconducting magnets are widely used in modern particle accelerators. Superconducting magnets are also used in magnetic resonance imaging (MRI) in medicine.

(vii) The superconducting quantum interference device (SQUID) may be configured as a magnetometer to detect incredibly small magnetic fields, i.e., small enough to measure the magnetic fields in living organisms.

(viii) Highly powerful strong field superconductor electromagnets are fabricated using liquid helium superconductors. These electromagnets are used in NMR spectrometers and NMR imaging that are employed in medical diagnosis. Electromagnets are also used to produce Josephson's devices, electromagnetic shields and magnetically levitating world's fastest trains.

(ix) Low temperature superconductors have been used to construct fractional wavelength antennas, leading to a significant improvement in radiation efficiency. However, the use of liquid helium as a cryogen limits the application of such antennas.

(x) It is seen that conventional metal guides at mm wavelength have attenuations of the order of 10 dB/m due to the high value of surface resistance of the metal walls at ~ 200 GHz. Therefore, potential application for superconductors is also in the construction of electromagnetic waveguides. The advantage over conventional metal waveguides would be at the higher frequencies.

(xi) Now it has been possible to design ceramic superconductors which can act at temperature > 77 K, i.e., it can act as hi-T_c superconductors. These superconductors have advantage over low T_c superconductors because liquid nitrogen can be used as coolant, which is cheaper and has better cooling due to its high thermal capacity.

(xii) Other industrial applications of superconductors are through magnets, sensors, transducers and magnetic shielding.

(xiii) Superconductors also have applications in power generation, energy storage, fusion, transformers and transducers.

SUMMARY

The topics covered in this chapter are summarised below.

- There are several factors that contribute to the electrical resistivity of a solid. For example, the deviations from a perfect lattice can scatter the electrons. Also, the vibrations of lattice ions (phonons) take place in normal modes. So, the electrons and phonon interaction scatters conduction electrons and hence causes more resistance. Therefore, it is clear that the electrical resistance of a solid will decrease if we cool the solid. It is amazing that electrical resistivity of some metals, alloys and compounds drops suddenly to zero when they are cooled below a certain temperature (called critical temperature). This phenomenon is called superconductivity.
- Properties of superconductors viz. electrical property, magnetic property and thermal property, were talked about. An important property of the superconducting phase is the repulsion of magnetic flux lines from the bulk of superconductor. It is called Meissner effect. When a specimen is placed in a magnetic field, the magnetic flux lines pass through it. Now, if the temperature is decreased below the transition temperature (T_c), it expels all the magnetic flux lines from inside of the specimen.
- Depending upon the magnetisation behaviour, the superconductors can be classified into two groups, namely type-I or soft superconductor and type-II or hard superconductor. Type-I superconductor expels all the magnetic field abruptly from the interior and becomes an ideal diamagnetic material. However, type-II superconductor loses magnetisation gradually rather than abruptly.
- The Maxwell's equations of electromagnetic waves were not able to explain the "zero resistance" and "ideal diamagnetism" of superconductors. However, London and London in the year 1935 derived two new equations to explain the superconducting state of matter. These equations are known as first London equation and second London equation, given below respectively

$$\frac{d\vec{J}}{dt} = \frac{ne^2}{m}\vec{E} \quad \text{and} \quad \vec{\nabla} \times \vec{J} = -\frac{ne^2}{m}\vec{B}$$

- Effects of isotopes and magnetic field on the superconductors were discussed.
- The second London equation $\vec{\nabla} \times \vec{J} = -\frac{ne^2}{m}\vec{B}$ explains well the Meissner effect. Then it was proved that this equation also predicts the penetration of supercurrent and magnetic flux in a superconductor.

  Calculations of this penetration depth were done and it was obtained that the penetration depth is given by $\lambda_L = \sqrt{\frac{m}{\mu_0 e^2 n}}$.
- The generation of Cooper pair was made clear based on the interaction of electrons with the phonons. Cooper pairs are formed due to the electron lattice interactions. The two electrons of a cooper pair have equal and opposite momenta. They also have opposite spins, i.e., one electron is spin up and another is spin down. Thus, the bound Cooper pair is a spin zero object and is a boson.
- Very late after the prediction of Einstein in 1924 that bosons could condense in unlimited numbers into a single ground state, anomalous behaviour of liquid helium was noticed at low temperatures. A remarkable discontinuity in heat capacity of helium was observed when it was cooled to critical temperature of 2.17 K. The liquid density dropped and a fraction of the liquid became a superfluid with zero viscosity. This superfluidity took place due to the fraction of helium atoms which condensed to the lowest possible energy. This was referred to as Bose Einstein condensation.
- The basis of a quantum theory of superconductivity was led by the classic 1957 paper of Bardeen, Cooper and Schrieffer, which is now called the BCS theory. The formulation of BCS theory is based on two experimental facts viz the isotope effect and the variation of specific heat of superconductors.
- The superconductivity occurs when an attractive interaction between two electrons due to phonon exchange dominates the usual repulsive interaction. This is the fundamental postulate of the BCS theory. The energy of the pair of electrons in bound state is less than the energy of the pair in free state. This difference of energy is called binding energy of the Cooper pairs. It means by applying this amount of energy we can break this pair. The pairing is complete at $T = 0$ K and completely broken at $T = T_c$. It was observed that the binding energy of the Cooper pair is maximum when electrons forming the pair have opposite momenta and spins. In addition, there is a BCS wavefunction composed of particle pairs.
- The concept of coherence was introduced. The coherence length is the maximum distance up to which the states of pairs of electrons are correlated to produce superconductivity. It is represented by ε_0. The ratio of London penetration depth to the coherence length, given by $k_k = \frac{\lambda_L}{\varepsilon_0}$, demarcates the two types of the superconductors. For type-I superconductors, $0 < k_k < 1/\sqrt{2}$ and for type-II superconductors, $k_k > 1/\sqrt{2}$.
- The high temperature (hi-T_c) superconductors represent a new class of materials, those bear extraordinary superconducting and magnetic properties. Bendroz and Muller discovered a class of superconductors with higher critical temperatures. The hi-T_c superconductors have transition temperature above 40 K or

may be even 90 K. Liquid nitrogen is better coolant with temperature (T_c > 77K) than helium because of its larger heat capacity. It is inexpensive also. Since it has a temperature of 77 K, the new materials can be maintained in their superconducting state relatively easily and cheaply.

- Finally the applications of superconductors in medical, electronics, industry, power generation, transportation etc. were talked about. It was felt that the superconductivity has diverse applications in different areas of science and engineering.

SOLVED EXAMPLES

EXAMPLE 1 The critical temperature of lead is 7.2 K. Determine the penetration depth in lead at 5.1 K if the penetration depth at 0 K is 380 Å.

SOLUTION Given T_c = 7.2 K *and* λ_0 = 380 Å, λ(5.1 K) = ?

Formula used is

$$\lambda(T) = \lambda_0\left[1-\left(\frac{T}{T_c}\right)^4\right]^{-1/2}$$

$$\lambda(5.1\,K) = 380\left[1-\left(\frac{5.1}{7.2}\right)^4\right]^{-1/2}$$

$$= \mathbf{439.29\,Å}$$

EXAMPLE 2 Determine the transition temperature and critical field at 4.2 K for a given specimen of a superconductor if the critical fields are 1.41 × 10^5 and 4.205 × 10^5 amp/m at 1.41 K and 12.9 K, respectively.

SOLUTION Given H_{C1} = 1.41 × 10^5 A/m at T_1 = 14.1 K and H_{C2} = 4.205 × 10^5 A/m at T_2 = 12.9 K

Formula used is

$$H_c = H_0\left[1-\left(\frac{T}{T_c}\right)^2\right]$$

Thus, the critical fields at temperatures T_1 and T_2 can be written

$$H_{C1} = H_0\left[1-\left(\frac{T_1}{T_c}\right)^2\right] = H_0\left[\frac{T_c^2-T_1^2}{T_c^2}\right]$$

$$H_{C2} = H_0\left[\frac{T_c^2-T_2^2}{T_c^2}\right]$$

From the above equations,we get

$$\frac{H_{C1}}{H_{C2}} = \frac{T_c^2-T_1^2}{T_c^2-T_2^2} \quad \text{or} \quad \frac{1.41\times10^5}{4.205\times10^5} = \frac{T_c^2-(14.1)^2}{T_c^2-(12.9)^2}$$

$$T_c = 14.67\,\text{K}$$

Using the values,

$$H_c = 1.41 \times 10^5 \text{ A/m}, T_1 = 14.1 \text{ K and } T_c = 14.67 \text{ K}$$

$$H_C = H_0\left[1-\left(\frac{T_1}{T_c}\right)^2\right]$$

$$1.41\times10^5 = H_0\left[1-\left(\frac{14.1}{14.67}\right)^2\right]$$

or $$H_0 = 18.504 \times 10^5 \text{ A/m}$$

The critical field at $T = 4.2$ K and $T_c = 14.67$ K

$$H_c = H_0\left[1-\left(\frac{T}{T_c}\right)^2\right] = 18.504\times10^5\times\left[1-\left(\frac{4.2}{14.67}\right)^2\right]$$

$$= \mathbf{16.99\times10^5\ A/m}$$

Example 3 Assuming that the critical magnetic field depends upon T, find the critical current density for 1.0 mm diameter wire of lead at 4.2 K. Take critical temperature for lead as 7.18 K and H_0 for lead as 6.51×10^4 A/m.

Solution Given $T = 4.2$ K, $H_0 = 6.51 \times 10^4$ A/m and $T_c = 7.18$ K

$$H_c = H_0\left[1-\left(\frac{T}{T_c}\right)^2\right] = 6.51\times10^4\times\left[1-\left(\frac{4.2}{7.18}\right)^2\right]$$

$$= 4.29\times10^4 \text{ A/m}$$

Critical current $I_c = 2\pi r H_c$, $r = \frac{d}{2} = \frac{1.0}{2}$ mm $= 0.5$ mm

Current density $J_c = \frac{I_c}{\pi r^2} = \frac{2\pi r H_c}{\pi r^2} = \frac{2H_c}{r}$

$$= \frac{2\times4.29\times10^4}{0.5\times10^{-3}} = \mathbf{1.716\times10^8\ A/m^2}$$

Example 4 The critical temperature T_c for Hg with isotopic mass 199.5 is 4.185 K. What will be its critical temperature when its isotopic mass is increased to 203.4.

Solution Given $T_{c_1} = 4.185$ K, $M_1 = 199.5$ and $M_2 = 203.4$

$$T_{c_2} = ?$$

Formula used is

$$T_c M^{1/2} = \text{constant}$$

Therefore, $T_{c_1} M_1^{1/2} = T_{c_2} M_2^{1/2}$

$$4.185\times(199.5)^{1/2} = T_{c_2}\times(203.4)^{1/2}$$

$\therefore$ $$T_{c_2} = \mathbf{4.1446\ K}$$

EXAMPLE 5 Determine the penetration depth in mercury at 0 K, if the critical temperature of mercury is 4.2 K and the penetration depth is 57 nm at 2.9 K.

SOLUTION Given $T_c = 4.2\,\text{K}$

$$\lambda(2.9) = 57\,\text{nm}$$

$$\lambda(0) = ?$$

Formula used is

$$\lambda(0) = \lambda(T)\left[1 - \left(\frac{T}{T_c}\right)^4\right]^{1/2}$$

$$\lambda(0) = \lambda(2.9)\left[1 - \left(\frac{T}{T_c}\right)^4\right]^{1/2}$$

$$\lambda(0) = 57 \times 10^{-9} \times \left[1 - \left(\frac{2.9}{4.2}\right)^4\right]^{1/2}$$

$$= 57 \times 10^{-9} \times [1 - (0.6905)^4]^{1/2}$$

$$= \mathbf{50.10\ nm}$$

EXAMPLE 6 Determine the critical temperature of aluminium if the penetration depth for aluminium is 16 nm and 96 nm at 2.18 K and 8.1 K, respectively.

SOLUTION Given $\lambda(2.18) = 16\,\text{nm}$ and $\lambda(8.1) = 96$ nm.

Formula used is

$$\lambda(T) = \lambda_0\left[1 - \left(\frac{T}{T_c}\right)^4\right]^{-1/2}$$

$$\frac{96}{16} = \frac{\lambda_0[1 - (8.1/T_c)^4]^{-1/2}}{\lambda_0[1 - (2.18/T_c)^4]^{-1.2}}$$

$$6 = \sqrt{\frac{1 - \left(\frac{2.18}{T_c}\right)^4}{1 - \left(\frac{8.10}{T_c}\right)^4}}$$

or

$$35 = \frac{1}{T_c^4}\{154968.19 - 22.58\}$$

$$T_c = \mathbf{8.16\ K}$$

EXAMPLE 7 The critical temperature of a given superconducting sample is 1.19 K with mass 26.91. Determine the critical temperature when the isotope mass changes to 32.13.

SOLUTION Given $Tc_1 = 1.19$ K, $M_1 = 26.91$ and $M_2 = 32.13$, $T_{c_2} = ?$

Formula used is

$$T_{c_1} M_1^{1/2} = T_{c_2} M_2^{1/2}$$

$$T_{c_2} = \frac{T_{c_1} M_1^{1/2}}{M_2^{1/2}} = \frac{1.19 \times (26.91)^{1/2}}{(32.13)^{1/2}}$$

$$= \frac{1.19 \times 5.127}{5.668} = \frac{6.173}{5.668}$$

$$T_{c_2} = \mathbf{1.089\ K}$$

EXAMPLE 8 Considering the critical temperature of mercury as 4.2 K, calculate the energy gap in eV at $T = 0$. Also find the wavelength of a photon whose energy is just sufficient to break up Cooper pairs in mercury at $T = 0$. Find the region of the electromagnetic spectrum where such photons may be observed.

SOLUTION Cooper pair binding energy or the energy gap is $E_g = 3kT_c$

$\Rightarrow$ $$E_g = 3 \times 1.38 \times 10^{-23} \times 4.2 = 1.74 \times 10^{-22}\ \text{J}$$

or $$E_g \text{ in eV} = \frac{1.74 \times 10^{-22}}{1.6 \times 10^{-19}} = 1.08 \times 10^{-3}\ \text{eV}$$

The wavelength λ of a photon of energy E_g

$$\lambda = \frac{hc}{E_g} \quad \left(\because E_g = \frac{hc}{\lambda}\right)$$

$$= \frac{6.6 \times 10^{-34} \times 3 \times 10^8}{1.74 \times 10^{-22}} = \mathbf{1.14 \times 10^{-3}\ m}$$

From the value of λ, it is clear that these photons are in the very short wavelength part of the microwave region.

EXAMPLE 9 In Example 8, does the metal look like a superconductor to electromagnetic waves having wavelengths shorter than 1.14×10^{-3} m?

SOLUTION Since the energy content of shorter wavelength photons is more than sufficient to break up the Cooper pairs or to excite the conduction electrons through the energy gap into the non-superconducting states above the gap, the metal would not work as superconductor to the said electromagnetic waves.

OBJECTIVE TYPE QUESTIONS

Q.1 The phenomenon of superconductivity was discovered by Kamerlingh Onnes in
(a) 1931 (b) 1911 (c) 1921 (d) 1811

Q.2 The temperature at which a normal material turns into a superconducting state is called the
(a) critical temperature (b) superconducting temperature
(c) high temperature (d) low temperature

Q.3 The transition temperature T_c of all superconducting materials satisfies the relation
(a) $T < T_c$ (b) $T > T_c$ (c) $T = T_c$ (d) none of these

Q.4 Resistance of a substance in superconducting state becomes
(a) infinite (b) finite (c) zero (d) arbitrary

Q.5 When a superconducting material is placed in an external magnetic field, it
(a) repels the magnetic field lines (b) does not influence magnetic field lines
(c) attracts the magnetic field lines (d) enhances the magnetic field lines

Q.6 In superconducting state, the material becomes
(a) ferromagnetic (b) perfect diamagnetic
(c) strong paramagnetic (d) ferrimagnetic

Q.7 If the bulk specimen passes through the normal state to the superconducting state, its entropy
(a) increases (b) decreases
(c) remains unaffected (d) becomes half of it

Q.8 The variation of critical temperature with increasing atomic mass is called
(a) isotope effect (b) isobar effect (c) isomeric effect (d) none of these

Q.9 State of superconducting is generally achieved at
(a) very low temperatures (b) low temperatures
(c) high temperature (d) superhigh temperatures

Q.10 The London penetration depth is given by
(a) $\lambda = \left(\dfrac{m}{n\mu_0 e^2}\right)^{1/2}$ (b) $\lambda = (nm\mu_0 e^2)^{1/2}$ (c) $\lambda = \left(\dfrac{m}{m\mu_0 e^2}\right)^{1/2}$ (d) none of these

Q.11 Superconductor is
(a) an ideal conductor (b) an ideal diamagnetic
(c) not an ideal diamagnetic (d) both (a) and (b)

Q.12 The magnetic susceptibility of a superconductor becomes
(a) positive (b) zero (c) negative (d) none of these

Q.13 Which of the following statement is correct interms of critical field strength of superconductor?
(a) It depends on mechanical stress (b) It varies with temperature
(c) It depends on the purity of metal (d) All of these

Q.14 Which of the following relation represents London's equation?
(a) $\text{curl } \vec{E} = -\mu_0 \dfrac{\partial \vec{H}}{\partial t}$ (b) $\text{curl } \vec{J}_s = -\dfrac{\vec{H}}{\lambda_L^2}$
(c) $\text{curl } \vec{J}_s = -\dfrac{\lambda_L^2}{\vec{H}}$ (d) both (a) and (b)

Q.15 Which of the following relation is correct?
(a) $H_c = H_0\left[1 + \left(\dfrac{T}{T_c}\right)^2\right]$ (b) $H_c = H_0\left[1 - \left(\dfrac{T}{T_c}\right)^2\right]$
(c) $H_c = H_0\left[1 - \left(\dfrac{T}{T_c}\right)\right]$ (d) None of these

Q.16 Which of the following relation is correct?
(a) $C_n(T) = \gamma T + \beta T^3$ (b) $\text{curl } \vec{J} = -\dfrac{ne^2}{m}\vec{B}$
(c) $\dfrac{\lambda(T)}{\lambda(0)} = \left[1 - \left(\dfrac{T}{T_c}\right)^4\right]^{-1/2}$ (d) all of these

Short-Answer Questions

Q.1 What do you understand by superconductivity?

Q.2 Explain whether a superconductor is a perfect diamagnet.

Q.3 What is Meissner effect?

Q.4 Mention some important changes that occur in substances when they change from normal state to superconducting state.

Q.5 Name the types of superconductors?

Q.6 What is isotopic effect in superconductors?

Q.7 What are high T_c superconductors? Give some examples.

Q.8 Discuss how Cooper pairs are formed?

Q.9 What do you understand by phonons?

Q.10 What is the binding energy of a Cooper pair?

Practice Problems

General Questions

Q.1 What is superconductivity? Give the main properties of a superconductor?

Q.2 What do you mean by critical field in superconductivity?

Q.3 Explain the distinction between type-I (soft) and type-II (hard) superconductors.

Q.4 Explain the difference between the type-I and type-II superconductors using Meissner effect.

Q.5 What do you mean by superconductivity? Describe the effect of (a) magnetic field (b) frequency (c) isotopes on superconductors.

Q.6 Define and explain the Meissner effect in superconductors.

Q.7 Describe the effect of an external magnetic field on the superconducting state of a material. What do you mean by flux exclusion and what is Meissner effect?

Q.8 What is the significance of critical temperature, critical magnetic field and critical current density for superconductors?

Q.9 Derive the London equations and discuss how they explain Meissner effect and flux penetration?

Q.10 Give brief outline of Bardeen, Cooper and Schrieffer (BCS) theory of superconductivity. Show that this theory provides adequate explanation of superconducting state.

Q.11 Describe major uses and potentialities of superconductors.

Q.12 Why superconductivity is a low temperature phenomenon?

Q.13 What do you mean by high temperature superconductivity?

Q.14 Write a note on

(i) Penetration of magnetic field in a superconductor and penetration depth.

(ii) Flux quantization.

Q.15 Discuss Bose Einstein condensation.

Q.16 Explain why the electrical resistivity of a solid goes down with the increase in temperature?

X-Rays

21

LEARNING OBJECTIVES

After reading this chapter you will be able to

LO 1 Know about origin of *X*-rays

LO 2 Learn about properties of *X*-rays, continuous X-ray spectrum, line spectrum

LO 3 Explain Moseley's Law on the basis of Bohr's theory and its importance

LO 4 Discuss practical applications of *X*-rays

Introduction

In the phenomenon of photoelectric effect, a photon can eject an electron from a metallic surface if its energy is greater than the threshold value. As is clear from this process, the photons of light can transfer energy to the electrons. However, in 1895 *Roentgen* observed the reverse process in which the kinetic energy of moving electrons was converted into photons under suitable conditions. He observed that when fast electrons impinge on the anode material in the Crooks discharge tube, some rays are produced that have highly penetrating power. These rays were named as X-rays. Actually in this process, the electrons passing near a nucleus in the target are decelerated and hence emit a continuous spectrum of radiation (Bremsstarhlung) ranging up from a minimum wavelength. In addition to this, the electrons may eject an electron from an inner shell of a target atom. Then the resulting transition of an electron of a higher energy level to this level produces radiation of specific wavelength, which is the characteristic X-ray spectrum of the target and is specific to the target element.

These rays are found to penetrate paper, thick wooden blocks, glass and thin metal sheet, etc. The penetrating power of these rays depends on the speed of the moving electrons, i.e., faster the moving electrons the greater the penetration power of the X-rays. The intensity of X-ray beam is found to be in direct proportion to the number of electrons, i.e., the intensity of X-ray beam is greater if the number of electrons is larger.

21.1 ORIGIN OF X-RAYS

LO1

X-rays are produced when fast moving electrons strike with a target of high melting point and high atomic number, for example, tungsten, platinum, molybdenum, copper and chromium. A Coolidge tube is generally used to produce *X*-rays, as shown in Fig. 21.1. It is evacuated at low pressure of 10^{-5} mm of Hg. This tube contains a cathode *C* and target *T*. The filament is heated with a low voltage source to emit electrons. A high voltage ($\approx 10{,}000 - 30{,}000$ volts) is applied between the filament and target to accelerate the electrons. When these high velocity electrons strike the target, *X*-rays are produced. The wavelength of *X*-rays produced depends on various factors like nature of target material, and initial and final energies of incident electrons.

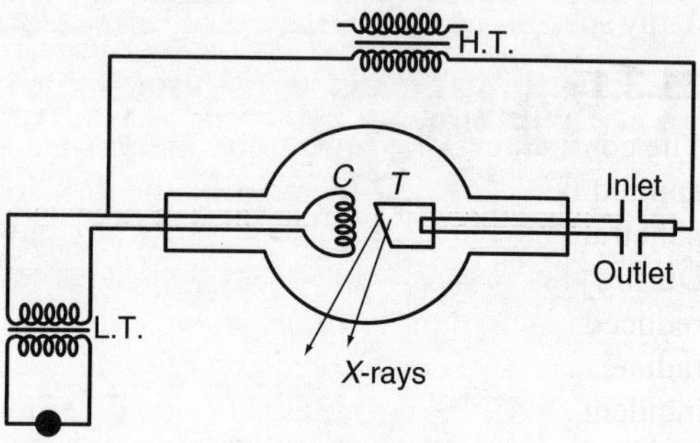

FIGURE 21.1

Only about 0.2% of the electron beam energy is used to produce *X*-rays, the rest of the energy transforms into heat. In order to remove the heat so produced by the beam, the target material is mounted on a hollow copper tube through which the cold water is flowing continuously.

21.1.1 Control of Intensity

The intensity of *X*-rays is controlled by controlling the intensity of incident electrons striking the target. As mentioned earlier, actually greater number of the incident electrons produce more intense *X*-rays. In other words, the intensity of *X*-rays can be controlled by controlling the current in the filament.

21.1.2 Control of Penetrating Power

The penetrating power of *X*-rays is controlled by controlling the potential difference between the filament and target. If the potential difference (V) is increased, the energy of incident electrons increases. This results in the *X*-rays of higher energy and hence their higher penetration power.

21.2 PROPERTIES OF X-RAYS

LO2

Properties of *X*-rays are listed below.

(i) *X*-rays are electromagnetic radiations of very short wavelength of the order of 10^{-10} metre and they travel with the velocity of light.

(ii) They are not deflected in the presence of electric and magnetic fields. It means they are not charged particles like electrons.

(iii) They produce fluorescence in some substances like barium platino-cyanide, zinc sulphide, etc.

(iv) They affect the photographic plate and the effect is more intense than light.

(v) *X*-rays ionize the gas when they are passing through it.

(vi) They produce photoelectric effect.

(vii) They can pass through many solids due to high penetrating power.

(viii) They show interference, diffraction and polarisation like light.

21.3 X-RAY SPECTRA LO2

X-ray spectra consist of two types of spectra, namely continuous spectrum and characteristic or line spectrum.

21.3.1 Continuous *X*-ray Spectrum

The continuous *X*-ray spectrum can be produced at sufficiently low potential applied between the target and filament. Some of the fast moving electrons penetrate the core of the atoms of the target material, as shown in Fig. 21.2. During this process, the electrons get decelerated and their velocities are reduced. According to the classical theory, the decelerated charge particle radiates energy in the form of *X*-rays. Due to different losses in the energy of incident electrons, the radiations of all possible wavelengths within a certain range are emitted which constitute a continuous spectrum.

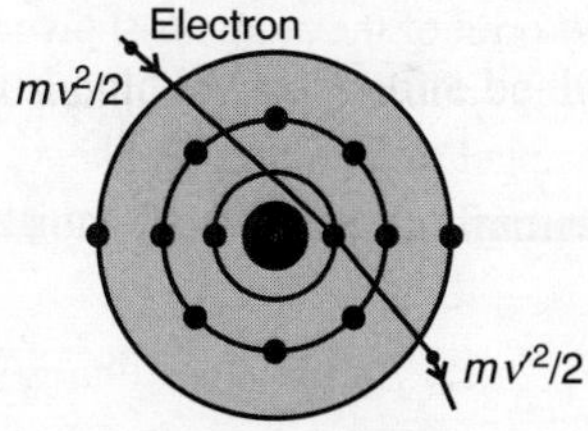

FIGURE 21.2

If the velocity of the incident electron reduces from v to v', then the frequency (ν) of the emitted *X*-ray photon is given by

$$\frac{1}{2}mv^2 - \frac{1}{2}mv'^2 = h\nu \tag{i}$$

If the incident electron is completely stopped, then the frequency of the emitted *X*-ray photon will be maximum, i.e.,

$$\frac{1}{2}mv^2 = h\nu_{\text{max}} = eV \tag{ii}$$

where, V is accelerating potential of electron,

$$\text{or} \quad \frac{hc}{\lambda_{\text{min}}} = eV \qquad \left(\because \nu_{\text{max}} = \frac{c}{\lambda_{\text{min}}}\right)$$

$$\text{or} \quad \lambda_{\text{min}} = \frac{hc}{eV} \tag{iii}$$

Eq. (iii) gives the minimum wavelength limit of the continuous *X*-ray spectrum. This is also called quantum limit. Putting the values of *h, c* and *e* in the equation, we get

$$\lambda_{\text{min}} = \frac{6.62 \times 10^{-34} \times 3 \times 10^8}{1.6 \times 10^{-19} \times V}$$

$$\lambda_{\text{min}} = \frac{12400}{V} \tag{iv}$$

Thus, Eq. (iv) shows that λ_{min} is inversely proportional to the voltage (V) applied between the cathode and target. If V is in volts, λ_{min} is obtained in Å.

21.3.1.1 *Features of Continuous X-rays Spectrum*

Some features of continuous *X*-ray spectrum are given below.

(i) *X*-rays are produced due to the deceleration of fast moving electrons.

(ii) The intensity of continuous spectrum increases as the potential is increased (Fig. 21.3).

(iii) The minimum wavelength limit shifts towards lower wavelength as the potential is increased.

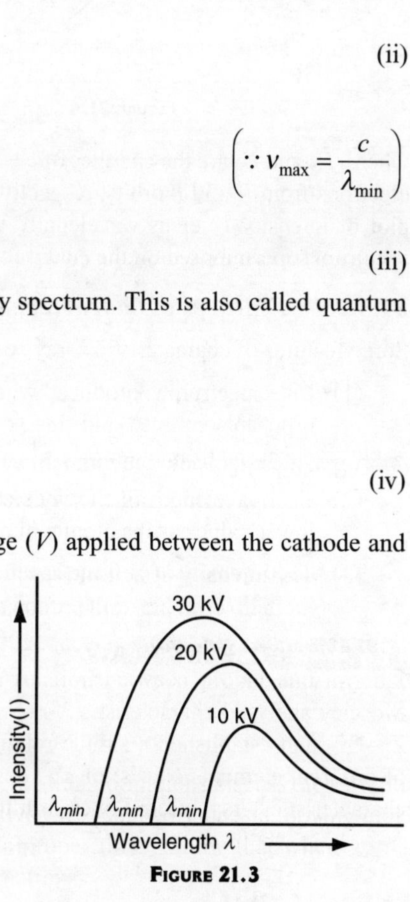

FIGURE 21.3

21.3.2 Characteristic *X*-ray Spectrum: Line spectrum

It consists of some well-defined wavelengths superimposed on the continuous spectrum. The spectrum lines generally occur in the form of small groups. The characteristic *X*-ray spectrum is produced when extremely high energetic electrons penetrate well inside the atoms of the target and collide with tightly bound electrons of innermost *K* orbit of the atom, as shown in Fig. 21.4. If this vacancy is filled up by the electron of the second orbit *L*, K_α line is produced and if it is filled up by the electron of the third orbit *M*, K_β line is produced and so on. In this manner we get different series like *K*, *L*, *M* etc. in the emission of *X*-ray spectra, as shown in Fig. 21.5. Therefore, the energy of the emitted *X*-ray photon corresponding to K_α line is given by

$$h\nu_{K_\alpha} = E_K - E_L \tag{v}$$

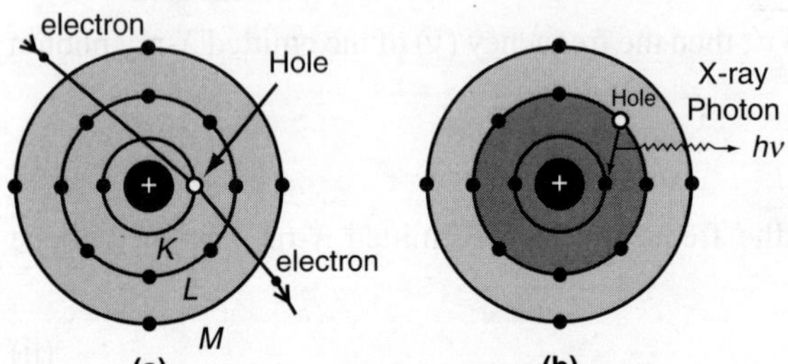

Figure 21.4

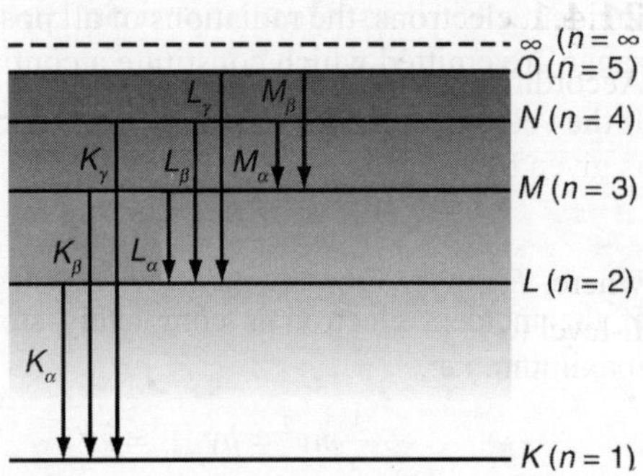

Figure 21.5

where, E_K and E_L are the energies of electrons required to remove electrons from *K* and *L* orbits, respectively. Fig. 21.6 depicts the plot of intensity *I* versus wavelength λ for characteristic *X*-ray spectrum superimposed on the continuous spectra.

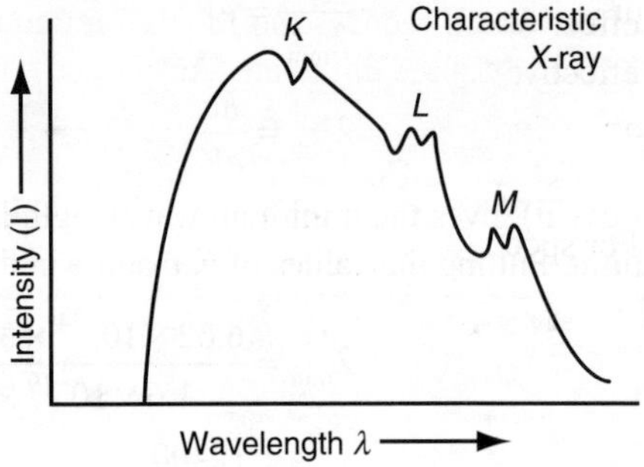

Figure 21.6

21.3.2.1 *Features of Characteristic X-ray Spectrum*

Some features of characteristic *X*-ray spectrum are given below

(1) Line spectrum is produced when electrons of innermost orbit are removed and this is followed by jumping of electrons from outer orbits.

(2) The wavelength of the lines of *K*-series shifts towards lower values as the atomic number of the target element increases.

(3) The intensity of definite spectral line depends on the probability of that particular transition.

21.4 MOSELEY'S LAW

LO3

Moseley analysed characteristic *X*-ray spectra emitted by targets of a number of heavy elements ($Z = 22$ to $Z = 30$) and established a simple relation between them. He observed that the characteristic *X*-ray spectra of different elements consist of *K*, *L*, *M* series. Also, he found that on increasing the atomic number *Z*, the characteristic *X*-ray spectral lines shift towards shorter wavelengths side or towards higher frequencies side.

Moseley plotted a graph of the *K*-series lines of the characteristic *X*-ray spectra for a number of elements between $\sqrt{\nu}$ (ν is the frequency) and *Z* of the target material and found almost a straight line (Fig. 21.7). From this he concluded that

$$\sqrt{\nu} \propto Z \text{ or } \nu \propto Z^2$$

This conclusion is known as Moseley's law. Mathematically

$$\nu = a(Z - b)^2$$

where, *a* and *b* are the constants for the given transition of the *K*-series.

The constant *b* is known as screening constant.

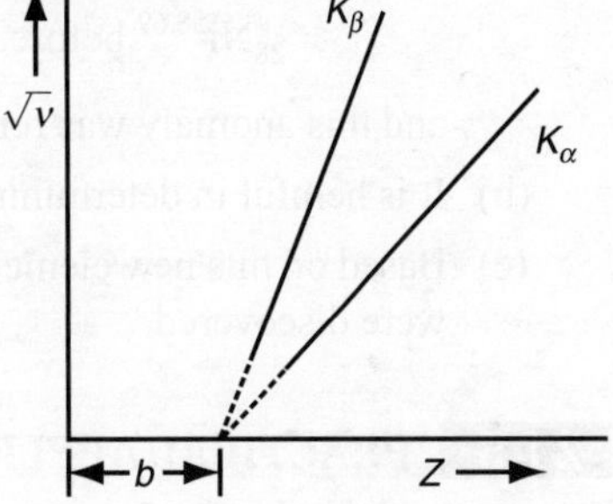

FIGURE 21.7

21.4.1 Explanation based on Bohr's Theory

According to the Bohr's theory,

$$\nu = \frac{c}{\lambda} = RcZ^2\left\{\frac{1}{n_1^2} - \frac{1}{n_2^2}\right\},$$

where, *R* is Rydberg constant and *Z* is the atomic number of the element. When an electron jumps from *L*-level to *K*-level, i.e., $n_2 = 2$ and $n_1 = 1$, the frequency of K_α line is given by

$$\nu = RcZ^2\left\{\frac{1}{1^2} - \frac{1}{2^2}\right\} = \frac{3}{4}RcZ^2$$

Now, if one of the two electrons of the *K*-shell of an atom is knocked off, an electron from *L*-shell would make a transition to the *K*-shell, thereby emitting K_α line. However, the remaining *K*-electrons produce a screening effect which reduces the force of attraction of the nucleus for the *L*-electron. Therefore, for *L*-electron the effective charge of the nucleus reduces to (*Z*–1)e. Hence, replacing *Z* by *Z*–1 in Bohr's formula, we get

$$\nu = \frac{3}{4}Rc(Z-1)^2$$

For spectral line L_α, $n_2 = 3$, $n_1 = 2$

$$\nu = Rc(Z - 7.4)^2\left(\frac{1}{2^2} - \frac{1}{3^2}\right)$$

$$= \frac{5}{36}Rc(Z - 7.4)^2$$

In general, we can write

$$\nu = Rc(Z - b)^2\left[\frac{1}{n_1^2} - \frac{1}{n_2^2}\right]$$

or $$\nu = a(Z - b)^2$$

or $$\sqrt{\nu} \propto (Z - b)$$

21.4.2 Importance of Moseley's Law

The importance of Moseley's law is listed below.

(a) It yields a unique method for identifying as well as for classifying different elements in the periodic table. According to Moseley's law, the atomic number *Z* of the element determines the physical and chemical properties of the element. Therefore, the elements in the periodic table must be arranged according to Z instead of atomic weight. Infact before Moseley, Mendeleef constructed periodic table by arranging elements in the order of increasing atomic weight and he put

$$_{28}Ni^{58.69} \text{ before } _{27}Co^{58.94}, \; _{19}K^{39} \text{ before } _{18}A^{40}$$

and this anomaly was removed by Moseley by putting them according to their atomic number.

(b) It is helpful in determining Z of rare earth elements and position in the periodic table.

(c) Based on this new elements like rhenium (75), hafnium (72), promethium (61), technetium (43) etc. were discovered.

21.5 PRACTICAL APPLICATIONS OF X-RAYS LO4

We mention below some of the important applications of X-rays.

(i) The X-rays are extensively used in surgery to detect fractures, diseased organ, etc. as these rays can pass through flesh and not through bones.

(ii) Since the X-rays kill the diseased tissues of the body, they are used to cure some skin diseases, malignant tumors, etc.

(iii) The X-rays are used to detect defects in motor tyres, golf and tennis balls, etc. and also for testing the uniformity of insulating materials.

(iv) People have explored the high penetrating power of the X-rays to investigate the structure of metals, crystalline solids, alloys, complex organic molecules, etc.

(v) The X-rays are used for the detection of explosive arms at customs posts.

(vi) X-ray spectrography and X-ray crystallography are the most recent applications of the X-rays.

SUMMARY

The topics covered in this chapter are summarised below.

- For X-ray generation, a process is used in which the kinetic energy of moving electrons is converted into photons under suitable conditions.
- Bremsstarhlung arises from the retardation of charged particles, commonly electrons, stopped by the target of high atomic number.
- The X-rays can pass through the opaque materials and cause phosphorescent substances to glow.
- The penetrating power of the X-rays depends on the speed of the moving electrons, i.e., faster the moving electrons greater the penetration power of the X-rays.
- The intensity of X-ray beam is found to be in direct proportion to the number of electrons, i.e., the intensity of X-ray beam is greater if the number of electrons is larger.
- The continuous X-ray spectrum is produced at sufficiently low potential applied between the target and filament of the Coolidge tube. Here, some of the fast moving electrons penetrate the core of the atoms of the target material, get decelerated and radiate energy in the form of X-rays.

- The minimum wavelength limit of the continuous spectrum, which is also called quantum limit, is given by $\lambda_{min} = \dfrac{hc}{eV}$, where V is the accelerating potential of the electron.
- The characteristic X-ray spectrum is produced when extremely high energetic electrons penetrate well inside the atoms of the target and collide with tightly bound electrons of innermost K orbit of the atom. If this vacancy is filled up by the electron of the second orbit L, K_α line is produced. If this vacancy is filled up electron of the third orbit M, K_β line is produced. Similarly L_α line is produced if the vacancy in L orbit is filled up by the electron of the third orbit M.
- Moseley analysed characteristic X-ray spectra emitted by targets of heavy elements whose atomic numbers ranged from 22 to 30. He achieved a conclusion that satisfies the expression $\sqrt{\nu} = a(Z - b)$. Here a and b are the constants for the given transition of the K series.
- Moseley's law was explained based on Bohr's theory that related the frequency of radiation with the atomic number of the element and the transitions between two energy levels.
- Moseley's law had removed the anomaly took place due to the arrangement of elements in periodic table as per the order of increasing atomic weight done by Mendeleef. Moseley suggested this arrangement based on the atomic number rather than the atomic weight.

SOLVED EXAMPLES

EXAMPLE 1 An X-ray tube operates at the voltage (i) 40 kV, (ii) 20 kV, and (iii) 100 kV. Find the maximum speed of electrons striking the anti-cathode and shortest wavelength of X-rays produced.

SOLUTION Given (i) $V = 40$ kV (ii) $V = 20$ kV and (iii) 100 kV. Formula used is

$$\frac{1}{2}mv^2 = eV \quad \text{and} \quad \lambda_{min} = \frac{12400}{V}$$

or

$$v = \sqrt{\frac{2eV}{m}} = \sqrt{V}\sqrt{\frac{2 \times 1.6 \times 10^{-19}}{9.1 \times 10^{-31}}}$$

$$v = 0.593 \times 10^6 (\sqrt{V})\,\text{m/sec}$$

(i) $V = 40 \times 10^3$ V

$$v = 0.593 \times 10^6 \sqrt{4 \times 10^4}, \quad \lambda_{min} = \frac{12400}{4 \times 10^4}$$

$$= \mathbf{1.186 \times 10^8\ m/sec} \qquad = \mathbf{0.31\text{Å}}$$

(ii) $V = 20 \times 10^3$ V

$$v = 0.593 \times 10^6 \sqrt{2 \times 10^4}, \quad \lambda_{min} = \frac{12400}{2.0 \times 10^4}$$

$$= \mathbf{8.39 \times 10^7\ m/sec} \qquad = \mathbf{0.62\text{Å}}$$

(iii) $V = 100 \times 10^3\ \text{V} = 10 \times 10^4$ V

$$v = 0.593 \times 10^6 \sqrt{10 \times 10^4}, \quad \lambda_{min} = \frac{12400}{10^5}$$

$$= \mathbf{1.875 \times 10^8\ m/sec} \qquad = \mathbf{0.124\text{Å}}$$

EXAMPLE 2 The short wavelength limit of the continuous X-ray spectrum emitted by an X-ray tube operated at 30 kV is 0.414 Å. Calculate the Planck's constant.

SOLUTION Given $V = 3.0 \times 10^4$ V, $\lambda_{min} = 0.414 Å$

$$e = 1.6 \times 10^{-19} \text{C} \quad \text{and} \quad c = 3 \times 10^8 \text{m/sec}$$

Formula used is

$$\lambda_{min} = \frac{hc}{eV} \quad \text{or} \quad h = \frac{eV}{c}\lambda_{min}$$

$$h = \frac{1.6 \times 10^{-19} \times 3.0 \times 10^4 \times 0.414 \times 10^{-10}}{3 \times 10^8}$$

$$= \mathbf{6.624 \times 10^{-34} \text{ J sec}}$$

EXAMPLE 3 An X-ray tube is operated at 25 kV. Calculate the minimum wavelength of X-rays emitted from it.

SOLUTION Given $V = 25 \times 10^3$ V

Formula used is

$$\lambda_{min} = \frac{12400}{V} = \frac{12400}{25 \times 10^3}$$

$$= \mathbf{0.496 \text{ Å}}$$

EXAMPLE 4 An X-ray tube operates at 13.6 kV. Find the maximum speed of electron striking the target.

SOLUTION Given $V = 13.6 \times 10^3$ V

Formula used for maximum kinetic energy is

$$\frac{1}{2}mv^2 = eV$$

$$v = \sqrt{\frac{2eV}{m}} = (0.593 \times 10^6)\sqrt{V} \text{ m/sec}$$

$$= 0.593 \times 10^6 \times \sqrt{13.6 \times 10^3}$$

$$= \mathbf{6.92 \times 10^7 \text{ m/sec}}$$

EXAMPLE 5 If the potential difference applied across an X-ray tube is 10 kV and current through it is 2.0 mA. Calculate the velocity of electrons at which they strike the target.

SOLUTION Given $V = 10 \times 10^3$ V and $I = 2 \times 10^{-3}$ A

Formula used is $v = \sqrt{\frac{2eV}{m}} = 0.593 \times 10^6 \times \sqrt{10 \times 10^3}$

$$= \mathbf{5.93 \times 10^7 \text{ m/sec}}$$

EXAMPLE 6 Electrons are accelerated in a television tube through a potential difference of 9.8 kV. Find the highest frequency and minimum wavelength of the electromagnetic waves emitted, when these strike on the screen of the tube. In which region of the spectrum will these waves lie?

Solution Given $V = 9.8 \times 10^3$V

Formula used is

$$\lambda_{min} = \frac{hc}{eV} = \frac{12400}{V}$$

$$= \frac{12400}{9.8 \times 10^3}$$

$$= 1.26 \text{ Å}$$

and

$$\nu_{max} = \frac{c}{\lambda_{min}} = \frac{3 \times 10^8}{1.26 \times 10^{-10}}$$

$$= \mathbf{2.38 \times 10^{18}\ Hz}$$

Example 7 If the potential difference applied across an *X*-ray tube is 12.4 kV and current through it is 2 mA, calculate (i) the number of electrons striking the target per second and (ii) the speed with which they strike it.

Solution Given $V = 12.4 \times 10^3$V and $I = 2 \times 10^{-3}$ A

(i) For current $I = ne$ or $n = \frac{I}{e} = \frac{2.0 \times 10^{-3}}{1.6 \times 10^{-19}}$ or $\mathbf{n = 1.25 \times 10^{16}}$ **electrons/sec**

(ii) $v = \sqrt{\frac{2eV}{m}} = 0.593 \times 10^6 \sqrt{V}$ m/sec

$$= 0.593 \times 10^6 \times \sqrt{12.4 \times 10^3}$$

$$= \mathbf{6.6 \times 10^7\ m/sec}$$

Example 8 An *X*-ray tube is operated at an anode potential of 10 kV and anode current of 15 mA. Calculate (i) number of electrons hitting the anode per second and (ii) the minimum wavelength produced by the *X*-ray tube.

Solution Given $V = 10 \times 10^3$ V and $I = 1.5 \times 10^{-3}$ A.

(i) For current, $I = ne$ or $n = \frac{I}{e} = \frac{15 \times 10^{-3}}{1.6 \times 10^{-19}}$

$$= \mathbf{9.38 \times 10^{16}}\ \textbf{electrons / sec}$$

(ii) $\lambda_{min} = \frac{12400}{V} \text{Å} = \frac{12400}{10 \times 10^3} \text{ Å}$

$$= \mathbf{1.24\ Å}$$

Example 9 An *X*-ray tube is operated at 50 kV and current through the tube is 1.0 mA. What is the number of electrons striking the target per second?

Solution Given $I = 1.0 \times 10^{-3}$ A and $V = 50 \times 10^3$ V.

(i) For current $I = ne$ or $n = \frac{I}{e}$

or

$$n = \frac{10^{-3}}{1.6 \times 10^{-19}}$$

$$= \mathbf{6.25 \times 10^{15}}\ \textbf{electrons/sec}$$

EXAMPLE 10 What voltage must be applied to an X-ray tube for it to emit X-rays with minimum wavelength of (i) 40 pm and (ii) 1.0 Å

SOLUTION Given wavelengths (i) 40×10^{-12} m (ii) 1.0×10^{-10} m

Formula used is $\lambda_{min} = \dfrac{12400}{V}$ Å or $V = \dfrac{12400}{\lambda_{min}}$

(i) $\lambda_{min} = 4.0 \times 10^{-11}$ m

$$V = \frac{12400 \times 10^{-10}}{4 \times 10^{-11}} = \mathbf{31\ kV}$$

(ii) $\lambda_{min} = 1.0 \text{Å} \times 10^{-10}$ m

$$V = \frac{12400 \times 10^{-10}}{10^{-10}} = \mathbf{12.4\ kV}$$

EXAMPLE 11 An X-ray tube operating at (i) 44 kV (ii) 50 kV emits a continuous spectrum with shortest wavelength (i) 0.284 Å and (ii) 0.248 Å, respectively. Calculate the Planck's constant.

SOLUTION Given (i) $V = 44 \times 10^3$ V, $\lambda = 0.284 \times 10^{-10}$ m and (ii) $V = 50 \times 10^3$ V, $\lambda = 0.284 \times 10^{-10}$ m

Formula used is

$$\lambda_{min} = \frac{hc}{eV} \quad \text{or} \quad h = \frac{eV}{c} \times \lambda_{min}$$

(i) $$h = \frac{1.6 \times 10^{-19} \times 44 \times 10^3 \times 0.284 \times 10^{-10}}{3 \times 10^8}$$

$$= \mathbf{6.66 \times 10^{-34}\ J\ sec}$$

(ii) $$h = \frac{1.6 \times 10^{-19} \times 50 \times 10^3 \times 0.248 \times 10^{-10}}{3 \times 10^8}$$

$$= \mathbf{6.61 \times 10^{-34}\ J\ sec}$$

EXAMPLE 12 The K-absorption limit for Uranium is 0.1 Å. What is the excitation potential of the tube to give this radiation?

SOLUTION Given K-absorption limit, means $\lambda_{min} = 1.0 \times 10^{-11}$ m

Formula used is $$\lambda_{min} = \frac{12400 \times 10^{-10}}{V} m$$

or $$V = \frac{12400 \times 10^{-10}}{\lambda_{min}} = \frac{12400 \times 10^{-10}}{10^{-11}} = \mathbf{124\ kV}$$

EXAMPLE 13 Given that K-absorption edge for lead is 0.14Å and the minimum voltage required for producing K-lines in lead is 88.6 kV. Determine the ratio of h/e.

SOLUTION Given K-absorption edge for lead, $\lambda_{min} = 0.14 \times 10^{-10}$ m and $V = 88.6 \times 10^3$ V

Formula used is

$$\lambda_{min} = \frac{hc}{eV} \quad \text{or} \quad \frac{h}{e} = \frac{V}{c} \lambda_{min}$$

or $$\frac{h}{e} = \frac{88.6 \times 10^3}{3 \times 10^8} \times 0.14 \times 10^{-10}$$

$$= \mathbf{4.134 \times 10^{-15}\ Jsec / C}$$

EXAMPLE 14 Calculate the wavelength of K_α line for an atom of atomic number 92 by using Moseley's law and considering Rydberg constant as $1.1 \times 10^5 cm^{-1}$

SOLUTION Given $Z = 92$, $R = 1.1 \times 10^5\ cm^{-1}$

Formula used is

$$\nu = cR(Z - b)^2 \left[\frac{1}{n_1^2} - \frac{1}{n_2^2} \right]$$

K_α line is obtained when an electron jumps from L-shell (n_2=2) to K-shell (n_1=1). Now b=1 for K-series. The wavelength λ is given by

$$\frac{1}{\lambda} = R(Z - b)^2 \left[\frac{1}{n_1^2} - \frac{1}{n_2^2} \right]$$

$$= 1.1 \times 10^5 \times (92 - 1)^2 \times \left[\frac{1}{1^2} - \frac{1}{2^2} \right] = 1.1 \times 10^5 \times (91)^2 \times \frac{3}{4} \qquad \left(\because\ \lambda = \frac{c}{\nu} \right)$$

or $$\lambda = \frac{4}{1.1 \times 3 \times (91)^2 \times 10^5}$$

$$= 1.464 \times 10^{-9}\ \text{cm}$$

$$= \mathbf{0.15\ Å}$$

EXAMPLE 15 If the K_α radiation of Mo ($Z = 42$) has a wavelength of 0.71Å, determine the wavelength of the corresponding radiation of Cu (Z = 29).

SOLUTION Given $\lambda_{Mo} = 0.71 \times 10^{-10}$ m corresponding to $Z = 42$. λ_{Cu} = ? corresponding to $Z = 29$

Formula used is

$$\nu = \frac{1}{\lambda} = a(Z - b)^2 \quad \text{For} \quad K_\alpha \text{ line } b = 1$$

$$\frac{l}{\lambda_{Mo}} = a(42 - 1)^2 \quad \text{and} \quad \frac{1}{\lambda_{Cu}} = a(29 - 1)^2$$

$$\frac{\lambda_{Cu}}{\lambda_{Mo}} = \frac{a(42 - 1)^2}{a(29 - 1)^2} = \frac{1681}{784}$$

$$\lambda_{Cu} = 1.52 \times 10^{-10}\ \text{m}$$

$$= \mathbf{1.52\ Å}$$

EXAMPLE 16 Determine the wavelength of K_α X-rays emitted by an element having $Z = 79$, $b = 1$ and $a = 2.468 \times 10^{15}\ sec^{-1}$

SOLUTION Given $Z = 79$, $b = 1$, $a = 2.468 \times 10^{15}\ sec^{-1}$

Formula used is

$$\nu = a(Z - b)^2$$

$$\nu = 2.468 \times 10^{15} \times (79 - 1)^2$$

$$= 1.502 \times 10^{19}\ \text{sec}^{-1}$$

$$\lambda = \frac{c}{\nu} = \frac{3 \times 10^8}{1.502 \times 10^{19}} = 1.997 \times 10^{-11}\ \text{m}$$

$$= \mathbf{0.1997\ Å}$$

EXAMPLE 17 Calculate the ionisation potential of K-shell electron of copper. Given that Z for copper is 29 and Rydberg's constant for hydrogen $R = 1.097 \times 10^7\ \text{m}^{-1}$.

SOLUTION Given $Z = 29$ and $R = 1.097 \times 10^7\ \text{m}^{-1}$

The frequency of K_α X-ray spectral line is given by Moseley's law

$$\nu = \frac{3}{4} Rc(Z - 1)^2$$

$$= \frac{3}{4} \times 1.097 \times 10^7 \times 3 \times 10^8 \times (29 - 1)^2$$

$$= \frac{3}{4} \times 1.097 \times 10^7 \times 3 \times 10^8 \times (28)^2$$

$$= \mathbf{1.936 \times 10^{18}\ Hz}$$

Energy of photon corresponding K_α line

$$E = h\nu = \frac{6.62 \times 10^{-34} \times 1.936 \times 10^{18}}{1.6 \times 10^{-19}}\ \text{eV}$$

$$= 8.01\ \text{keV}$$

Now

$E = E_K - E_L$, where E_K is the ionisation energy, i.e., energy of electrons required to remove electrons from K-orbit

$$\therefore \quad E_K = E + E_L$$

$$= 8.01 + 0.931\ \text{keV (Taking } E_L = 0.931\ \text{keV)}$$

$$= \mathbf{8.942\ keV}$$

EXAMPLE 18 Calculate the frequency of K_α line, when atomic number of the anti-cathode is 79. Given $R = 1.097 \times 10^7\ \text{m}^{-1}$.

SOLUTION Given $Z = 79$ and $R = 1.097 \times 10^7 \text{m}^{-1}$

Formula used is

$$\nu = \frac{3}{4} Rc(Z - 1)^2$$

$$= \frac{3}{4} \times 1.097 \times 10^7 \times 3 \times 10^8 \times (79 - 1)^2$$

$$= \mathbf{1.502 \times 10^{19}\ s^{-1}}$$

EXAMPLE 19 Find the energy in eV and wavelength of K_α X-ray line of Co ($Z = 27$).

SOLUTION Given $Z = 27$

Formula used for K_α line is

$$\nu = \frac{3}{4}Rc(Z-1)^2$$

$$= \frac{3}{4} \times 1.097 \times 10^7 \times 3 \times 10^8 \times (27-1)^2$$

$$= 1.67 \times 10^{18} \text{ Hz}$$

$$\text{Energy } E = h\nu = \frac{6.62 \times 10^{-34} \times 1.67 \times 10^{18}}{1.6 \times 10^{-19}} \text{ eV}$$

$$= 6.91 \text{ keV}$$

Now wavelength

$$\lambda = \frac{c}{\nu} = \frac{3 \times 10^8}{1.67 \times 10^{18}}$$

$$= \mathbf{1.79\ \text{Å}}$$

OBJECTIVE TYPE QUESTIONS

Q.1 X-rays are
(a) electromagnetic radiations (b) a stream of protons
(c) a stream of electrons (d) a stream of neutrons

Q.2 The speed of X-rays in vacuum is
(a) greater than that of the visible light (b) equal to that of the visible light
(c) smaller than that of the visible light (d) none of these

Q.3 In order to produce X-rays, an element of high atomic number is bombarded by
(a) photons (b) electrons (c) neutrons (d) protons

Q.4 Which one of the following has the lowest wavelength
(a) visible light (b) X-rays (c) infra-red rays (d) ultraviolet light

Q.5 What would be the energy of the X-ray photon when an X-ray tube is operated at 30 kV?
(a) 60 keV (b) 30 keV (c) 90 keV (d) 30 MeV

Q.6 The minimum wavelength limit of the continuous X-ray spectrum will be doubled if
(a) accelerating potential of electron is doubled.
(b) accelerating potential of electron is halved
(c) accelerating potential of electron is increased 4 times.
(d) accelerating potential of electron is reduced 1/4 times

Q.7 The minimum frequency limit of X-ray spectrum depends on
(a) nature of target material (b) degree of vacuum in Coolidge tube
(c) potential energy of incident electron (d) kinetic energy of incident electron

Q.8 In characteristic X-ray spectrum, K_{α} line is produced when the vacancy in K-orbit is filled by the electron of

(a) third orbit M (b) second orbit L (c) fourth orbit N (d) none of the above

Q.9 In continuous X-ray spectrum the intensity

(a) increases as the potential is increased
(b) decreases as the potential is increased
(c) increases if the number of fast moving electrons is increased
(d) increases if the number of fast moving electrons is decreased

Q.10 The frequency of any line in characteristic X-ray spectrum is directly proportional to

(a) square root of the atomic number of target element
(b) atomic number of target element
(c) square of atomic number of target element
(d) thickness of the target

Q11. X-rays were discovered by

(a) Roentgen (b) Curie (c) Bohr (d) none of these

Q12. Penetrating power of X-rays can be increased by

(a) increasing the potential between the anode and cathode
(b) increasing the cathode filament current
(c) both (a) and (b)
(d) none of these

Q13. The wavelength of X-rays is the order of

(a) 1 Å (b) 1 m (c) 1 mm (d) none of these

Q14. Which of the following relation is correct?

(a) $\lambda_{\min} = \dfrac{hc}{eV}$ (b) $\nu = Rcz^2\left[\dfrac{1}{n_1^2} - \dfrac{1}{n_2^2}\right]$

(c) $\nu = a(z-b)^2$ (d) all of these

Q15. The intensity of X-rays in Coolidge tube increases with

(a) increasing filament current
(b) increasing the potential difference between anode and cathode
(c) decreasing filament current
(d) none of these

Q16. The Bragg's law of diffraction is

(a) $2d \sin\theta = n\,\lambda/2$ (b) $2d \sin\theta = n\lambda$ (c) $\dfrac{2d}{\sin\theta} = n\lambda$ (d) none of these

SHORT-ANSWER QUESTIONS

Q.1 What are X-rays?

Q.2 How are the X-rays produced?

Q.3 How do you control the intensity of X-rays?

Q.4 Differentiate between characteristics X-ray spectrum and continuous X-ray spectrum.

Q.5 How will it affect the cut off wavelength of *X*-rays if separation between the cathode and target is doubled?

Q.6 What is Bremsstarhlung?

Q.7 Explain the difference in origin of *X*-rays and visible light.

Q.8 Is it appropriate to regard *X*-ray production as the inverse of photoelectric effect?

Q.9 What is the difference between optical spectra and *X*-ray spectra?

PRACTICE PROBLEMS

General Questions

Q.1 Discuss *X*-rays in view of their production and properties.

Q.2 Describe the construction and working of a Coolidge tube. How can you control (i) the intensity (ii) the quality of *X*-rays? What are hard and soft *X*-rays?

Q.3 Why should anti-cathode have high atomic number and high melting point?

Q.4 What are continuous and characteristics *X*-rays and how are they produced? What is the minimum wavelength limit and how it is related with the voltage applied across the *X*-ray tube?

Q.5 **(a)** Discuss the origin and mechanism of production of the continuous *X*-ray spectra. What is the source of energy of photon of continuous *X*-rays? Show that the lowest wavelength limit of continuos *X*-ray spectra is inversely proportional to accelerating potential of *X*-ray tube.

(b) Draw the graph of relative intensity of continuous spectra versus wavelength of *X*-rays and show that λ_{min} is proportional to $\frac{1}{V}$.

Q.6 The potential difference between the cathode and anode in *X*-ray tube is doubled and the separation between the cathode and target is also doubled. How will it affect the cut-off wavelength?

Q.7 Distinguish between continuous and characteristic *X*-ray spectra. Why is the characteristic spectra so called? How is the production of characteristic *X*-ray spectra accounted for? Discuss the transition for *K* and *L* series.

Q.8 What is Moseley's law? How can it be explained on the basis of Bohr's theory? What is its importance?

Q.9 **(a)** Describe Moseley's work on *X*-rays. State and explain Moseley's law. Show it graphically.

(b) Derive Moseley's law on the basis of Bohr's theory.

(c) Discuss the importance of Moseley's observations of *X*-ray spectra of different elements. What conclusions were drawn by him?

UNSOLVED QUESTIONS

Q.1 An *X*-ray tube operates at (i) 50 kV and (ii) 18 kV. Compute the shortest wavelength of *X*-rays produced and also find the maximum speed of electrons striking the target.

[Ans: (i) 0.248 Å, 1.33×10^8 m/sec (ii) 0.69 Å, 7.996×10^7 m/sec]

Q.2 Calculate the minimum wavelength when the potential difference applied to the *X*-ray tube is 98 kV.

[Ans: 0.125 Å]

Q.3 What is the shortest wavelength of *X*-ray produced in a tube when the applied voltage is 12.4kV?

[Ans: 1.0 Å]

Q.4 An *X*-ray tube operated at 40 kV emits a continuous *X*-ray spectrum with a short wavelength limit $\lambda_{min} = 0.31$Å. Calculate the Planck's constant. **[Ans: 6.61×10^{-34} J sec]**

Q.5 What voltage must be applied to an *X*-ray tube for it to emit *X*-rays with minimum wavelength of (i) 0.5 Å and (ii) 0.25 Å. **[Ans: (i) 24.84 kV (ii) 49.68 kV]**

Q.6 An *X*-ray tube is operated at an anode potential 12.4 kV and current 15mA. Calculate (i) minimum wavelength produced by *X*-ray tube and (ii) number of electrons hitting the anode per second.

[Ans: (i) 1.0 Å (ii) 9.4×10^{16} electrons/sec]

Q.7 An *X*-ray tube operating at (i) 30 kV and (ii) 200 kV emits a continuous spectrum with shortest wavelength (i) 0.414 Å and (ii) 6.2×10^{-12} m. Calculate the Planck's constant.

[Ans: (i) 6.62×10^{-34} sec (ii) 6.61×10^{-34} sec]

Q.8 For a platinum ($Z = 78$), the wavelength of L_α line is 1.32 Å. For an unknown element, the wavelength of L_α line is 4.174 Å. Determine the atomic number of unknown element. Take $b = 7.4$ for L_α line.

[Ans: 47.1]

Q.9 The wavelength of K_α line in copper is 1.54 Å. Calculate the ionisation potential of *K*-shell electron in copper. [Given the energy of *L*-shell as 0.923 keV.] **[Ans: 8.99 keV]**

Q.10 Calculate the frequency of K_α *X*-ray line of *Pb* atom. [Given *Z* for $Pb = 82$ and $R' = 3.289 \times 10^{15}$ sec^{-1} [Hint : $R' = Rc$]. **[Ans: 1.618×10^{19} Hz]**

Nanoscience and Nanotechnology

22

Learning Objectives

After reading this chapter you will be able to

LO 1 Differentiate between nanomaterials and bulk materials, and nanoscience and nanotechnology

LO 2 Understand quantum confinement, nanowires and their synthesis through top down and bottom up approaches, VLS and VS methods, and catalyst free growth mechanism, single walled and multi-walled carbon nanotubes and their fabrication methods, i.e., arc discharge method, laser ablation method, CVD technique, n-hexane pyrolysis, their properties, inorganic nanotubes and biopolymers

LO 3 Know nanoscales in 2D and 3D, nanoparticles and various methods of their synthesis including Ball milling, gas condensation, sputtering, CVD, CVC, Sol-gel and electrodeposition techniques, properties of nanoparticles, Bucky balls or fullerenes, their synthesis, properties and applications, QDs and their fabrication and characterization

LO 4 Explain applications of nanotechnology

LO 5 Evaluate limitations and disadvantages of nanotechnology

Introduction

The prefix nano in nanotechnology means a billionth ($1 \times 10^{-9} \equiv 1$ nm). The typical dimension of nanomaterials or nanostructures spans from subnanometer to several hundred nanometers. Generally the dimesion/length may be less than 100 nm. Figure 22.1 shows how things scale and how small a nanometer is. Depending on the number of dimension, the nanomaterials are classified as quantum dots (0D: zero dimension), quantum wires (1D: one dimension) and quantum wells (2D: two dimensions). Here the dimensionality, refers to the number of degrees of freedom in the electron momentum.*

In semiconductor industry, the efforts have been to reduce the size of the devices and the continued decrease in device dimensions has followed the well-known Moore's Law. The Moore's law was predicted in 1965, which states that the transistor size is decreased by a factor of 2 in every 18 months. In view of this, it is said that the study of materials in the nanometer scale is partly driven by the ever shrinking of devices in the semiconductor industry.

* If the number of degrees of freedom is 3, then the material is said to be in bulk form. Since the electrons are free to move in three dimensions, the extent of the confinement in bulk materials is zero. Similarly the extent of the confinement in quantum well is 1, in quantum wire it is 2 and in quantum dot it is 3.

22.1 HOW NANOMATERIALS ARE DIFFERENET FROM BULK MATERIALS? LO1

The difference in the properties of nanostructures and that of micro or larger scale systems has its origin in a very fundamental physical property of materials. In any material an atom on the surface is different from an atom in the bulk. This difference is primarily because of the different potentials that the two atoms are experiencing. Now consider as a simple case a sphere of radius r, the volume of this sphere is $V = \frac{4}{3}\pi r^3$ and the surface area is given by $S = 4\pi r^2$. So the ratio of the atoms on the surface to that in the bulk can roughly be approximated as,

$$\frac{N_b}{N_s} = \frac{r}{3}$$

where N_b is the number of atoms in the bulk and N_s is the number of atoms on the surface. Now this approximation gives us an idea that as the size of the particle decreases the number of atoms on the surface is a greater proportion of the total number of atoms. Hence, as the particle size becomes smaller and smaller the surface atoms start dominating the properties of the whole material. This is what gives rise to the difference in the properties of nanoparticles and their so varied applications.

At the nano-scale the quantum effects become important. For example, confinement of carriers, discrelization of energy level and modification in band gap take place. The high surface area to volume ratio enhances the ability of the structure to take part in chemicals reaction. Due to the absence of dislocations between separate crystals the nonomaterials carry high tensile strength. The ability of manipulating the dimensions of a structure allows one to tune the optical properties. Their novel optical properties enable them to absorb or emit precise frequencies of light. Finally the simple difference is the size effect which plays the important role in the applications of nanomaterials.

22.2 DIFFERENCE BETWEEN NANOSCIENCE AND NANOTECHNOLOGY LO1

Nanoscience is the study of objects having size less than hundred nanometers at least in one dimension, whereas nanotechnology is the engineering of these objects using different techniques. Nanotechnology is all about the techniques and tools to come up with a nanoscale design or system, which exploit the properties at molecular level to be more accurate and efficient. The difference between nanoscience and nanotechnology is just like the difference between theory and practice. If we say that science has developed it does not necessarily mean that the technology has developed. However, if we say that the technology has developed, it means science might have definitely developed. Nanotechnology deals with the ability to develop and use the technology to manipulate and observe at the nanoscale, whereas nanoscience is the study of phenomena and objects at the nanoscale. We can say that nanoscience is just the study of nanostructures, whereas nanotechnology is the application and implementation of understandings in different industries. We can take an example of carbon nanotubes which are a subject of nanoscience and we will have to work on small scale to create them. However, these carbon nanotubes will be of great use as a building material and the study about its usage can be referred to as nanotechnology. Nanotechnology is applied in many areas including information technology, automobile, textile, healthcare and agricultural industries.

22.3 QUANTUM CONFINEMENT AND CLASSIFICATION OF NANOSTRUCTURES LO2

Quantum confinement is the restricted motion of randomly moving electrons in specific energy levels when the dimensions of a material approach the deBroglie wavelength of electron. The length scale corresponding to quantum confinement regime usually ranges from 1-25 nm. Quantum confinement leads to change in optical and electrical properties. The change in properties occurs because of the energy levels becoming discrete and because of restricted motion of electrons.

Based on the number of dimensions that are confined, nanostructures are classified as 3D (in which no dimensions are confined); 2D (in which one dimension is confined, e.g. thin films); 1D (in which two dimensions are confined; e.g. nano wires) and OD (in which all the dimensions are confined e.g. Quantum Dots, Nano crystals).

22.4 NANOSCALE IN 1-D

LO2

One-dimensional nanomaterials are nanofiber, such as nano rod, and nano tubular materials with fibre, such as rod and tube, having length from 100 nm to tens of microns. Below we discuss in detail the nanowires, carbon nanotubes, inorganic nanotubes and bio-polymers falling within this category.

22.4.1 Nanowires and Their Different Kinds

A nanowire is a nanostructure having the diameter of the order of a nanometer. A structure with the ratio of the length to width greater than 1000 is also termed as nanowire. This ratio is known as aspect ratio. We can say that the nanowires are the structures which have a thickness (diameter) constrained to 10 nm or less and an unconstrained length. Since these structures have an unconstrained longitudinal size, these are referred to as one-dimensional (1-D) materials. At these scales or if the diameter of the nanowire is of the order of the de Broglie wavelength of the electron, then quantum effects become dominated. In view of this such wires are also known as quantum wires. Nanowires have many interesting properties which are not seen in bulk materials. The reason for this is that the electrons in nanowires are quantum confined laterally because of which they occupy energy levels different from the traditional continuum of energy levels or bands found in bulk materials. A quantum mechanical restraint on the number of electrons that can travel through the wire at the nanometer scale leads to the discrete values of the electrical conductance, which are often referred to as the quantum of conductance.

Nanowires can be superconducting, metallic, semiconducting and insulating in nature. For example, the wires prepared using YBCO are superconducting in nature, whereas the nanowires made from SiO_2 or TiO_2 are insulating. The nanowires of Ni, Pt and Au behave as metals, while the silicon nanowires (SiNWs), InP and GaN wires are semiconducting in nature. Other type of the nanowires is the molecular nanowires which are composed of repeating molecular units of either organic (DNA) or inorganic ($Mo_6S_{9-x}I_x$).

There exit many applications where nanowires may become important in electronic, optoelectronic and nanoelectromechanical devices. These are used as additives in advanced composites. These are used as field-emitters and as leads for biomolecular Nano sensors. The nanowires are also used in making metallic interconnects in nanoscale quantum devices.

22.4.1.1 Synthesis of Nanowires

A nanowire is an extremely thin wire. Two very simple processes by which nanowires can be manufactured are suspension and deposition. In the method of suspension, a suspended nanowire is held up by the ends in an evacuated chamber. Then, in order to reduce its diameter, the wire is chemically etched or bombarded with high speed atoms or molecules. Another method, called the deposition method, involves indenting the surface of a wire in the center of a suspended span, raising the temperature, and then stretching the wire while it is near its melting point. This way the deposited nanowire is fabricated on a surface consisting of some non-conducting substance such as glass or plastic. This process is similar to that by which semiconductor chips are grown. The only difference is that here we get a linear (one-dimensional) structure rather than a flat (two-dimensional) or solid (three-dimensional) structure.

In addition to the above processes, other two basic approaches to synthesizing nanowires are called top down technique and bottom up technique.

(i) Top-Down Approach

In top-down approach materials are reduced from two-dimensional thin films to a desired structure. In order to fabricate nanowires, lithographic techniques (photo lithography and electron lithography) and various etching techniques are used.

(ii) Bottom-Up Approach

Since nanowires are a result of anisotropic 1-D crystal growth on a nanometer scale, the key issue related to the growth of nanowires is how to induce 1-D crystal growth in a controlled manner. For accomplishing this many approaches have been employed, which include the metal catalyst assisted vapour liquid solid (VLS) mechanism, vapour solid (VS) mechanism and template assisted mechanism.

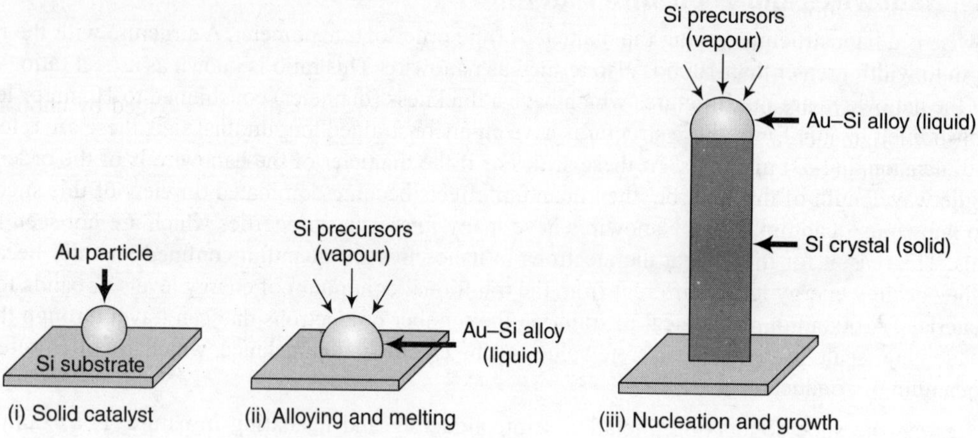

FIGURE 22.1

(a) Vapour Liquid Solid (VLS) Method

This is a common technique for creating a nanowire, which is used to grow single crystalline nanowires of semiconducting materials. It requires a catalyst particle. A typical VLS process starts with the dissolution of gaseous reactants into nano-sized liquid droplets of a catalyst metal, followed by nucleation and growth of single crystalline rods and then wires. In this mechanism, the grown species are supplied from vapour to metal catalyst that seat on the nanowire and remain on the top of the nanowire during growth. The one-dimensional growth is induced and dictated by the liquid droplets, whose sizes remain essentially unchanged during the entire process of wire growth. Here each liquid droplet serves as a virtual template to strictly limit the lateral growth of an individual wire. The growth is done at or above the eutectic point of the catalyst metal particle. A schematic diagram of VLS growth is shown in Figure 22.1.

(b) Vapour Solid (VS) Method

Vapour solid (VS) mechnism is generally referred to as a process of one-dimensional material growth from vapour phase precursors in the absence of the catalyst or obvious VLS evidence. In this technique vapour is first generated by evapouration, chemical reduction or gaseous reactions. Then vapour is subsequently transported and condensed on the substrate. This enables one to grow the

1-D nanostructures using the VS process, if one can control its nucleation and its subsequent growth process. In consideration of thermodyanamics and kinetics, the VS growth of nanowires could be possible via (i) a self catalytic VLS growth, (ii) an oxide assisted growth and (iii) Frank's screw dislocation mechnism.

(c) Catalyst Free Growth Mechnism

Through this mechanism the self assembles growth and the patterned growth of nanowires can be achieved. In self assembled growth, the nanowires grow by self assembly without any lithographic technique and without metal particle. In this mechnism, an oxide layer of SiO_2 is deposited over the substrate, usually Si. The SiO_2 is amorphous in nature and contains large number of pin holes which act as nucleation center for the growth of nanowires. The nanowires grow at an optimum thickness of the SiO_2. Then vapours of growing materials are supplied by various techniques such as molecular beam epitaxy (MBE), metal organic chemical vapour deposition (MOCVD) and pulse laser deposition (PLD). Each technique has its own advantages and some disadvantages. Since the grown nanowires contain the same orientation as the substrate in this technqiue, the growth is called epitaxial growth of nanowires. On the other hand, in Patterned growth, a pattern is made on the oxide grown substrate by electron beam lithography. Then vapours are provided by choosing the appropriate technique. Finally, the nanowires grow at the holes only (Figure 22.2). So by patterning the substrate we can grow an ordered array of nanowires which can be used in many applications.

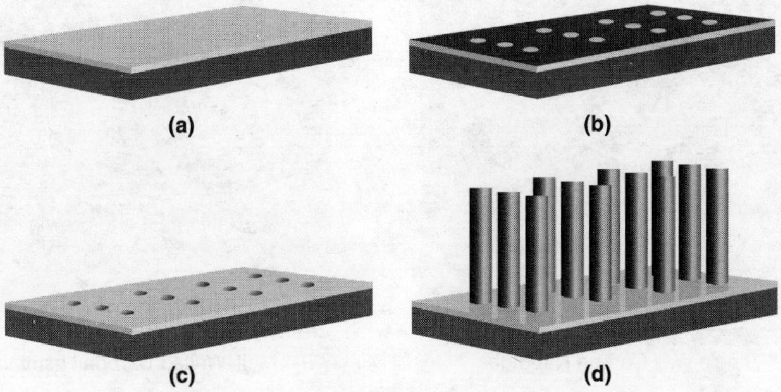

FIGURE 22.2

The nanowires can be characterized for their structural, transport and optical properties. The characterization techniques used for the structural properties are scanning electron microscopy (SEM), transmission electron microscopy (TEM), scanning tunneling probes (STPs), X-ray analysis and Raman spectroscopy. Their transport properties are characterized with the help of I-V measurement, temperature dependent resistance measurements and magnetoresistance. The optical properties of these nanowires can be investigated through photoluminescence.

22.4.2 Carbon Nanotubes

Carbon nanotubes were discovered in 1991 by S. Iijima. These opened up a new era in materials science. Carbon nanotubes, abbreviated as CNTs, are allotrope of carbon with a cylindrical nanostructure as thin as a few nanometers yet be as long as hundreds of microns. Owing to their extraordinary thermal conductivity and mechanical and electrical properties, carbon nanotubes find application as additives to various structural materials in addition to their importance in nanotechnology electronics and optics. Nanotubes are members of the fullerene structural family. The structure of a carbon nanotube is formed by a layer of carbon atoms that are bonded together in a hexagonal (honeycomb) mesh. This one atom thick layer of carbon is called graphene, which is wrapped in the shape of a cylinder and bonded together to form a carbon nanotube. The combination of the rolling (chiral) angle and radius decides the nanotubes properties; for example, whether the individual nanotube shell is a metal or semiconductor. Carbon nanotubes have a range of electric, thermal, and structural properties that can change based on the physical design of the nanotube. Nanotubes can have a single outer wall of carbon called single-walled nanotubes (SWNTs) or they can be made of multiple walls (cylinders inside other cylinders of carbon) called multiple-walled nanotubes (MWNTs). Figure 22.3 shows how various types of carbon nanotubes can be formed from graphene. Similar to those of graphite, the chemical bonding of nanotubes is composed entirely of sp^2 bonds. These bonds which are stronger than the sp^3 bonds found in alkenes and diamond provide nanotubes with their unique strength.

Carbon nanotubes have found applications in field emitters / emission, conductive or reinforced plastics, energy storage, molecular electronics with CNT based nonvolatile RAM and transistors. The CNT based ceramics, fibers and fabrics have also attracted the people for their enormous usage.

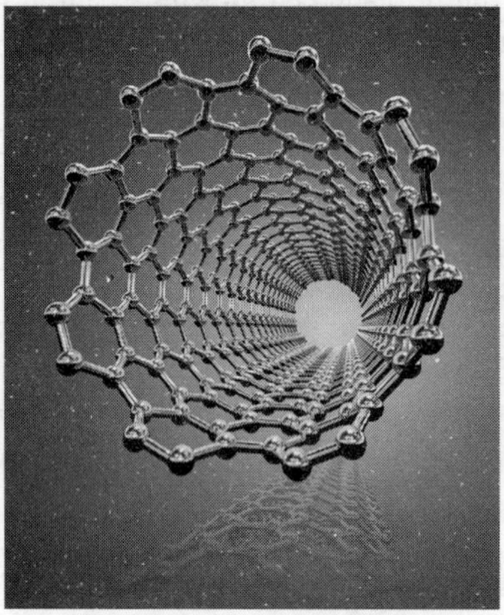

Single Wall Carbon Nanotube

Multi Walled Carbon Nanotube

Figure 22.3

22.4.2.1 *Kinds of Carbon Nanotubes*

A broad category of carbon nanotubes comprises single-walled carbon nanotubes and multi-walled carbon nanotubes.

(i) Single-walled Carbon Nanotubes (SWCNTs)

Single-walled carbon nanotubes can be formed in three different designs, namely, Armchair, Chiral and Zigzag (Figure 22.4). The design depends on the way the graphene is wrapped into a cylinder. For example, rolling a sheet of paper from its corner will assign one design, and rolling the paper from its edge will attain a different design. The structure of a SWCNT is represented by a pair of indices (n, m) called the chiral vector. The structural design has a direct impact on the electrical properties of the nanotube. When $n - m$ is a multiple of 3, then the nanotube is described as metallic, i.e., highly conducting, otherwise the nanotube is a said to behave as a semiconductor. The Armchair design is always metallic while other designs can make the nanotube a semiconductor.

(ii) Multi-walled Carbon Nanotubes (MWCNTs)

There are two structural models of multi-walled nanotubes, namely, Russian doll model and Parchment model. In the Russian doll model, a carbon nanotube contains another nanotube inside it such that the inner nanotube has a smaller diameter than the outer nanotube. In the Parchment model, a single graphene sheet is rolled around itself multiple times, resembling a rolled up scroll of paper. Multi-walled carbon nanotubes have similar properties to single-walled nanotubes. However, the outer walls of these nanotubes can protect the inner carbon nanotubes from chemical interactions with outside materials. These nanotubes also have a higher tensile strength than the single-walled nanotubes.

22.4.2.2 Fabrication of Carbon Nanotubes

There are several procedures that have been developed for fabricating carbon nanotube structures.

(a) Arc Discharge Method

Arc discharge method is used for the synthesis of carbon nanotubes where a DC arc voltage is applied across two graphite electrodes (anode and cathode) immersed in an inert gas (such as He) filling a chamber. The electrodes are vapourized by the passage of a DC current of about 100 A through the two high-purity graphite separated by about 1–2 mm in 400 mbar of He atmosphere. In this process a carbon rod is built up at the cathode after arc discharging for a period of time. This method can mostly produce multi-walled nanotubes but can also produce single-walled nanotubes with the addition of metal catalyst such as Fe, Ni, Co, Y or Mo, on either the anode or the cathode. Various parameters such as the metal concentration, inert gas pressure, type of gas, plasma arc, temperature, the current and system geometry are the deciding factors for the quantity and quality such as lengths, diameters and purity of the nanotubes. When pure graphite rods are used, fullerenes are deposited as soot inside the chamber and multi-walled carbon nanotubes are deposited on the cathode. However, single-walled carbon nanotubes are generated in the form of soot when a graphite anode containing a metal catalyst (Fe or Co) is used with a pure graphite cathode. The advantage of this method is that it produces a large quantity of nanotubes. But the main disadvantage is that there is relatively a little control over the alignment (i.e., chirality) of the produced nanotubes, which is critical to their characterization and role.

(b) Laser Ablation Method

Laser ablation process was developed by Richard Smalley and co-workers at Rice University, who were primarily involved in blasting metals with a laser to produce various metal molecules. However, they replaced the metals with graphite to create multi-walled carbon nanotubes when they heard of the existence of nanotubes. In laser ablation, a pulsed laser is used to vapourize a graphite target (kept in a quartz tube) in a high-temperature reactor, where a flow of argon gas is maintained throughout the reaction. Here the carbon vapourizes, which is carried away by the argon gas and is condensed downstream on the cooler walls of the quartz. This condensation leads to the formation of single-walled carbon nanotubes and metallic particles. Therefore, purification methods are employed to this mixture. The key to the proper formation of the condensed nanotubes is that the location where the carbon atoms begin to condense should be set up as a curved sheet of graphene with a catalyst metallic atom nearby. As carbon atoms begin to attach and form rings, the metallic atom, if it has the proper electronegativity properties, will preserve the open edge of the tube and prevent it from drawing to a close. This phenomenon is termed as the scooter effect. The laser ablation method yields around 70% and produces primarily single-walled carbon nanotubes with a controllable diameter, which can be determined by the reaction temperature. This technique includes a relatively low metallic impurities, since the metallic atoms involved tend to evaporate from the end of the tube once it is closed. However, it is more expensive than either arc discharge or chemical vapour deposition techniques. Moreover, the nanotubes produced from this method are not necessarily uniformly straight, but instead do contain some branching.

(c) Chemical Vapour Deposition Technique

The chemical vapour deposition (CVD) technique allows carbon nanotubes to grow on a variety of materials, which makes it more viable to integrate into already existent processes for synthesizing electronics. This process involves the chemical breakdown of a hydrocarbon on a substrate, which is due to the fact that a main way to grow carbon nanotubes is by exciting carbon atoms that are in contact with metallic catalyst particles. The CVD method extends this idea by embedding these metallic particles (for example, iron) in properly aligned holes in a substrate (say, silicon). Essentially, tubes are drilled into silicon and implanted with iron nanoparticles at the bottom. After that a hydrocarbon such as acetylene is heated and decomposed onto the substrate. The carbon comes into contact with the metal particles embedded in the holes and start to form nanotubes, which are templated from the shape of the tunnel. The advantages of this method are that the yield is very high and the size of the growth area is theoretically arbitrary. Moreover, the alignment of the nanotubes is consistent, which is crucial for creating particular types of nanotubes, e.g. semiconductor or metallic. However, the main disadvantage is that the large sized areas (several millimeters) tend to crack, shrink, and otherwise twist. Hence, the substrates need to be dried very thoroughly to prevent against this.

(d) n-hexane Pyrolysis

Researchers have developed a method to synthesize large, long single-walled nanotube bundles in a vertical furnace by pyrolyzing hexane molecules. These n-hexane molecules are mixed with certain other chemicals, which have been shown independently to help with growth of nanotubes. These are burned (pyrolyzed) at a very high temperature in a flow of hydrogen and other optional gases. The use of a different hydrocarbon or using a different gas has been shown to prevent the formation of long nanotubes. The primary advantage of this method is that it produces macroscopic nanotube bundles (micro tubes), whose their diameters are typically larger than that of human hair, and their length is several centimeters. However, the disadvantage is that the alignment is not as produced from other methods, making it viable for creating micro cables, but not nanotubes with precise electrical properties. Moreover, the elasticity of these nanotube bundles is not found to be as great as hoped, i.e., they are more brittle.

22.4.2.3 Properties of Carbon Nanotubes

The carbon nanotubes are known for their strength, electrical properties and thermal properties. These properties are discussed below.

(a) Strength

Carbon nanotubes have a higher tensile strength than steels, which comes from the sp^2 bonds between the individual carbon atoms. This bond is even stronger than the sp^3 bond found in diamond. Under high pressure, individual nanotubes can bond together and exchange some sp^2 bonds with sp^3 bonds. This way the possibility of producing long nanotube wires is enhanced. Carbon nanotubes are not only strong but they are also elastic. However, their elasticity does have a limit, it is possible to permanently deform to shape of a nanotube under very strong forces. The strength of a nanotube is weakened by defects in its structure, occurring from atomic vacancies or a rearrangement of the carbon bonds. This in turn causes the tensile strength of the entire nanotube to weaken.

(b) Electrical Properties

The structure of a carbon nanotube determines how conductive the tube is. A carbon nanotube is highly conductive under the situation when the structure of atoms in it minimizes the collisions between the conduction electrons and the atoms. The strong bonds between carbon atoms also allow carbon nanotubes to withstand higher electric currents than copper. The electron transport occurs only along the axis of the nanotube. Single-walled nanotubes can route electrical signals at the speeds up to 10 GHz when used as interconnects on semiconducting devices.

(c) Thermal Properties

The strength of the atomic bonds in carbon nanotubes allows them to withstand high temperatures, because of which carbon nanotubes have been shown to be very good thermal conductors. Compared with copper wires which are commonly used as thermal conductors, the carbon nanotubes can transmit over 15 times the amount of Watts/m/K. However, the thermal conductivity of carbon nanotubes is dependent on the temperature of the tubes and the outside environment.

22.4.3 Inorganic Nanotubes

An inorganic nanotube is a cylindrical molecule that is often composed of metal oxides or group III-Nitrides. These nanotubes are morphologically similar to carbon nanotubes, but are non-carbon nanotubes. The inorganic nanotubes have been found naturally in some mineral deposits the way carbon nanostructure are found naturally. Minerals such as white asbestos and imogolite have been shown to have a tubular structure. The inorganic nanotubes have been synthesized based on molybdenum disulphide (MoS_2) and tungsten disulphide (WS_2).

Inorganic nanotubes are the nanostructures which are distinct from pure inorganic nano wires and carbon nanotubes in two ways. First, the resultant physical properties and electronic structure show combined characteristics of both one- and two-dimensional materials. Second, these hollow nanotubes can serve as nanoscale containers or pipes to deliver fluids and molecular species. These are excellent building blocks for the construction of large scale nano fluidic systems. Inorganic nanotubes show easy synthetic access and high crystallinity, needle like morphology, good uniformity and dispersion, and good adhesion to a number of polymers. So these are promising candidates as fillers for polymer composites with enhanced thermal, mechanical and electrical properties. Inorganic nanotubes are heavier than carbon nanotubes and are not as strong under tensile stress, but these are particularly strong under compression, leading to potential applications in impact resistant applications such as bulletproof vests.

22.4.4 Biopolymers

Biopolymers are polymers which are produced by living organisms. It means these are the polymers that are biodegradable. In other words, we can say that these are polymeric biomolecules. The biopolymers represent the most abundant organic compounds in the biosphere and constitute the largest fraction of cells. Their main classes are distinguished according to their chemical structures. For example, there are four main types of biopolymer based on sugar, starch, cellulose and synthetic materials. The input materials to produce these polymers may be either synthetic or renewable, i.e., based on agricultural plant or animal products. Two main strategies may be followed in synthesizing a polymer. One is to build up the polymer structure from a monomer. This is done by a process of chemical polymerization. The alternative is to take a naturally occurring polymer and chemically modify it to give it the desired properties. However, the biodegradability of the polymer may be adversely affected by chemical modification. Therefore, it is often necessary to seek a compromise between the desired material properties and biodegradability.

Some biopolymers such as polylactic acid (PLA), naturally occurring zein, and poly-3-hydroxybutyrate can be used as plastics, which replace the need for polystyrene or polyethylene based plastics. Some plastics are now referred to as being degradable, oxydegradable or UV degradable, which means they break down when exposed to air or light. However, these plastics are still largely oil based.

For their use in the packaging industry as food trays, blown starch pellets for shipping fragile goods, thin films for wrapping, etc., biopolymers are produced from biomass which comes from crops such as sugar beet, wheat or potatoes. The conversion of these takes place in the following manner. For example, sugar beet is converted to glyconic acid which finally gives polyglyconic acid. The fermentation of starch is done in order to get lactic acid which is converted to polylactic acid. On the other hand, the fermentation of biomass leads to the bioethanol which gives ethene and the ethene is converted into polyethylene.

22.5 NANOSCALE IN 2-D LO3

Thin films with thickness ranging from atomic monolayer to a few atomic layers are 2D nanomaterials. They are characterized by a high aspect ratio and their small size perpendicular to the film gives rise to quantum confinement of electrons.

Synthesis of thin films is carried out by either physical or chemical techniques. The chemical techniques employed are chemical vapour deposition (CVD), Sol-Gel technique and electrochemical plating. Chemical techniques are based on decomposition of molecular precursors and subsequent deposition of the desired material on the target.

Physical techniques are based on the vaporization of material to be deposited and its condensation *n* the substrate surface as thin film. Based on the method used for vaporization, physical techniques are classified as Molecular Beam Epitaxi (MBE), sputtering, and Pulsed Laser Ablation and Deposition (PLD).

22.6 NANOSCALE IN 3-D LO3

If there is no restriction of carrier confinement or carriers are free to move in all 3 directions, the material is called bulk material (also termed as 3-D). In this case the property or the behavior of nanomaterials stand the same as that of the materials. It means the energy levels are continuous and density of state varies as $E^{3/2}$, where E is the energy of the state. Ingot GaAs, Si, Ge etc. fall within this category of 3-D nanomaterials.

22.6.1 Nanoparticles

Figure 22.4 shows how things scale and how small a nanometer actually is. Although it may seem that such structures have come into being in the very recent past, this is not true. Humans have been known to take advantage of the peculiar properties of nanoparticles as early as the 4th century A.D. Roman glassmakers were fabricating glasses containing nano sized metals. The great varieties of beautiful colors of the windows of medieval cathedrals are due to the presence of metal nanoparticles in the glass.

22.6.1.1 Synthesis of Nanoparticles

Nanoparticles can be synthesized by means of various techniques, some of which are discussed below.

22.6.1.1.1 Mechanical Method Ball milling is the best example for mechanical method. A ball mill (Figure 22.5), a type of grinder, is a cylindrical device used in grinding (or mixing) materials like ores, chemicals, ceramic raw materials and paints. Ball mills rotate around a horizontal axis, partially filled with the material to be ground plus the grinding medium. Different materials are used as media, including ceramic balls, flint pebbles and stainless steel balls.

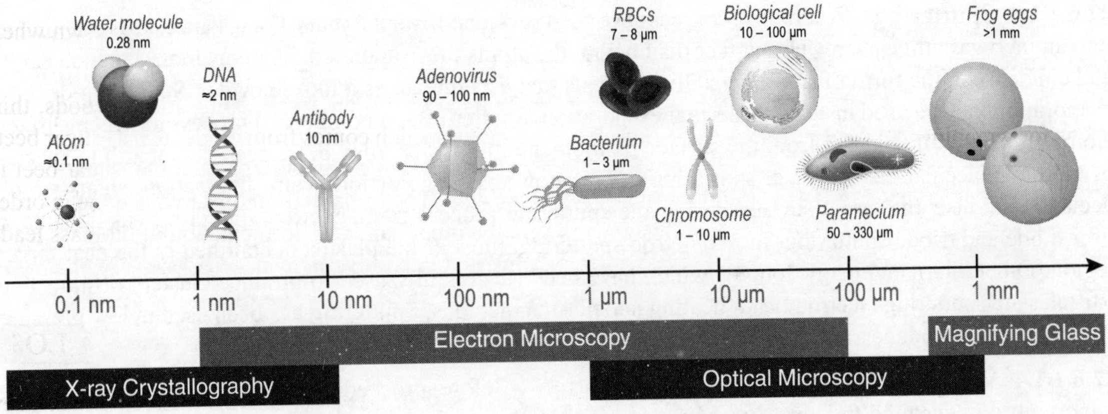

FIGURE 22.4

An internal cascading effect reduces the material to a fine powder. Industrial ball mills can operate continuously fed at one end and discharged at the other end. Large to medium-sized ball mills are mechanically rotated on their axis, but small ones normally consist of a cylindrical capped container that sits on two–drive shafts (pulleys and belts are used to transmit rotary motion).

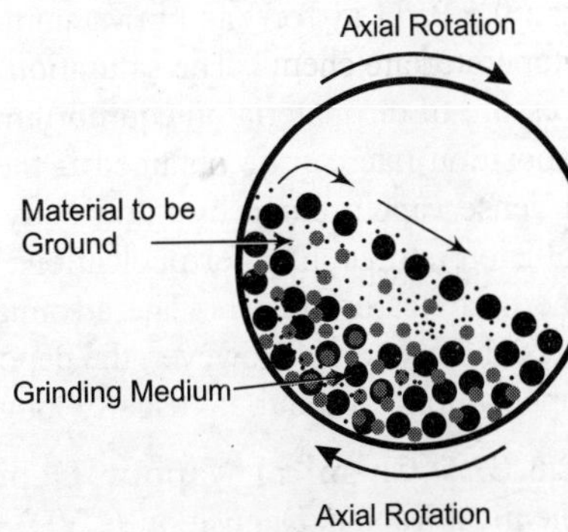

FIGURE 22.5

A rock tumbler functions on the same principle. Ball mills are also used in pyrotechnics and the manufacture of black powder, but cannot be used in the preparation of some pyrotechnic mixtures such as flash powder because of their sensitivity to impact. High-quality ball mills are potentially expensive and can grind mixture particles to as small as 0.0001 mm, enormously increasing surface area and reaction rates. The grinding works on principle of critical speed. The critical speed can be understood as that speed after which the steel balls (which are responsible for the grinding of particles) start rotating along the direction of the cylindrical device; thus cause no further grinding.

22.6.1.1.2 Gas Condensation Nanocrystalline metals and alloys were first fabricated using the gas condensation technique. In this technique, a metallic or inorganic material is vaporized using thermal evaporation sources such as a refractory crucibles or electron beam evaporation devices, under a pressure of 1–50 mbar. In gas evaporation, a high residual gas pressure causes the formation of ultra fine particles (100 nm) by gas phase collision. The ultra fine particles are formed by collision of evaporated atoms with residual gas molecules. The sources used for vaporization may be resistive heating, high energy electron beams, low energy electron beam and inducting heating. Clusters form in the vicinity of the source by homogenous nucleation in the gas phase that grows by incorporation by atoms in the gas phase. It comprises a ultra high vacuum system fitted evaporation source, a cluster collection device of liquid nitrogen filled cold finger scrapper assembly and a compaction device. During heating, atoms condense in the super saturation zone close to the heating device. The nanoparticles are removed by scrapper in the form of a metallic plate. Evaporation is to be done from W, Ta or Mo refractory metal crucibles. In case the metals react with crucibles, we can use the electron beam evaporation technique. The method is extremely slow. A few limitations of the method are source precursor incompatibility, temperature ranges and dissimilar evaporation rates in an alloy.

22.6.1.1.3 Sputtering Alternative sources have been developed over the years. For instance, Fe is evaporated into an inert gas atmosphere. Through collision with the atoms the evaporated Fe atoms loose kinetic energy and condense in the form of small crystallite crystals that accumulate as a loose powder. Sputtering or laser evaporation may be used instead of thermal evaporation. Sputtering is a nonthermal process in which surface atoms are physically ejected from the surface by momentum transfer from an energetic bombarding species of atomic/molecular size. Typical sputtering uses a glow discharge or ion beam. Interaction events which occur at and near the target surface during the sputtering process in magnetron sputtering has advantage over diode and triode sputtering. In magnetron sputtering, most of the plasma is confined to the near target region. Other alternate energy sources which have been successfully used to produce clusters or ultra fine particles are sputtering electron beam heating and plasma methods. Sputtering has been used in low pressure environment to produce a variety of clusters including Ag, Fe and Si.

22.6.1.1.4 Vacuum Deposition and Vaporization Before proceeding to the other methods, it is important to understand the terms vacuum deposition and vaporization or vacuum evaporation. In vacuum deposition process, elements, alloys or compounds are vaporized and deposited in a vacuum. The vaporization source is the one that vaporizes materials by thermal processes. The process is carried out at pressure of less than 0.1 Pa (1 m Torr) and in vacuum levels of 10 to 0.1 MPa. The substrate temperature ranges from room temperature to 500°C. The saturation or equilibrium vapour pressure of a material is defined as the vapour pressure of the material in equilibrium with the solid or liquid surface. For vacuum deposition, a reasonable deposition rate can be obtained if the vaporization rate is fairly high. Vapour phase nucleation can occur in dense vapour cloud by multi body collisions, the atoms are passed through a gas to provide necessary collision and cooling for nucleation. These particles are in the range of 1 to 100 nm and are called ultra fine particles or clusters. The advantages associated with vacuum deposition process are high deposition rates and economy. However, the deposition of many compounds is difficult. Nanoparticles produced from a supersaturated vapour are usually longer than the cluster.

22.6.1.1.5 Chemical Vapour Deposition (CVD) and Chemical Vapour Condensation (CVC) Chemical Vapour Deposition (CVD) is a well–known process in which a solid is deposited on a heated surface via a chemical reaction from the vapour or gas phase. CVD reaction requires activation energy to proceed. This energy can be provided by several methods. In thermal CVD, the reaction is activated by a high temperature above 900°C. A typical apparatus comprises gas supply system, deposition chamber and an exhaust system. In plasma CVD, the reaction is activated by plasma at temperatures between 300°C and 700°C. In laser CVD, pyrolysis occurs when laser thermal energy heats an absorbing substrate. In photo laser CVD, the chemical reaction is induced by ultra violet radiation which has sufficient photon energy, to break the chemical bond in the reactant molecules. In this process, the reaction is photon activated and deposition occurs at room temperature. Nano composite powders have been prepared by CVD. SiC/Si_3N composite powder was prepared using SiH_4, CH_4, WF_6 and H_2 as a source of gas at 1400°C.

Another process called Chemical Vapour Condensation (CVC) was developed in Germany in 1994. It involves pyrolysis of vapours of metal organic precursors in a reduced pressure atmosphere. Particles of ZrO_2, Y_2O_3 and nano whiskers have been produced by CVC method. A metal-organic precursor is introduced in the hot zone of the reactor using mass flow controller. For instance, hexamethyldisilazane $(CH_3)_3$ Si NHSi $(CH_3)_3$ was used to produce $SiC_xN_yO_z$ powder by CVC technique. The reactor allows synthesis of mixtures of nanoparticles of two phases or doped nanoparticles by supplying two precursors at the front end of reactor and coated nanoparticles, n-ZrO_2, coated with n-Al_2O_3 by supplying a second precursor in a second stage of reactor. The process yields quantities in excess of 20 g/hr. The yield can be further improved by enlarging the diameter of hot wall reactor and mass of fluid through the reactor.

22.6.1.1.6 Sol-gel Techniques In addition to techniques mentioned above, the sol-gel processing techniques have also been extensively used. Colloidal particles are much larger than normal molecules or nanoparticles. However, upon mixing with liquid colloids appear bulky whereas the nanosized molecules always look clear. It involves the evolution of networks through the formation of colloidal suspension (sol) and gelatin to form a network in continuous liquid phase (gel). The precursor for synthesizing these colloids consists of ions of metal alkoxides and aloxysilanes. The most widely used are tetramethoxysilane (TMOS), and tetraethoxysilanes (TEOS) which form silica gels. Alkoxides are immiscible in water. They are organo-metallic precursors for silica, aluminium, titanium, zirconium and many others. Mutual solvent alcohol is used. The sol-gel process involves initially a homogeneous solution of one or more selected alkoxides. These are organic precursors for silica, alumina, titania, zirconia, among others. A catalyst is used to start reaction and control pH. Sol-gel formation occurs in four stages: (i) Hydrolysis (ii) Condensation (iii) Growth of particles, and (iv) Agglomeration of particles.

(i) Hydrolysis

During hydrolysis, addition of water results in the replacement of [OR] group with [OH-] group. Hydrolysis occurs by attack of oxygen on silicon atoms in silica gel. Hydrolysis can be accelerated by adding a catalyst such as HCI and NH_3. Hydrolysis continues until all alkoxy groups are replaced by hydroxyl groups. Subsequent condensation involving silanol group (Si-OH) produced siloxane bonds (Si-O-Si) and alcohol and water. Hydrolysis occurs by attack of oxygen contained in the water on the silicon atom.

(ii) Condensation

Polymerization to form siloxane bond occurs by either a water producing or alcohol producing condensation reaction. The end result of condensation products is the formation of monomer, diamer, cyclic tetramer, and high order rings. The rate of hydrolysis is affected by pH, reagent concentration and H_2O/Si molar ratio (in case of silica gels). Also ageing and drying are important. By control of these factors, it is possible to vary the structure and properties of sol-gel derived inorganic networks.

(iii) Growth and Agglomeration

As the number of siloxane bonds increases, the molecules aggregate in the solution, where they form a network, a gel is formed upon drying. The water and alcohol are driven off and the network shrinks. At values of pH of greater than 7, and H_2O/Si value ranging from 7 to 5, spherical nanoparticles are formed. Polymerization to form siloxane bonds by either an alcohol producing or water producing condensate occurs.

$$2\ \text{HOSi (OR)}_3 \rightarrow \text{(OR)}_3\ \text{Si O Si (OR)}_3 + \text{H}_2\text{O}$$

or

$$2\ \text{HOSi (OR)}_3 \rightarrow \text{(OR)}_2\text{OH Si O Si (OR)}_3 + \text{H}_2\text{O}$$

Above pH of 7, silica is more soluble and silica particles grow in size. Growth stops when the difference in solubility between the smallest and largest particles becomes indistinguishable. Larger particles are formed at higher temperatures. Zirconium and yttrium gels can be similarly produced.

Despite improvements in both chemical and physical methods of synthesis, there remain some problems and limitations. Laser vaporization technique has offered several advantages over other heating techniques. A high energy pulsed laser with an intensity flux of $10^6 - 10^7$ W/cm^2 is forced on target material. The plasma causes high vaporization and high temperature (10,000°C). Typical yields are $10^{14} - 10^{15}$ atoms from the surface area of 0.01 cm^2 in a 10^{-8} s pulse. Thus a high density of vapour is produced in a very short time (10^{-8} s), which is useful for direct deposition of particles.

22.6.1.1.7 Electrodeposition Nanostructured materials can also be produced by electrodeposition. These films are mechanically strong, uniform and strong. Substantial progress has been made in nanostructured coatings applied either by DVD or CVD. Many other non-conventional processes such as hypersonic plasma particle deposition (HPPD) have been used to synthesize and deposit nanoparticles. The significant potential of nanômaterial synthesis and their applications is virtually unexplored. They offer numerous challenges to overcome. Understanding more of synthesis would help designing better materials. It has been shown that certain properties of nanostructured deposits such as hardness, wear resistance and electrical resistivity are strongly affected by grain size. A combination of increased hardness and wear resistance results in a superior coating performance.

22.6.1.2 Properties of Nanoparticles

Nanoparticles are generally considered to be a number of atoms or molecules bonded together such that the dimension of the bonded entity is of the order of 100 nm. And since 1 nm is 10 Å, hence the critical limit for nanoparticles size is 1000 Å. For the sake of completeness to this size based criterion for nanoparticles, we may say that particles that are large enough so that not almost all the atoms are at the surface are still nanoparticles. But it is quite evident that this size based scheme is arbitrary. For example, the heme molecule, $FeC_{34}H_{32}O_4N_4$ that is found in hamoglobin contains around 75 atoms. Thus, a more convincing definition of nanoparticles would be that they have sizes smaller than the critical lengths for certain physical phenomenon. This critical length can characterize processes like electrical conductivity or excitonic processes. So, one definition of nanoparticles of metals can be given by their scattering length, which is the distance that an electron moves between two successive collisions with the vibrating atoms or impurities in the material. It is below these critical lengths that the materials begin to demonstrate new physical or chemical phenomenon that is not observed in bulk.

Now that we know what nanoparticles are let us study how the physical and chemical properties of materials change when we enter the nanoparticles paradigm.

(i) Optical Properties

The optical properties of nanoparticles are markedly different from those of bulk. However, the changes that are observed are quite different for different materials. We would talk about the metals first and then go on to talk about the semiconducting materials.

(ii) Metals: Surface Plasmons

In the case of metals as the size of the particles decreases we start observing oscillations of electron gas on the surface of nanoparticles. These oscillations are called *surface plasmons*. So, if the nanoparticles are exposed to an electromagnetic wave (light) having a wavelength comparable to or greater than the size of the nanoparticles and the light has a frequency close to that of the surface plasmon then the surface plasmon would absorb energy. Thus nanoparticles start exhibiting different colors as their size changes and the frequency of the surface plasmon changes with it. This kind of a phenomenon is not observed strongly in bulk metals. This frequency of the surface plasmon absorption is a function of the dielectric constant of the material, size of the particles and also the specific geometrical shape that the particle has. This phenomenon of surface plasmon resonance and subsequent absorption was used to obtain different colors of the stained glasses used in the medieval cathedrals. Surface plasmons have been used to enhance the surface sensitivity of several spectroscopic measurements including fluorescence, Raman scattering, and second harmonic generation. However, in their simplest form, SPR reflectivity measurements can be used to detect DNA or proteins by the changes in the local index of refraction upon adsorption of the target molecule to the metal surface. If the surface is patterned with different biopolymers, the technique is called *Surface Plasmon Resonance Imaging* (SPRI).

(iii) Semiconductors

In the case of semiconducting nanoparticles the properties change in a different fashion. One of the important properties that changes as the size of the nanoparticles changes is the absorption spectrum of the material. In any semiconducting material there is a band gap between the conduction and valence bands in the solid. This band gap is specific to materials like bulk Si has a band gap of 1.1 eV. The band gap in any material affects its electrical as well as optical properties. Taking an optical perspective whenever light of any particular wavelength is passed through a material sample there is some degree of absorption by the material. The strength of absorption depends on the material and the wavelength passed. A plot of the absorption strength against the wavelength is what is referred to as the absorption spectrum. Figure 22.6 shows the absorption spectra of a direct as well as indirect band gap semiconductor. For a given material in its bulk state the absorption spectrum is unique. But when the material is in the form of nanoparticles then the absorption spectrum changes and undergoes a blue shift. By blue shift we mean that the whole spectrum shifts towards a smaller wavelength.

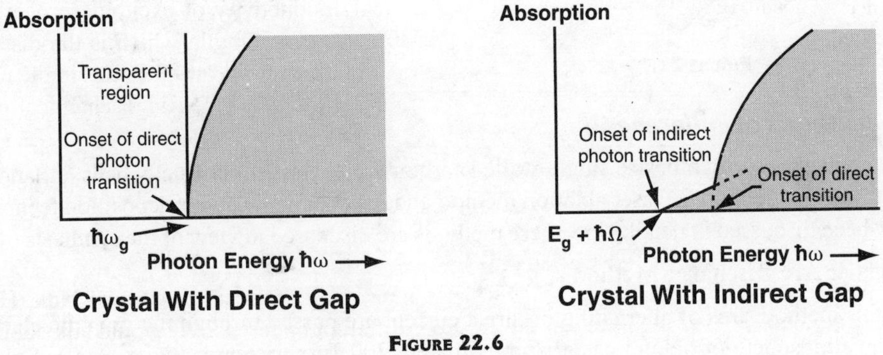

FIGURE 22.6

22.6.2 Carbon Allotropes, Bucky balls or Fullerenes

Fullerene is a molecule of carbon which could be in the form of a hollow sphere, ellipsoid, tube and other shapes. In 1985, a new allotrope of carbon was discovered in which the atoms are arranged in closed shells. The new form was found to have the structure of truncated icosahedrons and was named Buckminsterfullerene after the architect Buckminster Fuller who designed geodesic domes in the 1960's. Previously six crystalline forms of the element carbon were known, namely two kinds of graphite, two kinds of diamond, *chaoite and carbon (VI)*. Fullerenes are similar in structure to graphite, which are composed of stacked graphene sheets of linked hexagonal rings; they may also contain pentagonal or sometimes heptagonal rings. Spherical fullerenes are also known as Bucky balls, as they resemble the balls used in football, whereas the cylindrical ones are called carbon nanotubes or Bucky tubes. Figure 22.7 provides a graphical representation of the 60-carbon atoms called C_{60} fullerene, whereas Figure 22.8 illustrates the arrangement of 540-carbon atoms called icosahedral fullerene C_{540} which is another member of the family of fullerenes.

An infinite number of spherical fullerenes are believed to be able to exist, the known forms of which include C-60, C-70, C-76, C-84, C-240 and C-540. All fullerenes consist of 12 pentagonal faces and a varying number of hexagonal faces. In general, for a fullerene C-n, there will be 12 pentagonal faces and half of n minus 10, i.e., n/2-10, hexagonal faces. It means the C-60 fullerene, which is called Bucky ball, has 12 pentagonal faces (rings) and 20 hexagonal faces (rings), forming a spheroid shape with 60 vertices for 60 carbons. The pentagonal rings sit at the vertices of an icosahedron such that no two pentagonal rings are next to each other. The average C-C bond distance measured using nuclear magnetic resonance (NMR) is found to be 1.44 A°. A diameter of 7.09 A° is calculated for the C-60 based on the fact that the C-C distance is equal to 1.40 A° for the hexagon bonds and 1.46 A° for the pentagonal bonds length.

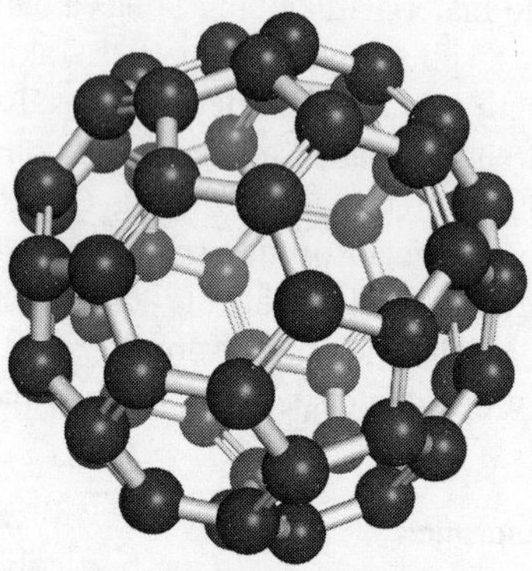

FIGURE 22.7

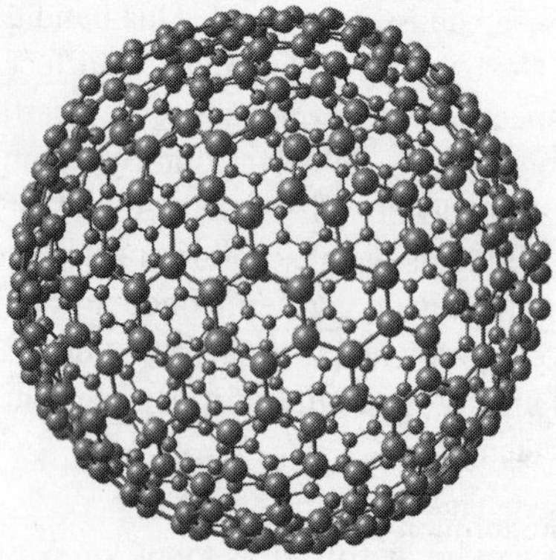

FIGURE 22.8

22.6.2.1 *Synthesis of Fullerenes*

The three main methods to synthesize single-walled carbon molecules, either Bucky balls or nanotubes are i) electric arc discharge method, ii) laser ablation method and iii) chemical vapour deposition technique. These have already been discussed in detail. Now these methods are discussed in view of the synthesis of fullerenes.

(i) Electric Arc Discharge Method

In this method, arcs of alternating or direct current are passed through the graphite electrodes kept in an atmosphere of helium gas at approximately 200 Torr pressure (Figure 22.9). The graphite is evaporated that takes the form of soot. This is dissolved in a nonpolar solvent. The solvent is dried away and the C-60 and C-70 fullerenes are separated from the residue. This method yields up to 70% of C-60 and 15% of C-70 at the optimal current, He pressure and flow rate.

(ii) Laser Ablation Method

Laser ablation method is one of the three methods of laboratory and industrial synthesis of Bucky balls in addition to single-walled and multi-walled nanotubes. Laser vaporization is also used for fullerene production. In a typical apparatus, a pulsed Nd:YAG laser is used as a laser source operating at 532 nm and 250 mJ of power and the graphite target is kept in a furnace at 1200 °C.

(iii) Chemical Vapour Deposition Technique

This technique is based on the thermal cracking of a carbon containing gas (e.g. a hydrocarbon or carbon-monoxide) in the presence of a catalyst. Hydrogen gas or an inert gas like Ar are used as vector gas. Sometimes metallocenes such as ferrocene, nickelocene or cobaltocenes are used whose cracking generates both the nanometric metallic catalysts as well as carbon for the formation of nanotubes.

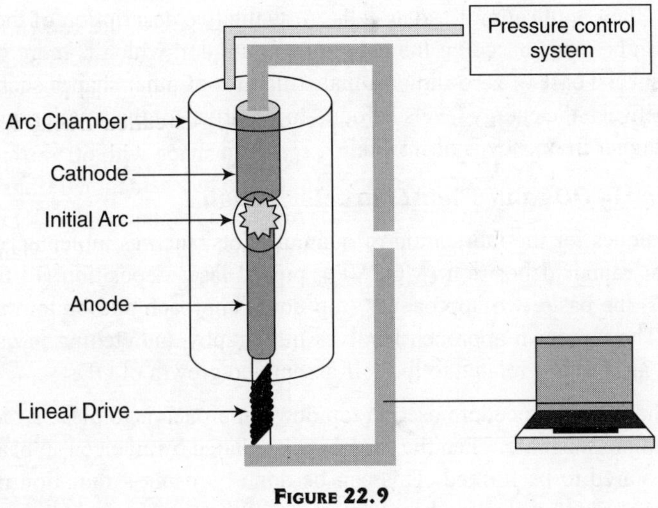

FIGURE 22.9

22.6.2.2 Properties of Fullerenes

The Bucky ball becomes more than twice as hard as its cousin, diamond, when compressed to 70 percent of its original size. These can withstand slamming into a stainless steel plate at 15,000 mph, merely bouncing back unharmed. This shows their high speed collisions property. The Bucky ball is the only known carbon allotrope that can be dissolved at room temperature and for which aromatics are the best solvents. Larger fullerenes (C-72) with trapped lanthanides have been found to have higher solubility.

22.6.2.3 Potential Applications of Fullerenes

Basic fullerenes and their functionalized derivatives have been suggested to have a large number of applications. Potential applications include organic hydrogen gas storage, sensors, polymer electronics, photovoltaic, as a molecular wire, as a precursor to diamond antioxidants, biopharmaceuticals, antibacterial, HIV inhibition, catalysts, water purification, MRI agents, optical devices, scanning tunneling microscopy, and atomic force microscopy. Fullerenes are being extensively investigated as carrier species for medical radionuclides in cancer therapy. On the other hand, Bucky papers are used in fire resistance and in television screens since these may be more efficient than CRT and LCD displays.

Pristine Fullerenes with no functional groups can have a positive effect and act as antioxidents. However, functionalized fullerenes or fulleres dissolved in organic solvents are hazardous to the environments.

22.6.3 Quantum Dots (QDs)

Quantum dots (QDs) are semiconductor nanocrystals that have a collection of electrons varying from a single electron to several thousands of electrons. Quantum dots are termed as zero- dimensional structured materials, as the carriers, i.e., the electrons, are spatially confined in all the three directions or the motion of the electrons is restricted in all the three dimensions. The energy levels of quantum dots are discrete and just as an atom their energy levels are quantized and isolated. In view of this, the quantum dots are sometimes referred to as artificial atoms. The density of state of quantum dots is a delta function of energy E. For this reason, many electrons can be accommodated in a single energy level. This property makes the quantum dots as an active component of high power and low threshold semiconductor LASERs and highly efficient photo detectors in addition to their application to solar cells. A qualitative description of the response of quantum dots of many shapes can be made based on the cuboid quantum dot, which is more often designated as the quantum box and is a special case of zero-dimensional structures of other shapes such as spherical quantum dots. The energy gap between the energy levels is found to be larger for the smaller quantum dots and hence, the emitted light with higher frequency is observed.

22.6.3.1 Fabrication of QDs and Their Characterization

There are several techniques for the fabrication of quantum dots, such as molecular beam epitaxy (MBE), metal organic chemical vapour deposition (MOCVD), pulsed laser deposition (PLD), etc. As in the case of other nanostructures, the basic two approaches (top down approach and bottom up approach) are used to fabricate the QDS. The top down approach involves lithography and etching in quantum well structure, whereas the bottom up approach is related to the self-assembled growth of QDs.

The first step of the lithographic procedure used in top down approach is to place a radiation-sensitive resist on the surface of the sample substrate. Then the sample is irradiated by an electron beam in the region where the nanostructure is required to be formed. This can be done by using a radiation mask that contains the nanostructure pattern or a scanning electron beam that strikes the surface only in the desired region. The next step is the application of the developer to remove the irradiated portions of the resist. After that an etching mask is inserted into the hole in the resist. Subsequently the remaining parts of the resist are lifted off. The areas of the quantum structure not covered by the etching mask are then chemically etched away in order to produce the quantum structure. Finally, the etching mask is removed.

Recent techniques for fabrication of quantum dots in bottom up approach involve strain induced self-assembly. The term self-assembly represents a process where a strained 2-D system reduces its energy by changing into a 3-D morphology. The $In_xGa_{1-x}As/GaAs$ system that offers a large lattice mismatch (7.2% between InAs and GaAs) is employed as the material combination most commonly used for this technique. Through this technique self-assembled InAs quantum dots can be grown on GaAs and the size, separation and height of the quantum dots can be controlled by the deposition parameters. As was the case with quantum nanowires, the random distribution of the quantum dots is, however, one of the drawbacks of this technique.

There are several techniques to characterize the quantum dots. These are atomic force microscope (AFM), scanning tunneling microscope (STM), transmission electron microscope (TEM), photoluminescence (PL) and in situ reflection high energy electron diffraction (RHEED).

22.7 APPLICATIONS OF NANOTECHNOLOGY LO4

Nanotechnology has been identified as the technology for the future and holds the promise to be the basic tenet of the next industrial revolution. Although there are still existing applications of nanotechnology in the world but the future holds much more promise. Some of the existing products that derive their functionality from nanotechnology are discussed below.

22.7.1 Self-Cleaning Glass

Nanoparticles are coated on the glass to make it photocatalytic and hydrophilic. The photocatalytic effect implies that when UV radiation from the light hits the glass the nanoparticles become energized and begin to break down the organic particles on the glass surface and due to its hydrophilic nature the glass attracts water particles, which then clean it.

22.7.2 Clothing

Scientists are using nanoparticles to enhance tour clothing. By coating the fabrics with a thin layer of Zinc Oxide nanoparticles clothes offering better protection from UV radiation can be created. Also, clothes can have nanowhiskers that can make them repel water and other materials thus making them stain resistant. Silver nanoparticles have been demonstrated to have an antibacterial effect on the clothes that they were coated on to.

22.7.3 Scratch Resistant Coating

Materials like glass are being coated with thin films of hard transparent material to make the glass scratch resistant. Anti–fog glasses with transparent nanostructures are also available, which conduct electricity and heat up the glass surface to keep it fog free.

22.7.4 Electronics

Carbon nanotubes have been used as conduits for electricity in very small electrical circuits due to their superior electrical properties and absence of electromigration.

22.7.5 Energy

The most advanced nanotechnology projects related to energy are: storage, conversion, manufacturing improvements by reducing materials and process rates, energy saving (by better thermal insulation, for example), and enhanced renewable energy sources. A reduction of energy consumption can be reached by better insulation systems, by the use of more efficient lighting or combustion systems, and by use of lighter and stronger materials in the transportation sector. Currently used light bulbs only convert approximately 5% of the electrical energy into light. Nanotechnological approaches like light emitting diodes (LEDs) or quantum caged atoms (QCAs) could lead to a strong reduction of energy consumption for illumination.

22.7.6 Space

Nanotechnology may hold the key to making space flight more practical. Advancements in nanomaterials make lightweight solar sails and a cable for the space elevator possible. By significantly reducing the amount of rocket fuel required, these advances could lower the cost of reaching orbit and travelling in space. Space science, as long, played a role in the research and development of advancing technologies. Spacecraft are being launched, with hulls that are composed of carbon fibres, a lightweight high strength material. Combine that with smaller on–board computers that perform hundreds of times faster than computers used on spacecraft just a decade ago, one can see the incredible advances in space exploration in just, past few years. The advancements in material science and computer science have allowed the building, launching and deploying of space exploration systems that continually do more and more as they become smaller and lighter.

22.7.6.1 Smart Materials

Some of the latest avenues being explored, which are more in the nano realm, in space science, include smart materials for the hulls of spacecraft. These would be materials primarily composed of nanotube fibres with nano sized computers integrated into them. These materials along with being even lighter will also be far stronger too. One idea is to create a surface that will help transfer the aerodynamic forces working on a spacecraft during launch. When the craft is launched the nano computers will flex the crafts hull to offset pressure differences in the hull caused by the crafts acceleration through the atmosphere.

Then the same nano computer network in the hull would go to work heating the shaded side of the craft and cooling the sun exposed side and to even create heat shielding for re-entry. To equalize the surface temperature, now a spacecraft must be kept rotating and although a slight spin is good in maintaining the attitude of a craft. Sometimes it interferes with the mission plan, like when a spacecraft is taking photographs or is in the process of docking with another craft.

22.7.6.2 Swarms

Another avenue being investigated is a concept of nano robotics called “Swarms”. Swarms are nano robots that act in unison like bees. They theoretically will act as a flexible cloth like material and are composed of what’s called Bucky tubes. This cloth will be as strong as diamond. Add to this cloth of nano machines nano computers and we have smart cloth. This smart cloth could be used to keep astronauts from bouncing around inside their spacecraft while they sleep, a problem that arises when the autopilot computer fires the course correction rockets. The cloth like material will be able to offset the sudden movements and slowly move the sleeping astronaut back into position. Still another application for the nano robot swarms, being considered, is that the smart cloth could be used in the astronauts’ space suits.

22.8 LIMITATIONS OF NANOTECHNOLOGY LO5

Any technology is generally known to solve a problem, but at the same time it may create a new problem or has a limitation. There are potential risks of nanotechnology as well. A few limitations of the nanotechnology are discussed below.

(i) The problem with nanotechnology is the scale it works on. Small and highly complex Nano machines are probably awesome at building things with carbon only. Moreover, nanites may not work well in the strong magnetic fields which are needed to produce permanent magnets. They may also not be able to meet or operate under the conditions necessary to manufacture superconductors.

(ii) In view of constant breaking and reforming bonds as well as the friction caused by millions of nanobots a large amount of heat is expected to be produced, which will lead problems in fabricating certain extremely temperature sensitive materials. For example, nitroglycerin would explode if it experienced even a one-degree shift in temperature. So large stations would likely be limited in their production by the amount of heat they are able to disperse. Also, because the ratio of surface area to internal space decreases as the size of an object increases, larger ships will have a harder time dispersing heat than smaller vehicles.

(iii) It will be difficult for nanites to build something in a hostile environment. The workforce may be billions of tiny robots strong but they will likely be unable to finish the job in a sandstorm or leave large chunks of partially completed product scattered over a large area.

22.9 DISADVANTAGES OF NANOTECHNOLOGY LO5

Nanotechnology has made significant impact in all fields of life and it has impressive applications in almost every area of work, but it also has some major disadvantages which need to be addressed.

(i) Nanotech devices and machines have taken place of human to work faster and accurately. This has lessened the importance of men power in the field of practical work, and reduced the employment in the fields of traditional farming and manufacturing and industrial sector.

(ii) Nanotechnology has made atomic weapons more powerful and more destructive. However, this has produced a big threat with regard to their easy accessibility. Unauthorized and criminal bodies can reach nuclear weapons easily and its formulation could be stolen.

(iii) Diamond is now being produced massively with the help of nanotechnology, which has reduced the value of diamond and increased the fall of diamond markets. Presence of alternative has decreased the demand because alternates are more efficient and do not require the use of fossil fuels. The manufacturer can now produce bulk of the products at molecular scale and decomposition is done to create new components.

(iv) The presence of nanomaterials which contain nanoparticles is not in itself a threat but their increased reactivity and mobility can make them risky. Nanotechnology has increased risk to the human health, as the nanoparticles due to their small size can cause inhalation problem and many other fatal diseases. Apart from what happens if non-degradable or slowly degradable nanoparticles accumulate in organs, another concern is their potential interaction with biological processes inside the body.

(v) Nanotechnology is the most expensive technologies, whose cost is increasing day by day due to the molecular structure and processing of the product. It has become difficult for the manufacturers to randomly produce dynamic products due to the huge pricing of nanotech machines. This is unaffordable for the common people.

(vi) Nanotechnology has raised the standard of living, but at the same time it has increased the pollution including water and air pollutions due to the wastes generated by Nano devices or during the nanomaterials manufacturing process. This pollution, called Nano pollution, may be very dangerous for living organisms. Most of the human made nanoparticles do not appear in nature, so living organism may not have appropriate means to deal with Nano waste. Hence, the whole life cycle of these particles needs to be evaluated with respect to their fabrication, storage, distribution, potential abuse and disposal. The impact on the environment may vary at different stages of the life cycle. Concerns are raised about Nano pollution as it is not currently possible to precisely predict or control ecological impacts of the release of these Nano products into the environment.

(vii) Finally, there are educational gap risk issues with regard to the nanotechnology, though it offers rapid advances across many areas of science and engineering which are crucial to the society. For example, the knowledge within scientific and industrial communities is not appropriately shared with the civil society, public and regulatory agencies. Because of this innovative opportunities may be lost and public confidence in transparency and accountability may wear away.

SUMMARY

Summary of the topics covered in this chapter is given below.

- Nanoscience is the study of objects having at least one dimension less than hundred nanometers; whereas nanotechnology is the engineering of these objects using different techniques.
- Nanotechnology deals with various structures of matter having dimensions of the order of a billionth of a meter. In any material, an atom on the surface is different from an atom in the bulk. This difference is primarily because of the different potentials that the two atoms are experiencing.
- The ratio of number of atoms in the bulk to the number of atoms on the surface reveals that as the size of the particle decreases the number of atoms on the surface is in a greater proportion of the total number

of atoms. Hence, as the particle size becomes smaller and smaller the surface atoms start dominating the properties of the whole material.

- Nanoparticles are generally considered to be a number of atoms or molecules bonded together such that the dimension of the bonded entity is of the order of 100 nm.
- The properties of nanomaterials are different from those of bulk because of two main reasons—the surface effect and the quantum effect. The varied applications of nanoparticles is a consequence of their varied properties.
- Synthesis of nanoparticles can be achieved by mechanical means such as ball milling; or by techniques such as Gas condensation, Sputtering, Vaccum Deposition and Vaporization, Chemical Vapour Deposition (CVD). Chemical Vapour Condensation (CVC), Sol-Gel Technique and Electrodeposition.
- Nanoparticles have properties different from those of bulk. Metallic nanoparticles exhibit Surface Plasmon Resonance (SPR) which enables their use in medical diagnostics. In case of semiconductor nanoparticles, the band gap increases with decrease in size leading to a change in absorption spectra compared to bulk material.
- Quantum confinement is the restricted motion of randomly moving electrons in specific energy levels on reduction of size. Quantum confinement leads to change in optical and electrical properties. Based on the number of free dimensions available; nanostructures are classified as 3D, 2D, 1D or 0D. 3D structures have none of its dimensions confined whereas 2D nanostructures have one dimension confined e.g. thin films. 1D nanostructures have two dimensions confined e.g. nanowires and 0D nanostructures have all their dimensions confined e.g. Quantum Dots.
- Nanowires are 1D structures and have an Aspect Ratio $>$ 1000. They have diameter $\leq$ 10 nm and unconstrained length. In Top down approach to Nanowire synthesis, etching and Lithographic techniques are employed. For bottom-up approach, the Vapour Liquid Solid (VLS) method, Vapour Solid (VS) Method and Catalyst free Template assisted methods are used.
- Carbon nanotubes, which are long and thin cylinders of carbon, were discovered in 1991 by S. Iijima. These are large macromolecules that are unique for their size, shape, and remarkable physical properties. They can be thought of as a sheet of graphite (a hexagonal lattice of carbon) rolled into a cylinder.
- Cartoon Nanotubes (CNTs) are single walled (SWCNT) or Multiwalled (MWCNT). MWCNTs have higher strength than SWCNTs.
- Procedures used for CNT synthesis are: Arc Discharge method, Laser Ablation Method, Chemical Vapour Deposition (CVD) method and n-hexane pyrolyris.
- CNTs have higher tensile strength than steels. They are highly elastic. The electrical conductivity of CNTs is structure-dependent as the structure influences the collisions between conductive electrons and atoms. The thermal conductivity of CNTs is 15 times superior to that of copper.
- Inorganic nanotubes are non-carbon nanotubes. They are cylindrical molecules that are often composed of metal oxides or Group-III Nitrides.
- Biopolymers are polymers produced by living organisms. They are biodegradable. Biopolymers are synthesized by chemical polymerization or by chemical modification of a naturally occurring polymer. They find application in the packaging industry as food trays, wrappings, plastic etc.
- Fullerene is a carbon-molecule which could be in the form of a holow sphere, ellipsoid or tube. C-60 spherical fullerenes are known as Bucky balls. C-60, C-70, C-76, C-84, C-240 and C-540 are some of the spherical fullerenes.

- The main methods of synthesizing fullerenes is by Electric Arc Discharge method, Laser Ablation method and chemical Vapour Deposition method.
- Fullerenes are twice as hard as diamond when compressed to 70% of their original size. They are soluble in organic solvents at room temperature.
- Quantum dots are 0-D nanostructures. The energy levels of Quantum dots are discrete, quantized and isolated as in atoms. They find application in high power, low threshold semiconductor lasers, high efficiency detectors and in solar cells.
- Commonly used techniques for fabrication of Quantum dots are Molecular Beam Epitaxy (MBE), Metal Organic Chemical Vapour Deposition (MOCVD), and Pulsed Laser Deposition (PLD).
- Nanotechnology has got diverse applications in self-cleaning glass, clothing, scratch resistant coating, electronics, energy, space and environment. Therefore, these applications of nanotechnology were talked about.

SOLVED EXAMPLES

EXAMPLE 1 Estimate the N_b/N_s ratio for a spherical particle of diameter 12 μm and compare it with the same for a nanoparticle of diameter 90 nm.

SOLUTION We have, $N_b/N_s = r/3$ where r is the radius of the particle. Hence, for the micrometer sized particle,

$$N_b/N_s = 6\times10^{-6}/3 = 2\times10^{-6}$$

Similarly for the nanometer sized particle,

$$N_b/N_s = 45\times10^{-9}/3 = 15\times10^{-9}$$

Hence, the ratio is smaller for the nanoparticle by a factor of $\mathbf{7.5\times10^{-3}}$.

EXAMPLE 2 How is it possible to obtain nanoparticles of the same material yet having different colours?

SOLUTION In the case of metallic nanoparticles the color of the particle depends on its surface plasmon resonance (SPR) frequency. Now it is known that the surface plasmon resonance frequency depends on the size as well as the shape of the nanoparticle at hand and shifts as any of these parameters are changed. Hence, even for the same material it is possible to have different SPR frequency for different particle sizes or shapes and thus a different colour.

EXAMPLE 3 How does a Single Electron Transistor (SET) work?

SOLUTION In a SET, the modulation is done by charge transfer on to a small island. This island has dimensions in the nanometer regime and even the transfer of a single charge e to it can change its potential substantially. So we can put a transistor in off state by transferring a single electron to the island and increasing its potential to a value so that the current transfer between the source and drain ceases. Thus, it is given the name of a Single Electron Transistor.

EXAMPLE 4 What gives the increased yield strength in the systems that have a grain size in nano regime?

SOLUTION The reason for the increase in yield strength as the grain size becomes smaller and smaller is that smaller grain sizes have more grain boundaries that offer resistance to the movement of dislocations. And since a material fails when the dislocations gather at a spot and yield to breakage, by limiting movement of dislocations the yield strength is increased.

Objective Type Questions

Q.1 A special molecule of carbon made up of 60 carbon atoms is understood as a structure that shows potential for a basic building block in the area of molecular manufacturing. The nontechnical name of these molecule is
(a) Fullerene (b) Nano rods (c) Bucky balls (d) Nanotubes

Q.2 Graphene is
(a) A one-atom thick sheet of carbon
(b) A new material made from carbon nanotubes
(c) Thin film made from fullerene
(d) A software tool to measure and graphically represent nanoparticles

Q.3 Single-walled carbon nanotubes (SWCNTs) are
(a) Poor conductor (b) Excellent conductor
(c) Poor conductor than MWCNTs (d) None of the above

Q.4 Diameter of Bucky ball is about
(a) 100 A° (b) 10 A° (c) 1 A° (d) 1000 A°

Q.5 Properties of nanoparticles
(a) Are significantly different from the properties of bulk materials
(b) Are little bit different from the properties of bulk materials
(c) May be the same as in bulk material
(d) Are none of the above

Q.6 Surface area per unit volume for nanoparticles is
(a) Higher than macro-sized particles (b) Same as macro-sized particles
(c) Lower than macro-sized particles (d) None of the above

Q.7 Starch and cellulose are both biopolymers of
(a) Glucose (b) Maltose (c) Starch (d) Fructose

Q.8 Self-healing pain can be cured through
(a) Biotechnology (b) Information technology
(c) Nanotechnology (d) None of the above

Q.9 Carbon nanotubes are
(a) Hollow cylinders made up of carbon atoms
(b) Circular tubes made up of graphite
(c) Nanotubes made of carbon sheet
(d) Nothing but simple carbon atoms

Q.10 Upon decreasing the dimension of a nanoparticle what kind of a shift is observed in the absorption spectra of a semiconducting particle?
(a) Red shift (b) Blue shift (c) Green shift (d) Violet shift

Q.11 What kind of a quantum mechanical system has a constant density of states?
(a) 1-D (b) 2-D (c) 3-D (d) 0-D

Q.12 A quantum mechanical system was found to have *spikes* in the plot of its density of states. Among the following physical systems which can represent such a system?
(a) A quantum well (b) A bulk system
(c) A quantum dot (d) A carbon nanotube

Q.13 Which of these pairs correctly represents the constituent particles of an exciton?
(a) Electron, positron (b) Electron, hole
(c) Electron, positronium (d) Hole, hole

Q.14 Which among these systems can show us excitonic effects?
(a) Low T, low purity, bulk (b) Low T, low purity, quantum dot
(c) Room T, high purity, bulk (d) Room T, high purity, quantum dot

Q.15 Which of the following methods are not currently employed for fabricating carbon nanotubes?
(a) Arc discharge (b) Laser ablation
(c) Chemical vapour deposition (d) Ball milling

Q.16 Which of these does not represent a type of carbon nanotube?
(a) Chiral (b) Zigzag (c) Wavy (d) Armchair

Q.17 Which of the following is not a stage of Sol-gel formation?
(a) Agglomeration (b) Condensation (c) Hydrolysis (d) Sputtering

Q.18 Which of the following elements is known for its anti-bacterial properties in its nanoparticle form?
(a) Ag (b) Fe (c) Pd (d) Cu

Q.19 The absence of which of the following phenomenon is most critical in making carbon nanotubes as a good conduit material in electronic circuits?
(a) Electromigration (b) Mechanical strain memory
(c) Poor mechanical strength (d) Thermal anisotropy

Q.20 Nanotechnology was brought into day light by delivering lectures by
(b) Max Planck (b) Einstein (c) Feynman (d) Lorentz

Q.21 Which of the following nano-particles are mostly used in paint industry
(a) Ag (b) Fe (b) TiO_2 (c) SiO_2

Q.22 Which of the following is correct for surface area to volume ratio in nanomaterials
(a) moderate (b) very less (c) very large (d) None of these

Q.23 The third known form of pure carbon is
(a) Diamond (b) Fullerene (c) Graphite (d) None of these

Q.24 When a bulk material is changed into nanoparticle which of the following will change state
(a) Physical (b) Chemical (c) both (a) & (b) (d) None of these

Q.25 Which of the following method is used to prepared carbon nanotube?
(a) Plasma arc-evaporation method (b) Chemical vapour deposition method
(c) Laser ablation method (d) All of these

Q.26 Bucky ball is the cluster of carbon atoms
(a) 10 (b) 60 (c) 15 (d) None of these

Q.27 Which of the following statement is correct for carbon nanotube
(a) almost 20 times stronger than steel (b) almost 6 times lighter than steel
(c) (a) & (b) (d) None of these

Q.28 Which of the nanocrystalline are used as separator plate in new generation batteries
(a) Nickel (b) Nickel hydrides (c) both (a) & (b) (d) None of these

Q.29 Carbon atom attain the shape in Bucky ball
(a) Hexagonal (b) pentagonal (c) trigonal (d) None of these

Q.30 Gold nanosphere of size 100 nm appears
(a) violet (b) red (c) orange (d) green

Practice Problems

General Questions

Q.1 Write a short note on nanotechnology.

Q.2 What do you understand by nanoparticles? Discuss their optical properties.

Q.3 Based on I-V characteristics, discuss the electrical properties of nanoparticles.

Q.4 In the light of mechanical properties of nanoparticles, explain how the yield strength varies with grain size?

Q.5 What do you understand by quantum confinement? Discuss density of states for different types of quantum confinements.

Q.6 What are carbon nanotubes? Discuss how various types of carbon nanotubes can be formed from graphene?

Q.7 Discuss in short various techniques for the synthesis of nanoparticles.

Q.8 List the difference between chemical vapour deposition (CVD) and chemical vapour condensation (CVC).

Q.9 Explain various steps involved in sol-gel technique used for the synthesis of nanoparticles.

Q.10 Discuss how nanotechnology is useful in environment and space?

Q.11 How nanomaterials are different from bulk materials?

Q.12 Discuss the difference between nanoscience and nanotechnology.

Q.13 Discuss the basic difference between 0D, 1D, 2D and 3D materials.

Q.14 Write a note on nanowires and their different kinds.

Q.15 Give a brief description of synthesis techniques of nanowires.

Q.16 What do you understand by carbon nanotubes? These structures fall within which category?

Q.17 Discuss single-walled and multi-walled carbon nanotubes along with their differences.

Q.18 What are the methods for fabrication of carbon nanotubes? Discuss in brief.

Q.19 Discuss properties of CNTs.

Q.20 Write a note on inorganic nanotubes.

Q.21 Write a note on biopolymers.

Q.22 Discuss in brief 2D nanomaterials with respect to their synthesis.

Q.23 Discuss sol-gel technique used for the synthesis of nanoparticles.

Q.24 Write down the details of Ball milling and gas condenstion techniques used for the synthesis of nano-particles.

Q.25 What are surface plasmons.

Q.26 What are Bucky balls or Fullerenes? How are these synthesized using electric arc discharge, laser ablation and CVD techniques?

Q.27 Discuss in short the properties of fullerenes and their potential applications.

Q.28 What do you understand by quantum dots? How are these fabricated and characterized?

Q.29 Discuss any five applications of nanotechnology.

Q.30 What are the limitations of nanotechnology?

Q.31 Discuss five disadvantages of nanotechnology.

APPENDICES

Appendix 1

Measurements and Errors

If you are asked to measure the same object two different times, there is always a possibility that the two measurements may not be exactly the same. Then the difference between these two measurements is called a *variation* or *error* in the measurements. In general, there are two types of errors in measurement. These are called *static error* and *dynamic error.*

A1.1 STATIC ERROR AND ITS CLASSIFICATION

Static error of a measuring instrument refers to the numerical difference between the true value of a quality and its measured value of quantity. This gives different indications when the same quantity is measured repeatedly. On the other hand, *dynamic error* is the difference between the true value of a quantity changing with time and the value indicated by the instrument.

Static errors are classified into three categories, namely, gross errors or human errors, systematic errors and random errors.

A1.1.1 Gross Errors

Gross errors include all the human mistakes while reading and recording. Mistakes carried out in calculating the errors also fall within this category. For example, while taking the reading from the meter of the instrument, a person may read 21 as 27 or 31. Gross errors can be avoided if proper care is taken in reading, recording the data and doing calculations accurately. We can also reduce such errors by increasing the number of experimenters and by taking the average of more readings.

A1.1.2 Systematic Errors

Systematic errors are the errors which tend to be in one direction (either positive or negative). Systematic errors include instrumental, environmental and personal errors. *Instrumental error* may be due to wrong construction or calibration of the measuring instruments. These errors also include the loading effect, misuse of the instruments and zero error in the instrument. *Environmental error* arises due to external conditions, which include temperature, pressure, humidity, external magnetic field, etc. We can minimize the environmental errors by maintaining the temperature and humidity of the laboratory constant through some arrangements, and ensuring that there is no external magnetic or electrostatic field around the instrument. On the other hand, *personal errors* are due to wrong observations, which may be due to lack of proper setting of the apparatus or individual carelessness in taking observations.

A1.1.3 Random Errors

After calculating all the systematic errors, it is observed that there are still some more errors in the measurement. These *random errors* are those errors which occur irregularly and are random with respect to their sign and size. Random and unpredictable fluctuations in temperature, voltage supply and mechanical vibrations of experimental set-up may lead to random errors. The important property of a random error is that it adds variability to the data but does not affect average performance for the group. For this reason, random error is sometimes referred to as *noise*.

A1.2 DYNAMIC ERROR

Dynamic errors are caused by the inertial properties of measuring instruments or equipment. Consider that a varying quantity is recorded with the help of a recording device. Then the difference between the obtained function and the actual process of change of the recorded quantity in time is called the dynamic error of the given dynamic instrument. It is clear that these errors are caused by the time variation in the measured quantity.

A1.3 SOURCES OF ERRORS

Sources of error, other than the inability of a piece of hardware to provide true measurements, are as under:

1. Insufficient knowledge of process parameters and design conditions
2. Errors caused by the person operating the instrument
3. Change in process parameters, irregularities, etc.
4. Poor maintenance
5. Poor design
6. Certain design limitations

A1.4 ACCURACY AND PRECISION

If you obtain a weight measurement of 6.5 kg for a given substance, and the actual or known weight is 10 kg, then your measurement is not accurate. It means your measurement is not close to the known value. The closeness of a measured value to a standard or known value is referred to as *accuracy*. On the other hand, *precision* refers to the closeness of two or more measurements to each other. The precision tells us to what resolution (or limit) the quantity is measured. In the above example, if you weigh a given substance six times and get 6.5 kg each time, then your measurement is very precise. It means the precision is independent of the accuracy. You can be very precise but inaccurate, and also you can be accurate but imprecise. Moreover, you can have accuracy without precision. For example, on an average your measurements for a given substance may be close to the known value, but the measurements may be far from each other.

A1.5 RESOLUTION

Resolution is the fineness to which an instrument can be read. We can take the example of two stopwatches, out of which one is analog and the other is digital. Both are manually actuated and are looked at for resolution.

The analog stopwatch has to be viewed on its dial. If we look closely, we can relate the big hand to the smallest tick mark on the big dial. That tick mark is a tenth of a second. It means the best a good eye can do is resolve a reading to 1/10 second. Hence, this is the resolution of the stopwatch. On the other hand, the digital stopwatch has two digits beyond the seconds. So it subdivides time in hundredths of a second. Since it is easy to read to 1/100 of a second, the resolution of the digital stopwatch is 1/100 second.

A1.6 MEASUREMENT UNCERTAINTY

Certainty is perfect knowledge which has total security from error. Certainty is also the mental state of being without doubt. Every experiment is approximate due to error in measurement. When a number of measurements are done to a stable voltage (or other parameter), the measured values will show a certain variation. This variation is caused by thermal noise in the measuring circuit of the measuring equipment and the measurement set-up. These variations or the uncertainties are shown in Figure 1.1.

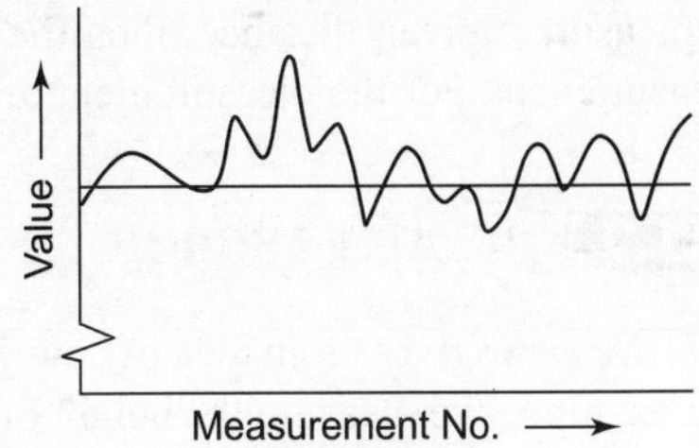

FIGURE 1.1

A1.7 STANDARD DEVIATION AND VARIANCES

The uncertainty is estimated by calculating the standard deviation. Let $x_1, x_2, x_3 \ldots x_N$ be the results of an experiment repeated N times. Then the *standard deviation* σ is defined as

$$\sigma = \sqrt{\frac{1}{N}\sum_{i=1}^{N}(x_i - \overline{x})^2}$$

where, $\overline{x}\left(\equiv \frac{1}{N}\sum_i x_i\right)$ represents the average of all the values of x. In this case, the uncertainty is of the order of $\pm\sigma$. The standard deviation is defined in terms of the square of the deviations from the mean, which is clear from the term $\sum_{i=1}^{N}(x_i - \overline{x})^2$ in the above formula. Moreover, σ^2 is known as the variance of the data.

The standard deviation σ is the root mean square deviation of the data, measured from the mean.

A1.8 ABSOLUTE ERROR

Absolute error is defined as the magnitude of difference between the actual and the approximated values of any quantity. For example, we measure a given quantity n times and $a_1, a_2, a_3, \ldots a_n$ are the individual values. Then the arithmetic mean (say a_m) can be found as

$$a_m = \frac{a_1 + a_2 + \cdots + a_n}{n} \qquad \text{(i)}$$

This can also be written as

$$a_m = \frac{\sum_{i=1}^{n} a_i}{n} \tag{ii}$$

The absolute error can now be obtained from the following formula:

$$\Delta a_n = a_m - a_n \tag{iii}$$

In simple words, absolute error is the amount of physical error in a measurement. If we use a metre stick to measure a given distance, then the error may be ±1 mm or ±0.001 m. This is the absolute error of the measurement. For the measurement of a quantity x, the absolute error is Δx.

A1.9 RELATIVE ERROR

Relative error gives us an idea of how good a measurement is relative to the size of the object being measured. For example, we measure the height of a room and the length of a small table by using a metre stick. We find the height of the room as 3.256 m ± 1 mm and the length of the table as 0.082 m ± 1 mm. Then

$$\text{Relative error} = \frac{\text{Absolute error}}{\text{Value of thing measured}} \tag{i}$$

∴ Relative error in measuring the height of the room

$$= \frac{0.001 \text{ m}}{3.256 \text{ m}} = 0.0003$$

Relative error in measuring the length of the table

$$= \frac{0.001 \text{ m}}{0.082 \text{ m}} = 0.0122$$

Here, it is clear that the relative error in measuring the length of the table is larger than the relative error in measuring the height of the room. In both the cases, however, the absolute error is the same.

A1.10 PERCENTAGE ERROR

If the relative error is represented in percent, then the error is called *percentage error*. For example, in the above example, the percentage error in measuring the height of the room is 0.0003 × 100 = 0.03%, while the percentage error in measuring the length of the table is 0.0122 × 100 = 1.22%.

The absolute error, relative error and percentage error can be summarized as follows. If we represent the given or actual value by a and its approximated value as a_{app}, then

$$\text{Absolute error} = \left|a - a_{\text{app}}\right|$$

$$\text{Relative error} = \frac{\left|a - a_{\text{app}}\right|}{a} = \left|1 - \frac{a_{\text{app}}}{a}\right|$$

$$\text{Percentage or percent error} = \left|1 - \frac{a_{\text{app}}}{a}\right| \times 100$$

The important point is that the actual value a talked about cannot be zero.

A1.11 ERRORS OCCURRING IN ARITHMETIC OPERATIONS

1.11.1 Addition and Subtraction

Let us consider two measured values $a \pm \delta a$ and $b \pm \delta b$ in which a and b are actual values, whereas δa and δb are absolute errors of a and b, respectively.

The error obtained in the sum of these quantities is given by

$$\begin{aligned} Q \pm \delta Q &= (a \pm \delta a) + (b \pm \delta b) \\ &= (a + b) + (\pm\delta a \pm \delta b) \\ \Rightarrow \quad \delta Q &= \delta a + \delta b \end{aligned} \tag{i}$$

Similarly, the error obtained in their difference is given by

$$\begin{aligned} Q \pm \delta Q &= (a \pm \delta a) - (b \pm \delta b) \\ &= (a - b) + (\pm\delta a \pm \delta b) \\ \Rightarrow \quad \delta Q &= \delta a + \delta b \end{aligned} \tag{ii}$$

Hence, in arithmetic operations of addition and subtraction, the absolute error in the resultant is the sum of the absolute errors of individual quantities. So errors are always added in these operations.

A1.11.2 Multiplication and Division

Let us consider the two measured quantities $a \pm \delta a$ and $b \pm \delta b$, where a and b are actual quantities and δa and δb are the absolute errors in a and b, respectively. The errors occurring in multiplication and division can be estimated as given below.

$$Q \pm \delta Q = (a \pm \delta a)\,(b \pm \delta b) = ab \pm b\delta a \pm a\delta b \pm \delta a\delta b \tag{iii}$$

On dividing by Q on the LHS and by ab on the RHS, we obtain

$$1 \pm \frac{\delta Q}{Q} = 1 \pm \frac{\delta a}{a} \pm \frac{\delta b}{b} \tag{iv}$$

In terms of percentage error, we have

$$\frac{\delta Q}{Q} \times 100 = \frac{\delta a}{a} \times 100 + \frac{\delta b}{b} \times 100$$

For general treatment, we can consider a quantity

$$Q = ka^x\, b^y c^z \tag{v}$$

Then error can be determined as

$$\frac{\delta Q}{Q} = x\frac{\delta a}{a} + y\frac{\delta b}{b} + z\frac{\delta c}{c} \tag{vi}$$

It means, if Q is a measure of x power of a, y power of b and z power of c, then the fractional error will be the sum of x times of fractional error in a, y times of fractional error in b and z times of fractional error in c. This also infers that maximum error will be encountered due to the quantity carrying highest power. Therefore, in experiments, that quantity should be determined with more precision in order to minimize the error.

SOLVED EXAMPLES

EXAMPLE 1 Find the standard deviation of the numbers 1, 5, 6, 7, 8, 10, 12.

SOLUTION The mean of these numbers is found to be 7. The deviations are –6, –2, –1, 0, 1, 3, 5, respectively. Now using the formula

$$\sigma = \sqrt{\frac{1}{N}\sum_{i=1}^{N}(x_i - \bar{x})^2}$$

$$= \sqrt{\frac{1}{7}[36+4+1+0+1+9+25]}$$

$$= \sqrt{\frac{76}{7}}$$

So, $$\sigma = \sqrt{10.86}$$

$$= \mathbf{3.295}$$

EXAMPLE 2 If two resistances given as $R_1 = (50 \pm 5)\ \Omega$ and $R_2 = (150 \pm 2)\ \Omega$ are connected in series, then find the equivalent resistance.

SOLUTION $R = (50 \pm 5) + (150 \pm 2)$

$$= (50 + 150) \pm (5 + 2)$$

$$= \mathbf{(200 \pm 7)\ \Omega}$$

EXAMPLE 3 If the mass of a bulb with air is 98.625 ± 0.002 g and the mass of an empty bulb is 98.305 ± 0.002 g, then find the mass of air.

SOLUTION Error in difference $= (a \pm \delta a) - (b \pm \delta b)$

$$= (a - b) + (\pm\delta a \pm \delta b)$$

$$= (98.625 - 98.305) \pm (0.002 + 0.002)$$

$$= \mathbf{0.320 \pm 0.004\ g}$$

EXAMPLE 4 If the capacity of a capacitor is $C = 2 \pm 0.4$ F and the applied voltage is $V = 20 \pm 0.2$ V, then find the charge on the capacitor.

SOLUTION Charge on capacitor, $Q = CV = 2 \times 20 = 40$ C

$$\text{Percentage error in } C = \frac{0.4}{2} \times 100 = 20\%$$

$$\text{Percentage error in } V = \frac{0.2}{20} \times 100 = 1\%$$

$$\therefore \text{ percentage of error in } Q = 20 + 1 = 21\%$$

$$\text{or error in } Q = 40 \times \frac{21}{100} = 8.4 \text{ C}$$

Hence, charge on the capacitor $Q = \mathbf{40 \pm 8.4\ C}$

EXAMPLE 5 The volumes of two bodies are measured to be $V_1 = (10.2 \pm 0.02)$ cm^3 and $V_2 = (6.4 \pm 0.01)$ cm^3. Calculate the sum and difference in volumes with error limits.

SOLUTION $V_1 = (10.2 \pm 0.02)$ cm^3

$$V_2 = (6.4 \pm 0.01) \text{ cm}^3$$

$$\Delta V = \pm(\Delta V_1 + \Delta V_2)$$
$$= \pm(0.02 + 0.01)\ \text{cm}^3$$
$$= \mathbf{0.03\ cm^3}$$

EXAMPLE 6 The mass and density of a solid sphere are measured to be (12.4 ± 0.1) kg and (4.6 ± 0.2) kg/m^3 respectively. Calculate the volume of the sphere with error limits.

SOLUTION Here, $m \pm \Delta m = (12.4 \pm 0.1)$ kg

$$\rho \pm \Delta\rho = (4.6 \pm 0.2)\ \text{kg/m}^3$$

$$\text{Volume}\quad V = \frac{m}{\rho} = \frac{12.4}{4.6} = 2.69\ \text{m}^3 = 2.7\ \text{m}^3$$

$$\frac{\Delta V}{V} = \pm\left(\frac{\Delta m}{m} + \frac{\Delta\rho}{\rho}\right)$$

$$\Delta V = \pm\left(\frac{\Delta m}{m} + \frac{\Delta\rho}{\rho}\right)V$$

$$= \pm\left(\frac{0.1}{12.4} + \frac{0.2}{4.6}\right)\times 2.7 = \pm 0.14$$

$$V \pm \Delta V = \mathbf{(2.7 \pm 0.14)\ m^3}$$

EXAMPLE 7 A current of 3.5 ± 0.5 A flows through a metallic conductor and a potential difference of 21 ± 1 volts is applied. Find the effective resistance of the wire.

SOLUTION Given $V = 21 \pm 1$ volts, $\Delta V = 1$, $I = 3.5 \pm 0.5$ A

$$\Delta I = 0.5\ \text{A}$$

$$\text{Resistance}\ R = \frac{V}{I} = \frac{(21 \pm 1)}{(3.5 \pm 0.5)} = 6.01 \pm \Delta R$$

$$\frac{\Delta R}{R} = \text{error in measurement} = \frac{\Delta V}{V} + \frac{\Delta I}{I}$$

$$= \frac{1}{21} + \frac{0.5}{3.5}$$

$$= 0.048 + 0.143 = 0.19$$

$$\Rightarrow \Delta R = 0.19 \times R = 0.19 \times 6 = \mathbf{1.14}$$

Effective resistance $R = \mathbf{6 \pm 1.14\ \Omega}$

EXAMPLE 8 A rectangular board is measured with a scale having an accuracy of 0.2 cm. The length and breadth are measured as 35.4 cm and 18.4 cm, respectively. Find the relative error and percentage error of the area calculated.

SOLUTION $l = 35.4$ cm, $\Delta l = 0.2$ cm

$w = 18.4$ cm and $\Delta w = 0.2$ cm

Area $(A) = l \times w = 35.4 \times 18.4 = 651.36\ \text{cm}^2$

$$\text{Relative error in area}\ (\delta A) = \frac{\Delta A}{A} = \frac{\Delta l}{l} + \frac{\Delta w}{w}$$

$$= \frac{0.2}{35.4} + \frac{0.2}{18.4} = 0.006 + 0.011 = 0.017$$

$$\text{Percentage error} = \frac{\Delta A}{A} \times 100 = 0.017 \times 100 = \mathbf{1.7\%}$$

EXAMPLE 9 A physical quantity Q is related to four observables a, b, c, d as follows:

$$Q = \frac{a^3 b^4}{d^2 \sqrt{c}}$$

The percentage errors of measurement in a, b, c and d are 1%, 3%, 4% and 3% respectively. What is the percentage error in the quantity Q? If the value of Q calculated using the given relation is 8.768, to what value should the result be rounded?

SOLUTION Given

$$Q = \frac{a^3 b^4}{d^2 \sqrt{c}}$$

Percentage error in Q is given by

$$\frac{\Delta Q}{Q} = 3\frac{\Delta a}{a} + 4\frac{\Delta b}{b} + \frac{1}{2}\frac{\Delta c}{c} + 2\frac{\Delta d}{d}$$

Since $\quad \frac{\Delta a}{a} = 1\%, \frac{\Delta b}{b} = 3\%, \frac{\Delta c}{c} = 4\%, \frac{\Delta d}{d} = 3\%$

$$\frac{\Delta Q}{Q} = 3 \times 1\% + 2 \times 3\% + \frac{1}{2} \times 4\% + 2 \times 3\%$$

$$= 3\% + 6\% + 2\% + 6\%$$

$$= \mathbf{17\%}$$

$\therefore$ percentage error in $Q = 17\%$

If the calculated value of Q is 8.768, the roundoff value is 8.8.

EXAMPLE 10 Find absolute error, relative error and percentage error of the approximation 3.14 to the value π.

SOLUTION $\quad \text{Absolute error} = |3.14 - \pi| = 0.0015926536$

$$\text{Relative error} = \frac{|3.14 - \pi|}{\pi} = 0.000506957383$$

$$\text{Percentage error} = \frac{|3.14 - \pi|}{\pi} \cdot 100\% = \mathbf{0.0506957383\%}$$

EXAMPLE 11 The refractive index (μ) of water is found to have the values 1.29, 1.33, 1.34, 1.35, 1.32, 1.36, 1.30 and 1.33. Calculate the mean value, absolute error, the relative error and percentage error.

SOLUTION $\quad \mu_{\text{mean}} = \frac{1.29 + 1.33 + 1.34 + 1.35 + 1.32 + 1.36 + 1.30 + 1.33}{8} = 1.3275 \simeq 1.33$

Absolute errors are

$$\Delta\mu_1 = \mu_{\text{mean}} - \mu_1 = 1.33 - 1.29 = 0.04$$
$$\Delta\mu_2 = \mu_{\text{mean}} - \mu_2 = 1.33 - 1.33 = 0.00$$
$$\Delta\mu_3 = \mu_{\text{mean}} - \mu_3 = 1.33 - 1.34 = -0.01$$
$$\Delta\mu_4 = \mu_{\text{mean}} - \mu_4 = 1.33 - 1.35 = -0.02$$
$$\Delta\mu_5 = \mu_{\text{mean}} - \mu_5 = 1.33 - 1.32 = 0.01$$
$$\Delta\mu_6 = \mu_{\text{mean}} - \mu_6 = 1.33 - 1.30 = 0.03$$
$$\Delta\mu_7 = \mu_{\text{mean}} - \mu_7 = 1.33 - 1.33 = 0.00$$

Mean absolute error,

$$\Delta\mu_{\text{mean}} = \frac{(|\Delta\mu_1| + |\Delta\mu_2| + |\Delta\mu_3| + |\Delta\mu_4| + |\Delta\mu_5| + |\Delta\mu_6| + |\Delta\mu_7| + |\Delta\mu_8|}{8}$$

$$= (0.04 + 0.00 + 0.01 + 0.02 + 0.01 + 0.03 + 0.03 + 0.00) \div 8$$

$$= 0.14 \div 8 = 0.0175 \simeq 0.02$$

Relative error ($\delta\mu$) = $\Delta\mu_{\text{mean}}/\mu_{\text{mean}} = 0.02 \div 1.33 = \pm 0.015 \simeq 0.02$

Percentage error = $\Delta\mu_{\text{mean}}/\mu_{\text{mean}} \times 100\% = \pm 0.015 \times 100 = \mathbf{\pm 1.5}$

Example 12 The radius of curvature of a concave mirror is given as $R = \frac{l^2}{6h} + \frac{h}{2}$, where l and h are given as 2 cm and 0.064 cm, respectively. Find the error in measuring the radius of curvature.

Solution $l = 2$ cm, $\Delta l = 0.1$ cm (LC of metre scale)

$h = 0.064$ cm, $\Delta h = 0.001$ cm (LC of spherometer)

$$R = \frac{l^2}{6h} + \frac{h}{2} \Rightarrow \frac{\Delta R}{R} = \left|\frac{2\Delta l}{l}\right| + \left|\frac{-\Delta h}{h}\right| + \left|\frac{\Delta h}{h}\right|$$

$$\frac{\Delta R}{R} = \frac{2\,\Delta l}{l} + \frac{2\,\Delta h}{h} = \frac{2\times 0.1}{2} + \frac{2\times 0.001}{0.064} = 00.1 + 0.03 = \mathbf{0.131}$$

Example 13 The time of 30 oscillations of a simple pendulum whose length is 90 cm was observed to be 60 s. According to given data, find the value of g and determine percentage error in the value of g.

Solution $T = 2\pi\sqrt{\frac{l}{g}} \Rightarrow g = 4\pi^2 \times \frac{l}{T^2},\ T = \frac{60}{30} = 2.00$ s

$$g = 4 \times 3.14^2 \times \frac{90}{2^2} = 887.364 \text{ cm/s}^2$$

Maximum error in the value of g,

$$g = 4\pi^2 \frac{2l}{T^2} = 4\pi^2 \frac{l}{(t/30)^2} = \frac{4\pi^2 l}{t^2} \times 30^2$$

Taking log on both sides,

$$\log_e g = \log_e 4 + 2\log_e \pi + \log_e l - 2\log_e t + 2\log_e 30$$

$$\log_e g = 1.386 + 2.289 - 0.105 - 8.188 + 6.8$$

$$\log_e g = 2.182 \text{ or } g = 8.86 \text{ m/s}$$

Differentiating both sides, we get

$$\frac{\Delta g}{g} = \left|\frac{\Delta l}{l}\right| + 2\left|\frac{\Delta t}{t}\right|$$

$l = 90$ cm, $\Delta l = 0.1$ cm (LC of metre scale)

$t = 60$ sec, $\Delta t = 0.1$ sec (LC of stopwatch)

$$\frac{\Delta g}{g} = \frac{0.1}{90} + \frac{2\times 0.1}{60} = 0.0011 + 0.0033 = 0.0044$$

$$= 0.0044 \times 100\% = \mathbf{0.4\%}$$

Example 14 In a measurement of the viscous drag force experienced by spherical particles in a liquid, the force is found to be proportional to $V^{1/3}$ where V is the measured volume of each particle. If V is measured to be 30 mm^3, with an uncertainty of 2.7 mm^3, what will be the resulting relative percentage uncertainty in the measured force?

Solution The relative percentage uncertainty in the measure of fore is $6F^2 = \left(\frac{\partial E}{\partial V}\right)^2 6V^2$

$$6F = \left(\frac{\partial E}{\partial V}\right) 6V, 6V \rightarrow \text{uncertainty in measurement of volume}$$

$$F \propto V^{1/3}$$

$$\frac{\partial F}{\partial V} \propto \frac{1}{3} V^{-2/3}$$

$$\Rightarrow \quad 6F = \frac{1}{3V^{2/3}} \times 6V = \frac{1}{3(30)^{2/3}} \times 2.7 = \frac{1}{3 \times 9.7} \times 2.7$$

$$\Rightarrow \quad \mathbf{6F = 0.09}$$

Example 15 One gram of salt is dissolved in water that is filled to a height of 5 cm in a beaker of 10 cm diameter. The accuracy of length measurement is 0.01 cm while that of mass measurement is 0.01 mg. When measuring the concentration c, what is the fractional error $\Delta c/c$?

Solution c = Mass/Volume

$$V = \pi r^2 h = \frac{\pi d^2}{4} h$$

$$\text{Fractional error} = \sqrt{\left(\frac{\Delta x}{x}\right)^2 + \left(\frac{\Delta y}{y}\right)^2}$$

$$\frac{\Delta V}{V} = \sqrt{\left(\frac{\Delta d}{d}\right)^2 + \left(\frac{\Delta h}{h}\right)^2} \qquad \frac{\Delta d}{d} = \frac{0.01}{10} = 10^{-3}$$

$$\frac{\Delta V}{V} = 2\sqrt{2} \times 10^{-3} \qquad \frac{\Delta h}{h} = \frac{0.01}{5} - 2 \times 10^{-3}$$

$$\frac{\Delta c}{c} = \sqrt{\left(\frac{\Delta m}{m}\right)^2 + \left(\frac{\Delta V}{V}\right)^2} = \sqrt{10^{-10} + (8 \times 10^{-6})} = .2\sqrt{2} \times 10^{-3} = \mathbf{0.28\%}$$

Example 16 A battery powers two circuits C_1 and C_2 as shown in Figure 1.2.

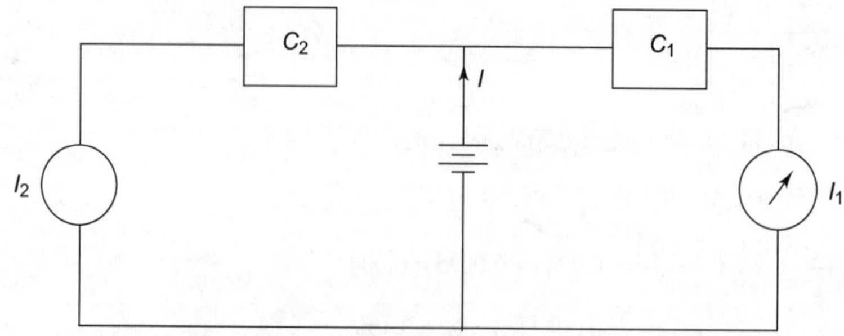

Figure 1.2

The total current I drawn from the battery is estimated by measuring the currents I_1 and I_2 through the individual circuits. If I_1 and I_2 are both 200 mA and the errors in the measurement are 3 mA and 4 mA respectively, what is the error in the estimate of I?

Solution

$$I_1 = (200 \pm 3) \text{ mA}$$

$$I_2 = (200 \pm 4) \text{ mA}$$

$$I = 400 \pm \Delta I$$

$$I = I_1 + I_2 = (400 \pm 7) \text{ mA}$$

$$\Delta I = \mathbf{7 \text{ mA}}$$

Example 17 A resistance is measured by passing a current through it and measuring the resulting voltage drop. If the voltmeter and ammeter have uncertainties of 3% and 4% respectively, then

(a) Find the uncertainty in resistance

(b) Find the uncertainty in the computed value of power dissipated across the resistance

Solution

(a) $V = IR$

Taking log on both sides and differentiating, we get

$$\frac{\delta V}{V} = \frac{\delta I}{I} + \frac{\delta R}{R}$$

$$\pm 0.03 = \pm 0.04 + \frac{\delta R}{R}$$

$$\frac{\delta R}{R} = \pm 0.07 \text{ (max.)} = \mathbf{7\ \%}$$

(b) $P = I^2R$

Taking log on both sides and differentiating, we have

$$\frac{\delta P}{P} = \frac{2\delta I}{I} + \frac{\delta R}{R}$$

$$= 2 \times 0.04 + 0.07$$

$$= 0.15 = \mathbf{15\%}$$

Optics

Appendix 2

A2.1 ELECTROMAGNETIC (EM) SPECTRUM

A large number of frequencies of electromagnetic waves visualized in numerical order constitutes an *electromagnetic (EM) spectrum*. Frequencies that are usable for radio communication occur near the lower end of the EM spectrum. With the increase of frequencies, the EM energy becomes dangerous to human beings. For example, a microwave oven can be a hazard if it is not shielded properly. Also, with the increase of frequencies, it becomes difficult to employ EM energy for communication.

The electromagnetic spectrum as per frequency range and usage of EM energy is given below in Table 2.1.

TABLE 2.1

Approximate Frequency Range	*EM Phenomena*	*Examples of Uses*
530–1600 kHz	Radio waves	AM radio
3–30 MHz		Shortwave radio
50–250 MHz		FM radio, VHF TV
450–800 MHz		UHF TV
3–300 GHz	Microwaves	Radar, Satellite communication
10^3–10^4 GHz	Infrared radiation	Photography
10^5–10^6 GHz	Visible light	Human vision
10^6–10^8 GHz	Ultraviolet radiation	Sterilization
10^8–10^9 GHz	X-rays	X-ray (medical)
10^{10}–10^{13} GHz	γ-rays	Cancer therapy
> 10^{14} GHz	Cosmic rays	Astronomy (Physics)

A2.2 LENS ABERRATIONS

The inability of a lens to form the white point image of an object is called *lens aberration*.

A2.2.1 Chromatic Aberration and its Kinds

When the rays of white light parallel to the principal axis are incident on a lens, the different colours of the white light are refracted by different amounts at different places. It can also be said that the rays are dispersed into various colours and are focused at different distances from the lens (see Figure 2.1). This happens due to the fact that the refractive index of the lens depends on the wavelength of light.

Actually, violet light gets refracted more than red light. Hence, the point at which violet light focuses is nearer the lens than the point at which the red light focuses. Thus, the image formed by a lens is usually coloured and blurred. This inability of a lens to form a single image of a white object is called *chromatic aberration*.

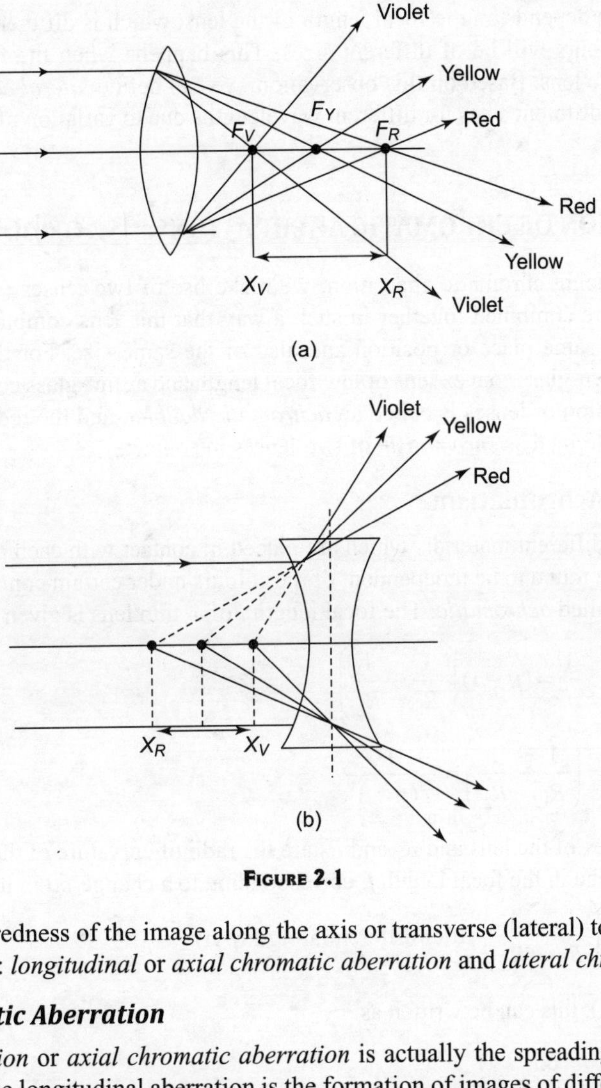

FIGURE 2.1

Depending upon the blurredness of the image along the axis or transverse (lateral) to the axis, the chromatic aberration is of two types: *longitudinal* or *axial chromatic aberration* and *lateral chromatic aberration*.

Longitudinal Chromatic Aberration

The *longitudinal aberration* or *axial chromatic aberration* is actually the spreading of an image along the principal axis. It means the longitudinal aberration is the formation of images of different colours at different positions along the axis. The axial distance between the positions of red and violet images is a measure of axial aberration. In Figure 2.1, $(X_R - X_V)$ is the measure of this aberration. Clearly, the longitudinal aberration is positive for the case of a convex lens and is negative for a concave lens.

Lateral Chromatic Aberration

If f is the focal length of the lens and u is the distance of an object from the lens, then the magnification is given by

$$m = \frac{f}{u + f}$$

Since the magnification depends on the focal length of the lens, which is different for different colours, the images of different colours will be of different sizes. This happens when the finite-sized white object is placed on the axis of the lens. Based on this observation, we can define *lateral chromatic aberration* as the formation of images of different sizes for different wavelengths due to variation of lateral magnification with the wavelength.

A2.3 CORRECTION OF CHROMATIC ABERRATIONS: USE OF DOUBLET

For minimizing or reducing chromatic aberration, we make use of two lenses (doublet). This can happen if two or more lenses are combined together in such a way that this lens combination produces images of different colours at the same place or position and also of the same size. For this, a combination can be prepared by using a crown-glass convex lens of low focal length and a flint-glass concave lens of greater focal length. Such a combination of lenses is called an *achromatic doublet*, and the reduction or minimization of chromatic aberration is termed as *chromatism* of two lenses in contact.

A2.3.1 Theory of Achromatism

Consider two lenses of different materials which are placed in contact with each other. Under this situation, if the focal length can be found to be independent of the colours under certain conditions, the combination of the two lenses will be called *achromatic*. The focal length f of a thin lens is given by

$$\frac{1}{f} = (\mu - 1)\left(\frac{1}{R_1} - \frac{1}{R_2}\right) \qquad \text{(i)}$$

or

$$\left(\frac{1}{R_1} - \frac{1}{R_2}\right) = \frac{1}{f(\mu - 1)} \qquad \text{(ii)}$$

Here, μ is refractive index of the lens and R_1 and R_2 are the radii of curvature of the two surfaces of the lens. If we take δf as the change in the focal length f, corresponding to a change $\delta\mu$ in the refractive index μ, then one can find $-\frac{\delta f}{f^2} = \delta\mu\left(\frac{1}{R_1} - \frac{1}{R_2}\right)$ after differentiating Eq. (i).

By making use of Eq. (ii), this can be written as

$$-\frac{\delta f}{f^2} = \frac{\delta\mu}{(\mu - 1)} \cdot \frac{1}{f} \qquad \text{(iii)}$$

The ratio $\frac{\delta\mu}{(\mu - 1)}$ $(= \omega)$ is the *dispersive power* of the lens between the two colours for which the difference in refractive index is $\delta\mu$ and the mean refractive index is μ. Hence,

$$-\frac{\delta f}{f^2}=\frac{\omega}{f} \tag{iv}$$

If f_1 and f_2 are taken as the mean focal lengths of two thin lenses of combination and ω_1 and ω_2 are the dispersive powers between two colours for which the combination is to be achromatized, then the focal length of the combination can be written as

$$\frac{1}{F}=\frac{1}{f_1}+\frac{1}{f_2} \tag{v}$$

From this, we get

$$-\frac{\delta F}{F^2}=-\frac{\delta f_1}{f_1^2}-\frac{\delta f_2}{f_2^2}$$

In view of Eq. (iv), we can write $-\frac{\delta f_1}{f_1^2}=\frac{\omega_1}{f_1}$ and $-\frac{\delta f_2}{f_2^2}=\frac{\omega_2}{f_2}$

$$\therefore \quad -\frac{\delta F}{F^2}=\frac{\omega_1}{f_1}+\frac{\omega_2}{f_2}$$

The lens combination will be said to be achromatic if F does not change with colour, i.e., $\delta F = 0$. This yields

$$\frac{\omega_1}{f_1}+\frac{\omega_2}{f_2}=0 \quad \text{or} \quad \frac{f_1}{f_2}=-\frac{\omega_1}{\omega_2} \tag{vi}$$

So this condition is the required condition for a doublet. The negative sign indicates that the combination should be a convex lens and a concave lens (Figure 2.2).

This condition also conveys that the ratio of the focal lengths of the two lenses is numerically equal to the ratio of dispersive powers of their materials. Since ω_1 and ω_2 are positive quantities, the focal lengths must carry opposite signs, which justifies the combination of a convex lens and a concave lens (Figure 2.2).

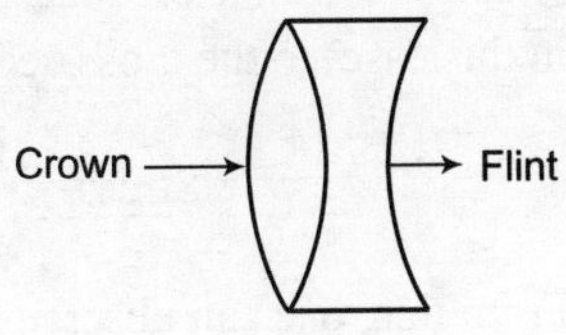

FIGURE 2.2

Appendix

3

Mechanical Properties of Materials

A3.1 ELASTICITY

Elasticity is a fundamental property of materials, and any material or body can be deformed by the application of an external force. If the body returns to its original shape after the removal of the force, it is said to be elastic. Springs of all kinds are examples of elastic bodies. Most substances are found to be elastic to some degree. In technical terms, a substance with high elasticity is the one that requires a large force to produce a distortion. For example, a steel sphere is a substance of high elasticity.

A3.2 STRESS AND STRAIN

We define certain terms such as stress and strain, for comparing the elasticity of materials. Consider a steel wire, which is held rigidly at the top end and has a load fastened to the lower end. The wire under this situation is said to be under *stress*, the magnitude of which is equal to the ratio of the applied force (the weight in this case) to the cross-sectional area, i.e.,

$$\text{Stress} = \frac{F}{A}$$

From this, one can observe that the SI unit of stress is N/m^2.

If the load is significantly enhanced, the wire will be stretched by an amount ΔL (for its length as L). Under this situation, we define another term called *strain*, which is a measure of the distortion of an object. Strain is defined as the change in a spatial variable divided by the original value of that variable. If we take this variable as the length, then

$$\text{Strain} = \frac{\Delta L}{L}$$

From this, one can observe that strain is a dimensionless quantity. It means it does not carry units.

There are three ways in which a body may change its dimensions under the action of an external force. Consider a solid cylinder, which is stretched by two equal forces applied normal to its cross-sectional area. The restoring force per unit area in this case is called *tensile stress*. Conversely, if the cylinder is compressed under the action of applied forces, the restoring force per unit area is called *compressive stress*. Since in both the cases, there is a change in the length of the cylinder, the tensile or compressive stress can also be termed as *longitudinal stress*. The change in the length ΔL to the original length L of the cylinder, in this case, is known as *longitudinal strain*. On the other hand. if two equal and opposite deforming forces are applied parallel to the cross-sectional area of the cylinder, there would be a relative displacement (say Δx) between the opposite faces of the cylinder. In this situation, the restoring force per unit area developed due to the applied tangential force is known as *tangential stress* or *shearing stress*. The strain so produced by the tangential force is known as *shearing strain*. This is defined as the ratio of relative displacement of the faces Δx to the length of the cylinder L, whereas the *volume strain* is defined as the ratio of change in volume (ΔV) to the original volume (V).

A3.3 HOOKE'S LAW

Hooke's law is related to the stress and strain produced in a body under the action of an external force. This law is an empirical law which is found to be valid for most materials. Hooke's law states that the stress and strain are proportional to each other for small deformations in the body. It means

$$\text{Stress} = k' \times \text{Strain}$$

where k' is the proportionality constant and is known as the *modulus of elasticity*.

A3.4 YOUNG'S MODULUS

Young's modulus is denoted by the symbol Y, and is defined as the ratio of tensile (or compressive) stress (σ) to the longitudinal strain (ε). It means

$$Y = \frac{\sigma}{\varepsilon}$$

Putting the values of σ and ε in the above formula, we get

$$Y = \frac{(F/A)}{(\Delta L/L)} = \frac{(F \times L)}{(A \times \Delta L)}$$

A3.5 SHEAR MODULUS

The *shear modulus* of a material is denoted by G, and is defined as the ratio of shearing stress to the corresponding shearing strain. The shear modulus is also called the *modulus of rigidity*, which is given by

$$G = \frac{(F/A)}{(\Delta x/L)}$$

For small values of Δx, the ratio $\Delta x/L$ can be defined as the angle θ (shown in Figure 3.1). Then

$$G = \frac{F}{(A \times \theta)}$$

Since F/A represents the shearing stress σ_s, we can have

$$\sigma_s = G \times \theta$$

It can be seen that the SI unit of shear modulus is N/m^2.

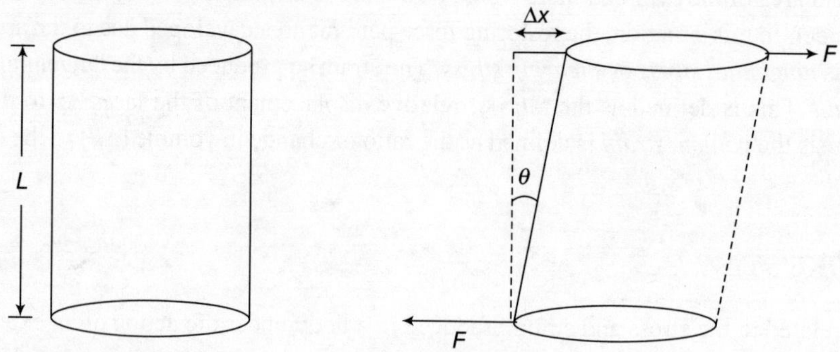

FIGURE 3.1

A3.6 BULK MODULUS

Bulk modulus is denoted by B, and is defined as the ratio of stress to the volumetric strain. In view of the unit of stress as N/m^2, the bulk modulus can be defined in terms of pressure, p, as

$$B = -\frac{p}{(\Delta V/V)}$$

Since there is a decrease in volume with the increase of pressure, if p is positive and ΔV is negative, the negative sign appears in the above formula. Thus, for a system in equilibrium, the value of the bulk modulus B is always positive. The SI unit of bulk modulus is the same as that of pressure, i.e., N/m^2 or Pa.

The reciprocal of the bulk modulus is known as *compressibility*, which is denoted by k. The compressibility is basically the fractional change in volume per unit increase in pressure, i.e.,

$$k = \left(\frac{1}{B}\right) = -\left(\frac{1}{\Delta p}\right) \times \left(\frac{\Delta V}{V}\right)$$

A3.7 RELATION BETWEEN ELASTIC CONSTANTS

Young's modulus (Y), bulk modulus (B) and modulus of rigidity (G) are the elastic constants. These are related to each other through Poisson's ratio (σ) as follows:

$$Y = 3B(1 - 2\sigma) = 2G(1 + \sigma)$$

These are also related to each other as per the relation

$$Y = \frac{9BG}{(G + 3B)}$$

Relation 1: $Y = 3B(1 - 2\sigma)$

We consider a cube $A_1B_1C_1D_1A_2B_2C_2D_2$ (Figure 3.2).

On each of its faces, an equal amount of force F is applied. Due to this action, the edges A_1B_1, B_1A_2 and B_1C_1 are elongated (in the direction of force), and are also contracted (in the direction perpendicular to the force).

If α represents the increase in length per unit length per unit tension along the direction of the force, and β represents the contraction produced per unit length per unit tension in a direction perpendicular to the force, then the Poisson's ratio $\sigma = \beta/\alpha$. In view of this, the elongation produced in the edges A_1B_1, B_1A_2 and B_1C_1 will be $F\alpha$ each. Similarly, the contraction produced in these edges will be $2F\beta$ each, as the two pairs of forces are acting perpendicular to each edge.

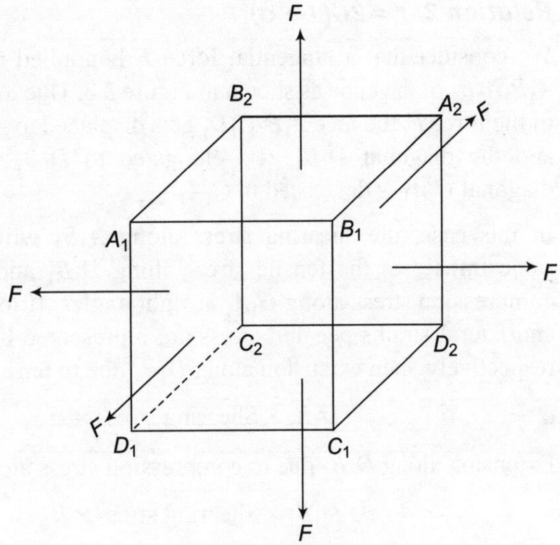

Figure 3.2

In view of the above, the lengths of the edges would be

$$A_1B_1 = 1 + F\alpha - 2F\beta = B_1A_2 = B_1C_1.$$

Hence, the new volume of the cube will be

$$= (1 + F\alpha - 2F\beta)^3$$

$$= (1 + F\alpha - 2F\beta)\,(1 + F^2\alpha^2 + 2F\alpha + 4F^2\beta^2 - 4F\beta - 4F^2\alpha\beta)$$

Since α and β are small quantities, the terms $F^2\alpha^2$, $F^2\beta^2$ and $F^2\alpha\beta$ can be neglected. Thus, the new volume becomes

$$(1 + F\alpha - 2F\beta)\,(1 + 2F\alpha - 4F\beta) = 1 + 2F\alpha - 4F\beta + F\alpha - 2F\beta$$
$$= 1 + 3F(\alpha - 2\beta)$$

From this, we find the change in volume ΔV as

$$\Delta V = 1 + 3F(\alpha - 2\beta) - 1 = 3F(\alpha - 2\beta) \quad \text{(i)}$$

If a pressure P is applied on all the faces to compress the cube, it can be found that

$$\Delta V = 3P(\alpha - 2\beta) \quad \text{(ii)}$$

From the above, the volumetric strain can be obtained as

$$\Delta V/V = 3P(\alpha - 2\beta) \qquad \because V = 1$$

Now, bulk modulus

$$B = \frac{\text{Stress}}{\text{Volumetric strain}} = \frac{P}{3P(\alpha - 2\beta)}$$

or $$B = \frac{1}{3(\alpha - 2\beta)}$$

Since Young's modulus $Y = 1/\alpha$ and Poisson's ratio $\sigma = \beta/\alpha$, we divide RHS of the above relation by α

$$\Rightarrow \quad B = \frac{1/\alpha}{3(1 - 2\beta/\alpha)} = \frac{Y}{3(1 - 2\sigma)}$$

This gives $\boxed{Y = 3B(1 - 2\sigma)}$

Relation 2: $Y = 2G(1 + \sigma)$

We consider that a tangential force F is applied to the face $A_1B_1A_2B_2$ of the cube as shown in Figure 3.3. Due to the action of the force F, the face $A_1B_1C_1D_1$ gets displaced to $A_1'B_1'C_1D_1$ and the diagonal D_1B_1 gets elongated to D_1B_1' while the diagonal C_1A_1 is decreased to C_1A_1'.

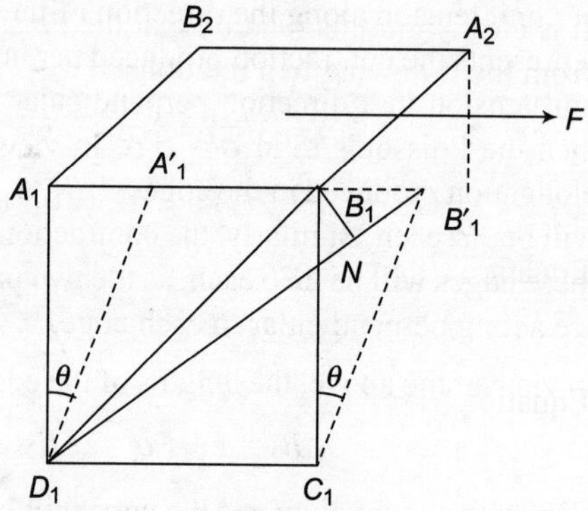

FIGURE 3.3

In this case, the shearing stress along A_1B_1 will have the same impact as the tensile stress along D_1B_1 and an equal compression stress along C_1A_1 at right angles. If longitudinal and lateral strains per unit stress are represented by α and β respectively, then extension along D_1B_1 due to tensile stress

$$= D_1B_1 \times \text{Shearing stress} \times \alpha \quad \text{(i)}$$

Extension along D_1B_1 due to compression stress along A_1C_1

$$= D_1B_1 \times \text{Shearing stress} \times \beta \quad \text{(ii)}$$

By adding Eqs (i) and (ii), we get the total extension along D_1B_1 as

$$= D_1B_1 \times \text{Shearing stress} \times (\alpha + \beta) \quad \text{(iii)}$$

Now, if θ is very small, then $\angle A_1B_1'C_1 \approx 90°$ and $\angle B_1B_1'N = 45°$

Hence, the increase in length of $D_1B_1 = B'_1N$

$$= B_1B'_1 \cdot \cos 45°$$

$$= \frac{B_1B_1'}{\sqrt{2}} \quad \text{(iv)}$$

From Eqs (iii) and (iv), we have

$$D_1B_1 \times \text{Shearing stress} \times (\alpha + \beta) = \frac{B_1B_1'}{\sqrt{2}}$$

or $$\text{Shearing stress} \times \frac{A_1\,B_1}{B_1B_1'} = \frac{1}{2(\alpha + \beta)} \qquad (\because\ D_1B_1 = B_1A_1\sqrt{2})$$

or $$\frac{\text{Shearing stress}}{(B_1B_1'/A_1B_1)} = \frac{1}{2(\alpha + \beta)}$$

or $$\frac{\text{Shearing stress}}{\text{Shearing strain}} = \frac{1}{2(\alpha + \beta)}$$

or $$G = \frac{1}{2(\alpha + \beta)} = \frac{1/\alpha}{2(1 + \beta/\alpha)}$$

or $$G = \frac{Y}{2(1 + \sigma)}$$

or $$\boxed{Y = 2G\,(1 + \sigma)}$$

Relation 3: Y = 9 BG/(G + 3B)

It is clear from this relation that the Poisson's ratio σ does not appear here. Hence, we intend to eliminate σ from the following two relations:

$$Y = 3B(1 - 2\sigma) \qquad \text{(i)}$$

$$Y = 2G(1 + \sigma) \qquad \text{(ii)}$$

Equation (i) gives

$$1 - 2\sigma = Y/3B \qquad \text{(iii)}$$

Equation (ii) gives

$$1 + \sigma = Y/2G$$

or $$2 + 2\sigma = Y/G \qquad \text{(iv)}$$

Adding Eqs (iii) and (iv), we obtain

$$\frac{Y}{3B} + \frac{Y}{G} = 3$$

or $$Y\frac{(G+3B)}{3BG} = 3$$

or $$\boxed{Y = \frac{9\,BG}{(G+3B)}}$$

A3.8 BEAMS AND THEIR BENDING

A beam is a structural member that is acted upon by a system of external loads at right angles to the axis. These loads acing perpendicular to the axis of the beam can deform the beam, which is called *bending of beam*. If the plane of loading passes through one of the principal centriodal axes of the cross section of the beam, the bending is said to be *plane*; otherwise the bending is said to be *oblique*. If the load can be considered to act at a point, then the load is called a *point load*. On the other hand, if the load is distributed or spread in some manner over the length of the beam, then it is called a *distributed load*. If the spread of the load is uniform, it is said to be *uniformly distributed*; otherwise it is said to be a *non-uniformly distributed load*.

A3.8.1 Classification of Beams

Depending upon the type of support, beams are classified as cantilever, simply supported beam, overhanging beam, fixed beam or continuous beam. A *cantilever* is a beam whose one end (say A) is fixed and the other end (say B) is free. The length between A and B is known as the length of the cantilever. If both the ends of a beam freely rest on a wall, columns or knife edges, the beam is called a *simply supported beam*. In all such cases, the reactions are always upwards. If the supports are not situated at the ends of the beam, i.e., one or both the ends project beyond the supports, then the beam is called an *overhanging beam*. On the other hand, if both the ends of the beam are rigidly fixed or built-in into its supporting walls, then the beam is known as a *fixed beam*. Finally, a *continuous beam* is defined as the beam that has more than two supports. In this case, the supports at the extremes are called *end supports* and all the other supports, except the extreme, are called *intermediate supports.*

A3.9 SHEARING FORCE (SF) AND BENDING MOMENT (BM)

With respect to the bending of beams, two terms, i.e., *shearing force* (SF) and *bending moment* (BM) are important to discuss. Consider a beam, which is in equilibrium under the action of a series of forces. When the beam is cut in some section *XX* (see Figure 3.4) and its left part to the section remains in equilibrium, then it is obvious that some force must act at the section. This force would be provided by the adjacent material prior to the cutting of the beam and would act tangentially to the section. Hence, this is clear that there will be shearing force at the section.

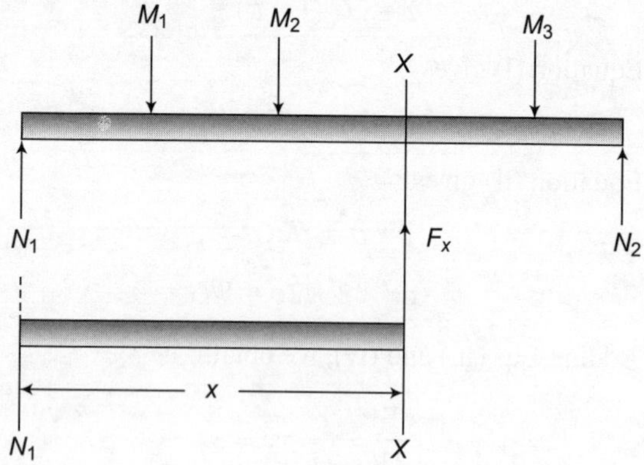

FIGURE 3.4

In view of the equilibrium of the material to the left of the section *XX*, there can be no resultant moment to the left of the section. It means any moment produced by the forces acting on the beam must be balanced by an equal and opposite moment produced by the internal forces acting in the beam at the section. This is called the *bending moment* at the section, which is the algebraic sum of moments to the left or right of the section. With respect to the equilibrium either for forces or moments, the resultant caused by the applied forces to one side of the section is balanced by the bending moment and shearing force acting at the section.

A3.9.1 Sign Convention

In general, a shearing force having an upward direction to the RHS of a section or downwards to the LHS of the section is taken to be +ve. The shearing force having a downward direction to the right of the section or upward direction to the left of the section will be –ve. Following the same terminology, a bending moment causing concavity upwards is taken to be +ve and called *sagging bending moment.* On the other hand, a bending moment causing convexity upwards is taken as –ve and called *hogging bending moment.*

A3.10 CANTILEVER WITH LOAD AT FREE END

As mentioned earlier, a cantilever is supported at one end only and being built-in at its support, it gives a fixed slope at that point. Now, we consider a section *XX* at a distance *x* from the free end *B* [see Figure 3.5(a)]. Then

shearing force at $X = S_x = -M$, and

bending moment at $X = B_x = -Mx$.

This shows that the shearing force is constant at all sections of the member between *A* and *B*. However, the bending moment at any section is proportional to the distance of the section from the free end. The SF diagram is shown in Figure 3.5(b) whereas the BM diagram is depicted in Figure 3.5(c).

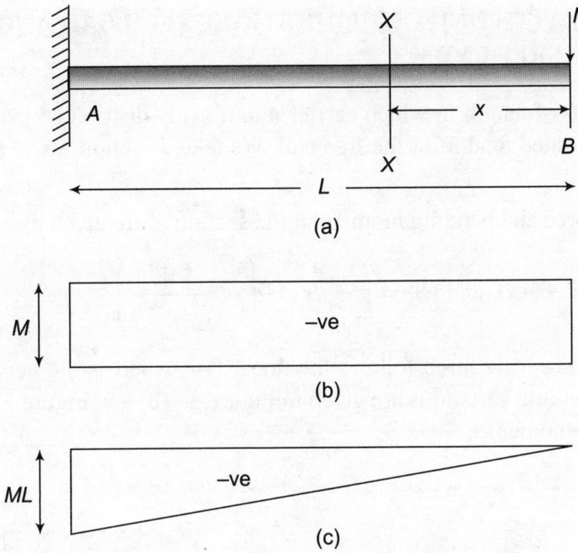

FIGURE 3.5

A3.11 CANTILEVER WITH UNIFORMLY DISTRIBUTED LOAD

As is clear, here we consider that the load is uniformly distributed over the whole length of the beam. So it is appropriate to take the loading as W per unit run [shown in Figure 3.6(a)].

Now, we consider the section XX at a distance x from the free end B. In this case, the shearing force at $X = S_x = -Wx$ whereas the bending moment at

$$X = B_x = \int S_x dx = -\frac{Wx^2}{2}$$

This shows that the variation of the shear force is according to the linear law. However, the variation of bending moment is parabolic, being proportional to x^2.

The shearing-force diagram is shown in Figure 3.6(b) and the bending-moment diagram is given in Figure 3.6(c).

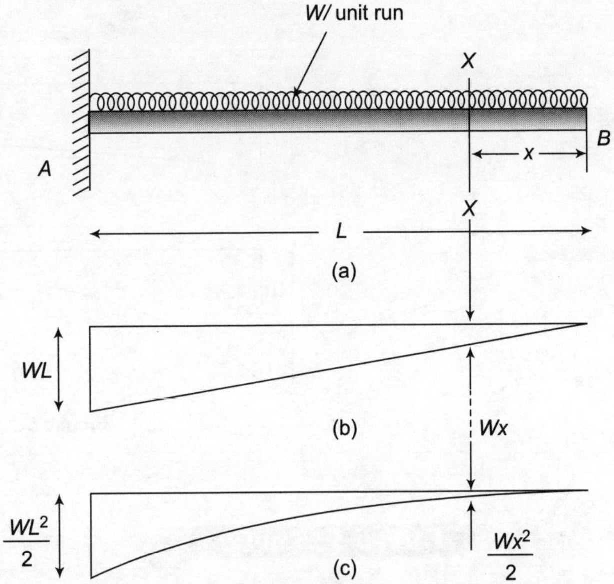

FIGURE 3.6

A3.12 CANTILEVER WITH UNIFORMLY DISTRIBUTED LOAD AND A CONCENTRATED LOAD AT THE FREE END

We consider a cantilever AB of length L, which carries a uniformly distributed load W per unit run over the whole length and a concentrated load M at the free end. We take a section XX at a distance x from the free end B [Figure 3.7(a)].

In this case, the shearing force and bending moment at the section X are given by

$$S_x = -(M + Wx) \text{ and } B_x = \int S_x\, dx = -\left(Mx + \frac{Wx^2}{2}\right)$$

Here also, we find that the shearing force follows the linear law, whereas the bending moment follows the parabolic law. The corresponding variations are given in Figure 3.7(b) and Figure 3.7(c), respectively, for the shearing force and bending moment.

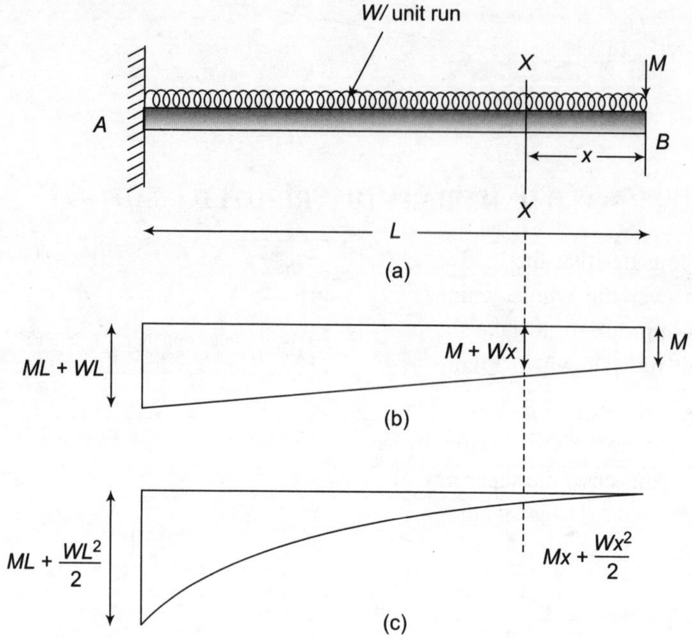

FIGURE 3.7

SOLVED EXAMPLES

EXAMPLE 1 A structural steel rod has a radius of 10 mm and a length of 1.0 m. A 100 kN force stretches it along its length. Calculate (a) stress, (b) elongation, and (c) strain on the rod. Young's modulus (Y) of structural steel is 2.0×10^{11} Nm^{-2}.

Solution We assume that the rod is held by a clamp at one end and the force F is applied at the other end, parallel to the length of the rod. Then the stress on the rod is given by

$$\text{Stress} = \frac{F}{A} = \frac{F}{\pi r^2} = \frac{100 \times 10^3 \text{ N}}{3.14 \times 10^{-4} \text{ m}^2} = 3.18 \times 10^8 \text{ Nm}^{-2}$$

The elongation

$$\Delta L = \frac{(F/A)L}{Y} = \frac{3.18 \times 10^8 \text{ Nm}^{-2} \times 1 \text{ m}}{2 \times 10^{11} \text{ Nm}^{-2}}$$
$$= 1.59 \times 10^{-3} \text{ m} = 1.59 \text{ mm}$$

The strain is given by

$$\text{Strain} = \Delta L/L = (1.59 \times 10^{-3} \text{ m})/1 \text{ m}$$
$$= 1.59 \times 10^{-3} = \mathbf{0.16\ \%}$$

Example 2 A square load slab of 50 cm side and 10 cm thickness is subject to a shearing force (on its narrow force) of 9.0×10^4 N. The lower edge is reverted to the floor (Figure 3.8). By how much will the upper edge be displaced?

Solution The load slab is fixed and the force is applied parallel to the narrow force. The area of the force parallel to which this force is applied is

$$A = 50 \text{ cm} \times 10 \text{ cm} = 0.5 \text{ m} \times 0.1 \text{ m}$$
$$= 0.05 \text{ m}^2$$

Therefore, the stress applied is

$$= (9.0 \times 10^4 \text{ N}/0.05 \text{ m}^2) = 1.80 \times 10^6 \text{ Nm}^{-2}$$

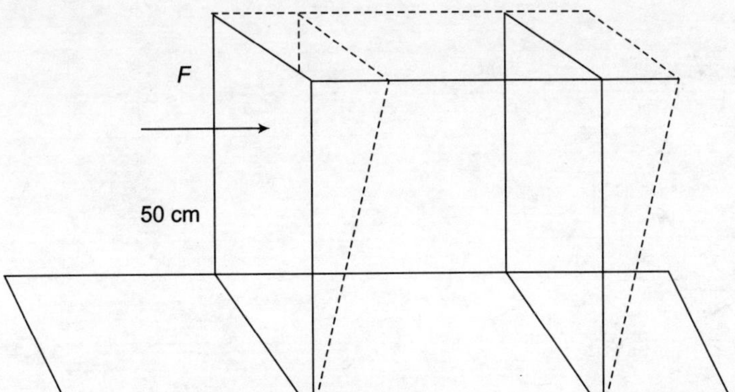

Figure 3.8

We know that

$$\text{Shearing strain} = \frac{\Delta x}{L} = \frac{\text{Stress}}{G}$$

Therefore, the displacement

$$\Delta x = (\text{Stress} \times L)/G$$

$$= (1.8 \times 10^6\ \text{Nm}^{-2} \times 0.5\ \text{m})/(5.6 \times 10^9\ \text{Nm}^{-2})$$

$$= 1.6 \times 10^{-4}\ \text{m} = \mathbf{0.16\ mm}$$

EXAMPLE 3 The average depth of the Indian Ocean is about 3000 m. Calculate the fractional compression $\Delta V/V$ of water at the bottom of the ocean, given that the bulk modulus of water is 2.2×10^9 Nm^{-2}. (Take $g = 10$ ms^{-2}).

SOLUTION The pressure exerted by a 3000 m column of water on the bottom layer

$$p = h\rho g = 3000\ \text{m} \times 1000\ \text{kg m}^{-3} \times 10\ \text{ms}^{-2}$$

$$= 3 \times 10^7\ \text{kg m}^{-1}\ \text{s}^{-2}$$

$$= 3 \times 10^7\ \text{N m}^{-2}$$

Fractional compression,

$$\frac{\Delta V}{V} = \frac{\text{Stress}}{B}$$

$$= \frac{3 \times 10^7}{2.2 \times 10^9}$$

$$= 1.36 \times 10^{-2}$$

or $= \mathbf{1.36\ \%}$

Chapter-wise Answers to Objective Type Questions

Chapter 1

Q1. (b) Q2. (c) Q3. (c) Q4. (c) Q5. (c) Q6. (a) Q7. (d) Q8. (c)
Q9. (b) Q10. (c) Q11. (c) Q12. (c) Q13. (d) Q14. (d) Q15. (b) Q16. (b)
Q17. (c) Q18. (a)

Chapter 2

Q1. (a) Q2. (c) Q3. (b) Q4. (b) Q5. (a) Q6. (d) Q7. (b) Q8. (b)
Q9. (d) Q10. (d) Q11. (b) Q12. (a, c) Q13. (d) Q14. (a) Q15. (d) Q16. (a)
Q17. (a) Q18. (d)

Chapter 3

Q1. (c) Q2. (a) Q3. (d) Q4. (a) Q5. (c) Q6. (a) Q7. (a) Q8. (d)
Q9. (c) Q10. (a) Q11. (a) Q12. (b) Q13. (d) Q14. (d) Q15. (a) Q16. (a)
Q17. (a) Q18. (a) Q19. (c)

Chapter 4

Q1. (b) Q2. (a) Q3. (d) Q4. (a) Q5. (c) Q6. (a) Q7. (a) Q8. (b)
Q9. (c) Q10. (b) Q11. (b) Q12. (b) Q13. (a) Q14. (c) Q15. (a) Q16. (d)
Q17. (a) Q18. (a) Q19. (c) Q20. (d) Q21. (a)

Chapter 5

Q1. (b) Q2. (b) Q3. (d) Q4. (a) Q5. (a) Q6. (a) Q7. (b) Q8. (d)
Q9. (d) Q10. (d) Q11. (a) Q12. (a)

Chapter 6

Q1. (c) Q2. (c) Q3. (a) Q4. (d) Q5. (b) Q6. (a) Q7. (a) Q8. (a)
Q9. (c) Q10. (b)

Chapter 7

Q1. (a) **Q2.** (d) **Q3.** (d) **Q4.** (d) **Q5.** (b) **Q6.** (b) **Q7.** (d) **Q8.** (b)
Q9. (b) **Q10.** (b)

Chapter 8

Q1. (a) **Q2.** (c) **Q3.** (b) **Q4.** (d) **Q5.** (c,d) **Q6.** (a,c) **Q7.** (b) **Q8.** (c)
Q9. (b) **Q10.** (c) **Q11.** (c) **Q12.** (a) **Q13.** (c) **Q14.** (b) **Q15.** (d) **Q16.** (b)
Q17. (a) **Q18.** (a) **Q19.** (b) **Q20.** (d)

Chapter 9

Q1. (d) **Q2.** (d) **Q3.** (b) **Q4.** (a) **Q5.** (a) **Q6.** (c) **Q7.** (b) **Q8.** (c)
Q9. (a) **Q10.** (b) **Q11.** (b) **Q12.** (a) **Q13.** (d) **Q14.** (a) **Q15.** (c) **Q16.** (c)
Q17. (b) **Q18.** (a) **Q19.** (a) **Q20.** (b) **Q21.** (b) **Q22.** (b) **Q23.** (c) **Q24.** (a)
Q25. (a) **Q26.** (d) **Q27.** (d) **Q28.** (d) **Q29.** (d) **Q30.** (b) **Q31.** (a) **Q32.** (d)
Q33. (d) **Q34.** (b) **Q35.** (c) **Q36.** (d) **Q37.** (d) **Q38.** (a) **Q39.** (b) **Q40.** (a)

Chapter 10

Q1. (b) **Q2.** (a) **Q3.** (c) **Q4.** (d) **Q5.** (a) **Q6.** (c) **Q7.** (a) **Q8.** (d)
Q9. (a) **Q10.** (a) **Q11.** (d) **Q12.** (a) **Q13.** (d) **Q14.** (b) **Q15.** (b) **Q16.** (b)
Q17. (a)

Chapter 11

Q1. (b) if the surface ϕ is constant
Q2. (c) **Q3.** (c) **Q4.** (c) **Q5.** (a) **Q6.** (b) **Q7.** (a) **Q8.** (d) **Q9.** (c)
Q10. (a) **Q11.** (c) **Q12.** (d) **Q13.** (a) **Q14.** (a) **Q15.** (c) **Q16.** (c) **Q17.** (a)
Q18. (c) **Q19.** (b) **Q20.** (a) **Q21.** (b) **Q22.** (c) **Q23.** (d) **Q24.** (b) **Q25.** (d)
Q26. (c) **Q27.** (c) **Q28.** (b)

Chapter 12

Q1. (a) **Q2.** (b) **Q3.** (a) **Q4.** (b) **Q5.** (d) **Q6.** (b) **Q7.** (d) **Q8.** (a)
Q9. (d) **Q10.** (a) **Q11.** (b) **Q12.** (b) **Q13.** (b) **Q14.** (b)

Chapter 13

Q1. (b) **Q2.** (b) **Q3.** (c) **Q4.** (a) **Q5.** (a) **Q6.** (b) **Q7.** (b) **Q8.** (a)
Q9. (b) **Q10.** (a) **Q11.** (d) **Q12.** (a) **Q13.** (d) **Q14.** (d) **Q15.** (c) **Q16.** (a)
Q17. (d) **Q18.** (b) **Q19.** (c) **Q20.** (c) **Q21.** (c) **Q22.** (b) **Q23.** (b) **Q24.** (d)

Q25. (d) Q26. (a) Q27. (c) Q28. (d) Q29. (c) Q30. (a) Q31. (c) Q32. (a)
Q33. (a) Q34. (a) Q35. (a) Q36. (b) Q37. (a) Q38. (a) Q39. (a) Q40. (b)
Q41. (d) Q42. (a) Q43. (b) Q44. (a) Q45. (a) Q46. (a) Q47. (a) Q48. (c)
Q49. (b) Q50. (c) Q51. (a) Q52. (c) Q53. (a) Q54. (c) Q55. (a)

Chapter 14

Q1. (d) Q2. (c) Q3. (b) Q4. (d) Q5. (d) Q6. (c) Q7. (d) Q8. (b)
Q9. (a) Q10. (c) Q11. (d) Q12. (b) Q13. (b) Q14. (b) Q15. (a) Q16. (b)
Q17. (c) Q18. (b) Q19. (d) Q20. (a) Q21. (a) Q22. (a) Q23. (a) Q24. (b)
Q25. (c) Q26. (a) Q27. (b) Q28 (b) Q29. (a)

Chapter 15

Q1. (d) Q2. (d) Q3. (a) Q4. (a) Q5. (d) Q6. (d) Q7. (a) Q8. (d)
Q9. (c) Q10. (b) Q11. (b) Q12. (a) Q13. (d) Q14. (a) Q15. (b) Q16. (b)
Q17. (a) Q18. (a) Q19. (b) Q20. (b)

Chapter 16

Q1. (d) Q2. (b) Q3. (a) Q4. (b) Q5. (a) Q6. (d) Q7. (a) Q8. (a)
Q9. (b) Q10. (a) Q11. (d) Q12. (d) Q13. (b) Q14. (a) Q15 .(c Q16. (a)
Q17. (c)

Chapter 17

Q1. (b) Q2. (d) Q3. (c) Q4. (a) Q5. (d) Q6. (c) Q7. (a) Q8. (a)
Q9. (a) Q10. (c) Q11. (a) Q12. (d) Q13. (c) Q14. (b) Q15. (c)

Chapter 18

Q1. (c) Q2. (d) Q3. (b) Q4. (c) Q5. (d) Q6. (b) Q7. (b) Q8. (b)
Q9. (d) Q10. (b) Q11 (a) Q12. (c) Q13. (d) Q14. (c) Q9. (b) Q15. (b)
Q16. (b) Q17. (c) Q11. (c) Q18. (b) Q19. (a) Q20. (a) Q21. () Q22. (a)
Q23. (b) Q24. (a) Q25. (b) Q26. (d) Q27. (d) Q28. (a) Q29. (d) Q30. (a)
Q31. (c) Q32. (c) Q33. (c) Q34. (a) Q35. (b) Q36. (c) Q37. (c)

Chapter 19

Q1. (a & d) Q2. (a) Q3. (b) Q4. (d) Q5. (b) Q6. (a) Q7. (c) Q8. (a)
Q9. (d) Q10. (d) Q11. (b) Q12. (c) Q13. (a) Q14. (b) Q15. (a) Q16. (b)
Q17. (a) Q18. (b) Q19. (a) Q20. (b) Q21. (d) Q22. (d)

Chapter 20

Q1. (b) **Q2.** (a) **Q3.** (a) **Q4.** (c) **Q5.** (a) **Q6.** (b) **Q7.** (b) **Q8.** (a)
Q9. (b) **Q10.** (a) **Q11.** (d) **Q12.** (c) **Q13.** (b) **Q14.** (b) **Q15.** (b) **Q16.** (d)

Chapter 21

Q1. (a) **Q2.** (b) **Q3.** (b) **Q4.** (b) **Q5.** (b) **Q6.** (b) **Q7.** (d) **Q8.** (b)
Q9. (a) **Q10.** (c) **Q11.** (a) **Q12.** (a) **Q13.** (a) **Q14.** (d) **Q15.** (a) **Q16.** (b)

Chapter-22:

Q1. (a) **Q2.** (a) **Q3.** (d) **Q4.** (b) **Q5.** (a) **Q6.** (a) **Q7.** (b) **Q8.** (d)
Q9. (c) **Q10.** (b) **Q11.** (b) **Q12.** (c) **Q13.** (b) **Q14.** (d) **Q15.** (d) **Q16.** (c)
Q17. (d) **Q18.** (a) **Q19.** (a) **Q20.** (c) **Q21.** (c) **Q22.** (c) **Q23.** (b) **Q24.** (c)
Q25. (d) **Q26.** (b) **Q27.** (c) **Q28.** (a) **Q29.** (d) **Q30.** (a)

Index

Symbols

A

B

C

D

R

S